CW00701583

A comprehensive reference book for those with interest in, or need to know, how operations in the world's factories work, and how common products, components, and materials are made.

Handbook of Manufacturing Processes

How Products, Components and Materials Are Made

James G. Bralla

With Contributions by a Distinguished Editorial Board

Industrial Press, Inc.

New York

Library of Congress Cataloging-in-Publication Data

Bralla, James G.
Handbook of manufacturing processes / James Brella.
p. cm.
ISBN 0-8311-3179-9
1. Manufacturing processes–Handbooks, manuals, etc. I. Title.

TS183.B73 2006
670–dc22

2006041834

First Edition

Handbook of Manufacturing Processes
How Products, Components and Materials are Made

Industrial Press
989 Avenue of the Americas
New York, New York 10018

Editor: John Carleo
Copy Editing: Robert E. Green
Art Director/Production Manager: Janet Romano
The line drawings in this book, not attributed to other sources, were drawn by the author.
They were enhanced or redrawn for publication by Lorraine Pawlewicz.

Cover photo of engine block provided by GE Fanuc Automation

10 9 8 7 6 5 4 3 2 1

TABLE OF CONTENTS IN BRIEF

DETAILED TABLE OF CONTENTS

Chapter 3—Machining Processes 87

Chapter 5—Glass and Ceramics Processes 211

Chapter 6—Woodworking Processes 255

Chapter 7—Assembly and Fusion (or Joining) Processes 295

Chapter 8—Finishing Processes (including Heat Treating) 329

Chapter 12—Food Processes 483

Chapter 13—Processes for Electronic Products 519

Chapter 14—Advanced Manufacturing Methods 585

Section II—How Products, Components and Materials Are Made 613
(listed in alphabetical order)

A 615

B 632

C 646

G 677

H 690

EDITORIAL BOARD—HANDBOOK OF MANUFACTURING PROCESSES

The editorial board is made up of experienced individuals with expertise in certain fields of manufacturing who have prepared some material for the handbook or edited that prepared by others. Their participation is for the purpose of insuring the accuracy and completeness of the material.

ABOUT THE AUTHOR

James G. Bralla has had a career spanning more than 50 years in manufacturing, as an engineer, consultant, and executive. He was Vice-President, Operations, for Alpha Metals, Inc., Director of Manufacturing, Asia, for the Singer Company, and Industry Professor at Polytechnic University. He holds a BS in Mechanical Engineering from Princeton University and an MS in Manufacturing Engineering from Polytechnic. He is a registered Professional Engineer, the editor of the *Design for Manufacturability Handbook* and the author of *Design for Excellence,* both published by McGraw-Hill.

DEDICATION

This book is dedicated to the thousands of people, worldwide, who keep all the manufacturing processes described in this book operating productively. These people come from all walks of life with varied amounts of education, ranging from the grammar school level to PhD's and even post-PhD's. All, however, share certain attributes. One is dedication to the task of keeping their process in operation, with willingness to stay with a problem—beyond normal working hours, if necessary—until it is solved. A second attribute is extensive self-education in the workings of the equipment for which they are responsible. A third is an innate knack for analyzing a process problem, finding the root-cause of the problem, and the ingenuity, when necessary, to devise a quick fix. They may realize, for example, that a certain linkage is sticking, that there is too much play in some moving parts, that a detector is not signaling the condition for which it is designed, that the workpiece material is out of spec, or whatever one of the thousands of things that can go wrong is causing the malfunction of the equipment. Then they have the energy to try different approaches, to disassemble a device to find out what is wrong, to research a problem with others, or from source documents, to find out what could be amiss. After all this, they have sufficient skill to do what has to be done to put the equipment back into productive working order. This may involve such skills as machining, to make or modify a critical part, to replace electronic devices or printed circuit boards, to add a simple sheet metal shim, to design or build a tool or fixture, or to have the ability to work with others who provide the specialist skills necessary. The net result of their efforts and skills is the continuing operation of the equipment that they care for, so that we all can benefit from the products and goods that they make.

This book is also dedicated to Steve Bralla, my son, who happens to be one of the gifted people noted above, except that his particular field is the operation of sophisticated earthquake detection apparatus, rather than production machinery. Steve was faced last year with a diagnosis of acute myeloid leukemia, a devastating and frequently fatal disease. But through his courage, strength and willingness to undergo the lengthy and debilitating ordeal that a cure involves, the support of his family, and with the guidance of talented and dedicated medical specialists, he now tests to be cancer-free. I salute him and all those who keep the world's production machinery in operation, and dedicate this book to them.

James G. Bralla
North Jackson, PA

PREFACE

This is a reference book. It was prepared to serve as a concise, easy-to-read, source for those who need to gain an intelligent insight into the workings of manufacturing processes. It is also for those who want or need to know how particular products, their components, or their raw materials are made.

Many books that are currently available give some very worthwhile instruction about the methods used in specific industries; others present good information over some range of industries, but these are textbooks rather than reference books and none have the breadth of coverage that is included here. This book gives descriptions of key operations in the major production industries: Metalworking, including Casting, Metal Forming and Machining, and the Plastics, Ceramics, and Woodworking Industries. There are chapters on Joining and Assembly, and on Product Finishing. The Paper and Printing Industries, Textiles, Garment-Making, Chemicals, Food Processing, and Electronics are all included in this book.

There are other books that describe how some products are made, but they are usually aimed at the general public, especially younger readers, and are quite limited in both scope and the depth of information provided. There is no reference book on this subject with engineering-level information. This book is intended to fill that void.

An objective of the book is to provide clear, easily readable and concise explanations, so that the reader can easily gain an understanding of what is involved and how each process works. Although the book includes much technical detail, we have tried to avoid including non-essential complexities of any process, but to explain it concisely in simple terms, so that the reader, even if not technically trained, can understand and, if necessary, explain the method to others. The text has been prepared to be explanatory, straightforward, to-the-point, and practical (rather than theoretical). To aid in this end, descriptions have been liberally supplemented with illustrations. The objective of each illustration is to present a clear, easily understood view of the workings of the method covered. To this end, most illustrations are schematic, concentrating on the basic principles of each process and stripped of unnecessary detail.

WHO SHOULD USE THE BOOK

People for whom the book was prepared include the following:

manufacturing engineers, those who design, build, plan, execute and maintain the equipment, tools, and processes that make the things that the public buys and uses.

process engineers, those who plan and engineer the manufacturing steps, equipment and tooling needed in production.

manufacturing executives, managers, and supervisors who need to know and understand what their employees are doing and why, and what new processes and equipment should be considered to improve their operations.
students interested in a career in manufacturing and especially those pursuing a career in manufacturing engineering, who can use this book for current instruction and for future reference.
product design engineers and draftsmen, who should have this book available for reference so that they understand how the products that they design are made.
government officials who are responsible for operational safety (OSHA), environmental conditions, and other regulatory matters. They can gain a better understanding, with this book, of the factory operations that they regulate.
consultants who have, or wish to have, manufacturing clients and want to be sure that they understand what is happening in their client's operations. These consultants should have this book available for reference.
salesmen and sales managers who deal with customers that are involved in manufacturing, and who need to know more about their operations.
faculty of engineering schools
engineering societies involved in manufacturing or related subjects should have a copy of this book in their libraries and should offer it to their members.
state, city, county, town and college libraries, for their constituencies.
purchasing people who buy manufactured components and products.
quality control managers and specialists who can gain, with this book, a better understanding of the processes, whose products they monitor.
maintenance and reliability managers and technicians who can similarly benefit from a better knowledge of the processes they are responsible for.

ACKNOWLEDGMENTS

I am indebted to the following people who provided valuable assistance in the preparation of this book:

Frank Andros, Mike DiPietro, Richard Redolphy and **Sandra Marsh** of the Integrated Electronics Engineering Center (IEEC) of Binghamton University—electronics manufacturing
David J. Aquilino, Bodine Assembly and Test Systems—automatic assembly
John Bartman, VP, Human Resources, Public and Media Affairs, Snyders of Hanover
Carolyn Boss—text editing
J. R. Casey Bralla, Manager of Manufacturing Engineering, Southco, Inc.—for much help in many areas, in addition to his Editorial Board participation
Matthew S. Bralla—fishing rods
Stephen J. Bralla, Scripps Institute, San Diego, CA.—lasers, sailplanes.
Andrew Broom, Eclipse Aviation—aircraft assembly, including friction stir welding
John Commander—Cookson Electronics Enthone—electronics
Sue Dean, CEO, Donald Dean and Sons, Montrose, PA—cabinet making
Victor DePhillips, President, Signature Building Systems, Inc., Moosic, PA 18507—prefabricated housing
Kate Dougherty, Cirrus Design Corp.—aircraft manufacture
Dave Duemler, CEO and Technical Director, and **Stephen Maund**, Director of Engineering/Manufacturing, Demco Automation
Renee' J. Fink, Executive Secretary, The Black and Decker Corporation
Paul Foster, retired plant manager, General Cable Corporation
Peter Frennborn, Alfa Laval, Inc.—plate type heat exchangers
Kristine Gable, Research Consultant, Corning Inc.
Ken Gilleo, ET-Trends, LLC—flexible circuit boards
Roger Glass, Emeritus Research Scientist, University of Michigan Aerospace Engineering
Ken Glover, Inland Paperboard and Packaging, Inc.—corrugated cartons
Angelo Gulino—Cookson Electronic Materials—solder and flux making
Charles A. Harper, Technology Seminars, Inc., Lutherville, MD
Steve Johnson and **Fran Borrego,** Universal Instruments—populating PC boards
Dawn Klehr, The Toro Company—lawn mowers
Jim Kessler, Service Forester, Montrose, PA—Applications of wood species
Don Lillig, Marley Cooling Technologies—cooling equipment
Roy Magnuson, PhD, Endicot Interconnect Technologies—electronics
Mark Martin PhD, president Design4X—aircraft
Jim McKee of McKee Button Co. Muscatine IA—button manufacture
Bob Papp—commercial designer and artist
Kelly Parke, Senior Designer, WMH Group—woodworking equipment

Amelia Paterno, Sharon Hoopes, and **Pam Bagnall** at the Susquehanna Borough Branch of Susquehanna County (Pennsylvania) library.
Mason and Michael Perryman, Rapport Composites—golf clubs
Anna Petrova—Bodine Assembly and Test Systems
Jerry Pinch, Pinch Heating and Cooling—air conditioners
Scott Plickerd, The McGraw-Hill Companies
Greg Pompea, Vice-president, Engineering and **Rob Horowitz,** Sales Manager, Contact Systems, Inc.
Sue Rehmus, General Motors—automobile manufacture
Bill Rollo, retired from The Singer Company.
Jim Rooney—retired from IBM.—electronics manufacture
Peter Schlotter, hunter—gun and bullet-making
Raymond P. Sharpe, Chief Executive Officer, Isola Group SARL
Wayne Smith and **Bob Hawley,** Smith-Lawton Mill Work, Montrose, PA—cabinet making
John Stein—Retired from the Singer Company—metal stamping
Gary Stitely, General Manager, and **Brian Shook,** Engineering, Landis Threading Systems
Brian Terski and **Jason Walden** of Pole-Kat—golf clubs
Carol S. Tower, librarian of Society of Manufacturing Engineers
Jason Tuttle, Dodds Company—woodworking
Jim Wilcox, PhD, of IBM, Endicott 607-429-3172
MaryAnn Wright—Engineering Supervisor, Metal Injection Molding and Powder Metallurgy, Remington Arms—gun manufacture

And, special thanks to my wife **Martha-Jane Bralla,** who has supported this project so well despite my physical and mental absences when doing the research, writing, sketching, editing and proof reading that this book required.

Readers are invited to call to my attention any errors that may have crept into the information presented in this book. Please address e-mail to info@industrialpress.com.

James G. Bralla

HOW TO USE THIS HANDBOOK

The book is in two sections: (I), the Process Section, in which common manufacturing processes in 17 key industries are described, and (II), the Products, Components and Materials Section, which explains how many of these are made.

Section I explains how each manufacturing process works, detailing what happens to the materials or workpieces that are being processed. Usually, these explanations are in general terms as they are not limited to a particular component or material. However, the description also identifies the normal components or materials to which the process applies.

Section II deals with specific products, components, and materials, outlining the manufacturing sequence and processes used for making each. It often refers the reader, using the designation of any applicable text entry, to where more detailed descriptions of the operations mentioned can be found in the book. This is done by showing, in parenthesis, the chapter and text entry designation from Section I, where the basic operation is described. For example, in the description of the manufacture of a metal part that requires case-hardening heat treatment, the description may include "(8G3b)" to tell the reader that the case-hardening heat treatment used on the part is described at greater length in entry G3b of Chapter 8.

The handbook text in Section I is organized in a typical outline structure to aid the reader in finding relevant information easily. Related processes are grouped together and sequential operations are covered in sequence when possible. Major topics are given an upper-case letter designation such as A, B, C, etc. Important sub-topics are designated with the capital letter and a number (for example, A1, B1, C1, etc.) Sub headings under these topics are indicated by adding a lower-case letter to the designation (e.g., A1a, A1b, A1c, etc.) The descriptions of further process variations may be given designations such A1a1, A1a2, A1a3, etc. For example, Chapter 1 is devoted to metal casting processes; section B in Chapter 1 covers sand-mold casting methods; entry B5 describes methods of making sand molds, B5e describes those methods that utilize a machine for the operation, and B5e1 describes one specific machine method, the jolt-squeeze method.

For ease of reference, the same designations used to identify text entries are also used to identify accompanying illustrations. For example, Figure 9B2 illustrates the process described in text entry B2 in Chapter 9. Figure 1B5e1 illustrates the jolt-squeeze machine described in entry B5e1 of Chapter 1.

Section II is simply arranged in alphabetical order by the name of the product, component, or material whose manufacturing method is described. Section II includes descriptions of manufacturing processes used in making each product, component, or material listed, though sometimes, if the process for that item has already been described in Section I, the Section II entry simply refers to the applicable entry in Section I. Thus, for the manufacture of gasoline, whose manufacturing process is described under "Petroleum Processing" in Chapter 11, the listing in Section II simply refers to (11H)—Petroleum Refining and Processing—where gasoline manufacture is described in considerable detail.

When an entry in Section II is referred to elsewhere in the book, the name is shown italicized to tell the reader that there is a description in Section II of how the item is manufactured. Thus, for example, if the reader sees a name such as "*detergents*" in italics, he or she knows that there is a description in Section II that tells how detergents are made. (Italics are also used in the text of the book to designate processes of particular importance.)

HOW TO FIND A HANDBOOK ENTRY

For a process description, if the usual process name is known, the reader can refer to the Index at the back of the book. If the reader is uncertain of the name of the process, he or she can refer to the table of contents, find the major heading where the kind of operation of interest is shown, and, by visually scanning the entries below the major one, find the listing and page number for the particular operation in mind.

For a product, component or material manufacturing description, the reader can refer directly to Section II of the book, where entries are arranged in alphabetical order or can refer to the Index to locate its page number. (Both Section II and the Index are arranged in alphabetical order, but Section II includes considerable descriptive material, and does not include listings of processes, equipment, methods, or operations by name, as does the Index.)

Section I

Manufacturing Processes

Chapter 1 - Casting Processes

A. Melting Metal for Casting

A1. ***cupola melting*** - Now much less common due to environmental factors, this method utilizes a furnace in stack form as shown in Fig. 1A1. Fuel and metal to be melted are in direct contact. The stack is lined with refractory material and alternate layers of coke and metal are placed in it. Some minerals, primarily limestone ($CaCO_3$), are included with the metals to be melted. Air is blown through the stack from the bottom through openings called tuyeres. The bottom layer of coke is ignited initially. Heat from the burning coke melts the metal, which flows to the bottom of the cupola from where it can be removed by opening a tap hole. Slag is also removed from the bottom, from an exit hole just above the one used to remove molten metal. As the coke is consumed and the metal charge melts, the burning gradually proceeds upward. The upper layers are preheated by the flow of hot gases. Additional metal, coke, and limestone can be added from a charging door in the upper part of the stack as the operation proceeds. Metal charges may consist of steel scrap, cast iron scrap or pig iron, or, more commonly, a combination of them. The molten metal absorbs carbon from the coke, so cupola melting is generally restricted to cast, malleable, and ductile iron (though the electric arc method is preferred for the latter).

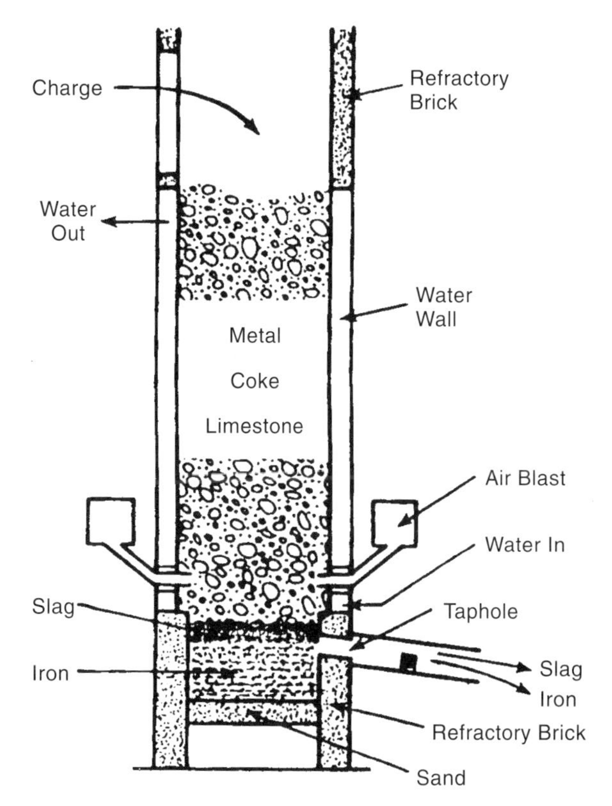

Fig. 1A1 Cross-sectional view of cupola melting cast iron. The metal charge (pig iron and scrap iron and steel) is in direct contact with burning coke. When the metal melts, it flows to the bottom of the cupola where it can be withdrawn. *(from Schey, Introduction to Manufacturing Processes, McGraw-Hill, New York, 1987)*

A2. ***electric arc melting*** - In this method, an electric arc similar to the one used in arc welding but much larger and more powerful, is used to provide the melting heat. In the *direct-arc* method, there are two arcs, one from an electrode to the metal and another from the metal to the second electrode. In the *indirect-arc* method, the arc extends from one electrode to another and the heat is transferred to the metal by radiation. Electrodes normally are made of carbon although, when molybdenum and other high-melting-temperature metals are processed, the electrodes may be of the same metal as that being melted. This variation is known as *consumable arc melting*. Fig. 1A2 illustrates the direct-arc method. This method can also be used with three electrodes and three-phase current. Electric arc melting is used extensively for in the production of alloy and carbon steels, and for malleable iron, ductile iron, tool steel, and high-strength cast iron. Control of environmentally undesirable emissions is easier with electric furnaces than with cupolas. An indirect arc is used in brass and bronze production.

A3. ***crucible melting*** - This method employs a cup-shaped, refractory-lined, metal furnace which is normally heated by gas or oil and sometimes by electrical resistance or induction. It has an inner crucible to hold the metal charge. The crucible is made of either a clay-silicon-carbide or a clay-graphite mixture. The furnace can either tilt for pouring or the crucible can be lifted out. Fig. 1A3 illustrates a tilting type with a lift-out crucible. The crucible method is used to melt brass, bronze, aluminum, and magnesium for sand castings. Except for induction heating, ferrous metals are not usually melted in this kind of furnace.

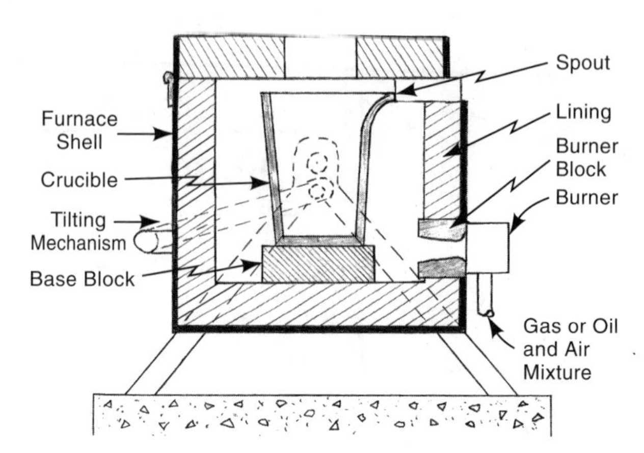

Fig. 1A3 Cross-section through a crucible furnace.

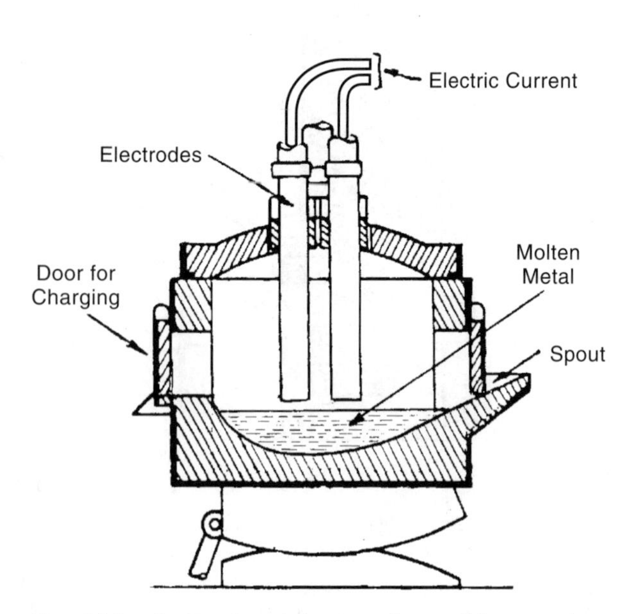

Fig. 1A2 A direct-arc furnace for melting steel or iron for castings. The arc passes from one electrode to the metal and back from the metal to the other electrode, providing heat that melts the metal.

A4. ***air furnace (reverberatory) melting*** - has similarities to open-hearth melting. Fig. 1A4 shows a typical air furnace. Oil or pulverized coal is burned in one chamber and the charge is placed in another. Heat from the burning fuel passes over and is absorbed by the charge, melting it. There is no direct contact between the metal and the fuel, allowing carbon content to be closely controlled. Oil or finely pulverized bituminous coal are used as fuels. Some smaller furnaces use natural gas. This type of furnace is used in the production of castings from malleable and gray cast iron, brass, and bronze.

A5. ***induction melting (high frequency and low frequency)*** - With this method, alternating electric current in a coil creates a magnetic field that induces corresponding secondary electrical currents in the metal charge. The resistance of the metal in the charge causes its temperature to rise to the melting point. Melting can be very rapid and there is no pollution or contamination from the heat source. Induction melting is used for steel, brass, bronze, aluminum, and magnesium.

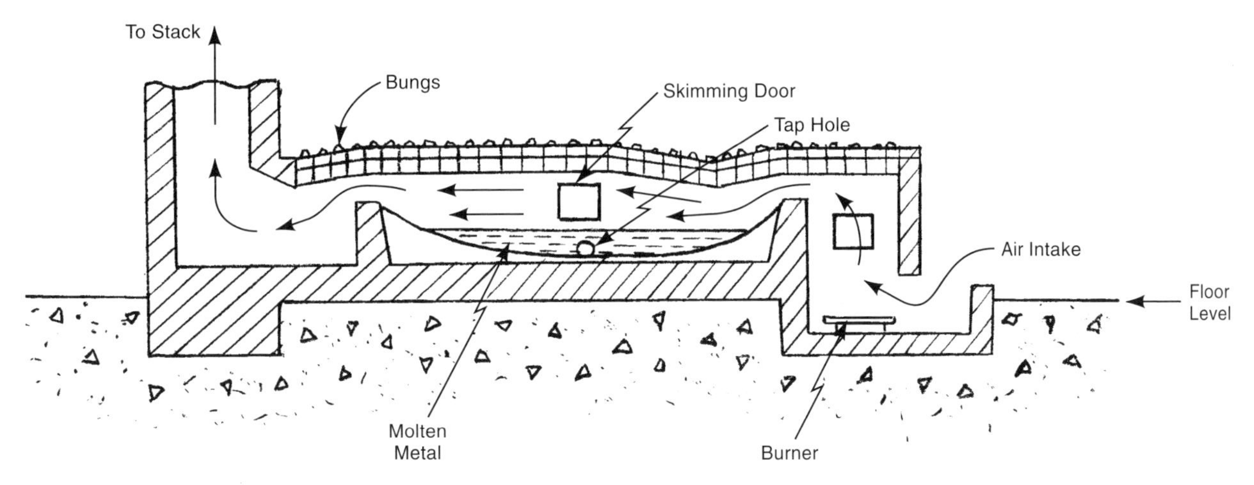

Fig. 1A4 An air reverberatory furnace.

With the *coreless* method, which commonly - but not always - is utilized at high frequencies, the coil surrounds a crucible containing the metal. The coil is made from copper tubing and water is circulated through the tubing to prevent the coils from overheating. Typical frequencies vary up to 10,000 Hz. but coreless furnaces can also operate at low frequencies (e.g. 60 Hz). The most common range is 250 to 3000 Hz. Melting is rapid. At the lower frequencies, the induction provides a stirring effect. At higher frequencies, higher power levels are possible. Brass, aluminum, cast iron, and steel are melted in coreless induction furnaces. Fig. 1A5 illustrates a typical coreless furnace.

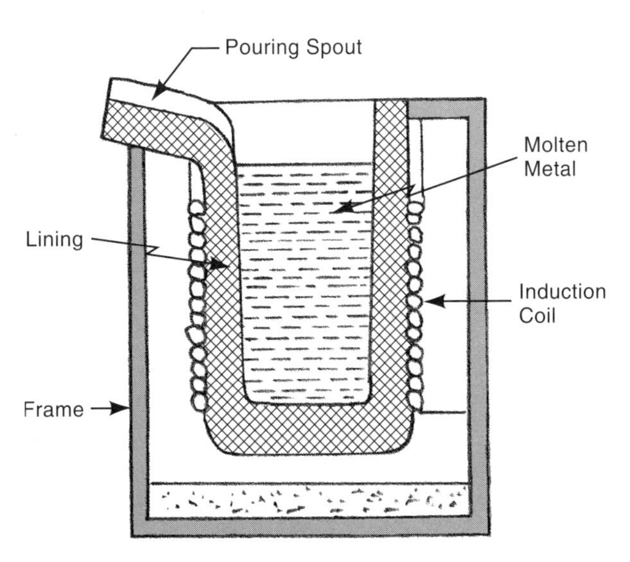

Fig. 1A5 A coreless induction furnace.

With the *channel* type of induction, the melting container itself forms a loop but only one portion of the loop is surrounded by the coil. The metal in this loop is heated by induction and the heat is transferred to the balance of the metal by convection and induction. The arrangement is shown in Fig. 1A5-1. Channel type furnaces operate at low frequencies. The melting rate is very high with this method and the temperature can be controlled accurately. However, there must be liquid metal in the channel for the induction effect to take effect, so an initial charge of enough melted metal to form a loop is required. Solid material can then be added. Low-frequency cored furnaces are often used as holding furnaces. Channel furnaces are used for brass and aluminum, and as duplex or secondary furnaces

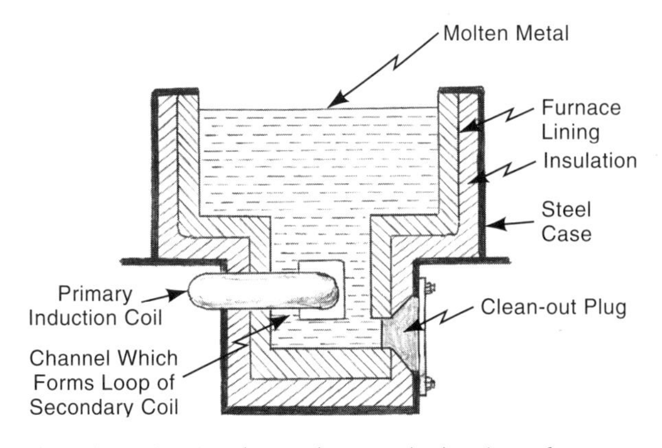

Fig. 1A5-1 A channel-type induction furnace. The molten metal in the furnace becomes the loop of a secondary induction coil.

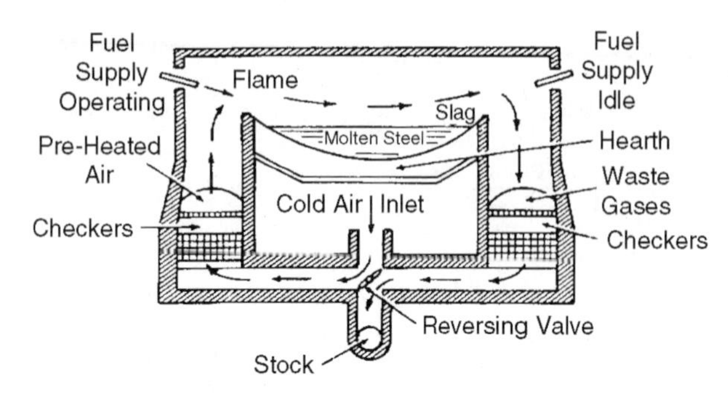

Fig. 1A6 Sectional-view of an open-hearth furnace. (*Courtesy Steel Founders' Society of America, Barrington, Illinois*)

for iron. In the latter case, molten cast iron from a cupola is fed to a channel induction furnace where the composition can be adjusted to meet specifications.

A6. ***open-hearth melting*** - This method, used in the production of steel and cast iron, is also used to supply molten metal for casting operations. Foundry open-hearth furnaces are usually smaller than those found in steel mills. Fig. 1A6 illustrates a typical open-hearth furnace which is both reverberatory and regenerative. Metal in the furnace is heated by a flame passing over the charge. The flame comes from the combustion of gas, oil, tar, or pulverized coal. The low roof of the furnace reflects heat downward to the metal in the furnace. Both fuel and air are fed from one side into the central area where the flame and heating take place. The chambers on the opposite side are heated by the flame and exhaust gases moving through them. The pool of molten metal in the furnace is shallow, which provides the maximum area for heat transfer per unit volume of metal. After a period of time, the direction of flow is reversed. The chambers heated from the previous cycle, in turn, heat the incoming fuel and air. Most open hearth furnaces are chemically basic (rather than acidic) as determined by the material of the brick furnace lining. The basic furnaces remove sulfur, silicon, carbon, and manganese from the charge metal. The charge used in making structural steel includes iron ore, limestone, scrap, and, later, molten pig iron. Additions can be made to the steel to produce the desired composition. Oxygen may be added to the furnace combustion area to reduce the process time and the amount of fuel required. Finished metal is removed from a hole in the rear of the furnace and transferred to a ladle.

A7. ***pouring*** - Metal is usually tapped from the melting furnace into either a ladle from which it is poured by gravity into the mold, or into one that is used to transfer a quantity of metal to a pouring ladle. Such transfer ladles are usually covered to reduce heat loss during transfer. Pouring ladle capacities range from about 60 lb (27 Kg) up. Ladles are frequently transported by overhead cranes. There are three basic types of ladles, as illustrated in Fig. 1A7: open-lip ladles that pour by

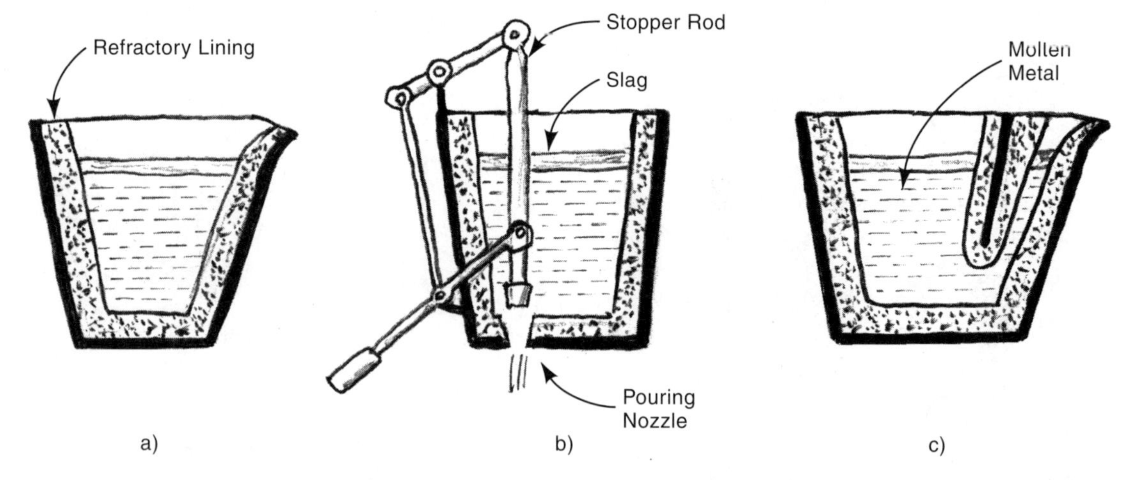

Fig. 1A7 Three different types of pouring ladles: a) a standard open lip-pour ladle. b) a bottom-pour ladle that avoids including slag in the metal poured. c) a "teapot" ladle that pours metal from near the bottom, avoiding the inclusion of slag in the casting.

tilting, "teapot" ladles that also pour by tilting but which avoid pouring slag, and bottom-pour ladles which also avoid pouring slag. Tilting ladles often utilize worm-gear tilting systems to provide better control and prevent the ladle from tipping too much or too fast. Numerous automatic pouring systems, designed to accurately meter the amount of molten metal poured, are also used. Some consist of mechanized or robotic dip-and-pour ladles. Others pour directly from a larger holding pot, using either stopper rods as shown in Fig. 1A7, or sliding gate valves. Some pouring vessels are fitted with electrical heating apparatus to maintain the metal at the proper pouring temperature. (The ideal pouring temperature involves a "superheat", a metal temperature sufficiently high to ensure that all parts of the mold are fully filled before solidification starts.) Other pouring systems include machine vision to sense when the mold is full, or weight controls to pour a prescribed amount, by weight, into the mold.

B. Sand-mold Casting

In sand mold casting the mold is made of packed sand. Molten metal is poured into a cavity in the sand. When the metal cools and solidifies, it has the shape of the cavity. The sand is removed, normally by a shaking action that is vigorous enough to cause the mold to break apart. The casting is then cleaned of sand; flashing and sprues are cut off and any jagged or sharp edges are ground smooth. (See snagging, B8g.)

The sand mold includes binders to hold the packed sand together and other additives. Bentonite clay is one of the most common binders. Organic materials and a certain amount of water are also used. The sand is either shoveled into the mold flask, dropped or blown from an overhead chute, or thrown by a sand slinging machine. The sand mixture is packed around a pattern which duplicates the shape wanted in the cast part. Various hand and machine approaches are used to compact the sand. Ramming, squeezing, slinging, and jolting are described below. After the sand has been compacted, the pattern is removed, leaving a cavity that retains the inverse of the pattern's shape. The sand is held together strongly enough so that it withstands the pressure and any eroding effects of the melted metal; is porous enough to allow gases to escape; yet it is weak enough to yield to shrinkage forces when the metal solidifies, and can be broken up and removed easily from the finished casting. The pattern can be of almost any material. In low quantity production situations, it may be made of wood. For repetitive manufacture, steel is more common. Plastics, aluminum, and other materials are also used. The pattern has the same shape as the desired cast part, but is slightly larger to provide a shrinkage allowance for the metal as it cools.

A typical sand mold is shown in Fig. 1B, and is normally made in two halves. The pattern is correspondingly split. The top half of the mold is called the "cope"; the bottom half the "drag". Both are held in a box-like container called "flasks". An entrance channel for the molten metal into the mold is provided by a basin and sprue formed in the cope half. Runners and gate are normally in the drag half. If the casting has some hollow or undercut elements, one or more additional sand pieces, called "cores" may be used. If a core is used, it is inserted in the mold cavity. The cope half of the mold is made similarly to the drag half and, after the pattern is removed, is inverted and placed over the drag. Pins in the flask insure alignment of the mold cavity. The two mold halves are held together with a clamp or weight. Sand mold casting can be used to make simple and complex parts from a wide variety of metals, though cast iron is the most common. Shapes with undercuts, contours, re-entrant

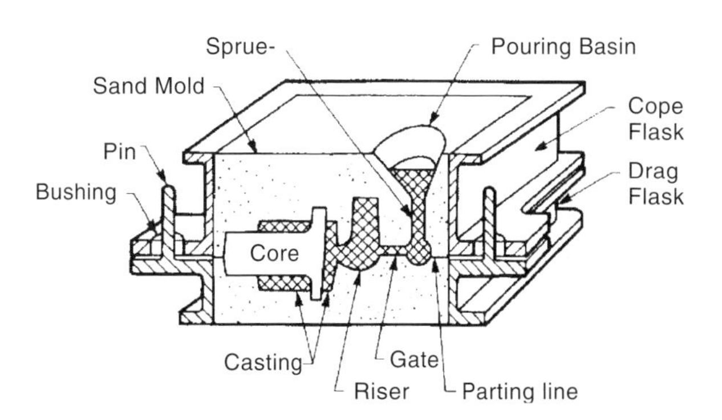

Fig. 1B A typical mold arrangement for sand-mold casting showing a typical core, pouring basin, riser, gate, and cope and drag flasks. (*From James G. Bralla, Design for Manufacturability Handbook, McGraw-Hill, New York, 1999*)

angles and other complications of shape can be cast. Castings weighing only one ounce to those of many tons can be cast with the process. Typical applications of sand mold casting are: automotive engine blocks, cylinder heads, connecting rods, crankshafts and transmission cases, machine tool bases and other mechanical components.

B1. ***green sand casting*** - is the basic sand-mold casting process described above. The green sand mixture consists of sand (usually silica), 6 to 8 percent clay binder (usually bentonite), 2 to 3 percent water and additives (sea-coal, starches, cellulose). ("Green" refers to the mixture, not the color.) It is moist and is not dried out before the molten metal is poured into the mold. The ingredients are mixed together thoroughly in a sand muller machine. The green sand process is inexpensive and has great versatility in regard to the metal that is cast and the size and shape of the castings made. Other sand-mold processes may give greater accuracy and smoother surface finish, however. The sand can be reclaimed and used again many times. The green sand method is the most common and least expensive sand casting process. Castings up to about 500 lb (230 kg) are commonly made with this approach. Automotive engine blocks, transmission cases, differential housings, railroad parts, and machinery components are all typical parts. Metals cast include gray cast iron, malleable and ductile iron, cast steel, aluminum, brass, bronze, and other non-ferrous metals.

B2. ***dry sand casting*** - In this process, the green sand mold is dried or baked before it is filled with molten metal. Typically, the mold is heated to 300°F (150°C) or higher, by baking or forced hot air until most of the moisture is evaporated. This approach produces a stronger mold and there is less gas (steam) generated when the molten metal is poured into the mold. One or more coatings of refractory material - silica, zircon, or graphite - are usually applied to the mold surface in a water or solvent carrier. Dry-sand molds can withstand more handling and longer storage, and have better resistance to the pressure of molten metal. The dry-sand process is normally used for medium-size to very large, multi-ton castings where greater mold strength is needed to withstand the mass of the molten metal. Dried sand gives a better surface finish but is more costly than green-sand molding because of energy, space, and equipment costs. With dry-sand casting, pitch is most commonly used to provide carbon instead of the sea coal used with green sand. Gilsonite, glutrin, corn flour, and molasses are other additives. These materials become thermoset at the baking temperature, adding to the strength and rigidity of the mold. Coarser sand is used to facilitate natural venting. Strong flasks and reinforcing bars, which extend into the sand, are also used to insure rigidity of the mold. Large castings in dry sand molds are cooled slowly to reduce internal stresses in the casting and the possibility of cracking.

B2a. ***skin-dried casting*** - To reduce the lengthy drying time of dry-sand casting, the drying is often limited to a depth of only about 1/4 to 1/2 in (6 to 13 mm). The patterns are usually first coated with a wash of refractory material. Heat for drying is supplied from a torch, from infra-red lamps, or from hot air. The mold is then referred to as a *skin-dried mold*. The dried skin is backed up with a mixture of green and dry sand. The approach is used extensively for the casting of steel, which involves higher pouring temperatures, and has also largely replaced dry sand molding for other applications. The casting surface is improved by elimination of the moisture in the facing sand, which could cause pin holes in the casting. Shake out is also facilitated. Additional binders such as linseed oil, corn flour or molasses may be used in the facing sand to improve the strength of the dried mold skin.

B3. ***other sand-mold casting processes***

B3a. ***shell mold casting*** - In this process, the mold sand is mixed with phenolic or another thermosetting plastic resin, either in liquid or solid form, with a catalyst. There is no clay binder but some other additives may be included for specific purposes. The mixture, when used, is dry and free-flowing. The pattern that forms the sand mold is heated to about 400 to 600°F (200 to 300°C), and clamped to a container of sand, which is then inverted, dropping the sand on to the heated pattern. When the sand is in contact with the heated pattern for a short period, the resin melts and then begins to polymerize, becoming hardened enough to bind the sand particles together. Coated sand particles that are not in contact with or near the pattern do not get heated sufficiently and do not bond together. Thus, a layer of bonded sand is formed next to the pattern. The pattern and sand are then re-inverted and the loose, non-bonded sand drops free. The pattern and the shell of bonded sand are heated additionally to complete the polymerization. The shell, normally from 0.2 to 0.4 in (5 to 10 mm) thick, is removed from the pattern by ejector pins and the flat faces of the two shell halves are fastened together with adhesive. Loose sand, gravel, or metal shot may then be used to support the mold during casting. This method is adaptable to large-scale production conditions and can produce castings of high complexity and accuracy but of limited size. The surface finish is smoother than that attainable with other sand casting methods because of the effect of the resin and the finer sand normally used.

The process is also used for making cores for molds used with other sand-mold processes. When shell cores are made, the core box is metal and is heated to cure the phenolic resin. Production is rapid. Cores can be made hollow by removing the sand from the center of the core before it has received enough heat to melt and cure the resin. The shell molding process is illustrated in Fig. 1B3a.

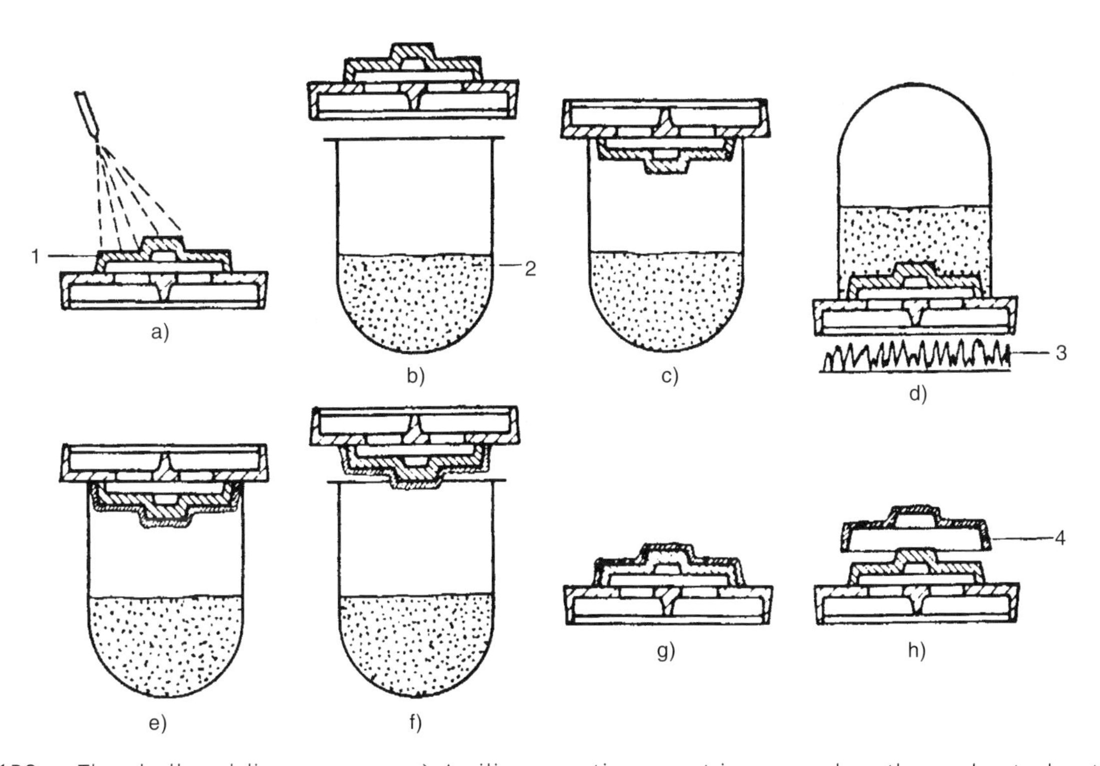

Fig. 1B3a The shell molding process. a) A silicon parting agent is sprayed on the preheated pattern. b) The preheated pattern is then brought to a dump box that contains a sand-resin mixture. c) The pattern is inverted and clamped to the top of a dump box. d) The dump box is inverted, sand falls against the pattern and is further heated for a short period. The resin in the sand mixture next to the hot pattern starts to set, causing sand particles to adhere together. e) The dump box is turned upright and the sand not in contact with the hot pattern or near to it, being unaffected, falls away. f) The pattern and the sand-resin shell are removed from the dump box. g) The pattern and shell are baked to fully cure the resin. h) The pattern is withdrawn from the shell. Two shells are fastened together, placed in a flask and supported with back up sand or metal shot. Molten metal is poured into the cavity formed by the two shells. Identified items: 1 - pattern, 2 - sand-resin mixture, 3 - burner, 4 - shell mold half. (*from Davidson, Handbook of Precision Engineering, Vol. 10, McGraw-Hill, New York*)

B3b. ***lost foam casting*** - This process is also called *full mold casting, evaporative casting* or *disposable pattern casting*. It differs from conventional sand mold casting in that the pattern is made of foamed polystyrene plastic (See 4C4.) instead of conventional pattern materials and is consumed in the process. The polystyrene pattern is coated with a slurry of permeable refractory material by dipping, spraying or brushing. This coating is dried before the pattern is placed. The mold is made by placing the pattern - which is made complete with sprues, risers, runners, and gates - in a flask and packing sand around it. The foam pattern remains in the mold during pouring and melts and vaporizes immediately on contact with the molten metal. Thus, it is not necessary to remove the pattern from the sand prior to pouring. The molten metal takes the place of the foam plastic pattern and fills the space it occupied in the sand. The sand mold is in only one piece; there is no cope or drag, hence the name, "full mold casting". Because gas is produced from the vaporization of the foam pattern, the sand mixture must be highly permeable to allow these gases to escape. Hence, there is no bonding agent for the sand. Loose sand is flowed into the flask to surround the pattern. Then, the filled flask is vibrated to compact the sand. If the pattern is of such a shape that the sand can fill the openings in it, no separate core pieces are required. Therefore, there is much less need in the process for separate core pieces. On the other hand, since the pattern is consumed during the process, an additional pattern has to be made for each casting in the lot. If the quantities are large, the foam pattern is molded by the normal methods used for polystyrene foam (See 4C4b). If quantities are limited, or if the casting shape is such that the pattern cannot be molded in one piece, the pattern may be fabricated by gluing together machined or molded polystyrene foam blocks. Pouring of a lost foam casting must be

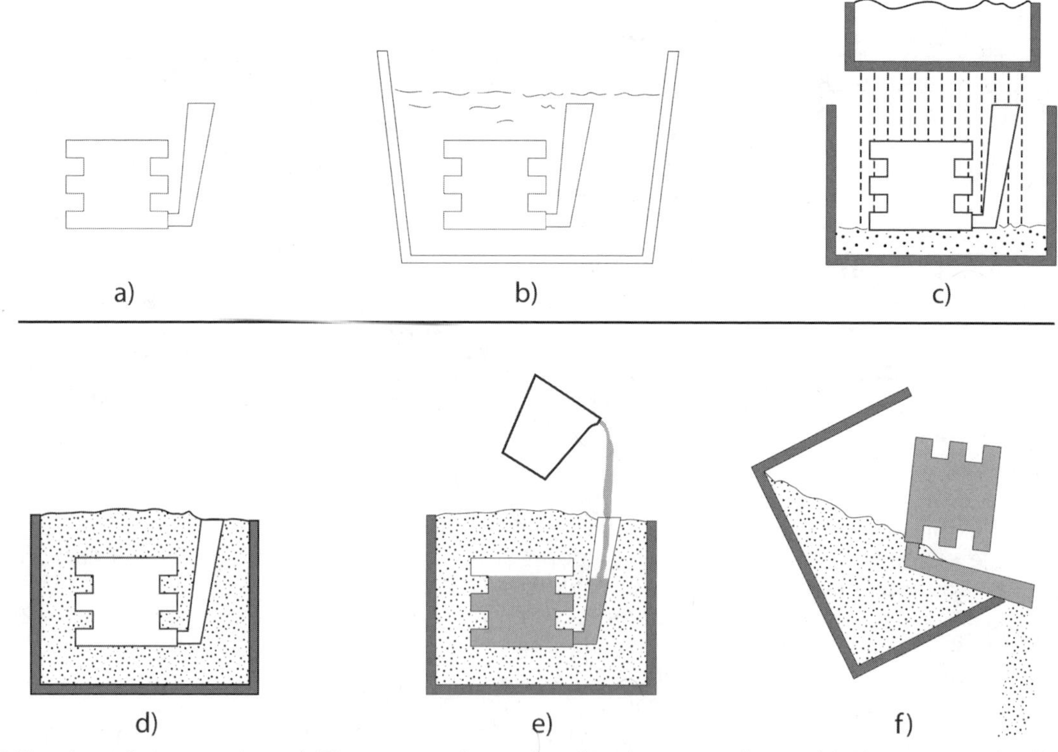

Fig. 1B3b Lost foam casting. a) The pattern is made of polystyrene foam. b) The pattern is dipped in a refractory slurry, c) The pattern is placed in a flask and surrounded with unbonded sand, d) The filled flask is vibrated to compact the sand. e) Molten metal is poured onto the pattern, vaporizing and replacing it. f) The solidified casting is removed from the flask; the sand is recycled.

controlled to maintain pressure in the unbonded mold so that it does not collapse. After pouring, the mold is allowed to cool and the metal to solidify. The flasks are then moved to a shakeout area where they are inverted to dump the loose sand. Castings are then cleaned by conventional methods. The lost foam process is suitable for complex shapes with undercuts and other irregularities. Complex castings can be made more easily than with other sand casting processes. Cleaning is also simpler than with other sand mold processes; there is no parting line, and the sand, being unbonded, is more easily recycled. The process has come into widespread use for automotive castings and is illustrated in Fig. 1B3b.

B3c. ***magnetic molding*** - is a variation of lost foam casting in that a polystyrene foam pattern is used and there is no cope or drag. However, instead of the sand used to back up the mold, iron powder with particles of 0.004 to 0.020 in (0.1 to 0.5 mm) in diameter is used. This iron is compacted by vibrating the flask and then is magnetized with a magnetic field to hold it in place during pouring. The magnetic field is turned off after the poured metal has cooled and solidified. The casting is then easily shaken out and the metal powder can be reprocessed. The heat conductivity of the iron powder results in a finer grain structure of the casting. The process has been used for casting steels, copper-based alloys and cast iron.

B3d. ***V-process casting (vacuum molding)*** - is another process that uses loose sand to back up a mold. In this case, the sand is contained between two plastic sheets, one of which is in the shape of the mold cavity. There are two such pairs of sheets filled with sand, one for each mold half. A vacuum is drawn between each pair of sheets to hold them against the sand. The outer plastic sheet, holding the sand, is flexible and essentially flat and does not require forming. The inner sheet is formed to cavity shape by conventional thermoforming processes as described in chapter 4, and this is the first step of the casting process.

Fig. 1B3d illustrates the operation sequence. Thermoforming and mold making take place at one location as a single operation sequence. The forming die for each inner plastic sheet stays in place beneath the sheet until a mold flask is put into position over it and is filled with sand. The mold flask has a vacuum connection. The equipment is vibrated to compact the sand, excess sand in the flask is removed, a pouring basin and runner are formed in the sand as needed, and the remaining sand is leveled and covered with the outer sheet of plastic film. A vacuum is applied to the sand so that atmospheric pressure, acting against the outer sheet and the formed sheet, holds the sand in position, producing a mold of high hardness. The flask is then lifted from the forming pattern while still under vacuum with the plastic sheets and is assembled with a similar flask for the other mold half. The mold halves are kept under vacuum during pouring and until the cast metal has solidified. The formed plastic sheets vaporize from contact with the poured molten metal. When the vacuum is released, the sand falls away and a clean casting remains. The process is used for production of medium-large castings at moderately high production levels.

B3e. ***cement-sand molding*** - simply involves the use of about 10 percent Portland cement (plus water) as a binder for the sand mold. This process variation is used for large parts/large molds where the improved strength of the cement in comparison with other binders is important. The sand-cement-water mixture is formed into mold halves immediately after mixing but full curing requires another 24 to 72 hours. The molds may be stored for extended periods. A disadvantage of the high strength of the cement-sand molds is that they are less apt to yield to shrinkage forces and some casting shapes and materials may be susceptible to tearing.

B3f. ***loam molding*** - is suited for large castings of circular shape. Patterns are not used; the mold cavity is made to approximate shape manually using a structure of bricks or wood to hold the sand. A slurry of coarse sand, with a high percentage of clay and water, is worked into the cavity over the structure to further form the approximate shape. Then, as shown in Fig. 1B3f, a profile board is swept from a central upright spindle through the rough cavity to scrape away excess sand and produce a cavity of the required shape. The mold is heated and thoroughly dried and a coating of refractory material is applied.

The cope half of the mold is made from a series of cores placed side by side around the central axis. Alternatively, if there is a flask of sufficient size

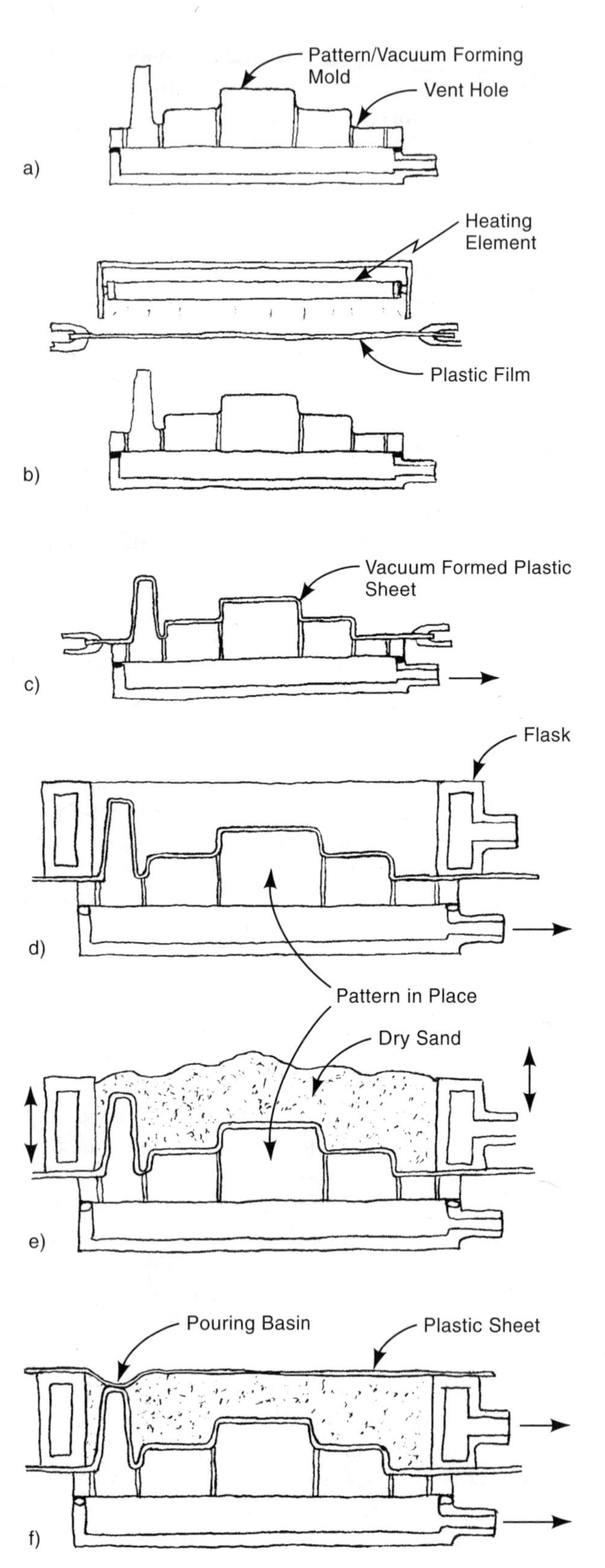

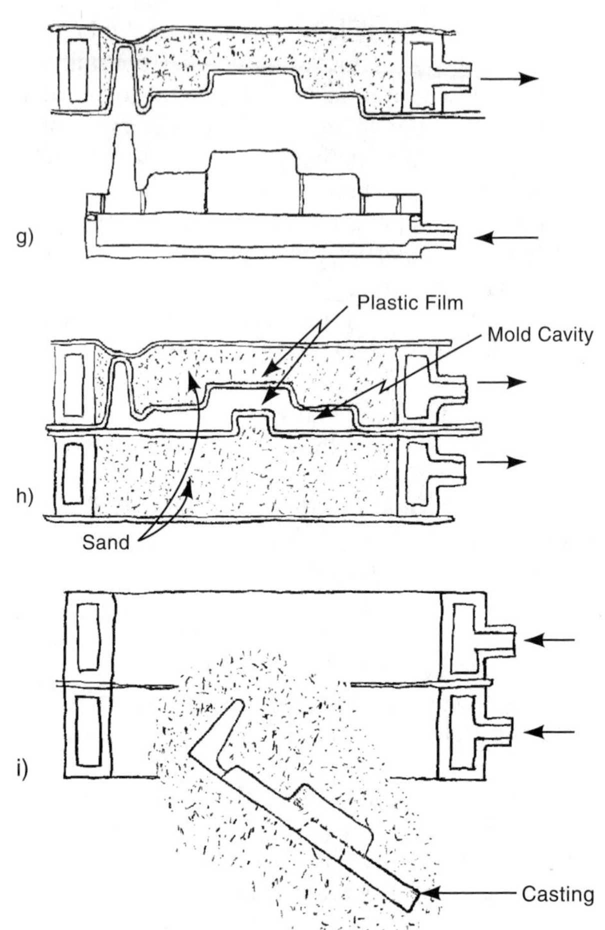

Fig. 1B3d The steps involved in V-process casting. a) The pattern, which also serves as a vacuum forming mold, is placed on a hollow carrier. b) A plastic sheet is put into position over the pattern and is heated to soften it. c) The plastic sheet is draped over the pattern and drawn against it by a vacuum in the carrier chamber drawn through vent holes in the pattern. d) A flask is set over the film-coated pattern. The flask has hollow walls and is connected to a vacuum pump. e) Dry sand is placed in the flask. Slight vibration is applied to compact the sand. f) A pouring basin is formed in the sand and another plastic sheet is placed over the mold. A vacuum is applied to the flask, causing the plastic sheets to be drawn tightly against the sand. g) The vacuum in the carrier is released and the flask, with vacuum still applied is lifted off the pattern. h) The cope and drag are assembled together with vacuum still applied to both flasks. The vacuum is retained during pouring. I) After the casting has solidified, the vacuums are released and the sand drops away from the flask and the casting.

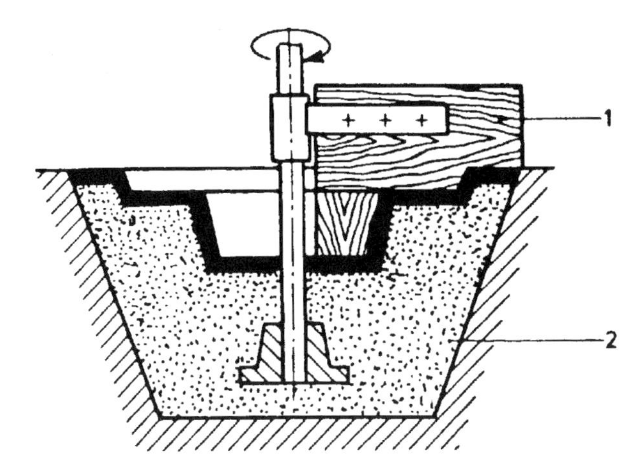

Fig. 1B3f Loam molding with a sweep, rotating about a central axis to form a round shape. 1) wooden sweep. 2) pit mold. *(from Davidson, Handbook of Precision Engineering, Vol. 10, McGraw-Hill, New York)*

and the necessary handling equipment is available, the cope can be made by methods similar to that for the drag, or by conventional methods and fitted in place over the drag. Skilled workers are required for satisfactory results. Equipment is not extensive: a bed plate, cover plate, spindle, and arbors for cores are needed. Straw, hay, cloth strips, or sawdust, may be used to reinforce the sand mixture. These materials burn off during casting. Large bells, cylinders, and rolls are cast with this method. Loam molding is currently in only limited use.

B3g. ***flaskless mold casting*** - is a high-production, automatic green-sand process. Casting rates of over five hundred pieces per hour are feasible. A typical mechanized sequence includes mold-making, core placement, pouring, cooling and shake-out, all automatic and all linked together by a conveyor in one production area. The flaskless process uses a four-sided mold chamber in which the molds are formed. The chamber is filled with a sand mixture from above by gravity and air pressure. The molds are formed and maintained with the parting line between mold halves in a vertical position. The pattern plates are supported vertically in the molding chamber with a plate for half of the mold cavity on each side of the chamber. The sand is compacted between the two plates by horizontal squeeze pressure. Each mold, then, has a half-cavity on each side, the cope half on one side and the drag half on the opposite. Cores, if needed, are inserted automatically.

One of the patterns swings out of the way and the molds are pushed together side-by-side, providing a complete cavity between them and, as the operation proceeds, forming a continuous line of molds. There are no flasks; the weight of the molds pushed together and the bonding of the sand is sufficient to resist any sideways thrusts. Sand is firmly compacted by air and hydraulic pressure to provide the holding strength needed without flasks. Pouring is automatic. The metal enters a basin and sprue, formed by the pattern, at the parting line of the two mold halves. As the operation proceeds and more molds are added to the side-by-side line of molds, the metal cools in those that have been filled, and the castings solidify. They are conveyed to a section where they are separated automatically from the sand and then to a cleaning area. The sand is recycled. Fig. 1B3g illustrates the mold making steps and how the molds are pushed together, side-by-side. Needless to say, a considerable investment is required for a complete flaskless molding operation. Gas stove grills, sewing machines, pipe fittings, and valve bodies are cast with this method.

B3h. ***Antioch process*** - In this process, gypsum plaster is used as a binder for the sand. Talc, sodium silicate, bentonite, Portland cement and terra alba may also be used to provide certain characteristics to the mixture. Water is added to the dry mixture and the resulting slurry is piped or poured around the pattern in the mold flask. A typical mixture is 50 percent sand, 40 percent gypsum, and 8 percent talc, plus water equal to half the weight of the dry mixture. After the material sets, it is air dried, heated with steam for 6 to 8 hours in an autoclave at 15 lbf/in^2 (103 kPa), air dried again for about 14 hours, and then oven heated to 450 to 475°F (230 to 250°C) for 12 to 20 hours. This sequence results in a mold permeability much greater than that achievable with regular plaster molds while retaining a smooth surfaces. Close tolerances can be maintained in the casting but the process is limited to non-ferrous metals. Alloys melting at 1900°F (1040°C) or less are suitable. Applications include the casting of aerospace parts and critical automatic transmission parts for automobiles.

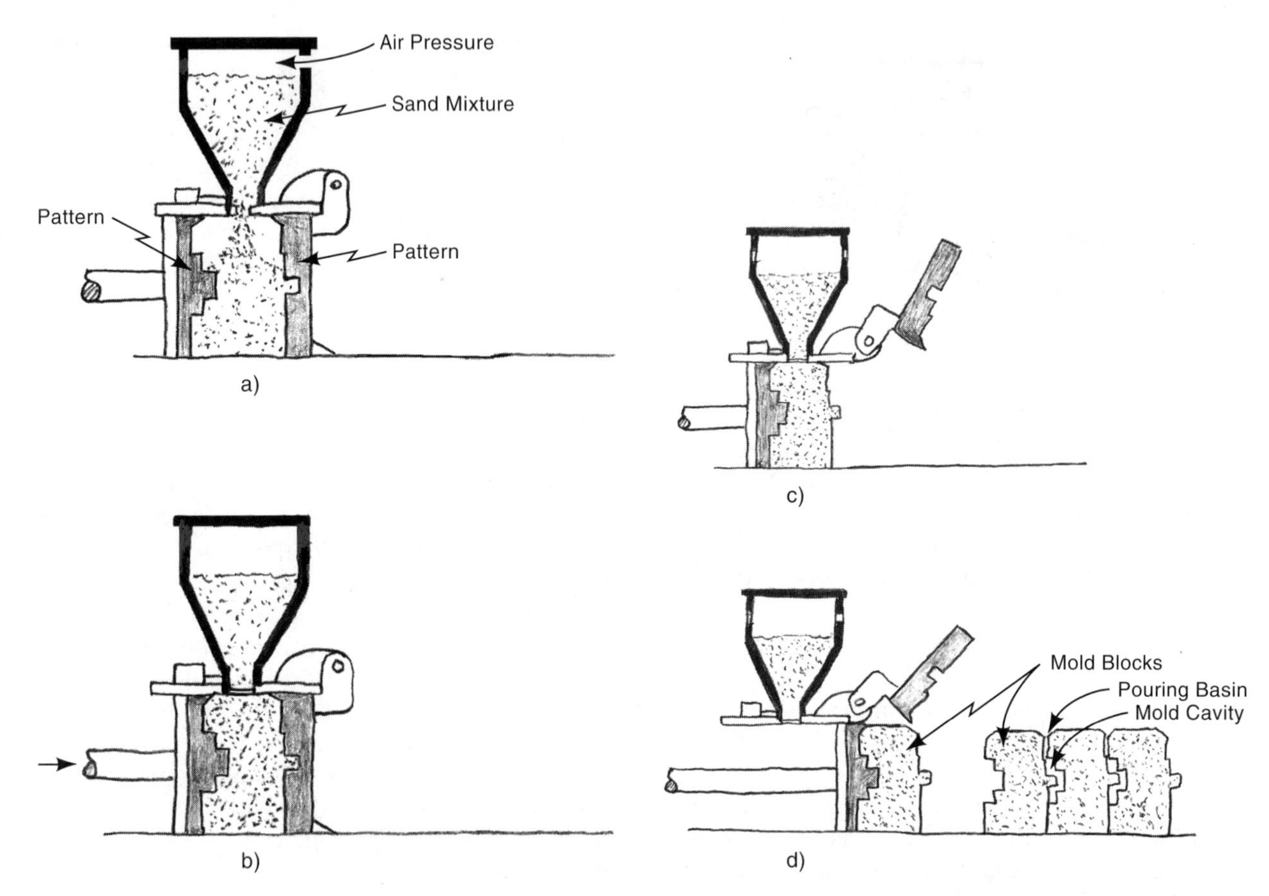

Fig. 1B3g Flaskless molding. a) Sand is blown between two pattern plates in a molding chamber. b) The sand mixture is squeezed to form a dense mold block which has half the mold cavity on each side. c) One of the patterns is withdrawn so that the newly formed mold block can be moved. d) The mold block is pushed tightly against previously made mold blocks. The pouring basin, sprue, and mold cavity lie at the vertical parting line between mold blocks.

B4. ***core making*** - Cores are used whenever there is an undercut, opening, or hollow area in the casting that cannot be made by the mold. Cores, sometimes called *dry-sand cores*, are separate pieces made to the shape of the opening, and are supported by the external mold. Fig.1B shows the placement of such a core in a sand mold. Cores are made from sand mixed with a binder to hold the sand grains together strongly enough to allow handling and placement of the core in the mold. One or more of a variety of binders may be used including linseed oil or other oils, phenolic resin, sodium silicate, bentonite, or other clay, and cereals. Core sand, with the binder, is compacted in a metal, wood, or plastic mold, usually called a "core box". After the core is removed from the core box, baking may be employed to set the binder. Some binders are activated by a reactive gas without additional heat. Another binder needs no external heat or reactive gas: the binder ingredients react and cure without external agents. Which process is selected depends on a number of factors: the size and shape of the core, the production quantity of the casting requiring the core, the metal being cast, and the mold-making process involved. Finished cores are placed in the mold before the two mold halves are placed together. *Core prints*, recesses in the mold, or *chaplets*, small metal pieces that will fuse with the molten metal and become part of the casting, are used to properly locate and hold the core in the mold. Sometimes, complex or large cores are made by gluing two or more simpler or smaller cores together. Typical cast parts requiring cores are: exhaust manifolds of internal combustion engines, and cylinder heads and engine blocks with cooling channels.

B4a. ***green sand cores*** - are not cores in the sense that a core is a separate piece inserted in the mold. They are appendages of the green sand mold that form openings or holes in the casting. When a center hole is needed in a casting for pulley, gear or other round part, the hole may be formed by a green sand element of the mold itself. The term, "green sand core" is sometimes used for such elements with "dry sand core" referring to those that are separate and made independently from the mold. Green sand cores are weaker than dry sand cores, require draft, and are suitable only for short, large, openings.

B4b. ***core blowing*** - is a pneumatic process for filling and compacting core sand in core boxes in production quantities. The premixed sand and binder are added to a chamber that is then sealed except for ports that are placed in line with inlet ports of the core boxes. The sand is "blown" with a sudden blast of high pressure air (40 to 100 lbf/in^2 (275 to 690 kPa) from the chamber and into a core box. The kinetic energy of the air/sand/binder mixture provides the force needed to pack the sand. The method used for hardening the binder depends on the chemical nature of the binder, but phenolic resin-coated sand is a common approach and the phenolic is cured with heat. All steps in the operation are automatic and production is rapid. Blown cores are used for applications requiring mass production of small or medium-sized cores.

B4c. ***core baking*** - The binder for the core sand is hardened by drying or polymerizing it in an oven of controlled temperature. Core-oil binders using linseed or vegetable oil are commonly oxidized and hardened by baking. Urea-formaldehyde and phenolic resins are also oven baked. These cores are formulated with the flour of corn or another cereal and/or clay and water, which coat the sand and provide sufficient green strength so that the cores can be handled. They are then removed from the core boxes before curing and placed in an oven. Ovens may be continuous, with a metal mesh conveyor, or stationary, the choice depending on the production level and size and complexity of the core. Baking is for one hour or longer at 400 to 500°F (200 to 260°C), the time depending on the thickness of the core section. When fully cured, the cores are removed from the oven and allowed to cool to room temperature.

B4d. ***oil-oxygen process*** - for core making. In this process, a combination of oils and additives is used with the sand. The oils polymerize when they contact oxygen-bearing activators (perborates, percarborates, permanganates, peroxides), and bond the sand particles together. Dry sand, additives, and activators are mixed together. The sand mixture flows easily into core boxes and does not have to be rammed. Depending on the amount and nature of the additives and activators, as well as the temperature and the presence or absence of metallic dryers, the additives begin to gel. When they are sufficiently hard, the cores can be removed from the core box, and are usually baked in ovens at 400 to 450°F (200 to 230°C) until polymerization is complete. The resulting cores are stable dimensionally, and sufficiently strong. After the casting has solidified, the core sand is easily removed. The process is used for a wide range of ferrous and nonferrous casting alloys. However, its principal use is in large castings made in small quantities. Large gears, steam turbine impeller wheels, and bridge construction components, are typical applications.

B4e. ***furan no-bake core process*** - Furans are a family of thermosetting resins that, when cured, act as binders for core sands. Curing takes place at ambient temperature and begins when two or more of the binder components are mixed together with the core sand. Curing is slow enough that the sand mix is flowable and workable for a period of time so that core boxes can be filled. After an additional period, curing has progressed enough so that the cores can be removed from the boxes. The curing then proceeds to completion. (Although baking is not strictly required, it may be used to speed curing and drying.) The time required for both preliminary and final curing can be varied from a few minutes to a number of hours, depending on the resin formulation. Both fural acid and phenolic acid resin systems can be used with this approach.

B4f. ***carbon-dioxide process***[1] - This process uses 3 to 6 percent sodium silicate as a binder for the core sand. Sand compaction in the core box is by blowing or other conventional methods. The compacted sand - mixed with sodium silicate - in a corebox is exposed to carbon dioxide gas for from 5 to 15 seconds. With small cores, the core box is double-walled or the pattern is hollow, to allow a closed conduit for the gas. With larger cores, lances are used to inject the gas into holes made in the core. In both methods, the gas permeates the core sand and causes the sodium silicate to gel and harden. Cores then are stored for 24 hours or more, while the hardening of the silicate continues, producing a core strength of about 100 to 200 lbf/in^2 (700 to 1400 kPa) so long as the atmosphere is not too humid. No baking is required; the operation can be performed at room temperature. [However, baking at 400°F (210°C) is sometimes done after the CO_2 treatment.] Surfaces of the cores are often coated with a graphite or zircon refractory coating using an alcohol wash. Most of the sand can be reclaimed for further use after pouring and shake-out. Iron, steel, copper, and aluminum alloys are cast with these cores for many applications.

B4g. ***shell process for cores*** - is essentially the same as the process used to make shell molds. See B3a above.

B5. ***sand mold methods***

B5a. ***ramming*** - is the packing of sand in a mold. The term usually refers to the hand operation performed with a hand tool, a *rammer*, which provides weight and a small, flat, packing surface to aid the operation. Pneumatic hand-operated rammers are also used. Hand ramming is labor-intensive, produces more variable results, and is slower than mechanized methods but is suitable when production quantities are too small to justify mechanized methods. Ramming is also useful to supplement sand slinging in large floor and pit molding operations.

B5b. ***bench molding*** - For small castings, the mold making and pouring may be done on a workbench where the work surface height is more convenient.

B5c. ***floor molding*** - When larger castings are to be made, it is common for the mold making and casting operations to take place on the factory floor. Sand compaction may be done by hand ramming, sand slinging, or large machines of other types. Sand is applied in layers and each layer is carefully knitted to the adjacent layers. The use of very hard sand compaction is necessary because of the weight of the casting and the amount of molten metal poured, especially if large flasks are not available to contain the sand. (When flasks are used they are normally so large that they require the use of an overhead crane. Otherwise, the cope may be made up of a series of side-by-side cores.) Dry-sand molding is often used. Cement-bonded sand or loam molding may also be employed.

B5d. ***pit molding*** - is used when the castings are too large to fit in flasks that would hold the cope and drag. Instead, a pit in the foundry floor takes the place of the lower half of the mold. The pit is filled with the sand to be used. An upright pattern is lowered and pressed into the sand, and additional sand is rammed tightly around it. Sets of cores, arranged side by side, are used to provide the cope half of the mold. These large castings are cooled slowly, sometimes over several days, before the mold is opened, in order to minimize internal stresses.

B5e. ***machine methods*** - In production situations, several machine methods are available to supplement or automate the sand mold-making operations, which would otherwise be manual. Mechanized methods typically provide more uniform sand densities than manual ramming. Often, the machines not only compact the sand, they invert the mold and remove the pattern from the sand. Mechanized methods include jolting, squeezing, jolting-squeezing, sand slinging, and flaskless casting.

B5e1. ***jolt-squeeze methods*** - use a machine that packs sand around a pattern by raising the pattern plate and flask with sand for a few inches and then allowing them to fall to an abrupt stop. The inertia of the falling sand packs it around and against the pattern. The operation is repeated until the desired density of compaction of the sand is achieved, but some hand or pneumatic ramming may also take place. Then, the machine uses pneumatic or hydraulic pressure to lift the table and force the packed sand in the flask against a flat plate that is fixed in position. The pressure further densifies the sand packing in the flask. In rollover machines that process both mold halves using a match-plate pattern (See B5e4 below), the double flask is inverted, sand is added, and the operation is repeated for the other mold half. The flasks are separated (The cope is lifted.), the pattern is removed, and the cope is lowered onto the drag to complete the mold. Fig. 1B5e1 shows some of the key steps of the sequence. The degree of mechanization and automation of the operations, other than jolting and squeezing, varies with the installation. High production volumes and large sizes of the molds make mechanization of lifting and inverting and other steps more economically justifiable. After the completed mold is assembled, it is normally placed on a conveyor that transports it to the foundry pouring area. Jolt-squeeze sand compaction works best when molds are not too deep and the casting shape is somewhat shallow and horizontal. Both *jolting* and *squeezing* may be performed as single operations on either jolt or squeeze machines, depending on the casting involved. Squeeze machines are limited to molds only a few inches thick. Some squeeze machines use a rubber diaphragm and air pressure to squeeze the sand.

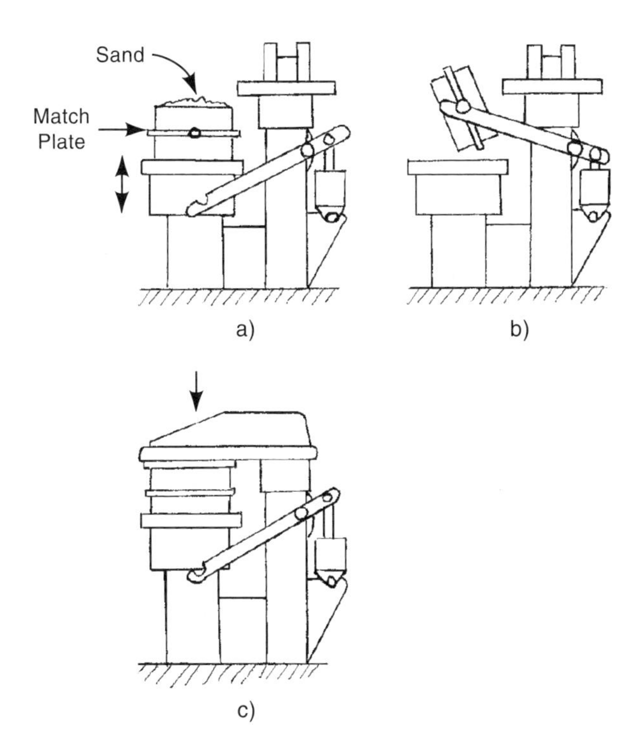

Fig. 1B5e1 A typical jolt-sqeeze-rollover machine with a match plate pattern. a) The sand in one mold half is jolted by rapid up and down motions of the machine table to compact it around the pattern. b) Rollover - The mold is inverted after the drag is completed so that the cope half can be filled. c) The head of the machine indexes over the mold and the squeezing cycle takes place to fully compact the sand.

B5e2. ***sand slinging*** - compacts the mold sand by slinging or throwing it at high velocity against the pattern in the flask. Centrifugal force from a rotating impeller provides the high velocity. The operation is fast and can produce uniform compaction. However, when sand slinging is used in jobbing foundries, considerable skill on the part of the machine operator may be necessary to insure consistent results and uniform mold density. Supplemental hand ramming may sometimes be employed but the process provides very dense and hard molds. For larger molds, the slinger machines are portable and are brought to the mold location.

B5e3. ***rap-jolt machines*** - are similar to jolt squeeze machines but, instead of jolting the pattern plate and flask, a weight strikes the underside with a controlled force at rapid intervals. The pattern plate and flask do not move up and down but the rapping causes the sand to densify. Squeezing is part of this method also and it may take place simultaneously with the rapping. The method provides dense, hard molds.

B5e4. ***match-plate molding*** - has patterns for the cope and the drag mounted on opposites sides of a single metal plate. Fig. 1B5e4 shows a typical match plate pattern. When these patterns are used, both the cope and drag halves of the mold are made in one operation, one on each side of the plate. Gravity fill, followed by a pressure squeeze, is one method used to supply and compact the sand. Sand fills the flask on one side of the plate, a plate closes the flask, the flask is rotated, and the other side is filled by gravity. Then the two sides are compacted by the same squeezing operation. Another approach is to blow the sand in

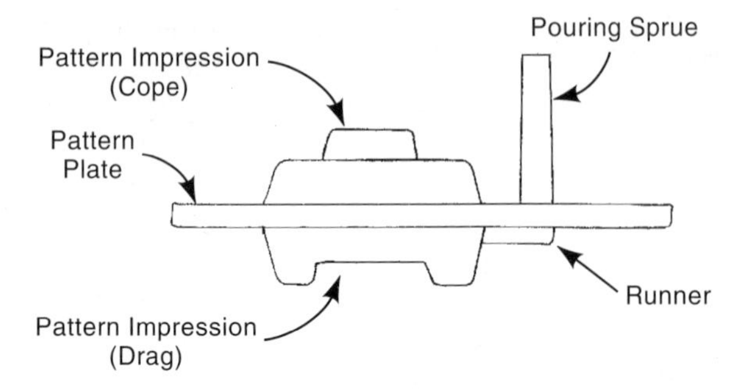

Fig. 1B5e4 A typical match plate pattern. Note that the two half patterns, for the cope and drag cavities, are placed on opposite sides of the plate. Sprues, runners and risers are also included in the pattern.

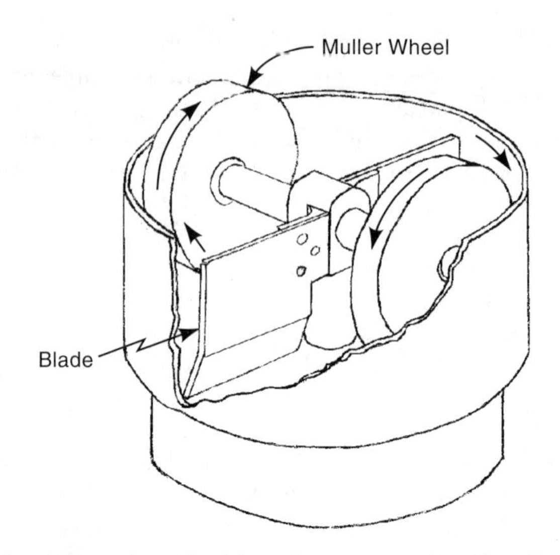

Fig. 1B6a A typical batch-type sand muller. Both the muller wheels and blades move and mix the sand, binder, and other ingredients.

on both sides of the pattern plate at the same time. Then the two halves are squeezed simultaneously. After squeezing, the mold halves are separated and the pattern is withdrawn, a core is placed, if applicable, and the cope and drag are assembled together in preparation for pouring. (Also see jolt-squeeze mold making above.) Match plate molding is suitable for high production situations. Gates and runners are normally included as part of the pattern.

B6. *sand processing*

B6a. ***sand mulling*** - or mixing is performed by a machine, a sand muller. Sand, and for green-sand molding, water, clay, carbonaceous material, and other additives are loaded into the machine. The muller has two rollers (muller wheels) at the ends of a horizontal arm that rotates. The rollers with two plow-like blades, break up agglomerates of sand and binder and distribute binder, water, and additives throughout the mixture. The machine operates at a low speed. Fig. 1B6a illustrates a typical batch type sand muller. There are also continuous mullers in which ingredients are fed into the equipment at one end and exit at the other end after being blended. One such continuous muller uses two connected side-by-side mullers similar to the one illustrated. When sand is processed after it has been used in casting, ingredients are replenished as needed and are blended. (Casting sand is re-used but is normally blended with a certain amount of fresh material.) After mixing, the sand is typically conveyed to an aerator that separates the sand grains and improves flowability and thence conveys it to hoppers serving the molding operations. Other types of equipment are available for sand mixing, some of which can operate with shorter mixing times than the conventional unit pictured in Fig. 1B6a.

B6b. ***reclamation of sand*** - is increasingly important in light of the increasing costs involved in the disposal of used sand. Reclamation is a feasible alternative, but is more difficult if several types of binder are mixed. Three reclamation methods are in current use: *mechanical reclamation, thermal reclamation* and *wet reclamation.* After these methods are used, additional sand, binder, and other additives are provided to insure that the sand mixture, including reclaimed material, meets specifications.

B6b1. ***mechanical reclamation*** - uses the sand muller described above to break up lumps so that only particles the size of sand grains remain. However, the process also involves several other steps such as the separation and removal of metal using magnetic or screening methods, and scrubbing, which hurls the sand against a target plate, either with high volume air flow ("pneumatic scrubbing") or centrifugal force ("mechanical scrubbing"). The sand grains impact against each other and against the target plate. This tends to remove the residue of bonding resin from the grains. A dust collection system removes the resin husks and various fine particles from the sand. If necessary, the sand is kept in the system for repeated action until the desired degree of removal of unwanted material has taken place. The processed sand is cooled, if necessary, and is screened or air classified to remove any further foreign or undersized material. Sometimes, both mechanical and pneumatic scrubbing is performed in sequence on the same lot of sand. Make-up sand is added as needed and is blended with the reclaimed sand. Typically, the finished product will be 80 percent reclaim and 20 percent new sand.[1]

B6b2. ***thermal reclamation*** - uses similar steps but includes heat processing. The sand is heated in a rotary kiln to a temperature high enough to ignite and burn off organic resins used in the shell molding process - about 1470°F (800°C). The burning of the resins contributes to the heat required for the operation. One method utilizes a refractory-lined rotary drum with an elevated feed end. Burners inside the drum at both ends are directed at the cascading sand flowing through the drum or, in other designs, external burners provide sufficient heat to ignite the resins. The drums are relatively inexpensive but have high heating costs and some weaknesses in control of material flow and air for combustion.

Fluidized bed machines are also used for thermal reclamation. Such machines, have a combustion chamber with provision for flow of air and hot gas from the bottom sufficient to fluidize the sand in the chamber. Burners are located below the chamber in the flow of incoming air and also in the chamber with the sand. Sand is introduced through a higher port of the chamber and withdrawn from a lower port. All resins and other hydrocarbons are burned off with this system. A second combustion zone above the chamber burns off any waste gases. This approach requires a larger investment than a rotary kiln but provides better energy efficiency and better control of the operation.

Clay-bonded sands can also be processed thermally. After the sand is crushed and metal residues are removed, it is fed into a calciner, which heats the mixture to a temperature sufficient to calcinate the clay but not high enough to cause the clay to fuse to the sand particles. Temperature control is critical. Following this step, the sand is cooled and processed with the scrubbing steps described above.

B6b3. ***wet reclamation*** - is used for sand mixtures bonded with silicates. A water wash replaces the heating cycle of thermal reclamation. It removes the silicate residues. The first step, however, is the breaking and crushing of lumps in the mixture. Then the water wash takes place, followed by sand and liquid separation. The sand is dewatered and dried. The liquid is treated to agglomerate the residues and allow separation by settlement. The water is then treated to permit safe disposal. When properly performed, all the sand processed can be re-used. The only new sand needed is to make up any that is lost.

B6c. ***metal separation from sand*** - It is necessary to remove smaller pieces of metal from the sand. These include sprues, runners, etc. that are separated from the casting in the shakeout operation. With iron and steel castings, these tramp metal pieces can be removed magnetically. A common approach is to use a belt conveyor for the sand and to place magnets both above and at the end of the conveyor. (See 11C8d.) With non-magnetic metals, screening is the most common method, using multiple screens of successively finer mesh. Other machines, whose method is based on the density difference between the metal and the sand, are also available.

B6d. ***cooling of sand*** - is necessary after use because hot sand causes moisture and other problems. Bentonite clay does not function as a binder if the temperature is above about 115°F (45°C). Sand does not conduct heat well and a mass of sand in storage will remain its heat for a long period. Cooling is accomplished by spraying the green sand with water and blowing air through the sand. Evaporation of 1 percent water content cools the sand by 45°F (25°C).[3] Water is sprayed on the sand, and air is blown through it in either a rotary drum or through a screen that retains the sand. In the latter method, the sand becomes a fluidized bed. With both the drum and fluidized bed methods, sand is fed from one side of the device and exits at the other, thus providing a continuous operation. Both air flow and moisture addition must be carefully controlled so that the sand mixture is not made too wet and that the amount of desirable fine materials carried away with the cooling air is not excessive. In some arrangements, many of the fine particles carried away by the cooling air are returned to the sand mixture.

B7. ***pattern making*** - Patterns are made to produce the proper sizes and shapes of cavities in the sand molds. Except for some very simple shapes, usually those that are circular, a pattern is necessary if the mold cavity is to have the correct shape and dimensions. Patterns are typically made from wood, plaster, plastics, various metals, and, for lost foam castings, from polystyrene foam. For investment castings, patterns are made from wax or plastics. Aluminum, brass, and cast iron are common metal pattern materials. The greater the production required, the harder and more wear-resistant the pattern material should be. Pattern-making may be costly because the process is lengthy and largely manual. Patterns must be slightly larger than the casting they will produce because of the shrinkage of metals as they change from the liquid to solid state and cool to room temperature. The initial operation for many pattern materials is contour milling. The pattern is typically rough contour milled with the machine under either hand, template, or computer control. It is then finished by hand with filing, sanding, and polishing operations.

Metal patterns are often produced by first making a wooden pattern and then casting a metal replica. Wood patterns are usually varnished to provide a moisture seal and smoother surface. Some of the available rapid prototyping methods are useful for making patterns suitable for various casting methods, chiefly for investment casting of smaller quantities. (See Chapter 14.) The laminated object method (LOM)(14A5a) is used to make wood-like casting patterns.

B8. ***post-molding operations***

B8a. ***shakeout*** - is the operation that separates castings from the sand mold. It is the first operation that takes place after castings solidify and cool in the mold. The basic method employed in the operation is to subject the mold and casting to a strong vibrating motion, which causes the sand mold to break up and fall from the casting. The sand falls through a grating that supports the casting but allows the sand to fall to a collection bin or conveyor below. Fine dust is collected by a cyclone-type dust collector. There are four prime methods by which shakeout is carried out: 1) *vibrating conveyor* - The molds containing castings are placed in a pan or trough that moves the casting with a vibratory motion as the sand falls away. 2) *shakeout table* - or deck, which processes the castings on a batch basis. This approach is used for lower quantity production. The frequency of vibration is lower, but with greater amplitude for larger, sturdier, castings and at a higher frequency and lesser amplitude for thin-walled or otherwise weaker castings. 3) *rotary shakeout* - This uses equipment that tumbles the molds and castings in a rotating cylinder. Castings are fed into the cylinder at one end and exit at the other. Sand also normally exits at or near the end but sand discharge openings can be at other points along the cylinder. Light castings with thin walls may not be suitable for this method. 4) *vibrating drum* - a large drum with a bottom grating allows the sand to fall as the drum vibrates. The drum does not rotate but the vibration imparts a rotary motion to the mass of sand and castings in the drum.

These devices all provide some degree of cooling to the sand and castings in the drum. Rotary shakeout machines may be equipped for water addition to provide increased cooling. Shakeout machines remove almost all the mold sand from the castings. Abrasive blast cleaning, which follows shakeout, removes whatever sand remains on the casting surfaces after the shakeout operation.

B8b. ***core knockout*** - Aluminum and other non-ferrous castings are poured at a lower temperature than iron or steel castings and the metal temperature may not be sufficient to burn out organic binders in molds and cores, making them less easily removed from the castings in shake out. This is particularly true with cores for hollow sections, which are protected by metal sections of the casting from shake out and tumbling forces. It is more common, in non-ferrous castings, to have a separate operation to remove core sand. One method involves the use of a manually operated chisel, which vibrates pneumatically. The impact of the vibrating chisel blade breaks up the core sand, which falls away. Another method uses a chisel or other tool that is mounted on a frame and pushes the core sand from the casting, or breaks it up. The casting is handled manually but the machine holds, advances, and retracts the vibrating tool. Other machines, sometimes designed or arranged for use on a specific casting, provide vibration to either tools or the whole casting to break up and remove the core sand. Holding fixtures are used to position the casting for maximum effect. Some machines use high-frequency mechanical vibration rather than pneumatic oscillation to create the forces needed. High pressure water jets, shot blasting, and vibrating media are used in other equipment. These vibration methods tend to generate high noise levels and some dedicated machines include acoustic enclosures.

B8c. ***blast cleaning*** - is a fast method for removing residual sand and scale from a casting after shake out. It also improves the surface finish of the casting. The blast medium is either sand, metal shot, grit, or glass beads, which are propelled against the casting by air, water nozzles, or mechanical means. Centrifugal wheels provide the most common propulsion method. See Fig. 1B8c-1. The shot, grit, or sand is fed to a rapidly rotating wheel, that has vanes to pick up, accelerate, and expel the media. The wheels can deliver larger quantities of media than nozzles, but nozzles can deliver the media more selectively to recesses and other areas needing special attention. The blasting operations produce dust, which must be contained except when water is used as the carrier for the media. The castings may be tumbled in the blast chamber to expose all surfaces, may be on a turntable, or may be carried on an overhead hook-type or other conveyor. Smaller quantity production can be processed in batch-type tumble blasting machines, as shown in Fig. 1B8c-2.

High production equipment utilizes a belt conveyor with means to tumble the castings as they pass under the blast nozzles. High production equipment may also have multiple blast sources arranged so that they reach all surfaces of the workpiece. Robots are now used to handle and position castings so that the blast medium strikes all necessary surfaces. For aluminum castings, the blast cleaning process can provide a more uniform surface finish, an improvement in appearance, and closure of surface porosity in addition to the cleaning effect.

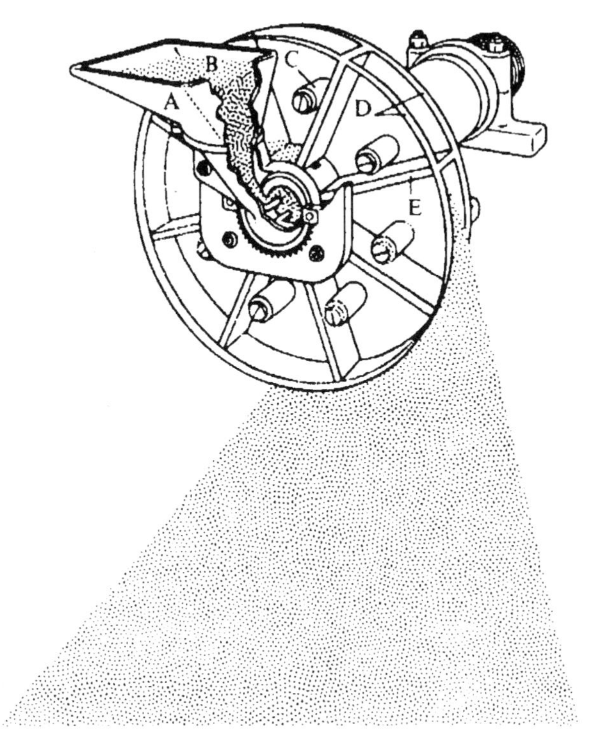

Fig. 1B8c-1 The wheel mechanism that uses centrifugal force to propel the grit, metal shot, glass beads, or sand in a blast cleaning machine. *(Courtesy Wheelabrator. Wheelabrator is a registered trademark of Wheelabrator Technologies, Inc.)*

B8d. ***tumble cleaning*** - Castings small enough to be placed as a group or lot in a rotating tumbling barrel can be cleaned of sand residues, scale, and fins by tumbling. Star shaped pieces of iron are included in the barrel with the castings and they abrade and burnish the castings as the barrel rotates, removing the sand and smoothing the surfaces. Sometimes water and a caustic are also included to hold down dust. Brass and bronze castings are tumbled with steel or iron balls, sand, or pumice, with water and detergents to improve their surface finish[2]. Tumbling can also be combined with blast cleaning as noted above for smaller castings. The tumbling action exposes all sides of the casting to the blast.

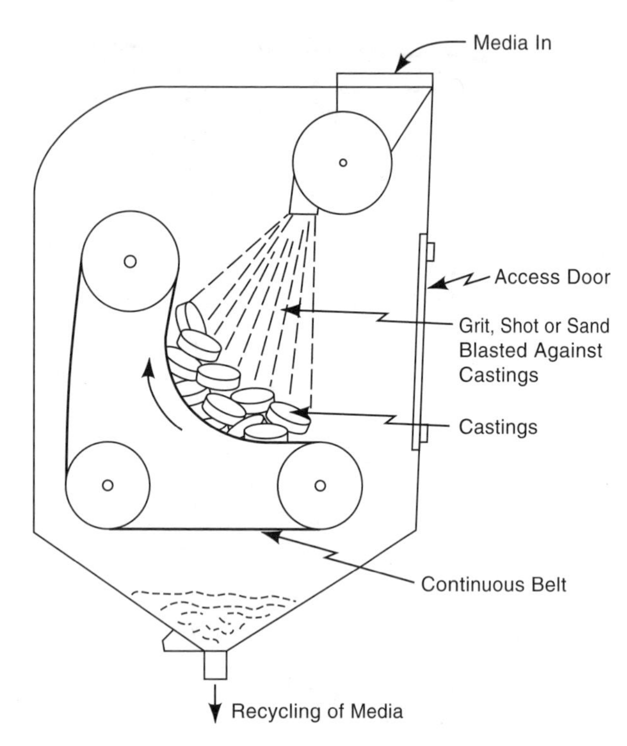

Fig. 1B8c-2 A tumble blasting machine for cleaning castings. The continuous conveyor causes the castings to tumble while being subjected to the blast of media, which removes sand and scale from the surfaces of the casting.

B8e. ***wire brush cleaning*** - is often used on aluminum, brass, and bronze alloy castings. It can produce a shiny surface on the casting. Various types and hardnesses of brushes can be selected, depending on the cast material, its shape, and the results desired. Power-driven rotating brushes are usually used. The operation may be performed robotically.

B8f. ***gate, riser and fin removal*** - This operation is sometimes referred to as *fettling*. It can be accomplished with any of a variety of metal cutting devices such as band saws, shears, abrasive cut-off wheels, flame cutting torches, and pneumatic hammers. Flame cutting is more common on steel castings if the gates and risers to be removed are large. Powder-assisted flame cutting (See 3H1 and 3H2) is sometimes used on alloys that are oxidation-resistant. The operation may be performed by robots.

B8g. ***snagging*** - utilizes portable and powered grinding tools, stationary stand-grinders, or swing-frame grinders, to remove excess metal from castings manually. The operation provides rough grinding only; further machining is required to produce accurate surfaces. Fins and flashing are often removed by snagging and the operation is also applicable to forgings. Hand chipping tools, including pneumatic hammers, are also used.

C. Other Expendable Mold Processes

C1. ***ceramic mold casting*** - The ceramic mold process has similarities with investment casting, sand mold casting, and plaster mold casting. It is similar to plaster mold and sand mold casting in that two-piece, cope and drag molds are used. The process is similar to investment casting in that it uses a ceramic mold material rather than sand or plaster, so that steel and alloys with high melting temperatures can be cast (whereas plaster molds are limited to non-ferrous materials having lower melting temperatures.). The pattern is not expended as in investment casting; normally a precision made, re-usable wood or metal pattern is used. The first step in the process is to pour a thick slurry of ceramic mold material around the pattern, which incorporates gates and risers and is mounted on a match plate. Fine-grain zircon and calcined, high-alumina mullite are commonly used refractory mold materials. The pattern is removed after the mold material gels but before it sets completely. This prevents bonding of the mold to the pattern. The other mold half is made similarly, and when both halves have set, they are assembled. The assembled mold is fired at approximately 1800°F (980°C) and is filled with molten metal while still hot. Castings can be considerably larger than typical investment castings but have the advantages of a fine surface finish, sharp detail, and high dimensional accuracy, except that dimensions across the parting line require somewhat greater tolerances. A disadvantage of the process is the high cost of the ceramic material, which is expended in the process. Ceramic molds are used for dies and die parts, cast from tool steel, to eliminate much of the machining that otherwise would be required of these difficult-to-machine alloys. Die casting and forging dies, thread rolling dies, injection molds for plastics, stamping dies, and cutting tools are cast in ceramic molds. Parts cast by this method include components for aircraft, chemical processing, and food equipment, using stainless steel, and many parts for marine and architectural applications made from copper alloys. Cast iron, ductile iron, aluminum, nickel, cobalt alloys, and titanium are also cast in ceramic molds.

C2. ***ceramic-shell process*** - To reduce the cost of the mold, with its loss of ceramic material, the ceramic mold process can be modified to make only a facing layer of ceramic around the pattern, with the balance of the mold made up of less expensive fireclay. This approach is used when the castings are large and a conventional ceramic mold would be too expensive. The Shaw and Unicast processes, described below, are two of several methods that can be used to make ceramic-shell molds.

C3. ***Shaw process*** - is a ceramic mold process that uses two different types of molds: 1), an all-ceramic mold with a process quite similar to that described above and used primarily for small castings and 2), a mold consisting of a ceramic facing 3/32 to 3/8 in (2.3 to 9.3 mm) thick, backed up by a larger amount of inexpensive fireclay. The economics of the use of ceramic material limits the size of castings in the first process; there are no size limitations to the second process. Both process variations produce castings with fine detail, smooth surface finishes, fine grain, and a high level of soundness and accuracy. Both approaches use a variety of mixes of refractory powders and a liquid carrier of hydrolyzed ethyl silicate in a proprietary formulation. These materials are mixed with a gelling agent to produce a slurry that can be poured over the pattern to produce a cope or drag.

In the all-ceramic approach, the slurry of refractory material with hydrolyzed ethyl silicate is poured over the mold pattern, which can be wood, metal or plaster. The mold is stripped from the pattern when the material has gelled but before it is fully set, as in regular ceramic molding. The mold, at this point, is quite rubbery, so some undercuts and back drafts are feasible. The volatile materials in the mold are burned off by torch heating before the mold is baked in a furnace for four to five hours. The burning operation produces microcrazing in the cavity surface, which allows the gases and air to escape without being large enough to allow the molten metal to penetrate. The resulting mold is strong enough to withstand the expansion forces from hot molten metal while still being sufficiently porous. The molds may be preheated before pouring. They can be filled gradually so that turbulence is avoided, improving the structure of the casting.

The composite mold process uses four patterns if the cavity extends into both the cope and the drag. Two of these patterns are oversized and are used to form the backup material for the cope and drag. The other two are sized for the casting dimensions and are used to produce the ceramic facing. The backup is formed by ramming or vibrating the fireclay mixture over the backup pattern. (In some arrangements, only one pair of patterns is needed; a felt or flexible plastic sheet is placed over the facing pattern, equal in thickness to the ceramic facing, so that the fireclay backup can be formed.) Both the facing and backup patterns include locating members so that both halves of both mold portions can be assembled in correct alignment. The backup mold portions are hardened by placing them in bell jars and diffusing carbon dioxide gas through them for about 20 seconds. The facing mold portion is formed by pouring the ceramic slurry through a pouring channel in the backup and over the casting pattern, in the gap between the pattern and the backup. Fig. 1C3 illustrates the arrangement. Gravity pouring of the slurry is used but a vacuum chamber can be employed to remove trapped air if the casting details are critical. When the material gels after 2 or 3 minutes, the mold is stripped from the pattern. As with the all-ceramic molds, the flexibility of the mold at that stage allows it to be stripped from the pattern even if there are mild undercuts. The composite mold is then subjected to torch heat, which ignites and burns off the volatile materials. The mold halves are then assembled.

C4. ***Unicast process***[1] - is similar to the Shaw process. It differs primarily in the treatment of the ceramic material (in both the all-ceramic and ceramic-facing alternatives) after it has gelled but before setting is complete. In the Shaw process, alcohol in the ceramic mold material is burned off before microscopic cracks in the mold surface, which are desirable if kept small, can become

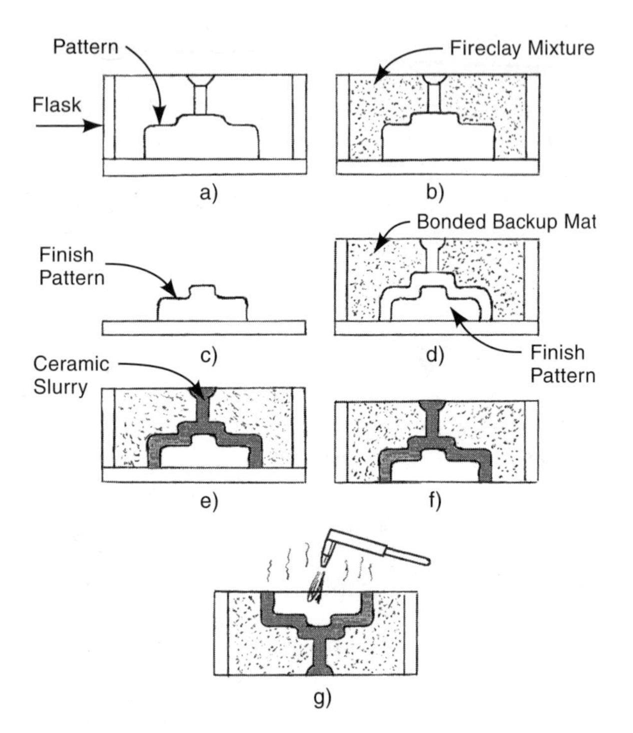

Fig. 1C3 The Shaw process for making composite molds. Two patterns are used, one to form the finish cavity, the other to form a somewhat larger cavity. The ceramic slurry poured between the larger cavity and the finish mold forms a ceramic shell suitable for accurate castings of many materials. a) The pattern for the backup material is placed in a flask. b) A fireclay mixture is introduced to the flask and is compacted by ramming or vibration. Containing sodium silicate, the mixture is bonded by gassing with carbon dioxide. c) The finish pattern on a pattern plate. d) The finish pattern and the bonded backup material are assembled in the same flask. e) Ceramic material in slurry form is poured into the opening between the backup material and the finish pattern. f) The finish pattern is removed. g) The mold half is inverted and heated by torch to burn off volatiles. Two mold halves, thus formed, are assembled together and are ready for pouring.

excessive or too large. In the Unicast process, the alcohol is dissolved in a liquid or vapor solvent, usually by spraying the solvent on the mold for 15 or 20 minutes. Application of the solvent limits the crazing and facilitates the hardening and stabilization of the ceramic material. Following this step, the hardening is completed by heating the mold in an oven at 1800°F (980°C). The cured mold is then ready for pouring. Another difference between the Unicast and the Shaw processes is the means of making composite molds. With the Unicast process, the ceramic facing slurry is applied to the pattern first. It gels almost immediately and a slurry of backing material is poured into the flask until it is full. Thus, only a single set of patterns is required. Very fine refractory material is used in the facing mixture and extremely sharp detail can be produced in the casting. The mold is also well vented with natural porosity permitting the casting of thin sections. Iron, steel, and a variety of non-ferrous alloys are cast with the process.

C5. ***plaster mold casting*** - This process is similar to sand-mold casting except that the molds and cores are made of a type of plaster of Paris rather than packed sand. The mold is made in a metal or wooden frame that contains the pattern. A liquid mixture of about 60 percent water and 40 percent metal-casting plaster is poured over the pattern. (Metal-casting plaster includes 20 to 30 percent talc and some other ingredients to speed setting.) The mold is vibrated lightly to ensure complete mold filling. After the plaster achieves an initial set, the frame and the pattern are removed, sometimes with a vacuum assist. Cores, if any, are added and the two mold halves are assembled together. The mold is then baked at a temperature ranging from 350 to 1600 °F (175 to 870 °C) to remove moisture, including the chemically-combined water, and to improve its permeability. Pouring of molten metal takes place with the mold hot and subject to a vacuum, to further remove the water of hydration. Fig. 1C5 illustrates a typical plaster mold before it is filled.

The process is used for non-ferrous metals with a melting temperatures of 2000°F (1100°C) or below. Aluminum, magnesium, zinc, and copper alloys are plaster cast. Cast parts can have an excellent surface finish, fine detail, and good dimensional accuracy. Intricate parts can be cast. However, production rates are lower and costs are higher than with sand mold processes. Typical parts include valves, pistons, cylinder heads, gears, cams, handles, pump parts, rubber tire molds, and plumbing fittings.

C5a. ***foamed plaster mold casting***[1] - is conventional plaster mold casting with a foaming

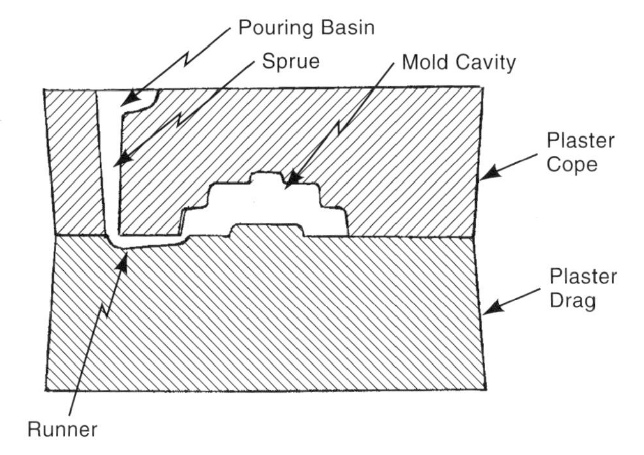

Fig. 1C5 A typical plaster mold before it is filled.

agent added to the plaster mix. The foaming, after the mold is completed, results in a permeable mold since the air cells tend to connect during the mold drying operation, providing an escape path for any gases generated during casting. The process differs from conventional plaster mold casting in that the foaming agent is added to either the dry plaster mix or to the slurry and there is intense mixing of the slurry. Proper mixing insures that the air cells are fine, no larger than about 0.01 in (0.25 mm) in diameter. Small air cells provide a mold surface that is smooth and strong enough to withstand the casting pressure. Otherwise the casting surface smoothness would suffer. The ideal amount of foaming produces 50 to 100 percent of volume increase. Drying of the foamed plaster molds before casting is done at the lower range of temperatures normally used for plaster molds because the insulating properties of the foam can result in cracking if the outer portions of the mold are overheated. Aluminum-magnesium alloys are most suitable for the process, but all-aluminum casting alloys can also be cast.

D. Permanent Mold Processes

D1. ***permanent mold casting*** - Permanent mold casting involves the use of reusable metal molds instead of the single-use molds of sand, ceramic, or plaster. The mold halves are usually hinged or mechanically guided together to permit quicker mold assembly. Filling with molten metal is by gravity. The process, then, falls between sand-mold and die casting. Molds are usually made of cast iron and are given a coating up to 0.12–0.03 in (0.30–0.75 mm) of refractory, suspended in liquid to prevent overly-fast cooling of the casting and to protect the mold surface. Powdered graphite is also sprinkled on the mold surface every few shots to facilitate release of the casting from the mold. The process is most practical for metals that melt at lower temperatures. Aluminum, magnesium, and copper-based alloys are the ones most frequently cast. Steel and iron can be cast in metal molds of suitable high-temperature alloys but this method is less common. Graphite molds are sometimes used, but they do not have a long life. Sand or metal cores may be used, depending on the shape required. (When sand cores are used with metal permanent molds, the process is known as *semi-permanent mold casting.*)

Molds are usually preheated before pouring and water cooled after pouring. Since the metal molds are not porous, small vent holes or channels to release displaced air and gases are usually incorporated in the molds. The process provides surface finishes and dimensional accuracy superior to those of sand-mold casting and denser structures than die casting. It is most suitable when shapes are not highly intricate. Since pressure is not used to force metal into the mold cavities, wall thickness must be greater than that used in die castings, typically 1/4 inch (6.3 mm) or greater. Fig. 1D1 illustrates a typical permanent mold ready for casting. Typical parts cast by this method are: automotive pistons,

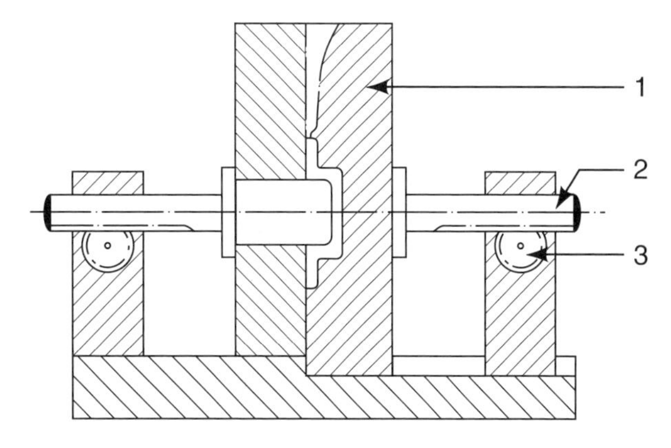

Fig. 1D1 A typical permanent mold. 1) movable mold half, 2) rack gear, 3) pinion gear. *(from Davidson, Handbook of Precision Engineering, McGraw-Hill, New York)*

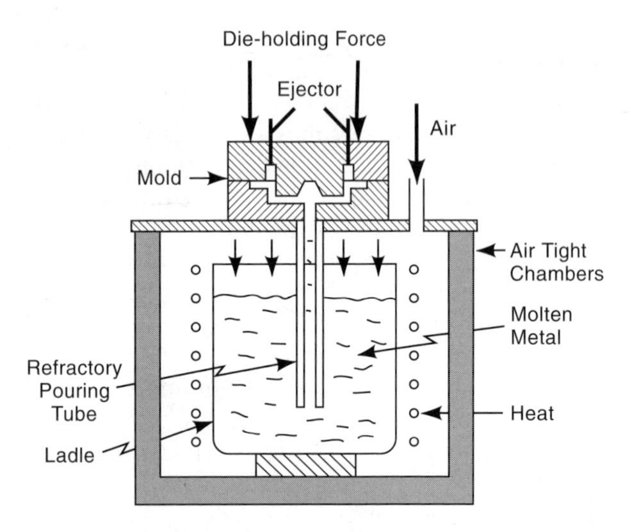

Fig. 1D2 Schematic view of low-pressure permanent mold casting. Air pressure forces the molten metal to flow upward against gravity and fill the mold. (*from Schey, Introduction to Manufacturing Processes, McGraw-Hill, New York, 1987*)

waffle irons, wheels, electric irons, pipe fittings, and gear housings.

D2. ***low pressure permanent mold casting*** - uses gas pressure to fill the mold and is a step more sophisticated than conventional permanent mold casting which uses gravity. Fig. 1D2 illustrates the low-pressure permanent mold process (LPPM). Gas pressure of about 5 to 15 lbf/in^2 (35 to 100 kPa) admitted to the chamber above the molten metal, forces the metal up through a filler tube and into the mold cavity. This approach provides a dross-free filling of the mold since the metal entering the mold comes from beneath the surface and is not in contact with the atmosphere. The method also can provide a highly controlled, non-turbulent, fill rate because the air pressure in the chamber can be controlled. A vacuum may be applied to the mold to ensure more complete filling. The process is suitable to higher production levels and produces high-quality castings. With graphite molds, the process can be used to produce ferrous castings.

D3. ***slush casting*** - This process is used for hollow objects of zinc-, lead- or tin-based alloys. It does not require a core. The molten metal is poured into a permanent mold and retained there long enough so that a shell of solidified metal coats the mold walls. The mold is inverted before all the metal has solidified. The still-liquid metal in the interior of the mold flows out and is returned to the melting pot. The mold is then opened and the casting removed. The process is suitable only for components whose interior dimensions are not critical, since the interior surface is irregular and rough. Statues, lamp bases, and other decorative objects are cast with this method.

D4. ***pressed casting*** - is similar to slush casting. The permanent mold is partly filled with a metered amount of molten metal. A closely-fitting core member is inserted into the molten metal, displacing it and pressurizing it enough to make it flow into the remainder of the mold cavity. When the molten metal has cooled and solidified, the core is immediately retracted, leaving a hollow casting. The process is used for components similar to those produced by slush casting but differs in that the inner cavity of the casting has controlled dimensions. As with slush casting, the process is used in the manufacture of ornamental objects.

D5. ***vacuum casting*** - is another permanent mold process, chiefly used for casting ingots, though parts can be cast also. Melting, reducing, and casting all take place in the same vacuum chamber, which contains the mechanisms necessary to transfer the molten metal from the melting pot to the mold. Centrifugal force may be applied to the molds if required by the nature of the part being cast. The vacuum prevents atmospheric contamination of the alloy and removes any entrapped gases. Induction is the most common heating method for melting but electric arc, electron beam, and plasma arc are also used. Vacuum chamber pressures are on the order of 10 µm. Vacuum casting is used for various alloys including those that are to be forged. Titanium requires a vacuum process when it is cast.

E. Centrifugal Casting

All centrifugal casting processes utilize rotation of the mold about a central axis and the resultant centrifugal force to drive the molten metal in the

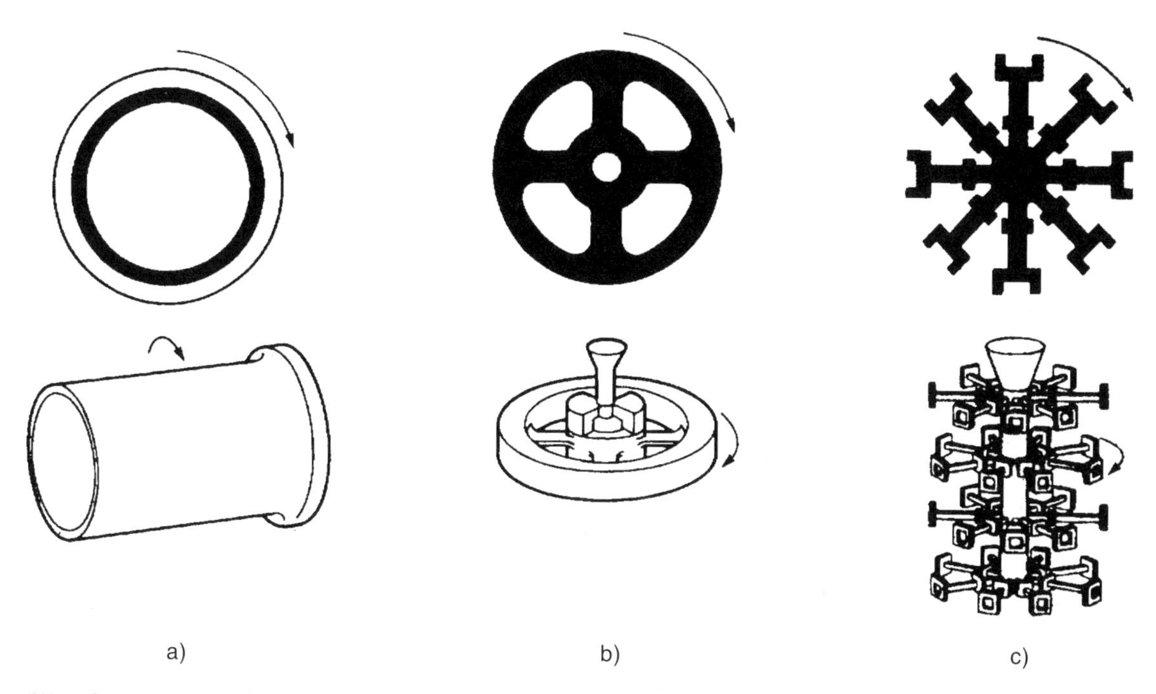

Fig. 1E Three types of centrifugal castings. (a) true centrifugal (b) semi-centrifugal, (c) centrifuging. *(from David C. Ekey and Wesley P. Winter, Introduction to Foundry Technology, McGraw-Hill, New York)*

desired portion of the mold. The cast metal is then relatively dense. Slag and other impurities cluster near the central axis and can be disposed of or removed more easily. Fig. 1E illustrates three common processes.

E1. ***true centrifugal casting*** - is used to manufacture piping, specialty tubing, cylinder liners, and other cast objects of cylindrical shape. Centrifugal force holds the molten metal against the outer walls of the rotating mold. The volume of metal in the mold determines the wall thickness and internal diameter of the part. No core is used except to form the bell ends of cast pipe. The mold is made of metal but may be sand-lined. If sand is employed, one of a number of binders may be utilized. Graphite is also used for some molds. The pouring spout may move axially as the mold rotates, providing a helical path to the flow of liquid metal. As the mold continues to rotate, the molten metal spreads evenly on the mold surface and solidifies. The rotation of the mold produces centrifugal force up to 100 times the force of gravity, the amount being partially dependent on the angle of orientation of the axis. The inner surface of the casting tends to collect dross and other impurities. Fig. 1E, view a), and Fig. 1E1 illustrate the process.

E2. ***semicentrifugal casting*** - In this process, parts that have a symmetrical shape around a central axis are cast, including wheels, gear blanks, nozzles, and similar parts. (See Fig. 1E, view b.) The rotational speed of the mold is less than that used in true centrifugal casting. Several molds may be stacked on the same axis and all filled from the same pouring source. A core may be used for the center hole, if any, or the sprue may be left in the center of the casting. In either case, the hole is machined to provide the necessary accuracy and surface finish. The outer portion of the casting has a dense structure.

E3. ***centrifuged casting*** - This process is used for smaller, intricate parts. Centrifugal force provides the pressure that ensures complete filling of the mold cavities. Molds are located radially about a central sprue or riser, which acts as the axis of rotation. Rotation can be about a vertical or horizontal

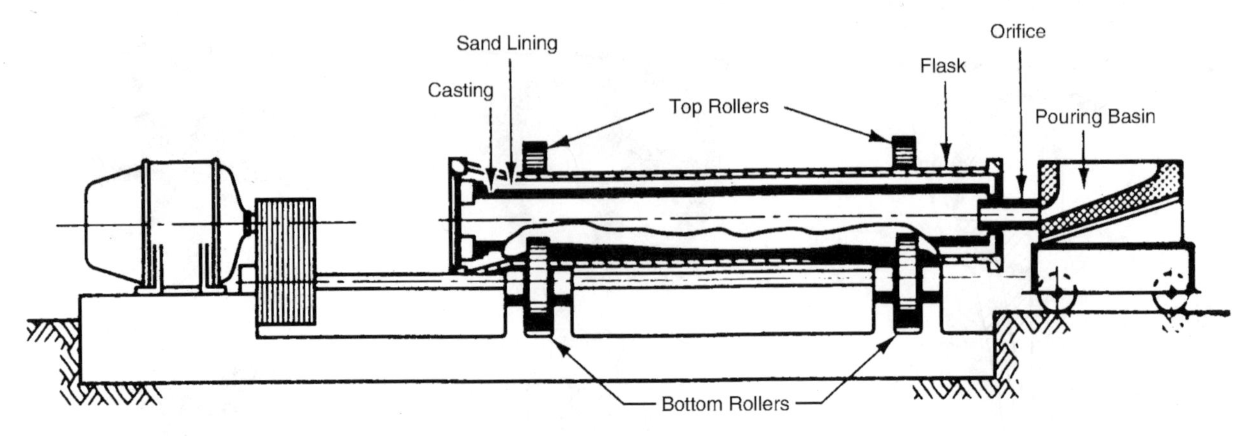

Fig. 1E1 A horizontal centrifugal casting machine equipped to make cast iron pipe. (*Courtesy American Cast Iron Pipe Company*)

axis and is at a relatively low rotational speed. Fluid pressure of the molten metal is proportional to its distance from the axis and the square of the speed of rotation. Molds may be stacked as shown in Fig. 1E, view c). Casting shapes that otherwise might present feeding problems can be cast with this process. Small caps and brackets, and dental inlays are among the components cast.

F. Die Casting

Die casting is a permanent mold casting process wherein the molten metal is forced into the mold at high velocity and under high pressure. That pressure is maintained until the metal solidifies. The molds, referred to as "dies", are water-cooled and are usually made from a hardened steel alloy that can withstand the pressure and heat of the process. Dies are in two halves and often have side cores if the part shape includes undercuts. These cores are actuated by cams, gears, or separate hydraulic cylinders. The dies are usually lubricated for each cycle prior to the clamping of the two halves together. The lubricant may be accompanied by water to cool the die and compressed air to remove any extraneous metal pieces that may remain on the die. Pneumatic, hydraulic, or mechanical force is used to close the die halves and to actuate the plunger that moves the molten metal into the die cavity. Because die filling is very rapid, some air may be trapped in the die cavity and result in some porosity in the casting. This porosity is normally limited to the interior of the casting, but surfaces tend to be dense. Vents and small overflow wells are incorporated in the die to allow trapped air and excess metal to escape. After the casting metal solidifies, the die halves part and ejector pins remove the workpiece. Cycle times range from just a few seconds for small zinc parts to 30 or 40 seconds for larger aluminum parts. The maximum practical casting size is about 40 lb (18 kg). A trimming operation is frequently needed after casting to remove flash, sprues, and overflow material. Although some ferrous die casting is being done, the process is primarily limited to non-ferrous alloys. Zinc, aluminum, magnesium, tin, and lead alloys are most commonly cast. Some die casting of copper-based alloys also takes place. The high-velocity filling of the dies permits the production of thin-walled and intricately shaped parts. Close tolerances can be held and post-casting machining is normally less than that required after other casting processes. Because the cost of equipment and tooling are high, the process is most suited to high-production conditions. Typical die cast parts are used in appliances, automobiles, hand tools, and builder's hardware.

F1. ***hot-chamber die casting*** - In this process, the plunger that forces the molten metal into the die is immersed in the molten metal at all times. Fig. 1F1 shows the "gooseneck" configuration of the plunger and injection channel. Prior to injection, molten metal is allowed to flow by gravity into the injection cylinder. The plunger is then actuated by either

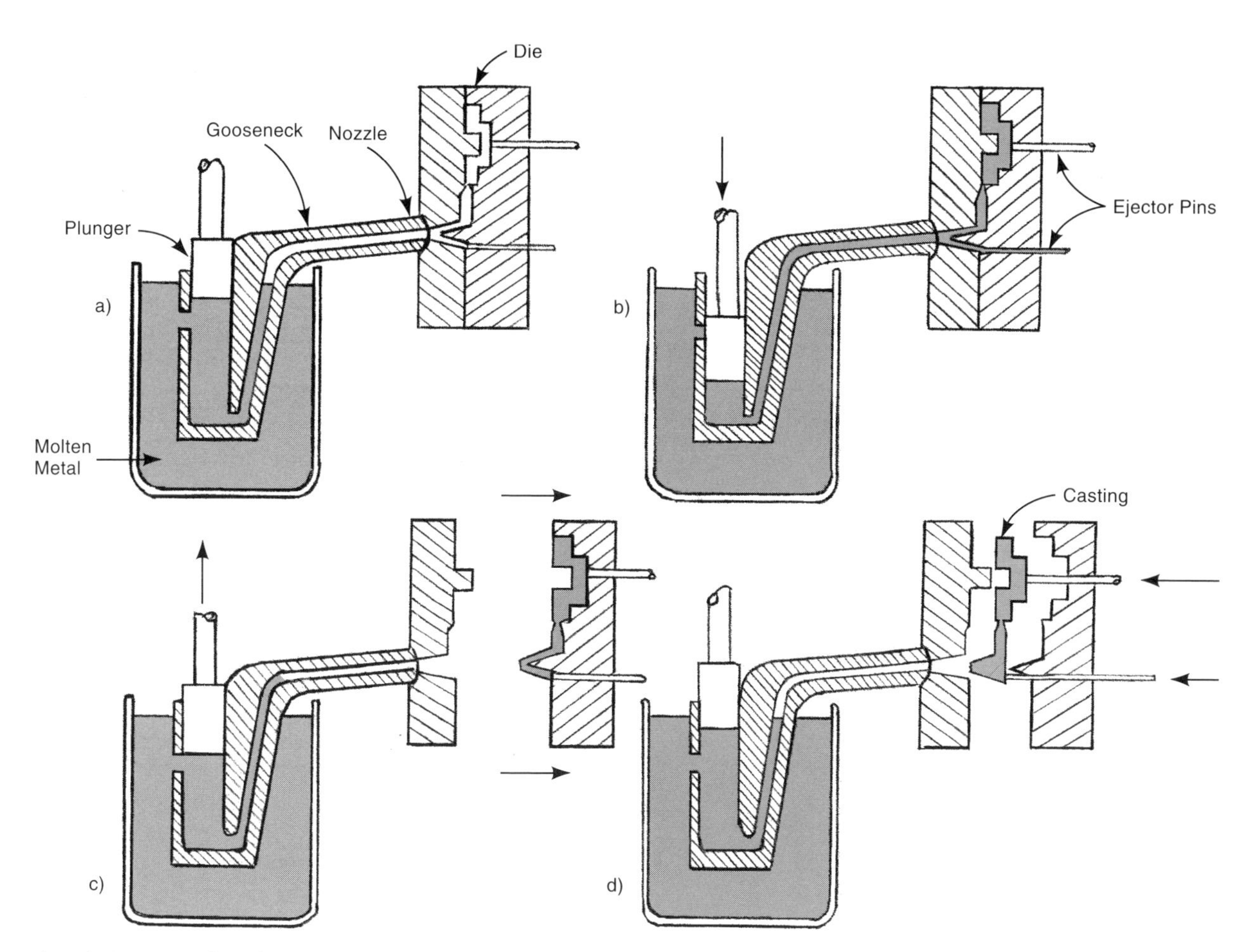

Fig. 1F1 Hot-chamber die-casting. a) The die is closed and the hot chamber (at the gooseneck) is filled with molten metal. b) The plunger descends and forces molten metal through the gooseneck and nozzle and into the die cavity. Metal is held under pressure until it solidifies. c) The die opens. The casting stays in the one half of the die. The plunger retracts, pulling molten metal back through the nozzle and gooseneck. d) Ejector pins push the casting out of the die. As the plunger uncovers the inlet, molten metal refills the hot chamber for the next cycle.

mechanical or pneumatic force to move the molten metal into the die. (There is no external transfer of molten material to the injection cylinder.) The cycle is automatic and, with small parts, can often repeat as fast as every 4 or 5 seconds. However, the constant contact of the injection equipment with molten metal causes difficulties with aluminum and other higher melting temperature alloys which attack the steel components of the injection system. As a result, hot-chamber die casting is used mainly with zinc, tin, lead, and, more recently, magnesium alloys.

F2. ***cold-chamber die casting*** - In this process, as illustrated in Fig. 1F2, molten metal is ladled into the horizontal injection cylinder from a separate melting and holding pot so that the metal contacts the injection mechanism only during the injection portion of the cycle. After ladling, the metal is immediately injected into the die and, as in the hot-chamber method, is held under pressure during solidification. This approach is not quite as rapid as hot-chamber die casting but nevertheless still provides high productivity. The process is used with aluminum, copper, and magnesium, which are not well suited to hot-chamber die casting because constant contact with these metals may result in short life of the injection equipment. The cold-chamber method limits the time

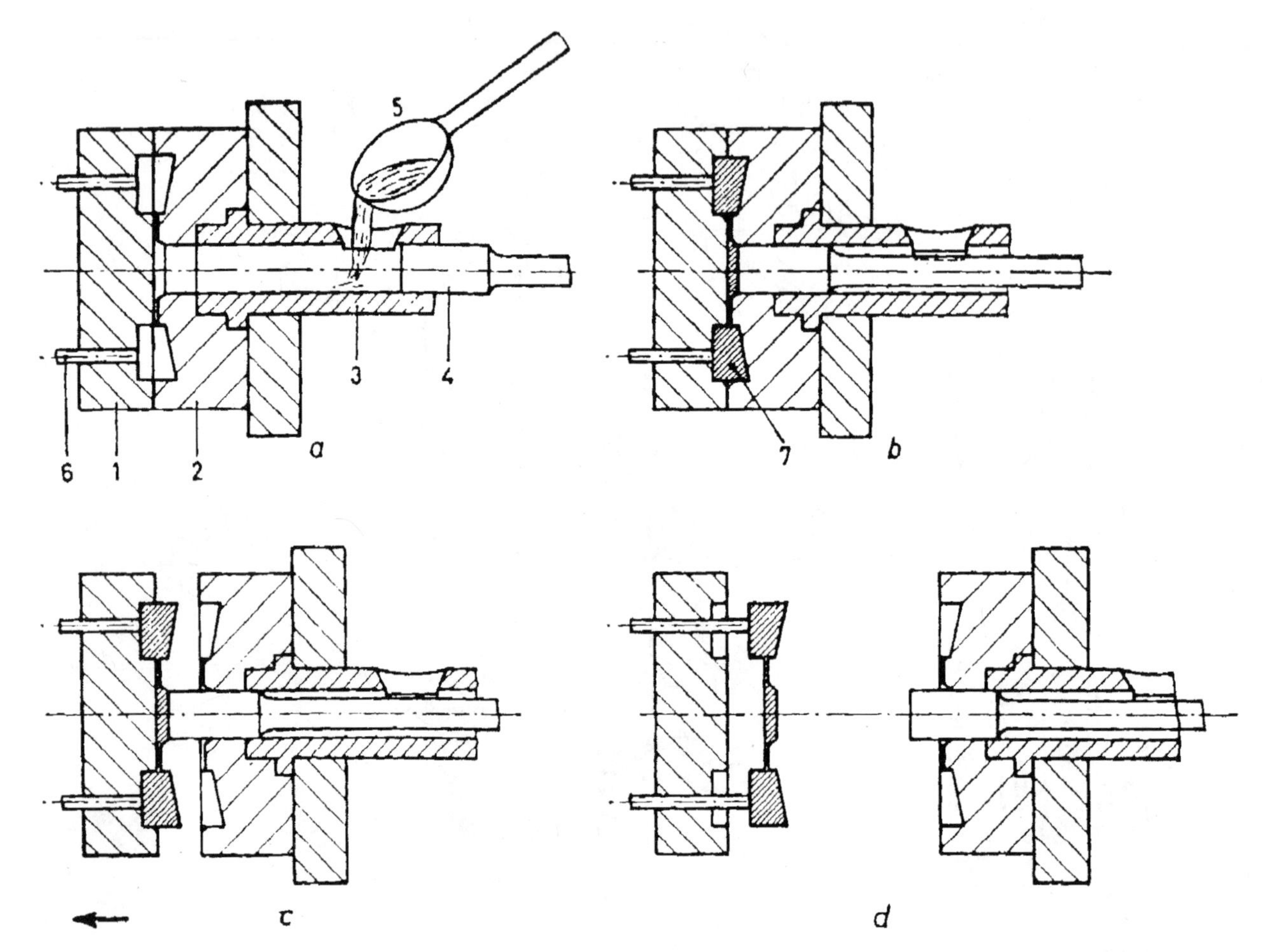

Fig. 1F2 Cold-chamber die casting. a) The mold is closed and molten metal is ladled into the injection cylinder ("cold chamber"). b) The plunger advances, forcing molten metal into the die cavity. c) The mold opens. d) The casting is ejected. Machine components: 1 - stationary die half, 2 - movable die half, 3) cold chamber, 4 - plunger, 5 - ladle, 6 - ejector pin, 7 - die casting. (from Davidson, Handbook of Precision Engineering, McGraw-Hill, New York)

of exposure of system components to the molten metal. The cold-chamber method is also used when ferrous alloys are die cast. Cold-chamber machines are of rugged construction because of the higher casting pressures needed with the alloys mentioned.

F3. ***trimming*** - of die-cast parts normally follows casting to remove flash, overflow material, gates and runners. The common method uses a hydraulic or mechanical press and a trimming die, made to fit the outline shape of the part. The operation is often performed immediately after casting when the workpiece is hot and more easily sheared. Trimming can also be a manual operation but the high production normally involved with die casting also justifies the use of dedicated tooling and production equipment. Grinding may also be employed in trimming operations.

F4. ***impregnation of die castings*** - is another operation that may follow die casting if the product application requires pressure tightness. The die casting is impregnated with organic material or sodium silicate to fill any pores that could affect the ability of the parts to hold pressure. The impregnation process involves the following steps: cleaning, placement of the workpieces in a suitable vacuum chamber, drawing of a vacuum in the chamber, introduction of the sealant to the chamber, pressure in the chamber to drive the sealant into any pores that exist on the workpiece, removal of the castings from the chamber, heating or baking to cure the impregnated material, if necessary, and washing and drying.[1]

G. Investment Casting (Lost Wax Process)

This process uses a one-piece mold made of ceramic material, the same material that is used in ceramic casting. The mold is made by surrounding an expendable wax, plastic, or frozen mercury pattern with the ceramic material in slurry form. When the mold material solidifies it is heated and the wax, plastic, or mercury replica of the part is melted out, leaving a cavity corresponding to the shape of the desired part. The mold is baked to remove all residues of the pattern and to fuse the ceramic. Molten metal is poured into this cavity. When the metal solidifies, the ceramic mold is broken free and removed from the casting. There are two basic variations of this process, the *flask, solid mold* or *monolithic method* which uses a metal flask to hold a solid mass of the ceramic mold material, and the *shell method* in which a thin (about 1/4 in - 6 mm) shell of ceramic without a flask surrounds the pattern. Investment casting is used extensively for intricate, usually small, precise parts of high strength alloys. Turbine blades, sewing machine and gun parts, valve bodies, wrench sockets, and gears are typical parts. Larger parts are also now produced weighing up to about 80 lb (35 Kg) per casting.

G1. ***flask method*** - The sequence for the simpler, but much less used, flask method is as follows:

1. The wax or plastic patterns are injection molded, often in a low-cost nonferrous mold.
2. A number of patterns are assembled together on a "tree" of the same wax or plastic so that a number of parts can be cast at one time. The tree consists of a central sprue and branch runners to which are attached the individual patterns.
3. The "tree" is precoated by dipping it in a slurry of refractory material and then is dried thoroughly.
4. The tree is inserted in a flask and investment material is poured around it. The worktable is usually vibrated to settle the investment material and drive out air pockets. The investment material is allowed to set and then air dry for six to eight hours.
5. The flask is inverted and placed in an oven to melt out the wax or plastic. Typically temperatures of about 375°F (190°C) are used for about 12 hours.
6. The flask is heated gradually, then held at about 1800°F (980°C) for about 4 hours to melt and "burn out" the wax or plastic residue and to fuse the investment material.
7. Molten metal is poured into the mold cavity. Gravity is the prime filling method but air pressure and centrifugal force may also be used. The metal cools and solidifies.
8. Shake-out takes place. The casting in "tree" form is removed from the flask and the investment material is broken away from it.
9. The tree is descaled by immersion for 10 to 15 min in a molten salt bath at 1100°F (600°C) followed immediately by a cold water dip and cleaning and neutralizing immersions.
10. The parts are cut from the tree.
11. Gates and runners are removed from the castings by abrasive or other machining methods. Fig. 1G1 illustrates the mold making portion of this process.

G2. ***shell method*** - This method has a very similar sequence to that of the flask method. The chief difference is that, instead of pouring a solid mass of investment material around the thinly-coated tree pattern, it is coated with a series of dip coats of ceramic slurry until the desired thickness of shell is built up. This thickness is usually about 1/4 in (6.3mm), and no flask is then required. About six or seven dips may be required, depending on the size of the part. The sequence of operations is illustrated in Fig. 1G2. The shell method has become the predominant method, though the flask method may give somewhat better surface definition to the casting since the molten metal cools more slowly. Less surface decarburization occurs with the shell process. It also has a shorter cycle time.

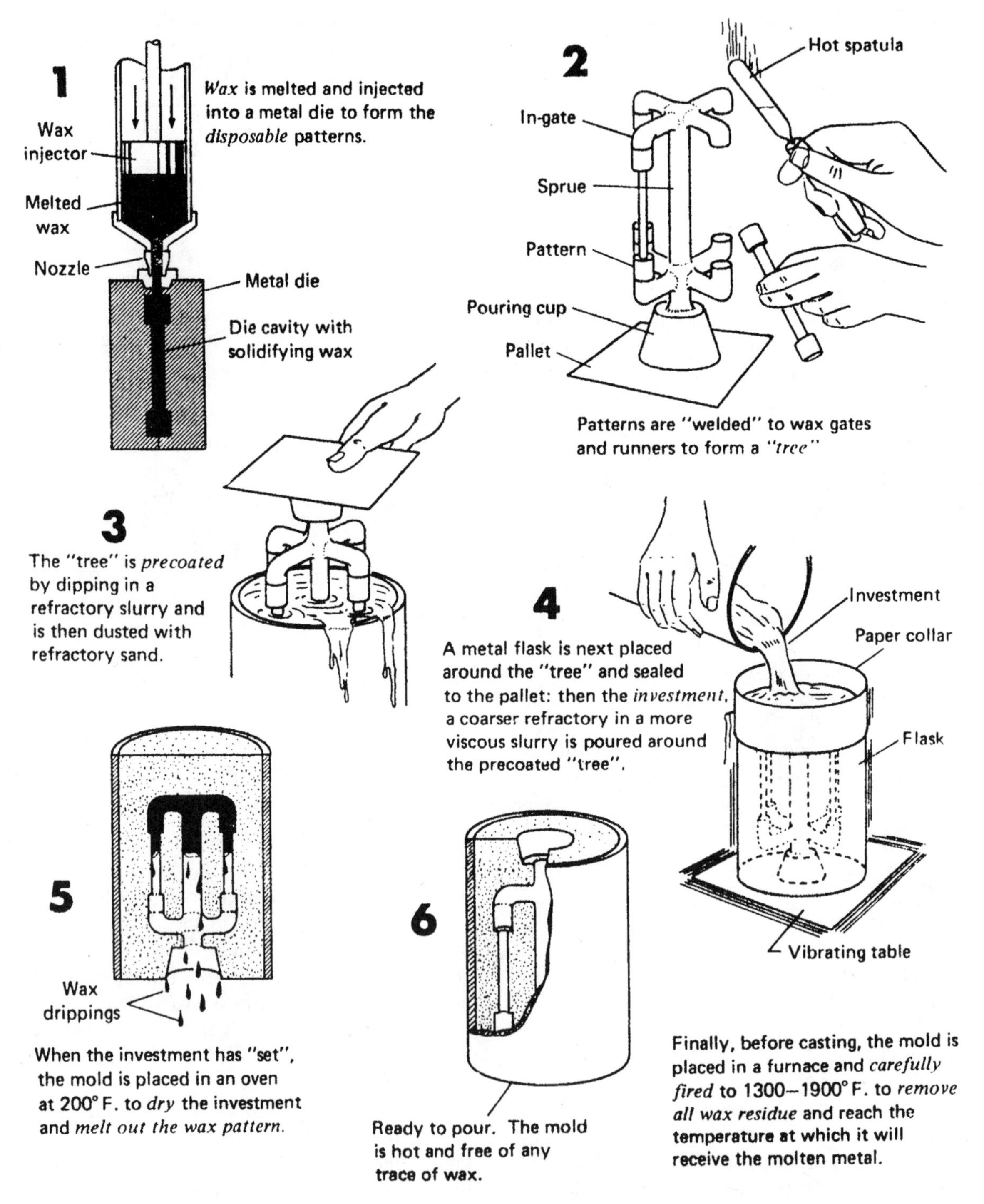

Fig. 1G1 The process for making a mold for the flask method of investment casting. (*from Niebel and Draper, Product Design and Process Engineering, McGraw-Hill, 1974, New York*)

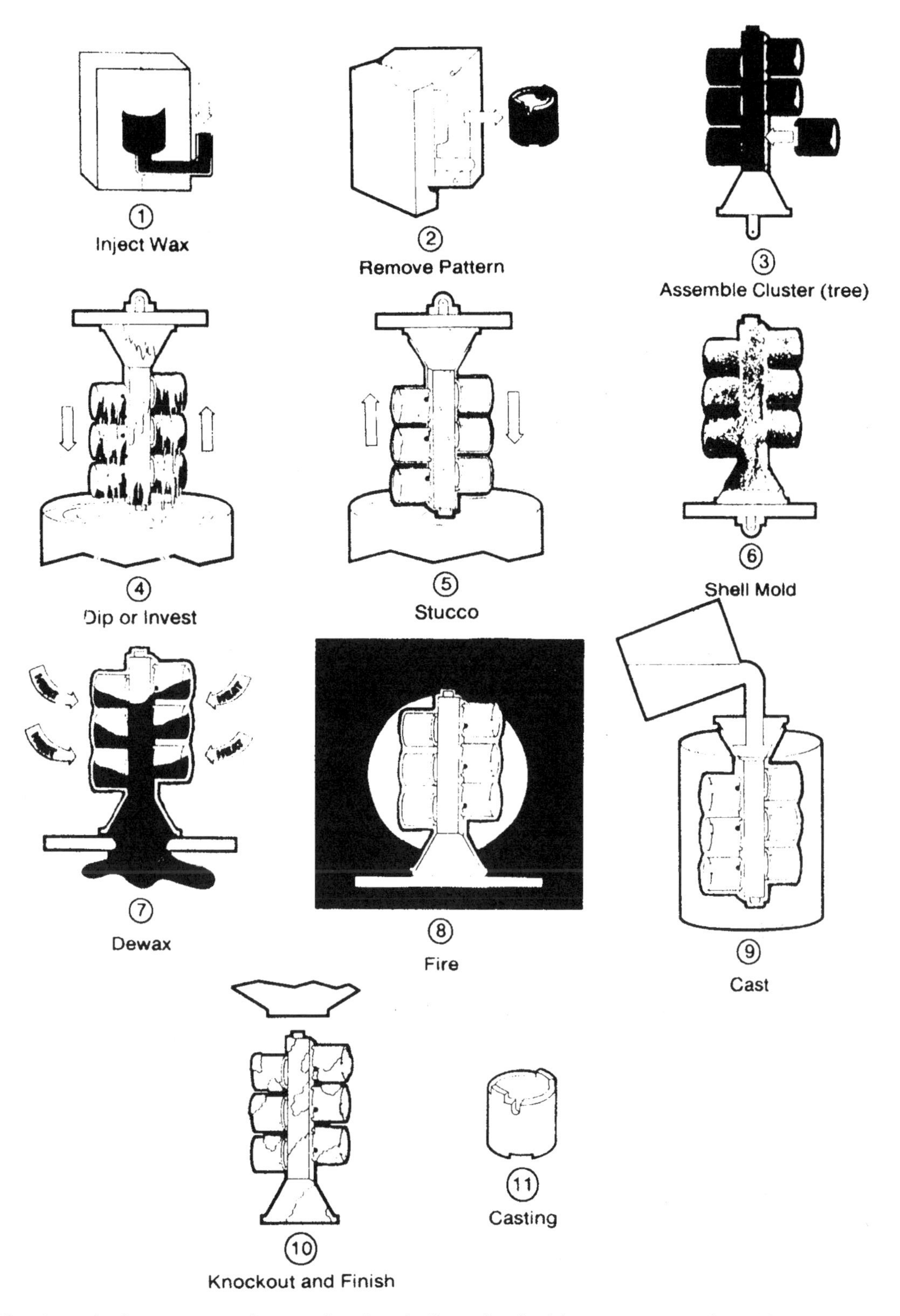

Fig. 1G2 A typical sequence of steps for the shell method of investment casting. (*Courtesy Investment Casting Institute, Dallas*)

H. Continuous Casting

Continuous casting is primarily a mill method for producing ingots, rather than discrete cast parts, but it is used for producing some long parts of constant cross section such as piping or tubing. The method solves certain quality difficulties that are more prevalent with conventional ingot casting and is more cost effective for mill-quantity production. The molten metal is continuously solidified as it is poured and its length is not determined by the length of the mold but by the length of time that the pouring and solidifying continues. The mold position is most commonly vertical with an open top and bottom. The material flows downward through the mold, which shapes it. As it starts to solidify and exit the bottom of the mold, space is created for additional molten metal at the top. Fig. 1H illustrates a typical arrangement.

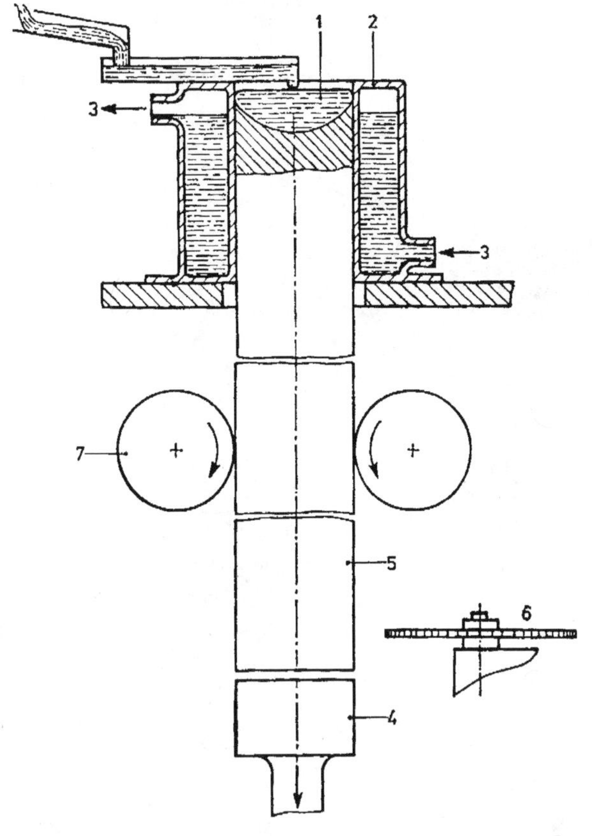

Fig. 1H Continuous casting of a solid bar with a vertical continuous casting machine. 1) molten metal, 2) water-cooled mold, 3) cooling water inlet and outlet, 4) (not shown) movable mold base which descends at a controlled rate to start the operation, 5) cast bar, 6) flying cutoff saw, 7) rollers to control movement of the cast bar. (from Davidson, Handbook of Precision Engineering, McGraw-Hill, New York)

The casting sequence is as follows:[1] 1) molten metal is delivered to the casting equipment. 2) The metal flows from a tundish (temporary container) into the mold; flow is controlled by a suitable stopper or slide gate valve. 3) The molten metal cools and partially solidifies as it passes through the water-cooled mold (which is often made from copper); the mold oscillates and is lubricated to prevent the casting from adhering to it. (When casting aluminum, air pressure and an electromagnetic field may be used to keep the metal from contacting and adhering to the mold.) 4) The section is supported and guided below the mold by opposed rolls that contact the cast section whose surface is solid but whose interior may still be molten; water spraying provides further cooling action. 5) The opposed rollers withdraw the solid cast section to a station where it is cut to length by flame cutting or a shear. It may, alternatively, be reheated and rolled before being cut to length. First utilized in the production of primary non-ferrous metals, this method is now used extensively in steel production as well. Non-ferrous metals are continuously cast as tubing and other special shapes; steel is more difficult to cast into special shapes.

References

1. *ASM Handbook, Vol. 15, 9th edition,* ASM International, Metals Park, OH, 1988.
2. *Cast Metals Technology,* J. G. Sylvia, Addison-Wesley, Reading, MA, 1972.
3. *The Foseco Foundryman's Handbook, 10th ed.,* J.R. Brown, ed., Butterworth-Heinemann Ltd., Oxford, England, 1994.
4. *Design for Manufacturability Handbook, 2nd ed.,* J.G. Bralla ed., McGraw-Hill, New York, 1998.
5. *Manufacturing Processes and Systems (9th edition),* Phillip F. Ostwald, Jairo Munoz, John Wiley and Sons, New York, 1997.
6. *Materials and Processes in Manufacturing (8th edition),* E. Paul DeGarmo, J.T. Black, Ronald A. Kohser, Prentice Hall, Upper Saddle River, NJ, 1997.
7. *Tool and Manufacturing Engineers Handbook, 4th edition,* Society of Manufacturing Engineers, Dearborn, MI, 1984.

Chapter 2 - Metal Forming Processes

A. Hot and Warm Forming Methods

(Note: Many metal forming operations can be performed with the workpiece metal either hot or cold. The operations discussed in this section are normally - but not always - performed on workpiece material that has been heated to make it more malleable for the operation involved. Metals that are to be hot formed are heated above their recrystallization temperature, one that varies with each material but is normally about 0.6 times the melting temperature on the Kelvin (absolute temperature) scale. For example, steels require a temperature above about 1800°F (980°C). Warm forming involves heating to a temperature 30 to 60 percent of the melting point, while cold forming takes place when the metal temperature is below 30% of the melting temperature[2].)

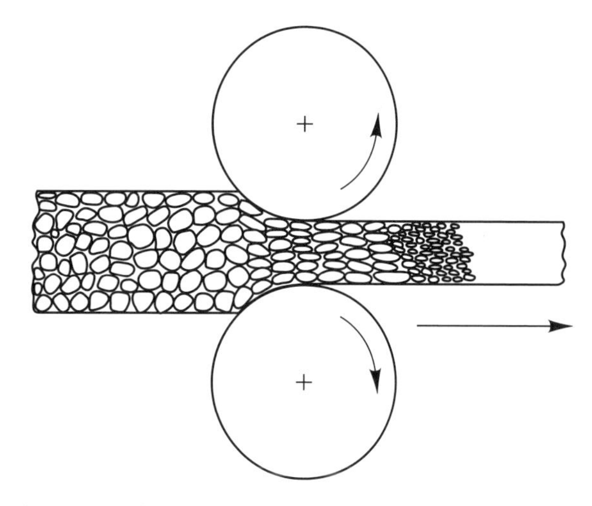

Fig. 2A1 The hot rolling process. The grain structure of the metal is deformed and then recrystallized.

A1. ***hot rolling*** - is commonly applied to convert steel ingots to blooms, billets, or slabs, and to make these shapes into salable forms. In the process, heated metal is passed between two rollers whose spacing is less than the thickness of the metal. The rotation of the rollers moves the metal forward, squeezing and elongating it. Fig. 2A1 illustrates the process. The process extends and refines the grain structure of the rolled material. A number of passes may be required, depending on the thickness desired and the thickness of the entering material. Reversing rollers are often used to facilitate multiple passes. Thin sheet or foil is best rolled with small-diameter rollers that are backed up with larger rollers to provide the necessary rolling force. As many as twelve rollers in a cluster may be used. Shaped rollers can produce material with various cross sections including those of structural shapes or special cross sections. Low-alloy or plain-carbon steel is heated to about 2200°F. (1200°C) before rolling and after being preheated in a soaking pit. In addition to ferrous metals, aluminum, copper and copper alloys, magnesium, nickel, titanium, and zinc alloys are hot rolled.

A2. ***hot drawing or cupping*** - is a process for making cup shapes of some depth (more than several times stock thickness) and thick and seamless tubes and cylinders from blooms, flat plate, or sheet. The process is similar to cold deep drawing of sheet metal (except that the material may be thinned during the operation whereas in deep drawing the material flows into the die and tends to thicken). Tubular parts can be made when the cup formed in one such operation is reheated and redrawn to a narrower diameter; then reheated and pushed through a series of draw bench dies that further reduce its diameter and extend its length. Fig. 2A2 illustrates both the cup-forming and redrawing operations. In addition to the redrawing that produces cylinders and seamless tubing, the process is used for forming relatively simple shapes, usually cylindrical, in thick material.

A3. ***extrusion*** - In this process, metal is forced through a die opening that gives it a uniform cross-sectional shape. Although the operation can be performed with many cold metals, the usual procedure is to preheat the metal to the plastic range to ease the transition. A heated ingot or billet is inserted in a chamber called a "container". A ram, normally hydraulically powered, forces the material through the die opening. As it flows through the die, the metal takes the shape of the die opening and closely conforms to its dimensions. It is quickly cooled as the metal exits the die so that the shape is maintained. Fig. 2A3 illustrates the process.

Aluminum, copper, magnesium, tin, lead, and their alloys are commonly extruded. Steel, including alloy and stainless varieties, and nickel alloys, are more difficult to extrude but can be processed by the Sejournet and related processes (See A3b below.)

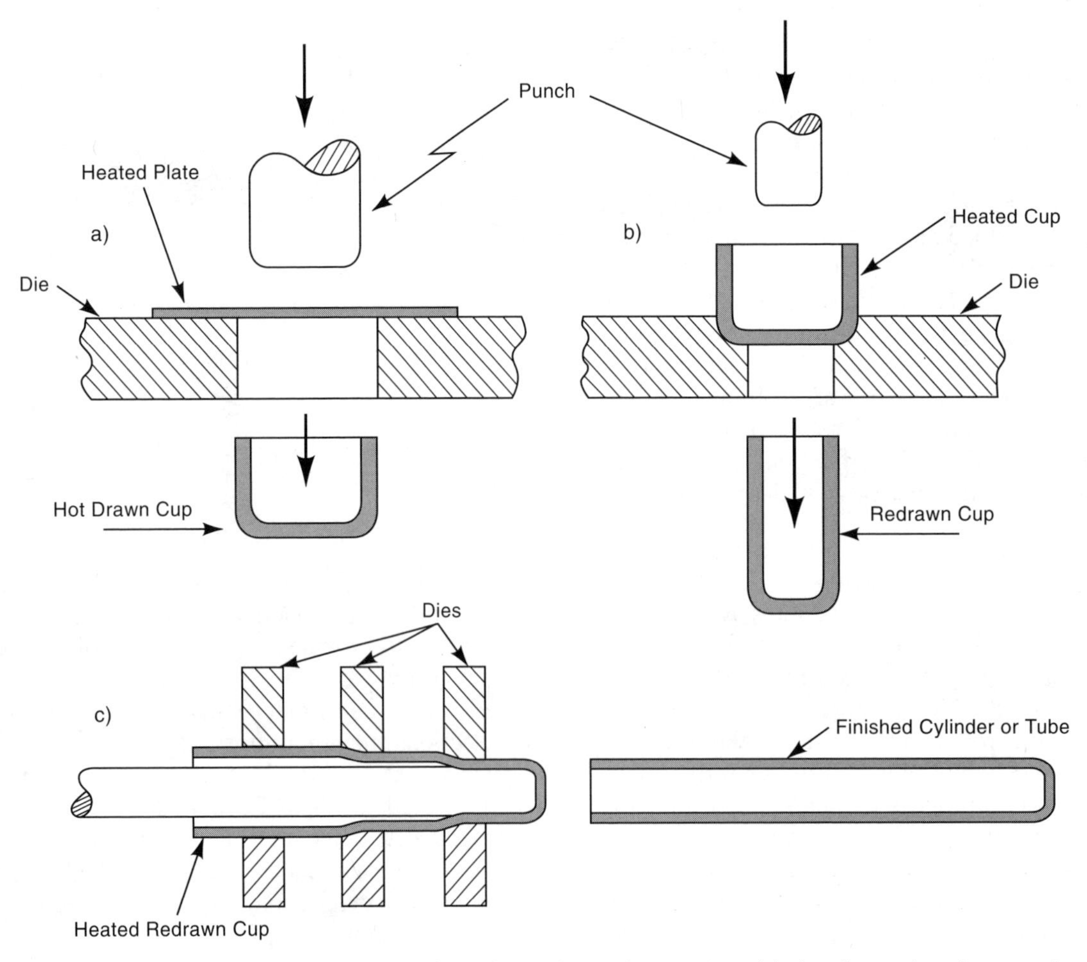

Fig. 2A2 Hot drawing or cupping. a) first draw, b) redraw, c) multiple die redrawing on drawbench.

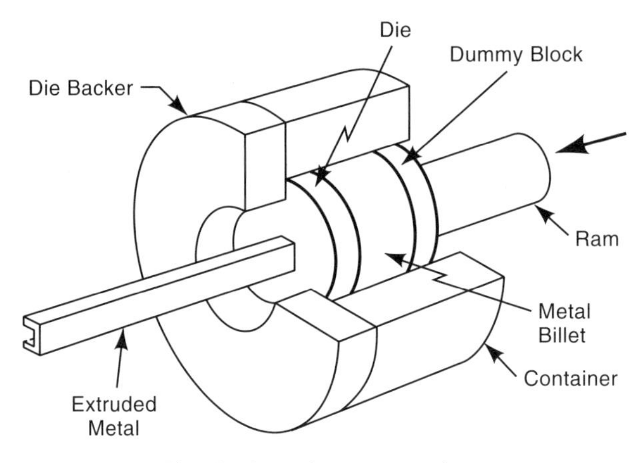

Fig. 2A3 Direct extrusion.

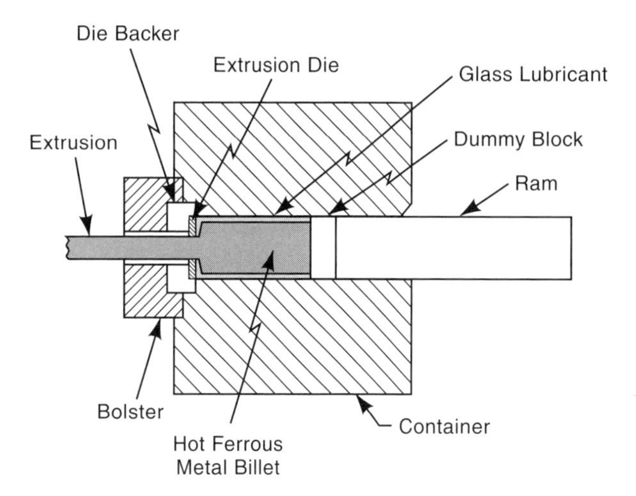

Fig. 2A3b The Sejournet process for extruding ferrous metals uses molten glass as a lubricant and insulator.

After extrusion, sections may be stretch-straightened to remove twist and camber that may exist. Extrusion is suitable for almost any part that has a constant cross section and is made from the above materials. A wide variety of complex shapes such as tubing and other hollow objects, door and window frame elements, ladder members, and structural sections can be extruded. A similar process is used to make like-shaped components in plastics or other materials (See chapter 4, section 4I.)

A3a. ***indirect extrusion*** - In this process variation, the ram is hollow and the metal is forced backward through a die and into the ram. Friction is reduced because the metal does not have to flow along the extrusion chamber walls, but the difficulty in supporting and removing the extrusion makes the process awkward. (See Fig. 2A3a.)

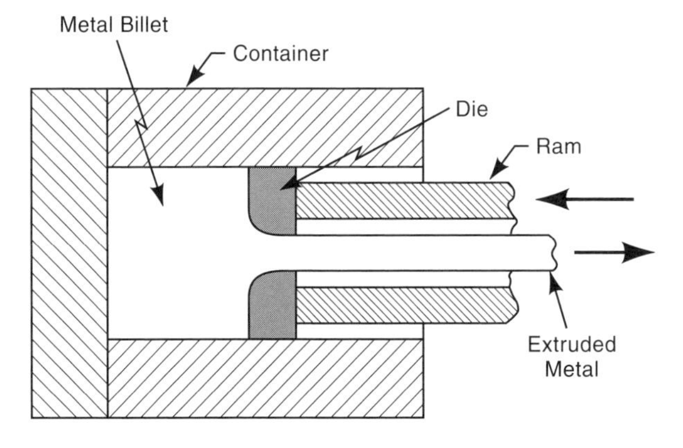

Fig. 2A3a Indirect extrusion.

A3b. ***Sejournet process*** - is used for the extrusion of ferrous metals. The operation is the same as that for other metals except that the steel billet to be extruded is heated to the recrystallization temperature range and then coated with molten glass that lubricates and insulates the metal as it is pushed through the die. See Fig. 2A3b. The workpiece is usually stretch-straightened afterwards to remove camber and twist that may result from the process.

A4. ***forging*** - is a process in which a metal slug undergoes plastic deformation into a useful shape, usually at an elevated temperature. Repeated strokes or extended pressure may be used. The process refines the grain structure of the material and improves the physical properties of the part. Equipment capable of applying compressive forces is required. Dies are often heated to a temperature between 300 and 400°F (150 and 200°C) or higher. Metals commonly forged include steel, aluminum, magnesium, brass, bronze, copper, stainless and alloy steel. Aluminum is typically forged isothermally at a temperature of 600 to 850°F (320 to 455°C). Many machine components that require high strength are produced by forging. Examples are aircraft landing gear components and other aircraft and aircraft-engine parts, connecting rods, spindles, couplers, parts for earth moving and agricultural equipment, valve bodies, ordnance parts, gears, turbine parts, and levers.

A4a. ***open-die drop-hammer forging (hammer, flat-die, or smith forging)*** - involves repeated blows with a powered flat-faced hammer or tools of simple shape. The workpiece is not confined. Except for the powered press strokes, the operation is the same as that traditionally performed by a blacksmith in that the operator positions and orients the workpiece for each blow. The process is often a preliminary one, used to provide an initial shape prior to further forging and other operations. The process is also used for forgings that are too large for impression dies, and for small quantities and situations where time schedules do not permit the fabrication of dies. It is illustrated in Fig. 2A4a. Simple shapes such as discs, shafts, or rings, are commonly produced. Steam, compressed air, or gravity ("drop forging"), provide the necessary force. The operation can be computer controlled to reduce dependency on operator skill and to ensure more accurate workpiece dimensions.

A4b. ***impression die forging (closed die forging) (die forging)*** - differs from hammer or smith forging in that the dies are closed. There are two die halves, one of which is fastened to the ram of the forging press and the other to the machine bed. Repeated press strokes or heavy hydraulic or mechanical pressure cause the workpiece metal to flow into the die cavities, taking their shape, as shown in Fig. 2A4b. If the part is complex, several different die sets may be used before the final shape is attained. The operation produces surplus metal called "flash" around the edges of the forging. The flash cools and hardens before the balance of the forging and aids in filling the unfilled portions of the die. The flash is removed in a secondary trimming operation.

Typical metal temperatures for impression die forging per Ostwald[1] are: steel - 2000 to 2300°F, (1100 to 1250°C), copper alloys - 1400 to 1700°F (750 to 925°C), aluminum alloys - 700 to 850°F (370 to 455°C), and magnesium - 600°F (315°C). The process results in particularly strong parts per unit of weight because it produces favorable orientation of the metal grains.

A4b1 - ***drop forging*** - is impression die forging using a drop hammer. A drop hammer is a machine that uses gravity, air, or steam pressure to make repeated blows against a workpiece. With gravity drop hammers, a heavily weighted ram is lifted above the forging die and then released.

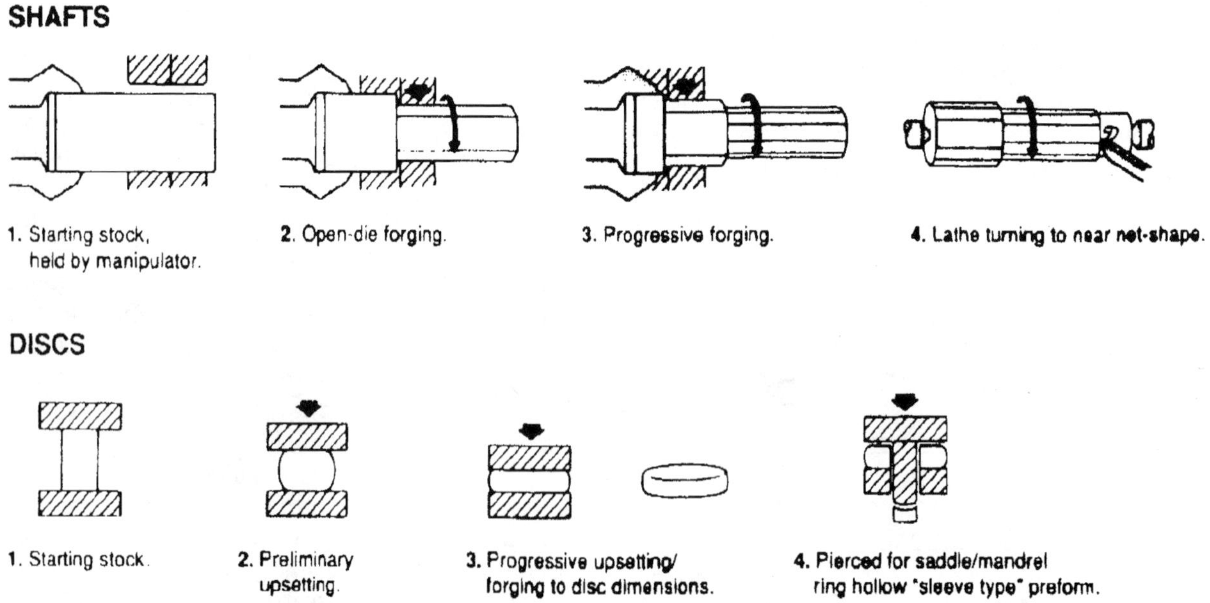

SHAFTS

1. Starting stock, held by manipulator.
2. Open-die forging.
3. Progressive forging.
4. Lathe turning to near net-shape.

DISCS

1. Starting stock.
2. Preliminary upsetting.
3. Progressive upsetting/ forging to disc dimensions.
4. Pierced for saddle/mandrel ring hollow "sleeve type" preform.

Fig. 2A4a Open-die forging of shafts and discs. Repeated press strokes with a flat or simple-shape die form the workpiece. (*Courtesy Forging Industry Association.*)

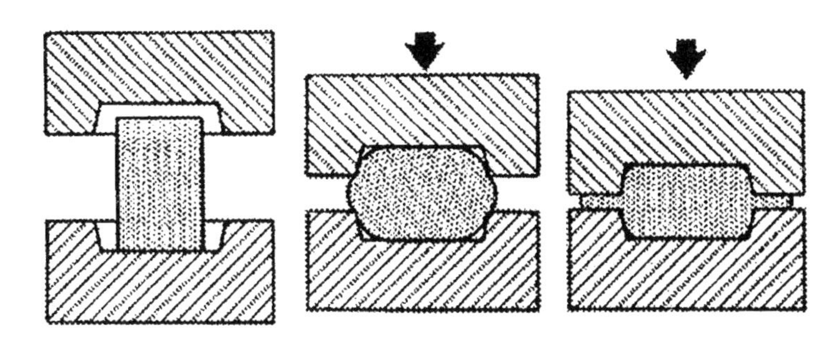

Fig. 2A4b Conventional impression-die (closed-die) forging sequence. (*Courtesy Forging Industry Association.*)

The press action is then repeated until the workpiece has attained the desired shape. Impact force aids the forging operation. Powered drop hammers use air, steam, or hydraulic pressure to add to the downward force.

A4b2 - ***precision forging*** - is a variation of impression die forging with a closely-controlled, more extensive forging process, accurately controlled blank sizes, and carefully designed dies, so that the forgings produced are close to the net shape required. The process greatly reduces the amount of post-forging machining required. This is because it can reduce side-wall draft to from 0 to 1° and permit thinner forging walls, smaller radii, and smoother surfaces. Since there is little or no machining, the grain flow patterns are not disturbed, extra metal does not have to be added to compensate, and strength-to-weight ratios are improved. A key approach in the process is to provide forging blanks of just the right size and shape with accurate and consistent dimensions. A certain amount of trial and error is usually required to develop a blank that will just fill the die completely without requiring excess metal in any area. A preformed (preforged or pre-machined) blank may be used. One method sometimes used to provide accurate blanks is the use of powder metal preforms since the powder metal process can provide accurate and consistent forging blanks. Very close attention to all process details is required: workpiece temperature throughout the billet, descaling, die temperature (dies are heated), press pressure and stroke, and lubrication, are all of vital importance.

The precision forging process is often used to produce forged gear blanks. Aluminum is a metal commonly precision forged. The metal is heated to about 800°F (425°C) and the die temperature is maintained at 700 to 800°F (370 to 425°C) to insure proper metal flow[9]. It should be noted that precision forging usually requires higher pressures and stronger dies than other forging methods, and, because the dies are typically run hot, they have a shorter die life.

A4b3. ***flashless forging*** - is characterized by an absence of the excess metal (flash) that normally escapes between the die halves in closed-die forging. Excess metal is often necessary to insure proper filling of all portions of the die cavity, particularly when the operation requires high deformation of the forging billet. One approach used to eliminate flash, although not all excess metal, when the part has a center hole (eg., a gear blank) is simply to confine the excess metal to that center hole. There, it is more easily removed with a punch-press operation. The principal approach in making flashless forgings, however, is to carefully size the billet and control its weight, so that there is no excess metal. There is an overlap between precision and flashless forging methods, and the process refinements and controls used in precision forging may also help to eliminate flash. Precision forging usually minimizes and sometimes eliminates flash[3]. However, precision forgings are not necessarily flashless, and flashless forgings are not necessarily precision forgings. The advantage of flashless forging, of course is the reduction of metal required and the elimination of the post-forging trimming operation.

A4c. ***press forging*** - differs from drop forging in that the forging action results from a slow squeezing action rather than a hammer-like impact. This squeezing action produces deformation more uniformly throughout the workpiece, resulting in greater dimensional accuracy and less need for sidewall draft. The dies may need to be preheated to reduce heat loss from the billet, and to promote finer surface detail. The presses used for press forging are normally hydraulic. This process is most suitable for more complex shapes; less intricate, simpler shapes may be more advantageously produced by drop forging.

A4d. ***upset forging*** - is an operation that increases the diameter of a workpiece while shortening its length. It is the hot-material equivalent of cold heading. The workpiece material is normally in the form of a bar, on which a head or other larger-diameter portion is produced by the process. One part of the die holds the bar while another part of the die is forced against it axially, causing the metal to flow and fill the die cavity. Normally, only the part of the bar to be upset needs to be heated. Making bolt heads is the most common application.

A4e. ***roll forging*** - involves placing heated bar stock between two cylindrical or semi-cylindrical powered rollers that have shaped cavities on their surface. The workpiece is reduced in diameter or thickness, increased in length and often changed in shape. Die rollers can have multiple cavities, side-by-side, when the forging operation requires successive passes with different dies. Fig. 2A4e shows typical shaped rollers. Levers, shafts, and axle blanks are typical roll-forged parts.

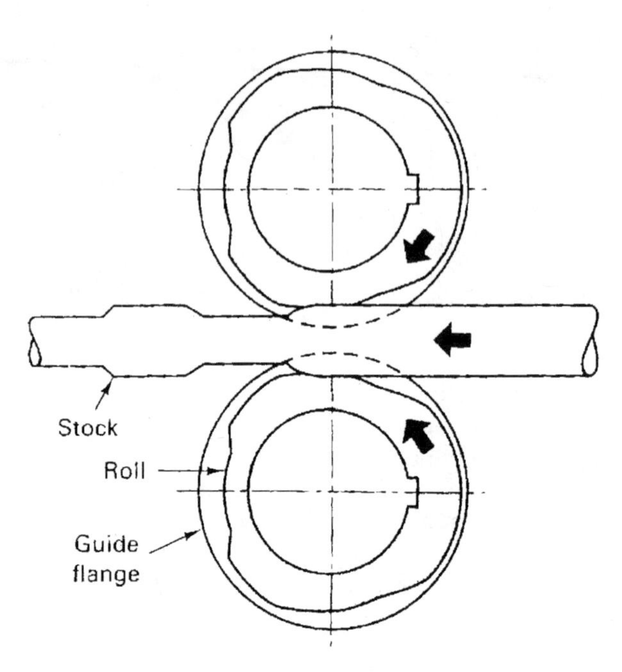

Fig. 2A4e The roll forging process. The stock is formed between two shaped rolls. (*Courtesy Forging Industry Association.*)

A4f. ***isothermal forging*** - is a forging operation during which the workpiece metal temperature remains uniform and constant. This is achieved by heating the dies to the same a temperature as that of the forging material. The dies then do not chill the forging, allowing it to be forged with very slow strain rates. The approach overlaps and has the same purpose as precision forging - to produce forgings that are closer to the net shape of the finished parts. Aluminum, nickel, and titanium are the most common materials to be forged by this process, and are used particularly in the manufacture of aerospace components.

A4g. ***swaging*** - is described in sections 2D12 and 2F1 and is normally a cold-forming operation. However, in some situations, it is performed on heated workpieces. The surface finish then is inferior to that achieved in cold swaging but some work-hardenable materials perform better when swaged hot.

A4h. ***ring rolling (mandrel forging)*** - is a process for making ring-shaped parts of particular cross sections by rolling a ring-shaped blank between rollers that control the diameter, width, height, and cross-sectional shape. The ring cross sections may have rectangular or contoured elements. See Fig. 2A4h.

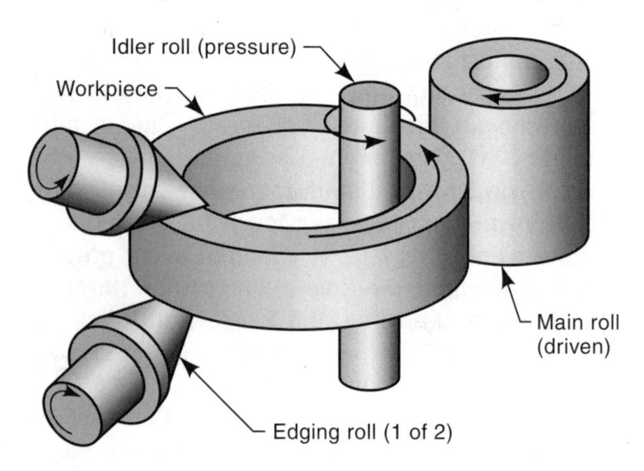

Fig. 2A4h Horizontal ring rolling to make a ring-shaped part with a rectangular cross section. (*Courtesy Forging Industry Association.*)

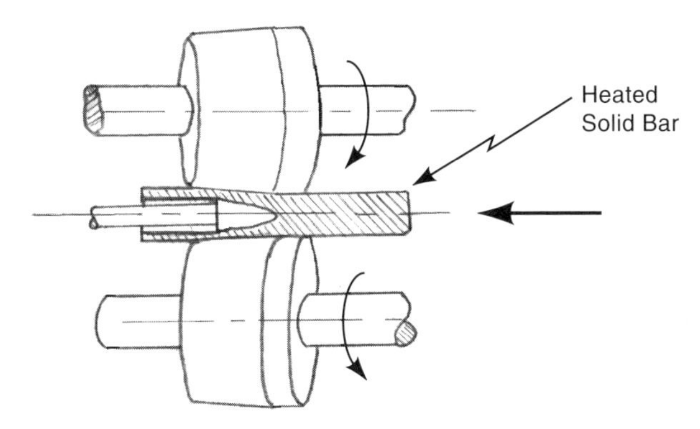

Fig. 2A5 Hot piercing, the first step in a method for making seamless tubing.

A5. ***piercing*** - as a hot working operation is different from that performed on cold sheet metal (section 3C5). Piercing, sometimes called ***rotary piercing***, is a process for manufacturing thick-walled seamless tubing. The process is illustrated in Fig. 2A5. In this process, a heated bar is passed between two tapered rollers that are canted at opposing angles of about 6 degrees from the axis of the heated bar. The bar is pushed forward against a pointed mandrel. The pressure and taper of the rollers tends to flatten the bar and cause a small crack to develop in the center. The pointed mandrel enters this crack and forces the metal to flow into a tubular shape. As the heated bar advances, a length of seamless tubing is formed. Various secondary operations are performed to size and straighten the tubing and to change the wall thickness. Seamless tubing up to about 150 mm (6 in) in diameter is made with this process but, when the piercing is used as a secondary operation on existing tubing, diameters up to 610 mm (24 in) are feasible[1].

A6. ***pipe welding*** - involves making pipe from heated strip (called "skelp" in this process) by roll forming it into cylindrical shape and pressing the heated edges together with enough force so that they fuse together to form a butt weld. The operation can also be performed by pulling the skelp through a tapered cylindrical die. The skelp can also be roll formed so that the edges overlap and form a lap joint. The rolls force the overlapping material against an internal mandrel where it fuses together. In all cases, additional rollers size and shape the welded pipe. Diameters for butt-welded pipe range from 3 mm (1/8 in) to 75 mm (3 in) and in lap welding from 50 mm (2 in) to 400 mm (16 in). Product length in the lap method is limited to about 7 meters (23 ft) because of the need for an internal mandrel[2].

A7. ***hot spinning*** - is metal spinning (section F6) with heated instead of cold material. Two common applications are the closing of the ends of tubing or cylindrical containers and the spinning of heavy plate stock.

A8. ***creep forming*** - relies on the property of some metals to flow at a slow rate when put under stress and heated. The stress level is below the yield point and the temperature is below that which anneals or otherwise heat treats the material. The process is used in the aerospace industry for large, thin components that require shallow contours. The manufacture of airplane wing skins is a common application. Aluminum sheet material is clamped in a fixture. Then the fixture with the clamped part is placed in an oven for a number of hours, typically between 4 and 8 hours. An alternative is to supply heat directly to the fixture. The sheet metal gradually assumes the shape of the tooling without significant internal stresses and retains the shape after the cycle is completed. Titanium and beryllium are also processed by this method with shorter holding times.

A9. ***warm heading*** - is the same as cold heading, described in section I2 below, except that the metal is heated to increase its ductility. The amount of upsetting possible per press stroke is increased. The work metal is heated to between 300 and 1000°F (150 and 540°C)[5]. The temperature depends on the metal to be headed, but is below the recrystalization temperature. The process is used chiefly for materials such as austenitic stainless steels, higher carbon steels, and other metals that are difficult to cold head because of their rapid work hardening characteristic[3].

B. Primary Cold-working Operations

B1. ***rolling*** - Cold rolling is a common process in the production of sheet, strip, bar, and rod forms of steel and other materials. It is similar to the hot rolling described in section A1 except that the metal is at room temperature. The material is passed between opposing rollers, sometimes with back-up rollers or in clusters of rollers, that compress the material and reduce its thickness - up to about 50%. Improved surface finish and dimensional accuracy are provided by the operation. With shaped rollers, various cross-sectional shapes can be produced and the operation can be a substitute for extrusion or machining. Cold-rolled material may exhibit improved yield strength because of the work hardening effect of the rolling operation. All malleable metals can be cold rolled.

B2. ***cold drawing*** - is a common method for reducing the size of wire, bar, tubing, and other shapes and is a different operation than the drawing (or deep drawing) of sheet metal. It may also be performed in conjunction with cold rolling or extrusion to produce special cross-sectional shapes. The material is forced through a die whose opening is smaller than the size of the entering metal. Except at the start, a pulling force is used to move the material through the die. The operation, illustrated in Fig. 2B2, improves the surface finish and dimensional accuracy of the material processed as well as reducing its cross sectional dimensions and sometimes, slightly changing its shape. Several dies are usually used in series, each progressively smaller than the preceding one. With some materials and depending on the size reduction and shape, intermediate annealing may be required between draws. Scale and dirt, if any, must be removed from the metal before drawing, and lubricants are normally employed to facilitate the operation. Cold finished steel usually is processed by cold rolling and drawing, but turning, grinding, polishing, and straightening operations may also be involved. The operation is performed on both solid wire (Fig 2B2, view a) and tubular material (Fig. 2B2, view b).

Fig. 2B2 Cold drawing. a) wire, b) tubing. The material is pulled through the die to reduce the diameter.

C. Sheet Metal Cutting Operations

C1. ***shearing*** - often refers to the cutting of sheet metal in a straight line without the generation of chips and without melting or oxidation. However, the cut may be curved as well as in a straight line. In many sheet metal operations such as blanking, punching (piercing), notching, parting, slitting, trimming, and lancing, the cutting action consists of shearing. The basic process is illustrated in Fig. 2C1. In the usual shearing process, the sheet metal lies on a lower tooling member, the die. The other tooling

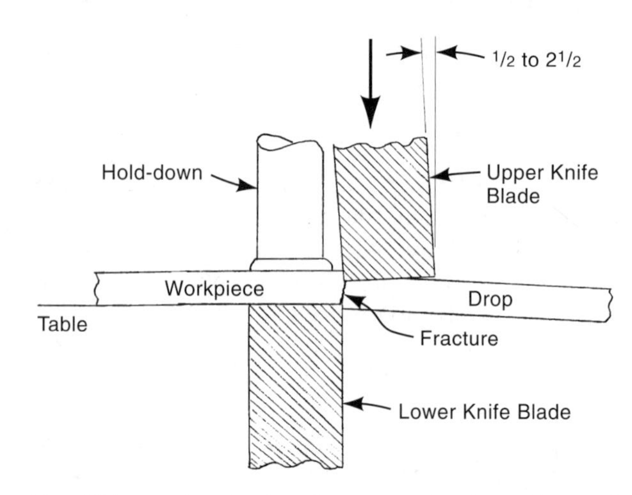

Fig. 2C1 A simple shearing operation is performed on a squaring shear.

member, the punch, has a sharp edge that is approximately opposite the sharp edge of the die. (There is a slight clearance between the two). Cutting takes place when the punch descends downward into the sheet metal workpiece. The sheet is cut and also deformed in a downward direction. When the force of the punch entering the workpiece exceeds the strength of the metal, the workpiece fractures. Fig. 2C4 shows the shearing action of a hole punch or blanking die. The edge of the sheared metal shows a slight depression at the one edge, where the punch or die has deformed it. There is an area of a smooth metal edge, the sheared portion, adjacent to which is an area of rough surface, the "break-away" portion. Fig. 2C4 shows the edge of a typical sheared or blanked sheet metal piece. Note the three portions of the edge.

C1a. ***squaring shears*** - are machines that perform the shearing operation. Fig. 2C1a illustrates a hydraulic squaring shear. The sheet to be cut is positioned on the machine table under the ram, which has a blade attached. The blade and the cutting edge of the machine table are in close alignment. When the ram descends, a set of clamping fingers or a clamping bar descend with it and hold the workpiece sheet firmly. The blade then moves down further to sever the workpiece. Usually, the blade is fastened to the ram at a slight angle to the horizontal so that the shearing action is progressive from one side to the other as the blade descends. The ram movement is usually machine-powered except for small machines, which may utilize human arm or leg power to lower the ram. The method makes accurate straight cuts on sheet. Sheet metal up to a maximum of about 1.5 in (38 mm) in thickness, and up to 20 ft in width, can be processed on the largest machines.

C1b. ***alligator shears*** - get their name because the pivoting action of the blades is reminiscent of the motion of an alligator's jaws. The shearing action is similar to that of scissors. These machines have somewhat shorter cutting blades than squaring shears and therefore are best adapted and most used for shearing rods, bars, and sections, rather than sheet, though they can cut sheet and plate material.

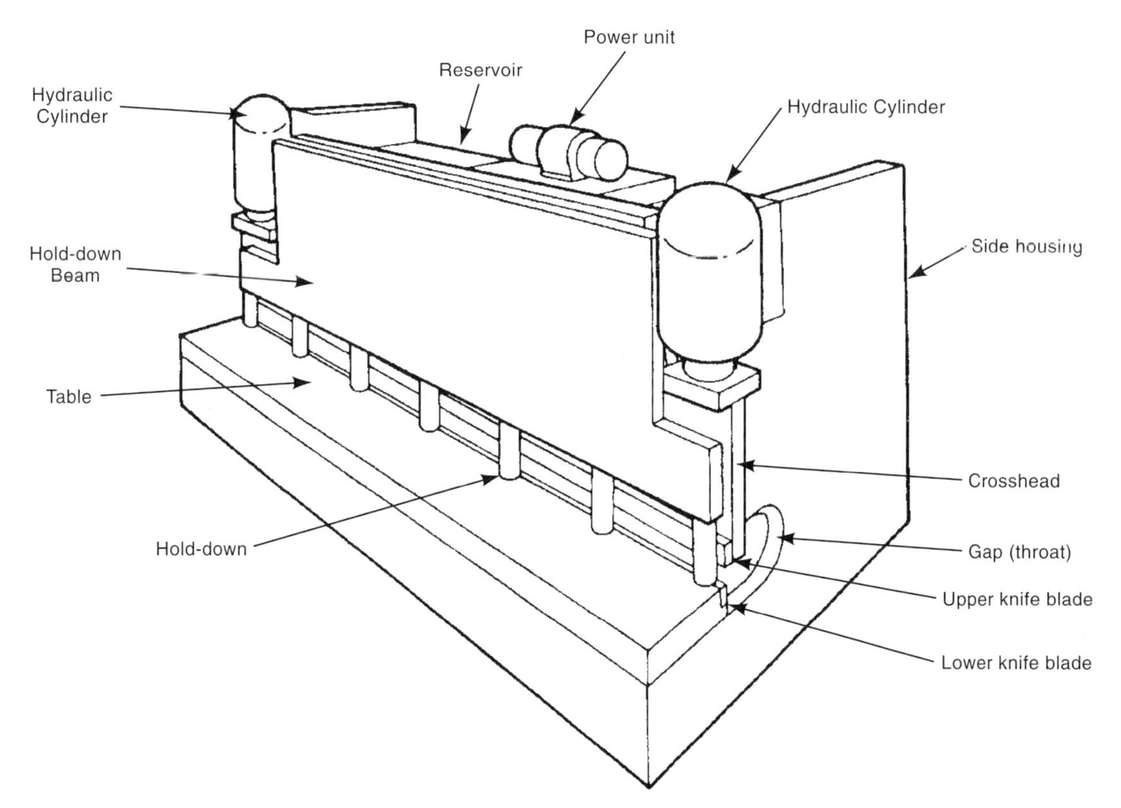

Fig. 2C1a A hydraulically-powered squaring shear. (*Courtesy Pacific Press Technologies.*)

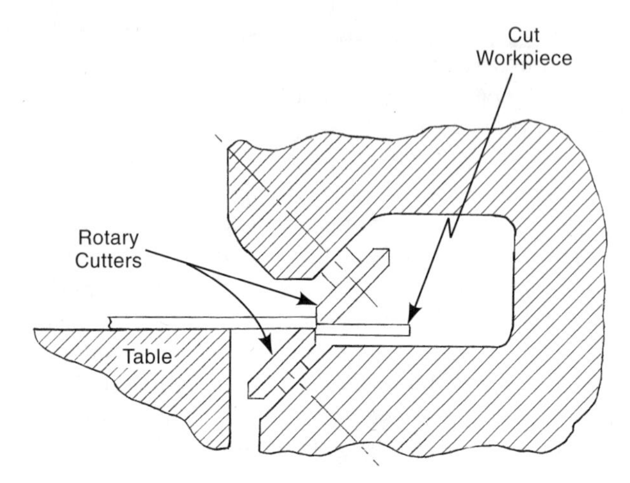

Fig. 2C1c A rotary shear used to make curved cuts in sheet material.

C1c. ***rotary shears*** - use two opposed and canted rollers, ground at an angle, that roll across sheet or plate and produce a shearing action. The arrangement is shown in Fig. 2C1c. The upper cutter is powered for both rotation and position. Steel plate up to about 1 in (25 mm) in thickness can be cut this way. Shearing can take place along a straight or curved path, and irregular parts can be cut out with this method. It is adaptable to low-quantity production.

C2. ***nibbling*** - is a sheet metal operation in which a long cut is made by punching a series of overlapping holes or slits in the workpiece. The nibbling press typically operates at 300 to 900 strokes per minute[3]. This approach permits complex cuts to be made, including blanking with simple standard tooling. The method works best when controlled in a CNC turret punching machine but can also be controlled manually, and by less sophisticated automatic equipment.

C3. ***slitting*** - is a shearing operation that divides sheet or coiled sheet metal into narrower widths. Circular shearing tools in a slitting machine, illustrated in Fig. 2C3, are mated cutters. The raised circular cutters on one arbor match a space between circular cutters on another arbor. Sharp edges perform the slitting operation. The operation is normally performed in a line of machines that feed coiled sheet material to the slitting machine and re-coil the stock after slitting. Edges are commonly slit to insure accurate width of the slit material.

C4. ***blanking*** - is a presssworking operation used in the production of parts from sheet metal. In the operation, a flat workpiece of the desired size and shape is cut out completely around the periphery in one press stroke with a punch and die set. The punch's cross-section and the die opening match the shape of the desired blank and differ in size only to the extent of the clearance between them. The punch descends and shears the sheet stock to sever the blank from the sheet, leaving stock entirely around the blanked workpiece. Fig. 2C4 illustrates the process.

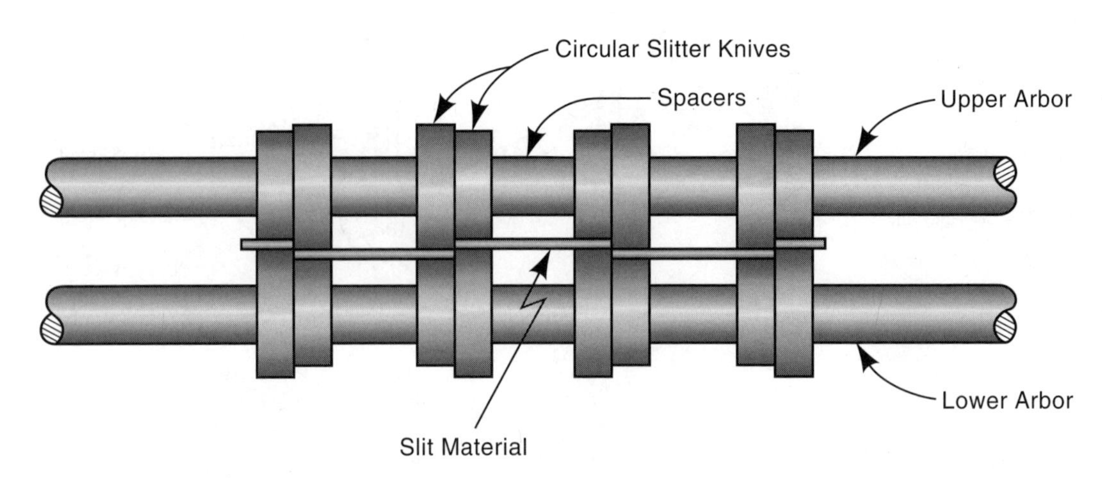

Fig. 2C3 Rotary slitting. Sharp edges of the circular slitting knives shear the sheet material.

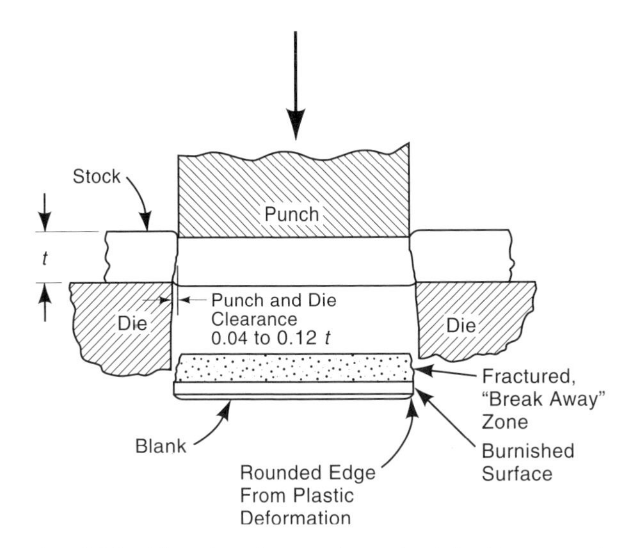

Fig. 2C4 The shearing action of a blanking punch and die. Note the edge of the blank which is typical of all sheared, blanked, or punched parts. There are three different edge areas: 1) the area of plastic deformation where the die edge first impinges the sheet, 2) the burnished area where pure shearing takes place, and 3) the breakaway area where the metal sheet fractures.

C4a. ***steel rule die blanking*** - This method is particularly suited to lower-quantity production of parts made of softer materials such as leather, rubber, plastic sheet, fibre, and paperboard. Dies are somewhat similar to cookie cutters. They consist of a thin strip of hardened steel bent to the shape of the blanked part and held in a base of wood or other material. The exposed edge of the strip is sharpened. The die is mounted on a press ram. On the opposing surface on the press bed, there is a hardened steel punch made from flat stock and cut to the shape of the blank. When the press ram descends, the steel cutting edge penetrates the work material and there is a shearing action between the steel rules and the punch. Although the materials noted above are most suitable, this method has allowed steels up to 3/8 inch (10 mm) and aluminum to 0.55 in (15 mm) to be blanked in short runs. Blocks of rubber in the die serve to strip the blanked workpiece from the die. Fig.2C4a shows the elements of a typical steel rule die.

C4b. ***dinking*** - is similar to steel rule die blanking in that the cutting is performed by a knife-like

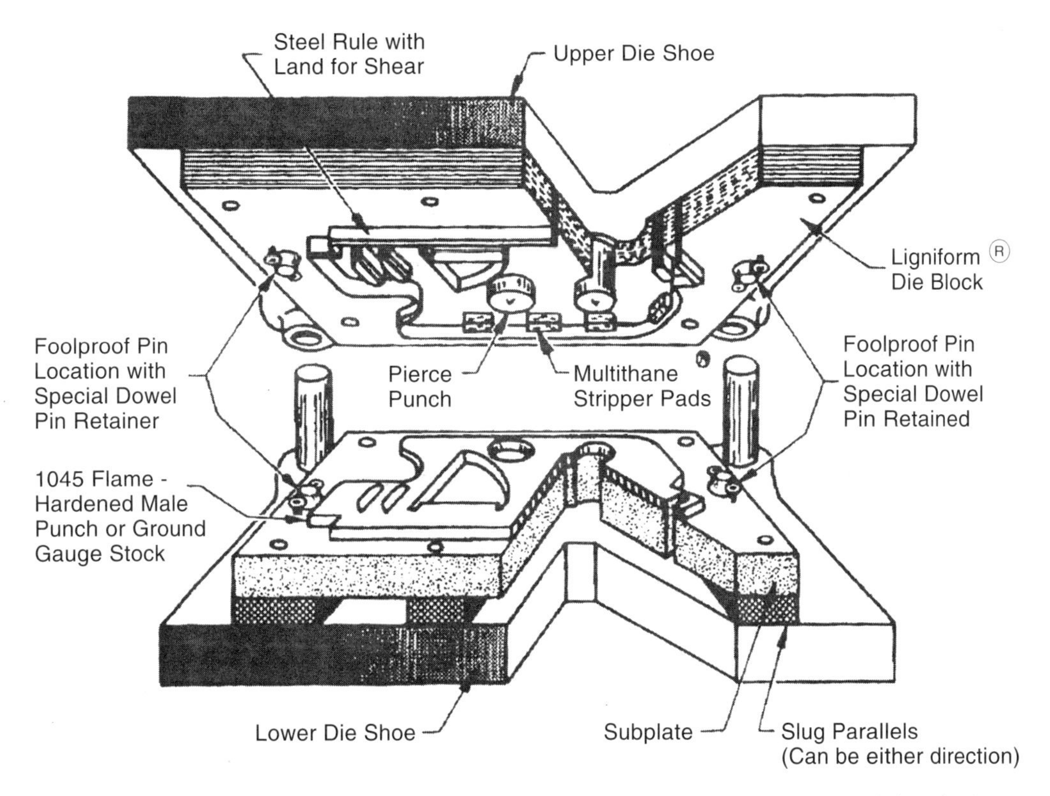

Fig. 2C4a Cross-section of a steel-rule blanking die. (*Courtesy J.A. Richards Co.*)

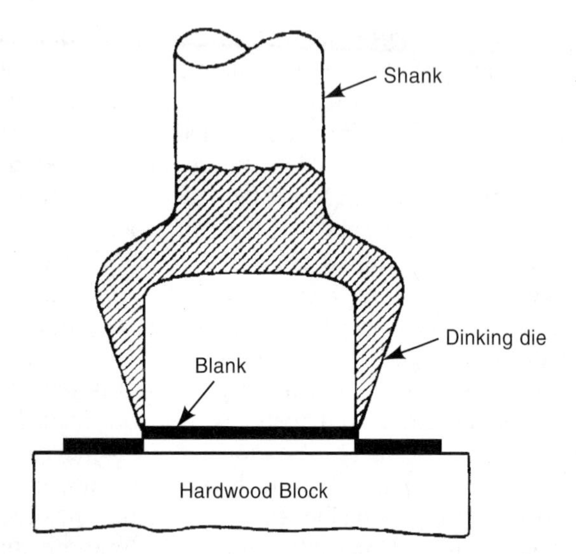

Fig. 2C4b Cross-section of a typical dinking die. The work material is cut when the die presses it against the hardwood block. (*from Niebel, Product Design and Process Engineering, McGraw-Hill, NY 1974*)

edge thast can bear against a flat surface. In this variation, the knife-like punch and its holder are all made from one piece of steel, hardened and sharpened at the edge. The process is illustrated in Fig. 2C4b. The process, like most steel rule die blanking, is used for cloth, fibre, rubber, and other soft materials.

C4c. ***cutoff*** - a stamping operation to sever a part from a sheet or strip of stock, normally with a single line cut. The operation is frequently the last one of a progressive-die sequence where a number of blanking, piercing, forming, and other operations are best performed on a work piece while it is still attached to the stock and the cutoff operation is the final one to create a discrete part.

C4d. ***parting*** - is similar to cutoff in that a part is severed from strip material except that the term, ***parting***, implies that some material is removed in the operation whereas cutoff implies a cut only along a single line. Fig. 2C4d illustrates the operation.

C4e. ***photochemical blanking*** - See chapter 3, paragraph S4.

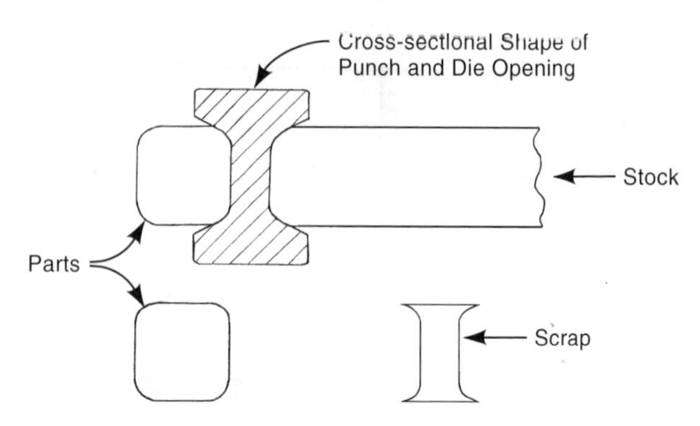

Fig. 2C4d A parting punch is used to make separate pieces from strip material when the end edges do not mate.

C5. ***punching*** - often called ***piercing***, is a press-working operation used to produce holes in sheet materials. The operation is a variation of blanking (2C4), being identical in that a separate piece (or pieces if multiple holes are punched in one press stroke) is sheared from another piece in one press stroke with a punch and die set, but different in that the purpose of the operation is to create a hole rather than a new part. The piece that is removed is scrap rather than a useful part. The hole punched can be circular or any other shape, according to the shape of the punch and the matching die opening.

C5a. ***turret punching*** - is performed on a machine that has two synchronized indexing tables, one containing punches and the other containing matching dies. The machine also has a free floating table to which the work piece is attached. The flat work piece can be moved, either manually, often with the aid of a template, or by computer-numerical control to position the work piece under the operating punch. When the press is actuated, the punch descends and punches a hole in the work piece. When there are several holes to be produced, the work piece is moved in succession to the designated locations and the press is tripped at each position. When there are holes of several sizes, the turret mechanism can index to present a different size punch and die. Notching, nibbling, lancing, louver forming, slotting, embossing, and other operations are also feasible with the appropriate tooling. This equipment is used for work pieces that are too large to be punched with a single die

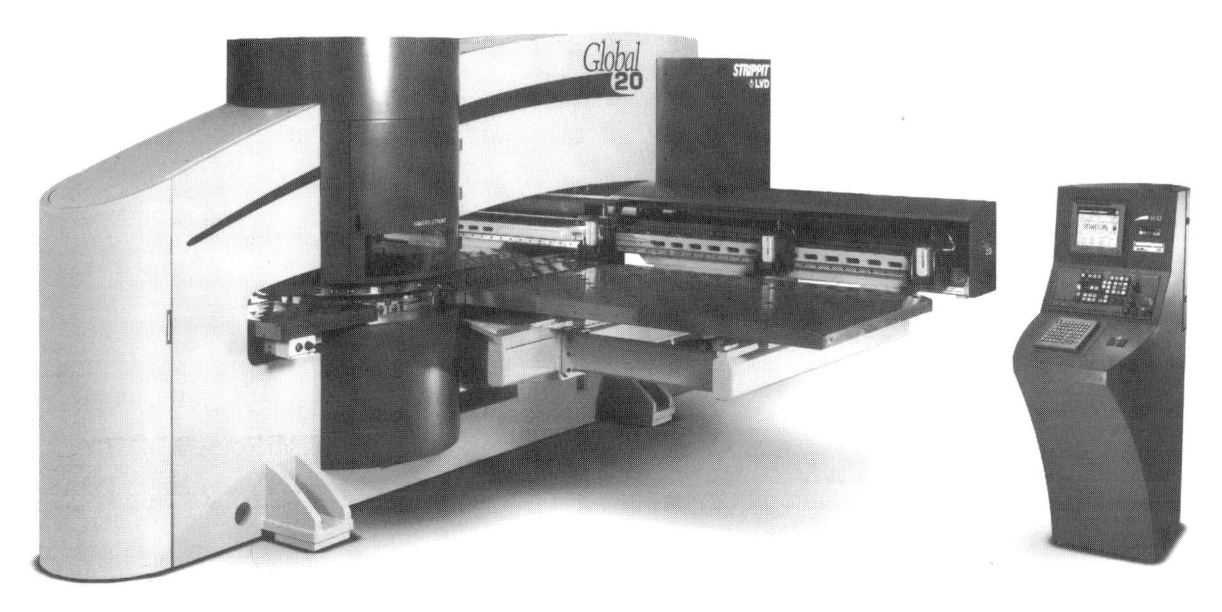

Fig. 2C5a A CNC turret punching machine. Movement of the sheet, changes in punches and dies through rotation of the turret, and the punching action, are all controlled automatically. (*Courtesy Strippit LVD, Akron, New York.*)

set or parts made in such low quantities that they do not justify an investment in dedicated tooling. See Fig. 2C5a.

C5b. ***notching*** - is punching or piercing performed at the edge of the work piece. The edge of the strip or blank becomes part of the perimeter of the piece that is removed. The operation is performed when the shape of a blank is too complex to fully incorporate in a blanking die, for low-quantity work when complex tools are not justifiable, to free material for a subsequent forming or drawing operation, or to remove material that would otherwise be distorted in a subsequent operation. A work piece with several notching operations performed is illustrated in Fig. 2C5b.

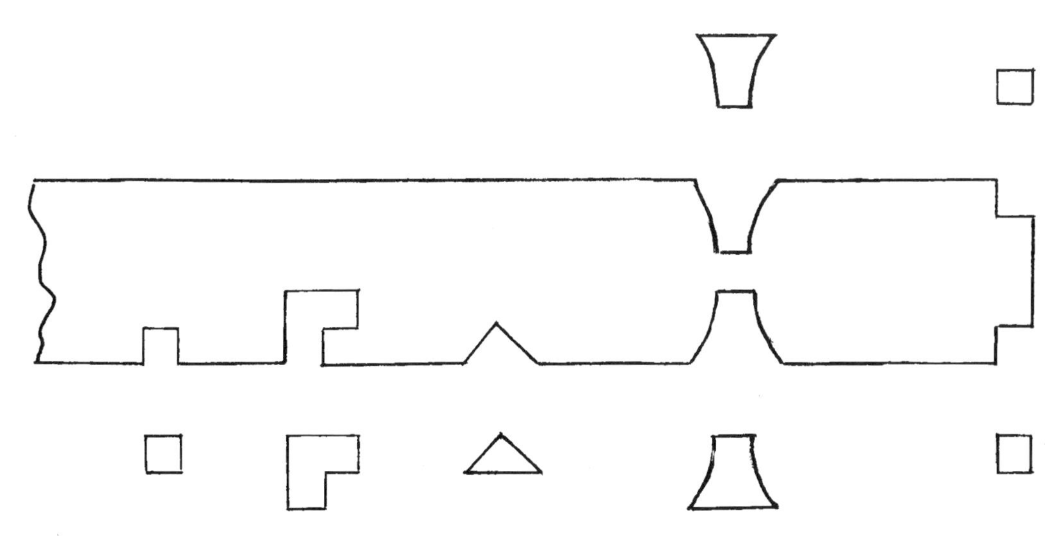

Fig. 2C5b Metal strip with various notches made to provide the desired workpiece shape or to facilitate a subsequent bending or forming operation.

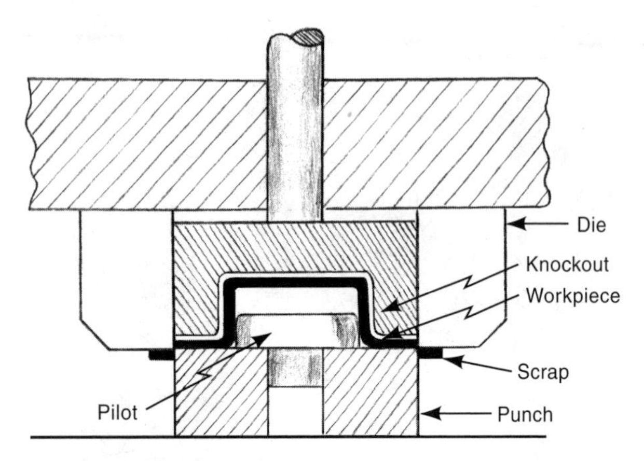

Fig. 2C6 Trimming a drawn cup. This method is intended to leave a flange at the top of the cup.

C6. ***trimming*** - is a stamping operation used to remove unwanted material that occurs as a result of some other process such as casting, forging, forming, or deep drawing. The operation is similar to blanking except that the workpiece is a semi-finished part instead of a flat piece of sheet stock, and the trimming die is shaped to contain the shape of the part. In die casting, trimming removes sprues, flash, and runners. In deep-drawing and some forming operations, trimming removes flanges left where the material was gripped or held. A workpiece with this trimming operation is illustrated in Fig. 2C6.

C7. ***shaving*** - is a secondary operation performed on sheet metal workpieces after blanking or piercing. Its purpose is to refine the edge finish or the dimensions of the blank. The process is similar to blanking except that the fit between the punch and die is very close. Very little metal, only a few thousandths of an inch, is removed from the part. The edge of the workpiece, after the operation, is sheared smooth through about 75% of the stock thickness, with reduced rounding or "pull-down" and very little rough breakaway area. Sometimes, a second shaving operation is performed on the workpiece to further improve the straightness and length of the smooth portion. Before the advent of fineblanking, the process was used to produce smooth-edged stampings. Stamped gears are a typical application.

C8. ***lancing*** - is a sheet metal operation that makes a slit or cut that is not long enough to produce a separate piece. No scrap is produced because the cut does not extend from edge to edge. The usual purpose is to facilitate a subsequent forming operation on the workpiece by allowing the material to flow more easily. A workpiece with this kind of cut is illustrated in Fig. 2C8.

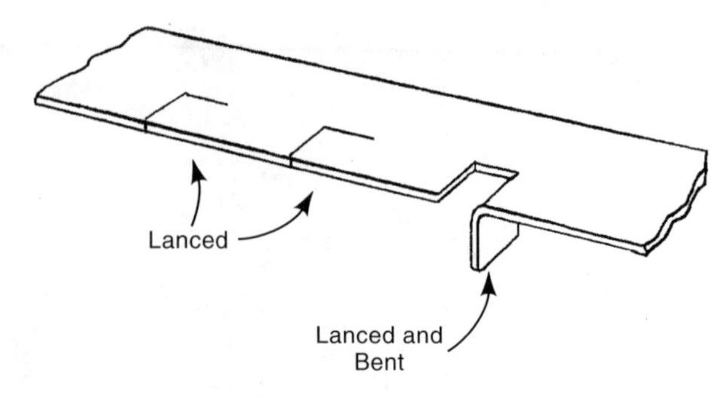

Fig. 2C8 A workpiece with lancing cuts, and one bending operation.

C9. ***fineblanking*** - is a method for producing smooth edges and other improved features on stamped metal parts. It involves both blanking and forming. The method requires closer control and several features in presses and tooling compared to those in conventional stamping. These features include a more precisely controlled press stroke, a closer fit between punches and dies, the use of a shaped pressure plate with a v-shaped impingement ring surrounding the die to hold the workmaterial and provide compressive stress in the part, a floating ejector pin to provide pressure on the material and against the punch force, and somewhat slower press speeds. Another difference is that the close-fitting punch does not enter the die. A typical fineblanking operation is shown in Fig. 2C9-1. The operation requires a press with the following characteristics: separate actions for the punch, v-ring pressure, and counterpressure; capability for a fast approach stroke, a slow shearing speed, and a fast ram retraction rate; control of the shut-height setting within 0.0004 in (0.01 mm) to insure that the punch stops at precisely the correct point; and sufficiently accurate guideways that the punch-die clearance can be maintained within half percent 2% of the stock thickness. Fig. 2C9-2 illustrates the difference between the edges of conventionally blanked and fineblanked parts. Fineblanking is utilized in the manufacture of precision parts that otherwise may require machining or other secondary

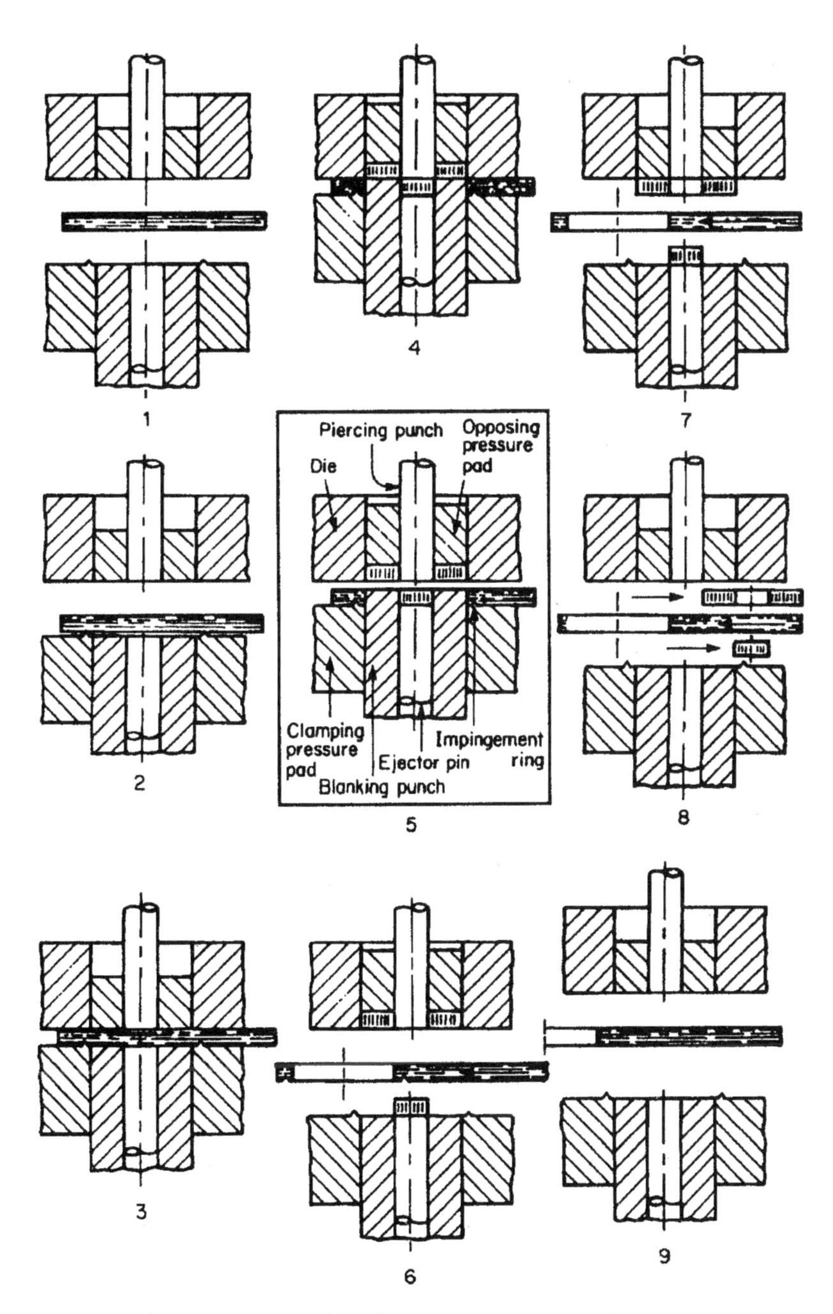

Fig. 2C9-1 The sequence of steps in a typical fine-blanking operation with a compound die that both blanks and pierces the sheet material: 1) The material is fed into the die. 2) The ram lifts the table, die set, and material to the die face. 3) The tool closes and the v-ring is embedded in the material. The counterpunch clamps the material against the blanking punch face inside the shear periphery. 4) The v-ring pressure and counterpressure are held constant while the punch moves upward, shearing the part into the die and the inner slug from the punch. At the top dead center position, all pressures are shut off. 5) The ram retracts and the die opens. 6) The v-ring elements moves upward, stripping the punch from the skeleton material and pushing the inner slugs up out of the punch. The material feed begins. 7) The counterpressure is reimposed, stripping the part from the die. 8) The part and the slugs are ejected from the die by either air jets or by a removal arm. 9) The cycle is complete and ready to repeat. (*from Bralla, Design for Manufacturability Handbook, McGraw-Hill, New York, 1999*)

Fig. 2C9-2 Comparison of the edge surface produced by conventional stamping (left) and fine blanking (right). (*Courtesy Feintool New York, Inc.*)

operations after blanking, and for making parts that may not be feasible by conventional stamping methods. Stacked gears, cams, sewing machine and business machine parts are examples. The method is not limited to blanking: forming, piercing, coining, embossing, etc. may also be performed on the workpiece with a separate operation or with compound and progressive dies, when required.

C9a. ***semi-piercing*** - is piercing with a shortened punch stroke so that the piece that would otherwise be removed from the part is offset rather than severed. If done properly, there is no fracture in the workpiece metal. In fact, the metal joining the offset and the base material is strengthened by the cold working it receives. The offset part can function as a cam, a stop, a rivet, or a locating pin. Fine-blanking is particularly suited for producing strong and accurate semi-piercings as illustrated in Fig. 2C9a.

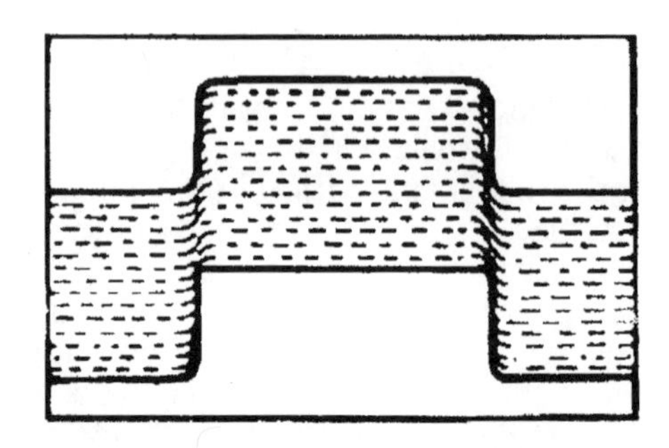

Fig. 2C9a Cross section of semi-piercing in sheet metal. (*from Bralla, Design for Manufacturability Handbook, McGraw-Hill, New York, 1999*)

D. Press Bending and Forming Operations of Sheet Metal

D1. ***bending*** - is the deformation of material about a straight axis. There is little or no change in surface area of the workpiece. Material on the outside of the bend tends to be stretched and material on the inside tends to be compressed. Bending is a very common sheet metal operation. When performed as an individual operation, usually in smaller quantity production, it can be done on a bending brake (sometimes called a bar folder), as shown in Fig. 2D1-1, if the sheet is about 1.5 mm (0.060 in) or less in thickness. Press brakes are used for thicker stock. Bending as a punch press operation is often carried out with wiper dies as illustrated in Fig. 2D1-2.

D1a. ***press brake bending*** - Press brakes, as illustrated in Fig. 2D1a, are mechanical or hydraulic presses with long, narrow, stationary

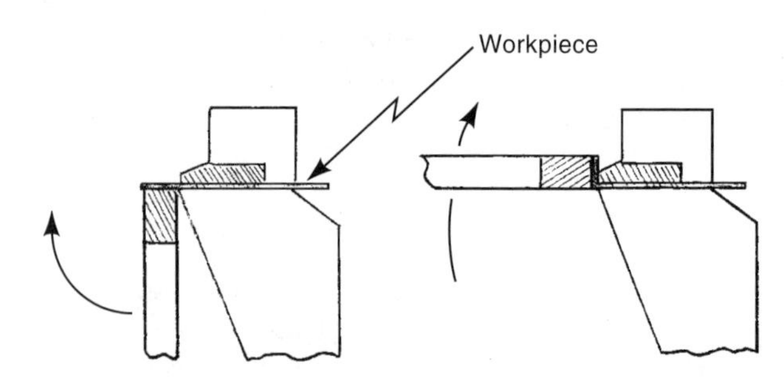

Fig. 2D1-1 The bending action of a manually-operated bending brake (bar folder).

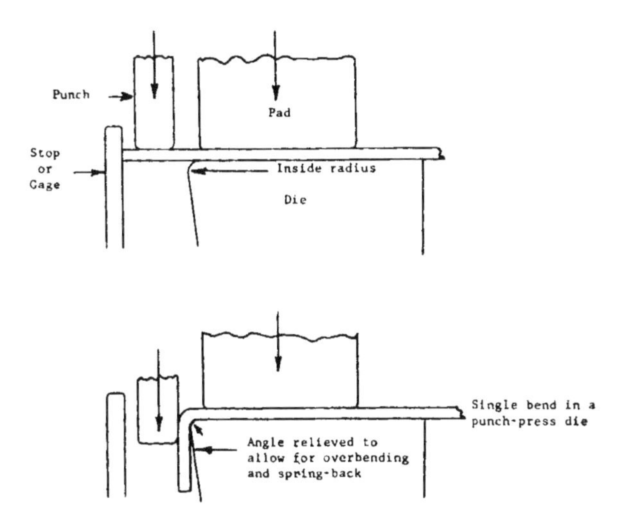

Fig. 2D1-2 Bending sheet metal with a punch press wiper die. *(from Niebel, Product Design and Process Engineering, McGraw-Hill, NY 1974)*

beds. Bed lengths range from 2 or 3 ft to 30 ft and press tonnages from 10 to several thousand. The ram stroke is short but adjustable. Dies are long, narrow, and often simple V-dies. Both sharp and gentle bends can be made, depending on the shape of the dies. Piercing, notching, forming, shearing, edge curling, beading, hemming, corrugating, and tube forming can be performed with suitable dies. In bending, sheet metal, placed between the bed and the ram is most commonly bent once with each press stroke. The bend occurs when a shaped punch, attached to the press brake ram, descends against the workpiece, forcing it into a suitably-shaped die, fastened to the press brake bed. Multiple bends are made by repositioning the workpiece sheet between press strokes. Press brakes are used in the bending of long, narrow workpieces and for other workpieces made in small quantities where standard press brake tooling can be employed.

D1b. ***V-die bending*** - V-block dies are commonly used in press brakes. The die half has a V-shaped cavity and the punch has a corresponding V-shape of the same or somewhat lesser angle. The punch and die extend over the entire width of the press. The angle of the bend is the same as that provided by the V-block less any spring back.

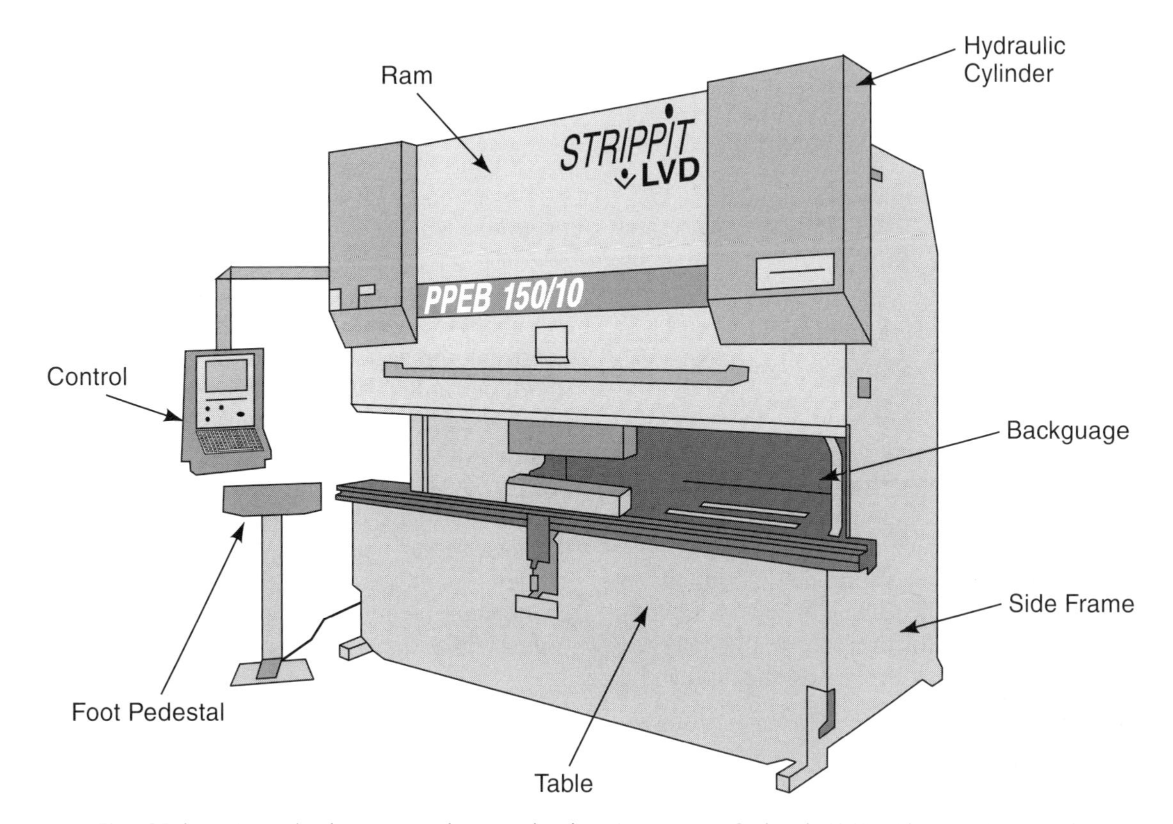

Fig. 2D1a A typical powered press brake. *(courtesy Strippit LVD, Akron, New York)*

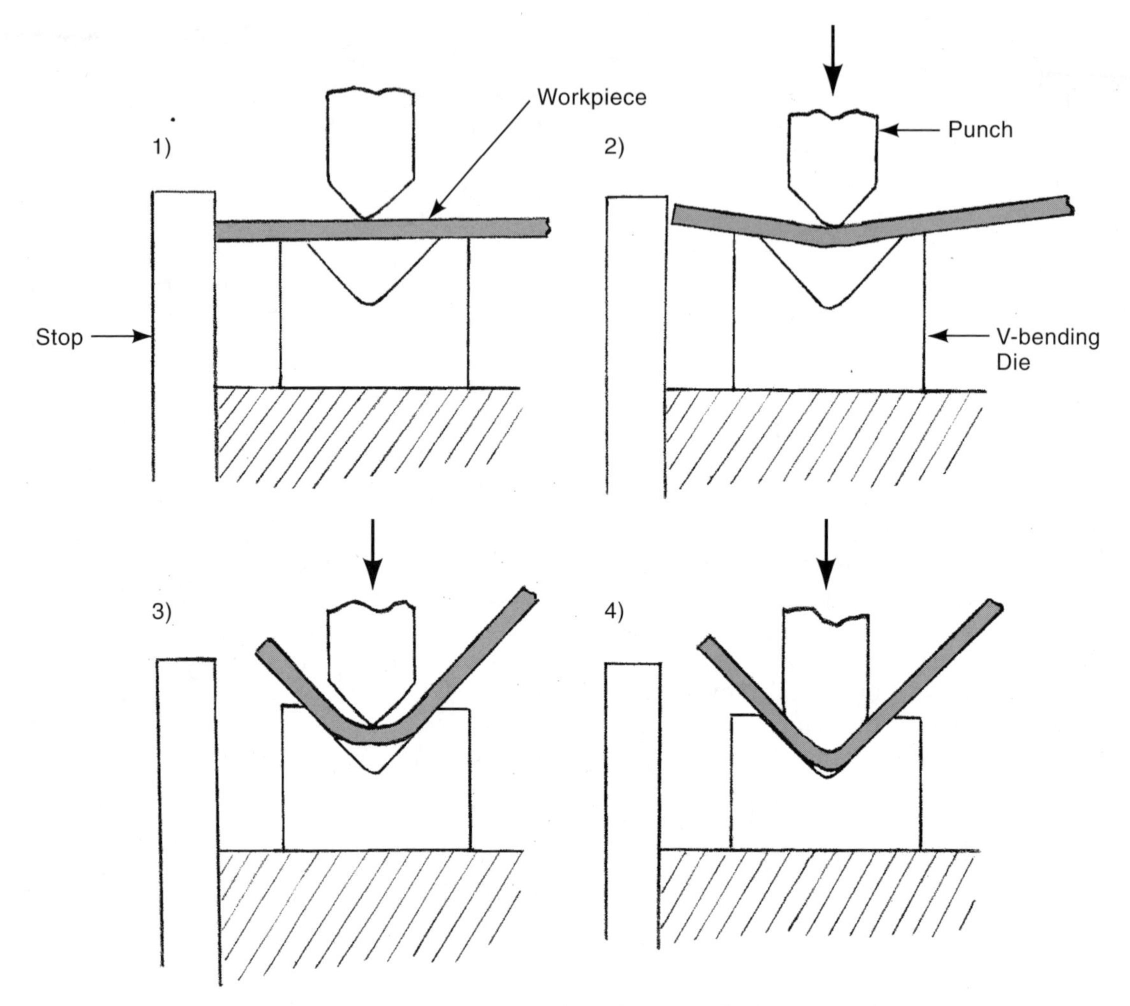

Fig. 2D1b Making a 90 degree bend in a V-die in a press brake.

Typical examples of V-dies are shown in Figs. 2D1b, 2D1c, and 2D11.

D1c ***air bending*** - In air bending, the upper die in a press brake set does not bottom in the v-block. The degree of bend depends on the depth of movement of the upper die, as shown in Fig. 2D1c. This approach has the advantage that adjustments can be made in the process to accommodate variations in materials and that one die set can be used for bends of a variety of angles. However, the results may not be as consistent as they are when the upper die bottoms in the V-block.

D1d. ***punch press (bending die) bending*** - all the bending operations feasible with press brakes can also be made in regular punch presses, though not over as great a length. Dies similar to the V-block dies commonly used in press brakes are sometimes used but other die configurations are more common. Dies dedicated to a particular part permit high-production bending of one or more bends, often in combination with other press operations. Again, the descending stroke of the press ram typically carries a punch that pushes the workpiece against an adjacent form block, bending one or more sides of the workpiece.

D2. ***forming*** - is any operation that changes the shape of a solid-state workpiece by force or pressure. Forming operations have one thing in common: permanent useful deformation or plastic

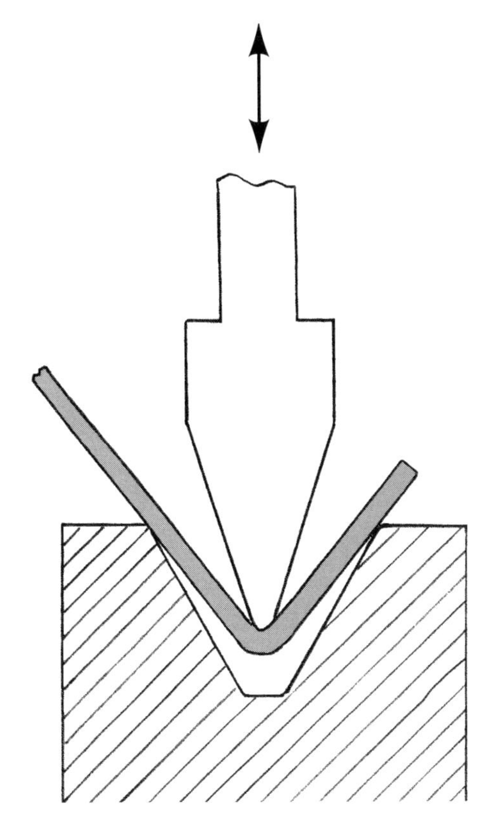

Fig. 2D1c Air bending sheet metal, using a V-die in a press brake.

flow of the workpiece material as a result of forces applied to it. This definition covers operations such as bending, rolling, forging, cold heading, extrusion, metal spinning, swaging, coining, and deep drawing. It applies to sheet metal and material of other shapes. Casting and machining, however, are not be forming operations. In sheet metal operations, however, the term, forming, is often applied to operations that involve somewhat more deformation than simple bending but that do not fall into another clear-cut category such as deep drawing, coining, embossing, or rotary swaging, especially if the operation is intended to change the thickness of the part. Included in the term, forming, then would be the production of dish-shaped parts, flanges or other bends along curves or at corners and other miscellaneous sheet metal operations. Fig. 2D2 illustrates a simple tooling arrangement for forming a sheet metal workpiece.

D3. ***forming with rubber tooling***

D3a. ***rubber tool forming (rubber pad forming)*** - is a method of bending and forming that simplifies the tooling required by substituting a confined rubber pad for one half of the die. The other half of the die consists only of a simple block around which or against which the part is formed. The confined rubber pad acts like hydraulic fluid exerting force in all directions as it is compressed. The tooling tends to be quite simple and the forming blocks can be made of low-cost materials including non-metals. The process is suited for short-run production since tooling costs are low. The process is commonly used in making aluminum parts in the aircraft industry.

D3b. ***Guerin process*** - is a common rubber-tool forming process that also can be used to blank parts. A rubber pad of fairly soft durometer, usually 6 or more inches thick, is fastened to the press ram and is surrounded on the sides with steel or cast iron walls strong enough to contain the rubber against the pressure of forming. The forming block is fastened to the press platen. When the press ram descends, the rubber pad forces the workpiece against the forming block and, as the workpiece bends, rubber surrounds the workpiece and applies horizontal as well as vertical pressure to force it against the forming block. The process is illustrated in Fig. 2D3b. The maximum feasible depth of the formed portion is about 1.5 in (37 mm).

When blanking is required, the edges of the lower blocks are hardened and have sharp edges. The blocks can be as little as 3/8 in (10 mm) thick. The pressure of the rubber pad causes the workpiece metal to be sheared by the sharp edge. Rubber of higher durometer is used when blanking is included in the operation. Aluminum sheet up to about 0.050 in (1.3 mm) thick can be blanked.

D3c. ***Marform process*** - is a rubber-tool process that uses a deeper rubber pad in the press ram and adds a flat steel holder plate, supported hydraulically or by springs to act as a blankholder during the downward movement of the rubber pad. The arrangement permits deeper forming and drawing of irregular parts. Drawing depths equal to the workpiece diameter are feasible.

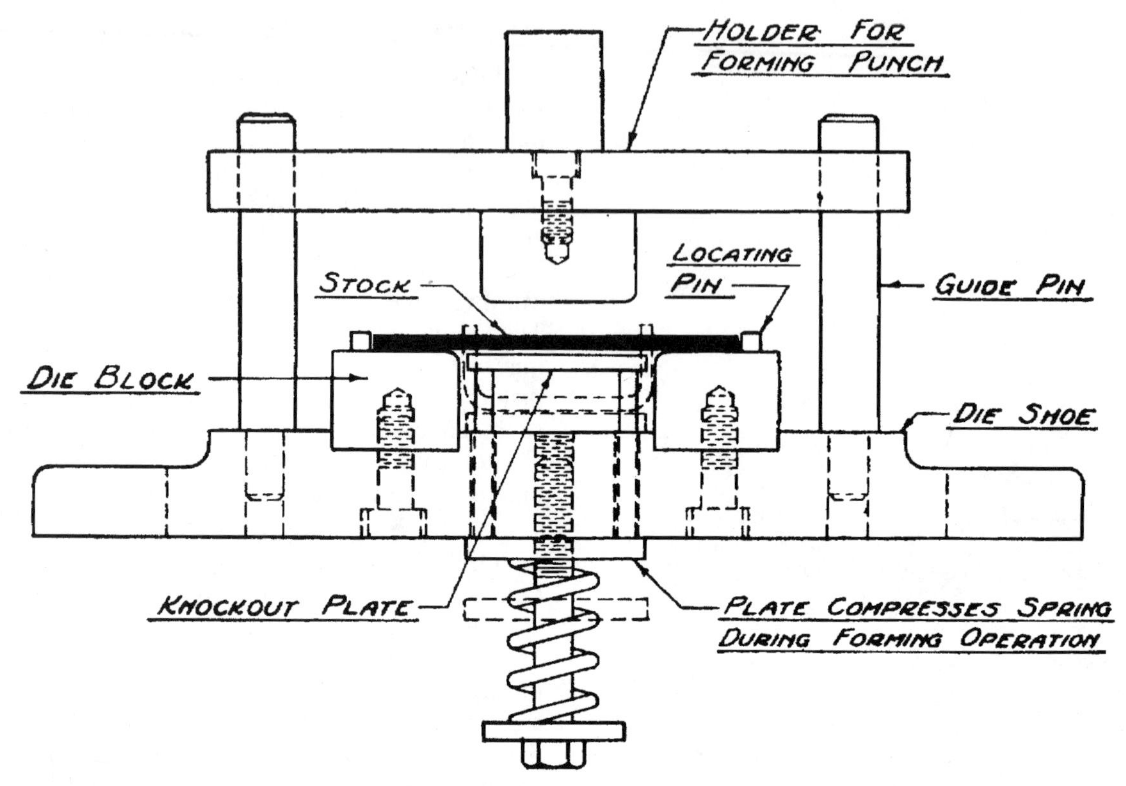

Fig. 2D2 A punch and die made for a forming operation.

D3d. ***rubber diaphragm forming (hydroform process) (fluid forming)*** - in this process, the rubber pad is replaced by a hollow flexible rubber diaphragm which contains hydraulic fluid. This fluid provides improved sideways pressing capabilities so the process can be used to form deeper parts. The forming block is attached to a hydraulic cylinder that forces it upward against the diaphragm, raising

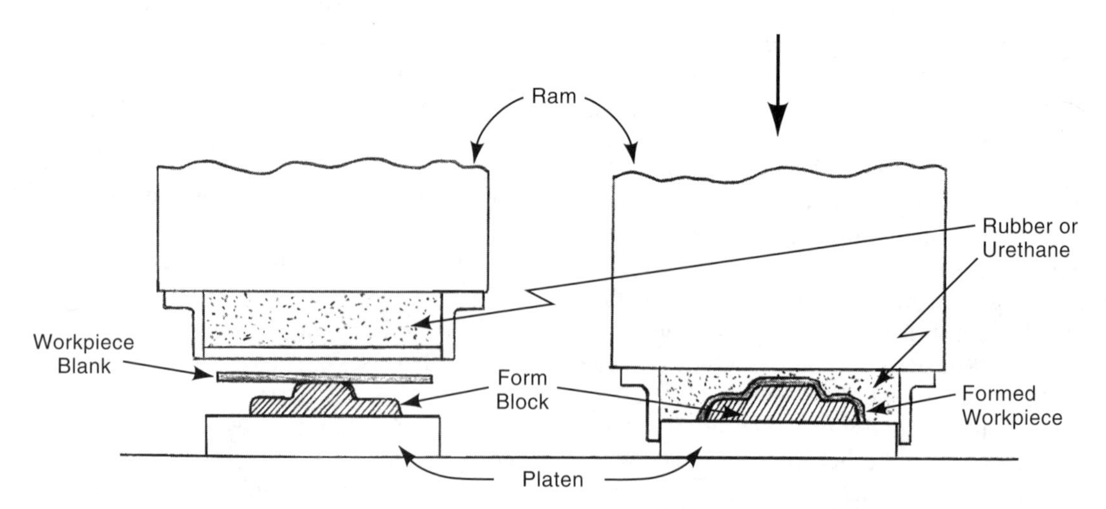

Fig. 2D3b The Guerin process for rubber tool forming. The rubber takes the place of one half of the forming die that would otherwise be required.

the pressure in the system. Hydraulic pressures up to 15,000 psi (100 MPa) may be used. Fig. 2D3d illustrates the process.

D3e. ***Verson-Wheelon process*** - is another rubber-diaphragm process. There is no press stroke. Instead, hydraulic pressure is applied directly to a diaphragm that forces a rubber pad against the workpiece. The process is applicable only for relatively shallow formed parts, similar to those processed with the Guerin process, but can produce somewhat greater detail and variety of forms. Fig. 2D3e shows how the process works.

D4. ***drop hammer forming*** - uses repetitive strokes of a drop-hammer press to form sheet metal parts. (See drop forging, A4a.) The process is particularly suitable for parts with shallow, smooth contours and generous radii, including parts with double curvature. Beaded parts can be produced.

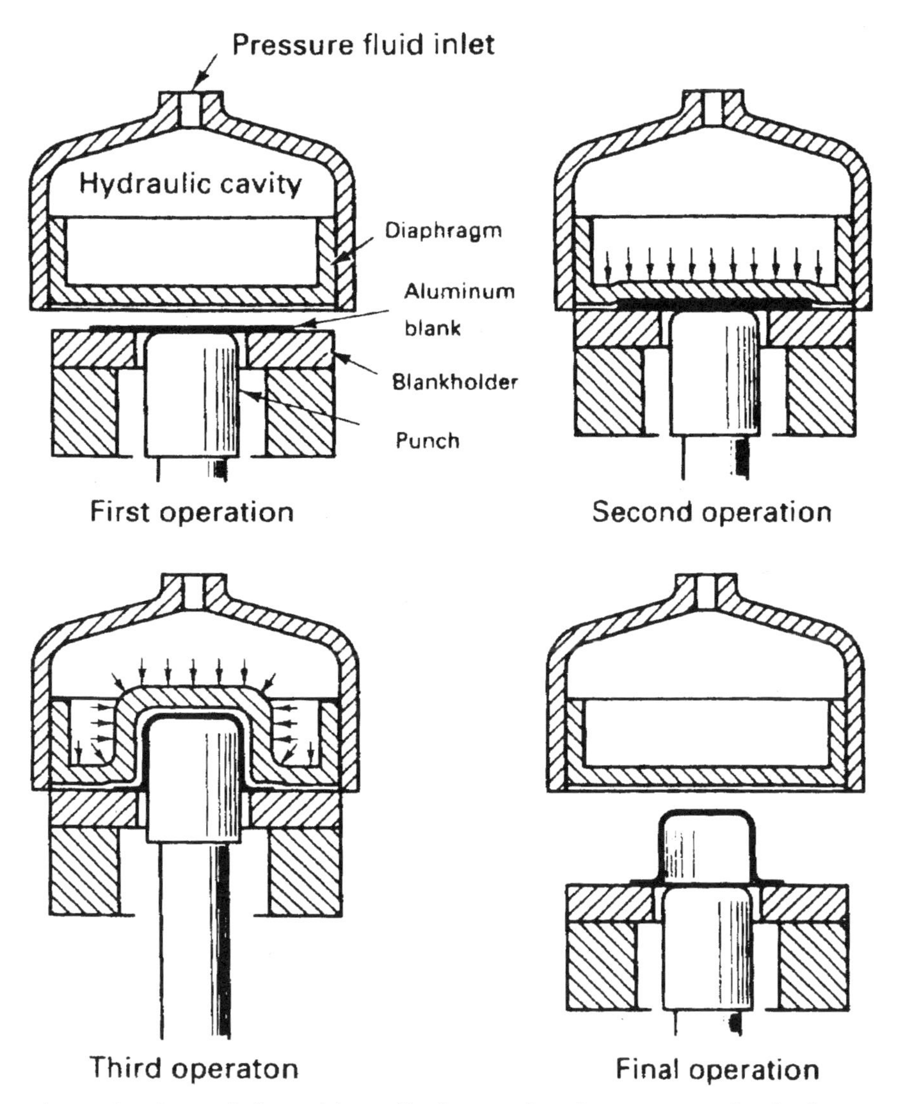

Fig. 2D3d A schematic view of the rubber diaphragm forming process (hydroform or fluid-forming process) showing: 1) the blank in place ready for forming, 2) the press closed and the cavity pressurized, 3) the ram advanced into the cavity and, 4) the pressure released and the ram and press retracted. (*Courtesy Aluminum Association, Washington, DC*)

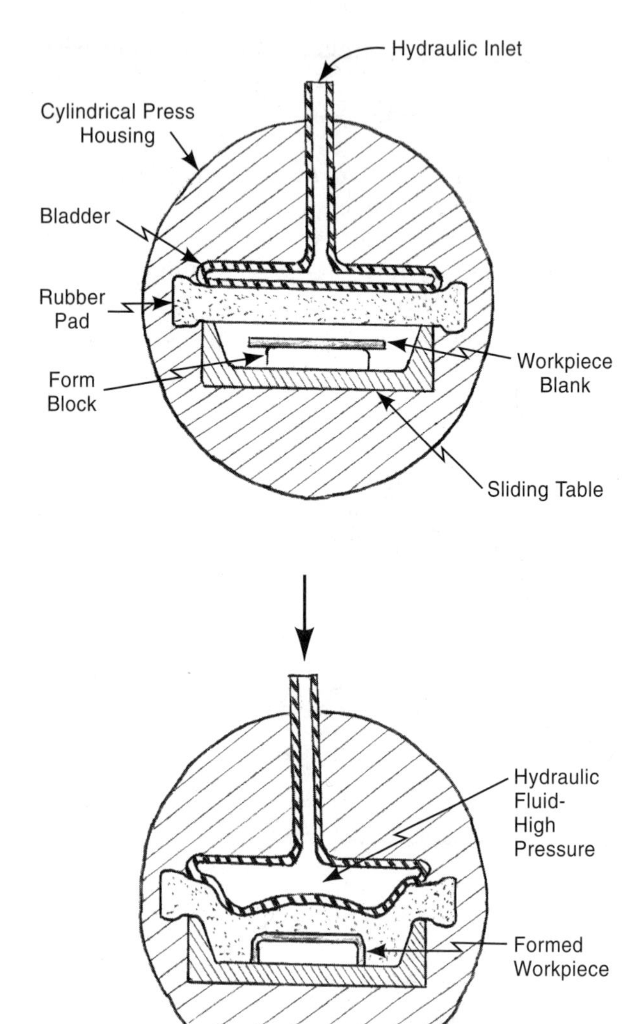

Fig. 2D3e The Verson-Wheelon process which uses a rubber bladder and hydraulic pressure to provide the forming force.

The tooling is simple and can be cast from low-melting-temperature alloys. Since tooling costs are relatively low, the process can be economical for limited production quantities. Applications are is normally limited to sheet metal thinner than 0.064 in (1.6 mm). Dimensional tolerances are wide.

D5. ***drawing (sheet metal parts)*** - is a process that forms a recess in a flat sheet-metal workpiece. The process can produce cup, cylinder, and box-shaped parts with a depth considerably greater than the diameter. (It is then called. "deep drawing".) When the press ram descends, a punch attached to the ram pulls or draws the metal into a die cavity over the die cavity's rounded edge. The operation involves flowing rather than stretching the sheet material. To control the operation, the drawing die or press includes a blank holding device. The blank holder (pressure pad) restricts the movement of the workpiece and keeps it from wrinkling but allows it to flow toward the punch and die cavity. Blankholder pressure is typically about one third of the pressure required for drawing. Thinner metal requires greater blankholder pressure[4]. The thickness of the sidewalls of the part normally increase during the operation. A close fit between the die cavity and the punch can be used to "iron" the sidewalls, maintaining their shape and reducing the thickness. Fig. 2D5 illustrates the operation. When the depth of draw is more than three-quarters of the diameter or width of the part, two or more draws, with annealing between draws, are usually required. The need to control the blank movement by holding it, usually

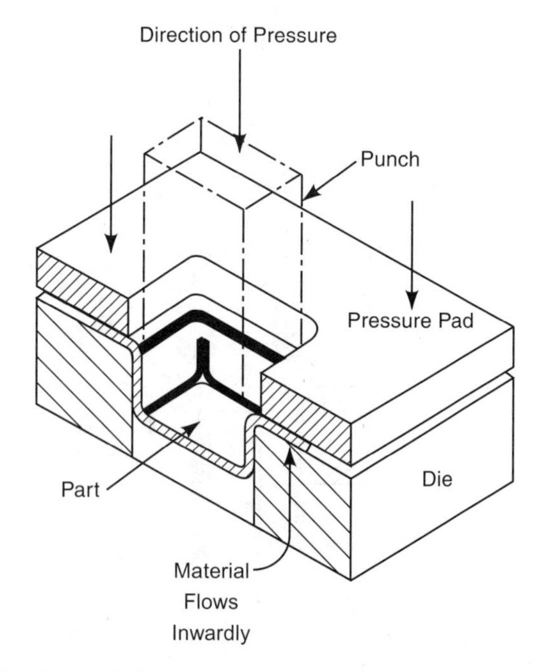

Fig. 2D5 Schematic of deep drawing showing the blank holder (pressure pad). (*from Bralla, Design for Manufacturability Handbook, McGraw-Hill, New York, 1999*)

results in excess material that needs to be trimmed off in a secondary operation. Typical parts made by drawing are: beverage cans, artillery shells, cartridge cases. auto body panels, and pots, pans, and other containers.

D5a. ***shallow drawing*** - commonly refers to a drawing operation when the depth of the part is less than one half its diameter.

D5b. ***deep drawing*** - commonly refers to a drawing operation when the depth of the part is greater than one half its diameter. One or more redrawing operations and, with some materials, annealing before redrawing, may be required.

D5c. ***redrawing, direct and reverse*** - Sometimes a drawn part is run through a second drawing operation to increase the depth and reduce the diameter of the part. The drawn part is placed in a die recess and a punch, smaller in diameter than that used in the previous drawing operation, pulls it into the die. The blank-holding function is also involved to control the metal flow. Annealing may be required between the original and the redraw operations if the workpiece metal is the type that work hardens.

In direct redrawing, the operation proceeds in the same direction as the original drawing. The punch is made in two parts, an inner punch that determines the inner diameter of the part after the operation, and a sleeve that surrounds this punch. Thc outer diameter of the sleeve is about the same diameter as that of the drawing punch in the first operation. This sleeve enters the workpiece first and serves the same function as the blank holder in the first drawing operation. As the inner punch strikes the workpiece, it pulls it downward around the sleeve into a smaller diameter. See Fig. 2D5c, view a). A double-action press provides both the holding action and the drawing action.

Reverse redrawing performs a similar operation, but the workpiece is inverted in the die and the die again is configured with a two-piece punch which provides both holding and drawing operation. The advantage of reverse drawing is that the workpiece is not subjected to severe stress in one direction and then the other. This allows a greater percentage diameter reduction in the operation and lessens the need for annealing between operations. Fig. 2D5c, view b), illustrates reverse redrawing.

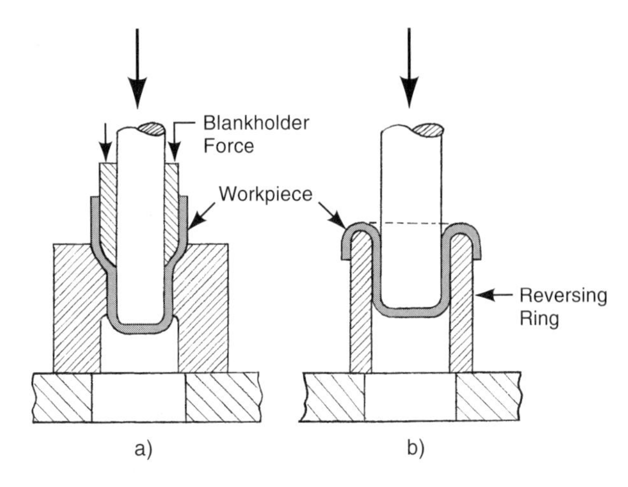

Fig. 2D5c Redrawing a drawn part: a) direct redrawing, b) reverse redrawing.

D6. ***coining*** - is an operation that changes the thickness of a sheet metal part, usually to create visible designs, markings, or other configurations on the surface. Very high compressive forces, and dies that confine the workpiece so that it doesn't flow in the lateral direction, are required. The workpiece thickness changes locally and its surfaces are formed to match accurately whatever design exists on the die surfaces. Fig. 2D6 illustrates the effect of the process and contrasts it with embossing. One difference is that coining permits the opposite surface of the part to have a different design. Coining is used to produce coins, jewelry, tableware, medals, and other parts where markings on both sides, or thickness changes, are required. Pressures as high as 100 tons per sq in (1400 MPa) may be required. Knuckle-joint mechanical presses are frequently used for the operation because they can provide the high pressure with a slow squeeze and dwell at the bottom of the stroke. Ductile materials, including low-carbon mild steel, are suitable for the operation. (Note: Also see roll coining, F4.)

D6a. ***drop hammer coining*** - is a process in which the high pressures required for coining are provided by the impact force of multiple blows of a drop hammer rather than by a high-force, slower-acting,

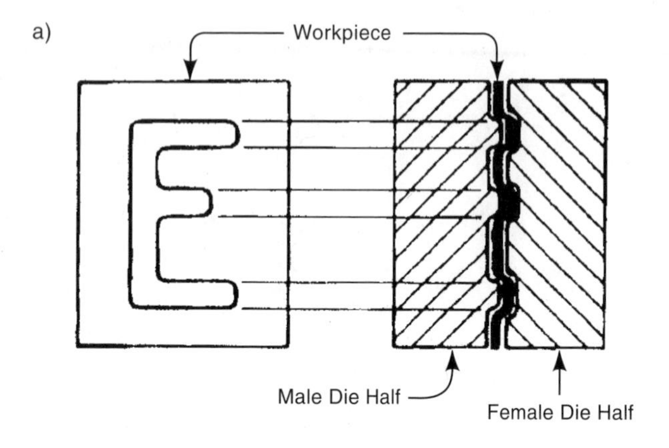

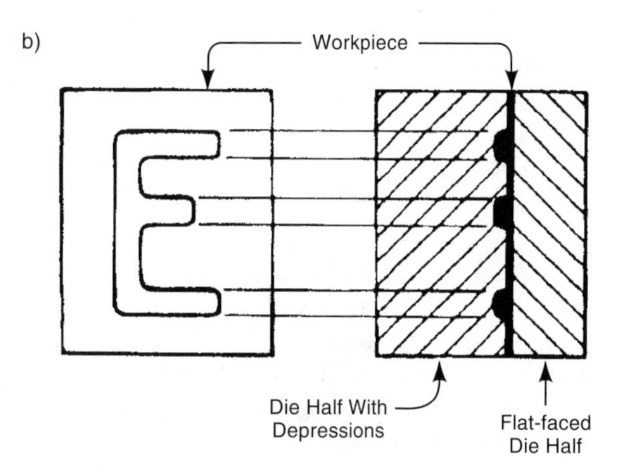

Fig. 2D6 A comparison of embossing (a) and coining (b) to imprint a surface design on a sheet metal part. (*from Bralla, Design for Manufacturability Handbook, McGraw-Hill, New York, 1999*)

punch press. Tableware is often coined with drop hammers.

D7. ***embossing*** - is a process, usually performed on sheet metal, that results in raised areas on one side of the sheet and depressed areas opposite them on the other side. The operation is basically a shallow drawing operation that does not significantly change the thickness of the sheet It is frequently performed to create a pattern in the surface of sheet metal workpieces. Reinforcing ribs and identification markings on stamped parts, tags, jewelry, and nameplates are typical applications. The height of the raised area is normally no more than three times the material thickness. Fig. 2D6 illustrates the process, and contrasts it with coining. (Note: Also see rotary embossing, F5.)

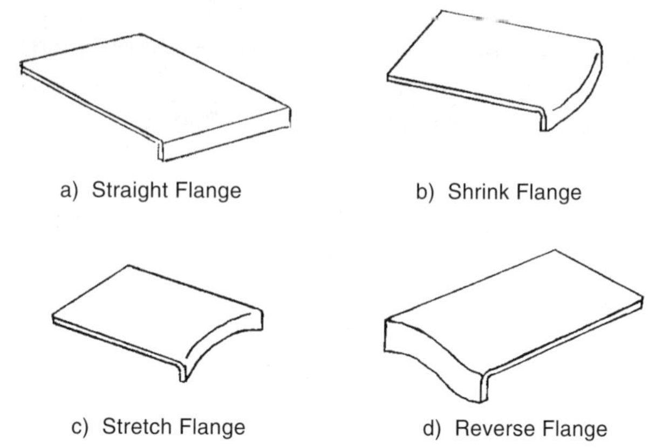

Fig. 2D8 Various types of flanges: a) straight flange, b) shrink flange, c) stretch flange, d) reverse flange.

D8. ***flanging*** - is a bend of approximately 90 degrees near the edge of a sheet metal part, usually for reinforcement. If the bend is along a straight line, a simple bending operation is all that is required; if there is an inside or outside curvature to the bend, as shown in Fig. 2D8 b), c), and d), more extensive forming is required because metal will be forced to flow as the bend is made.

D9. ***beading*** - forms a shallow round trough of uniform width in a sheet metal workpiece. The trough can be depressed or raised, and can be in a straight line or can be curved or circular. Its purpose is to provide stiffening or decoration in the part produced. The thickness of the sheet does not change. The process is similar to embossing except that embossing implies a more complex pattern of formed material. A simple stiffening bead is illustrated in Fig. 2D9, where view a) shows the cross section of a press brake die for making a simple stiffening bead and view b) shows the bead itself in section and plan views.

D10. ***hemming and seaming*** - Hemming is the bending of a sheet edge back 180 degrees on itself to provide a rounded edge or reinforcement. The bend can be produced by roll forming, or by two bending operations, or in one operation with a

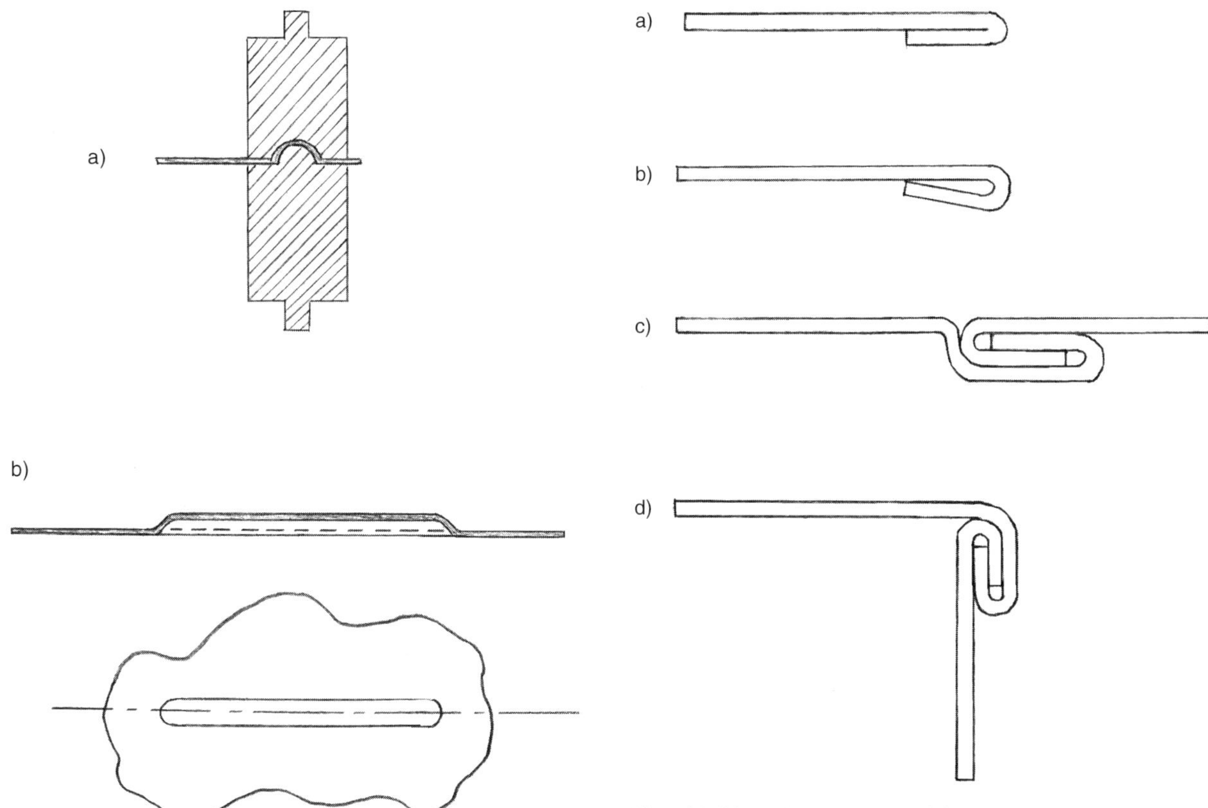

Fig. 2D9 A simple stiffening bead in a sheet metal part. The operation used to form the bead is very similar to that used in embossing.

Fig. 2D10 Hems, a) and b), and seams, c) and d), on sheet metal parts. Seams are formed by interlocking hems on the edges of two sheets.

cam-operated or compound die. Seaming is the joining of two opposite edges of a workpiece with interlocking hems or other interlocking bends. It is frequently used in the manufacture of containers such as pails, drums, and cans. Special machines are often used for the operation, particularly in high production applications. Fig. 2D10 illustrates both hems and seams.

D11. ***edge curling*** - provides a coiled or partially-coiled edge on the workpiece for reinforcement or to provide a smooth edge. On straight edges, curling can be performed with two press brake dies as illustrated in Fig. 2D11. On curved edges the operation is more complicated because curling reduces the diameter of the curved section and provision must be made for removal of the punch. Punch removal may be provided for by using a segmented punch. When the punch is to be removed, a segment is retracted so that the punch diameter is reduced.

D12. ***swaging*** - has several meanings. As a press operation, sometimes called *flat swaging*, it compresses a workpiece severely to reduce the thickness of all or part of it, to flatten it, or otherwise changes its shape. The process is similar to coining, but does not form the workpiece as completely as coining does. However, the term *swaging*, is most commonly used to refer to rotary swaging described in F1.

D13. ***sizing*** - is a press operation that improves the dimensional precision of a workpiece by subjecting it to high force in an accurate die. Perhaps the most common example is the re-pressing operation

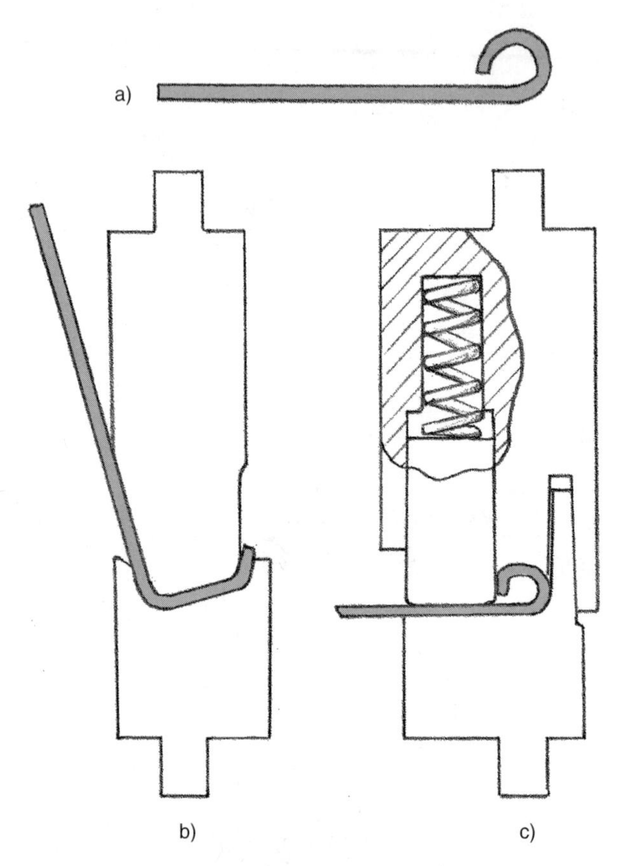

Fig. 2D11 Edge curling. a) cross-section of a sheet with a curled edge. b) and c) show the two-step sequence of forming a curled edge with press brake dies. The second die includes a member to hold the sheet from shifting during the final curling step.

performed on powder metal parts (See below.) Sizing is also performed on forgings, castings, and various cold-formed parts. With sheet metal parts, the operation is similar to coining in that the thickness of the workpiece may be reduced, except that the part is not confined in the die as it is in coining. The operation is used to flatten and sharpen the edges of stampings.

D14. ***ironing*** - is an operation that smooths and thins the walls of a cup-shaped part by forcing the part through a die with a punch. The operation works the material severely, so that annealing may be required afterward. It is illustrated in Fig. 2D14.

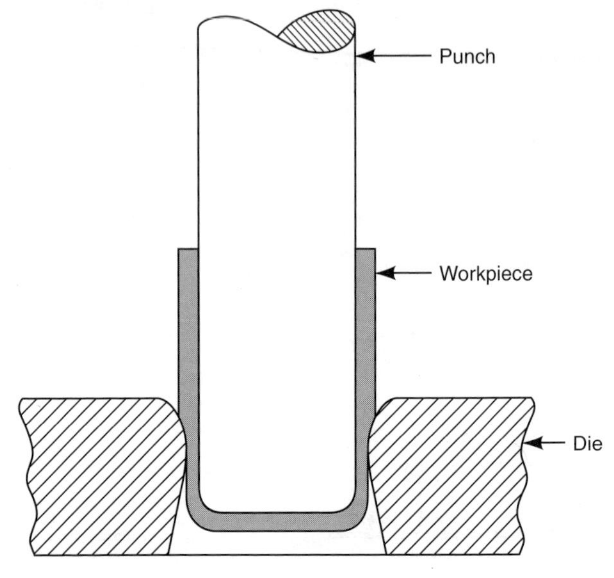

Fig. 2D14 Ironing a drawn part smooths the side walls, reduces their thickness, and lengthens the workpiece. *(from Schey, Introduction to Manufacturing Processes, McGraw-Hill, New York, 1987)*

E. Multiple Die Stamping Press Operations

E1. ***progressive die operations*** - In this metal stamping method, the work material moves from station to station through the die in step-by-step increments. There can be two or more stations in the die and, often, ten or more. After each movement of the work material, the press ram makes a stroke and a stamping action takes place. Once the front end of the work material has moved all the way through the die, a finished piece is produced with each press stroke. Stamping operations take place at all die stations with each press stroke (except for die stations left blank to facilitate material flow or for other reasons). Complex stamped parts can be produced with this method at rapid rates. The approach is widely applicable but sufficient production quantity is needed to provide economic justification for the cost of the tooling. Operations at each die station can run the full gamut of stamping including piercing, notching, lancing, blanking, forming, drawing, and trimming. Pilot holes to assure alignment at subsequent stations are typically punched at the first station. Typical parts produced with this approach include a wide variety of mass-produced products and components such as laminations for electric motors, metal kitchen tool parts, automotive parts, aluminum cans, and all kinds of mass-produced sheet metal parts. In the last station, the part is cut away from the strip or sheet material. Fig. 2E1 illustrates a typical, but simple, progressive die operation.

E2. ***transfer die (transfer press) operations*** - Transfer dies consist of several separate die stations for sequential operations on a particular part, all positioned on one press bed and are all actuated with the same press stroke. The workpiece can be manually moved from die to die between press strokes, but the major benefits of the system occur when the transfer press is equipped to move the workpiece from die-station to die-station automatically. The operation is similar to that of progressive dies except that the parts are severed from the strip stock before all the operations on it are complete, and the transfer mechanism then moves the workpieces. The transfer mechanism consists of parallel rails with grippers, fingers, cam-actuated slides or levers, that move and position the workpieces. Each die-station is independent and can be adjusted separately from the others, but all perform an operation with each press stroke, unless deliberately made idle. Blanking, piercing, shearing, bending and forming operations can be performed. An advantage of the transfer die approach is that it can be used to perform secondary stamping operations on parts already formed and may reduce material loss since

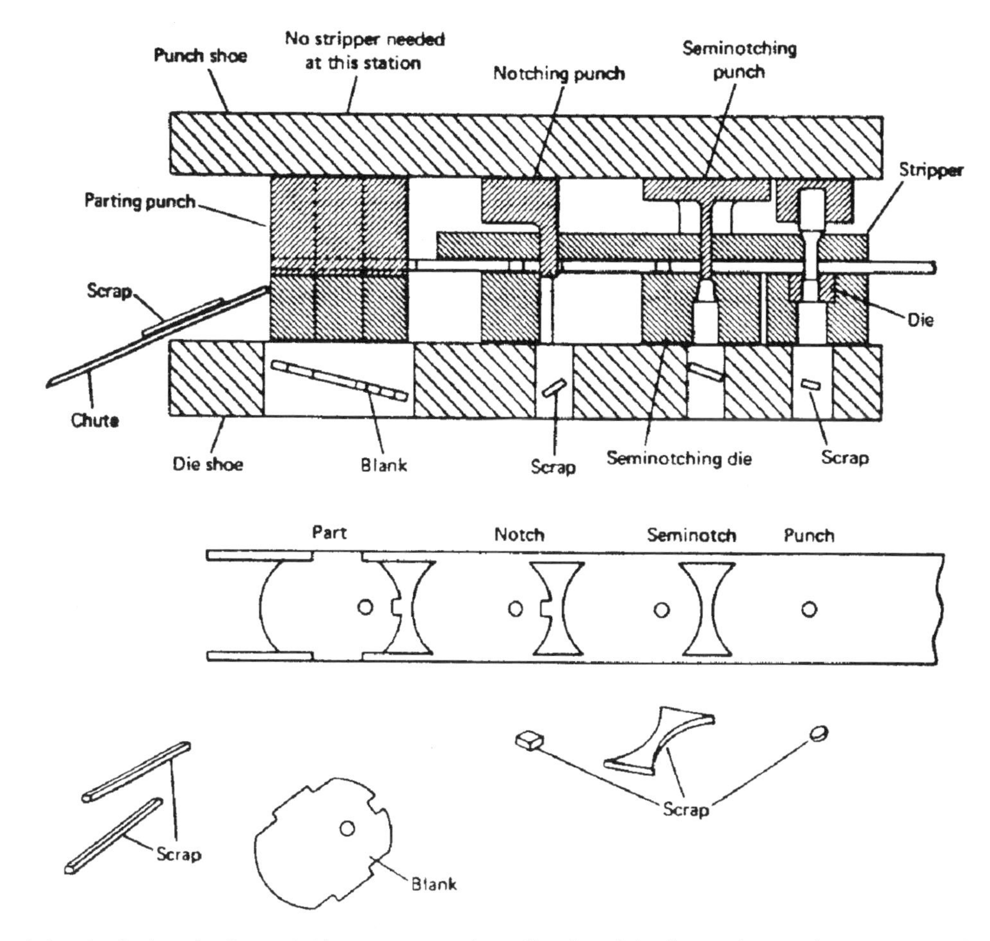

Fig. 2E1 A typical simple four-station progressive die that blanks and punches a part. (*from Niebel, Product Design and Process Engineering, McGraw-Hill, NY 1974*)

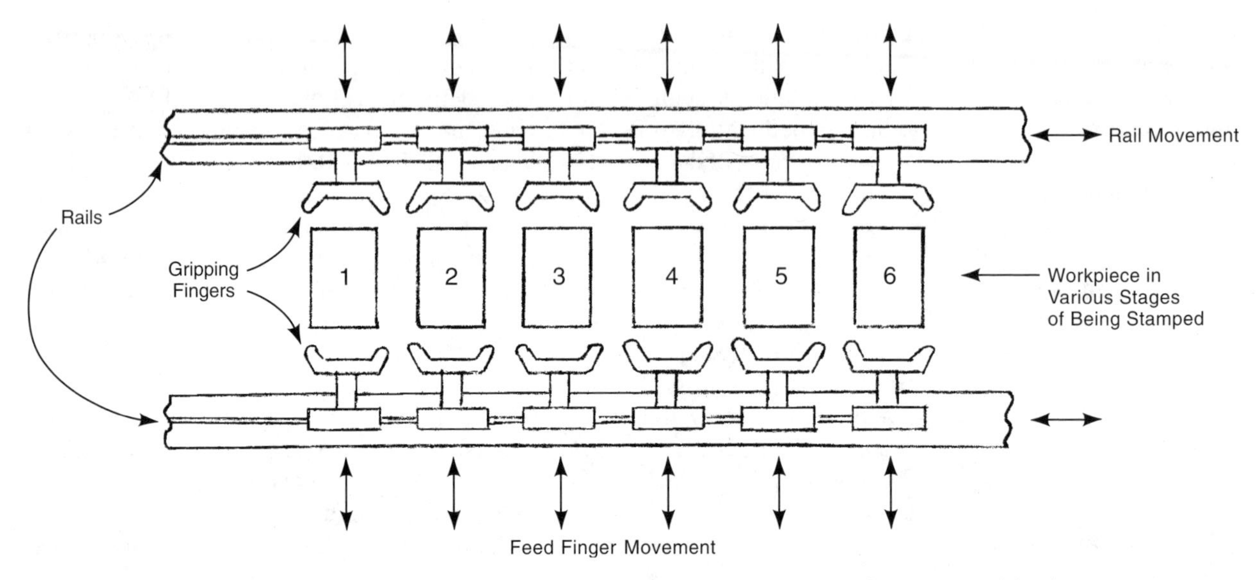

Fig. 2E2 A schematic view of the transfer mechanism used in a transfer press with six stamping stations. Gripping fingers move forward to grasp the workpiece at each station; the rails then move to the right to the next station; fingers then retract and the press makes another stroke. The cycle is then repeated. After station no. 6, the workpiece is deposited in a take-away chute; new workpieces are loaded at station 1 after each press stroke.

the alternative method, progressive die stamping, requires the use of some stock material to hold the workpieces as they are moved from station to station. The transfer material then becomes scrap. The transfer die process is useful in making parts that are difficult to form when they are connected to the stock, as in progressive die stamping. Making small rings, cylinders, and cups, from strip stock are common applications. The approach depends on high quantity production to amortize the high costs of tooling, equipment, and set-up costs involved. Fig. 2E2 shows a typical transfer die operation.

E3. ***compound die operations*** - Compound dies perform several operations on the workpiece at one die location and in one press stroke. Typical combinations are blank and bend, blank and form, blank and pierce. An example would be an operation in which the part is first blanked and then, as the press ram continues to descend, die elements perform a bending operation on a portion of the blank that was just created by blanking. Another common use is to blank a part and simultaneously punch an opening in it, as illustrated in Fig. 2E3. The advantage of a compound die operation is that it eliminates the

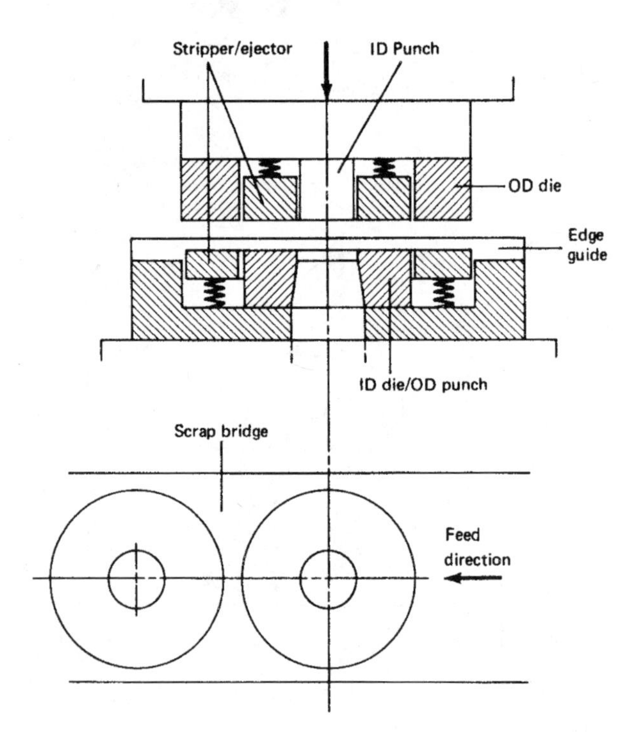

Fig. 2E3 A simple compound die that blanks and punches a part in one press stroke. The blanking punch contains the die for punching. (*from Schey, Introduction to Manufacturing Processes, McGraw-Hill, New York, 1987*)

need for two or more separate press operations and is normally more accurate than two operations on separate dies. Compound die operations are also usually better from the cost and accuracy standpoints than operations with progressive dies. However, since compound dies are more complex and usually more expensive than several single-operation dies, a greater levels of production is usually required to amortize the die expense.

F. Sheet Metal Operations Performed on Equipment Other Than Presses

F1. ***rotary swaging*** - is an operation to reduce the cross-sections of tubes, bars, rods, and wires. It is accomplished by a rapid series of controlled blows from the die elements of a rotary-swaging machine. The machine, illustrated in Fig. 2F1, contains one

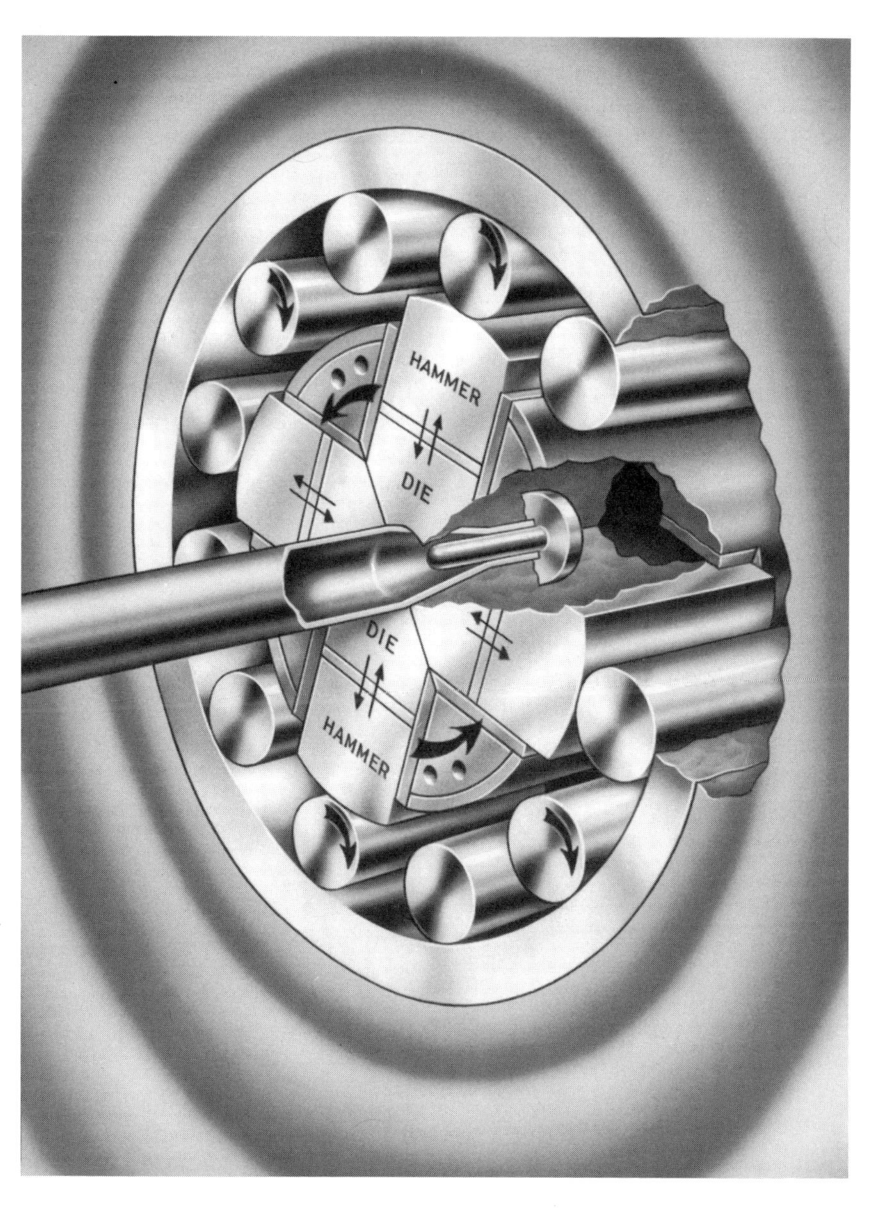

Fig. 2F1 A conventional rotary-swaging die reducing the diameter of a tubular workpiece. (*Courtesy Fenn Manufacturing Co.*)

or more pairs of opposed dies that strike the workpiece repeatedly as the machine spindle and the dies rotate. Centrifugal force throws the die members outward, where they encounter a series of rollers that force them to move inward and strike the workpiece. A common operating speed is 1000 blows per minute. The workpiece is inserted axially between the dies, held in place during the die blows, and then removed. For some workpieces it is necessary to have a mechanical assist or mechanical feed to move the workpiece against the taper of the dies. An internal mandrel may be used with tubular workpieces to control internal dimensions, to control wall thickness when it is reduced, to prevent collapse of thin walled tubing, and to form internal shapes.

The operation can produce tapers, pointed ends, and other shape changes, as well as diameter reductions. The method is also used to fasten fittings to cable, wire, and hose. Sewing machine needles, golf club shafts, and automotive torque tubes, are typical rotary-swaged parts. Tubing up to 14 in (375 mm) and bars up to 4 in (100 mm) in diameter have been rotary swaged[3].

F1a. ***stationary die swaging*** - is a variation in rotary swaging used when the workpiece does not have a round cross section. The operation is the same as conventional rotary swaging except that the dies do not rotate and are configured to conform to a non-round workpiece shape. The machine head does rotate, along with the rack of rollers, so that the rollers strike the back ends of the dies, driving them into the workpiece. The process has the same applications as conventional rotary swaging except that the workpiece develops a square, rectangular, or other non-round cross section.

F2. ***three-roll forming*** - See roll bending, section H2f.

F3. ***stretch forming*** - involves the stretching of a length of sheet metal (the workpiece) as it is wrapped around a form block. Stretching is achieved with two or more pairs of gripping jaws equipped to provide tension in the sheet. The combination of movement of the forming block and stretching of the material causes the material to be stressed beyond the elastic limit (typically, 2 to 4 percent total elongation). Simple tooling is involved because only a form block is required and there is no need for close alignment and fit of two die halves. Form blocks can be made of wood, plastics, and cast iron, as well as steel. The method is particularly applicable to large parts with small or modest amounts of forming, but it is not feasible for sharp contours. Tooling is inexpensive, so the method is justifiable for low-quantity production, despite the need to trim off material held in the gripping jaws. The method is best suited for producing shallow contours. Sheet aluminum components for aircraft are a common application, as are sheet steel body parts used in the truck and automobile industries. The method can also be used to make longitudinal bends in roll-formed or extruded sections.It is illustrated in Fig. 2F3.

F3a. ***stretch draw forming*** - is stretch forming when a mating female die also engages the workpiece on the form block. This permits more severe forming to take place. Automotive panels and door posts are made with the process. Stainless steel and titanium sheet, as well as the more common steel sheet can be processed. There is material loss and secondary trimming since the material in the grippers is trimmed from the workpiece, as in conventional stretch forming.

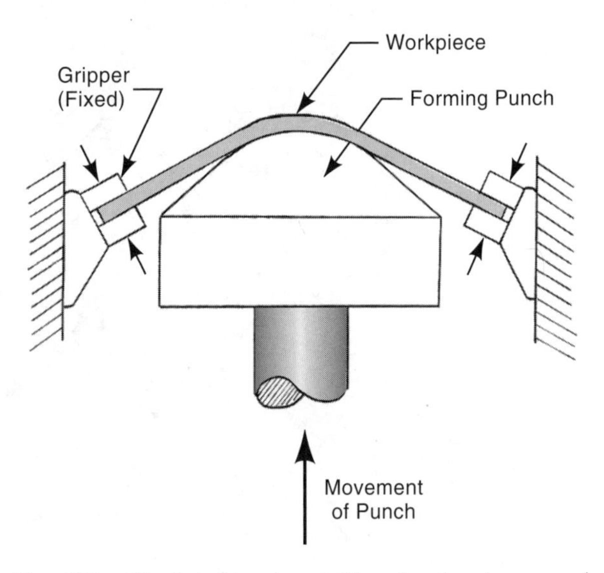

Fig. 2F3 Stretch forming with gripping jaws and a form block. Relative motion between the form block and jaws stretches the material just beyond its yield strength. *(from Schey, Introduction to Manufacturing Processes, McGraw-Hill, New York, 1987)*

F3b. ***stretch wrapping (stretch wrap forming)*** - uses the stretch principle and a rotary table to make stretch-formed parts without scuff marks on the workpiece surface. The rotary table allows the workpiece to be wrapped around the form block without sliding motion between the two. When reverse bends are needed, additional form blocks on the table are utilized. The process with three form blocks is illustrated in Fig. 2F3b. In this illustration, after initial stretching, the table rotates, creating a first bend in the workpiece. Additional movable blocks advance, and the table rotates in either direction, depending on the configuration to be made.

F3c. ***compression forming*** - is one of the family of stretch forming operations, but the force, instead of being used to stretch the material, is used to press it against the forming block. A roller or shoe, machined to match the cross-sectional shape of the workpiece, is used. The workpiece, which can be a metal strip or part with some cross-section, is pressed firmly, with hydraulic force, against the rotating form block. The compression forming method enables small radius bends to be made without exceeding the elongation limits of the outer portion of the workpiece. It also helps maintain the cross-sectional dimensions of the workpiece. The method is similar to stretch forming with a rotating form block except that only one end of the workpiece is gripped and the only workpiece stretching is that which results from the forming. Fig. 2F3c illustrates the process, which is used for vehicle bumpers and various structural members.

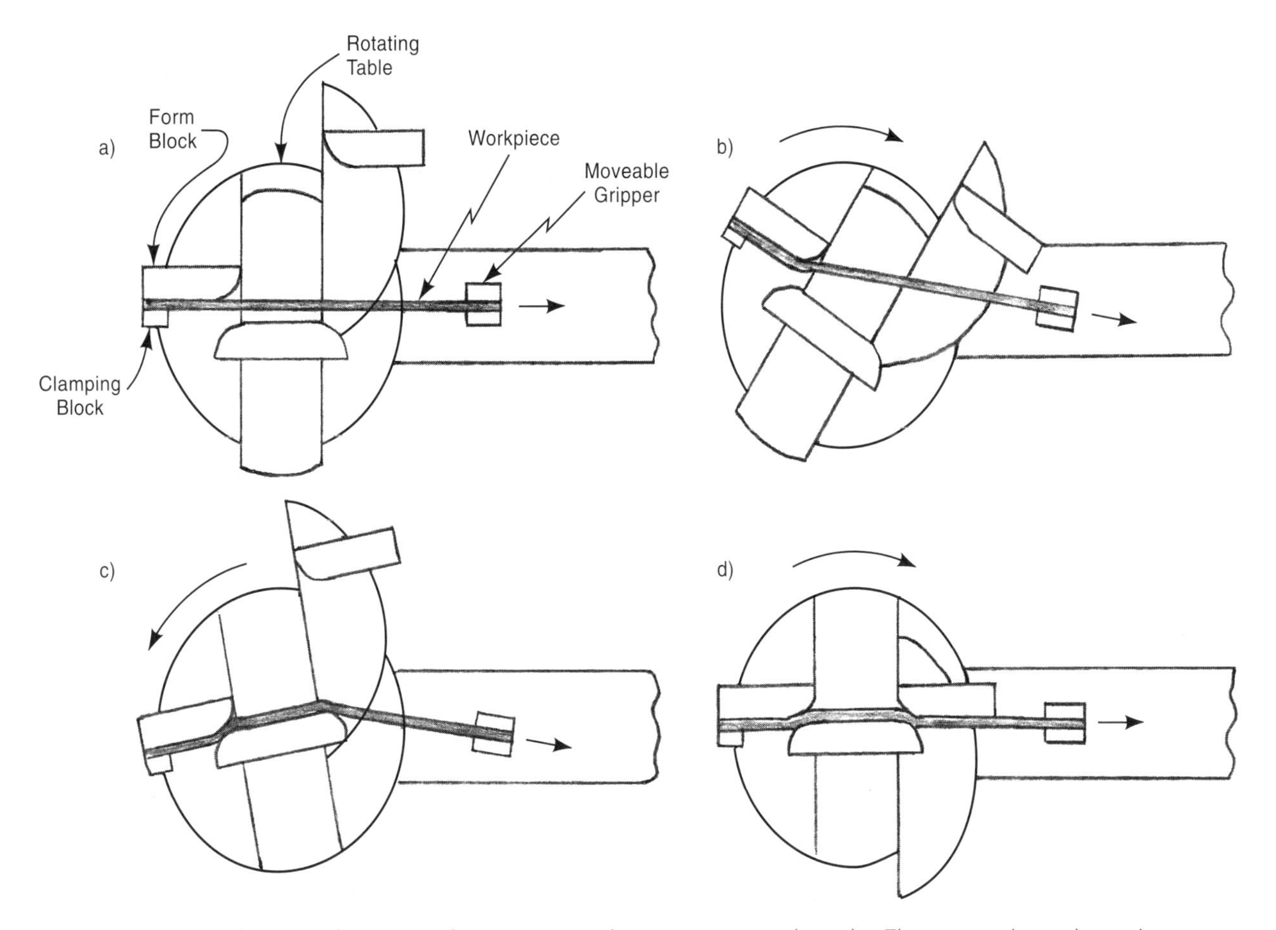

Fig. 2F3b Stretch-wrap forming of a part to make two reverse bends. The operation takes place on a rotary table with form blocks on slides. a) starting position, b) first forming operation, c) second forming operation, d) third forming operation.

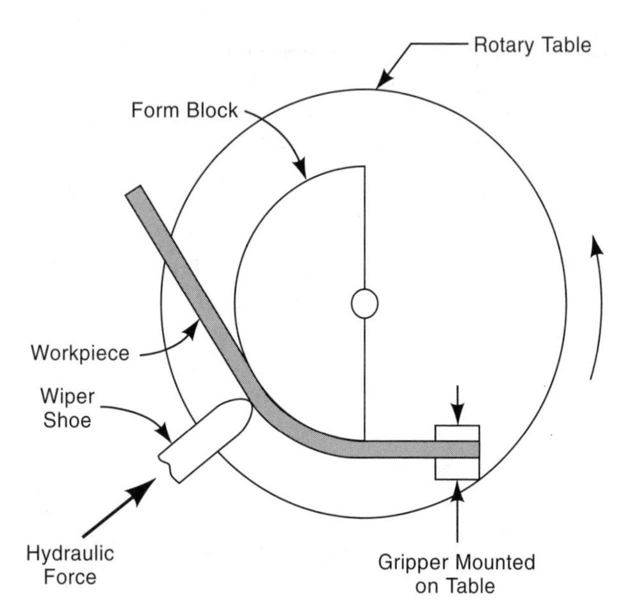

Fig. 2F3c Compression forming. The wiper shoe presses the workpiece against the form block. As the block rotates, bends are made in the workpiece.

F3d. ***radial-draw forming*** - is another stretch forming operation. It is similar to compression forming described above except that the workpiece is kept in tension. The method is used in bending extrusions and other workpieces with non-flat cross-sectional shapes. There are three points of application of hydraulic force: on the two ends of the workpiece and on the wiper shoe that presses the workpiece against the form block. Parts with a twist as well as a bend can be formed with this method by incorporating a twisting motion in one of the grippers, at the appropriate time during bending. See Fig. 2F3d.

F4. ***roll coining*** - is similar to rotary embossing, described below, in that the work is fed between two opposed rollers that have the desired surface shape. The process can be more rapid than press coining but normally is limited to small parts produced in large quantities from metals of lower yield strength. Interlocking fastener strips, made from jewelry bronze, are an example.[3]

F5. ***rotary embossing*** - Embossing, normally a press operation, can also be performed by passing the work material through a pair of opposed rollers that

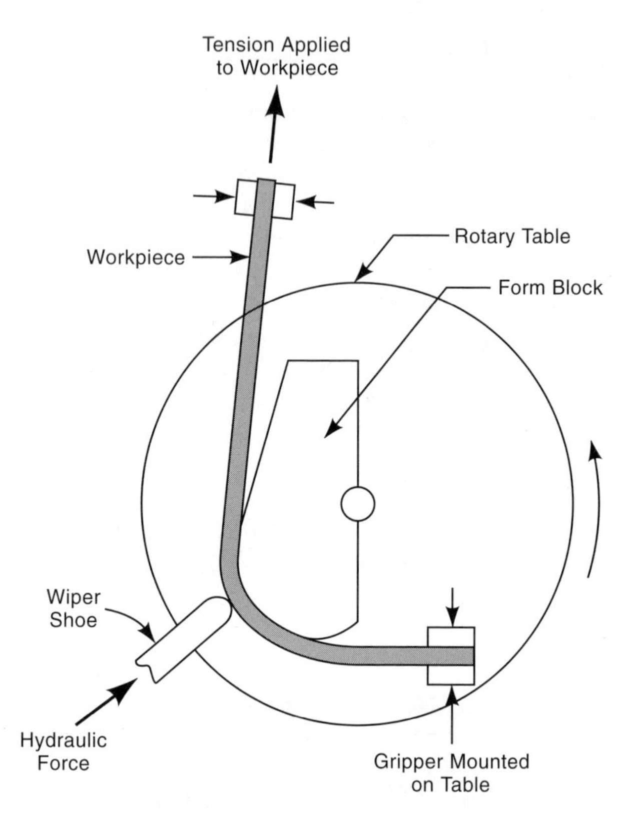

Fig. 2F3d Radial-draw forming. Similar to compression forming but the workpiece is stretched as it is pressed against the rotating form block.

have the desired surface shape. The roller approach is particularly applicable with thinner metals and foils at high production levels, when continuous strip is fed to the rollers and the embossed parts are separated from the strip in a subsequent operation.

F6. ***metal spinning*** - is an operation that forms a disc of sheet metal into various seamless circular shapes by pressing it against a form while it rotates. The disc is clamped against a form block of circular cross section, held in the headstock of a lathe, so that the two rotate together. As the disc rotates it is pushed with localized pressure against the form block, with suitably shaped spinning tools, and gradually assumes the shape of the form block. The motion and pressure of the pressing tool can be provided manually or with mechanical power. This method is used in the production of various circular sheet metal objects such as reflectors, lamp bases, bowls, bells, cooking pans, funnels, metal drinking glasses, and conical parts. Tooling is very simple; form blocks, called "chucks," can be made from wood, wood fibre, cast iron, steel, or other metals. The operation is therefore economical for prototypes and low-quantity production, particularly when manually powered. Some deeply-formed parts may require several passes, sometimes with blocks of different shapes, before the part is completed. With some metals, annealing between successive spinning operations is required. Intermediate form chucks are sometimes referred to as "breakdown chucks."

For parts that are narrower at the open end than at the closed end, and other parts with re-entrant contours, a one-piece chuck would be trapped in the part after forming. These shapes can often be produced, however, if a segmented chuck is used, ie., one that can be removed, piece-by-piece, after the spinning is completed. Another approach, more common with higher production levels, is to use a smaller diameter, off-center, contoured roller, that can be withdrawn from the workpiece after spinning.

Spinning may be combined with deep drawing if the part has considerable depth. Deep drawing takes place first and the part is then spun to its final shape.

F6a. ***manual spinning*** - Metal spinning can be performed manually as long as the metal used is thin and ductile enough to respond to manually-applied pressure. Pressure is applied to the workpiece by the rounded end of a wooden or metal (eg., aluminum-bronze) lever held by the machine operator but supported by a tool rest on the lathe's cross slide. The operator often also uses a "back stick", a second tool to guide the workpiece material and prevent wrinkling. Lubricants such as: grease, petroleum jelly, beeswax, or brown laundry soap, facilitate the operation. Tooling costs are very low, but skill is required in order to assure uniform results. Sheet steel up to about 1/8 in (3 mm) in thickness can be formed by manual spinning. Fig. 2F6a illustrates the process schematically.

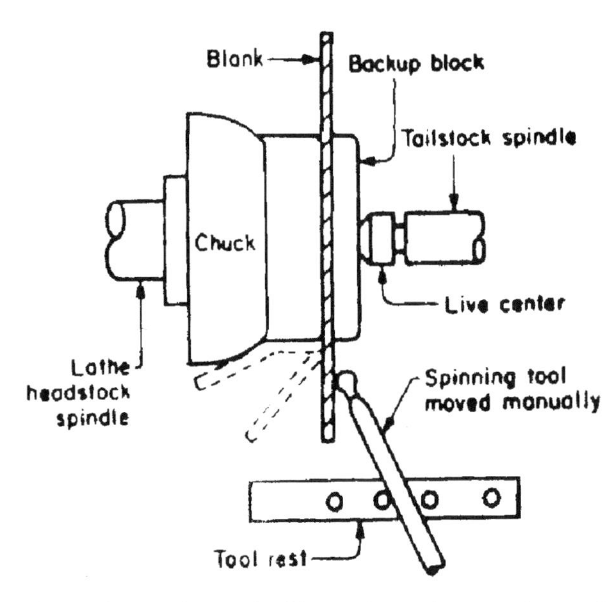

Fig. 2F6a Schematic illustration of manual metal spinning. (*from Bralla, Design for Manufacturability Handbook, McGraw-Hill, New York, 1999*)

F6b. ***power spinning , "flow turning" or "shear spinning"*** - refer to a variety of methods that apply powered forming pressure to the spinning process. Rollers are usually used to contact and shape the workpiece. These methods work the material more severely than hand spinning and reduce its thickness. Heavier parts can be produced in this way and the operation, with a metal form block, can be applicable to high production levels. However, shapes may be more limited than those attainable with manual spinning. With repeated operations, closed-end cylindrical parts can be produced. The method has been applied to workpieces as thick as 1 in (25 mm) and 240 in (6 m) in diameter[3]. Typical parts produced are: conical light fixtures, air deflectors, and domed tank ends. In shear spinning, the workpiece metal is subjected to shearing deformation, being pushed ahead of the forming roller and compressed between the roller and the forming block. The operation is illustrated in Fig. 2F6b.

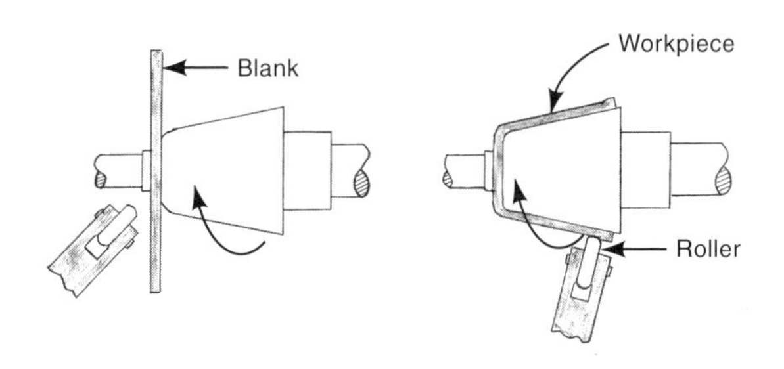

Fig. 2F6b Schematic illustration of power spinning, also known as "shear spinning" or "flow turning".

F7. ***roll forming (contour roll forming)*** - is a method for forming tubing, channels, pipe, roof gutters, siding and roof panels, metal joists and studs, metal picture frames, curtain rods, toy train tracks, decorative strips on railroad cars and trucks, and other shapes of constant cross-section from strip material. The strip is guided through a series of contoured mating rollers, mounted in tandem, which progressively make longitudinal bends in the strip as it passes through the sets of rollers, called "stands". Quite complex shapes can be produced in very long lengths. Only bending takes place; the stock thickness does not change. Typically, from one to 40 stands are employed. The process is an alternative to extrusion in some applications, and is illustrated in Fig. 2F7. Steel and other sheet metals are processed with the method including prepainted and preplated metals in thicknesses from about 0.005 in (0.13 mm) to 3/4 in (19 mm) but standard machines are limited to about 5/32 in (4 mm) thick steel. All metals that can be bent by other methods can be roll formed. The operation is rapid [about 100 ft (30 m) per min and sometimes much faster], and suited to high-volume production. Roll sets must be changed when the section to be produced is changed. When a particular section requires welding, a resistance welding station can be incorporated at the end of the forming rolls. Notching, piercing, embossing, and other operations can be performed if suitable equipment is added to the roll forming line, though these usually slow the roll-forming operation. Straightening rollers or guides may be placed at the exit end of the machine to correct for any twist distortion that may occur in the operation. Flying cut-off equipment can be placed at the end of the line of rollers to cut the workpieces to the desired length.

F7a. ***roll forming of tubing and pipe*** - When tubing and pipe are made by roll forming, a welding unit is incorporated in the process. The metal strip, as it moves through the machine, is formed into a circular cross section. The edges of the strip are brought together as they exit from the forming rolls and are arc or resistance welded together. Typically, 10 pairs of rollers are required to bend, finish, and size the tubing. The welding station normally limits the speed of the operation. An interlocking seam joint can also be produced (with thinner sheet) by the rolling operation. Welded tubing of from 1/4 to 24 in (6.3 to 625 mm) in diameter can be produced[4]. Straightening and cutoff equipment are normally part of the production line.

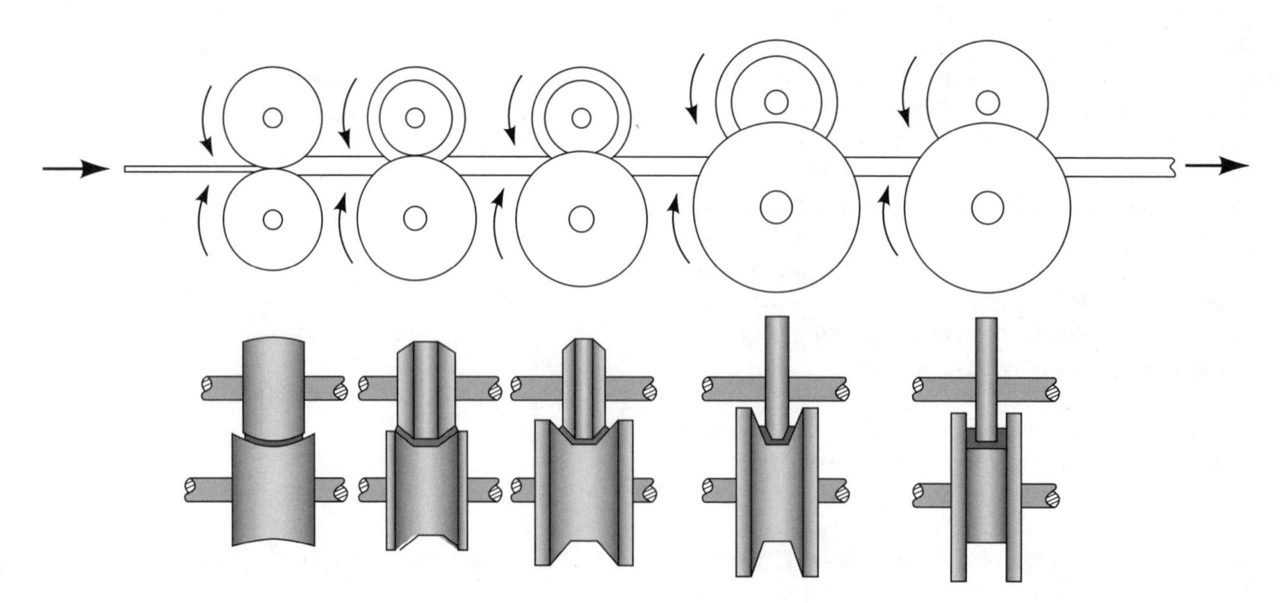

Fig. 2F7 The stages of contour roll-forming a U-shaped channel.

G. Wire Forming Operations

G1. ***wire forming*** - Components made from bent and formed wire are useful when an open configuration, such as those used in baskets and fan or machine guards, is required. In many other applications, wire can provide an economical configuration, sometimes with a spring effect incorporated and sometimes as a rigid member. Bending methods used for sheet metal, rod, tube and other cross-sections are normally also applicable when the workpiece material is wire. Other operations that can be performed on wire include swaging, cold heading, resistance, arc or gas welding and threading.

G1a. ***manual forming of wire parts*** - Wire can be bent with hand-powered bending brakes, kick presses, or various fixtures with hinged or pivoted elements, since forces required are low. Bending tools similar to tube benders can be used. Hand-bending may be applicable when quantities are limited.

G1b. ***wire forming in power presses and special machines*** - All bending operations performed by presses, press brakes, four-slide machines, and other equipment can be performed, with suitable tooling, on wire workpieces. See sections C1, D1, D2, D12, H2, I2 and K. Additionally, wire is formed into springs of various shapes. (See *springs*.) In high production situations, the wire is fed from continuous coils and is formed and cut off automatically.

G2. ***forming in four-slide machines*** - is applicable to parts formed from wire or sheet metal strip. Four-slide machines have a forming area with four press slides set 90 degrees apart and driven by cams. There is also a center post that can be shaped to facilitate forming. Tools on the slides are designed to blank, pierce, notch, bend, form, emboss, and cut off the wire or strip as it progresses through the machine. Mechanisms feed the material (from coil stock) and eject the workpiece where appropriate. Rollers are often positioned on the machine to straighten the stock prior to the blanking or forming operations. One or two small, horizontal punch presses can also be mounted on the machine upstream from the four-slide area to perform additional operations.

Fig. 2G2 shows the layout of a typical four-slide machine for forming one typical part. Very complex

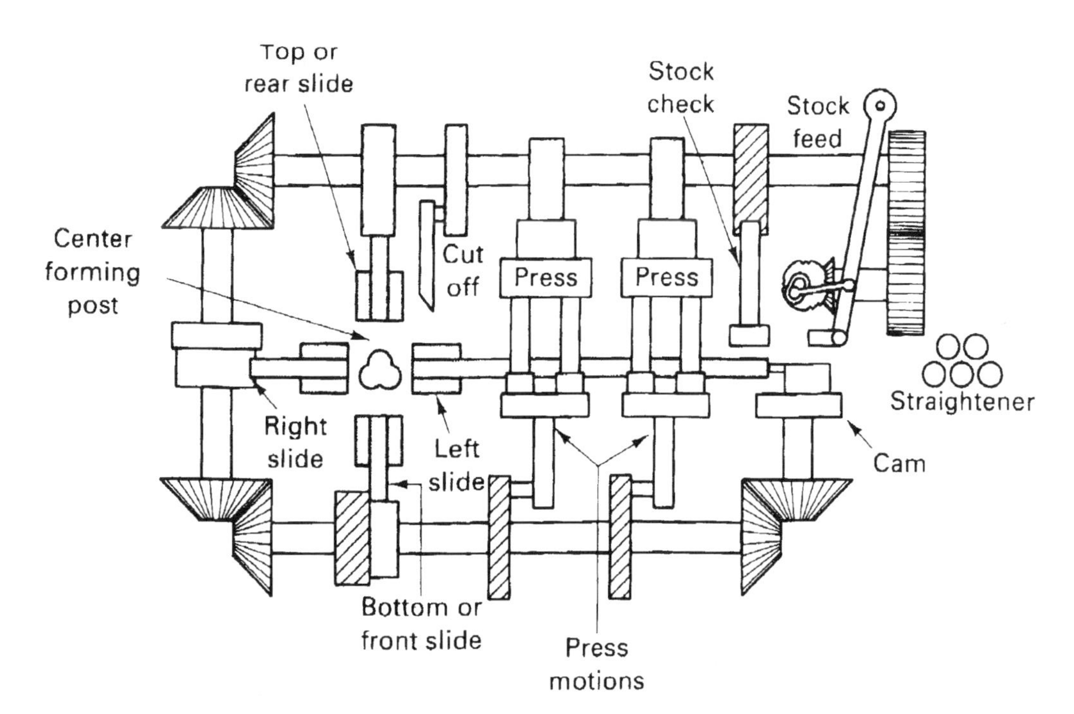

Fig. 2G2 Schematic view of a four-slide machine. Various blanking, punching, forming, and cut-off operations are performed as the stock passes through the machine (from right to left in this illustration). (*Courtesy U.S. Baird Corporation, Stratford, CT.*)

parts can be produced with this process and they are often completely fabricated in one four-slide operation. Production is rapid, but large lot sizes are needed because set up is lengthy. Resistance welding, and drilling and tapping, are sometimes added to the operation by mounting the appropriate head on the machine. Four-slide parts, though often of quite a complex configuration, are generally on the small size because of machine limitations. Deep draws and severe coining are not normally found. Clips, rings, hooks, bobby pins and electrical contact and switch parts are made with the process.

G3 ***spring forming*** - see *springs*

G4. ***Turk's-head rolling of wire*** - is used to change the cross-section of round wire. Rectangular, square and special shapes can be imparted to the wire by passing through opposed rollers arranged so that the opening between them is the shape of the cross-section desired. Fig. 2G4 illustrates the roller arrangement used to produce square wire. The operation is rapid, up to about 600 ft (180 m) per minute. Contoured cross-sections can be produced so long as the shape desired can be ground into the rollers.

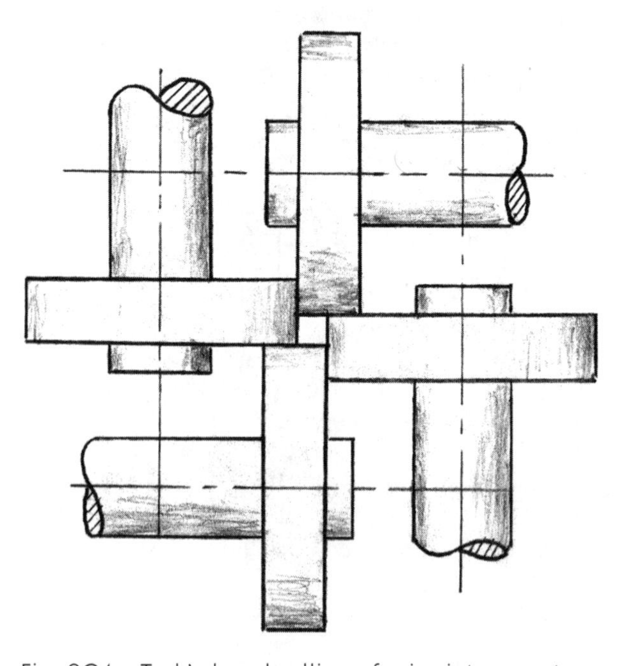

Fig. 2G4 Turk's-head rolling of wire into a rectangular cross-section. By moving the rollers to slightly different positions, rectangular wire can be rolled. Contoured rollers can be used to make wire of other cross-sections.

H. Tubing and Section Operations

H1. ***tube spinning*** - Tubes can be spun to provide tapers or flanges by using metal spinning methods similar to those used in spinning sheet metal. Other uses are to thin the walls and increase the strength of tubular members. Internal mandrels are normally used to provide support for the workpiece metal. Hollow mandrels external to the tubing can be used when the tubing is expanded or flared at the end. All ductile metals can be processed with this method. Two different methods can be employed: forward tube spinning and backward tube spinning.

H1a. ***forward tube spinning*** - sometimes called tube stretching, is a means for increasing the depth of cooking pots or the length of cylindrical parts or tubing. The diameter is constant. The power-spinning method is equivalent to shear spinning, and forces the metal to flow axially along a mandrel, increasing the length of the workpiece while decreasing the thickness of the sidewalls. The forming tool moves in the same direction as the lengthening of the workpiece, away from the headstock of the spinning lathe, as illustrated in Fig. 2H1a.

H1b. ***backward tube spinning*** - is the same as forward tube spinning except that the forming tool moves in the direction opposite to that of the expansion of the workpiece. ie., it moves toward the headstock of the spinning lathe and the workpiece metal flows away from the headstock as shown in Fig. 2H1b. This arrangement enables the tool travel distance to be less than the eventual length of the workpiece because the workpiece material flows along the mandrel beyond the forming roller. However, distortion can take place in the workpiece.

H2. ***tube and section bending***

H2a. ***draw bending*** - is used in bending tubing, sections, and bars. The workpiece is held against a bending form by a shaped clamp. As the bending

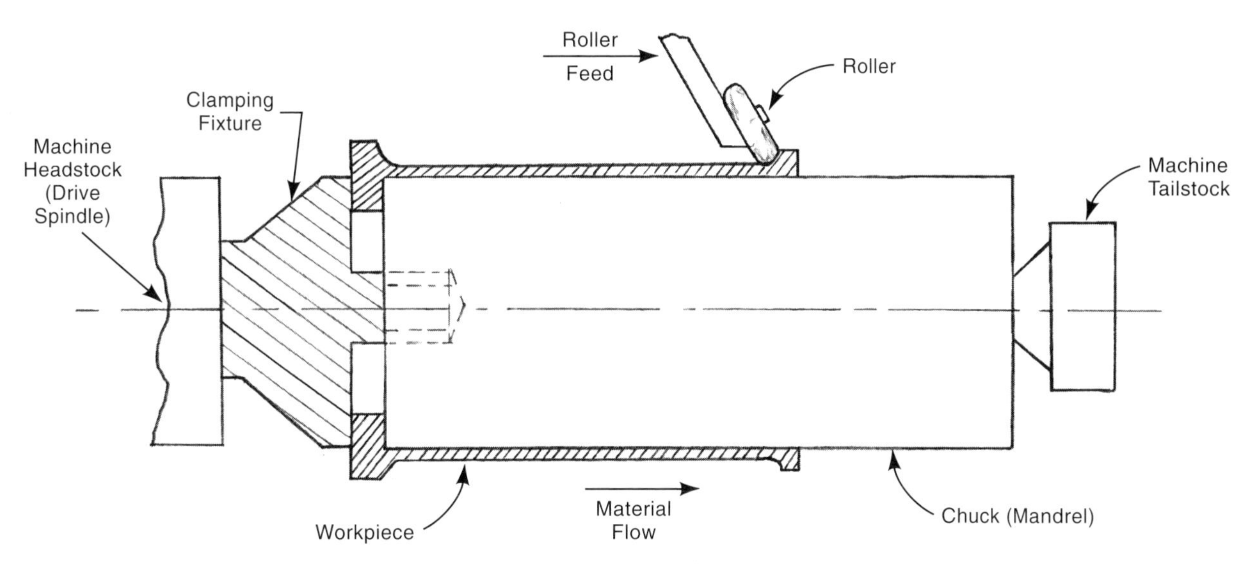

Fig. 2H1a Forward tube spinning. The illustration shows the second or third pass of several required, the number depending on the metal used and the length and wall thickness desired. An opposing roller used to balance forces is not shown in this illustration.

form and clamp rotate together, the workpiece is pulled or drawn around the form and against a pressure die whose form also matches the cross section of the workpiece. The pressure die can be either stationary or movable along its longitudinal axis. An accurate bend is produced by this method, which is illustrated in Fig. 2H2a. Internal mandrels may be used to prevent or minimize flattening. Typical flexible mandrels are illustrated in Fig. 2H2a-1. The operation is often power driven, and ultrasonic

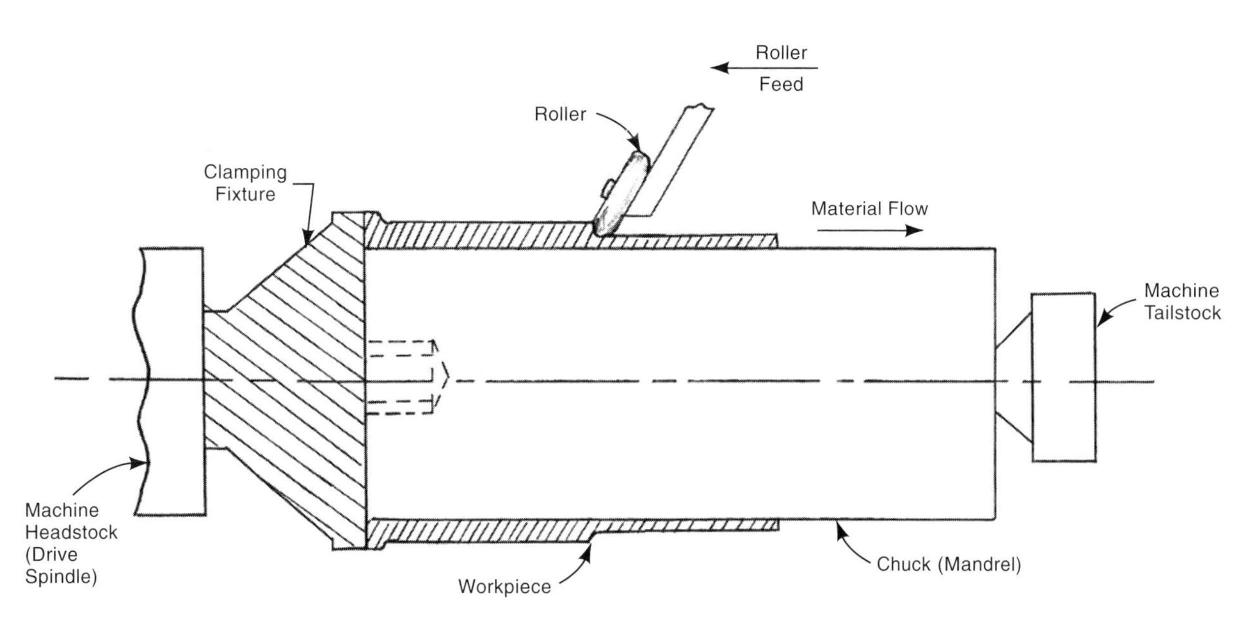

Fig. 2H1b Backward tube spinning. The material flows in the opposite direction from that of the pressure roller. The illustration shows the second pass of the roller. An opposing roller used to balance forces is not shown in this illustration.

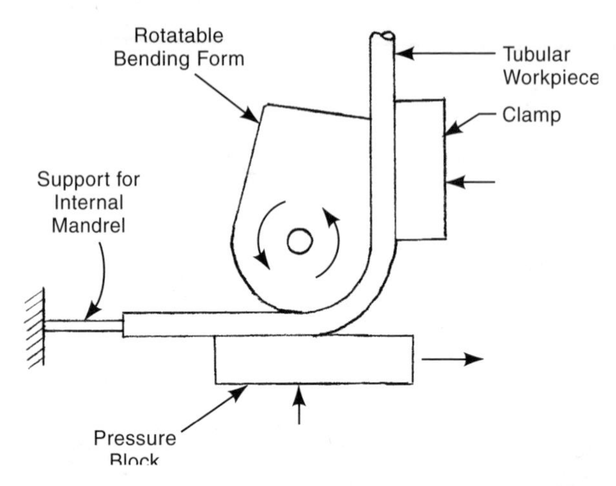

Fig. 2H2a Draw bending of tubing. The form block rotates with tubing clamped to it. A mandrel is often used to prevent collapse of the tubing.

vibration may be applied to the tooling or workpiece to reduce friction. The method is applicable to tubing of 1/2 inch to 10 in (12 to 250 mm) in diameter and bends to 180 degrees. Draw bending provides better control over the workpiece shape than other tube bending methods. With use of a mandrel and care in the operation, thin-walled tubing can be bent with little wall collapse to a center line radius as small as one tube diameter.

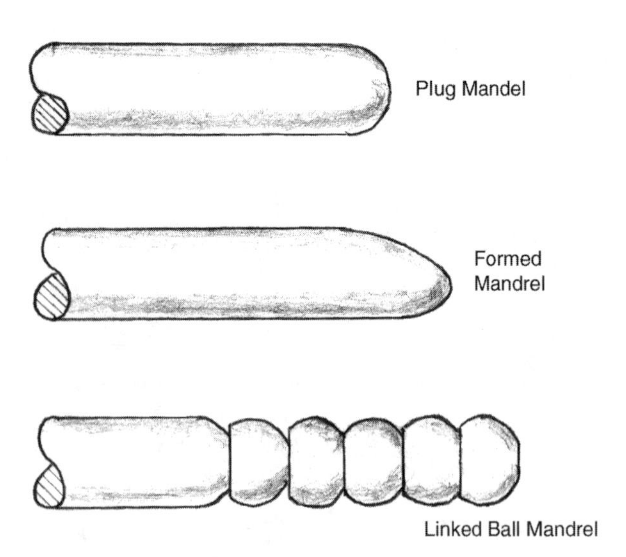

Fig. 2H2a-1 Mandrels used in tube bending to prevent collapse of the tubing during bending. The linked balls in the mandrel shown at the bottom can pivot to provide support to the tubing wall throughout a tight precision bend.

H2b. ***compression bending*** - is similar to draw bending, but the forming block and clamp do not rotate. Instead, the pressure die is replaced by a wiper shoe or fitted roller which moves along the periphery of the stationary bending form, causing progressive bending of the workpiece. No internal mandrel is used. Except for the outer surface, the tubing is subjected to some compressive stress. The operation is normally manually powered. The method is less applicable to thin-walled tubing than draw bending.

Center-line bend radii as small as about 2.5 times the tube diameter are possible, but 4 times is a more common minimum. Bends in tubing to 170 degrees are feasible. The method is useful when multiple bends in a workpiece are close together, but bends may have more distortion than draw bends made with a mandrel. Painted tubing can be compression bent because there is little stretching of the outer surface. Fig. 2H2b illustrates the process.

H2c. ***ram-and-press bending*** - is a method applicable to tubing, bar, and other sections, and is illustrated in Fig. 2H2c. A forming block is attached to the ram of a hydraulic press. Wing dies below the ram hold the tubing and pivot to wrap it

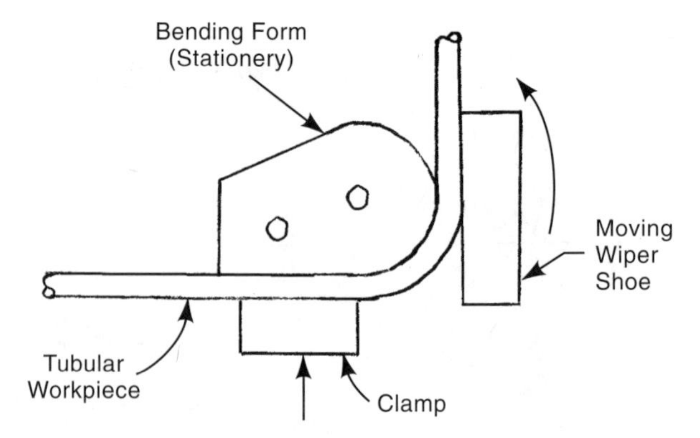

Fig. 2H2b Compression bending of tubing. A moving tool compresses the workpiece against a stationary form. (*from Schey, Introduction to Manufacturing Processes, McGraw-Hill, New York, 1987*)

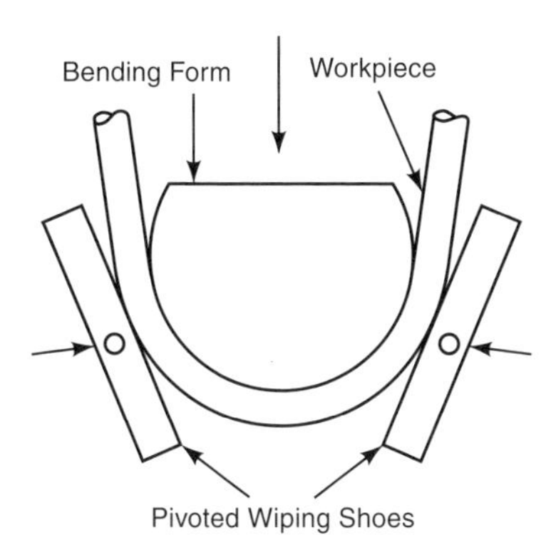

Fig. 2H2c Ram-and-press bending. (*from Bralla, Design for Manufacturability Handbook, McGraw-Hill, New York, 1999*)

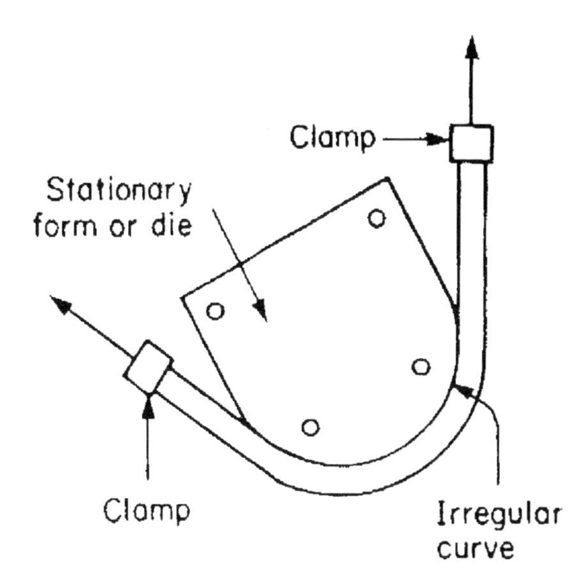

Fig. 2H2d Stretch bending of tubing. (*from Bralla, Design for Manufacturability Handbook, McGraw-Hill, New York, 1999*)

around the forming block as the press ram descends. A wiping action of the wing dies controls the flow of metal and provides a compression bend. The length of the press stroke determines the angle of bend. Typically, hydraulic presses, in which the press stroke can be stopped at any point, are employed. No mandrel is used. There is less control over the workpiece than that provided by draw or compression bending, but the operation is relatively rapid. This method can be used to make successive bends if the tubing is repositioned between press strokes. Bends to 165 degrees can be made. Automobile exhaust pipes and other tubing components used in large quantities are commonly bent by this method.

H2d. ***stretch bending*** - The stretch forming method for sheet metal, described above in section F3, also is applicable, with variations, for bending of tubing and other sections. The workpiece is stretched longitudinally to the yield point and then is wrapped around a bending die or form. A mandrel is not needed, but the method is not rapid. It is applicable to bends of non-uniform radius. More than one low-angle bend can be made in one operation. The ends of the workpiece may have to be trimmed off after the bending operation because of distortion from the gripping jaws that provide the stretching tension. One common application of the process is the bending of structural members made from angle or channel sections when such members require a curved shape. The frames of rockets and other aerospace vehicles are bent this way. See Fig. 2H2d.

H2e. ***wrinkle bending*** - is a method for bending large, heavy-wall, tubing or pipe. It is applicable to field conditions since it can be done by hand with no special tooling. One side of the tubing is heated locally by gas torch to the point where the tubing wall softens. A compressive force is applied to the pipe, causing the soft area to wrinkle and the tubing to shorten on that side, producing a shallow bend. The operation is repeated at another point a short distance from the first wrinkle and then successively until the desired degree of bend has been achieved.

H2f. ***roll bending*** - is a means of putting gentle bends in tubing, pipe, bars, rolled or extruded shapes, plates and sheets. It is illustrated in Fig. 2H2f. Usually, three parallel rollers are provided in a triangular arrangement with the lower two being mechanically driven. The third is placed above and between them at a height that can be varied, depending on the diameter or thickness of the workpiece and the desired degree of curvature. This upper roller usually is not power driven. Sometimes, more than three rollers are employed. When workpieces

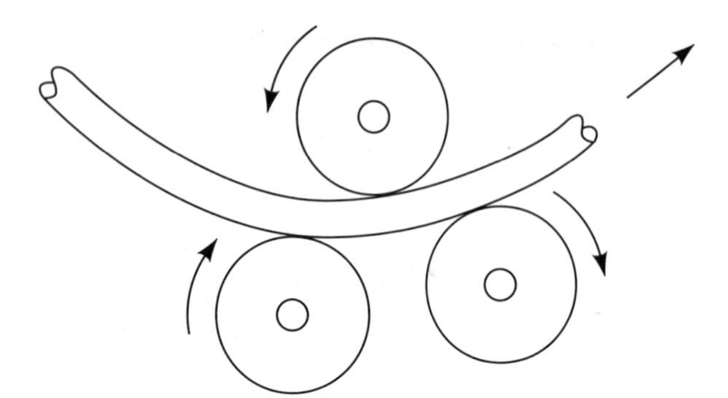

Fig. 2H2f Roll bending of tubing.

other than sheets or plates are bent, the rollers are grooved to fit the cross section of the workpiece. When used for pipe or tubing, the process is limited to heavier-walled workpieces with bend radii usually not tighter than 6 times the diameter though 4 times is possible. Pipe up to 8 in (200 mm) in diameter can be bent with this method. Rings and coils can be produced. When the process is used with sheet metal, a common application is the production of cylindrical parts.

H2g. ***roll extrusion bending*** - This method is used for large, heavy-walled pipe. One wall of the pipe is swaged from the inside, causing it to elongate and the pipe to bend. Pipes of 5 to 12 in (125 to 300 mm) can be bent with this method to a minimum bend radius of 3 times the diameter. Successive bends in different planes can be made with this method but a straight section must be allowed between bends.

H2h. ***bulging, mechanical*** - is an operation that expands a portion of a tubular or cylindrical part. When done mechanically, a segmented die - with segments held together by springs - is inserted into the tubular part. During the press stroke, a tapered punch pushes the segments apart and they, in turn, push out the walls of the workpiece. The process produces a patterned expansion (flat spots around the tubing) because of open spaces between the die segments when they are expanded. These flat spots can be minimized by rotating the segmented die and repeating the operation. However, the method is otherwise straightforward and well suited to production conditions.

H2i. ***bulging, hydraulic*** - expands tubing, pipe, or a cylindrical part, by applying internal force with a pressurized liquid or an elastomer (low-durometer rubber or polyurethane) punch. The workpiece is contained by a die that is split so that the bulged part can be removed after the operation.

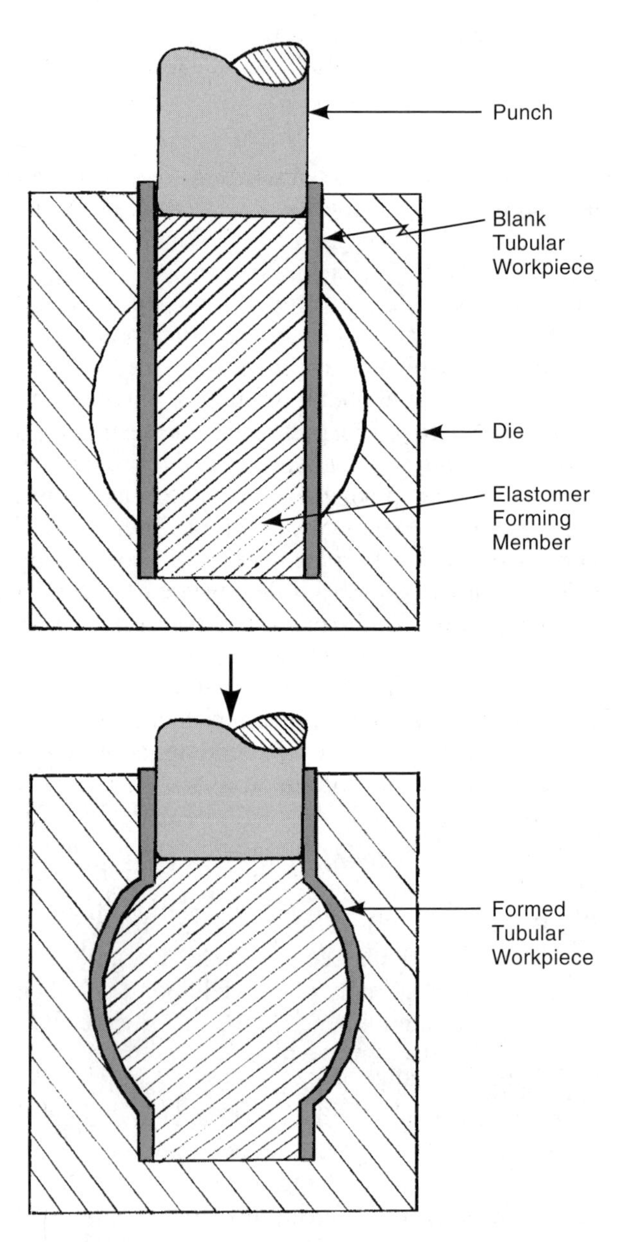

Fig. 2H2i Hydraulic bulging of tubing with an elastomer. The elastomer is deformed by the descending punch and forces the tubular workpiece outward into the die.

With direct hydraulic forming, the tubing is filled with water, sealed at the ends, placed in the die and pressurized. Water pressure forces the tubing walls outward to conform to the shape of the external die. With the other variation, an elastomer punch is commonly used since there are no leakage problems with it and it is wear resistant. As the press ram bears against it, the elastomer deforms but does not compress significantly; the compressive force of the press is transferred outward, forming the workpiece. When the press ram withdraws, the material springs back to its original shape and can be removed from the workpiece. Fig. 2H2i illustrates hydraulic bulging with rubber or polyurethane.

H2j. ***other methods of bulging tubing*** - Most of the high-energy-rate forming methods described in section J can be employed to expand tubular workpieces.

I. Non-sheet Forming Operations

I1. ***shearing of bars and other non-flat shapes*** - Alligator shears (see Section C1b), are commonly used for shearing bars and other sections to length. Guillotine shears, vertical presses, permanently fitted with short shear blades, are another alternative in shops where there is much bar shearing. The blades of guillotine shears may be notched or grooved to aid in maintaining the location of the workpieces during cutting.

I2. ***cold heading*** - a process for ***upsetting*** (enlarging and shaping) the end of a rod, wire, or bar. It is commonly used for the production of bolts, rivets, nails, and screws, and is illustrated in Fig. 2I2. One or more blows of a heading tool against the end of a rod or bar displace some portion to change its shape and enlarge its diameter. Material is usually fed from coil stock. The material upset is that portion of the workpiece that extends from a stationary die. A series of blows with different punches may be required in order to produce the desired head shape. As can be seen in Fig. 2I2, upsetting can take place in the punch, in the die, in both the punch and die and between the punch and die. Production rates can exceed 500 parts per minute. The part is cut off by shearing either before or after heading, and other operations may be performed after heading, often automatically when the production quantities are large. Machining, bending, and flattening are examples, but thread rolling is probably the most common accompanying operation. No material is wasted in the operation, burrs are not produced and grain flow provides improved mechanical properties. Although fasteners are the major application, a wide variety of parts of different shape can be formed in this process. Valves, knobs, rollers, shafts, spark-plug bodies, gear blanks, and hose fittings are examples. Upsetting within the length of the part rather than the ends is also possible. Cold heading is particularly economical when production quantities are large. The operation can be performed upon heated workpieces if the material involved is difficult to form or work hardens quickly. The operation then is referred to as warm heading. High carbon steels and austenitic stainless steels may be processed with this approach.

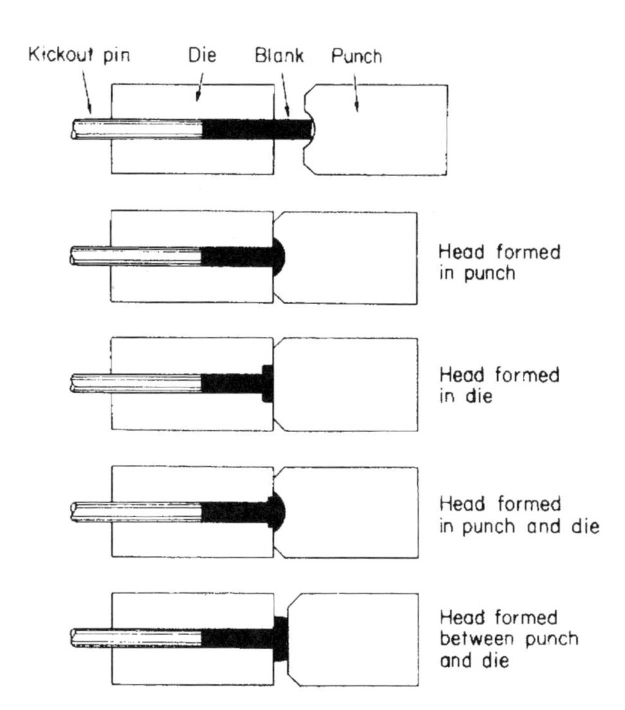

Fig. 2I2 Cold heading. Upsetting the end of the workpiece material can be achieved by a moving punch, in the stationary die, or in both punch and die and between the punch and the die. *(Courtesty National Machinery Company)*

I3. ***thread rolling*** - See screw threads.

I4. ***impact/cold extrusion*** - forms a part of some length by plastic flow of metal under compressive stress into an opening of limited size. The process is normally referred to as *impact extrusion* when non-ferrous metals are involved and *cold extrusion* when the workpiece is ferrous. A blank is placed in the die and a punch puts pressure on the blank, causing metal to flow in the desired direction. The press used can be either mechanical or hydraulic. With the rapid force of a mechanical press, the term, "impact extrusion" is appropriate. There are three varieties of the process: forward extrusion, backward extrusion and combined extrusion. Impact/cold extrusion produces parts with smooth surfaces, with no loss of material in the form of machining chips. The process is fairly rapid and is suitable for mass-produced components used in quantities of 100,000 pieces per year or more. Cylindrical and near-cylindrical shapes are typical parts made with the process. Collapsible metal tubes are made with the process. Aluminum, copper, lead, and magnesium alloys are the most commonly extruded, along with low and medium-carbon steels.

I4a. ***backward extrusion*** - In this process, shown in Fig. 2I4a, a metal blank is placed in the die, and the metal is compressed by the punch and forced to flow backwards around the punch to form a hollow object. The side wall thickness of the part depends on the amount of clearance between the punch and the die. Depending on the shape of the bottom of the punch and the bottom of the die, various shapes at the closed end of the part can be produced. The thickness of the base depends on the press stroke and is independent of the wall

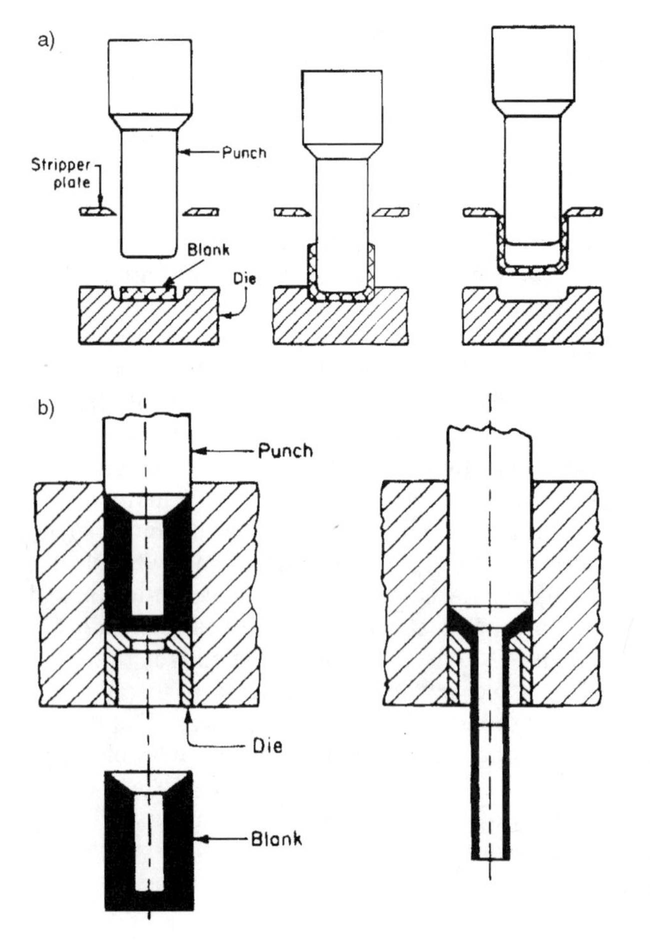

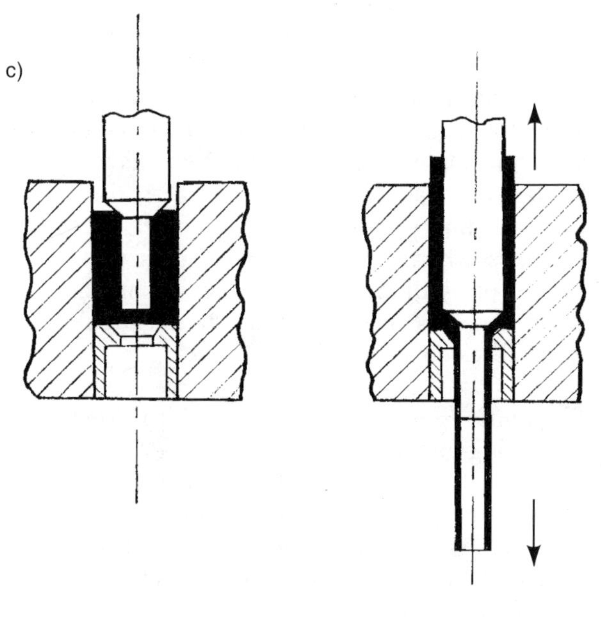

Fig. 2I4 Impact extrusion methods, a) backward extrusion, b) forward extrusion, c) combined backward and forward extrusion. (The method is usually called, "cold extrusion" when ferrous metals are processed.) *(from Bralla, Design for Manufacturability Handbook, McGraw-Hill, New York, 1999)*

thickness. The process is most frequently used with softer non-ferrous metals such as zinc, lead, tin, and aluminum to form cans and collapsible tubes for toothpaste, paint pigments, shaving cream, and other materials.

I4b. ***forward extrusion*** - In this process, sometimes called the ***Hooker process***, the workpiece metal flows forward (normally downward) through an orifice rather than upward around the punch. A close fit between the upper portion of the punch and the die prevents the upward flow. Hollow parts are produced by providing clearance between lower portion of the punch and the die orifice. The process differs from conventional extrusion described above in section A3 in that, with this method, a discrete part is produced that has one end closed or is of a particular shape. However, open- or closed-end tubes can be produced with the process. The operation is also sometimes part of a sequence that also involves cold heading. Preformed blanks are often used to provide a desired shape in the finished part. Fig. 2I4 (view b) shows the process.

I4c. ***combined extrusion*** - is both forward and backward impact extrusion at the same time on one part, as shown in Fig. 2I4c. The backward extrusion can have a different solid or hollow shape than the forward extrusion. Typical parts have a central flange with different cross sectional shapes above and below it. Fig. 2I4 (view c) illustrates the process.

J. High-energy-rate Forming Methods

HERF methods, sometimes referred to as HVF (*high velocity forming*) methods are useful in the forming of large workpieces and difficult-to-form metals through the application of large amounts of energy in a very short time period. These processes are characterized by the high velocity of the forming action (more than 50 feet per second[4]) and the use of a pressure wave rather than a forming punch to effect the change in the workpiece (except for pneumatic-mechanical forming described below). Energy is supplied by explosive charges, detonation or combustion of gas mixtures, spark discharge, exploding bridge wire, electromagnetic pulses, and the sudden release of compressed gas. Water may be used to transmit the shock wave. Except for pneumatic-mechanical forming, tooling and equipment are relatively simple and less expensive compared to that required with more conventional forming methods. The processes are often advantageous for prototype and limited quantity production. However, safety aspects of the process used must be dealt with.

J1. ***explosive forming*** - uses an explosive charge to provide the energy necessary to effect the forming operation. Normally, the explosion takes place in water and the water pressure then forces the workmetal against the die walls, but rubber, sand, glass beads, oil, molten salts, and air can also be used as a pressure transfer medium. (Air is an inefficient medium.) The pressure wave travels at several hundred feet per second. Only a female half die is required to form depressed shapes. For short-run production, dies made of concrete, fiberglass/plastic, sheet metal, or cast epoxies can be used. Another advantage is that there is little or no springback after the operation. Fig. 2J1 illustrates explosive forming, a) with a water medium and free forming and b), with a confined system. One common application is the forming of steel plate components for large storage tanks. Blanks to 13 ft (4 m) in diameter have been formed with this method. The method is useful when the workpiece material, eg., stainless steels, have work hardening properties. Most such operations are performed outdoors. The tank used to hold the medium must be strong enough to withstand the force of the explosions. The explosive charge is sometimes affixed to the workpiece ("contact operation") and sometimes separated from the workpiece ("standoff operation") as illustrated in Fig. 2J1.

With smaller parts, a confined system can be used. This is one which has a die that completely encloses the workpiece. The method is practical for forming and sizing tubular parts and other smaller parts where the cost of a fully-enclosing die is not too expensive. Applications for explosive forming include aircraft and jet-engine components, ducts, panels, and housings[5].

J2. ***combustible gas forming*** - With this HERF process, hydrogen, methane, or natural gas, mixed with air, oxygen, or ozone, when ignited, provide the pressure wave that forms the sheet metal workpiece.

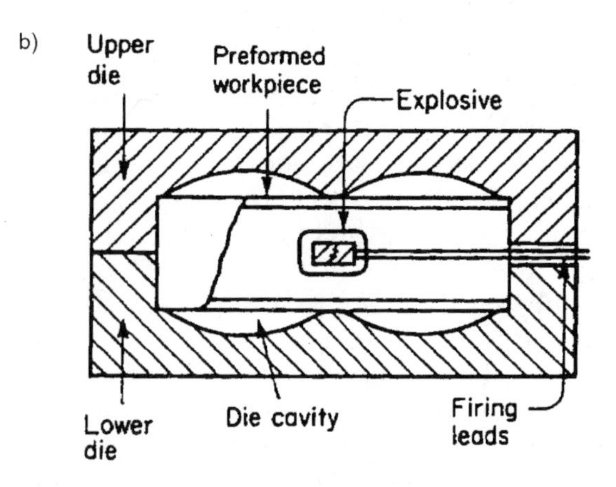

Fig. 2J1 Typical explosive forming arrangements: a) free forming with a water medium and b) with a confined system. (*from Bralla, Design for Manufacturability Handbook, McGraw-Hill, New York, 1999*)

The speed of the pressure wave is considerably less than when high explosives are used, and its duration is shorter but still measured in milliseconds. The pressure wave occurs in the ignited gas rather than in water or some other medium. The process is more suited for in-plant operation and for larger production quantities than high-explosive forming. More complex shapes can be produced. Large parts of sheet metal are feasible. Fig. 2J2 illustrates the process.

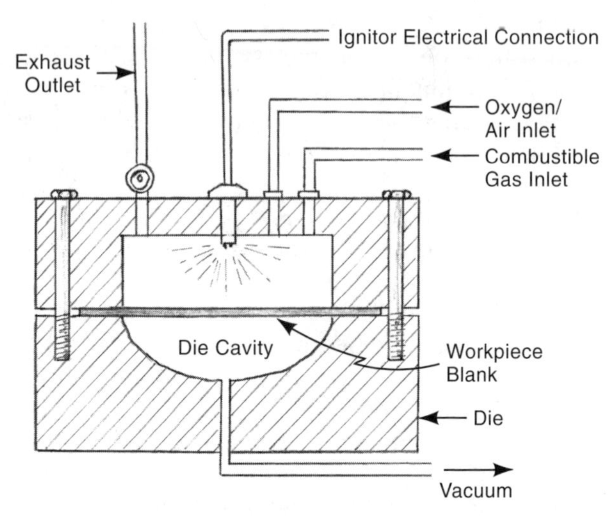

Fig. 2J2 Typical arrangement for combustible gas forming.

J3. ***electromagnetic forming (EMF)*** (also known as ***magnetic-pulse forming)*** - uses the electromotive force generated by a sudden electromagnetic pulse to form conductive sheet metal workpieces. A magnetic coil, placed adjacent to the workpiece, is subjected to sudden electrical current flow from the discharge of a bank of capacitors. The magnetic field thus produced in the coil induces a secondary current in the workpiece and there is a repulsive force between that current and the primary current in the coil. This repulsive force drives the workpiece metal against the die or other part and stresses it far beyond its yield point. There is no contact between the coil and the workpiece, and the magnetic energy will pass through non-conductive materials such as fibre, rubber, plastics, etc. The operation works best with highly conductive workpiece materials such as copper, brass and aluminum. Low-carbon steel can also be processed. Materials of low electrical conductivity can be formed if a "driver", a sheet of conductive material, is placed between the coil and the workpiece. Equipment consists of a storage capacitor, a power supply to charge the capacitor, and switches. The magnetic pulse lasts only 10 to 100 μs. One common application is to contract (swage) tubing around fittings or tubular fittings around pipe, cable, or hose. This requires a coil around the part to be swaged. In all electromotive forming, the coil must be physically strong enough to withstand the repulsive force, but otherwise, tooling is simple and inexpensive. The operation is illustrated in Fig. 2J3. It can be used for embossing, forming, expanding tubing, piercing, and blanking, as well as swaging, but there are limitations to the complexity of shapes that can be produced.

J4. ***spark-discharge forming (electrospark forming)*** - Both this method and electrohydraulic forming, described below, use electrical energy stored in a bank of capacitors to create a shockwave, which is transferred through a liquid medium. In spark-discharge forming, the energy released is in the form of a spark across two electrodes. Switching circuits instantaneously discharge that energy to the two electrodes. The intense energy emanating from the spark sends a shock wave that forces the workpiece material against the die walls. The equipment set up is similar to that depicted in Fig.2J1 a) and b) except that the explosive charge is replaced with spark discharge. The liquid medium used must be non-conductive because the electrodes generating the spark are in the same liquid. The use of a spark discharge permits careful control of the energy intensity. Repetitive cycles can be employed without removal of the workpiece when successive blows can aid the forming operation. Tooling and equipment costs are low so the process can be economical even with small production quantities.

J5. ***electrohydraulic forming*** - is very similar to spark-discharge forming except that the energy is released from an exploding bridgewire connecting

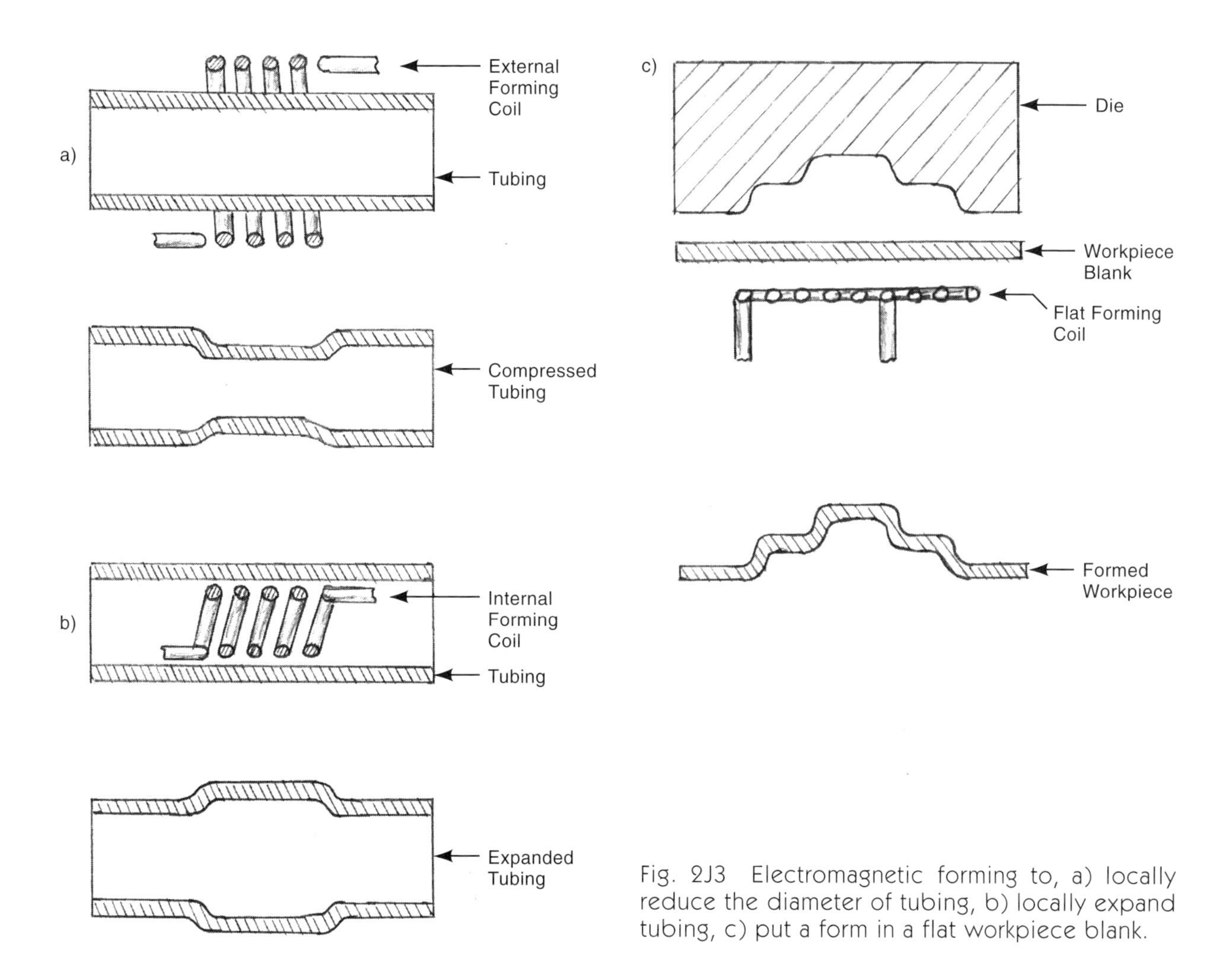

Fig. 2J3 Electromagnetic forming to, a) locally reduce the diameter of tubing, b) locally expand tubing, c) put a form in a flat workpiece blank.

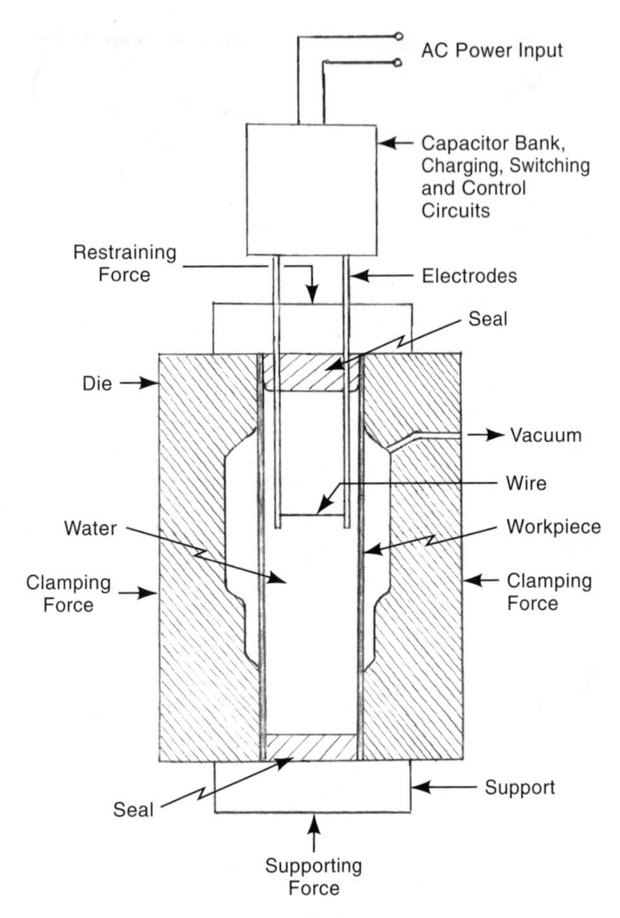

Fig. 2J5 Tooling arrangement for use of electro-hydraulic forming to change the shape of a tubular workpiece.

two electrodes. Water transfers the force that forms the workpiece. The operation cannot be repeated as easily as spark-discharge forming but it does provide somewhat greater forming force. The process, illustrated in Fig. 2J5, is most suited to expansion forming of tubular parts. The process is effective; however, if the operation can be performed by more conventional methods, they are usually more economical.

J6. ***pneumatic-mechanical forming*** - can be considered a variation of drop hammer forging. Compressed gas stored in a pressure tank, or the energy released in the form of gas from combustion of a fuel-oxidizer mixture, is used to drive the press ram to a high velocity, two to ten times that of conventional drop hammers[4]. The best application of the process is the forging of parts having thin vertical dimensions and a high level of detail. The process is advantageous with workpiece materials of high unit cost when the forging can be produced in sufficient detail so that material wastage is minimized. Otherwise, conventional forging will probably be more economical.

J7. ***peen forming*** - is a forming process that does not require any dies and can be performed at room temperature. Sheet metal is impinged on one surface with a stream of metal shot as illustrated in Fig. 2J7. Each shot piece acts as a very small hammer stroke, compressing the surface vertically but elongating it horizontally. The shot is delivered by either nozzle or centrifugal wheels. The total effect is the generation of a gentle compound convex curve. The process is useful when the form required does not involve abrupt curvature changes. It is used in the aircraft industry to produce wing panels. An advantage is that both the top and bottom surfaces have residual compressive stresses, which aid in providing fatigue strength. Workpieces with some reinforcing structure can be processed. Sometimes a preload is applied to the surface to assist in producing the desired final shape. Aluminum sheet from 0.050 to 0.200 in (1.25 to 0.5 mm) and steel from 0.016 to 1.00 in (0.40 to 25 mm) thickness can be processed.

K. Straightening

Straightening is required when the workpiece material is not in the form required for subsequent operations (eg., coiled rather than straight) or has undergone some deformation from stresses induced by prior operations. Several methods are available.

K1. ***manual straightening*** - covers a variety of operations performed with surface plates, levers, vises, anvils, hammers, twisting tools, grooved blocks, and heating torches. Typically, the workpiece is clamped at one end or where there is no deformation, and the deformed portion is moved in the opposite direction of the distortion. The workpiece is overbent so that it springs back to a straight orientation. Hammer blows or heat sometimes aid in the process.

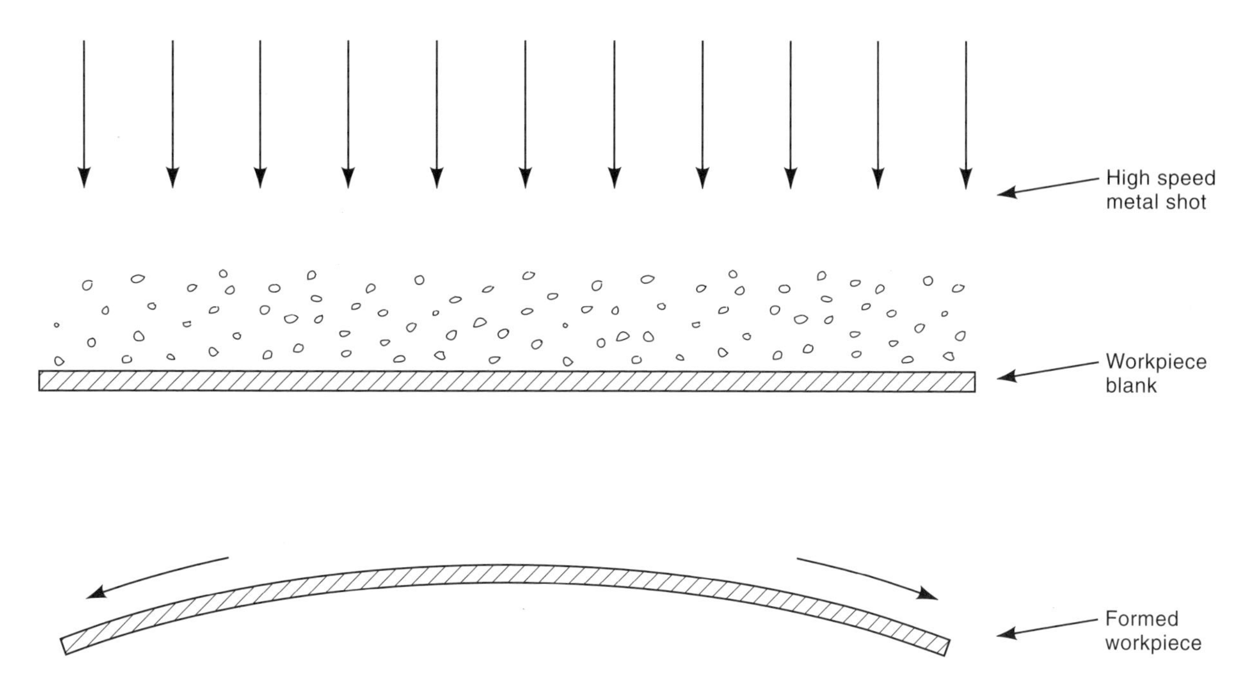

Fig. 2J7 Peen forming. High velocity shot striking the upper surface elongates the metal grains of a sheet and creates a convex shape.

K2. ***press straightening*** - uses an arbor or hydraulic press to make the corrective bends. Typically, the deformed workpiece is placed on support blocks in the press bed so that the convex area is on the upward side. The ram descends and bends the workpiece in the opposite direction far enough so that the elastic limit of the material is exceeded. If done properly, the workpiece will spring back to a straight position. Indicators and blocks may be used to find the high points in the deformed areas and to measure the effect of press strokes. Heat may also be applied to the area of deformation. The process works best with materials of Rockwell C hardness less than 40[3].

K3. ***parallel roll straightening*** - illustrated in Fig. 2K3, is used on sheet, rod or wire. A series of parallel rollers on opposite sides of the workpiece subjects the workpiece to progressively decreasing reverse bends as it passes through the rollers. As the metal is bent back and forth, it is stressed slightly beyond the yield point, exiting from the rollers with a reasonably flat or straight shape. The operation is rapid since it is continuous.

K4. ***rotary straightening*** - is a method for straightening round rods and bars. It uses two or more rollers with axes at an angle to each other and at an angle to the axis of the part. One of the

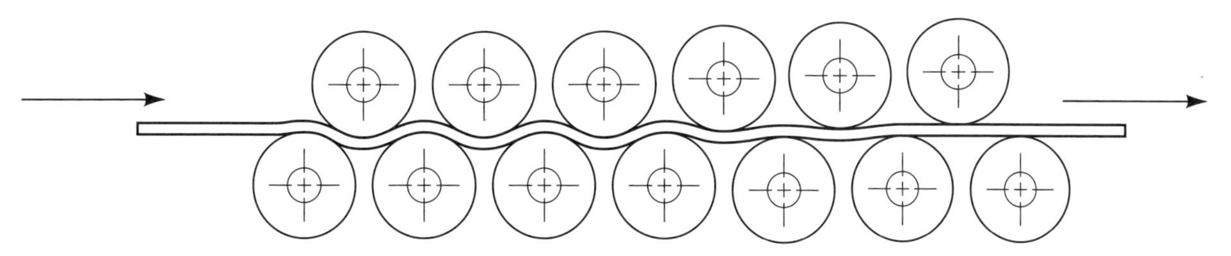

Fig. 2K3 Straightening of workpiece material by passing it through a set of straightening rolls that reverse-bend it a gradually decreasing amount as the material passes through.

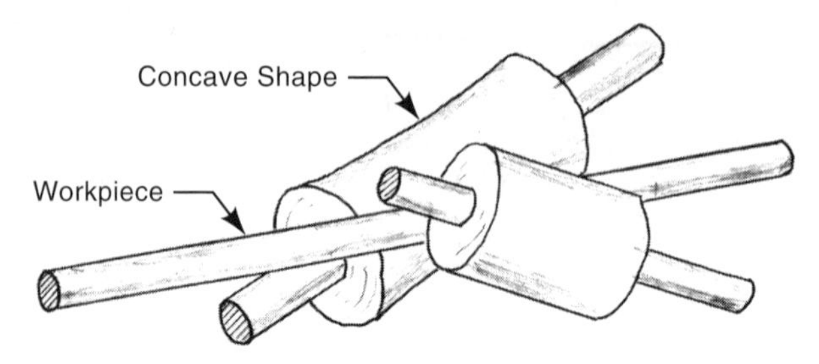

Fig. 2K4 Rotary straightening of bar material. The workpiece moves axially. Alternating compressive and tensile stresses straighten the workpiece. (Guides for the workpiece are not shown.)

rollers has a concave shape, as shown in Fig. 2K4, while the other is cylindrical. There are guides to position the direction of the rod to be straightened. Both rollers are driven and they rotate the workpiece and cause it to move axially. As the workpiece moves, the curvature of the concave roller(s) puts the workpiece surface into alternate stresses of compression and tension. The angles of the rollers and the pressures that they apply are both adjustable. The method is particularly applicable to short workpieces that are more difficult to straighten by other methods. Workpieces of from 1/16 to 10 in (1.6 to 250 mm) long have been straightened with this method[3].

K5. ***epicyclic straightening*** - is a proprietary process for straightening axles, tubular driveshafts, and propeller shafts, I-beam and other sections, and symmetrical forgings. The workpiece is supported on its ends and is rotated. A powered arm, attached to the center of the workpiece, deflects the center of the workpiece so that it is stressed in all directions beyond its elastic limit. The powered arm reduces the magnitude of deflection gradually so that the neutral axis of the part describes either a circular or elliptical spiral toward the center.

K6. ***stretch straightening*** - The workpiece is gripped mechanically and stretched in a straight line beyond the yield point. Twist deformation can also be corrected; when this is needed, the workpiece is reverse-twisted when it is stretched. The process is normally limited to workpieces of constant cross section. It also results in some material loss since the ends damaged by gripping must be removed.

L. Other Forming Processes

L1. ***powder metallurgy (P/M) processes*** - metal powders mixed with certain solid lubricants are compacted under pressure to form the desired shape (See L1c.); they then are heated to a temperature sufficient to bond the particles strongly together (See L1d.). The sintered part may be pressed a second time to improve dimensional accuracy and surface finish and, sometimes, to modify the shape. (See L1e.) Typical powder metal parts are rather small (About 3 square inches in cross-sectional area is a typical maximum because of the high press forces involved.). The process is used for a variety of precision mechanical parts, but bearings, and other parts with surfaces that have are metal-to-metal sliding contact are good applications because the porosity inherent in these parts provides space for a reservoir of lubricating oil. Almost all metals can be processed by this method. Iron, steel, bronze, copper, brass, nickel alloys, and stainless steels, are the most commonly used. Small gears, cams, small levers, sliding blocks, sprockets and pawls are other applications. Further applications are parts made from alloys that are otherwise difficult to machine or fabricate such as carbide cutting tool inserts, and tungsten lamp filaments. Parts requiring dual materials such as graphite-carbon motor brushes, copper or silver and tungsten electrical contacts, and tin-copper bearings are other examples. The process tends to be limited to high-production situations since the tooling required is not inexpensive. However, the labor content is low, particularly in comparison with applications that otherwise would require machining. The process works best with parts having straight sides, although tapers and curvature of sidewalls are possible over short distances if properly designed. Undercuts and holes in sidewalls are not feasible and must be produced by secondary machining operations.

L1a. ***metal powder manufacture*** - See *powders, metal.*

L1b. ***powder blending and mixing*** - The ideal powder for parts making may contain a mixture of powders of different metals or alloys, nonmetals, different particle sizes, lubricants, and binders.

Lubricants, eg., stearic acid, lithium stearate, and graphite, provide better compressibility and flow characteristics. Binders improve the "green strength" of the unsintered part. Both binders and lubricants are burned off or volatilized in the sintering process. Blending of these ingredients may take place either wet or dry. Water, or other solvents facilitate the mixing and reduce dust. Mixing usually is performed as a batch operation. The mixing cycle should be short to prevent damage to or work-hardening of the powder particles. Drum, double-cone, cubical-shaped, V-shaped, and conical mixers, with rotating screws are commonly used. Fig. 2L1b shows the shape of several common mixers.

L1c. ***pressing (compacting)*** - Metal powders, normally at room temperature, are introduced to, and fill, a die cavity of precise shape, with sidewalls that are usually straight. A punch of the same cross-sectional shape as the die, and with a close fit to the die cavity, descends into the die and presses the metal particles together. A lower punch, which rises to aid in the compaction and later ejection of the part, is common. Parts with stepped shapes or flanges may have more than one lower and upper punch. (Multiple punches are required because the powder does not behave like a liquid; friction between particles and die sidewalls reduces the density of areas away from the punch face. Almost all compaction is vertical, and pressures of up to about 1600 MPa (230,000 psi) are used. The pressing operation forms a "green" part of sufficient strength for further handling (but not enough strength for functional use.) The die and press motions are summarized in the caption accompanying Fig. 2L1c. Note that there is an upward force and motion applied from the bottom punch. This is advisable to provide more uniform compaction, overcoming die wall friction. Multiple-action presses with multiple-motion punches may be used to provide the necessary compaction in all portions of more complex workpieces. In one pressing method variation, the basic die is not fixed; instead, its motion during the operation provides the equivalent effect of a moving bottom punch. This variation, called ***withdrawal-die pressing*** , is shown in view (b) of Fig. 2L1c.

L1d. ***sintering*** - The "green" compact is heated to an elevated temperature in a controlled atmosphere.

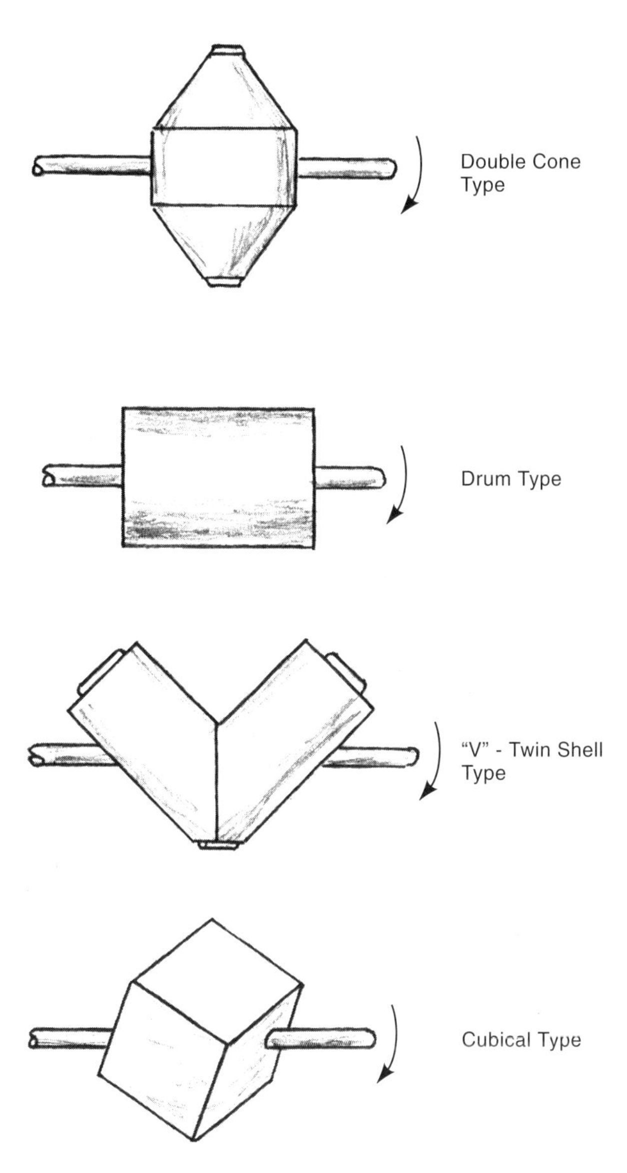

Fig. 2L1b Four types of rotating mixers used to blend metal powders.

Either a batch-type or continuous conveyor furnace may be used. The furnace temperature is well below the melting point of the alloy involved but high enough and maintained for a long enough period so that diffusion bonding of the metal particles takes place. The metal particles bond securely together and there may be some increase in density but the resulting part is not 100 percent dense. Densities of 75 to 90 percent are typical.

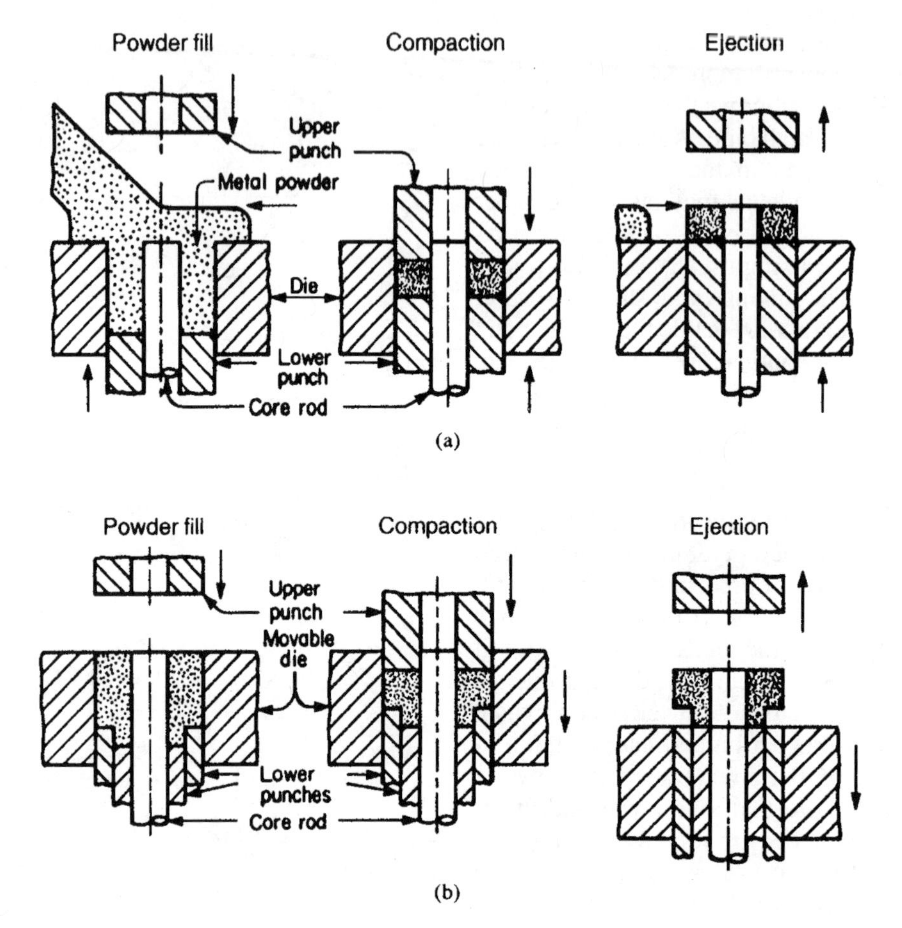

Fig. 2L1c The pressing sequence for powder metal parts, (also applicable to ceramic parts. See chapter 5.) (a) **Fixed-die system.** The die is fixed in the press table. The lower punch is withdrawn when the feed shoe (not shown) is over the die cavity. The excess metal powder is pushed aside. The upper punch lowers the die. The upper and lower punches move simultaneously. Powder is compacted. The upper punch is then withdrawn. The lower punch continues to rise, ejecting the newly compacted part from the die. The feed shoe moves across the face of the die, pushing the part to a collecting station. The cycle is repeated. (b) **Withdrawal die pressing system.** The main lower punch is fixed; the die is movable. The die is filled. The upper punch enters the die. The die is withdrawn at half the speed of travel of the upper punch. When two lower punches are employed, one lower punch is fixed and one is movable. When the movable punch has completed its compaction motion, it is allowed to move to enable compaction of the second level to occur. When compaction has been completed, the upper punch is raised. The die is further withdrawn to effect ejection by stripping the die from the component. The feed shoe moves across the face of the die, pushing the part to the collection station. The die rises to allow it to fill from the feed shoe. The cycle is repeated. (*from Bralla, Design for Manufacturability Handbook, McGraw-Hill, New York, 1999*)

The sintering process actually has three phases:

1. burn off, which removes air and volatilizes binders and lubricants as the temperature of the compacted part gradually increases. The volatilized materials must be removed from the furnace atmosphere before the second phase begins.
2. a high-temperature stage during which the diffusion bonding takes place. This can require a period of minutes or several hours. For iron-based parts, the temperature ranges from

1850 to 2100°F (1010 to 1150°C); copper, bronze, and brass parts allow somewhat lower temperatures[4].

3. cooling phase, which lowers the temperature of the sintered workpieces while maintaining the controlled atmosphere.

The atmosphere used is commonly one with oxide-reducing properties. Dissociated ammonia, hydrogen, or cracked hydrocarbons, are most often used, though the atmosphere may also be inert, especially with nitrogen. Vacuum sintering is utilized with stainless steel, refractory alloys, and titanium. Fig. 2L1d shows a typical continuous type furnace.

L1e. ***repressing*** - The sintering process results in parts with some shrinkage from the green state as binders and lubricants are removed from the parts. There is also some inevitable distortion from the thermal expansion and contraction that takes place during sintering. To improve the dimensional accuracy of the parts, to improve surface finish, to sharpen certain details, and sometimes, to reduce the porosity of the parts, they may be subjected to an additional pressing (or "calibrating") operation. With some parts, repressing can take place in the same die that was used for compacting; for other parts, separate dies are used. The parts are pressed severely enough to cause plastic flow of the material. In addition to the improved dimensions and surface finish, the parts gain some strength from being cold worked and densified.

L1f. ***secondary operations*** - Powder metallurgy parts can be machined, thread rolled, heat treated, and surface finished after fabrication. Machining takes place when undercuts, side holes, screw threads, and other features not feasible with the powder-metallurgy process, are needed. Techniques are straightforward but the effect of porosity must be considered. It is usually not recommended to machine bearing surfaces if impregnated lubricants are planned to be used, since the machining will tend to close the surface pores of the workpiece. Heat treating with liquid-cyanide is not recommended for PM parts because the liquid salts can be trapped in the pore structure, leading to corrosion. Similarly, electroplating and other surface treatment chemicals can be trapped in these pores; pretreatment with materials that close the pores permits electroplating and other chemical treatments.

L1g. ***powder metal forging*** - In this process, the sintered powder metallurgy part becomes a blank for forging. The operation involves more drastic deformation than conventional repressing, and the P/M part is heated to forging temperature beforehand. This process has the advantage that the forging blank can be made to optimum size and shape, eliminating or greatly reducing wasted material, and permitting the production of more complex forged shapes with reduced necessity for machining after forging.

Forgings made from P/M blanks can be more precise than conventional forgings. No flash is produced

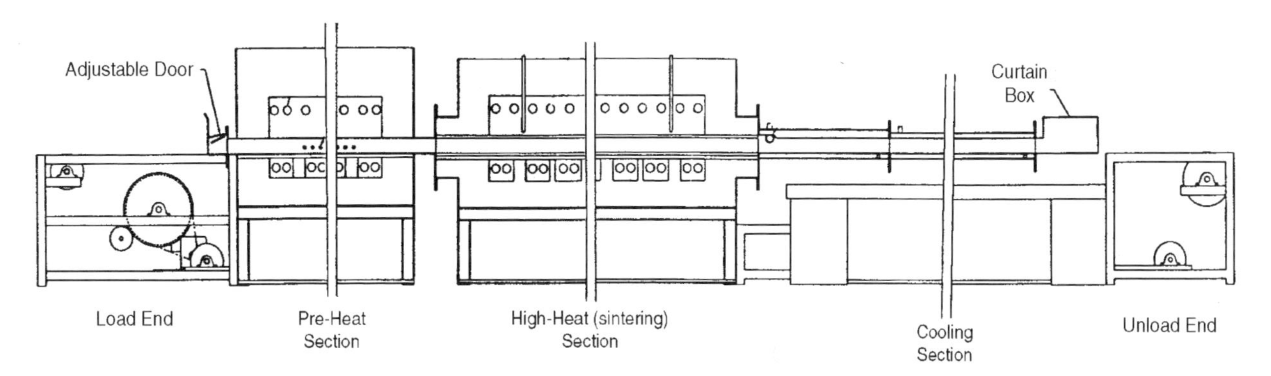

Fig. 2L1d Conveyorized, continuous-type, sintering furnace for powder metal parts. The furnace includes a preheat section, a high-heat sintering section and a cooling section. (Note: The lengths of the sections are considerably longer than shown here. The drawing was shortened to fit the page.) (*Drawing courtesy Abbott Furnace Co.*)

and draft is not necessary. The part is stronger than a conventional repressed P/M part because it gains the benefit of the improved grain structure that forging provides, and the density is increased to as much as 99 percent. Cams, gears, splines and connecting rods are typical parts made with this process. To prevent oxidation, protective atmospheres are used during the heating and forging cycles or the workpiece is coated with graphite.

L2. ***electroforming*** - Electroforming is a process that utilizes electroplating (See 8C.) techniques to make a formed sheet component. Whereas, in electroplating, the deposited material is very thin and is left on the surface of the plated workpiece, in electroforming, the deposited material is much thicker and is separated from the substrate after plating to produce a separate part. There are three steps, then, in the process:

1. Preparation of a mandrel,
2. Electroplating of the mandrel to the thickness needed, and
3. Separation of the electroplated material from the mandrel.

The process is particularly suited to complex shapes where very high accuracy and great detail are needed. One common application is dies used in production of compact audio and video discs. The accuracy and detail produced depend on the precision of the pattern or mandrel onto which the plated material is deposited. The electroformed part takes on the detail and accuracy of the mandrel. Therefore, it is necessary for the mandrel to be made with the accuracy and surface smoothness wanted in the electroformed part. Casting and molding methods can sometimes be used in making the mandrel. This may have to be repeated for production of multiple parts if the mandrel is expendable, that is, if it is made of materials that are melted or dissolved to separate the electroformed part and the mandrel.

Typically, the wall thickness of the electroformed material will range from the minimum needed to maintain the part's integrity to as much as 5/8 in (15 mm). Nickel, copper, and silver are metals most easily used for the electroformed parts, with nickel probably being most common. Iron can also be processed. The mandrel can be made of

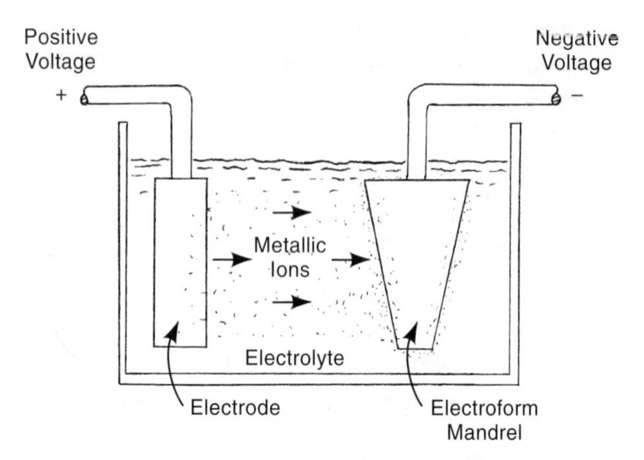

Fig. 2L2 Schematic representation of electroforming. The part is formed from a build up of electroplated coating on a mandrel.

either non-metals or metal, though the former require a conductive coating so that electrodeposition can take place. Plastics, glass, and even wax, can be used for mandrels in addition to metals such as aluminum or stainless steel. Surface finishes down to 2μin are feasible on the interior surface of the part. (The other surface may be quite irregular, but this condition is seldom critical because compensations can be made on the mating parts, if necessary.) Fig. 2L2 shows the process schematically and Fig. 2L2-1 shows a typical electroformed part.

Electroforming can be used to "weld" several pieces together. The weld is stress-free because it is made at room temperature; melting of the fillet or the base materials is not involved. The fillet is developed by electrodeposition of metal on plating the joint area to a sufficient thickness.

L3. ***metal injection molding (MIM)*** - is a process somewhere between standard powder metal forming and injection molding, as used with many plastics (See 4C.). A mixture of fine metal particles and binders that usually include a thermoplastic, wax, plasticizers, and dispersants, is heated enough to provide a paste-like consistency and is injected under high pressure into a mold cavity. The mixture cools to form a molded workpiece that is ejected in the same way as are plastic parts. The molding equipment used is very similar to that used for injection molded plastics. Molds are of hardened tool steel. The molded part is in the "green"

Fig. 2L2-1 A copper electroformed microwave guide (shown on the right) and the aluminum mandrel on which it was formed (shown on the left). The mandrel is dissolved chemically after the plated coating is of the desired thickness. The flanges in the finished part were "welded" to the guide by plating additional metal at the joint after assembly. Electroforming is particularly suitable for complex parts like this, particularly if surface details must be accurately produced. In this example, the interior dimensions are the most critical. *(Courtesy A. J. Tuck Company)*

state immediately after molding, strong enough for handling in the factory but not for product use. It is then subjected to a treatment that removes the binder material, either by solvent extraction, catalytic action, or by high temperature vaporization, or a combination of these methods. This may take from one to 24 hours, depending on the size of the parts and the method used. Metal density is then typically about 60% and the workpiece is in the "brown" state, still strong enough for handling.

The workpiece is then sintered in a vacuum or hydrogen/nitrogen atmosphere with methods that are similar to those used in conventional PM sintering. (See description in L1d above.) The sintering temperature usually ranges from 2200 to 2500°F (1200 to 1400°C) for a period of 3/4 to 4 hours. Sintering bonds the metal particles together to form a strong, usable part. The parts are 95 to 99% dense but have shrunk to a size 15% to 25% smaller than the molded dimensions.

The MIM process is carried out with low alloy steels, stainless steels, soft magnetic materials, copper, and other non-ferrous metals. Complex parts can be made with the process to good dimensional tolerances. A final punch press sizing operation may be employed when required tolerances are particularly close. Because of the fine particle size required in the metal powder (less than 20μm in diameter - finer than that used for conventional PM parts), the raw material tends to be expensive. The time-consuming binder removal process is also expensive and must be closely controlled. The process is most applicable for high production quantities of small parts (under golf ball size). Intricate parts, not feasible with conventional PM forming often can be made with the process. Parts produced by the process include pistol, rifle, and shotgun parts, stepper motor rotors, fuel injector components, and automotive parts.

References

1. *Manufacturing Processes and Systems, (9th edition),* Phillip F. Ostwald, Jairo Munoz, John Wiley and Sons, New York, 1997.
2. *Materials and Processes in Manufacturing, (8th edition),* E. Paul DeGarmo, J.T. Black, Ronald A. Kohser, Prentice Hall, Upper Saddle River, NJ, 1997.
3. *Metals Handbook, Volume 4, Forming, (9th edition),* ASM International, Metals Park, OH.
4. *Tool and Manufacturing Engineers Handbook, Vol. 2 Forming, (4th edition),* Society of Manufacturing Engineers, Dearborn, MI, 1984.
5. *Design for Manufacturability Handbook, (2nd edition),* James G. Bralla (ed.), McGraw-Hill, New York, 1998.
6. *Schuler Metal Forming Handbook,* Springer, Berlin, New York, 1998.
7. *Handbook of Metalforming Processes,* H.E.Theis (ed.), Marcel Dekker, New York, 1999.
8. *Introduction to Manufacturing Processes,* J.A. Schey, McGraw-Hill, New York, 1987.
9. *Processes and Materials of Manufacture, (4th edition),* Roy Lindberg, Alllyn and Bacon, New York, 1990.

Chapter 3 - Machining Processes

A. Lathe and Other Turning Operations

A1. ***lathe operations (general description)*** - produce, with a cutting action, surfaces of rotation (surfaces having a round or partly-round cross section), both external and internal, in a workpiece. The workpiece is rotated in a lathe, screw machine, or chucking machine. It is held between centers or in a chuck or collet, or fastened to a face place. The cutting tool is fed into the work or along the work, or both, to produce a part of the desired shape. There are several basic types of lathes and related machines as described below and many varieties of tools that can be fed against the workpiece. These machines are used extensively in the production of parts that contain surfaces of rotation. The basic operations performed on lathes are the following:

A1a. ***turning*** - is the most prevalent lathe operation. In its most common form, a single-point cutting tool is moved on a precise path with respect to a rotating workpiece. When the tool moves parallel to the axis of rotation, straight turning takes place and the surface machined is cylindrical or part of a cylinder. When the cutting tool moves uniformly closer or farther from the axis of rotation as it moves longitudinally, a tapered surface is generated. (This is often accomplished in engine lathe by moving the tailstock supporting center for the work to an off-center position, out of alignment with the headstock axis of rotation.) Fig. 3A1 shows examples of straight and tapered turning. Engine lathes, turret lathes, screw machines, and chucking machines all perform turning operations, to produce all kinds of shafts, axles, spindles, pins, etc. with straight or tapered turned surfaces. Parts as small as wristwatch shafts and as large as ocean liner propulsion components are turned on lathes. Producing a turned surface of considerably smaller diameter than that of the original workpiece may involve one or more roughing cuts, usually with a large depth of cut. Finishing cuts to produce greater accuracy or smoother surfaces normally have a small depth of cut.

A1b. ***form turning*** - occurs when a single-point cutting tool moves longitudinally in a path other than a straight line, or when a cutting tool ground to a particular curved or otherwise irregular edge is fed radially into the rotating workpiece. The profile of the round workpiece then takes the shape of the cutting tool path or the inverse of the form tool edge. The lengths of surfaces produced by form tools on a lathe, shown in Fig. 3A1b, are limited by the width of the tool.

A1c. ***tracer turning*** - is a process for form turning. The inward and outward motions of the cutter as it moves longitudinally along the length of the part are controlled by the motion of a stylus as it bears against a template or master part. Several passes may be used, after which the part duplicates the shape of the template. A variety of tracing mechanisms are available, powered hydraulically, pneumatically, or electrically. The tracing operation is one of those shown schematically in Fig. 3A1. (The movement may also be controlled electronically

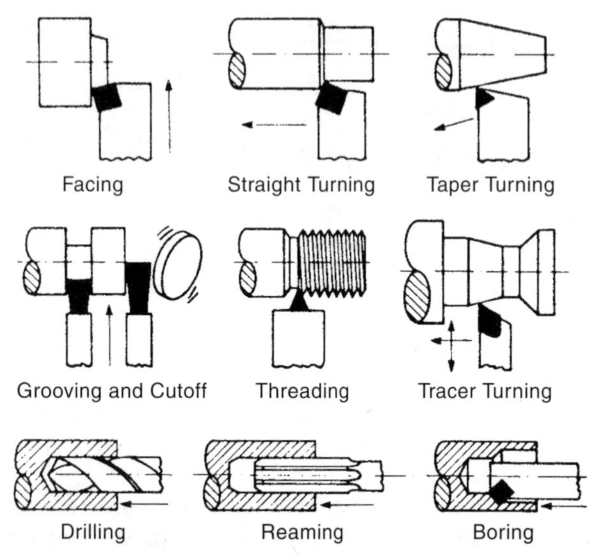

Fig. 3A1 Facing, turning, grooving, thread cutting, drilling, reaming, and boring performed on turning equipment. In facing, the tool moves perpendicularly to the axis of rotation of the spindle and a flat surface is produced. In straight turning, the tool moves parallel to the axis of rotation and a cylindrical surface is generated. In tapered turning, the tool moves at an angle to the axis of rotation and a tapered surface is produced. In tracer turning, the single point cutting tool moves inward and outward as it moves longitudinally, creating a contoured or other shaped surface on the rotating workpiece. The tool's path can be controlled by a template or by computer numerical control (CNC). In grooving, the tool, ground to the width of the groove desired, is plunged, or fed inwardly, creating a groove in the outer surface of the workpiece. Grooving can also take place internally when a cutter inserted in the center hole of a workpiece is fed outwardly. In threading, the cutting tool, shaped to the flank angle of screw threads, moves longitudinally in a fixed relationship with the rotation of the workpiece. *(from Bralla, Design for Manufacturability Handbook, 1998, McGraw-Hill Companies. Reproduced with permission.)*

without tracing, by using numerical or computer-numerical control, CNC). Some tracer lathes have an additional tracer-controlled cross slide at the rear of the work to facilitate the machining of grooves, undercuts, and chamfers.

A1d. ***facing*** - produces flat surfaces whose plane is at right angles to the axis of rotation of the part.

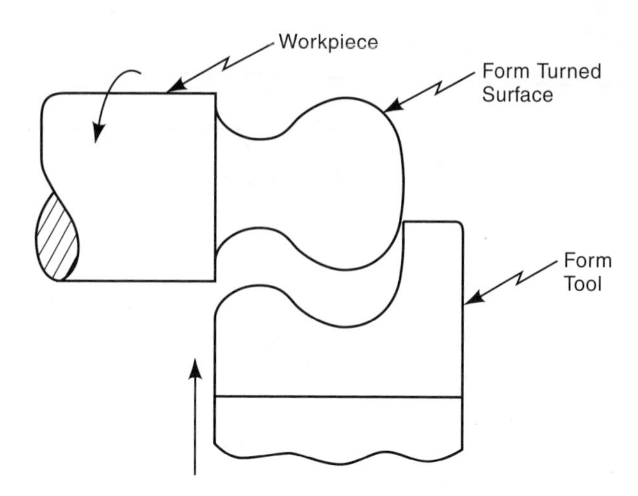

Fig. 3A1b Form turning - The tool is ground to the shape desired and is fed into the workpiece as it rotates, cutting a surface whose shape is the inverse of that of the form tool.

As the part rotates in the lathe, the cutting tool moves radially toward the axis of rotation, removing material as it advances. (It can also move outward from the center, but the other direction is much more common.) The operation is used to produce flat surfaces on castings and other parts that usually also require turning or some other lathe operation. The process is illustrated in Fig. 3A1.

A1e. ***grooving*** - The cutting tool, usually ground to the width and bottom shape required, is fed into the work, cutting a groove of the desired dimensions. Groves can be cut into any external or internal surface that such a cutting tool can reach. (Internal grooves are usually called recesses.) Fig. 3A1 includes a view of the grooving operation.

A1f. ***knurling*** - is not really a machining (cutting) operation because the knurl is formed, not cut, in the workpiece. Knurling is a common lathe or screw machine operation. The hardened knurling tool rolls against the cylindrical surface of the rotating workpiece with high pressure, causing the surface material of the workpiece to flow into peaks and valleys according to the pattern of the knurling tool. The result is a surface in the finished part that is roughened to a particular pattern, useful to improve the grip if the part must be held or rotated by hand when it is used. Several different patterns are possible. Other uses for the operation are for decoration and to increase the diameter of the part slightly to facilitate a press fit.

A1g. ***cutting off (parting)*** - When parts are made in lathes and screw machines from bar stock, the final operation is to sever the part from the remaining bar material. This is accomplished by advancing the cutoff tool, a narrow grooving tool, radially into the work. When the cutting edge advances to the axis of rotation of the part, the part is severed and falls to the bed of the machine. Some machines, which make blanks for further machining or other operations, are designed to perform only cut off operations and other simple ones on bar and tubular stock. The operation is shown schematically along with grooving in Fig. 3A1.

A2. *lathes and other turning machines*

A2a. ***engine lathes*** - are general purpose machines that provide the most basic means of performing turning, facing, grooving, knurling, and threading operations (Fig 3A1). These machines can also drill, ream, and bore holes at the center of rotation. A typical engine lathe has a chuck or collet to hold the workpiece in a powered rotating spindle, and a bed that normally consists of two ways. A tailstock holds the end of the workpiece opposite the spindle if the workpiece is long enough to require support at the end. A drill, reamer, or other tool can also be mounted in the tailstock so that holes can be machined at the axis of rotation of the workpiece. Other cutting tools (usually single point) are positioned in a compound rest and cross-slide that are mounted on a carriage that can move in an axial direction along the bed of the lathe. The ways of the bed are machined to be smooth and precisely in line with the spindle. Transverse movement of the tool is provided by the cross slide or the compound rest. The compound rest can be placed at an angle for short taper turning or for angle facing cuts. In all cases, the movement of the cutting tool can be precisely controlled with machine screws, mostly with manual crank actuation. Axial movement of the tool carriage, however, can be automatic through rotation of a lead screw. With the lead screw, the movement of the carriage also can be geared to the rotation of the spindle so that screw threads can be machined. Fig. 3A2a illustrates a typical engine

Fig. 3A2a A typical turning operation being performed on an engine lathe. The bar workpiece is held in the three-jaw chuck and is supported by a center in the tailstock. (*Photo courtesy Clausing Industrial, Inc.*)

lathe. These machines are very versatile and are especially applicable for performing turning and related operations when production quantities are limited.

A2b. ***turret lathes*** - differ from engine lathes in that they have a slide carrying a usually hexagonal tool holder, a "turret", in place of the tailstock. The turret can hold cutting tools in each of its sides and can be fed longitudinally and, on some machines, transversely. Turret lathes are, therefore, lathes adapted for production work. They are particularly suited for moderate production levels when automatic screw or chucking machines may not be justifiable, but when the turret lathe's tooling arrangement permits more rapid production than is feasible with engine lathes. Spindles of machines intended for machining of bar stock are equipped with collet chucks. When other components are to be machined, the spindles are fitted with chuck jaws suitable for gripping the workpieces.

A typical ram-type turret lathe is shown in Fig. 3A2b. When the turret lathe is set up for a particular operation, retraction of the turret automatically indexes it and positions the cutting tool required so that it is in place for the next operation. Feeds can be automatic or manually operated to pre-set stops. Turret lathes generally have a four-sided turret on the cross slide of the machine. Some machines also have a fixed tool at the back of the cross slide. Tools on the cross slide turret can also be indexed and fed in the sequence that the particular workpiece requires. Once the turret lathe has been set up with the required tools, the component to be produced can usually be fully machined without tool changes or adjustments.

Semi-automatic and automatic operation of turret lathes can be achieved by incorporating various modifications including automatic headstock control, power feed to the turrets, and automatic turret indexing. The use of some of these features may enable one operator to tend more than one machine. In the fullest degree of incorporation of these devices, the turret lathe becomes an automatic screw machine or chucking machine as described below. Then an operator is needed only to load and unload, and monitor the operation. Current practice, however, instead uses computer numerical control to achieve automatic operation of turning machines.

A2c. ***screw machines*** - are automatic lathes, originally developed for the mass production of screws from bar stock but long since used for a wide variety of parts from bar stock over a broad size range. Once set up, these machines run completely automatically. It is common practice for one operator

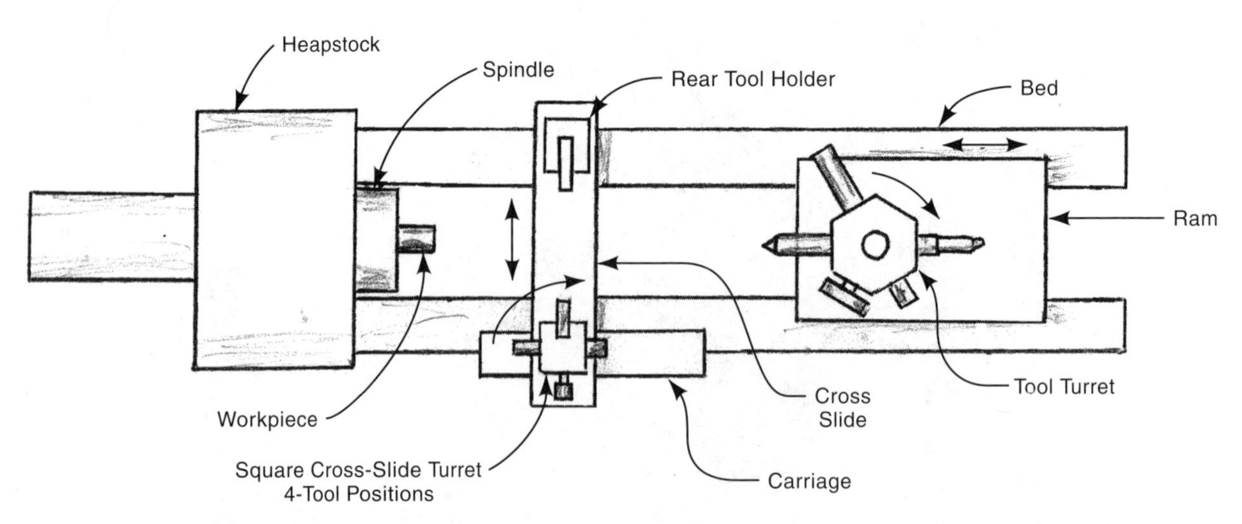

Fig. 3A2b Overhead view of a ram-type turret lathe. The turret on the ram holds up to 6 cutting tools. Four tools can also be held in the turret on the cross slide and an additional one on the rear tool holder.

to run a bank of such machines. Machines are available for producing parts from bar stock as large as 8 in (200 mm) in diameter but most work is performed on machines capable of processing bar of 2 in (50 mm) diameter or less.

A2c1. ***single spindle screw machines*** - There are two common varieties of single-spindle screw machines: the Brown and Sharpe or turret type and the Swiss type. The Brown and Sharpe type is essentially an automatic turret lathe. Typically, it has a six-sided turret, mounted vertically on a ram, a cross slide on which two tool-holders can be placed and one or two upper tool slides mounted near the spindle. Cutting tools on these holders are moved to machine the bar as it rotates. The bar is held in a spring collet in the spindle. All motions of these devices are controlled by cams. With the proper cams and set ups, the machine will operate fully automatically. Bar stock is fed automatically through the hollow spindle after each piece is cut off; the operation, therefore, continually repeats. Additional bars can be fed automatically from a magazine attachment. Some machines are equipped with pick-off attachments that grasp the turned part and hold it for secondary operations that require the workpiece to be non-rotating. Screw driver slots, flats, and cross holes can thus be machined with this attachment. Fig. 3A2c1 illustrates

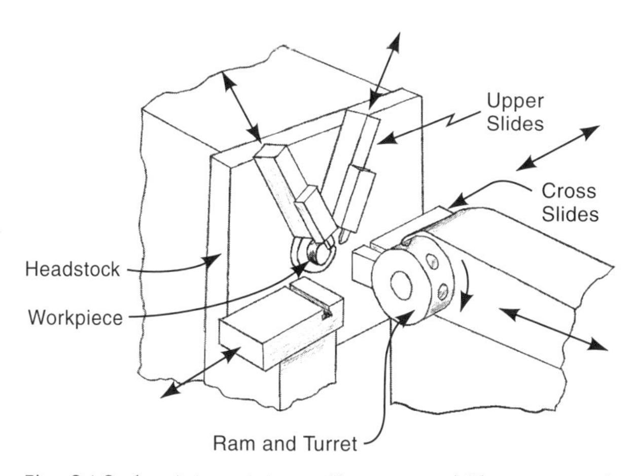

Fig. 3A2c1 A turret-type (Brown and Sharpe type) single spindle screw machine. The two upper slides, the two cross slides and the turret all can hold cutting tools.

the Brown and Sharpe type of machine. These machines are no longer manufactured, having been replaced by machines whose tool and stock movements are actuated by computer numerically controlled servo motors rather than cams. However, very many of these cam-operated machines are still in use throughout the world.

A2c2. ***Swiss-type screw machines*** - In the Swiss-type machines, tools are mounted on two cross slides and three upper slides arranged radially around the spindle. The headstock, holding the rotating bar, can slide longitudinally on the machine's ways. Tool movements are controlled by cams. Tools feed radially to the center of rotation of the bar. Rocker arms, driven by cams, provide the approach and infeed motions for the upper tools. Longitudinal feeds are made by shifting the headstock rather than tool slides, again by cam control. Single-point cutting tools are normally used on all slides. Swiss-type machines were originally developed for clock and watch manufacture and are particularly suited for small shafts and other very small parts. Fig. 3A2c2 illustrates the cutting tool and feed arrangements of these machines. Most current designs of Swiss-type screw machines utilize CNC with servo or stepping motors instead of cams to provide tool and headstock motions.

A2c3. ***multiple spindle screw machines*** - have 4, 6, or 8 spindles instead of the single spindle of the Brown and Sharpe and Swiss-type machines. The spindles are held in an indexing drum in the headstock of the machine. Each spindle has a collet that holds a separate piece of bar stock. This arrangement permits the cutting action to be divided so that a portion takes place simultaneously on several different workpieces. The spindles all rotate and then index to a new position after each cutting sequence. There is a tool slide in line with each spindle and, often, a cross slide for each spindle position. All spindle and tool slide motions are automatic, controlled by cams or servo motors. Bar stock is fed at one spindle position. Some machines are equipped to stop the rotation of one spindle and to perform milling and transverse drilling operations on the workpiece. The production rate of multiple-spindle machines is considerably faster than that on single-spindle machines because the operations

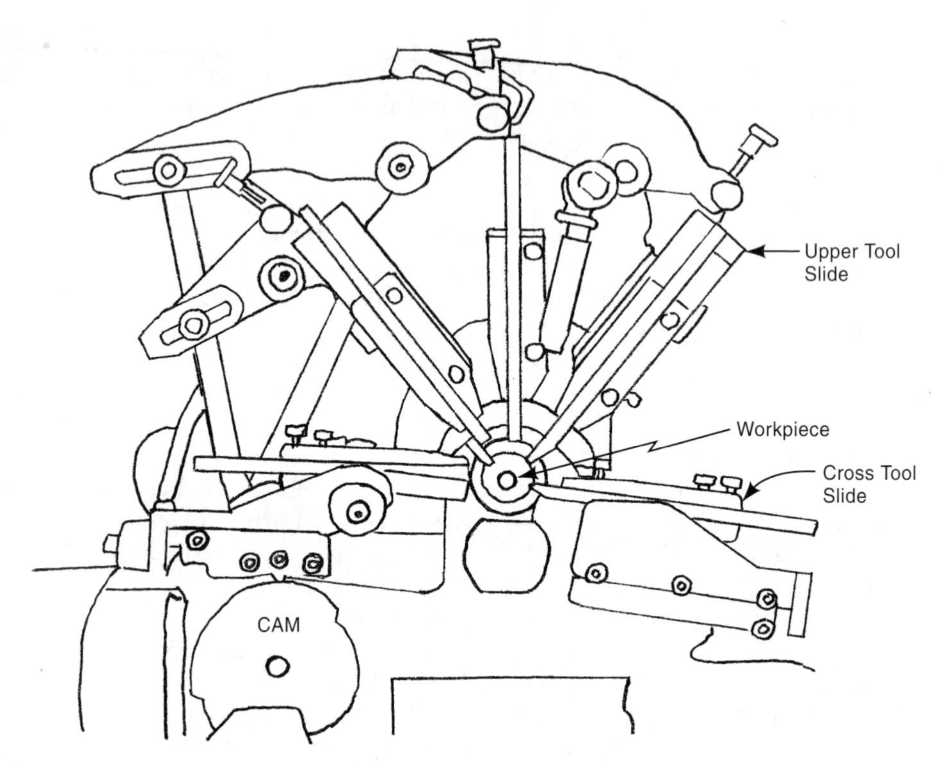

Fig. 3A2c2 Swiss-type screw machine. Cutting tools are arranged radially around the machine spindle. Tool infeed is controlled by cams. Longitudinal feed is provided by moving the spindle head.

are almost all performed simultaneously. Cycle time for a particular part is equivalent to the time required for the longest cutting operation plus the time required to index the spindles. These machines are ideal for high production situations. One operator typically tends a group of such machines. Fig. 3A2c3 illustrates a typical multiple spindle headstock and tool arrangement.

A similar operation can take place with discrete parts instead of bar stock. Each spindle then has chuck jaws made to fit the workpiece. One spindle position is used for loading and unloading workpieces. At each of the other positions, one or more operations are performed on the workpiece. The machines are then designated *multiple-spindle automatic chucking machines.* (See following paragraph.)

A2d. ***chucking machines*** - are automatic lathes designed for operations on castings, forgings, and other parts, rather than on bar-stock. Mechanisms and other features are similar to those of screw machines except for the bar feeding

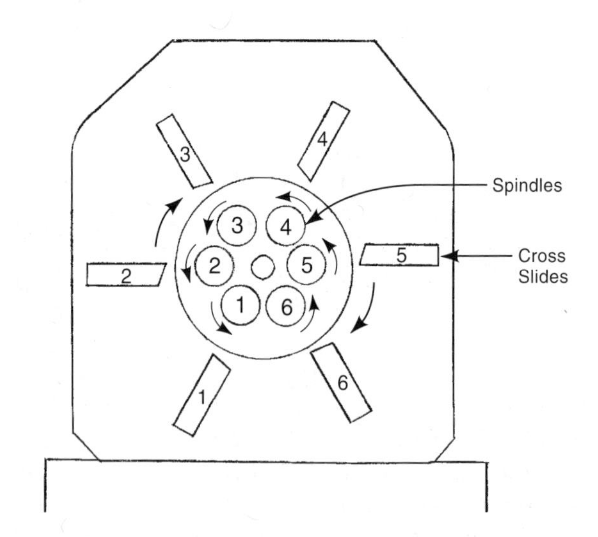

Fig. 3A2c3 The headstock of a typical six-spindle, multiple-spindle screw machine. All six spindles carry bar stock and rotate continuously. Periodically, each spindle is indexed to a new position. Cutting operations at all indexed positions take place simultaneously. There are also as many as six cutting tools located on the tailstock of the machine.

system. Instead, chucking machines have means at the spindle for holding the part to be machined. Special chuck jaws usually must be made to hold the workpiece if its shape is at all irregular. Most chucking machines are single-spindle but multiple-spindle machines also are widely used. Multiple-spindle chucking machines have one or more spindle locations where the spindle rotation stops for loading and unloading workpieces. Chucking machines are made for various sizes of workpieces ranging up to above 50 lb (23 kg) for vertical-spindle machines. Most horizontal-spindle machines are for workpieces under 10 lb (4.5 kg).

A2e. ***turning centers*** - (See T1.)

B. Round-Hole-Making Methods

B1. ***drilling*** - The most common tool for drilling, a twist drill, is a rod with helical flutes and two or more cutting edges at the end. It is rotated about its axis and fed axially into the work. As it advances, it produces or enlarges a round hole in the workpiece. The chips are carried away from the hole by the flutes in the drill. (When drilling an axial hole with a lathe, the workpiece rotates rather than the drill.) There are other types of drills that may not have helical flutes. Others may have only one cutting edge. The drilling process is very common and is used with a wide variety of machines ranging from the most sophisticated computer-controlled or multiple-spindle machines to hand-held electric or crank-driven drills. The most common diameter range for drilled holes is about 1/8 in (3 mm) to $1^1/_2$ in (38 mm) although diameters from 0.001 (0.025 mm) to 6 in (150 mm) can be drilled with commercially available special drills. Fig. 3A1 includes an illustration of drilling as performed on turning equipment. Fig. 3B1 shows some typical drills.

B2. ***counterboring*** - enlarges a hole for part of its depth and usually machines a flat bottom in the

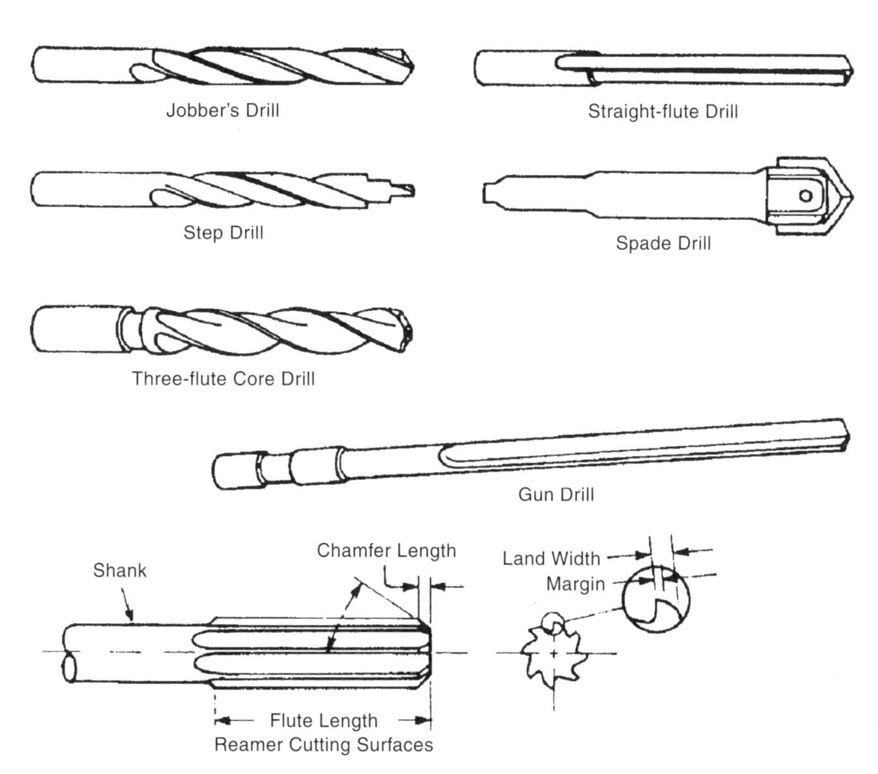

Fig. 3B1 A series of drills and, at the bottom of the group, a typical reamer. *(from Bralla, Design for Manufacturability Handbook, 1998, McGraw-Hill Companies. Reproduced with permission.)*

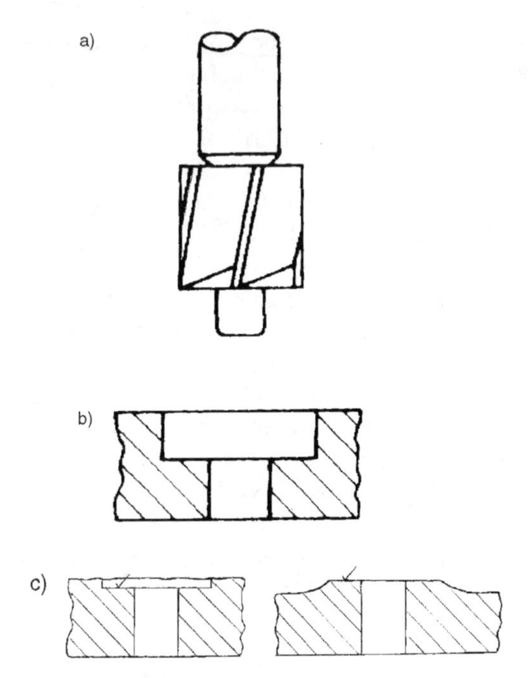

Fig. 3B2 A counterboring/spotfacing tool (view a) and a cross-sectional view of the counterbored hole it produces (view b). View c) shows cross-sections of two slightly different spotfacings produced by the same tool. The purpose of counterboring is to produce a recess of prescribed depth while spotfacing is performed to provide a smooth and perpendicular flat surface for a fastener or other object.

enlarged portion. The operation is most often performed to provide clearance for a bolt head or multi-diameter part. The rotating cutter is guided by a pilot that fits into the existing hole, so that the counterbored surface is concentric with the original hole. A multi-diameter counterboring tool can produce stepped counterbores. Fig. 3B2 illustrates a counterboring tool in view a) and the counterbored hole it produces in view b). (The tool also can be produce spotfacing, as shown in view c).

B3. ***countersinking*** - is an operation that adds a chamfer at the entry end of a hole. A rotating cutting tool, with the edge set to the angle of chamfer desired, is fed into the hole and removes material at the edge. The tool is centered by the hole; therefore the chamfer is concentric with the hole's axis. The operation is typically used to remove burrs or a sharp edge at the end of a hole, or to provide space for a tapered screw head or other tapered object.

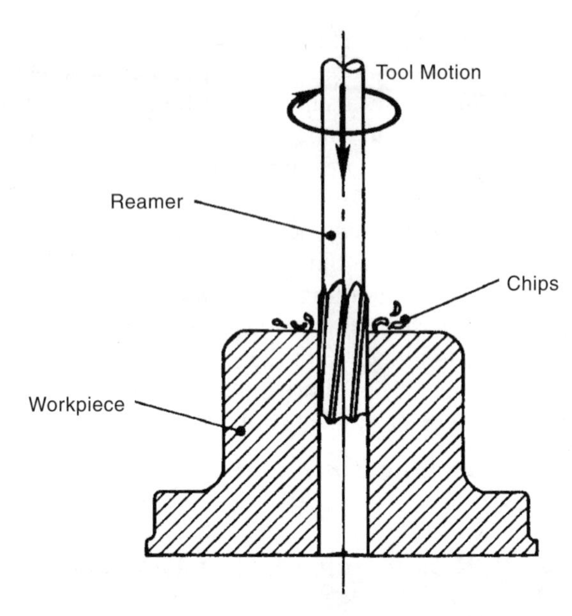

Fig. 3B4 Reaming is used to improve the accuracy, surface finish and straightness of round holes. *(from Todd, Allen, and Alting, Manufacturing Processes Reference Guide, Industrial Press, 1994.)*

B4. ***reaming*** - is a secondary machining operation for existing holes. It can provide a more accurate diameter, improved straightness, and a smoother surface finish as it slightly enlarges the hole. A rotating tool, a reamer, is used. The operation can be performed on a drill press or other drilling machine and is sometimes done by hand. Reamers normally remove 0.005 to 0.015 in (0.13 to 0.38 mm) of diameter. Reamers normally float, that is they follow the direction and location of the existing hole, but they can also be guided by bushings to slightly improve the hole's direction or location. The operation is most common with holes from 1/8 to $1^1/_4$ in (3 to 32 mm) in diameter but both smaller and larger holes can be reamed. A typical reamer is illustrated in Fig. 3B1 and the reaming operation on a lathe is shown schematically in Fig. 3A1 and Fig. 3B4. Taper reamers are used for finishing tapered holes.

B5. ***boring*** - is an operation that enlarges and improves the accuracy of an existing hole. Either the work or the cutting tool rotates about the center axis of the hole. The single point tool describes a circle, removing material from the surface of the existing hole as it advances, enlarging the hole, normally increasing the precision of any of a number of

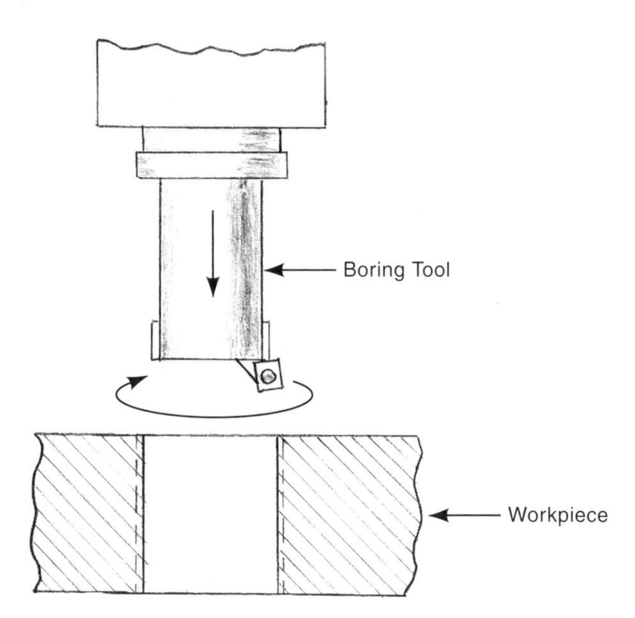

Fig. 3B5 Boring operations slightly enlarge and improve the precision of an existing hole.

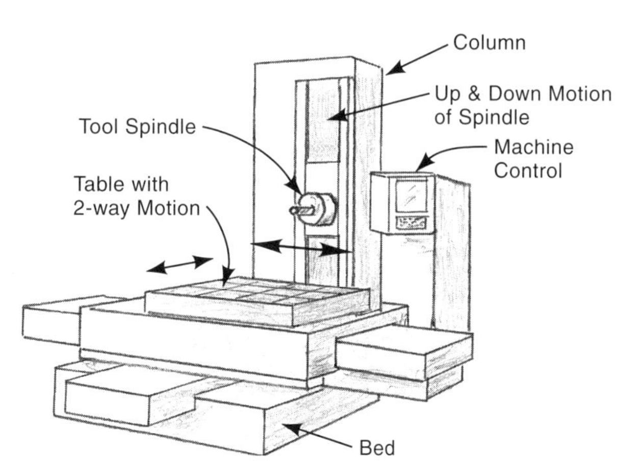

Fig. 3B5b A horizontal boring mill. This machine can perform boring, milling, and drilling operations.

factors. They are: its location, diameter, direction, cylindricity, and finish. When this operation is performed on a boring machine, the workpiece is stationary and the cutting tool rotates; when performed on a lathe, the workpiece rotates. On a lathe, the operation can then be considered to be internal turning. The tool spindle and the workpiece holder must be rigid enough to provide the desired accuracy in the bored hole. The operation is performed on holes from about 1/4 in (6 mm) in diameter and larger but is more common on larger holes, especially those too large to be drilled accurately, and for the machining of cast or forged large holes. Fig. 3B5 illustrates the process, and Fig. 3A1 shows it as one of a series of lathe operations.

B5a. ***jig boring*** - is performed on jig boring machines, which are vertical boring machines of very high accuracy. The table movement is extremely accurate and the spindle and spindle bearings are very precisely made. The machines are mainly used for making jigs, gages, dies, and fixtures, especially where accurate layout and hole location are essential.

B5b. ***horizontal boring mills*** - are basically large horizontal milling machines capable of performing boring, milling, and other machining operations on large and often complex parts. These units are sometimes called, horizontal boring and milling machines. The table can move in x and y directions. (Some machines have a table that also swivels.) The headstock that holds the spindle can be raised or lowered. The tool-holding spindle can move inward or outward. These machines are used in the machining of large components that have horizontal holes requiring the precision that boring provides. The machines normally include an end support column, opposite the spindle, for long boring bars. Tolerances with the machines can be as low as one or two ten thousandths of an inch (0.003 to 0.005 mm). Fig. 3B5b illustrates a horizontal boring mill.

B5c. ***vertical boring mills*** - are machines with a horizontal table rotating on a vertical axis, and a precision tool head (often two tool heads) capable of movements up and down and side to side (in and out radially). There may be more than one cross slide with tool-holding capability. These machines can be considered to be large lathes turned on end. They are especially suited to boring and other operations on parts too large for a conventional lathe. Workpieces are typically round and heavy with large diameters and shorter lengths. The workpiece is clamped to the rotating table, which can be as large as 40 ft. (12 m) in diameter. Both boring and facing are possible. There is no spindle for milling cutters; all cutting is by single point tools. Fig. 3B5c illustrates a typical vertical boring mill.

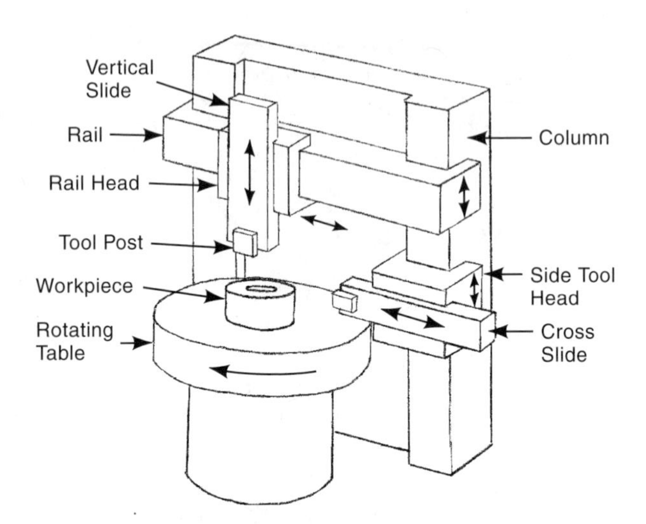

Fig. 3B5c A vertical boring mill. (vertical boring and turning machine.)

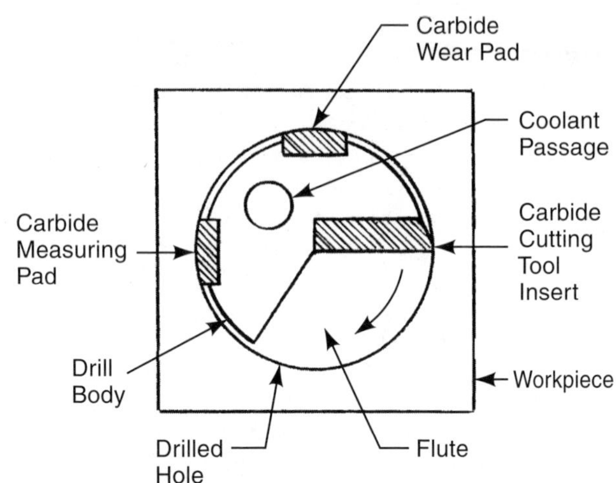

Fig. 3B6 A typical gun drill, viewed from the cutting end.

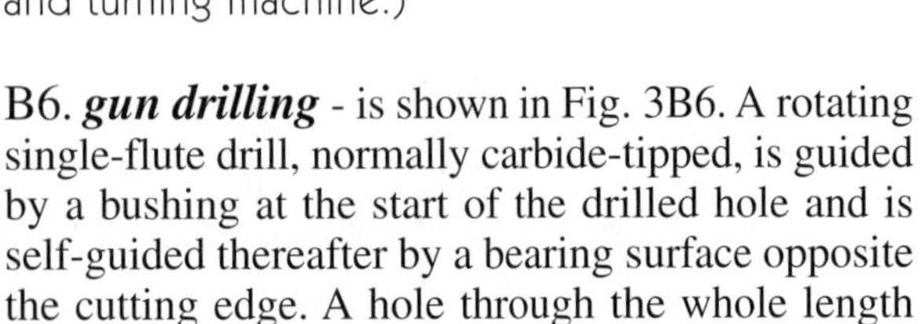

B6. ***gun drilling*** - is shown in Fig. 3B6. A rotating single-flute drill, normally carbide-tipped, is guided by a bushing at the start of the drilled hole and is self-guided thereafter by a bearing surface opposite the cutting edge. A hole through the whole length of the drill provides a means for oil coolant to flow at high pressure to the cutting edge and to flush chips from the hole. Deep, straight, holes are possible with the process which was originally developed for manufacture of gun barrels. Hole depths of over 250 times diameter are possible. Fig. 3B6-1 shows

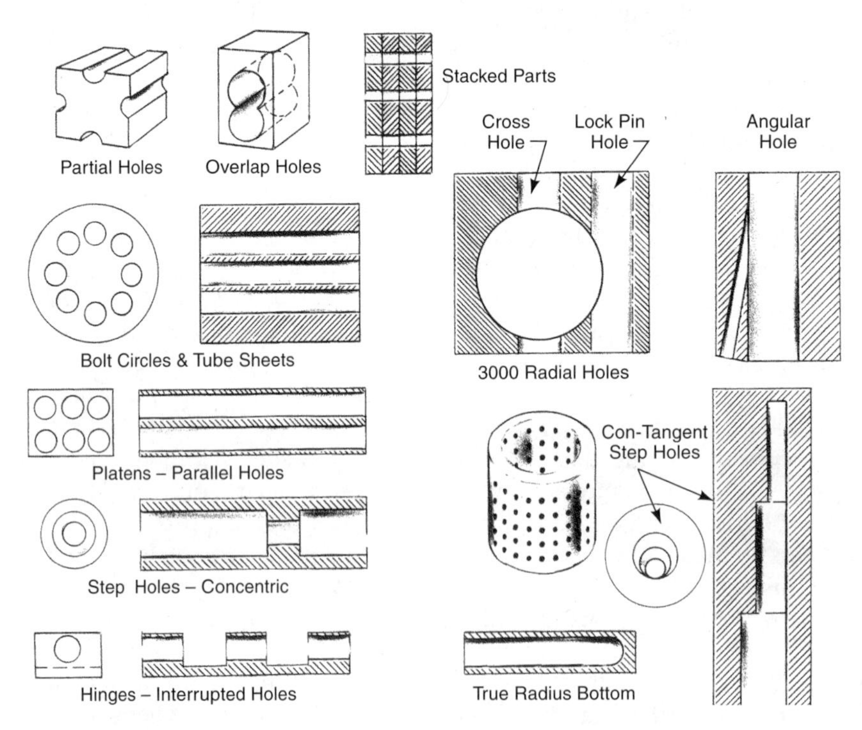

Fig. 3B6-1 The range of hole drilling applications, in addition to gun barrels, for which gun drills are used. *(Courtesy Eldorado PCC Specialty Products)*

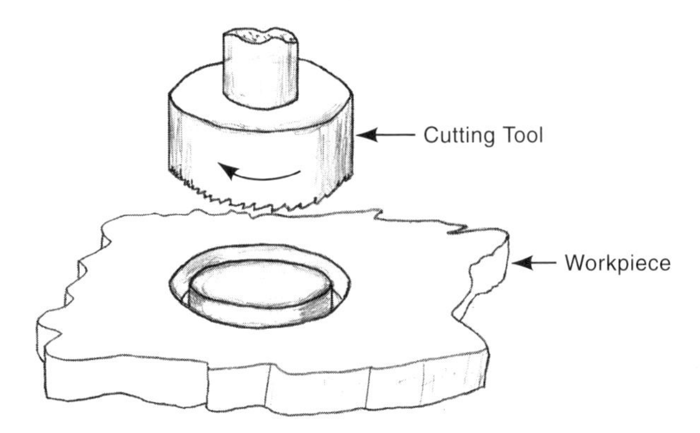

Fig. 3B7 Trepanning. In this example, a hole saw type of milling cutter is being used.

various applications of gun drilling. Also see *guns* and Fig. G9.

B7. ***trepanning*** - makes a circular groove in a workpiece through the use of one or more cutters or cutting teeth rotating about a central axis. If the grooves are cut all the way through the workpiece, a hole is created and a circular center piece (called a *slug*) is produced. The process is used primarily for large, shallow holes. It is also used to machine round disks from flat stock and is illustrated in Fig. 3B7.

Deep-hole trepanning is similar to gun drilling in that forced lubrication is used, the drill is self-piloting, and special drilling machines are employed. It differs in that a center slug is produced.

B8. ***multiple-spindle drilling*** - When production quantities are sufficiently large, it may be justifiable to construct a drilling head with a number of drills, all of which are driven from the same power source and make contact with the workpiece at the same time, drilling a number of holes simultaneously. There are three basic types of multiple-spindle drill heads: adjustable, geared, and gearless. The adjustable variety uses universal joints so that the drill positions can be varied, and are desirable for moderate size lots. For higher production and greater precision of hole location, the geared variety are preferable. The gearless type allows close spacing of the drilled holes. Fig. 3B8 shows a typical multiple-spindle, adjustable drilling head.

Fig. 3B8 A typical multiple spindle drilling head with provision for adjustment of the position of individual drills. (*Courtesy RMT Technology, Bellwood, IL.*)

C. Grinding and Abrasive Machining

At the point where the cutting takes place, grinding is very similar to other machining operations, the difference being that the workpiece is cut by the sharp edges of small pieces of abrasive material, rather than the edge of a hardened steel or carbide cutting tool. The irregularly-shaped abrasive particles may be bonded to a wheel or coated belt, or may be used loose. The particles commonly consist of aluminum oxide, silicon carbide, cubic boron nitride, diamond, or other hard materials. The individual abrasive grains are each smaller than a conventional metalworking cutting tool, and the grains on a typical wheel make a multitude of minute cuts. Fig. 3C illustrates the grinding process schematically.

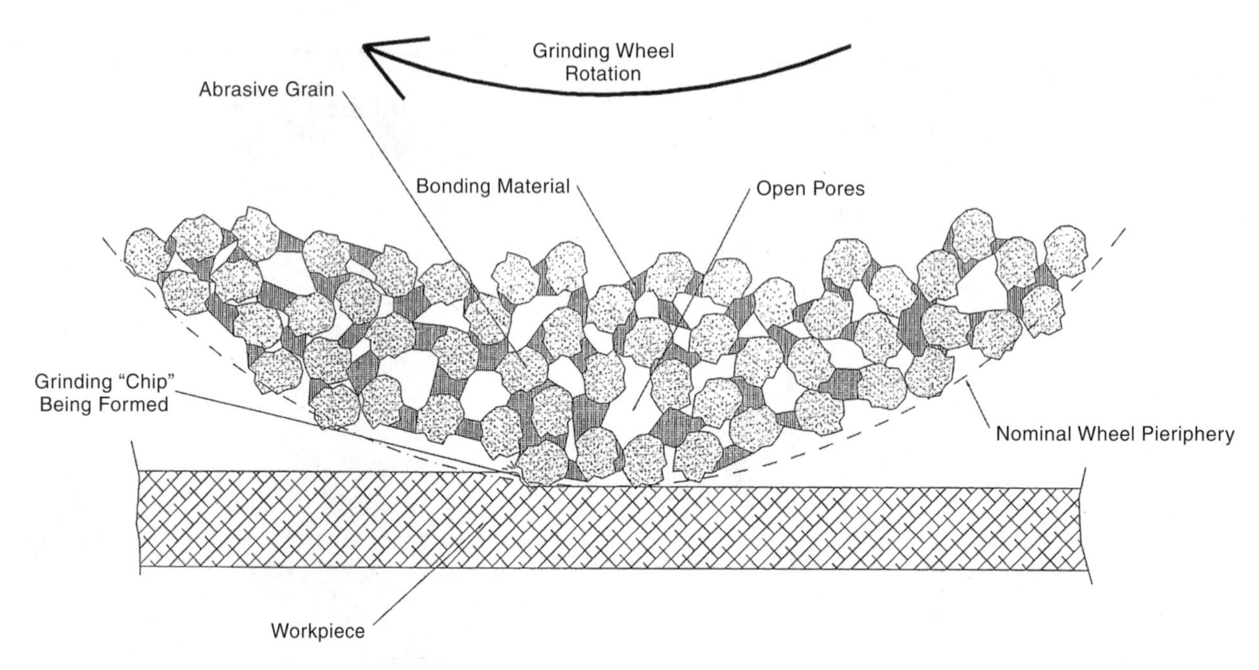

Fig. 3C The grinding process. Sharp edges of individual abrasive grains act as minute cutting tools, removing small amounts of material from the workpiece. (*Courtesy J. R. Casey Bralla*)

(Some grains, depending on their shape, do not cut but instead rub or slightly deform the surface of the workpiece.) Cutting speeds are high but the depth of cut from each grain is shallow. A water or water-oil emulsion is often sprayed on the wheel and workpiece to control the dust that otherwise arises and to overcome the heating effect of the operation. Grinding wheels are often porous, especially those designed for use with softer materials.

As the wheel cuts, it wears, causing some abrasive particles to become smooth but causing others to fracture, exposing new sharp edges. New wheels, and those that have become worn, are dressed with a diamond tool that removes some of the abrasive material and bonding agent, exposing sharp edges of new abrasive grains and providing a straighter, more uniform, cutting surface. Grinding is most commonly a finish-machining operation to provide a smoother surface or greater dimensional accuracy, particularly with hardened materials. When used as the primary metal removal method, the term, *abrasive machining* is often used.

C1. ***cylindrical grinding*** - is used to produce external cylindrical surfaces by removing material, creating smoother surfaces, and providing more precise dimensions. In all such operations, both the grinding wheel and the work rotate. The grinding wheel moves toward the workpiece to contact it and away from the work after the grinding is completed. However, in many cases, the wheel also traverses the work or vice versa. There are two basic methods for grinding the surfaces of components such as shafts, axles, cylinders, and rolls: center-type grinding and centerless grinding.

C1a. ***center-type cylindrical grinding*** - is performed on lathe-like machines. The workpiece is usually held at each end on pointed centers and is rotated about these centers. (It may also be held by a chuck or other holding device.) The grinding wheel normally rotates on an axis parallel to the axis of the workpiece. The wheel and the workpiece contra-rotate so that the contacting surfaces move in opposite directions. After the wheel and the work have made contact, there usually is axial motion between the wheel and the work for the full length of the surface to be ground, plus some overrun. The wheel may also be fed only transversely into the workpiece as it rotates. In this case, the wheel has either

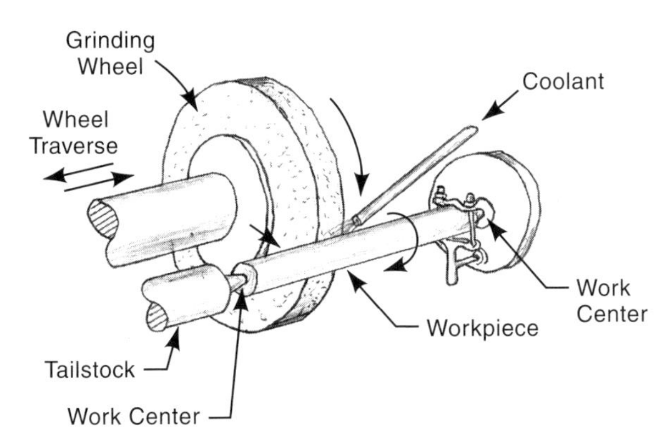

Fig. 3C1a Center-type grinding. Note that the grinding wheel rotates, has transverse motion across the whole length to be ground, and is fed into the workpiece as material is removed. The workpiece, held in centers, rotates against the grinding wheel.

a flat face or have a form dressed into it. The ground surface of the workpiece, then, can have contours, grooves, or whatever shape is dressed into the face of the wheel. If tapered surfaces are desired, the machine is set so that the axes of rotation of the work and the wheel are not parallel. Long, slender parts, and others subject to deflection or vibration during grinding may be supported by a steady rest. Fig. 3C1a illustrates the process.

C1b. ***centerless grinding*** - is a process for machining cylindrical surfaces wherein the workpiece is not held between centers or in a chuck. Instead, the work is supported by a work-rest blade at the correct height and contained between two wheels, as shown in Fig. 3C1b. One wheel is the grinding wheel; the other is a regulating wheel. The regulating wheel does not grind; it rotates the workpiece at a constant rate of speed. The process produces accurate diameters and roundness, with smooth surfaces in parts such as pins, shafts, and rings. Both throughfeed and infeed methods can be used, and production can be quite rapid. The method is normally not applicable if there are flats, keyways, or other interruptions in the workpiece's cylindrical surface. Conventional centerless grinders can accommodate solid parts up to about 7 in (18 cm) in diameter and rings and tubing up to about 10 in (25 cm) in diameter.

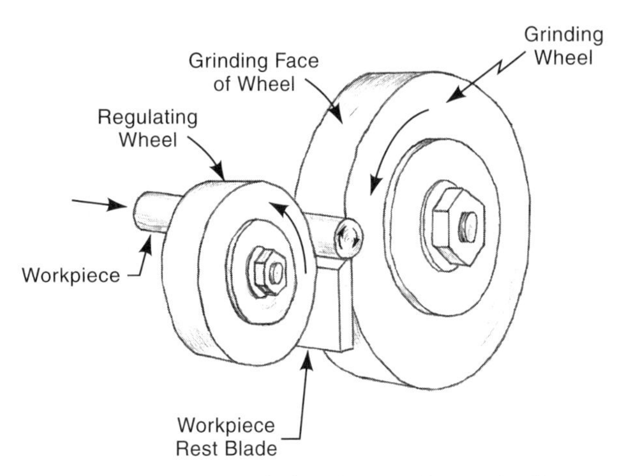

Fig. 3C1b Through-feed centerless grinding.

Larger sizes are too heavy for smooth operation and require special equipment.

C1b1. ***through-feed centerless grinding*** - In this method, the regulating wheel is canted slightly, as shown in Fig. 3C1b1, causing the work-

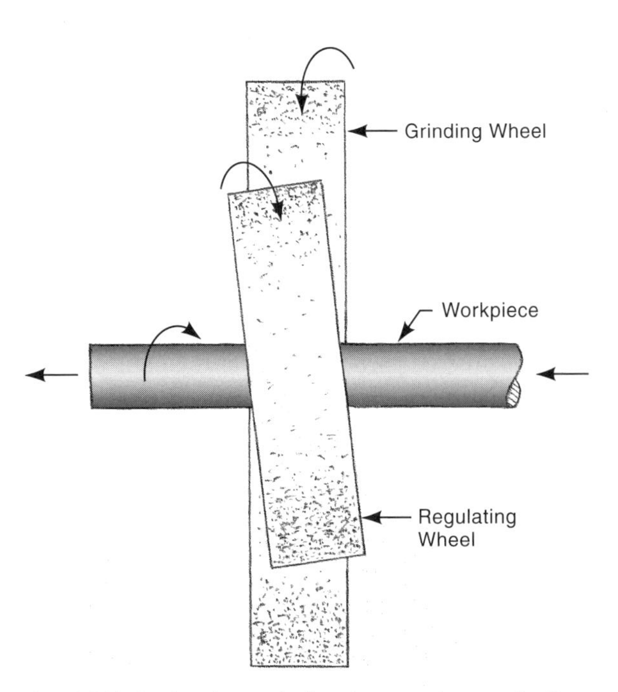

Fig. 3C1b1 In through-feed centerless grinding, the inclination of the regulating wheel induces a sideways thrust on the workpiece.

piece to move axially across the grinding wheel as the grinding takes place. This arrangement provides a rapid operation that can be made automatic if production quantities are sufficient to justify the use of an automatic feed for workpieces to be ground, feeding them end-to-end. The process is used for grinding pins, shafts, and similar parts of constant diameter. Piston pin grinding is a notable application. Parts that have heads, projections, or shoulders that would block the movement of the workpiece through the machine, cannot be through-feed centerless ground.

C1b2. ***infeed centerless grinding*** - differs from the through-feed process in that the part does not move axially. Stops in the machine prevent axial movement until the part is ejected. Instead, the grinding wheel is fed into the work once the workpiece is positioned. After grinding, the wheel retracts, the part is removed or ejected, and another is inserted. If the grinding wheel is dressed with a form, that form is ground into the workpiece. Tapers, multiple diameters, grooves, and other irregular shapes can be ground so long as the grinding wheel is appropriately dressed. The process is also useful when a portion of the workpiece is larger than the surface to be ground. It is illustrated schematically in Fig. 3C1b2. The process is fast but not quite as rapid as the through-feed method since workpieces must be inserted and removed individually. The process is used in the manufacture of ball bearings and parts not having a uniform diameter.

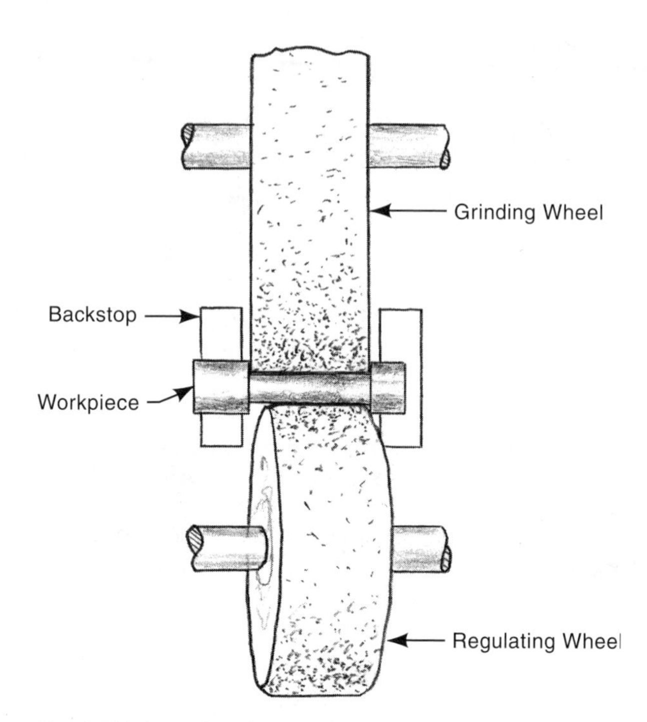

Fig. 3C1b2 Infeed centerless grinding. The method is used when a portion of the workpiece is larger than the ground surface. The workpiece is placed in position; the grinding wheel is fed into the work; the wheel retracts after grinding and the workpiece is removed or ejected.

C1b3. ***end-feed centerless grinding*** - is used for tapered parts. Either the regulating wheel or the grinding wheel, or both, are dressed to provide the desired taper angle. The two wheels and the workrest blade are set in fixed relationship to each other. There is a stop to control the position of the workpiece, which is fed and removed axially. Fig. 3C1b3 shows the process.

C2. ***internal grinding*** - is a process for finish machining existing round holes. It normally is performed on a lathe-like machine. The workpiece is held in a chuck or faceplate holding fixture and rotated about the axis of the hole to be ground. The rotating spindle carrying the small grinding wheel is inserted into the hole and then fed radially to contact the surface. The wheel can be cylindrical or dressed with a form. If the internal surface to be ground is cylindrical, the wheel is also traversed axially. Diameter control can be maintained by dressing the grinding wheel with a diamond tool in a fixed position with respect to the axis of the hole. Internal grinding is most commonly used to finish machine precision holes in hardened workpieces. Large, heavy parts are internally ground on vertical spindle machines since loading and positioning such parts is easier if the rotary surface is horizontal. If the large workpiece is too bulky or unbalanced to rotate, it is held in a fixed position and the grinding wheel is moved in a planetary motion around the hole's axis, as in jig boring.

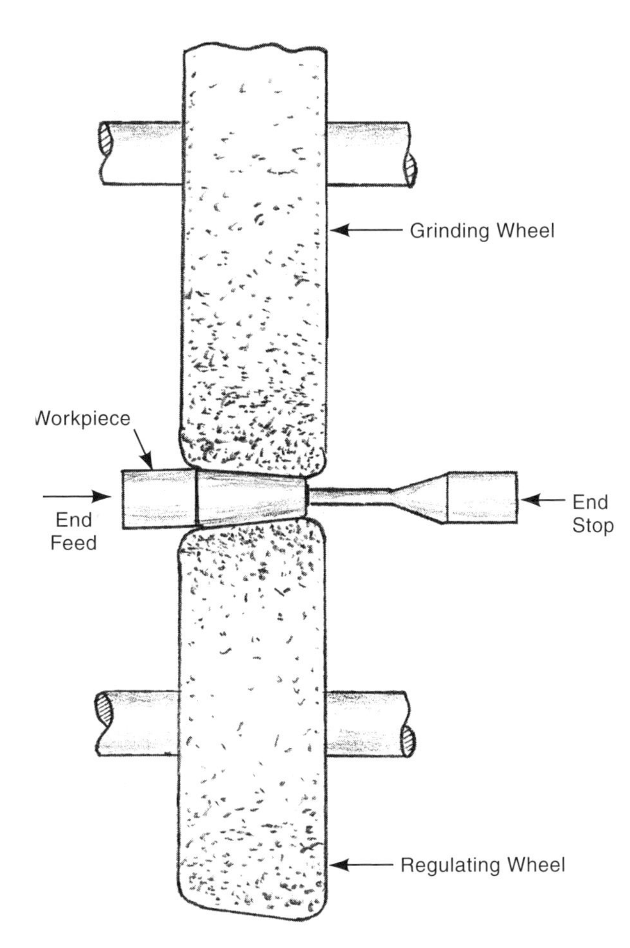

Fig. 3C1b3 End feed centerless grinding is useful when the surface to be ground is tapered. The workpiece is positioned and removed axially.

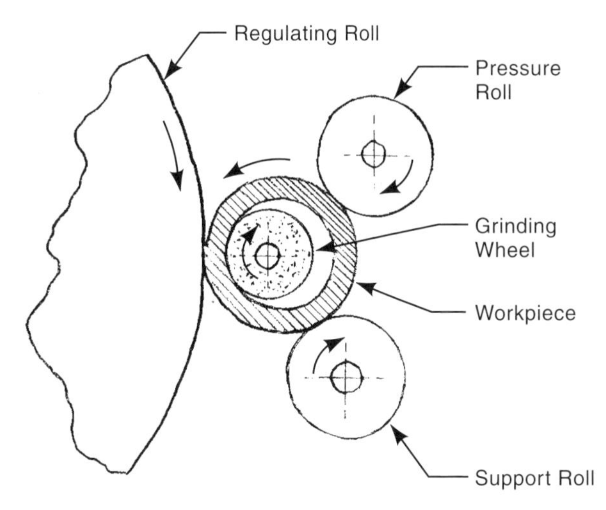

Fig. 3C2a Internal centerless grinding.

C2a. ***internal centerless grinding*** - can be used to finish machine sleeves, rings, and similar parts. As illustrated in Fig. 3C2a, the workpiece is held between three rolls, which locate it and provide rotation. The grinding wheel is inserted into the center hole and fed radially to contact the inner surface of the hole as in regular internal grinding. The process insures concentricity between the outer diameter of the part and its center hole . It is suited to mass production situations because the part does not have to be placed in a chuck, and handling is simplified. The grinding of bearing raceways is a common application. Sleeves and cylinder liners are also ground by this method. Tapered holes can be ground on some machines that allow the axis of the grinding wheel to be set at an angle to the axis of rotation of the workpiece.

C3. ***surface grinding*** - moves the workpiece in a horizontal plane so that it passes under a revolving grinding wheel which contacts it and removes surface material. The result is a part with a flat, smooth surface and an accurate thickness or height. Various arrangements of grinding wheels and various table movement methods can be used. Major arrangements are shown in Fig. 3C3. The table movement is most often reciprocating but rotary tables are also common. The wheel spindle may be horizontal or vertical. The latter requires a cup-shaped or segmented wheel, which cuts on its face.

The most common method for holding the workpiece is with a magnetic chuck, although holding fixtures and other methods may also be used. Surface grinding is also used to sharpen cutting tools. For some applications, especially when heavier metal removal rates are appropriate, the operation may be carried out with an abrasive belt (passing over a roller at the point of contact) rather than a bonded wheel. When the wheel is dressed with a form, the workpiece surface can be ground with that shape instead of a flat surface.

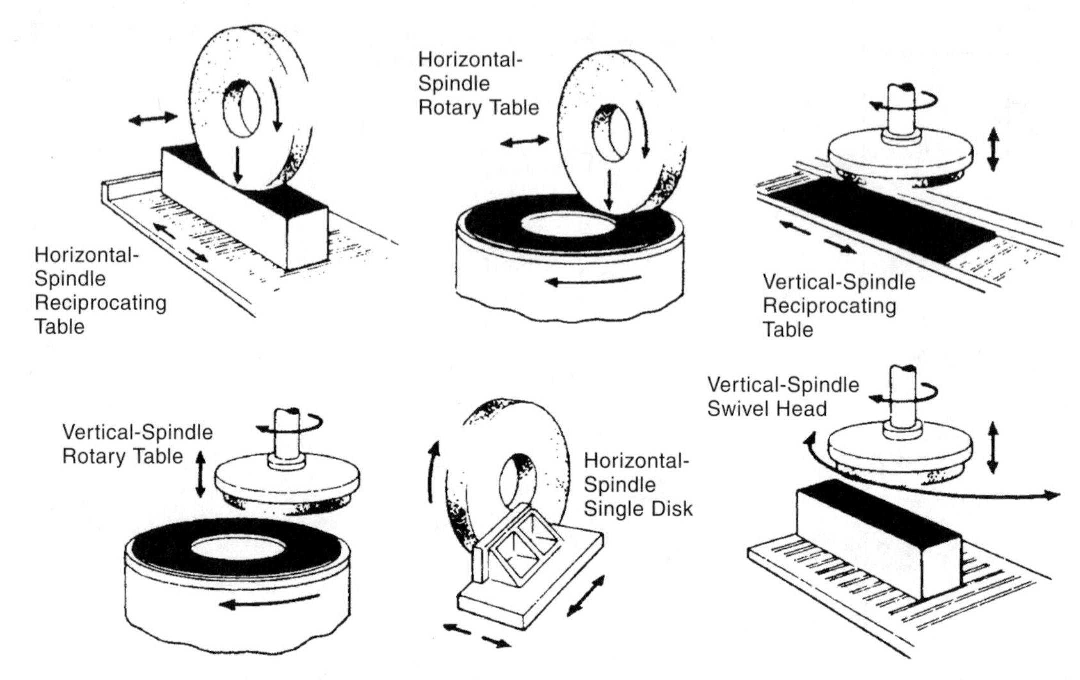

Fig. 3C3 Various equipment arrangements for grinding flat surfaces. (*Courtesy American Machinist, Penton Media, Inc.*)

C3a. ***horizontal spindle surface grinding*** - is surface grinding with a horizontal grinding wheel spindle. The wheel cuts on its periphery. The table may be either circular with a rotational motion or, more commonly, rectangular with a reciprocating motion. In both arrangements, the wheel advances perpendicularly to the basic direction of motion with each cycle of table movement, and thus generates a flat surface. See Fig. 3C3. This approach is used extensively in tool shops to make and recondition dies and to produce machine surfaces on which there are moving parts. With a reciprocating table, horizontal-spindle surface grinding can be used to machine slots in hardened workpieces. By dressing a form on the grinding wheel, the ground surface can be given a form rather than a flat surface.

C3b. ***vertical spindle surface grinding*** - uses a grinding wheel mounted on a vertical spindle with either a rotary or reciprocating table. Various methods are used to move the wheel across the work, as shown in Fig. 3C3. The cup-shaped, segmented, or cylindrical wheel covers a larger area than the side of the wheel does in a horizontal spindle arrangement, so cross feed of the worktable or wheel may not be necessary. These machines are typically used in production applications, while horizontal-spindle machines are more common in toolroom or jobbing work. Vertical spindle machines are used in the manufacture of cylinder head surfaces, gear and pulley faces, and other moving parts that require flat, smooth, surfaces. Weldments, castings, and forgings are ground on these machines. These parts usually have somewhat liberal tolerances for dimensions and surface finish; however, vertical spindle machines have the capability of producing the fine finishes and close tolerances typical of toolroom work. Some machines have more than one spindle and thus can do rough and finish grinding in one pass, or can grind several surfaces of the workpiece simultaneously.

C3c. ***creep-feed grinding*** - is an abrasive machining process capable of heavy stock removal.

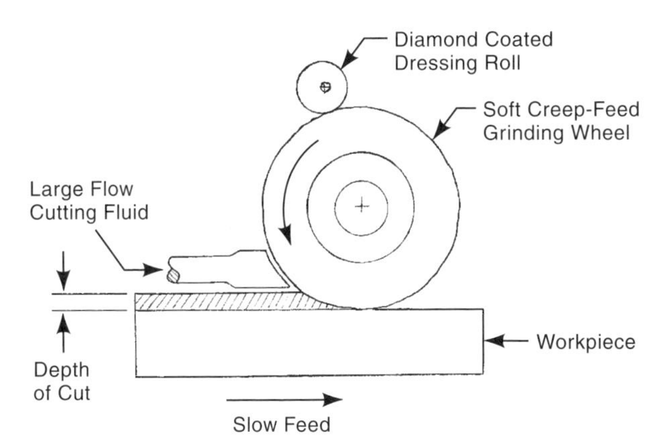

Fig. 3C3c Creep feed grinding. Note the slow feed, large depth of cut, soft wheel, and heavy flow of cutting fluid.

It is particularly applicable to hard materials that would be difficult or unfeasible to machine with a milling cutter. The depth of cut normally ranges from 0.20 to 2 in (5 to 50 mm), much more than in conventional grinding, but the table speed is low, 0.4 to 20 in (1 to 50 cm) per minute. The process thus resembles milling, but the grinding wheel takes the place of the milling cutter. The grinding wheel is soft and made with a very open porous structure to provide space for flow of cutting fluid. The wheel spindle is lowered gradually as the wheel wears. Continuous crush-dressing is normally used, and wheels can be dressed with a form if desired. Cutting forces are high and the machine must be rigid and capable of closely-controlled table speeds; otherwise wheel breakage would occur. Fig. 3C3c illustrates the process. Note the "climb milling" direction of the grinding wheel rotation.

C4. ***jig grinding*** - A jig grinder is a precision vertical-spindle machine, similar to a jig borer. It is used for internal grinding. It has a horizontal table that can be moved and located very accurately. The spindle head contains a high-speed grinding head that rotates the grinding wheel about its axis but that axis can also revolve in a planetary motion. The head can also move vertically with reciprocating motion. Thus, holes can be ground to very accurate locations and diameters and, as the head reciprocates vertically, to high levels of cylindricity. The machine is used for finish grinding of hardened dies, molds, jigs, and fixtures.

C4a. ***tool post grinding*** - involves the mounting of a power-driven grinding wheel for either external or internal grinding on the tool post of a lathe. The wheel then performs conventional grinding operations on parts held in the lathe. This method is a useful means to provide cylindrical grinding in shops not equipped with the conventional cylindrical grinding machines. Care must be taken to avoid damage to the lathe components from trapped grinding dust from the operation.

C5. ***low-stress grinding*** - is a variation of conventional grinding in which the objective is to minimize stresses developed in the workpiece by the grinding operation. It differs from conventional grinding in that very low wheel infeed or downfeed rates, frequently dressed coarse, softer, open-grain wheels, and liberal flow of coolant are used. Wheel speed also may be low. Abrasive grains in the grinding wheel should be sharp, and wheels should not be allowed to become loaded. The process is applicable when the part to be ground is subject to high stresses in use or is susceptible to damage from heat. Parts made from some materials such as high-strength steels, cobalt alloys, titanium alloys and high-temperature nickel alloys that are particularly sensitive to surface cracking and residual stresses from conventional grinding, can be finish-ground with this approach[2].

C6. ***plunge grinding*** - is simply a center-type grinding operation where there is no transverse feed of the grinding wheel. Instead, the wheel is fed directly into the work. If the wheel is dressed with a form, that form is ground into the workpiece. If the width of ground area is no more than the width of the wheel, the operation can be referred to as plunge grinding. The term "in-feed grinding" is also used.

C7. ***disc grinding*** - is a means for producing flat surfaces. The workpiece is held against the flat

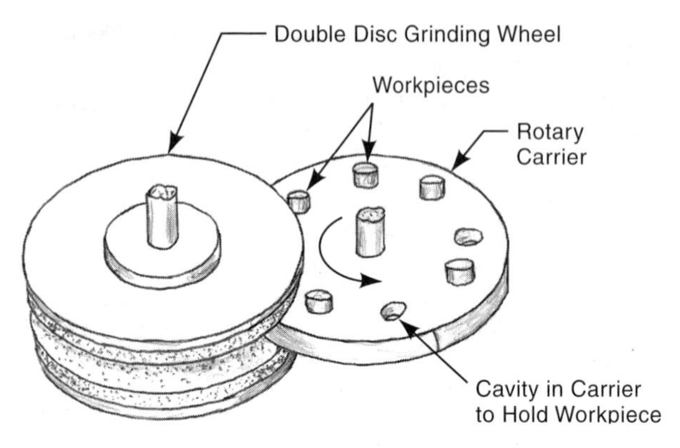

Fig. 3C7 Double disc grinding. Both the top and bottom surfaces of the workpiece are ground at the same time.

side (face) of a large rotating abrasive disc. (See Fig. 3C3.) The operation can be performed manually when dimensional requirements for the part are not severe. In production situations, *double disc grinding,* as illustrated in Fig. 3C7, is sometimes employed. The workpiece is fed between two abrasive discs which grind the opposite surfaces of the workpiece at the same time, thus controlling flatness, thickness, and parallelism in the same operation. Parallelism and flatness of surfaces is particularly good with these machines because there is no magnetic chuck to pull down non-flat parts, only to have them spring back to a non-flat condition after the operation. Double-disc grinders are used in the production of automotive connecting rods, disc brake rotors, compressor vanes, and cast-iron rocker arms.

C8. ***abrasive belt grinding*** - uses an abrasive-coated cloth belt to remove metal. The endless belt runs between a drive wheel and a contact wheel, and the workpiece bears against the belt at the contact wheel. The abrasive grains on the belt are arranged with orientation and spacing to optimize metal removal rates. The process can provide faster metal removal than grinding with a wheel and is then often referred to as *abrasive belt machining*. Metal-removal rates of 30 in^3/min/in (193 cm^3/min/cm) of belt width are feasible with standard belts.[4] This rate is faster than those attainable with milling machines, even under the fastest metal removal conditions. Belts as wide as 10 ft (3 m) allow the entire surface to be ground in one pass. The process can provide lower heat levels than grinding with an abrasive wheel because the belt carries away heat effectively. Lubricants may be used to facilitate the operation. The approach is used for rough metal removal from castings, forgings, and other shapes, especially when the workpiece is large. Surface, cylindrical, and centerless grinding can be performed with abrasive belts. Abrasive belts can also be used for hand grinding and polishing.

C9. ***abrasive jet machining*** - provides cutting action from the effect of finely-powdered abrasive in a high-velocity stream of gas. With the gas carrier, the abrasive and gas are raised to a pressure between 30 and 120 lbf/in^2) (200 and 830 kPa). The stream is passed through a nozzle of 0.005 to 0.032 in (0.13 to 0.81 mm), and reaches a velocity of 500 to 1000 ft/s (150 to 300 m/s). The stream is directed at the desired place on the workpiece. This can be done by hand for rough cuts, stripping, or deburring. However, precision cuts require the nozzle to be mounted on suitable equipment. Masks of rubber, glass, or copper may be used to help limit the abrasive action. The process can be used for drilling, cutting, etching, trimming, cleaning, and deburring operations on a variety of metals and non-metals, especially hard and fragile materials. (The process is less suitable for softer materials that may trap the abrasive.) Ceramics, glass, silicon, and mica are machined with this method. Sheets of these materials can be cut and drilled, and abrasive jet can produce intricate holes that would not be easily made by other methods. Other applications are marking identification on workpieces, trimming, and cleaning electronic components, and removing surface coatings including plating. The process is illustrated schematically in Fig. 3C9. Cutting rates are slower than those achieved with more conventional processes but there is no heat damage to the workpiece.

C10. ***abrasive flow machining (AFM)*** - uses a viscous, putty-like abrasive medium which is pumped so that it flows against the workpiece. The abrasive particles in the medium rub against the workpiece, gently removing surface material.

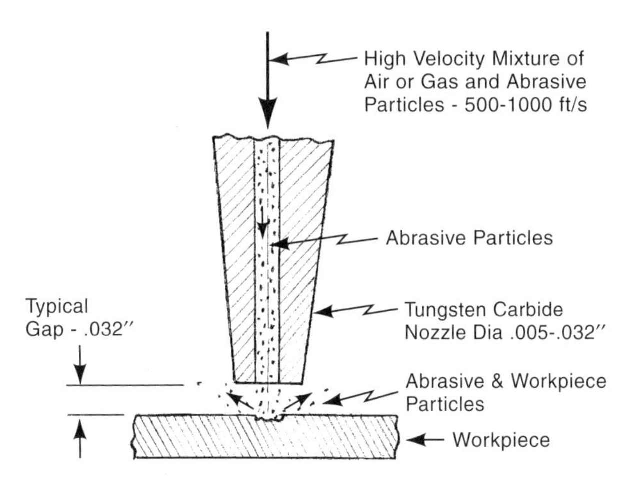

Fig. 3C9 Abrasive jet machining - A high velocity stream of air or gas and abrasive particles impinge on the workpiece and remove workpiece material. The process is particularly useful in drilling, cutting, or engraving hard, brittle materials including non-metallic materials. (*Courtesy Omax Corporation.*)

The process is used to round sharp corners, finish edges, remove recast layers from EDM or laser machining, polish surfaces, and remove burrs (See K1.) AFM is particularly useful when the area to be machined is not accessible for other methods. In practice, two cylinders on opposite sides of the workpiece are used to contain and pump the medium, which flows back and forth repeatedly (through as many as one hundred cycles) until sufficient material is removed from the workpiece.

C11. ***ultrasonic machining*** - utilizes an abrasive in a water slurry and a shaped tool. With the abrasive between the tool and the workpiece, the tool is vibrated at ultrasonic frequency. The vibration of the tool drives the abrasive particles against the workpiece, cutting a cavity that has the same shape as the tool. The frequency of vibration is usually between 19,000 and 25,000 Hz with a low amplitude - 0.0005 to 0.0025 in (0.013 to 0.063 mm). The abrasive slurry is pumped through the gap between the tool and the work. The gap ranges from 0.001 to 0.004 in (0.025 to 0.1 mm). The stainless or carbon steel tool is attached to an ultrasonic generator through a "horn", usually of Monel metal. Fig. 3C11 illustrates the process schematically. It is most advantageous for shallow cuts in hard, brittle, non-conductive materials that are difficult or unfeasible to cut by other methods. Slots, holes, and cavities of various shapes can be produced in ceramics, glass, carbide, tool steels, honeycomb material, and gem stones. Holes and cavities can be non-round and curved. Material removal rates are low, but there are no heat effects, no burrs, and surface stresses are low. Holes have some inherent sidewall taper.

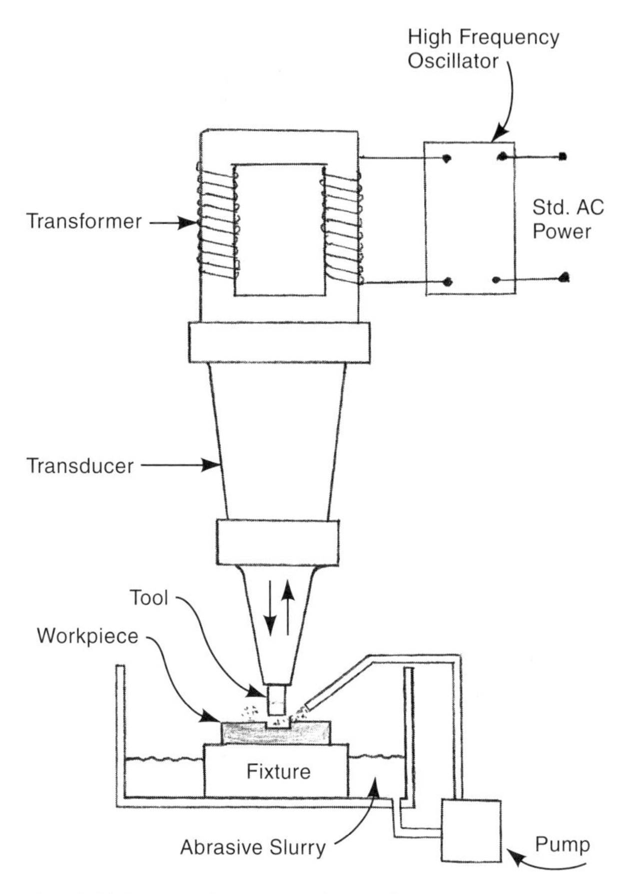

Fig. 3C11 A schematic view of a typical arrangement for ultrasonic machining.

C11a. ***rotary ultrasonic machining*** - uses a rotating tool, coated with diamond abrasive and ultrasonically vibrated, to machine hard and brittle materials. There is no abrasive slurry but a liquid coolant, usually water, is used. Milling, drilling, and threading operations can be performed, but drilling is the primary application. Non-metals: glass, alumina, ceramic, quartz, sapphire, and composite materials are normally processed.

D. Milling

Milling is a means of creating a desired surface with a rotating multi-toothed cutter. Each tooth of the cutter removes material as the workpiece advances against it. The axis of rotation of the cutter may be either horizontal or vertical. The cutter can provide cutting action on its side or at its end (face), or both. The cutter rotates rather rapidly and its position is normally stationary; the work moves past the cutter with a suitable depth of cut at a relatively slow feed rate. Milling is the most common machining operation for producing flat surfaces, but slots, and contoured or stepped surfaces and screw threads can also be produced. A variety of milling operations and the cutters used are illustrated in Fig. 3D. (Also see ***machining centers***, T.)

D1. ***face milling*** - shown in Fig. 3D1, produces a flat surface at a right angle to the axis of rotation of the cutter. Depending on the depth of cut, some machining also takes place on the periphery of the cutter. For flat surfaces, face milling is generally preferable to peripheral milling from the standpoints of tool economy, simplicity of set-up, and cutter rigidity. However, the operation is limited to flat surfaces.

D2. ***peripheral milling*** - The milled surface, if flat, is parallel to the axis of rotation of the cutter, and is produced by cutting teeth located on the periphery of the cutter body. The operation is usually performed on horizontal-spindle machines. The milling cutter or cutters are mounted on an arbor that has outboard support. The surface may be flat or contoured, depending on the profile of the cutter. Flat and contoured surfaces, slots, and key-ways, are machined by this method. (Fig. 3D, in views a), b), f), and g), shows peripheral milling. Views c), d), and e) show both peripheral and face milling.)

D3. ***end milling*** - uses a cutter, commonly of smaller diameter, with teeth on both the end (face) and periphery. Fig. 3D, in views n) and o), illustrates the operation. The approach is versatile in that slots, recesses and profiles can be machined. Machining can also be carried out in areas not accessible to other types of cutters. However, the length-to-diameter ratio of end mills is high and they can be supported only at one end, so they are less rigid than cutters for other milling methods. Lighter feeds may be required to reduce cutter deflection. Material removal rates are less than with other milling methods and accuracy may not be as great.

D4. ***slab milling*** - is peripheral milling with cutters that produce a flat surface over a wide area. The axis of rotation of the cutter is parallel to the machined surface. The cutter often removes large amounts of material. Sometimes, two or more cutters are used per arbor with opposing helixes to balance cutting forces. See view b) of Fig. 3D.

D5. ***form milling*** - When the peripheral cutting edges of the milling cutter are ground with a form rather than in a straight line, that form is transferred to the workpiece as the milling operation proceeds. The operation is called form milling and is illustrated schematically in Fig. 3D, views i), k), l) and m). Milling of gear teeth is a common application of this approach.

D6. ***gang milling*** - is simply milling with more than one cutter on the arbor of the milling machine. This produces multiple surfaces on the workpiece with one pass of the cutters. Also see straddle milling, as follows.

D7. ***straddle milling*** - involves the use of two cutters on one arbor with a space between them. Two surfaces are cut in one pass, but the area between them is not machined, as illustrated in Fig. 3D, view Q).

D8. ***fly cutter milling*** - involves the use of a single-point cutter rather than a multiple-tooth cutter to perform a milling operation. It is face milling with only one cutting tooth. The method is useful for producing

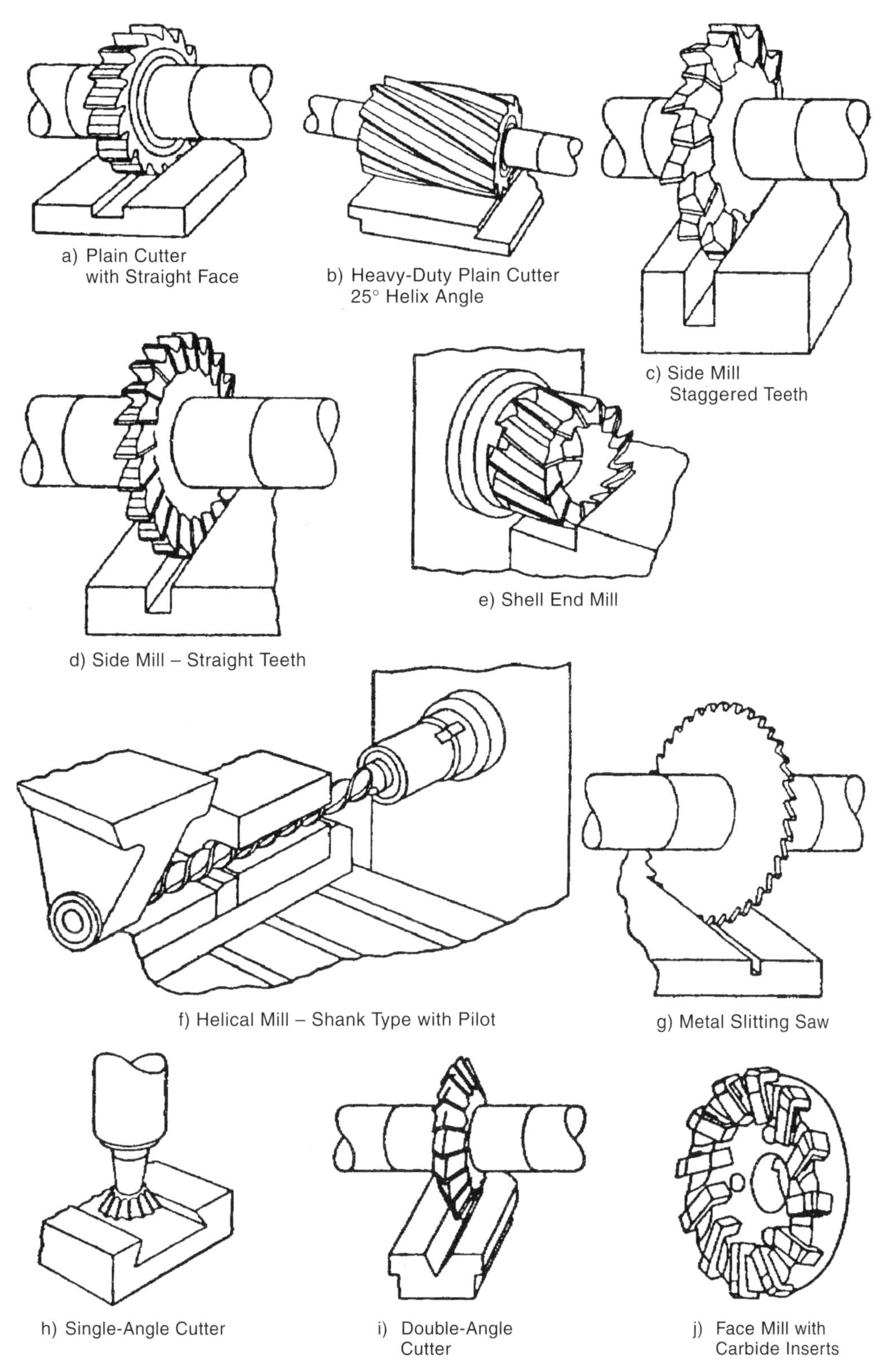

Fig. 3D A collection of milling cutters and the operations that they perform. (*from LeGrand, American Machinist's Handbook, 1955, McGraw-Hill Companies.*)

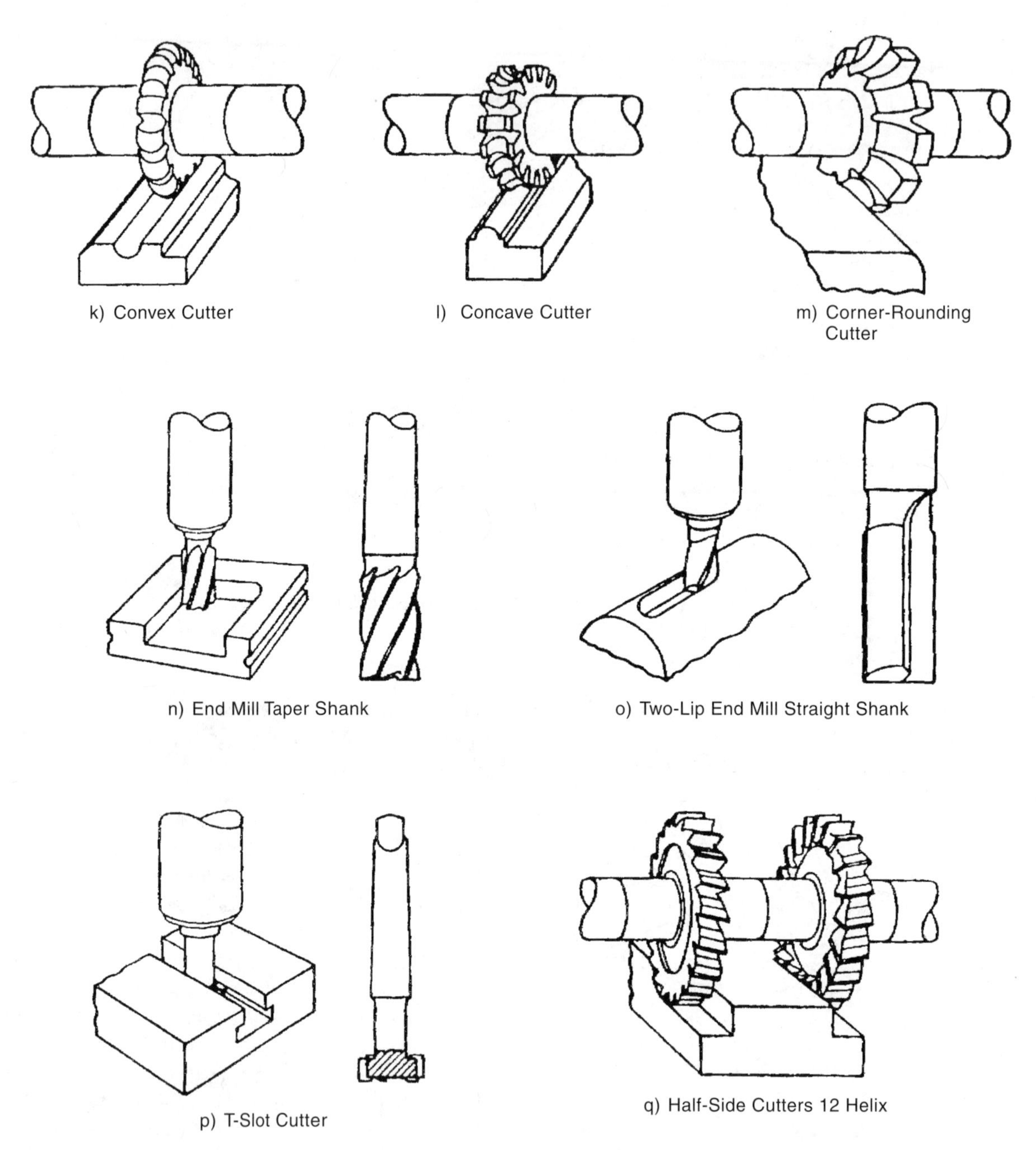

Fig. 3D *(Continued)*.

flat surfaces in a tool room situation where the optimum multiple-toothed cutter may not be available. Obviously, cutting feed rates are much less than with face mills but may be satisfactory when flycutters are the only tools available and requirements are for only one piece or a small quantity.

D9. ***pin routing*** - involves the use of a template to guide the movement of a high speed routing cutter (small diameter end mill). Typically, the process is used to blank flat stock of sheet metal or other materials. Stacks of thin material can be cut by this method, to produce multiple parts.

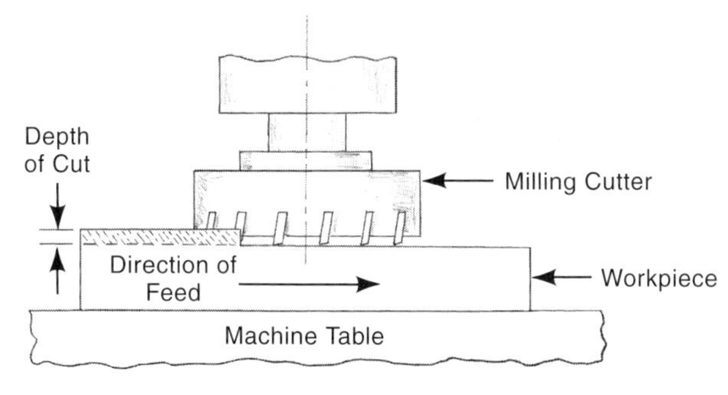

Fig. 3D1 Face milling.

D10. ***spotfacing*** - is a simple operation, shown in Fig. 3B2 view c) that is normally used to provide a small flat bearing surface, perpendicular to the axis of a hole, for a bolt head or nut. An end-cutting rotating tool is fed into the workpiece along the axis of the bolt hole, often with a drill press rather than a milling machine. Depth of cut is often not critical as long as the surface machined is flat and perpendicular to the axis of the bolt hole. The operation is the same as counterboring except that the depth of cut is shallow, only enough to create a flat machined surface. It is most commonly performed on castings and forgings where the surface prior to the operation has some irregularity.

E. Screw Threads

Screw threads are made with a large number of different methods involving both machining (cutting) and forming processes. Cutting methods include the use of hand-operated taps and dies for internal and external threading, and machines for single-point screw-thread cutting, thread-cutting die heads, thread milling, and thread grinding. Forming methods include rolling of external threads and cold-forming of internal threads.

E1. ***hand-die external threading*** - A typical external hand-threading die is illustrated in Fig. 3E1. These are called button or acorn dies and are useful for cutting screw threads in prototype and limited-quantity production. The die is placed on the end of the screw blank and is rotated with a wrench-like

Fig. 3E1 A hand die for cutting external threads.

tool. It feeds itself forward at the lead rate of the screw as it cuts the threads.

E2. ***internal thread tapping*** - utilizes taps such as that shown in Fig. 3E2. The tap is rotated like a drill or reamer and is self-fed axially into the hole in the workpiece to create an internal thread. Taps can be positioned and fed manually, and rotated with a hand tool, or can be used with a drill press, lathe, automatic screw machine with a tapping attachment, or a special tapping machine. Tapping machines and attachments have the capability of reversing the rotation of the tap to remove it from the work after the thread cutting is

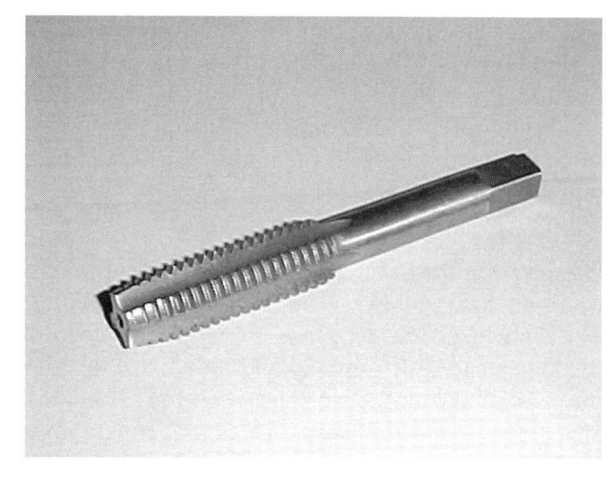

Fig. 3E2 A solid tap for machining internal screw threads.

Fig. 3E2-1 A collapsible tap. (*Courtesy Landis Threading Systems.*)

completed. These solid taps are used for internal threads of 0.050 in (1.2 mm) to 6 in (150 mm) diameter[5].

Collapsible taps automatically retract the cutters to clear the new threads and do not have to be reversed for removal, but are not available for diameters less than about 1 1/4 in (32 mm). They are usable for diameters up to about 24 in (600 mm)[5]. See Fig. 3E2-1.

E3. ***single-point screw-thread cutting*** - is a lathe operation. A form cutting tool that has a profile corresponding to that of the space or spaces between the required screw threads makes repeated passes along the workpiece. The cutter is attached to the lathe carriage and is moved longitudinally along the bed by the lathe's lead screw, which is geared to provide the rotational rate needed to produce a screw of the proper pitch. Both external and internal threads can be generated by this method, as shown schematically in Fig. 3E3. The method is used for large screws, for which die heads are not available or impractical, when quantities are small, or when the material is difficult to machine.

E4. ***thread cutting die heads*** - are efficient, accurate, and widely used in production threading. They are used on all kinds of lathes and screw machines, on drill presses, and as part of special threading machines. As illustrated in Fig. 3E4, they have sets of insertable multi-toothed cutters that are changeable when different pitches of threads are to be machined or when the cutters need re-sharpening. Like button dies, die heads are

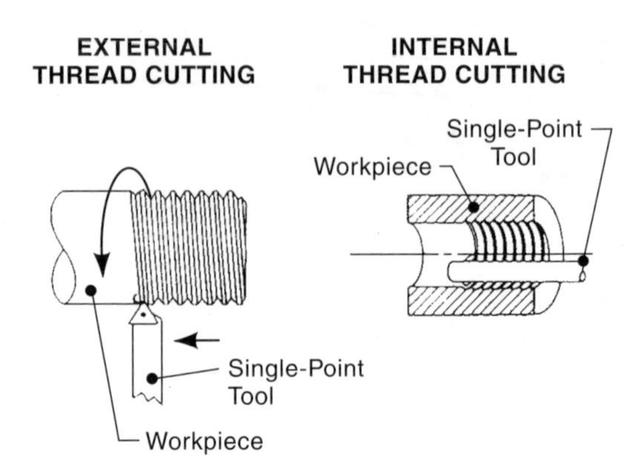

Fig. 3E3 Single point machining of external and internal screw threads. (*from Manufacturing Processes Reference Guide, R.H. Todd, D. K. Allen, and L. Alting, Industrial Press, New York, 1994.*)

self-feeding at the lead rate of the thread. At the end of the cut, the cutters retract automatically for rapid withdrawal. Perhaps the largest single application is pipe threading, on special small machines that can be transported to building construction sites.

E5. ***thread milling***[5] - utilizes a form-milling cutter. Both the workpiece and cutter rotate. Most cutters are the multiple-rib type shown in Fig. 3E5, but single-rib cutters are also used, though cutting one thread at a time requires more time. The method is applicable to both internal and external threads. (In internal thread milling, the cutter diameter should not exceed 1/3 of the hole diameter.) Thread

Fig. 3E4 A die head for external threads with inserted multi-toothed cutters. (*Courtesy Landis Threading Systems.*)

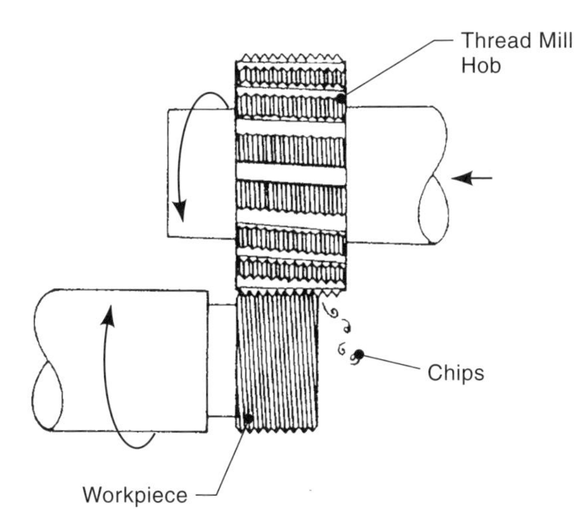

Fig. 3E5 Thread milling of external threads. (*from Manufacturing Processes Reference Guide, R.H. Todd, D. K. Allen, and L. Alting, Industrial Press, New York, 1994.*)

milling is used in situations where die-head cutting or tapping is difficult, for example, in difficult-to-machine materials, long threaded lengths, high helix-angle threads, or large, coarse-pitch threads. However, milling is not suitable for square threads or others with a flank angle nearly 90 degrees from the axis.

E6. ***thread grinding***[6] - is used when particular accuracy and surface smoothness are required, when heat-treated workpieces are threaded, or when the material is otherwise difficult to machine. Threads are ground in stainless and tool steels and in sintered iron components, particularly when they have been heat treated for hardness or are used in precision applications such as adjustment screws and feed screws. Both center-type (3C1a) and centerless cylindrical grinding machines (3C1b) are used. The wheels may be dressed to grind multiple ribs or a single rib. In either mode, the workpiece moves axially across the wheel as it rotates.

With center-type grinding, several passes are usually required. Center-type grinding can be used on short runs, but is not limited to small-quantity production.

With centerless grinding, the wheel is always dressed for multiple-rib grinding. The workpiece is first ground to the correct diameter and then the threads are made as the work passes across the wheel. Centerless thread grinding is more of a high production method, used when quantities are 10 or 15 thousand per lot or more. Process times are very short but setups require more time. Fig. 3E6 shows centerless thread grinding.

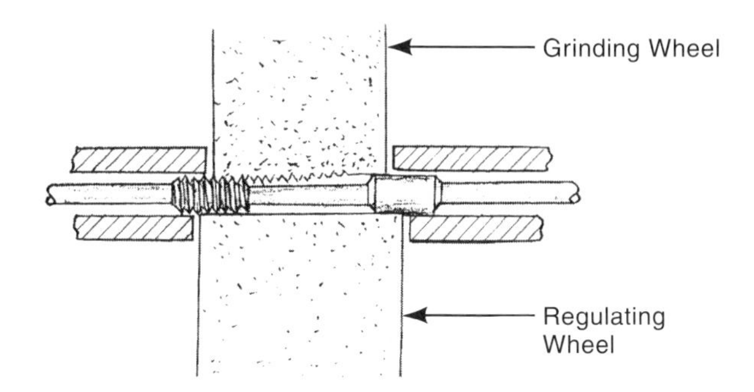

Fig. 3E6 Centerless thread grinding of headless screws.

E7. ***thread rolling*** - is a cold-forming method for making external screw threads. The blank workpiece is rolled between opposing dies which have a negative screw thread profile. The dies may be flat or cylindrical. Die heads, somewhat similar in appearance to cutting die heads, are often used, but these units have hardened rolling tools in place of the cutters. A third method allows several blanks to be run at one time between a cylindrical die and a concave die in planetary fashion. With all these process variations, the screw thread produced is of high quality with an accurate thread and smooth surface finish. No material is wasted and there are strength advantages because of the cold-working of the workpiece material. However, the diameter of the blank to be threaded must be accurate for the finished part to have the correct pitch diameter. The operation is rapid and is particularly suited to mass production of threaded fasteners. The flat-die method is illustrated in Fig. 3E7.

E8. ***cold-form tapping of internal threads*** - differs from tapping with cutting taps as described above, in that the workpiece metal flows rather than being cut and removed to form the threads. Both kinds of taps are used in essentially the same

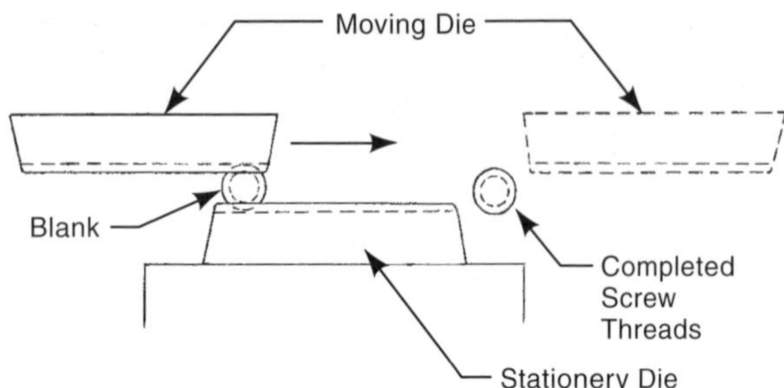

Fig. 3E7 Thread rolling with flat dies.

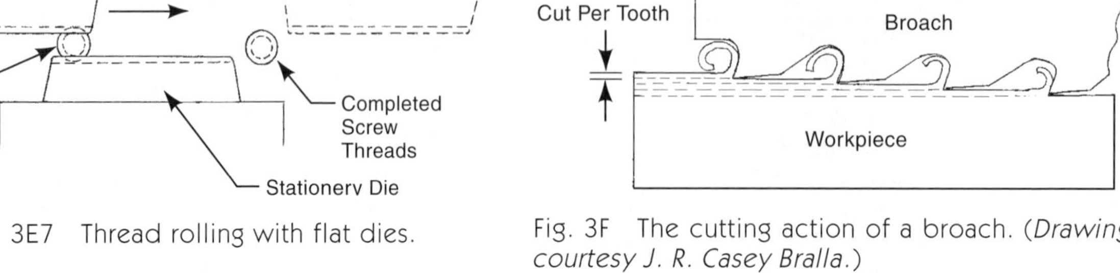

Fig. 3F The cutting action of a broach. (*Drawing courtesy J. R. Casey Bralla.*)

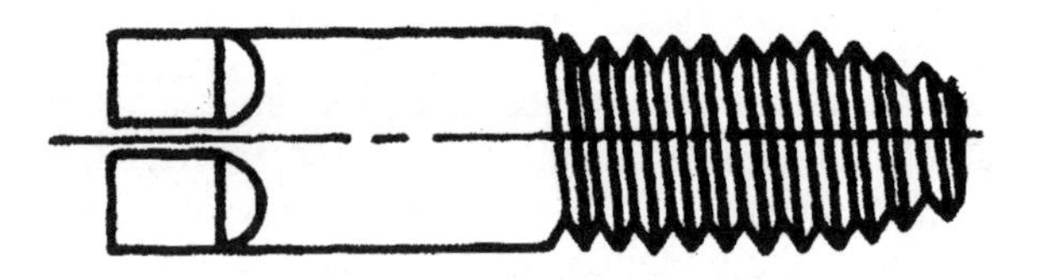

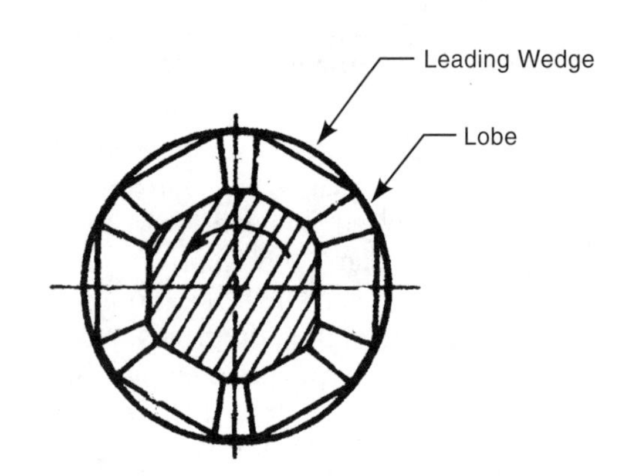

Fig. 3E8 A typical thread forming tap.

way and they look somewhat similar. A cold-forming tap is illustrated in Fig. 3E8. Cold-forming taps produce strong screw threads but the threads formed fill only 65 percent or less of the space of full threads. Soft and ductile materials must be used. Tapping speeds are higher than with cutting taps.

F. Broaching

Broaching is a high-production machining operation that involves the one-way movement of a broach, a cutting tool with a series of progressively-stepped teeth. The teeth move parallel to the surface being machined and each one removes a precise amount of material. The action is similar to that of a saw except that each tooth is set slightly higher than the one that precedes it and is wide enough to machine the entire surface. As with a saw, the spaces between the teeth hold the chips until the teeth pass from the workpiece. One pass of the broach results in the completion of the operation. The broach can be pulled or pushed across the work. Fig. 3F illustrates the cutting action schematically. Broaching is used to machine holes, (particularly non-round holes), flat surfaces, splines and slots. One common application is the machining of keyways in pulleys and gears, as shown in Fig. 3F1. The initial teeth to contact the workpiece are often designed to provide a roughing cut, while the final teeth provide finishing to the dimension desired. In that way, the operation can provide both high

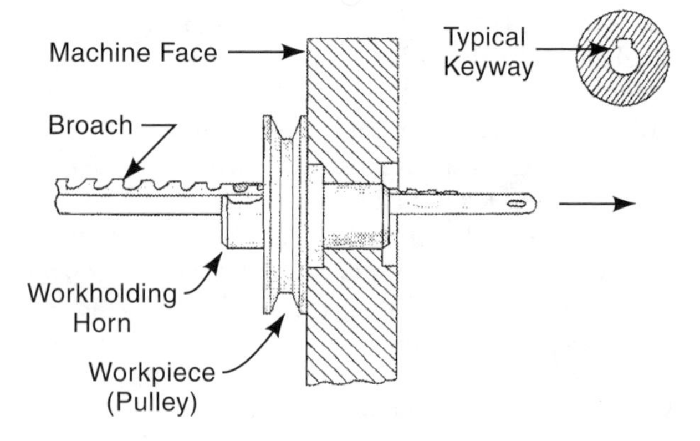

Fig. 3F1 Internal broaching of a keyway in a pulley and the cross-section of a typical broached keyway. The pitch of broach teeth is exaggerated to illustrate the process.

accuracy and excellent surface finish. The operation is rapid and particularly suited to mass production conditions. Operator skill requirements are not high because the needed precision is incorporated into the broaching tool. However, broaches are expensive and this is why broaching is chiefly found in mass production situations where the tooling cost can be amortized. However, some internal shapes are not practical to machine by other methods and that is why broaching is commonly used for keyways, splined-shaped holes, dovetail slots, and "fir tree" shaped slots for turbine blades. The equipment required for broaching ranges from simple arbor presses for keyway broaching, to huge special machines used in the automotive industry for surface-machining engine blocks and other large components. Holes as small as 0.050 in (1.3 mm) and surfaces as large as 20 in (0.5 m) wide, have been machined with this method.[4]

F1. ***internal broaching*** - is machining by broaching the inside surface of a hole or other opening in the workpiece. The most common broaching operations produce keyways, key slot openings in locks, spline-shaped openings, and other non-round enlargements of holes in pulleys, gears, or other parts that are to be fastened to shafts. There must be an initial hole or opening in the workpiece to provide room for entry of the broach. Helical surfaces can be cut with helical broaches if the workpiece is rotated as the broach advances. Fig. 3F1 illustrates internal broaching of a keyway.

F2. ***external broaching*** - takes place when the broaching tool machines an outside surface of a workpiece. Examples are the machining of flat surfaces, cam surfaces, and gear or ratchet teeth.

F3. ***pot broaching*** - uses a ring- or tubular-shaped broaching tool with the cutting teeth on the inside. Normally, the workpiece is pushed through the tool with hydraulic force. The tool cuts the entire workpiece periphery in one pass. Machining of gear teeth is a typical application.

G. Sawing

Sawing is the parting of material through the use of a narrow cutter, a saw, which contains a series of cutting edges that pass against the work in a continuous or reciprocating motion. As the cutter is advanced into the work, material is removed by each tooth, and a slot is formed, eventually extending through the entire thickness of the workpiece, and severing it into two pieces. The chip produced by each tooth is carried in the space between the teeth until the teeth exit the workpiece. The cutter can be in disk, band, or reciprocating blade form. Cutting teeth are typically set, i.e., offset slightly and alternately from both sides of the saw blade to provide a slightly wider cut (kerf) than the thickness of the blade so that there is room for its passage. The operation is used to cut billets, extrusions, castings, forgings, and various other shapes into blanks for further operations. Bars of various cross sections, rods, angles, and various other structural sections are cut to length by sawing.

G1. ***circular sawing*** - uses a saw in the form of a circular disk, with cutting teeth on the periphery. As the circular saw rotates, it is fed against the workpiece, machining a narrow slot in the workpiece and eventually severing it. Circular saws for metal cutting are sometimes called *cold saws* because they don't significantly heat the workpiece as friction saws do. They often have inserted cutter teeth of carbide rather than teeth formed of the blade material, or have cutter segments fastened to a center disc. Blades are sometimes large in diameter to permit the sawing of bulky workpieces. Kerfs are considerably wider than those on band or hack-saws because the circular blade must be thick enough to provide rigidity. Accurate and smooth cut surfaces are feasible with this method. The circular saw process is used to make blanks for subsequent operations or to cut structural members to the desired length. Fig. 3G1 illustrates a circular cut-off saw (coldsaw).

G2. ***band sawing*** - is most commonly a cut-off operation. Instead of a circular disk, the saw is an endless steel band with cutting teeth on one edge. The blade moves as it cuts in one direction. Cutting is continuous and blade wear is uniform over the whole length of the blade. Since the blade is normally thin, little material is lost to chip waste and power requirements are modest. The workpiece or the blade can be fed manually or mechanically.

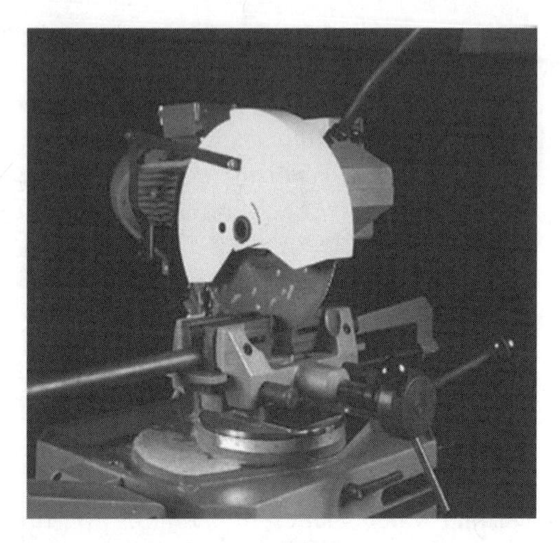

Fig. 3G1 A circular cutoff saw. (cold saw). *(Courtesy Clausing Industrial, Inc.)*

For cutting off, the blade is normally horizontal and is fed mechanically or by gravity. Band saws are also used for contour sawing. (See G3 below.) Band sawing is suitable for thin-walled tubing and other somewhat fragile parts because cutting forces are lower than with other sawing processes.

G3. ***contour sawing*** - uses a band saw. The powered band normally moves vertically downward and the workpiece, resting on a horizontal table, is fed into the moving blade either manually, with or without power assist, or automatically, into the moving blade. The cut can be in a straight line or a curved path, depending on the shape of the part to be produced. Contour work is not uncommon in tool shops for templates, cams, dies, and fixtures. Flat stock is the most common shape cut by this method but material of other shapes is also processed. A quantity of parts of thinner material can be cut in one operation by stacking a number of blank sheets. Views of some typical applications of contour sawing are shown in Fig. 3G3.

G4. ***hacksawing*** - uses a reciprocating saw, either manually or mechanically operated, to perform a cut-off operation. Cutting takes place only on the forward stroke. Feed of the blade is commonly by gravity but hydraulic and other mechanical feeds are often used on production hacksaws. Production hacksaws also incorporate a rapid return stroke to minimize the cutting time. The short, straight blade is less expensive than a bandsaw or circular saw blade but the cutting is less rapid. Once loaded, power hacksaw machines complete the operation automatically, stopping when a limit switch is tripped. Blades and equipment are less costly than those used with other sawing methods.

G5. ***abrasive sawing*** - uses a circular, band, or reciprocating saw blade, with a cutting edge that consists of abrasive particles rather than teeth formed in the edge of the metal blade itself. Thin, circular, cutters of bonded abrasive are most common. Surface speeds are much higher than with conventional sawing. Abrasive sawing is used when the workpiece material is hard or otherwise difficult to cut with metal cutting teeth. Hot metal billets are cut with this method. Cutting is quite rapid with large-wheel, high-horsepower machines. Coolant may be used to improve the finish of the cut surface, reduce burrs, and minimize heat effects. Smaller machines are usually operated manually. Equipment costs are low for this approach but the wheel cost is high compared with other sawing methods.

G6. ***diamond-edge sawing*** - is a kind of abrasive sawing that uses diamond abrasive particles because of particular hardness of the workpiece material. Cutting of ceramics, glass, stone, carbide, hardened die steel, nickel, and cobalt alloys are typical applications. Sufficient coolant and accurate feed control are necessary.

G7. ***friction sawing*** - uses an extremely high speed saw blade that softens or melts the workpiece material with frictional heat. The cutting teeth then remove the softened material with a cutting action. Either a bandsaw or large circular saw is used. The movement of the blade prevents it from overheating since any one part of it is in contact with the hot work for only an instant. However, the blade may also be water cooled. The blade is heavier and not necessarily as sharp as conventional blades. The process is suitable for hardened

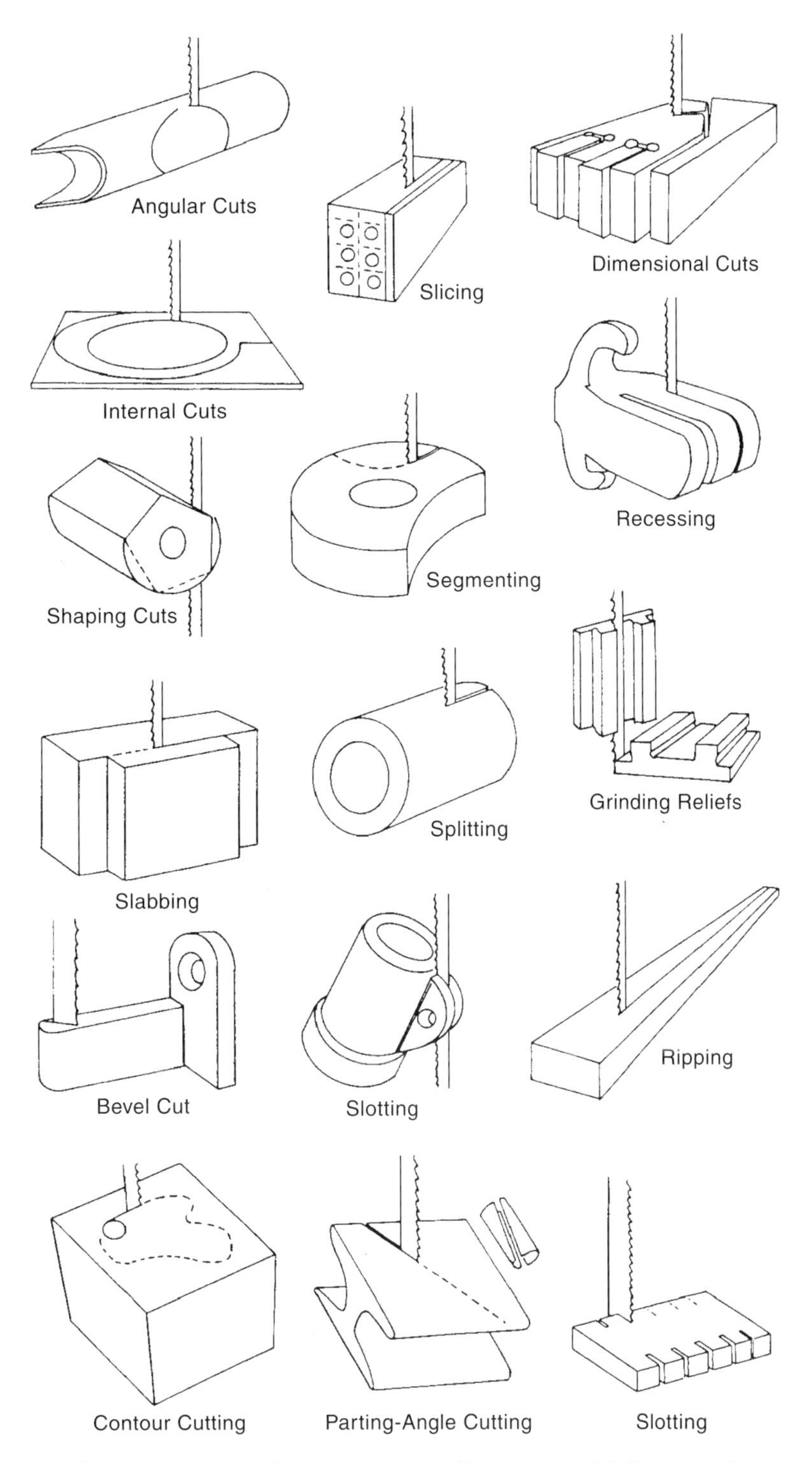

Fig. 3G3 Various contour sawing operations (*Courtesy DoAll Company*).

materials (Rockwell C 42 or more) which can not be cut well with conventional sawing. Material adjacent to the kerf is heat-affected. The method is slightly less accurate and has a poorer surface finish than that produced by other sawing methods. Copper, aluminum, and cast iron are not suitable for the process.

H. Flame Cutting (Thermal Cutting)

Flame cutting involves one of two underlying processes: 1) the severing or cutting of metal pieces by melting a narrow slit. 2) severing or cutting by chemical action wherein the metal oxidizes at high heat. The variety of methods used is discussed below. Flame cutting is most widely used for cutting heavy plate - as thick as 30 in (75 cm) - and structural shapes and sheet where tolerances are not so strict. The cutting torch is often manually guided, especially in field work or when tolerances can be liberal. However, flame cutting machines guide the torch more accurately along straight lines or curved paths. Template tracing, optical tracing of part drawings, numerical- or computer-numerical control are methods used to guide the torch tip along whatever path is required to cut the desired workpiece from the stock to be cut. The process is most often used to cut parts from flat stock, However workpieces of many other shapes can be cut with the process. It is also commonly used in dismantling various structures or equipment. Most flame cutting torches can be portable, so field dismantling is quite feasible.

H1. ***oxy-fuel gas cutting (OFC)*** - uses acetylene, natural gas, propane, or hydrogen in combination with oxygen to fuel the process. With ferrous metals, a small area of the workpiece to be cut is first brought to a temperature of about 1600°F (870°C) with a flame of pure gas or a gas/oxygen mixture. When the metal reaches the proper temperature, oxygen is added to the fuel mixture and the workpiece metal in a narrow section is oxidized (burned) with the aid of the oxygen. Iron oxide, in liquid form, is generated and is blown away from the workpiece by the force of the gas stream exiting the torch nozzle. As the torch moves, a narrow slit is made in the workpiece. With non-ferrous metals, oxidizing may not take place, and the metal is removed by simple melting. The torch is commonly held and controlled manually, particularly for work not requiring great dimensional accuracy, but mechanically-held torches controlled by templates, optical tracing, or computer numerical control are also used. The process works best with carbon and low alloy steels. A typical cutting speed for 1 in (25 mm) thick steel plate is 18 in (0.46 m) per minute. When cutting stainless steels, metal powders or chemical additives are incorporated in the stream to improve the cutting action. Some machines have multiple torches and can cut a number of parts from one piece of sheet or plate material at the same time.

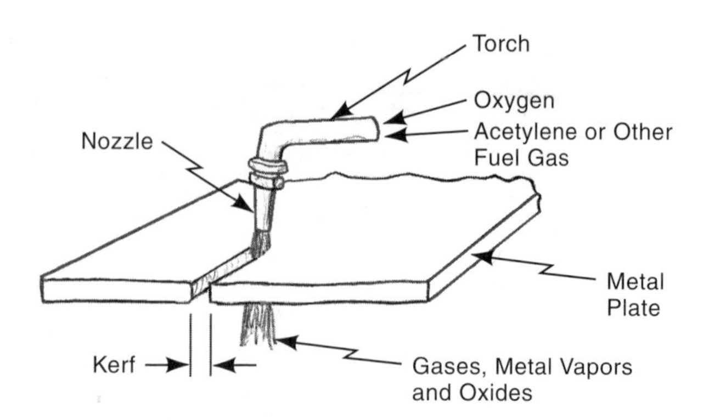

Fig. 3H1 Schematic view of oxy-fuel gas cutting (OFC).

Parts are most commonly cut from flat plate. Material from 1/8 to 60 inches in thickness can be cut. Ship building, construction, tank, pressure vessel, and heavy equipment manufacture, are industries that use the process .It is illustrated in Fig. 3H1. Straight or beveled cuts can be made, and welding is a common subsequent operation. There is a heat-affected zone adjacent to the cut , and this can have adverse effects with some materials and applications.

H2. ***metal powder cutting (POC)*** - is a method applicable to metals that do not flame cut very easily. Preheated iron powder is injected into the flame to raise the cutting temperature. Oxidation of the iron provides an aid to cutting metals that are highly resistant to oxidation. Chromium-nickel stainless steels, high-alloy steels, and cast iron are cut with this method. It is also used to remove gates and risers from iron and stainless steel castings, and to cut stacked sheets or plates more easily.

H3. ***chemical flux cutting (FOC)*** - adds flux to the cutting gases to make the metal oxide more fluid and more easily removed from the gap of the cut. Powdered chemicals such as sodium carbonate, and other salts of sodium are used. The process is sometimes called *flux injection*. Its use is relatively minor.

H4. ***arc cutting*** - involves a group of processes that use the heat of an arc to melt a metal workpiece to severe it or remove metal. All common arc welding processes (See chapter 7.) can be adapted to metal cutting as well as metal fusion. When arc processes are used to cut steel and other easily oxidized metals, oxygen may be used in conjunction with the arc to support the oxidation. In metals that do not oxidize, the cutting action is mainly from pure melting and air or a shielding gas is used to provide the force needed to help expel the molten metal from the gap. The most significant processes in this group are plasma arc cutting, air carbon arc cutting, and oxygen arc cutting (oxygen lance cutting). Gas metal arc cutting, gas tungsten arc cutting, shielded metal arc cutting, and carbon arc cutting, are other arc cutting processes that are not widely used in industrial production situations.

H4a. ***plasma-arc cutting (PAC)*** - uses the extremely high temperature (20,000 to 50,000°F (11,000 to 28,000°C) of the plasma arc to melt the workpiece metal. Severing is a result of melting rather than oxidation. A direct current arc between a tungsten electrode and the torch body creates the required high temperature, and partially ionizes a stream of nitrogen, hydrogen, argon, air or a mixture thereof. (Partially ionized gases are a mixture of positively-charged ions, free electrons, and neutral atoms.) As the stream of gas leaves the nozzle, the separated electrons recombine with the gas atoms and release additional energy. The constricting effect of the nozzle further increases the arc temperature, and quickly heats and melts the workpiece metal. A shielding gas is often used to protect the ionized stream, and sometimes a surrounding stream of water is also used to help confine the stream so that the melting takes place in only a narrow kerf. The force of the stream, the shielding gas, and water, all expel the melted metal from the cut.

The process provides very high cutting speeds, 4 to 5 times higher than those achievable with standard oxy-fuel cutting. It works equally well with metals that do not oxidize because melting rather than oxidation is the means by which the metal is removed. There is a heat-affected zone but it is narrower than that resulting from oxy-fuel-gas cutting. Stainless steel and aluminum are major applications of the process, although mild and alloy steel, titanium, bronze, copper, and magnesium are also processed. Smooth cuts in plates up to about 6 in (150 mm) thick are feasible[2]. Ductwork, over-the-road tanks of aluminum or stainless steel, food and chemical processing equipment, ships and barges, are all manufactured with the use of this process. Fig. 3H4a illustrates plasma cutting.

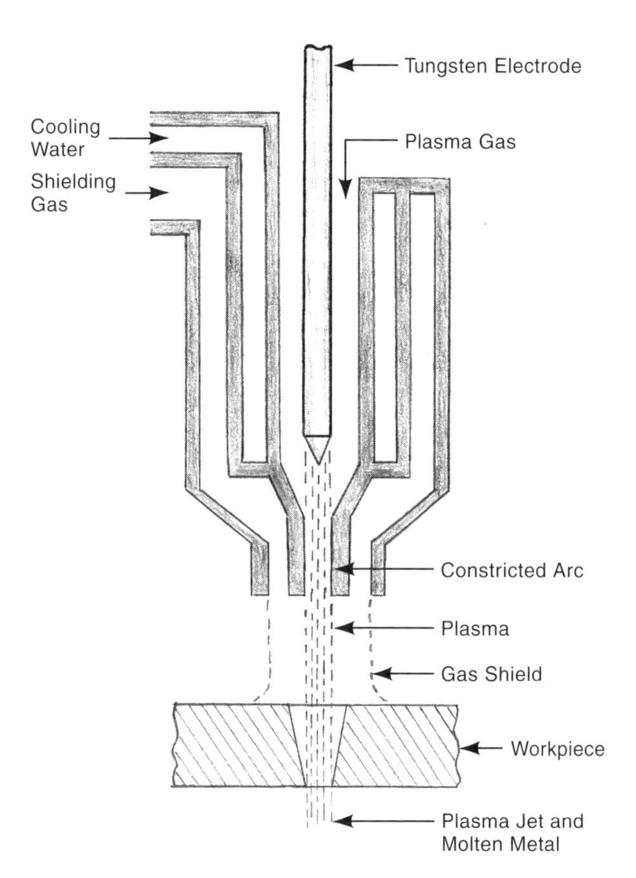

Fig. 3H4a The principles of plasma arc cutting.

H4b. ***air-carbon arc cutting and grooving*** - In this process, intense heat from an electric arc between the tip of a carbon electrode and the metal workpiece melts workpiece material. A high-velocity air stream passing through the arc blows the molten metal away from the area of the cut. Though it can be used for through-cutting, the method is commonly used for gouging, an application not feasible with oxy-fuel gas cutting. When used for gouging, it leaves a clean groove. One common application is the preparation of square-edged plate for welding. Another is to remove defective welds so that the joint can be reworked. The process is also used to remove sprues, runners, risers, and fins from castings. Cast iron, steel, stainless steel, ductile and malleable iron, nickel, aluminum, copper alloys, and other non-ferrous metals have been processed with this method. Equipment is simple, but the operation is noisy and the scattering of molten metal may be a problem.

H4c. ***oxygen lance cutting (LOC)*** also known as ***oxygen arc cutting*** - is used in steel mills to sever material that is already hot from the steelmaking process. A small pipe (lance) carries oxygen to the workpiece, which is hot enough to start the oxidizing reaction without the need for further heating by fuel. The heat generated by the oxidation, combined with the oxygen flow from the pipe, are sufficient to continue the oxidizing reaction and oxide removal until cutting is complete. The lance is usually a simple length of pipe or tubing with a diameter of 1/8 to 1/4 inch (3 to 6 mm) coated with non-conductive mineral material and connected to a valve and oxygen supply hose. The lance is consumed as the operation proceeds. When cool workpieces are to be cut, the surface of the metal is heated with a separate torch to a temperature high enough to start the oxidation reaction with the oxygen lance. The operation can be performed under water.

In some process variations, aluminum or magnesium wire in the lance develops additional heat of oxidation enabling the lance to cut concrete, bricks, and other non-metal objects.

H4d. ***gas metal arc cutting*** - is the metal cutting version of GMAW, gas metal arc welding. In both processes, the heat is provided by an electric arc between the workpiece and a continuously-fed metal wire, shrouded by the flow of an inert shielding gas. In the case of metal arc cutting , the purpose of the operation is to sever parts of the workpiece rather than to fuse it to another part. The molten metal is evacuated from the area of the cut by a combination of the force of the shielding gas, the arc, and vapor pressure from materials vaporized by the arc. This method is not highly significant in industrial production situations.

H4e. ***gas tungsten arc cutting*** - is the cutting variation of GTAW, gas tungsten arc welding. For cutting, higher amperages and increased shielding gas flow are employed. The arc melts workpiece metal, which is forced out of the cut by gas pressure from the shielding gas and the effects of the arc. Gas tungsten-arc cutting is used for cutting stainless steels and various non-ferrous metals including aluminum, copper, nickel, magnesium, silicon-bronze, and copper nickel. An argon-hydrogen gas mixture is commonly used. As with gas metal cutting, this method is not widely used in production situations.

H4f. ***shielded metal arc cutting*** - is similar to "stick" welding when that method is used to cut the workpiece rather than to fuse it to another member. A metal electrode in rod form, coated with flux, is used without any gas jet or blanket. The heat of the arc of high current density melts the workpiece, and gravity is used to carry away the molten metal. This method is limited to situations where it is inconvenient to use more effective flame cutting methods.

H4g. ***carbon arc cutting*** - is not a significant industrial process but is found in small shops that lack more-sophisticated equipment. The arc between a carbon-graphite electrode and the workpiece melts the workpiece where the arc contacts it. Gravity and pressure of the arc, rather than the air jet used with air-carbon-arc cutting, removes the molten metal.

H5. ***laser cutting*** - is an alternative process for cutting plate and sheet materials. See O below.

I. Electrical Machining Processes

I1. ***electrical discharge machining or EDM*** - is sometimes called, *spark erosion machining*. It uses a series of fine, electrical discharges or sparks to erode the workpiece material. The discharges pass from the tool (cathode) to the workpiece (anode) at a rate greater than 20,000 times per second. The workpiece material, which must be electrically conductive, melts or vaporizes at the point where it is touched by the spark. A dielectric fluid, usually kerosene, circulates between the electrode and the work. The fluid confines the spark, cools and solidifies molten material, and carries away the residue. The process is advantageous for hard, conductive material including hardened steel and carbide. There is no significant cutting force, so delicate shapes can be produced.

The gap between the anode and the work is only about 0.002 in (0.05 mm) and is servo controlled. There is an overcut equivalent to the length of the spark from the electrode to the work. The cutting rate is slow compared with conventional machining but is still advantageous for materials too hard or otherwise not easily machined. The rate of cutting and surface finish are controllable by varying the frequency, voltage, and current in the electrical pulses. Higher energy levels in the pulses produce faster erosion of the workpiece but a rougher surface finish. Typically, the rate of cutting is reduced toward the end of the operation, in order to provide smoother surfaces. Spark erosion produces small craters in the workpiece and these craters are manifested as a matte finish on the workpiece. The tool also is eroded since some spark action is in the reverse direction. Additionally, some secondary operations may be required to remove a hard, thin, re-cast surface layer, or fine surface cracks caused by thermal stresses, depending on the material used and the function of the part produced.

I1a. ***ram EDM*** - is sometimes called "die-sinker" EDM, and the principle is illustrated in Fig. 3I1a. The electrode is shaped to fit the desired

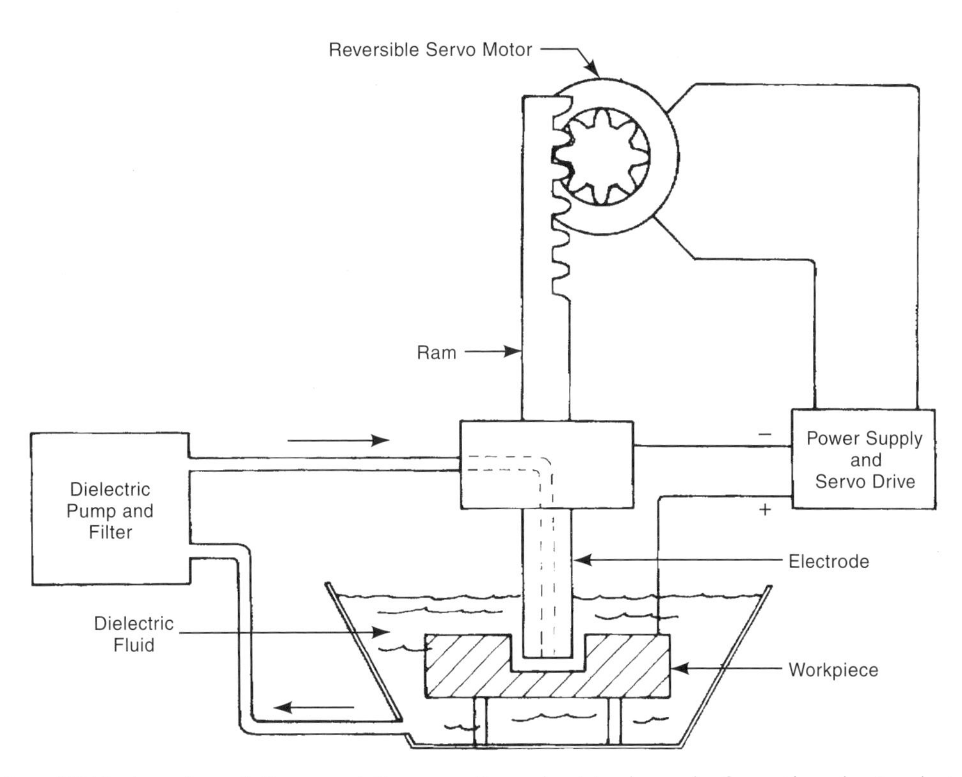

Fig. 3I1a Ram EDM. A series of fine, rapidly repeating electrical sparks from the electrode to the workpiece erode a cavity in the workpiece. The cavity matches the shape of the electrode. (*Courtesy American Machinist, Penton Media, Inc.*)

cavity and is fed into the workpiece, which is eroded to match the electrode shape. The electrode is made undersize to allow for the expected overcut, commonly 0.0005 to 0.020 in (0.013 to 0.5 mm). Feed is downward into the work, but CNC controls can also provide transverse movement of the electrode to produce special shapes in the cavity. Since the tool also erodes and becomes tapered, extra tools are often made. Graphite is a common electrode material, but copper, brass, aluminum, copper-tungsten, zinc-tin, and other alloys are also used. The process is mainly used to machine cavities in hardened steel dies and molds. It is also used in machining carbide, and in salvaging hardened parts or tools such as broken taps. Slots, non-round holes, small deep holes, and the machining of honeycomb and other fragile parts, are additional applications.

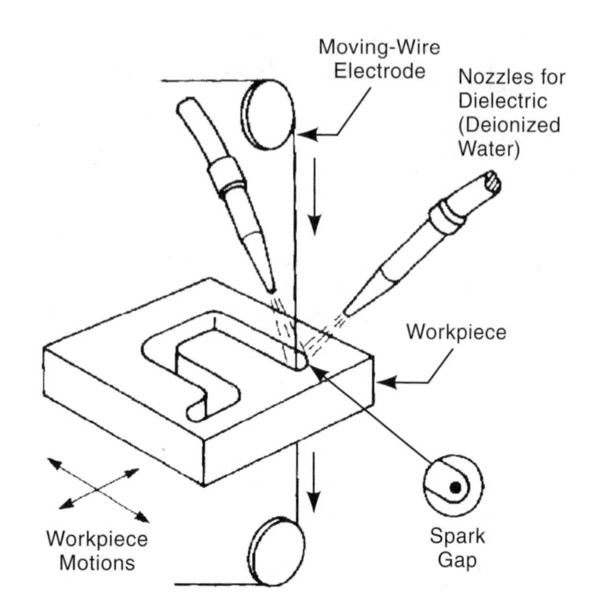

Fig. 3I1b Wire EDM. Electrical sparks from the wire to the workpiece cut a contoured slit in the workpiece as the wire advances. *(Courtesy American Machinist, Penton Media, Inc.)*

I1b. ***wire EDM*** - uses a constantly-moving wire instead of a shaped electrode. The wire, of 0.001 to 0.013 in (0.025 to 0.33 mm) diameter, passes through the work, with a vertical axis, (though it may be set at an angle when required when cutting apertures for stamping tools.) Tungsten, copper, and brass are common wire materials. The wire or the work, is fed horizontally as the cut progresses, to cut a slit or shaped through-hole in the workpiece. Different wire material is constantly exposed to the spark, so wear of the wire is widely distributed and is not a problem. The process, shown in Fig. 3I1b, is often used to cut die openings in hardened stock to produce dies and die components. A high level of accuracy and fine detail can be achieved.

I1c ***electrical discharge grinding (EDG)*** - is similar to electrical discharge machining (See I1), except that the electrode is a rotating wheel instead of an electrode that is stationary except for its downfeed. The workpiece and wheel are immersed in dielectric fluid, and metal is removed by the same kind of spark erosion as that which occurs with EDM. The wiping action of the wheel produces better surface finishes than with ram-type EDM. The graphite wheel is dressed as necessary to compensate for its wear in the process. The volume of material lost by the wheel from wear averages about 1/3 of that removed from the workpiece but is considerably less in many instances. Since the wheel wear is spread over its circumference, the amount of reduction of wheel diameter from dressing is normally small. The process is used in shaping carbide form tools and in grinding fragile or brittle materials. Material to be ground by EDG must be electrically conductive. There is a thin heat-affected layer from the process. The layer varies from 0.0001 to 0.0015 in (0.0025 to 0.038 mm) in depth. Fig. 3I1c illustrates the process.

I2. ***electrochemical machining (ECM)*** - removes metal by a reverse electroplating process. The workpiece becomes the anode and the tool is the cathode of the electrolytic process. A highly conductive electrolytic fluid is pumped into the space between the workpiece and the tool, and high-amperage current removes workpiece material by anodic dissolution. The workpiece takes a mirror-image shape of the tool. The gap between the electrode (tool) and workpiece is a small as 0.001 in (0.025 mm) and pressure of the electrolyte must be high to insure adequate flow. Temperature control

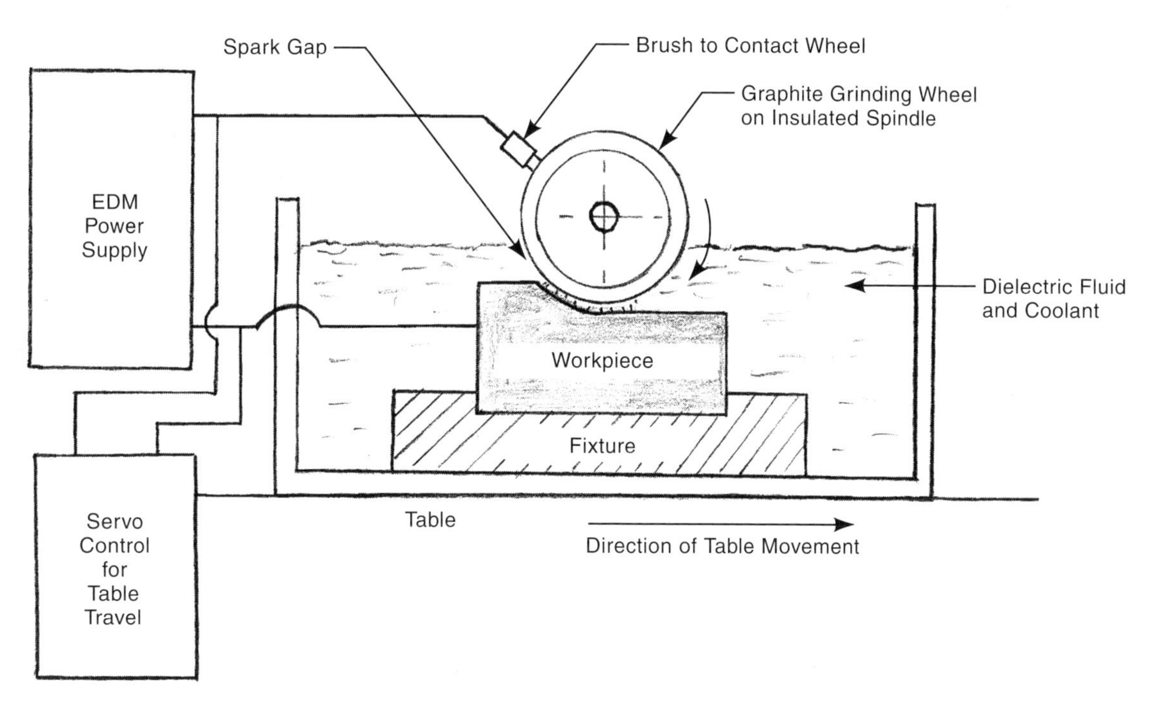

Fig. 3I1c Electrical discharge grinding (EDG) - is EDM with a rotating wheel electrode. Electric sparks from the electrode erode the workpiece. The shape of the periphery of the wheel is transferred to the ground surface of the workpiece.

of the electrolyte, control of the gap and feed rate of the tool and maintenance of electrolyte cleanliness are important elements in the process. The electrolyte is circulated through equipment that removes the operational debris from the fluid. The process is particularly suited to hard materials and others that are difficult to machine by conventional methods. Workpiece hardness does not affect its machinability by the process, but the workpiece material must be electrically conductive. The tool does not wear and no stresses are induced in the workpiece by the operation. Common tool materials are copper, brass, bronze, stainless steel, and copper-tungsten. Metal removal rates are good compared to a number of other non-traditional machining processes, and average 1 in^3 (16 cm^3) per minute per 10,000 amperes of current. The process is used for die sinking, manufacture of jet engine parts, cam profiling, and the machining of small, deep holes. ECM is most advantageous for materials that are difficult to machine by conventional methods. A typical electrochemical machining set up is illustrated in Fig. 3I2.

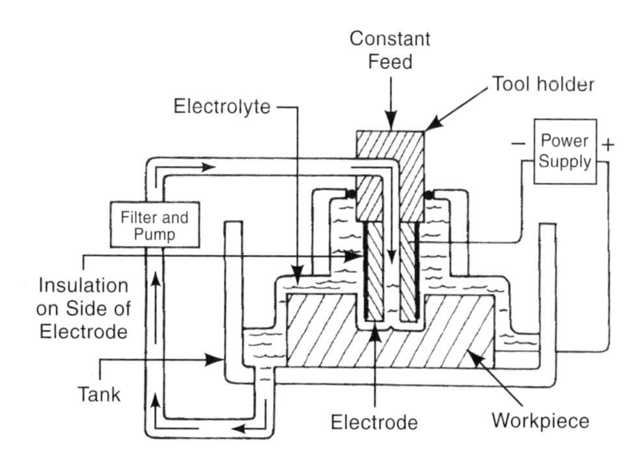

Fig. 3I2 Electrochemical machining (ECM). Heavy electrical current passing through the electrolyte between the electrode and the workpiece causes anodic dissolution of the workpiece material. (*from Bralla, Design for Manufacturability Handbook, 1998, McGraw-Hill Companies. Reproduced with permission.*)

I2a. ***electrochemical grinding (ECG)*** - is similar to electrochemical machining but replaces the relatively stationary tool with a rotating conductive grinding wheel. The wheel normally consists of aluminum oxide abrasive bonded to a metal wheel, which acts as the cathode of the electrolytic circuit. Electrolytic fluid circulates in the area where the abrasive contacts the work. A combination of electrolytic and mechanical action removes material from the workpiece but the electrolytic action predominates, accounting for about 90 percent of metal removal. Anodic dissolution of the workpiece metal leaves surface metal oxides. In conventional ECM, the flushing action of the electrolyte removes these oxides. In electrochemical grinding, the abrasive mainly functions to remove the oxide film, exposing a new metal surface to the electrolyte. The abrasive also separates the metal wheel from the work, preserving a fine (0.001 in or 0.025 mm) gap between the two. It also carries the electrolyte solution to the gap.

The process has the advantage of relatively high metal removal rates for hard metals, freedom from heat damage to the workpiece, and the ability to grind fragile parts. Plunge, surface, cylindrical, and internal grinding are all feasible with the process. However, capital costs are high and the electrolyte can be corrosive to the equipment and workpiece. The process is commonly used in sharpening carbide cutting tools, avoiding the high wear rates of expensive diamond-abrasive wheels that would otherwise be required. It is also used for grinding surgical needles, honeycomb structures, and other fragile parts. ECG is illustrated by Fig. 3I2a.

I2b. ***electrochemical turning (ECT)*** - is another application of electrochemical machining. The workpiece rotates as in conventional turning but the cutting tool is replaced by an electrode. Electrolytic fluid is directed to the gap between the tool and the work, and material is removed by electrolytic action between the workpiece (anode) and the electrode (cathode). Facing and turning cuts can be made. Disc forgings and bearing races are machined by the process.

I3. ***electrochemical discharge grinding (ECDG)*** - is sometimes called, electrochemical discharge machining. It is a combination of electrochemical grinding (ECG) (See I2a) and electrical discharge grinding (EDG) (See I1c). Stock removal is primarily by ECG but oxides from ECG are then removed by intermittent spark discharges instead of

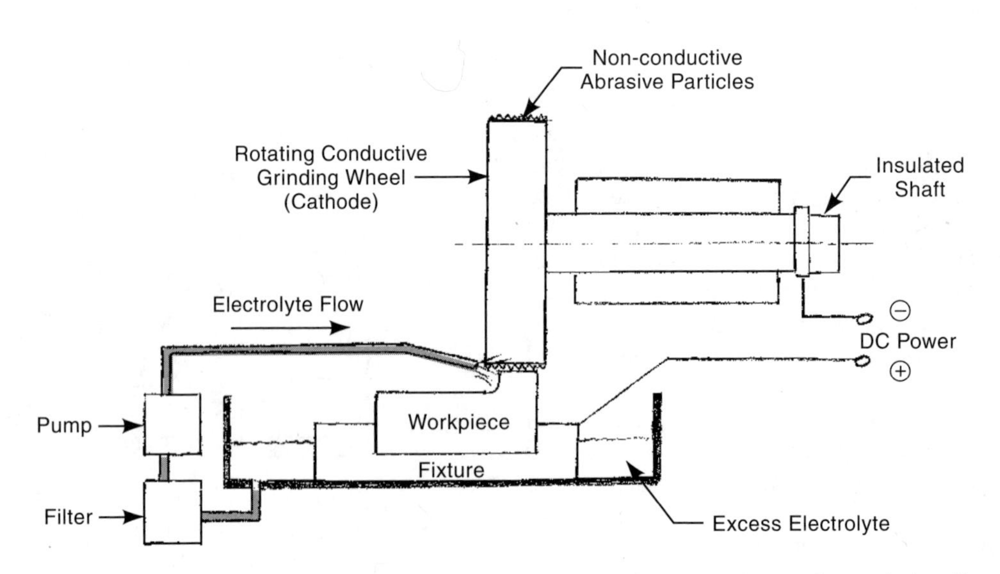

Fig. 3I2a Electrochemical grinding. Most of the metal removal results from electrolytic dissolution of the workpiece caused by flow of electrical current in the electrolyte between the conductive grinding wheel and the workpiece. Abrasive grains on the wheel remove surface oxides and expose more metal to the electrolyte.

by abrasive particles. The wheel is conductive (normally made of graphite) and has no added abrasive. The fluid used is a highly-conductive electrolyte in contrast with EDG, which uses a dielectric fluid. The wheel rotates rapidly (4000 to 6000 ft/min - 1200 to 1800 m/min), bringing fresh electrolyte between the workpiece (the anode) and the wheel (the cathode). Spark discharges occur randomly when the breakdown voltage of the oxide film is exceeded. Alternating current or pulsating direct current are used. Current densities are less than in ECG (to avoid cratering of the workpiece and wheel), and metal-removal rates are considerably slower. However, the wheel cost for the graphite-only wheels is less than that required for abrasive wheels used with ECG. The process is used in the grinding of carbide cutters. It can also be used with hardened tool steels, nickel alloys, and parts that are heat-sensitive or fragile. Honeycomb and other fragile parts are advantageous because they are finished free from burrs and stresses. Form grinding can be used if the wheel has a profile.

I4. ***electrochemical honing (ECH)*** - is electrolytic action added to conventional honing (J1), and is quite similar to electrochemical grinding. Material is removed from the workpiece by a combination of electrolytic action and mechanical abrasion. The electrolytic action removes most of the material, and the abrasive action of the honing stones removes the oxides produced by the electrolytic action. The equipment provides both reciprocating and rotational movement of the honing stones, as in conventional honing. However, electrolytic fluid handling is added. The tool that holds the honing stones is hollow and fluid passes from the tool into the gap between the stones and the workpiece. The gap between the tool and the work is approximately 0.003 to 0.005 in (0.08 to 0.13 mm) at the start of the operation, increasing to about 0.020 in (0.5 mm) at the conclusion, but the stones remain in contact with the workpiece, normally the surface of a machined hole. (The wedging action of a conical piece in the tool forces the stones outward as the honing takes place.) The operation provides metal removal three to five times as fast as regular honing, while the wear of the honing stones is considerably less. Deburring action is also superior and the work is more apt to be free from stress-induced or heat damage. However, the operation requires more complex equipment due to the need for control of the electrolytic action and the corrosiveness of the electrolyte. Therefore it is not necessarily more economical than conventional honing. The process is used to refine the bores of hardened parts such as gears and pump components, particularly when production quantities are large.

I5. ***electro etching*** - is a method for marking workpieces that are electrically conductive. A stencil is placed on the workpiece and a pad containing an electrolyte is placed or dabbed on the stencil. The workpiece and the pad are connected to sources of direct electrical current. The current removes workpiece material electrolytically in the areas corresponding to openings in the stencil. The method is useful for placing identification markings on metal parts with fairly smooth surfaces. Electro etching is normally performed manually and is a suitable method when production quantities are modest.

J. Finish Machining Operations

J1. ***honing*** - is a low-velocity abrasive machining process. One or more bonded abrasive stones or "sticks" are put in area contact with the surface to be machined (in contrast with grinding processes where the abrasive and the work are in contact essentially on a line). Slow movement of the sticks in several directions abrades the high spots on the workpiece surface and makes the surface smoother and more true (i.e., with greater dimensional and geometric accuracy). Because of the slow movement, there are no heat effects to the metal surface. The process, illustrated in Fig. 3J1, removes only a small amount of material - less than 0.005 in (0.13 mm)[2] - but improves the dimensional accuracy and surface finish of the surface being honed. The operation can be performed by hand but, for production situations, machines that impart reciprocating motion in several directions, or a combined reciprocating and rotary motion, are used. Cutting fluids are normally employed. The most common application is the finish machining of bored holes to remove tool marks, waviness or taper. The stones are allowed to "float" and follow

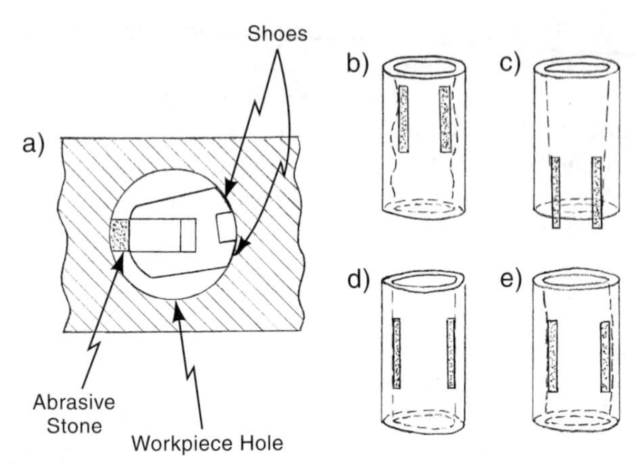

Fig. 3J1 Honing. a) end section view of typical single-stone abrasive honing tool in a hole. Views b) through e) show, greatly exaggerated, hole discrepancies that are improved by honing: b) waviness, c) taper, d) out-of-roundness, e) bowed shape.

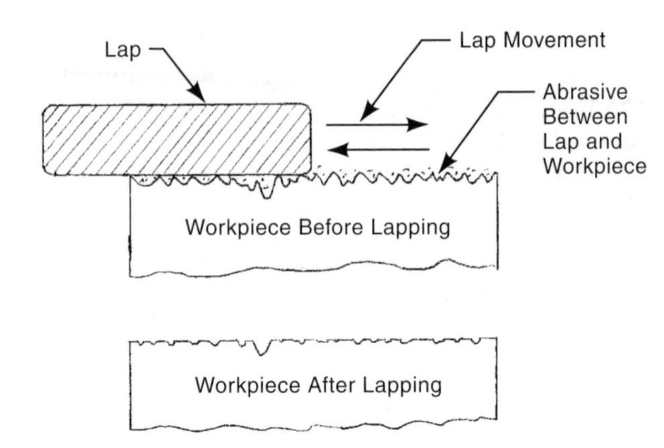

Fig. 3J2 Lapping. The abrasive particles held in place by the soft lap, remove peaks from the workpiece surface. Note: Peaks and valleys in the workpiece are shown greatly exaggerated.

the direction of the hole's axis. The cylinder walls of internal combustion engines are typically finished by honing. Gear teeth and bearing races are also honed.

J2. ***lapping*** - is another low-velocity abrasive machining process. It utilizes a fine abrasive in paste or powder form, which is contained between the workpiece and the lapping tool. The lapping tool is made of material softer than the workpiece and is shaped to mate with the surface being machined. It has the function of holding the abrasive. A random, reciprocating motion of the lap and the abrasive, with light pressure applied, refines the dimensions and smoothness of the workpiece surface to a close tolerance. Unlike honing, the motion is not necessarily in more than one direction. Very little material is removed, seldom as much as 0.001 in (0.025 mm)[2]. The purpose is to remove fine scratch marks or to create very flat or otherwise smooth and precise surfaces. The process is also used to provide a very close fit between mating surfaces. Flat, cylindrical, or spherical surfaces, or those of special shape can be processed with this method. The operation can be performed by hand or with machines designed for the purpose. In some cases, abrasive coated paper, bonded abrasives, or cloth laps are used. The relative motion between the lap and the workpiece is always slow. Typical applications are finishing of gage blocks, plug gages, ball bearing races, valve seats, roller bearings, and optical parts. Parts requiring metal-to-metal sealing surfaces are frequently lapped. The process is illustrated in Fig. 3J2.

J3. ***superfinishing*** - is a third low-velocity surface-refinement process. A solid abrasive is used, but it is loosely bonded so that it wears to the shape of the workpiece surface. Thus, a large surface area is in contact with the abrasive at all times during the operation. A controlled, bidirectional movement of the workpiece and the abrasive under light pressure removes high spots and smooths the surface, removing surface material that has been smeared or distorted by prior grinding or machining. Typical speeds of movement between the work and the abrasive are 12 to 50 ft (4 to 15 m) per min at a pressure of 10 to 40 lbf/in^2 (70 to 275 kPa). A liberal amount of oil-based cutting fluid normally accompanies the operation. Typically, from 0.0002 to 0.001 in (0.005 to 0.025 mm) of material is removed. Mirror-like surface finishes can be produced. Fig. 3J3 illustrates the process which is used to smooth flat, curved, or spherical bearing surfaces. Automotive crankshafts, camshaft bearings, brake drums, and distributor shafts are typical superfinished parts.

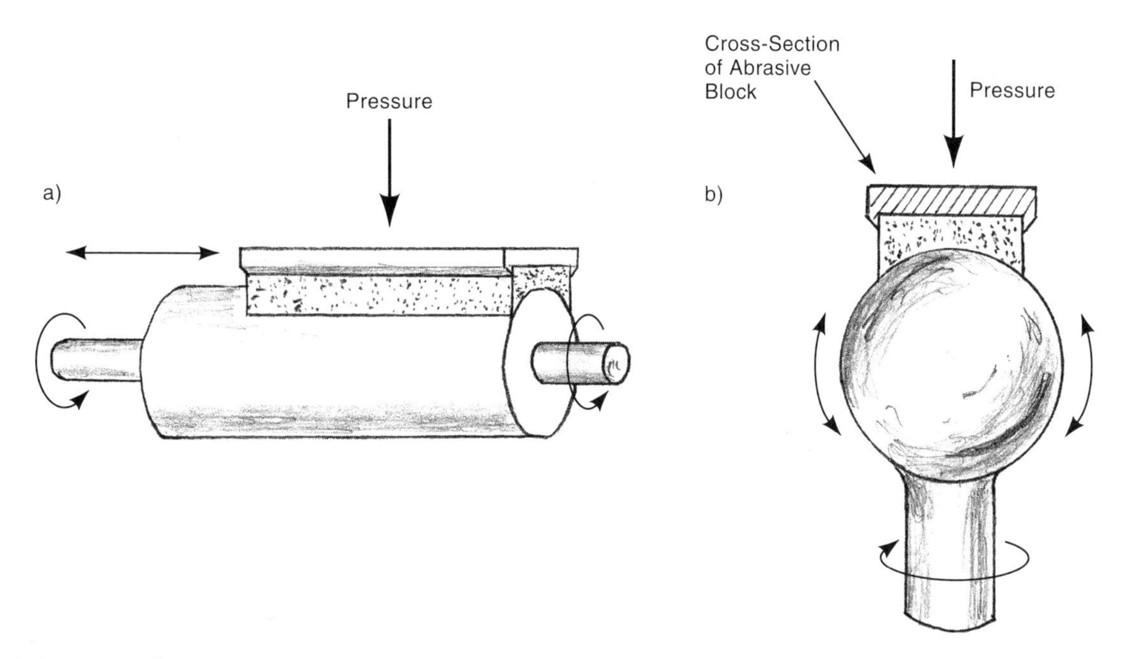

Fig. 3J3 Superfinishing. a) a cylindrical surface. The workpiece revolves while the abrasive block is moved back and forth along the axis. b) a spherical surface. The abrasive block moves around the spherical workpiece as it rotates. In all applications, the abrasive conforms to the shape of the workpiece.

J4. ***burnishing*** - is a means of smoothing machined or other surfaces by rubbing a smooth hard object against the surface with considerable pressure. Burnishing is not a machining operation - no workpiece material is removed - but instead deforms and presses down localized high spots from cutting tools, and thereby smooths rough areas.

J5. ***roller burnishing*** - refines the surface of a workpiece by pressure rolling rather than by removing metal. The burnishing tool incorporates one or more hardened, finely polished rollers, which bear against the workpiece at high pressure. Each roller achieves the desired effect by deforming the surface material of the workpiece as it rolls against it, compressing the minute peaks of surface roughness into the valleys. The force applied to the burnishing tool depends on the amount of pressure required to exceed the yield point of the workpiece material. Multiple rollers may be used, depending on the shape of the surface being roller-burnished. Gear-tooth finishing is one common burnishing operation, performed by rolling the gear workpiece against three smooth, hardened, burnishing gears. Fig. 3J5 shows the principles of the operation and Fig. 3J5a, b, and c illustrates typical configurations that are roller burnished and the burnishing tool that is used. Compressive stresses, left in the surface after the operation, and surface work

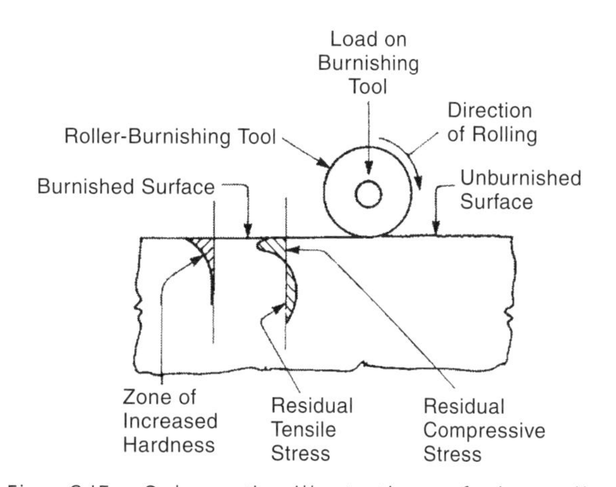

Fig. 3J5 Schematic illustration of the roller burnishing process. *(from Bralla, Design for Manufacturability Handbook, 1998, McGraw-Hill Companies. Reproduced with permission.)*

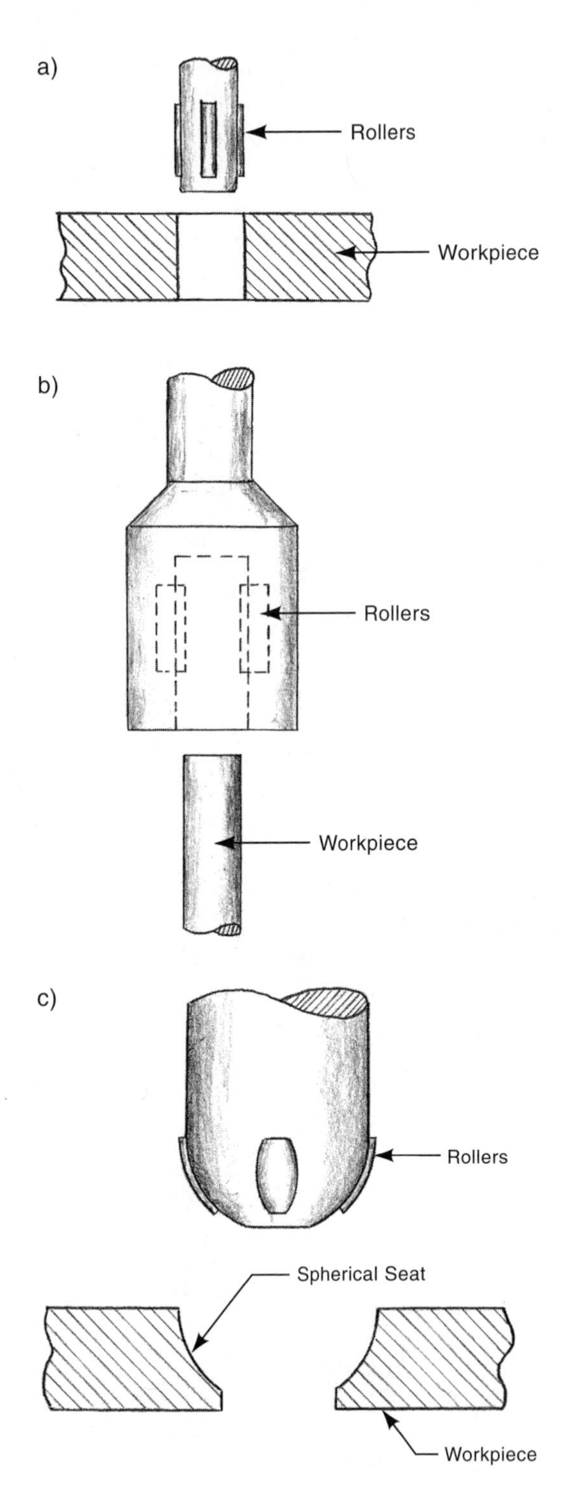

Fig. 3J5a Typical surfaces that can be roller burnished and the burnishing tools used. (*from Bralla, Design for Manufacturability Handbook,* 1998, McGraw-Hill Companies. Reproduced with permission.)

hardening improve resistance to wear and fatigue failure. The process is limited to workpieces up to about Rockwell 40C in hardness and with walls thick enough to withstand the forces involved. Roller burnishing of holes can be a substitute for, or a supplement to, reaming and boring. Typical parts that undergo the operation are cylinder bores, valve stems, piston rods, turbine shafts, pump plungers, and rolls for the plastics and paper industries.

K. Deburring

The thin, sharp, ragged metal edge that forms whenever a metal part is machined or stamped, normally must be removed for reasons of safety, accuracy of dimensions, fit of mating parts, and appearance. Typical burrs are illustrated in Fig. 3K. Burrs form on edges and corners of surfaces, holes, and slots. Similarly, flash, the thin edges of metal (or whatever material is used) that often accompany casting, forging and molding operations, must be removed or controlled. The following is a summary of the many methods that can be employed to remove burrs and flash.

K1. ***abrasive flow deburring*** - is an application of abrasive flow machining. (See C10.) A putty-like material containing abrasive is flowed by hydraulic pressure over the workpiece edges that have burrs. The force of the material helps break off the burrs and the abrasive particles provide cutting action that smooths the edges. The process is

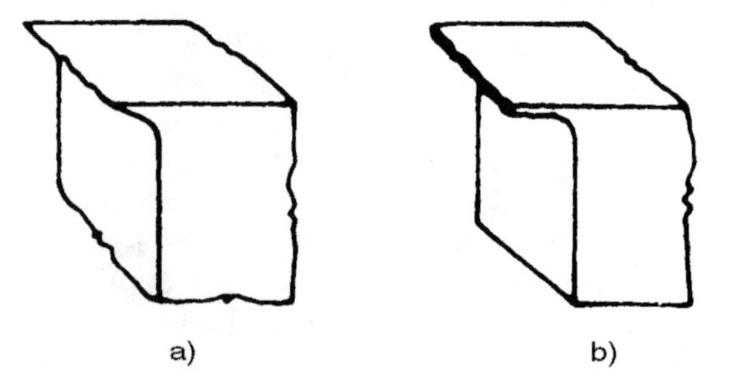

Fig. 3K Typical burrs. View (a) shows the typical shape of the most common burrs. View (b) illustrates a more pronounced burr. (*From Bralla, Design for Manufacturability Handbook, McGraw-Hill, NY.*)

repeated back and forth for a number of cycles until the desired results are achieved. It is particularly applicable to removing burrs in places that are not accessible to more conventional deburring methods. Surfaces can also be refined by this method. Ultrasonic motion from a graphite tool is sometimes used to provide more aggressive abrasive action. An orbiting tool may provide similar benefits. Abrasive flow deburring is illustrated in Fig. 3K1.

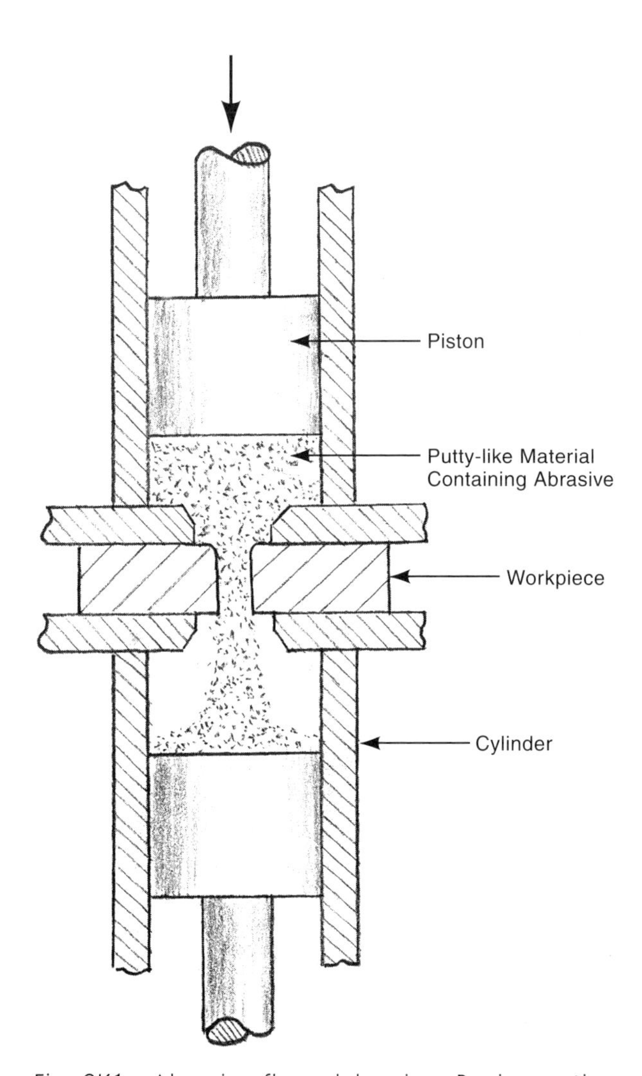

Fig. 3K1 Abrasive flow deburring. Reciprocating movement of the pistons forces the putty-like material containing abrasive particles, back and forth against the workpiece, removing burrs and smoothing the surface.

K2. ***abrasive jet deburring*** - uses a high-velocity stream of abrasive particles or glass beads that are directed against the burrs. The burrs are either broken off, peened over by the force of the abrasive stream, or are worn away. The abrasive particles can be carried by a gas stream, or mechanically thrown against the workpiece. Plastic and rubber workpieces may be cryogenically cooled just prior to the operation. The low temperature makes these materials more brittle and facilitates the fracture of the burrs. (See C9.)

K3. ***barrel tumbling*** - A group of parts to be deburred is placed in a rotating barrel that contains abrasive powder, small, pebble-size stones, and water. As the barrel rotates, the mixture continually slides to the lowest point. During this sliding, the breaking and peening action of the stones, combined with the cutting action of the abrasive, removes the burrs. (Also see 8B2.) This method is economical since workpieces do not have to be handled individually and little operator attention is required during the operation. Fig. 3K3 illustrates the process. Barrels rotating on a horizontal axis are most suitable for deburring. Machined parts, stampings, small sand-mold castings, die castings, forgings, and powder metal parts are all deburred with this approach. Plastic, glass, and rubber parts

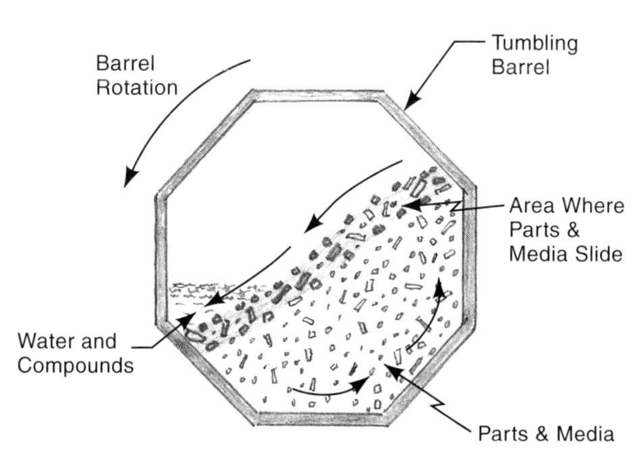

Fig. 3K3 barrel tumbling - As the barrel rotates, about 25 percent of the mixture of parts and media form a layer that slides downward, providing abrasion that breaks off and wears down burrs. The operation is slow but labor costs are low because a number of barrels can run at the same time with almost no operator attention.

can also be processed. After tumbling, the parts and media are separated by screening.

K4. ***chemical deburring*** - uses the same process as that used in chemical machining (See S.). A buffered-acid solution dissolves workpiece material including, and especially, burrs. Large groups of parts can be deburred simultaneously by immersion in the chemical solution, making the method economical. However, since the chemical solution acts on all surfaces of the workpieces immersed in it, critical-dimension surfaces may have to be masked. The process can be used for both ferrous and non-ferrous metals. Stampings and machined parts are commonly processed.

K5. ***electropolish deburring*** - is an application of electropolishing (8B3). Electrolytic action removes stock from all surfaces, especially the burrs, which are more prominent and draw a more dense electric current. Machined parts and stampings, including stamped gears, are deburred by electropolishing. Thin or fragile sections can be processed since there are no forces exerted on the workpiece by the operation. Excellent surface finishes can result from the process. It is limited to electrically-conductive workpiece materials.

K6. ***electrochemical deburring*** - is similar to electropolish deburring except that shaped electrodes are used. As such, the process is a special application of electrochemical machining (See I2.). The electrodes are positioned to concentrate the electrolytic action at the burrs. Surface residue is removed after the operation. The method is particularly suitable for removing burrs from inside surfaces that are not accessible for other methods. As long as an electrode can be positioned near the burr, the electrochemical dissolution will take place.

K7. ***liquid hone deburring*** - uses an abrasive (typically 60 grit) suspended in water. The mixture is then forced over the burred edges of the workpiece. The method is applicable to very fine burrs and very little edge radiusing takes place.

K8. ***manual deburring*** - is hand deburring with any of a variety of tools such as scrapers, knives, files, and emery cloth, to cut burrs from the workpiece. Powered rotary files or burrs, belt sanders, or other powered hand tools may be used. These methods are appropriate for small-quantity production where it does not pay to establish a more automatic approach, for burrs inaccessible to automatic devices, for burrs of variable size, and for fragile parts.

K9. ***thermal energy deburring*** - involves placing the workpiece in a closed container into which a charge of natural gas is introduced. The gas is ignited and the high temperature wavefront thus generated burns off and vaporizes the burrs. The workpiece is not significantly heated but the thin, prominent burrs are subjected to an instant of very high temperature. Surfaces of the workpiece are not dimensionally changed, or otherwise affected by the operation, though the material where the burrs were located may have a heat-affected zone. The method is applicable to ferrous, nonferrous and some plastic workpieces. Plastic moldings can have thin flashing removed by this process. One common application is the deflashing of zinc die-cast carburetor bodies.

K10. ***ultrasonic deburring*** - is used to remove minute burrs such as those developed during honing operations. A mixture of buffered acid and fine abrasive powder in a tank is agitated by ultrasonic vibration. This causes the burr to be both abraded and chemically dissolved. (The process is actually closer to ultrasonic cleaning - described in section 8A2b - than the ultrasonic machining described above, because the vibration is over a wide area in the tank instead of immediately below a particular tool.) A delicate balance must be achieved between three factors: burr size, acid, and abrasive, if the process is to give best results.

K11. ***vibratory deburring*** - shown in Fig. 3K11, is similar to barrel tumbling except that the agitation is produced by mechanical vibration rather than tumbling action, and the vibrations provide a somewhat more aggressive cutting action. (See 8B2.) Large quantities of parts can be processed at one

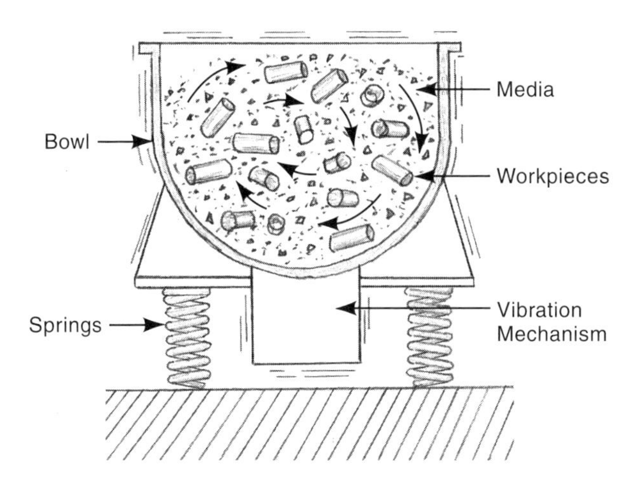

Fig. 3K11 Vibratory deburring. An eccentrically-mounted weight on a rotating shaft provides the vibration.

time. The method is faster than conventional barrel tumbling but not necessarily more economical since both processes proceed automatically once the parts are loaded.

K12. ***water-jet deburring*** - The force of a jet of water of high velocity is used to break burrs from the workpiece. The process can be used for all metals, with higher pressures used for harder metals. For hard steel, deburring pressures in the area of 40,000 psi (275 MPa) are used. (See M, hydrodynamic machining.) The process is also useful for removing loose chips and dirt from machined parts, and some CNC milling machines are equipped with water jets for that purpose.

K13. ***centrifugal barrel tumbling*** - is similar to barrel tumbling except that centrifugal force is used to increase the cutting and breaking forces. In this process, the tumbling barrel is placed at the end of a rotating arm. The rotation provides centrifugal force that may be as much as 25 times the force of gravity, greatly accelerating the deburring action.

K14. ***spindle finishing*** - is a variation of barrel tumbling. Instead of being placed loosely in the barrel along with the media and other parts, the part to be finished is clamped to a spindle and is then immersed in the moving media where it is rotated. Deburring action on the part then be quite rapid, although only one part is normally deburred at a time. By a suitable design of the holding fixture for the workpiece, the abrasive action can be concentrated where it is needed and other areas can be protected. The process can provide improved surface finishes. Although the operation is effective, labor costs may be higher than with other methods unless loading and unloading are automated. The operation is frequently used for finishing cylindrical automotive and appliance parts.

K15. ***powered brush deburring*** - uses the force of a powered, rotating wire brush to break off and remove burrs from the workpiece. Machined gears are often deburred with this method, which, with the proper brushes, can also round sharp corners. Cleaning, descaling, and surface finish improvement can also be accomplished. Splines, tubing, screw threads, and stampings are also deburred by this method. It is also used with nylon-abrasive brushes. These are brushes of nylon filaments that are impregnated with abrasive grits, and are useful for removing small burrs, for radiusing corners, and improving surface finish.

K16. ***abrasive belt sanding*** - is a manually-controlled operation aided with belt sanders or flap wheels. For a manual operation, high output rates are possible. The workpiece is held in such a way that the abrasive action of the belt is directed against the burr to be removed. The process can remove sizable burrs but can also produce a very small burr itself. Flashing from castings, forgings, and cold heading can be removed. Flap wheels provide flexibility needed to reach depressed areas. De-scaling and cleaning are also accomplished with these wheels.

K17. ***laser deburring*** - utilizes laser energy, directed by CNC equipment, to remove burrs. (See O.)

K18. ***plasma glow deburring*** - is a method used on plastics. Heat from a low-temperature argon-oxygen plasma removes flashing and thin burrs.

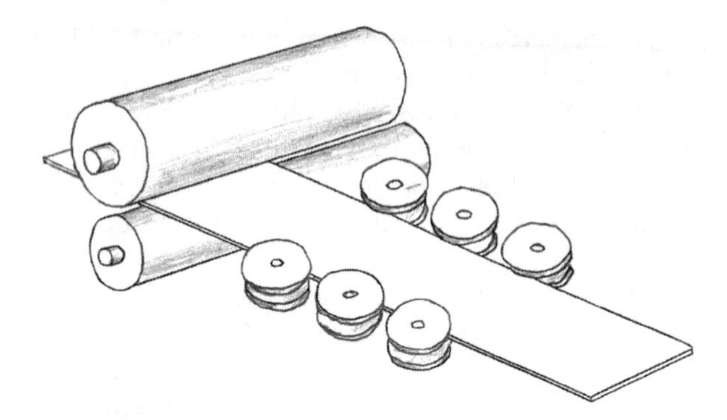

Fig. 3K20 Edge rolling uses pressure rollers at the edges of a strip of sheet metal to force burrs down against the sheet after slitting or shearing operations.

K19. ***skiving*** - is used when sheet metal is slit. A cutting tool is set to scrape off burrs as the strip passes the tool. The operation is thus automatic and does not affect other workpiece surfaces.

K20. ***edge rolling*** - is another method used when sheet metal is slit. Instead of scraping off the burr, it is pressed down against the sheet by the force of metal rollers, which are set to bear against the slit edge. The method is illustrated in Fig. 3K20.

K21. ***burnish deburring*** - is similar to edge rolling except that the slit edge is run against a smooth metal surface instead of a roller. The burr is broken off or pressed down, and the edge is smoothed.

K22. ***edge coining*** - is used with stamped sheet metal parts. Press tooling, designed to fit the part, coins the burred edges, forcing the burr into the base metal.

K23. ***robotic deburring*** - is an automatic, robotic version of manual deburring. The deburring tool, a file, rotating brush, or other tool, is manipulated by a robotic arm that is programmed to maintain the tool in contact with the burr-containing edge. If the tool used is compliant, as a brush, exact location of the tool is not critical and robotic control is relatively easy. Sometimes the robot is programmed to handle the workpiece rather than the deburring tool, picking up and moving the workpiece against a brush, bur, or abrasive tool.

K24. ***CNC machining center deburring*** - Tool-changing machining centers can be programmed and equipped to use a rotating wire brush instead of a rotating cutting tool. The machined moves the brush along the burred edges of the part.

K25. ***cryogenic deflashing*** - Plastic and rubber workpieces can have molding flash removed by tumbling them in either a rotating barrel or vibratory equipment at cryogenic temperatures. {The temperature used depends on the workpiece material but ranges from about −210 to −300°F (−135 to −180°C.)} Liquid nitrogen is introduced to the tumbling chamber to provide the required low temperature. The entire workpiece tends to be chilled to the cryogenic temperature but the thin flash, made brittle at the low temperature, is more vulnerable and it tends to break off from the workpiece. Flash formed at the junction of the mold halves, from gates, and from mold cores or ejector pins, are all removed by this method. Suitable media: steel balls, ceramic cylinders, walnut shells, nails, etc. and, sometimes, only the workpieces themselves, provide the striking force that breaks the flash. In another approach, the workpieces can be cryogenically cooled and placed in an abrasive stream that breaks off and abrades the burrs.

L. Filing

Filing involves the removal of material from a workpiece with a file, a tool that has cutting teeth arranged in succession along a surface. A file is somewhat like a saw that is very wide, wide enough to cut a surface rather than a slit. The cutting teeth of files are typically quite small, providing slow and easily-controlled cutting action. A variety of cutting tooth arrangements are available, the choice of which depends on the use to which the file will be put. Files are usually used to remove only a small amount of material from workpieces. Filing is most often a completely manual operation with the motion and force of the file provided by hand. However, filing machines are available that provide either a reciprocation motion like that of a jig saw, a continuous motion similar to that of a band saw, or a continuous motion provided by a rotating disc-shaped file. With these machines, the workpiece is still manually held and controlled. Filing is performed for many purposes including burr or flash removal, size and shape adjustments, and surface smoothing. A skilled machinist can make accurate parts by filing in conjunction with other machining operations, though such an approach involves high labor costs.

L1. ***burring (rotary filing)*** - Burrs and rotary files are small filing tools of round cross-section that are power-rotated. They are normally utilized with hand control of portable electric, pneumatic, or flexible-shaft-driven tools. They are used for such operations as burr and flash removal, chamfering corners, elongating holes and slots, and making other minor shape changes in workpieces. Fig. 3L1 illustrates several typical burrs and rotary files.

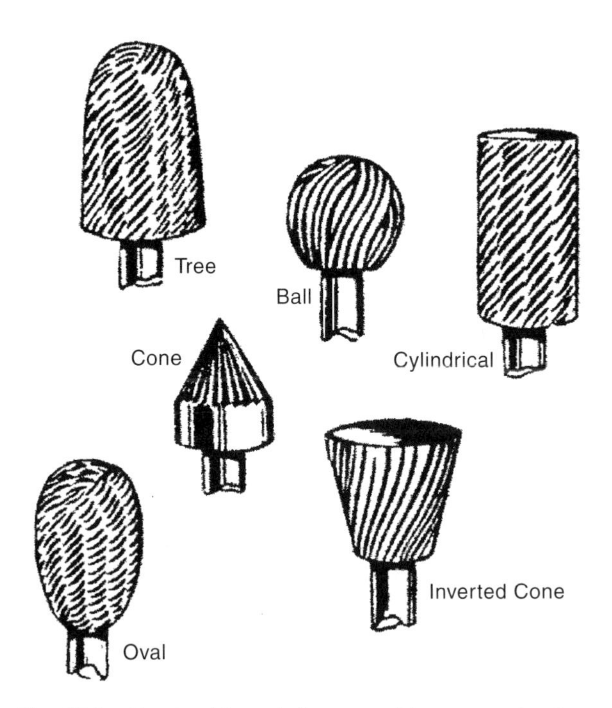

Fig. 3L1 Typical head shapes of burrs and rotary files. (Both rotary files and burrs have cutting teeth. The teeth of rotary files are made by upsetting the tool surface with sharp forming tools; the teeth on burrs are made by grinding.) *(from LeGrand, American Machinists Handbook, 1955, McGraw-Hill Companies. Reproduced with permission.)*

M. Water Jet Machining (Hydrodynamic Machining)

This process uses a narrow, high-velocity jet of liquid as a cutting agent. The jet travels at a speed of up to 3000 ft/s (900 m/s) and is from 0.002 to 0.040 in (0.05 to 1.0 mm) wide. The liquid is primarily water, but polyethylene oxide or other polymers may be added to keep the stream coherent. Cutting occurs where the jet strikes the work material. Although thin soft metals can be cut with the process, it is best adapted to non-metallic materials such as wood, rubber, plastics, fabric, gypsum board, leather, acoustic tile, paperboard, and various food products. The process is mostly used to cut out parts from materials in sheet form or to slit web materials. Cutting disposable diapers is a significant application, as are gasket, shoe sole, and carpet cutting. Water jet is also used for wire stripping, cutting of foods, separating printed circuit boards, paint stripping, and cleaning. Noise from the jet is a disadvantage of the process which is illustrated in Fig. 3M.

M1. ***abrasive water jet machining (AWJ)*** - adds abrasive particles to the water jet to aid the cutting action. The abrasive is added after the stream has left the orifice. This enables the process to be used for cutting of a wide range of ferrous and non-ferrous metals and non-metallics. Because the process is sensitive to variations in process parameters such as the type and amount of abrasive, water pressure and flow rates, tool traverse rate, and material thickness, automatic computer control is vital. Cutting is normally carried out under water, which eliminates objectionable noise that would otherwise accompany the process. Many but not all composite materials can be cut with the process without delamination of the material. The process is also used to blank parts from sheet and plate materials. Materials from 1/16 to 3 in (1.6 to 75 mm) thick are

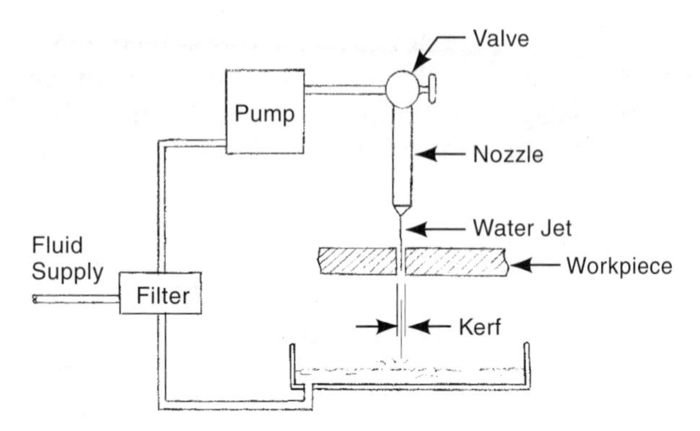

Fig. 3M A schematic illustration of water jet (hydrodynamic) machining.

commonly cut. Removal of sprues, risers, and gates from castings, other trimming of castings, forgings, and other parts and beveling of edges prior to welding are other applications. The process is slower but more accurate than plasma cutting and much faster but not as accurate as wire-EDM cutting. There is no heat affected zone and the cut edge is smooth. Fig. 3M1 illustrates the process. Some plastics that give off toxic fumes when heated, can be cut with this method without ill-effects. Aluminum, which can give problems with laser cutting because of its reflective surface, can also be cut advantageously. Stainless steel, tool steel, Inconel, brass, titanium, glass, ceramics, marble, and carbon fiber reinforced materials are also cut with abrasive water jet.

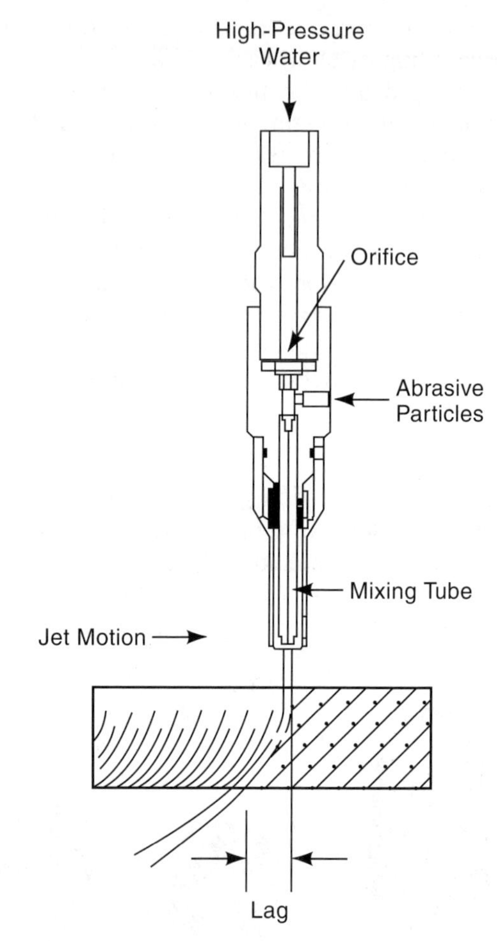

Fig. 3M1 Schematic view of abrasive water jet machining. (*Courtesy Omax Corporation.*)

N. Electron Beam Machining (EBM)

EBM is essentially the same process as the more common electron beam welding, except that the beam is used to cut instead of to fuse. The beam size, power, and dwell time are set to provide a cutting action. Magnetic coils focus and direct the high-energy electron beam. The electrons striking the workpiece melt and vaporize the workpiece material. Any material can be processed. Very narrow slits and small holes can be machined, as narrow as 0.0005 to 0.001 in (0.013 to 0.025 mm). Depth-to-diameter ratios of 100 to 1 are possible. Holes can be drilled in steel up to about 0.3 in (7.5 mm) thick. When holes and slits are machined, a backing material is placed under the workpiece. When the beam penetrates the workpiece, this backing material vaporizes, expelling the molten workpiece material from the hole. The process takes place in a vacuum of 10^{-5} mm/Hg.[2] The machining part of the operation is very rapid but the part must be placed in a sealed chamber and a vacuum must be drawn. The size of the workpiece that can be machined is limited by the size of the chamber. Equipment costs for this process are high and there must be protection against the x-rays that are generated when the electron beam strikes the workpiece. The operation

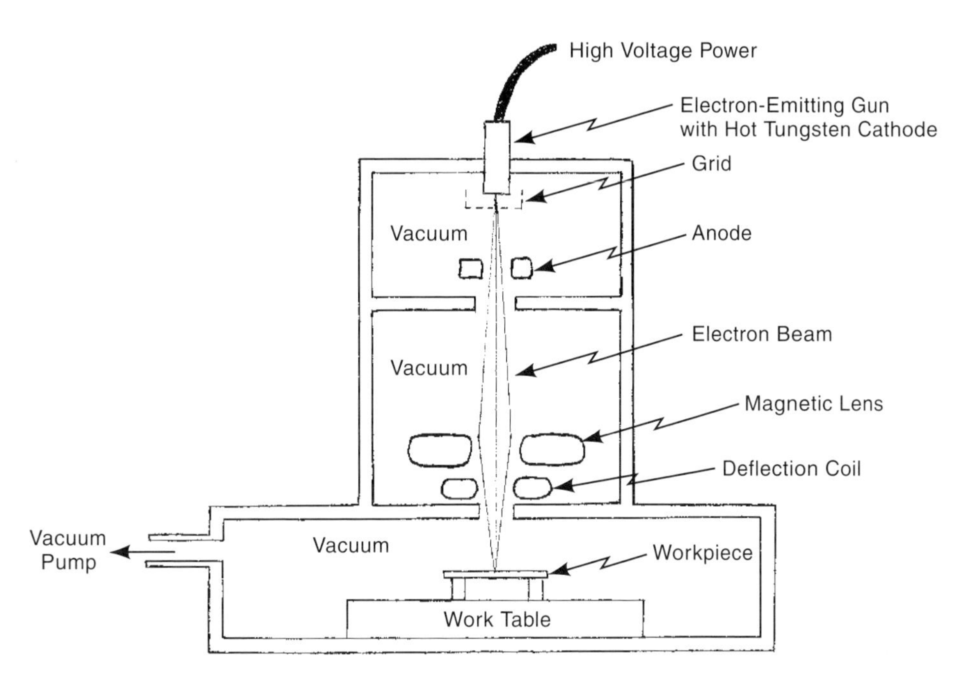

Fig. 3N Schematic illustration of electron beam machining. A high speed stream of electrons, emitted from a cathode, is accelerated and focused. It strikes the workpiece and heats and vaporizes the material at the point of contact. A vacuum chamber is necessary to generate and focus the beam and the workpiece is normally also in a vacuum. The workpiece table is controlled to be movable in two directions.

is normally computer controlled. There is a small heat-affected zone and a thin recast layer at the cut. Machining of semiconductors and sapphire bearings are two commercial applications. Filters and screens are made by drilling multiple holes in sheet material.

O. Laser Beam Machining

The laser beam, used for welding, is also adaptable to machining (cutting) operations. The heat generated by a powerful beam of coherent light, melts and vaporizes workpiece material. The process is best suited for drilling very small [0.005 in (0.13mm)] holes, but is also used increasingly for cutting flat stock. Depth-to-diameter ratios of 10 are feasible in laser-machined holes.[2] The process works best with materials less than 0.2 in (5 mm) thick but can be used with greatly reduced cutting speeds with metals up to about 0.5 in (13 mm) in thickness and non-metals up to about 1 in (25 mm) thick. With thin materials, the process is faster than mechanical cutting. Holes can also be drilled rapidly.

CO_2 or another gas may be blown from the nozzle to assist in removal of melted material from the cut. There is a thin recast layer and zone of heat effect where the cut takes place. The surface of the cut can be irregular, and there tends to be some taper in it. The depth of cut in blind holes and grooves is difficult to control. Typical applications of laser cutting are the machining of carbides and diamond drawing dies. Scribing, engraving, perforating, slitting, trimming and deburring are other operations. Turning, threading and milling of difficult-to-machine materials are feasible. Machining silicon wafers and other electronics components is a common application.

The process has grown as a desirable method for cutting out sheet material components, particularly

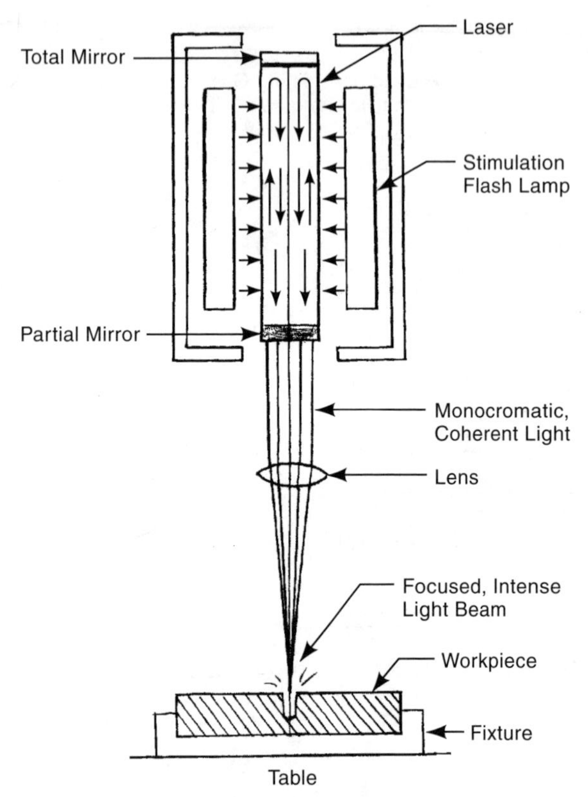

Fig. 3O Laser beam machining. The focused beam of coherent monochromatic light provides intense energy that melts and vaporizes workpiece material in a very small spot. By moving the beam and/or the workpiece, grooving, slitting, engraving, and other machining operations can be performed.

when quantities are too low to justify the expense of blanking dies. Steel, Inconel, Hastalloy, titanium, and stainless steel, and various plastics and other non-metallic materials are cut with the process. The degree of light reflectance of the material affects the type of laser that is best to use. Reflective metals such as copper, aluminum, gold, and silver do not work as well. Fig. 3O illustrates the process.

O1. ***laser-assisted hot machining (LAM)*** - uses laser processing combined with milling or turning to machine materials that may not be easy to machine by normal methods. The laser and cutter work together; laser energy heats the workpiece material locally at the point of cutting, softening the material and facilitating the cutting action. Tool wear is reduced substantially, cutting forces are lessened, and higher speeds can be used. Titanium, cast iron, steel, and silicon carbide ceramic have been cut experimentally with this approach.[7]

P. Shaping

Shaping is no longer a common machining operation, having been largely replaced by milling. Shaping produces flat or contoured surfaces by the reciprocating action of a single-point cutting tool that moves in a straight line. The stroke of the tool is normally horizontal. (Vertical shapers are called *slotters*. See section R below.) (Also see *gear shaping*.) The tool is hinged so that it doesn't affect the workpiece on the return stroke. The only movement of the workpiece is its gradual feed across the path of the cutter between strokes. The reciprocation motion and feed are automatic. If the vertical position of the tool is changed between strokes, slots, contours, gear teeth, or other shaped surfaces can be produced. Shapers are versatile machines, simple to set up, and use inexpensive cutting tools. They are still useful for toolroom and maintenance machining. Fig. 3P illustrates a typical horizontal shaper.

Q. Planing

Planing is similar to shaping except that the workpiece rather than the cutting tool has the reciprocating motion. The workpiece is clamped to a table that moves back and forth in a straight line against a single-point cutting tool. Planing is used for parts that are too large for it to be practical to machine them with a shaper. Machining ways of machine tools is an example. The tool is fed across the path of the work between strokes. Multiple tools can be used at one time and can be fed in different directions to produce several machined surfaces on the part at one time. Some planers have been equipped with milling cutters in place of the single point cutting tools in order to increase their

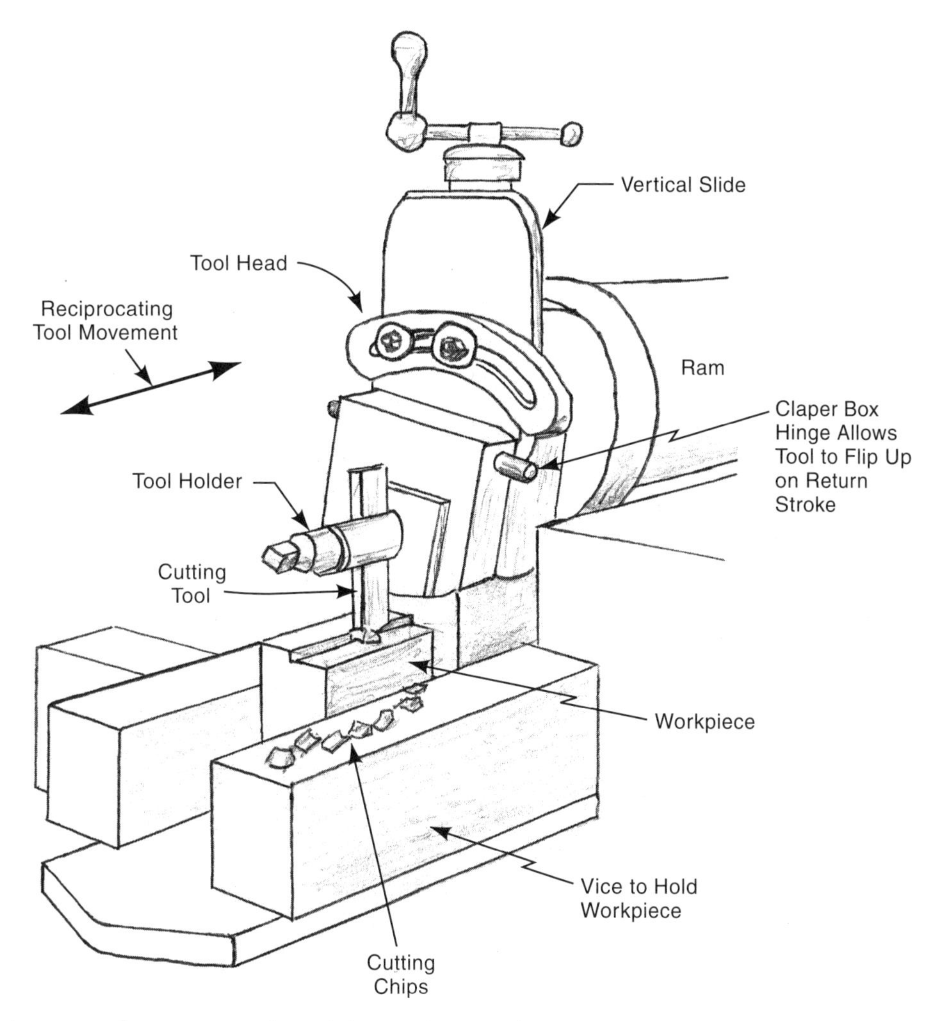

Fig. 3P A shaper makes repeated straight cutting strokes with a single point cutting tool to generate surfaces, slots, and other shapes in a workpiece.

metal removal rates, and are then called plano-milling machines or planer-mills.

R. Slotting

This process is essentially the same as shaping except that the movement of the cutter is most often vertical, and the process is usually used to make a keyway or other slot in the part, rather than a flat or contoured surface. Gear shapers are special machines similar to slotters, used for machining spur and helical gears.

S. Chemical Machining

In all chemical machining processes, metal is removed by the etching action of an alkaline or acid solution that is in contact with exposed surfaces of the workpiece or blank material. There is no external electrical circuit. There are several steps in the process: 1) Metals to be processed are cleaned of oil, dirt, scale, and other contaminants that will interfere with the exposure of the surface to the etching solution. Alkaline cleaning, solvent wiping, and vapor degreasing are the most common cleaning methods. 2) Areas not to be chemically machined may be masked to prevent exposure to the chemical agent. 3) The workpiece is etched (machined) by contact with the chemical solution, usually by immersion. Circulation of the solution or agitation of the workpiece are necessary to ensure uniform metal removal rates. 4) The mask is removed, the workpiece is cleaned of the etching solution, rinsed, and dried.

Aluminum, steel, titanium, and magnesium are the most common materials chemically machined but the process is applicable to almost any metal. Honeycomb structures and other delicate parts can be processed because there are no cutting forces. The process has also been used to salvage oversize parts. It creates no residual stresses in the workpiece. Equipment costs are lower than for conventional machining.

S1. ***chemical milling*** - is the term used to designate chemical machining when the purpose of the operation is to change the shape or some dimensions of the part, or to remove a significant amount of material. Maskants are normally used to control the location of the stock removal. Usually the entire workpiece is covered by the maskant by spraying or dipping. The maskant is then removed from the areas to be machined. Machining takes place when the workpiece is immersed in the chemical solution. Agitation of the solution ensures that it makes equal contact with all surfaces of the workpiece. The depth of cut is controlled by regulating the immersion time in the solution. The process is most suitable for shallow cuts, which can take place over large areas. One common area of application is to machine pockets or cavities in aerospace parts for weight reduction. Other applications are tapering, removal of undesirable surfaces, and overall size reduction of workpieces. Aluminum is the material most commonly chemically milled. Advantages of the process are the fact that it can be applied to hardened or difficult-to-machine metals, and difficult-to-machine shapes, and no burrs or internal stresses are produced. About 0.5 in (13 mm) is about the maximum depth of cut. Weld joints and other non-homogeneous material structures may yield satisfactory results.

S2. ***chemical engraving*** - is the process name used when the purpose of the operation is to make patterned surface depressions in the workpiece so as to produce lettering, other nomenclature, figures, or decorations. The pattern can be either depressed or raised. Masks are applied either as a photoresist or by screen printing. A variety of metals can be processed including stainless steels. Brass, aluminum, and copper are also commonly treated. Chemical engraving of printing plates is one of the earliest applications of chemical machining. The making of nameplates and instrument panels are other applications. Depressed patterns may be filled with contrasting paint for improved readability. The process is more suitable than pantograph machine engraving when fine detail is required.

S3. ***chemical blanking*** - uses chemical machining to create blanks from sheet material. It is most often used to make parts from thin stock using the photochemical method for configuring the maskant. Manual methods of scribing and stripping the maskant from around the desired part, are also feasible for larger, simpler parts, and those requiring less stringent dimensional accuracy. Manual methods can also be used when a photoresist mask may not have adequate chemical resistance to the solution used, especially when quantities are low and the cost of making a blanking die is not economically justifiable. The mask can be applied by screen printing or offset printing. It is normally duplicated on both sides of the sheet in accurate register so that machining can take place from both sides at the same time.

Etching takes place either by immersing the workpiece in the chemical solution or by spraying the solution on the workpiece. Typical components produced by chemical blanking include electric motor laminations, templates, magnetic recording heads, disk springs, and gaskets.

S4. ***photochemical blanking*** - is illustrated step-by-step in Fig. 3S4. This process uses a photosensitive resist, which is a light-sensitive material, as the maskant. in chemical machining. The resist material is applied by dipping, spraying, roller coating, or flow-coating. The material is hardened in the areas wanted by exposing it to light through a photographic negative whose image corresponds to the shape of the part to be produced. This photo negative is prepared in advance of the operation.

The usual procedure is to draw the blanked part, greatly enlarged, photograph it, and create a negative the same size as the part to be produced.

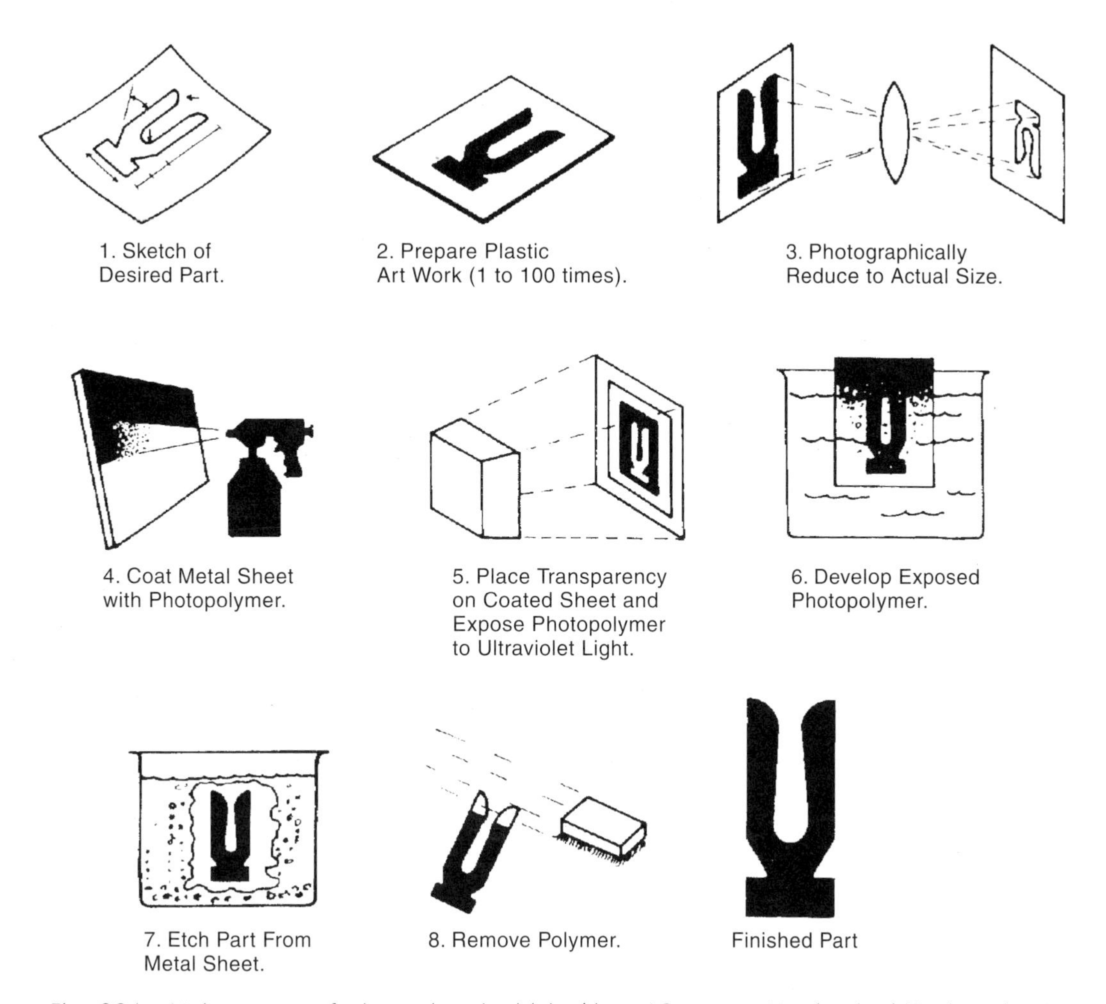

Fig. 3S4 Major steps of photochemical blanking. (*Courtesy Mechanical Engineering.*)

The negative is attached to the workpiece after the workpiece has been coated with the masking material and it has dried. Normally, for blanking, a similar negative is attached to the workpiece on the opposite side but in precise registration with the first one. Blue light shone through the negatives hardens (polymerizes) the resist material in the areas exposed to the light. The negatives are removed and the workpiece is processed to remove the unhardened portions of the maskant. The workpiece is then immersed in the chemical reagent, that removes the workpiece material that is not masked. The process is particularly adapted to small, complex, and precise parts, made from very thin materials that are not suitable or feasible for conventional blanking. Photochemically blanked parts are burr free. The electronics industry is an extensive user of photochemical blanking and machining in the production of integrated circuits, circuit boards, and other components. Shadow masks for television sets and screens for various purposes are made with the process.

T. Machining Centers

Machining centers are machines that can perform varied machining operations in sequence, automatically, under computer numerical control and in only one set-up. Milling (including contour milling), drilling, boring, reaming, and tapping are common operations performed by such centers, which are outgrowths of the conventional milling machine. The keys to their operation are twofold; 1) The machines can be programmed for operation by computer numerical control (See below.) to make whatever cuts the workpiece requires with preselected feeds and speeds. 2) They include automatic tool changers that permit the use of a wide variety of cutters, each suited to the operation they perform. The tool changers store whatever tools are needed for the complete series of operations on the part and insert and remove them from the machine spindle as needed. Fig. 3T illustrates a well-equipped horizontal machining center.

Horizontal-spindle machines are most common, but vertical-spindle machines are also available. Machines designed for larger parts may have a traveling spindle column instead of a traveling table. Current state-of-the-art machines have five or more axes and five kinds of controlled movement of either the workpiece or tool position and the feed. The axes controlled may vary with the machine's purpose and design, but can include table movement in and out and side to side, table rotation, spindle head movement up, down, and rotative, and

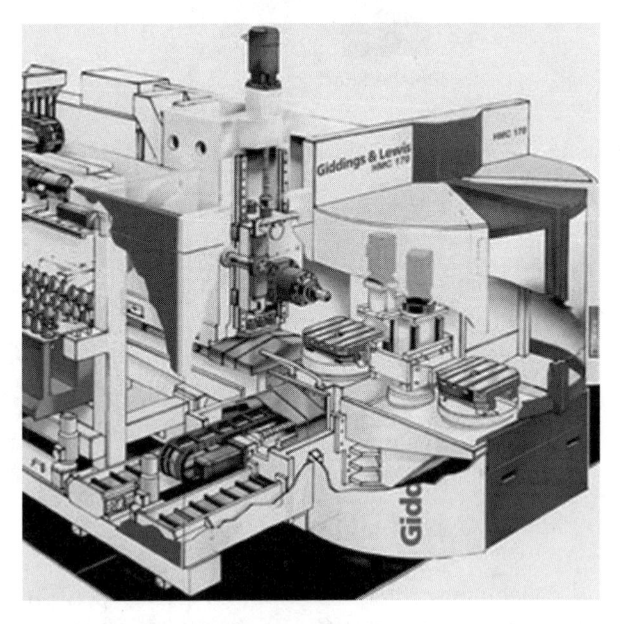

Fig. 3T A horizontal machining center. This CNC machine has a large-capacity tool changer and dual work tables so that one workpiece can be machined while another is unloaded or loaded. (*Courtesy Giddings and Lewis Machine Tools.*)

axial movement of the spindle. Fig. 3U illustrates one example of 5-axis control. Tool rotational speed, tool feed, tool dimensional offsets, coolant flow and tool changing can all also be performed under the computer control. For example, table and tool movements can be at the slow speeds needed for feeding the cut, or can involve rapid traverse to new cutting positions; the spindle can stop, start, and change speeds, as appropriate for the operation; and cutting fluid can be turned on and off, as needed. Operations can be performed on different sides of the workpiece in sequence, because the table can move or rotate to present a new surface to the cutters. All these operations are automatic. The machines are engineered to provide precise, close-tolerance positioning of the cutters and workpiece. They are particularly applicable to low- and moderate quantity production that otherwise would require a sizable number of separate machining operations on single-purpose equipment. Since machining centers operate automatically and don't require nearly as much operator skill as is needed to operate traditional single-purpose equipment, labor costs are low. However, programming requires special skills, and the investment required for the equipment is sizable.

Some machines are equipped with multiple work-positioning pallets and mechanisms to transfer the pallets to and from the machine table. The pallets and table are designed so that, when assembled, they are in a fixed relationship with one another. Given sufficient memory in the CNC system, as many as 20 or more pallets can be loaded with workpieces, that may or may not be identical. The workpieces are then processed automatically by the machining center, which loads each pallet, machines its workpiece in accordance with the program, and then discharges the pallet and workpiece. The machine then repeats the process, assembling another pallet to the table, machining its workpiece, discharging it, and so on. In this way, machines can be run on a night shift or at weekends, unattended, when power costs are usually lower. Machining centers equipped for this type of operation have sensors included that can detect tool wear, tool breakage, or some other problem that would render the machined part defective. When such a problem is encountered, an alarm may be sounded and the machine stopped automatically.

The types of parts most advantageously processed on machining centers are those that are machined on several surfaces and those with a large number of operations, particularly those made in smaller quantities, where special-purpose equipment is not justifiable economically.

T1. ***turning centers*** - While the typical machining center was based on earlier boring mills or milling machines, the turning center has its roots in the turret lathe. Turning centers differ from turret lathes in having automatic tool changers that can remove and install cutting tools in the machine's turret, in having the tailstock replaced by a live powered spindle and chuck, and in having NC or CNC control. These machines provide the same benefits and have similar applications to those of machining centers but deal primarily with parts that have surfaces of revolution.

T2. ***multiple operation machines*** - are part of a trend to provide more complex CNC machines that can perform the operations of both the turning center and the machining center. By adding additional cutter spindles and feeds, and additional tool changing capacity, both turning and milling as well as drilling, tapping, boring, and other capability, more operations can be performed on a workpiece without moving it to another machine. This capability can provide greater dimensional precision from cut to cut, less possibility of marring the workpiece during handling and storage between operations, and faster throughput times. As this trend continues, the distinction between milling, turning, and other operations on separate machines becomes more blurred. Such developments are made possible by the development of the controls and machine-element positioning devices that are part of CNC machining.

U. Numerical and Computer Control

These controls provide a means for operating machine tools and other equipment automatically. The represent a difference from earlier automation in that the machine elements are controlled by electronic pulses rather than mechanical devices. The pulses activate drive motors and other devices of the machine tool. Setting up a machine for a particular operation involves the entry of coded numerical data rather than the fabrication and installation of cams or other mechanical apparatus. The coded data are in the form of numbers, letters, and special symbols. The complete set of such data for any operation is called the *part program*. Data are stored on various magnetic media: magnetic tape, computer storage disks, etc. Perforated paper or "Mylar"plastic tape, widely used for data storage in earlier numerical control systems, is seldom seen any longer. The part program is stored on a CD-ROM or hard drive. When the program is read by the control unit, it is converted to a series of electrical pulses sent to the machine's motors. These electrical pulses provide the power to move the machine's spindle, tool slides, worktable, and other elements. They control the amount, direction, and speed of each moving machine element, as needed to carry out the operation.

There are three elements to a numerical control system: 1) the part program - a planned sequence of commands acted upon by the machine controller and then by the machine itself. 2) the controller - the numerical control system, computer-numerical control system, or programmable controller that reads the program and directs the machine tool, and 3) the machine tool, equipped with precise electrical power drives (servo motors or stepper motors) for the machine movements to be controlled. This approach provides great flexibility and ease of changeover so that automatic operation is economically feasible, even with one-of-a-kind or other limited production. There are various degrees of sophistication of this approach. They reflect developments as electronic devices and computers have become more powerful and more affordable. The development of numerical control has coincided with and has been an essential factor in the development of machining centers. The advantage of numerical control is that it enables the machine to run automatically in accordance with an optimized program; a machine operator is not needed to manipulate handwheels, levers, or other mechanical devices. If a number of parts are to be produced, all are made with a consistent machine sequence. With NC and CNC, automatic operation is affordable even with limited production quantities.

The simplest arrangement may have only two axes of movement and position to control, the longest table slide (usually the X-axis), and the machine

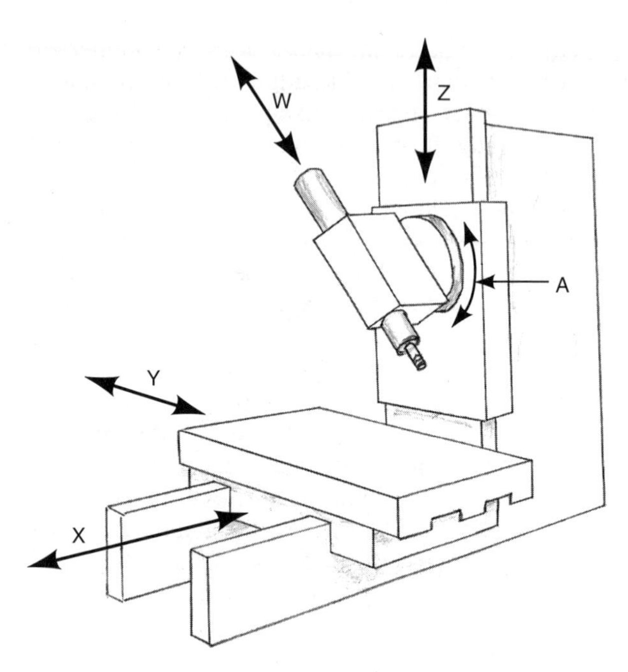

Fig. 3U A representation of the machine motions of a 5-axis milling machine.

table slide at right angles to it (usually the Y-axis). To these can be added the advance, feed and retraction of the machine spindle (the Z-axis) and the arrangement then would constitute three-axis control. In more elaborate systems, up and down movement of the worktable or machine head, rotation of the machine table, and the movement of the spindle mounting may all be added. Fig. 3U illustrates movements on a typical machine with 5-axis control.

Several degrees of sophistication of machine movement can be offered by the control system: *Point-to-point* positioning controls the position of the table but not necessarily its path from one location to another. This facility may be useful for hole-making operations, turret punching, and spot welding. Straight line control adds the control of the tool or table movement from one position to another so that straight milling cuts can be made. With earlier equipment, movement was usually only along the X or Y axis. Straight line motion where more than one axis is involved is typically called, *linear interpolation. Contouring* or *continuous-path* control, also called *circular interpolation*, allows curved surfaces to be machined, controlling both the direction and velocity of motion of machine elements.

Another important aspect is the degree to which information is fed back to the controller or computer during the operation. *Open-loop systems* do not have any feed back. The machine simply follows the instructions specified by the controller. This arrangement may be quite acceptable for simple two- or three-axis situations where tolerances are less stringent. *Closed loop systems* include sensors that transmit information back to the computer about the instantaneous location of the machine elements and the computer then may adjust the position of the machine element in accordance with that information. In more sophisticated installations, sensors may detect whether the tool has broken. They may measure the changing dimensions of the workpiece and adjust the cutting tool position accordingly, and may measure torque or actual cutting speeds. Compensations can be made to adjust for material variations, or dulling or wear of the cutters. This approach is sometimes referred to as *automatic adaptive control (AAC).* Fig. 3U-1 illustrates both closed-loop and open-loop systems schematically.

U1. ***NC, numerical control*** - is an earlier term for electronic control of machine tools. In the system, tool movements are instigated and controlled from numerical data that has been entered and stored. The numerical data are in the form of numbers, letters, and special symbols. The complete set of such data for any operation is called the *part program.* In the earliest systems, numerical data were stored in the form of holes punched in paper or Mylar plastic tape. The program was read by a tape reader/controller which then fed electrical pulses to the NC machine. Numerical processing power and sophistication were very limited. The term *numerical control* is still widely used to designate control systems that may be more sophisticated than earlier systems, but the more proper term for almost all current systems is *computer-numerical control, CNC.* Fig. 3U1 illustrates a typical traditional NC system.

U2. ***computer numerical control (CNC)*** - introduced in the 1970's, uses a computer in addition to a controller (and, more recently, in place of most of the functions of the machine control unit) as the heart of the system. Earlier NC systems used hard-wired controllers with a fixed logic. CNC,

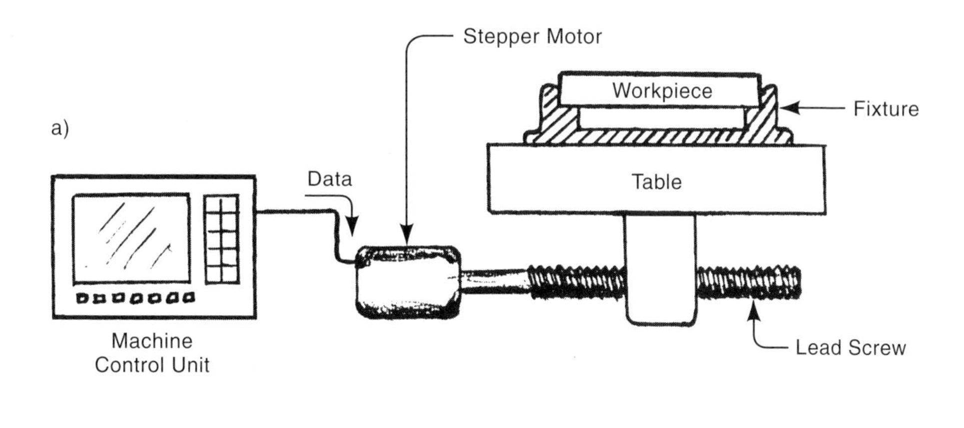

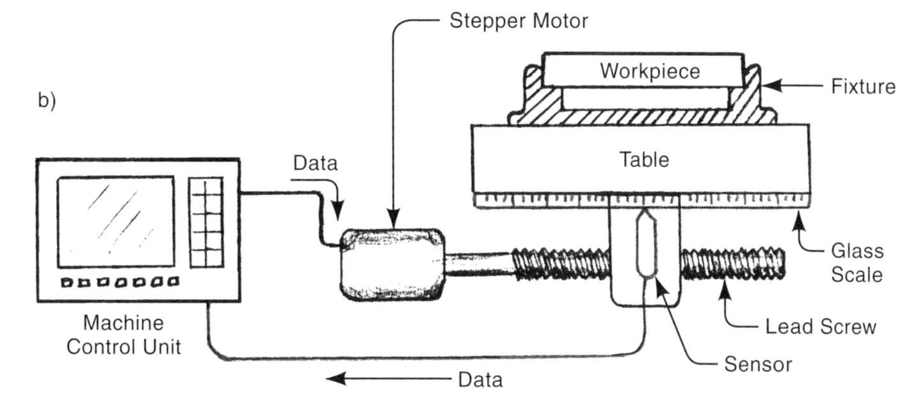

Fig. 3U-1 Open and closed loop control systems. An open-loop system in view a) shows data being transmitted to one of the machine tool's stepper motors causing it to move the machine table a stipulated amount. There is no feed back to verify the table's position. In view b) sensors detect the table's position with respect to a glass scale and this information is fed back to the machine control unit, which may send further signals to the stepper motor to adjust the position of the table. In some closed-loop systems, additional sensors may similarly transmit data of other conditions back to the machine control unit.

with a "soft" wired computer as part of the machine control unit (MCU), provides much greater flexibility and vastly increased functions in the control system. The MCU embodies circuitry that translates computer signals into movements of the machine elements. The computer simplifies the tasks of modifying and storing instructions for particular parts since the computer's data storage system rather than perforated tape can be used for this function. An alphanumeric keyboard provides a means of entering the program manually, and a visual monitor screen displays the program or an illustration of the movement path of the tooling on the workpiece. The addition of a computer to the system also increases the ability of the system to process feed-back information from sensors. This can greatly improve the accuracy and efficiency of the operation. With sophisticated programs involving feed-back and, increasingly, fuzzy logic and artificial intelligence, the computer can optimize the operation, achieving maximum cutting speeds and feeds and responding to problems revealed by sensors on the machine. Almost all CNC systems also provide the more usable contouring/continuous-path machining.

The sequence of steps in programming and operating a CNC machine are shown in Fig. 3U2 and are as follows: 1) The part specification, with any essential manufacturing notes, is prepared. 2) A NC part program for performing one operation (or several combined operations necessary to make the part) on a particular machine tool is prepared from

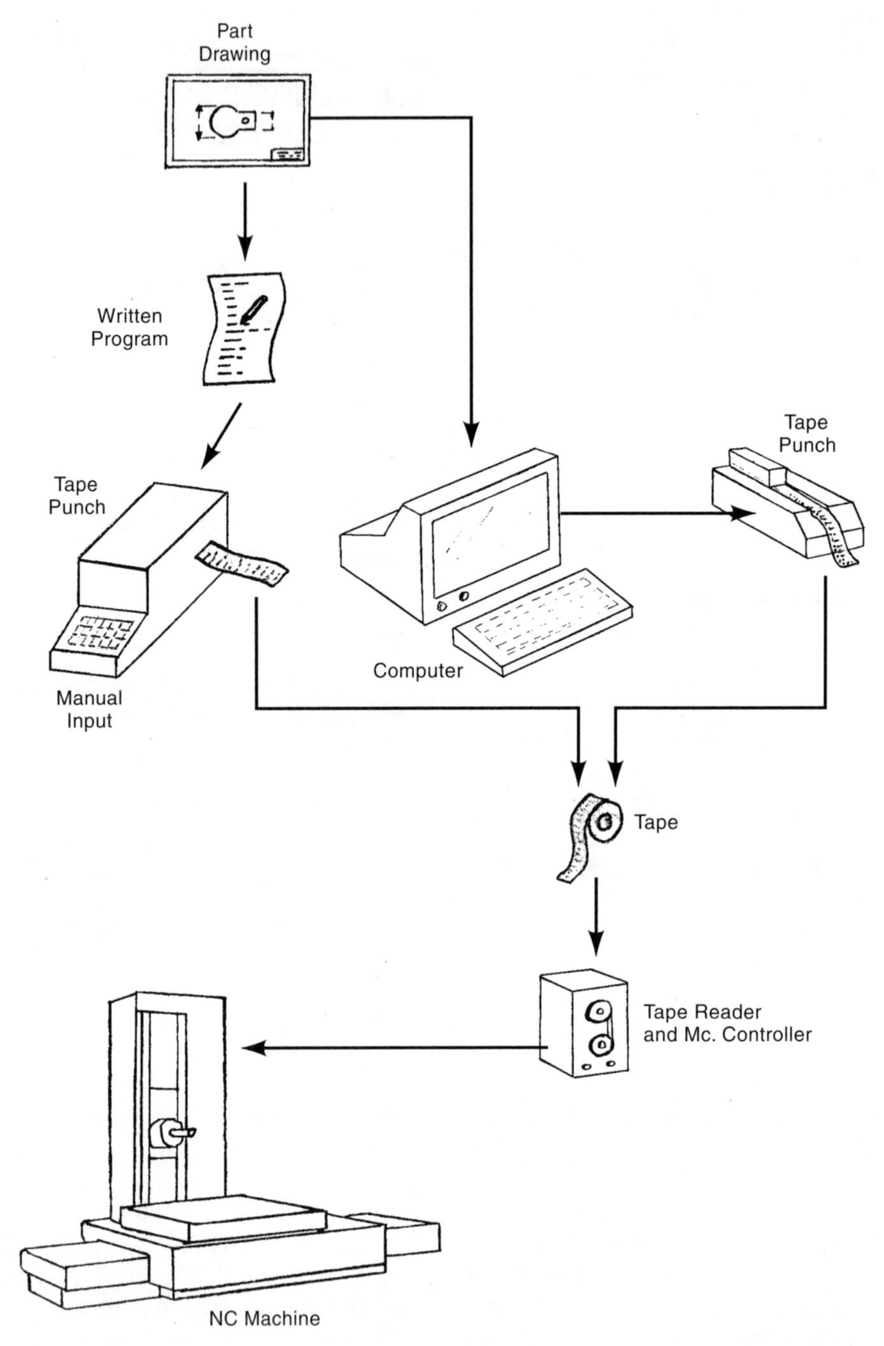

Fig. 3U1 The elements of a typical earlier NC system. The NC machine is controlled by perforated tape, which is punched with either a manual punch or a device connected to a personal computer.

the part specification. This work is typically done on a computer that can send the processing program for the part to the machine control unit (MCU) by one of a number of methods: a) directly by wire, b) in older systems, by means of a tape punching device connected to the computer and then a tape reader connected to the MCU, c) by magnetic tape which is then fed to the MCU, d) by

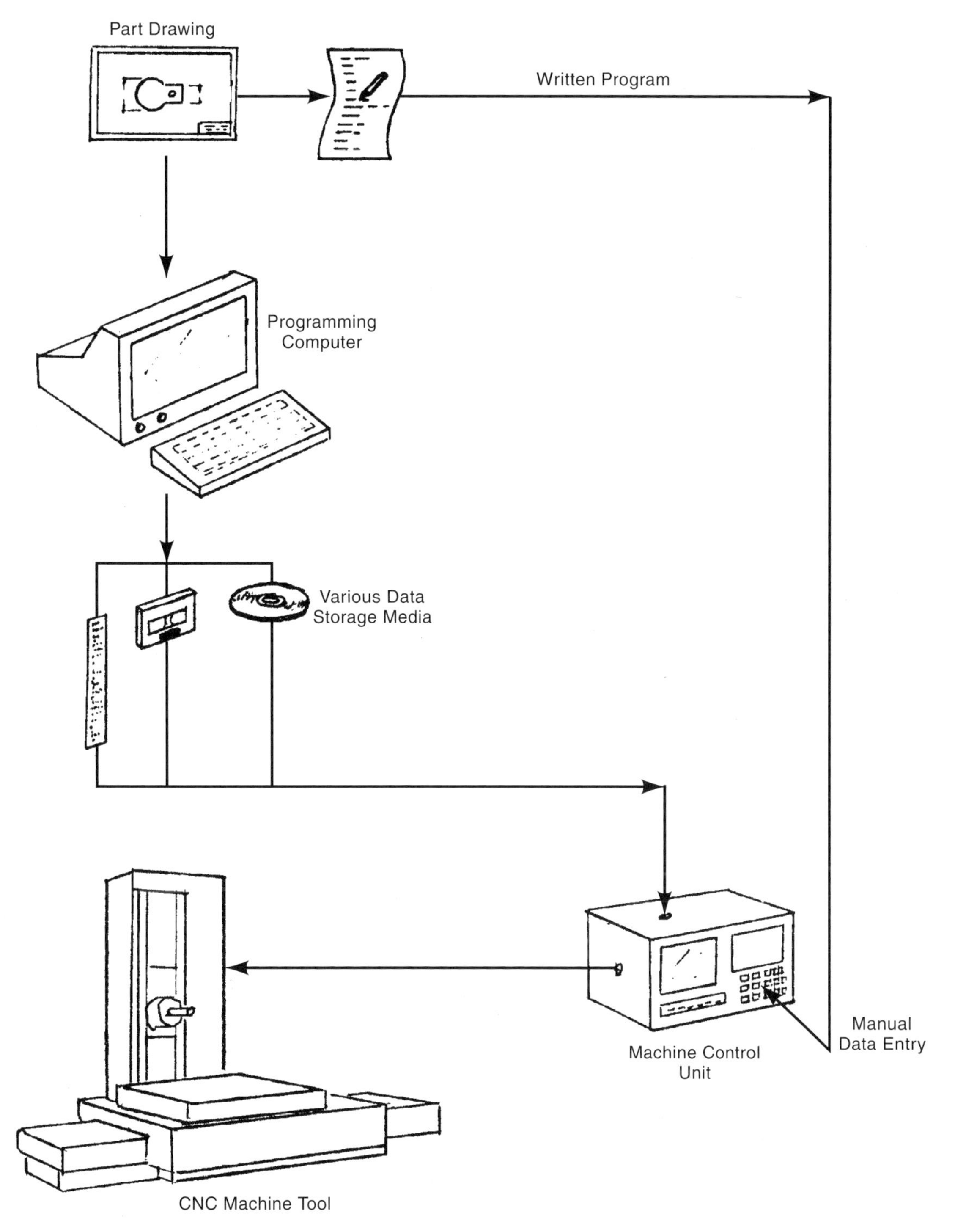

Fig. 3U2 A typical CNC system. The machine control unit (MCU) responds to a program entered on perforated tape, on one of a variety of magnetic storage media, or from a program manually entered directly into the MCU.

a computer disc or disc pack, which is inserted in a reader in the MCU. 3) The MCU reads the program and sends electrical pulses to the NC machine which, following the electrical pulses, performs the operational steps required.

U2a. ***DNC (direct numerical control)*** - is an earlier term for *computer numerical control, CNC.* It was used when the numerical program was fed to the controller directly from a computer rather than from punched tape. Direct numerical control now refers to the situation in which one central computer controls a number of machines at the same time, downloading the NC data as it is needed by each machine. *DNC* now also refers to *distributive numerical control,* a more complex and larger arrangement where one central host computer feeds programs to several satellite DNC computers, which, in turn, feed the programs to CNC units they control. Each of The machine tools each have a microprocessor that processes data from the central computer. The whole program may be downloaded at one time into the memory unit of each machine's controller or may be downloaded piecemeal as needed to keep ahead of the operation. The latter approach is referred to as "drip feed". The programs may reside in the central computer's memory system or may be loaded into the central computer from a floppy disc, punched tape, or other data storage device. Machines connected to a central computer in this way are part of a *local area network (LAN).*

U3. ***programmable controllers*** - This term dates from the advent of numerical control. The earliest programmable controllers were devices that were very simple, by present day standards, and were first used primarily in process equipment such as mixers, providing time and speed control for a number of steps of automatic operation. The term *programmable logic controller* (PLC) was also used. The terms then referred to machine tool controllers that were hard wired for the particular machine functions they would control. They accepted numerical data from punched tape but also had capability for program modification or entry from a keyboard. The term is still used to designate the electronic device at the machine tool that can perform the entire CNC function but which is now more commonly referred to as the *machine control unit (MCU).* As computers have replaced the hard-wired circuits, and computer power has become more economical, these units have been developed to perform more and more functions. Now, in addition to providing the basic program, programmable controllers include capability for closed loop control and adaptive control, utilizing feed-back from various sensors on the machine. Panel displays, machine interlocks, and other auxiliary control functions are incorporated. The basic program can come from the unit's own permanent storage, from a disc reader in the unit, or from a central computer.

U4. ***CAD-CAM - Computer Aided Design - Computer Aided Manufacturing*** - involve the creation of a part's design in digital form and then the translation of that design into a program for CNC operation of a machine tool. In the CAD phase, the designer develops the part's design by computer instead of by making a drawing on paper or film. CAD equipment includes a personal computer or workstation, a design program, and storage media for the completed design. The designer enters coordinates and dimensions with the computer keyboard, mouse, or light pen. The design is pictorially displayed on the computer monitor and can be printed, if desired. In the ideal situation, not yet realized, it would be possible to feed the computerized design automatically to a CAM computer or computer circuits, that would transform it into a control program for a machine tool. The program could then be fed to the machine tool to produce the part just designed. Thus, the task of programming the CNC system to make the part would be performed by the computer, and no human intervention would be required. Differences in machine control systems and between design objectives and manufacturing objectives make the complete achievement of this ideal quite difficult. Existing CAD-CAM systems simplify but do not eliminate the need for human participation in the programming of the machine tool except in some very simple 2-axis applications. In present practice, the production engineer operates a CAM computer, enters design data and specifies the sequence of operations, starting and ending points, intermediate dimensions, and process parameters such as feeds and speeds. The CAM program puts all these data in a form that can be read by the machine control unit. The CAM computer also may display tool movements, and the appearance of the workpiece at various stages of the operation. (A complete CAM system also may provide various administrative control information to monitor such items as work-in-process, cycle time, inventory, machine and labor efficiencies, and schedule compliance, etc.)

U5. ***digital readouts in machining*** - These are electronic devices that display for the machine operator the precise position of the machine table or other machine elements. As such, they are an aid to the manual operation of machine tools, simplifying for the operator the task of moving machine tables and cutters to the correct position for machining or other operations.

U6. ***automatic tracing*** - is used in some machining operations to control the position and movement of a cutting tool so that the operation can be performed automatically. Various devices are used. Tracing attachments for lathes and milling machines typically follow a fabricated template using electrical, pneumatic, or hydraulic systems. Contoured surfaces can thus be machined. These devices improve the productivity and versatility of some machine tools, because tracing often permits the machining of surfaces that would be difficult or impracticable to machine manually. The approach is common in the machining of injection molds and die casting dies. When parts are to be blanked from flat stock by flame cutting, routing, water jet, and abrasive water jet machining, optical tracing is sometimes employed. With this method, the cutter duplicates the path of lines on a drawing of the part's outline.

U7. ***robots and robotic operations*** - See chapter 14 of this handbook.

V. Trimming

Many manufacturing processes for shaping parts leave some unwanted material that must be removed from around the desired shape. Deep drawing, die casting, forging, plastic sheet molding, compression molding, reaction injection molding, and rotational molding are examples of processes that may leave flash or other excess material that must be removed. Various processes are used for such an operation, including the following: die cutting with dies like blanking dies but with suitable clearances for the part to be trimmed, pin routing (vertical milling), laser beam machining, abrasive jet machining, hydrodynamic (water jet) machining, and abrasive water jet machining.

W. High-Speed Machining

High speed machining is most common in milling operations but is also used in drilling and boring. The procedure is characterized by very high spindle speeds which, with normal feed rates per revolution, provide very high feed rates per minute. Surface cutting speeds may also be faster than normal. The method is heavily used in the aircraft industry to machine aluminum structural parts. Aluminum is the predominant material machined by high speed methods, although steel is also processed. One particularly attractive field of application is called *unitization*, the machining of a large, complex single part from a large material plate or block instead of assembling a series of small parts. The method requires careful balancing of tools and tool holders, CNC control of the operation, and more expensive machine tools with less-massive but high-powered spindles. Adaptive control systems are advisable. Spindle speed maximums range from 8000 to 40,000 rpm. Feed rates are as high as 800 ipm.

Another application of this approach is to the machining of dies and molds from tool steels. Small cutters are used, rotating at high rpm's but with normal surface speeds. The higher speeds permit more close passes to be made without time penalty, providing better precision and smoother contours, and greatly reducing the need for hand finishing benchwork, thereby significantly reducing cycle time and finishing costs.

X. Special Purpose Machines

When production quantities are very large, it is often advantageous to design and build machine tools that are dedicated to the production of specific parts. Such machines can provide both labor and quality advantages because the necessary precision can be built in, and the machines can incorporate spindles and tooling to perform several machining operations on the part simultaneously. Handling of the part is greatly reduced. Advantages of specialization can be gained and machines do not need apparatus and structure for a variety of parts. Examples of special purpose machines are those built to drill, bore, ream, or tap a number of holes simultaneously, to drill holes from more than one side, or to mill or broach one surface while performing another operation on another surface. Another method for accomplishing the same objectives is to construct a machine having an indexing table. The table rotates a fixed amount on a timed sequence and its position is accurately located after each movement. Several different machining heads can be located in fixed positions next to the table. The workpieces are positioned in fixtures on the table. After each operation, the table indexes to move each workpiece to the next station, with one station used for loading and unloading. Each time the machine indexes, one part is completed through a series of operations.

Y. Transfer Lines

Transfer lines move the workpieces automatically from one machine to another. In such arrangements, a whole series of machine tools are dedicated to the production of one part, or sometimes a family of similar parts. The workpieces are held on fixtured pallets and the transfer mechanism moves the pallets from machine to machine between operations. Devices on each machine automatically position and clamp the pallet and fixture to the precise location for the necessary operations. The machine tools at each station are automatic and are usually dedicated to specific operations on each part. Automotive engine block manufacturing is typically carried out on transfer machining lines with automatic equipment at each station.

References

1. *Manufacturing Processes and Systems, 9th ed.*, Phillip F. Ostwald, Jairo Munoz, John Wiley and Sons, New York, 1997.
2. *Materials and Processes in Manufacturing, 8th ed.*, E. Paul DeGarmo, J.T. Black, Ronald A. Kohser, Prentice Hall, Upper Saddle River, NJ, 1997.
3. *Metals Handbook, Volume 4, Forming, 8th ed.*, ASM International, Metals Park, OH.
4. *Tool and Manufacturing Engineers Handbook, 4th ed.*, Society of Manufacturing Engineers, Dearborn, MI, 1984.
5. *Design for Manufacturability Handbook, 2nd ed.*, James G. Bralla ed., McGraw-Hill, New York, 1998.
6. *Deburring and Edge Finishing Handbook*, LaRoux K. Gillespie, Society of Manufacturing Engineers, Dearborn, and American Society of Mechanical Engineers, New York, 1999.
7. *Lasers Claim New Roles in Manufacturing*, Thomas Begs, Manufacturing Engineering, October, 1998.
8. *Welding Handbook, 7th ed.*, American Welding Society, 1976.
9. *Introduction to Computer Numerical Control*, James Valentino, Joseph Goldenberg, Regents/ Prentice Hall, Englewood Cliffs, NJ, 1993.

Chapter 4 - Processes for Plastics

A. How Plastics Are Made

A1. ***definition - Plastics*** are organic chemical compounds of high molecular weight that have a structure resulting from the repeated linking together of small molecules, called monomers. The resulting long-chain molecules are polymers. Plastics are one type of polymer; elastomers (rubber-like materials) are another. Polymers can be either natural or synthetic. The molecular weight of the polymer is a multiple of the molecular weight of the monomer. Usually, at least 100 monomer molecules and up to many thousands of monomer molecules may be incorporated in one molecule of a polymer or elastomer.

There are two basic types of plastics, *thermoplastics,* which soften and melt when heated (melting takes place over a wide temperature range) and *thermosetting* plastics, whose long molecules are crosslinked together. Thermosetting plastics do not melt upon heating but remain rigid until they reach the charring or burning temperature.

A2. ***polymerization reactions*** - are the chemical reactions in which the individual, relatively small, monomer molecules link together into very long chains, called polymers. The two basic polymerization reactions are *addition polymerization* and *condensation polymerization.*

A2a. ***addition polymerization*** - is a simple combination of molecules without the release of any by-products. In one addition polymerization reaction, a chemical activation of the molecules with the aid of an initiator (a chemical catalyst or other substance), causes the atoms of the monomer molecules to bond together, creating chains of molecules. These chains may have as many as several hundred thousand repeating molecular units and have a molecular weight from tens of thousands to several million. In another process variation, the atoms within the molecules are rearranged. This causes the molecules to link. In still a third addition polymerization method, monomer molecules are composed of rings of atoms. A suitable catalyst causes the rings to open up and connect with the rings of other similar molecules. The reaction is exothermic; heat generated must be removed from the reactor. Polyethylene, polystyrene, polypropylene, and polytetrafluorethylene ("Teflon") are made by addition polymerization. Fig. 4A2a shows the addition polymerization reaction for polyethylene.

```
Ethylene      Catalyst
 H  H                          H   H
 C = C    +      X     ──►  X - C - C - *
 H  H                          H   H

   H   H     H   H             H   H   H   H
X - C - C - * + C = C  ──►  X - C - C - C - C - *
   H   H     H   H             H   H   H   H
                                 Polyethylene
```

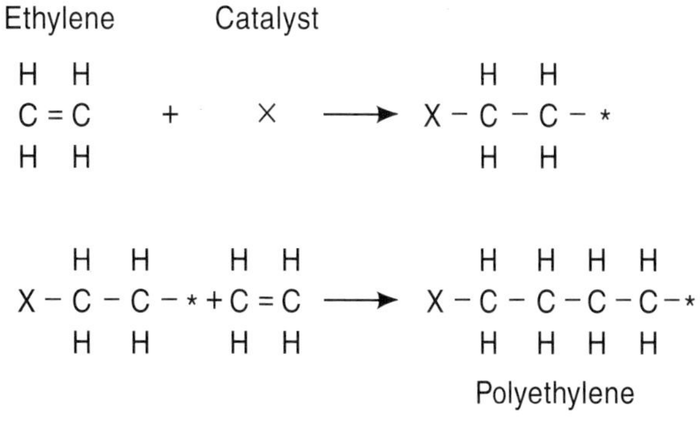

Fig. 4A2a The addition polymerization reaction for polyethylene.

A2b. ***condensation polymerization*** - differs from addition polymerization in that the monomer molecules are modified to cause the linking process to take place. A polymer is formed by a reaction between two functional groups attached to a monomer core. The modification also creates atoms, which form small molecules of condensation, most commonly of water, which is removed from the mixture. The repeating unit of the condensation polymer is different from the monomer because of the elimination of the condensation molecules. The condensation compound must be removed immediately. Otherwise it would interfere with further polymerization and could constitute undesirable contamination of the finished polymer. The condensation reaction requires the addition of heat. Some nylons, phenolics, and polyester pre-polymers, are made by condensation polymerization. Fig. 4A2b illustrates the condensation polymerization reaction involved in the production of phenolic plastics.

A3. ***polymerization methods*** - There are a number of approaches for effecting the polymerization: bulk, solution, suspension, emulsion, and gas-phase methods.

A3a. ***bulk polymerization*** - is a widely used approach. It takes place with the monomers and reactants in the reaction vessel but with no other materials present except initiators. The monomer is a solvent of the polymerized material. As polymerization proceeds, the polymerized plastic dissolves in the monomer, increasing the viscosity of the mixture. Polymerization continues until all the monomer has been polymerized. Exothermic heat must be removed to keep the reaction under control and prevent it from becoming explosive or developing local hot spots. These may result in degradation and discoloration of the polymer. Heat removal is critical because the reaction is accelerated at higher temperatures. The mixture must also be agitated, even though this can become difficult as the viscosity rises. One method of preventing overheating is to carry out the polymerization reaction in two steps, with special arrangements in the second step to aid in the dissipation of heat. The special arrangements include using thin layers of reacting material, finishing the polymerization in small-diameter tubing or in a free fall in open space or along the walls of the container. In another method, the final step is deferred until the monomer-polymer mixture is cast in molds into rods, bars, tubes, or other useful components. Cast acrylic shapes are made with this method.

In another procedure, the polymer is not soluble in the monomer but separates from it as polymerization proceeds. Nevertheless, plastics of very high molecular weight are produced with this procedure.

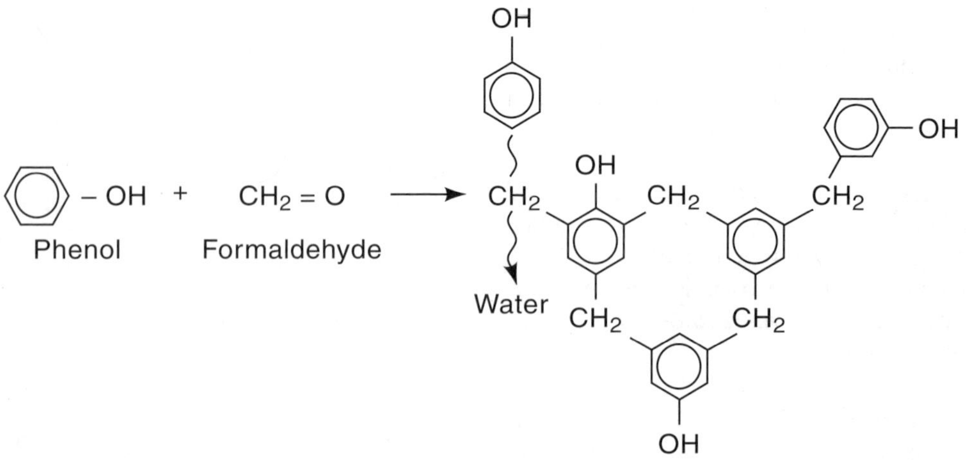

Fig. 4A2b The condensation polymerization reaction, which combines phenol and formaldehyde to produce polymers of phenolic thermosetting plastic.

Bulk polymerization is used in the production of most step-growth polymers and many chain-growth polymers.

A3b. ***solution polymerization*** - In this procedure, used when the monomer is not a solvent for the polymerized material, a separate inert solvent is used. This solvent may dissolve the monomer and the polymer as well as the initiator. However, it may dissolve the monomer without dissolving the polymer or may dissolve the polymer only partially, depending on the materials involved. The technique used varies with the solubility of the solvent as does the rate of polymerization and the molecular weight of the finished product. The solvent has the function of absorbing and conducting away exothermic heat and making the mixture easier to stir. It also reduces the rate of polymerization. These effects all virtually eliminate the problem of heat removal that is inherent in bulk polymerization. A solvent is chosen that does not react with other compounds in the reactor, although complete inertness to the reaction is not always possible.

Solvent removal after polymerization may be difficult. For this reason, the solution method is well adapted to the production of polymers that are used in liquid form as adhesives, coatings, impregnating fluids, and laminating resins. The method is also useful when the monomers are gaseous.

A3c. ***suspension polymerization*** - In this approach, the initiator is dissolved in the monomer and the monomer is dispersed in a liquid in which it is insoluble, usually water. The polymerized material is also insoluble in this liquid. A dispersing agent is incorporated in the mixture to stabilize this suspension of the monomer and polymer. The polymerized material and the monomer from which it is formed remain dispersed in the non-solvent liquid as small beads or "pearls" of 0.004 to 0.040 in (0.1 to 1.0 mm) diameter. The polymerization proceeds in each individual bead in a manner similar to that of bulk polymerization. However, the water or other liquid is valuable in conducting away the exothermic heat of the reaction and in other respects aiding in the control of the process. Other advantages are the lower cost of water compared to organic solvents and the fact that it does not react with other compounds in the reactor. Another benefit of suspension polymerization is that the finished plastic exits the process in smaller-size chunks, eliminating the need for later pelletization. The small chunks are easily washed, filtered, and dried for later molding.

A3d. ***emulsion polymerization*** - has similarities to suspension polymerization. It uses water, soap (an emulsifier) and a water-soluble free-radical initiator to carry out addition polymerization of the monomer. The monomer, which is insoluble or only slightly soluble in water, is dispersed in an aqueous continuous phase. The reaction tends to take place within small hollow spheres formed by the soap molecules. The monomer diffuses into these spheres. The reaction starts when the initiator reacts with the monomer in the small spheres. The control of the conditions that form the spheres also controls the polymerization reaction. A colloidal dispersion of the polymer, a latex, results. The polymerization is much faster than with other procedures including bulk and solution methods and the polymers produced can have very high molecular weights. The process is widely used. Polystyrene, acrylics, polyvinyl chloride (PVC), polyvinyl acetate, and ABS plastics are made with this method. Synthetic rubbers are also made by emulsion polymerization.

A3e. ***gas-phase polymerization*** - is a process in which a gaseous monomer such as ethylene, vinyl chloride or tetrafluoroethylene is polymerized. The gas-phase monomer is introduced under pressure into a reactor along with a catalyst. A reaction takes place that results in the formation of a solid polymer in granular form. Polymer particles can be removed from the reactor and it is not necessary to separate the catalyst. For polyethylene, the reaction takes place at low temperatures [170 to 212°F (75 to 100°C)] and a low pressure of 290 psi (20 bar). Some processes use a fluidized bed of the polymerized material and the solid powder catalyst; others use a stirred bed. Linear low-density polyethylene, high-density polyethylene, and polypropylene are commonly polymerized by a gas-phase process.

A4. ***compounding plastics*** - The typical plastic molding material contains a number of ingredients in addition to a plastic resin. Depending on the application and the resin involved, the molding compound may contain any of the following: fillers, reinforcements, colorants, stabilizers, antioxidants, ultraviolet light absorbers, antistatic agents, flame retardants, blowing agents, lubricants, fragrances, mold release agents, smoke suppressants, antifogging agents, antimicrobials, and plasticizers. Properly blending these ingredients involves a thorough mixing operation. The objective is that any unit of volume of the material has the same distribution of components as that of any other sample and of the entire lot. Mixing methods currently in use are described in the following paragraphs.

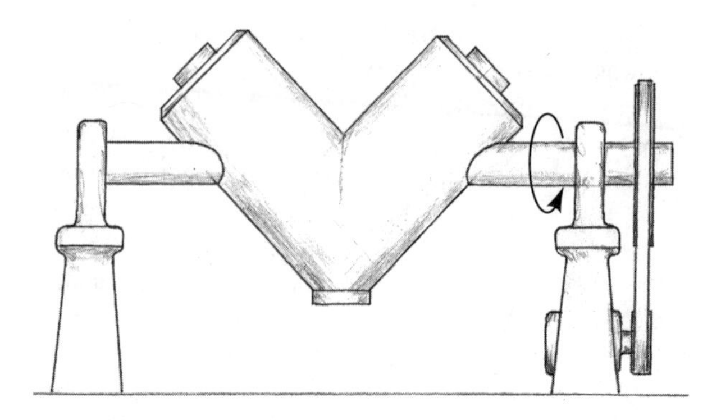

Fig. 4A4a Mixing by tumbling in a V-shaped barrel.

A4a. ***mixing by tumbling*** - Tumbling barrels are effective for mixing dry, solid materials. The barrel is rotated about a central axis and the material it contains falls from the top to the bottom, gradually dispersing additives in the base material. Intake of ingredients and discharge of the mixture after tumbling take place through an opening in the barrel. Several different shapes of barrel are used. The V-shaped barrel, shown in Fig. 4A4a, is common. When the "V" is upright, the materials in each side fall together; when the "V" is inverted, the materials fall into each side approximately equally. The repeated combining and dividing in half as the barrel rotates, gradually disperses each material until a uniform mixture is obtained. Granules, crystalline materials, and dry and partly-dry powders can be processed.

A4b. ***intensive dry mixing*** - involves a container such as that in Fig. 4A4b, with a high speed

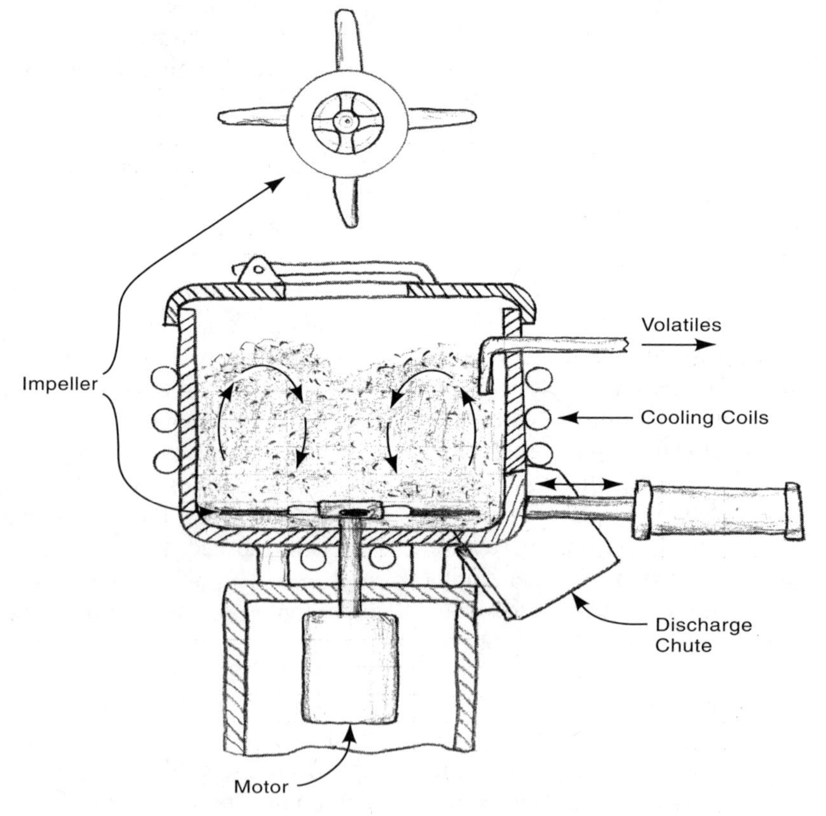

Fig. 4A4b Intensive dry mixing - propeller type.

rotating propeller-like blade at the bottom. It is used for blending powdered resins with various additives. PVC, blended with plasticizer and other additives, is a common application. This type of mixer is equipped with cooling apparatus to remove frictional heat. Volatilized moisture is also drawn off to a vortex and vented.

A4c. ***internal intensive batch mixing*** - utilizes mixers of the Banbury type, illustrated in Fig. 4A4c. They normally consist of two horizontal cylindrical chambers, side by side, each with a shaft centered on the cylindrical axis. Each shaft holds rotors of various shapes. The rotor blades move near but do not touch the cylindrical walls of the chambers, producing a strong shearing action. The blades rotate in opposite directions and interact with each other. Some blades are also helical and impart an axial movement to the material being mixed. There are other blade varieties, elliptical or fishtail in shape. The effect of all these shapes is to severely fold, shear, and mix the material in the mixer. The rotors and chambers contain channels

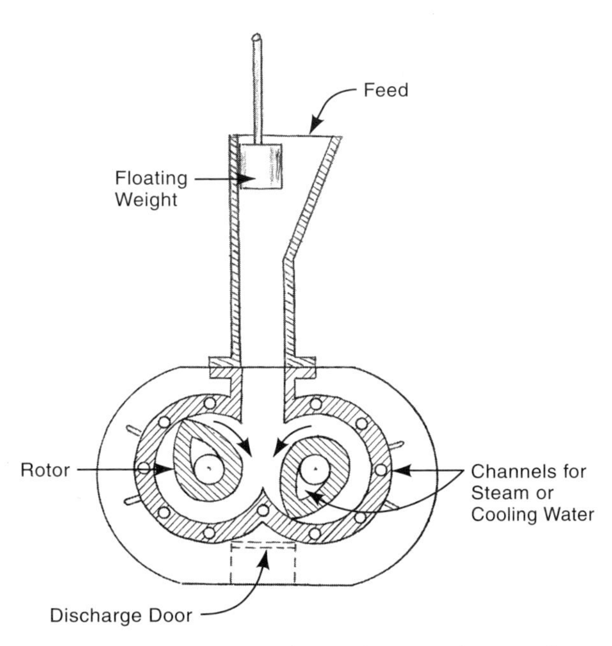

Fig. 4A4c Cross section of a Banbury mixer. Rotors within the two parallel cylindrical chambers knead, fold, shear, and mix the material thoroughly. A weighted piston holds the material in the chambers. Rotors can be of several shapes that may vary along their lengths.

for cooling or heating liquids so that the proper mixing temperature for the particular polymer involved can be maintained. High power levels may be required but mixing of the batch normally can be completed in 2 to 4 minutes and a highly homogenous mixture results. This equipment is used in the mixing of polyolefins, vinyls, ABS, polystyrene, ureas, and melamines.

A4d. ***continuous mixing*** - is quite similar to the internal intensive batch mixer (Banbury mixer) except that it is designed for continuous rather than batch operation. Like the batch-type Banbury mixer, there are two overlapping cylindrical chambers, each with a rotating mixer. There is a feed hopper at one end and a discharge chute at the other end. Near the feed end, the mixers are shaped like screw conveyors to move the material to the more aggressive mixing portion of the chambers. This mixing portion, near the discharge end, contains rotor shapes that shear and knead the material, and interchange it between the rotors. The machines are designed to provide high-quality mixing over a wide range of throughput rates. Even when operating well below maximum capacity, a homogenous mixture results. Fig. 4A4d illustrates this kind of mixer.

A4e. ***single screw extruders*** - Standard single screw extruders, as shown in Fig. 4I, have the capability of mixing plastic materials fed through them. Mixing is a desirable function during an extruding operation although it is not the prime function of the machine. Although these machines do not have the blending power of continuous mixing equipment or other dedicated mixing machines, they have proven useful for modest mixing tasks. Common operations are the mixing involved when fillers, antioxidants, stabilizers, color concentrates, or other ingredients are added to the resin compound to be extruded. These machines are also capable of blending reground scrap material. The back pressure of an extrusion die develops some back flow in the material in the barrel, further adding to the mixing. In all such machines, the added ingredient must be metered properly into the feed hopper.

A4f. ***compounder-extruder mixing*** - uses a machine similar to an single-screw extruder, but whose prime purpose is mixing rather than extruding.

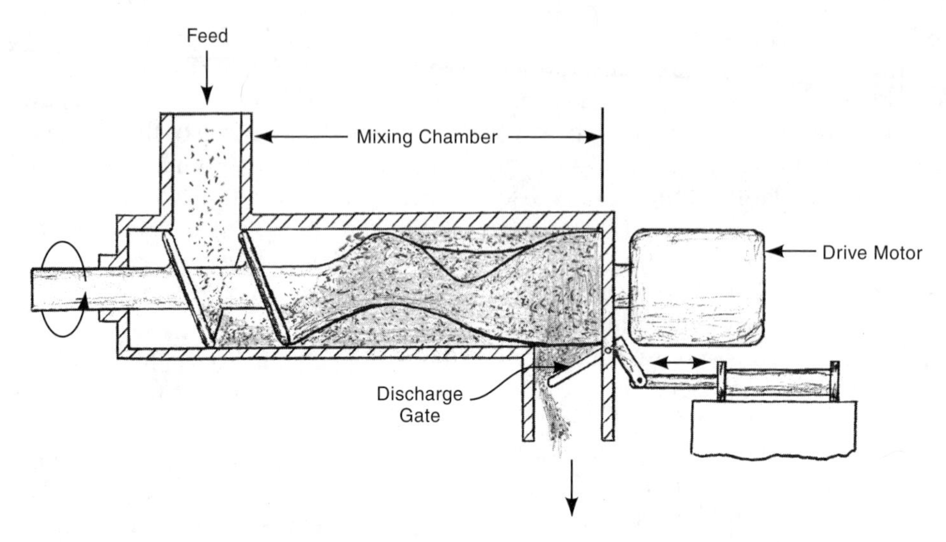

Fig. 4A4d Continuous mixing with a mixer similar to a Banbury type.

The barrel of the machine contains a series of stators outside the flutes of the extruder screw. The stators are opposite handed from the screw and are of alternatingly different heights. Torpedo and screw modifications are also made in the machine. These involve varied pitch and diameter of the screw and spiral grooves in the barrel. The screw is configured to provide an extensive mixing section and a high intensity variable-shear section. In the central area, the screw and barrel are shaped to permit the escape of trapped air and volatile materials. A vacuum pump draws them off.

A4g. ***twin screw extruder mixing*** - is performed on machines with two feed screws side-by-side in a double barrel. The screw flights usually overlap and intermesh. The screws usually rotate in the same direction, which means that there is a wiping and shearing action in the area of overlap. The material follows a figure-8 path as it moves through the machine and is thoroughly mixed. Fig. 4A4g shows the cross-section on one twin-screw mixer.

A5. ***pelletizing and dicing of plastics*** - Commercial plastics intended for molding and extruding operations are most often supplied in pellet form. The advantage of pellets is that they can be easily handled, accurately weighed, and conveniently stored. Machines to create pellets from compounded material are of two basic types:

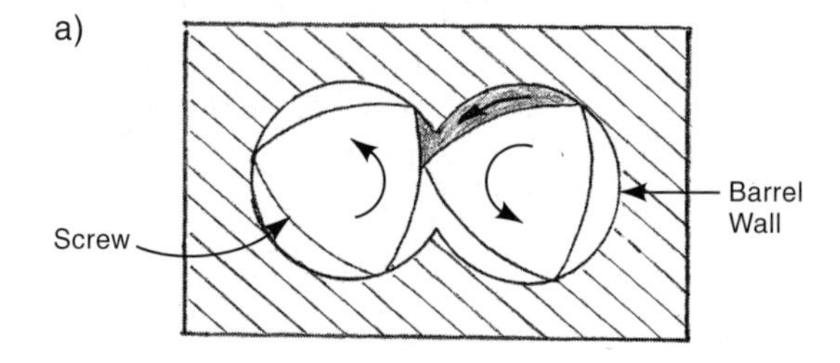

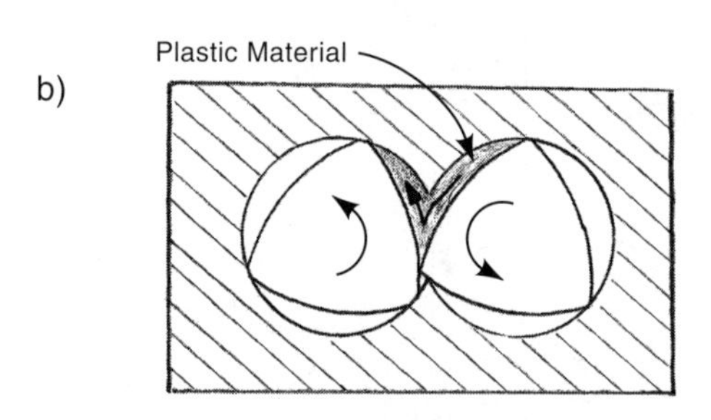

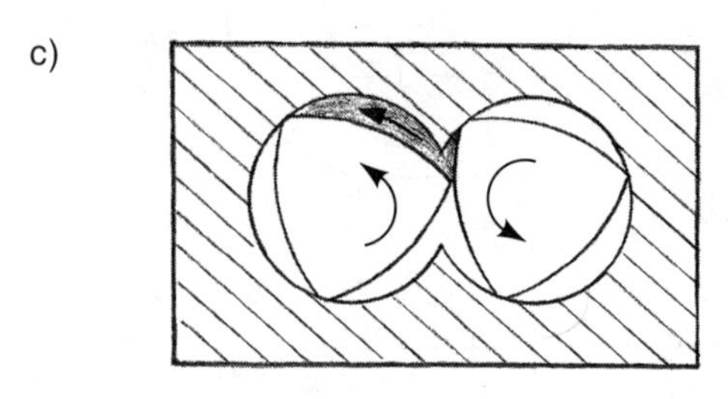

Fig. 4A4g The cross-section of a twin screw mixer showing how the material is mixed as it is transferred from one screw to the other.

1) rolls that mix the material and convert it to a sheet that is slit and diced and, 2) strand pelletizers, or extruders with attachments at the outlet end to cut extruded rod-like shapes into small pieces. The final pelletizing may be carried out with the material either cold and solid or still hot as it emerges from a die.

A5a. ***mixing and dicing with two-roll mills*** - Two-roll mills can be used as a final mixing step to introduce and blend plasticizers and fine particles of solid additives. Typically, the horizontally-shafted rolls rotate in opposite directions and they pull and nip the material through the space between them, providing good shearing action. The exiting material is slit by strip cutters into ribbons that are fed to a cooling tank and then to a dicer. Fig. 4A5a illustrates the two-roll mixing action.

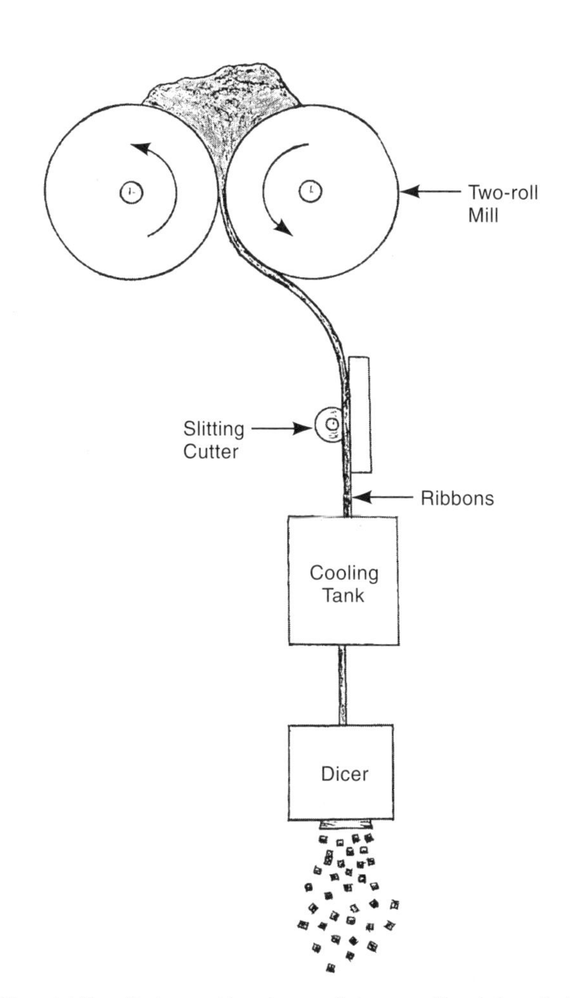

Fig. 4A5a Schematic view of two-roll mixing followed by slitting and dicing.

A5b. ***strand pelletizers*** - each consist of a screw or gear-pump extruder and die, and a rotating cutter that works against stationary blades. Each machine also has provision for cooling: by air, air-vacuum, or water, either before or after cutting. The machines include a drying system if water is used, and a means to collect the pellets. Hot-face cutters cut the extruded strands into pellets while they are still soft. Cold-cutting systems cut the strands after they have cooled. With cold-cutting systems, the strands may be drawn and pulled through a water bath before cutting. With hot-face cutting systems the cutters act upon the strands before they are cooled by air, fluidized bed, water spray, and/or water stream. The method chosen depends on the properties of the plastic, particularly its melt strength and sensitivity to temperature and its ability to withstand a residence period at a high temperature.

A5c. ***underwater pelletizing*** - Material exiting from a mixer-extruder flows through heated multiple-opening extrusion dies into a water chamber where the strands are sheared into pellets by a rotating, multi-bladed cutter moving across the die face. Water circulated through the chamber cools the material and conveys the pellets through a discharge port, away from the cutting area, and to a dryer. Water provides a convenient handling medium. The dryer then removes the water from the plastic pellets by centrifugal force. Polyethylene, polypropylene, PET, polystyrene, ABS, SAN, and thermoplastic elastomers are pelletized with this method.

B. Compression and Transfer Molding

B1. ***compression molding*** - Fig. 4B1 illustrates the process. In the most common approach, a preheated preform or tablet, normally of thermosetting plastic, is placed in an open, heated, mold half that is held in a molding press, which usually operates vertically. The press ram descends with the force half of the mold, compressing the preform, so that the material flows to fill the entire mold cavity.

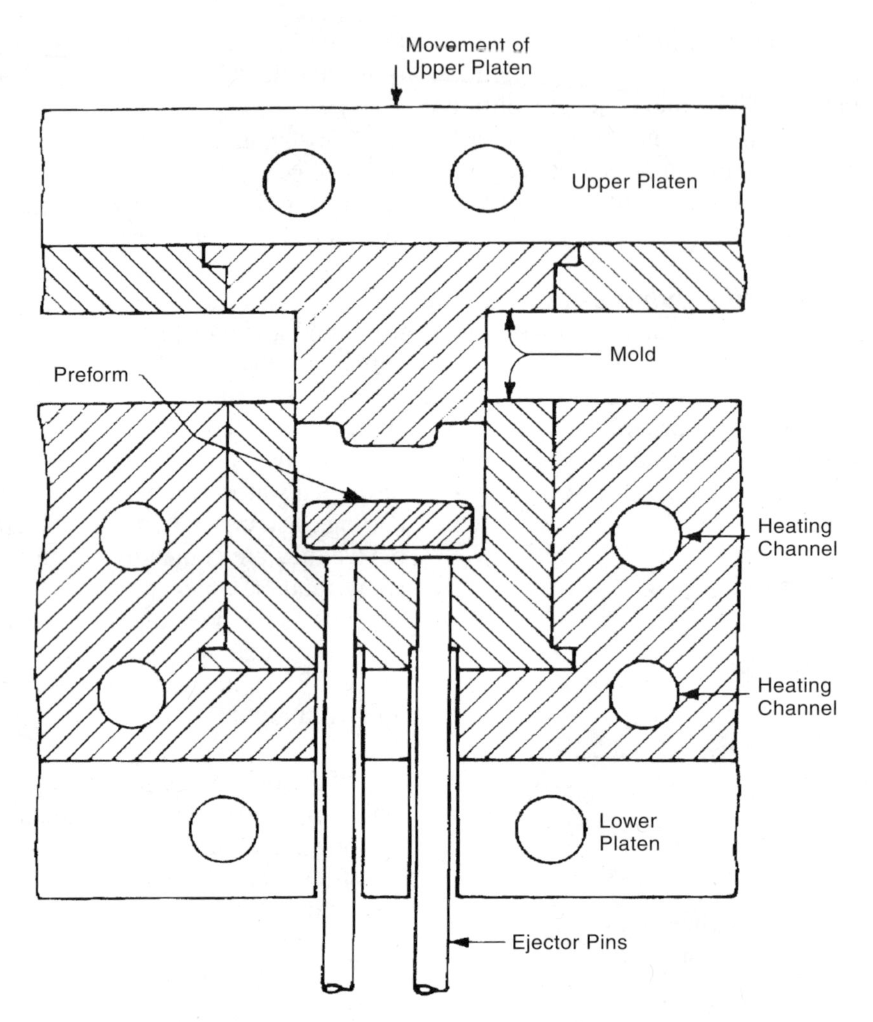

Fig. 4B1 Schematic illustration of compression molding. A preform of thermosetting plastic material is placed in a mold cavity. Heat softens the preform and the ram pressure forces the material into all portions of the mold cavity where it takes the shape of the cavity. The material then sets into a solid part that is ejected from the mold. *(from Bralla, Design for Manufacturability Handbook[6].)*

The heat of the mold, and that added by the friction of the molding process causes the thermosetting material to polymerize, changing from a somewhat pasty state to a strong, solid state. The mold can be opened and the part removed. There are no sprues or runners in compression molding but the process often produces flash on the molded part at the parting line of the mold halves. This flash is removed manually with a knife, by tumbling (See 3K3.) or, in some cases, with a press containing a trimming die that fits the contour of the part. The use of a preform of developed size minimizes the amount of flash. Preforms are made by compressing granules of plastic material in a die of suitable size. Instead of preforms, the molding compound may be in the form of powder, paste, or paper or various fibers impregnated with plastic material. Ejector pins are provided in the mold to ease removal of parts. Typical compression molded parts are automobile distributor caps, kitchen utensil handles, and various rubber items.

The use of fillers and reinforcing materials is common with compression-molded thermosetting plastic parts. Section G below describes methods particularly applicable to reinforced plastics molding.

B1a. ***automatic compression molding*** - is simply the automation of the otherwise manual compression molding process. Automatic compression molding equipment usually provides: storage of the molding material, metering of the charge of material to be deposited in the mold, transfer of this quantity of material to the mold cavities, operation of the compression molding press, ejection or removal of the molded workpiece from the die cavity, and removal of any extraneous molding material that may remain in the mold cavities after the operation is completed. Although injection molding of thermoset materials is now quite common, automation of the traditional compression-molding operation is often a viable alternative as a labor-saving approach.

B2. ***transfer molding*** - This process is a cross between injection and compression molding. It normally applies only to thermosetting materials. The material is first heated in a transfer chamber, which usually forms part of the mold. When the mold closes, the material is forced by a plunger, which is also part of the mold, through sprues and a gate into the heated mold cavity. More than one mold cavity can be fed from a central chamber or "pot". The material polymerizes in the mold cavity, and a solid part is removed or ejected. Usually a preheated preform or measured amount of material is placed in the transfer chamber at the start of each molding cycle. Transfer molding permits the production of more intricate parts with thinner sections than are possible with the compression-molding process. Fig. 4B2 illustrates pot-or sprue-type transfer molding. This process is economicall for molding thermosetting plastic parts when the quantities involved are somewhat higher than are economical for regular compression molding.

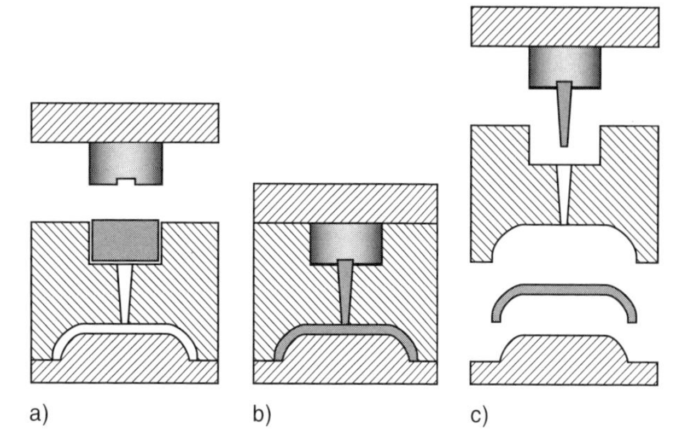

Fig. 4B2 Simplified sectional view of transfer molding. a) A heated preform of thermosetting material is placed in the transfer pot. b) The mold closes, forcing the material into the sprue channel, through the gate, and into the mold cavity. c) After the material is cured, the mold opens, pulling the sprue away from the part, and the part is ejected from the cavity.

B2a. ***plunger molding*** - is a variety of transfer molding. Instead of using a fixed element of the mold to force the material from the pot into the mold cavity, force is supplied by an auxiliary ram, after the mold is closed. This arrangement allows the speed and pressure of the transfer to be independent of the mold-clamping operation, thereby providing better control. Fig 4B2a illustrates a typical plunger mold.

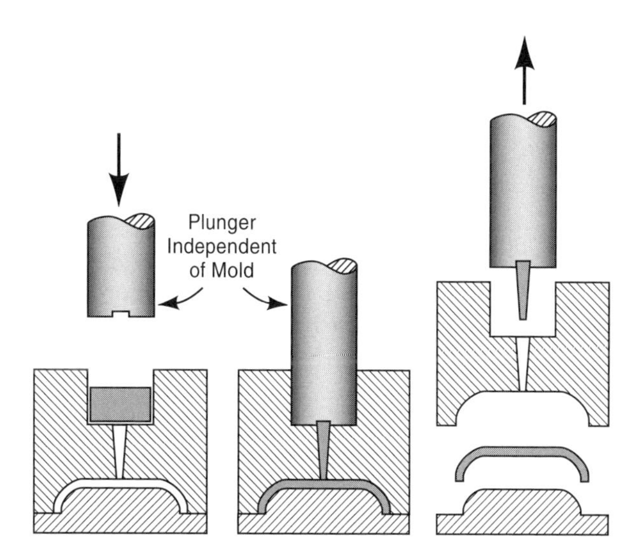

Fig. 4B2a Simplified section view of plunger molding. The process is a variation of transfer molding. It differs in that the plunger is not part of the mold but can be advanced and retracted independently.

B2b. ***screw transfer molding*** - is a further refinement of the transfer molding process. The material is heated and mixed with a plasticizing screw in the molding machine and is dropped in the pot of the mold. A plunger then forces the material into the mold cavity. Fig. 4B2b illustrates the process, which is useful when the molding material is difficult to preform.

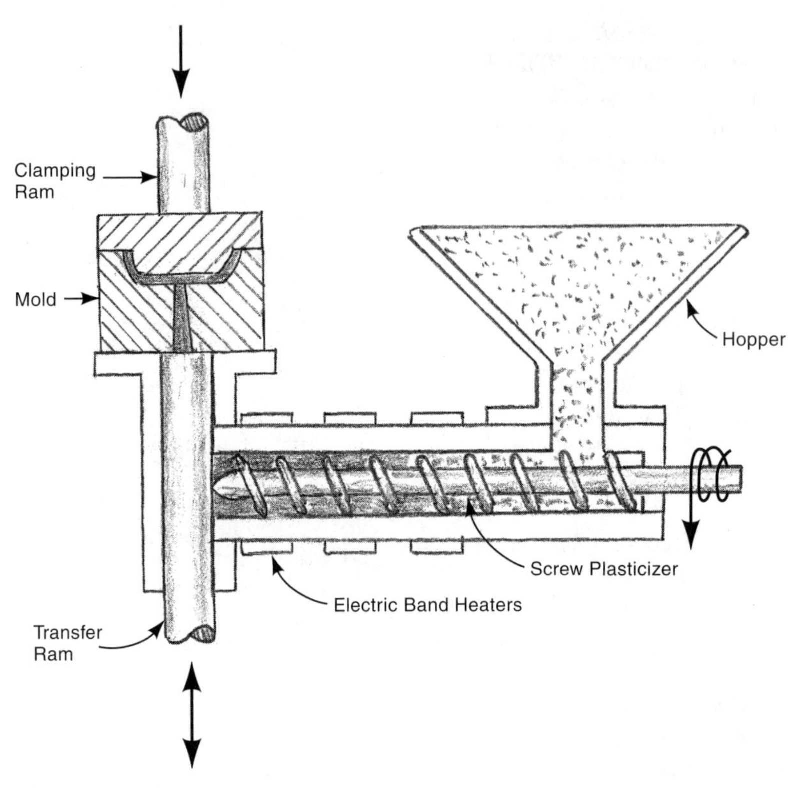

Fig. 4B2b Screw transfer molding.

B3. ***cold molding*** - is regular compression molding with no use of heat during the compression phase. The thermosetting material, including binders, is pressed to shape in the mold, then removed, and cured in a separate oven. The pressing operation is thus quicker than with hot molding, and the full molding sequence is more economical. However, the dimensional accuracy and surface finish of the molded part are usually inferior to that achieved with conventional compression molding. A number of parts can be cured simultaneously in the oven. The process can be useful for limited quantity production. The term *cold molding* is also used to identify the molding of reinforced thermosetting plastics that have been formulated to polymerize and cross-link at room temperature.

C. Injection Molding

C1. ***conventional injection molding*** - Injection molded parts begin with a plastic material in granular form, most commonly a thermoplastic. The granulated material passes from a hopper to a heated cylinder, where it changes from the heat into a pasty mass. (Most plastics, when heated, do not melt into a full liquid, but change into a mass with a consistency similar to that of peanut butter.) The pasty mass is moved forward with a screw feed, by plunger, or both, and, under pressure, is injected into a mold. (Rotation of the screw mixes the material and provides frictional heating; axial movement of the screw forces the material into the mold.) The mold is normally a steel cavity, made from at least two pieces, tightly clamped together. The cavity has the same shape as that of the final part. Multiple cavities per mold are very common. With thermoplastics, the mold is cooler than the cylinder from which the plastic was injected, cool enough so that the plastic, after it enters and fills the mold, cools and solidifies. A typical temperature of the material in the machine cylinder is 400°F. (200°C) and, in the mold, 180°F (80°C). The machine cylinder is heated electrically and the

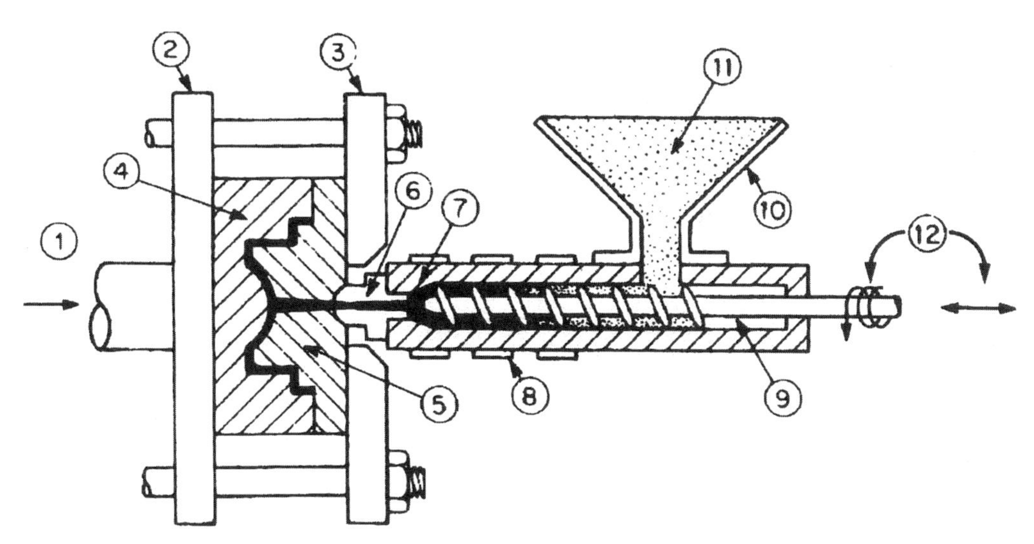

Fig. 4C1 Schematic illustration of injection molding. (1) Mold clamping force. (2) Movable mold platen. (3) Fixed platen. (4) Cavity half of mold. (5) Force half of mold. (6) Nozzle. (7) Cylinder. (8) Electric band heaters. (9) Reciprocating screw. (10) Hopper. (11) Granulated-plastic material. (12) Rotary and reciprocating motion of screw. *(from Bralla, Design of Manufacturability Handbook[6].)*

mold temperature is regulated with circulating water. The process is normally fully automatic. Cycle times range most commonly from about 15 to 45 seconds, the time being dependent on the wall thickness and size of the part and many other process factors. Figure 4C1 illustrates the injection molding process.

Injection molding is by far the most common process used in the manufacture of parts from plastics. It is almost an ideal mass production method. Parts can be complex and can have color and surface texture molded in (with no need for costly secondary operations). They can have such features as hinges, springs, screw threads, and bearings incorporated fairly easily. Many companies that have made significant DFM (Design for Manufacturability) improvements in their products have done so by combining a number of separate parts and fasteners in a fewer number of injection-molded plastic parts that snap together without external fasteners.

Some typical injection molded parts are the following: housings for various products such as computers, home entertainment products, telephones, caps for containers, pails, handles, laundry baskets, toys, and parts for automobiles and refrigerators.

C1a. ***hot runner molding (heated runnerless molding)*** - Injection molds have channels that convey the molten plastic from the molding machine nozzle to the gate of each mold cavity. Normally, the material in these channels cools and solidifies along with the part, forming runners, which are ejected from the mold with the part. The runners are cut from the part, usually by the molding machine operator during the next machine cycle. Although the runner material, usually can be reground and used, it is advisable to avoid the need for these steps by the use of hot runner molds. These molds incorporate electrical units to heat the runners so that the material in them remains in the molten state between shots and is not ejected from the mold with the part. Trimming of the runners from the part is eliminated, as is the need for scraping or regrinding and reprocessing the runner material. Less material needs to be heated for each shot; material is not subjected to repeated heating, which may degrade it, and it is kept at a more uniform temperature as it enters the mold. Another advantage is that the molding cycle is often speeded up because thin-walled parts do not have to wait for the thicker sprues and runners to solidify, and less material needs to be heated for each cycle. Some plastics have a narrow range of processing temperatures and are better used in a hot runner system that does not force them to be heated to counteract a cooling environment before they reach the mold cavity.

C2. ***injection molding of thermosetting plastics*** - This process is very similar to the injection molding of thermoplastic materials. A major difference, however, is that the mold is heated rather than cooled. Another difference is that the injection nozzle may be alternately heated and cooled - heated during injection to facilitate polymerization, but cooled between shots to retard the polymerization of material not yet injected. The thermosetting material is preheated in the cylinder of the machine, but the temperature is not high enough to cause polymerization to take place during the limited time the material is in the machine barrel. Heating reduces the viscosity of the material, however, which facilitates injection and mold filling. Polymerization does take place after the material is subjected to the heated nozzle, the frictional heat of injection, and the heat of the mold. After polymerization, a solid piece can be removed from the mold. Injection pressure is up to about 28,000lbf/in^2 (190MPa). The cycle is normally longer than that required for thermoplastic materials but shorter than that required for compression or transfer molding. The process is sometimes called, automatic transfer molding.

C3. ***structural foam molding processes*** - provide parts with a wall structure that includes a solid skin and a core that is cellular. The cellular core comes from the expansion, during molding, of a blowing agent, either an injected gas, usually nitrogen, or a chemical that decomposes into a gas at the molding temperature. A nucleating agent, a fine dry powder, is also added to the resin to control the size of the cells. Structural foam components have a high degree of stiffness in comparison to their weight. Large parts like containers, pallets, and housings, are molded by this method because it provides a means to maintain stiffness without using excessive material or slowing the molding process unduly. The process can also be used with flexible materials to produce components with cushioning or sealing properties.

C3a. ***low-pressure injection molding of structural foam plastics*** - In this process, the amount of material injected into the mold is carefully controlled and is less than that required to fill the mold. A blowing agent in the plastic material causes it to expand and fill the mold cavity. The portion of the material that contacts the cool surface of the mold cavity forms a dense skin, while the interior portion becomes cellular. Since the mold is not packed with plastic material, pressures are much lower than with conventional injection molding. Mold costs are low because the molds do not have to withstand high pressures. Large parts - up to about 120 lb (55 Kg) - can be molded with the process. Typical applications are containers, pallets, lawn furniture, outdoor equipment components, and shutters.

C3b. ***reaction injection molding*** - This process involves the injection, into a mold, of highly-reactive liquid components, usually of thermosetting material. The components (Polyurethane plastic is most common.), are fed to a mixing chamber just prior to injection. Heating is not required. The materials react, polymerize, and foam in the mold, forming a part with a somewhat smooth and dense skin and a cellular core. Pressures are high in the mixing chamber but low (50 lbf/in^2 or 340 KPa) in the mold to allow foaming to take place. This permits the use of lighter, lower-cost molds. After molding, parts are post-cured at about 250°F (120°C) for up to 1 hour. The process is suitable for large parts, including appliance components, doors, furniture frames, and vehicle body parts, as well as smaller parts such as gears, wheels, rollers, and sporting goods, Both rigid and flexible material formulations can be used. Fabric and strand reinforcement may be mixed in to provide added strength and stiffness. Epoxy, nylon, and polyester are also molded with this method. See Fig. 4C3b for an illustration of the process. With flexible formulations of urethane, reaction injection molding is used to produce furniture cushions and mattresses.

C3c. ***high-pressure injection molding of structural foam plastics*** - In this process, normal injection-molding action and pressure, fill a mold completely with a mixture of a polymer and a chemical blowing agent. The size of the mold cavity

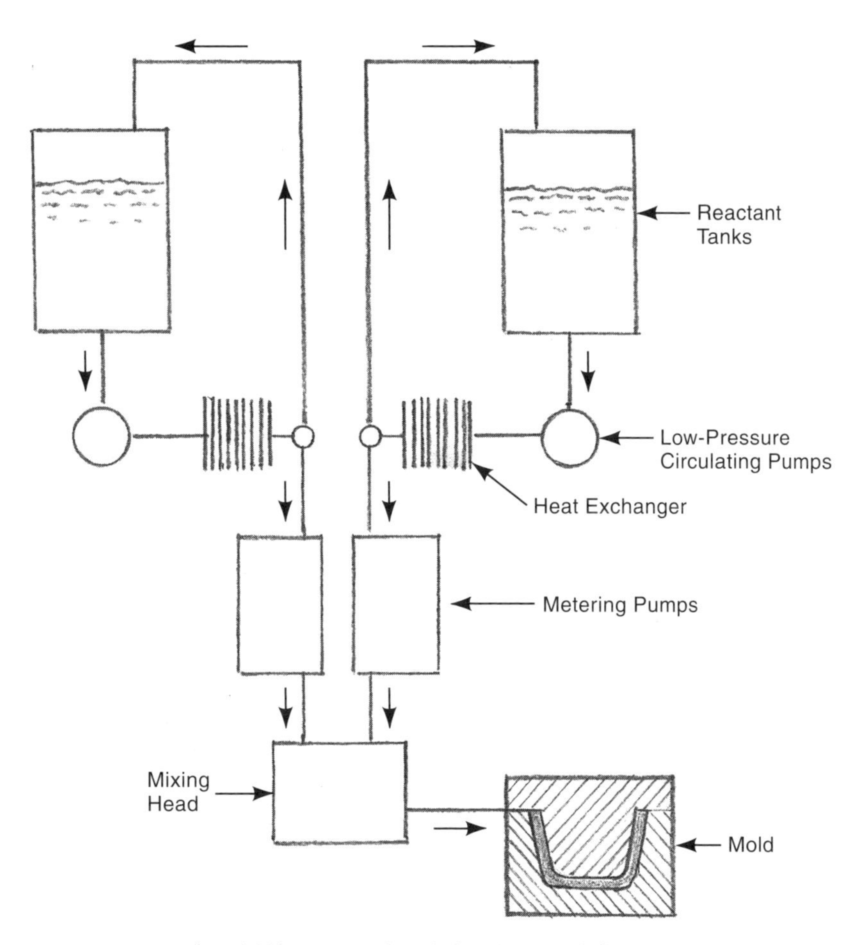

Fig. 4C3b Reaction injection molding.

is then increased by opening the mold platens or by retracting a core element. The material in the mold then foams and fills the expanded space. The high-pressure process uses more expensive and complex tooling but provides a smoother part surface than that obtained with low-pressure injection molding or by reaction injection-molding of structural foam. This process is used for items such as business machine housings where the swirled appearance of the surface of low pressure molded parts may not be acceptable.

C3d. ***gas counterpressure molding*** - The principle of this process is to delay the foaming until after a solid skin has formed on the plastic entering the mold, insuring that the molded part has a smooth surface. This is accomplished by pressurizing the mold with nitrogen just prior to injection. The pressure is slightly above the expansion pressure of the blowing agent in the injected plastic. At the proper instant, before the mold is completely filled, the pressure is released and the foaming action commences. However, foam does not break through the surface of the injected material. Therefore, no swirling pattern appears on the part's surface.

C3e. ***co-injection or sandwich molding*** - Two materials are injected in sequence and partly simultaneously into the mold through a special valve. One material forms the skin of the part and the other forms the core. The core material contains a blowing agent to provide foaming action. Two different plastics can be used. The core material can consist of a lower cost material or even recycled plastics. Surface details can be produced accurately

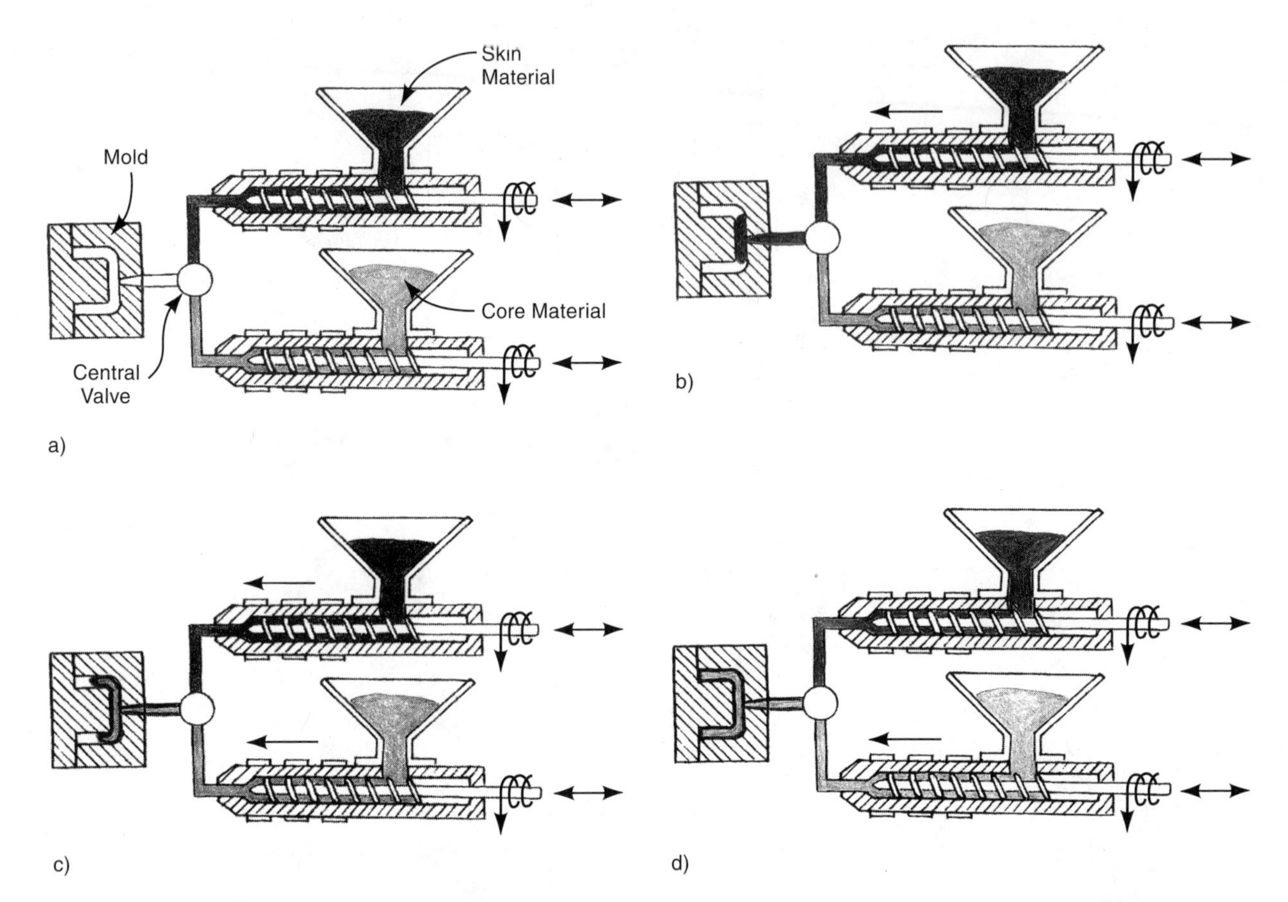

Fig. 4C3e Co-injection or sandwich molding: a) The machine arrangement. b) The skin material (shown in black) is injected first and a thin layer solidifies as it meets the cool walls of the mold. c) Both materials are then injected until all mold walls are contacted by the skin material. d) Filling of the mold is completed with further injection of the core material (shown in gray).

and there is no swirl pattern on the surface. See Fig. 4C3e for an illustration of the process.

C3f. ***gas-assisted injection molding*** - Parts made with this process do not have a foam core but instead have a hollow interior due to the injection of nitrogen into the plastic charge as it enters the mold. The gas does not break through the surface of the plastic but does form channels in the hotter, less-viscous material in the interior of the charge. The result, when the mold is filled, is a part with a smooth surface that matches the shape and finish of the mold, but is not solid. Less material is used and the part has similar stiffness advantages, per unit of weight, as a structural foam part. Fig. 4C3f provides a schematic illustration of the process.

C3g. ***casting of structural foam plastics*** - is described under *casting of plastics*. See H2.

C3h. ***extrusion of structural foam plastics*** - Extruded profiles can be made with a solid skin

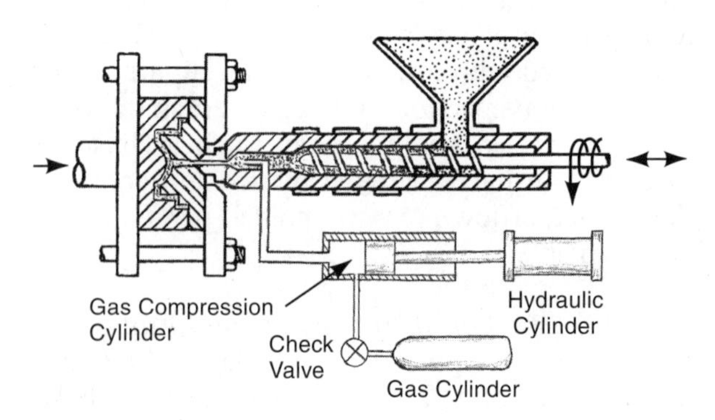

Fig. 4C3f Gas-assisted injection molding creates a part with properties similar to those of a part molded with structural foam.

and foam interior, similar to the structure of injection-molded structural-foam parts. The means of achieving the cellular structure is the same as that used with structural-foam injection molding. The extrusion process is also quite similar to that used for solid profiles. When the part is to have a cellular core, one method, called *free-foaming*, extrudes the material - containing blowing and nucleating agents - through a conventional die. Pressure in the extrusion barrel prevents the blowing agent from foaming until the material passes through the die. The extrudate is allowed to expand to approximately its final dimensions after it exits the die. It is then passed through a cold sizing die, which produces the final dimensions. In another method, the patented Celuka process, the extrusion die is of approximately the final dimensions desired. However, the die contains a centered mandrel that tapers to a point some distance from the die face. This mandrel allows room for the polymer to foam inwardly toward the center, thus producing a low-density foam core. A second die, in line with the first, sizes the extrudate and cools the surface to form the desired dense skin. Structural foam extrusions are used for window and door frames, building panels, and other extrusion applications where the lighter weight and better stiffness per unit weight of structural foam are desired. Extruded-foam polyethylene is used as an insulator for coaxial cable. (Note: Also see extrusion of non-structural foam in paragraph I3.)

C3i. ***slabstock foam process*** - might be considered to be reaction injection molding of polyurethane foam without a mold. Instead of being injected into a mold, the reactive polyurethane material is deposited on a wide, moving conveyor belt. There it continues to polymerize and foam, expanding to a mass as much as 8 feet (2.4 m) wide and 4 ft (1.2 m) high. Thereafter the mass is cut into long lengths and stored for additional curing before it is processed further. Further processing usually involves straight or curved cuts to shape the foam slab into such products as seat cushions, mattresses, carpet underlayments, and components of textile-based products. Thin sheets are sliced from the slab and rolled for later processing.

C4. ***expanded polystyrene foam processes*** - Expandable polystyrene (EPS) is supplied by manufacturers in bead form. The beads contain a blowing agent (usually pentane) that expands them from 2 to 50 times upon exposure to heat. By controlling time and temperature, the amount of expansion can be controlled. Expansion of the beads during molding produces a component with excellent insulating and flotation properties. Processing usually involves a pre-expansion of the beads as a first operation.

C4a. ***pre-expansion of EPS beads*** - can be accomplished with any of several batch methods involving ovens, steam chambers, or hot water baths, and several continuous methods involving hot air, radiant heat, or steam heat. All these methods expand the beads to approximately the level of expansion specified for the eventual molded part. The continuous steam-heating approach is by far the most common. It involves a continuous feed of beads to a steam chamber where the beads are mechanically agitated and mixed with steam. As they are heated by the steam, they expand from the vaporization of the blowing agent. Additional expansion comes from the absorption of steam. Agitation prevents the beads from fusing together. The amount of expansion is controllable and depends on the temperature of the steam, the feed rate of the beads and the amount of air introduced to the expansion chamber. As the beads expand, they tend to rise to the top of the chamber and overflow it and are conveyed by air flow to an open storage bin. In the storage bin, the beads cool and their expansion subsides somewhat and gradually stabilizes. The bins are open and subjected to additional air flow to dry the beads. After several hours (typically from about 3 to 12), the beads are dry and ready for molding.

C4b. ***shape molding of EPS beads*** - Fig. 4C4b illustrates this process. Pre-expanded beads are conveyed by air to the mold where a measured amount is introduced to the mold. The mold closes and steam is introduced through small holes, heating the beads and causing them to soften, fuse together, fully expand, and fill the entire mold cavity. The expanded beads block the small steam holes, preventing further introduction of steam. The mold is then cooled with water until the part stabilizes. The mold is opened and the part is ejected. Drinking cups for hot beverages, picnic coolers, fast-food containers, fitted support blocks in shipping

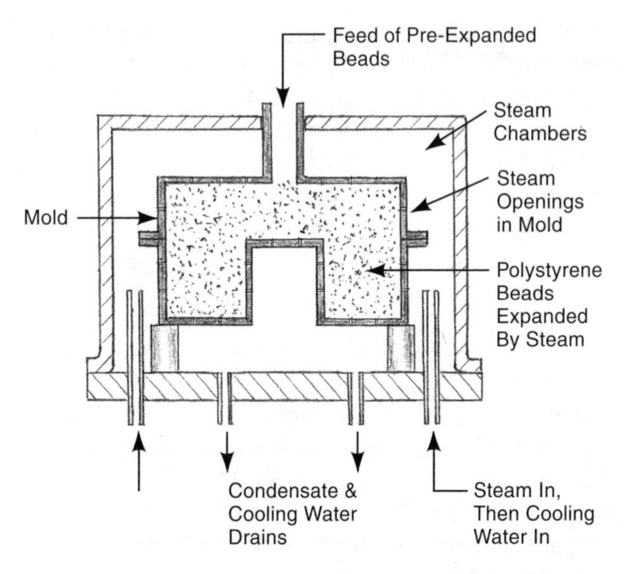

Fig. 4C4b Shape molding with EPS beads.

cartons, and display figures, are typical products molded from EPS with this method. Drinking cups are produced from small beads with typical densities of 2 to 5 lb/ft^3 (32 to 80 Kg/M^3). Sometimes, a vacuum is drawn on the mold cavity to provide room for the steam and to facilitate the expansion of the beads. Other heating methods - hot air or conduction through the mold - may be used to heat the beads.

C4c. ***block molding of EPS*** - to produce sheets or slabs of foam material proceeds quite similarly to shape molding. In this method, the mold cavity is rectangular and can be rather large - as large as 4 × 16 × 3 feet. Larger beads are also used for such components. Vacuum assistance may be employed to remove air from the mold. After molding, the blocks produced are sliced into thin slabs or other shapes by hot wire or band saw. Building insulation panels and flotation blocks are products made by this approach. Typical densities of insulating board are 1 to 2 lb/ft^3 (16 to 32 Kg/M^3).

C4d. ***expanded polyolefin foam process*** - Recently, polyolefins have been produced in a fashion similar but not identical to EPS. Special polypropylenes (EPP) have been manufactured to meet this market. Unlike EPS beads, EPP beads do not retain a blowing agent very well, so, EPP beads are usually pre-expanded by the resin manufacturer. The converter fills aluminum molds with the pre-expanded beads and applies high-temperature steam, heated air, and other gases to heat the beads until the surfaces are tacky. The mold is then cooled under pressure until the beads are fused together. EPP foams are very soft and ductile. They are used in packaging for shock mitigation and in vehicle bumpers. EPP foams tend to be porous and so are not used in liquid containers.

C5. ***two-color injection molding*** - This method is used for computer keyboard keys, two-color automobile tail lights and similar parts with inlaid color effects. The part is molded in two operations using two different mold cavities. The part molded in the first operation is used as an insert when the final part is molded in the second operation using a different color plastic. Fig. 4C5 illustrates the process sequence.

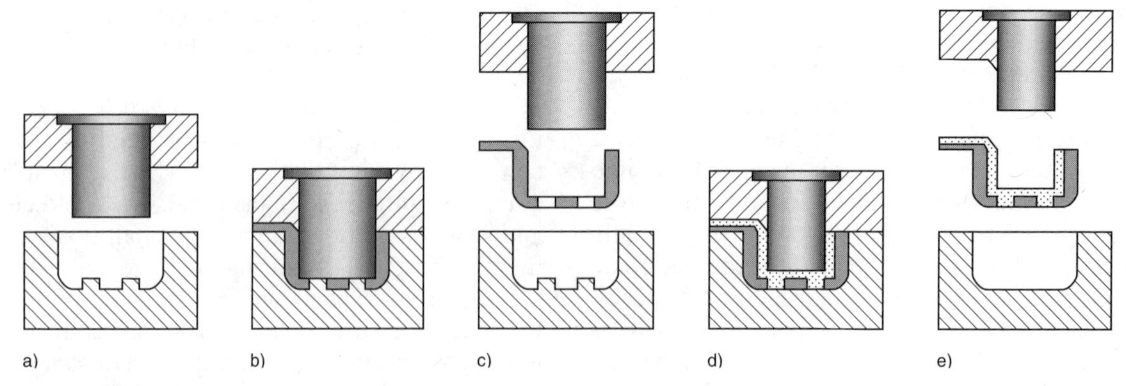

Fig. 4C5 Two-color injection molding: a) The mold for the outer shell. b) Plastic is injected in this mold to make the outer shell. c) The molded outer shell ejected from the mold. d) The second color is injected into a second mold that contains the outer shell as an insert. e) The second mold opens and the two-color part is ejected.

C6. ***insert molding*** - simply involves injection (or compression) molding in which other components are placed in the mold before the mold closes and the plastic material is introduced. Plastic material flows around the inserted components and holds them in place. This is the oldest method for combining metal and plastic parts and is extensively used. It provides excellent holding power for the insert. However, placement of inserts slows the molding cycle, adds the risk of damage to the mold from a fallen insert and has the possibility of allowing plastic to enter an unwanted area of the inserted part, necessitating rework. Metal inserts are sometimes used when screw threads, studs, or other elements, having strength or wear resistance greater than that of the plastic material, are required. Another application is the production of electrical plugs, sockets, and switches, when the metal electrically conductive elements are encapsulated in an insulating plastic. One common molding method uses duplicate mold halves on a turntable so that the inserts can be loaded in one mold cavity on the turntable while molding takes place at another turntable location. Inserts are most commonly metal, but inserts of other plastics, paper (e.g., labels), ceramics, or other materials can also be used. (Other methods for combining plastics and other materials in a single component are discussed in section N1 below.)

D. Thermoforming (Vacuum Forming)

Thermoforming, as the term is normally used, involves the shaping of a thermoplastic sheet by heating it to the softening point and then bringing it into contact with a cooler mold whose shape it takes. The process is often called vacuum forming because the most common method involves the use of a vacuum to draw the sheet against the mold. However, air pressure, a mating die, or some combination of methods, can be used to force the sheet to conform to the mold. The sheet cools after a period of contact with the mold and stiffens into the desired shape. Trimming of the formed sheet usually follows forming. Since thermoforming involves modest forces, tooling can be made from nonmetallic materials such as wood or plaster for prototype and low-quantity applications. Molds for commercial-quantity production are usually made of either cast or machined aluminum and are temperature controlled. Thermoforming equipment tends to be less expensive than that used for injection molding or extrusion, especially when larger parts are involved. Holes, cutouts, and slots in thermoformed parts are produced by secondary drilling, punching, or routing operations.

Applications for thermoforming can be classified into two branches: 1) disposable parts made in high quantities from thin-walled material. This kind of thermoforming is sometimes called *roll-feed thermoforming*, since the initial sheet thickness is usually less than 0.060 in (1.5 mm) and is delivered to the machine in a roll. Thin-gauge thermoforming is a high-speed process used extensively for blister packaging and other packaging for food or medical items, hardware, and many other products. Frozen food trays are typical examples. 2) heavy-walled parts used in permanent applications, made from sheets precut to the approximate dimensions of the finished part and delivered to the machine stacked on a pallet. This kind of thermoforming is sometimes called *cut-sheet thermoforming*. Heavy-gauge thermoforming is substantially slower than thin-gauge thermoforming primarily because of longer sheet-heating times. Lighting panels, equipment housings, and interior panels and liners for automobiles are examples.

The several process variations described below produce either a better or more intricate form or a more uniform wall thickness after forming than straight vacuum forming.

D1. ***straight vacuum forming*** - is probably the most common thermoforming process. It usually involves the use of a female mold. The heated sheet sags and then is drawn by vacuum against the mold where it cools and becomes more rigid. The process is useful when the depth to width ratio of the formed portion is 0.5 or less. This method is illustrated by Fig. 4D1.

D2. ***pressure forming*** - is very similar to vacuum forming. However, instead of relying on atmospheric pressure on one side of the sheet and a vacuum on the other side, higher pressure from compressed air is applied on the top side to force the sheet against the mold. The mold is vented and, additionally, a vacuum may be pulled on the underside of the sheet.

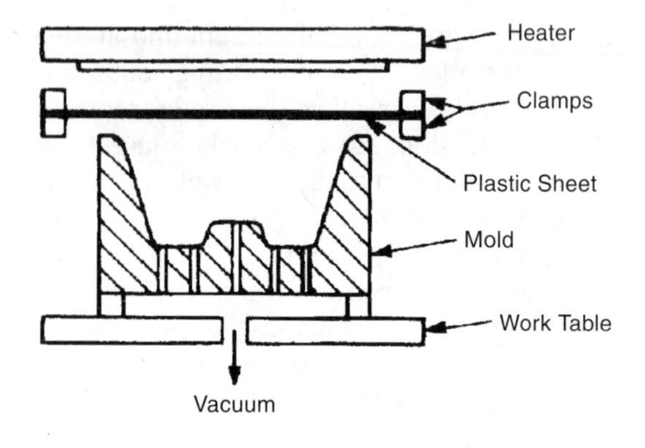

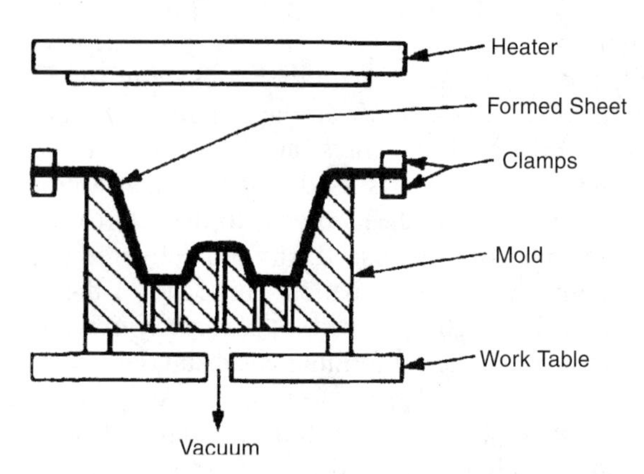

Fig. 4D1 Straight vacuum forming. Thermoplastic sheet is clamped, heated to the softening point, and placed against a one-piece mold cavity. A vacuum pump evacuates the air below the sheet, and atmospheric pressure forces the sheet against the mold so that it assumes the shape of the mold. The sheet cools and stiffens and can be removed and, if necessary, trimmed. (*from Bralla, Design for Manufacturability Handbook*[6].)

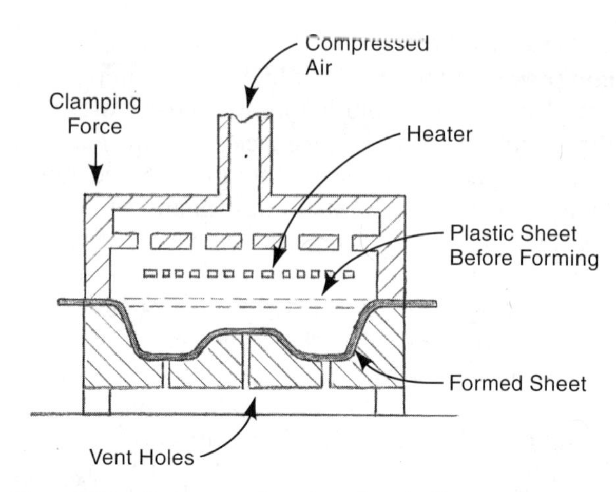

Fig. 4D2 Pressure forming is similar to vacuum forming except that the pressure differential across the sheet is achieved by positive pressure rather than a vacuum.

The advantage of this approach is that a somewhat higher pressure differential can be obtained between the upper and lower sides of the sheet, making it possible to form thicker sheet - up to about 0.375 in (10 mm) - and to produce more sharply-defined forms in the sheet. Fig. 4D2 shows this approach.

D3. ***drape vacuum forming*** - uses a male mold. The heated sheet is pulled down and "draped" over the male mold, prestretching the sheet. When the frame holding the sheet contacts the mold and seals the edges, a vacuum is applied, pulling the sheet firmly around the mold. When the sheet has cooled, it is removed from the mold. This approach is useful for deep-formed parts and permits reentrant shapes to be formed. See Fig. 4D3.

D4. ***plug-assist forming*** - uses a male plug that approximately conforms to the mold cavity shape to pre-stretch the material before it is drawn into the female mold. This sequence, shown in Fig. 4D4, minimizes thinning of the sheet at the bottom of the formed portion and permits easy removal of the formed part from the mold. The heated sheet is placed over the mold and the plug advances to push the sheet into the mold cavity. As the plug advances, the air under the sheet is compressed, forcing the sheet up around the plug. The prevents the sheet from touching and being cooled by the sidewalls of the cavity as the sheet is stretched. The movement of the plug stops before it makes contact with the cavity. The air below the sheet is evacuated and the vacuum draws the sheet off the plug and against the cavity. Then the sheet cools and takes its permanent shape. (The process can also be carried out with positive pressure above the sheet and no vacuum, with only a vent below the sheet.) The plug occupies 70 to 90% of the volume of the cavity. It has the approximate shape of the cavity, and a smooth surface, It is heated to a temperature just below that of the heated sheet. This process is useful for making cup- and box-shaped parts.

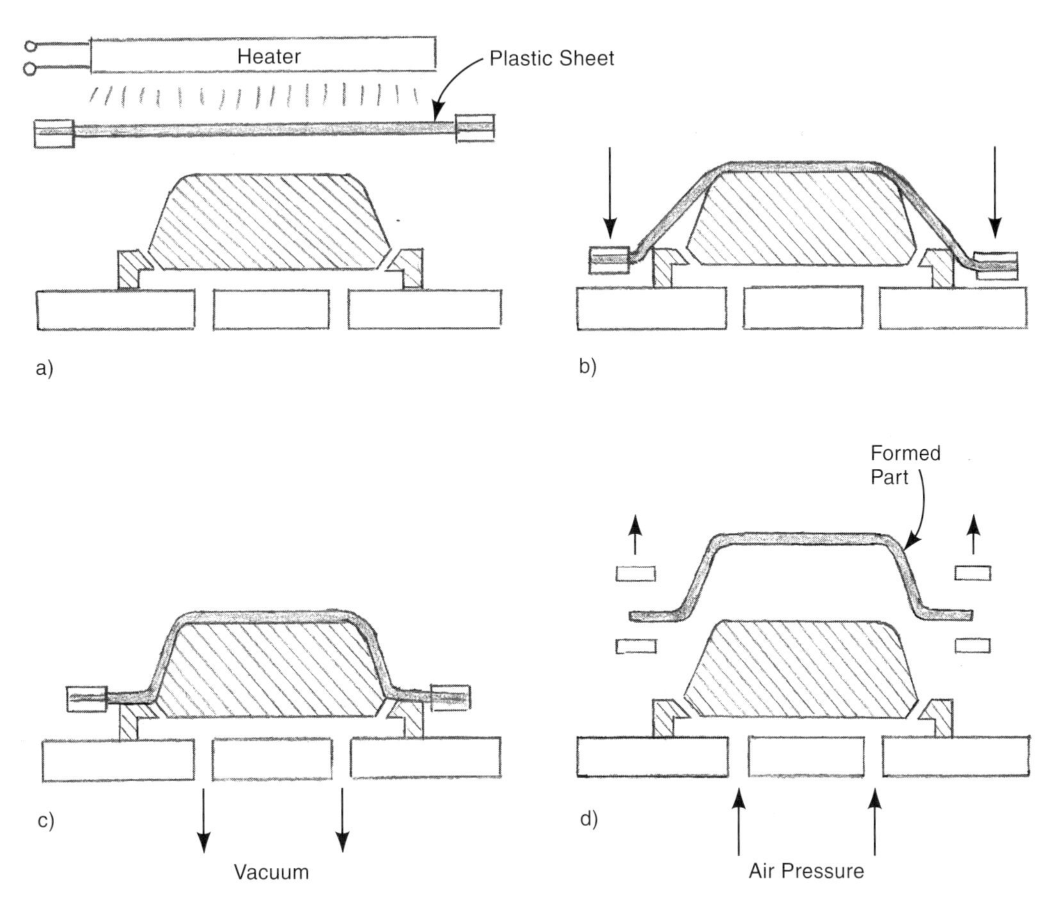

Fig. 4D3 Drape vacuum forming. a) The plastic sheet is placed above the male mold and heated. b) The heater is retracted and the sheet is draped over the mold. c) A vacuum is applied to the underside of the sheet, drawing it tightly against the mold. d) When the sheet has cooled and hardened, the vacuum and grippers are released and positive air pressure forces the sheet off the mold.

D5. ***vacuum snap-back forming*** - In this method, a sheet is heated and sealed across the top of an open chamber. A vacuum pulls it into a concave shape. A male mold is then introduced to the cavity from above. The vacuum under the sheet is gradually reduced and a vacuum is created on the opposite side of the sheet. It pulls the sheet tightly against the male mold where it cools and hardens. This procedure aids in maintaining a more uniform sheet thickness, provides some other quality advantages, and can reduce the necessary size of the starting sheet. Luggage components, automobile parts, and computer housings, which have a textured surface on the convex side, are formed with this method. Fig 4D5 illustrates the process.

D6. ***slip-ring forming (slip forming)*** - is a process similar to deep drawing of sheet metal (2D5), and provides a more uniform wall thickness for thermoformed shapes. The method is used when the plastic at its forming temperature is too stiff to be stretched into complex shapes. The primary applications are with highly-filled and fiber-reinforced thermoplastics. It is used for making such products as firemen's helmets and military aircraft structural supports. The sequence of steps in the operation is as follows: An oversize sheet is held loosely by a ring whose shape is similar to that of the mold. The sheet is heated and descends against a male die. (In some arrangements, the die descends against the sheet.) The sheet slips through the ring as it is drawn into

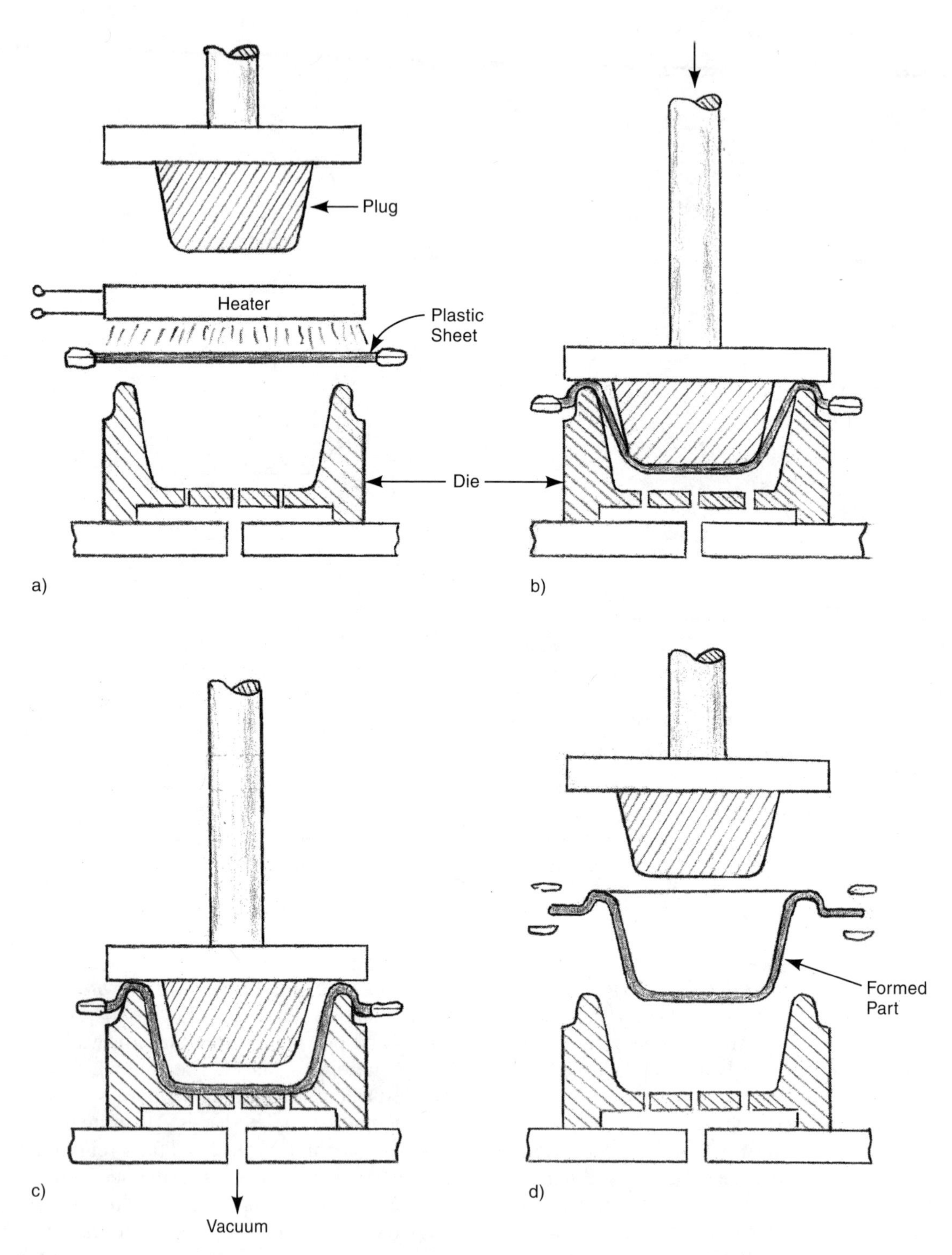

Fig. 4D4 Plug-assist forming: a) The plastic sheet is placed above the mold and heated. (The plug is also heated but to a slightly lower temperature.) b) The heater is removed and the plug descends to move the sheet into the mold cavity. c) A vacuum draws the sheet against the walls of the mold cavity, where the sheet cools and sets to its formed shape. d) The plug retracts and the formed part is removed from the mold.

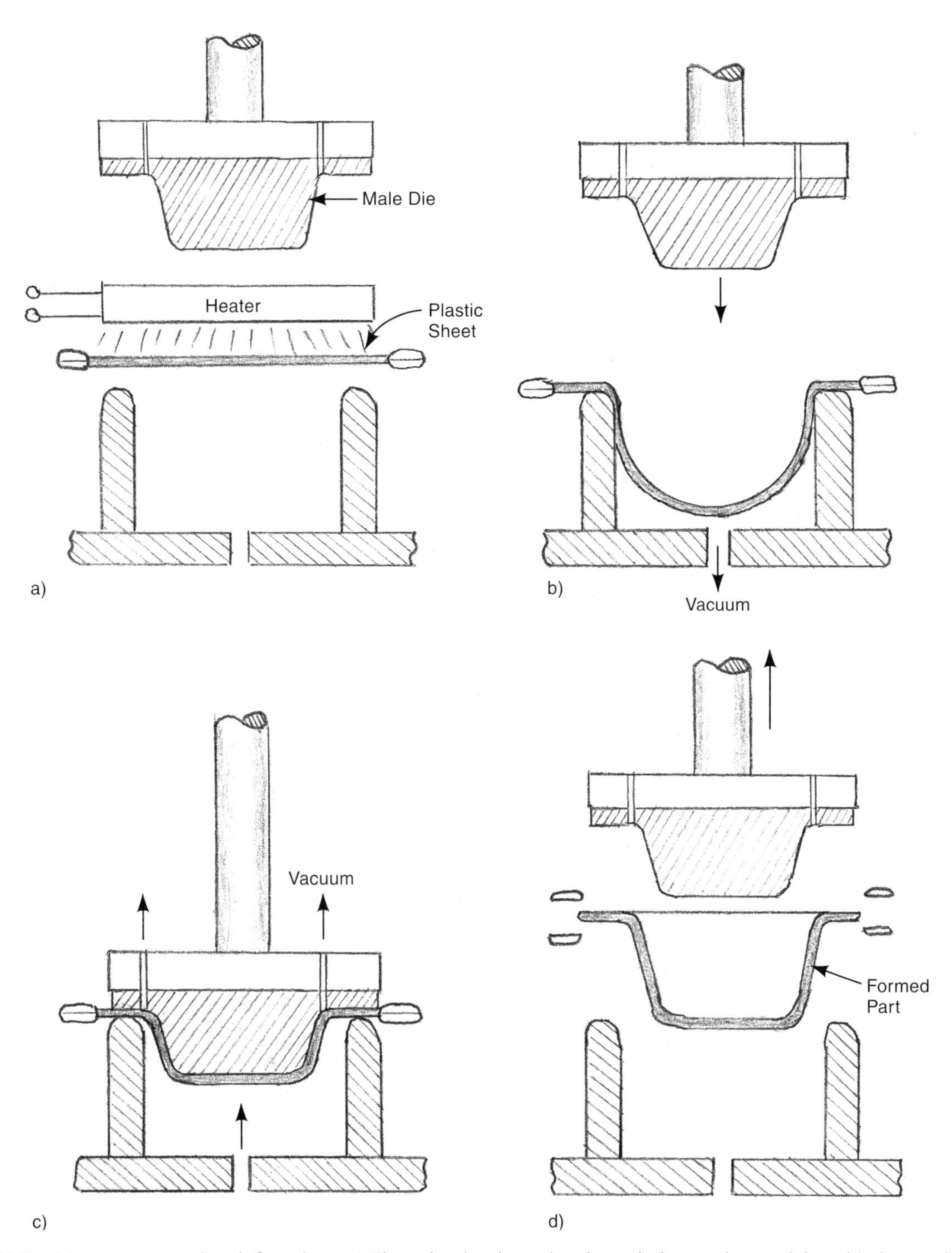

Fig. 4D5 Vacuum snap-back forming. a) The plastic sheet is placed above the mold and is heated. The heater is then retracted. b) The plastic sheet is lowered and sags into the mold, assisted by a vacuum drawn from below. c) The male die descends against the sagged sheet. The vacuum is reversed, drawing the sheet against the male die. d) When the sheet cools and sets to its formed shape, the mold is opened and the formed part is removed.

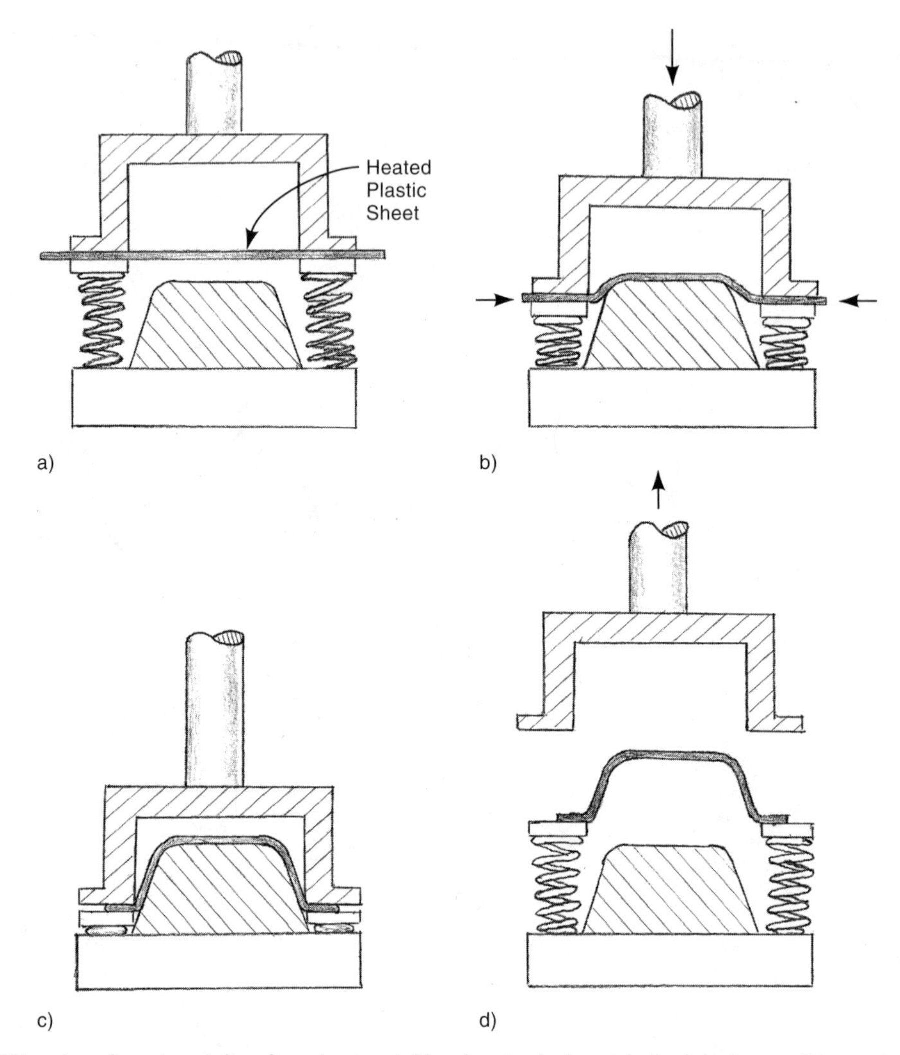

Fig. 4D6 Slip ring forming (slip forming): a) The heated sheet is held above the male mold. Holding force is only moderate. b) The sheet is pulled over the male die and partially slips through the spring-loaded holders. c) When the ram has completed its downstroke, the sheet is drawn over the male die. d) When the sheet has cooled, the ram retracts and the formed, drawn part is pushed upward and can be removed from the die.

the desired shape, instead of simply being stretched. There is no vacuum or air pressure on the sheet. Sheet temperature, the hot strength of the plastic involved, and spring (or air) clamping pressure against the sheet are critical factors to ensure that the sheet material flows without stretching or being scored. Fig. 4D6 illustrates this method.

D7. ***matched mold forming*** - uses mating forming dies between which the heated and softened sheet is placed. The forming dies are mounted in a conventional press. When the dies close, the sheet is pressed into the desired shape and when it further cools (aided by the cooler mold) the dies are opened and the formed part is removed. There is no air pressure or vacuum. Vent holes allow any trapped air to escape. Fig. 4D7 illustrates this method which is comparable to that used for forming sheet metal parts except that dies can be made from various metals and, sometimes, of non-metallic

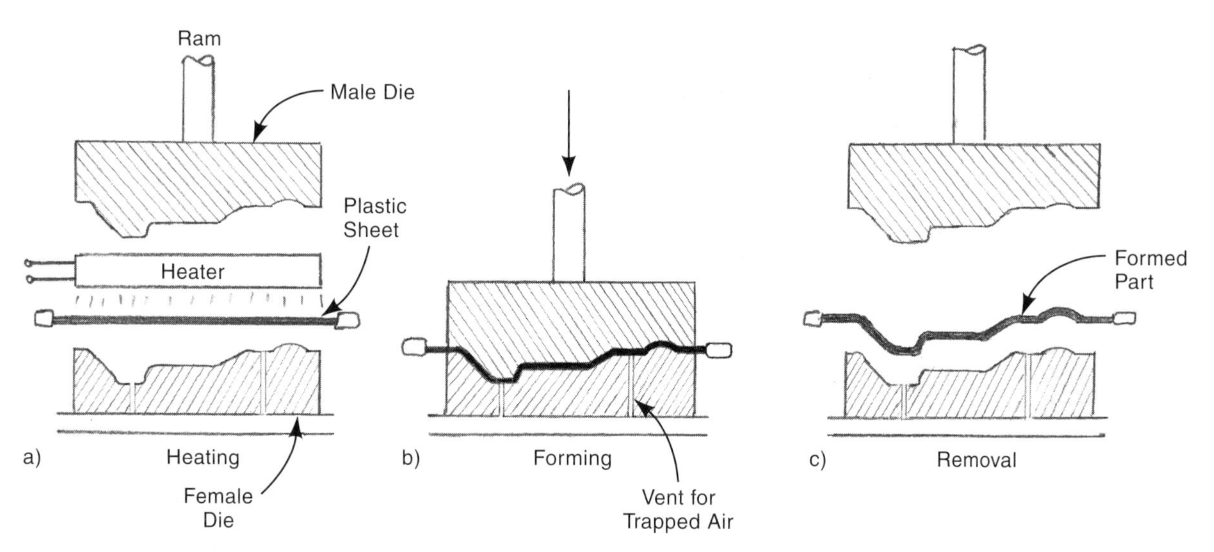

Fig. 4D7 Matched mold forming. a) The plastic sheet is placed between the die halves and heated. The heater is then retracted. b) The upper die descends, pressing the softened sheet between the die halves. The sheet cools and solidifies. c) The die opens and the formed sheet can be removed.

materials, instead of hardened steel. The die halves are kept in close alignment by the press or a die set unless the form is such that one die half can be made from rubber or another soft material that can be used to press the sheet against the other die half. The shape of the rubber die then needs only to approximate the shape of the part. Matched mold forming can produce special surface effects, lettering and other fine detail, and accurate forms. It is also used when the plastic sheet, at its forming temperature, is too fragile to be stretched into complex shapes. The sheet is then formed at a lower temperature using matched tooling. The process is used extensively for forming low-density foam containers such as meat trays, egg cartons, and for substructures in vehicles. (Also see paragraph G9 which refers to the forming of fiber reinforced plastic sheets.)

D8. ***pressure-bubble plug-assist forming*** (also called ***reverse-draw plug-assist forming, pressure-bubble plug-assist forming, billow-up plug-assist forming*** or ***reverse-draw with plug assist***) - In this process, the sheet is clamped, heated, and sealed against the top of the female mold cavity. Air is introduced into the cavity, blowing the sheet upward to a dome-like shape and stretching it evenly. A male plug, whose shape approximates that of the mold cavity but which is only about 85% as large, descends into the dome-shaped sheet. The plug is preheated to a temperature slightly less than that of the sheet. When the plug achieves its full descent, the space below the sheet is connected to a vacuum, drawing the sheet against the female mold cavity. Sometimes the space above the sheet is also pressurized. An advantage of the process is that wall thickness can be quite uniform and well controlled. Fig. 4D8 illustrates the process which is useful for deeply-formed, large-area parts.

D9. ***pressure-bubble vacuum-snapback forming*** or ***billow-up vacuum snap-back forming*** - is quite similar to pressure-bubble plug-assist forming except that the plug is shaped exactly rather than approximately to the final part shape wanted, and the sheet is drawn against the plug instead of against the female cavity. The process sequence is as follows: The sheet is clamped, heated, and sealed against the top edges of the box-like cavity. Air is introduced to the cavity, blowing the sheet upward to a dome-like shape. A male plug of the shape wanted in the part descends into the dome-shaped sheet. As the descent continues, the pressure below the sheet is released and a vacuum above the sheet draws the sheet tightly against the plug where it cools and hardens. This process variation is useful when there is a textured finish on

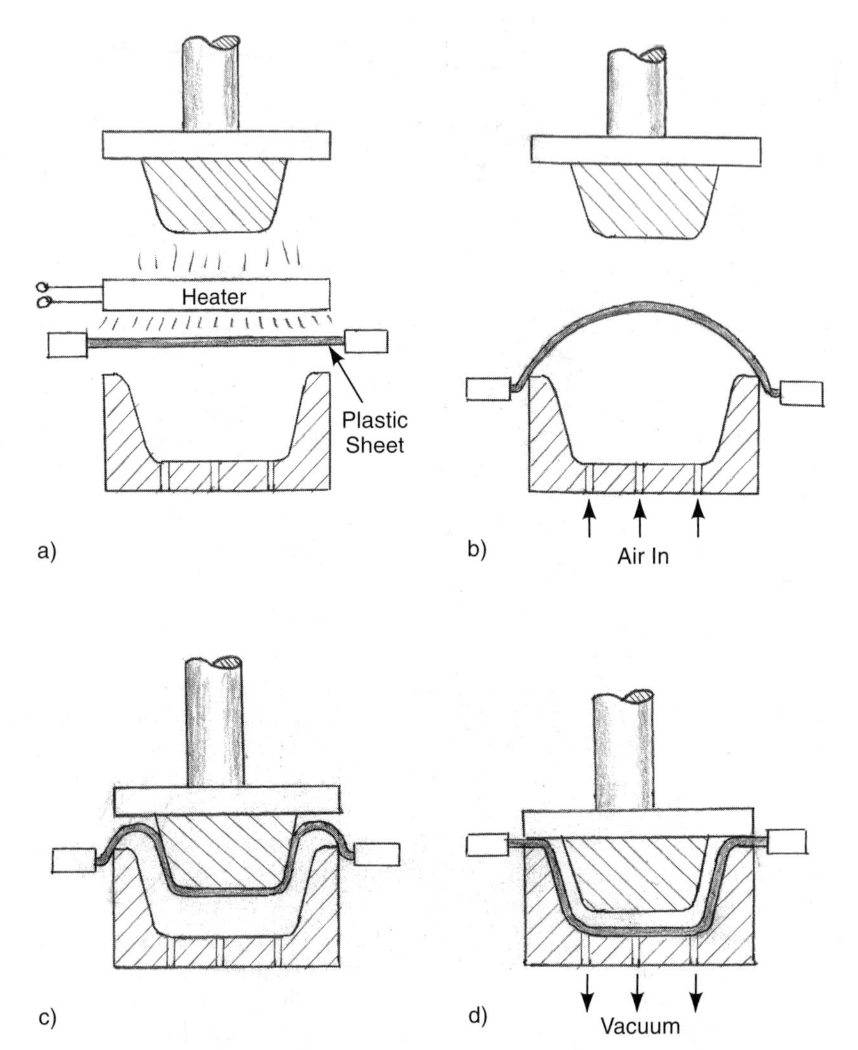

Fig. 4D8 Pressure bubble plug assist forming. a) The plastic sheet is placed above the mold and is heated. The plug above the sheet is also heated. The heater is then retracted. b) Air pressure from below causes the sheet to form a bubble, uniformly stretching the sheet. c) The plug descends against the bubble, bringing the sheet into close proximity to the female mold. d) A vacuum from below the mold draws the sheet against the mold where it cools and solidifies. The mold and plug then open and the formed workpiece is removed, similarly as illustrated in Fig. 4D4.

the sheet that must be incorporated in the outside of the formed part.

D10. ***trapped sheet, contact heat, pressure forming*** - This method is similar to straight vacuum or pressure forming except that the plastic sheet is heated by direct contact with a hot plate. The hot plate is porous so that air can be blown or drawn through it. The plastic sheet to be formed is placed between the mold cavity and a hot plate. The mold and hot plate are then brought together, trapping the plastic sheet between them and sealing the line of contact. A vacuum, applied above the hot plate, draws the plastic sheet to it, where it is heated. Air pressure may also be applied from the mold cavity to ensure close contact between the sheet and the hot plate. When the sheet is sufficiently heated, the air flow is reversed, creating pressure above the trapped sheet. The pressure forces it down against the mold cavity. Air trapped between the sheet and mold is vented, and a vacuum may be drawn below the mold to further force the sheet against the mold. When the sheet has cooled, the resulting formed part can be removed or ejected. Sometimes steel knives are used along the line where the sheet is trapped by the mold. Then, when extra force is applied, the knives trim the formed part to the desired outline. Contact heating is used extensively when the sheet thickness is about 0.010 in (0.25 mm). The primary application is as one operation in a continuous process called, *form, fill, and seal or FFS*. Rigid or semi-rigid containers are formed, filled with either solids or liquids, and then sealed, in a continuous operation. FFS is used extensively in pharmaceuticals for unit dose drugs and in food packaging for single-servings of cheeses or juices.

D11. ***air-slip forming*** - is a variation of snap-back forming. A sheet is clamped, heated, and sealed to the top of a forming box. Pressure is applied below the sheet and the sheet billows up. At the same time, the male mold in the forming box moves upward. (Gaskets at the sides of the mold form a sliding seal at the chamber wall.) Air pressure keeps the sheet from contacting the male mold. When the mold is fully in the up position, the pressure below the sheet is vented and pressure is applied above the sheet to force it against the male mold, where it cools and hardens.

D12. ***free forming (free blowing)*** - This method uses no mold but is useful for forming dome-shaped parts. The sheet is clamped, heated, and blown with air pressure or drawn with vacuum to the desired degree. There is no contact with any forming elements, except at the edges where the sheet is clamped. Acrylic sheet parts whose application needs optical clarity comprise the prime application.

D13. ***dual sheet forming (twin sheet forming)*** - is a method for producing hollow objects from two plastic sheets. Fig. 4D13 illustrates one method for achieving such a result. Two sheets are fed to the machine together, slightly spaced apart. Both are clamped and heated and moved between two halves of a mold. The process requires longer heating time than that required for a single sheet. This is sometimes compensated for by using a rotary table system that includes two heating stations so that each sheet goes through two heating cycles before it is formed. After the sheets are placed for forming, an inflation pin enters the space between them and the mold closes. Air pressure is introduced between the sheets and vacuums are drawn from the two opposing mold cavities. Pressure on one side and vacuum on the other causes the sheets to press against the walls of the two mold cavities. The pressure of the mold closure also bonds the softened plastic sheets together. The formed sheets and the joint between them cool and harden. The inflation pin is withdrawn, the mold opens, and a hollow part is ejected or removed.

Sometimes, the bottom sheet is formed first and an insert is placed on it before the other half of the part is formed and assembled. Another variation of the process introduces urethane foam instead of air pressure between the sheets. The foam adheres to both sheets, to make a strong sandwich construction. Foam-filled boat hulls are produced with this method.

Dual sheet forming is used extensively in Europe to produce such items as phone booth roofs and gaming table tops from PVC or ABS sheet. Shipping pallets and other dunnage products are produced from high-density polyethylene heated to above its melt temperature.

D14. ***solid phase pressure forming*** - is a technique used with several thermoforming processes rather than a particular forming method. It was first used in the thermoforming of thin gauge sheet polypropylene homopolymer. Some sheets tended to split when thermoformed at the temperature that would normally be indicated for the operation. Instead, an approach using a slightly lower temperature for the material was developed. The temperature was such that the material would, technically, still be in the solid phase. Higher forming pressures were used. This approach is still in use although new copolymer polypropylene formulations are thermoformable at higher temperatures. The technique is applicable to various crystalline plastics as well as polypropylene. Because of the reduced formability at the temperatures used, an increased pressure of 50 to 100 psi (345 to 690 kPa) is required. Vacuum forming does not provide sufficient pressure. Careful monitoring and control of the heating process is also necessary. The major

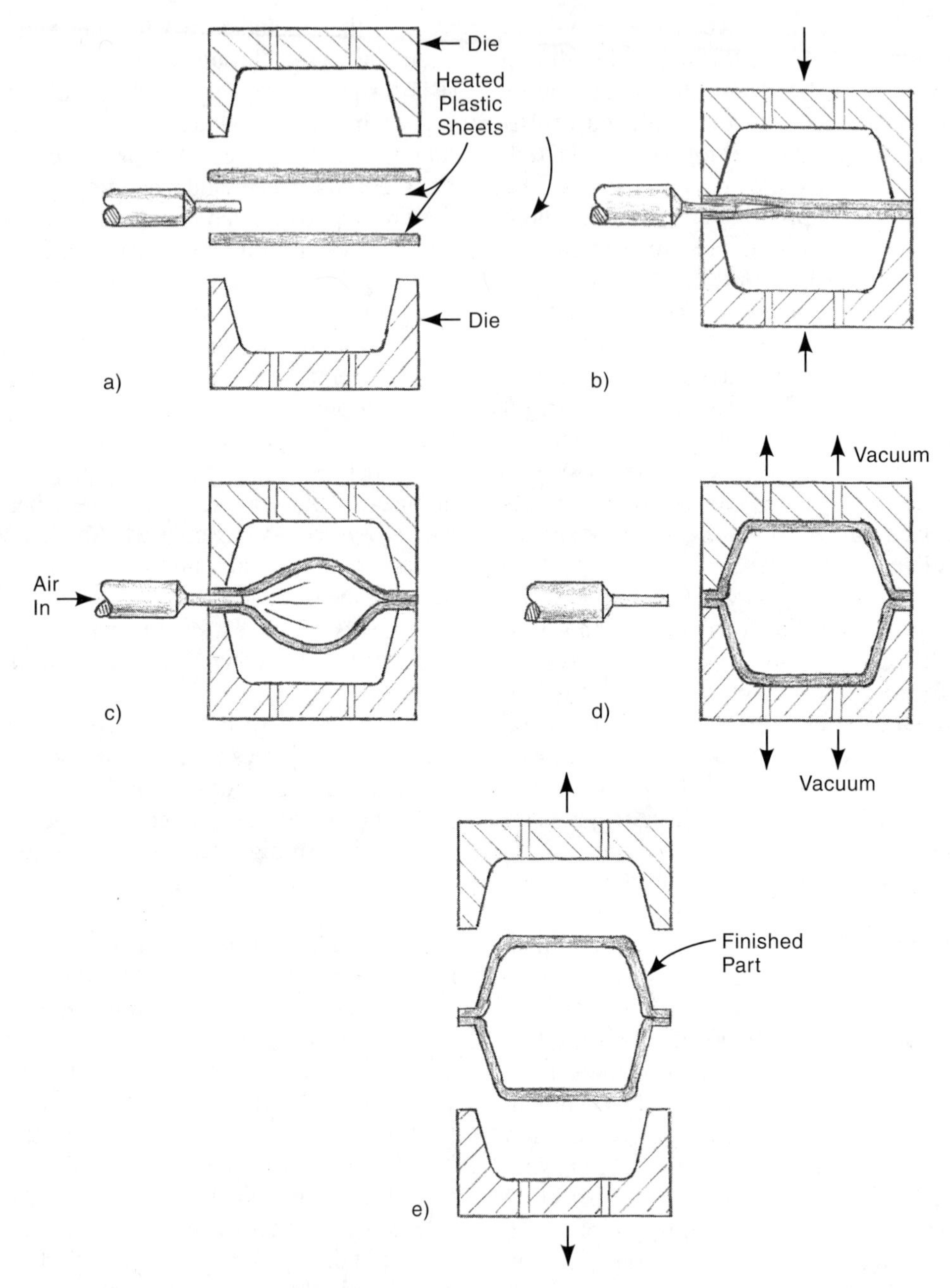

Fig. 4D13 Dual sheet (twin sheet) forming: a) Two heated plastic sheets are placed between the halves of the die. A thin air-inflation tube is also positioned between the die halves. b) The die closes, clamping the heated sheets and the inflation tube. c) Compressed air is pumped into the space between the sheets, driving the sheets against the die cavities. d) Vacuums draw the sheets tightly against the die walls. The air-inflation tube is withdrawn. Pressure of the die halves seals the two sheets together. e) The formed sheets cool, creating a sealed hollow component. The dies open and the finished part is removed.

application is the forming of smaller food cups such as those used for single portion servings.

E. Rotational Molding

Rotational molding, sometimes called, rotational casting, is a means for producing components that are thin-walled, hollow, seamless, and often large. It utilizes the two-axis rotation of a heated, clamshell-like, thin-walled, metal mold. A measured amount of liquid or powdered thermoplastic resin is charged to the mold. The mold is heated as it rotates in two planes. The resin continuously falls by gravity to the lowest point, and the heated mold walls become coated with the resin, which fuses together. The mold is then subjected to cooling by water, cold air, or a sprayed water-air mixture. This cools the plastic, causing it to solidify. The mold is then opened and the hollow part is removed. The equipment commonly provides three stations for the mold: 1) a loading-unloading station where the mold does not rotate, 2) a heating station where the mold has entered a hot-air oven, and 3) a cooling station. The mold is mounted on an arm, which carries it sequentially to these three stations. However, many other machine configurations are in use, including those with straight line and batch-type arrangements. Large containers, tanks, and outdoor play equipment, are made from polyethylene powder by this method. Gaskets, syringe bulbs, beach balls, hollow doll parts and other toys, are other typical applications and are made from liquid polyvinyl chloride (PVC)(vinyl plastisol). Fig. 4E illustrates the process.

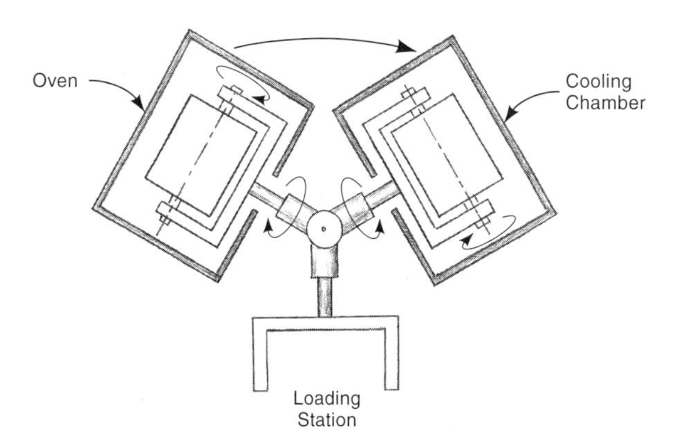

Fig. 4E Three-station equipment layout for rotational molding.

F. Blow Molding

Blow molding is a high-production method for making thin-walled, hollow, one-piece, objects of thermoplastics. Air pressure applied inside a small hollow and heated plastic piece (called a parison), expands it like a balloon and forces it against the walls of a mold cavity, whose shape it assumes. There it cools and hardens. The mold opens and the part is ejected. Flash, if any, is trimmed off and recycled. Normally, all these operations, including the forming of the parison, are part of an automatic sequence.

F1. ***extrusion blow molding*** - In this process, the parison is extruded as a tube with essentially the same method as is used for other plastic extrusions. The tube is then inserted in a blow-molding die with one end engaging a blow pin or needle. As the die is closed, the tube is pinched at both ends. The pinched-off tube is expanded by air pressure against the cooled walls of the die. The pressure is held for a brief period while the part cools and the material hardens. After the die opens and the part is ejected, the surplus material adjacent to the pinched-off areas is removed. All these operations are automatic. Fig. 4F1 illustrates the process which is used for about 75% of current blow-molded products. These include all kinds of containers, especially of larger blow moldings, for a variety of industries. Typical products besides containers are components such as automobile tanks, bumpers, seat backs and center consoles, housings, enclosures, toys, balls, and plastic duct sections.

F2. ***injection blow molding*** - The operation for injection blow molding is similar to extrusion blow molding except that the parison is made by injection molding instead of by extrusion. The parison is molded over a mandrel to provide the hollow shape, and this mandrel transfers the hot parison to the blow-molding die, and then functions as the blow nozzle. Air entering the blow nozzle expands the parison against the cool walls of the blow mold. Trimming of the molded part is normally not required. In the usual arrangement, a three-station, horizontal, indexing table is an

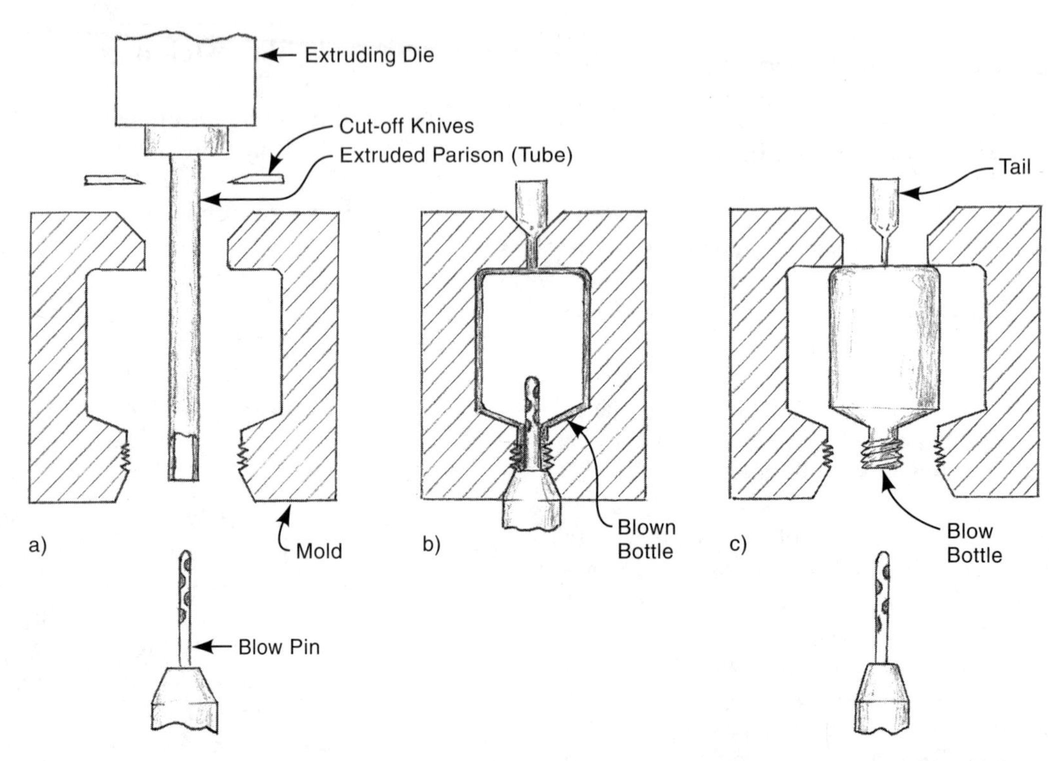

Fig. 4F1 Extrusion blow molding. a) The die is open and the parison tube is extruded between mold halves. b) The tube is cut off as the mold closes and a blow pin is inserted in the other end of the tube. The tube is pinched off at the top (bottom of bottle). The tube is expanded by air pressure to line the interior surface of the mold, forming a bottle. c) The bottle cools and the plastic solidifies. The blow pin retracts and the mold opens.

essential part of the equipment. The injection molding of the parison takes place at one station, inflation of the parison at the second station, and ejection of the finished part at the third station. (Some machines have a fourth station for pre-inflation of the part, or for a post-molding operation such as label attachment.) The process is adaptable to hollow parts that have some special shaped portion. The neck and opening of bottles, including screw threads for the cap, are produced in the injection mold as part of the parison. They can be made to closer dimensional tolerances than with extrusion blow molding and the wall thickness can be set as needed and more accurately controlled. The injection blow molding process is used extensively for smaller bottles of household products. It is illustrated in Fig. 4F2.

F3. ***In stretch blow molding*** - a center rod stretches the parison to about two times its length. This axial stretching, plus the circumferential stretching action of the inflation, produces a biaxial orientation of the molecules in the walls of the part, improving the strength, barrier properties, and clarity of the walls. The process has some complexities. The temperature of the workpiece during the stretching operation is critical, and that temperature must be essentially uniform throughout the wall of the part; the inflation air pressure must be somewhat high in order to achieve the benefits of stretching. There are two basic stretch blowing methods: the continuous or single-stage process in which the temperature conditioning and stretching take place immediately after the parison is molded, and the two-stage process in which temperature conditioning and stretching take place later.

The single-stage method involves the following steps: 1) injection molding of the parison 2) temperature conditioning, in which the parison is

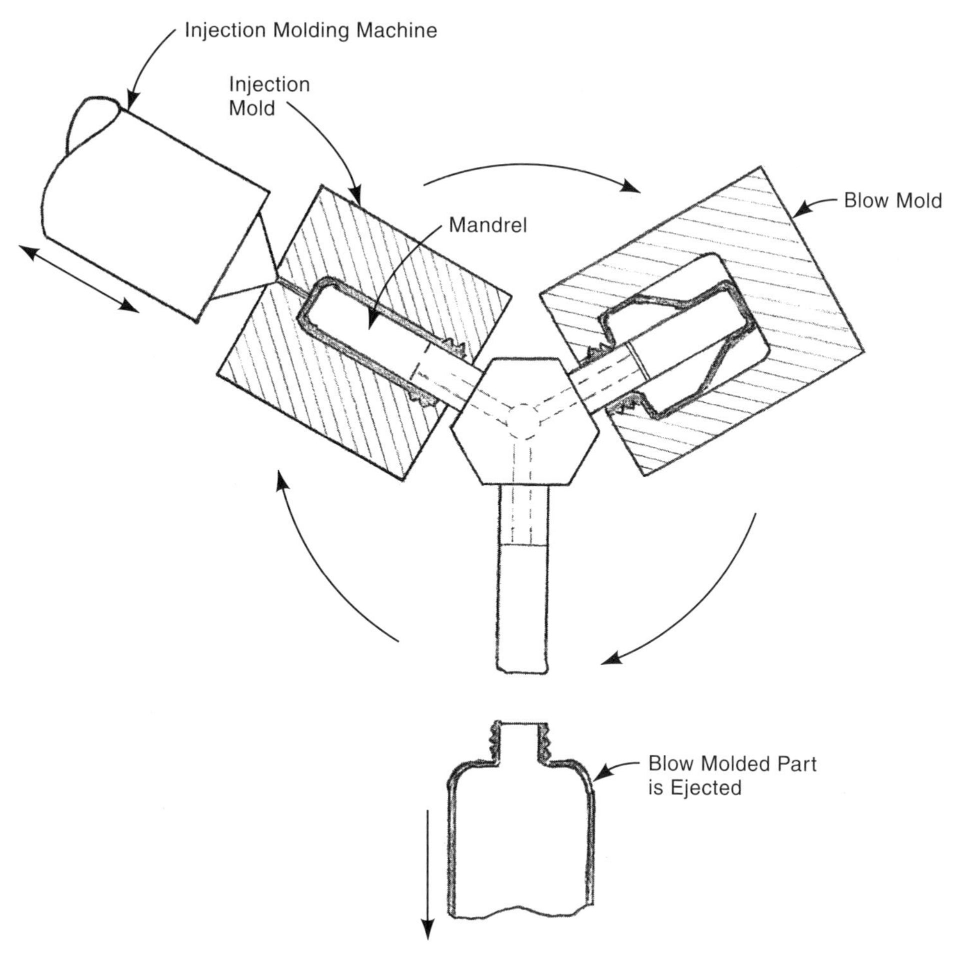

Fig. 4F2 Injection blow molding with a three-station index table arrangement.

brought to the best temperature for stretching, with uniformity throughout, 3) stretching and inflation in the mold and, 4) cooling and ejection of the finished part. Though the single stage process has some simplicity, it is not as rapid, overall, since the individual operations do not require the same amount of time. The two-stage method requires the parisons to be reheated, but the slower operations can be done in multiples to balance the flow. The two-stage approach, though more capital intensive, is most appropriate for the mass production levels required for most applications. Plastic soft drink bottles constitute the major application. Small bottles for pills and vitamins is another important use. PET, used for soft drink bottles, is the prime material, though other applications often involve different thermoplastics.

F4. ***multilayer blow molding*** - is a blow-molding operation that utilizes co-extrusion (see 4I2) - or co-injection-molding (see 4C3e) - to provide two or more layers in the parisons and in the final blow molded products. ***Coinjection blow molding*** and ***coextrusion blow molding*** are terms also applied to this approach. These processes are used in the production of containers when it is important to provide barriers against permeation and odor escape, and when the container is to be used for solvents, gasoline, herbicides, cosmetics, or pharmaceuticals. Stretching operations, as described above, are also common in the production of multiple-layer blow-molded bottles.

F5. ***dip blow molding*** - uses plastic resin adhering to a core rod, instead of an injected or extruded piece, as a parison. The core rod, whose diameter is the same as the inside diameter of the finished part's neck opening, is inserted through a narrow opening into a chamber holding molten plastic. The core rod is then withdrawn while, at the same time, a piston advances into the chamber from the opposite end, maintaining pressure in the chamber and insuring that material remains on the rod. The core rod, when withdrawn, then has a coating of hot plastic. The coated rod is transferred to the blow molding station where air is blown through it. The air expands the plastic coating into contact with the mold cavity walls. The product is then cooled and hardens into a hollow part.

F6. ***other blow molding processes*** - Labels are sometimes placed in the mold before the inflation phase to provide better adhesion and protection of the label. In high-production situations, when labels are inserted, automatic equipment picks up each label by vacuum and positions it in the blow-mold cavity. A vacuum source, drawing air through small perforations in the mold cavity, holds the label in place as the parison is inflated in the mold. This method provides very good label adhesion because the plastic material it contacts is almost in the molten state.

Some extrusion blow-molding dies are equipped with de-flashing jaws inside the mold to grab and tear off the bottom flash as the finished bottle is ejected.

G. Processes for Reinforced Thermosetting Plastics

This section deals with composites of reinforcing materials such as glass or carbon, and a matrix of thermosetting plastic such as polyester or epoxy. Other reinforcements can be used with other polymers but this section refers to processes where the reinforcement and thermosetting materials predominate. The reinforcing fibers in the composite material provide much greater strength and rigidity than is possible with unreinforced plastics. Several processes have been developed to manufacture useful components and products from these reinforced materials.

G1. ***hand lay-up*** - In this method, an open (one piece) mold is used. Its cavity surface is coated with wax or another release agent. Normally, a "gel coat" of the resin, which is a layer of resin without reinforcing fibers, is first applied to the mold with a spray gun, and allowed to set. This is in order to ensure that the reinforcement will not show through the surface of the molded part and to ensure that the surface will be smooth. Fiberglass or other reinforcement in the form of a mat of unwoven fibers or a fabric, or both mat and fabric, is then placed manually in the mold. Liquid thermosetting resin is poured, brushed, or sprayed onto the reinforcing material and is spread to a uniform distribution with hand rollers or by other methods. (The mat or fabric may also be pre-impregnated with resin.)

It is important to ensure that the reinforcing fibers are properly wetted by the resin and that the materials are compacted into a solid mass. Often, several layers of resin and reinforcing material are applied. Normally, the plastic is catalyzed to cure at room temperature. When the plastic has polymerized, the part is solid and can be removed from the mold. Trimming is usually required at the edges and is performed with hand trimming tools. Depending on which surface must have the superior finish, the mold will be of female or male shape (female shapes are used for fiberglass boats, machine covers, or housings). The mold itself may be made from fiberglass reinforced plastics, but wood, sheet metal, plaster, and other inexpensive materials may be used. The process is adapted to large parts, especially those made in small quantities. Truck wind deflectors, aircraft parts, and vehicle panels are typical components made with the process. Fig. 4G1 illustrates hand lay-up. Robotic and

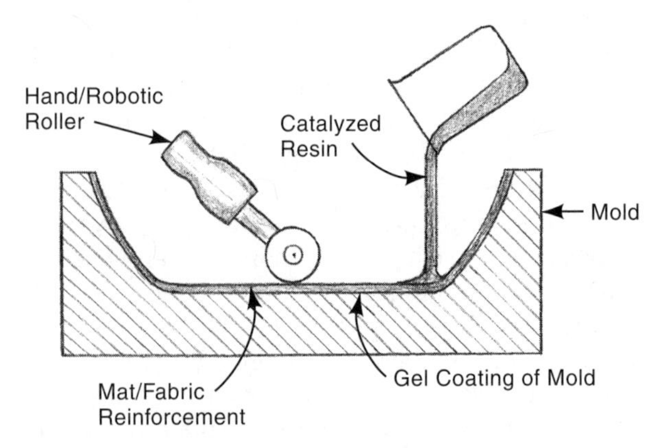

Fig. 4G1 Hand lay-up of a reinforced thermosetting plastic part.

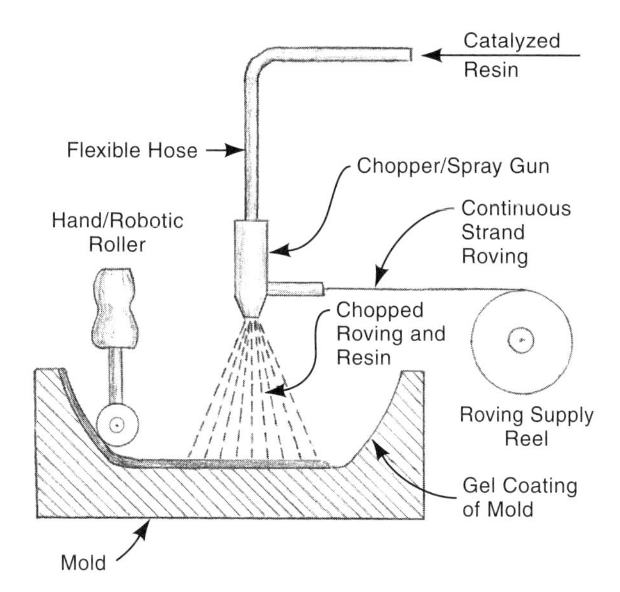

Fig. 4G2 Spray-up molding of a reinforced thermosetting plastic part.

other mechanical assistance is increasingly being employed to reduce the labor required to apply the reinforcements and resin and to properly distribute the resin.

G2. ***spray-up*** - In this method, illustrated in Fig. 4G2, an open mold is used, as with hand lay-up. The resin and reinforcing material are sprayed together into the mold from a gun that chops the reinforcing fibers into short lengths and mixes them with the catalyzed resin. The mixture then must be rolled to ensure that there is a dense, fully-wetted mat of reinforcement material. The process is used to make plastic shower stalls, bathtubs, and other products at higher production rates than are possible by hand lay-up. The method is less labor intensive than hand lay-up. Robotically-controlled guns and rollers are used for some components. The adverse effect of overspray on air quality in the workplace is a factor that must be dealt with when this method is used. As with hand lay-up, room temperature curing is usually used, though ovens or other supplementary heating is sometimes employed. A gel coat is commonly applied to the mold before the spray-up commences.

G3. ***vacuum-bag molding*** - is an augmentation of lay-up or spray-up molding. It involves the placement of a plastic film on top of the laminate in the open mold, sealing the edges, and then using a vacuum from the bottom of the mold to draw the film tightly against the laminate. Atmospheric pressure then helps provide a smoother interior surface and tends to close any voids that exist.

G4. ***pressure-bag molding*** - is another augmentation for lay-up or spray-up molding. In this case, a bag above the plastic film on top of the laminate is clamped to the mold and inflated, providing pressure to force the film tightly against the laminate. Pressures normally used are 30 to 50 psi (210 to 345 KPa). Venting of the mold allows any excessive air in the laminate to escape.

G5. ***autoclave molding*** - is a third means of providing pressure to smooth the surface of the laminate and close any voids that exist. Film is laid on top of the laminate in the open mold and the filled mold is placed in an autoclave. Pressure inside the autoclave to about 80 psi (560 kPa), forces the film against the laminate. This pressure is higher than can be achieved with an inflated bag, and provides even stronger smoothing and compression of the laminate. The autoclave is normally heated to accelerate the cure of the thermosetting plastic in the laminate and permit the use of resins that require elevated temperatures for curing. This approach is used in the aerospace industry, where quality requirements for reinforced plastic parts are highly critical.

G6. ***centrifugal casting*** - is a process used in the production of fiber-reinforced plastic piping, cylindrical containers, and tanks. The operation takes place inside a hollow, cylindrical mold. A reinforcing mat is placed inside the mold to cover the full inner surface. Then, resin is sprayed on the mat material as the mold is rotated slowly. (An alternative method is to spray a mixture of chopped reinforcing fibers and resin as is done with conventional spray-up molding.) When the required amount of reinforcing mat and resin are in place, the speed of rotation of the mold is increased. The resulting centrifugal force, then ensures that the fibers are fully wetted and that the resin-fiber mix is sufficiently dense. The mold may be heated to accelerate the curing of the resin.

G7. ***filament winding*** - is a reinforced-plastics process that uses continuous lengths of reinforcing fiber instead of chopped strands, mats, or fabric. The continuous lengths provide superior strength. The filaments are wound, along with a thermosetting plastic resin, over a mandrel that has the diameter of the part to be produced. The process was originally used in the fabrication of pressure or storage containers where the extra strength was particularly important. It has been developed to be applied to other shapes, including non-round cross sections. Helicopter rotors and windmill blades are examples. Other products include pipes, tanks, lighting poles, rocket components, drive shafts, aircraft fuselage sections, ski poles, golf poles, and tennis rackets. The resin can be applied to the filament before or during the winding operation and, occasionally, after winding. Most commonly, the filament is either precoated or passed through a resin bath during winding. The winding direction can be circumferential, helical, polar, or some combination of these, depending on which direction is most important from the strength standpoint of the finished component. Cris-cross patterns, and others that provide strength in multiple directions, are common. Mandrels can be inflatable, or mechanically collapsible if the end shape of the workpiece requires a size reduction of the mandrel so that it can be removed. Fig. 4G7 illustrates the process. Polyester and epoxy are the most common resins used. Glass fibers are most common but carbon, aramid and other fiber materials are also employed.

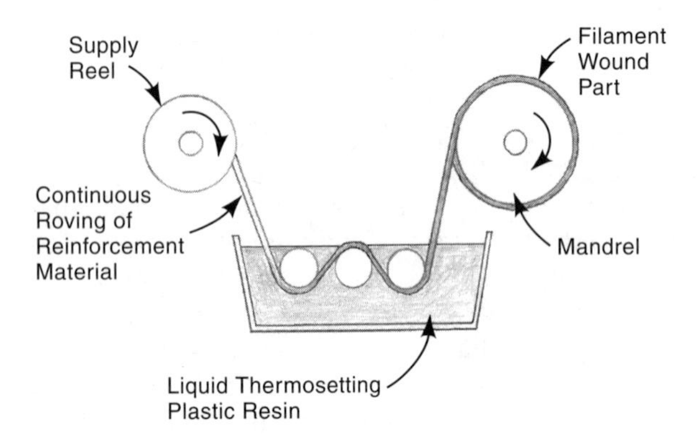

Fig. 4G7 Filament winding is useful for making tanks, pipes, and various round and non-round hollow parts of reinforced plastics.

G8. ***continuous laminating*** - is an operation in which sheets of reinforced plastics are produced. The term applies whether the operation is performed with thermoplastics or thermosets. If it involves a thermoplastic, the operation is usually is a preparatory one; the sheet produced is intended for later forming. When thermosetting plastics are used, the purpose of the operation may be either preparatory, to produce a sheet molding compound for later molding, or final, to produce a rigid sheet to be is used for such applications as skylights, building walls, greenhouse glazing or printed electronic-circuit boards. Note: Also see I4 below, *extrusion coating and laminating*.

G8a. ***continuous laminating with a thermosetting plastic*** - in this process, shown in Fig. 4G8a, reinforcing fibers, usually chopped lengths of glass fiber, are placed to cover a plastic film carrier which lies on the surface of a belt conveyor. A liquid resin, most commonly polyester, is applied to the reinforcing material. The resins used usually incorporate several additives: colorants, UV stabilizers, flame retardants, and fillers. A top film is applied. The resin-fiber mix on the film is conveyed to a pressure roller that bears against it, kneading it and ensuring that the fibers are fully wetted. Additional rollers remove any trapped air and control the thickness of the laminate. If the laminate is to be used for later forming (see matched metal mold forming below), it is coiled or cut to length at this point. If the laminate is to be made into rigid sheet, it is conveyed to a curing area where the resin cross-links, producing a solid sheet. There may be other rollers that put corrugations in the sheet before the resin is cured. After curing, the plastic films are stripped from the top and bottom of the laminate, edges are trimmed, and pieces are cut to length.

G8b. ***continuous laminating with a thermoplastic*** - Polypropylene or another thermoplastic, in sheet, extrudate, or even powder form, is fed both above and below a fabric or mat of reinforcing material into a double-belt machine. A layer of metal foil may also be included. The belts bear against the top and bottom of the material and supply

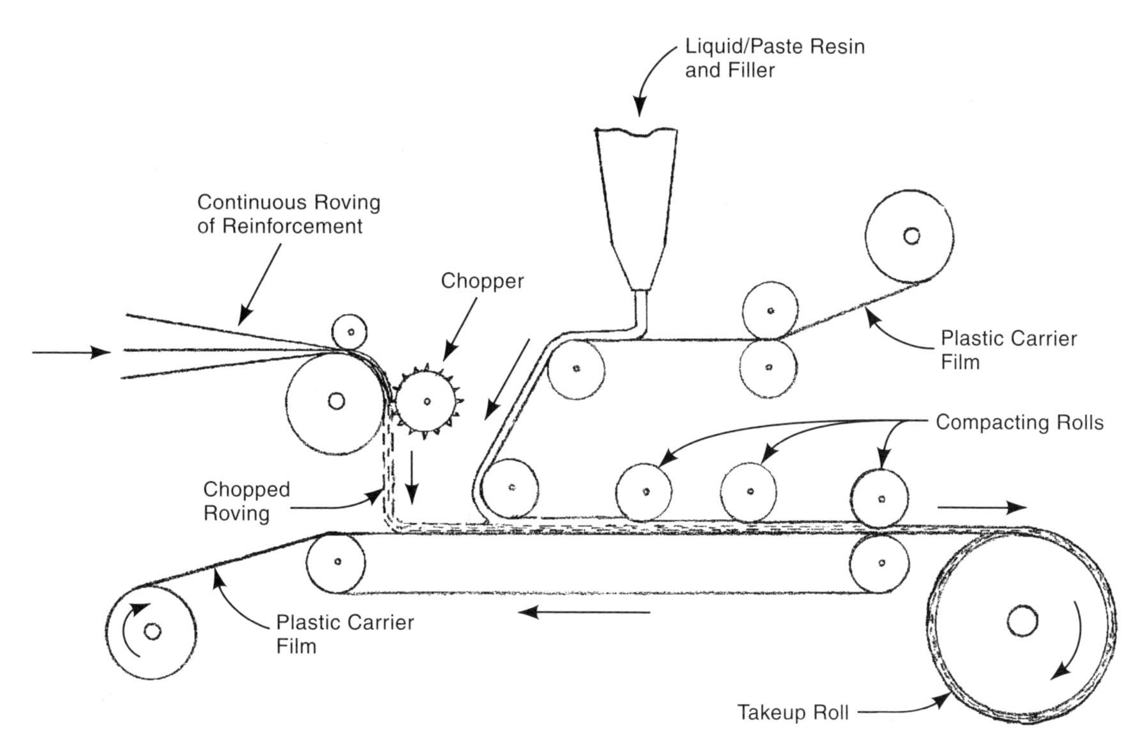

Fig. 4G8a Making a continuous laminate of reinforced thermosetting plastic for later molding. Edge trimming and cutting to length usually follow the operation shown.

heat, which softens the thermoplastic, causing it to flow around the reinforcing material. As the material proceeds along between the belts, it enters a cooling zone where the plastic re-solidifies. Edges may be trimmed, and the resulting sheet material may be either coiled or sheared into sheets. The laminated material that is produced is suitable for a variety of forming operations, the most common one being the matched metal-mold forming operation described below.

G9. ***matched metal mold forming (cold stamping)*** - is the same as the matched mold forming method described in paragraph D7 except that it applies to fiber-reinforced thermoplastic sheets (which are sometimes called, STC for "stampable thermoplastic composites"). The term, "stamping" is used because the softness of the heated composite sheet permits a quite-rapid die closure motion. The process, for which the illustration in Fig. 4D7 is also applicable, is used to make automotive components, where the combination of resilience (from the thermoplastics), and strength (from the reinforcing fiber), is important. Bumper bars, engine covers, and battery trays are common applications.

G10. ***matched metal mold forming of reinforced thermosetting material*** - Sheet molding compound (SMC), a combination of thermosetting plastic and reinforcing mat or fabric, is often formed into useful components by a compression molding process. The process is used to form machine housings, garden and farm equipment parts, trays, and other parts that can be formed from flat sheet. The sheet is placed in the heated mold, which consists of two accurately-machined halves. The top half descends, pressing the sheet between the two halves, where the heat from the molds cures and hardens the thermosetting plastic matrix. For applications requiring a smooth surface that does not show the reinforcing fibers, an in-mold additional resin coating may be applied. Fig. 4G10 illustrates the process.

G11. ***pultrusion*** - is a process used for making parts of reinforced plastics that have constant cross sections of any length. Such sections are similar in

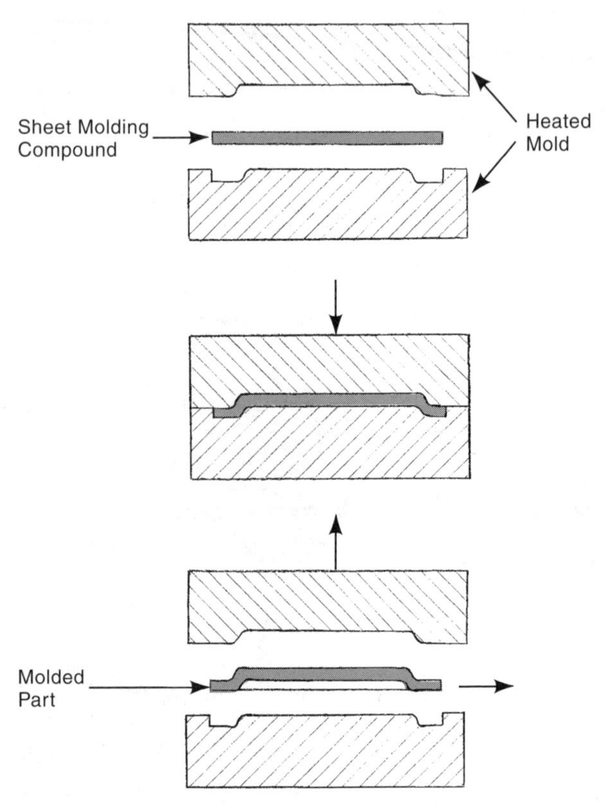

Fig. 4G10 Matched metal mold forming of reinforced thermosetting sheet molding compound (SMC).

shape to profile extrusions, but contain reinforcing fibers, which are arranged to lie in an axial direction in the part. Bundles of long reinforcements, usually glass fibers, are guided through a bath of liquid resin (normally thermosetting) so that each fiber is fully wetted. These reinforcements are then guided together and preformed into the approximate profile desired. This preform is then pulled through a heated die of the exact profile desired. The heat of the die changes the thermosetting material from liquid to a soft semi-solid, and then into a cured rigid plastic. A part with full fiber reinforcement is created. The term, pultrusion, results from the fact that the material is pulled through the die rather than pushed, as is done with conventional profile extrusions. See Fig. 4G11. After the pultrusion, parts are cut to the length desired, typically by cutoff saw. Pultrusions are used in applications requiring light weight, corrosion resistance, electrical insulation, and strength. Commonly pultruded shapes are I-beams, channels, angles, bars, and rods. Ladder parts, fishing poles, ski poles, and golf club shafts are among common applications.

G12. ***pulforming*** - is similar to pultrusion (see Fig. 4G11) except that it produces parts of non-uniform cross section. Tool handles and plastic leaf springs are two common products produced.

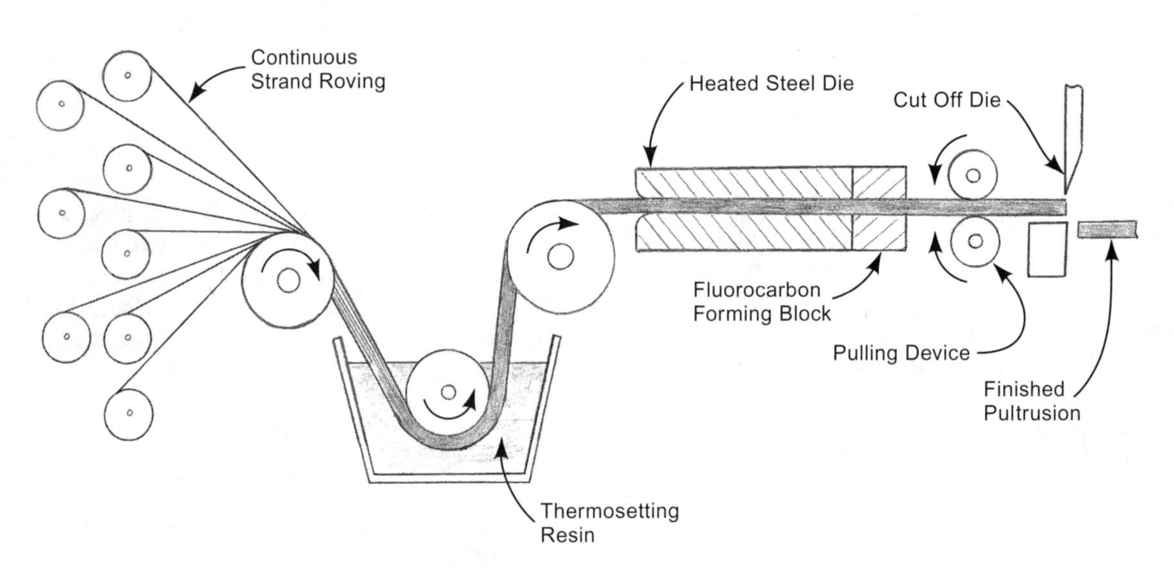

Fig. 4G11 Pultrusion shown schematically.

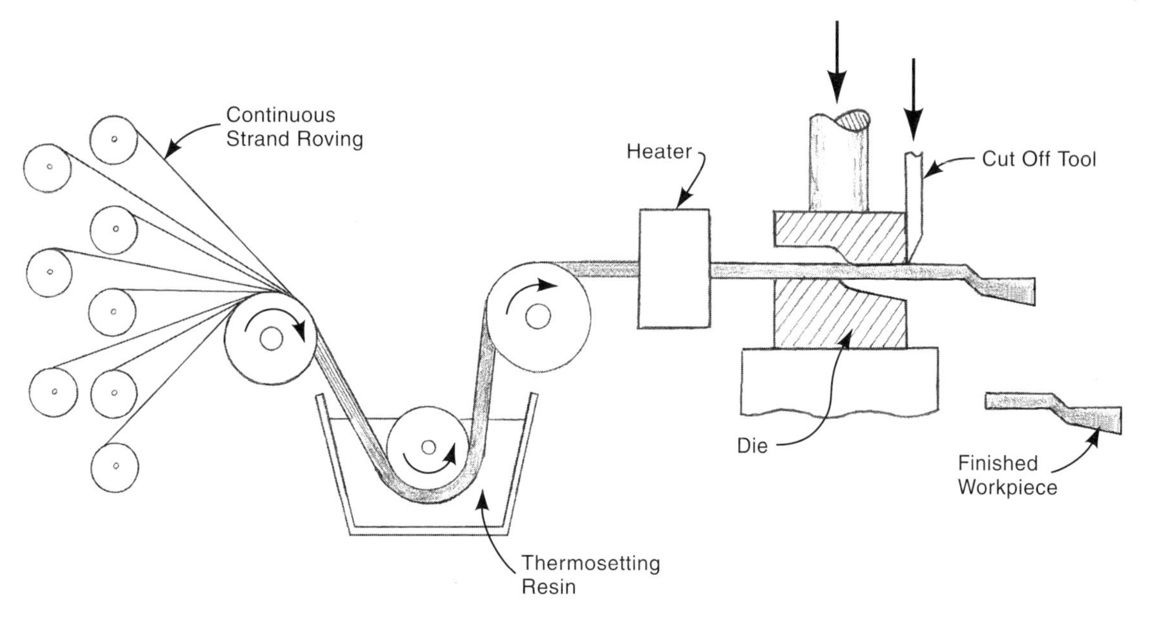

Fig. 4G12 Pulforming shown schematically. Before the die closes, additional uncured material may be added to accommodate the shape of the part, which can be tapered or otherwise non-uniform in cross section.

Instead of pulling the preformed material through an extrusion-like die to create a constant cross section, it is placed in a mold. Additional material may be added where needed. The heated mold is closed, curing the resin-reinforcement mix to the desired shape. Although the process does not produce constant cross sectional shapes, parts produced often have only minor changes in shape along their length. Fig. 4G12 presents an example of the process.

G13. ***resin transfer molding (RTM) (liquid resin molding)*** - In this process, reinforcing material, in mat or woven form, is placed in an open two-piece mold. The mold is then closed and low-viscosity liquid thermosetting material from two metering pumps - one for the resin and the other for the catalyst - is fed into a motionless mixing device and then into the mold. The material cures in the mold. Injection pressures are low, from 5 to 50 psi (35 to 345 kPa), so fragile inserts can be incorporated in the mold. There is only minimum mold wear. Inexpensive molds with shorter lead times can be employed. Typically, the mold gate is at the bottom and the mold fills with the liquid resin mixture from the bottom, as air bleeds from vent holes at the top of the mold. The material cures in the mold, either with the aid of heat or with a resin formulation that cures at room temperature. The mold then opens, allowing the solid part to be ejected. Both sides of the part can have the surface finish imparted by the mold, so there is not the rough finish on one side of spray-up or hand lay-up molding. The process is economic for medium-quantity production. Polyesters, epoxies, acrylics, vinylesters and phenolic resins are used. Reinforcements of glass, carbon, boron or Kevlar fibers can be incorporated. Gel coats can be sprayed on both mold halves before the reinforcement is placed. The process can be used for rather large parts including vehicle panels, bathroom and shower units, truck air deflectors, chairs, and antenna dishes. High-temperature aerospace parts are also made by this method. Fig. 4G13 shows the process.

G14. ***other processes for reinforced thermosetting plastics*** - Other processes include casting (See 4H), compression molding (4B), injection molding (4C), reaction injection molding (4C3b), extrusion (4I), and rotational molding (4E), that are used for unreinforced materials, can be used with reinforced thermosetting materials as well. Most of these processes are quite workable when the reinforcing fibers are short, so that the resin compound

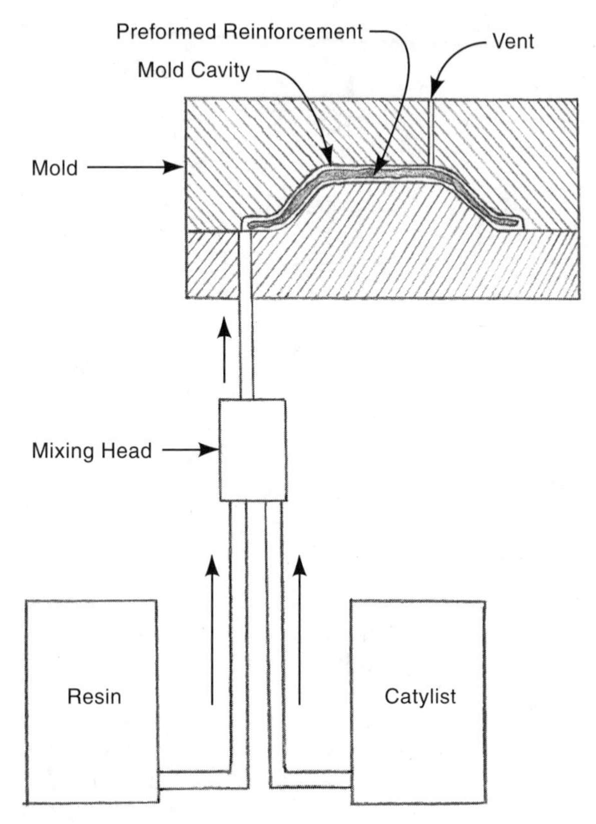

Fig. 4G13 Schematic illustration of the resin transfer molding (RTM) process.

can be worked in a manner similar to the way it would be worked without reinforcement. However, even short fibers can provide significant property enhancement to plastic parts.

H. Casting of Plastics

When plastics are cast, the resin, in liquid form, is poured into a mold cavity. The viscosity of the resin is low enough so that pressure is not required to fill the mold cavity. However, a vacuum may be applied to the liquid in the mold, in some cases, to remove trapped air or other gases. Manual mixing and casting are common, but automatic mixing, metering, and handling equipment may be utilized, if the production quantities warrant it. Thermosetting materials are the ones most commonly cast. They include epoxies, polyesters and silicones, including silicone rubber. They are cast as monomers and polymerization occurs in the mold, and heat may be applied to the mold to induce polymerization, depending on which thermosetting plastic is used. Some thermoplastics are also cast, but they also are dispensed into the mold as monomers and then are polymerized. Nylon, acrylic, and polyurethane are the thermoplastics most commonly cast as monomers. Vinyl dispersions are also cast. See sections K, K3, and K4.

Molds for casting plastics may be made of many materials since forces involved are very low. Lead and its alloys, aluminum, plaster, silicone rubber, and rubber latex are among the materials used to make molds. Casting is also used in the production of sheet materials.

H1. *casting of sheet*

H1a. ***cell casting of sheet*** - is a batch process for making sheet stock by the casting method. The major material is *acrylic* and the following description refers to the method used with that material. The molds or "cells" for casting acrylic sheet are made from polished plate glass, plus end and side members. The molds are filled with methyl methacrylate monomer mixed with some partly-polymerized resin in liquid form plus desired additives such as catalysts, ultra-violet absorbers, and colorants, etc. The filled mold is sealed and air is evacuated before it is moved to an oven. There, the mold is slowly heated to a temperature of about 200°F (93°C) for an extended period (up to 16 hours for 3/4 in thick sheets, longer for thicker sheets), during which time the monomer polymerizes. The mold is then cooled and opened, and the cast sheet is trimmed as necessary. Sheets of thicknesses from 0.125 to 4 in (3 to 100 mm) are cast by this method. Cell cast sheets are used for glazing, skylights, outdoor signs, and in plumbing and spa products. Opaque and colored sheets can be produced by the process, but its prime advantage is the superior optical properties the sheets possess compared to continuously-cast acrylic. Fig. 4H1a shows a typical mold arrangement used for cell casting.

H1b. ***continuous casting of sheet*** - uses a pair of moving, parallel, highly-polished, and endless, stainless steel belts, between which is poured or pumped an acrylic syrup similar, but not identical, to that used for cell casting of acrylic as described above. (The material mixture may be varied in

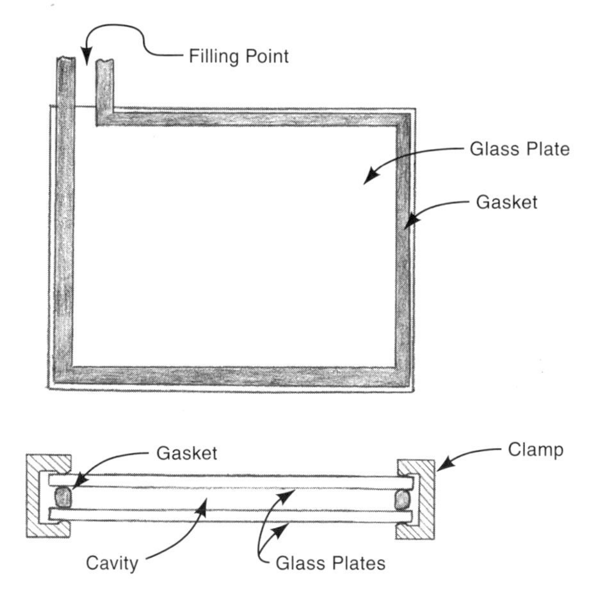

Fig. 4H1a Apparatus for cell casting of acrylic sheet.

order to provide the proper viscosity for the method used. The final sheet, though a thermoplastic, will contain some cross-linked material to improve chemical and stain resistance and thermo-formability.)

Flexible gaskets at the edges of the sheets prevent leakage. The stainless steel belts convey the material through a heating phase to induce polymerization, followed by a quick cooling phase to prevent bubbling of the material. Another heating phase up to about 260°F (125°C) is then used to complete the polymerization to the desired level. Then the solid product exits the belt system, the surfaces are covered with a protective film and sheets are cut to the length desired. Fig. 4H1b illustrates the process. The equipment required is rather large and expensive but output rates are high. Most acrylic sheet is produced by the continuous casting method. Sheet thicknesses can range from about 0.080 to 0.500 in (2 to 12.5 mm). Applications are the same as those for cell cast sheet except that optical properties are not as good as those from cell casting. Skylights, sign components, and glazing are all applications for continuous cast acrylic sheet.

H2. ***casting structural foam parts*** - is simply reaction injection molding (as discussed above in 4C3b) performed on a manual basis. Liquid components of thermosetting resins (usually polyurethane) are mixed and poured into a mold. Polymerization and foaming take place in the mold. Heated molds are

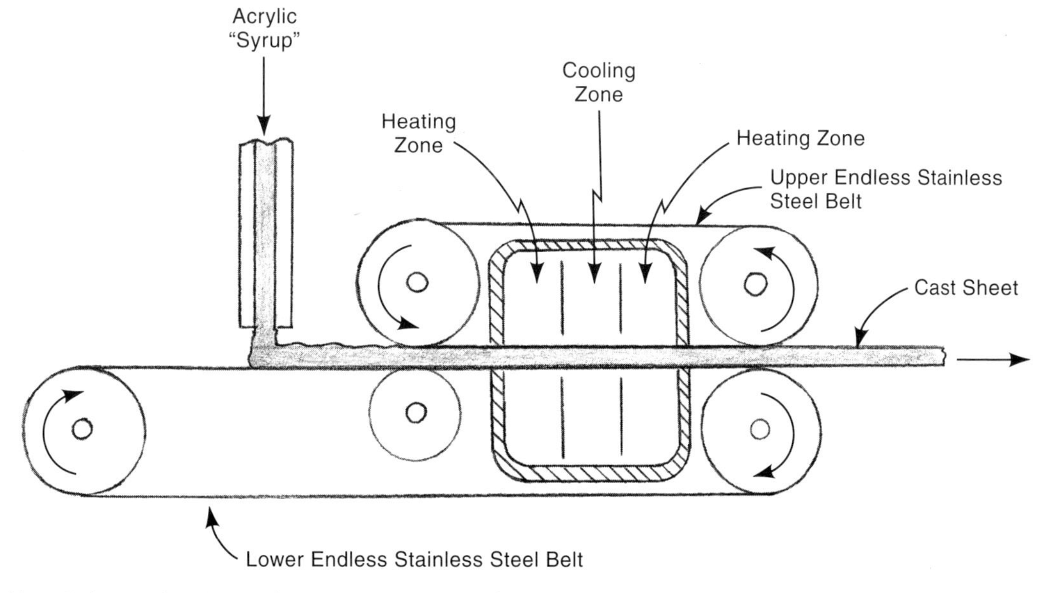

Fig. 4H1b Schematic view of equipment used for the continuous casting of acrylic sheet. The acrylic syrup is spread and compressed between two polished stainless steel belts, which carry it through a series of ovens that polymerize the acrylic by heating, cooling, and reheating to produce a solid cast sheet. (Note: For convenience of illustration, the heating and cooling zones are shown much shorter than those in actual production equipment.)

used to accelerate polymerization. Oven curing usually takes place after casting. This process is particularly applicable to prototypes and limited quantity production, and can be used effectively for large plastic parts.

H3. ***casting nylon parts*** - Cast nylon has superior machinability, stiffness and heat deflection properties than nylon 6/6 which is injection molded or extruded. The nylon casting process involves use of the monomer; polymerization takes place in the mold. The monomer must be heated and melted before casting, and must be protected against moisture absorption during the casting sequence since it is highly hygroscopic. Sealed containers and inert atmospheres may be used to prevent moisture from coming in contact with the material. Molds can be inexpensive since the forces they are subjected to are low. Aluminum, epoxy, and silicone rubber, as well as steel, have been used as mold materials. The casting operation often takes place in heated ovens. (The monomer material is heated to 390°F (200°C). For smaller parts, centrifugal casting may be employed, i.e., the molds are spun to provide sufficient force to fully fill the mold cavities. Cooling of finished parts is slow to permit relief of internal stresses from shrinkage (15%). Large parts (up to 400 lbs – 180 Kg) can be produced with the process. Gears, sheaves, cams, various machine parts, bushings, and bearings are common applications. The casting process lends itself well to low-quantity production levels.

H4. ***casting acrylic parts*** - The same kind of monomer-polymer mixture used in cell casting of acrylic sheets (See H1a above) can produce shaped parts if a suitable mold is used. The process is useful at low production quantities when the cost of an injection mold and molding machine cannot be justified. The optical clarity of acrylic makes it useful for embedding an object or biological specimen in a clear protective or decorative block. Embedded objects are pre-positioned in the mold before casting. Molds can be made of almost any material because the operation does not require elevated temperatures. Art statuary, and marble-like kitchen and bathroom counters and sinks are other applications. When marble-like objects are cast, the acrylic is mixed with up to 60% of inorganic filler which gives the appearance of marble.

H5. ***encapsulation and potting*** - various electronic and electrical devices are encapsulated with plastics to provide insulation, and mechanical and environmental protection. Potting involves partly surrounding a device with plastic to fix it in place in some component. Both encapsulation and potting involve casting, usually with thermosets. Epoxy potting of small transformers for electronic devices is a common example of potting. High voltage transformers are frequently potted with silicone rubber. As with regular casting, simple molds can be used. Often, immediately after casting, the filled mold is placed in a vacuum chamber so that any trapped gases or air will bubble out of the liquid casting resin before it sets. This procedure eliminates or minimizes voids in the cured material.

I. Plastics Extrusion

Extrusion is the process of forcing a heated, semi-solid plastic through a die whose cross-sectional shape it retains when it cools and solidifies. The operation is normally continuous. Thermoplastic material in granular form is fed from a hopper to the heated barrel of the extruder. A rotating screw transports the material through the barrel and mixes it as it melts, providing a uniform flow rate. The temperature of the material, as it reaches the extruding die, is uniform throughout. The die is essentially a plate with an opening of the shape desired in the cross section of the extrudate. The material exiting from the die is cooled by air blast, water spray, or water trough, so that it hardens, forming a product of constant cross section and indefinite length. Some further sizing operations may be performed after the material exits the extrusion die and before it fully cools. One method often employed is to pull the material by conveyor at a slightly faster rate than it leaves the extrusion machine. This reduces the size of the cross section and aids in keeping the extrusion straight. Fig. 4I illustrates a typical profile extruder.

I1. ***profile extrusion*** - The extrusion process is used to produce profiles of many different cross-sectional or profile shapes. Tubing, pipe, rods, sheet, fiber, and film can be extruded as well as many complex shapes. Coating of wire and cable is another common application. Typical profile shapes

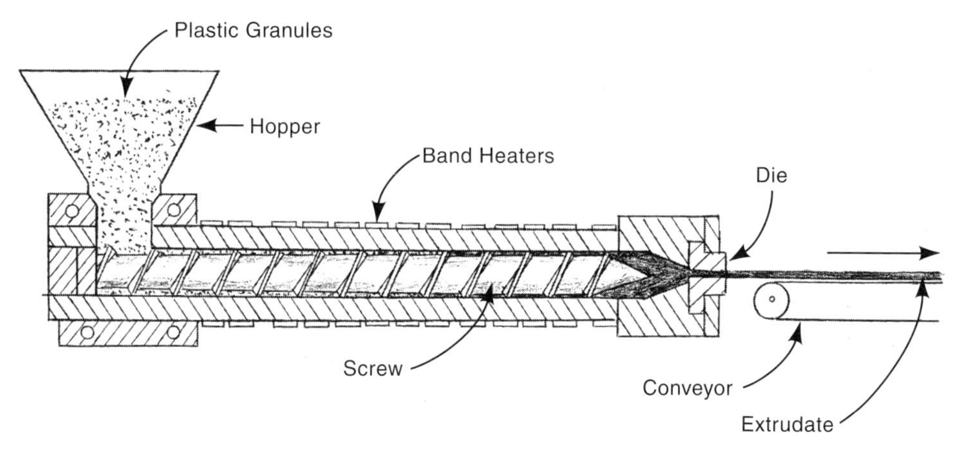

Fig. 41 Schematic section view of the extrusion of plastics.

produced are those required for vehicle trim, house siding, picture frames, door and window components, moldings, and edge trim. Sheet material, after extruding, is commonly rolled between rolls having smooth polished surfaces - or embossed surfaces - depending on the surface finish wanted on the finished sheet. Fiber is usually extruded in multiple strands and then is stretched or drawn down to the diameter wanted. Pipes and tubing are made from dies that have a mandrel or other core to form the hollow interior. These products are pressurized internally after extrusion, and may then be cooled in a vacuum chamber filled with water and a sizing sleeve. The internal pressure in the pipe, and the external vacuum, if used, cause the outer walls of the pipe to bear against the sleeve as the pipe cools, helping maintain consistency of diameter of the pipe or tubing.

I2. ***coextrusion (dual extrusion)*** - is the extrusion of two different plastics at the same time. This procedure may be carried out to create properties in the extruded part that are not feasible with one material, for example, color, strength, barrier, or other properties. The process involves the use of two separate extrusion barrels feeding material into a complex die. If the two materials have melting temperatures too far apart, the operation may be done in two stages. The higher-melting-point material is extruded first and is then treated as an insert around which another material is extruded. The second extrusion follows a method very similar to that used in coating metal wire with plastic insulation. Applications include window component profiles with both rigid elements, for stiffness, and flexible portions, for sealing, and pipe with both insulating foam and conventional walls.

I3. ***foam extrusion*** - (Note: Also see structural foam extrusion, C3h.) Extrusion of foam plastics is used to produce low-density thermoplastic sheet and board material, which is used for a variety of applications. Polystyrene and polyolefins in densities of 2 to 10 lb/ft^3 (32 to 160 kg/m^3) are used. Major applications are thermoformed foam sheet for meat and produce trays and egg cartons. Foam extrusions are used for automotive products such as carpet underlayment and indoor insulation. Tubular insulation for water pipes, and slabs for building insulation and for flotation, are other applications. In flexible formulations, foam blocks and slabs are used for shock absorbers in heavy-duty shipment packaging.

The extrusion process involves the use of two screw extruders, mounted end-to-end. The first extruder heats and mixes the plastic material, particularly mixing the blowing agent (usually in liquid form, but based on carbon dioxide gas). The material is also premixed with a nucleating agent (in small proportion) to insure uniformity and dispersion of the gas bubbles. Proper dispersion of the blowing agent also requires that the polymer mixture be sufficiently heated. The blowing agent is normally injected into the mixture during its passage

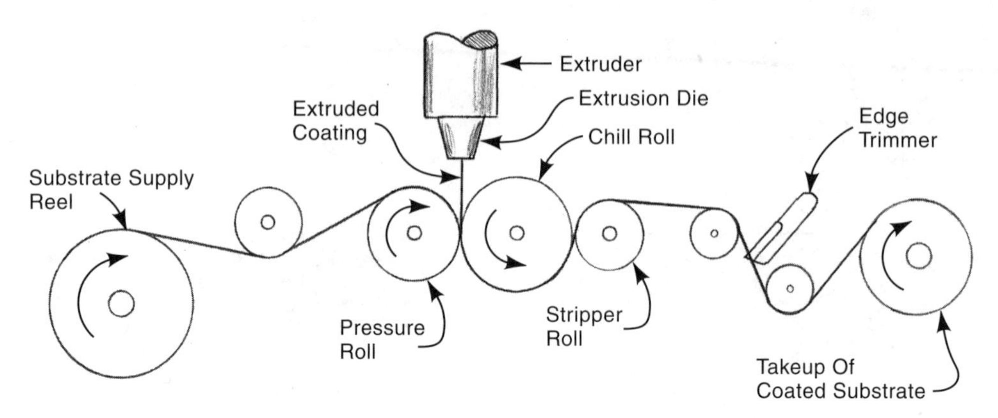

Fig. 4I4 Plastic extrusion coating and laminating another sheet material.

through the first extruder. After thorough mixing, the mixed material moves to the second screw extruder. This second machine maintains the pressure necessary to keep the blowing agent from causing preliminary expansion, while cooling the mixture so that its viscosity is increased when it is forced through the extrusion die. The extrudate then is cool enough to retain the shape imparted by the die. Pressure on the mixture is released as it passes through the die, and the blowing agent expands, producing a series of gas bubbles in the extrudate. The extrudate is further cooled after leaving the die by either water or air or both. It is then cut and slit as desired. Both open-cell (permeable) and closed-cell (fluid tight) materials can be produced. It is possible to produce a fine skin on the extrudate, which can be printed on for product identification. Polyethylene and polystyrene are the two plastics that are most frequently extruded as foams.

I4. ***extrusion coating and laminating*** - In this process, material is extruded through a slot die. The resulting plastic sheet is applied to a moving web of paper, fiberboard, film, foil, or other substrate material, fed from rolls. When two layers of substrate are used, the plastic is extruded between the two layers and the process is known as *laminating*. The substrate material is normally treated beforehand to provide a surface to which the plastic can adhere. The plastic coating is drawn to a thickness of from 0.0005 to 0.002 in (0.013 to 0.05 mm) although it is about 0.020 in (0.5 mm) thick at the die. To ensure good flow properties and adhesion of the plastic coating, the extrusion is run at a very high temperature and the viscosity of the plastic is very low. Because of the low viscosity, the extrudate normally exits the die in a downward direction. Chill rolls cool the material after the layers are joined. The finished material is edge trimmed and sometimes slit after the extrusion operation. The process is illustrated in Fig. 4I4.

The substrate material provides strength and/or stiffness and a surface that is printable, while the plastic layer provides a barrier to liquids or gases. Coated material is used for containers for milk, juice, sauces, puddings and other processed foods. Flexible packaging is another application for various liquids or solids when a barrier is needed. Other applications are credit card stock, wallpaper, insulation, automotive carpet backing, multiwall paper bags, composite drums and cans, oil proof papers, and product wrapping.

I4a. ***wire coating*** - is an extrusion operation similar to that used in profile extrusion and laminated sheet material in that a rotating screw drives the molten plastic through a special die. The process differs from laminating extrusion in that the material to be coated moves at a right angle to the direction of plastic resin flow from the barrel, and the die cavity is circular, surrounding the wire to be coated. This arrangement provides for the flow of material on all sides of the wire. The wire is normally uncoiled and straightened before passing through the die. The coated wire passes through a water trough for cooling, is tested electrically, and is then recoiled. Fig. 4I4a illustrates a typical wire coating die.

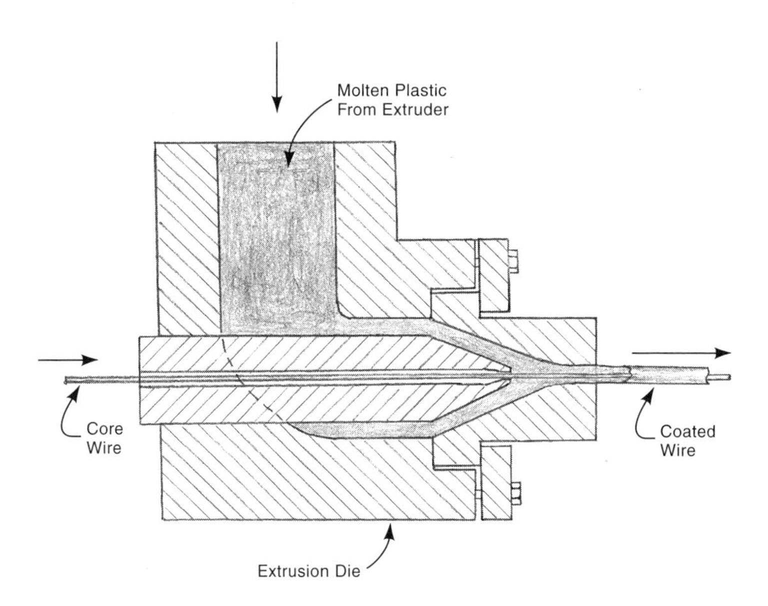

Fig. 4I4a An operating extrusion die for coating wire, shown in cross-section.

I5. *extrusion of film and sheet*

I5a. ***blown film extrusion*** - is a common method for manufacturing plastic films. The process, shown in Fig. 4I5a, involves the extrusion of very large but thin-walled plastic tubing, usually directed upward. The tube is simultaneously expanded from internal air pressure, to up to four times the die diameter and stretched axially, thinning the wall to the desired film thickness and improving its tensile strength. Bi-axial orientation of the film molecules, achieved by stretching the film in two directions, greatly increases its tensile strength in both directions. Cooling air on the outside, supplied by a ring of outlets, also cools the material. Sometimes there is internal cooling as well. The tube is closed by a set of "nip" rolls some distance above the die. The cooled and solidified material can be slit and spread open, and wound on rolls as thin sheet or film. It may also, if used for plastic bags, be cut to length instead of slit. The whole process is continuous and rapid. Polyethylene, both low and high density, is the most commonly-used material but polypropylene, polycarbonate, ethylene copolymers, and other materials are also used. Finished film thickness is 0.0001 to 0.050 in (0.0025 to 1.25 mm). Typical applications are all kinds of food bags, garbage bags, trash bags, agricultural and construction sheet, stretch wrapping material, and water barriers. Film for some food

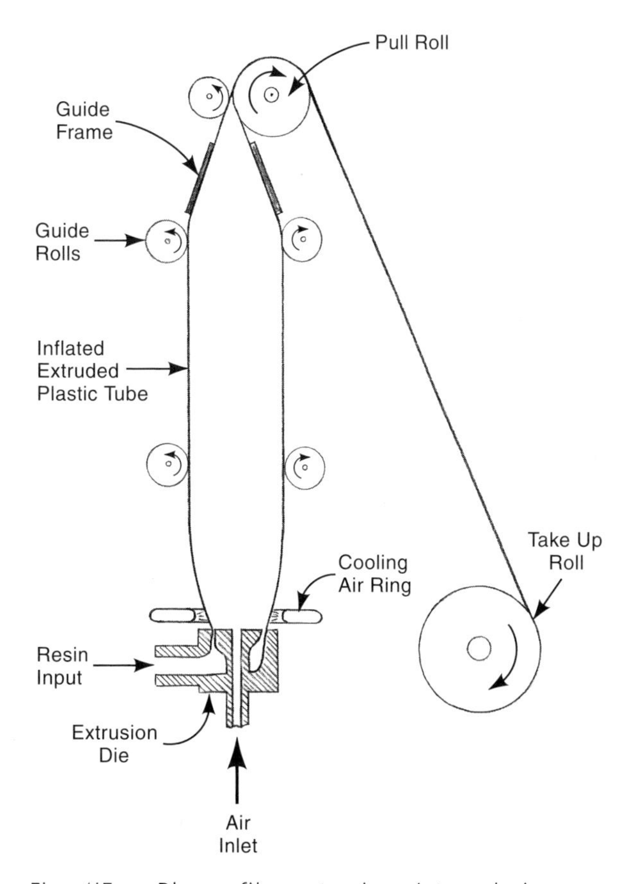

Fig. 4I5a Blown film extrusion. Internal air pressure expands a tube of extruded film and, combined with upward pulling, thins the walls and improves the strength of the film.

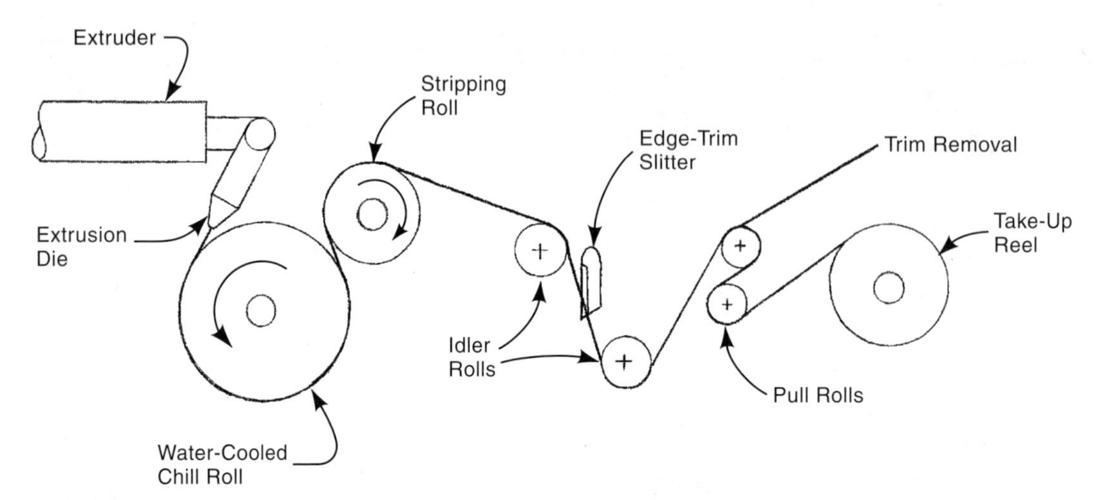

Fig. 415b Cost film extrusion shown schematically. The highly polished and plated chill roll imparts a smooth, glossy finish to the film.

applications is coextruded with as many as five layers of material.

15b. ***cast film extrusion*** - involves extrusion as the first step. The process runs at high output rates - up to about 1500 ft/min (450 m/min). Dies have slot openings as wide as 12 ft (4 m), with typical thickness openings of 0.010 to 0.020 in (0.25 to 0.5 mm). The still-molten extruded sheet is contacted by a chill roll, on which it cools and solidifies. However, in the process, it is drawn down to a thickness as little as 0.002 or 0.003 in (0.05 to 0.076 mm). The chill roll is highly polished and plated, and imparts a smooth glossy finish to the film. The film is then edge trimmed and wound on reels for further processing. For some applications, the as-extruded single axis orientation is modified, using a biaxial orientation process. Common applications are the production of film for coating and laminating uses, for stretch wrapping, and for other packaging. Polyethylene is the prime material processed with this method but polypropylene, nylon, polyester, and PVC are sometimes used. See the process illustration in Fig. 415b.

16. ***operations after extruding*** - In all plastics extrusion, cooling of the extrudate, after the material leaves the extrusion die, is an essential operation. Sizing of the extrudate, by some method, also is very often required. Cooling with air frequently is sufficient but some situations require water cooling. When water immersion is used, the extrudate enters the long water tank through a die in the end wall. This die has approximately the same diameter as the extrusion die and thereby has some sizing benefit. When such a die is used, a pulling device (a pair of powered rollers or belts that gently grip the material) may also aid in sizing. Pressure or vacuum may additionally be involved as a sizing aid. Other post-extrusion operations, often in line with extruding are: roller and spray coating, cut off, edge trimming, slitting, printing or otherwise marking, drilling or piercing, milling, blanking, welding, heat sealing, and assembly to other parts. In many high production situations, these operations can be performed on the extrudate as it moves. “Flying” or rotary tooling, if feasible, is then used for the operations.

J. Calendering

Calendering is a method for producing plastic sheet and film. A heated, softened, plastic is forced between two heated rollers (“nip” rollers), with fixed spacing. The rollers form the plastic into a thick, continuous sheet. Additional rollers reduce the thickness and, if wanted, emboss the sheet. The sheet may be further reduced in thickness by stretching it. The process is used to make various sheet plastic components such as flooring and tape, and to provide material for further operations. Sheet thickness

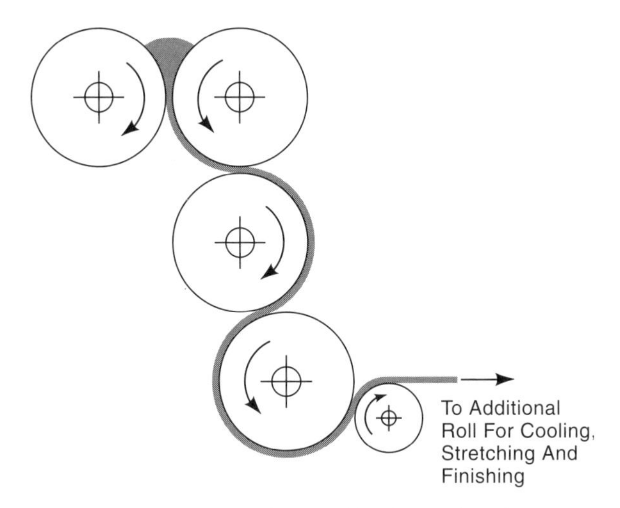

Fig. 4J Forming plastic film by calendering.

ranges from 0.002 to 0.060 in (0.05 to 1.5 mm). ABS and PVC are the most frequently processed plastics. Upholstery sheet, rainwear, shower curtains, and tape are made with the process. By calendering two sheets of plastic with one or more layers of paper or other materials, such objects as credit cards, wallpaper, and playing cards can be made. Fig. 4J illustrates the calendering operation.

K. Vinyl Dispersion Processes

Polyvinyl chloride (PVC), in fine particles, can be formulated as a suspension in a liquid plasticizer, often with some diluents. The viscosity of the suspension can range from that of a paste to that of a pourable liquid. When this dispersion (with copolymers) is heated to a temperature of 260 to 350°F (125 to 175°C), the plastic particles absorb plasticizer and swell, forming a gel. With some further heating, the dispersion is converted to hot melt. When cooled, it turns to solid, plasticized vinyl. When there are no significant diluents or thinners, the material is referred to as a *plastisol*. When diluents and thinners are included, the dispersion is known as an *organosol*. Several methods are available to make use of these materials. A major application is as a protective and insulating coating on other products.

K1. ***spread coating*** - is a general term denoting the application and distribution of a dispersion over the surface of some other material. Knife coating, roll coating, and curtain coating, are all spread coating methods. After spreading, the substrate material with its coating is passed through an oven where the heat causes the dispersed vinyl particles to fuse together. Typical products produced by spread coating are automotive padding, vinyl roll flooring, building siding, coated fabric used in apparel manufacture and soft foamed upholstery materials.

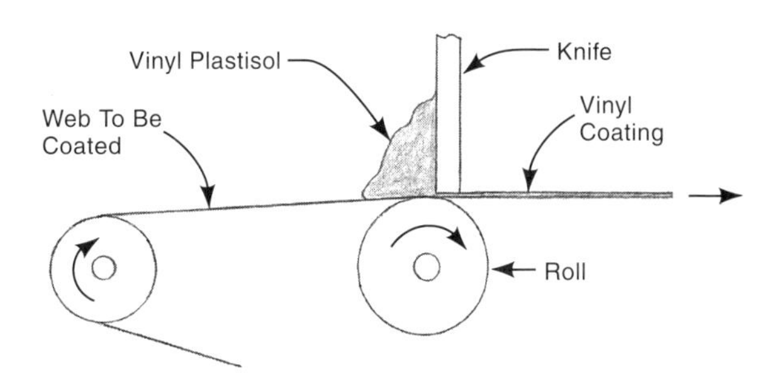

Fig. 4K1a Schematic of knife coating (knife-over-roll coating) in which vinyl plastisol is distributed evenly over the web to be coated. A baking operation to fuse the coating follows the coating operation.

K1a. ***knife coating (knife-over-roll coating)*** - methods use a horizontal straight edge, slightly above the web surface to be coated, to smooth, spread, and control the thickness of the vinyl coating material. The production arrangement is shown in Fig. 4K1a.

K1b. ***roll coating*** - has speed and quality advantages over knife coating in many cases. Fig. 4K1b illustrates two roll coating methods.

K2. ***dip coating (hot dipping) and dip molding*** - When a heated object is immersed in plastisol and withdrawn, and the material remaining on the object fuses, a useful coating can be provided. The object is heated to a temperature high enough to cause the plastisol to gel. The thickness of the coating depends on the temperature of the part to be dipped, its shape, the length of time it remains in the plastisol, the characteristics and temperature of the plastisol and the rate of withdrawal. The coating can be as thin as 0.005 in (0.13 mm) and as thick as 0.25 in (6 mm). The process constitutes an easy method for coating tool handles, glass bottles and other objects to provide easier grasping, electrical insulation or cushioning. Dish drying racks are another application.

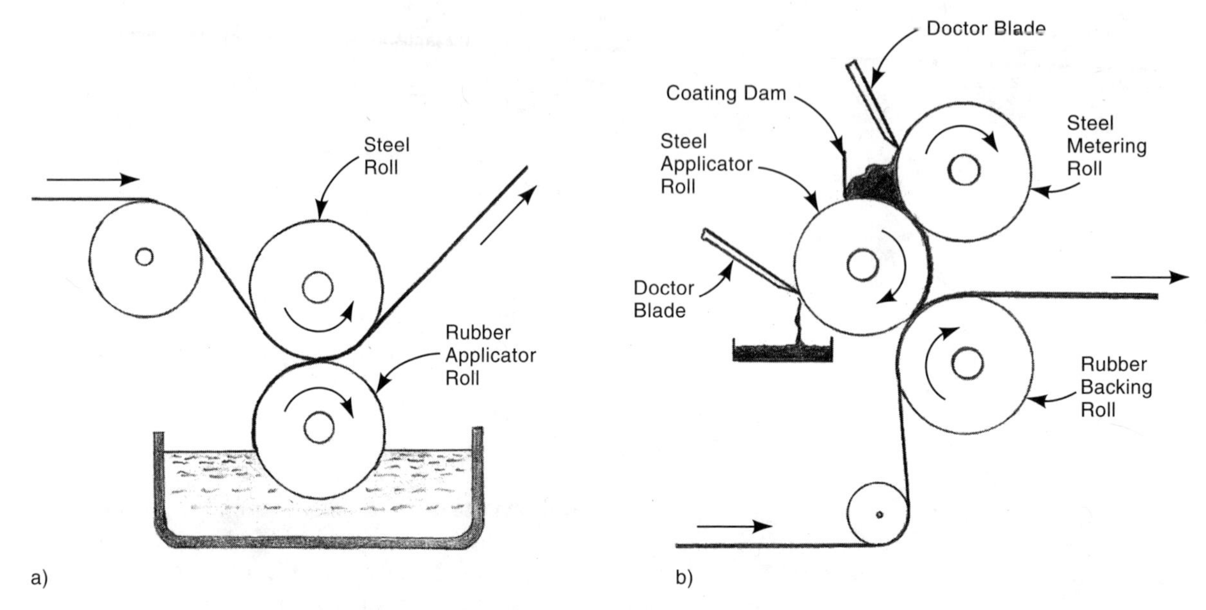

Fig. 4K1b Two roll-coating methods: a) Direct roll coating, used only when the coating material is of low viscosity. The coating liquid is picked up by the rubber applicator roll and is transferred to the web. b) Three-roll nip reverse coating. The metering roll rotates slowly in the opposite direction from that of the applicator roll. The nip between them governs the amount of coating on the applicator roll. The applicator roll rotates in the opposite direction from the web travel and the coating is wiped from the applicator roll and deposited on the web. Coating thickness can be controlled by varying the nip and relative speed of the web and the applicator roll. Coatings having a wide range of viscosities can be deposited with this method.

A typical sequence involves priming the part to provide improved adhesion, reheating the part, immersing it in the plastisol, withdrawal, additional heating to complete fusion, and then cooling. The operation can be done on a batch basis or can be conveyorized to be continuous and automatic.

If this operation is performed with a shaped mandrel or mold, from which the coating is stripped after the coating cools, the operation is referred to as *dip molding.* One common application is the manufacture of medical gloves as illustrated in Fig. 4K2. Another application, using a collapsible mold, is the molding of flexible bellows. Closures and caps are also made with this method. Although vinyl plastisol is the prime material used in dip molding, it is possible to dip mold objects of nylon, silicone and polyurethane.

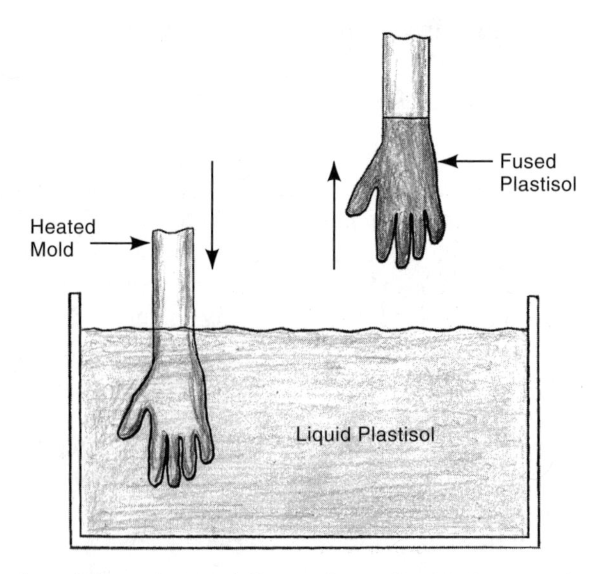

Fig. 4K2 Dip molding of medical gloves. The plastisol coats the heated mold and fuses. After the mold is withdrawn, the solidified coating is stripped from the mold. The thickness of the coating depends on the time of immersion, the temperature of the mold, and the rate of withdrawal.

K2a. ***cold dipping*** - with the part or mold to be dipped at room temperature, is sometimes used when the part to be coated cannot retain enough heat to gel the plastisol or has limited resistance to high heat levels. It may also be used when the surface detail desired may not be achievable when the plastisol gells rapidly on a hot mold. Cold dipping is somewhat slower than hot dipping. The procedure is as follows: The cleaned part is dipped into the liquid plastisol at a uniform rate of immersion. The immersion is held for a few seconds and the part is withdrawn at the same slow and steady rate. The part is then suspended in an oven at 365 to 500°F (185 to 260°C) until the plastisol fuses. Coating cloth work gloves is a common application.

K3. ***slush molding*** - is a casting process for vinyl plastisol and is identical to dip molding except that the plastisol material contacts the inside of a hollow mold instead of the outside of a male mold. The hollow mold is filled with plastisol and heated sufficiently to gel the material in contact with the inner surface of the mold. The liquid material in the mold is poured back into the source container, leaving a shell of gelled material in the mold. The higher the mold temperature and the longer the plastisol is contained in the mold, the greater the thickness of the fused material. Further heat applied to the mold fully fuses the plastisol. The mold is cooled and the hollow finished part is removed. The process is useful for making products such as beach balls, dolls, boots, hollow toys, and the surface of automotive head and arm rests. The process is suitable for hand operation for making prototypes and for short run production but can be automated for mass production situations. For products with fine surface detail, two or more fills of the mold may be employed, the first one with the mold cold to allow the plastisol to contact all surface details before gelling. Excess material from the first fill is poured out and the mold is heated to gel the first-coat material. The mold is then filled again while hot, with the same plastisol or sometimes with a cellular plastisol in order to provide a soft, thicker lining to the product.

K4. ***cavity, in-place, and low-pressure molding*** - are all essentially casting operations to produce solid parts from plastisol. With *cavity molding,* a mold is filled with plastisol and heated until the part has fused. The mold is then cooled so that the part can be removed. *In place molding* is simply the casting of a soft plasticised vinyl as a seal and adhesive in some other part. Examples are the seals for jar and bottle caps where plastisol is metered into the inverted cap which is then heated to fuse the plastisol. A similar approach is used in the manufacture of automotive air cleaner filters, to bond and seal the filter material to the steel frame of the device.Clay sewer pipe sections are also sealed together with this approach. *Low-pressure molding* is a somewhat mechanized version of cavity molding, wherein metered amounts of plastisol are pumped into closed molds which are then heated to fuse the plastisol. The molds are then cooled and opened. Shoe soles and printing plates are made with this method. The process is also used to encapsulate electronic components.

K5. ***strand coating, using plastisols and organisols*** - Wire, filaments and cords can be coated with vinyl plastisol. The operation is performed by running the strand through a plastisol reservoir or by pumping the plastisol over the strands which then may pass through a circular small-diameter die to control the amount of material covering the strand. The die wipes off any excess material that may be on the strand. In some cases, no die is needed and a low-viscosity organisol leaves only a thin layer of coating on the strand. Multiple passes may then be made. With a die, there are two approaches. In the floating die method, the die is loosely held and it tends to center itself on the strand. Low-viscosity organisol or plastisol is used with this approach. In the set-die method, the die is securely mounted and the strand is guided to pass through the center of the die to insure a concentric application. The strand must be under some tension if concentricity is to be controlled. In all these process variations, the coated wire is subjected to heat after coating, to fuse the plastisol. Coated strands are used in the manufacture of fiberglass screening, electrical wire, rope, thread, and woven cords.

K6. ***spray coating*** - Plastisols and organisols can be sprayed to apply decorative or protective coatings to various objects. The process is used for outdoor furniture, appliances and building components, and is particularly useful if the item to be coated is irregular in shape and too large in size or otherwise unsuitable for dip coating. Essentially conventional spray painting equipment can be used. After the coating has been allowed to level, oven baking at a temperature of 350 to 400°F (175 to 200°C) follows to fuse the coating.

Organisols are more commonly sprayed than plastisols since spraying works best with low viscosity fluids. The process is used for tank and drum linings and automotive anti-corrosion sealants.

K7. ***extrusion of plastisol*** - is another coating method. The liquid plastisol is fed to a heated barrel of the extruder where its PVC particles absorb plasticizer and fuse. The resulting compound is extruded as film through a die that has a slot opening. It is fed onto and adheres to, a fabric or other substrate. One application, in addition to coated fabrics, is the manufacture of battery separators, where the extrudate is fed to a fiberglass sheet where it forms spacer ribs on the fabric.

L. Welding and Adhesive Bonding of Plastics

Welding of plastics is carried out by heating, in the joint area, the thermoplastic pieces to be joined to the point where the material softens or melts and fuses together. There are a number of different methods used to provide the heat, several using friction and several using heat from external sources. Welding or bonding of plastic components takes place when large or complex components are desired and it is not practical or convenient to make them with primary processes. It is also used when parts of different colors or materials are to be joined or when some other component is to be enclosed.

L1. ***friction or spin welding*** - is a method useful when the joint is circular. One part is held stationary and the other, while being pressed against it, is rotated or spun rapidly while the surfaces to be welded are in contact. The friction from the rubbing of the two surfaces develops heat, which softens the plastic at the joint. The rotation is stopped and the two parts are continued to be pressed together. When the joint cools, a strong bond is created. Rotational speeds are typically 10 to 40 fps (3 to 12 mps) and pressures typically 300 to 700 lbf/in^2 (2100 to 4800 kPa). Hermetic seals can be produced with this approach. The process is applicable to nearly all rigid thermoplastics; low-density polyethylene and other soft materials may not process as well. As long as one of the parts can be put in a chuck and rotated and a circular joint is acceptable, the process can be used. See Fig. 4L1 for an illustration. Containers for cosmetics and other products,

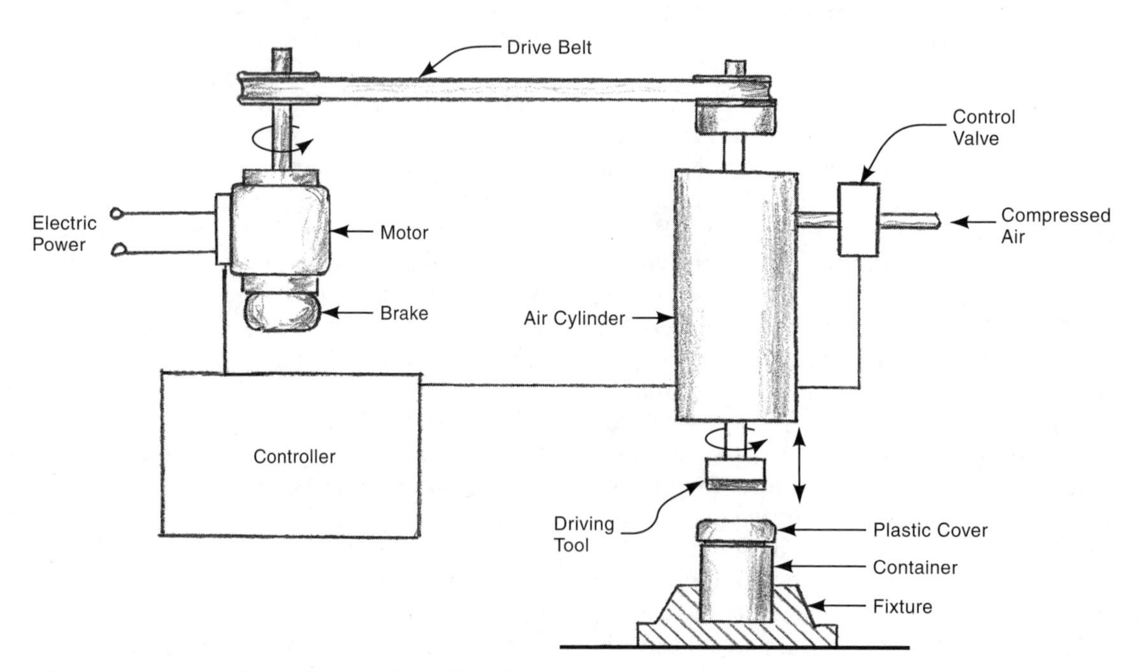

Fig. 4L1 Arrangement for spin welding. The driving tool descends and rotates the plastic cover against the plastic workpiece. Frictional heat fuses the two together. Speed of rotation, timing, and pressure are controlled automatically.

bottles and gasoline filters are welded with this process.

L2. ***hot plate welding*** - uses a heated metal platen to soften the surfaces of two parts to be welded together. In production situations, the full operational sequence is as follows: 1) the mating parts are placed in holding fixtures, which align them for assembly. 2) an electrically- heated plate, whose surfaces match the mating surfaces of the parts, is placed between them. 3) the parts and fixtures are brought together so that the heated plate heats the surfaces to be joined. 4) when the surfaces are sufficiently hot, i.e., when the surface material has softened, the plate is withdrawn. 5) the mating parts are brought together with some pressure and are held in contact until the mating surfaces, which have fused together, are cool. The fixtures are withdrawn and the welded part is removed from the lower holding fixture. The holding fixture and plate contain stops to control the amount of deformation and the final height of the assembly.

Hot plate welding is used on a variety of automotive parts including tail light assemblies, vehicle fuel tanks and tanks for other liquids, pipe fittings, storage battery cases, and appliance parts. The process is well suited to large assemblies. Liquid and gas-tight seals can be made. Surfaces of the mating parts do not have to be flat so long as the surfaces of the heating plate can be contoured to match them. A considerable variety of plastics can be welded by this method including many dissimilar pairs. See Fig. 4L2 for an illustration of the process.

L3. ***vibration welding*** - uses mechanical oscillating or orbital movement between the surfaces to be welded to create frictional heat. The process differs from ultrasonic welding in that the frequency of

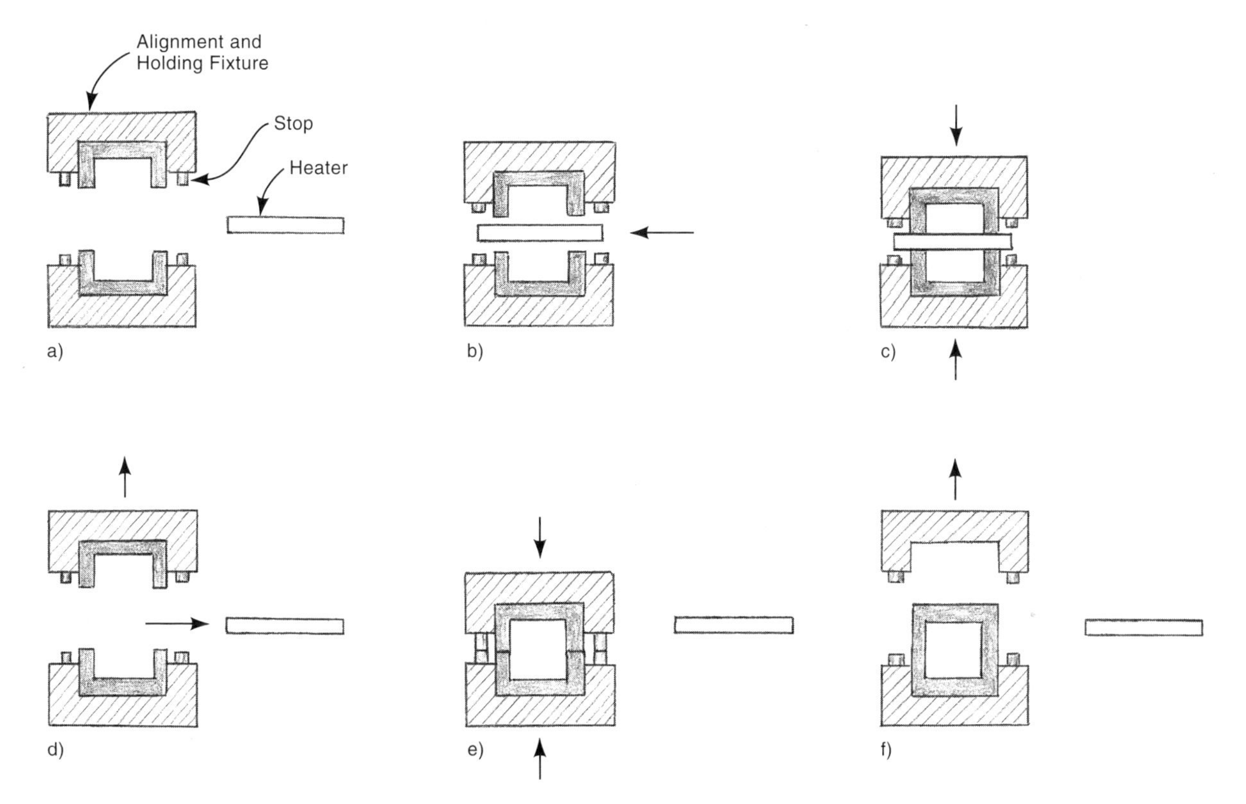

Fig. 4L2 Hot plate welding: a) Parts to be joined are inserted in the alignment and holding fixture; heater is nearby, b) Heater is placed between the two parts, c) Fixture closes to bring parts and heater in contact, d) After part surfaces are sufficiently heated, fixture opens and heater is withdrawn, e) Fixture closes, bringing the softened surfaces together. Stop pins prevent excessive force against parts. f) After part surfaces cool, fixture opens so that welded part can be removed.

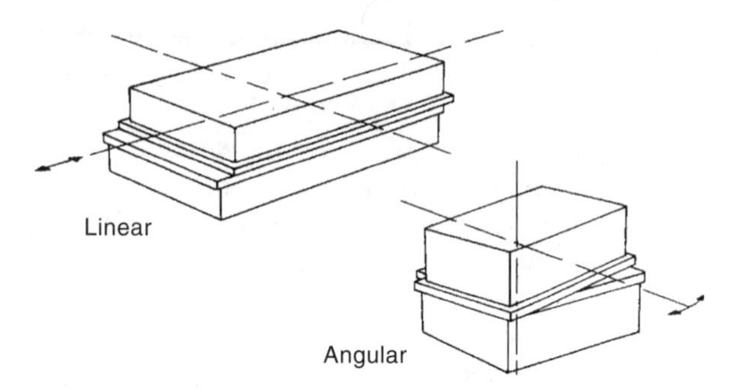

Fig. 4L3 The movement of parts undergoing vibration welding can be either linear or angular. (*from Bralla, Design for Manufacturability Handbook*[6].)

oscillation is much less, the amplitude is greater and the direction of oscillation is parallel rather than perpendicular to the plane of the surfaces being joined. Oscillating motions can be linear or angular. Typical frequencies of vibration are 100 to 300 Hz with amplitudes of 0.015 to 0.100 in (0.4 to 2.5 mm). Pressures of 200 to 250 lbf/in^2 (1400 to 1700 kPa) are exerted on the surfaces during the operation. Movements are guided by holding fixtures for the parts to be joined. After the vibration has softened or melted the joint surfaces sufficiently for bonding to take place, the vibration stops and the parts are held together briefly until the joint cools and solidifies. The equipment is designed to align the parts properly when the vibration is stopped. The operation is quick; typical vibration periods are 1 to 3 seconds and weld solidification times are usually less than one second. The operation is feasible for large and small parts but is particularly applicable for large parts of crystalline resins. Joint surfaces are usually flat, but curved and stepped surfaces also can be welded. The process is used extensively in the automobile industry. Plastic bumper assemblies, dashboard assemblies, air conditioning ductwork, fuel tanks, tail lights, fluid reservoirs, and wall panels are all joined with this method. A typical vibration welding operation is shown schematically in Fig. 4L3.

L4. ***ultrasonic welding and sealing*** - is similar in principle to vibration welding, but there are significant differences. With ultrasonic welding, the frequency of vibration is in the range of 20 to 40 kHz, far higher than with vibration welding. The magnitude of vibration is correspondingly smaller, typically being less than 0.001 in (0.025 mm) at 20 kHz. The direction of movement is at right angles to the joint surface instead of parallel to it and the area of weld is also typically much smaller. The vibration energy is directed to the joint through the parts to be welded; there is no visible movement of the parts. The equipment converts high frequency electrical oscillations to mechanical vibratory energy at ultrasonic frequencies. A tuned metal horn tool transmits this energy to the workpieces when it makes contact with them. The energy passes to the joint area where the vibration causes friction. Frictional heat melts the plastic at the joint and the surfaces fuse together as the energy is cut off. When the joint area cools, a strong bond results. Electronic systems in the welding equipment have made it possible to accurately control the amount of material melted, the joint and part dimensions, and the amount of flash produced.

Dissimilar plastics can be welded as long as their melting points are within about 30°F (17°C). Except for spot welding and sealing, rigid plastics work somewhat better than the softer plastics for most ultrasonic welding operations. The process is usable for joining plastic parts, inserting metal parts into plastic components, staking (upsetting ends of bosses to form a rivet-like fastener), spot-welding, and sealing sheet and film material. Typical applications of ultrasonic welding include all kinds of small appliance parts, medical devices, double-wall insulated drinking cups and mugs, film cartridges, electronic calculators, clock frames, vehicle taillight assemblies, decorative panels, and toys. Fig. 4L4 illustrates a typical ultrasonic welding operation.

L5. ***ultrasonic spot welding*** - Ultrasonic welding equipment is well suited to making spot welds in plastic components. A standard shape of horn tip is used. The tip has a point long enough to fully penetrate the top layer of material and part way into the underlying layer. Vibration of the tool heats and melts the material it contacts. The process is useful for joining large parts, making welds in areas that may be difficult to join with other methods, and for fastening parts that do not require a continuous weld seam. Fig. 4L5 illustrates a typical ultrasonic spot welding operation.

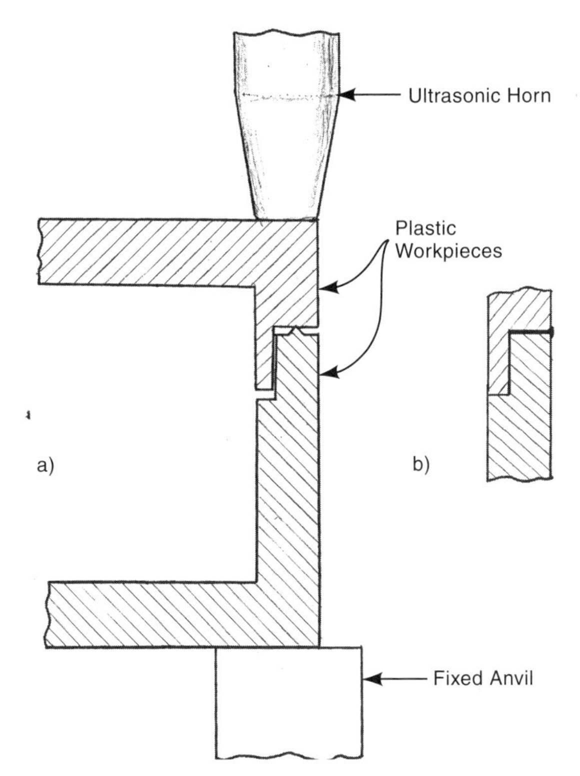

Fig. 4L4 Ultrasonic welding of two plastic parts: a) The joint before welding. b) The joint after welding.

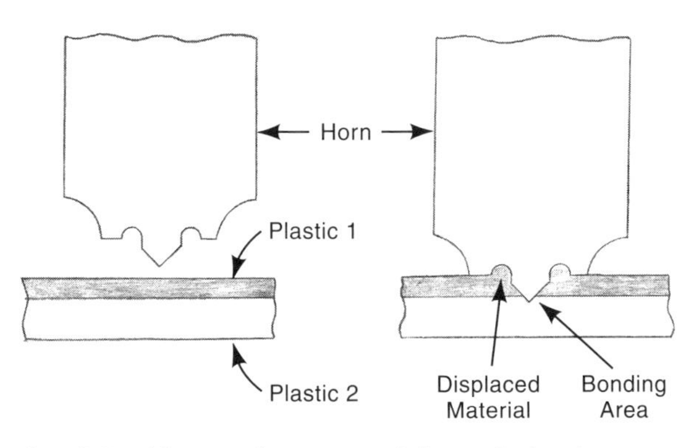

Fig. 4L5 Ultrasonic spot welding of plastic materials.

L6. ***ultrasonic film and fabric welding***[2] - Layers of film and fabric can be joined in continuous seams by ultrasonic welding. The ultrasonic welding horn is equipped with a smooth rounded tip and the underlying anvil has either a rotating wheel or a smooth surface. The material to be joined is pulled under the horn tip at the proper rate for seam welding. The rotating anvil can be engraved in a stitch pattern which is transferred to the fabric or sheet, simulating the appearance of a sewn seam. The process is used to join film sheet, and woven or non woven plastic fabrics. The sealing of polyester film is a common application.

L7. ***adhesive bonding of plastics*** - Information on adhesive bonding processes can be found in Chapter 7, section D. The material that follows applies only to the special steps and procedures involved when the parts to be joined are made from plastics.

L7a. ***solvent cementing of plastics*** - Some plastics can be bonded with strong joints by softening the surfaces in contact with a solvent and then pressing the two surfaces together for a short period. The plastics most suited to this approach are the amorphous type: polystyrene, acrylic, cellulosics, vinyl, polycarbonate, and copolymers of polyphenylene ether. Polyethylene and polypropylene are not bondable with this approach. The solvent used is one whose solubility is close to that of the material to be bonded. A solvent that is suitable for use with one plastic may not be suitable for another. Best results are often obtained if a small amount (up to 15%) of the base material is dissolved in the solvent beforehand. As with all bonding methods, the surfaces must be thoroughly cleaned before the bonding operation. The solvent can be applied by brushing, dipping, spraying, flooding, or capillary action. After some softening, the surfaces are pressed together under a moderate pressure - 100 to 200 psi (700 to 1400 KPa) - is sometimes recommended, but pressure must not be so high that the parts are distorted. Post heating in an oven to a temperature below the softening point of the base plastics involved may also be carried out. Objects fabricated from acrylic are commonly solvent-cemented but perhaps the major use of this method is for installation and assembly of PVC and CPVC piping.

L7b. ***pretreatment of plastic surfaces for bonding*** - See section M1 on page 197.

L7c. ***electromagnetic adhesive bonding*** - uses induction heating of an adhesive to join two plastics or other non-metallic components. The adhesive

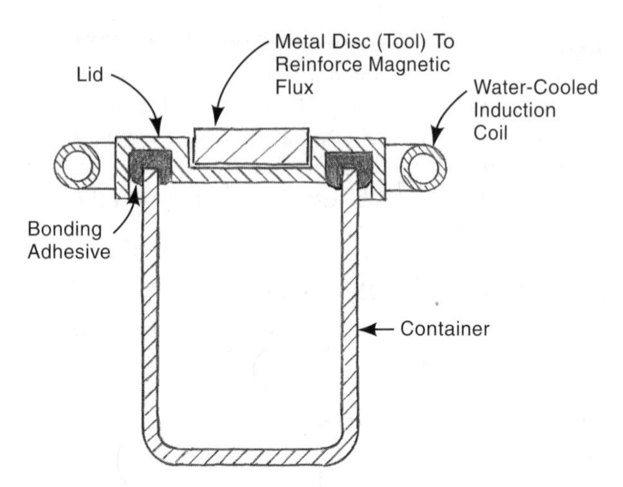

Fig. 4L7c Electromagnetic adhesive bonding in section view. Electromagnetic induction from the coil heats and agitates ferromagnetic particles in the adhesive, heating and softening the adhesive so that it adheres to the parts to be joined.

consists of a mixture of resin in which is dispersed very small ferromagnetic particles. The adhesive can be in the form of a paste, or a preform molded or cut from tape, sheet, or strands. When the induction generator is activated, radio-frequency electrical oscillations in a coil near the joint emit electromagnetic energy that causes the ferromagnetic particles to oscillate. Eddy currents generated in the ferromagnetic particles cause them to become heated. The heat from the particles and the friction from their movement quickly softens or melts the adhesive so that it wets and bonds with the surfaces to be joined. Thermosetting adhesives polymerize and harden from the heat; thermoplastic adhesives cool and solidify after the energy is shut off. The process provides good seals as well as strong joints. Joints need not be flat or regular, so long as the adhesive can fill any gaps. The process is most suited to moderate- or high-production levels, since fixtures, coils, operating settings, and possibly adhesive preforms, have to be developed for each application. Automotive parts are a major application, including such items as heating and air conditioning ducts, and seat backs. Medical devices and filters are other applications. Fig. 4L7c shows an example of electromagnetic adhesive bonding.

L8. ***induction welding*** - is similar, in many respects, to electromagnetic adhesive bonding. However, the workpieces themselves, rather than an adhesive, are melted by the induction heating. The induction effect, which requires an electrical conductor, is achieved by inserting an open-grid metallic piece between the opposing surfaces of the joint. The metallic piece may be a screen, foil, wire, or some other conductive part that will absorb the electromagnetic energy, get hot, and heat the surrounding plastics. When the power is shut off, light pressure is maintained across the joint and the melted joint materials fuse together and solidify. The metallic insert remains in the joint. The process is very fast; only a few seconds are required to weld the joint surfaces together. The equipment and tooling required are quite similar to those used to induction braze or solder joints in metal parts. See descriptions and illustration in sections 7A2h and 7B4.

L9. ***radio frequency sealing (dielectric sealing)*** - is a method for joining plastic film and sheet parts. Energy in the form of electromagnetic radiation at radio frequencies is directed at the joint, and it agitates the plastic molecules. A radio frequency of approximately 27 MHz is commonly used. Polar molecules of the plastic sheets vibrate at that frequency, and experience friction with other molecules, producing heat. The material in the joint area is thereby softened. Under pressure, the joint surfaces fuse together and, when the power is shut off and the material cools, a permanent seal is achieved. The process is used most frequently with flexible and rigid PVC but many other plastics can be processed including thermoplastic polyurethane, ABS, polyester film, EVA, acetate, and acrylic. Polystyrene, polyethylene, and polypropylene, however, are not suitable for the process. The equipment involves, in addition to a RF electrical oscillator, a press and dies for the particular application Typical products assembled by this operation include inflatable toys, swimming pool liners, shower curtains, rainwear, medical bags, looseleaf notebooks, packaging, and automobile interior components.

L10. ***thermal sealing (heat sealing) of sheet*** - In this method, the heat is applied externally and travels through the sheet sufficiently that the sheet materials in contact are soft enough to fuse together. The heat source is a bar, knife edge, metal band, wheel or roller, that is heated by electrical resistance or radio frequency energy while in contact with the top or bottom sheet. This tool is coated with PTFE ("Teflon") to prevent the softened sheet from sticking to it. Some high-production machines use heated sealing rolls or wheels, followed by pressure wheels, followed by cooling wheels as the sheet materials pass through them. Polyethylene sheet and film are frequently bonded with this method.

L11. ***hot gas welding*** - is a method used to join plastic sheets, normally in thicknesses from 1/16 to 3/8 in (1.5 to 10 mm). The process is very similar to gas welding of metals, except for the lower temperature and the absence of direct flame. Hot gases from a hand held gun are directed to melt (soften) the edges of the parts to be joined enough that they can fuse together. A welding rod, of the same plastic material as the parts being joined, is also softened and added to the material at the joint to provide any additional material needed for a fillet. The welding rod is often of a triangular cross-section and is pressed by the operator into a v-shaped space at the joint. When the heat is withdrawn, the material cools and solidifies into an integral joint. The hot gas is not supplied by a flame at the nozzle but is simply air or, in some cases, nitrogen, heated electrically to a temperature of 400 to 570°F (200 to 300°C). Edges of plates to be welded are beveled mechanically before welding. No flux is needed and there is no slag to be chipped away.

Polyvinyl chloride (PVC), polyethylene, polypropylene, acrylic, polystyrene, polycarbonate, and ABS are the plastics most commonly used for welding applications. Tanks, piping, large outdoor signs, ducting, and structural assemblies are common applications. The process is particularly useful in making components too big to mold or cast by primary methods.

M. Surface Finishing and Decorating Processes for Plastics

Many methods are available to change the surfaces of plastic products to improve their appearance and make them more useful. Surface property and color changes, the use of contrasting colors, trademarks, and other identification, product use information, and special effects all can be achieved with the processes described below.

M1. ***surface treatments for plastic parts*** - Various chemical and physical surface treatments may be applied to plastics to prepare them for further operations, usually to facilitate adhesion of coatings or bonding agents but sometimes to improve appearance or functional usefulness. These treatments include cleaning, chemical etching, flame treatment, corona discharge, plasma treatments, and priming.

M1a. ***washing and cleaning*** - methods are described in Chapter 8 of this handbook for parts made of metals and various other materials, often including plastics. Due caution must be observed with cleaning methods such as abrasive and brush cleaning which can damage surfaces of plastics. Steam or other high temperature methods may involve temperatures that soften surfaces of plastic parts. Some solvents may cause crazing or other adverse effects on some plastics, so it is desirable to test proposed solvent processes on a sample of the material before using them. Plastic parts, however, are less apt to require removal of scale, grease, and other soils that characterize metals.

M1b. ***chemical etching*** - is done on some plastics to make the surfaces more bondable or plateable. Fluorocarbon plastics are notable examples. Naturally very slippery, these plastics are treated with a sodium naphthalene solution prior to adhesive bonding. Other plastics may be immersed in, or have an application of, sulfuric or chromic acid. The resulting surface has enough micro-roughness to improve its wettability.[2] Epoxy printed circuit boards are treated with a potassium permanganate solution prior to electroless copper plating. Acrylic, polycarbonate, and PVC window panels are treated to reduce reflections.

M1c. ***corona discharge*** - consists of a series of small, uniformly-distributed electric sparks, which cause changes in the atomic structure of the surface being treated. The process, which is represented in Fig. 4M1c, is used most often on plastic film to increase the bonding strength of the surface. The process is often used to prepare surfaces for printing and labeling. Inks, paints, and adhesives

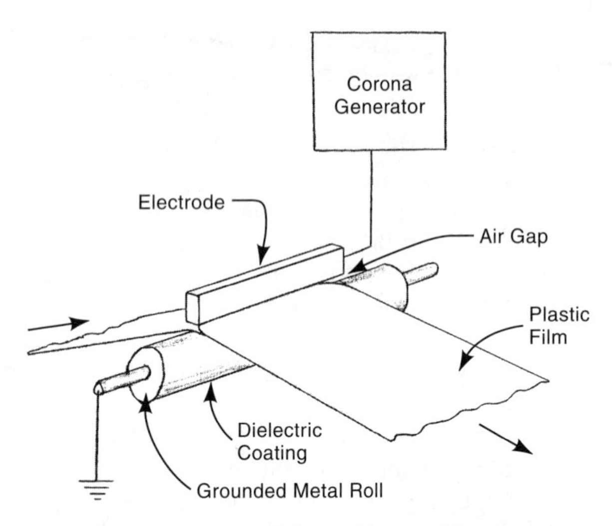

Fig. 4M1c Corona discharge treatment to prepare a plastic film surface for printing, painting, or adhesive bonding. The corona discharge from the electrode in the rectangular ceramic tube increases the surface energy of the film that passes beneath it.

adhere to the treated surface due to improved intermolecular bonding resulting from the treatment.

The equipment required consists of a power supply and a treater. The power supply provides a high-frequency oscillation (10 to 30kHz) at high potential. The treater subjects the workpiece to a corona from this electrical charge. For film, the treater consists of a rectangular ceramic tube containing an electrode that carries the high-potential output of the power supply, and a grounded metallic roll, coated with a dielectric material. The roll supports the film as it passes under the electrode. The corona discharge affects the surface that faces the high voltage; the surface facing the ground is not affected. The electrostatic discharge also produces ozone, which must be removed from the apparatus. Polypropylene, high density polyethylene, PET, and PVC are processed in sheet or woven fabric form. Products processed by corona discharge include combs, trays, cups, bottles, and other containers.

M1d. ***flame treatment*** - is a very common process for making the surfaces of molded plastic parts more receptive to ink, paint, and other coatings. It is used on blow-molded plastic bottles to prepare them for label printing. An oxidizing flame at a temperature of 2000 to 5000°F (1100 to 2800°C) impinges on the surface for less than a second, but causes enough oxidation so that inks, paints, lacquers, and adhesives will adhere. The usual method in high-production situations, as shown in Fig. 4M1d, is for the bottles on a conveyor or rotating machine table to briefly pass through a stationary flame. Polyethylene is the material most notable for being flame treated.

M1e. ***plasma treatment*** - is a newer surface treating process for plastic parts to make their surfaces more wettable and more suitable for bonding or coating. In a chamber containing the parts to be treated, a gas or gas mixture, with its molecules dissociated, contacts the surfaces of the parts, micro-etching them and making them cleaner and

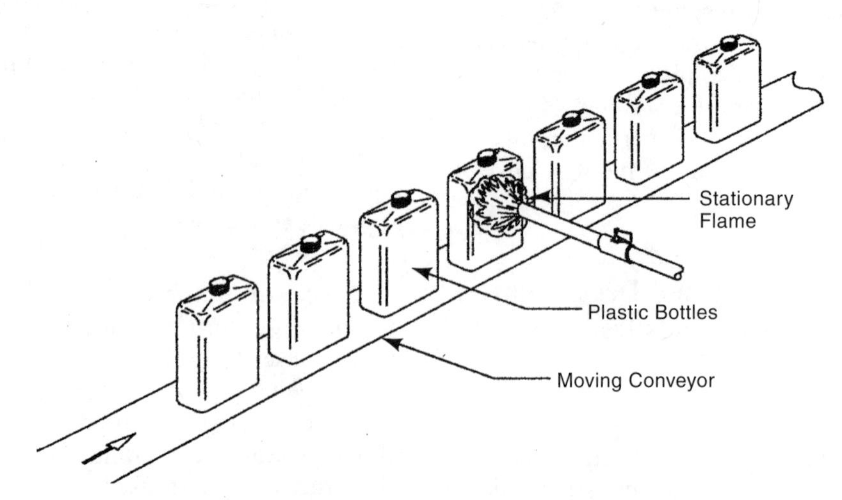

Fig. 4M1d flame treatment of plastic bottles facilitates adhesion of printing on the surface.

more active. The process works on a batch basis. Parts are placed in a chamber. A low vacuum is drawn. The gas to be ionized is then introduced into the chamber under a partial vacuum, and is subjected to an electromagnetic field at a radio frequency. This partially ionizes the gas, which then consists of ions, free electrons, and neutral particles, excited at a broad range of energy levels. Electrons, ions, free radicals, and photons, react with the surface molecules of the plastic workpieces and cause a chemical and physical modification. Gases or gas mixtures used may include: nitrogen, argon, air, oxygen, nitrous oxide, helium, water vapor, methane, tetrafluoromethane, carbon dioxide, and ammonia. Each gas yields a specific plasma composition and produces particular surface effects. Despite the high energy of some particles in the plasma, the overall plasma is not hot enough to affect the workpieces except at the very surface.

Plasma treatment is used to provide adhesion for coatings including graphics, applied by silk screening, other printing methods, or painting, to allow adhesive bonding, to etch microcircuits, and for medical applications, including improvement of membranes used in dialysis treatment.

M2. ***hot stamping*** - is a common method for applying decorations, printed material, and other surface changes to plastic products. The process involves a selective transfer of coating material to the surface of the workpiece. A multi-layer plastic foil is placed against the surface to be decorated. Heat from a hot pad, roller, or die, which is pressed against the foil, causes the coating material to be transferred to the workpiece. The foil typically consists of the following layers: 1) a carrier film, usually polyester, 2) a layer of a thin release agent that is activated when heated, 3) a layer of the decorative coating to be transferred to the workpiece. This coating may be metallic, or an ink, dye, or paint, often in several colors, 4) a protective layer for the decorative coating, and 5) an adhesive, which is formulated to bond the desired coating to the workpiece. The rubber or metal hot pad, is hot enough to melt the release coating of the foil, and to activate the adhesive coating. Often an engraved die is used. It is made to the form of the decorative design or lettering to be transferred. Then the design transferred is in the form of the die, rather than from a printed design on the foil. When either the die or the pad is briefly held against the foil and workpiece, the decorative material is transferred. Upon cooling, the coating is firmly bonded to the workpiece surface, which is slightly depressed in the coated area. The protective layer is also transferred and serves to shield the decorative coating of the part from abrasion or chemical attack.

All kinds of plastic parts are hot stamped to provide nomenclature, decorations, and different surface effects, usually in localized areas. Wide area hot stamping is also feasible and is employed to produce a simulated wood grain on TV cabinets and other plastic objects. One common approach for localized areas is to mold the part so that the area to be decorated is raised and formed to the lettering or decoration wanted. Then the foil is stamped on the raised surface with a flat pad or roller, and the decoration takes its shape from the workpiece shape. This procedure works well when raised lettering or decoration is desired. Fig. 4M2 illustrates hot stamping with both an engraved metal die and a flat silicone pad.

M2a. ***transfer coating (hot transfer coating)*** - is similar to hot stamping, but is primarily a decal process as described in section 8I9. Decals with information or decoration, printed on a plastic film and backed with paper or plastic are transferred to a workpiece. The printing is usually by the gravure method and is multicolored. The transfer is effected with heated rubber pads or rollers but at lower pressures than those used in hot stamping. The film is released from its backing and adheres to the workpiece. The method is quick and economical and is used for many kinds of labels and other decorative effects such as wood graining.

M3. ***painting*** - Although coloring can be incorporated into plastic parts as they are molded, it is common for plastic parts to be painted as well. Plastic parts are painted when an area of contrasting color is desired, when the finished appearance desired is not feasible with molded-in colors, to match the color of other parts that are painted and to cover surface imperfections. Selective painting, i.e., painting only a portion of the part involved, is common. Many painting processes are available for painting plastic parts. These are described in Chapter 8 since they are also applicable to workpieces made from materials other than plastics.

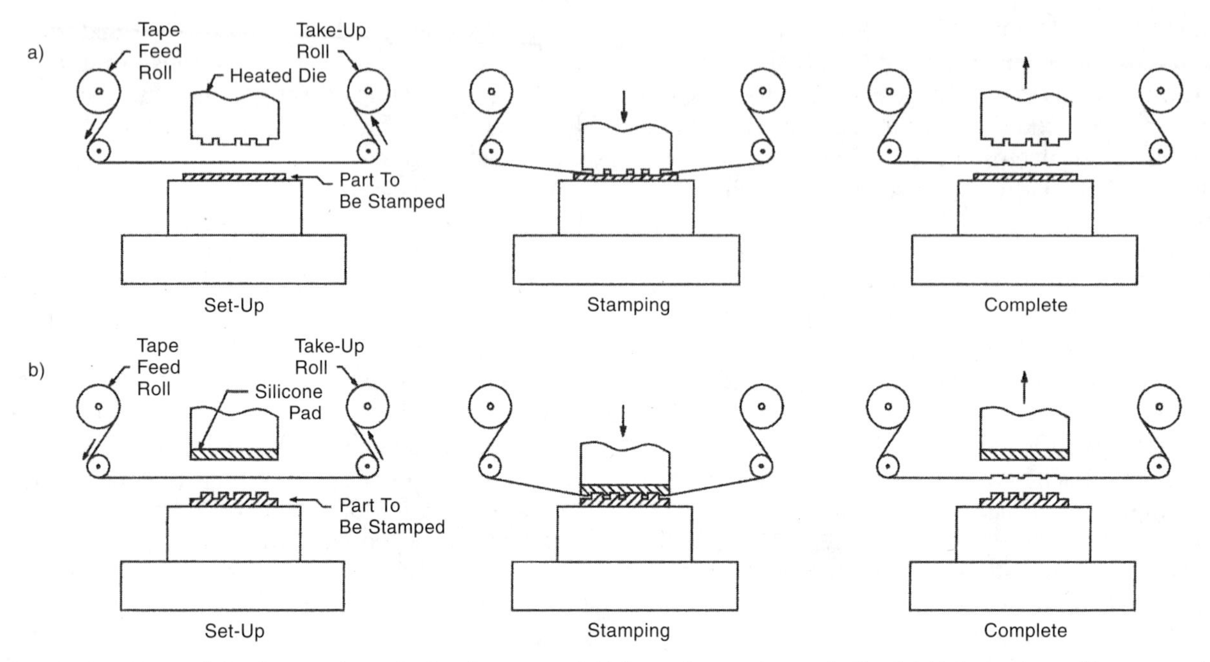

Fig. 4M2 Hot roll-leaf stamping of plastic parts: a) Using a heated metal die, b) Using a hot silicone pad.

They include spray coating (in several variations), dip, curtain, flow and roller coating. Paint material used with plastics are usually plastics-based and the paint material and solvents used, if any, are ones that are compatible with the substrate plastic. Fig. 4M3 illustrates a typical paint mask used in selective spray painting a plastic part.

M3a. ***spray and wipe*** - is used to put recessed areas of decorations or marking in a different color. The process simply involves spray painting the part and wiping off the paint from the non-recessed areas.

M3b. ***powder painting of plastics*** - Application methods for powder paints are summarized in Chapter 8 of this handbook. Thermosetting plastic substrate materials are coated with methods parallel to those involved with other non-conductive materials. Electrostatic spray, fluidized bed, electrostatic fluidized bed, and friction static spraying are applicable methods. Pretreatment of plastic surfaces to ensure adequate adhesion and electrostatic attraction may also be required. The elevated temperature required to fuse the powder, often about 400°F (200°C), has prevented significant commercial use of powder coating for thermoplastics.

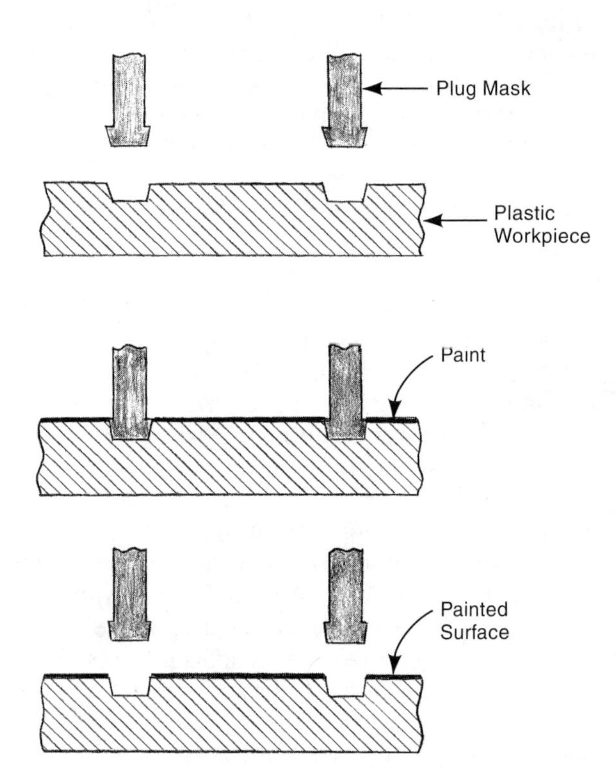

Fig. 4M3 Painting a plastic part with a plug mask. Recessed areas are kept free of paint while the surface of the part is painted.

M3c. ***manufacture of powder coatings*** - Three different methods are in use for the manufacture of powder paints: dry blending, melt mixing, and solution.

M3c1. ***dry blending*** - is used to mix PVC and thermosetting powders for thick film applications. A high-intensity mixer (See 4A4b.) or ball mill mixer (See 11D3.) is used. The ingredients include the basic resin, pigments and other solid additives. In the case of PVC, plasticizer and other liquid additives are also added. Temperature is controlled during the operation. For PVC, a temperature of about 212°F (100°C) is maintained. Sieving is performed after the mixing operation using a 60 to 100 mesh screen.

When a ball mill is used, the operation requires much more time. Because heat is generated, the catalyst is usually added toward the end of the mixing process. Ball-mill dry blending is used to make powders for fluidized bed and electrostatic applications. Ball-milled powder is also screened after mixing.

M3c2. ***melt mixing*** - involves the following steps: 1) premixing of ingredients. (One of a variety of mixers may be employed: cone mixer, high intensity mixer, ball mill, ribbon blender.) 2) melt mixing. This involves heat and continuous mixing with equipment such as that described in section 4A4c. 3) cooling. The material discharged after mixing is fed to a cooling conveyor or other cooling device where its temperature is lowered to about 100°F (38°C). 4) crushing. The material is crushed into flake or kibble (large particles). 5) pulverizing. The coarse particles produced are then pulverized by hammer mills or other methods. 6) sieving. The pulverized material is classified by sieving in a screen of 80 to 140 mesh.

M3c3. ***solution method*** - involves manufacture of the coating as a liquid paint and then removing the solvent by either spray drying, devolatizing or flocculation in water.

M4. ***decorating plastic parts with processes that are also common to non-plastics*** - Silk screening, roll coating, electroplating, labeling (except in-mold labeling described below), laser marking, flocking, dyeing, vacuum metalizing, sputtering, polishing and buffing are all described in Chapter 8 of this handbook. Printing (various processes) is described in chapter 9. Polishing and buffing operations for thermoplastics must be carried out with softer wheels, slower speeds and lighter pressures than with metal parts, to avoid overheating the surface of the workpiece.

M4a. ***electroplating of plastics*** - A number of plastics can be electroplated by being treated to make the surface conductive before the plating operation. One way to do this is to plate the part by the electroless method with a thin layer; electroless plating can be used with non-conductive materials. Another approach is to add carbon to the plastic when the part is molded. A third method is to coat the workpiece with a conductive paint. The electroless method involves several steps: 1) acid etching to create a microporous surface, 2) a neutralizer bath to reduce any residual acid, 3) a catalyst bath to deposit palladium in the surface micropores, 4) an accelerator bath to prepare the palladium for electroless plating and, 5) electroless plating with either copper or a nickel-phosphorous alloy. The conductive part is then plated by conventional methods. ABS and PEC (polyphenylene ether copolymer) are the most frequently plated plastics, but polystyrene, ABS/polycarbonate, nylon and polysulfone are also processed. Automotive hardware and trim are often made from plated plastic moldings. A major application is the plating of the circuit paths on printed circuit boards. Epoxy/glass, polyimide, phenolic and Teflon/glass are common substrates. The plating of household faucets, knobs, marine hardware, hospital equipment and kitchenware are other uses.

M5. ***in-mold decorating*** - It is quite feasible to place decorative or other material in a mold so that the item, after molding, is integral with the molded part. Several alternative methods are possible: 1) The mold cavity walls can be coated with another material prior to molding to provide a surface finish on the molded part different from that which would otherwise be achieved. Both liquid and powder coatings are used. 2) product labels can be inserted in a mold before the molding operation and, 3) decorative foils or other objects can also be placed in the mold so that they become part of the finished product. This approach is feasible with injection molding, compression molding, RIM and other molding processes, including those carried out with reinforced plastics. The advantage with coatings is that they eliminate surface porosity, and provide a durable, smooth surface for the part. The advantage with inserted foils and labels is that the inserted object is securely held and, in some cases, increases the strength of the finished part while reducing the amount of material required. The need for later finishing operations is often reduced. Foils with a brushed aluminum or wood grained appearance are common inserts. The inserting operation at the molding machine can be manual or, in many cases, robotic. One early and still common in-mold decorating operation is the use of preprinted, resin-impregnated paper in the molding of melamine dishes and urea and phenolic products. Thermoplastics can also be processed with decorative inserts. Polypropylene, polyethylene, polystyrene, polycarbonate, ABS and acrylics are suitable for in-mold label and foil decorating.

M6. ***sheet and film embossing*** - Several methods are used to produce textures in the surfaces of plastic film and sheet. Such textures change the appearance of the sheet, provide a different feel and can hide other imperfections. One common example of a changed surface of sheet plastic is the modification of flexible vinyl sheets to provide a leather-like appearance. The common method for producing this effect is to run the heated sheet through a pair of cooled forming rolls. At least one of the rolls is engraved with the reverse of the pattern desired on the plastic sheet. When passed through the rollers, the sheet is formed and then cooled so that the texture becomes permanent. Depending on the results wanted, the roll used to back up the engraved roll can be rubber or another resilient material. It can be flat steel or can be engraved to better emboss the sheet or provide some other pattern on the reverse side. Although it is most common to have a heated sheet and cool rolls, sometimes, the sheet is initially cool and is formed by rolls that are heated. Still another method uses formed paper, with a release agent on its surface, in a film casting operation so that the film or sheet as initially made has the texture in it. Sometimes, when deep embossing is desired, the roll is porous and a vacuum is drawn in it. Other methods involve dielectric or frictional heating in small areas of the sheet to aid in forming.

Forming or embossing is sometimes combined with printing a pattern on the surface to enhance the appearance of the texture. *Valley printing*, that is, printing in the depressed parts of the surface, is illustrated in Fig. 4M6 which shows equipment that

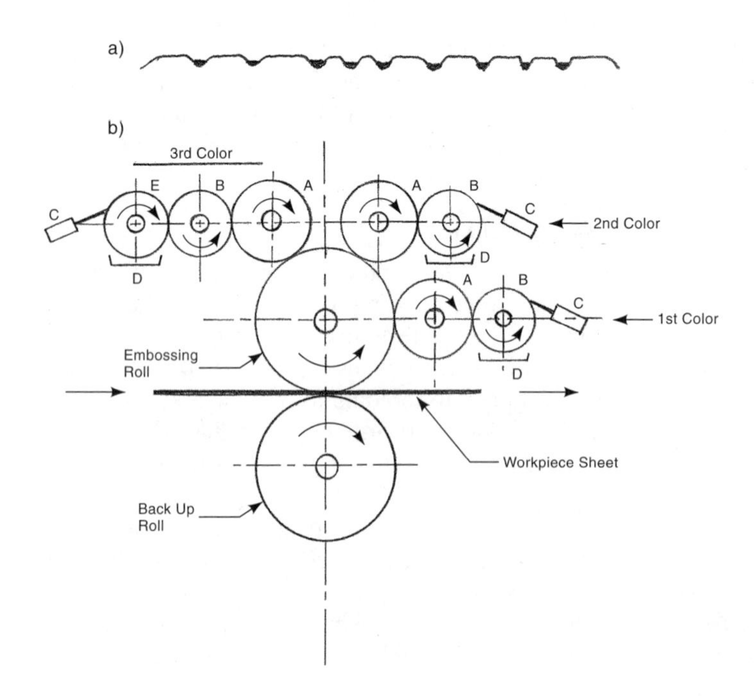

Fig. 4M6 Valley embossing and printing of film or sheet.[2] a) Valley printing - printing in the depressed areas of an embossed sheet. b) Schematic view of combination embossing and valley printing equipment: A - patterned applicator rolls, B - analox rolls, C - doctor blade to remove excess ink, D - ink pan, E - furnisher roll.

embosses and prints the surface in the same operation. This approach is useful in imparting wood grain, textile or leather effects on vinyl, urethane or other plastic sheets.

N. Other Plastics Processes

N1. ***insert assembly*** - Often it is desirable to incorporate metal or other non-plastic components in a plastic part. This is done when strength, hardness, dimensional precision, wear resistance, appearance or some other characteristic can be supplied better by some material other than the base plastic of the workpiece. Perhaps the most common example is the use of metal screw threaded inserts in plastic parts that have other parts attached to them with screw threads. Threaded inserts are particularly advantageous when the part is expected to be disassembled periodically. Inserts can be designed for various methods of insertion and attachment to the base part. Common methods are the following:

N1a. ***molding-in inserts*** - involves placement of the insert in the mold from which the part is made. Plastic material flows around the insert during molding and holds it in place. This method is described above in paragraph C6.

N1b. ***ultrasonic insertion*** - Ultrasonic welding of plastics is described above. (See L4, L5 and L6.) A similar approach can be used to insert a part into a slightly undersize hole in a plastic component. Ultrasonic vibration directed to the insert causes it to vibrate ultrasonically as it is positioned at a hole. Frictional heat softens the plastic and allows the insert to be pressed in. When the vibration stops, the plastic cools and firmly holds the insert in place. The process is for thermoplastics though some thermosetting plastics can be processed. Fig. 4N1b illustrates the method.

N1c. ***expansion installation*** - The insert is designed to be expandable. After it is placed in a molded or drilled hole in the workpiece, it is expanded with a suitable tool or by threading a screw into it. In either case, the walls of the insert spread and bear tightly against the walls of the hole. This is an inexpensive approach but its holding power is not as great as the molded-in or tapped-hole methods.

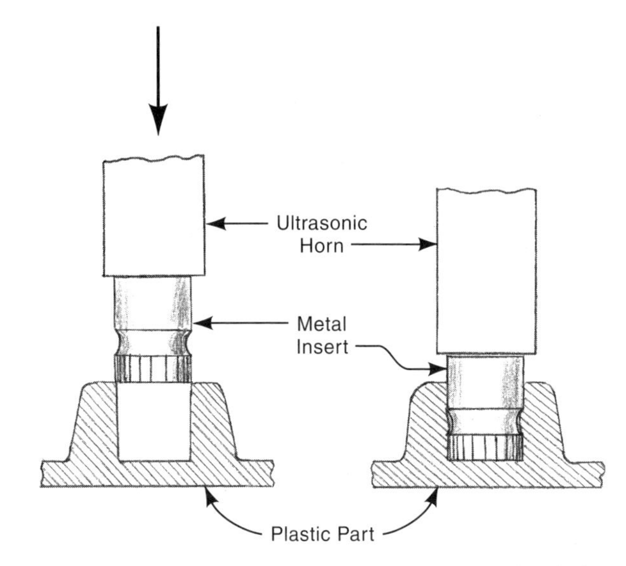

Fig. 4N1b The ultrasonic insertion method for assembling metal inserts to plastic parts.

N1d. ***threaded hole installation*** - The workpiece holes that are to receive the inserts have screw threads molded or tapped into them. The insert - which has external threads - is screwed into place. This method produces an assembly with high holding power.

N1e. ***press insertion*** - With this method, the outer diameter of the insert and the inner diameter of the molded or drilled hole create a press fit between the insert and base piece. Pressing the insert in place is a quick operation but the holding power is inferior to that of molded-in inserts. Some inserts have a spiral track on the outer surface so that the insert travels a rotational path when it is pressed into place. This method is used with harder, more brittle, thermoset plastics.

N1f. ***self tapping*** - Inserts with thread-cutting or thread-forming external screw threads are screwed into unthreaded holes. This approach is a less costly than the use of pre-threaded holes. Holding power is good for pull-out but limited for torque-out resistance.

N1g. ***cemented installation*** - utilizes an adhesive that is compatible with both the base plastic and insert materials to bond the insert into a straight walled hole. Holding power tends to be lower than when inserts are molded in.

N1h. ***hot insertion*** - In this method, the inset is heated to a temperature high enough to soften the plastic surrounding the hole. The insert is pressed into place and the softened plastic flows around knurls on the insert walls, producing good holding power when the plastic cools.

N2. ***other mechanical assembly of plastic parts*** - For assembly operations that are common to plastic and non-plastic parts, see Chapter 7.

N3. ***granulating plastics*** - is the operation that reduces molded sprues, runners and scrap pieces to a size and shape that can be blended and processed easily with virgin material. Then the blended material can be fed to injection molding machines, extruders and other processing equipment. Granulating machines have intermeshing stationary and rotating cutters that sever and recut the pieces fed into the machine until they are small enough to pass through a sizing screen. Pieces to be cut are fed either manually or by conveyor to the granulating machine. Parts too large for feeding to the machine are processed first in a shredder, a large machine with similar cutters but which runs at a lower rate of rotation. Granulating machines are also used to prepare scrap material, including thermoset plastics, for disposal since the granulated material is more compact and easily handled.

N4. ***coloring and blending of plastics*** - The method used for blending colorants into plastic resins varies with the type of colorant used. Colorant blending can be difficult with some materials, particularly organic pigments which are often finer and less dense than the plastic. Dyes - colorants that dissolve in the resin when it melts - are relatively easy to blend. The usual method is to use a tumbling barrel or rotating drum to mix measured quantities of both resin and dye until the dye is well dispersed. Another simple method for the molder is to utilize color concentrates, resin pellets containing dispersed or dissolved colorant. These can be blended in the barrel of the extruder or injection molding machine or tumble mixed beforehand. Blending of other additives and more difficult colorants is described more fully in section A4 above.

N5. ***drying of plastics*** - It is necessary to insure that hygroscopic plastics, notably nylon, PET (polyethylene terephthalate) and polycarbonate, are dried before they are processed in molding or another operation. Other plastics that may gain surface moisture from ambient air are also often dried before further processing. There are two basic approaches in common use: 1) Hot air systems that blow heated air through a silo or other container of the plastic material. This approach, illustrated in Fig. 4N5-1, is limited to plastics that have only some surface moisture. 2) For the hygroscopic plastics mentioned above, a typical dryer uses two or more cylinders of desiccant. While one cylinder is in use, the other is being regenerated or is undergoing temperature change. Drying is effected by blowing warm air through both the resin pellets and the desiccant in a closed, pressurized system. (Pressure is needed to get sufficient air flow through the containers of plastic pellets and desiccant.) Warm air blowing is continued until the desiccant reaches its full moisture content. While this is taking place, another cylinder is being regenerated by blowing much hotter air through it. This hot air, containing moisture, is then exhausted. When sufficient moisture has been removed from the desiccant, it and the cylinder are cooled to normal operating

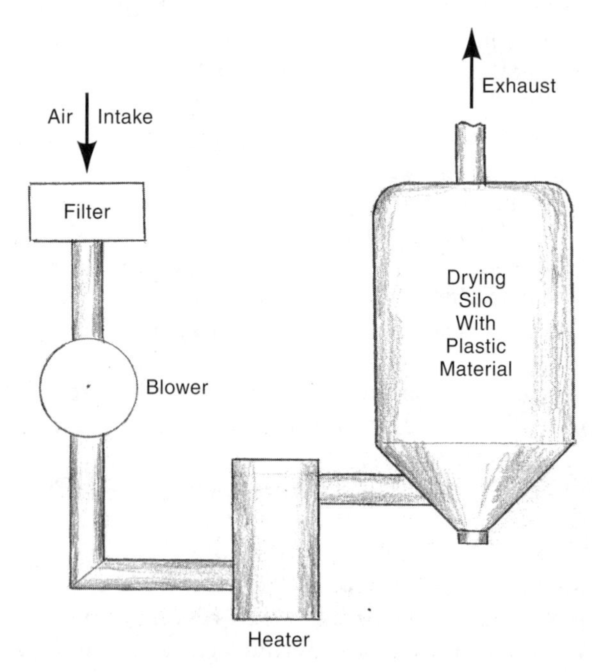

Fig. 4N5-1 A hot-air drying system for removing surface moisture from plastic molding material.

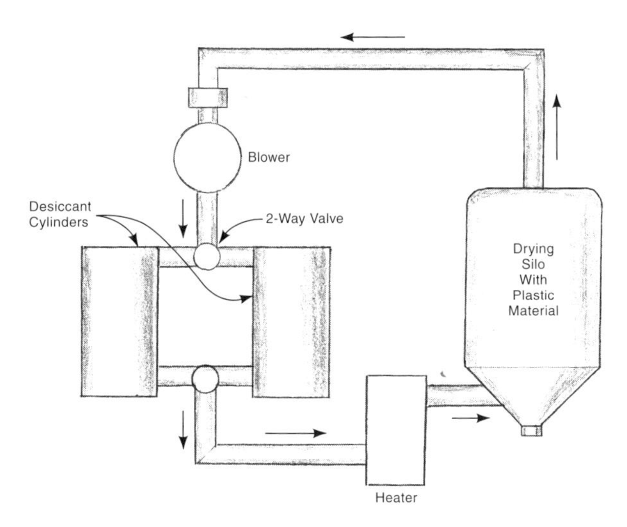

Fig. 4N5-2 A pressurized drying system for hygroscopic plastic molding materials. The desiccant cylinders are used alternately. When the desiccant in one cylinder is saturated with moisture, the system switches to the other cylinder and the first cylinder is subjected to a flow of heated air (not shown) to redry the desiccant it contains.

temperature and are ready for reuse. Fig. 4N5-2 shows this method.

N6. ***machining of plastics*** - Both thermosetting plastics and thermoplastics are machinable with conventional metal-working equipment (See chapter 3). However, the special properties of plastics require different machining techniques. Additionally, not all plastics are alike. What works well for one material may not give good results with another. The following are differences in properties of plastics, compared to metals, that must be reckoned with: 1) Thermoplastics are normally softer, more difficult to clamp securely and more apt to deflect from cutting pressures. Workpieces must be well supported. 2) Plastics do not conduct the heat of cutting friction well and may overheat at the point where cutting takes place. 3) They have a thermal coefficient of expansion about 10 times that of most metals and therefore are more apt to distort from the heat of cutting. The result of most of these factors is a less accurate final machined workpiece. The following steps are usually taken to get best results: 1) Tools are kept very sharp 2) Rake angles of tools are positive and larger than for most metals 3) Tool surfaces touching the workpiece are fully polished to reduce friction. 4) Where possible, chips are long and continuous. Filled and reinforced thermosetting plastics may be abrasive to cutting tools and may create dust that can damage machine ways and be a health hazard.

N7. ***trimming and die cutting of plastics*** - trimming is a common operation after various plastic molding and forming operations. Fig. 4N7 illustrates three different die designs for such operations. The die in Fig. View a) is a simplified view of the type of die used for blanking and

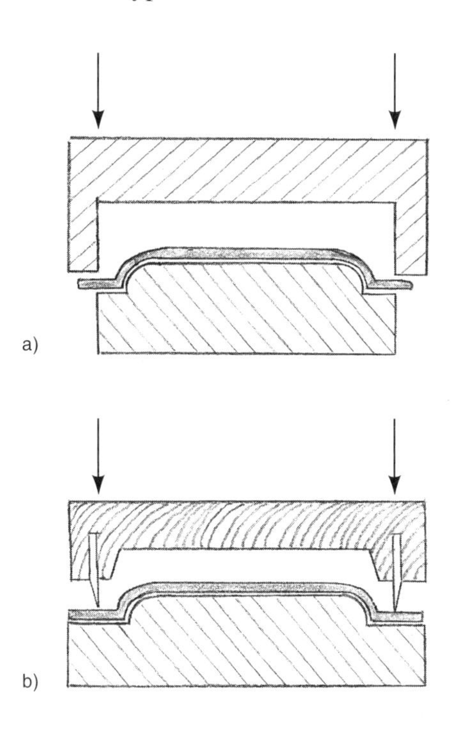

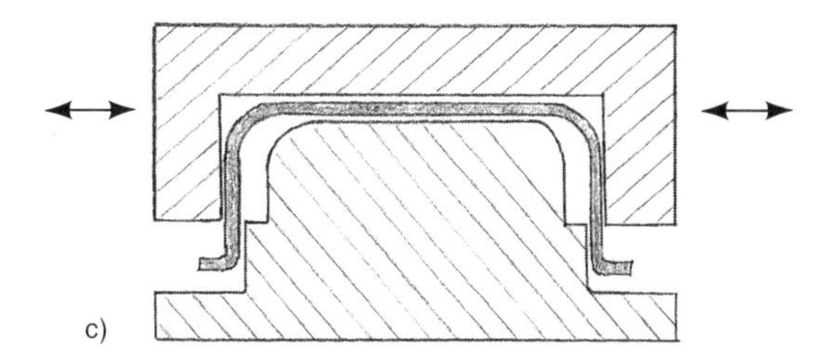

Fig. 4N7 Three different dies used to trim formed and molded plastics parts: a) A simplified view of a production blanking or trimming die, b) A steel rule die, c) The upper half of the die has a planetary movement in a side-to-side direction, to trim the side walls of the workpiece.

trimming metal workpieces and is described further in sections 2C4 and 2C6. Steel rule dies, shown in view b), are described more fully in section 2C4a. Dies of this type are used extensively in blanking and trimming softer materials. The die shown in view c), which oscillates horizontally, is called a planetary die and is useful for trimming vertical walls.

N8. ***radiation processing***[2] - is being increasingly used in plastics processing for curing thermosetting materials, cross-linking thermoplastics and making graft copolymers. Sterilization of medical products is another application. Two basic methods are available for providing ionizing radiation that can produce these desirable effects: 1) gamma ray and other radiation from radioactive isotopes, and 2) electron beam radiation from equipment-produced particle acceleration. In all cases, the production sequence is established so that, at some stage of the sequence, usually a production line in large-quantity situations, the workpieces are subjected to the radiation which produces a change in the properties and/or state of the workpiece material.

Crosslinking thermoplastics (Polyethylene is the most common thermoplastic given this treatment.) gives these materials improved properties: improved tensile strength, higher temperature resistance, improved chemical and weather resistance and reduced dielectric losses. Wire coatings, including foam insulation of coaxial cable, are given gamma ray or electron-beam radiation to produce crosslinking with its better strength. Crosslinked material also gains elastic memory. Components stretched after crosslinking will return to their pre-stretched state upon heating. Heat shrinkable tubing and film are made this way.

Graft copolymers are made by radiation of a homopolymer in contact with another homopolymer or monomer. The linkage is with the side chains of the material's backbone, providing a retention of the homopolymer properties plus the enhancement that the other polymer provides. Permeable film used in desalinization equipment is made by this method.

Electron beam radiation is used to cure paint coatings without the need for heating. This method is often advantageous with thermoplastic substrates that may not be able to withstand the heat of a paint curing oven. Painted plastic automobile body components are often treated with this approach.

Wood and concrete components are impregnated with plastics to provide improved properties. They are harder, stronger and more wear resistant than the unimpregnated equivalent components. The base materials are dried thoroughly with the aid of vacuum and heat, are impregnated with a plastic in liquid monomer form, and are exposed to either radioactive or electron-beam radiation.

N8a. ***non-ionizing radiation processing*** - involves radiation from ultraviolet, infrared, dielectric, microwave and induction sources. It is usually accompanied by some degree of heating which may, in itself, promote reactions in the plastic material. The process is used for crosslinking thermoplastics and curing of thermosets. Induction, microwave and dielectric methods are heat producing and are convenient means of applying heat selectively or quickly for bonding, localized curing, preheating for molding, sealing, and other molding, forming, or joining applications.

N9. ***vacuum handling and loading of plastic materials*** - is the most common method for short-distance transportation of plastic materials in pellet or powder form from storage containers to machine hoppers. This kind of movement can be accomplished with relatively low-power, compact equipment at each machine. Portable vacuum/pump units convey material at the molding machines from shipping containers directly to machine hoppers and do so automatically as material is needed. Some systems have intakes from both regrind and virgin material containers, and automatically feed the prescribed portion of each material. Similarly, color concentrates and other colorants can be metered into the material fed to the machine hopper. In some equipment, a color-virgin material blender is incorporated at the machine hopper.

For longer distance movement of plastics, vacuum equipment with positive displacement pumps to provide the vacuum is also used. For high production applications and a common material, a central vacuum pump with valves and piping for all machines is used. A central computer interprets sensor reading at each machine and, by controlling the vacuum pump and valves, delivers the needed amount of material to each machine. Material is also unloaded from incoming rail car or truck shipments in bulk using vacuum conveying equipment.

N10. ***robotic handling*** - is becoming more common in injection molding operations. Robots are used for removal of finished parts, sprues and runners from injection molds, to position parts for subsequent operations, and to perform some secondary operations. For some operations such as sprue removal, simple mechanical robots are often adequate. For more complex operations involving obtaining and placing inserts or handling, removing, stacking or packing finished pieces, more sophisticated computer-controlled robots are required. (Removing parts often requires a more complex operation than sprue removal because more extensive motions after removal are included.) The most complex robotic operations involve sensors, continuous path computer control and sophisticated programs. They can provide fully automatic operation of an injection molding machine. See Fig. 14G5.

N11. ***deflashing*** - is necessary after a number of molding operations for both thermoplastic and thermosetting plastic parts. The deburring and deflashing methods described in Chapter 3 of this handbook are generally applicable to plastic and elastomer parts. Of particularly interest is cryogenic deflashing, described in 3K25.

N12. ***reel-to-reel molding (continuous strip molding)*** - is an automatic insert molding process. Prior to the molding operation, the insert is blanked and formed from metal strip but is not severed from the strip, which is wound on a reel. The strip is then fed from the reel into the injection molding machine, which molds plastic material around the insert. The metal strip is precisely indexed in the injection molding machine. The individual molded parts are not separated from the strip during or immediately after molding. Instead, the strip of molded parts is wound onto a take-up reel. The major use of the process is for small electronic components which are not separated from the strip until they are automatically assembled to circuit boards or other components. The process requires high volume production to amortize the costs of the equipment required. Connectors, dip switches, shunts and other devices are the most common applications but medical devices, toys, and other products may also make use of the process. The molded part may include several stamped components. Wire may be used instead of strip in some components.

O. Rubber and Elastomers

O1. ***natural rubber***[4] - is produced from latex, the sap of the rubber tree. (See *latex*) To make rubber from latex, the rubber is separated from the latex fluid by coagulation. This involves the addition of formic acid or other acids or salts to the latex. The rubber particles coagulate into a dough-like material which is then easily separated. This material is then milled into sheets to remove contaminants and to facilitate drying. The sheets are subjected to wood smoke to kill bacteria and become the crude rubber that is shipped to processing plants.

The rubber sheets are cut into granules by a series of shear and rotating knife cuts and are dried in mechanical drying equipment over a period of several hours. The rubber is then softened so that various additive ingredients can be blended in. The most common softening method is the use of large mixers with eccentrically shaped blades that work the rubber against the mixer walls. Typical mixing lot sizes are one quarter ton. After mixing has softened the rubber, carbon black (which functions as a filler), vulcanizing chemicals (sulfur or sulfur compounds and accelerators), oil, a vulcanization accelerator and a protective antioxidant are added and mixed into the batch, either in the same mixer or with mixing rolls. The rubber is then ready for molding or extrusion into useful shapes.

Vulcanization of rubber takes place from the heat of the molding or forming operation when it is made into the desired product. Vulcanization temperatures range from 285°F to 360°F (140°C to 180°C). Vulcanization with 3 to 6% sulfur cross-links the molecules so that the rubber no longer will melt when heated and will maintain its flexibility over a wide temperature range.

Rubber is made into a wide variety of consumer and industrial products. Most of these use both natural and synthetic rubber. Vehicle tires are a major rubber product. Other uses are: hoses, balls for various games, electrical insulation, shoe soles, seals for building panels and for doors and windows of buildings and vehicles, seals and belting for machinery, floor tile, coatings, and adhesives.

O2. ***rubber, synthetic***[4] - Synthetic rubbers, often referred to as *elastomers*, are plastic materials, similar in structure to other plastic materials. However, they have the ability to stretch to at least double their length and to return almost immediately to the original length or close to it.[2] Much more synthetic rubber is produced in the world at this time than natural rubber. Styrene-butadiene rubber (SBR) is, by far, the most common synthetic rubber. It is also referred to as Buna-S and GR-S.[2] Other synthetics are: butadiene, ethylene-propylene, butyl, neoprene, nitrile and polyisoprene rubbers.

The manufacturing processes for these rubbers have much in common. The monomers that are reacted all can be handled similarly and the process equipment for one particular rubber is usable for several others. The process illustrated in Fig. 4O2 for making styrene-butadiene copolymer rubber is therefore quite typical for all the synthetic rubbers. The sequence starts with two separate sub-processes, one which produces styrene monomer and the other, butadiene monomer. Butadiene monomer is shown being made with the one-step Howdry process from *n*-butane with aluminum and chromium oxide catalysts. The butadiene thus produced is then purified by adsorption with cuprous ammonium acetate. The styrene is made from ethylbenzene

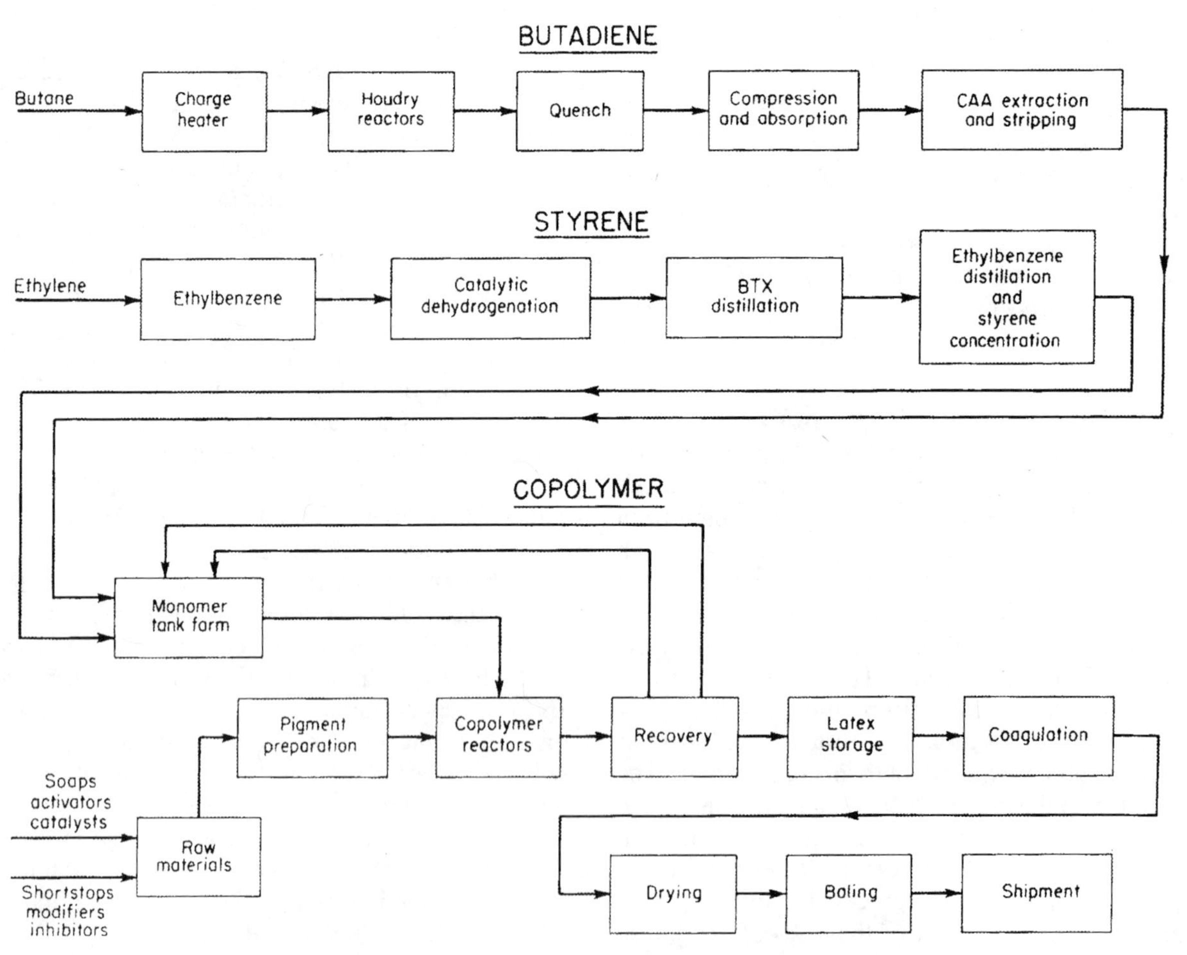

Fig. 4O2 The manufacturing sequence for styrene-butadiene rubber (SBR) showed in a simplified flow chart.

which results from alkylating benzene with the butane raw material. It is then dehydrogenated to styrene over an aluminum chloride, solid phosphoric acid, or silica-alumina catalyst. Then the monomers are fed into a polymerization reactor along with some additives and a catalyst. The proportion of each monomer varies somewhat, depending on the planned application, but 70 to 75 parts butadiene and 25 to 30 parts of styrene is typical. The emulsion polymerization reaction takes place for 8 to 12 hours at a temperature of 41°F (5°C). The heat of polymerization is removed from the reactors with cooling coils. Following polymerization, which is partial, unpolymerized monomers are returned to the reactor. Polymerized material proceeds through several physical steps to put it into usable form.

Neoprene is a product of coal, limestone, salt and water. Calcium carbide, from coal and limestone is reacted with water, forming acetylene gas (C_2H_2). The gas is reacted with hydrogen chloride to form chloroprene which is then polymerized to make neoprene.

Urethane, polysulfide, chlorinated polyethylene and silicone elastomers have superior properties for some applications where rubber-like material is needed. Urethane is used for forming pads for press forming of metal, solid tires, rollers and shock absorbing pads and bumpers. SBR rubbers are used extensively for tires and also for shoe soles, floor tile, in mechanical applications and as latex which becomes adhesives and coatings. Nitrile rubbers have particular resistance to oils, water, salts, soaps and most foods and are used in equipment where such resistance is important. Neoprene is used for automotive parts, adhesives, sealants, shoe soles, o-rings, bellows, conveyor belts, printing rolls and coatings. Butyl rubber is used for linings of tubeless tires, for innertubes, steam hose, tank lining and weatherstripping. Silicone rubbers are used for o-rings and seals for high temperature and corrosive conditions.

O3. ***rubber compounding***[4] - Both natural and synthetic rubbers are seldom used without additives as part of their formulation. The additives are needed to impart the necessary strength, elasticity, toughness, degree of hardness and abrasion resistance. Additionally, all natural rubbers require vulcanization, usually with sulfur compounds and many synthetic rubbers are similarly processed. Accelerators speed up the vulcanization. Antioxidants are added to improve the life of the rubber product. Fine powder fillers are added to reduce overall cost and improve hardness and shape retention. Carbon black and silica fillers, however, actually provide greater strength and improved abrasion resistance and resilience. Pigments may be included to provide the desired color of the rubber product.

The first step in compounding is *mastication.* The operation is normally performed in a Banbury-type mixer (Fig. 4A4c). The rubber is sheared repeatedly, breaking down molecules, and providing easier flow. Mixing of the additives and rubber then follows.

O4. ***rubber fabrication methods*** - are very similar to those used with plastics, particularly thermosetting plastics. Molding is usually by compression or transfer molding techniques. Injection molding is also used, particularly for thermoplastic elastomers. Calendering is used to provide rubber coatings on fabrics. Extruding is used extensively in the manufacture of weatherstripping, hose, innertubes and tire components. Another processing method used with rubbers is dipping wherein a master form is immersed in a liquid formulation of rubber or elastomer (natural rubber latex, neoprene, silicone or vinyl plastisol). The form is removed and the liquid that adheres is permitted to dry. Dipping and drying are repeated to build the coating to the needed thickness. The operation may be aided by using electrostatic charges, speeding the attraction of the liquid and providing thicker coatings with each dip. The dipping method produces uniform wall thicknesses and is used for boots, gloves and fairings[5]. Some flat rubber parts for seals and pads are made by die cutting sheet rubber material with steel rule or other simple dies. (See sections above for descriptions of these processes. Also see *tires, rubber.*)

References

1. *Plastics Handbook,* edited by Modern Plastics Magazine, McGraw-Hil, New York, 1994.
2. *Plastics Engineering Handbook of the Society of the Plastics Industry,* Inc. 5th ed., Michael L. Berins, editor, Van Nostrand Reinhold, New York, 1991. ISBN 0-442-31799-9.

3. *Encyclopedia of Polymer Science and Engineering*, John Wiley and Sons, New York, 1986.
4. *Encyclopedia of Chemical Technology*, Vol. 16, 4th ed., John Wiley and Sons, New York, 1995.
5. *Materials and Processes in Manufacturing*, 8th ed., E.P. DeGarmo, J. Black, R. Kohser, Prentice Hall, Upper Saddle River, NJ, 1997.
6. *Design for Manufacturability Handbook,* 2nd ed., J.G. Bralla, ed., McGraw-Hill, New York, 1999.

Chapter 5 - Glass and Ceramic Processes

"Glass is an inorganic product of fusion which has cooled to a rigid condition without crystallizing." - American Society for Testing Materials Committee C-14[1]

A. Glass Processes

A1. ***basic glassmaking*** - Glassmaking involves three basic steps: batching, melting, and forming. Batching is the preparation of a mixture of sand and stabilizing oxides, all in fine granular form. Melting involves the heating of the mixture to change it into a liquid and to further homogenize the various ingredients. Forming is the creation of useful objects or products from the molten mixture before it has completely solidified. The process can be carried out on either a batch or continuous-flow basis, the latter being used in mass production situations. Normally, forming operations take place immediately after the basic glassmaking, with the molten glass being cooled to increase its viscosity for forming.

There are many different kinds of glass. Soda lime glass is used for bottles, window panes and drinking glasses. Lead-alkali silicate glass has lead oxide in place of much of the calcined lime and is used for highly worked shapes including decorative glassware ("lead crystal") that is engraved. Borosilicate glasses, which contain boric oxide, are used when chemical and temperature change resistance is important, for example, in pharmaceutical containers, chemical process components and lamp envelopes. Aluminosilicate glasses are used where high temperature conditions exist. Several other mixtures may be used when optical properties are important.

A1a. ***raw materials*** - Silica sand (SiO_2) is the most common glass ingredient and has excellent resistance to attack, low thermal expansion, and resistance to devitrification (crystallization which impairs the optical and mechanical properties). However, in its unalloyed state, silica sand is difficult to process because of its high melting temperature and high viscosity when melted. Various other oxides are added to silica to improve its processibility and modify the properties of the finished glass. When soda-lime glass, the most common variety, is made, the ingredients consist of about 73 percent sand (SiO_2), about 14 percent soda ash or sodium carbonate (Na_2CO_2) and about 13 percent limestone ($CaCO_3$). Sodium oxide, (Na_2O) is an effective fluxing agent, i.e., a means for reducing the melting temperature, but too much can produce glass that is water soluble. Calcium oxide, calcined lime (CaO), increases the hardness and resistance of the glass to moisture. Alumina (Al_2O_3) improves durability and reduces thermal expansion. Potassium oxide (K_2O) from potash, increases durability and helps prevent devitrification, which has adverse effects. Other glass ingredients include borax or boric acid for boric oxide (B_2O_3), fluorspar (CaF_2), litharge or lead oxide (PbO), barium carbonate ($BaCO_3$), magnesium oxide (MgO), zinc oxide (ZnO), and other inorganic materials, some of which are colorants. Glass cullet (factory scrap or recycled glass), may be added to the mixture. It provides fluxing action and reduces the energy required for melting. About 30 to 40% cullet provides the maximum furnace efficiency[2]. In some mixtures, cullet content can reach 66 percent[3]. A typical commercial mixture has from 7 to about 12 different minerals, 4 to 6 of which are major ingredients[3].

A1a1. ***coloring materials*** - Glass is colored by adding small quantities (usually less than 0.5 percent) of certain metal oxides or other metallic compounds to the glass batch. Copper produces light blue; chromium - green and yellow; iron - bluish green or yellowish brown; cobalt - intense blue; nickel - grayish brown, yellow, green, blue or violet depending on the glass matrix; neodymium - reddish violet; manganese - violet; vanadium - green or brown[2].

A1b. ***batching*** - involves weighing, milling as necessary, and mixing to produce the glass furnace charge, a blend that can be melted to provide the composition desired. Quality control, including chemical analyses, must precede these steps to insure that each raw material is of the proper composition with impurities within limits. and of the proper grain size. Grain or particle size is important and must be controlled so that materials do not segregate during mixing, storage, and handling and so that they melt properly. Overly-fine particles of some materials may retard the elimination of gas bubbles from the melted charge. Milling and screening of raw materials may be required for some mixtures, though the common practice is to have suppliers of raw materials provide them with the desired grain size and size distribution. Milling and crushing methods are described in Chapter 11. Water may be added to the batch to the extent of 2 to 4 percent to prevent segregation prior to melting.

More recently, methods have been developed to consolidate the batch material in a form that more easily preserves the uniformity of the batch mixture, provides easier handling, improved melting, and better uniformity of the glass mixture during melting. These consolidation methods usually involve the following steps:

1. reducing and controlling the grain size of the batch materials by various milling operations and screening,
2. adding wetting and binding agents to the mixture,
3. thoroughly mixing the mixture and additives,
4. consolidation - briquetting, pelletizing or other means of holding the mixture into a stable but easily handled form, and
5. preheating the consolidation before melting.

A1c. ***melting*** - Melting the glass materials, known as the batch, enables the ingredients to be completely blended to produce glass of the desired properties and puts the glass in condition for forming. Typical melting temperatures are approximately 2640 to 2900° F (1450 to 1600° C). Heat is provided by gas, oil or electricity. Natural gas is the major fuel; propane is used as a standby. When quantities are small, melting is performed on a batch basis in pot furnaces or day tanks. High production melting is done in continuous furnaces that have output levels ranging to several hundred tons per day. Pots are made of refractory clay and are heated in brick furnaces. Day tanks are larger pots for batch production and are typically run on a one-day cycle, with melting at night and production and refilling the next day. Ten tons is a typical daily production quantity. Pots are typically round crucibles made of one piece of refractory material with individual capacities of one to two tons of glass. Several pots may occupy one furnace. Day tanks are made from refractory blocks.

Continuous furnaces are used for flat glass and for mass-produced containers and other high-production items. They are lined with refractory ceramics and are divided into a large melting section and a small refining section called a forehearth. The forehearth is used to cool glass from the melt temperature to a suitable temperature for whatever forming operation follows. Daily production levels are on the order of 100 to 400 tons of glass. The glass charge is fed from one end of the melting area. Temperatures in the melting area are as high as the glass mixture can tolerate in order to drive off carbon dioxide, steam, trapped air, and other gases, which could cause bubbles in the glass. Convection currents in the molten glass, which result from natural unevenness of heating and cooling from side walls, provide stirring that helps the glass mixture to become homogeneous. The molten glass that passes to the refining section does so through an opening below the surface of the melt, thus preventing any surface foam or scum from entering the forehearth. The temperature in the forehearth is typically cooler than that in the melting section by 180 to 360°F (100 to 200°C).

Furnaces may operate continuously for approximately a year before rebuilding is necessary. With gas and oil furnaces the glass is heated by exhaust gases that travel above the molten glass. Air for combustion is preheated by either a preheating chamber in the furnace or by regeneration where the cold air and cold gas are made to flow through brickwork that shortly before carried hot exhaust gases from the furnace. The flow is typically reversed at half-hour intervals. Immersed electrodes are used when heat is provided by electrical resistance. This resistance is that of the glass when current is passed through the molten glass from electrode to electrode. Electric heating is sometimes used as a booster in gas- or oil-fired furnaces. Electrical heating has quality and environmental advantages and is more common for batch production of specialty glasses, particularly those with a volatile component. Electrical induction heating is used for small quantity work.

Fig. 5A1c illustrates a typical glass pot furnace (a), a melting tank (b), and an electric heating system (c). Fig. 5A1c-1 shows a typical continuous furnace.

A2. *primary forming processes*

A2a. ***pressing*** - A "gob" of molten glass is placed in a mold by an automatic gob-feeding machine. A plunger descends and presses against the gob of glass which flows upward around the plunger and outward to fill the mold cavity. When the glass cools and solidifies, the plunger is withdrawn, the mold is opened, if necessary (because of undercuts in the part), and the part is removed. In some cases, excess glass may have to be trimmed from the part. The process is illustrated by Fig. 5A2a. In production situations, a turntable is used to carry the molds and may have as many as twenty. As the turntable indexes to new positions, each mold proceeds step by step through the full cycle of loading, pressing, cooling, trimming, and ejection or removal. Pressing is used to make drinking glasses and other household glassware, lenses, lamp globes, and TV tube parts.

A2b. ***blowing*** - is similar to blow molding of plastics. As in blow molding of plastics, there are two operations, one to make the parison and the other to make the hollow glass object from the parison. The operation can be manual, with or without a mold to control the shape of the finished part, or automatic with a number of process variations. Hand blowing into the open air, without a mold, but with shaping of the bubble with the aid of hand tools, has been practiced for centuries. The basic process with molds is illustrated in Fig. 5A2b. Blowing is used extensively in the production of glass bottles, containers, and vases and jars.

A2b1. ***manual blowing*** - The skilled artisan uses a glassworker's blowpipe consisting of a metal tube with a wooden handle and mouthpiece at one end and a nose or gathering head at the other end. Making a container, vase, drinking glass, etc. with purely manual methods, involves the following steps:

1. Gathering - The nose end of the blowpipe is immersed in melted glass and is rotated slowly. The viscous glass sticks to the end of the blowpipe. For large objects, several repeats of gathering may be required.
2. The blowpipe is continually rotated to keep the gob of glass centered and the artisan blows a small amount of air from the mouth through the pipe, making a bubble in the center of the glass gob and thereby creating a parison.
3. Marvering - The parison (hollow gob) is rolled against a surface of metal or stone or wet wood, cooling the surface and imparting a straight or curved side to the object.
4. The parison is enlarged by further blowing.

Further contact of the parison with the work surface and with hand-held shaping tools in a series of steps, gradually produces the desired shape. Some reheating may be required and continual rotation of the blowpipe is carried out to keep the workpiece circular and centered. Selective cooling or heating, cutting with shears, and attachment of glass handles or other elements may be carried out before the object is completed. Fig. 5A2b1 illustrates a typical sequence in making a glass pitcher. Fig. 5A2b1-1 shows a collection of glassblower's hand tools.

For repetitive blowing of some particular object, the glassworker may blow the parison into a mold made from water-soaked wood (beechwood has been traditionally used), graphite, or cast iron. This reduces or eliminates much of the tool and workpiece manipulation required, speeds the operation, and reduces the skill required by the glassworker. Because of the high level of skill required, the use of manual methods of blowing has declined in favor

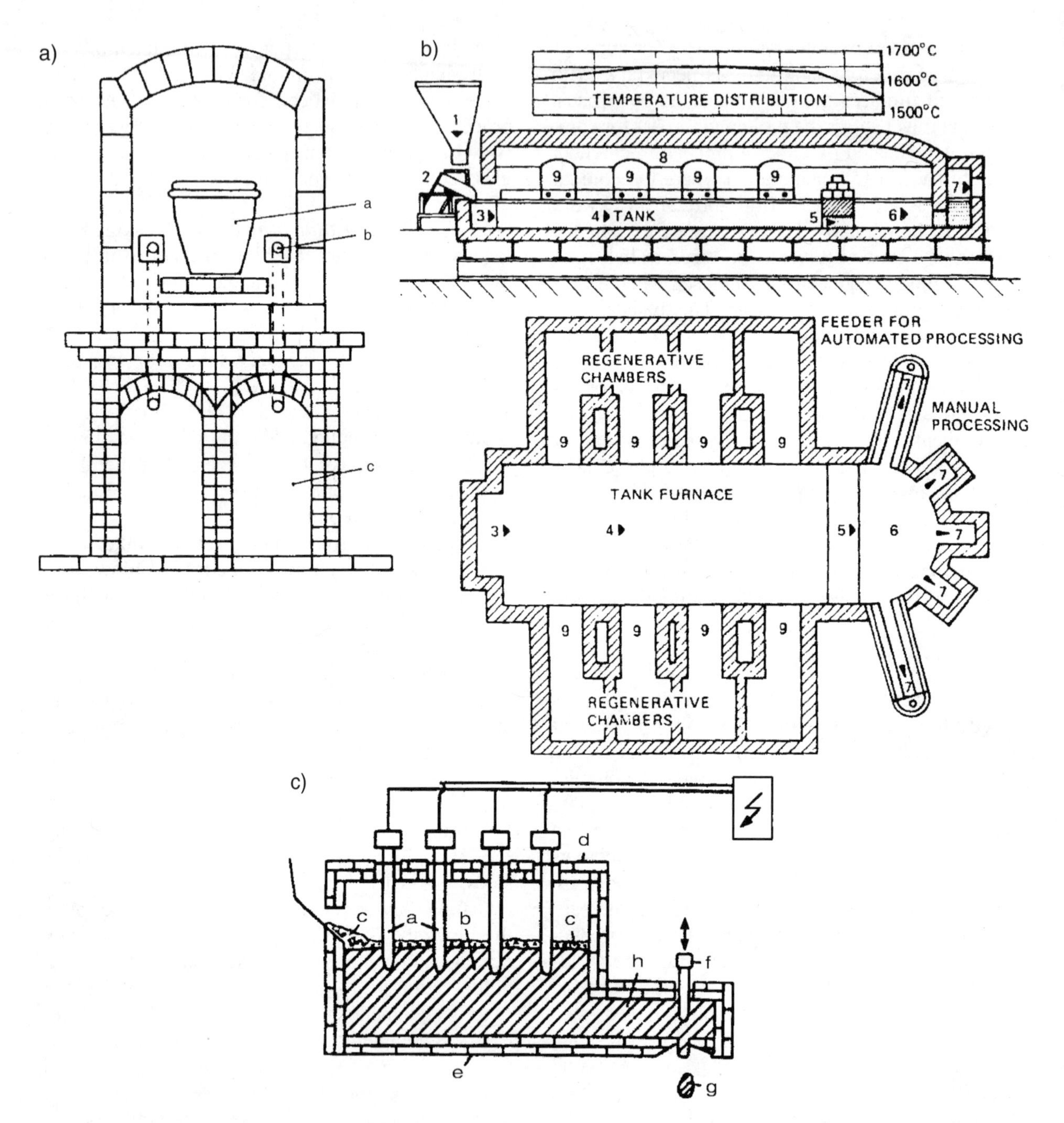

Fig. 5A1c Illustration of glass melting processes:
(a) schematic view of a pot furnace: a - melting pot, b - burners, c - regenerative chambers for heat recovery.
(b) schematic of cross section and floor plan of a melting tank: 1 - hopper for glass batch, 2 - feed chute, 3 - batch feeding compartment, 4 - melting and refining tank, 5 - "doghole" (tank throat), 6 - forehearth, 7 - molten glass feeder for either manual or automatic processing, 8 - crown or roof of the melting furnace, 9 - burner ports in pairs for combustion gas and flue gas.
(c) schematic of electrically heated tank: a - platinum electrodes, b - molten glass, c - batch of unmelted materials, d - crown, e - tank bottom, f - plunger for gob feeding, g - gob of molten glass, h - forehearth. (Courtesy Schott Glas.)

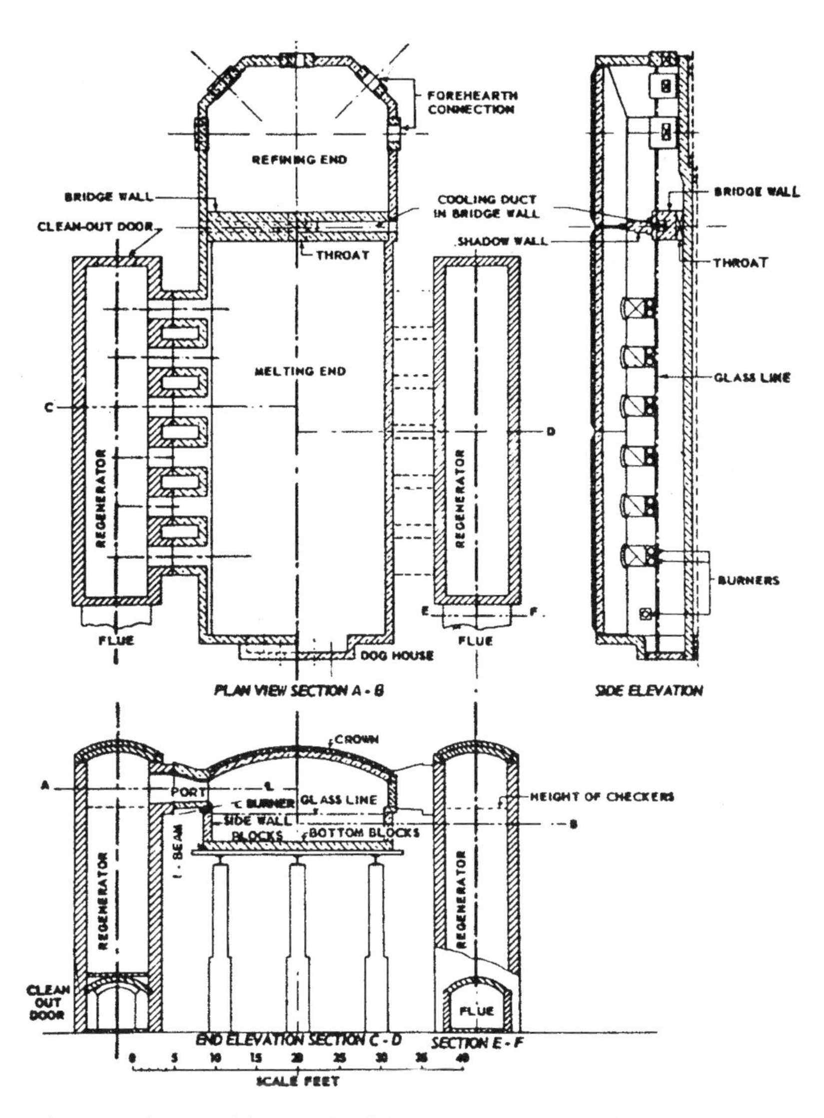

Fig. 5A1c-1 A continuous glass melting tank of 350 tons capacity. (from Glass Engineering Handbook, E.B. Shand, McGraw-Hill, New York, 1958.)

of machine blowing except for artistic work. The method is still used for art work, prototypes, and small quantity production of bottles, containers, laboratory vessels, and other specialty glassware.

A2b2. ***lampworking (lamp blowing, and scientific glass blowing)*** - is the forming of glass articles from tubing and rods by heating in a gas flame ("lamp"). The operation is essentially manual, but differs from the manual glass blowing described above in that it starts with a tube or rod rather than a gob of molten glass. Its primary application is the fabrication of laboratory apparatus and instruments. Medical, veterinary, food processing, and chemical industries require apparatus that use glassware made with this approach. The tubing or rod is heated by a gas flame and then formed by any of a variety of manual operations including blowing, bending, flaring, cutting, sealing, joining, and working with a large number of hand tools.

In higher production situations, the end of a glass rod is heated and placed in a die that presses the softened material into a small part and severs it from the rod. Tubing can be similarly heated at the end, which

Fig. 5A2a Glass pressing with three mold variations: A - with a block mold, B - with a split mold, C - with a font mold. In C, excess glass on the part must be trimmed after pressing. (from Glass Engineering Handbook, E.B. Shand, McGraw-Hill, New York, 1958.)

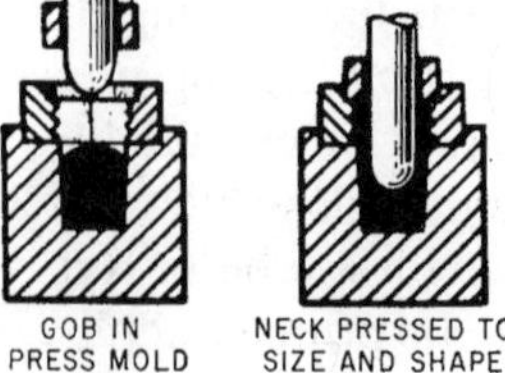
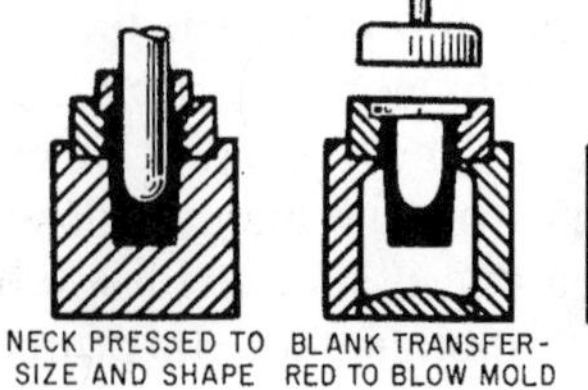

Fig. 5A2b Glass blowing with three process variations: A - with a paste mold, B - with a hot-iron mold, C - the press and blow method (also illustrated in Fig. 5A2b3e.) (from Glass Engineering Handbook, E.B. Shand, McGraw-Hill, New York, 1958.)

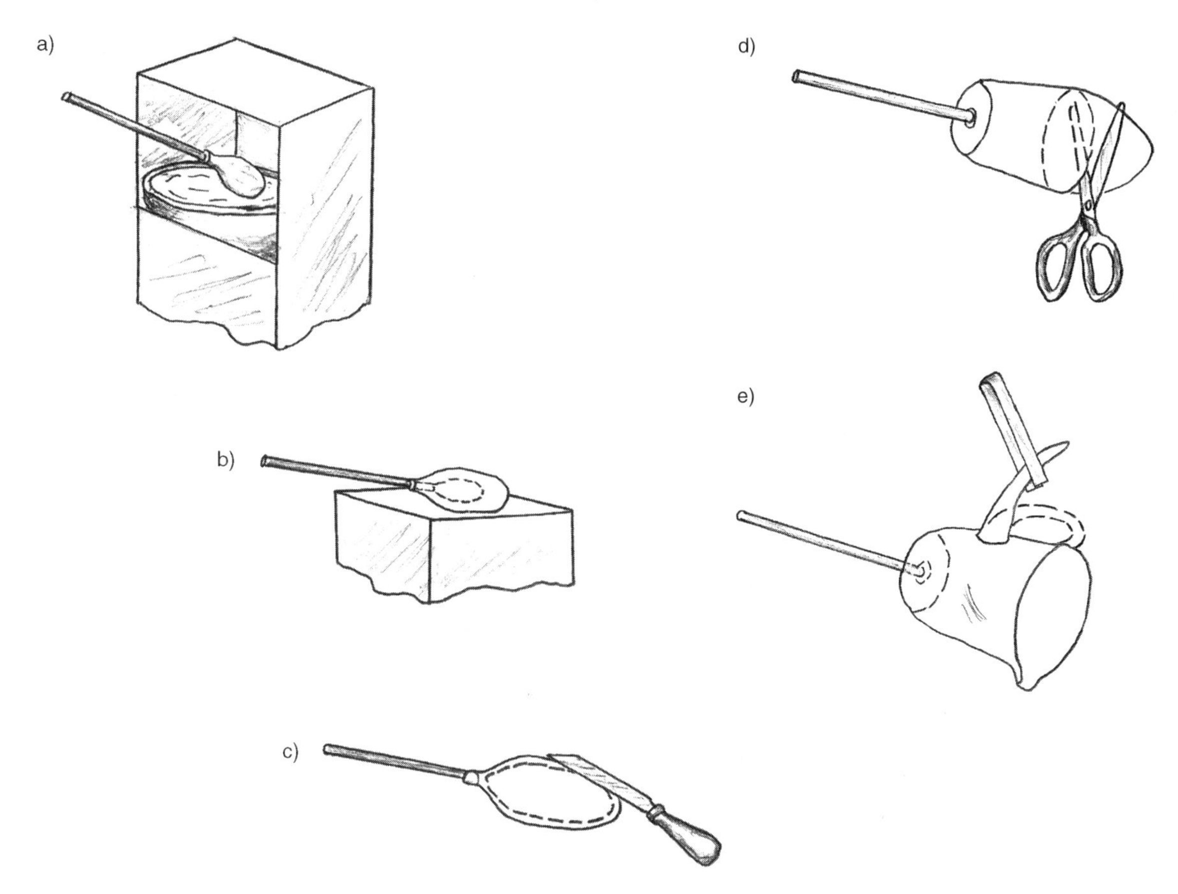

Fig. 5A2b1 forming a glass pitcher using manual methods. a) A gob of molten glass is gathered at the end of the blowing iron. b) An internal bubble is blown and the hollow gob is shaped by "marvering" or rolling on a flat surface. c) The gob is further blown to uniform thickness while being shaped with hand tools. d) The blowing iron is removed and the glass is fastened to a "punty iron" at the other end. Additional shaping with hand tools takes place and excess material is removed with hand shears. e) A pouring lip is formed and a second small gob of glass is formed to be a handle and is attached. The punty iron is removed.

can then be formed by blowing in a suitable mold. Scientific glass blowing has broadened in recent years to include working with flat and powdered glass as well as tubing and rods, and working with a variety of glass types and surface treatments.

A2b3. ***machine blowing*** - Automatic machine blowing is used in the production of glass bottles, jars, drinking glasses, and other glass containers that are manufactured in mass-production quantities. Machine blowing methods have the following elements:

1. equipment for feeding a gob of melted glass to the machine,
2. a means for converting the gob into a parison, i.e., introducing a hollow in the gob for later blowing,
3. inflation of the hollow gob (parison) against the inner surfaces of a mold,
4. a means for forming the elements at the open end of the object molded,
5. a means for trimming any excess material from the finished object, and,
6. annealing the finished product.

Material in process may be reheated during the operation sequence. Notable machine methods are the "press-blow", "blow-blow", "suck-blow" and "rotary-mold (paste mold)" processes.

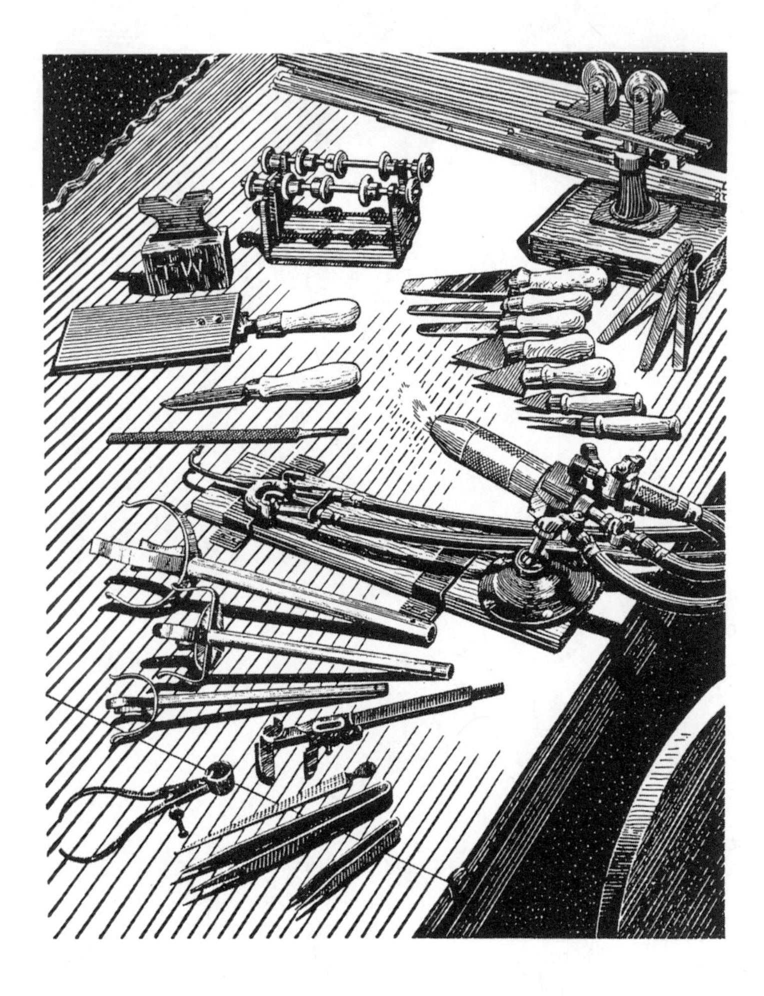

Fig. 5A2b1-1 Hand tools commonly used by glass blowers.(Courtesy Corning, Incorporated, corporate archives.)

A2b3a. ***gob feeding machine*** - In this machine, the feeder is integral with the forehearth of the melting furnace. A typical machine has the following elements to provide suitable gobs of molten glass for the blowing machine:

1. an orifice in the bottom of the forehearth,
2. a plunger or needle to push the gob through the orifice,
3. a rotating tube around the plunger to control the amount of molten glass in the gob,
4. shears to sever the gob from the other material that passes through the orifice.

The gobs fall by gravity directly into the molds of a rotary blowing machine or into a chute that carries them to the machine molds. Operation of the gob feeder, which is timed to the speed of the blowing machine, is illustrated in Fig. 5A2b3a. Gob feeding machines are also used to provide material for pressing machines.

A2b3b. ***Owens bottle machine (the suck-blow process)*** - The original machine developed by M. J. Owens was put into production around 1904 but has been much further developed since then. The glass is brought into the parison mold by suction, hence the name, suck-blow process. The work is performed on a large rotary table. Motion of molds and other elements is controlled by cams. The operating sequence is as follows:

1. The parison mold, with an open bottom, is lowered into the surface of molten glass. Suction applied to the top of the mold draws glass into the mold cavity. A pin with a rounded end puts

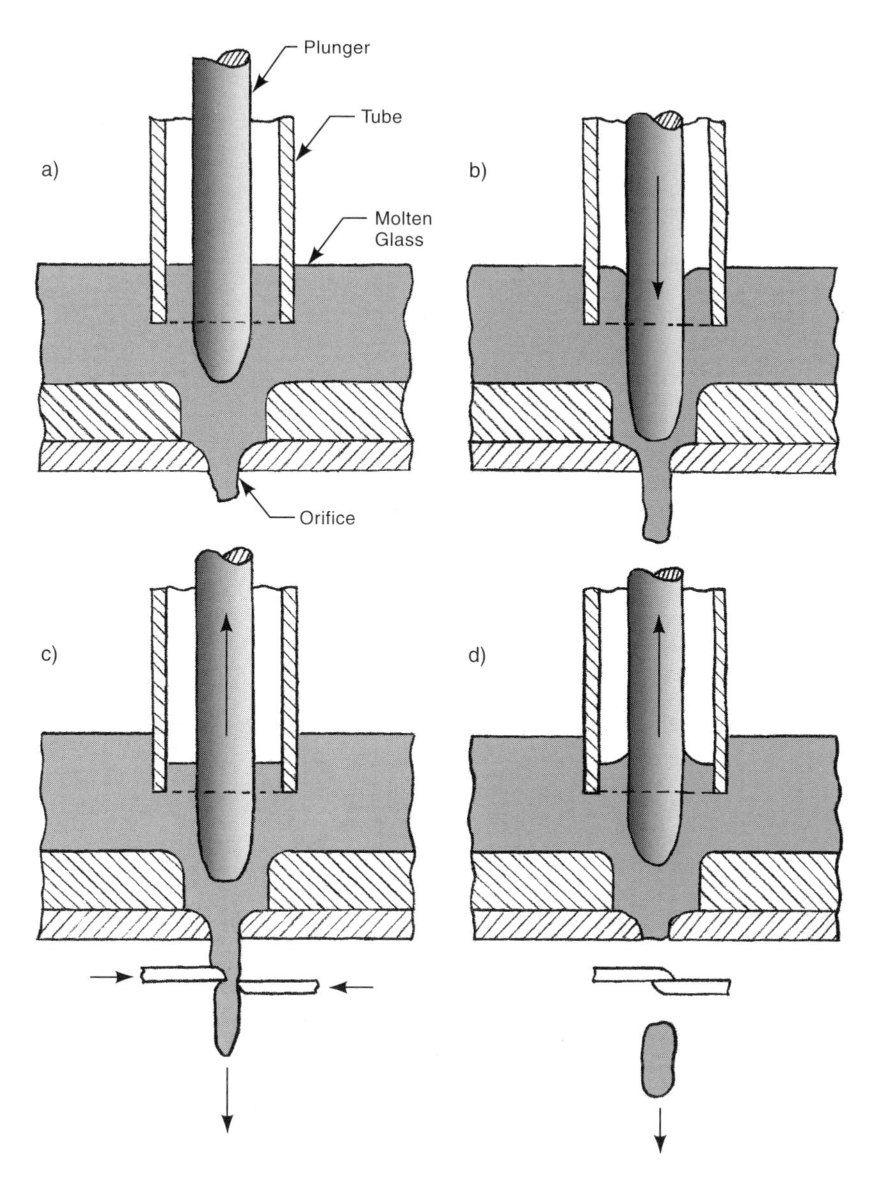

Fig. 5A2b3a gob feeder cycle: a) Molten glass starts to flow through orifice, b) The plunger descends and forces more molten glass through the orifice, c) Shear blades advance to cut gob from stream of molten glass; plunger starts to retract, d) Gob falls as retracting plunger pulls glass back from shears.

a depression in the top of the glass in the mold. The neck portion of the bottle is also formed in this mold.

2. The parison mold is lifted and a knife passes across the bottom of the mold, severing any excess glass from the parison. At the same time, the rounded pin at the top is withdrawn, and air pressure in the resulting opening enlarges the top of the parison, forming a "bubble".
3. The mold opens, freeing the parison which is held by neck rings at its upper end. The parison is out in the open as the machine table rotates. The parison elongates from the effects of gravity and from several puffs of air into the bubble.
4. The parison enters the blow mold, which closes around it. Air is blown into the bubble, expanding the parison against the mold walls.

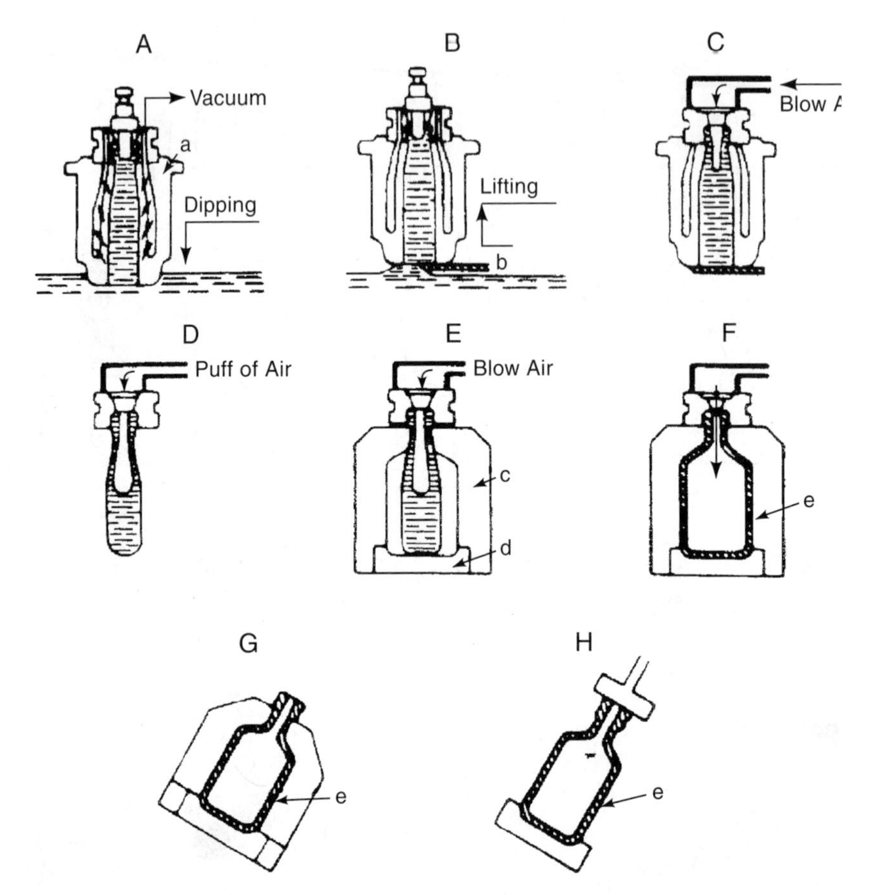

Fig. 5A2b3b The suck-blow molding process for holloware: A - preform mold being filled by a dipping-sucking process, B - bottom of preform of sucked glass is sheared off and the mold starts its upward movement, C - connection to compressed air supply, D - puff of air partly inflates the preform, E - preform is inserted into the finishing mold, F - the finishing blow, G - removal of the mold and workpiece from the blowing mechanism, H - the finished workpiece. Identified components: a - preform mold, b - shears, c - finishing mold, d - bottom-forming mold section, e - finished glass bottle. (Courtesy Schott Glas.)

5. The neck rings open and the mold with the bottle inside drops below the pot as the table continues to rotate. The mold and bottle cool.
6. Upon further rotation of the table, the mold opens and the bottle is discharged. Fig. 5A2b3b illustrates the molding action schematically. The machine is used for large-scale production. With smaller bottles, double and triple molds are used so that each cycle of the machine produces two or three bottles.

A2b3c. ***the blow-blow process*** - uses a gob feeder instead of drawing molten glass into the parison mold by suction. The parison mold is in an inverted position so that the part of the parison that will form the bottle opening is at the bottom. The neck ring and a center pin are in place. The gob is dropped into the mold and is forced by air pressure from above to settle into the bottom of the mold. As in suck-blow molding, a rounded pin forms the start of the bubble. The pin withdraws and air pressure from below forms a bubble in the parison. The mold opens and the parison is removed, turned back to a right-side-up position, and placed into a nearby blow mold. The parison is reheated and elongates slightly. The final blow then takes place. After cooling, the bottle is removed from the blow mold. See Fig. 5A2b3c for an illustration of the process sequence.

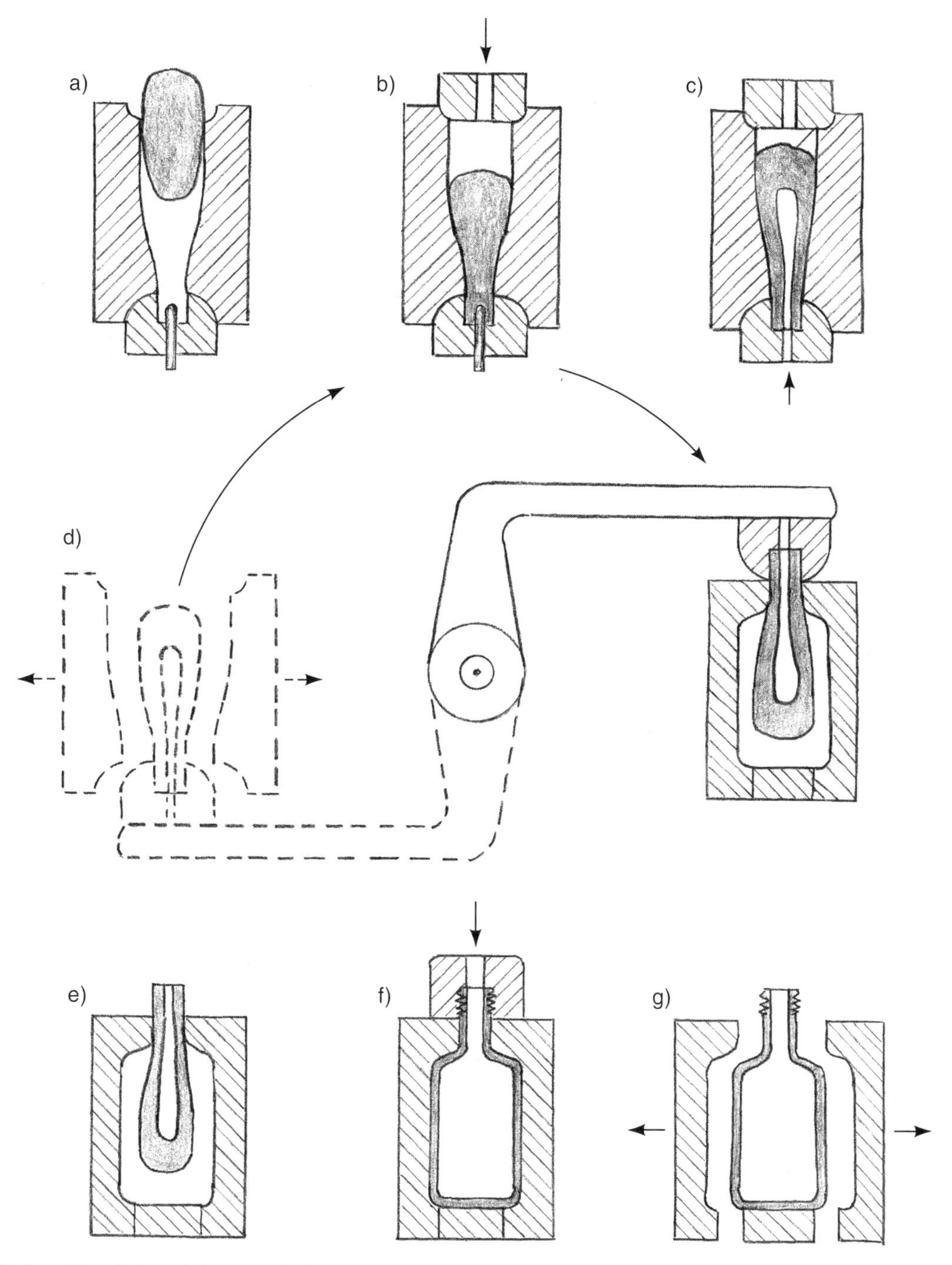

Fig. 5A2b3c The blow-blow molding process. a) Gob feeder places gob of molten glass in the parison mold. b) air pressure from the top of the mold settles the gob at the bottom of the mold. c) Counter blow - a brief blast of air from the bottom of the mold forms a bubble in the parison. d) The parison mold opens and the parison is removed, inverted and transferred to a bottle mold. e) The parison, in the bottle mold, is reheated. f) A blowing head is placed at the top of the mold and, with a final blow, inflates the glass bubble to a bottle shape. g) After some cooling, the mold opens and the cooler bottle is removed from the mold.

A2b3d. ***the press-blow process*** - is similar to the blow-blow process except that the parison is not blown but is pressed between the parison mold walls and a central plunger extending from the neck area of the mold. The plunger provides a internal hollow space for later blowing. Use of pressing with fixed tooling gives improved glass distribution in the parison. After pressing, the parison is transferred to a blow mold and blown to final shape. Because this process gives better control of wall thickness, it is used in the production of thin-walled containers including disposable glass bottles. The bottom line (item C) of Fig. 5A2b illustrates the approach in simplified form.

A2b3e. ***the rotary-mold (paste mold) process*** - is suitable only for parts of a circular cross section. The molds are also always circular and are lined with an absorbent coating that is soaked with water before blowing. During blowing, the water produces steam from the heat of the molten glass. This provides a cushion between the mold and the glass that permits the glass to be rotated during blowing and ensures that mold seams will not be seen on the blow molded articles.

The parison for rotary mold blowing can be made by any of the methods described above. One commercial machine uses a gathering device and blowing to make the initial parison; another uses pressing. However, the significant difference in the process is that, during the final blowing, the glass or the mold rotate against one another and the mold surface has a saturated water-absorbent coating. Water is applied to the mold surface for every blowing cycle by dipping or spraying the mold. The parison is held by a metal ring that can be driven to provide the rotation. The moil, solidified glass that contacts the retaining ring and air nozzle, is not usable, and is separated from the finished article. Applications include medium and high-grade tumblers, cereal bowls, and lamp bulbs. The process is illustrated in Fig. 5A2b3e.

The mold coating is cork, other carbonaceous material, graphite, or some proprietary material. It is adhered to the mold surface with a drying oil followed by baking at a high temperature. Roughness in the coating smooths out from the first few usages but some rotational lines may be visible on the final article. For limited quantities and artwork, the rotary-mold approach may be carried out manually.

A2b3f. ***ribbon machine blowing process*** - This variation of the paste mold process is used to produce light bulbs. The machine is provided with a stream of molten glass that flows continuously from a forehearth. Two steel rolls flatten the stream into a "ribbon". The rolls are shaped so that the ribbon has regular patties of greater thickness connected by a thin web. Beneath the ribbon, as it moves through the machine, is a conveyor with steel plates that support the ribbon. Holes in the steel plates match the positions of the patties in the ribbon. The molten glass sags into these holes, forming a shallow bubble. Blowing nozzles on an upper conveyor, movement of which is synchronized with the steel plate conveyor, supply puffs of air that expand the bubbles. At the same time, a series of wet paste molds traveling under the steel plate conveyor and synchronized with it, open, envelope each bubble, and close.

The wet paste molds rotate as additional air is blown into the bubbles, converting them to glass bulbs. After the bulb is blown, the mold opens and the bulb is separated from the ribbon by a light mechanical hammer blow, falling to another conveyor where it is moved to an annealing lehr. The conveyor moves rapidly and many bulbs are processed simultaneously. As many as 10,000 small light bulbs per minute can be made with the ribbon machine process.[5] Fig. 5A2b3f illustrates the operation of the machine.

A2c. ***glass tubing manufacture*** - The Danner process starts with a continuous strand of molten glass, which flows from a forehearth onto a slowly rotating mandrel. The mandrel is hollow, made from fire clay and is angled slightly downward, as shown in Fig. 5A2c. The rotation of the mandrel causes the glass to wrap around it. As it slides down and off the mandrel, the mandrel leaves a hollow in the center. Air blown through the mandrel maintains the hollow core in the glass. The glass leaving the mandrel is drawn away from the mandrel, forming tubing of smaller diameter and thinner wall than that which first forms on the mandrel. The operation is continuous until the tubing is cut to length. The tubing is made into fluorescent

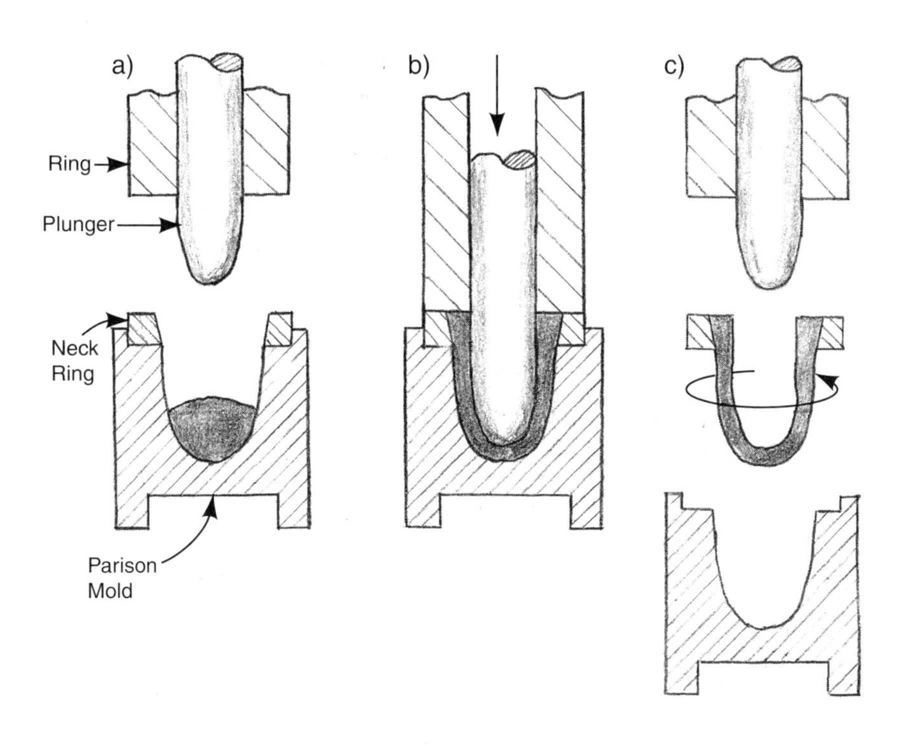

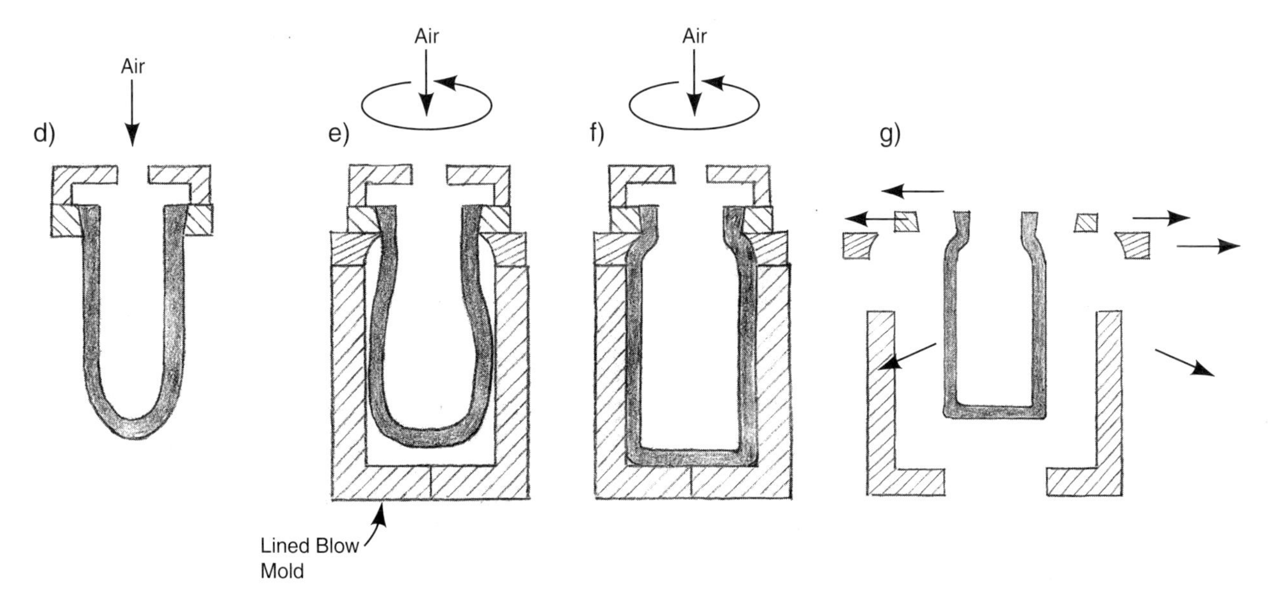

Fig. 5A2b3e The rotary mold (paste mold) process: a) Gob of molten glass is placed in parison mold and plunger is poised to descend. b) Plunger descends, presses the glass gob to a form a parison and retracts. c) The parison is reheated as it is rotated. d) A puff of air into the parison elongates it. e) The parison is inserted in the lined blow mold. Another puff of air opens it further as it rotates. f) The final air blast fully forms the glass into the desired shape as it rotates in the mold. Steam from the heated mold lining prevents the glass from contacting the mold surface during the rotation. g) As the formed glass part cools, the mold opens and the part is removed.

light tubes, vials, ampules, industrial glass piping, and containers.

Another tube drawing method, the Vello process, uses a ring-shaped orifice exiting from the forehearth. The glass flows downward through the orifice and has a hollow center formed by the opening and maintained by a pipe in the center of the ring. The glass tube, while still soft, is then drawn off along a horizontal roller track, cooled, and cut to length.

A2d. ***centrifugal casting*** - A gob of hot glass is dropped into a rapidly rotating steel mold. Centrifugal force causes the glass to flow outward and coat the surface of the mold, forming a wall of uniform thickness. Excess glass is trimmed with a sharp-edged wheel or other cutter while the glass is still in the plastic state. When the glass has cooled sufficiently to have hardened, the cast part is removed from the mold. This approach has been used to form the funnel-shaped bodies of television picture tubes, and for column sections in chemical plants.

A3. *flat glass processes*

A3a. ***manual methods for producing flat glass*** - Two methods were used to make flat glass sheets prior to the advent of machine methods. In the crown method, a round bubble was first blown. An iron rod called a "punty", was attached opposite the blowpipe and the blowpipe was removed. The open bubble was spun on the axis of the punty so that the glass flattened into a circular disk. After cooling, the smoothest areas of the circular disk were cut into rectangular panes. A more productive and better quality method involved manual blowing of a bubble that was worked into a cylindrical shape. The two ends of the cylinder were cut off and an axial cut was made in the resulting tube for its whole length. The curved rectangular piece that resulted was placed on a flattening stone in an oven. Where the glass had softened, it was straightened and ironed flat against the stone. As with the crown method, the quality of the window panes produced was poor and quite variable from piece to piece, but larger panes could be produced. These methods are no longer used except for art work, reproductions of antique window panes, and some other specialty, low-quantity production.

A3b. ***drawing sheet glass (the Fourcault process)*** - The Fourcault process was the first successful mechanized production method for drawing sheet glass directly from a tank. It was first

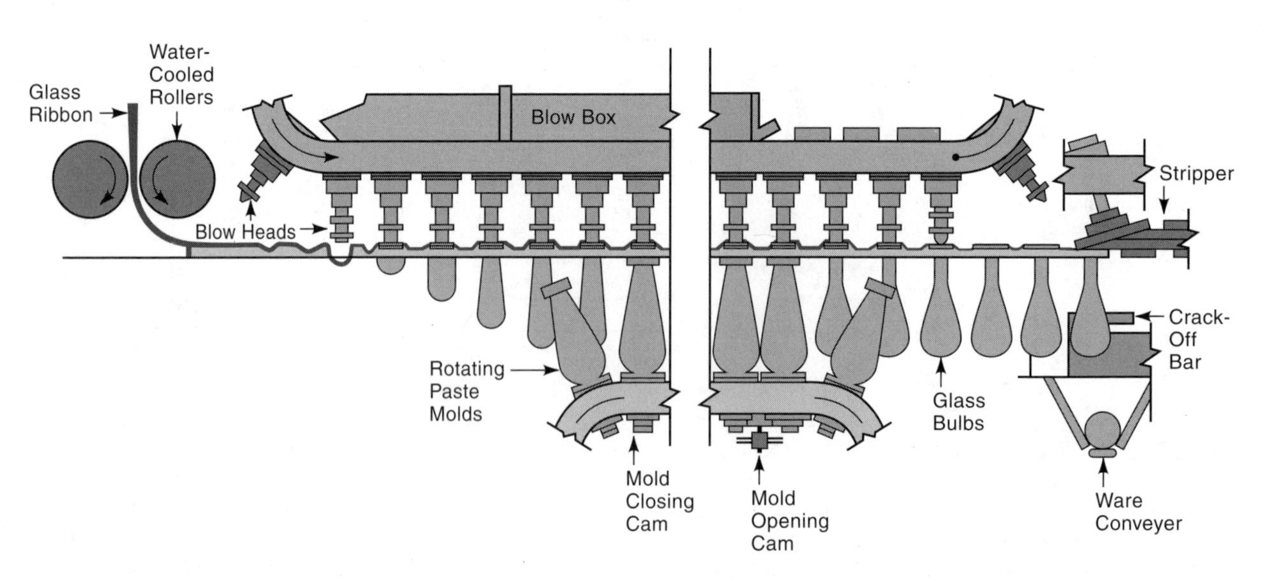

Fig. 5A2b3f The operation of a ribbon machine producing lamp bulbs at a high production rate. (Courtesy of Corning Incorporated, corporate archives.)

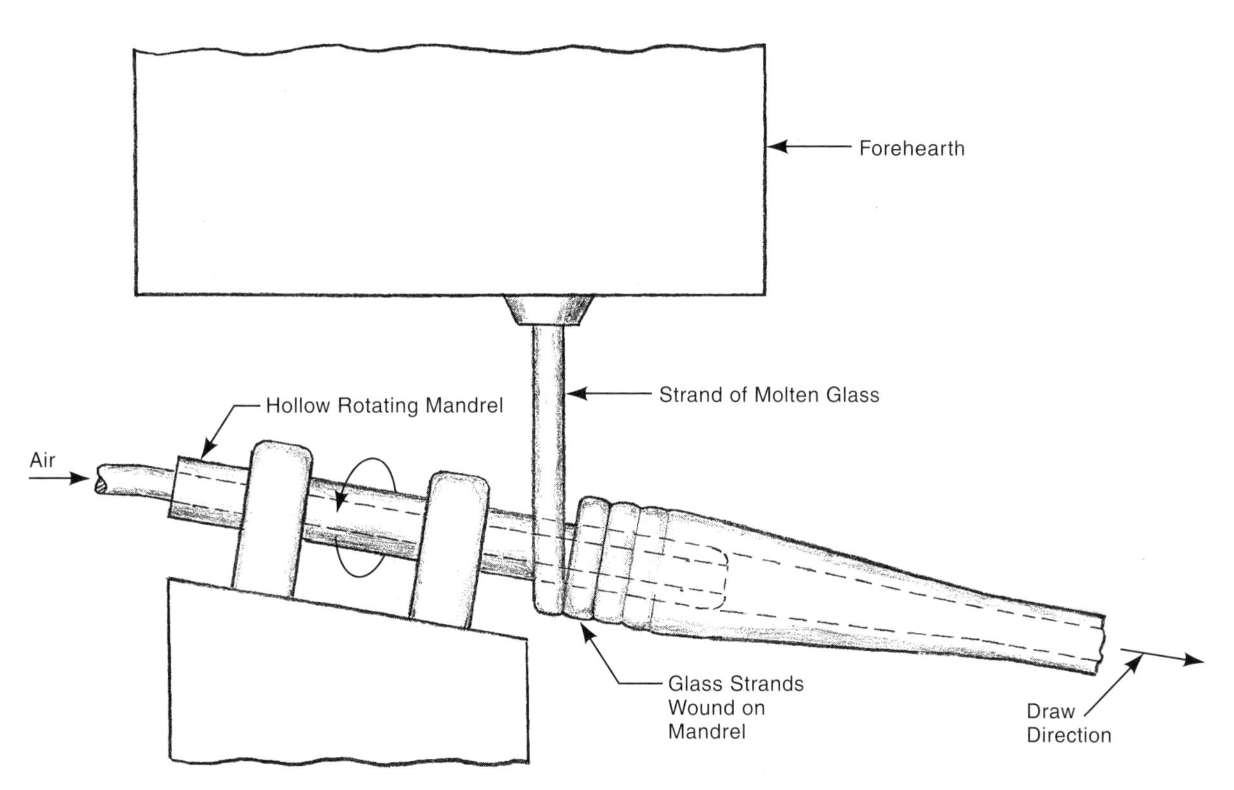

Fig. 5A2c The Danner process for making glass tubing from a strand of molten glass wrapped around a rotating mandrel.

carried out on a production basis in 1914. Prior to that, the production of flat glass was at least partly a manual operation. The method is keyed to the "debiteuse", a long clay block with a lengthwise slot. The block floats on the molten glass but, when it is pressed slightly down, into the molten glass, some glass rises out of the slot. This glass is grasped by an iron "bait" and is pulled upward past a cooling station and into an annealing tower. The tower contains rollers that draw the glass upward and the operation is thereafter continuous.

The rate of drawing, among other factors, determines the thickness of the glass. (Slower drawing yields greater thickness.) The length of the slot in the debiteuse determines the width of the ribbon. Width is maintained by pairs of knurled rollers at the edges that maintain a constant side pull on the ribbon. The drawback of the process is a tendency toward a small amount of waviness in the sheet, which cannot be avoided. There may also be fine marks on the glass surface left by the rollers and some tendency to devitrification caused by the refractory material from which the debiteuse is made[2]. Fig. 5A3b illustrates the process.

A3c. ***drawing sheet glass (the Colburn or Libby-Owens process)*** - This process, seen in Fig. 5A3c, is similar to the Fourcault process but does not use the debiteuse. Instead, the initial ribbon of glass is picked up from the tank with a metal "bait" and immediately controlled by chilled rollers at the edges. It also is diverted into a horizontal direction by a polished roller after traveling upward only about 27 in (70 cm). It is stretched, flattened, and supported by transporting rollers as it moves into a 200 ft (60 m) annealing lehr. The drawing speed with the Colburn process is twice that used with Fourcault.

A3d. ***drawing sheet glass (the Pittsburgh process)*** - In use since about 1928, this process is

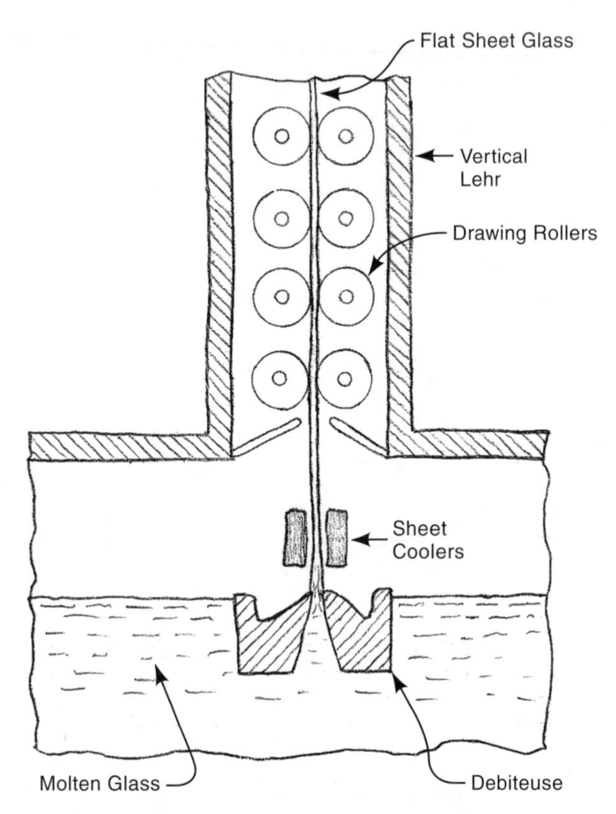

Fig. 5A3b Drawing sheet glass with the Fourcault process. A continuous flat sheet of glass is drawn vertically upward.

similar to the Fourcault and Colburn processes and combines the best features of both. Glass is drawn vertically through a cooling zone and annealing lehr. The process can be run at high speed, gives higher quality than the other drawing methods, and allows quick changes of sheet thickness. The quality advantages are most significant for thinner sheets. Thicknesses down to 0.050 in (1.25 mm) are of high quality. Instead of the debiteuse of the Fourcault process, a refractory guide or "draw bar" for the glass to be drawn is positioned several inches below the surface in the plane where drawing takes place. This bar improves the flow currents in the tank and conditions the glass. The initial end of the glass ribbon is held by cooled grippers that are shaped like hollow plates. The path of the glass sheet and arrangement of the equipment is illustrated in Fig. 5A3d.

A3e. ***plate glass manufacture*** - involves the rough and finish grinding of rolled glass. The full operation has two phases: the production of rough glass by rolling, and then the grinding and polishing. However, in the highest mass-production conditions, these phases are combined in one continuous sequence requiring a factory production line

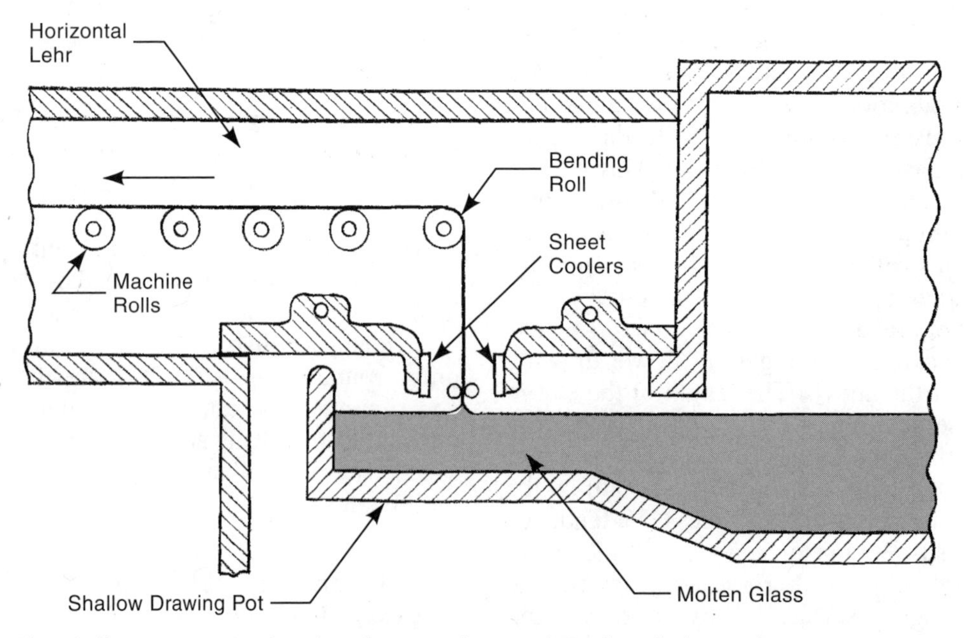

Fig. 5A3c The Colburn process for drawing continuous lengths of sheet glass.

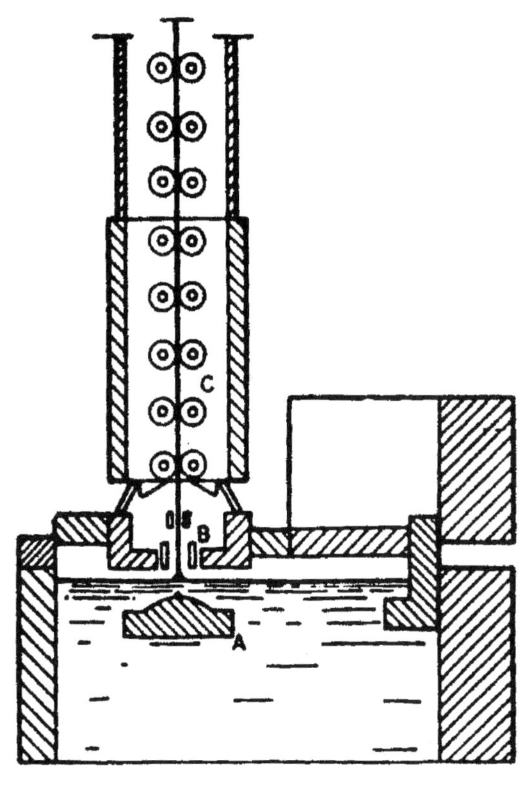

Fig. 5A3d The Pittsburgh process for drawing flat sheet glass: A - drawbar submerged in molten glass, B - sheet coolers, C - vertical drawing and lehr area. (Courtesy PPG Industries Corp.)

almost 2000 feet long. Rolling involves the following steps:

1. allowing the molten glass to flow over a weir or through a slot from the melting tank,
2. passing the resultant ribbon of glass between a pair of water-cooled rolls to give it accurate thickness and width,
3. stretching it slightly to improve flatness, and
4. passing it through an annealing lehr.

In the continuous process, the cooled glass ribbon, still uncut, moves to a grinding section where vertical-spindle grinding machines above and below the ribbon remove material from both the top and bottom surfaces. The operation is fully automatic. Abrasive compounds and large flat rotating disks work against the glass surfaces. Sand, garnet, and emery, with water, are the abrasives used. The top and bottom grinding and polishing disks are directly opposite one another and rotate in opposite directions to help keep the glass ribbon in line.

Following grinding, the ribbon, still uncut, moves to a unit where both top and bottom surfaces are polished with iron oxide or cerium oxide, again by large rotating disks. Many grinding and polishing disks, in line, work on the ribbon before it is finished. The glass ribbon is then cut into separate sheets for warehousing and further cutting. Fig. 5A3e lists the full manufacturing sequence. In earlier, and somewhat less high-production arrangements, the glass was cut after rolling, and the grinding and polishing was performed on separate sheets with the operations done only on the top surface until the sheet was inverted and the grinding-polishing sequence repeated on the other surfrace. In both methods, with plate glass, the surface flatness, parallelism, thickness, and optical qualities are superior to those achieved with other flat glass processes. However, the quality of float glass is almost as high and far less costly to produce, and float glass has therefore replaced plate glass in almost all applications. Common applications for plate glass were mirrors, automotive safety glass, windows, especially for commercial establishments, and as a starting material for glass finishing operations. The thickness range was from 0.125 to 1.25 in (3.1 to 31 mm).

A3f. ***float glass process*** - The molten glass is fed onto a shallow tank containing molten tin. The glass, being lighter than the tin, floats on the surface. The tank is typically 13 to 30 ft (4 to 9 m) wide and 200 ft (60 m) long. The glass spreads to form a sheet with parallel surfaces and a natural thickness of slightly over 1/4 in (slightly under 7 mm). The operation is carried out in a mildly reducing atmosphere to prevent oxidation of the molten tin. The glass enters the tank at a temperature of about 1920°F (1050°C). It cools as it flows along the tank and, at the exit end, is about 1200°F (650°C) and is solid enough to be lifted off the bath of molten tin. The glass is carried to an annealing lehr before being cut to size.

Glass sheet from 0.060 in (1.5 mm) to 0.80 in (2.0 mm) in thickness can be made with the float process. In producing thin glass, rollers are used to stretch the sheet to a lesser thickness and to

Manufacturing Sequence for Plate glass

1) Raw materials are received and stored
2) Materials are weighed and mixed
3) Batch is fed to furnace and melted
4) Molten glass flows from the furnace through forming rolls to form a rough, continuous ribbon
5) Ribbon is stretched slightly to improve flatness
6) Ribbon travels through annealing lehr
7) Both top and bottom surfaces are ground flat by a series of vertical spindle, large-disc grinding machines as the ribbon travels past them
8) Both surfaces are polished with similar machines using finer abrasive
9) An acid wash removes grinding residue
10) Sheets are cut from the ribbon
11) Sheets are inspected and sent to storage for later final cutting and shipment

Fig. 5A3e The manufacturing sequence for plate glass.

control the speed and width of the glass ribbon. Thicker glass is made by partly damming the flow of glass on the tin surface. The float method has replaced almost all plate glass production because manufacturing costs are considerably lower and the dimensional quality is virtually as good. Float glass also has a more brilliant fire-polished surface.

The float glass process has also replaced sheet glass drawing methods because float glass quality is superior to that of drawn glass. Float glass is used for window glazing, vehicle safety glass, mirrors, visual displays, and other applications where transparency and a flat surface are required. Fig. 5A3f illustrates the process.

A3g. ***rolling (casting) flat glass*** - Rolling is carried out for two major purposes:

1. as the first operation in the production of plate glass, and
2. to produce a flat glass which is not totally transparent.

Glasses with patterned surfaces are made by rolling. The process for higher production quantities of rolled glass is the same as described in paragraph A3e (plate glass) above, except that, for rolled glass, a patterned or textured roller is used instead of one with a smooth surface and there is no grinding and polishing. The glass sheet will have the roller pattern embossed on its surface. Usually, only one of the two surfaces is patterned; the other is essentially smooth. The spacing between the rollers determines the thickness of the rolled sheet. The method of feeding the molten glass from the forehearth to the rollers is sometimes referred to as

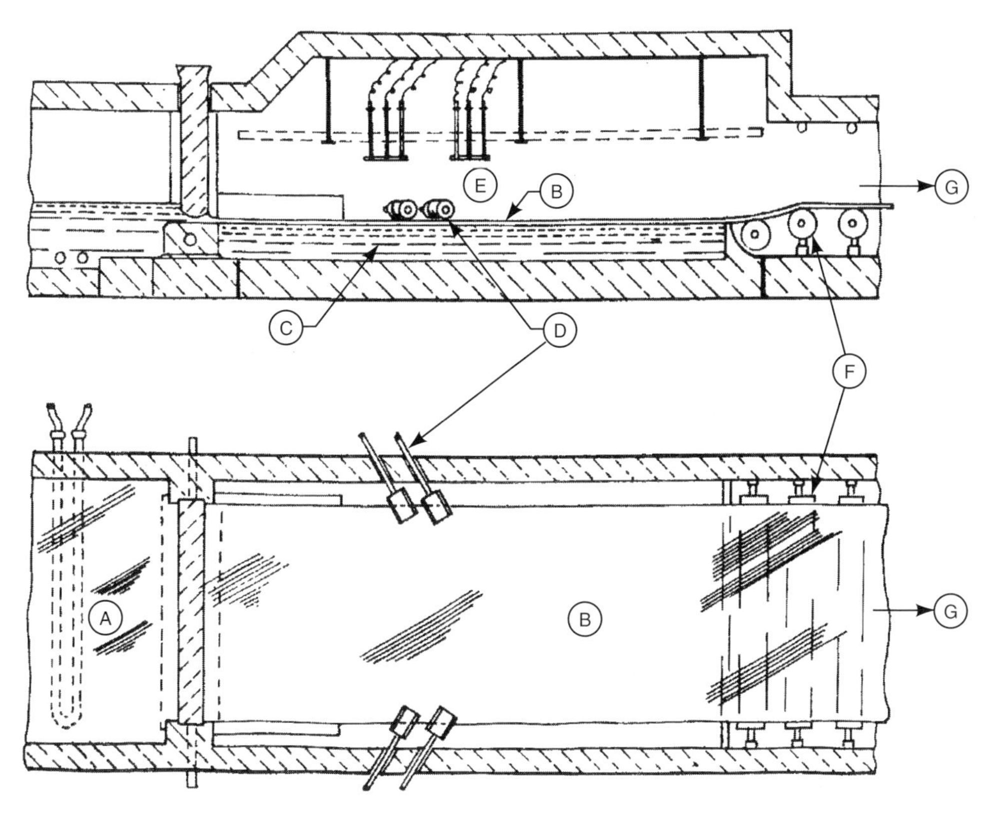

Fig. 5A3f The float glass process: (A) molten glass, (B) layer of glass floating on molten tin, (C) molten tin, (D) guide rollers, (E) heated chamber (fire polish zone), (F) transport rollers, (G) to annealing lehr. (Courtesy PPG Industries Corp.)

"casting" and rolled glass is sometimes referred to as cast glass.

For smaller-quantity production, the glass may be cast or poured onto a flat iron table, after which a roller is passed over it. With somewhat higher quantities, a patterned surface on a movable rolling table is used. The molten glass is ladled onto the surface and the moving table, on a track, then passes under a set of rolls that produce a sheet of the desired thickness. Wire glass, that is flat glass with a wire mesh incorporated in it, is also made with this moving table method. First, enough molten glass to produce half the thickness of the sheet is ladled onto the movable table; the table is moved under a roller and the wire mesh is then fed onto the sheet. Then, another ladle of molten glass is poured onto the sheet; the molten glass spreads and the table is moved under a second roller, which provides the desired surface and thickness for the final sheet containing the wire mesh in the middle.

A wide variety of surface textures can be embossed on the surface of the rolled glass sheet, depending on the application involved. Rolled glass is used for skylights, light fixtures, table tops, greenhouses, office partitions, bathroom doors, office doors, and industrial and commercial glazing where transparency is not wanted. Greenhouse glass has a special surface designed to scatter light. Partition glass can have many different surface designs: ribbed, "hammered" (like metal), or faceted surfaces.

A4. *heat and chemical treating operations*

A4a. ***annealing*** - is a heating procedure intended to reduce residual stresses and strains in the glass object or sheet. These stresses and strains could lead to fracture of the glass and, in optical applications, could result in unacceptable variations in optical properties. The stresses arise when the glass cools unevenly from its initial molten state due to the shape of the object or ambient conditions. In annealing, the glass is first heated to a temperature high enough that existing stresses are relieved and then is cooled slowly enough that stresses induced are within an acceptable range. The annealing temperature is considerably lower than the forming temperature for the same glass composition. Heating to the annealing temperature can be rapid so long as the glass does not fracture.

Annealing temperatures for window glass range from about 815 to 1100°F (435 to 595°C) depending on the type of glass. This temperature is maintained for sufficient time to relieve existing strains and stabilize the glass. Cooling then begins at a slow rate and in a controlled manner down to a strain point (for window glasses) of about 670 to 930°F (355 to 500°C), at which point rapid cooling does not induce additional permanent stresses. Heating and cooling times depend on the thickness of the glass, its expansion coefficient, the shape of the object being annealed and the amount of permanent stress and strain allowable for the application. Thick sections, and objects with both thin and thick sections, take more time.

Optical applications necessitate much lower levels of residual stress and correspondingly longer annealing cycles. Sheet glass normally takes considerably less time. A heavy section for optical use may require several weeks for fine annealing, while thinner sheet may require less than half-hour for the annealing cycle.

A4b. ***tempering*** - is a means of making glass tougher, by inducing a compressive stress in the surface of the tempered workpiece. Failures with glass almost always occur because of tensile stress failures at the surface. (Internal tensile stresses are less critical). Residual compressive stresses in the surface of the glass counterbalance tensile stress loads on the product. Tempering, by inducing these compressive stresses, can increase the stress required to break glass parts by a factor of $2^1/_2$ to $3^1/_2$.[4] The glass object does not break until the surface compression is overcome by sufficient applied force and the surface is put into tension (or until the interior tension becomes large enough to cause failure). Another advantage of tempered glass is that, when it breaks, it shatters into many small pieces of somewhat regular shape instead of into sharp-pointed or sharp-edged shards.

The tempering process involves heating the glass to approximately the softening point, a temperature about 270°F (130°C) above the transformation temperature of the glass. Immediately afterwards, the glass is subjected to a blast of cold air from a suitably arranged array of jets. Alternative methods involve plunging the glass workpiece into a bath of oil or fused salts. This chilling quickly quenches the glass surface while the interior glass remains at a higher temperature. As the interior glass slowly cools, it shrinks and pushes the adhering surface material into compression. The interior glass remains in tension.

Tempered glass is used in glass doors, ship portholes, shelves, store fronts, oven doors, arc lights, and radiant heaters. It is also used for automotive side windows.

A4c. ***chemical toughening*** - also induces a compressive stress in the surface of the glass and tension in the center. The glass workpiece, after otherwise being finished, is immersed in a molten salt bath. The temperature is below that which would soften the glass. An ion exchange takes place. Some ions at or near the surface of the workpiece migrate into the salt bath. Larger ions from the bath diffuse back into the glass to fill the spaces left by the migrating ions. Since the larger ions normally would require more space, they become compressed upon entering the glass surface. This produces a compressive stress at the surface. The advantage of chemical treatment is that all surfaces of the glass workpiece are treated equally and the object treated does not suffer any deformation. It is therefore useful for formed objects. Tubing, containers, and pressed objects can be chemically treated. The depth of toughening can be controlled by the length of time in the salt bath and its temperature. Thinner sections, down to about 0.040 in (1 mm) can be treated. Glass compositions vary in their suitability for chemical toughening. Soda-lime-silica glasses, used for sheet applications, are less adaptable to the process. The process is used for toughening eyeglass lenses, and for glass components used in aircraft and processing industries, and for lighting components.

A5. *secondary, finishing, and decorating operations*

A5a. ***bending and sagging*** - Flat glass is bent to provide curvature for architectural and automotive applications. One method is gravity bending, which uses only a one-piece mold. The flat sheet is placed on the surface of the open mold that has the shape wanted for the finished part. Both the mold and the glass resting on it are transported on a roller conveyor through a heating tunnel that heats the workpiece to the desired temperature. The heated glass sags to conform to the shape of the mold. The glass and mold then proceed together through quenching or annealing. Sometimes, spot heaters (gas burners or electric heating elements), are placed to provide extra heat to areas requiring a sharper bends. This method is used for production of dished shapes and cylindrical segments from flat glass, and in bending rods and tubing. Another approach, after the glass has been heated to the suitable bending temperature, is to place the sheet between two matched ceramic dies, which press it to the desired shape. Previously this method produced unacceptable distortion in the glass but process improvements have reduced the distortion to acceptable levels. If the glass is to be tempered, it is quenched immediately after bending. Some allowance must be made in the bending dies for shape changes during quenching.

A more recent development uses a flow of heated air to support the glass sheet as it is heated and as it moves from a heating area to a forming area, and then to a quenching area. A ceramic bed with numerous small holes, or a series of adjacent, nested, small cup-shaped holes, as shown in Fig. 5A5a, supply the heated air. Return holes and channels remove the excess air. The return holes are necessary to balance air that escapes at the edges. Without them, there would be too much air at the center that would create a bulge in the sheet. Air pressure is enough that the glass sheet virtually floats about 0.010 to 0.020 in (0.25 to 0.50 mm) above the bed. The bed is inclined slightly to make the glass sheet slide along the bed. The initial bed is flat but the glass sheet slides along, controlled by drive wheels, to an area where the supporting bed is curved to the desired shape and then, after it conforms to the curvature of the bed, to a quenching area.

A5b. ***grinding*** - Glass is normally machined with abrasive processes. Grinding with bonded abrasive wheels and with loose abrasives driven by metal wheels are both used. Water is used as a cutting fluid to prevent overheating of the glass, to increase the grinding rate, and to prevent glazing of the grinding wheel.

Cutting of glass objects can be done with one of several methods: Score line cutting involves scribing a line in the glass surface and then applying a bending force across the line to break the glass apart. The scribe line can be made with a small sharp wheel that is pressed into the surface as the wheel is rolled across the glass or with a file, knife blade, or carbide- or diamond-pointed tool.

Another method, particularly applicable to glass tubing, uses a hot wire. The Nichrome wire loop is resistance heated to the point of being red hot. Holding the wire against the tube makes a score line and suddenly chilling of the line with water, or manual bending at the score line, breaks the tubing along the line. A third method, more suitable to larger quantity production but also suitable for laboratory or prototype conditions is the use of a wet abrasive cut off wheel. Silicon carbide is the abrasive used, commonly of 120 grit, with substantial water flow. Another cut-off method uses a dry steel disk or stone wheel pressed against the tubing as it rotates. Frictional heat creates thermal expansion that causes the glass to part.

Decorative cutting involves the grinding of grooves in the surface of the workpiece with rotating grinding wheels whose edges are dressed to the shape of the groove. Much cutting is artisan work. A design is first drawn on the glass surface with red lead, rosin, and turpentine, and the cutting action takes place on the scribed lines. Wide, narrow, shallow, or deep cuts are made by guiding the workpiece by hand against a suitable wheel. Various standard cuts referred to as leaf and stem,

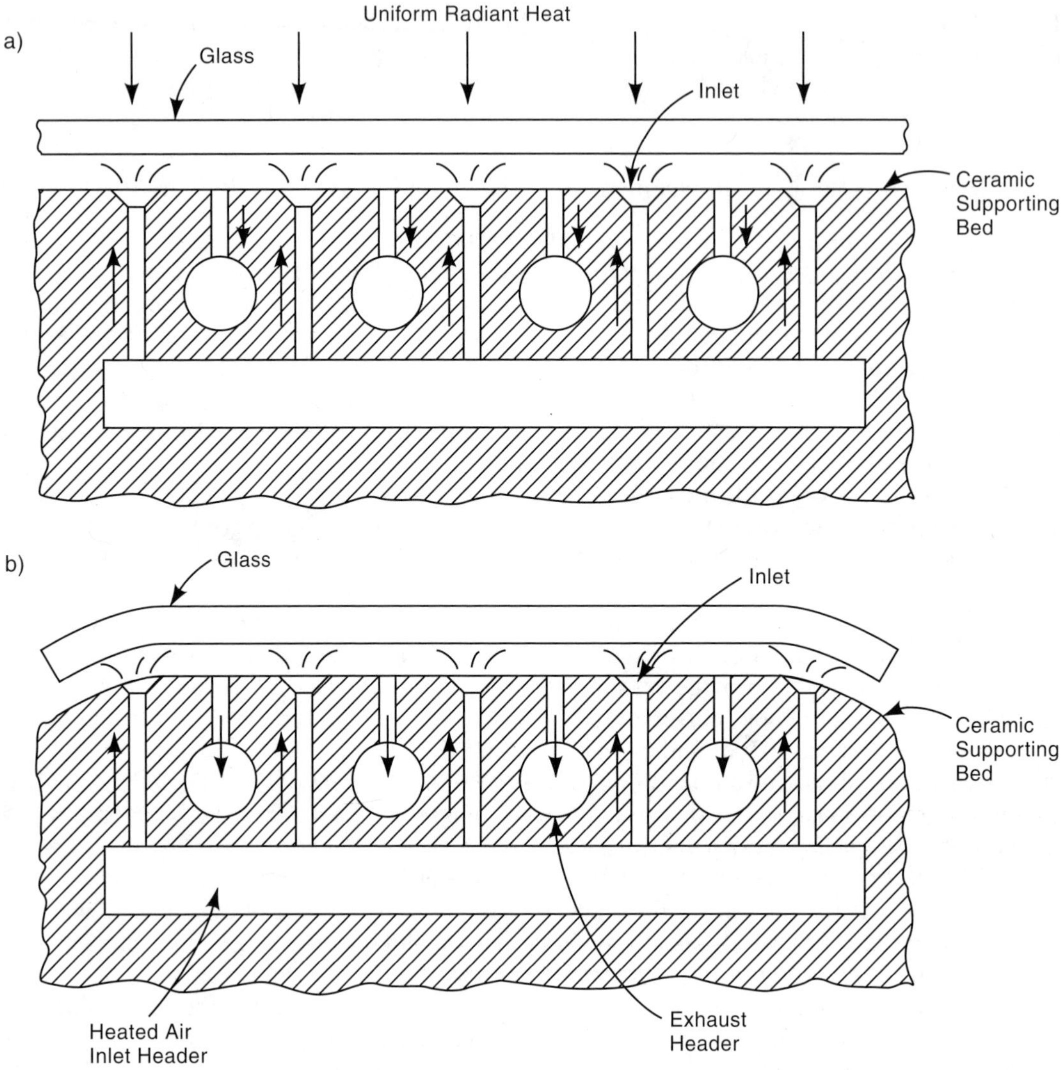

Fig. 5A5a Cross-sectional view of air float bending and tempering of sheet glass. a) heating section. The glass sheet floats above a ceramic bed on streams of hot air. The sheet is heated as it slides on the air cushion (guided by drive wheels) to the forming section. b) At the forming section, the shape of the supporting bed is changed and the glass sheet sags and bends. The sheet then continues to slide to a tempering or annealing section.

flutes, bands, spikes and punties are used. The hardness of the stone cutting wheel is selected to suit the hardness of the glass.

Another artisan technique for engraving designs is to use copper wheels, abrasive grit, and motor oil. The designs are applied manually by the wheel, which may be as small as 1/8 in (3 mm) in diameter. The engraved surfaces have a frosty appearance.

Drilling small holes - less than 3/8 in (10 mm) diameter - is done with hardened steel or tungsten carbide-tipped drills. The drills are rotated slowly and turpentine is used as a coolant. Chipping at the exit side of the glass part is avoided by countersinking that side first or having a back up sheet attached to the underside of the workpiece. Larger holes are drilled with brass tubes having vertical slots at the cutting end, charged with abrasive powder and water. Rotating the tube and applying pressure produces a cutting action from the abrasive. A glass disk, corresponding in diameter to the interior of the tubing, is produced. Tube drills are also commercially available with diamond dust bonded to the cutting edges, and do not need abrasive. Diamond abrasive saws and wheels of various types are also used for cutting and surface grinding.

Cylindrical and internal grinding are used to produce mating surfaces for stopcocks in laboratory glassware. In large-lot production, centerless grinders may be used. Several passes may be made with progressively finer grit sizes starting with 120 grit metal-bonded diamond wheels and ending with 400 or 600 grit silicon carbide wheels. Flat surfaces for bevels at the edges of sheet glass are ground with surface grinders. Loose abrasives, water, and large-diameter horizontally-rotating steel wheels are used to put flat surfaces on apparatus bases and bell jars. When done manually, the glass workpiece is held and moved in a figure-8 pattern against the wheel. The wheel is cleaned and the operation is repeated with a finer grit until the desired finish is obtained. Wet abrasive belt machines are used in glass-blowing shops to perform many rough grinding operations.

A5c. ***polishing*** - is often the final stage of a grinding operation whose purpose is to further smooth the ground surface of the workpiece. A fine, loose, abrasive is used with water as a coolant. The polishing compound is applied with a powered rotating buff, which may be cloth fabric, a bristle brush, leather, felt, poplar wood or one of the softer metals. The workpiece is held against the rotating buff or vice-versa. For final polishing, water and a soft powder such as cerium oxide (optical powder) or rouge (ferric oxide) is used. Very little material is removed by such an operation, but minute irregularities in the service are leveled and a high degree of smoothness can be attained. For optical applications, a very fine and accurate surface is possible.

Acid polishing is an alternative method. It can produce a brilliant surface (rock crystal). Workpieces are placed in Monel metal baskets and immersed alternatively in an acid bath and water until polishing is complete, typically for a total time of 10 to 15 minutes. The time for each immersion will vary from 10 to 30 seconds depending on the type of glass involved and the acid strength. The acid is usually 3:1 or 3:2 concentrated sulfuric and 60 percent hydrofluoric acids. Too much time in the acid in at any immersion can result in an uneven surface.

A third polishing method is fire polishing. With this method, the glass is heated to a temperature of 900 to 1300°F (500 to 700°C) at the surface by directing a flame to the workpiece. This heating causes the surface glass to flow and, because of surface tension, become more smooth.

A5d. ***fusion sealing*** - Glass-to-metal seals are important in glassware made for scientific use. They are made by heating both the glass and the metal surface. The metal should have a thin oxide surface. The glass "wets" this surface and absorbs the metal oxide so that part of it dissolves in the glass. This may change the color of the glass at the interface and thereby can provide an indication that the seal is sound. Proper oxide thickness for the operation is facilitated by heating the metal part sufficiently, in an oven, to oxidize the surface. With an oven, time and temperature can be controlled accurately, but sometimes flame oxidizing is used instead. Machined parts are typically degreased and baked in a hydrogen atmosphere prior to the surface oxidizing step. Glass-metal seals are made with copper, steel, nickel-iron alloys, platinum, chrome-nickel-iron alloy, high chrome iron alloys, and nickel-cobalt-iron alloys.

Glass-to-glass seals require glasses of similar thermal expansion coefficients; otherwise stresses will develop at the seal that cannot be relieved by annealing the joint. The same is true of glass-to-metal seals except when the metal involved is relatively soft. Glass-to-glass seals are made by heating the ends to be joined, often with gas burners, pressing the ends together, and blowing, pulling, or smoothing the joint. Sealing lathes, which rotate both parts in unison, may be used to hold the workpieces and bring them together. Electric arc heating is another method. The glass workpieces are first heated with gas burners. When the glass has heated enough so that it is conductive, an arc is drawn for a few seconds, heating the ends to the fusion temperature. The parts are pressed together and the joint is dressed as necessary. Heavier sections, and those of irregular shape, can be joined with this method. Rectangular TV picture tube sealing is an example.

Diffusion sealing is another method for making glass-to-glass assemblies. The surfaces to be joined are polished to a flatness within 30 millionths of an inch (3 fringes of yellow light), kept free of any dust, and assembled together under pressure from a light weight. The assembly is then heated to the annealing temperature. The surfaces in contact will diffuse together.

A5e. ***grit blasting*** - Abrasive particles, driven against the surface of glassware by a jet of high velocity air, remove small amounts of material from the surface. The technique is useful in permanently marking, decorating, or modifying the surface of glass workpieces. Both sand and aluminum oxide (Al_2O_3) are used. By masking the workpiece, the surface effect can be confined to narrow areas, so various designs can be marked on the glass surface. The technique is also called sand blasting or sand carving. Variable cuts, including deep sections, can be made, if desired. The process is also used to create a matt surface in the workpiece. Grooves and holes can also be made. The operation is rapid. Masks are made from rubber, lacquer coatings, lead foil and heavy paper masking tape. The process is often used for decorating vases, bowls, and art glass.

A5f. ***acid etching*** - uses an acid solution to etch the glass. One approach for acid etching designs in glassware, called needle etching, is to coat the entire glass object with a coating of wax by dipping or brushing, and then scribing lines in the wax. Scribing the wax coating may be manual or may be done with an engraving machine. The workpiece is then immersed in the acid bath. Another approach, called plate etching, is useful for fine detailed etchings such as those including floral patterns. With this method, a wax mixture is transferred from an etched metal plate to pottery printing tissue and thence to the glass surface before immersing the workpiece in acid. A 60 percent hydrofluoric acid bath is used. The glassware is protected in other areas not covered by the transfer by brushing on or dipping in the wax resist to coat areas that are not to be acid etched. After engraving, the glass workpiece is rinsed and the wax is removed with hot water or by vapor degreasing. The etched area is typically filled with a paste or glaze of contrasting color. Graduations of scientific glassware as well as decorations of other glass products are etched by this method. Acid etching is also used to put a matt finish (or frosting) over a large area of a glass workpiece, and on some incandescent light bulbs.

Another method for acid etching is to use a paste of ammonium bifluoride, hydrofluoric acid, and barium sulfate, and to brush, screen, or stencil the paste on the surface of the glass workpiece. The paste acid etches the surface where it makes contact.

A5g. ***shrinking*** - is a method for making accurate internal diameters in glass tubing. The process uses a vacuum to reduce the tubing diameter against a mandrel. The mandrel is machined from alloy steel to the dimension desired with allowance for the thermal expansion coefficients of both the glass and the mandrel material. The mandrel is placed in the tubing; the tubing is heated to the softening point; the ends are sealed and a vacuum is applied to the space between the mandrel and the tubing bore. The tubing collapses against the mandrel. When the tubing has cooled, the mandrel is withdrawn. Accurate bores are produced in this manner. The method is used to make tubes for fluid flow meters[4].

A5h. ***staining*** - Glass can be stained for decorative reasons or, in laboratory glass, to filter out light rays that may be harmful to biological solutions. Oxides or salts of copper and silver are mixed with an inert material, commonly, kaolin (clay) or ocher. Linseed oil, glycerine, alcohol, turpentine, water, or a combination of them is blended into the mixture, which is then applied to the glass. When the glass is heated almost to the transformation (softening) temperature, the ions of copper or silver migrate into the rigid glass solution. The glass takes on a permanent transparent color, usually red, amber, or yellow. A black tint can be produced after silver migration by subjecting the glass to hydrogen or another hot reducing gas. Vases, goblets, and other drinking glasses and blown ware are sometimes stained.

A5i. ***decorating with vitrifiable colors (vitreous enamels)*** - These materials are glasses that have low melting temperatures and contain inorganic pigments. Various glass formulations may be used. Their melting temperatures are as low as 900°F (480°C) and as high as 1400°F (760°C). The lower melting temperature glasses used are softer and have limited resistance to acids or alkalis or other aggressive or corrosive chemicals. They are also less apt to be able to provide opaque coloring. These vitrifiable colors consist of the pulverized glass in a liquid vehicle such as turpentine. They are applied to glass workpieces by various methods: spraying, silk screening or decals.

The workpiece is then fired at the temperature required to fuse the colors and make them adhere to the workpiece. This temperature is below the level that would deform the workpiece. Applications include coloring the entire workpiece or an expanded area, or to specific portions of the glassware. Colors applied to limited areas may be for decorative or artistic purposes, or to provide functional markings such as graduations, scales, ruled lines, or numbers. Housewares, architectural spandrels and other components, thermometers and other laboratory devices, signs, containers, and lamps are colored or decorated with vitrifiable colors. (See *enamel, vitreous* in the products section.)

A5j. ***metallic coating*** - Metallic coatings for glass are primarily used for decorative purposes and are made chiefly from platinum, palladium, and silver- and platinum-gold alloys. They are applied to the glass in liquid or paste form, the liquid being a suspension of metal particles in oil. When a small amount of glass flux is added to the liquid to aid bonding or dispensing, the resulting material is a paste that can be spread on the glass surface. Printing, silk screening, stenciling, and spraying may be used to apply the metallic material, depending on the effect wanted. When the coated glass workpiece is fired, the metals diffuse into the glass and coat its surface. An oxidizing furnace atmosphere is used. After firing, the metal can be polished or burnished to a bright finish. This approach is used for various decorative effects, for example, to coat the rims of higher-priced drinking glasses. It is also used to provide conductive, printed electronic circuits.

Other metallic coatings for glass involve the spraying of molten metal on hot glass (as described for metal substrates in 8F4), vacuum deposition (8F3), and the spray coating of tin oxide with other ingredients that form an iridescent film with properties of improved alkali resistance, electrical conductivity, and infrared ray reflection. Mirrors are coated with silver nitrate, which is converted to silver metal by the action of glucose. A stannous chloride rinse is also involved. Metal coatings on glass are used in the production of some small trimmer capacitors and glass inductors for high-frequency electronic circuits.

A5k. ***organic decorating*** - Organic enamels are used to decorate and mark glass products, and have the advantage of providing a wider variety of colors than are available with vitrifiable coatings. Adherence may be a problem, however, because glass is a difficult substrate for paint adherence over a long period, particularly if there are temperature changes, moisture, and handling stresses on the coating. Organic enamels are oven cured, typically at 450°F (230°C) for 15 min. Application can be by any painting method, silk screening, stenciling, offset printing, or dipping. Both one-part and two-part enamels are used. Common applications are the marking of bottles for soft drinks, wine, beer, and other beverages, food containers, cosmetic jars, Christmas ornaments, lighting fixtures, and tableware.

A5l. ***silk screening (screen printing)*** - is a printing method described in 9D4b, Fig. 9D4b and 8I7b. It can be utilized to put markings or decorations on glass surfaces.

A6. ***glass fiber manufacturing*** - Glass fibers are made for several major applications including textiles, plastics reinforcement, glass wool for thermal insulation, and optical fibers for two classes of use:

1. traditional fiber optics, the transmission of illumination or images, and,
2. communications fiber optics, the transmission of information such as computer data or telephone messages with pulses of light.

The methods used to produce fibers for these applications are similar but have some distinctive differences. Except for the communications fiber, all the methods start with molten glass flowing through small orifices to form strands of glass. All the methods have some means of attenuating the strands, reducing their diameters to those of useful fibers. Fibers for optical data transmission and most for textile applications are drawn into continuous strands, while those for insulation are made into discontinuous strands.

Optical fibers are composed of two different glasses with different properties. The inner core of the fiber is highly refractive; the glass surrounding it, the sheath, is of lower refractive index. Textile fiber applications include fabrics for draperies and curtains, and tire and plastics reinforcements. With many of the fiber manufacturing methods, the molten glass may first be formed into glass "marbles", which are later fed into a melting pot and drawn into fiber.

A6a. ***mechanical drawing of continuous textile fibers*** - One production method is shown in Fig. 5A6a. Molten glass flows from many small holes in a platinum bushing in the bottom of a forehearth and is drawn into fibers of 4 to 20 microns, μ, (0.00016 to 0.00079 in) diameter. (The glass may also come from glass marbles remelted in an electric furnace.) To achieve the small diameters, the fibers are drawn at speeds of over 5000 ft (1500 m) per min. An organic sizing is applied to be a protective coating, and the fibers are gathered into one strand and wound on a tube for later processing.

In another variation, shown in Fig. 5A6a-1, the glass flowing from the bushing is blown immediately with air to attenuate it into long, discontinuous fibers. These drop onto a rotating collecting drum, from which they are withdrawn and gathered together into one continuous multi-filament strand. The strand is wound on a spooler that runs at a higher peripheral speed than the web, so that the strand is elongated into a textile sliver. This sliver is later twisted and plied with others to form a coarse yarn. It may then be drafted and twisted again to form a fine yarn.

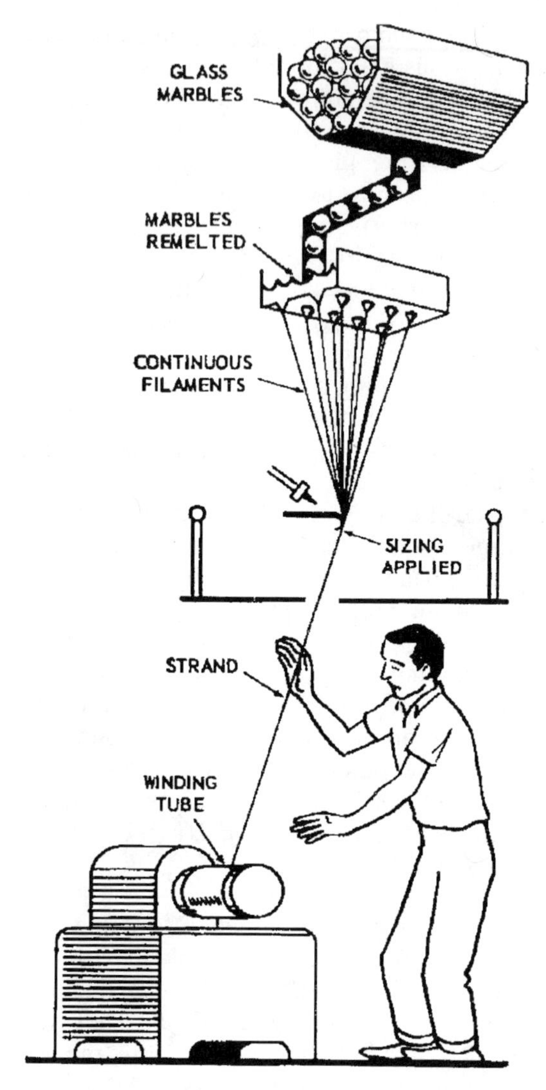

Fig. 5A6a Mechanical drawing of continuous textile fibers. (Courtesy of Owens-Corning.)

A6b. ***steam blowing*** - Streams of molten glass flowing from a melting tank through sieve-like platinum bushings are impinged upon by jets of steam. The jets approach the glass streams at a small included angle, and push them at a faster rate, causing them to draw into finer fibers. If the steam jets are strong enough, the fibers will break into shorter lengths. The strength of the steam jets determines how discontinuous and fine the final fibers will be.

The fibers can be processed in several ways. When used to make thick pads or "wool", they are sprayed with a binder and fall on to a conveyor where they pile up into wool. The binder is dried and the wool is cut into discrete batts of fibrous

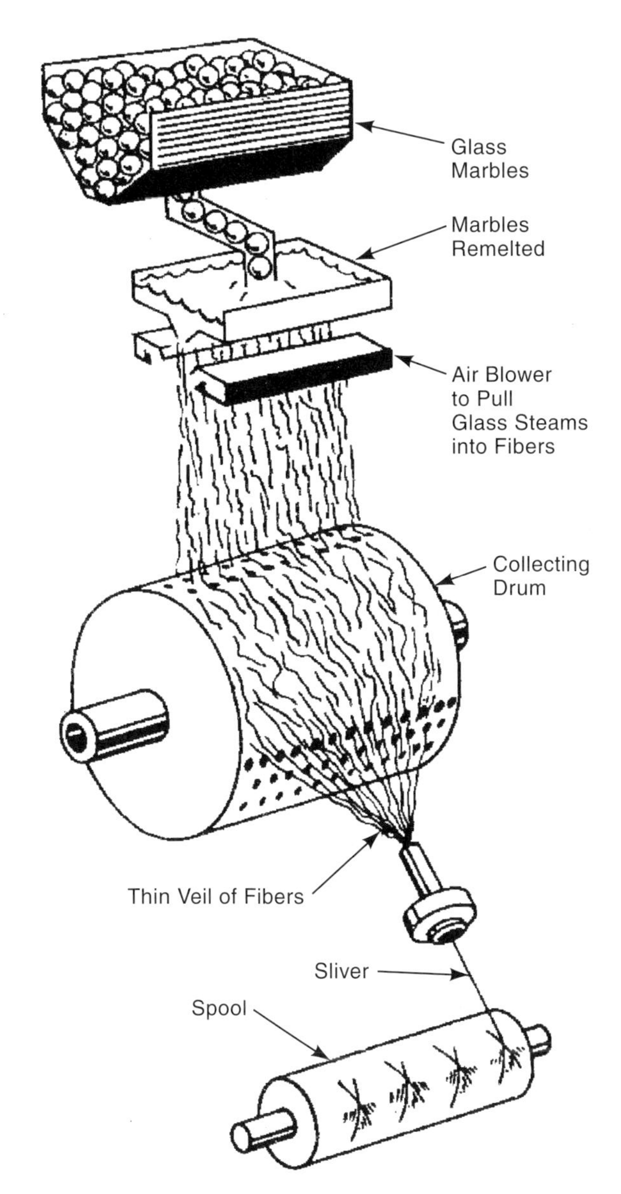

Fig. 5A6a-1 Manufacture of staple textile fibers. Air blows molten glass into fine streams that are gathered from a collecting drum to become a multi-filament strand. The strand is drawn into a textile sliver because the surface speed of the spooler is faster than that of the drum. (Courtesy of Owens-Corning.)

glass. Fig. 5A6b illustrates this process. When made into glass mat, the fibers are allowed to fall into a thin, web-like mat that is then immersed into a bath of binder and then passed through a drying oven. A major use for such mat is the reinforcement of fiber glass-reinforced plastic products. This process variation is shown in Fig. 5A6b-1.

A6c. ***flame blowing (superfine process)*** - involves two major steps: 1) large diameter fibers are drawn from the melting tank through multiple platinum bushings. 2) the fibers move to a high velocity chamber where they are reheated to the melting point and propelled by a high velocity burner. This treatment attenuates the fibers into fine diameter discontinuous filaments. They may be also coated with binder in the same operation. The operation is shown in Fig. 5a6c. The process is used to make ultrafine fibers for papermaking, and for fine fiber-glass wool.

A6d. ***rotary wool forming process*** - Rotary forming is the prime method used for the production of fiberglass wool insulation. Molten glass is fed into a rapidly rotating cylindrical container that has a large number of small holes (approximately 20,000) in the bottom. The molten glass flows through these holes and is thrown outward by centrifugal force. The glass filaments encounter a high velocity stream of gas that attenuates them into discontinuous fibers. The gas may be steam, air, or a combustion gas. The fibers are sprayed with a binder as they fall to a conveyor below. The binder preserves their open structure. Fig. 5A6d shows the process. The air spaces between the fibers account for the excellent insulation properties of the wool. The weight of the loose wool ranges from about 2 to 12 lb/ft^3 (32 to 192 kg/m^3). Since slag, or lower-quality raw materials can be used in the process, (without adverse affect on insulation properties and with a possible improvement in chemical resistance compared to window glass) the term, mineral wool or rock wool is sometimes used for the resulting product.

A6e. ***methods for production of traditional optical glass fibers*** - These fibers have two portions, a central core that transmits the light, and a sheath that, due to a difference in its properties, helps ensure that the light does not escape from the core. The sheath also protects the central core from scratches and other surface damage, which would allow the light to scatter and leak from the core.

One method of producing this kind of optical glass fibers is the rod-tube process. A rod of

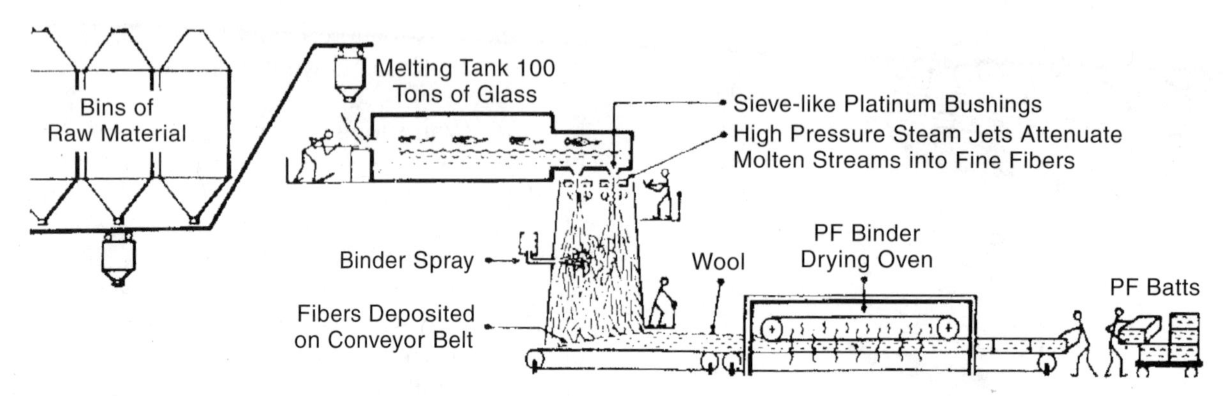

Fig. 5A6b Steam blowing of glass fiber to form glass wool. (Courtesy of Owens-Corning.)

highly-refractive glass is placed inside a tube of lower-refractive glass. The pair are heated in a furnace so that both components become soft. They are then drawn together into a single thin fiber with core and sheath of different materials. The fiber is then wound on spools or drums.

Another method is the two-crucible process in which the two glasses are melted separately but drawn through dual concentric orifices to form a fiber. Fig. 5A6e illustrates both the rod-tube and two-crucible processes schematically.

Optical glass fibers are also normally bundled together so that sufficient illumination is transmitted or, when an image is to be transmitted, so that each fiber transmits a portion of the total image. In some applications, the bundled fibers are fused

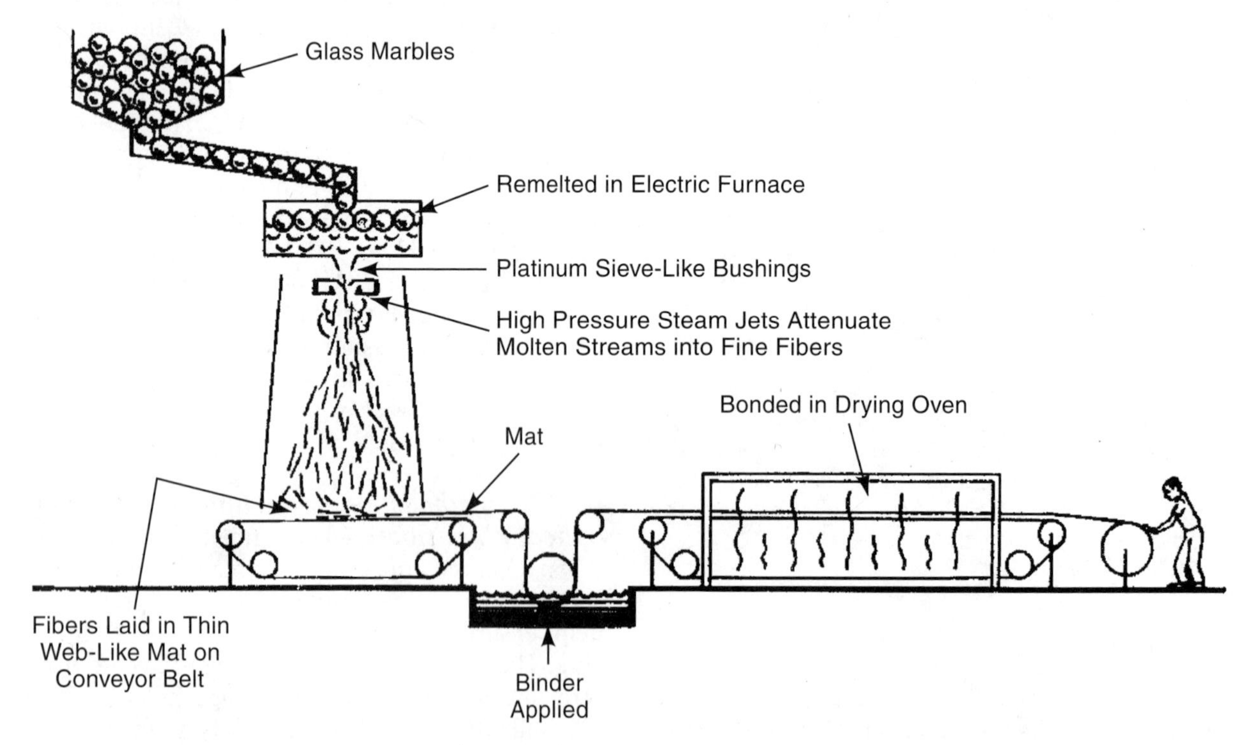

Fig. 5A6b-1 Steam blowing of glass fiber to produce matting for reinforcement of plastics components (Courtesy of Owens-Corning.)

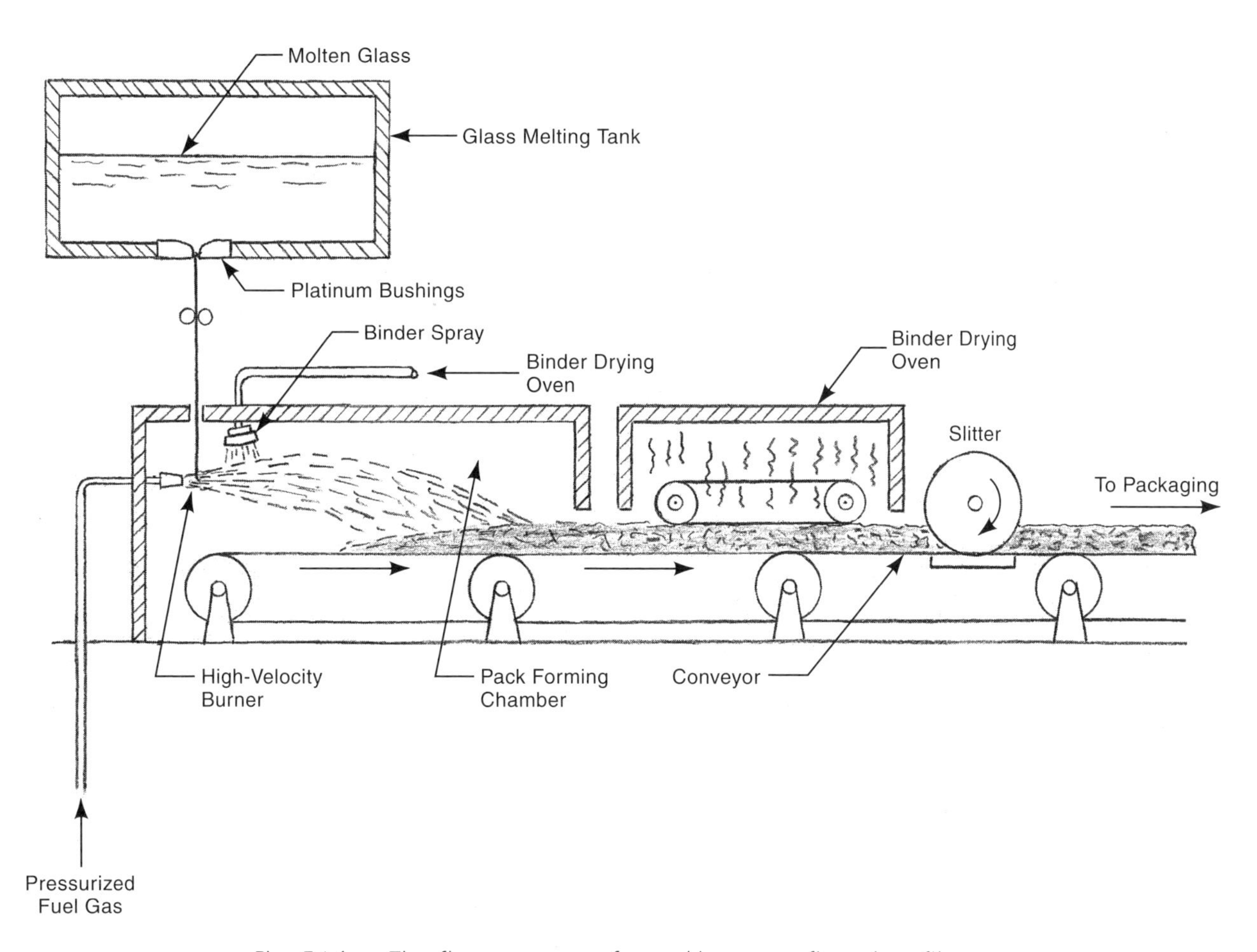

Fig. 5A6c The flame process for making superfine glass fibers.

together to provide greater strength. Sometimes, the bundle of fibers is reheated and redrawn to a smaller cross section, and then bundled with other bundled stacks of fibers. The redrawn bundles may be reheated, redrawn, and rebundled, until there is a sufficient number of fibers in the bundle and the bundle is small enough for the application planned. Light absorbing strands may be incorporated in the bundles to insure that light does not leak from fiber to fiber. Bundled optical glass fibers are used in medical applications to view internal bodily organs for diagnosis and surgery. The bundles are also used for lighting of instrument panels of aircraft, automobiles, and scientific equipment, and for remote inspection and sensing in process equipment[6].

A6f. ***methods for production of optical communications fibers*** - These fibers, even more strongly than traditional optical fibers, need the property of low-loss transmission of digital light pulses and the property of containing the light within the fiber. They normally differ from much traditional optical fiber in that, instead of having a distinct core and sheath with different refractive indexes, there is a gradual transition in refractive index from the core material to the outer cladding. This arrangement provides a clearer signal at the receiving end of the fiber. Low-loss transmission over long distances also requires a high level of purity in the glass. Silica glass is used because it can be made to the level of purity required. However, silica glass is more difficult to process by conventional methods than other glass; silica has high viscosity and is volatile at the temperature needed to melt it when it is in a crystalline phase[6].

One common process used for optical communications fibers starts with $SiCl_4$ in liquid form. The liquid also contains certain dopants (germania and

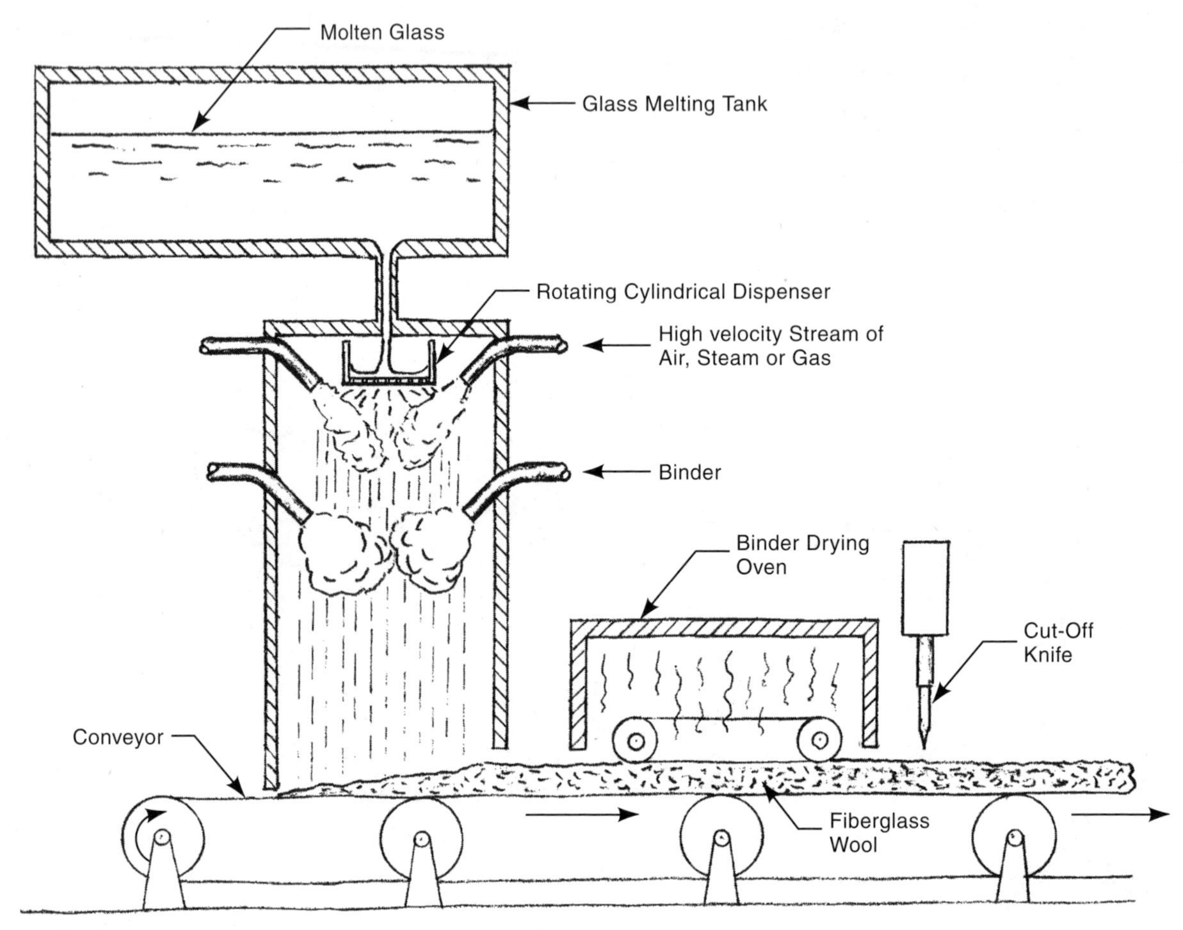

Fig. 5A6d The rotary wool forming process.

fluorine), which control the refractive index of the glass. A carrier gas (oxygen or helium) is bubbled through the liquid at a controlled temperature. This converts the $SiCl_4$ to a vapor. The vapor is fed into a special burning chamber containing a high-temperature gas-oxygen flame. The gas is either natural gas or hydrogen. A hydrolysis reaction takes place that yields silicon oxide and oxides of the dopants. These oxides occur as molten droplets of transparent "soot".

There are two basic process variations in how the soot is collected and processed further. In one variation, the outside process, the soot is collected on a rotating mandrel rod. The first soot collected on the rod is of glass that will form the core of the fiber; then glass material that will form the sheath is processed and collected. There can be a gradual transition from core to cladding material. The preform of soot particles is removed from the mandrel rod and heated to fuse the particles together more strongly. It is treated, first with chlorine gas and then with helium to remove trapped water and air, both of which would impair light transmission. This process variation is sometimes referred to as OVD for outside vapor deposition.

In the other major process variation, the inside process, the $SiCl_4$ and dopants are fed into a fused silica glass tube. Heat from a burner outside the tube causes a reaction that results in the deposit of soot on the inside of the tube. The first soot deposited is of glass material that will form the outer portions of the fiber. Then the formulation is progressively modified so that the final glass deposited on the tubing walls is of glass that will form the fiber's core. No moisture is contained in the soot because the source of the moisture, the gas-oxygen flame, is

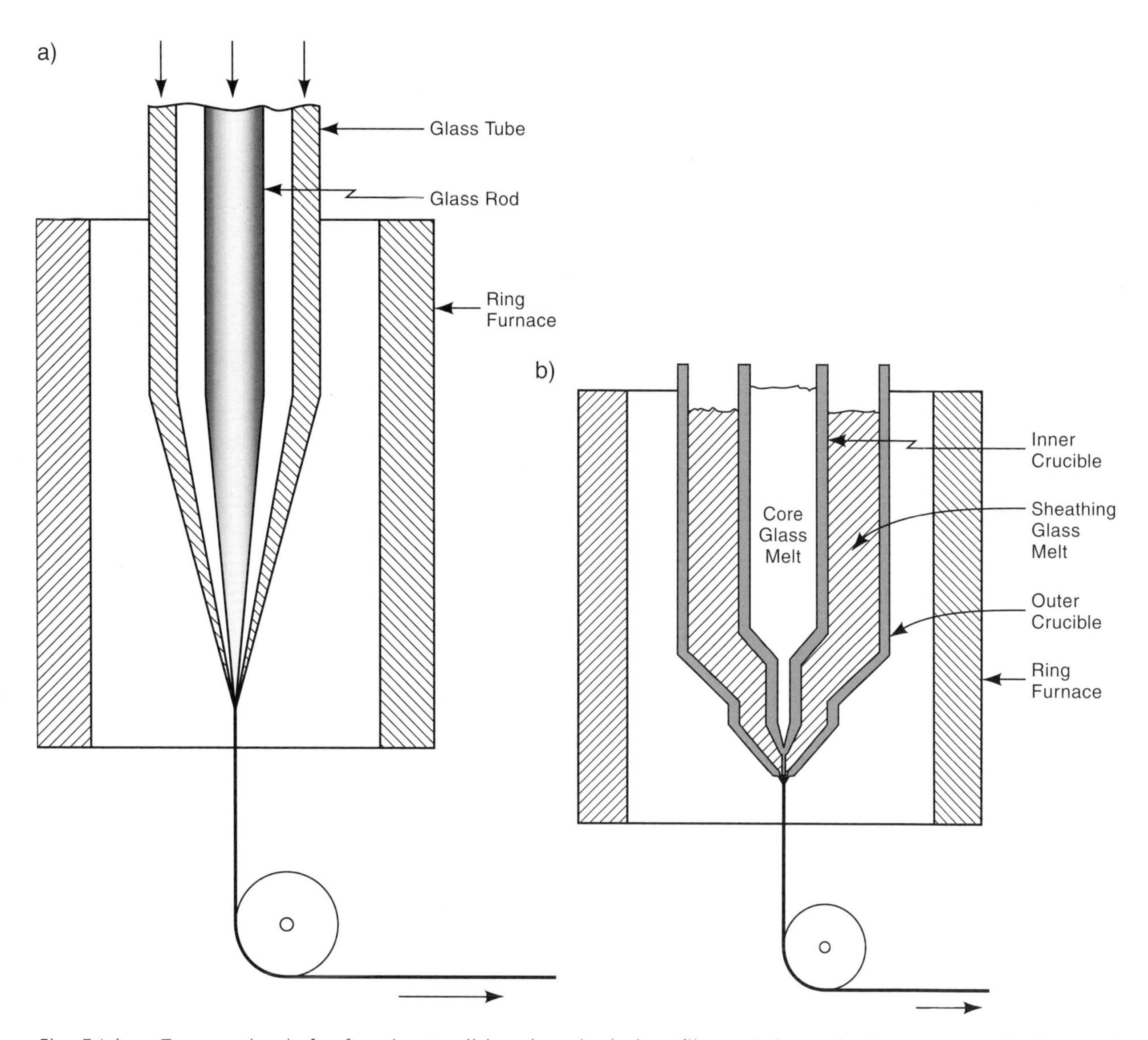

Fig. 5A6e Two methods for forming traditional optical glass fiber: a) the rod-tube process. A glass rod and glass tube arranged concentrically are melted together and drawn into fiber. b) Two concentric crucibles, one containing melted core glass and the other melted sheathing glass, discharge into a common opening from which the fiber is drawn.

prevented by the glass tube from contacting the soot. The soot particles are hot enough to fuse together inside the tube, layer by layer. After sufficient glass soot is deposited in the tube, the tube and its contents are collapsed with the aid of heat and an internal vacuum to form a solid rod. The fused silica glass tube will form the outer layer of the eventual fiber. This process variation is sometimes referred to as MCVD (modified chemical vapor deposition).

The changed glass material in different layers for both process variations provides an "index gradient" that will contain light in the core of the fiber. Fig. 5A6f shows both these process variations for collecting the silica glass soot. The next step in the process for both variations is to draw the preformed glass rod into fiber. The drawing process is the same in principal as other fiber drawing but incorporates significantly higher levels of cleanliness and precision. Commercial drawing

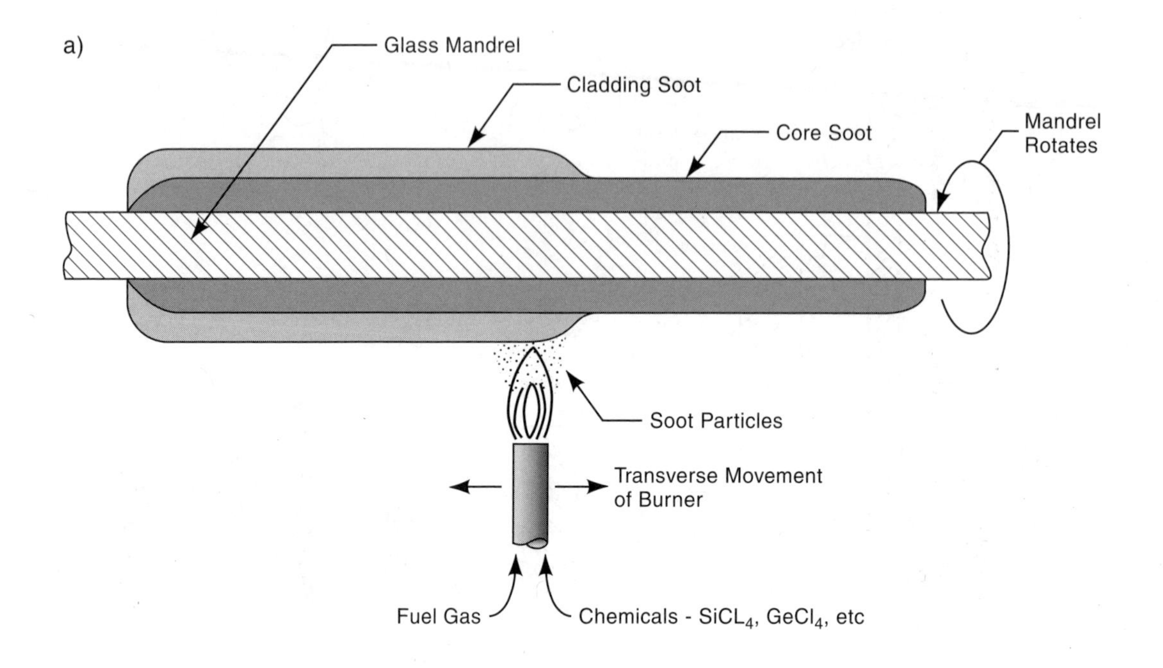

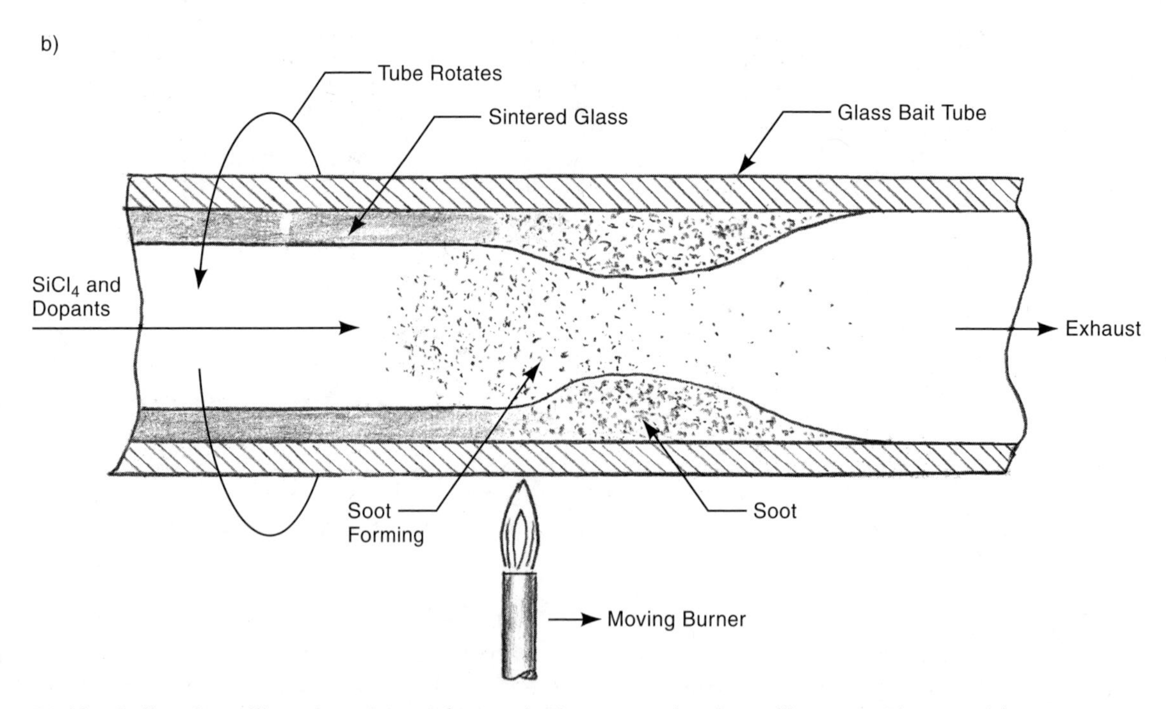

Fig. 5A6f Collecting silica glass "soot" for optical communications fibers: a) The outside process (DVD process) - $SiCl_4$ vapor and dopants fed into a high temperature gas-oxygen flame yield transparent soot, which is collected on a rotating glass mandrel. The mandrel is later removed and the soot is sintered and then drawn into a long fiber. b) The inside process (MCVD process) - The chemical ingredients are fed into a glass tube and do not directly contact the flame. Soot is deposited on the inner surface of the tube and is sintered by the heat of the moving burner. The glass tube becomes the outer sheath of the fiber when the tube and its soot contents are further heated and drawn into fiber.

towers are large, and drawing speeds for the final fiber are rapid. The communications fiber produced may have a core diameter of only 0.0003 in (8 microns) though the fiber diameter, including cladding, is normally about 0.005 in (125 microns). Light intensity loss in such fibers is less than 1 percent in one kilometer (0.6 mile) of transmission distance.

The final step in the manufacturing process, prior to testing, is to coat the fiber with an organic material, sometimes in several layers, for protection of the glass. UV or heat-curing polymers are used. The fiber is tested for strength before spooling. Tests are also made of the fiber's light transmission qualities against a series of specified characteristics.

A7. *manufacture of other types of glass*

A7a. ***glass ceramics manufacture*** - Where conventional glass is a supercooled liquid (non-crystalline), ceramics consist of crystalline particles bonded together. It is possible to convert glass to ceramic form by controlled devitrification, the generation of crystals in the glass structure. This creates products with properties significantly different from those of products made from conventional glass. The process involves the incorporation of nucleating agents in the glass batch, and an operation sequence that uses some specific time-temperature cycles to create crystals.

Glassware to be converted to the ceramic form is first fabricated by pressing, blowing, rolling, or casting. The special nucleating agents in the workpiece are then activated to form small crystals in the structure. This nucleating process can be effected by a special heat treatment, that follows a particular time-temperature curve. Another method that may be used, depending on the glass composition, exposes the workpiece to ultraviolet light followed by a heat treatment. Crystals are then grown by raising the workpiece temperature into the devitrification range for the glass composition involved. The resulting crystals are smaller and more uniform that those of conventional ceramics and, in total, they occupy from 50 to 90% of the volume. The glass ceramic product is normally opaque and has greatly increased strength and hardness compared to conventional glass, greater rigidity at high temperatures, and higher thermal conductivity. Glass ceramics can be formulated with almost no thermal expansion, providing superior thermal shock resistance. Because of these properties, glass ceramics are used for ovenware, cook tops, mirror supports for astronomical telescopes, radomes of missiles, and length standards.

A7b. ***photosensitive glass manufacture*** - is related to the process described above for manufacture of glass ceramics. Some glass formulations are photosensitive. Slight amounts of silver compounds and cerium oxide are key ingredients in the photosensitive effect. Exposure to ultraviolet radiation activates the glass so that, under heat treatment, crystals form. The UV light causes electrons from cerium oxide to migrate, but the electrons are trapped by silver ions when the glass is heated. This process forms metal atoms, which immediately transform into metal colloid particles that act as nuclei for devitrification. Further heating causes devitrification, creating a brown or yellow coloration. If the areas exposed to UV radiation are limited with a mask or photographic film, only the exposed areas will later exhibit the color. If bromine and chlorine are present in the glass, different colors can be created. Glass components for signs, and clock and radio faces can thus be produced.

Another effect can be created because of the fact that the devitrified areas are more easily etched by hydrofluoric acid. This provides the basis for selective precision etching of special shapes and perforations. Printing dies and display components can be made with this approach.

A7c. ***cellular glass (foam glass) manufacture*** - starts with a mixture of pulverized glass and a gas-generating material such as pulverized carbon. This mixture is loaded into metal trays and covers are positioned. The trays are placed in a furnace where the glass is melted. The gasifying material forms many small bubbles in the glass, which expands to fill the trays. A closed cell structure is formed. After cooling, the foamed glass is removed from the trays and cut into blocks or other shapes. Typical densities are on the order of 10 lb/ft^3 (160 kg/m^3). The finished material is frequently used as a self-supporting thermal and acoustical insulation. Buildings, cold storage structures, and piping, are particular applications. Another application is use in flotation devices.

A7d. ***glass microsphere manufacture*** - Spheres of glass with a diameter of 0.008 in (0.2 mm), or less, are used in making reflective surfaces, for reinforcement of plastics, and for micro-blasting metal and other surfaces for polishing. There are several methods for manufacture of such spheres[2]: One method is similar to the shot-tower operation used in the manufacture of the metal pellets in shot gun shells. Crushed glass, screened for the proper particle size, is fed into the top of a heated vertical tubular furnace. The glass particles melt and, because of surface tension, assume a spherical shape as they fall to the bottom of the furnace. The lower part of the furnace is cooler, and the spheres solidify before they reach the bottom. Another method uses a blast of upward flame. The crushed glass particles are fed into the furnace, melt, and are blown upward into a cooler zone where they solidify and are collected. Another approach sprays a thin stream of molten glass into a chamber where the stream breaks it up into small spherical droplets, which then cool and solidify, fall to the bottom and are collected.

Microspheres are used in reflective highway signs and projection screens, and in the finish coatings of reflective foils. The spheres act as optical lenses to direct light so that it is more strongly reflected. Microspheres are also used as a reinforcement in molded plastic parts to provide them with better dimensional stability, surface hardness, rigidity, and sliding properties. Glass spheres are used in shot peening to provide a more glossy surface, for fatigue-failure resistance, and for sheet metal forming, as described in sections 2J7 and 8H of this handbook.

A8. ***powdered glass processes***

A8a. ***dry pressing and sintering*** - Glass in powder form, with a binder, can be pressed in a die of the desired shape and then sintered to fuse the powder particles together. This process makes a solid part, albeit one with many small trapped air bubbles. The process is similar to that used to make metal parts from metal powders (See 2L1). To make the powder, crushed glass is further pulverized in ball mills. It is then screened so that the resulting powder has the proper particle size and distribution. The binder, mixed in before pressing, holds the pressed workpiece together while it is in the "green" state. When sintered in an oven, the binder burns off and the particles then fuse together. The finished object is opaque white in color because of the trapped air. Sintered glass is used to make small parts with holes, parts that may be difficult to make by pressing or other methods.

A8b. ***slip casting of glass*** - uses essentially the same ball-milled glass powder as that used in dry pressed and sintered parts but, in this method, the powder is mixed with water to form a liquid slip. This slip is poured into a plaster of Paris mold. The mold absorbs the water from the slip, providing a casting that is firm enough to handle. The casting is additionally dried after it is removed from the mold and before it is fired. Firing fuses the glass particles together. Slip casting is used to make larger parts in very low quantities, in shapes that would be difficult or unfeasible with other methods. This process is also used for ceramics, as described below in paragraph B8.

A8c. ***fritted filter manufacture*** - uses methods that are similar to those used in dry pressing and sintering. However, the glass powder, though from crushed glass, is not ball milled. Instead, it is screened to a particular particle size distribution, depending on the pore size desired in the filter to be made. Sometimes glass fibers are used. A binder is added and the material is then pressed and sintered. The desired porosity is achieved by controlling the particle size and the pressing and sintering process. The filters, in disc or tubular form, are joined to solid glass tubes or crucibles. Glass filters are used in laboratories for gas washing as well as filtering.

B. Ceramics Processes

B1. ***the nature of ceramics*** - Ceramic parts are hard, highly chemical and corrosion resistant, strong in compression, non-flammable, usually dielectric, and capable of use at high temperatures. They are normally brittle and limited in tensile strength. Most have crystalline structures. Ceramics can be classified into the following groups:

1. whitewares which include china, tiles, earthenware, electrical insulators and other components, mechanical parts, and porcelain,
2. refractories - heat-resistant and insulating bricks and other items,
3. structural clay products such as bricks, clay pipe, and tiles,
4. vitreous enamels to coat steel, cast iron and other metals, and
5. glass.

Ceramics can be made from a wide variety of material formulations although many have clay as a basic raw material. They can also be classified into two basic groups: Traditional ceramics (natural ceramics) include standard products in the above five groups. Advanced ceramics are premium materials made to more exacting specifications and used for applications such as engine components, cutting tools, medical implants, bearings, valves, and various electronic components. They are also sometimes referred to as high-technology ceramics, modern ceramics or fine ceramics. They include oxides including Al_2O_3, MgO, BeO, ZrO_2, ThO_2, and $MgA_{12}O_4$, magnetic ceramics ($PbFe_{12}O_{19}$, $ZnFe_2O_4$, and $Y_6Fe_{10}O_{24}$), silicon carbide (SiC), nuclear fuels, (uranium oxide and uranium nitride) and other nitrides, carbides and borides[7]. Except for glass and the glass binders used in many ceramics, all ceramics have a crystalline rather than an amorphous structure.

B2. ***ceramic materials*** - are inorganic and nonmetallic. The traditional ceramic raw material, clay, consists chiefly of silica (SiO_2) and alumina (Al_2O_3). Clay is the prime ingredient in pottery, bricks, tiles, pipe, sanitary ware, dinnerware, china, and pottery. Quartz and feldspar are often added to clay. Water is added as an aid in forming. Ceramics are made from oxides, carbides, nitrides, silicates, fused cordierite and titania. Refractories are made with magnesium oxide and chromite (Cr_2O_3) as well as silica and alumina. Electronics parts are made from clay and other standard materials, beryllium oxide, zirconia (ZrO_2), boron nitride, and barium titanate. Structural ceramics include silicon nitride, silicon carbide, zirconia, alumina, sialon (a combination of silicon nitride, silica, alumina and aluminum nitride), boron carbide, boron nitride, titanium diboride, and composites of ceramics. These materials are used in high-stress, high temperature-resistant applications, including turbine blades.

Some ceramic materials are manufactured rather than mined from the earth. These, and more refined materials made from mined raw materials, are sometimes referred to as advanced ceramics or modern ceramics. Silicon carbide and silicon nitride, and other carbides, nitrides, and borides are made by chemical reactions. Alumina and magnesia (MgO) occur in nature but, for industrial applications requiring particular properties, are made from hydroxides. Cubic boron nitride is used as an abrasive. Cermets are combinations of ceramics (oxides, carbides, nitrides, or carbonitrides), and metals.

B3. ***ceramics operation sequence*** - The operation sequence for making ceramic parts, can vary significantly, depending on cost factors, the materials involved, and the shape, complexity, and application, of the parts being fabricated. However, most ceramics are made from powdered material. A typical sequence for making such parts involves the following:

1. preparation of the powder: grinding, classifying, and mixing, and
2. a forming operation such as pressing, jiggering, casting, or extruding, etc.,
3. drying,
4. firing and
5. finishing.

Fig. 5B3 is a block diagram of the typical operation sequence.

B4. ***ceramic material preparation*** - Mined clays and other ceramic materials are crushed and ground to a fine powder. The material then is screened or undergoes flotation, filtering, or magnetic separation to remove out-of-size or undesirable components. The resulting powder is mixed with various other materials, including various metallic oxides as fluxes, organic binders, plasticizers, and, often, water. Ball milling or hammer milling equipment may be used to remove any agglomerated material and ensure thorough mixing.

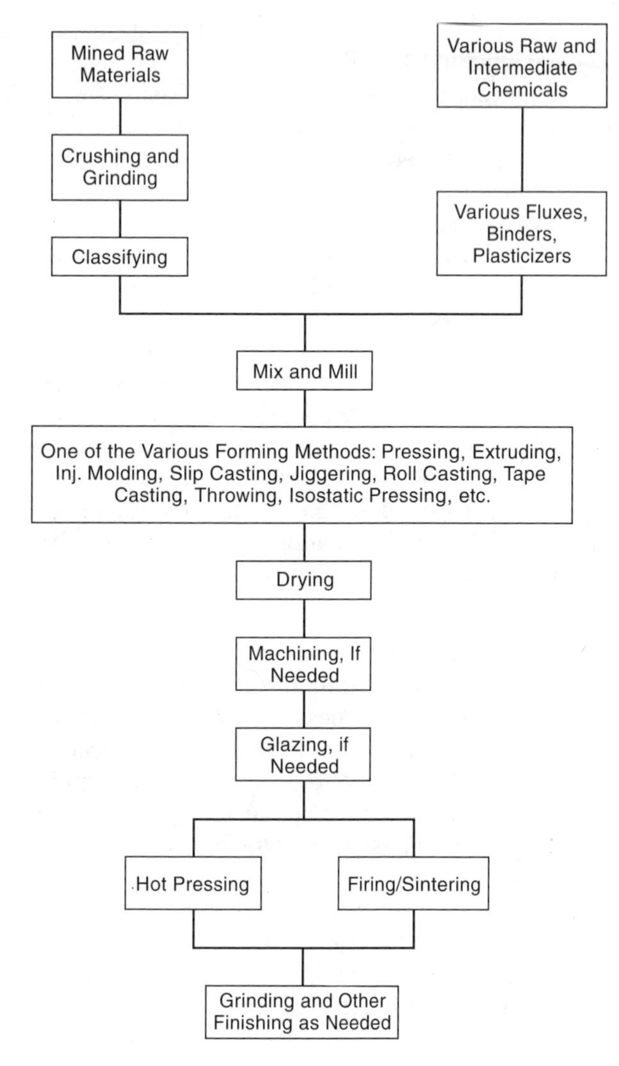

Fig. 5B3 Chart of the sequence of common manufacturing operations for ceramic products.

B4a. ***material preparation for advanced ceramics*** - Powders for advanced ceramics may be made with the same processes indicated for standard ceramics but with stricter standards for material purity, particle size, and mixing. Advanced ceramics are also prepared by several sophisticated processes that provide a still higher purity, better particle size distribution, and more uniformity of characteristics from lot to lot. Raw materials may undergo chemical reactions that create intermediate compounds that are more easily purified and then chemically converted to the oxide or other final compound. Processes that may be used include vapor deposition, precipitation, and calcination.

Vapor-phase production uses reactant gases that are mixed and heated in a suitable chamber. Particles nucleate and grow from the gas phase.

Precipitation may be carried out with one of several process variations. In addition to conventional precipitation, precipitation under pressure (hydrothermal precipitation), and coprecipitation may be used. The latter approach provides better mixing if several oxides or salts are included in the powder mix. Precipitation is a key procedure in the production of high-purity alumina (Al_2O_3) from bauxite.

The aluminum hydroxide in bauxite is first dissolved in caustic soda. The liquid is then separated from non-soluble material by filtering, and the resulting aluminum trihydrate is precipitated by the addition of seed crystals and by changing the pH. Very fine alumina powders result and they have highly desirable properties when used in ceramics[7].

Calcination is common in the production of advanced ceramic powders. With this process, the raw material is heated to a high temperature to decompose salts or hydrates or to remove volatile ingredients, chiefly water. The process is controlled to provide the desired material properties. After the heating operation (calcination), the powders are ball milled to break the bonds between crystallites, which are formed by diffusion during the calcination. A very fine powder can be produced. Magnesium oxide, MgO is often made by extracting magnesium hydroxide from seawater and calcining it to convert it to the oxide. Magnesium oxide is used to make high-temperature electrical insulators and refractory brick.[7].

B5. ***pressing*** - is the most common ceramics forming operation and is the preferred method, if it is feasible. Most pressing is similar to pressing of glass or compression molding of plastics in that the material is placed in a metal mold and then compressed with strong force from a descending punch. Mechanical or hydraulic presses are used. Lubricants and binders may be included in the material. Binders may be organic or inorganic. If the ceramic mixture is relatively dry (0 to 5% water), the method is known as dry pressing, and the operation is very similar to the compaction of powder metal parts. High pressures and precision dies are used. Floor and wall tiles, spark plug insulators, ceramic capacitor components and enclosures are dry pressed.

If the ceramic mixture is relatively wet (wet pressing), the material contains more water and is capable of greater flow when pressed. The operation becomes more similar to the compression molding of plastics where the plastic material flows into remote areas of the mold. Wet pressing is used for somewhat more complex parts than dry pressing but the greater water content of the material leads to greater dimensional variations and larger tolerances than those needed for dry-pressed parts.

Generally, pressing, especially dry pressing, is used when the part's shape is relatively simple. After pressing, the part may be further shaped by additional forming or machining. Fig. 2L1c (in chapt. 2) shows press forming of powder metals, and the methods shown are also applicable to the pressing of ceramic powders.

B5a. ***isostatic pressing*** - Some pressing is isostatic, using a mold made of a flexible elastomer, and enclosed in a pressure vessel. Pressure in the vessel, transferred by a medium of water, oil or glycerine, compacts the powder in the mold. This approach is used with spark plug insulators, which have a central metal mandrel to form the center hole and flexible elements to rough form the outer surfaces. They are then finish-machined in the green state before the part is fired. However, in many other applications of isostatic pressing, the part can be finished to final shape without machining or further forming. The advantage of isostatic pressing is that pressing occurs in all directions, providing more uniform compaction. Fig. 5B5a shows the process schematically.

B6. ***injection molding*** - Injection molding machines, similar to those used for plastics injection molding but without the band heaters needed for plastics, are often used to mold green parts from ceramic paste. The process is useful for large-quantity production of parts with greater complexity than those produced by pressing. Relatively wet material mixtures with organic binders are used, but the binders must be removed before final firing.

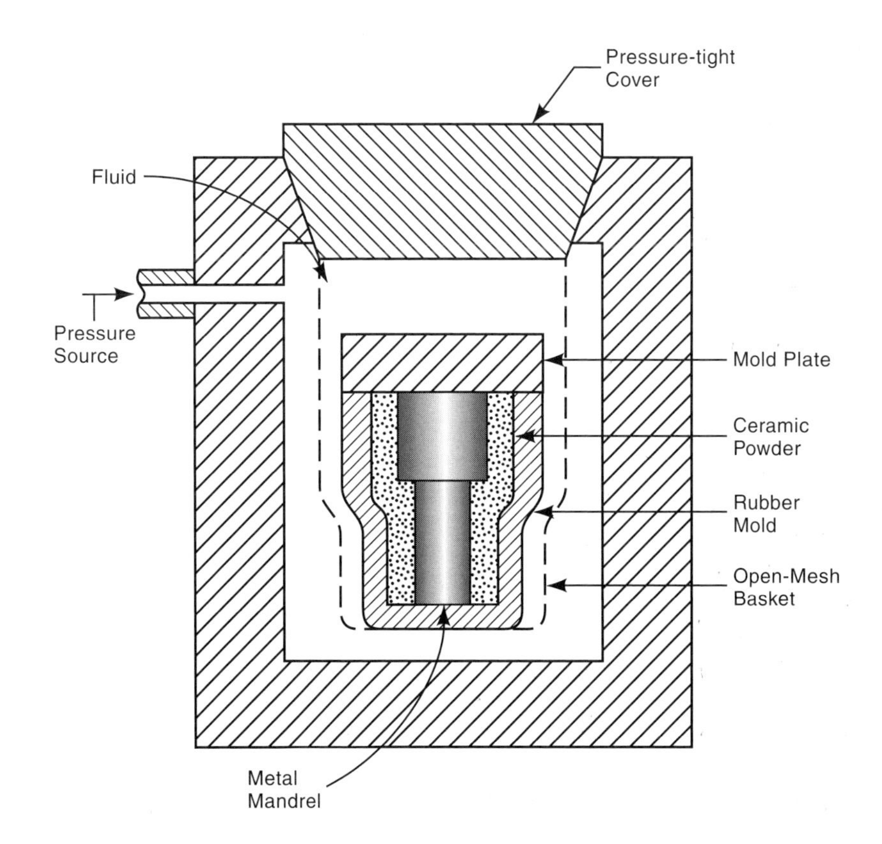

Fig. 5B5a Isostatic press forming.

Tolerances are broader than those for dry-pressed parts. Ceramic gas-turbine blades and thread guides are made using injection molding.

B7. ***extruding*** - is similar to extrusion of plastics as described in 4I1 in that a screw auger drives the material through the extrusion die. It differs in that there is no need to heat the material to plasticize it. Prior to extrusion, the material is prepared for the operation by mixing in enough water to produce a stiff paste. Trapped air or gases may be removed during material preparation, or the ceramics extruder may have a vacuum chamber for this purpose. If the extruded shape is to have a hollow interior, a mandrel of the desired shape is incorporated in the die. After extrusion, the green part is cut to length. Since extruded sections are of a uniform cross section, they are sometimes machined or shaped by other methods prior to firing, to provide the desired external shape. Bricks, clay pipe, thermocouple insulation tubes, and automotive emission catalyst supports, are among products made by extrusion.

B8. ***slip casting of ceramics*** - uses a liquid ceramic material mixture and plaster of Paris molds to cast green parts for later firing. The mixture contains fine particles with deflocculants and dispersants to insure uniform suspension of the ceramic particles. The mold is filled by gravity and the plaster of Paris, through capillary action, absorbs water from the mixture. The part gradually assumes a leathery consistency as the water is absorbed by the mold. After a suitable time, the part can be removed from the mold, handled and finished, prior to drying and firing. Finishing usually involves careful wiping with wet sponges to remove mold flash, and to smooth any areas where needed. If the part is to be hollow, excess material can be poured from the mold after the material in contact with the mold walls has solidified to a desired thickness. This approach is sometimes called drain casting and is used in the production of artware and hollow sanitary ware components. Refractory parts, including ceramic molds for metal casting, are also made with this method. Fig. 5B8 illustrates drain casting.

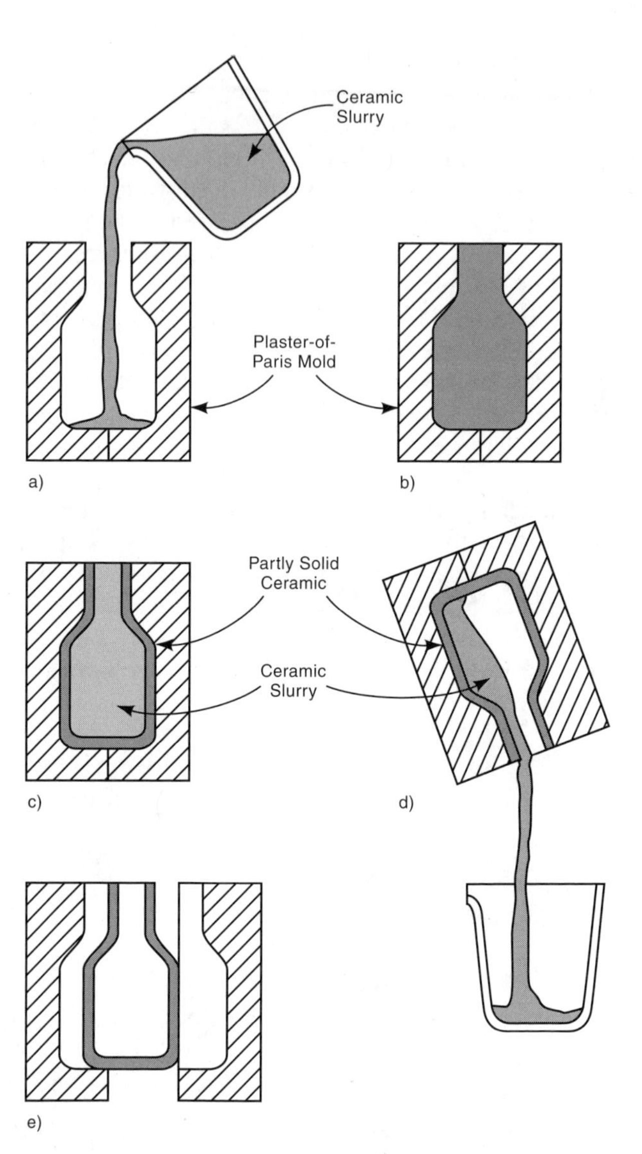

Fig. 5B8 Drain casting. a) Ceramic slurry is poured into a plaster-of-Paris mold. b) The mold starts to absorb moisture from the slurry by capillary attraction. c) The dryer ceramic material on the mold walls forms semi-solid walls of the ceramic workpiece. d) Excess slurry is poured from the mold, leaving a semi-solid workpiece in the mold. e) The mold is opened and the workpiece is removed for trimming, if needed, and further drying and firing.

B8a. ***pressure casting*** - applies pressure to the slip in the slip casting process to partly or completely replace the capillary action of the porous mold. This procedure enables the casting operation to proceed more quickly. More importantly, it results in a better part. Dewatering of the cast material is greater. There is less post-casting shrinkage, and the dimensions of the finished piece are more accurate. The part has an improved surface finish and greater strength. The process also allows some materials that are less suitable for slip casting to be processed. Pressures up to about 580 lbf/in^2 (4 MPa) are used. Molds are made from porous plastic polyelectrolytes, which can withstand the pressures involved.

B9. ***jiggering*** - is an advancement of the manual process used to form bowl-shaped or other hollow workpieces on a potter's wheel. In the current process, a heavy plaster of Paris mold is set up to rotate on its axis. A glob of paste material in placed in the mold, which is of the shape needed to form the outside surface of the part to be produced. Centrifugal force causes the material to flow to the mold walls. The process is aided by a tool, a jigger shoe, that descends either automatically or manually into the mold and shapes the inside surface of the desired part. The process, therefore, differs from slip casting since the wall thickness is controlled by the mold and jigger shoe and the inner surface of the ceramic part takes a more controlled shape. Fig. 5B9 shows the cross section of a porcelain fitting, and the mold and jigger shoe used to produce it by jiggering. Dinner plates are often made with this method. Sometimes, several parts, made by jiggering, are assembled together before drying and firing to produce a more complex part.

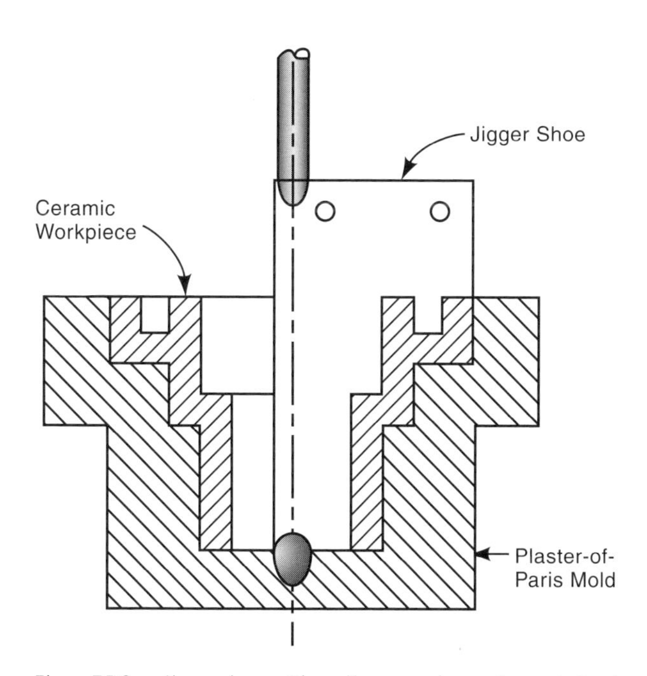

Fig. 5B9 Jiggering. The jigger shoe is rotated about the central axis to form the internal shape of the workpiece.

B10. ***roll compaction*** - uses a slurry of fine ground powders, fluxing agents, organic binders, and plasticizers. The slurry is sprayed onto a flat surface, and is partially dried into a putty-like material in sheet form. The sheet is fed into a pair of large parallel rollers that compact the sheet to a uniform thickness. The sheet is then still flexible enough that it can be reeled. In subsequent operations, the sheet is dried, blanked, and punched to the shape desired and then fired. Roll compaction, like tape casting, described below, is used for making small components for the electronics industry such as substrates for integrated circuits.

B11. ***tape casting*** - is another sheet production method. Slurry, made with an easily-evaporated organic liquid, is dispensed onto a moving plastic belt. The belt carries the ceramic material under a doctor blade in a position parallel to the belt surface. Excess material is held back by the blade while the belt carries the other material in a sheet of uniform thickness. The sheet is heated to remove part of the organic liquid and a carrier plastic sheet is applied. The sheet is then taken up on a reel with for later processing. Further drying, blanking, and punching to the shape desired, and firing then take place. The process is used to make ceramic substrates for integrated electronic circuits. Fig. 5B11 shows the operation.

B12. ***throwing*** - is a manual method for forming ceramic parts with circular cross sections. A glob or blank of ceramic material in paste form is placed on a revolving table or disk, called a potter's wheel. While it is turning, the glob is shaped with the hands and a variety of hand tools. The process is used for art work, craft items, and limited quantity production. For precision parts, shaping may be only to a rough shape, followed by machining, before firing, to provide more accurate surfaces and dimensions.

B13. ***drying*** - Several methods are available to remove moisture and other additives from ceramic workpieces before they are sintered. Room-temperature drying is

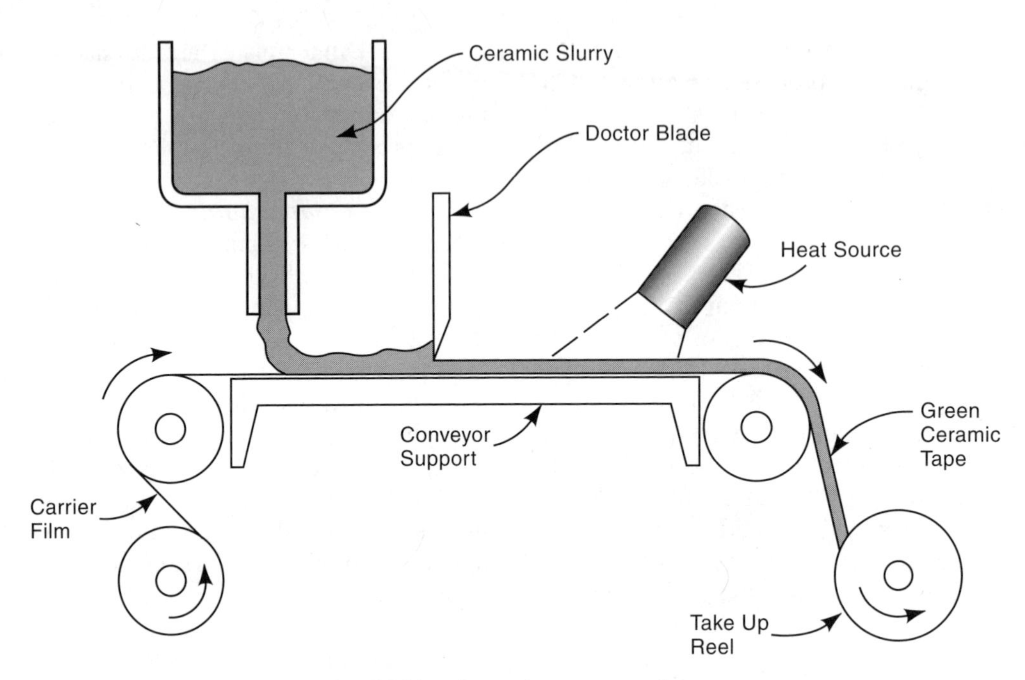

Fig. 5B11 Ceramic tape casting.

the simplest. After forming, some free water may be removed from workpieces that have been made with wet forming methods, by allowing them to stand for a period at room temperature. For further drying, workpieces may be allowed to dry for several days in a drying room or chamber. These facilities are kept at a high humidity or the workpieces are wrapped with plastic film so that the drying is gradual and even.

Under more elaborate production conditions, the workpieces are subjected to a drying operation under conditions of controlled heat and humidity. The ideal cycle in a batch drying chamber starts with high-humidity, cool air being blown onto the workpieces and ends with high-temperature, dry air after a gradual change over a period of time. High velocity of airflow is needed throughout the cycle. Various methods may be used to provide heat to facilitate the operation while still controlling the moisture removal rate. 180 to 210°F (80 to 100°C) is a typical temperature range for such a device when used to dry clay-based ceramics.

If the drying is a continuous operation, a drying tunnel is used and the workpieces proceed through it on a conveyor or on a series of material-handling buggies. The starting high temperature air enters the tunnel at the exit end. As the air flows toward the entrance end, it gradually gets more moist and cooler, so the first air blast hitting the entering workpieces is cool and moist and the air blast at the end is warm and dry.

Drying must be carried out in a controlled manner to avoid cracking, distortion, or excess internal stresses. The control is needed because ceramics shrink when moisture is removed and, at a certain point, are brittle enough for differential shrinkage to cause cracking. Drying is most important for workpieces formed with one of the wet forming methods such as casting, extruding, injection molding, and wet pressing. Drying helps remove organic binders and plasticizers as well as moisture from the workpiece. The degree of drying depends on the accuracy required in the final dimensions and what subsequent forming or shaping operations, if any, are required. Controlled drying is sometimes the first stage of the firing process.

B13a. ***partial in-process drying*** - In some cases, drying of wet-formed ceramic parts is performed in two stages. The first is a preliminary partial drying which puts the workpieces in the green state so that they can be handled and trimmed, or otherwise processed, before the final drying. Workpieces requiring trimming are often partially dried before the trimming operation.

B13b. ***final drying*** - Organic materials as well as moisture, must be removed prior to sintering. Otherwise, gases formed in the workpiece from the decomposition of organic materials or from steam may cause voids or other problems with the workpieces. Final drying, whether a separate operation or the first stage of sintering, is intended to remove these materials completely.

B13c. ***microwave drying (also called dielectric or radio frequency drying)*** - is used in production shops to speed the drying process. The process is similar to heating and cooking food in a microwave oven in that radio-frequency energy provides the heat that dries the workpieces. The process is much quicker than other drying methods because heat is generated within the workpiece and does not have to penetrate from the surface as it would if radiant, convection, or conduction heat sources were used. Greater heat is generated in those portions of the workpiece that are wettest. Hence uniformity of drying is promoted. Because the operation is faster than other methods, it minimizes the quantity of work in process in the drying cycle. Both batch and through-feed ovens are available.

B14. ***machining and grinding*** - Forming methods may not always produce the final shape wanted in ceramic parts, and the precision required in some dimensions may not be achievable from the forming operation. Machining of ceramics is possible and may be required in order to bring the workpiece to the shape and dimensions needed. Machining can involve turning, milling drilling, boring, threading, tapping, and other operations described in Chapter 3. These operations are normally carried out before the part is sintered, when it is in the "leather-hard" condition, after drying.

Carbide cutting tools are used because of the abrasive nature of the ceramic material. Such machining is made difficult by the weak and fragile nature of unsintered ceramics. There are also dimensional changes from firing which may make it difficult to achieve close dimensions of machined surfaces after sintering. If dimensional adjustments are needed after sintering, they may be effected by grinding or lapping with diamond abrasives. Drilling and cutting with diamond-tipped drills and saws is sometimes employed. Ultrasonic machining, laser- and electron-beam machining, and chemical machining, are alternative processes. Some ceramic materials have been developed to provide less difficult machining conditions, when machining of sintered material is necessary.

B15. ***glazing*** - Glazing material can be added to the workpiece before firing to provide a smooth, glossy surface. It is sometimes applied after firing and, if so, there is a second, lower-temperature firing. Glazing coatings are made up of glassy material or partially crystalline material in slurry form. Clay or organic binders may be added to the mixture to improve its adhesion to vertical workpiece surfaces.

B16. ***sintering (firing)*** - In sintering, the dried ceramic workpiece is heated to a high temperature for a specified period. Typical temperatures are around the 2500 to 3000°F (1400 to 1650°C) range. Some ceramic materials may require a temperature over 3600°F (2000°C) Alumina without glass materials included in the mixture, is fused at 3500°F (1930°C)[6].

There are two prime sintering mechanisms: solid state diffusion, when particles fuse together without melting (as with the pure alumina) and liquid-phase sintering. When glasses are included in the mixture, liquid-phase sintering takes place. The glasses melt and fuse the particles together at a lower temperature. Some surface melting of other particles in the material may also occur, providing a bond between the particles. Also, chemical reactions at the elevated temperature may create liquids that provide liquid-phase sintering. When glasses are not incorporated in the materials, diffusion is the usual means of sintering .

When sintering is completed with either mechanism, the particles fuse into a hard, strong, homogeneous, and dense state. Both heating and cooling phases take place very slowly and, though the actual sintering does not require an extended period, the total sintering process can require several days or longer. Careful control of temperature and temperature changes over time is important to insure proper sintering and to prevent adverse effects on the workpiece during the process.

As the material is slowly heated, several changes take place: The first is drying, if the workpiece is not already dried. Then, if the material contains water of crystallization, that is driven off at

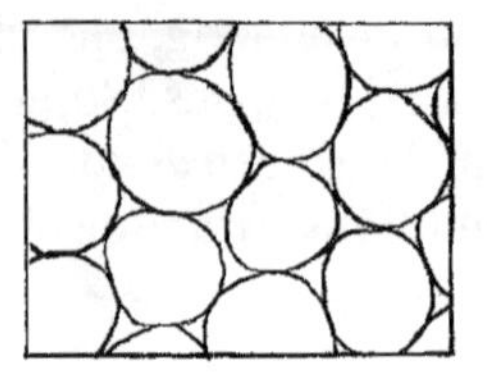

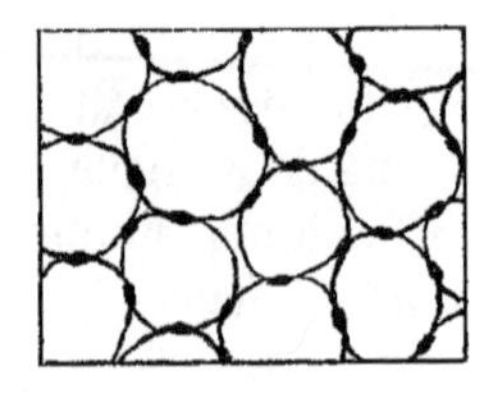

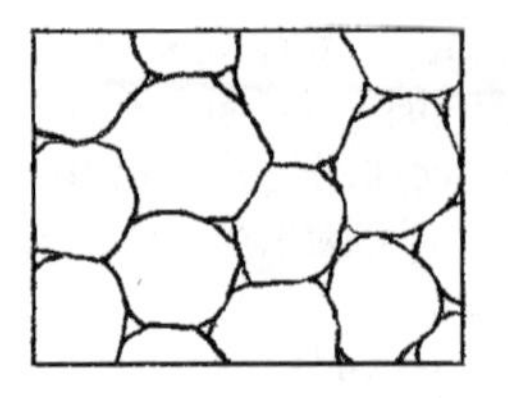

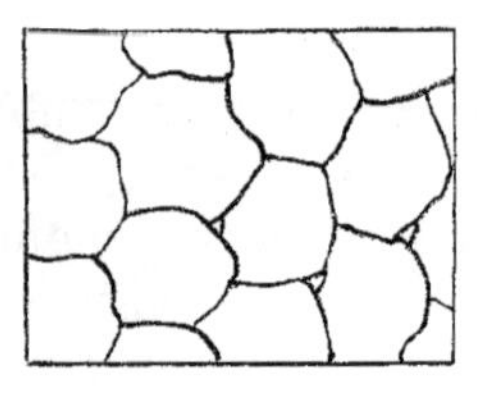

Fig. 5B16 The sintering process for ceramics. The shrinkage of the workpiece during sintering is due to the elimination of most of the pores between particles.

temperatures of 660 to 1100°F (350 to 600°C). The sintering is initiated by several different reactions or physical changes. Typical sintering action on the material grains is shown in Fig. 5B16. Considerable shrinkage may take place during sintering as material moves to fill the open spaces between grains.

B17. ***hot pressing*** - is used to make cutting tool inserts and other ceramic parts of simple shape. Dry, fine, ceramic powder or a powder preform is placed in a graphite die supported by ceramic parts. Pressure is applied at the same time that the ceramic workpiece is heated to the sintering temperature. However, the temperature is lower than that required for regular sintering because the combination of pressure and temperature aids in bonding the particles. The diagram in Fig. 5B17 shows a typical tooling arrangement. The shapes of parts that can be processed in this way are limited, but the parts, after pressing, are finished or nearly so. The process is used with some materials that cannot be densified by sintering without pressure. The process is costly. Microcrystalline ceramics with high strength properties can be produced, and the parts made are used in higher technology, structural applications.

B17a. ***hot isostatic pressing (HIP)*** - involves isostatic compaction of powders (and other materials) at an elevated temperature. A water-cooled pressure vessel with an internal high-temperature furnace is employed. Pressures reach about 45,000 lbf/in^2 (310 MPa) and temperatures about 3600°F (2000°C). Argon, nitrogen, or helium gas is pressurized and acts against the surfaces of the workpiece through a hermetically sealed glass or metal encapsulation. Because of the high temperatures involved, sheet metal, if used for encapsulation, must be refractory. Glass envelopes soften at the temperatures involved but still transmit the pressure to the ceramic material. Pressure and temperature are closely controlled. Electrical resistance heating is usually used. The method is advantageous for producing more complex shapes of parts than by regular hot pressing. Improved, more-uniform compaction for critical parts is an

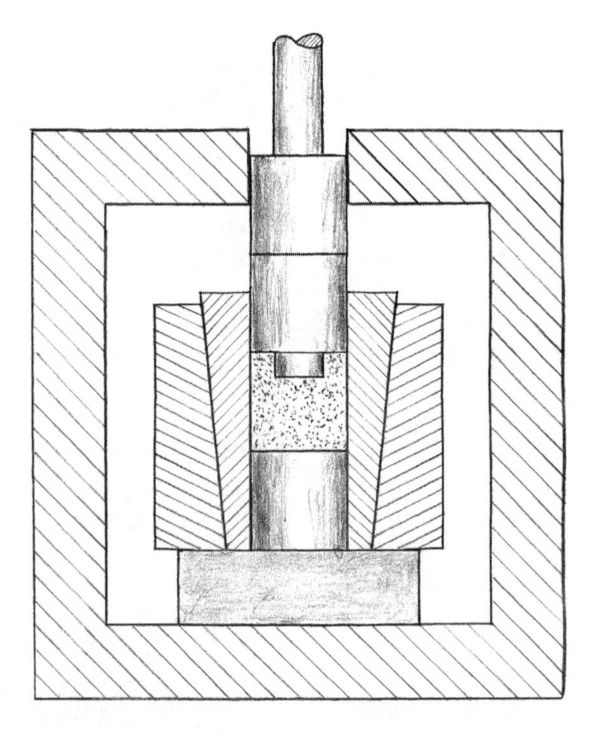

Fig. 5B17 Hot pressing of a part from ceramic powder.

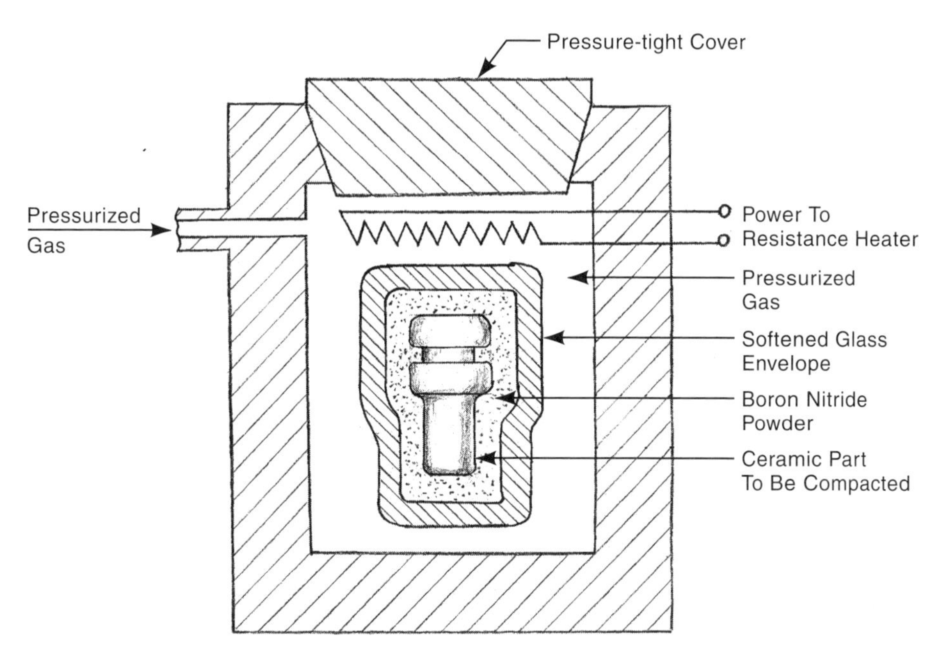

Fig. 5B17a Hot isostatic pressing.

important advantage. The process is applicable to powder metals and cermets as well as ceramic powders and is used to remove voids in castings for critical parts such as turbine blades, to compact powder metal parts to almost 100 percent density of the metal involved and to bond dissimilar materials together. Fig. 5B17a shows the process schematically.

References

1. *Encyclopedia of Physical Science and Technology, Vol. 6*, R.A. Meyers, ed., Academic Press, London, 1987.
2. *Schott Guide to Glass, second edition*, H.G. Pfaender, Chapman and Hall, London, 1996.
3. *The Handbook of Glass Manufacture, 3rd ed.*, F.V. Tooley, Ashlee Publishing, New York, 1984.
4. *Glass Engineering Handbook, 2nd ed.*, E.B. Shand, McGraw-Hill, New York, 1958.
5. *The McGraw-Hill Encyclopedia of Science and Technology, vol. 9*, McGraw-Hill, New York, 2002.
6. *Handbook of Ceramics, Glasses and Diamonds*, Charles A. Harper, ed., McGraw-Hill, New York, 2001.
7. *Modern Ceramic Engineering*, D.W. Richerson, Marcel Dekker, New York, 1982.

Chapter 6 - Woodworking Processes

A. Lumber Making, Including Saw Mill Operations

Logs are delivered to the saw mill and are sorted and stored according to species, length, diameter, and expected end use. Handling at the saw mill is by cranes, heavy lift-trucks, or derricks. Bark is removed and the logs are sliced into slabs, which become boards when their edges are trimmed and their ends are cut square. The boards are seasoned (dried) and given a smooth surface.

A1. ***debarking*** - is carried out by several different methods. One common approach utilizes self-tumbling of the logs in a revolving drum. The tumbling action of the logs against one another breaks the bark free from the logs. Other methods are the Cambio ring or Rosser head debarkers that chip the bark away using blunt knives or metal teeth mounted on a rotating ring or wheel. Other debarking machines have two or three debarking tools that rotate against the log as it is fed axially through the machine. Another method uses chain flails as the logs rotate and move through the debarking chamber. High pressure water is used in some sawmills, particularly when logs are irregular.

The Rosser head debarker rotates each log while a rotating tool, similar in appearance to a gear or wide circular saw (but not sharp enough to cut the wood significantly) bears against the log surface with hydro-pneumatic pressure. This debarking tool breaks the bark free as the log and the tool rotate. The log is fed axially so that the tool passes over the whole surface of the log. In modern sawmills, the operation is highly automatic but one operator oversees the operation and controls the conveyor that moves logs to and from the machine. Fig. 6A1 illustrates this machine in operation.

The purpose of debarking is to remove stones, grit, or other foreign objects that dull cutting saws, and to eliminate the bark from pieces that will be made into particle board and similar products. Another advantage of removing bark is that bark-free logs can be evaluated better for cutting into boards. Logs are fed from a feed chute and discharged after debarking to a conveyor that moves them to the next

Fig. 6A1 A Rosser head debarking machine in operation. (*Courtesy Connerstone Forest Products, Kingsley PA.*)

operation. Sometimes, logs are washed after debarking to remove any remaining sand, dirt, or other foreign material. The bark removed from the logs is utilized as landscaping mulch.

A2. ***headsaw operation (breakdown sawing)*** - is the first saw cutting of the log. In well-equipped saw mills, this is a semi-automatic operation, performed on a headsaw and controlled remotely by a skilled operator who is assisted by a computer. With computer assistance, the operator determines how each log will be cut. The particular pattern of the cut depends on the condition and size of the log and the need for boards with particular widths, thicknesses, and grain patterns. The headsaw includes a long moving carriage that holds the log and conveys it axially through the cutting blade and back past the blade for additional cuts.

Cutting is commonly done with a bandsaw blade in the larger production mills. (Some saw mills use a large circular blade or, for large logs, two circular saws, one above the other, arranged so that both cut in the same plane. Bandsaw blades are less costly than circular saws and have a narrower kerf which wastes less material as sawdust.) Sensors feed data on the log's shape to the computer. The computer calculates an optimum first cut to provide a flat surface on the log with a minimum amount of material loss. A red light, projecting a line on the log, shows the operator the location of the cut. The operator may overrule the cut suggested by the computer. Fig. 6A2 illustrates a present-day headsaw in operation.

Fig. 6A2 A head saw performing a squaring operation on a log. (*Courtesy Connerstone Forest Products, Kingsley PA.*)

There are three basic types of saw cuts, each of which produces a distinctive grain pattern. They are illustrated in Fig. 6A2-1. *Plain sawing* is the simplest, quickest and most common cutting method. All cuts are tangent to a growth ring of the tree and are parallel to one another. This system has the highest yield of boards per log, and provides boards that are easiest to kiln dry, but the boards are more susceptible to warping. The growth ring lines in the boards extend wide and have a V-pattern[2]. See view (a) of Fig. 6A2-1. *Quarter sawing* produces boards with mostly straight grain lines, fewer splits and checks and less warpage after drying than plain-sawn boards. However, these boards are usually narrower than with plain sawing and the yield of boards per logs is less. See view (b) of Fig. 6A2-1. *Rift sawing*, shown in view (c) of Fig. 6A2-1 is similar to quarter sawing and has a distinctive grain pattern.

If plain sawing is to take place, the log is squared. The slab produced by the first cut is conveyed away, and the log, which now has a flat surface along one side, is returned to the starting position. (A slab is a board with one flat surface and one curved surface from the near-cylindrical

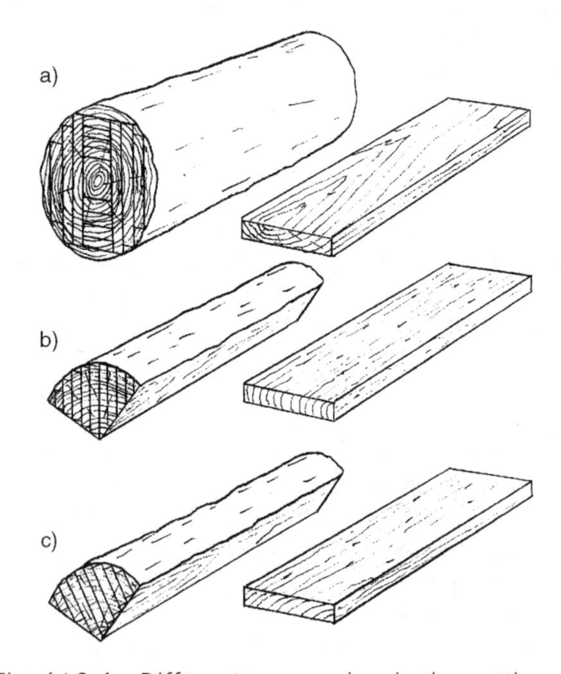

Fig. 6A2-1 Different approaches in the cutting of logs into boards yield differences in grain appearance and properties of the board: a) plain sawing, b) quarter sawing, c) rift sawing.

surface of the log.) The log is rotated a quarter of a turn about its axis and is moved to the saw blade for another cut. This action is repeated so that, after four cuts, the log is square or rectangular in cross section. The square is not so small that all surfaces are flat for their full width, because to do so would waste too much of the wood in the tree trunk. In most mills, the cant (squared log) is conveyed to a secondary saw for cutting into boards.

If the log is to be cut by quarter sawing or rift sawing, it is cut into quarters with one cut through the center of the log. Center cuts are then made to separate the two halves into quarters.

A3. ***plain sawing of boards*** - (view a of Fig. 6A2-1) may be done at the headsaw or on a secondary saw. In plain sawing, the cant (squared log) remains on the cutting table of the headsaw. The table is shifted sideways one board thickness and the table again traverses into and across the saw blade, cutting one board from the cant. This operation is repeated, perhaps after the cant is rotated about its axis to present a better face for cutting the next board. After each board is cut, it drops on a conveyor and is moved to the next operation. After a number of boards have been cut, the cutting reaches the area near the center of the log, which is apt to have more knots and other defects which yield boards of lower quality. This center portion of the log may be cut into lower-grade boards for use in pallets or crating, or use as timber for railroad ties[5].

Many sawmills limit the headsaw to preliminary cuts and do the cutting of cants or log quarters into boards on a secondary cutting saw. Such saws are similar to headsaws in having a carriage that carries the cant or log quarter through a saw blade. Fig. 6A3 shows a secondary saw and cants waiting to be sawed into boards. In high production mills, usually for softwood construction lumber or other large-quantity applications, gang saws with multiple blades that are spaced one board thickness apart may be used to cut two or more boards at a time. In some mills, especially for smaller and lower-grade boards, gang saws are used to cut cants completely in one pass.

A4. ***quarter sawing*** - uses larger logs. They are first sawn into two half-round logs and then into four quarter-round logs as shown in Fig. 6A2-1, view (b). Each quarter is then sawed into boards by making cuts at an angle from about 65 to 90 degrees from the growth rings. These cuts may be made successively with the headsaw or secondary saw or, if production volume is high, they may all be sawn at once with a gang saw.

Fig. 6A3 Cants (squared logs) being cut into boards at a secondary saw. *(Courtesy Connerstone Forest Products, Kingsley PA.)*

A5. ***rift sawing*** - Boards cut by this method are also cut from quarters, but the cutting starts at an angle of about 45 degrees from the growth rings as illustrated in Fig. 6A2-1, view (c). The properties are essentially the same as those of quarter-sawn boards with a slightly different grain appearance.

A6. ***edging*** - Since the edges of many sawn boards show the irregularities of a tree trunk, their edges must be trimmed. *Flitches* (boards with an irregular edge or edges) from the board cutting operation are conveyed automatically to a double band or double circular saw called an *edger*. One blade on each side cuts the edge of the board as it passes through the machine. The amount of cut on each edge is determined by a machine operator; the position of the saws is adjustable. The operator moves the saws to maximize the width of the boards. The cut edges are disposed of as wood scrap or ground into chips and used in the manufacture of oriented strand board, wafer board, or particle board, unless there is some portion that can be trimmed into a shorter narrow board as shown in Fig. 6A6.

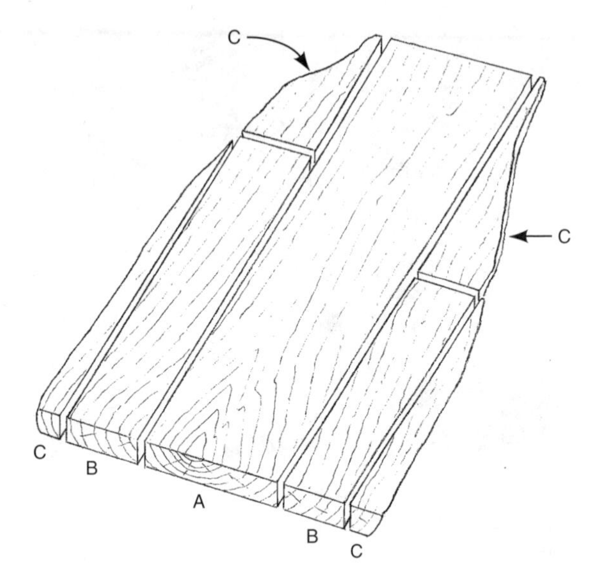

Fig. 6A6 Edge trimming of boards cut from logs. Part A is the maximum size board that could be cut from the flitch. Parts B are shorter and narrower pieces that could also be cut where the log was larger. Parts C are scrap which may be used as material for particle board or other wood-fiber products.

A7. ***trimming (cutting to length)*** - The ends may not be square since they result from the chainsaw cuts made when each log was cut from a tree. The full length of the board may not be usable because of defects in the wood or an irregular edge. Therefore, it is necessary to trim the boards to have square ends. The boards are crosscut to length at a *trimmer* saw by an operator who gets the longest board possible. Multiple-bladed saws may be used. Fig. 6A7 shows a multiple blade trimming saw in operation.

A8. ***inspecting and grading boards*** - After edging and trimming, boards are conveyed to a "green chain" inspection area where they are sorted by size (thickness, length, width), grade (quality, which involves freedom from knots and other defects, nature of the grain, and overall appearance), and species. In some saw mills, sorting is partly performed with mechanical equipment. Grading and inspection standards vary considerably, depending on the expected end use of the boards. Softwood and hardwood standards are considerably different because of their different end uses. Fig. 6A8 illustrates this operation.

Fig. 6A7 The trimming operation with a multiple blade saw. The ends of the boards are trimmed square and defective or unfinished sections are removed at this saw. The operator determines, with computer assistance, for each board, where the cuts are made. (*Courtesy Cornerstone Forest Products, Kingsley, PA.*)

Fig. 6A8 Inspecting and grading green lumber. (*Courtesy Cornerstone Forest Products, Kingsley, PA.*)

A9. ***drying (seasoning)*** - Harvested wood normally has a much higher moisture content than is desirable in wood products. If milled and used when "green" (freshly cut), wood pieces will shrink considerably and often warp, check, crack, or split. A typical desired moisture content for furniture making is 6 to 11 percent by weight, whereas green lumber will have values of 50 percent or higher. Seasoning involves either air or kiln drying of the cut lumber or a combination of both. The operation must be carefully controlled in order to dry each board uniformly throughout its thickness, and to minimize shrinkage stresses that create these defects. Drying ensures better dimensional stability of the wood, better color and strength, and lower transportation costs because of reduced weight.

A9a. ***air drying***[1] - is performed by stacking cut boards out of doors. Ends of the boards may first be coated with wax or paint by brushing, spraying, or dipping, to protect them against fungi or insect attack and to limit too-rapid drying at the ends, which can lead to splitting or checking. The bottom of the stack is elevated above ground about one foot. Each layer of boards is separated from the layer below by spacer sticks 1/2 to 1 in thick but of uniform thickness, spaced 12 to 2 feet apart. (These dimensions depend on the thickness of the boards and the season.) The separations allow free air movement around each board. The spacer sticks are placed so that the boards are kept straight and well supported to minimize warpage. The stack is usually protected from direct rain and sunlight, ideally with an overhanging roof. Drying must be slow so that the outer portions of each board do not dry too much while the inner portions are still wet, otherwise, splitting will occur. Drying times vary with the climate and type of wood. Hardwoods may require 9 or more months; softwoods may be satisfactorily dried in three to six months. Pines and other light woods dry faster. (Drying is faster in the warmer months of the year.) Typical moisture content after air drying is about 20 percent. The target level for construction lumber is 19 percent[2]. Ideally, for cabinet work, hardwood moisture content should be about 8 percent for most areas in United States, but as low as 6 percent in the arid Southwest and up to 11 percent in humid areas.

A9b. ***kiln drying*** - takes place in a large, heated, and well-insulated chamber or building. The operation is much faster than air drying but must be carried out very carefully in several steps to avoid distortion and splitting of the boards. The boards are neatly stacked, similarly to the system used in air drying. The operation may be carried out on a batch or continuous basis where the stacked boards proceed through the length of the kiln on trucks. The heated air in the chamber is controlled for temperature, humidity, and circulation. Large fans are employed. At the start of drying, the air has a higher humidity, achieved by injecting live steam into the system, but only a slightly higher temperature. This arrangement prevents rapid drying of the surface wood, which could cause the boards to crack and split. The humidity is gradually lowered and the temperature raised as the wood dries. Sample boards are checked for moisture content to provide information to aid control of the rate of change in temperature and humidity. The kiln operator also cuts samples to evaluate the amount of stress in the wood. As a last step, some additional humidity is added to prevent the surface of the wood from becoming too dry and "case hardened". A successful kiln operation provides consistent moisture content in both the interiors and surfaces of the boards. Drying to 10 per cent moisture content is normally achieved with hardwoods in 3 to 12 weeks. Often kiln drying takes place after air drying and, in such cases, the kiln cycle time is normally 1 to 4 weeks. After kiln drying, the boards are usually stored in a warm area; otherwise, the moisture content will revert to a normal equilibrium value of 15 to 22 per cent[1]. Wood scrap is commonly used to fuel the kiln. Fig. 6A9b illustrates a typical kiln drying operation.

A9c. ***radio frequency drying*** - In this process, stacks of boards to be dried are placed between two metal plate electrodes. Power is applied by high-frequency current that induces molecular activity in the wooden boards, generating internal heat. The amount of heat generated depends on the level of moisture in the wood. The operation is usually performed in a vacuum chamber, which reduces the temperature level needed. Lumber and wood components can be dried in much less time than with other techniques and quite evenly with this method. When processing lumber, separation strips between layers of boards are not needed. The approach is also used in plywood production, providing more rapid curing of the adhesives that bond the veneer layers.

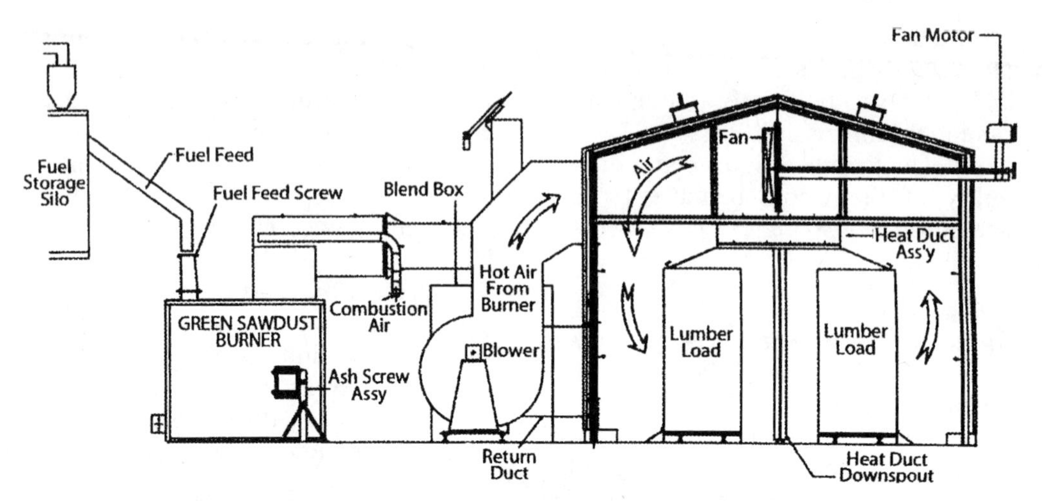

Fig. 6A9b Kiln for removing excess moisture from lumber. This particular kiln was designed for softwood, but hardwood kilns are quite similar. Note the sawdust burner, which provides heat for the operation. (*Drawing courtesy of USNR.*)

B. Making Wooden Components

Fig. 6B shows an assortment of equipment used to machine wooden components for cabinets, furniture, and other wood products.

B1. ***cutting boards to size*** - usually involves a sequence of ripping and cross-cutting operations. However, when CNC routers or laser cutters are used, especially with smaller pieces, and with fiber board or plywood, the operation may be done in one computer-controlled step.

B1a. ***ripping*** - cutting of board to a desired width along the grain, is performed in low-quantity shops as a manual operation, commonly with a table saw set for ripping. The board is fed along a rip fence, which is set parallel to the circular saw blade, a prescribed distance from the blade. For greater quantities, two or more parallel cutters may be used in a gang saw so that boards fed to it are trimmed at the edges and the boards exiting from between the cutters are of the desired width. Higher production ripping is more sophisticated. Fig. 6B1a illustrates a computer-controlled, high production rip saw, part of a system that adds computer-controlled processing to the operation. The five-cutter saw includes a scanning system that inspects each board before it is ripped. From data on the length and width needed, and quality inspection criteria, all entered beforehand, the computer program maximizes the yield from each board. It plans the finished cuts to allow for the removal of defects and even aids in matching ripped boards of similar color if instructed to do so.

B1b. ***cutting to length (crosscutting)*** - cuts a board across the grain to the length desired. The operation may be performed under manual control on a table saw, a radial arm saw, a miter saw (See Fig. 6B.) or a chop saw. A semi-automatic chop saw is illustrated in Fig. 6B1b-1. Fig. 6B1-2 shows a highly automatic machine that reads an inspector's markings to cut out defects, cuts the board to lengths that optimize lumber utilization and marks each cut board to identify it. In high-production situations, when boards have untrimmed ends, they are conveyed against two parallel circular saw blades spaced so as to cut the board to the desired length. Such an operation is shown in Fig. 6B1b-3.

B2. ***surfacing with planers and jointers*** - These operations remove the surface roughness that remains when boards are cut from logs, produce lumber of standard thickness and squareness, and correct for some warpage (at the expense of some loss of thickness).

B2a. ***planing (surfacing)*** - is the operation that smooths and flattens the wide surface of

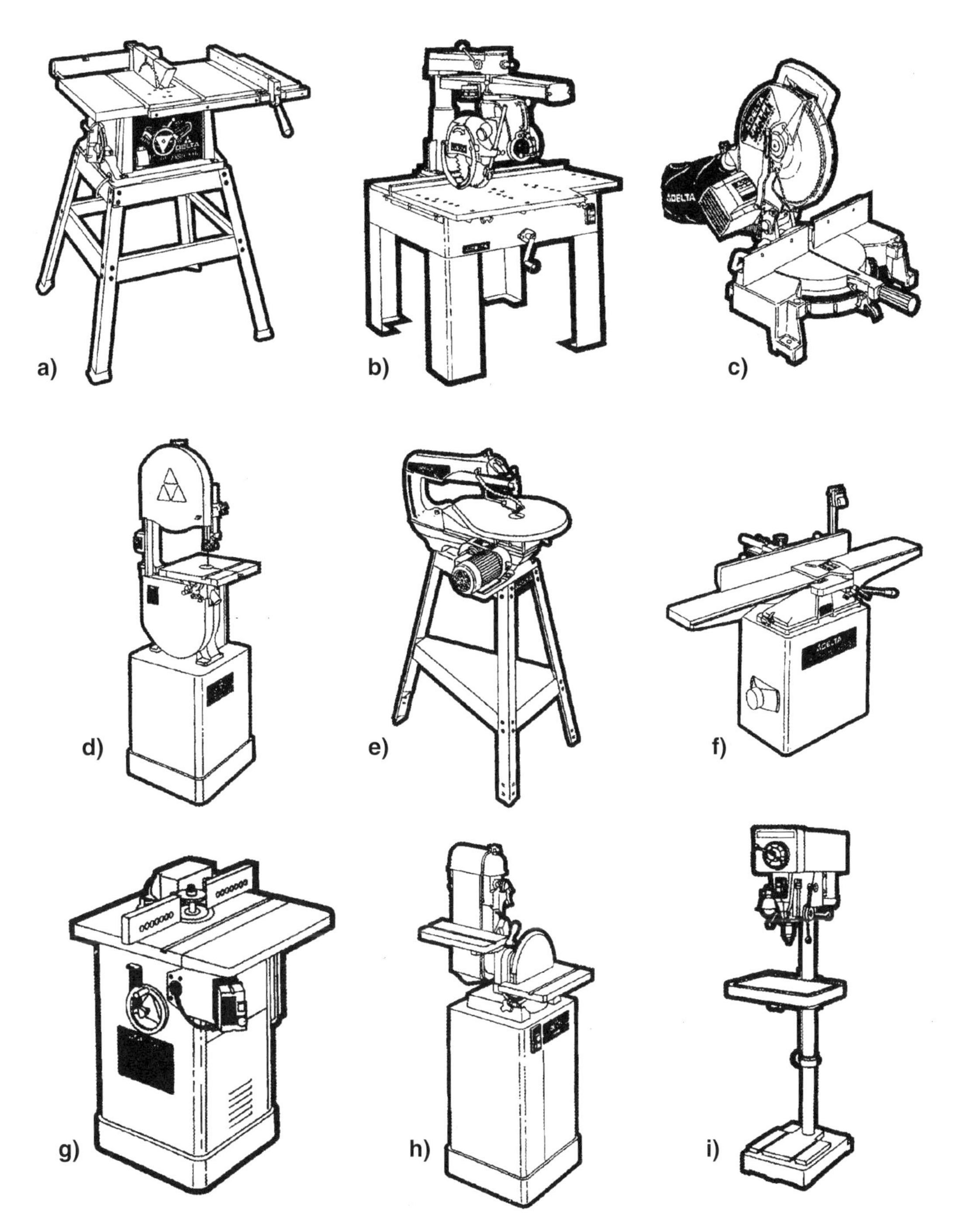

Fig. 6B Common equipment used in machining wooden components: a) table saw, b) radial arm saw, c) miter saw, d) band saw, e) scroll saw, f) jointer, g) shaper, h) combination belt and disc sander, i) drill press. (*Courtesy Delta Machinery.*)

Fig. 6B1a Ripped boards exiting a multi-blade, computer-controlled, rip saw. (Computer functions are described in the text.) (*Courtesy Donald Dean and Sons, Inc., Montrose, PA.*)

Fig. 6B1b-1 A chop saw. The operator keys in the workpiece length wanted in the key pad, and the machine then moves the work stop to the proper position. The circular cutting blade moves against the workpiece from underneath. (*Courtesy Donald Dean and Sons, Inc., Montrose, PA.*)

Fig. 6B1b-2 A computerized cut off saw. The machine reads an inspector's markings on the piece to be cut and cuts out defects. The length to be cut is then chosen by the computer from a number of preset possibilities to maximize the utilization of the lumber. Each piece cut is marked automatically with its length and identification. (*Courtesy Donald Dean and Sons, Inc., Montrose, PA.*)

Fig. 6B1b-3 A dual cut-off saw. The machine cuts both ends of a workpiece to provide a precise length and square, smooth, ends. (*Courtesy Donald Dean and Sons, Inc., Montrose, PA.*)

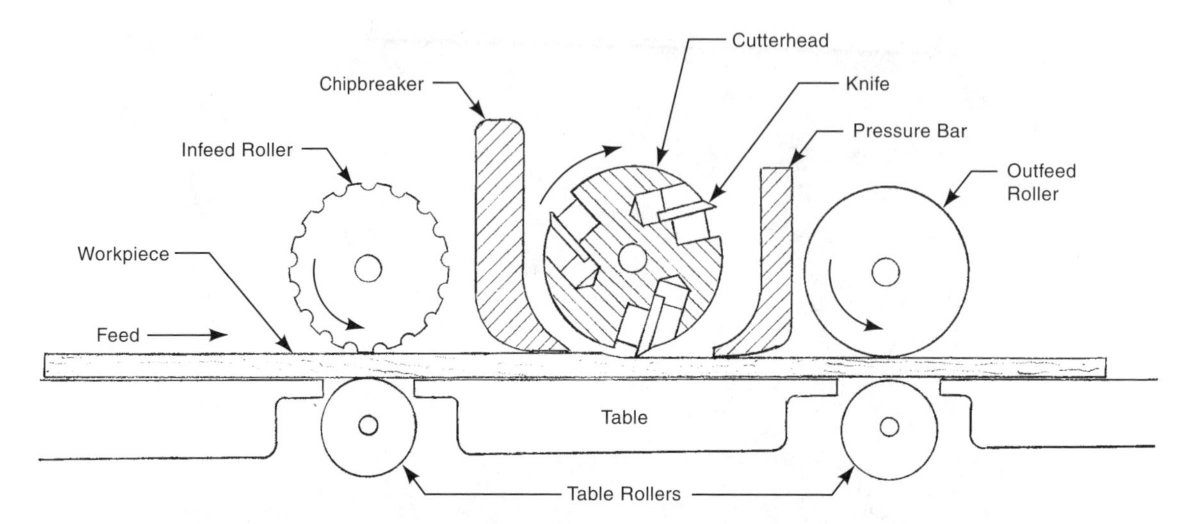

Fig. 6B2a Drawing showing the planing operation that leaves a smooth surface on a board and machines it to a prescribed thickness.

boards. A wide, multi-tooth cutter removes material across the surface of the board, eliminating the surface roughness and producing a board of standard thickness. Fig. 6B2a illustrates the principle of planer operation. Boards are fed into one end of the planer and are thereafter mechanically pulled through by the machine and discharged at the other end. Devices in the planer hold the board securely against the machine table at the point of cutting. Cutting heads of planers are as wide as 48 in (1.2 m). The depth of cut is normally 1/16 in (1.6 mm) or less. Boards as thin as 3/8 in (10 mm) can be planed as is and thinner boards can be processed if attached to a backing board. Fig. 6B2a-1 shows a simple portable planer used for craft work and Fig. 6B2a-2 shows a high-production planer system.

Fig. 6B2a-1 A planer used for craft work. (*Courtesy DeWalt.*)

B2b. ***jointing*** - A jointer is a machine similar to a planer, but is more versatile. It can make smoothing and squaring cuts on the edges and ends of boards as well as the wide surface. It is normally manually operated. The board is guided through the machine manually, in contrast to the planer which moves the board automatically. The jointer has a narrower cutting head than a planer, normally 4 to 12 in (0.1 to 0.3 m). The cutter head is below the table, which is divided at the cutter into two sections, the infeed table and the outfeed table. The outfeed table is slightly higher than the infeed table because of the depth of the cut. A fence provides side guidance and support for the workpiece. Work is fed into the jointer by sliding it along the infeed table and against the fence. Jointing is illustrated schematically in Fig. 6B2b, and Fig. 6B(f) shows a jointer. The fence is usually set at a 90-degree vertical position so that jointed edges or ends are square to the board's surface, but it can be tilted for angle cuts. Jointers, like planers, normally remove 1/16 in (1.6 mm) or less

a)

b)

Fig. 6B2a-2 A high production planer processing boards coming from the drying kiln: a) Feeding mechanism. b) Planed boards exit this end of the machine. (*Courtesy Donald Dean and Sons, Inc., Montrose PA.*)

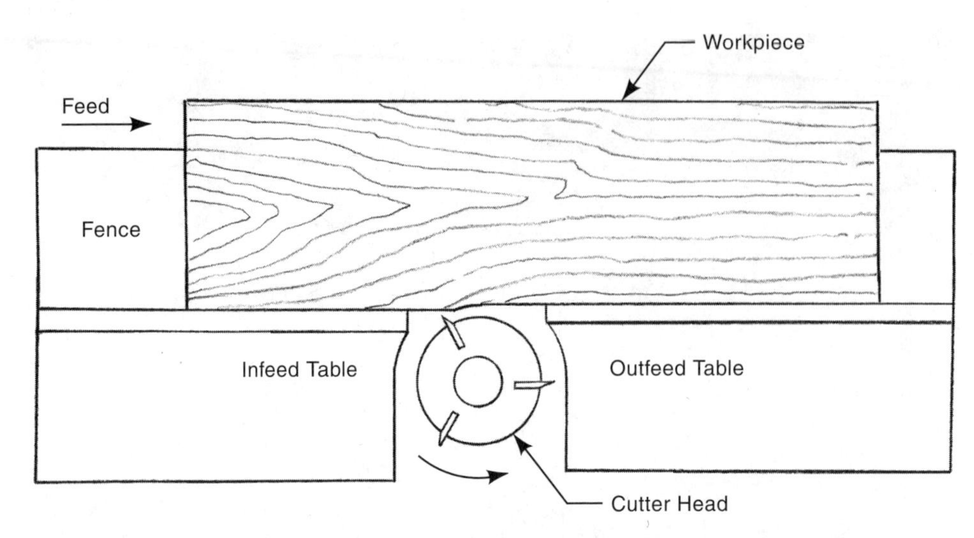

Fig. 6B2b Jointing puts a smooth edge on a workpiece. The operation is commonly manually controlled.

but, when a board end is jointed, the depth of cut is usually 1/32 in (0.8 mm) or less. Jointers are better for removing warp such as bow, twist, or cup, in a board, because the planer's feed mechanism presses the board flat for cutting, after which it can spring back to its warped condition, while the jointer can be fed without pressing the board and can more easily remove the high areas from a warped surface.

B3. ***contour (curved line) sawing*** - is traditionally performed on a band saw or scroll saw. Both machines have narrow blades. The workpiece is moved manually so that the blade follows a curved line traced on the workpiece. A typical woodworking band saw, which uses a continuous thin steel blade in the form of a loop that passes over two wheels, is shown in Fig. 6B(d). A scroll saw, as shown in Fig. 6B(e) and, in operation in Fig. 6B3, uses a very narrow reciprocating blade. Band sawing is further discussed in sections 3G2 and 3G3. Other methods of cutting curved lines involve routers, which follow a template or computer program, or laser cutters, which are also computer controlled. Machines of these types are particularly well adapted to cutting out grid and filigree parts. With both band and scroll saws, the narrower the blade, the sharper the curves that can be cut. Scroll saws are used for intricate, decorative cuts; band saws, with wider blades, for more gentle contours. Although intended primarily for contour cutting, the band saw can be used for ripping and cutoff operations. Both band and scroll saws can make bevel cuts if the worktable is tilted.

AU: Please update figure 6B3.

Fig. 6B3 A scroll saw is used for intricate decorative and other curved cuts in thin stock. (*Courtesy DeWalt.*)

B4. ***turning*** - is the operation used for making parts with round cross sections, and is performed on a lathe. Woodworking lathes are available for both high-production applications and for use in small cabinet shops

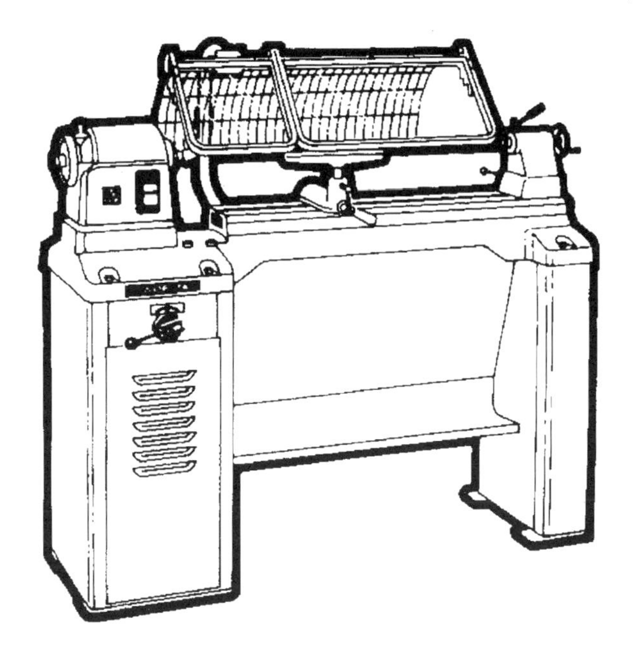

Fig. 6B4 A lathe for manually-controlled turning operations. (*Courtesy Delta Machinery.*)

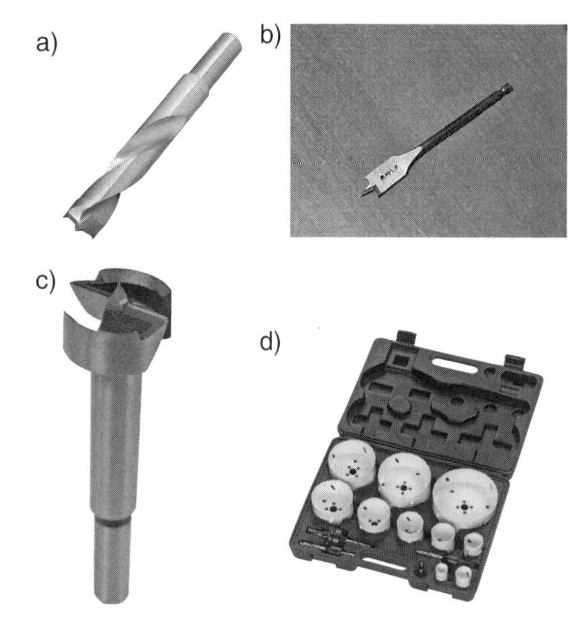

Fig. 6B5 Various wood drills: a) a carbide-tipped Brad point drill, b) spade drill, c) Forstner or power bit, d) hole saws. (*Note: Views a), c) and d) copyright WMH Tool Group, Inc. All rights reserved.*)

and home workshops. As the workpiece rotates, held and driven by the machine's spindle, cutting tools make contact with the surface and remove material to leave a round cross-section. Furniture legs, baseball bats, lamps, bowls, round tables, and other round wooden objects are made on lathes.

Production lathes may be controlled with servo mechanisms, which guide the cutter to follow a template, but current state of the art utilizes computer control to guide the cutter along whatever path is needed to produce the desired shape of part. In manual lathes, the operator may follow a template with a hand-held cutting or shaping tool, or may operate completely free hand to produce an original shape. Fig. 6B4 illustrates a woodworking lathe used for manually controlled turning.

B5. ***drilling and boring*** - In principle and in basic method, making round holes in wood components involves the same methods as described in section 3B (Machining-Drilling and Boring). Wooden workpieces can be drilled or bored with metalworking equipment. In practice, however, the easier cutting of wooden components compared to metal, and the lower precision normally needed, makes the methods somewhat different, particularly with respect to the type of cutting tools used. Fig. 6B5 illustrates a number of cutting tools used to produce holes in wooden components. The ease of drilling wood enables hand electric drills and even manually-powered drills to be used in low-quantity craft woodworking.

B6. ***routers and shapers*** - are machines for cutting shaped edges on wooden components. Usually the shaped edge is for decoration but it may also be part of the means used to join two pieces. One common use of shapers is to cut moldings in strip material. Another purpose is to engrave designs into surfaces. Fig. 6B6 illustrates a number of edge forms and the cutters that make them. The machines used to make these cuts are spindle shapers and routers. The terms are somewhat interchangeable but, usually, shapers are stationary machines with vertical spindles, normally driven from below a horizontal table. The workpiece is moved against the cutter. Machines or powered hand tools in which the cutter is above the workpiece, with the workpiece fixed and the cutter movable, are most commonly referred to as routers. Both types of machines have high speed spindles (4000 to 10,000 rpm) to drive the cutters.

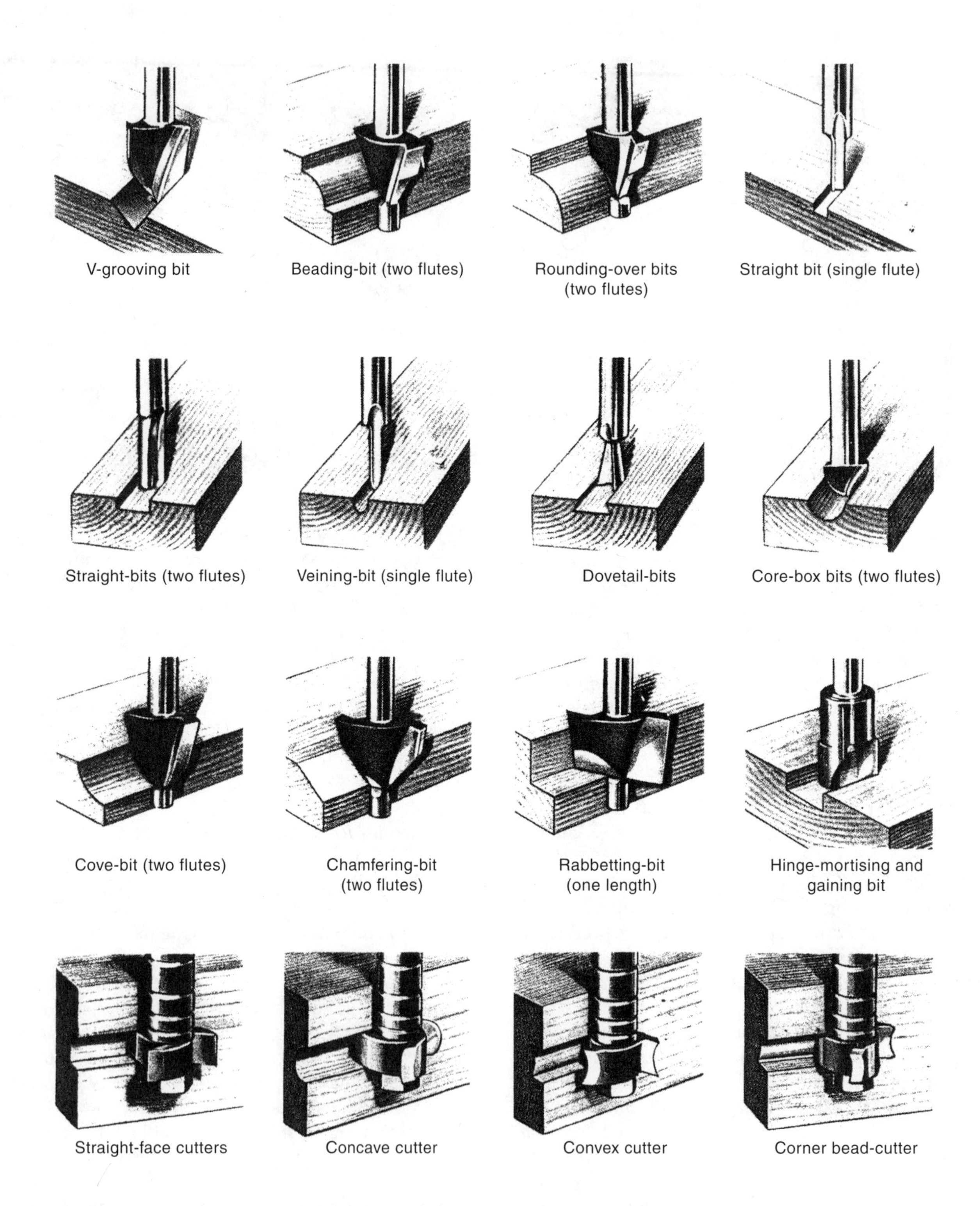

Fig. 6B6 Various router and shaper form cutters and the edge shapes they produce. (*Courtesy Black and Decker.*)

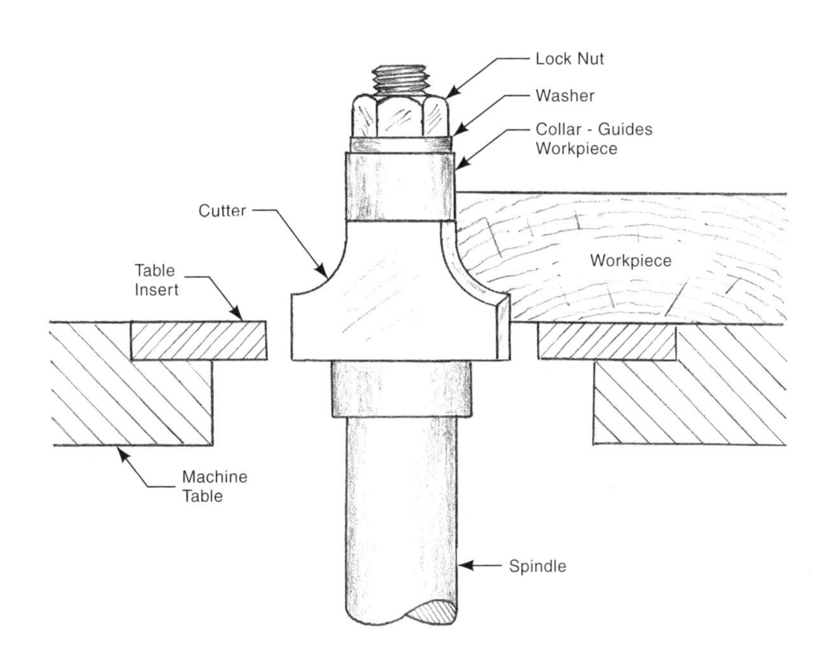

Fig. 6B6-1 A typical router cutter with a collar to guide the cutter along the edge of the workpiece. The collar may alternatively bear against a template instead of the workpiece.

The high speed of the cutters permits smooth surfaces to be produced with little sanding[2].

When edges are to be shaped, the tool is provided with a non-cutting collar, that guides the cutter by bearing against the edge of the workpiece. It prevents the cutter from cutting too deeply and changing the outline of the workpiece. Fig. 6B6-1 illustrates how the collar is placed on a typical cutter in a shaper. For straight-line or circular edges, fences on the machine can be used to guide the cutter. Raising or lowering the cutter may change the form that is machined. Sometimes several passes are made with different cutters or with the same cutter in different positions, to produce intricate or special edge forms. Sometimes a template, a separate-piece pattern, is used to guide the shaper. When a template is used, the edge of the workpiece does not have to be finished smooth or to the final dimension before shaping. The template guides the shaper to machine the part to the finished size and shape. More recently-developed equipment uses computer control to guide router cutters in straight or contoured paths. It also controls the height of the router cutter and thereby the shape of the edge it produces. Fig. 6B6-2 illustrates such a machine, designed specifically for cabinet doors.

Fig. 6B6-2 The workpiece left on the machine table illustrates one cut made with this computer-controlled shaper. Any one of three cutters on the spindle can be selected and set at the desired height to rout the edge or engrave a design in the workpiece. The path of the tool can be straight or contoured as the sample piece indicates. The operator loads and unloads the machine, but the operation is otherwise automat-ic.(*Courtesy Donald Dean and Sons, Inc., Montrose PA.*)

Another machine that is used for engraving and for edge cutting is called the *overarm router*. This machine has the cutter spindle mounted vertically above the table. A pin protrudes slightly from below the table surface, and in alignment with the cutter spindle and, when used, engages a groove in the workpiece or workpiece holder to guide the workpiece is during the cutting. The cutter descends into the work to make the cut and the workpiece is moved in the lower groove to create the design in the surface. The pin is not used if a fence is used to guide the cutter, when free-hand engraving takes place, or when the edge of the workpiece is used to guide the operation.

Another method used to produce these edge forms is with a *hand router*, a manually-operated power tool that brings the cutter to the work. The router has a bottom surface that bears against the workpiece. A spindle and cutter and, for edging, a collar, extend from the router surface. The workpiece is clamped to a worktable and the operator moves the router along the edge of the workpiece to shape its edge. Various guides are used to make straight and circular cuts. Templates can also be used.

Table saws, radial arm saws, and drill presses can be used to make moldings and formed edges of workpieces if the machine used is equipped with the proper form cutter and the necessary guiding devices are employed.

B7. ***other form, and special cutting and joint-making operations*** - Special machines are also used to make moldings used in furniture and as architectural treatments. Fig. 6B7 illustrates one such machine suitable for machining modest quantities of straight or curved moldings of various cross sections. Fig. 6B7-1 shows a high-production machine capable of machining 100 ft of finished molding per minute.

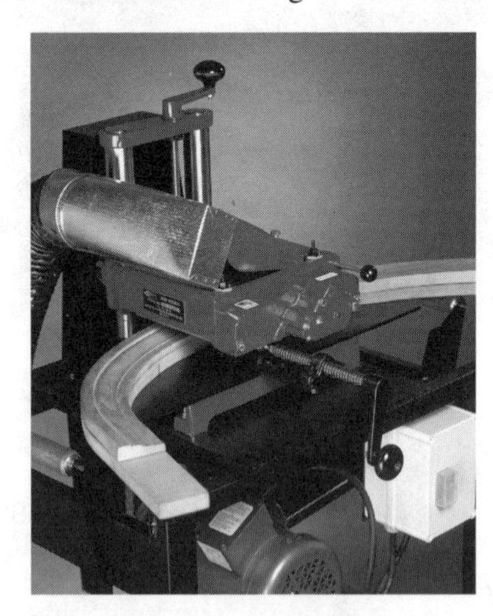

Fig. 6B7 A special machine for making wood moldings used in furniture and in architectural applications. The illustration shows a variable-radius curved molding but other curves and straight moldings can also be produced. This machine uses two flat form tools on the same spindle to produce the desired cross section. (*Courtesy Williams and Hussey Machine Co., Inc.*)

B7a. ***computer-controlled routering*** - Computer control is particularly useful for cutting curved components, those of other irregular shapes, and those that include engraving. It is used to cut cabinet and furniture parts from large sheets of plywood, particle board, or glued-up panels of solid lumber. With computer control (or CNC for "computer numerical control"), the pattern to be cut is stored in the computer's memory or in other devices that store digital data. No templates or other guidance devices are needed. Automatic production is feasible for production of only one piece or for a lot of thousands. The typical computer-controlled router utilizes a fixed table, which may be as small as 8 × 8 inches (20 × 20 cm) or as large as 5 × 12 ft (1.5 × 3.6 m). Mounted above the table is a bridge-like support that holds a powered router with a vertical spindle. The bridge-like support can move a controlled distance along the length of the table. It carries a horizontal slide so that the router can move a controlled amount across the table. The router is mounted on a vertical slide and moves upward and downward under computer control.

Fig. 6B7-1 A high-production molding machine. The machine has an automatic feed and is capable of machining 100 feet of finished form moldings per minute. (*Courtesy Donald Dean and Sons, Inc., Montrose PA.*)

All three motions can occur simultaneously if the shape of the part requires it. Such machines are similar to metalworking milling machines or machining centers as described in Chapter 3, but they are of somewhat lighter, less rigid construction because of lower cutting forces and somewhat less stringent dimensional tolerances that characterize the usual wooden components. As a result, these routering machines are also less costly than metal-working machines of comparable size. Three-axis control is standard, but machines can be equipped with additional axes by making the router mounting adjustable, so that the router can be swivelled and fed at an angle under computer control. The router bit can be of any shape used with manually- or template-controlled routers. Personal computers provide the computer-control and store the programs used. A typical computer-controlled routering machine is shown in Fig. 6B7a cutting out cabinet parts from a plywood sheet.

These machines are used on softwoods and hardwoods, plywood, particle board and on some non-ferrous metals. Typical components machined are cabinet doors, curved moldings, paneling, staircase parts, decorative shelf supports, all kinds of curved furniture parts including those used in upholstered furniture, signs, plaques, wood carvings, and bas relief objects.

B7b. ***laser machining of wooden parts*** - (See section 3O and Fig. 3O.) Computer-controlled lasers are primarily used for contour cutting of thin stock but are also applicable to straight line cutting. CO_2 lasers are used because other types do not provide a beam that is sufficiently absorbed by wood. A high level of accuracy is feasible with laser cutting but the heat of the laser discolors the cut edge and lasers are not normally used to cut out the usual cabinet and furniture parts. Most applications involve the production of decorative pieces of thin stock and veneer, especially when shapes are intricate. Veneer inlays, puzzles, craft pieces, and guitar body parts are laser cut. Mounting boards of fiber board or maple for steel-rule blanking dies are cut with lasers. Another application is the engraving of designs and lettering into wood plaques, trophies, gifts, and panels. Very detailed engravings can be made. Cutting speeds depend on the power of the laser and the type and thickness of the wood, and range from about 15 in to 13 ft (0.4 to 4 m) per minute. A thickness of 1 in (25 mm) is a practical maximum for laser cuts in wood.

Fig. 6B7a A CNC router cutting cabinet parts from a panel of plywood. (*Courtesy Thermwood Corp.*)

B7c. ***dedicated special machines and multi-operation machines*** - When production quantities are large, special machines are often developed and used to produce common wooden parts. Operations like cutting off and drilling, or cutting off and automatic machining of mortises, tenons, and biscuit slots, may be performed on multiple-cutter machines. Since cutting forces for wooden components are modest, these machines do not have to be as rigid as metalworking equipment and are more easily developed. Air cylinders or mechanical linkages and gearing may power the movement of cutting heads. Products and components made with special machines include tool handles, coat hangers, clothespins, chair legs, picture frames, and pencils.

Some machines are made to perform several operations on families of similar parts. Fig. 6B7c shows one such machine. It machines a form on the edge of cabinet components and finish sands the surface at the edge to remove any burrs or sharp edges. Other multi-operation equipment machines the pockets for hinges and drills screw holes to fasten them, in one automatic operation. Sockets for clip fasteners used in ready-to-assemble furniture are made with similar machines.

B8. ***filing and sanding*** - are operations for smoothing or otherwise modifying the surfaces of wood components after they have been cut to size. Sanding

Fig. 6B7c A combination shaper and sanding machine that cuts a desired shape in the edge of the workpiece and then sands the area near the cut to remove burrs and sharp edges. *(Courtesy Donald Dean and Sons, Inc., Montrose PA.)*

is an abrasive machining operation that utilizes paper or cloth coated with abrasive grains. The operation is normally performed with the aid of powered equipment, which may be hand operated, bench-mounted, or highly automatic. The purpose of sanding is to remove saw tooth and other cutting tool marks, ripple effects from jointers and planers, grooves, dents, etc. Bench-mounted abrasive machines most often utilize continuous abrasive belts backed up by flat or cylindrical supporting surfaces. Exceptions are narrow belt and flap sanders, which rely on belt tension or centrifugal force to provide pressure for the moving abrasives against the wood surface. Flat belt, drum, and disc sanders all may be used. Sanding motion is always along the grain except when board ends are sanded; otherwise, the cutting marks left by the abrasive grains may be visible. Sanding often involves several passes, first with a larger grained abrasive, then with finer abrasive.

C. Making Wood Joints

Several types of joints are used in making cabinets, other wood furniture, and other wood products. Joints of some complexity are relatively easy to produce with wood products because of the ease of machining wood and the precision of production woodworking equipment. The simplest wood-to-wood joint, a *non-positioned joint*, however, is when two wood surfaces come together face to face, with no additional elements on either part to position the parts or hold them together. These simple butt joints include edge-to-edge and surface-to-surface joints, or plain miter joints. They are held together with adhesive or some type of mechanical fastener. The strongest non-positioned joints are those where both the mating surfaces are along the grain of the mating wooden parts, i.e., surfaces or edges. Board ends involved in one or both parts make the joint considerably weaker.

A second joint type, the *positioned joint*, has components that have been machined to fit together in a particular way and may have shapes that lock together. The mortise and tenon, dovetail, and locked miter joints are examples. These joints are stronger than the non-positioned joints because of the holding elements and the larger area of adhesive bonding that is inherent in their shapes. They often also provide more bonding surface along the grain. They aid in proper alignment of the parts to be joined, and hold them together during assembly and gluing. Interlocking joints hold the parts together even if the glue joint should fail.

The third type of joint is the reinforced joint, which uses some extra component for support, in addition to the joined components. Dowels, splines, plates, glue blocks, and "biscuits" are examples of such wooden reinforcing parts, but metal fasteners such as nails, screws, plates, staples, and corrugated fasteners also fit this category.

Most of these joints can be made with the equipment discussed above, though special attachments or apparatus may be advisable in some cases, particularly when interlocking of well-reinforced joints is involved.

C1. ***butt joints*** - The edges of boards to be assembled edge-to-edge or edge-to-surface are first planed or jointed to make them smooth and straight. (See planing and jointing above). Simple glued butt joints can be strong if joined edge to edge, face-to-face or edge-to-face along the same grain direction. The effectiveness of this type of joint is demonstrated by the great number of table tops, carving boards, and other surfaces made from a series of edge-glued boards. Joints involving board ends are far weaker than edge- or face-glued, and it is usually advisable to use some reinforcement or fastener in addition to adhesive. Joints utilizing the grain direction, but with grain directions at right angles or near right angles to one another, may have good initial strength but may weaken over time because of differences in the amount of movement when the members expand or contract from moisture changes. Fig. 6C1 illustrates the typical butt joints.

After adhesives are applied to the surfaces to be joined, butt joint components are firmly clamped until the adhesive has set. Screw, pneumatic, or hydraulic clamps hold the joint surfaces tightly together during this period.

C2. ***rabbet joints*** - are corner joints that aid in positioning cabinet and case members during assembly, provide a more secure joint, and greater strength than simple butt joints. Their greater strength results in part from a larger glue area. The rabbet is a step at the end or edge of a wooden component. Two kinds of rabbet joint are illustrated in Fig. 6C2. Rabbets are machined by one of several methods: two saw cuts at right angles, use of a dado cutter (see below), or by routing or shaping. In small-quantity craft work, special hand planes can be used. Rabbet joints are commonly used in drawers and in the attachment of the back panels of cabinets, bookcases, and similar wood products.

C3. ***dado and groove joints*** - Dado joints include a slot machined across the wood grain; groove joints include a slot machined with the

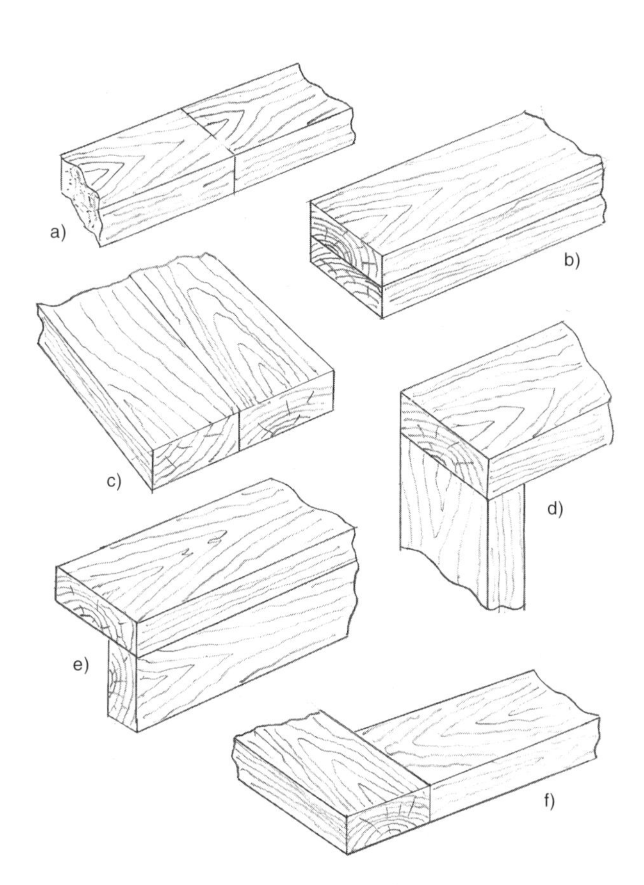

Fig. 6C1 Several butt joints: a) end-to-end, b) face-to-face, c) edge-to-edge, d) end-to-face, e) edge-to-face, f) end-to-edge.

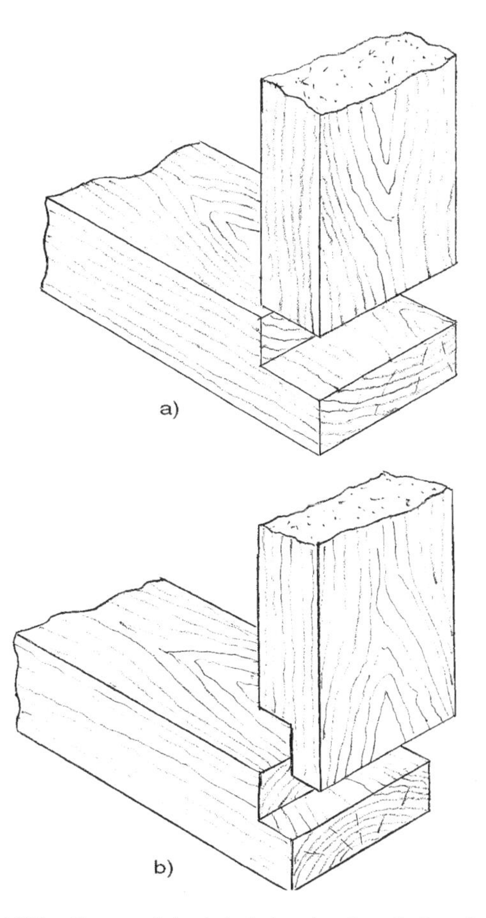

Fig. 6C2 Two rabbet joints. In the full rabbet joint (a), the step on one piece is the same width as the thickness of the mating board. In the half rabbet joint (b), the widths of the steps on both pieces are usually one-half the board thickness.

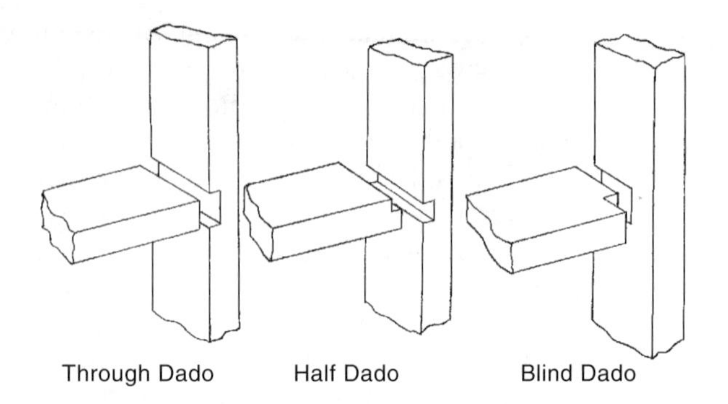

Fig. 6C3 Three common dado joints.

wood grain[2]. The mating parts fits snugly in the slot. Fig. 6C3 illustrates several varieties of dado joints including those that have features of rabbet, and tongue and groove, as well as dado. The slots can be machined in several ways. The most common method employs a circular table or radial arm saw with a dado cutter. Dado cutters are set to cut a slot of the desired width in one pass. This is achieved by incorporating an adjustable amount of wobble so that the blade cuts a wide swath. The other method utilizes two circular cutters with several chipper blades positioned between them. The combined width of the cutters and chipper blades equals the slot width. Shapers and routers can also be used to cut the necessary slots, using a straight bit. When the width of the slot is greater than that of the bit, two or more passes are necessary. The dado joint is common for mounting cabinet and bookcase shelves. The double dado (also called the dado tongue and rabbet) is useful for drawer fronts because it hides the end grain of the drawer sides.

C4. ***tongue-and-groove joints*** - are butt joints with added alignment and strength. The added surface area in the joint provides additional glue area and holding power. Fig. 6C4 illustrates some typical tongue and groove joints. The tongue is typically centered on the edge and its thickness is one-third of the board thickness, as is the width of the mating groove. Tongue-and-groove joints are machined in production on shapers with special cutters that have matching shapes for tongue and groove. The operations can also be performed on table or radial arm saws using dado cutters of the proper width.

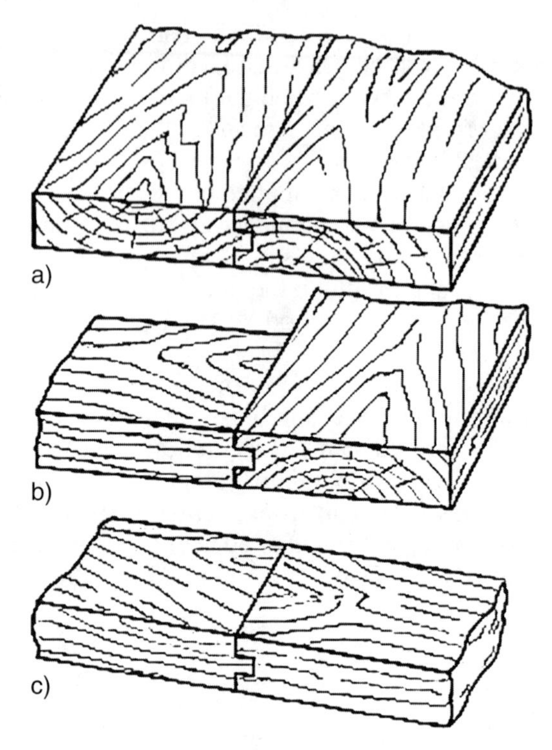

Fig. 6C4 Typical tongue and groove joints: a) edge-to-edge, b) end-to-edge, c) end-to-end.

The tongue can also be a separate machined piece and the mating boards then will both have grooves. Joints of this kind are called splined joints. The spline, a thin piece of wood, hardboard, or plywood, is assembled so that the grain of the spline is at right angles to the grain of the boards being joined. This arrangement produces a joint stronger than a standard tongue and groove.

C5. ***mortise and tenon joints*** - are very strong joints found on quality furniture and cabinets. They are characterized by a projecting tab (the tenon) on one piece that is inserted in a rectangular opening (the mortise) on the mating piece. Fig. 6C5 illustrates a variety of mortise and tenon joints. In many cases the mortise and tenon are hidden and the joint appears to be a simple butt joint. This is the case with the most common blind mortise and tenon joint, where the mortise and tenon extend only part way through the joint. In production, the mortise is made with special machines, similar in appearance to drill presses. A drill encased in a square, non-rotating chisel, makes square holes in a workpiece. The square chisel plunges into the work along with

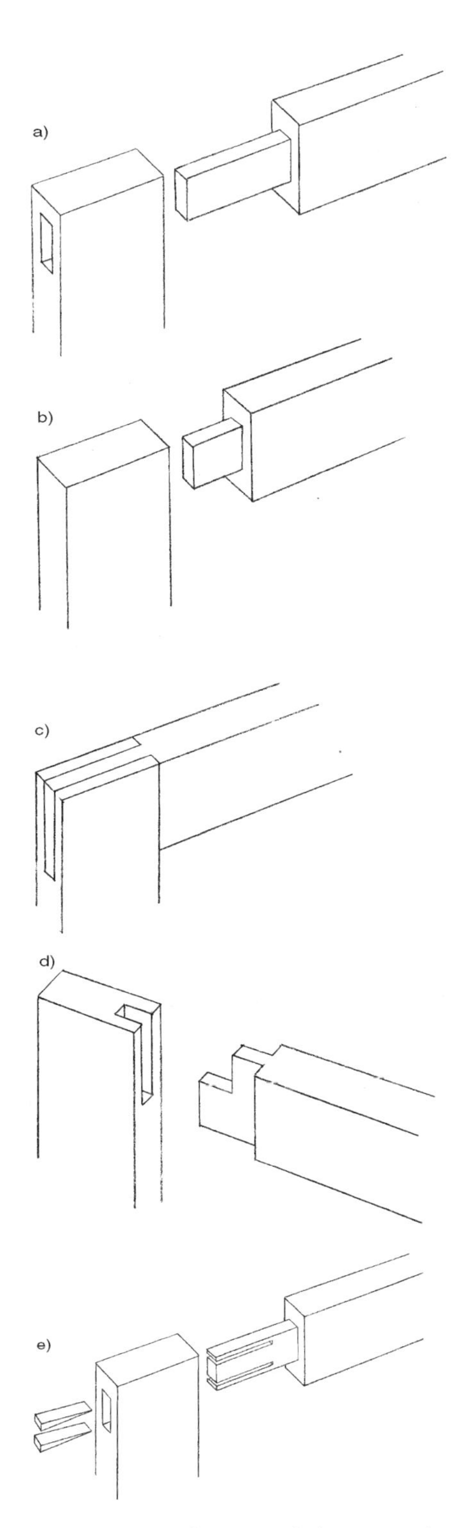

Fig. 6C5 Mortise and tenon joints: a) through mortise and tenon, b) blind mortise and tenon, c) open mortise and tenon, d) mortise and haunched tenon, e) mortise and wedged tenon.

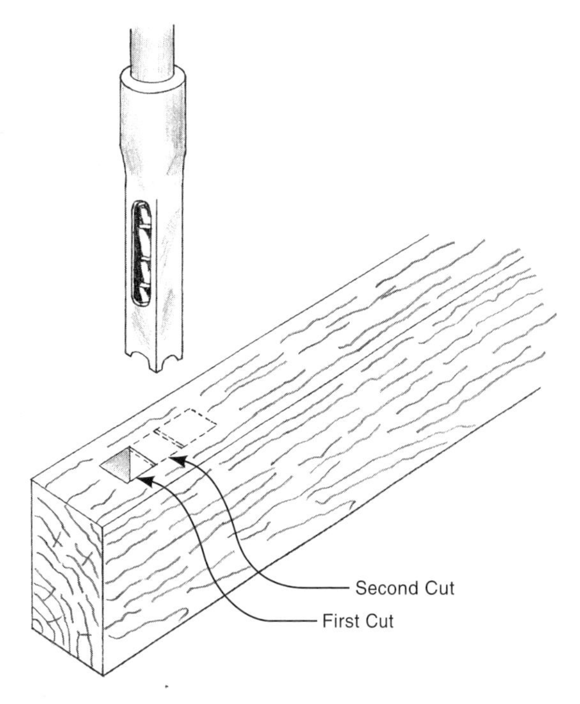

Fig. 6C5-1 Successive cuts with a mortising chisel assembly produces a rectangular mortise in the workpiece.

the drill, converting the round drilled hole into a square shape. A series of square holes, side-by-side and slightly overlapping, make a rectangular slot into which the tenon is assembled, as in Fig. 6C5-1.

In production situations, the mortising machine makes the series of slightly overlapping square holes automatically. In low quantity production, the work is done under manual control in a drill press, with a fence to guide the workpiece. The tenon is also produced by specialized machines when production quantities make it justifiable. Some machines of this type have four cutters to shape the tenon on one end of a workpiece. Some similar machines are double ended so that a tenon is machined from both ends of the workpiece. In limited production situations, the tenon can be cut from the workpiece using a series of table saw cuts, often with suitable fixtures to control the movement of the workpiece against the saw. Tenons are also machined with dado and router cutters.

C6. ***dovetail joints*** - Fig. 6C6 illustrates two common dovetail joints, both of which are sturdy. The appearance of dovetail joints is attractive and the reverse tapers of the dovetails provide a strong

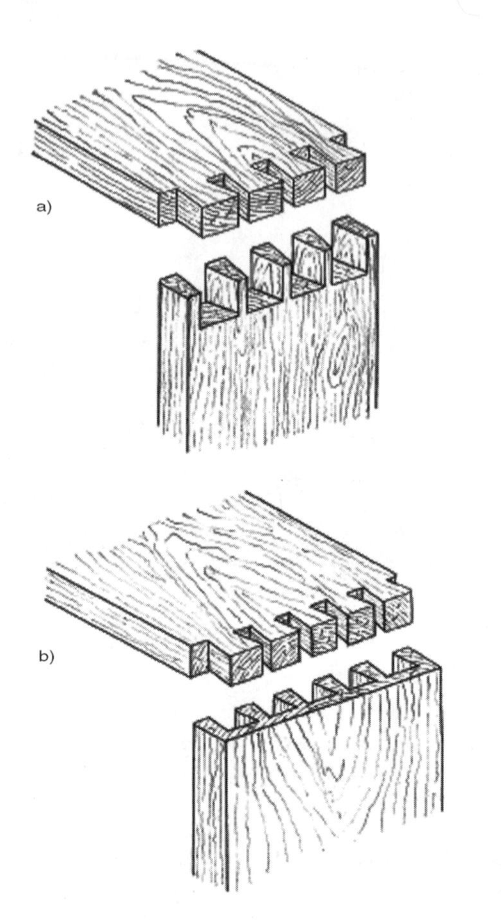

Fig. 6C6 Dovetail joints: a) through dovetail, b) half-blind dovetail

locking effect. Fig. 6C6-1 shows a dovetailing fixture with finished workpieces. A hand router with a dovetail-shaped router cutter is guided by the fixture to machine both pieces. The dovetails are the same size. and are uniformly spaced, where hand-cut dovetails may have different patterns. Fig. 6C6-2 shows dovetail pin and socket shapes produced by the fixture illustrated in Fig.6C6-1. Fig. 6C6-3 shows a machine that makes dovetail joint cuts in both pieces in each automatic cycle.

C7. ***dowelled and biscuit (joining plate) joints*** - provide reinforcement and positioning assistance in the assembly of butt joints. These kinds of joints are useful when curved or angled pieces, which are less easily clamped, are joined. Dowels are cylindrical pins normally made from hardwood or plastic. They are often made with grooves for adhesive and air

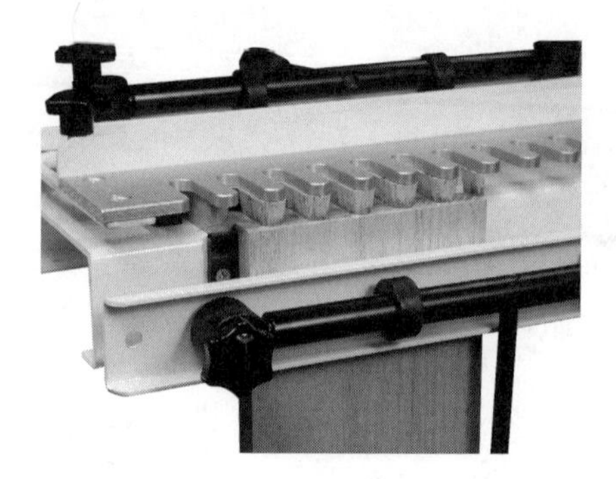

Fig. 6C6-1 Fixture for machining dovetail joints with a hand router. The dovetail cutter on the router follows the path controlled by the metal plate and machines both pieces to be joined in the same operation. Both pieces fit perfectly together with a line-to-line fit. (*8 WMH Tool Group, Inc. All rights reserved.*)

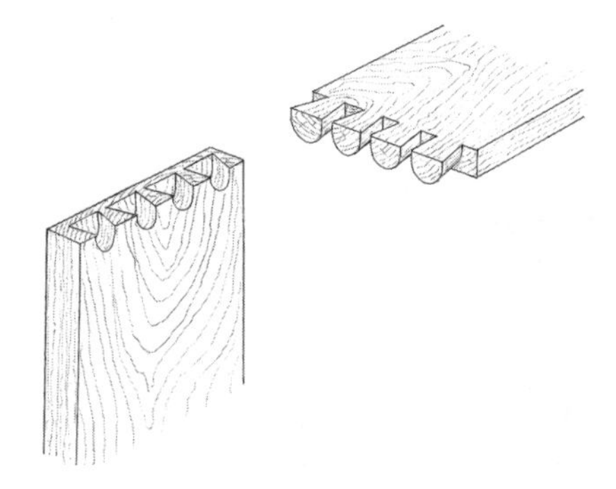

Fig. 6C6-2 Half-blind dovetail cut with a router and dovetailing jig. or with a special dovetailing machine.

Fig. 6C6-3 This semi-automatic machine makes the dovetails of Fig. 6C6-2 with computer-numerical cutter control. The operator loads the two pieces to be joined. The machine cuts all the dovetails simultaneously, using as many of its 25 spindles as necessary. (*Courtesy Dodds Co.*)

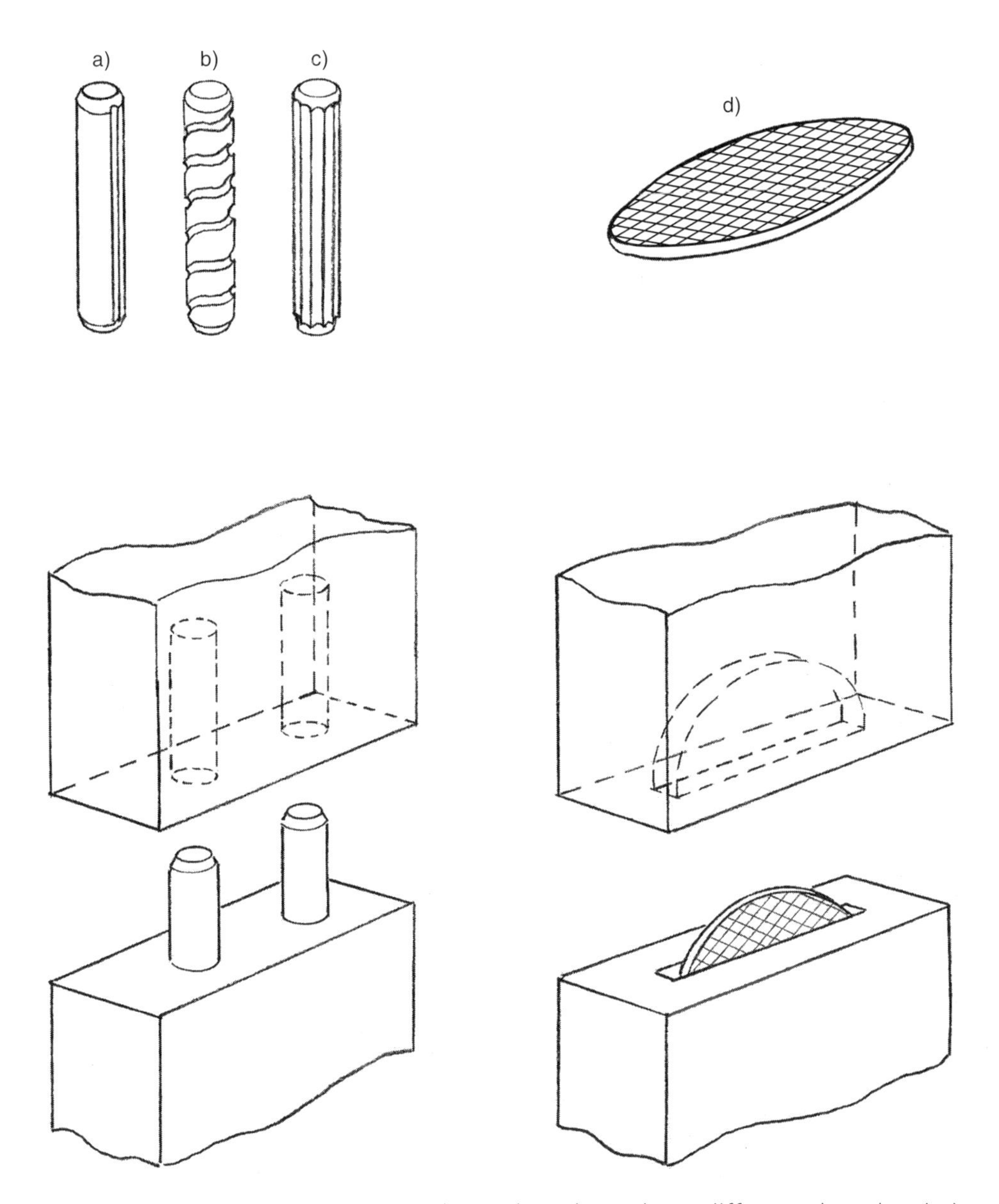

Fig. 6C7 Dowel and biscuit joints. Views a), b) and c) show three different dowel variations, each of which has grooves to contain adhesive and allow air to escape. View d) shows a typical biscuit. The biscuit is compressed but expands in the joint from the moisture in the adhesive to provide a snug fit. The lower views show typical butt joints that incorporate dowels (left view), or biscuits (right view), to provide additional reinforcement, greater glue area, and joint alignment.

flow when the joint is assembled. Biscuits (joining plates or wafers) are small, oval-shaped, compressed plates, usually of birch. Fig. 6C7 illustrates both these approaches. Dowelled joints are not common in production work but are still used in custom or hobby work and can be made with the aid of suitable drill fixtures to locate the dowel holes in the mating pieces. Dowel holes and biscuit slots must be accurately located to insure proper alignment of the parts to be assembled, but current production drilling and slotting equipment will provide the requisite precision. Biscuits are purchased items, made to standard size and thickness. Biscuit slots are made with special machines that provide the necessary alignment accuracy. When glued in place with water-bearing adhesives, the biscuits expand from their compressed state in the joint to form a tight fit. Normally two or more biscuits are used per joint.

C8. ***lap (halved and bridle), scarf and bevelled joints*** - Fig. 6C8 illustrates examples of these joints. Bevelled and scarf joints are made by angling a table, radial-arm, or miter saw, or by using angled guide blocks with a table saw. These joints may be used in assemblies having angled or curved members. Special gluing fixtures may be required for the assembly and bonding of such assemblies. Scarfed joints have greatly increased joint areas compared to butt joints, and allow the joints to be essentially along the grain. They are much stronger than butt joints.

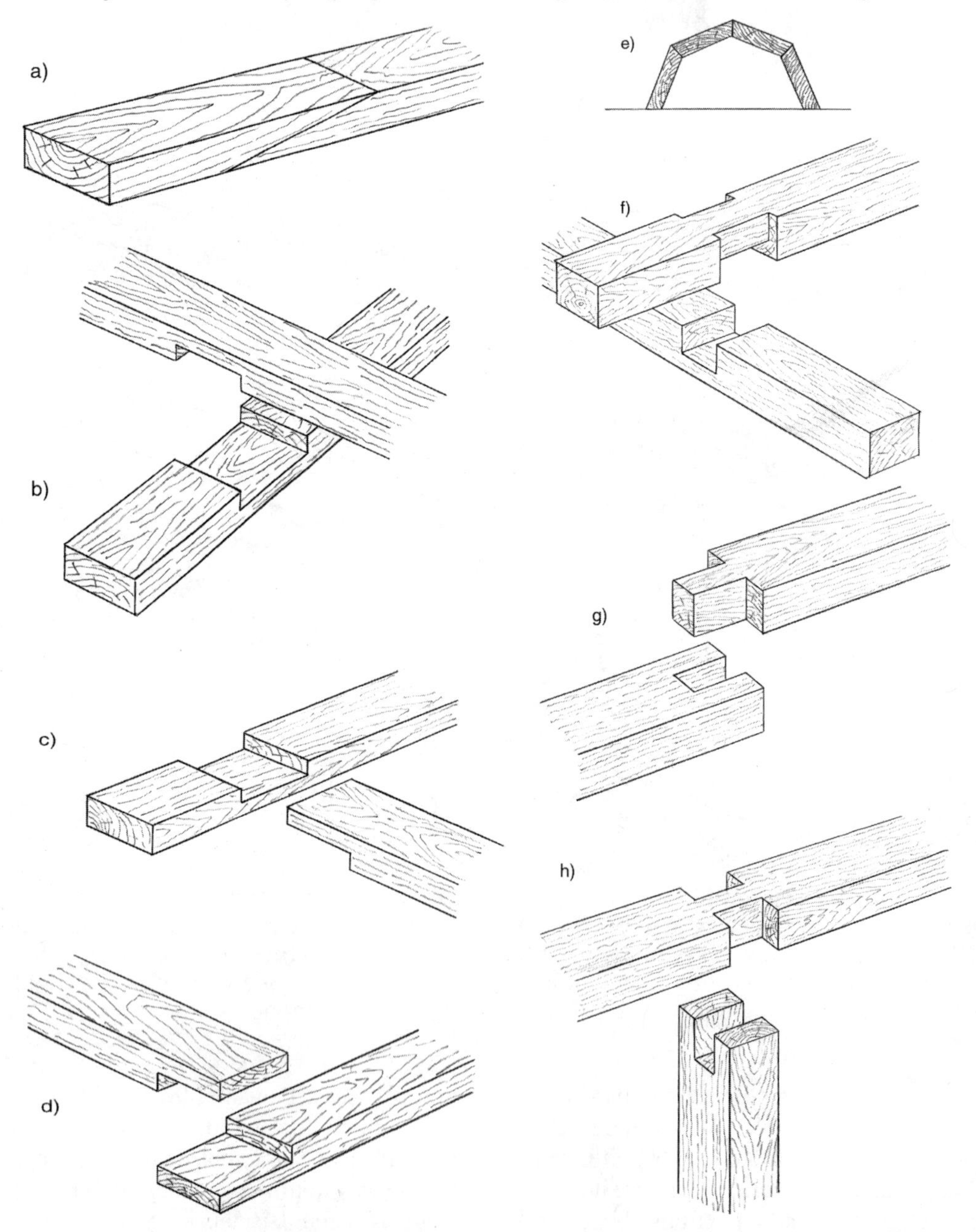

Fig. 6C8 Scarf, lap (halved), bevel, and bridle joints: a), scarf joint, b), middle lap (halved) joint, c), T-lap (halved) joint, d), end lap (halved) joint, e), bevel joint, f), g), and h), bridle joints.

Lapped (halved), joints, unlike butt joints, are normally self-fixturing. Joint cuts for them are normally made with table or radial-arm saws and dado cutters.

D. Making Bent Wooden Components

There are two common methods for making bent wood pieces: wet bending and fabrication of laminated parts. A third method, kerf bending, is less frequently used.

D1. ***wet bending*** - Wood bends much more easily if the workpiece has a high moisture content and is heated. One wet bending method involves heating the wood to 212°F (100°C) in a steam chamber for approximately 3/4 hour per inch of thickness.[1] This treatment softens the fibers so that the wood is in a semi-plastic state. Ends of the workpiece are sealed with varnish beforehand to prevent absorption of excessive moisture. The ideal moisture content is between 20 and 30 percent.[2] Bends made in the wood workpiece, if it is held until the piece cools and dries, will be retained with essentially unchanged curvature.

Another heating method involves immersion of the workpiece in boiling water, but this may increase the moisture content of the wood excessively. The workpiece is bent by hand or in a press, immediately upon removal from the steam chest, and is left in a bending fixture until cooled and dried. Drying may take several days. The fixtures may, themselves, be wooden, but a common element is a flexible spring steel strap, bearing against the outside surface of the bend. This arrangement guards against bruising the wood at pressure points in the fixture and insures a smooth curvature. More importantly, it also shifts the center of curvature so that less of the wood is in tension, which it cannot withstand well, and more of it is in compression, which the wet wood can more easily absorb. See Fig. 6D1.

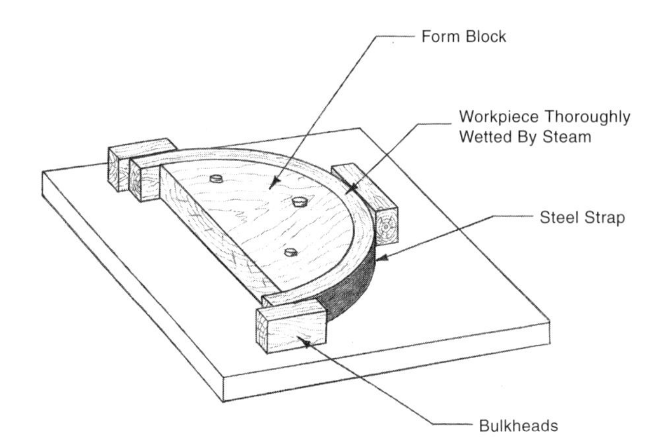

Fig. 6D1 Steam bending a solid wooden component. The steel strap applies compressive force to the outer layers of the workpiece, preventing stretching that would adversely affect the piece.

D2. ***making curved laminations*** - Curved laminations can be made by draping veneers of 1/8 in (3 mm) or less in thickness, coated with glue on mating surfaces, over a male die half. Plies all have the same grain direction. A gap-filling synthetic resin glue is used. The laminated part is kept under the pressure of the closed die until the glue has set. Dies may be heated to accelerate the setting. Sometimes only a male die is used, with a vacuum bag over it to provide the necessary pressure during curing. In other cases, a metal strip or sheet is clamped over the work in the male die. Wax paper may be used between the wood and the steel strip to prevent glue from bonding the strip to the workpiece. An example of a curved, laminated component of high quality, is the cabinet of a grand piano which has a side with an S-shaped curvature. Because of the acoustic requirements, the shape, solidity, and moisture content of these curved laminated pieces are quite critical.[1] Other common applications for curved laminations are chair backs, skis, and tennis rackets. Laminated parts are stronger than solid lumber of the same thickness and, for this reason, lamination is used for wood beams and other highly stressed components. Laminated parts are also less likely to warp than solid pieces.

D3. ***kerf bending*** - uses dry solid pieces. A series of shallow cuts is made in one surface of the workpiece to a depth about 1/16 in (1.5 mm) from the opposite surface. Then the workpiece can be bent relatively easily. Fig. 6D3 illustrates how the kerf cuts react to the bend. If the kerf-cut side of the board is the side that is visible in the final product, a layer of veneer is glued over it. The depth, width, and number of kerf cuts, are critical and usually have to be developed with trial pieces before production takes place. Grand piano bodies are also given curves with this method.

a)

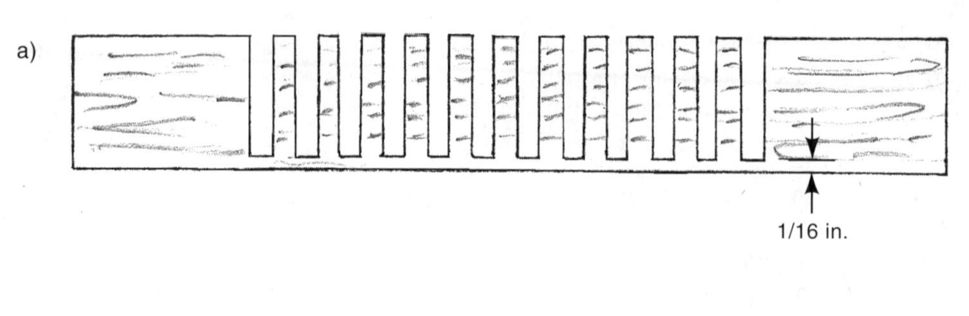

b)

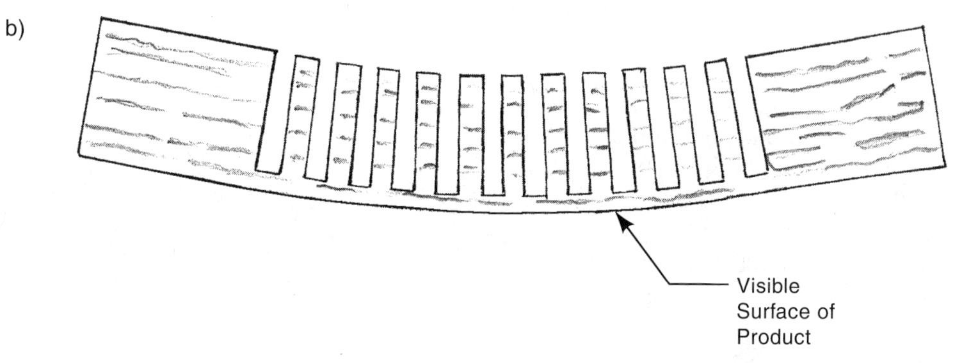

c)

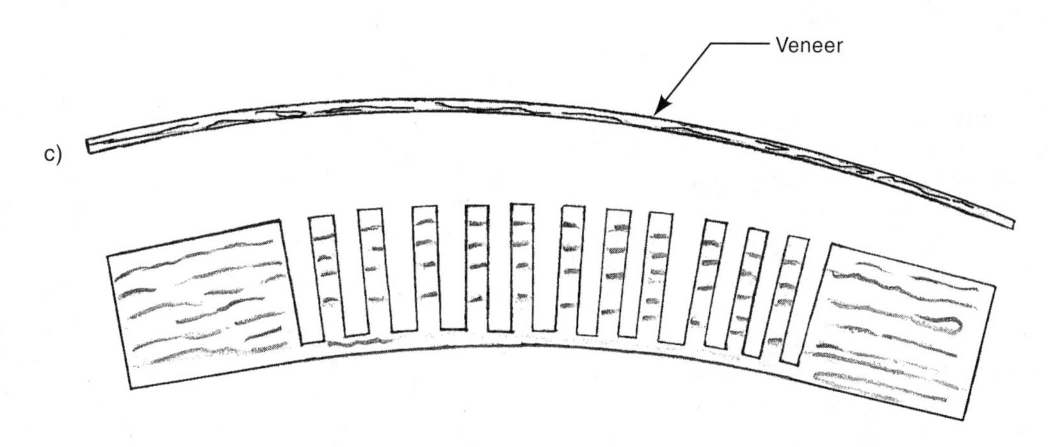

Fig. 6D3 Kerf bending. a) A series of parallel cuts are made from one side of the board to within about 1/16 in (1.5 mm) of the opposite surface. b) The uncut surface then becomes the visible surface of the product. c) If the cut surface is on the exterior of the product, veneer is added to hide the cuts.

E. Assembly and Fastening of Wood Products

Assembly and fastening may involve any of the major classes of joints described above, including simple edge-to-edge, side-to-side, end-to-end connections, or combinations of them. Sometimes these joints are made with only adhesives or metal fasteners, or they may be connections made with the aid of wooden components - dowels or biscuits - or blocks, to aid in locating and holding the separate components together. Or they may be interlocking joints, where the pieces are cut so that they locate

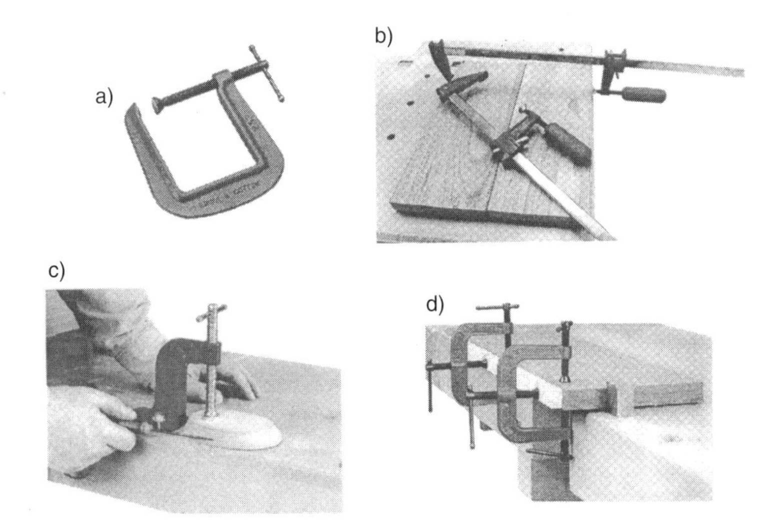

Fig. 6E1 Woodworking clamps: a) C-clamp, b) bar clamp, c) hold-fast clamp, d) edge clamp. (© *WMH Tool Group, Inc. All rights reserved.*)

with one another more positively or are held together, at least partially, by their shapes. Another type of assembly is based on the use of "knock-up" or "knock-down" fittings of metal or plastic, which hold wood parts together. Often, these fittings are used in such a way that the assembly can be taken apart by the user when necessary. A fourth type of assembly is one that permits movement between the wooden pieces. A hinged door and door jamb is a prime example of such an assembly.

Correct alignment of wooden components in an assembly is achieved with one of two methods: 1) fixturing so that the pieces being joined are held in the correct position, and 2) interlocking joints so that the parts are self-locating. Wooden locating fixtures are easily fabricated and generally have sufficient accuracy for parts placement. When interlocking parts are employed, current production woodworking equipment and processes provide the precision necessary for accurate alignment and snug fit.

Wood components are fastened by two prime methods: metal fasteners and adhesives. Sometimes, wooden (or plastic) fasteners are used. During assembly and some machining operations, a variety of holding devices, clamps, vises or fixtures may be used.

E1. ***clamps and fixtures*** - of various kinds are useful in holding wood workpieces for gluing or other fastening, and during some drilling, routing, or cutting operation. Several useful types of clamps are illustrated in Fig. 6E1. Two different fixtures used in assembly of wooden components are shown in Fig. 6E1-1, and 6E1-2. Fixtures are employed in production situations where a number of identical components are assembled. They provide assurance of accurate alignment of parts and consistent and accurate dimensions of the finished assemblies.

E2. ***mechanical fastening*** - The variety of metal fasteners used in woodworking is extremely broad, ranging from common household nails to special fas-

Fig. 6E1-1 Gluing fixture for edge-bonded panels. Individual boards have glue applied from an applicator roller on the roller conveyor. The operator positions each board on the fixture, which holds it securely under sufficient clamping pressure until the glue sets. (*Courtesy Donald Dean and Sons, Inc., Montrose PA.*)

Fig. 6E1-2 Gluing fixture for assembly of cabinet doors. Pneumatic clamps provide pressure to seat all components and to square the assembly. The fixture can hold the panel assembly together until the glue sets, but most commonly is used only temporarily. After clamping, headless pins are driven into the panel to hold it together until the glue sets. (*Courtesy Donald Dean and Sons, Inc., Montrose PA.*)

teners that draw parts together when tightened. The latter are particularly important with ready-to-assemble furniture that is sold fully disassembled. The parts of such furniture usually come with fasteners that utilize screw threads, cam surfaces and eccentrics to drive components tightly together. Tightening one screw fastener draws against the angled surface of another fastener component to push or pull the mating part into place. Ready-to-assemble furniture kits also often include plastic elements attached to the wooden components that allow two parts to be pressed together or driven together with hammer strokes. Components normally come with pre-drilled holes for screws, dowels, and other fasteners, and pre-machined pockets for hinges and special fasteners. Common metal fasteners may be hidden in woodwork with wood putty or wooden plugs.

E3. ***adhesive bonding*** - See Section 7D of this handbook for coverage of adhesive bonding in general, including wood products. Also see *adhesives*. Adhesives that have a history of use with wood products include glues made from animal and vegetable materials. Synthetic glues made from thermosetting plastics have gained ascendancy for use in commercial woodworking, however, because of their superior properties and ease of use. The most common woodworking adhesives, include the following which are available in ready-to-use liquid form: hide glue, polyvinyl acetate glue, and alphatic resin glue. Water-mixed glues for woodworking include casein glue and plastic resin glue. Resorcinol glue is a two-part woodworking adhesive. Table 6E3 summarizes properties and working arrangements for a number of glues used in wood product assembly.

The techniques used in bonding wood parts parallel those used when bonding other materials. The surfaces to be bonded should be clean, especially free of greases and oils but also of dust, moisture, and other foreign materials. The best joints for adhesive bonding result when the joint is along the grain of the wood, face to face, face to edge, or edge to edge. End grain butt joints do not hold adhesive well. The adhesive may be applied by rollers, spray guns, brushes, toothed spreaders, or spatulas. Some wood glues work best if the gluing takes place soon after the surface has been machined. Woods with moisture content of 20 percent or more can present bonding problems, especially with water-based glues.

Glue joints, especially butt joints, are normally clamped after they are assembled. This ensures that the glue is distributed in a uniform, thin film and the clamps hold the parts until the glue has set. Optimum gluing pressures are 100 to 250 psi (700 to 1700 KPa).[5]

The curing of wood glues, both the types that require evaporation of a solvent and those that cure by polymerization of plastic resins, are accelerated by application of heat. Heat is commonly used to cure glues used in wood products. Heating methods include various oven methods using forced heated air, or radiant heat sources; resistance-heated platens, or metal strips in contact with wooden joint members; and induction/microwave heating of the glue itself.

Fig. 6E3 illustrates a special machine for applying glue for the assembly of dovetail joints.

E4. ***Assembly of veneer and inlaid surfaces*** - is inherently a manual operation, requiring considerable operator skill to insure a tight bond, to get a good fit of pieces in marquetry, and to eliminate splitting of veneer and other defects. (Marquetry is

Table 6E3 Common Adhesives for Woodworking

Name or Type	Characteristics Properties	Working Arrangements	Applications
Hide glue	liquid strong no moisture resistance	indefinite pot life short assembly time clamp for 2–3 hrs	fine furniture
Casein glue	strong good moisture resistance	mix powder with water short pot life and assembly time clamp for 2–3 hrs	oily woods
polyvinyl acetate (PVA)	white liquid strong little moisture resistance	long pot life fairly short assembly time clamp for 1–2 hrs	interior cabinets and furniture
aliphatic resin	cream-colored liquid stronger and more solvent resistant than PVA	short setting time clamp for 1–2 hrs	general woodwork
epoxy	paste or viscous liquid excellent moisture resistance	mix resin and catalyst, then short pot life	metal to wood glass to wood
contact cement	viscous liquid no clamping needed	apply to both surfaces	fasten veneers or laminates to wood

the art of fitting together pieces of veneer to create a design.) The traditional hide glues have been replaced with water-based synthetic resin adhesives, which somewhat reduce the skill required. These glues have a longer tack time and allow more time for aligning the veneer. (Hide glues also penetrate the thin veneer and can interfere with later filling or staining.) Veneer pre-coated on the back with hot-melt glue is another alternative. After this veneer is positioned and aligned, the assembly is heated to melt the glue. Applied adhesive should be in a thin layer, which must be uniform in thickness.

When veneer sheets are to be joined on one surface, or when inlays are used, the veneer pieces can be fitted and taped together, and laid on the surface as one piece. The synthetic resin adhesive must be of a uniform thickness; the veneer is pressed against the surface in a hydraulic, pneumatic, or mechanical press until the adhesive has set. Veneer is usually laid to overlap all edges of a surface. After the adhesive thoroughly sets, the excess can be trimmed off with a hand trimming tool, which resembles a plane but has a sharp vertical knife that cuts the veneer both with and across the grain.

Another technique with inlays or joined sheets of veneer is to overlap them and make a cut with a sharp knife midway in the overlap area. The trimmed pieces from each sheet are then removed

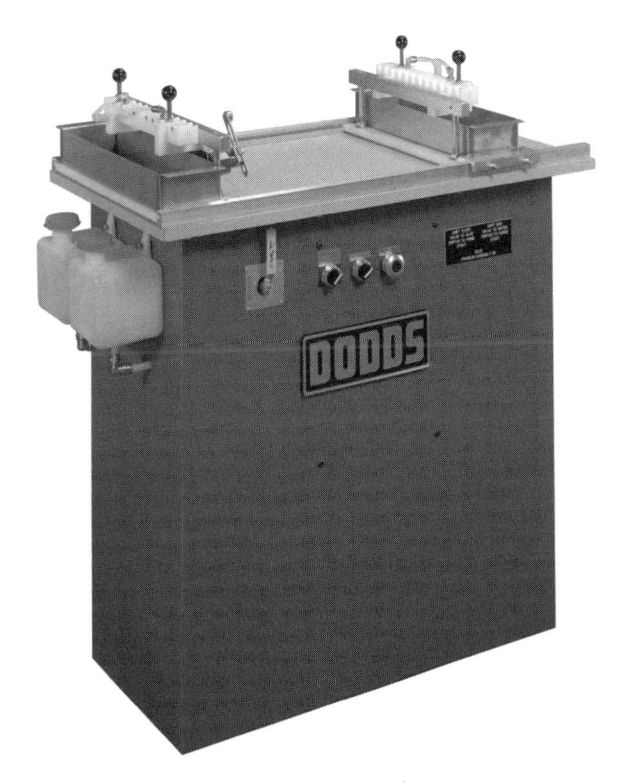

Fig. 6E3 A glue applicator machine for dovetail joints. When the board with dovetails is inserted, the machine dispenses a measured amount of glue from a manifold to each dovetail. (*Courtesy Dodds Co.*)

and the edges pressed against the substrate surface. The grain of a veneer laid on a board should be in the same direction as the grain of the board; with plywood, the veneer grain should be at right angles to the grain of the top layer of plywood.

F. Manufacture of Plywood and Other Panel Materials

Plywood and other panel materials are often advantageous for use in wood products because of increased stability, strength properties, or lower cost, or a combination of these properties. Stability, or freedom from warping or splitting, is an important factor. Cost is another. They are also available in much wider dimensions than can be obtained with natural lumber. All panel products used in woodworking are made from wood or wood fibers, and utilize adhesives to bond smaller or thinner pieces of wood or wood fiber together. In cabinet work, they are more apt to be used for internal, less visible, components because of appearance factors, but they can be finished with veneer or printed laminations to provide the appearance of solid lumber. Manufactured panels, other than plywood, are sometimes known as *composite panels*.[5]

F1. ***making veneer*** - Wood veneers are thin wood sheets used primarily as structural material in plywood or as visible surface material in cabinetry and architectural panels. Veneers are almost always cut by slicing a thin sheet of wood from a log or flitch (part of a log) although saw cutting can still be used. (Saw cutting can produce very good veneers but results in a loss of about half of the wood from the log as sawdust.) There are two prime cutting methods: rotary peeling and flat slicing. In both methods, the logs are selected for their freedom from knots and other defects and for a desirable grain pattern. The logs are cut to length, debarked, trimmed as necessary to cylindrical shape, and immersed in boiling water for a prolonged period to soften and heat the wood grain. While still hot from boiling, logs for rotary peeling are mounted on a lathe that includes a fixed horizontal knife at least as long as the log. As the log rotates, the knife is fed forward slowly to slice a thin sheet of wood from the log surface. The operation proceeds, peeling the wood from the log in one thin continuous web. The web is then sheared into sheets, graded, and sorted, repaired as necessary when there are knots and other defects, dried in drying equipment, and stacked for later use. Drying may be performed on a batch basis, or by feeding the veneer through the length of a drying oven on a roller system or a belt conveyor. Rotary-cut veneers are typically from 0.013 to 0.375 in (0.33 to 9.5 mm) in thickness. 1/28 in (0.9 mm) and 1/40 in (0.6 mm) are common thicknesses. The distinctive grain pattern of rotary cut veneer is somewhat different than the grain pattern of solid lumber. This prevents its use in furniture and panel surfacing where authentic solid lumber appearance is needed. It is, however, the prime veneer type used in structural plywood.

Flat-sliced veneers are produced in special machines. Logs are first rip cut in half or in quarters to make flitches. Boiled and heated flitches are attached to a flitch table. The table moves vertically up and down against a sturdy slicing knife, which is at least as long as the flitch. With each downstroke, the knife cuts along the grain to produce a wide ribbon of wood from the length of the log. After each downstroke, the slicing knife is advanced the thickness of the desired veneer and, with the next downstroke, another sheet is produced. Each sheet is as long as the flitch and as wide as the section being cut. Typical flat-sliced face veneers are from 0.010 to 0.035 in (0.25 to 0.88 mm) thick. 1/60 inch (0.4 mm) is a common thickness. Fig. 6F1 shows rotary slicing and two approaches to flat slicing.

Veneer techniques are also used to make thin wood pieces for other applications such as coffee stirrers, ice cream sticks, and basket weaving material, and for forming with suitable adhesives into trays or other formed objects.

F2. ***making plywood*** - Plywood consists of panels of wood made from an odd number of layers of veneer glued together. The grain direction of each layer is at right angles to that of the adjacent layers. Plywood, especially that intended for construction use, is most commonly made from Douglas fir and other softwoods that can be rotary peeled with little waste (See veneers). Plywood for cabinet and furniture use normally has hardwood veneer on at least one surface and may have hardwood in the core veneers. Particle board may also be used as a core material. Fig. 6F2 illustrates some common plywood construction. The veneer is cut into convenient

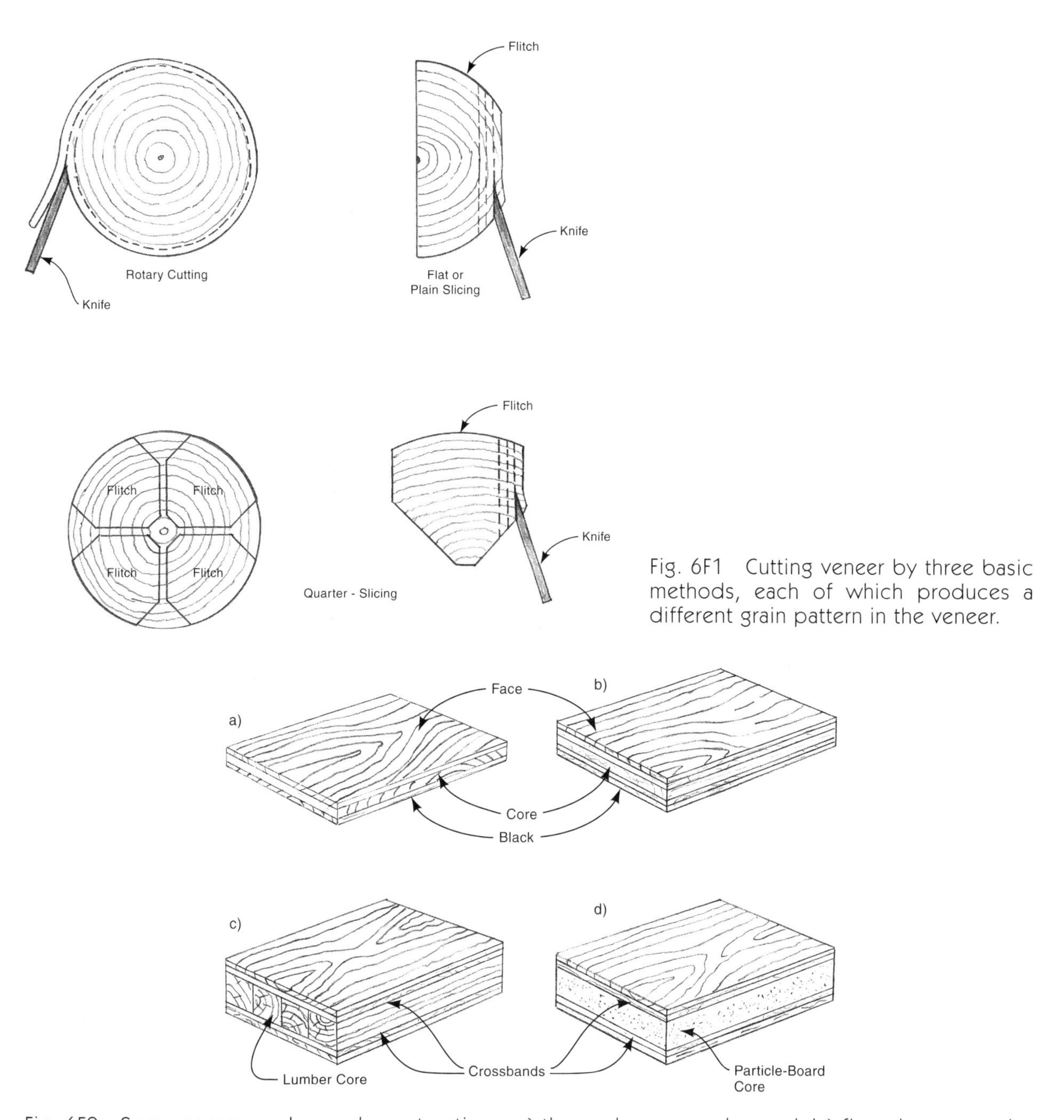

Fig. 6F1 Cutting veneer by three basic methods, each of which produces a different grain pattern in the veneer.

Fig. 6F2 Some common plywood constructions: a) three-ply veneer plywood, b) five-ply veneer plywood, c) five-ply lumber-core plywood, d) five-ply plywood with particle board core.

lengths and is dried to a moisture content of from 5 to 14 percent (depending on the adhesive used). Defects are repaired by cutting them out and inserting plugs of acceptable material. Splits may be taped. If necessary, narrow pieces are edge glued together. (This is especially so with face veneers, where good appearance may be necessary. Face veneers may not be rotary cut and individual pieces, therefore, will be narrow.) Glue is commonly applied by roller applicators, but also by spraying or curtain coating, to each sheet. Some odd number of sheets, usually 5 or 7, are stacked together with each sheet having the grain direction at right angles to the direction of the adjoining sheets. However, some plywood has as little as 3 layers with a central core of solid lumber.

Lumber core plywood is more common in thicker grades or when the edges are intended to hold hardware or be exposed. The adhesive is cured with the sandwich of sheets under heat and pressure, in heated hydraulic presses. These presses typically have from 5 to 25 openings for the assembled plywood. Platen temperatures are typically 175 to 350°F (80 to 180°C). The platens are heated by hot water, steam, or hot oil. (Assemblies may be lightly cold pressed to facilitate handling and loading before they are inserted into the heated hydraulic presses.) Waterproof adhesives are used for exterior grade plywood. The plywood sheet is then trimmed to size and the surfaces are machine-sanded by belt or drum sander for smoothness and final panel thickness. The final moisture content is normally 12 percent or less.

There are two basic applications for plywood. Softwood plywood is used as panels in buildings and in other commercial construction. Hardwood plywood is used in the manufacture of furniture, for interior wall panelling, for sporting goods, for industrial fixtures, and for equipment bases and parts. The standard size for construction plywood is 4 by 8 feet (1.2 by 2.4 m) in thicknesses up to 1 in (25 mm). Hardwood plywood is produced in these and other sizes.

F3. ***making wafer board (chipboard)*** - Wood chips, approximately 1 to 2 in (25 to 50 mm) long and of various widths and thicknesses, are machined from debarked logs and scrap lumber with special machines. Sawdust and other residues are not used. The chips are inspected and selected, mixed with a plastic resin adhesive (usually phenolic) and spread in random orientation on heated press platens. The presses force the chips into a board of uniform thickness and the heat cures the adhesive. The boards are trimmed to size, normally 4 by 8 ft (1.2 by 2.4 m) and the surfaces are sanded. These boards are less expensive than plywood to produce, and find numerous applications in the building industry where sheet material is needed. An average breaking strength in bending is 2400 psi (17 MPa).[5] Common uses are as an underlayment for exterior walls and roofs.

F4. ***making oriented strand board***[2] - uses methods similar to those used for making wafer board except that the chips are oriented in specific directions in layers instead of being allowed random orientation. Typically, the outer layers are oriented in the direction of the long dimension of the panel and are separated by a central layer in the perpendicular direction. In a this sense, oriented strand board is similar to plywood. The three layers and the strands of wood are all bonded together with phenolic resin, with heat and pressure. The board has a high resin content and high strength. An average breaking strength in bending is 8000 to 10,000 psi (55 to 69 MPa).[5] Thinner board, from 1/4 in (6.3 mm) thick is used in cabinet work for cabinet backs and drawer bottoms. Sheets from 3/8 to 1/2 in (10 to 13 mm) thick are used for subfloor sheeting and roof and wall underlayment. Sheets thicker than 1/2 in (13 mm) and up to 23/32 in (18 mm) are used for single-layer flooring. Surfaces of oriented strand board may or may not be sanded smooth, depending on the intended application.

F5. ***making particle board*** - Particle board is similar to wafer board but the chips are much smaller. The manufacturing steps are very similar, but sawdust and other small particles of wood are used, as well as green logs. The grinding of wood material into particles is more extensive, a richer mixture of resin (urea, phenolic or melamine) is used on surface layers, and wax may also be added to surface layers. The increased resin adds to the strength of the boards and the wax provides water repellency[2]. The material is subjected to pressure and heat to compact the particles and cure the resin. The process is also very similar to that described in section F6 below for fiberboard. Particle board is used in furniture and in construction. It may be laminated with veneer, or finished with paint, film, or paper, and printed with a wood grain pattern.

Particle board is made in densities ranging from 30 to 70 lb/cu ft (480 to 1120 kg/cu m). Although standards differ with manufacturers, low density particle board, used for door cores, generally has a density of from 30 to 40 lb/cu ft (480 to 640 kg/cu m). Medium-density particle board, used extensively in furniture, generally has a density of 40 to 50 lb/cu ft (400 to 800 kg/cu m). High density board, used for core material in plywood, generally has a density of 50 to 70 lb/cu ft (800 to 1120 kg/cu m).

Thicknesses range from 1/4 in (6.3 mm) to 1 $^{7}/_{8}$ in (48 mm).

Medium-density particle board has the following manufacturing operation sequence: 1) Bark is removed from wood logs, both hardwood and softwood. 2) Logs and wood scrap are reduced to chips. Chippers, hammer mills, disc cutters, or knife-ring-flakers. all may be used. 3) The chips are ground into fine particles. 4) The mixture is then dried. This is a continuous process using rotating cylindrical dryers in which the particles are suspended in hot air while moving from end to end. Steam, hot water, or hot oil, circulating in tubes, supplies the heat. 5) Particles are then sieved with vibrating screens or air classifiers to remove oversize particles, that can later be reduced in size, and to remove undersize particles that would necessitate using excessive adhesive. 6) Adhesive, commonly urea (Phenolic or melamine is also used.), in liquid form, is added along with wax and other additives. Adhesive comprises 3 to 10 percent of the weight of the mixture. The wood particles and adhesive are blended in mixing vessels. 7) The mixture is spread as a mat on a belt conveyor by equipment that concentrates the finer particles at the surfaces and the larger particles in the core of the mat. 8) The resulting mat is pre-pressed between flexible metal webs to partially compress it. 9) The mat is cut into separate pieces. 10) Cut pieces are loaded and stacked in a heated platen press with steel plates separating the layers. 11) The pieces are compressed by hydraulic pressure and subjected to heat of 280 to 400°F (140 to 200°C) to cure the thermosetting adhesive, forming panels that are most often 3/4 in (19 mm) thick. 12) Panels are rough trimmed to squareness within ±1/2 inch. Scrap wood is used as a fuel to provide the heat needed by the process 13) Panels are sanded on both the top and bottom surfaces with belt or drum sanders. 14) Panels are cut to the desired size, commonly the same standard 4 × 8 ft (1.2 × 2.4 m) size of wafer board and plywood. Panels may also be cut into boards at this point. 15) Groups of panels or boards are strapped together and moved into stock.

Particle board is used in furniture and in construction. It may be laminated with veneer or finished with paint, film, or paper, and printed and impressed with a wood grain pattern. Floor underlayment is a major application for particle board.

Some particle board is made by extruding the mixture of wood particles and resin through heated dies. Surface laminations are bonded to the board as part of the same operation. Board is made by this method in the high-volume production of bathroom and kitchen counter tops, particularly when there is a curved edge.

F6. ***making fiberboard*** - Fiberboard is a heavy sheet material made from wood fibers and other organic fibers pressed together. The fibers are deposited on a conveyor in sufficient depth to form into mats. They are in random orientation, often with small amounts of synthetic resin adhesive. The mats are then cut into sheets called wetlaps. The wetlaps are subjected to pressure and heat, which bonds the fibers together. Heat may be provided by radio frequency induction. The board panels thus made are quite strong. Three different density grades are produced: *hardboard, medium density fiberboard,* and *insulation board.*

Hardboard, high density fiberboard, is produced using the highest pressure and is available in densities of 50 to 80 lb/cu ft (800 to 1280 kg/cu m) but, most commonly, 60 to 65 lb/cu ft (960 to 1040 kg/cu m). Common thicknesses are 1/8, 3/16, and 1/4 in (3.1, 4.7 and 6.3 mm). Standard widths are 4 ft (1.2 m) and lengths are 8, 12 or 16 ft (2.4, 3.6 or 4.8 m). The brand name, Masonite, has become rather generic for these hardboards, which are quite strong and abrasive resistant. They are used extensively in wall paneling, drawer bottoms, and hidden back panels of cabinets. The boards are made from long wood cellulose fibers and lignin of the wood serves as a bonding agent. The finished boards may have a smooth surface on one or both sides. They are available as peg board, panels with regularly-spaced perforations that can be used to mount metal hooks for tools, household implements, etc.

Tempered hardboard is regular hardboard given a secondary treatment with liquid resin at the surface followed by heat curing. This treatment increases hardness, wear, and moisture resistance, strength, and stiffness further, and raises the density to 60 to 80 lb/cu ft (960 to 1280 kg/cu m).[5]

Medium density fiberboard (MDF) is available in thicker panels than high density, typically from 3/8 to 1 in (10 to 44 mm) thickness and in the same lengths and widths as high-density board. Urea is typically used as an adhesive and wax may be added.

The mats are stacked in a press with steel sheets as separators. Hydraulic pressure with heat cures the thermosetting adhesive. The board has a density of 32 to 50 lb/cu ft (512 to 800 kg/cu m). Medium density fiberboard is a plywood or particle board substitute. Cabinet tops, drawer fronts, and shelving are typical applications. It is also used for moldings and other millwork. It can be stained, painted, laminated, or printed with wood grain. It is equally strong in all directions and can be cut with moderately smooth edges without finish sanding. It is not as strong but much more dimensionally stable than lumber.

Low density fiberboard, insulation board, is light in weight with densities of 10 to 30 lb/cu ft (160 to 480 kg/cu m). It is typically made in 1 in (25 mm) thick, 4 by 8 ft (1.2 x 2.4 m) panels. The manufacturing process involves the cold compression of mats of wood fibers without bonding resin, followed by drying and trimming. The natural cohesion of wood fibers provides the bonding force. It is used as a backing for exterior siding in building construction, as board material in upholstered furniture, bulletin boards with a cork or burlap surface, and in other fabrication situations where high strength is not needed.

F7. ***making engineered lumber, prefabricated wooden beams, and joists***[5] - The term, "engineered lumber" refers to composite wood components which have size, quality, or effectiveness that is superior to those of conventional one-piece lumber. Hoadley[5] recognizes four classes of such components: finger-jointed lumber, glued-laminated components, structural composite lumber, and composite I-joists.

Finger-jointed lumber incorporates joints similar to that in Fig. 6F7 to produce long lengths of knot-free material for such items as door and window parts, stair railings, strip moldings, closet poles, and trim pieces.[5] The finger joint provides a moderately high-strength scarf joint without wasting a large amount of material. The joint is made automatically in special equipment that machines the ends of the pieces to be joined to the proper mating shapes, applies adhesive, clamps the pieces together, and applies microwave heat to cure the adhesive. Joints have approximately 75 to 85 percent of the strength of knot-free wood.

Glued-laminate components, "glulam", consist of structural members made from layers of thinner boards glued together face-to-face with their grain in the same direction. Beams and arches are common applications. Long components can be made from smaller pieces. The resulting member is stronger that an identically-sized member made from one piece of lumber and its strength is more predictable. Most such members are manufactured in one-of-a-kind or limited quantities for specific architectural applications. However, some standard beams and joists made by this method are available.

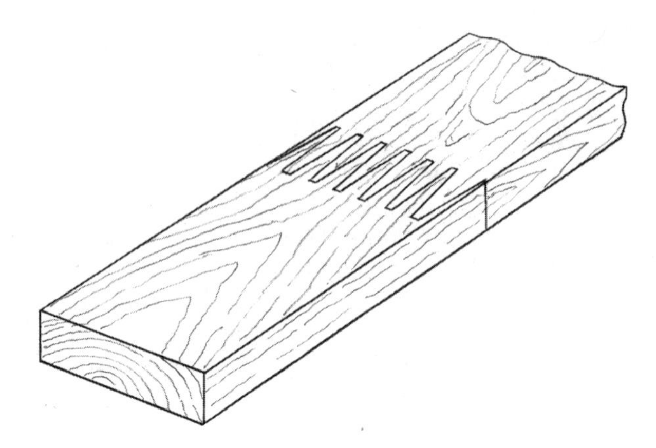

Fig. 6F7 Finger joints, like this one, provide ample glue area and mating surfaces along the grain, both of which add to the strength of the joint.

The term, *structural composite lumber (SCL),* refers to a family of products made from veneer, strands, chips, and particles glued together to form lumber or timbers. They are of the same sizes available in single, solid pieces, except that longer lengths are feasible. This lumber can then be machined and assembled with the same or similar methods used for conventional lumber, but it has superior strength and stability characteristics. Depending on its construction, this material is made with veneer, strands, and/or particles glued and pressed together and trimmed to standard sizes. It is used in applications where greater strength or length are required.

I-joists are wooden structural members with a cross-section similar to that of steel I-beams. A thin central web connects thicker and wider elements at the top and bottom. The central web can be made of plywood or oriented strand board. The top and bottom flanges are made of structural composite lumber or high-quality finger-jointed conventional lumber. See Fig. 6F7-1. Waterproof adhesives are

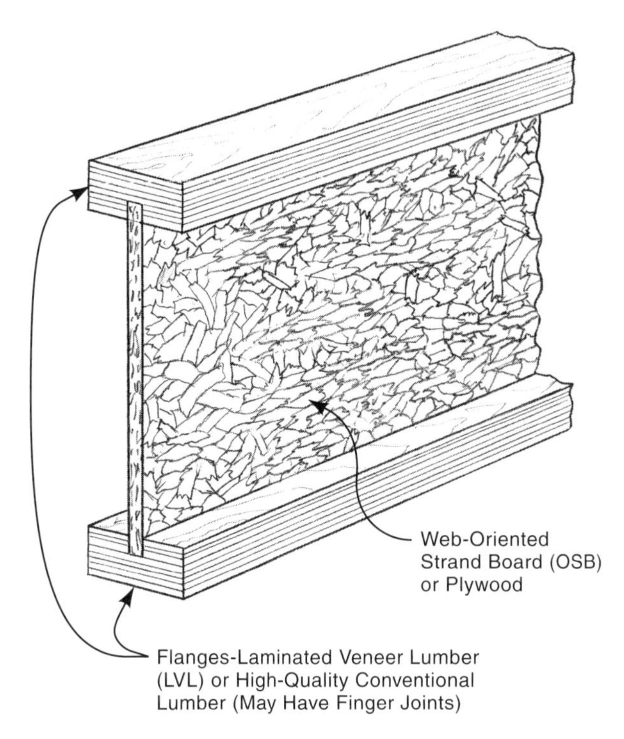

Fig. 6F7-1 An I-joist made from engineered lumber.

used in all components and in the assembly of the joists. These joists have superior stiffness compared to the solid lumber joists used in house construction and can span greater distances.

F8. ***making rigid plastic laminates (high pressure laminates)*** - Decorative surface materials such as Formica and Micarta are normally made from layers of kraft paper impregnated and bonded with phenolic resins. A cover paper is printed with whatever decorative pattern or design is wanted and is impregnated with melamine resin. The top sheet is coated with melamine resin for protection against scratches, dents, heat, and attack from moisture or chemicals. The multi-layered material is placed in a heated press with highly polished platens. Heavy pressure and heat polymerize the phenolic and melamine resins to produce a laminate of about 1/16 in (1.5 mm) in thickness.

The top melamine coating may have a glossy, satin, or other finish, including textures, depending on the intended use and nature of the decorative pattern. Wood grain laminates are textured to simulate the feel of real wood grain. Fig. 6F8 illustrates the composition of a rigid plastic laminate. The laminated board is often bonded to plywood or particle board for support. The back of the bottom sheet is kept somewhat rough to aid in the adherence of the laminate to a substrate. Several grades and thicknesses of rigid laminate are available to permit post forming to make curved surfaces and use in wall panels and cabinet interiors.

F9. ***making synthetic lumber (composite lumber) and components from plastics*** - Two fairly important synthetic wood products made from plastics are available: One consists of injection-molded polystyrene parts that are molded with a wood grain appearance. They often have shapes that replicate carved wooden parts like drawer pulls, cabinet panels with carving, and other decorative components. These moldings can be finished to have the appearance of expensive carved wooden parts. The other significant synthetic lumber products are boards that are extruded from recycled plastics, usually with a wood filler. There are numerous formulations for the synthetic boards. Most utilize high density polyethylene derived from waste bottles and jugs, with ample ground-wood fill material including sawdust and wood waste. This combination makes very good deck boards that are virtually immune to warping, rotting, or splitting. They are, however, more costly

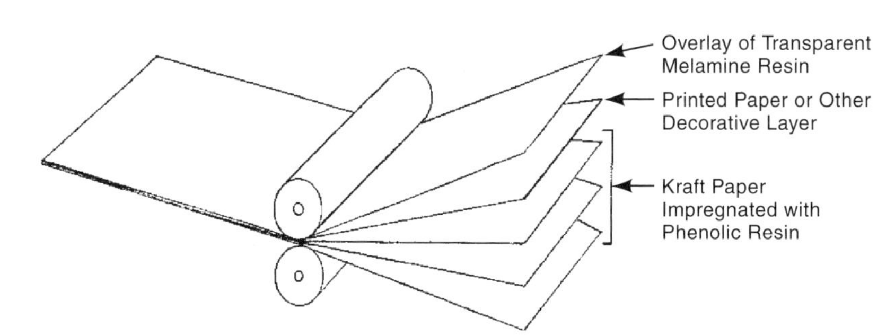

Fig. 6F8 Construction of rigid plastic laminate. Phenolic and melamine thermosetting plastics bond layers of paper-and sometimes other materials-into one rigid sheet.

than treated lumber, with which they compete for use in outdoor decks, porches, docks, boardwalks, and similar applications. Railings and posts are also made of this material. However, the materials are not usually recommended for structural applications such as joists, studs, beams, stringers, or columns. They can be nailed and cut or drilled with woodworking equipment utilizing carbide cutting tools. Boards are typically 50 percent wood fiber and 50 percent recycled plastics, though wood fiber content of 60 percent is not unusual. Boards may be made with other fill materials in addition, or in place of, wood fibers. Coal fly ash, waste glass, textile fibers, rice hulls, and sand have also been used as filler material. Some boards are extruded with foaming agents to produce a cellular structure.

G. Wood Finishing

Finishing is undertaken to protect the surfaces of wood products, to enhance the appearance, usually of the wood grain, to seal the pores of the wood, to provide a different or matching color, and to provide resistance to wear, liquids and chemicals, and to provide easier cleanability. Natural oils, dyes, and waxes, used in earlier times as finishing materials, are still in use, but production woodworking now relies primarily on synthetic materials. These materials are easier to use and have improved properties. Mass-production finishing methods are used in much of the woodworking industry, but the fact that appearance and style are important in furniture, leads to decorations, such as antiquing, stencilling, or application of decals that require manual operations.

G1. ***preparing surfaces for finishing*** - can involve any of the following steps: visual inspection to identify defects and decide which prefinishing steps are needed; sanding the surfaces for better smoothness, removing cutter marks and slivers and slightly rounding edges and corners; bleaching to remove any unwanted coloring; repairing of dents, scratches, cracks, and other defects; removal of excess adhesive, if any; and distressing the wood with marks, dents, etc., if the objective is to give an antique appearance. Sanding prior to finishing normally is done with a progression of abrasive sizes ending with the finest-grained abrasives and is performed both manually and automatically. See B8 above. Some open-grained woods such as walnut and oak are given a coating of a paste filler prior to sanding in order to fill indentations and produce a smooth surface.

Bleaching is used to remove stains or to change natural wood colors for some desired finishing effect. Bleaches are applied in liquid form by brush, rag, sponge, or spray. Some bleaches require a rinse or neutralization application, or both, after the bleach has acted. Oxalic acid, dissolved in water from powder or granules, is a common commercial wood bleach. Chlorine laundry bleach is also used for many woods and stains. Different woods and stains vary considerably in their ability to be bleached. Oak is easy to bleach; cherry is difficult. Penetrating oil stains do not always respond to water-based bleaches, which tend to work only at the surface. Application of mineral spirits or paint removers may also be needed.

G2. ***staining*** - Stains are dyes or pigments or both, carried in a liquid. The liquid may be water, "spirit" (alcohol or acetone), or oil. Stains are absorbed in the wood but do not obscure the grain pattern. They are applied to enhance or alter the grain appearance or modify the color of a wood. Sometimes their purpose is to hide an unattractive grain. One other purpose is to make a lower-priced wood appear to be a more expensive species. Stains are applied by spraying, brushing, wiping, rolling, or dipping. Excess stain is often wiped off after the wood has absorbed the color desired. Water-based stains raise the grain of the wood and light sanding usually follows the stain application. Some woods receive a wash coat before staining. This is a highly diluted coating of a sealer such as shellac or lacquer. Its purpose is to partially seal the grain of woods of different density so that the stain does not overemphasize the grain. Board ends, which absorb stain readily, may need extra wash coating.

G3. ***varnishing, lacquering, and painting*** - with woodworking, follow the same methods outlined in section 8D (painting). The term, top coating, is often used in the woodworking industry to differentiate between such a final coating and the staining, sealing, or filler coats. The major differences between woodwork finishing and other product finishing covered in section 8D, is that clear finishes are the norm with woodwork because of the desire to show the wood grain. Wood is also non-conductive, so that electrostatic methods of application require that the surface be made conductive with a preliminary treatment. In production situations, spraying is the primary finish application method. Both air-atomized and airless systems are used. Curtain and dip coating are also used in volume production. Roller, pad, and brush applications are used in one or few-of-a-kind custom shop situations.

Varnishes used include formulations based on polyurethane, urea formaldehyde, melamine, alkyds, phenolic, acrylic, polyester, and epoxy resins. All consist of the resin, a solvent and other ingredients. These include drying agents, plasticizers and, sometimes, colorants or "flatterers", which provide a satin finish. Others have catalysts in a separate container. Lacquers dry by evaporation of the solvent; varnishes dry partially by solvent evaporation but primarily by chemical reaction, often with the oxygen in the air or with an included chemical. Two-part varnishes, such as epoxies, have two separate liquids that are mixed just before application. These dry by polymerization of the mixed liquids. They provide tough top coatings with properties of chemical, heat, and abrasion resistance, including freedom from discoloration when water and alcohol are spilled on them. Natural varnishes using linseed or tung oil, or shellac, can make beautiful finishes but require more finishing labor and provide less protection to the underlying wood.

Often, multiple finishing coats are used to build up a thick top coating. When this is done, the dried coats are usually sanded with a fine abrasive or treated with a liquid de-glosser between coats to correct any irregularities and insure good bonding of subsequent coats.

G4. ***polishing*** - Wax polishing sometimes follows the varnish or lacquer coating to provide a deeper, richer finish. Carnuba wax, made from a Brazilian palm tree, provides very good hardness and gloss[1]. Polishes and waxes come in solid or liquid form and, in the liquid form, can be sprayed. Buffing follows the application. Manually-held electric buffing machines, robotically-controlled machines, and dedicated special machines may be used, depending primarily on the level of production of the product.

When eggshell, matt or semi-matt finishes are wanted, the required effect is often achieved by using varnishes with additives that provide the desired surface finish. Such finishes are also produced by buffing a gloss finish with a suitable compound. The best results with eggshell finishes is achieved with this method. Pumice is a typical abrasive.

H. Upholstery

Production of cushioned furniture, is largely by assembly operations that add cushioned or other special surfaces to wooden or metal furniture. The work is most commonly performed manually with hand tools. The prime purpose of upholstery is to provide more comfortable contact surfaces for the user of the furniture. A secondary purpose is to provide greater attractiveness from a decorative or textured surface. Upholstery elements normally include the following:

1. A frame structure of the furniture. This supports the seat or other upholstered surfaces. Frames are most often made of wood, using the methods described above. Other frames are made from metal "angle iron", other structural pieces, or tubing. Frame members are welded or bolted together. (See chapters 2, 3 and 7 of this handbook.) Sometimes, upholstery frames are molded from structural foam plastics.
2. Webbing to provide a semi-resilient support for springs and cushioning.
3. Springs or other resilient materials that support soft cushioning. (See *springs* and chapter 2.) Sometimes, some organic materials with the required resiliency are used without cushioning. Examples are chair seats made from cane, reed, or rush. These materials are obtained from palm trees and other plants, primarily from the tropics, and are made into strands that are usually machine woven into webbing of an attractive pattern. The webbing is then attached to a chair frame or other surface. Such webbing is also used as a decoration on paneling or furniture doors.

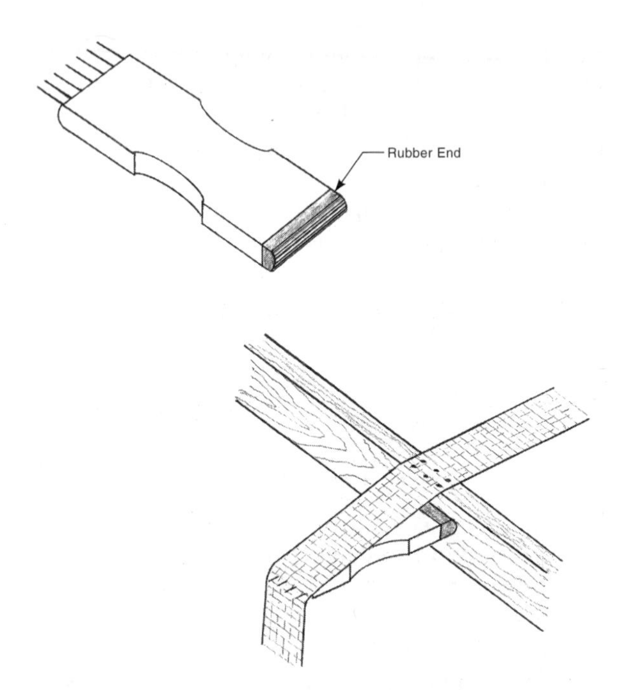

Fig. 6H1 Installing jute webbing with a webbing stretcher.

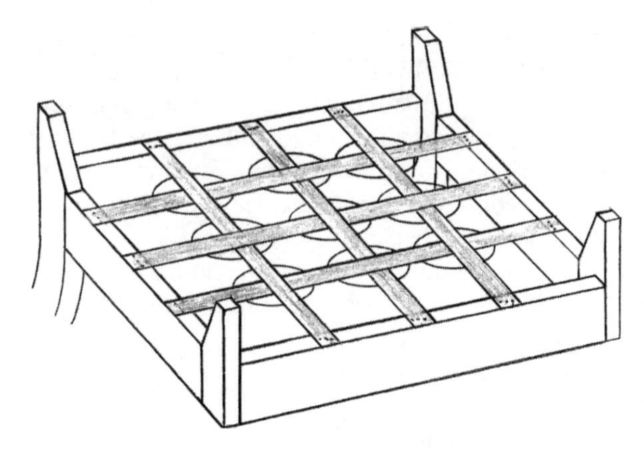

Fig. 6H1-1 The normal webbing pattern of an upholstered chair seat. Steel or fabric webbing is installed in a criss-cross weave pattern and coil springs are supported from where the webbing crosses.

4. Padding and cushions made from a collection of organic fibers or foam rubber, or foam plastic pieces, or molded from a flexible foam plastic (See 4C3b for reaction injection molding).
5. An outer covering, sewn from an attractive woven fabric. (See chapter 10).
6. Tufting, channelling, or other shaping of the upholstery to provide a special decorative effect.

H1. ***installing webbing*** - Webbing is made from perforated or corrugated steel strip, wire mesh, or narrow strips of woven fabric. Jute is a common fiber used in woven fabric webbing. These strips are stretched tight and fastened to a wooden frame by tacking or nailing. Fig. 6H1 shows how a stretching tool is used to put tension on fabric webbing. Webbing, under the seats of upholstered furniture, is usually installed in a crisscross woven pattern as shown in Fig. 6H1-1.

H2. ***installing springs*** - Three types of springs are common in upholstered furniture: coil springs, sinuous (sagless or zig-zag) springs, and marshall units (pocket springs), which are coil springs in canvas or burlap bags. The coil springs are used for seats, seat backs, and cushions. Sinuous springs are used when a low profile is desired. Marshall units are pre-assembled as complete rows or sets, ready for installation in seats, seat backs, or cushions. Fig. 6H2 illustrates the three types of springs. Coil springs and marshall units are fastened to fabric webbing with stitching twine. Hand needles are used to stitch the twine into the webbing and to tie knots to hold the coil spring. With metal webbing, the springs are held with twisted wire or corrugations, and slots in the webbing. Coil springs are positioned at the intersection of two strips of webbing. The coil springs are tied together at their other ends with additional twine and standard knots. The tying arrangements are made to insure that the coils do not fall over and that the spring action is in the right direction. A layer of burlap may be placed over the coil springs and attached by stitching.

Sinuous springs are attached to furniture frames with metal stampings that retain the spring wire. They include holes or slots through which nails or other fasteners to the frame can be driven. Sinuous springs are placed with a slight upwardly-curved surface. Small coil extension springs connect spring strips to one another and to side rails of the furniture frame. Marshall units are attached by stitching their burlap or muslin covering to the webbing and frame. Wire hog rings are sometimes used instead of stitching twine.

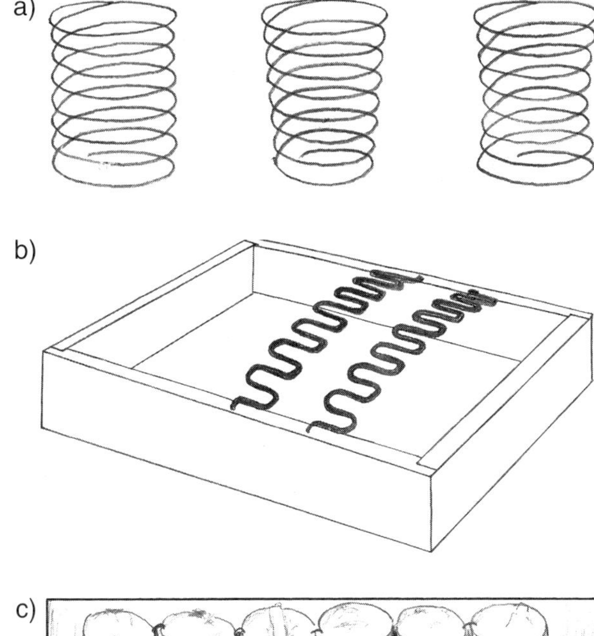

Fig. 6H2 Three types of upholstery springs: a) coil springs, of three different shapes, b) sinuous springs, being installed in a chair seat, c) marshall units, attached coil springs covered with burlap or other fabric as a subassembly.

H3. ***installing padding and cushioning*** - Padding and cushioning materials include sisal (fibers from the leaves of the hemp plant, sometimes rubberized); animal hair, curled by soaking in hot water and then rubberized; foam plastics or rubber; and cotton and polyester fibers. Fiber materials are gathered together in masses of the desired size and thickness. They are usually covered with fabric and fastened in place by tacking, stitching, or adhesive bonding.

H4. ***installing covering*** - Some furniture has two layers of covering over the padding, springs, and structure. An unbleached muslin layer provides strength to hold the cushioning materials in place and the final cover provides decoration and a durable surface. These covers are sewn with typical needle-trade techniques (as described in Chapter 10) with piping, pleats, and blind stitches suited to the styling intended. The pre-sewn cover is slipped over the furniture piece and may be tacked to the frame or stitched in place in certain areas so that it is held as needed.

H5. ***cushions, channeling, and tufting*** - The typical construction of an inner spring cushion is shown in Fig. 6H5. Channels (fluting or piping) and tufting are created by incorporating padding within two layers of the sewn final cover.

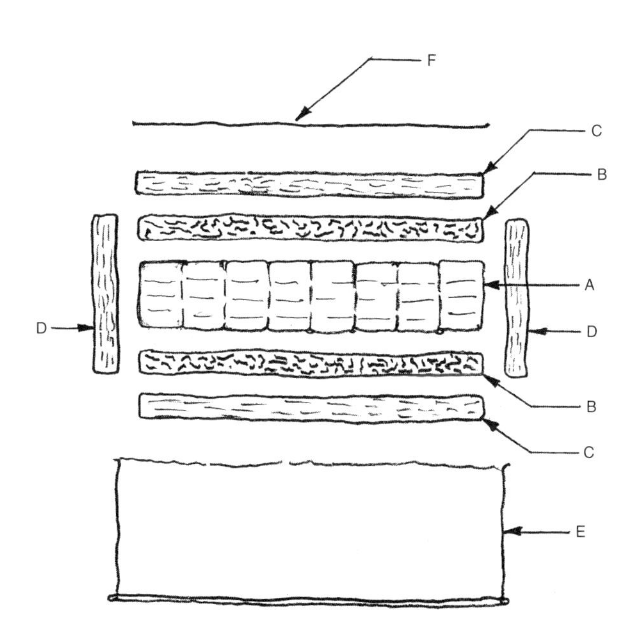

Fig. 6H5 Components of an innerspring cushion: A - marshall unit - coil springs fastened together and covered with burlap or other fabric, B - rubberized hair mat, above and below the marshall unit, C - resilient mat of polyester fibers, above and below the rubberized hair mats, D - resilient mats of polyester fibers on all four sides, E - bottom and side covers of upholstery fabric, sewn into a box shape, F - top cover of upholstery fabric to be sewn to complete the fabric cover of the cushion.1

Bibliography

1. *The Encyclopedia of Furniture Making,* Ernest Joyce, Drake Publications, London and New York, 1970.
2. *Modern Cabinetmaking,* William D. Umstattd, Goodheart-Willcox Co., Inc., South Holland, IL, 1990.
3. *The Art of Woodworking—Cabinet Making,* Time-Life Books, Alexandria, VA, 1992.
4. *Upholstering methods,* Fred W. Zimmerman, The Goodheart-Wilcox Company, Inc., South Holland, IL, 1992.
5. *Understanding Wood,* R. Bruce Hoadley, The Taunton Press, Inc., Newtown, CT, 2000.
6. *Woodworker's Handbook,* Roger Cliffe, Sterling Publishing Co., Inc., New York NY, 1990.

Chapter 7 - Assembly and Fusion (or Joining) Processes

A. Soldering and Brazing

Soldering and brazing are closely related methods for joining separate components. In both cases, a filler metal that melts at a lower temperature than the melting or maximum exposure temperature of the pieces to be joined, "wets" the surfaces to be joined and, when it solidifies, provides a solid mechanical or metallurgical bond between the pieces. In soldering, the filler metal has a liquidus (melting point) below 800°F (425°C). Common solders are alloys of tin and lead. Antimony and silver are also included in some solders in lesser percentages. Relative solderability of base metals in descending order is as follows: tin, gold, silver, copper, brass or bronze, lead, nickel, zinc, iron, steel, stainless steel, chromium, and aluminum. Solder joints are made to provide an electrical connection (the prime current application), to provide a seal, to provide a mechanical joint between parts (although the strength of soldered joints is usually inferior to those that are brazed or welded), or to aid in heat transfer between the parts being joined.

In brazing, the liquidus of the filler metal is above 800°F (425°C). Common brazing alloys utilize silver or copper as the major element. Phosphorus, silicon and aluminum are other alloying ingredients. Brazed joints are made for the same purposes as soldered joints but the prime application is to provide a strong mechanical assembly of separate pieces. Brazing may be an economical method for fabricating complex or bulky components including those composed of parts made with different processes or of dissimilar materials.

Soldering and brazing operations consist of six basic steps: 1) A cleaning operation almost always must precede soldering or brazing. The purpose is to remove oils, dirt, and other contaminants that would prevent the surfaces from being wetted by the filler alloy. (2) A flux is applied to the surfaces to be joined or an inert or reducing atmosphere is made to surround the joint to prevent oxidation, which would inhibit or prevent the surface wetting. 3) The joint members are both heated to a temperature above the melting temperature of the filler metal. 4) Filler metal is introduced to the joint where it melts and flows into the joint interstices by capillary action. 5) The joint cools and the filler metal solidifies. 6) Exccss flux, if used, is cleaned from the joint area since it could otherwise cause corrosion or other adverse effects.

A1. *solder application methods*

A1a. ***wire or rod soldering*** - Solder, in wire form, provides easy application in repair, touch-up and low-quantity production operations. The end of the wire solder is touched to the heated joint at the opening between the pieces to be joined. The solder melts and capillary action draws it into the joint opening. The wire solder frequently has a hollow core containing flux. For heavier work, such as plumbing joints, the solder may be in rod rather than wire form. In these cases, flux is not contained in the rod and is applied as a separate step before the workpieces are heated.

A1b. ***preform soldering*** - Solder preforms are parts made of solder alloy in particular shapes for insertion in the joints to be soldered. For production applications, it may be advisable to make solder preforms to provide exactly the amount of solder needed and to preposition them in the ideal locations in the joints being soldered. Rings, washers, spheres, and tubes are the most common preform shapes but special shapes can be made by processing solder alloys with conventional metal forming methods. Some preforms are made with flux incorporated. The preform is placed in the joint and, when the joint is heated, it melts and flows into the space within the joint.

A1b1. ***how solder preforms are made*** - Solder preforms are made with the standard methods for formed metal parts described in chapter 2. The exact sequence depends on the shape, size, and flux coverage of the particular preform. However, the following can be considered a typical operation sequence: 1) the solder alloy required is cast into an extrusion slug. 2) the slug as extruded as a round rod or wire or with whatever cross section is needed. If the shape is round, it may be extruded with a flux core. 3) If the desired preform is small, the rod or wire may be drawn to a smaller diameter. The flux core, if any, is correspondingly reduced in size. 4) Many preforms are made from flat stock. If so, the rod or wire is passed between paired pressure rollers, which change the shape from round to flat. 5) A common shape required is that of a round washer. A punch press operation then blanks a washer shape from the flattened stock. If the wire used is flux cored, the washer will also contain the necessary flux. Blanking, then, may be the final operation. However, any shape desired may be blanked from the flat stock. 6) If the preform is not flux cored, but a flux coating is wanted, the next operation is to spray a batch of the preforms with liquid flux while the preforms tumble in a barrel. When the sprayed flux coating has dried, the preforms are ready for inspection and shipment to the customer. Other preform shapes are made with common metal forming operations.

A1c. ***solder paste soldering*** - Solder pastes are homogeneous mixtures of finely powdered solder alloy, flux, and other ingredients in paste form. Plumbers' paste is used in joining copper tubing but, by far, the most common applications for solder paste involve the attachment of electronic devices to printed circuit boards. (See 13B2, 13B2b, and 13C6 for printed circuit board use of solder paste.) In plumbing applications, the paste is applied by brush or other dispensers to the tubing or fittings in the areas where they join. When the parts are assembled and heated, usually by torch, the solder in the paste melts and wets the joint.

A1d. ***Dip soldering (DS)*** - involves immersion of the joint into a bath of molten solder. The joint is cleaned and fluxed prior to immersion. The operation is rapid; heating and solder application take place at the same time. Solder flows into the joint by capillary action but does not wet unfluxed areas. The workpieces to be joined may be held by fixtures, which insure correct final dimensions and maintain proper joint clearances. Larger workpieces may be preheated prior to the operation to provide faster soldering and to avoid overcooling the solder bath. Common applications are the soldering of electronic assemblies including printed circuit boards (wave and drag soldering described below are dip soldering methods), electrical wires twisted together, or the "tinning" of wire ends prior to their being soldered to other components. Tinning facilitates the soldering operation. Another major application is the soldering of automotive radiators. The process is also useful for assembling small parts and can be an economical method when production quantities are limited.

A1e. ***wave soldering*** - is similar to dip soldering except that the liquid solder is lifted as a standing wave that contacts the joints to be soldered. Wave soldering is used extensively in the electronics industry to make connections on printed circuit boards. It is explained and illustrated in 13C5.

A1f. ***drag soldering*** - is another method used to solder circuit board connections in the electronics industry. It is an automatic soldering method wherein the workpiece joint is pulled through a static bath of molten solder. Circuit boards are held so that just the underside contacts the molten solder. Dross is automatically skimmed from the surface of the solder periodically, or as each circuit board is processed. Section 13C4 describes and illustrates this method.

A1g. ***ultrasonic soldering*** - is primarily a variation of dip soldering. The solder bath is subjected to high frequency vibration from ultrasonic transducers. The vibration removes oxides from the surface of non-ferrous metals and promotes metallurgical bonding of the solder. Flux is not required, so cleaning after soldering is also not necessary. This approach is particularly useful with aluminum and also with stainless steel, glass and ceramics. Wire tinning is a common application. Ultrasonic energy can also be applied to a soldering iron when hand soldering aluminum, or can be applied to a surface being rubbed while solder is flowed onto it. In both these methods, the ultrasonic vibration breaks up surface oxides and allows the solder to wet the surface of the workpiece. Fig. 7A1g shows ultrasonic soldering of aluminum heat-exchanger tubing.

A2. ***workpiece heating methods***

A2a. ***with soldering iron (INS)*** - Soldering irons are solid copper hand tools, nickel or iron plated, heated electrically or by immersion in a gas flame. Contact between the iron and the base materials conducts heat to the latter so that solder in contact with them will melt and flow into the joint.

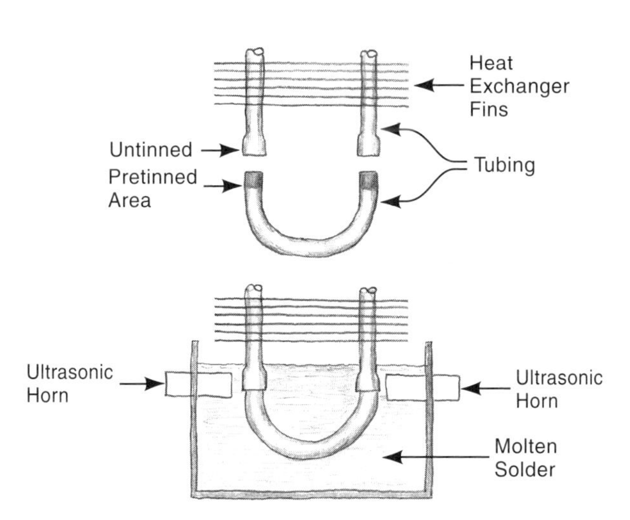

Fig. 7A1g Ultrasonic soldering of return bends of an aluminum heat-exchanger assembly. Ultrasonic vibrations from the horns, transmitted to the molten solder, break up oxide coatings on the surface of the aluminum tubing and enable the solder to wet the surface without the use of flux.

(Proper technique involves application of heat to the base materials rather than to the solder. The base material must be above the melting point of the solder if the solder is to flow properly into the joint and wet the surfaces.) Soldering irons are useful for prototypes and limited production, and light work such as wiring connections and printed circuit board touch up or repair. Large irons are used for sheet metal work.

A2b. ***with gas torch*** - An oxygas torch can be used to heat the workpieces to be joined. One common application is the joining of copper tubing and fittings. The tubing and fittings are heated by torch and the solder then melts and flows into the joint spaces by capillary attraction.

A2c. ***oven or furnace heating*** - Several methods are used: circulated air, infrared lamps, electric elements. An inert or reducing atmosphere may be introduced into the oven to facilitate the operation. Conveyorized pass-through ovens can be used for high production applications. Infrared oven heating is widely used for reflow soldering of electronic printed circuit boards and is covered in section 13C6a.

A2d. ***selective infrared heating*** - Infrared light can be focused onto very small areas, so the heated area can be limited to the area of the joint. It is not necessary to place the assembly in an oven.

A2e. ***vapor-phase heating*** - is a method used to "reflow" solder paste deposits on electronic printed circuit boards. See section 13C6b for a description of the process.

A2f. ***Resistance heating (RS)*** - uses the electrical resistance of the workpieces themselves to provide the heat required for soldering. Metal or carbon electrodes, carrying a low-voltage, high amperage current, are attached or clamped to the workpieces or brought in contact with them. The workpieces are usually preassembled with a solder preform and flux or solder paste prior to heating. One common method is to have one part of the assembly clamped to a grounded fixture. The other electrode then can be manually brought in contact with the other part. The heating operation is usually quick. The manually held electrode is lifted from the workpiece by the operator when the parts become hot and the solder starts to flow. Other systems use clamped parts and a manual, timed, or automatic switch. Shut-off can be made automatic with a circuit that senses the drop in electrical resistance when the melting solder wets the parts and improves the conductivity between them. Resistance heating is suitable for higher production levels and is particularly useable when other methods of heating are less workable due to inaccessibility of the joint or the need to avoid overheating other parts of the assembly. If the dimensions of the parts to be joined and their resulting electrical resistance vary somewhat, the process will be more difficult to control. In such situations, the process is limited to less critical applications.

A2g. ***laser heating*** - provides very rapid, extremely localized heating. Its principal soldering use is in the reflow soldering of electronic printed circuit boards. The method is described in section 13C6c.

A2h. ***induction heating*** - generates heat in the joint from its electrical resistance to induced eddy currents. Eddy currents are induced in the joint by high-frequency alternating current in a coil adjacent to or surrounding the joint. The amount of heat generated depends on the amount of electrical power in the coil, the magnetic properties of the workpiece material and how well the coil and workpiece are electromagnetically "coupled". The advantages of induction heating are its speed and the fact that the heat is localized in the joint area. The joint is preassembled prior to induction heating, and solder paste or preforms and flux are included in the assembly. Fixturing, either external or by self-fixturing parts, is required to hold the joint in the desired arrangement and the induction coil must be fabricated to fit the shape of the joint. A certain amount of development may be needed to insure good coupling between the coil and the joint to insure proper energy transfer. Because of the cost of these factors, moderate or high levels of production are required for economic operation. However the operation, once established provides very good repeatability and low unit costs. Robotic and other automatic operation can be developed if quantities are sufficiently large to justify the investment required. Typical soldering applications are: fittings for tubing and hose, container seals, and complex machine components where the strength requirements are modest enough to allow use of solder as a filler metal. Induction heating is more commonly used as a heating means for brazing than for soldering. Heating for localized heat treating is another application. Fig. 7A2h shows a typical arrangement of induction coil and workpieces.

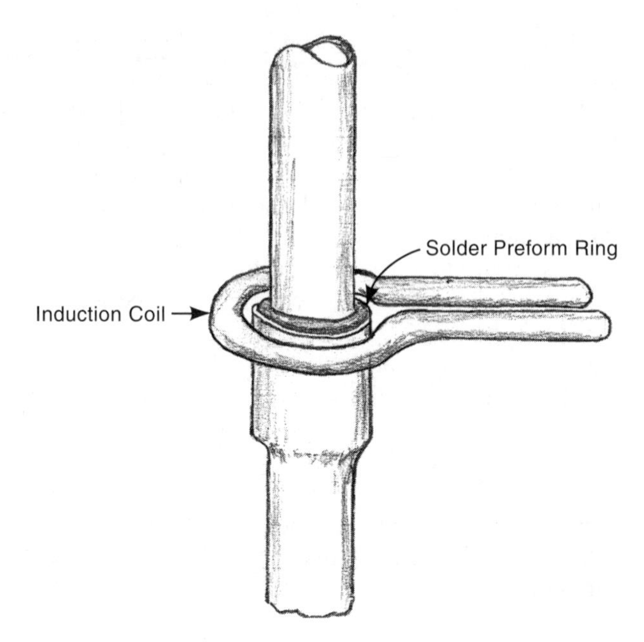

Fig. 7A2h Arrangement for induction heating to melt a solder preform in the joining of two pieces of metal tubing.

A2i. ***hot gas*** - Hot gas soldering is a technique used in the electronics industry for repair operations. It uses a stream of hot gas to provide localized heating of joints that require rework. See section 13C7.

A3. ***fluxes*** - All soldering, brazing and welding requires a clean metal-to-metal contact in order to produce a satisfactory joint. Oxides and other coatings on metal surfaces prevent this contact. Oxidation and other surface tarnishing are very likely at the elevated temperatures required in these metal joining operations. Fluxes are used to prevent and remove such surface coatings and to shield the metal surfaces so tarnishing does not reform. Fluxes also have another function; they provide greater fluidity (spreading and surface wetting) of the filler metal used. Fluxes may be in gaseous, liquid, or solid form, but solids, pastes, and liquids are the common forms.

Welding fluxes are included in the electrode coating or core of flux-bearing welding wire and include calcium carbonate, fluorspar, dolomite, and sodium silicate that produce shielding gas or shielding slag as well as fluxing action. Welding processes may use inert gases - usually argon - to shield the molten metal from formation of oxides and other unwanted compounds.

In brazing, combinations of borax, borates, boric acid, fluorides, chlorides, and fluoborates are used as fluxes. The most active soldering fluxes include inorganic acids and inorganic chlorides used in structural soldering applications, and hydrogen and hydrogen-chloride gases used in transistor manufacture[6]. Less active soldering fluxes are organic acids and halogens. Rosin, a derivative of pine sap, is a natural mild flux. It is inactive at normal temperatures but provides fluxing action when heated to above its melting temperature. Rosin is used in electronic soldering applications, though less extensively than previously. Almost all commercial fluxes are mixtures of several materials to provide the desired properties. Solvents, viscosity modifiers, combinations of active ingredients, surfactants, and other additives may be included. These ingredients are processed to a blended mixture in mixing equipment of the type described in 11G with the choice of mixing equipment dependent on whether the materials are solid or liquid and, if liquid, the viscosity involved. (See 13C1 for flux application in electronics manufacture.)

B. Brazing

B1. ***application of filler metal*** - While brazing alloy in wire or rod form can be fed into the joint by hand - and is commonly done so in repair, prototype, and limited-quantity production - in production situations, the filler metal is more often pre-placed in the assembly before heating takes place. Unless an atmosphere furnace is used for the heating phase, flux is also applied during the preassembly. The preassembled filler metal can be a preform - a ring, washer, shim, or other shape that fits the joint to be brazed - or can be in the form of a clad or electroplated coating of brazing alloy, or a paste applied to the joint. If paste is used, it contains both the brazing alloy in powdered form and the flux. Proper joint design is important so that the filler metal can fill the joint by capillary action. It is desirable to stake, press fit, or otherwise fasten the parts together before brazing, with sufficient strength so that the heat of brazing or handling prior to heating does not cause the parts to separate. If fastening is not feasible, a fixture to hold the parts during the brazing operation may be required.

B2. ***torch brazing*** - as in gas torch soldering, the flame from the combustion of a fuel gas provides the necessary heat. Either oxygen or air is mixed with the fuel gas (propane, acetylene, or natural gas). A torch, often hand-held, directs the flame and heat to the joint to be brazed. This approach is versatile, being applicable to assemblies of various sizes and to various production quantities, especially lower-volume production. The heat is directed at the joint and the entire assembly does not need to be heated. Brazing alloy is either prepositioned before heating or is fed in the form of rod or wire as the joint members reach the melting point of the brazing alloy. As in soldering, the process involves heating the joint members rather than the brazing alloy. When the joint members reach the desired temperature, the brazing alloy (filler metal) is heated by conduction. Because the operation takes place in regular atmosphere, flux must be applied to the joint area to prevent oxidation, and the flux must be cleaned from the assembly after the brazing operation is completed. Equipment costs are very low but operator skill is required for best results. Sometimes, multiple torches can be used and the torch movement and operation can be made automatic. Torch brazing is used to join copper and steel tubing in the refrigeration, air conditioning, and heating industries. Bicycle frames, furniture, carbide cutting tool inserts, and automotive components, are examples of components that are frequently torch brazed.

B3. ***furnace brazing*** - Atmosphere and vacuum furnaces can be used to supply the heat necessary for brazing. Parts are cleaned beforehand and assembled with the filler metal before entering the furnace. A fixture is often needed, but some parts are self-fixturing. The process is suitable for mass production, particularly when a conveyorized arrangement is employed. Box (batch) furnaces are also common and may incorporate the use of a retort to insure that the atmosphere is correct. The atmosphere, if not a vacuum, may be reducing or inert. Common gases are dry hydrogen, dissociated ammonia, nitrogen and argon. The furnace temperature is typically 100 to 150°F (55 to 85°C) above the melting point of the brazing alloy. The heating rate, temperature, time, and cooling rate can all be controlled closely with excellent repeatability. Little operator skill is needed once the set up is correct. A reducing or neutral atmosphere eliminates the need for flux. Other advantages are that multiple joints can be brazed simultaneously and distortion is at a minimum since the whole part is heated. There is also no flux entrapment since none is used. However the initial investment is higher and the heating cost is higher than with other processes. The process is used to braze jet engine parts, vacuum devices and automotive components.

B4. ***Induction brazing (IB)*** - is the same process described above in paragraph A2h but applied to brazing alloys rather than solders. It is used when the assembly has a shape to lend itself to placement of induction coils and when high-strength, heat-resistant joints are needed. The technique is widely used in brazing and is well adapted to brazing alloys except those that are in the high range of melting temperatures. Filler metal can be fed by the operator but is more commonly prepositioned before the induction heating operation. Flux is normally required though some induction brazing is done in a protective atmosphere. The process has the same advantages for brazed joints as for soldered joints: localized heating, fast process time, accurate control of heat, and uniform results with less need for operator skill. However, unique coils are usually needed for each assembly and development of the right coupling between the coil and workpiece may take some development. Induction brazing is used for aerospace components, appliance assemblies, industrial equipment, hand and machine tools, and hose and tubing fittings. Fig. 7A2h is also illustrative of brazing applications.

B5. ***dip brazing*** - similar to dip soldering. The assembly or the joint is immersed in molten filler metal, which is covered by a layer of molten flux. Filler metal flows into the joint but also will coat other portions of the workpieces immersed in the bath. Primarily for this reason, the process is not widely used. The method normally is restricted to small assemblies and only some brazing alloys.

B6. ***salt bath brazing*** - The parts to be brazed are immersed in a bath of molten salt that is maintained at a temperature slightly above the melting temperature of the filler metal. The method has a number of advantages: heating is rapid but overheating can be avoided; the bath provides protection against oxidation. Because of this, fluxing is often not required. However, fluxing agents may also be part of the salt bath. The salt bath also prevents decarburization of the workpiece (though it can occur after the workpiece is removed from the salt bath). Preheating is advisable to prevent the salt from freezing around the joint area of a cold workpiece. Multiple joints can be salt-bath brazed in one operation. The process is also useful for hidden joints. Carburizing and cyaniding can be performed in the same salt bath. Parts must be held in fixtures or held together by other means and the filler metal must be assembled to the joint beforehand. The brazed assembly is washed afterward to remove the salts, since trapped salts cause corrosion. Salt bath brazing is used with aluminum, copper, and ferrous alloys but is especially suited to aluminum.

B7. ***Resistance brazing (RB)*** - uses resistance to electrical current in both the workpieces and the electrodes that contact them to provide the heat necessary to melt the brazing filler alloys. The parts are pressed together between two electrodes and the current flows through both the electrodes and the parts. The parts heat up and additional heat is conducted from the electrodes to the parts. The method is best adapted for lower-melting-temperature silver brazing alloys. Carbon electrodes are often used. Regular resistance welding machines can be utilized for the operation. As with other brazing operations, a flux is needed. The method can provide very rapid brazing with precise, repeatable, high quality results. Equipment is economical. However, the process is best suited for small workpieces or small joints in larger assemblies. It is used for electrical components such as cable connectors and contacts. It is not suitable for large or complex assemblies. Fig. 7B7 illustrates the process.

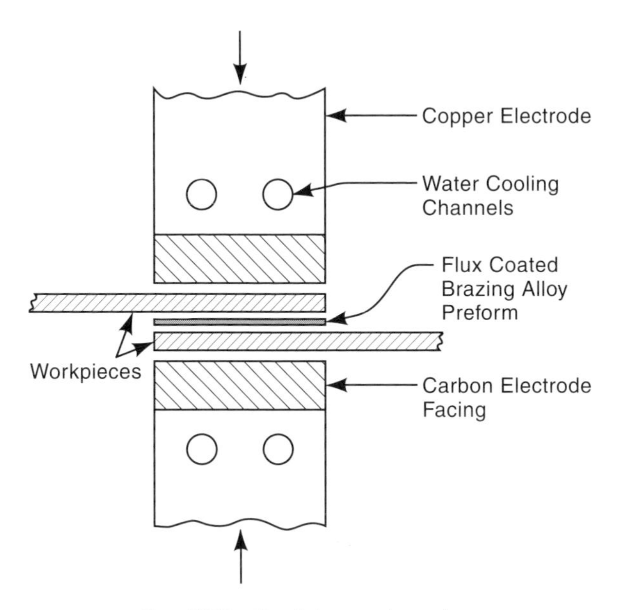

Fig. 7B7 Resistance brazing.

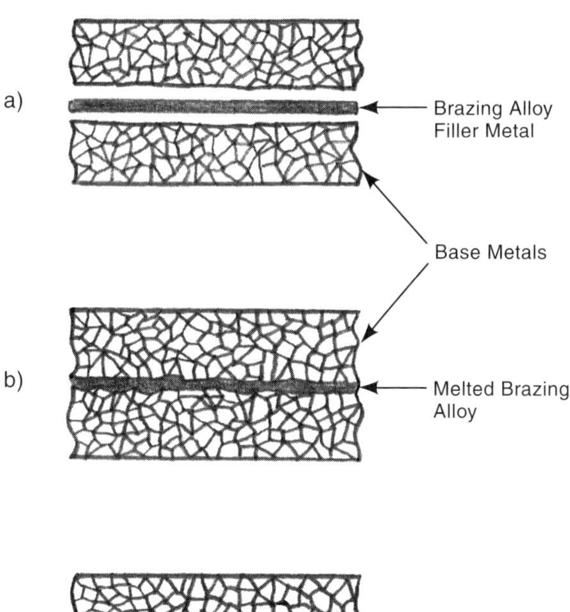

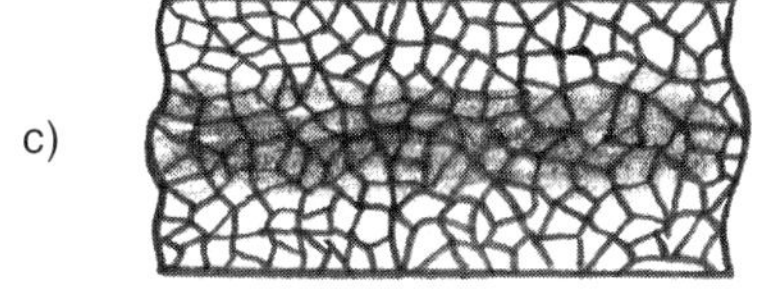

Fig. 7B8 Diffusion brazing with extensive diffusion of the brazing alloy: a) The brazing alloy is placed between two base-metal workpieces. b) Heat causes the brazing alloy to melt. c) Further heating causes the brazing alloy to diffuse into the base metals. d) After prolonged heating, the brazing alloy is fully diffused and the original junction of the two workpieces is no longer visible.

B8. ***diffusion brazing*** - is a variation of brazing in which the filler metal not only wets the surfaces to be joined but actually diffuses into them. The term *diffusion bonding* is sometimes used, although that term applies to diffusion welding also, when there is no filler metal. The process is the similar to diffusion welding as described in paragraph C13g but a foil of different material (a brazing alloy) is placed between the two surfaces of the joint. The brazing alloy melts and, under prolonged heating, diffuses extensively into the base metal. Diffusion brazing is most common when dissimilar metals are to be joined, and in the aerospace industry where it is used to bond titanium, nickel, cobalt, and aluminum alloy components. The furnace processing cycle can require from $^1/_2$ to 80 hours or more[1]. Fig. 7B8 illustrates the progression of a diffusion-brazed joint where the diffusion is extensive enough that the identity of the original joint is lost.

C. Welding

C1. *Arc welding*

C1a. ***Shielded-metal arc (SMAW)*** or ***stick welding*** - In this manual process, the electrode is in rod form and is covered with flux. The end of the electrode is drawn across the work, striking an arc, and then is held slightly above the work surface. The arc extends between the workpiece and the end of the electrode, and its heat melts both the electrode and the portion of the workpiece touched by the arc. As the electrode is consumed, its metal is added to the molten metal of the workpiece, forming the weld fillet. When the fillet cools and solidifies, the workpiece materials are joined. The process is illustrated by Fig. 7C1a.

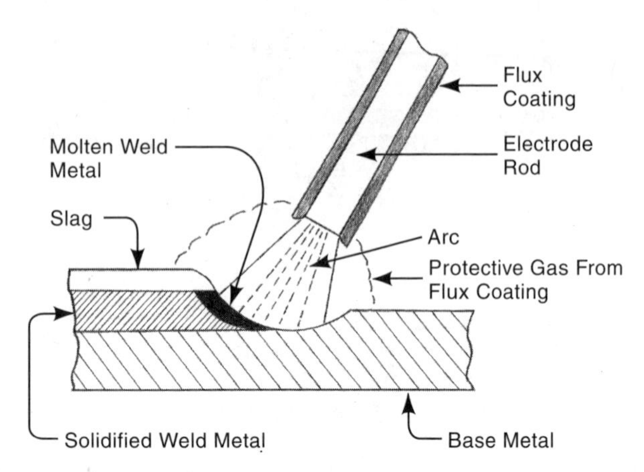

Fig. 7C1a Shielded metal arc or "stick" welding (SMAW).

C1b. ***Submerged arc welding (SAW)*** - uses granulated solid flux instead of a flux coating on the welding electrode. A thick layer of the flux covers or "submerges" the end of the electrode, the arc and the molten metal. The electrode is in wire form and is fed automatically. The dispensing of flux and movement of the welding gun are also automatic, as is the vacuum pick up of the flux granules after the weld. The process is used mainly for in-plant welding of products such as pipe or storage tanks, which have long weld seams. See Fig. 7C1b.

C1c. ***Flux-cored arc welding (FCAW)*** - In this process, the electrode is tubular and the flux is contained inside the electrode. The process thus differs from shielded-metal arc welding, which utilizes externally-coated electrodes. Otherwise, the two processes are almost the same. However, having the normally brittle flux contained inside a tubular electrode enables a coiled, continuously-fed electrode to be used, saving the time required to change welding rods.

C1d. ***Gas-metal arc welding (GMAW)*** - uses a shield of inert or non-reactive gas to prevent contamination of the workpiece and filler material. Filler material is supplied by a consumable, bare, solid wire electrode, that is fed continuously by the

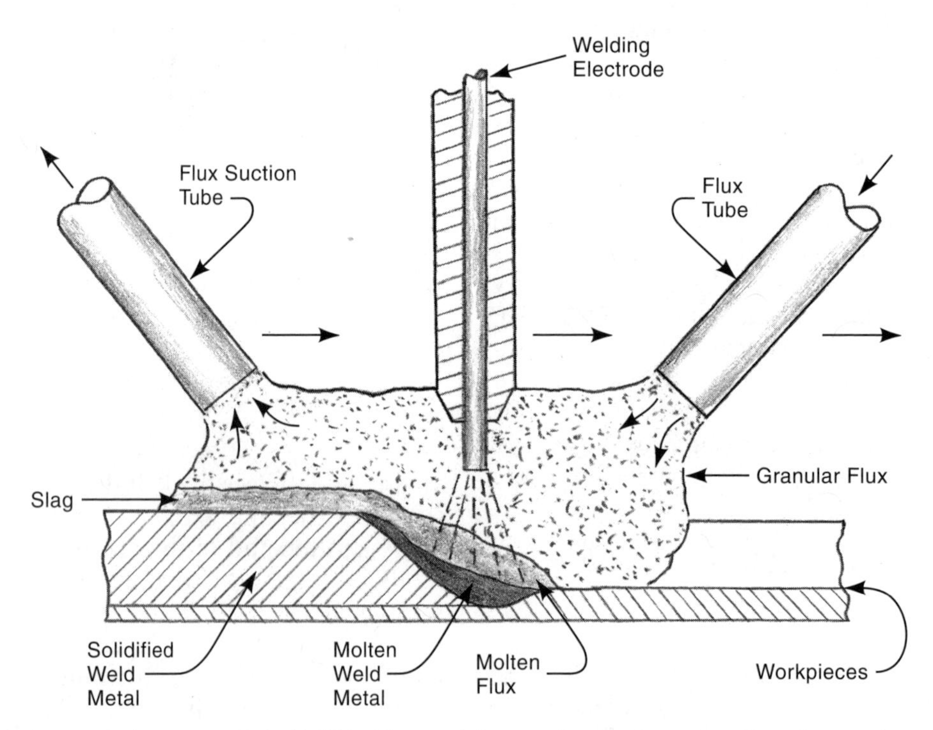

Fig. 7C1b Submerged arc welding (SAW). The weld seam is produced as the electrode and flux tubes move from left to right.

welding gun from a reel or coil as the operation proceeds. An arc between the electrode and the work provides the necessary welding heat. This method was formerly designated as metal-inert-gas (MIG) welding. Argon, helium or carbon dioxide or a mixture of them, are the most commonly used shielding gases. The shielding gas is also fed through the welding gun. The process is illustrated in Fig. 7C1d. The GMAW process is less labor intensive and faster than stick welding because welding-rod changes are not required and there is no slag to be chipped away. GMAW is used with material of 0.5 in (12 mm) or less thickness. Other processes are normally used with thicker stock. The process is widely used in production work but is less common outdoors because wind may interrupt the inert gas envelope. High quality welds on horizontal, vertical, and underside locations are feasible. Fig. 7C1d-1 shows the production GMAW welding of the mower deck for a professional riding lawn mower.

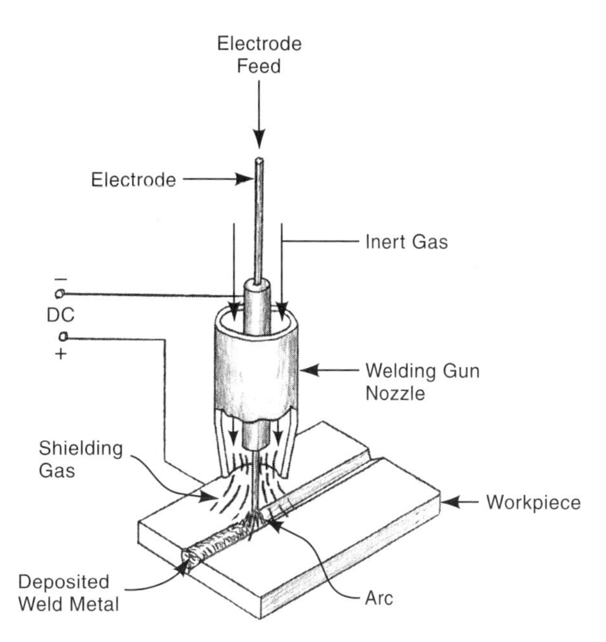

Fig. 7C1d Gas metal arc welding (GMAW).

Fig. 7C1d-1 The production (GMAW) welding of the mower deck of a professional riding lawn mower. (*Courtesy the Toro Company.*)

C1e. ***Gas-tungsten arc welding (GTAW)*** - was formerly designated *tungsten-Inert gas (TIG) welding*. It is quite similar to the GMAW process. However, the tungsten electrode in the GTAW process is not consumed and does not provide filler metal. An inert gas shield (argon, helium or a mixture of the two) is used. The electrode holder is water cooled. An auxiliary rod is used if filler metal is required. High-quality, clean, slag-free welds can be produced. The process is particularly applicable to welding of closely fitting sheet metal. Sheet thicknesses down to 0.005 in (0.12 mm) can be welded. Thicknesses over 0.25 in (6 mm) usually are better welded with other methods. All metals and alloys can be welded including reactive metals and high-temperature refractory metals. See Fig. 7C1e.

C1f. ***Plasma arc welding (PAW)*** - is similar to GTAW in that a non-consumable electrode is used and there is a plasma, a zone of ionized gas. However, in plasma arc welding, the amount of ionized gas is greatly increased and the heat of welding is supplied from it more than from the arc itself. The plasma is formed by constricting the orifice of the shielding gas at the location of the arc. The plasma consists of free electrons, positive ions, and neutral particles, and has a temperature higher than that of a normal arc. The arc may exist entirely within the welding gun rather than from the electrode to the workpiece. Fig. 7C1f illustrates the process, which is used for thin-walled materials that require high-quality welds.

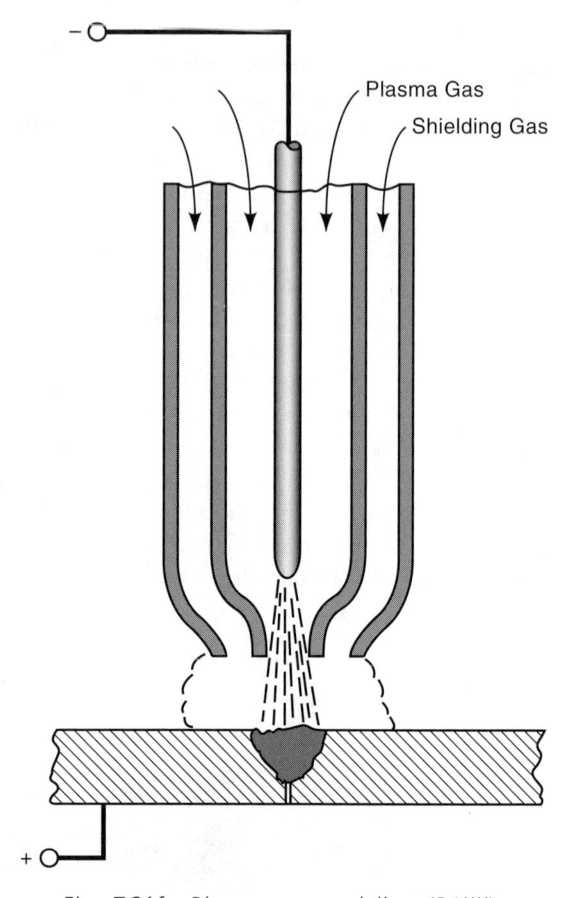

Fig. 7C1f Plasma arc welding (PAW).

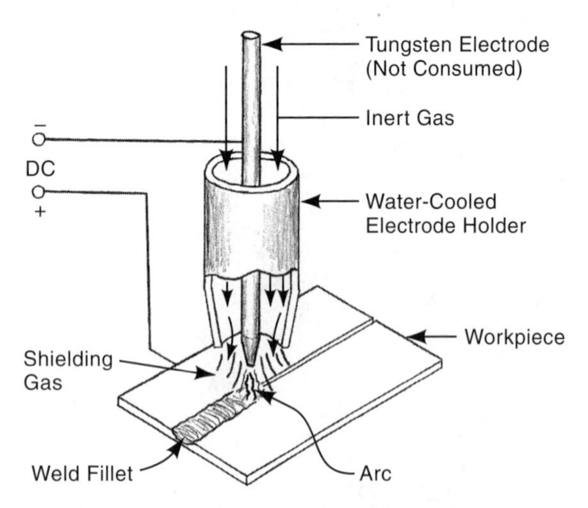

Fig. 7C1e Gas tungsten arc welding (GTAW). (Water cooling of the electrode holder is not shown.)

C1g. ***Electroslag welding (ESW)*** - is applicable to vertical joints in thick material. After an arc is initially drawn, granular flux melts and the consumable electrode is submerged in a pool of molten flux (slag) in the space between the two parts to be joined . The resistance of the slag to the electrical current between the electrode and the workpieces heats the slag and, in-turn, the joint area. Although classified as an arc welding process, the heat for fusion really comes from the resistance of the slag to the current flow. Water-cooled copper shoes dam the molten slag and metal. The copper shoes are gradually moved upward as the welding

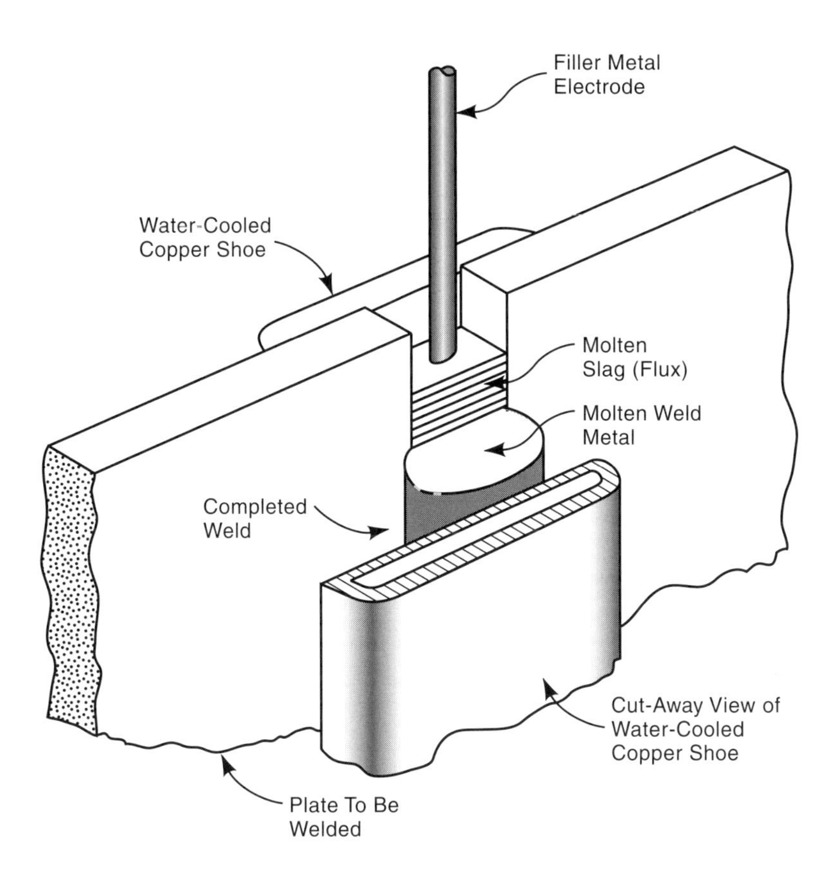

Fig. 7C1g Electroslag welding (ESW). Resistance of the molten slag (flux), to the current flowing from the electrode, melts the electrode filler metal and the plate to be welded. The operation moves upward as the metal melts and then solidifies. The copper cooling shoes move upward with the pool of molten slag.

progresses and the location of the electrode end also must be controlled. A steel tube may aid in guiding the electrode and to provide filler metal as it is consumed. Sometimes several electrodes are used to provide sufficient filler metal. The process is used extensively for welding plate 1 in (25 mm) thick and thicker, and heavy structural members. Shipbuilding, machinery, and heavy pressure vessel manufacture, and building construction are common applications. Fig. 7C1g illustrates the process.

C2. ***induction welding (electromagnetic welding)*** - Induction heating as described above (Paragraphs A2h and B4) can also be used to create sufficient heat at the joint of two workpieces to weld them together. Pressure is also often used to force the heated pieces together. Coils are designed to concentrate the heat at the edges or surfaces to be joined. Typical frequencies are 400 to 450 cps.[2] A number of ferrous and non-ferrous alloys and dissimilar metal combinations can be induction welded. Fig. 7C2 illustrates the principle when used to butt weld two sections of bar together. Other applications are the sealing of containers and the fabricating of structural sections from flat stock. Tubing and pipe, made from strip stock is also induction welded, but direct contact may be used rather than induction from a coil, to produce a high-frequency current (200,000 to 500,000 Hz) in the material. The strip is first roll formed to produce a round cross section.

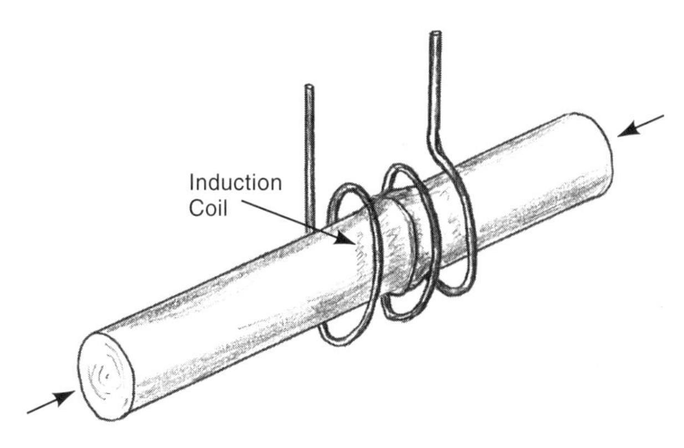

Fig. 7C2 Induction welding the ends of two metal bars together.

The edges are then heated to the high temperature to fuse the mating edges together, forming a longitudinal weld seam.

C3. ***Oxyfuel gas welding (OFW)*** - commonly known as gas welding or oxyacetylene welding, uses a stream of acetylene gas, the combustion of which, in a stream of oxygen gas, provides the heat necessary for welding. Both gases are supplied to the torch through flexible hoses. The temperature and heat concentration of OFW is less than that of arc welding processes. OFW is used for welding sheet metal parts together for such components as trays and tanks.

C4. ***Electron beam welding (EBW)*** - utilizes a narrow stream of high-velocity electrons, directed at the joint, to provide the heat of fusion. The electrons are emitted by a heated cathode and are focused and accelerated by electrostatic and magnetic elements, all in a high-vacuum chamber. The energy of the electrons striking the workpiece provides intense localized heat. The process works best when the workpiece is also in a vacuum of 10^{-4} torr, or lower. However, a less severe vacuum or even atmospheric pressure can be also used if the application permits a less focused electron beam. The process can be used for welding foils and thin sheets to heavy plates. Deep weld joint penetration is possible but the weldment must be small enough to fit into the vacuum chamber. Fig. 3N in chapter 3 illustrates an electron beam machining arrangement. When used for welding, the arrangement is very similar, differing only in that the energy is used to melt and fuse the workpiece metal rather than to remove it.

C5. ***Laser-beam welding (LBW)*** - This process uses the energy of a laser beam, an extremely concentrated beam of coherent, monochromatic (single wavelength) light, to melt the metal at a joint. A gas or solid-state laser emits pulses of coherent light that is focused to concentrate it within a very small area, providing an extreme concentration of energy (e.g., 60 kW/in^2) which melts metal in a narrow area. Penetration can be deep with a very narrow heat-affected zone and without heat distortion of the base metal. All metals can be laser-beam welded. The process is quite similar to laser-beam machining, which is discussed in chapter. 3 and illustrated in Fig. 3O. It is used for welding thin gauge parts, heat sensitive parts, and inaccessible areas. The pool of melted metal that forms the joint is very narrow. Edges to be joined must be carefully prepared to insure a narrow gap. Laser welding is extensively used to connect wire leads to small electronic devices. However, stainless steel plate up to about 3/4 in thick can be laser-beam welded.[2] The cost of laser equipment is somewhat high but much cheaper than electron beam welding equipment. A vacuum chamber is not required. Care must be taken to avoid eye injury from the scattered or reflected laser beam.

C6. ***resistance welding*** - achieves fusion from the heat generated by the resistance of the metal workpieces and the joint between them to the flow of a heavy electric current. The process is applicable when one or both of the parts to be welded is sheet metal. Electrodes supplying the current contact the work where the weld joint is to be located and apply pressure. Heat is generated in the area of contact of the workpieces; there is no external heat source. Pressure is increased when the metal softens. Fluxes and filler metals are not required. Resistance welding is rapid and the equipment required is not expensive.

C6a. ***Spot welding (RSW)*** - This resistance welding method is illustrated schematically in Fig. 7C6a. The process is widely used to join sheet metal components, usually with a series of localized "spot" welds. High electric current at low voltage is conducted by electrodes contacting and pressing on the work on opposite sides of the joint area. The current flows through the workpieces. Heat is generated by their electrical resistance, especially at the area. where they contact. The workpiece metal at the point of contact of the two pieces melts. When the current stops, the metal solidifies, forming a weld spot.

C6b. ***Seam welding (RSEW)*** - The electrodes are in the form of wheels. They press the parts together while rolling and conduct heavy current in a series of pulses. The pulses produce overlapping spot welds that provide a weld joint in the form of a seam. Vehicle gasoline tanks and other containers or ductwork where pressure tightness or fluid-carrying ability are required are a typical examples. Fig. 7C6b shows the process.

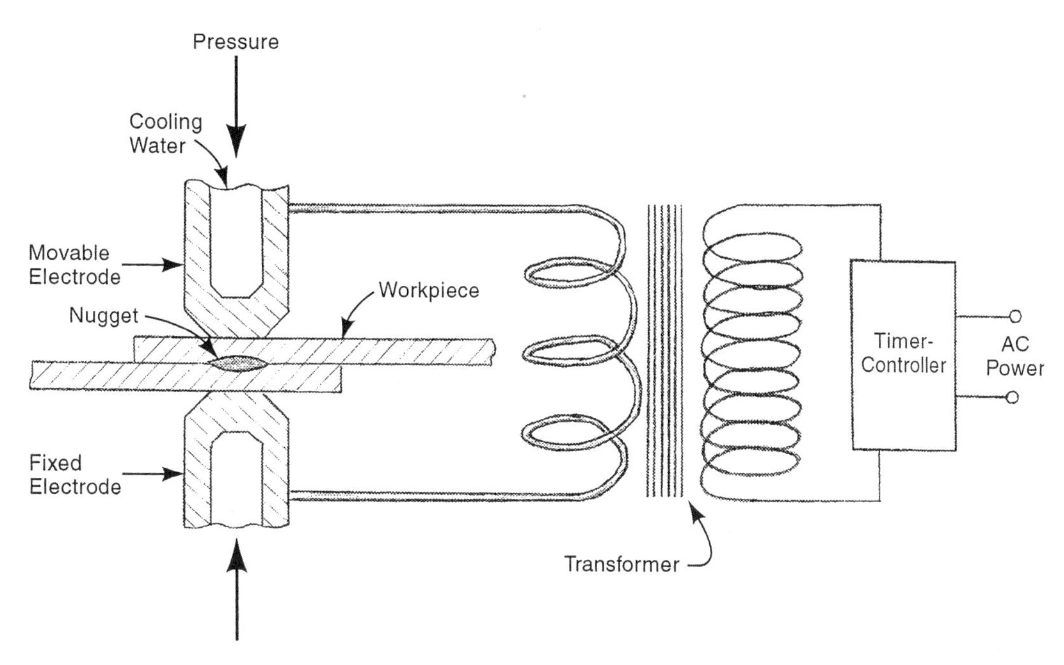

Fig. 7C6a Spot welding (RSW).

C6c. ***Projection welding (RPW)*** - uses raised elements on the workpiece surface to concentrate the electrical current and heat of resistance welding. The workpieces are pressed together by the welding equipment as in conventional resistance welding. The raised areas, projections, on one or both of the workpieces, perform the same function of concentrating the electrical current as does the localized pressure from electrodes in conventional resistance welding. Projections can consist of embossments of sheet metal parts or small raised lumps in the surfaces of castings or forgings. The projections can normally be incorporated in the tooling of the parts-making operation and thus involve little or no additional cost. They can be shaped to fit the requirements of the welded assembly and multiple projections can usually be welded at one time. An advantage of the process is that there are no surface depressions from the electrode pressure as there are in conventional spot welding. Additionally, lower current levels, less pressure, better consistency from piece to piece, and shorter welding times normally are feasible than if there were no projections and conventional spot welding were employed. Fig. 7C6c illustrates projection welding. The process is particularly applicable when production quantities are large. The process is also applicable when parts are too thick for conventional spot welding.

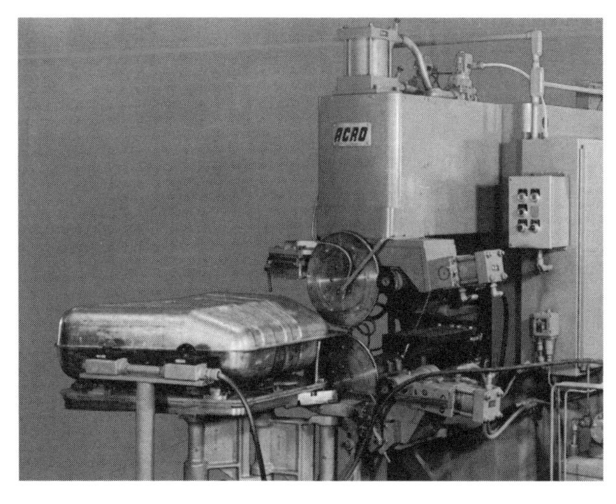

Fig. 7C6b Seam welding (RSEW) a vehicle fuel tank. (*Photo courtesy Acro Automation Systems, Inc.*)

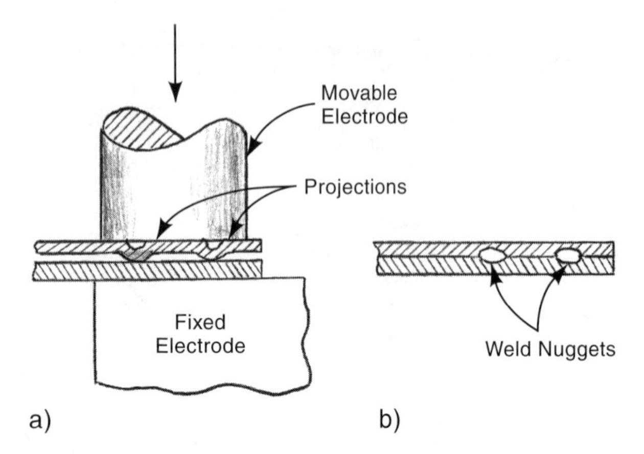

Fig. 7C6c Projection welding (RPW): a) prior to the application of pressure and electrical current. b) the completed weld.

C7. ***stitch welding*** - is seam welding with a non-continuous seam. It is used when pressure tightness is not required and has the advantage of producing less warpage than a full seam weld. The "stitches" are produced by switching the electrical current alternately on and off as the wheel-shaped welding electrode rolls along the weld area.

C8. ***stud welding*** - is an arc welding process used to attach fasteners or studs to the surface of a metal part. The stud to be welded acts as an electrode; a direct current arc is struck between the workpiece and the end of the stud. After the arc has melted and softened sufficient metal, the parts are brought and pressed together until the weld joint has solidified. A hand gun is used to hold the stud and provide the arc and pressure required. The arc duration and current, and the pressure, are automatically controlled. The stud withdrawal to create an arc and its subsequent pressure against the mating surface are effected by mechanisms in the welding gun. The gun also includes a ceramic ferrule to confine the heat and molten metal, and to shield the operator from the arc. Shielding gas also is used for some work. Stud welding can be employed whenever some object must be fastened to a surface. Typical examples are securing fasteners for liners in boxcars, trucks and tanks, concrete anchors for structures, electrical panel covers, and fastening legs and feet to appliances and machines. Fig. 7C8 illustrates the process.

C9. ***Friction welding (FRW)*** - The frictional heat developed when one part rotates and is pressed

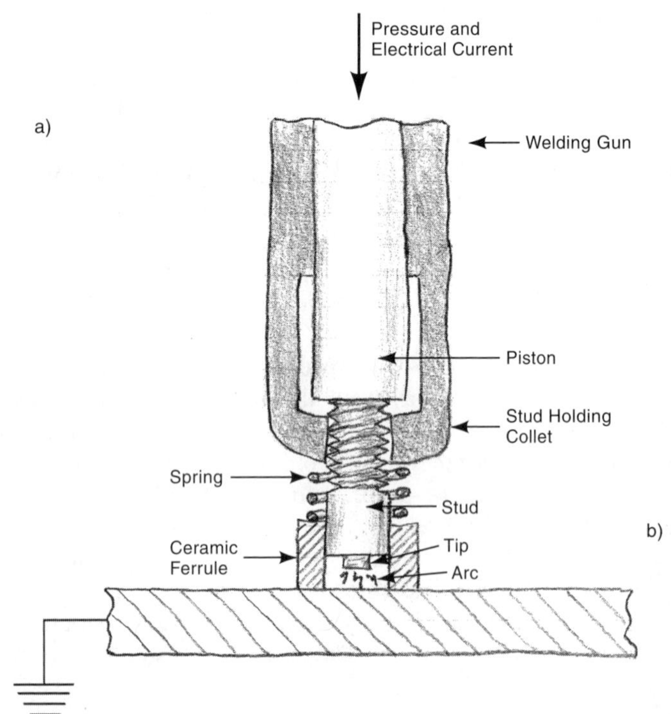

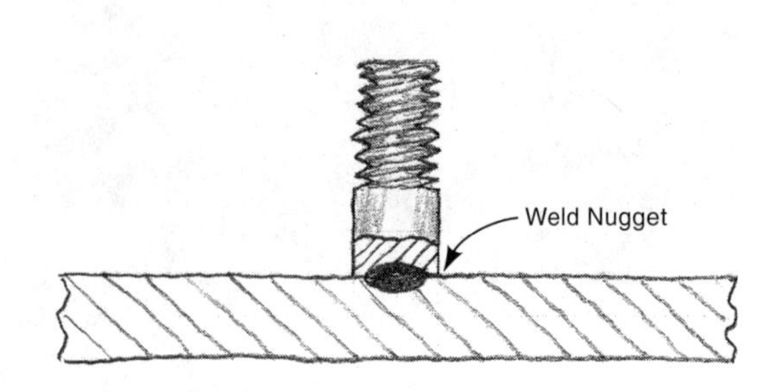

Fig. 7C8 Stud welding: a) the operation, b) section view of stud welded part.

against a stationary part can be sufficient to weld the two parts together. Typically, a lathe-like machine is used to rotate one part against a stationary part until the temperature of the interface is sufficiently high. Then the rotation is stopped and the parts are pressed more firmly together. The molten metal at the interface fuses together. When the joint has cooled, a strong weld can result. Either a strong electric motor or a flywheel is used to provide the rotational force. If a flywheel is used, the process may be called, ***inertia welding***. Rotational speed, time of rotation, and axial force must be developed for each application. The process is used to make joints between parts when at least one part can be rotated about a central axis. Pipe, bars, and other parts of circular or near-circular cross section are welded with this process.

C10. ***Flash welding (FW)*** - sometimes called ***flash butt welding*** - uses an arc between two workpieces, rather than an arc between an electrode and the workpieces, to provide the necessary heat for welding. The parts to be welded are both tightly clamped to electrical current sources. The parts are brought together with light pressure and the current is switched on. The parts are moved slightly apart so that an arc passes between them, softening and melting the metal at the interface. The parts are then pressed together with strong force. This creates the weld and upsets the metal in the joint area. Impurities are expelled in the flash. The current is shut off and the holding force is maintained until the joint metal solidifies. The process is illustrated schematically in Fig. 7C10. Preheating may be achieved in some cases by holding the parts together for a period with current flowing before the arc is struck. Extra material at the upset joint may be machined off as the application requires. The process is useful for welding tubular or solid workpieces. The process is also used in making pipe and tubing from strip stock. Non-ferrous and dissimilar metals can be flash welded. However, flash welding is not recommended for alloys containing high percentages of zinc, lead, tin, or copper.[2]

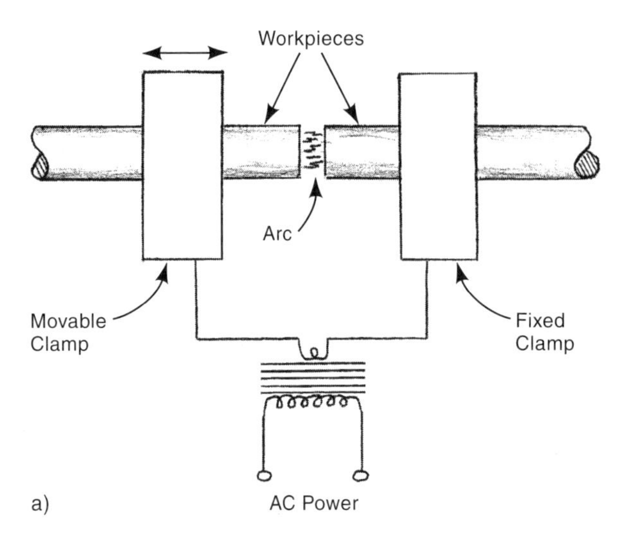

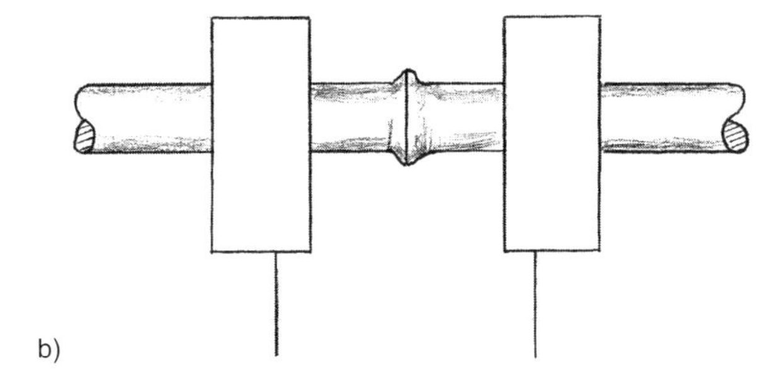

Fig. 7C10 Flash welding (FW) to join two pieces of metal rod: a) The workpieces have been moved together to touch and drawn slightly apart to create an arc which melts the metal at the ends. b) The workpieces have been pressed together; the heated metal at the joint has cooled and a finished weld results.

C11. ***percussive welding*** or ***percussion welding*** - is a variation of flash welding. Parts to be welded are securely held some distance apart against heavy spring pressure. When the hold is released, they move together rapidly. When the parts are about 0.060 in (1.5 mm) apart, a sudden release of electrical energy from a bank of capacitors, or from a collapsing electromagnetic field, causes a large, intensive arc to form. The arc duration is very short, only 1 to 10 milliseconds, until the parts contact one another, extinguishing the arc. The parts are then held together by the spring pressure. (In some set-ups, devices other than springs may be used to provide the movement and percussive force.) Only a small amount of metal is melted so there is little upsetting when the parts are forced together. The heat-affected zone is very small. Heat treated parts can be welded with this process without annealing.

However, workpieces must have relatively regular joint areas of 0.5 in^2 (320 mm^2) or less. The process is used to weld electrical connections, silver contacts to copper components, carbide tool bits to tool holders and other joints of dissimilar metals. Dissimilar metals not weldable by other processes can be welded by percussive welding. Because the melting cycle is so rapid, nearby heat sensitive components are not adversely affected.

C12. ***Thermit welding (TW)*** - is an old process that can be carried out in the field without the use of conventional welding equipment on a joint up to about 10 in^2 (60 cm^2) in area. The joint is made by casting molten metal around the workpiece edges to be joined. To weld steel or iron parts, iron oxide powder is mixed with aluminum powder. The following reaction takes place after the mixture is ignited:

$$3Fe_3O_4 + 8Al \Rightarrow 4Al_2O_3 \text{ (slag)} + 9Fe$$

The iron liberated by the reaction is in liquid form and can be directed with a sand mold to act as a filler metal in a joint. Iron alloy pellets may be added to reduce the temperature of the reaction and to provide additional molten metal. The molten metal, since it is very hot, readily fuses with the workpieces. After the molten iron solidifies, the sand mold can be broken and removed. Any excess iron in the joint area can be removed by chiseling while it is still hot and soft. The process has been used to join sections of railing and reinforcing rods, and to repair large castings, in locations where conventional welding equipment is not available. However, it is no longer in widespread use, having been largely replaced by more recently developed processes. With suitable oxides, a number of nonferrous alloys can be joined by thermit welding.

C13. ***Solid-state welding (SSW)*** - There are a number of processes that metallurgically join two workpieces without creation of a liquid phase:

C13a. ***forge welding*** - is an old process, originally performed by blacksmiths who, with borax, fluxed the joint areas of the parts to be welded, heated them to forging temperature and, with a series of hammer strokes, forced them together. Hammer strokes were normally started in the center of the joint and moved outward to force any scale out of the joint. When done correctly, the metals of each part would weld together, forming a strong bond. The process is currently little used today except in the forge-seam welding of pipe.

C13b. ***forge-seam welding*** - The forge welding process is now used to make pipe from strip steel material. The material is heated, formed into a cylinder and, while the edges are at forging temperature, they are forced together under the pressure of rolls or a die. See *pipe, metal* and Fig. P2.

C13c. ***cold welding*** - is, by definition, the solid state joining of metal components through the application of pressure without the application of heat from external sources. The operation takes place at or near room temperature. Surfaces must be reasonably clean before the operation; wire brushing is one cleaning method used. The joint area is subjected to localized pressure. In sheet materials, when the joint thickness is reduced by about 50 percent from the external pressure, welding takes place. One or both of the materials being joined must be ductile for the process to work satisfactorily. The process is most frequently used for small parts and electrical connections. Copper, aluminum, lead, nickel, zinc, and Monel are suitable for the method. The process is shown schematically in Fig. 7C13c.

C13d - ***metal cladding*** - is the joining of two or more sheets of different materials to produce one sheet with improved properties. It is carried out to provide corrosion resistance, improved appearance, or other properties not achievable with only one metal. U.S. coins, previously solid silver alloy, are now made with nickel cladding on a copper substrate. They have similar properties to the silver coins they replace, particularly for coin-operated vending machines that are engineered to reject slugs, but at much lower cost than the silver coins. The surface metal is bonded metallurgically to the substrate material. The two prime methods for cladding, described below, are roll welding and explosive welding.

C13e. ***Roll welding (roll bonding) (ROW)*** - is a solid-state process used chiefly for metal cladding. Two or more sheets of metal are joined by

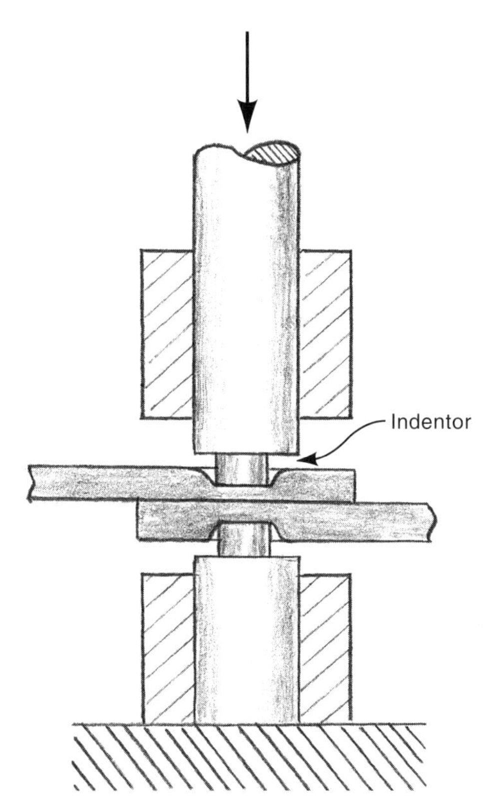

Fig. 7C13c Cold welding. Pressure from the opposing indentors deforms the workpiece metal and causes the mating surfaces to fuse together without application of heat.

passing them through pressure rollers. (Sheets must be clean before the operation.) If pressure is sufficient to reduce the thickness of the sheets, bonding between the sheets will take place. Heating may be used for some materials but is not necessary if at least one of the metals to be joined is ductile. Roll welding is applicable to sheets of the same metal and to dissimilar metals. Fig. 7C13e provides an illustration of the process. Sheets can be selectively bonded if the areas that are not to be bonded are coated with a material that prevents the bonding. Refrigerator cooling panels are made in this way. Before bonding, one sheet is coated with anti-bonding material in a pattern corresponding to the desired channels. The sheets are roll bonded together and then are heated to vaporize the anti-bonding material, The internal pressure thus generated expands the unbonded metal areas, creating channels for refrigerant to flow.

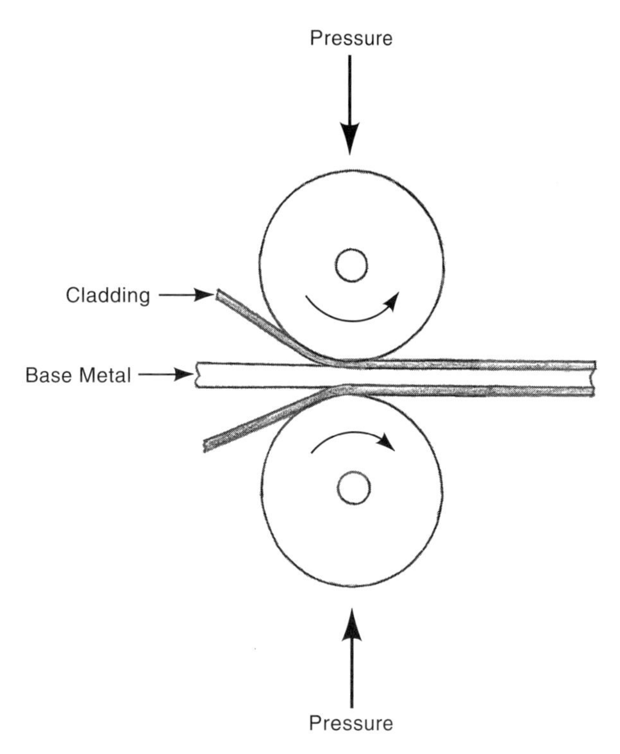

Fig. 7C13e Roll welding (roll bonding) (ROW).

C13f. ***Explosive welding (EXW)*** - is another process that relies on pressure rather than heat to join two workpieces. Explosive welding utilizes the enormous pressure resulting from the ignition of an explosive charge to cause the workpieces to bond together. There is no heat-affected zone at the joint. The process is principally used as a *metal cladding* method, to bond sheets of corrosion-resistant metal to plate material, particularly when large areas are involved. The bottom plate is placed on a solid support and the top sheet-the one to be welded to it - is placed above it at a small angle. (See Fig. 7C13f.)

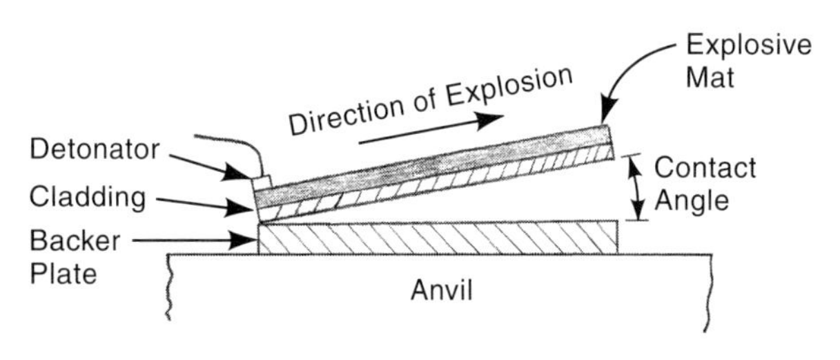

Fig. 7C13f Explosive welding (EXW). The extreme force of the explosion bonds the cladding to the plate.

An explosive charge, in sheet form, is placed on top of the sheet material. The explosion takes place in progressively, moving across the joint from the point where the metals are in initial contact. The compressive force from the explosion amounts to several hundred thousand pounds per square inch, forcing the metals together tightly. The sweeping nature of the compressive force aids in expelling any oxides or other foreign materials on the surfaces of the joint materials. The process is well suited to the welding of dissimilar metals. The joint is normally fully as strong as the weaker of the two metals being joined. Sheets as large as 7 × 20 ft (2 × 6 m) can be welded with this method. The process is also used for the internal cladding of tubes and pressure vessels.

C13g. ***diffusion welding (diffusion bonding)*** - This process involves placing the clean, highly smooth mating surfaces of the joint in contact under moderate pressure and at elevated temperature (but below melting temperature) for a prolonged period of time. The process temperature is typically about 70% of the melting temperature - on an absolute scale - of the lower melting temperature workpiece material. The pressure causes some plastic deformation of the workpieces at the joint area, but only at a microscopic level, chiefly to ensure close metal-to-metal contact. Gas pressure is often used in the process; another method is to confine the workpieces in a container having low thermal expansion properties so that, when the workpieces are heated and expand, they are forced together. Atomic diffusion and grain migration complete the metallurgical bond. Continuous, leakproof joints can be produced without deterioration of the properties of the workpiece materials. The process is most frequently used when dissimilar metals are to be joined. Sometimes, with dissimilar metals, a thin foil of a third metal is used in between the two workpieces. When furnaces with inert or protective atmospheres are used, reactive metals - titanium, zirconium, beryllium, and high temperature refractory metals can be welded. The operation is slow, so multiple parts or joints are often welded in the same furnace batch. The operation is most applicable to low-quantity production situations. The welding of titanium airframe components is one application. Other applications are in the electronic and atomic energy industries.

C13h. ***ultrasonic welding*** (of metals) - is another solid-state process. The joint surfaces are held together under light pressure and are vibrated at a high frequency (10,000 to 200,000 Hz). The rapid vibrations, parallel to the plane of the joint surfaces, break up oxide films, bringing the metal into intimate contact. There is some heating from the friction but not enough to melt any metal. The process, which is limited to thin materials, is illustrated in Fig. 7C13h. Sheet, wire, and foil can be ultrasonically welded. Lap joints are most common. The thinner of the two sheets is limited to a thickness of about 0.125 in (3 mm) for aluminum and 0.040 in (1 mm) for steel. The process can be used to weld dissimilar metals and even metals to glass and other non-metals. Temperature-sensitive components can be welded because of the low temperatures attendant to the process. Ultrasonic welding is used to make connections to electronic microcircuits, in sealing and packaging applications, especially with foil, and in bonding refractory and reactive metals. (Note: Ultrasonic welding is also used with plastics. See 4L4 for a process description.)

C13i. ***Friction stir welding (FSW)*** - is a relatively recent development for making solid-state welds in sheet or plate material. The joint metal softens but does not melt from the operation. The method utilizes a rotating cylindrical tool that bears against the workpiece to create frictional heat. The tool incorporates both a vertical pin and a horizontal shoulder. When the tool rotates, it is pressed into the joint to be welded. The friction from the pin softens the joint metal so that the pin can penetrate. Thereafter, the tool moves horizontally as it rotates. The shoulder of the tool then generates most of the frictional heat. As the tool rotates and moves, the joint metal is deposited behind the pin where it cools to form a bond in a continuous seam. The workpieces must be securely fixtured before the operation begins so they can withstand the separation force of the tool. The process is primarily used with aluminum, but can join dissimilar alloys, some of which are not normally weldable. The heat-affected zone and workpiece distortion are less than with conventional welding because the maximum temperature is only about 80% of the

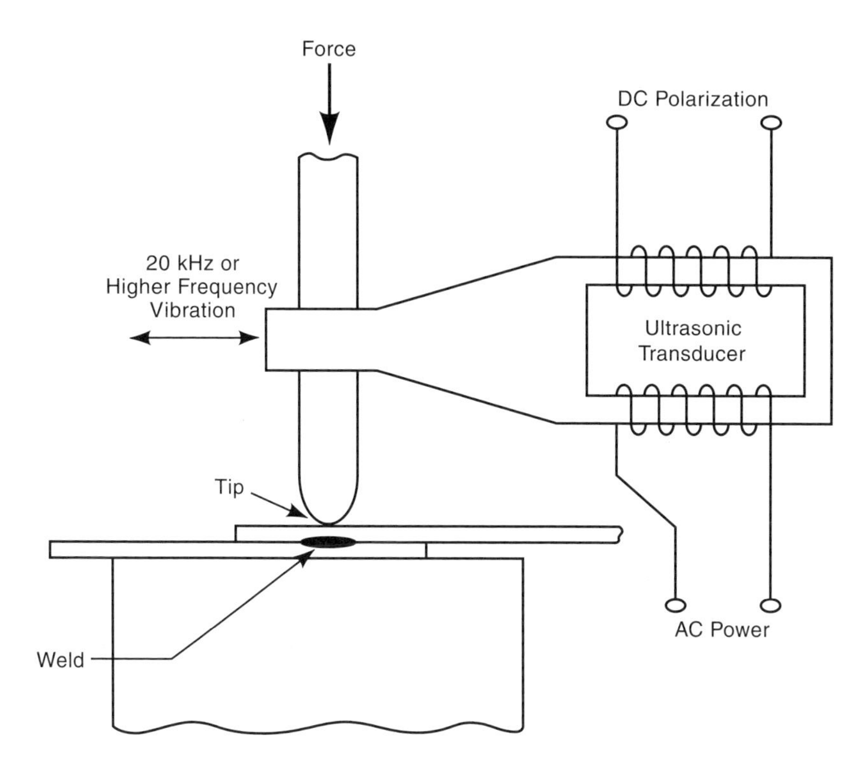

Fig. 7C13h Ultrasonic welding of sheet metal.

melting temperature of the metal. Friction stir welds are fine-grained, and superior in density (no entrapped oxides or gas), strength and fatigue resistance to arc welded joints in the same material. Normally, no filler metal or flux is consumed in the operation. Fig. 7C13i illustrates the principle and

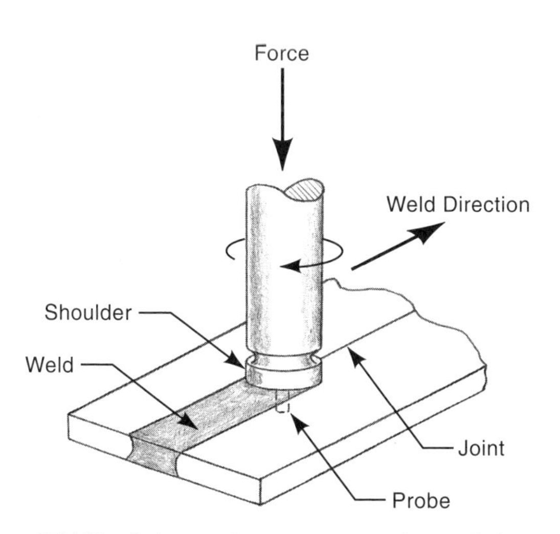

Fig. 7C13i Schematic representation of friction stir welding (FSW).

Fig. 7C13i-1 shows equipment for welding aluminum aircraft skin to frame members. Figs A4 and A5 in Chapter A show aircraft components welded by FSW. The process works best with lower-melting temperature metals and with butt joints or other simple joints. Steel can be welded by FSW with the use of shielding gas but with high tool wear.

D. Adhesive Bonding

Adhesives are materials that can hold other objects together by surface attraction between the adhesive and the workpiece material (adherend). All adhesive bonding processes have some characteristics in common, although the adhesive used and its application method may vary. Typical sequences are as follows: 1) Surface preparation: cleaning - The surfaces to be joined must be free of oils, dirt, and other contaminants. Some kind of cleaning operation is sometimes required. (See section 8 for cleaning methods.) Some surface roughening may also be advisable, and plastic adherends may require a surface ionization pretreatment. 2) Application - The adhesive is applied to one or both

Fig. 7C13i-1 Friction stir welding of aluminum aircraft skin to frame members. Note the clamping devices used to withstand the separation force of the probe of the welding tool. (*Photo courtesy Eclipse Aviation.*)

surfaces to be joined. 3) Assembly - The workpieces to be joined are assembled together and immobilized. Fixtures may be used in production situations to insure accurate placement and holding of the workpieces. Clamping may be used to hold the joined pieces together. 4) Curing - The liquid or semi-liquid adhesive becomes a solid bonding material. (Dry film or other solid adhesive first must be brought to a semi-liquid state in order to wet the joint surfaces). To be cured, the adhesive may be subjected to heat, ultraviolet, or other radiation, or may cure at room-temperature. Polymerization, some other chemical reaction, evaporation of solvent, or simple cooling may be involved in solidifying the adhesive. Clamping, if any, is then removed. 5) Excess adhesive, if any, is removed and the assembly is ready for use or further operations.

Adhesive bonding is useful when fragile or heat-sensitive materials are joined, when the weight of the assembly must be minimized, when the parts to be joined must be electrically insulated from each other, when materials dissimilar in composition, thickness, or stiffness are joined, when sealing as well as bonding are needed, and when vibration or sound dampening is desirable. However, joint strength of adhesive bonded joints may be less than with other methods, unless a large-area joint is possible. Adhesive joints are generally difficult to disassemble and operating temperatures of the joint may be limited. Some common applications are vehicle brake disc and brake band assemblies, plywood and wood furniture, helicopter blades, nameplate attachment, floor tile installation, automotive rear-view mirrors attached to windshields, skis, automotive headlight and taillight assemblies, building wall coverings and insulation, and even structural elements of aircraft. Also see *adhesives* (Chapter. A).

D1. ***surface preparation methods for adhesive bonding*** - are performed to improve or make possible the adhesion of the adhesive to the surfaces to be joined. They include various cleaning methods, and physical and chemical treatments of the surfaces.

D1a. ***cleaning*** - is performed to remove oil, grease, mold release agents, dust and other soils. Complete removal of these contaminants is necessary for best bonding results. Solvent, alkaline, and other cleaning methods described in chapter 8, section A may be used. Vapor degreasing is a preferred solvent cleaning method to avoid contamination of the solvent with oils and grease from previous operations on the workpieces.

D1b. ***surface roughening*** - Abrasive methods are used: hand or power sanding, grit blasting, grinding, or chemical etching. These convert a smooth surface to one with minute peaks and valleys and increased surface area, both of which improve the holding power of the adhesive. They also remove surface oxides that could impair bonding and are not removed by a normal cleaning process. Wire brushing may be another useful method to remove these oxides. Degreasing or other cleaning is often advisable after surface roughening to remove residual grit.

D1c. ***surface ionization pretreatment*** - Flame treatment, corona treatment, chemical etching, and plasma treatment are used to make the surface of plastic parts more adherent for bonding materials. All these processes reduce the slipperiness of plastic surfaces from their as-molded or as-extruded condition. See M1b through M1e in chapter. 4 for process descriptions. Flame treatment is used mostly with polyethylene and polypropylene materials. Corona and plasma treatments are used on plastics with low surface energy. Polyethylene and polypropylene can often be treated by applying a suitable primer instead of using flame treatment. PTFE ("Teflon"[1]), silicones and some thermoplastic elastomers also may need primer treatment to achieve satisfactory bondability.

D2. ***adhesive application methods*** - may involve any of the following: brush, spray, roller or spatula application for liquid and semi-liquid adhesives; assembly of dry film; use of hand-powered pumps or mechanical dispensing devices for application of hot-melt, other viscous adhesives and, sometimes, powdered or granulated adhesives. Dispensing devices can be integrated with assembly lines in high-production situations.

D2a. ***brush, spray, dipping, roller*** or ***spatula application*** - Liquid and semi-liquid adhesives can be applied by these methods to one or both of the cleaned adherend surfaces. *Pumps* and *dispensers* or *dispensing containers* can also be used. Applicable adhesives are epoxies, phenolics, polyesters, solvent cements, urethanes, vinyls, acrylics, and anaerobic adhesives. Roller application is common for bonding objects or materials of large surface area to other surfaces. Common applications are bonding of fabrics, fibers, elastomeric parts to steel surfaces in appliances and automobiles, and wall paper installation after the adhesive is applied to the back side of the paper, by roller or brush.

When a *contact cement* is used, it is normally applied as a liquid to the two surfaces to be joined. The adhesive is allowed to dry and then the two surfaces are brought together. The pressure sensitive adhesive "grabs" the two surfaces, once they touch, and pulls them together. For this reason, careful alignment before contact is essential. Contact cement is used to bond wood, plastic, metal, rubber, leather and cloth. A common application is the bonding of laminates to wood surfaces.

D2b. ***dry film application*** - A plastic film is placed, at room temperature, between the parts to be joined. The parts are held or pressed together while heat is applied to soften the film so that it adheres to the mating surfaces. The film can be cut beforehand to fit the joint area. The dry film approach can be less messy than bonding with liquid adhesives. Both thermoplastic and thermosetting plastic films are used. Heat is provided by ovens or other heating methods. If the part to be bonded is thin (e.g., a sheet of veneer), a contact heating source can be used. The method is used for making laminated items when the adhesive film is fed from a roll. Other applications involve the bonding of glass, leather, fabrics, and various metal parts. Aircraft structures and brake linings are bonded with this method. Epoxies, phenolics, and polyamide adhesives may be in the dry film form.

D2c. ***hot melt adhesives application*** - These adhesives are thermoplastics that are applied from either a dispenser or hand gun that melts the plastic or by spatula from a pot of melted adhesive. Beads or webs of the hot adhesive are applied to one of the joint surfaces. Parts are immediately brought together before the adhesive cools. When it does cool, a solid bond exists between the parts. This approach has the advantage of quickness, simplicity, and little need for operator skill. It is used primarily with polyolefins in household applications and with EVA (ethylene vinyl acetate) and polyamide adhesives industrially. Typical applications are: carton sealing, bookbinding, veneer and edge gluing to wood and particle board substrates, labeling, laminating, footwear manufacture, and bonding construction materials.

D2d. ***pressure sensitive and contact adhesive application*** - These kinds of adhesives are useful for commercial products that are intended to be adhered to other products, for example name tags, labels, small signs, etc. In these applications, the pressure-sensitive adhesive is sprayed on the back side of the object to be attached and a thin shield of plastic film is temporarily placed over the adhesive coating. When the device is used, the plastic film is stripped from the back and the sticky surface that is exposed will adhere to other surfaces. Urethane and silicone adhesives are utilized for this approach.

D2e. ***pump/ pressure application of solid adhesives*** - may be accomplished with powdered or granulated adhesives, which are then heated to activate them. Thermosetting adhesives, based on epoxies, silicones, and polyesters can be dispensed this way.

D3. ***joint assembly methods*** - Assembly methods for parts to be adhesively bonded parallel those for the assembly of parts to be fastened by other methods. The methods discussed below in section F are applicable. Adhesive dispensing usually is part of the assembly operation. Dispensing is facilitated by equipment that is operated manually, with a foot pedal, or with some degree of automation. For the kinds of adhesives that require some curing step such as the application of heat, radiation energy, or placement of the assembly in an oven, the operation and control of such apparatus is usually part of the assembly operation. It sometimes is necessary to hold the parts to be bonded during curing, and the fixtures to do this are manipulated as necessary by the assembly operator. When the assembly is automated to some degree, with robots or dedicated equipment, the dispensing and curing steps are similarly automated. Robots are capable of being set to spray, dispense, or otherwise apply an adhesive, and then assemble the adhesive-coated part to other parts, or to a fixture. Dedicated mechanical equipment often provides for dispensing and curing of adhesive as well as mechanical assembly operations.

D4. ***curing methods*** - All adhesives (except pressure sensitive types) require some physical or chemical change after application and assembly in order to achieve holding power. Curing normally involves a change from the liquid or semi-liquid state to the solid state. The change is brought about by one of several mechanisms: cooling, evaporation of solvents, or polymerization. Polymerization, as exemplified by epoxy and silicone adhesives, is usually brought about by heating, but with the proper catalysts, it can take place at room temperature.

In other situations, with thermosetting adhesives, polymerization is initiated by some other factor: Anaerobic adhesives polymerize at room temperature in the absence of air (oxygen). Contact with some metals yields metallic ions that act as catalysts for rapid curing. (These adhesives are used frequently by machinists to lock metal screw fasteners in place, or lock pulleys or gears to shafts).

Other adhesives, (acrylics), polymerize when exposed to ultraviolet (UV) light. One application is for dental fillings, particularly those that are visible and must match the tooth color. These can be cured in a short time at body temperature with a directed exposure to UV light. Another adhesive, a modified acrylic, cures at room temperature when an activator is applied to one of the bonding surfaces and the adhesive is applied to the other surface. No mixing is required. Other radiation sources that may be used for some adhesives are visible light, infra red light, electron-beam or microwave.

The cyanoacrylate adhesives ("super glues") are single component liquids that start to polymerize in only a few seconds from ambient humidity in the air or moisture on the adherend surface. Almost any surface has sufficient moisture to initiate polymerization. The cure takes place at room temperature. These adhesives cure more quickly if the workpiece surfaces are slightly alkaline. They are applied sparingly to only one surface of the joint. They are most useful for bonding small assemblies and are employed in adhering plastic parts.

Silicone and urethane adhesives and sealants polymerize from reaction with ambient moisture. They are used when flexibility is required in the joint. Silicone is particularly advantageous when sealing is required in wet environments subject to temperature changes.

Hot melt adhesives change from solid to liquid or semi-liquid when heated. They then flow to wet the surfaces of the joint and, when cooled, change to solids and gain sufficient holding strength.

Adhesives that harden from evaporation are often either water emulsions such as polyvinyl acetate (white glue) or hydrocarbon-solvent-based (rubber cement and household glue). Phenolics, and polyurethanes can also be solvent-based. Porous substrates such as paper or wood aid in evaporation of the solvent.

E. Welding of Plastics

(For welding of plastics, see sections L though L6 in chapter 4.)

F. Mechanical Assembly Processes

All products with more than one part require some kind of assembly operation and, if packaging is considered, even the one-piece products undergo assembly. Assembly can be defined as the act of placing and fastening two or more parts together to form a useful component. It has been, historically, a manual operation. Increasingly, however, it is being performed by automatic equipment, particularly if production volumes are large. When it is manual, the work is often assisted and augmented by a variety of tools, fixtures, and powered apparatus. Fixtures may be used to hold one part or the semi-finished assembly while other parts are being added, or to guide the added parts to a precise relationship to the base part and those assembled to it. Gages and other measuring devices help the assembler achieve the necessary precision. A wide variety of fastening methods is available to hold assembled components together. Hand tools and powered tools, including powered screw drivers, rivet setting presses, and similar machines are often used to facilitate the fastening of the parts. Sometimes, foot-operated devices provide the extra force needed for press fits, crimping, or other fastening methods. Often and increasingly, parts are designed to snap together with light manual pressure. These designs speed the assembly operation and eliminate the need for separate fasteners. Sometimes, it is advantageous to assemble only a portion of the parts comprising a product and to hold that partial assembly (*subassembly*) for later processing, often for use in a number of different products. Assembly operations commonly comprise the largest item of labor cost in the manufacture of a product.

F1. ***bench assembly*** - In lower quantity production, assembly of an entire product or subassembly may take place at one workbench and be performed by one skilled assembler. In the ideal situation, parts, fixtures, and tools are arranged in a "motion economy" arrangement on the workbench, so that the distances reached and movements made by the assembler are a minimum, and items needed are easily grasped and disposed of. Fixtures and other holding devices and gages may be fastened to the workbench, or otherwise provided, to ensure correct and accurate placement of parts. Hand and bench-mounted power tools are on hand to facilitate the operation. Some products are assembled with a bench arrangement with only one person performing the entire assembly operation, because the sensitivity or precision of the product requires assembly by an expert. Rifle assembly is one example of this approach. When large products such as machines are assembled in one workplace, it may be on the floor rather than on a bench, but with a similar arrangement for various parts and tools. These items are arranged around the work area in an organized fashion from nearby benches and racks. Fig. 7F1 illustrates a bench assembly operation and Fig. 7F2 shows some bench assembly workstations in addition to line assembly.

F2. ***assembly lines*** - are common in mass production industries. The assembly task is divided into a number of short-cycle operations, each

Fig. 7F1 Bench assembly. Note stacked bins for small parts, and devices mounted on the workbench for pressing component parts together. (*Courtesy The Toro Company.*)

Fig. 7F2 An assembly line for the production of a powered hand tool. The operation starts at the workstation to the left of the painted line on the floor (near the bottom of the photograph). The line is U-shaped. As the assembly progresses, the product is handed from station to station and is complete when it reaches the last workstation on the right side of the painted line. There, it is packed in an individual carton. The worker at the base of the "U" (with his back to the camera) operates test equipment. This operation verifies the function, speed and safety of the tool. The operators off-line in the left side of the photo are doing bench subassembly of components that are incorporated into the product on the line. (*Courtesy The Black and Decker Corporation.*)

performed by a different person, who specializes in one operation. A conveyor normally moves the assembly from workstation to workstation located according to the assembly sequence. At each workstation, only a small portion of the total assembly operation may be performed but, when the assembled workpiece reaches the end of the conveyor, the operation is complete. This approach has the advantage of allowing the individual operations to be short and simple enough so that training of operators is simplified and worker skill can be developed quickly. The materials and tools required for each operation are limited enough that they can be arranged for optimum efficiency. Additionally, the workpiece is moved automatically rather than by human labor. (Note: For smaller and lighter products, the workpiece may be moved from station to station simply by having each operator hand it to the next station. Fig. 7F2 illustrates an assembly line where this approach is used.) When conveyors are used, the type of conveyor depends on the size and shape of the product to be assembled and the degree of sophistication of the method. Belt conveyors, roller conveyors, or mechanical transfer devices may be used. Some lines utilize a pallet to hold the product being assembled and the pallet may have fixture elements to guide assembled components to the proper position. Some assembly line systems allow facilities for storage of several assemblies at each workstation so that there is a buffer between stations. This helps prevent a delay at one station from stopping the whole line. At some workstations, the assembly can be performed by robotic equipment or by special dedicated equipment. Test and inspection operations can be incorporated on the assembly line. To be effective, line operations must be balanced, i.e., the time required for the work performed at each station must be approximately equal to that of the other stations; otherwise, time will be wasted at the faster operations because the pace of the entire line is limited to that of the slowest operation. If one workstation has more work content than the others or has a less skilled, slower operator, the output of the line will be limited to the output rate of that station.

Assembly lines are used in the final assembly of almost all common products: automobiles, household appliances, power tools, electronic equipment, computers and computer accessories, clocks, watches, toys, hand tools, stereo and TV sets, furniture, production machinery, cameras, bicycles, and many other products. Fig. 7F2-1 and Fig. 7F2-2 show other typical assembly lines, one for the assembly of professional lawn mowers, the other for room air conditioners. Also see Figs. A2, A3, A9 and A10.

F3. ***automatic assembly*** - Assembly, historically, has been a high-labor-content operation and therefore a costly one. Cost-reduction activity among manufacturers has often focused on the development and use of automatic assembly equipment in order to reduce the amount of labor required. Earlier mechanization of assembly operations involved the use of specialized and dedicated equipment suitable for only one particular product configuration. Such equipment is still preferable in many situations, but the use of robotics, sensors, standardized module

Fig. 7F2-1 The assembly line for professional riding lawn mowers used for golf course putting greens. The mowers being assembled are conveyed from work station to work station by an overhead rail system. (*Courtesy The Toro Company.*)

Fig. 7F2-2 A well engineered assembly line for the production of room air conditioners. (*Courtesy Carrier Corporation.*)

Fig. 7F3a Multiple vibratory parts feeders bring small parts to the assembly station and put them in a consistent orientation. They can then be quickly picked up with short motions by a human assembler or a robot, or a special purpose pick and place mechanism. (*Photo courtesy Service Engineering, Inc., Greenfield, Indiana.*)

mechanisms, and computer control has introduced significant flexibility of application. This makes an automatic approach increasingly justifiable for more moderate production quantities.

F3a - ***parts feeding equipment*** - Automatic devices to sort and orient small parts prior to their assembly to a product have been used for many years to simplify hand assembly and as part of mechanized assembly equipment. These devices are sometimes referred to as bowl feeders. Using rotary, reciprocating, or vibratory motion, they take jumbled small parts from a bowl-shaped container and move them onto "tracks" (chutes or slides) where they move to a point where they are easily grasped and handled. Parts are in random orientation in the tracks but the tracks are designed so that parts not oriented in the desired direction drop off, or are forced off, and fall back into the bowl. Those parts that are properly oriented do not fall off but move on to the assembly operation. Parts that are slightly misoriented are straightened in position by guides along the track. Some feeders use a reciprocating pick-up device, that is shaped so that parts that are not properly oriented fall off as they are being picked up. Bowl feeders are used for all kinds of fasteners, pins, buttons, knobs, plugs, screw machine parts, stampings, and other small parts. Fig. 7F3a illustrates a group of typical vibratory feeders for several such parts.

F3b - ***high speed assembly with dedicated equipment*** - Products such as ball point pens, marking pens, razors, flashlight batteries, toothbrushes, electrical outlets, switches and plugs, lipsticks and other cosmetics, light bulbs, hinges, keys with plastic handles, and similar mass-produced items are assembled with automatic equipment that is designed and fabricated specifically for the product being assembled. The large production quantities of such products makes it economic to undertake the considerable development that dedicated equipment may involve. Such equipment includes the following elements: 1) a means to hold the product being assembled, 2) a means to move the product from work station to work station, 3) a means to deliver the parts and fasteners to the assembly stations and to orient them so that they can be inserted or positioned on the product. 4) a means to insert or position each part on the product assembly. 5) a means to fasten the parts to the product assembly, 6) often, a means to inspect or test that the part is installed properly and that it functions or that the product functions after all parts are assembled to it,

7) Often, a means to attach a label or print nomenclature on the product, 8) some kind of packaging for the product, for example a blister pack, a box, card, or bag.

Holding the assembly: The assembly machine includes a series of fixtures or pallets to hold the base part and the product as it moves through the machine.

Moving the assembly: Since it is normally not feasible to do all the loading, assembly and discharge of a product at one machine station, there are a number of workstations and the assembly equipment includes a means to move the assembly from station to station. The workpiece in process is held in a pocket or holding fixture as it moves. The movement can be accomplished on a turntable or carrousel, or by a linear conveyor of some kind. In either case, the motion normally includes both a moving phase and a dwell phase. The assemblies move from station to station and then remain there until the assembling operations at the station are completed, whereupon they move to the next station. The movements and dwell periods are simultaneous for all stations, although some linear systems are equipped to provide a storage bank of assemblies before each workstation, and more independent timing of the movement between stations, so that a delay at one station does not shut down the entire machine. This buffering is more common if there is a combination of automatic and manual assembly on the conveyor line. Movement and equipment actuation on these machines can be mechanical, hydraulic, pneumatic, or electrical. Rotary table machines generally are limited to simple products because of space limitations.

Parts feeding: (See F3a above.) Parts not arriving in a jumbled storage may be placed in magazines and may be inserted in them in the earlier fabrication operations, often automatically. Other parts are supplied in strip, bandolier, or reel form when the strip-making or reeling can be incorporated in the automatic parts-making sequence. For such parts, the orientation of the part from a mold or die is preserved for later handling in the assembly machine.

Parts insertion: "Pick and place" mechanisms of assembly machines grasp parts at the end of the feeding track, magazine, or bandolier, move them to the insertion point or positioning location on the product and, if applicable, insert them into the assembly. Movements may be controlled by cams, linkages, powered cylinders, stepping motors, or other mechanisms, or a robot may be used.

Parts fastening: In the case of threaded fasteners, many pick and place mechanisms also rotate the fastener after it is positioned. In other arrangements, the rotation and torque is applied at a subsequent station. Press and snap fits are handled similarly. Where applicable, machines are equipped with rivet setting, welding heads (including plastics welding), soldering heads, or adhesive dispensers. Labels and some parts are precoated with a self-sticking adhesive and are pressed after positioning.

Inspection and testing: Vision systems and other sensors on the machine can verify that a part is assembled to the product and that it is properly placed. Functional testers are also incorporated in some machines to verify that the product works satisfactorily. These units also test such factors as electrical characteristics, pressure-holding capability, and freedom of movement of some component.

Marking and labeling: In addition to label attachment equipment, machines can be equipped with various kinds of printing and marking devices using various methods including laser marking.

Packaging: Packaging equipment is often incorporated as part of the assembly machine.

Fig. 7F3b shows an indexing-table arrangement suitable for the assembly of small components. Fig. 7F3b-1 illustrates a much larger machine engineered to assemble and weld rear axle assemblies for pick-up trucks.

F3c. ***robotic assembly*** - Robots are "reprogrammable multifunctional manipulators designed to move materials, parts, tools, or other specialized devices through variable programmed motions for the performance of a variety of tasks" according to the Robot Institute of America.[2] They are described and illustrated in chapter. 14 of this handbook. One common task performed by robots is the moving of parts from a source (feed magazine, tray, vibratory or other orienting device), to an assembly, and positioning or inserting the part into the assembly. However, a very wide range of operations is possible, depending on the production conditions, the product being assembled, and the type of robot available. Robots are also used to apply tools in the assembly operations. Spot and arc welding can be done robotically in some cases. Robots can apply heat from a gas torch or other heat source to carry

Fig. 7F3b An indexing table arrangement for small assemblies. At each stopping point for the table, as it rotates, a mechanism places a part, fastens it or performs a different operation until the assembly is completed and ejected at another station. Using standard mechanisms, such a machine can be engineered to assemble various different products. (*Courtesy Demco Automation.*)

Fig. 7F3b-1 A special machine for the automatic assembly and welding of rear axle assemblies for pick-up trucks. (*Courtesy Acro Automation Systems, Inc.*)

out various kinds of joining operations. Vision systems may aid in directing the robot to the part to be grasped, to identify it, and to verify its correct position. Such systems also aid in directing the part or tool held by the robot to the exact location. Tactile, force, and sound sensors may aid in performing the operations successfully.

The fastest assembly robots are *pick and place machines*, robots with limited motion flexibility, relatively simple programmability, high speed movements, small size, and positioning accuracy to about 0.010 in (0.25 mm). One notable use is in the electronics industry to populate printed-circuit boards.

F4. *mechanical fastening methods*

F4a. ***assembly with threaded fasteners*** - Threaded fasteners (screws, bolts, nuts, machine screws, cap screws, set screws, and drive screws) are probably the most common type of fasteners. With the simplest methods, applicable to repair situations or individual or low-quantity production, they can be started manually and tightened with a hand screwdriver or wrench. In high-production situations they can be started and driven with special equipment that feeds the fastener from a magazine or a vibratory hopper, inserts it in the appropriate opening, and rotates and tightens it with automatic torque control. When production quantities are intermediate, insertion may be manual followed by the use of a powered driver, positioned either manually or automatically. Powered drivers can be electrical or pneumatic. Impact, impulse, or shutoff wrenches can be adjusted to apply the proper torque to the threaded fastener. Robotic assembly is possible with robots that obtain or contain the fastener, position it, and drive it. These more sophisticated methods require sufficient production quantities so that the cost of obtaining or developing and setting up the equipment can be amortized.

Some screw fasteners are self tapping. They either cut or form the threads in the hole. They are useful in applications where the workpiece material is soft enough to be cut or formed. Similar screws for wood are both self-drilling and self-tapping so that even a pilot hole is not necessary.

Fig. 7F4a illustrates a variety of threaded fasteners.

F4b. ***riveting*** - is a common and long-used method for permanently fastening parts together. The rivet, a fastener with a head on one end and a smooth shank, is inserted through aligned holes in the parts to be joined. The shank end is upset (a head is formed) from force against a die, to lock it in place and clamp the parts together. A steel shank may be heated to red heat to facilitate upsetting if the rivet is large and solid, particularly in field structural applications where only a limited amount of force can be applied. (In factory situations, upsetting is almost always a cold forming operation.) Upsetting is performed with press force, repeated strokes of a shaped hammer, or orbital motion of the upsetting tool. Tubular and semi-tubular rivets, split (bifurcated) rivets, and eyelets, are typically hopper-fed, inserted, and clinched in inexpensive equipment. Traditionally, such equipment has been manually operated with the clinching powered by foot force, air pressure, or electricity. For high production situations, the operations can be made fully automatic.

Self-piercing rivets are those that punch a hole and are inserted in it, all in one operation. These rivets are used with non-metals and metals to a hardness up to R_b50. The maximum metal thickness for self-piercing is about 0.15 in (4 mm).

In the aircraft industry, special machines punch the holes, insert the rivet, and upset the shank end, all in one operation.[4] Tubular and semi-tubular rivets are clinched in place by flaring the tubular shank end. Blind rivets are useful when it is not possible to have access to the back side of the riveted assembly. Although holding power is reduced compared to that of conventional rivets, blind rivets can be inserted and clinched easily with hand tools that pull a shaped center plug that expands the tubular shank. Explosive rivets are blind rivets that carry an explosive charge in the shank end. When heat is applied to the rivet, the explosive charge in the shank end is activated, expanding the tubular walls and clinching the rivet. There is no need for heavy axial force.

Fig. 7F4b illustrates the upsetting of several types of rivets.

F4c. ***stitching/stapling*** - is useful for fastening sheet materials, primarily non-metals but can be used with sheets of softer metals. The fasteners are made from either coiled wire or wire preformed into a U-shape and fed from a magazine. With both types, the wire ends are mechanically driven through the sheets to be joined without any holes

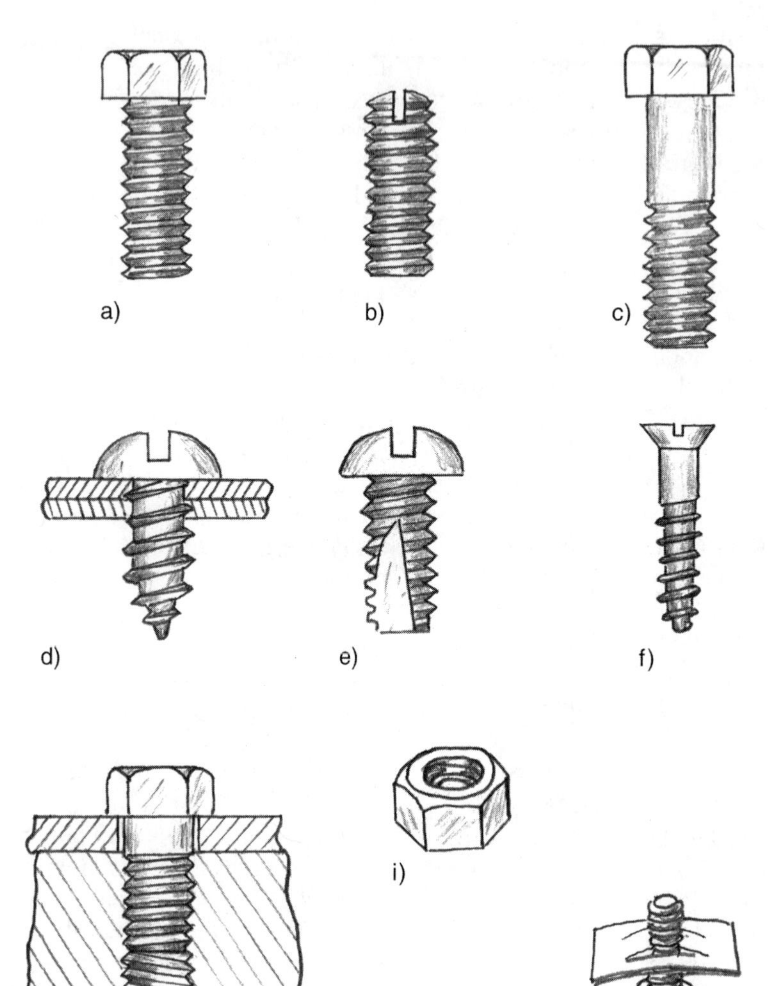

Fig. 7F4a A collection of threaded fasteners: a) hex-headed machine screw b) headless set screw, c) hex-headed bolt, d) thread-forming sheet metal screw, e) thread-cutting screw for plastics and soft metals, f) flat-head wood screw, g) cap screw assembly (in threaded hole), h) hexagonal nut, i) speed nut and screw.

being made beforehand. A forming die, held below the sheets, clinches the wire. The operation, on a semi-automatic basis, has been in existence for many years with operators using stitching or stapling machines. These machines are operated by either foot pressure ("kick press"), pneumatic, or electrical force. Fully automatic equipment is used if production quantities are sufficient. Stitching and stapling are rapid and low in materials costs. The process can be used with fiber, paper, wood, plastics, leather, fabric, and some metal parts. Stitching is commonly used to fasten a sheet to a wood backing. It can also be used to fasten wires, tubes, or rods up to about 1/4 inch (6.3 mm) in diameter to a sheet with a suitable die to clinch the wire around the item fastened. Cold rolled steel sheets to 14 gage (0.080 in)(2 mm) have been stitched together. Fig. 7F4c illustrates typical stitched or stapled joints.

F4d. ***snap fit fastening*** - Snap fits occur when the parts to be assembled are designed so that the act of assembly engages elements on the parts that cause them to be held together. These elements normally incorporate some flexibility of one or both of the parts, creating a snap or click when the parts engage. The snap fit approach is most common in plastics, which have some flexibility, and where it is relatively easy to mold catches

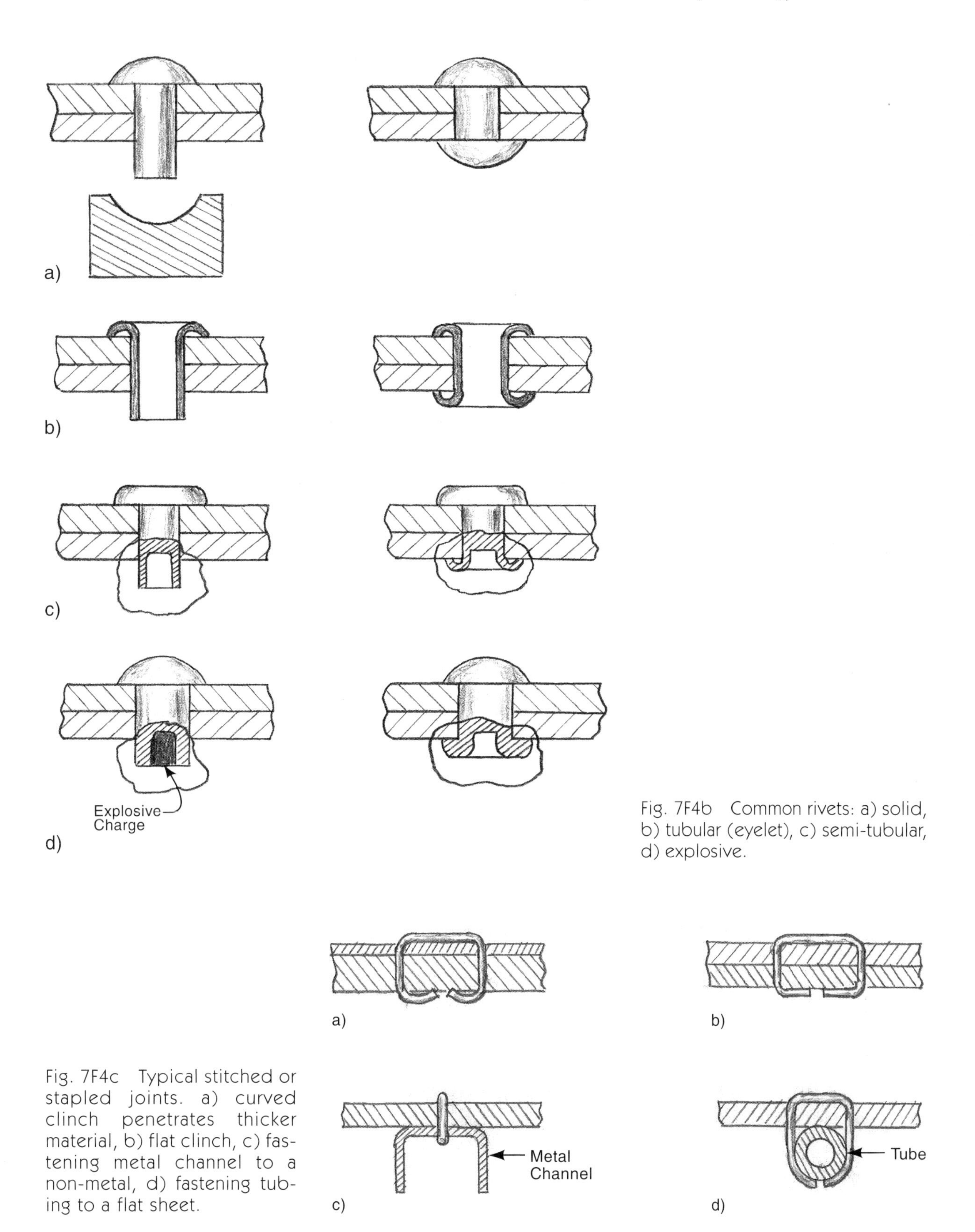

Fig. 7F4b Common rivets: a) solid, b) tubular (eyelet), c) semi-tubular, d) explosive.

Fig. 7F4c Typical stitched or stapled joints. a) curved clinch penetrates thicker material, b) flat clinch, c) fastening metal channel to a non-metal, d) fastening tubing to a flat sheet.

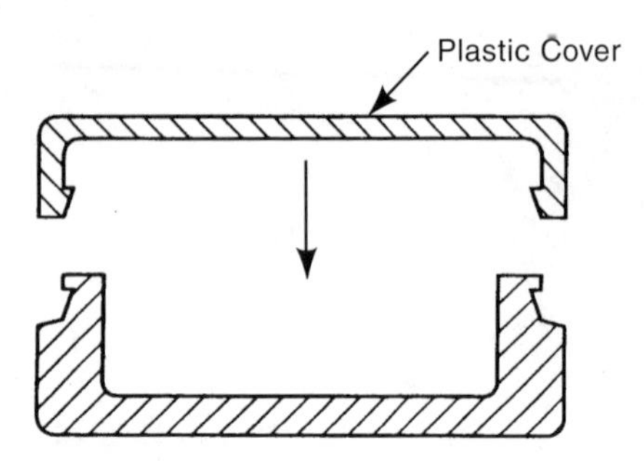

Fig. 7F4d The principle of snap-fit components illustrated with a plastic container cover. (from James G. Bralla, Design for Excellence, McGraw-Hill, New York, 1996.)

or hook-like elements in the parts. Normally, molding the part involves the inclusion in the mold of some kind of side core, an element of the mold that moves at right angles to the direction of the mold opening and closing. When this core is advanced before the mold fills, it creates an undercut in the molded part that provides the holding action. When the core retracts, the hooked or socketed part is free to be ejected from the mold when it opens. Snap fits can also be incorporated in many sheet metal parts. The forming die for such parts can have a cammed punch that pierces and bends the sheet metal to form a catch or hook. The advantage of snap fits is that they eliminate the need for screw fasteners, welds, or other means of attachment that involve additional parts and additional operations. Snap fits not only save labor and eliminate some factory operations, but reduce the need to purchase and inventory fasteners that otherwise would be needed. Usually, a simple direct engagement of the parts also engages the snap-fit elements. Fig. 7F4d illustrates the principle with a plastic container and lid.

F4e. ***press and shrink fit fastening*** - can be a low-cost method for permanently fastening parts together. The method involves the use of heavy force to drive one part, usually a pin, shaft, stud, or other round part, into a hole where the fit is tight or where there actually may be an interference fit. In such a fit, the diameter of the male part slightly exceeds the diameter of the female part. The disadvantage of press and shrink fits is that the dimensions of the mating parts must be closely controlled. Pins to be inserted are often centerless ground to provide an accurate diameter, and the holes to receive them are normally reamed or bored to insure an accurate internal diameter. Often, the end of the part to be inserted is tapered slightly or the hole is beveled slightly to permit easier initial insertion. In a typical situation, the part to be inserted is manually positioned in the hole and then driven into position with a hand, foot, or powered press. In a shrink fit, the receiving part is heated sufficiently to expand it so that the two parts can go together. When it cools, the outer part shrinks around the inserted part, holding it securely. Press and shrink fits are more common in heavier machinery. Fig. 7F4e illustrates a simple press fit of a small pin and two alternative designs that lessen the amount of precision needed to insure a satisfactory fit.

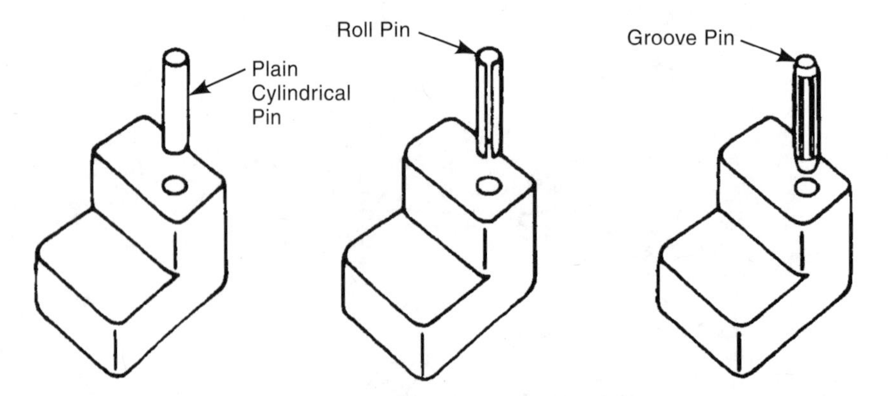

Fig. 7F4e Three varieties of pins press to fit into a metal component. The roll pin at the center and the groove pin at the right allow a lesser degree of precision in the diameters of the hole and the pin; for these pins, reaming of the hole after drilling is not normally required. (*from Design for Manufacturability Handbook, James G. Bralla, ed., McGraw-Hill, New York, 1998.*)

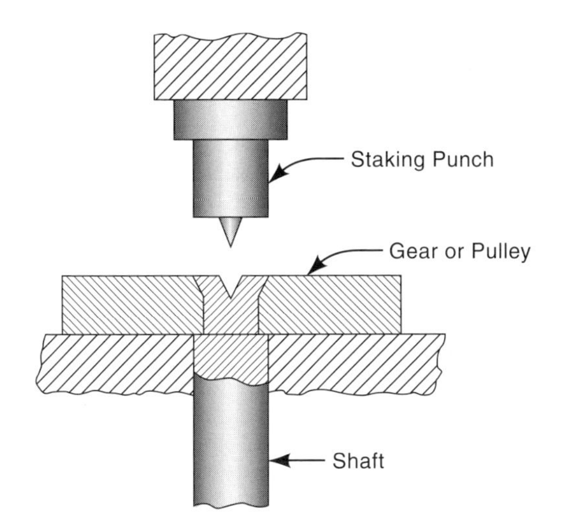

Fig. 7F4f Sectional view of a staked assembly.

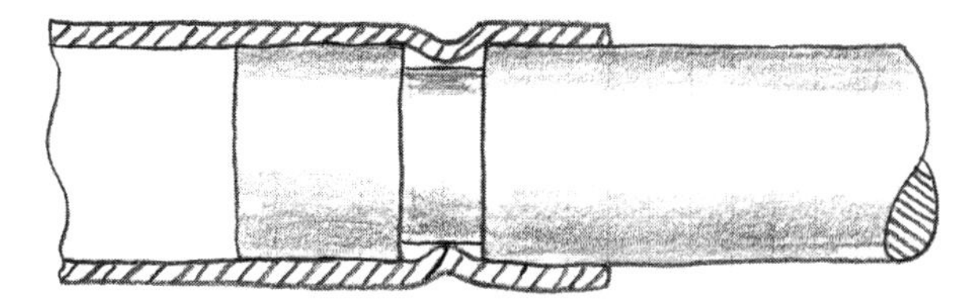

Fig. 7F4f-2 Crimping used to fasten a length of tubing to the end of a rod.

F4f. ***staking, seaming and crimping*** - are other methods of fastening parts together.

Staking is similar to riveting except that, instead of using a separate rivet, one of the parts is configured to fit into a hole in the other, and is upset to hold it in place and thereby hold the parts together. Fig. 7F4f illustrates a typical staking operation.

Seaming is a means for fastening sheet metal parts together at their edges. Fig. 7F4f-1 illustrates cross sections of a group of typical seam joints. These joints can be made by a series of operations on a press brake or, in high production situations, by dedicated tooling that fits the parts involved. (See chapter 2, sections C and D for sheet metal shearing and bending processes.) Containers such as drums, cans, and pails are routinely fastened by seaming. The approach is also common on all kinds of sheet metal work including the manufacture of ducting for buildings for heating and air conditioning systems.

Crimping, when involved in assembly, involves the bending of sheet metal parts to lock them into place. The term usually applies to cylindrical parts like caps, which fit over smaller, more rigid, cylindrical parts. The diameter of the cap is reduced where there is a circumferential groove in the inner part. Crimping is usually a fairly simple operation, performed with a hand or foot-operated lever tool or a light punch press. Electromagnetic forming is also sometimes used. (See 2J3.) Crimping is often less costly than using fasteners to hold the parts together, because it avoids the need for holes and screw threads and eliminates the need to maintain a stock of some kind of fastener. The most common applications of crimping are for the attachment of connectors to electrical wires, fittings to the ends of mechanical wires and cables, the attachment of hose and tubing to end fittings, and shells to bullets. Fig. 7F4f-2 illustrates a typical crimped assembly.

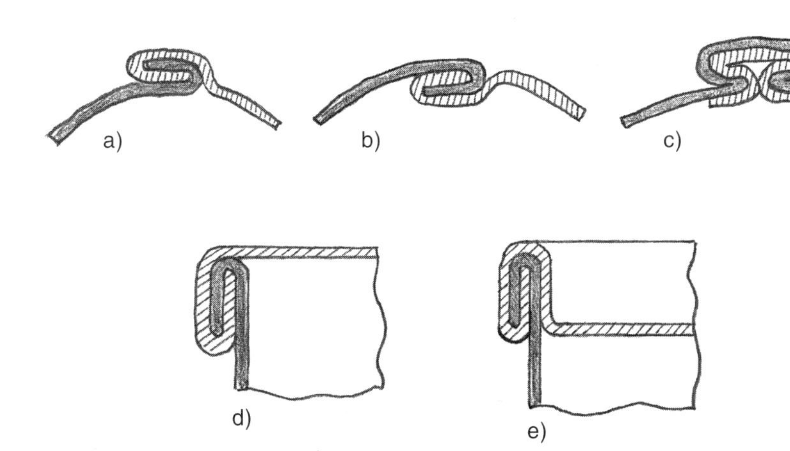

Fig. 7F4f-1 Sheet metal seam joints. a) outside seam, b) inside seam, c) compound seam, d) double seam for containers, e) double seam for containers with recessed end.

References

1. *ASM Handbook, vol. 6, Welding, Brazing and Soldering*, ASM International, 1993.
2. *Manufacturing Processes and Systems, 9th edition*, Ostwald, Munoz, John Wiley, New York, 1997.
3. *Loctite Worldwide Design Handbook, 1996/1997 edition*, Loctite Corporation, Rocky Hill CT.
4. *Materials and Processes in Manufacturing, 8th edition*, DeGarmo, Black and Kohser, Prentice Hall, Upper Saddle River, NJ, 1997.
5. *McGraw-Hill Encyclopedia of Science and Technology, 8th ed., Vol. 6*, New York, 1997.
6. *Solders and Soldering, 4th ed.*, Howard Manko, McGraw-Hill, New York, 2001.

Chapter 8 - Finishing Processes (including Heat Treating)

A. Cleaning Processes

Most industrial cleaning processes use a combination of the operations listed below. The particular cleaning process chosen depends on the nature of the soil to be removed, how heavily the workpiece is soiled, how clean it must be after the operation, and the size, shape, and material of the workpiece. Cleaning operations are performed for aesthetic reasons but most importantly are done to prepare the surface for some coating or finishing operation or assembly. Most cleaning operations are followed by rinsing and drying, particularly if a liquid cleaning agent is used.

A1. ***mechanical cleaning processes***

A1a. ***brushing*** - A fiber or wire brush, normally power driven in industrial applications, is moved against the surface of the workpiece, removing solid material such as rust, caked dirt and loose paint. The stiffness of the bristles depends on the bristle material and the bristle length and thickness. The brush configuration can be adapted to the strength and degree of adhesion of the soil. Brushing is usually only the first operation of a cleaning sequence.

A1b. ***abrasive blasting*** - Abrasive particles, at high velocity, are driven against the workpiece surface. The process is used to remove scale, rust, dry surface dirt and paint but is not effective in removing grease. Various abrasives can be used, from hard silicon carbide, aluminum oxide and steel shot, to softer materials such as plastic beads, corn cobs, nut shells and rice hulls. In addition to cleaning, abrasive blasting may be used for deburring, surface strength improvement or surface roughening.

A1b1. ***wet blast cleaning*** - is abrasive blasting when water is mixed with the abrasive. This permits finer abrasives to be used and provides better results when the workpiece has irregular surfaces. The water also minimizes dust.

A1b2. ***dry blast cleaning*** - is abrasive blasting without the use of a water carrier for the abrasive. The process can be made automatic for high production situations somewhat more easily than wet blast cleaning.

A1c. ***steam jet cleaning*** - High pressure steam is directed at the workpiece. The process can be manual or automatic. Grease, oils and dirt can be effectively removed

A1d. ***tumbling*** - can be an effective means for removing scale and rust from smaller parts as part of a cleaning sequence as well as for polishing and deburring. See B2 below for further information.

A2. ***chemical cleaning processes***

A2a. ***solvent cleaning*** - utilizes liquid hydrocarbons as cleaning solvents. There are several basic methods, described below, for applying the solvent to the workpiece. Common hydrocarbons used are petroleum solvents (mineral spirits, Stoddard solvent and kerosene), chlorinated hydrocarbons (trichloroethylene and perchloroethylene) and Freon.

A2a1. ***immersion cleaning*** - is used to remove grease, oil, and oil-bearing dirt. The part is submerged and sometimes soaked in the solvent. The solvent bath may be agitated to aid in soil removal. The weakness of the process is that the solvent becomes contaminated quickly and loses its cleaning power. The process is nevertheless used as an early stage of a more lengthy cleaning process.

A2a2. ***spray degreasing*** - is sometimes used as one step in a cleaning sequence. The workpieces are sprayed with a liquid hydrocarbon to remove oils and oily contamination. As in immersion cleaning, the solvent tends to get contaminated rather quickly.

A2a3. ***vapor degreasing*** - uses a hydrocarbon vapor as a cleaning agent. The use of vapor solves the problem of solvent contamination because the solvent touching the workpiece is clean since, when it boils off from the reservoir of the equipment, it leaves the contaminants behind. In a typical vapor degreaser, illustrated in Fig. 8A2a3, cooling coils in the vapor chamber walls contain the vapors so that they do not escape. The part may be briefly immersed in the boiling or heated liquid solvent, or both in sequence, and/or sprayed with solvent before being raised to the part of the chamber that contains the vapor. The part, if not immersed, is normally cooler than the vapor; the vapor condenses on the part and drips into the reservoir below, carrying away oils and contaminants from the workpiece. As the workpiece warms, the condensation stops and the workpiece can be removed from the vapor, dry and very clean. The cleaning action of the vapor stage of the process for heavy soils is limited because the amount of flushing provided by the condensing vapor is not great. That is the reason for the preliminary immersion and spraying. Some degreasers include another tank (not shown in Fig. 8A2a3) of cooled, clean, distilled solvent. The workpiece is power-sprayed with enough of this solvent to cool it as well as further clean it. Immersion again in the vapor, until the workpiece heats up, provides further flushing and self-drying.

A2b. ***ultrasonic cleaning*** - is immersion cleaning with the addition of agitation of the cleaning solution at an ultrasonic frequency. Most systems agitate at 10 to 40 kHz. Such agitation produces small momentary cavities in the fluid. These tend to pull soils from the workpiece. The process can be

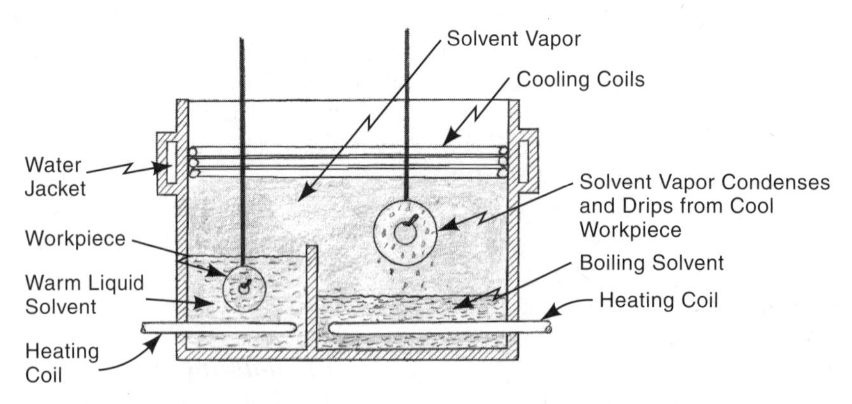

Fig. 8A2a3 A typical vapor degreaser shown in section view. Hot clean vapor condenses on room-temperature workpieces that have been lowered into the vapor, flushing off oils and dirt. More seriously soiled workpieces may first be immersed in the warm liquid solvent. Some degreasers also have a liquid solvent sprayer or a third tank of liquid solvent to aid in cleaning.

carried out with hydrocarbon solvents or water-based cleaners. It is primarily used for intricate smaller parts that can be contained in the ultrasonic tank.

A2c. ***emulsion cleaning*** - uses a mixture of a hydrocarbon solvent, water and emulsifying agents. The solvent, usually petroleum-based, comprises from 1% to 10% of the total and is dispersed in the emulsion as fine globules with the help of an emulsifying agent (soap, glycerol, polyether or polyalcohol). Workpieces are sprayed with the emulsion or are immersed in it with agitation. The emulsion typically is heated to a temperature around 60°C (140°F). The process is useful for removing heavy deposits of soils from the workpiece, for example, caked buffing compounds or mixtures of grease and solid material. It usually leaves a thin coating of oil on the workpiece. A hot-water rinse normally follows the cleaning operation.

The method is economical and safe because of the high water content. It is commonly used as an in-process operation.

A2d. ***alkaline cleaning*** - uses an aqueous solution of various alkaline salts with a wetting agent and detergents as the cleaning medium. Trisodium phosphate, caustic soda and silicates are common choices. The workpiece is immersed in the solution with agitation or the solution is pressure sprayed on the workpiece. The solution is heated to a temperature of from 140 to 200°F (50 to 90°C). This process is more economical than solvent or emulsion cleaning and is widely used, especially in mass production applications. It is effective in removing a wide variety of soils including grease, oil and shop dirt, and, at a slower rate, rust, light scale and carbon smut. Rinsing in water usually follows the alkaline cleaning operation. Safety requirements must be considered when highly alkaline solutions are used.

A2e. ***acid cleaning*** - is very similar to alkaline cleaning except that the alkaline salt is replaced by an acid or acid salt. Otherwise the method is the same; a detergent and wetting agent are also part of the solution. Acid cleaning is superior to alkaline cleaning in removing light rust or other metal oxides, scale, tarnish and similar deposits. For heavy coatings of grease and oil, acid cleaning is not the best process. The process may slightly etch the workpiece surface, but this is desirable for paint adhesion. Aluminum, steel, iron and copper are often cleaned with this approach.

A2f. ***pickling*** - is similar to acid cleaning but is a more aggressive procedure. It involves the removal of oxides - in the form of scale and other surface films - from metal by chemical etching with an acid solution. Any of a variety of acids and acid combinations may be used depending on the metal and the nature of the scale and other oxides on the surface. A stronger acid than that used for acid cleaning and a wetting agent are used. For steel, hydrochloric, sulfuric, hydrofluoric or phosphoric acids may be involved. The acid attacks the surface of the metal workpiece, destroying the bond between the scale or soil and the workpiece. Immersion of the workpiece in the acid solution is the usual method. After the operation, the part is immersed in an alkaline rinse to neutralize the pickling acid.

Pickling is an effective method for removing scale, dirt and oxides from metals. The scale that is removed results from hot forming, welding, or heat treating operations and from corrosion. The operation is performed prior to plating, painting, phosphating, or coating with vitreous enamel. Aluminum, copper, stainless steel, magnesium, and nickel alloys are cleaned by this method but steel is the most commonly processed metal.

A2g. ***salt bath cleaning*** - This is a multi-step method for removing scale, carbon and oxides from metals. The first step involves the immersion of the workpiece in a bath of molten salt which normally is at a temperature between 825 and 975°F (440 to 525°C). The salt solution partially loosens the soil and, if it is an oxide scale, tends to chemically reduce it. The workpiece is then immersed in water. Because the workpiece is now hot, it turns water that it contacts into steam. The expansion of the water as it turns to steam blasts the soil from the surface of the workpiece. The part is then dipped into a neutralizing acid solution and a water rinse. The process is useful for parts with irregular surfaces that may be difficult to clean with other methods. A wide variety of metals are suitable for the process.

A3. ***electrochemical cleaning processes***

A3a. ***electrolytic cleaning*** - Alkaline cleaning, acid cleaning, pickling and salt bath cleaning can all be performed with the assistance of electric current. When alkaline cleaning is made electrolytic, the alkaline solution becomes an electrolyte, the workpiece becomes one electrode and the tank or a separate steel plate is used as the other electrode. With electric current, oxygen is released at the anode and hydrogen at the cathode. The bubbles of these gases provide a scrubbing action to the workpiece. Additionally, the electrical charges that are developed on the workpiece and on the soil cause them to separate. The net effect of these phenomena is an effective cleaning operation. However, as the operation proceeds, the soils contaminate the electrolyte, reducing the efficiency of the operation. Because of this, electrolytic cleaning is normally used only as the final operation in a sequence after mechanical and chemical processes.

A3b. ***electrolytic pickling*** - uses the acid pickling solution as an electrolyte, the workpiece as the cathode and the tank or a plate as the anode. Gas released at the workpiece aides in loosening and removing scale. The process requires close control to avoid pitting the workpiece or changing its dimensions.

B. Polishing Processes

B1. ***conventional polishing*** - is a surface smoothing operation performed through the cutting action of abrasive particles glued to or impregnated in a flexible wheel or belt. The wheel may have varying degrees of stiffness, depending on the part to be polished and the desired degree of metal removal. Coarser abrasives and harder wheels remove material more rapidly. Wheels can be made of stacked layers of fabric or leather or of wood or rubber. Wax, fatty acids and tallow are used as lubricants during the process. The work is normally held manually against a powered wheel but the operation can be automated. The workpiece is moved as necessary to apply the cutting action against the entire surface to be polished. Belt polishing can be used if the surface to be polished is relatively flat or regular. Rough polishing is sometimes referred to as "roughing", intermediate polishing as "fining" and finer polishing as "oiling" The purpose of polishing operations, in addition to improving appearance, is to aid fluid flow, to remove burrs, to provide clearance for assembly, and to remove surface indentations that could be stress raisers or sites for entrapment of corrosive substances.

B1a. ***buffing*** - is similar to polishing except that the abrasive is normally finer or milder and is not bonded to the wheel, but instead is loosely held. Charging the wheel or belt with abrasive is effected by holding a bar of abrasive against the wheel or belt. Buffing is usually a secondary operation which follows polishing to further smooth the workpiece surface.

B2. ***barrel polishing (tumbling)*** - usually used for deburring, can also be an effective polishing method. The process involves the use of a rotating barrel or vibrating hopper charged with an abrasive compound (a fine abrasive plus detergents or other cleaners), water, a medium (stones or chunks of ceramic or metal) and the parts to be polished. The rubbing of medium and trapped abrasive compound against the parts, as the barrel moves, provides a polishing operation for the parts' surfaces. The operation is automatic and can provide surfaces comparable to those produced by conventional polishing and buffing if the proper materials and conditions are used. See Fig. 3K3 (deburring by barrel tumbling) and Fig. 3K11 (vibratory deburring).

B3. ***electropolishing*** - is an electrolytic process, the reverse of electroplating. The workpiece is connected to the positive (anode) side of the power supply while a cathode is connected to the negative side. When both workpiece and cathode are immersed into a conductive solution, metal is removed from the workpiece and deposited on the cathode. The electrolytic action tends to concentrate on removal of the high spots, including those of minute imperfections. The result is a gradual leveling and smoothing of the surface and an improvement in its glossiness.

B4. ***burnishing*** - smooths the surface of a workpiece by compressive deformation, normally with pressure rollers, rather than by metal removal. See 3J4 and 3J5.

C. Plating Processes

C1. ***conventional electroplating*** - deposits a metallic coating on the surface of a workpiece. The workpiece is connected to the negative terminal of a direct current source. A piece of the metal to be deposited, is connected to the positive terminal. Both the workpiece and piece of metal are immersed in a solution (normally aqueous) containing ions of the metal to be deposited. The workpiece becomes the cathode in an electrolytic circuit and the piece of metal, the anode. When electric current flows, metal dissolves at the anode and is plated on the workpiece. The workpiece must be cleaned to a high level for the process to be successful. Electroplating is performed to improve the appearance of the workpiece, to improve its corrosion, abrasion or wear resistance, to improve electrical conductivity, to change a dimension or for a combination of reasons. Fig. 8C1 illustrates the process.

C2. ***electroless plating*** - is plating without an electric current. Instead, a chemical reducing agent in solution reduces a metallic salt. The metal then deposits on a catalytic surface of the workpiece. In nickel electroless plating, the most common application, nickel is supplied in the form of nickel chloride which is put in an aqueous solution with sodium hypophosphate as a reducing agent. The solution is heated to a temperature above 160°F (70°C). The workpiece is catalytic, as is the nickel deposit on the workpiece, so the process can continue indefinitely. The plated material is an alloy of nickel and phosphorus with from 4 to 12 percent phosphorus. The electroless process is particularly advantageous when it is necessary to plate recesses and other irregularities that do not cover well with conventional electroplating. Some plastics, when pretreated properly, can be plated with the electroless method.

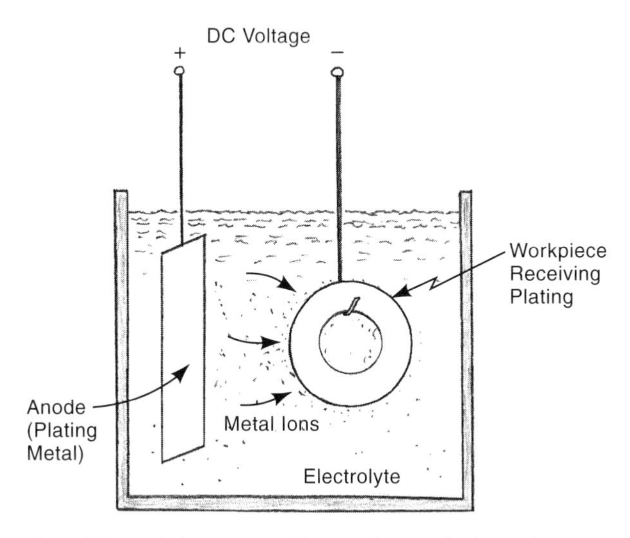

Fig. 8C1 Schematic illustration of the electroplating process.

C3. ***barrel plating*** - Small parts can often be placed in bulk in a porous metal barrel that is immersed in the plating solution. The barrel is electrically conductive and the parts connect to the electric power source from contact with the barrel and each other. Tumbling action of the rotating barrel ensures that all the parts have the opportunity, during the operation, to be in position to receive the metal deposit.

C4. ***mechanical plating*** - involves tumbling the parts to be plated in a mixture of powder or dust of the metal to be plated, glass beads, certain promoter chemicals and water. The force of the glass beads striking the parts peens or hammers the powder particles against the surfaces of the parts, causing them to "cold weld" and tightly adhere to the surface. The resulting coating does not have the sheen of electroplating but can provide galvanic protection for the plated part. The process is also known as *peen plating*, *impact plating,* or *mechanical galvanizing*.

C5. ***brush plating*** - is a mean for plating only a portion of the workpiece. The process is also known as *selective plating*, *swab plating* or *contact plating*. Instead of immersing the workpiece in a plating solution, the solution is applied to the workpiece by a tool. The tool is shaped to fit the surface to be plated and is coated with absorbent material that is saturated with the plating solution. A direct-current electrical circuit, in which the tool is the anode and the workpiece the cathode, carries metal ions to the surface of the workpiece. The electrolytic action therefore is identical to that of conventional immersion plating. The shaped tool is usually made from graphite and has an insulated handle. The process is used to plate large parts that do not require plating on their entire surface or are too large for immersion in a standard tank. It is also used for repair work. A common application is the plating of contact points on printed circuit boards, circuit breakers and other electrical or electronic devices where low electrical contact resistance is needed.

C6. ***electroplating of plastics*** - See 4M4a.

D. Painting (Organic Finishing)

D1. ***brushing*** - the common household method for applying paint. A soft bristle brush is dipped in the paint, touched to the workpiece to be painted and moved across it. The paint flows from the brush to the workpiece and, if done with care, covers it with a fair degree of uniformity. However, because of surface variations in the paint coating from the bristles, the process is most satisfactory, from an appearance standpoint, with non-glossy finishes.

D2. ***roller coating*** - is similar to brushing except that a cylindrical roller, normally with a soft, fluffy surface, is used instead of a brush to apply the paint. The finished surface can have a light texture from the surface texture of the roller.

Two other roller coating methods, primarily for continuous sheet or film, are illustrated in Fig. 4K1b (Chapter 4). They are suitable for the application of thin coatings.

D3. ***curtain coating*** - The paint is placed in a horizontal trough that has a slit of narrow but controllable width in the bottom. The trough may be pressurized. Paint flows through the slit in a "curtain". The part to be painted is moved beneath the trough at a fixed rate and passes through the curtain which coats it with the flowing paint. Excess paint is captured and recycled through the system. The process is economical for high-production situations. See Fig. 8D3 for an illustration of curtain coating equipment.

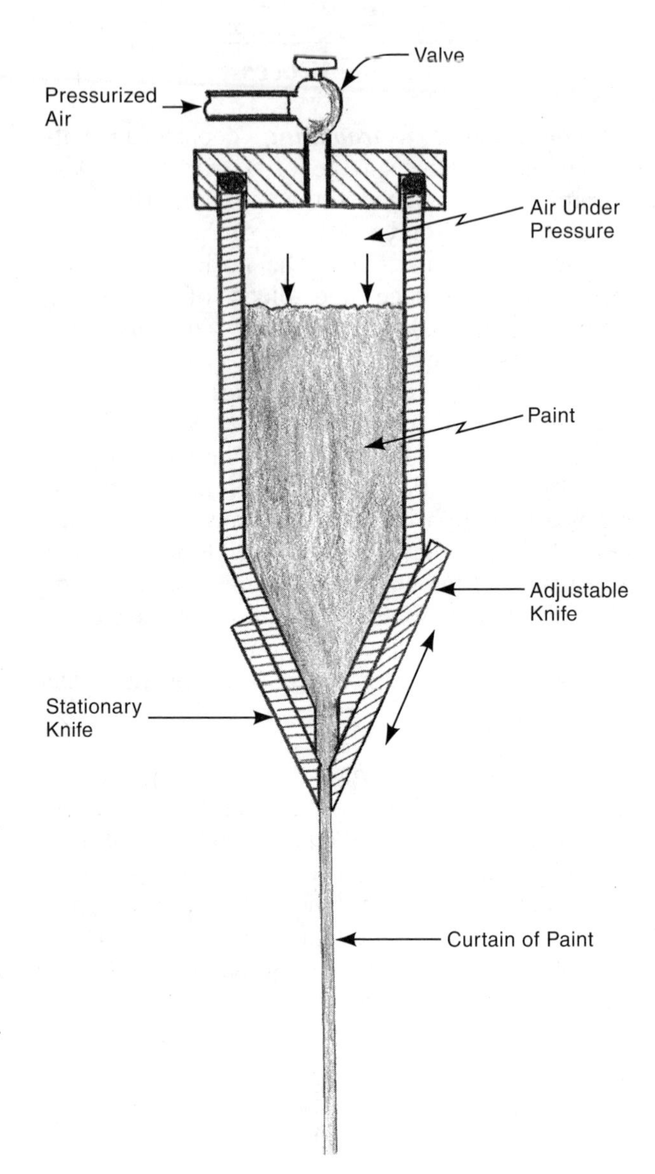

Fig. 8D3 Sectional view of curtain coating equipment. Workpieces are painted as they pass through the falling curtain of paint.

D4. ***dip painting*** - involves immersion of the workpiece in a container of paint. Excess paint is allowed to drain away after the part is lifted from the container. The process is applicable to parts with irregular configurations that may not be feasible for coverage by other painting methods. However, parts must be of such shape or must have drain holes to allow the excess paint to drain away.

D5. ***flow coating*** - is similar to curtain coating except that the paint flows through one or more nozzles that are directed at the workpiece. Excess paint is drained from the workpiece, captured and recycled.

D6. ***spray painting*** - In this process, liquid paint is atomized and directed at the workpiece. Spray painting is widely utilized and, when properly applied, produces a smooth, uniform paint coating. Portions of the workpiece that are not to be painted are suitably masked. Different spray atomization methods in use are: air, airless and, with some electrostatic painting, centrifugal.

D6a. ***air atomized spray*** - Compressed air atomizes the liquid paint and propels it from the nozzle of the spray gun. This method allows good control of the spraying rate during the operation.

D6b. ***airless spray*** - Hydraulic pressure forces the liquid paint through a nozzle, where it breaks up into fine droplets that are propelled toward the workpiece. This approach avoids the tendency with air spray of having some of the paint carried by the air flow away from the workpiece. However, control of the spray and varying the amount sprayed is not as feasible as it is with air atomization.

D7. ***electrostatic painting*** - involves the use of an electrical charge on the droplets of atomized paint (or particles of powdered paint). The workpiece is grounded and the paint particles are then attracted to it. This greatly reduces overspray and correspondingly improves the yield per volume of paint consumed. The workpiece must have some electrical conductivity for the process to work effectively, but non-conductive parts can be dip-coated with a conductive coating as a preliminary operation.

D7a. with ***centrifugal spraying head*** - Some electrostatic painting is done with a rotating disc or bell-shaped dispenser to atomize the paint. The paint is introduced at the device's center of rotation and moves to the periphery and into the air in fine droplets by centrifugal force. The disc or bell rotates at a speed of 900 to 1800 rpm and is charged with a high negative potential. The paint droplets are thus electrostatically charged and are attracted to the grounded workpiece. Discs are normally mounted horizontally (i.e., with a vertical axis) and the workpieces to be painted are conveyed in a circular path around the disc. The disc may also be oscillated upward and downward, particularly if the workpieces are long. Bell shaped atomizers usually have the bell opening positioned to face the surface of the workpiece to be coated. These atomizers may be mounted next to a workpiece conveyor or may be hand operated.

D8. ***powdered paint coating*** - Coating parts with powdered material rather than liquid paint has a number of advantages: No solvent is required, so the pollution, venting and air make-up factors are virtually eliminated. Powder not utilized can be recovered and used, so yields are higher and the problems of clean up and disposal of dried overspray are minimized. Thicker coatings can be made in one application. However, the powdered material, when suspended in air, can constitute an explosion hazard. Paint in powdered form consists of finely ground plastic resin with coloring agents. Epoxy, nylon, acrylic, vinyl, polyester, and polyethylene are commonly used materials. They are applied to the workpiece with several methods. Normally, several cycles of workpiece cleaning and, often, a phosphating treatment precede the coating operation. After the coating application, the coated workpiece is oven heated, causing the paint particles to melt, flow and fuse together. The resulting coating is tough and well bonded to the part's surface. Typical applications are appliances, office furniture, farm equipment, industrial machinery, automotive parts, and architectural components.

D8a. ***electrostatic spray powder coating*** - Dry powdered plastic resin is pneumatically fed from a supply source to the spray gun where it is electrically charged. The charged particles from the spray gun are attracted to the electrically grounded workpiece by electrostatic attraction and adhere to it. Oversprayed powder settles to the bottom of the spray booth where it is recovered and reused. The powder coating on the workpiece adheres sufficiently so that the workpiece can be moved to the next operation. This involves oven heating the workpiece so that the plastic particles in its powder coating soften, fuse together and bond to the workpiece surface. Typical coating thicknesses are 0.001 to 0.004 in (0.025 to 0.1 mm) though thicker coatings can be applied. Because of the electrostatic attraction, the process can be automated relatively easily. See Fig. 8D8a.

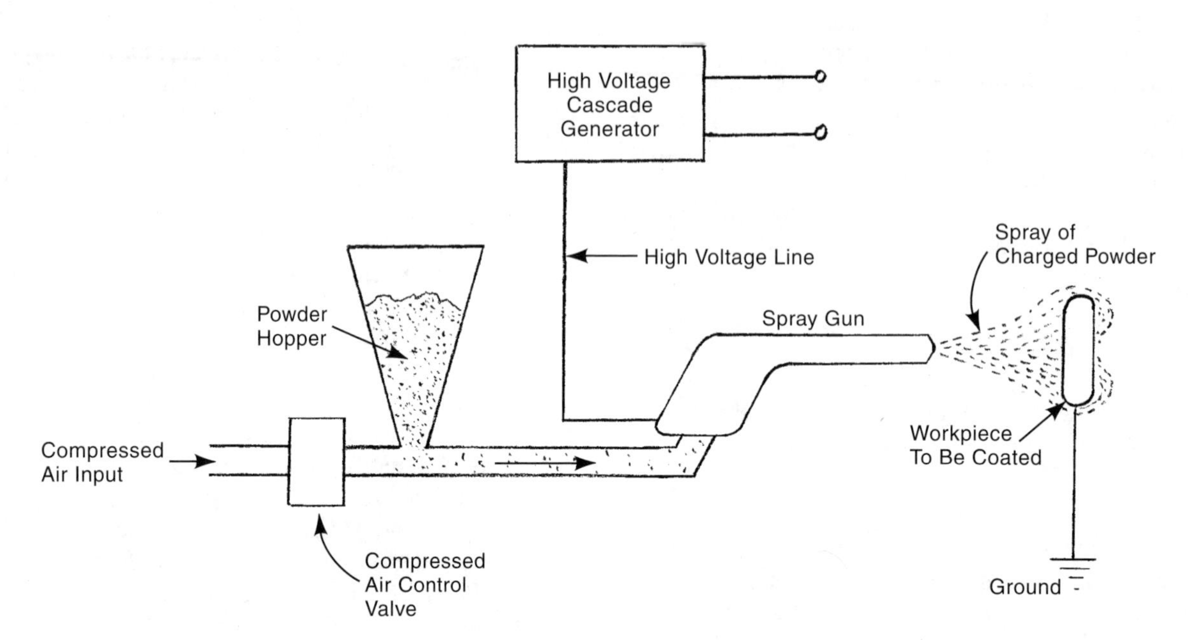

Fig. 8D8a Electrostatic powder spraying - apparatus for applying powder coating to a product with the aid of electrostatic attraction.

D8b. ***fluidized bed powder coating*** - utilizes a tank-like container containing dry plastic powder that is aerated from below with low-pressure air flow through a porous membrane. The air flow causes the particles to be suspended in a dense cloud; the mass behaves like a fluid. The workpiece, preheated to 400 to 600°F (200 to 315°C), is lowered into the tank. The plastic particles contact the workpiece, soften and melt from its heat and adhere to the workpiece surface. The workpiece then is oven heated to further fuse the particles into a uniform film. When the workpiece is removed and cooled, the powder becomes a tough, solid, adherent coating that also is decorative. Standard coating thicknesses are somewhat thicker than those from spraying liquid paint, averaging from 0.007 to 0.015 in (0.18 to 0.38 mm) in thickness. Because of the greater thickness, the process is particularly applicable to parts that require resistance to abrasion or impact. Epoxies and polyvinyl chloride are common powder materials but acrylics, polyesters, polyimides and silicones are also used. Fig. 8D8b shows a typical fluidized bed.

D8c. ***electrostatic fluidized bed powder coating*** - In this process variation, the workpiece is held just above the surface of the fluidized bed. The particles are electrically charged by a grid positioned just below the top of the fluidized bed or by a pre-charged air flow. The workpiece may be heated or not, but those particles floating just above the fluidized bed are attracted to the grounded workpiece and adhere to it. This approach provides somewhat better control over coating thickness than conventional fluidized bed coating, but is limited to parts

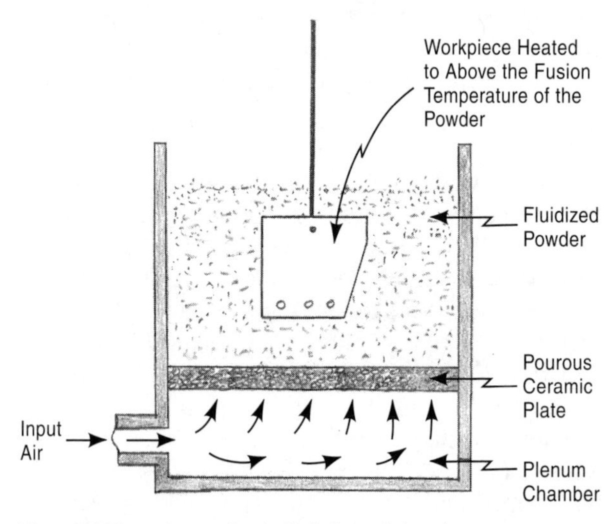

Fig. 8D8b A typical fluidized bed arrangement for powder coating workpieces.

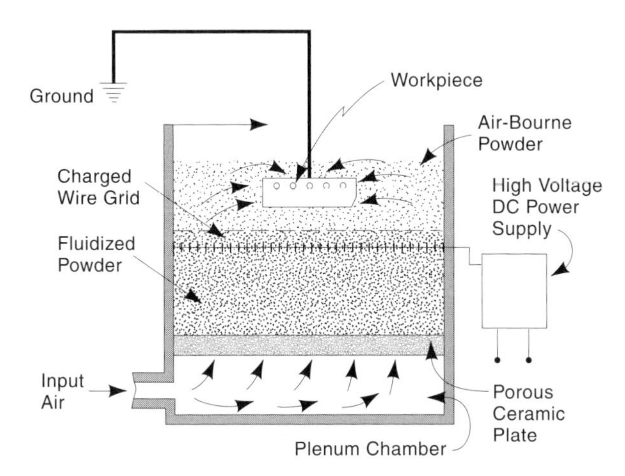

Fig. 8D8c The arrangement used for electrostatic fluidized bed powder coating. The arrangement is very similar to that shown for a regular fluidized bed except that the powder is given an electrostatic charge which causes the particles to be attracted to the grounded workpiece.

that are low enough in height to be able to be immersed in the cloud of powder particles just above the fluidized bed - approximately 2 to 4 inches (50 to 100 mm). The method has both the advantages and disadvantages of electrostatic spraying in that the attraction of the particles to the workpiece may result in less coverage in deeply recessed areas in the workpiece and there may be uneven coverage of vertical surfaces. Fusing the coating by oven heating follows the coating operation. Fig. 8D8c illustrates the process schematically.

D8d. ***friction static spraying*** - In this method, friction between the powder particles as they collide with each other and the spray apparatus generates a static electrical charge on the particles. The charge may be either positive or negative. The powder spraying apparatus removes either the positive or negative charges, leaving a powder with uniformly charged particles when it leaves the spray gun. The spray has the ability to enter narrow openings and the space between narrowly angled surfaces more easily than with conventional electrostatically-charged powder. However, powder must be sprayed at a lower rate with this technique so its use is best limited to the coating of components that are difficult to cover with conventional electrostatic methods.

D8e. ***flame spraying of powdered paint*** - The flame spraying method for applying metallic coatings from powder (See F4b below.) also can be used to apply powdered plastics. The powder is fed to a fueled spray gun where it is melted and blown against the surface of the workpiece. The heat for melting is supplied by the combustion of gases, usually propane and oxygen. The powder is fed in a stream of air through the center of the ring-shaped flame. The liquid or semi-solid plastic particles adhere strongly to the heated workpiece and fuse together. Fluidized-bed-grade powder is used. The process is not widely used because it is a manual process requiring a skilled operator and the coating is difficult to apply in a uniform thickness. Undercut areas that are too small for gun access cannot be coated. Also, the method is somewhat slow and overspray cannot be recovered since molten droplets fuse together. However, it is useful for coating objects too large for a fluidized bed or those that are in the field, since the spraying equipment can be portable.

D9. ***electrocoating*** (also referred to as ***electropainting, electrophoresis, electrophoretic coating or electrodeposition of paint***) - is a method of application of water-based paints to electrically conductive parts. The workpiece is electrically charged and immersed in a tank containing the aqueous-based paint and deionized water. The paint particles are given an opposite charge, causing them to migrate to an immersed workpiece and adhere to its surface. The workpiece is removed from the tank and flushed with water to remove excess paint. The remaining paint film on the part is level and tightly adhering. Oven baking completes the process. Epoxies and acrylic-based paints are commonly used. Polyesters, vinyl-based plastics, phenolics and alkyds are also employed. The process is fast, provides uniform film thickness, good coverage of recesses and high paint utilization. It can be fully automatic. However, color changes are more difficult, surface defects may show through the coating and process conditions must be closely controlled. The process has been used on automobile bodies.

D10. ***vinyl plastisol coatings*** - These are put on tool handles for ease of handling and electrical insulation. They are also added to woven steel wire fencing and other items for appearance and corrosion protection. The normal method of application is dipping. The part to be coated is heated and immersed into the liquid plastisol which is a suspension of vinyl plastic and plasticizer. The heated liquid that touches the workpiece "gels", forming a coating up to about 1/8 in (3 mm) in thickness, and sometimes up to about 1/4 in (6 mm). As the workpiece is withdrawn, the plastisol cools and becomes a flexible solid. (See 4K and 4K2.)

E. Chemical Surface Treatments

E1. ***anodizing*** - is an electrolytic process that produces an oxide coating on aluminum and some other metals. Aluminum forms a natural oxide surface without anodizing, but such a surface is much thinner and has much more limited properties than those that can be provided by anodizing. The workpiece, connected as the anode of a DC power circuit, is immersed in an acid solution. The current flow liberates oxygen at the surface of the workpiece and the oxygen reacts with the aluminum to form aluminum oxide. Chromic, sulfuric or oxalic acid are ones used most commonly in the electrolyte solution. Thorough cleaning and etching or treatment with a brightening solution are performed before anodizing.

The anodized coating is hard, smooth, wear and corrosion resistant, and easily colored, but somewhat porous. Colors are applied by dipping the workpiece in a liquid dye after anodizing. Coloring can also be produced by incorporating organic acids in the anodizing solution. Anodized surfaces are also usable as a base for painting. Sealing of the surface is a common subsequent operation that closes surface pores and makes the surface stain resistant.

Magnesium, titanium and zinc can be anodized but these metals represent only a small portion of the materials processed. Anodized surfaces normally range from 0.0002 to about 0.0007 in (0.005 to 0.018 mm) in thickness although hardcoat anodizing can produce thicker surfaces.

E1a. ***hardcoat anodizing*** - uses one of a number of proprietary processes to produce a thicker, harder anodized surface. In one process, a sulfuric and oxalic acid mixture is used at somewhat lower temperatures and higher current densities than with conventional anodizing. Surfaces as thick as 0.004 in (0.10 mm) are feasible. This approach provides superior wear and corrosion resistance.

E2. ***phosphating*** - Iron or steel workpieces are phosphated by immersing them in a dilute solution of phosphoric acid into which a metallic phosphate has previously been dissolved. (Iron, zinc, or manganese are the usual phosphates.) The operation is preceded by a cleaning sequence, the nature of which depends on the type of soil on the workpiece. With large workpieces, the phosphating solution may be applied by spraying instead of immersion. Small parts may be tumbled as they are immersed. The phosphating solution is normally heated to a temperature of 90 to 210°F (32 to 99°C). Immersion times vary with the thickness desired, the bath temperature and the type of phosphating, but typically range from one or two to about 39 minutes. Spraying is a somewhat faster operation. Phosphating provides a thin coating which penetrates into the base material. Phosphated surfaces provide a good base for painting or bonding by improving adhesion and helping to prevent corrosion. Wax and oil can be retained also, providing another means of corrosion resistance. Phosphate coatings also facilitate drawing and other forming operations. Galvanized steel can be phosphated as a base for painting. Phosphating of stainless steel and some alloy steels is difficult.

E3. ***chromate conversion coatings*** - The workpiece is immersed, sprayed or brushed with an acidic solution of hexavalent chromium compounds. Chromic acid, sodium or potassium chromate or dichromate, hydrofluoric acid or hydrofluoric acid salts, phosphoric acid or other mineral acids may be employed. The chemical reaction of the solution with the workpiece metal forms a protective film. The film is comprised of complex chromium compounds. It is very soft when first formed, but when dried and allowed to age, becomes more abrasion resistant. Chromate conversion coatings are applied to workpieces made or coated with zinc, cadmium, magnesium, aluminum, copper or silver. The process is used to provide a protective barrier against corrosion. It also can be used to improve the appearance of the workpiece by providing better brightness or color to the coating. Chromate conversion coatings also provide excellent bonding surfaces for painting or lacquering. Most processes are proprietary.

E4. ***black oxide coating*** - is a chemical treatment for iron and steel parts. The procedure involves the following steps: 1) Soak the parts in an alkaline cleaner at 180°F (80°C). 2) Rinse the parts in water at 150°F (65°C). 3) Place the parts in a solution of caustic soda (sodium hydroxide), sodium nitrate/sodium nitrite with wetting agents and stabilizers at 290°F (143°C) for 15 to 30 minutes. 4) Rinse the parts in cold water. 5) Immerse the parts in oil or molten wax. The treatment provides a semi-porous surface without changing dimensions. After oil or wax impregnation, a modest amount of corrosion protection is provided and the black surface has a pleasing appearance. Black oxide treatment is used for firearms parts, spark plugs, tools, gears, and sprockets. The procedure is somewhat hazardous because of the caustic solution and high temperature. Cold blackening systems with proprietary solutions are available but their results are not quite as satisfactory. Most apply a copper-selenium coating.

F. Other Coatings

F1. ***porcelain enameling*** - provides a vitreous coating on metal products. Steel, cast iron and aluminum are the metals most commonly processed. Porcelain enamel coatings provide corrosion and weather resistance, improved appearance, electrical insulation, and a more easily cleaned surface. Stove tops, cooking containers, kitchen sinks, bathtubs, architectural components, signs, reflectors and water heater tanks are products that commonly receive this treatment. Porcelain enameling requires several steps: 1) chemical cleaning of the workpiece to remove oils, scale, and dirt. 2) mechanical treatment of the surface by grit blasting or other methods to provide a better surface for enamel adherence. 3) coating of the surface with the frit material. There are several methods available: fluidized bed or spraying of dry powder, and spraying, dipping or flow coating of liquid slips. If a liquid slip is used, coating is followed by drying. 4) firing of the coated workpiece at temperatures from 900 to 1500°F (480 to 800°C) to fuse and bond the enamel coating. The process parameters vary somewhat depending on the material to be coated. Often, several coats of enamel are applied. The enamel frits are prepared beforehand in a series of operations. Up to about 15 ingredients (silicate glass, metal oxides, ceramics, pigments and other additives) are mixed and melted together, then made into thin sheets which are broken up and milled into a fine powder. The powder then is usually mixed with water to form a slip. (Also see 5A5i and *enamel, vitreous.*)

F2. ***hot dip coating (galvanizing)*** - is achieved by immersing the workpiece in a molten bath of the coating metal. The sequence of operations is as follows: 1) Thoroughly clean the part to be coated of oil, rust, scale and other contaminants. 2) Dip the part in a solution of aqueous flux. (An alternative is to have a layer of liquid flux on the bath of molten metal.) 3) Immerse the part in the bath of molten coating metal. The bath temperature for zinc coating (galvanizing) is about 840°F (450°C) and for aluminum, about 1290°F (700°C). 4) Perform whatever post-coating operations are required by the application for the coated part. Operations include slow cooling, quenching, and conversion coating.

F3. ***vacuum metalizing*** - puts a thin coating of a metal or metallic compound on a workpiece in a vacuum chamber. In the chamber, under a vacuum, metals, alloys or chemical compounds are vaporized with heat and are deposited on the surface of the workpiece. The full sequence of operations is as follows: 1) The part is thoroughly cleaned. Vapor degreasing is a common cleaning method. 2) If necessary to seal the part's surface or provide a smoother substrate, the part is coated with a precoat material. 3) Vacuum coating takes place. The parts to be coated are placed in the chamber that is evacuated to 10^{-3}–10^{-5} mbar. A source of coating material is placed in a central location in the chamber. After the vacuum is achieved, in the most common approach, the source metal is heated by electrical resistance, induction or other methods to the temperature at which the metal vaporizes. The metal vapors condense on the workpiece surfaces that face the source of metal. (The metal vapors travel only in a straight line, so the parts must be arranged and held in the proper position.) Parts may have to

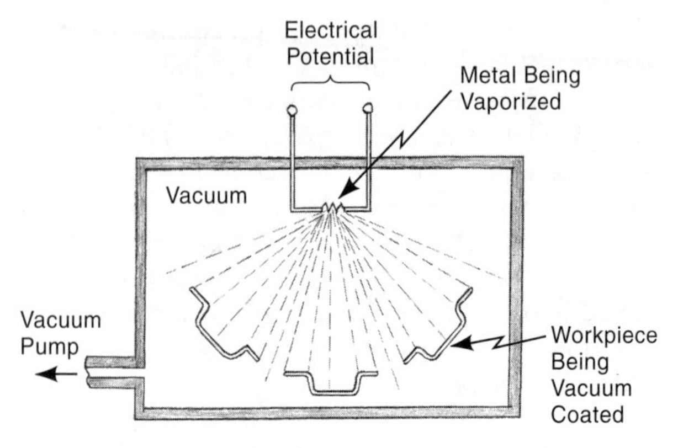

Fig. 8F3 Vacuum metalizing. In this example, electrical resistance heating vaporizes the coating metal. The metal vapors travel in a straight line and condense on the surface of the workpieces.

be rotated during the operation if more than one surface or a curved surface is to be coated. 4) After coating, a clear lacquer protective coating may be applied to the surface of the workpiece to protect the coating. Fig. 8F3 provides a schematic illustration of the process which is used for appearance improvement, reflectance or to provide greater wear or friction resistance to the workpiece. Vacuum metalizing is also an important method for applying thin films to substrates in the manufacture of integrated circuits (13K3a4).

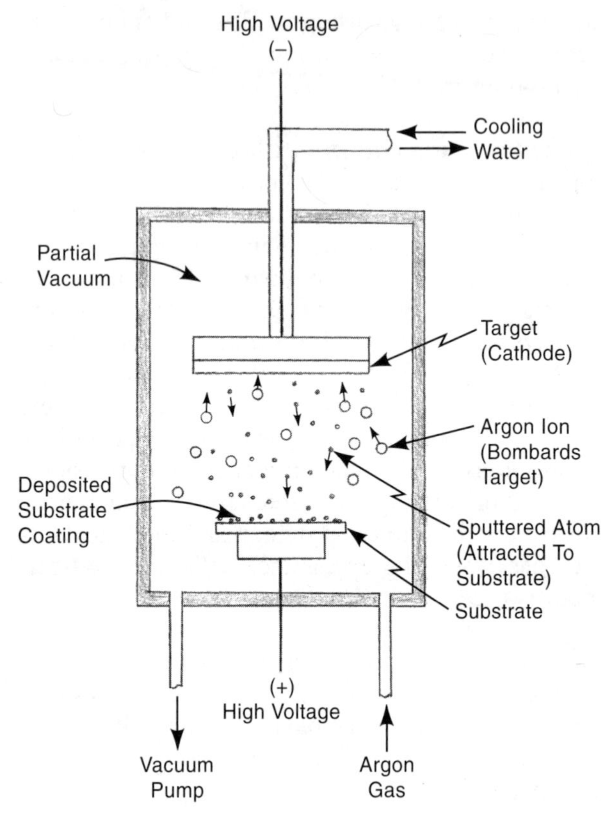

Fig. 8F3a An arrangement for sputtering to apply any of various materials to a substrate (workpiece) surface.

F3a. ***sputtering*** - can deposit any material to any substrate. The process uses ionic (plasma) bombardment rather than heating to vaporize the coating material for vacuum deposition. The source material, in a partial-vacuum atmosphere containing argon, has a negative electrical charge and is subjected to a stream of ions against its surface from either a glow discharge or ion beam. High voltage (2 to 6 kV) ionizes the argon. A magnetron may also be used as the source of ionizing energy. In any case, atoms of the source material are dislodged from the surface. They are attracted to the workpiece surface, which has a positive charge, causing the atoms to adhere to it. The process has advantages in thickness uniformity, adherence of the coating, and ability to coat the workpiece with nonmetallic as well as metallic materials. Semiconductors, metal alloys, insulators, and various compounds can be deposited with this method. Deposition rates, however, are low and the process is normally limited to sheet and tubular workpieces. Three dimensional workpieces may not be suitable because material transfer is line-of-sight. Fig. 8F3a illustrates the process. Applications of sputtering include deposition of thin films in integrated circuit manufacture, production of thin film resistors, capacitors and lasers, deposition of optical coatings on eyeglasses and lenses, and the coating of costume jewelry. The process can also be used for engraving if the target material is the workpiece rather than the source of material to be deposited on another workpiece.

F3b. ***chemical vapor deposition (CVD)*** - is a coating process that uses a reactant gas or vapor in a chamber with a heated workpiece. The gas decomposes at the heated surface of the workpiece, depositing a solid element or compound. The process has similarities to carbonitriding or gas carburization. The material is either simply deposited on the surface or absorbed into it. A second material in gaseous form also results from the reaction and is drawn off the work chamber. Sometimes, provisions are made to introduce various reacting compounds in succession. Rates of deposit are high and the coatings may be corrosion resistant and durable.

The process is used with a variety of materials and applications: One application is to provide hard cutting tool coatings of titanium carbide, titanium nitride, aluminum oxide and other materials. It is also used to provide anti-oxidation coatings of refractory metals and other alloys that are subjected to high temperatures, semiconductor coatings in the production of integrated circuits and carbon-carbon coatings for rocket and space components. (Also see paragraph 13K3a3.)

F3b1. ***plasma-enhanced (or plasma-assisted) chemical vapor deposition***[5] - is a process variation of chemical vapor deposition, (CVD). This approach, (sometimes referred to as PECVD), enables the CVD to take place at considerably lower substrate temperatures and with higher deposition rates. The reactant gas is subjected to an electrical field at a frequency of 50 kHz to 13.5 MHz or, in some cases, at microwave frequencies. The electrical field causes collisions of the gas molecules, producing ions, free radicals, excited neutrals and electrons which are more reactive. The plasma promotes the reaction of the chemical deposition. The process is used to deposit thin films in the fabrication of semiconductor wafers where the higher temperature of thermal CVD would be detrimental. It is also used in depositing optical coatings and wear-resistant coatings on cutting tools.

F3c. ***physical vapor deposition (PVD)*** - is a general term to denote coating processes whereby individual atoms or molecules of the coating material are deposited on the workpiece surface. Sputtering and vacuum metalizing (see above) are physical vapor deposition processes.

F3d. ***ion implantation*** - Ionized atoms of a material are bombarded against a workpiece surface in a high vacuum, causing the alloying of the workpiece material and the material that strikes it. The alloying occurs only at the very surface; penetration is shallow. A frequent application for cutting tools is the implanting of nitrogen to provide improved wear resistance. The process is used in the semiconductor industry to introduce dopant atoms into silicon wafers. Another important application is the bombardment of titanium or cobalt-chromium artificial joint bearing surfaces with nitrogen ions for wear resistance. There are no significant dimensional changes resulting from the process. However, the equipment is costly and a line-of-sight must exist between the source of the ions and the surface to be treated.

F4. ***thermal spray coating*** - sometimes known as ***flame spray coating***, is a process for applying a coating of a high-performance material on a workpiece. Liquid metal, or other material, in spray form, is directed at the workpiece and coats it. Metals, alloys, intermetallics, carbides, ceramics, cermets and plastics can be applied by this method. The process is used to salvage worn or undersized parts, to provide coatings for wear resistance, for corrosion or heat-oxidation protection and for electrical conductivity. It can also be used to produce parts with superior properties though they are fabricated from lower cost, more easily processed, materials. Coating thicknesses range from about 0.004 to 0.5 inches (0.1 to 12 mm). In most of these processes, substrate temperatures do not rise to very high levels so the operation can be performed with a number of non-metallic materials as the substrate. In all thermal spray coating processes, the cleaning and degreasing of the workpiece is a prerequisite. Scraping, wire brushing, grit blasting, chemical cleaning or machining can be used to remove scale and other foreign materials. Vapor degreasing and aqueous methods are used to remove grease and oils. Often it is desirable to roughen the workpiece surface before spraying to improve the bonding of the coating.

F4a. ***wire metalizing*** - is the earliest thermal spraying process. Wire (or rod) is fed through a dispensing gun through a ring of flame which melts the tip of the wire. Compressed air or gas atomizes the melted metal and propels the fine droplets at 300 to 800 ft/sec (90 to 240 m/s) to the workpiece. The droplets, 0.0004 to 0.004 inches (10 to 100 μm) in diameter, strike the workpiece surface, flatten and solidify. Oxyacetylene, propane or other fuel gases can be used to provide the heat for melting.

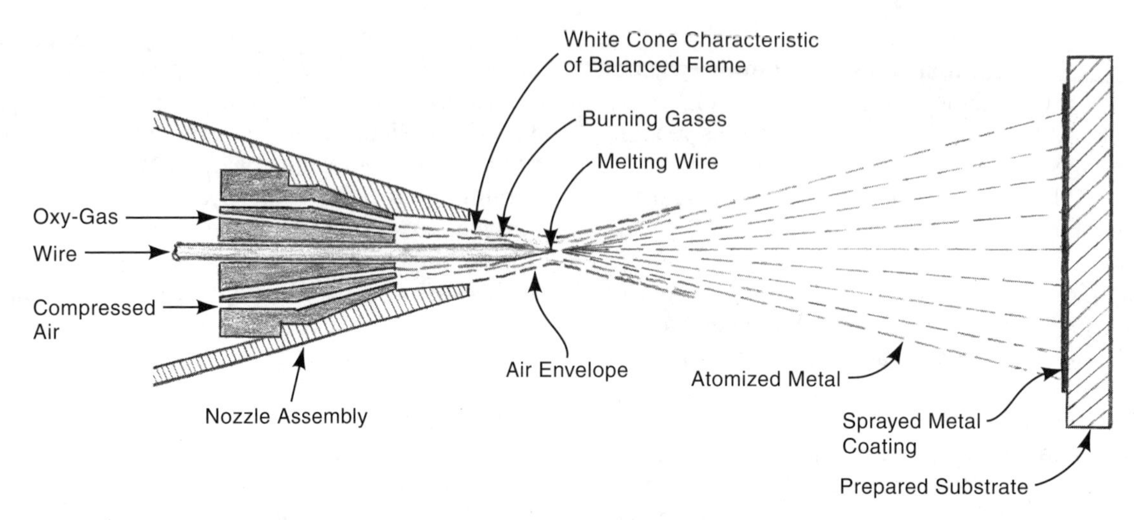

Fig. 8F4a Wire metalizing to spray liquid metal on the surface of another metal part.

The surface to be coated is often grit blasted to clean and roughen it. Occasionally, it is machined to provide a still rougher, interlocking surface. Additionally, steel workpieces are sometimes plated with nickel, chromium, nickel-aluminum or molybdenum to provide better adhesion of the sprayed-on material. There is some porosity in the coatings, usually about 6 to 13 percent. One drawback of the process is the limited bonding strength of the coating, typically about 5000 psi (35 MPa). Fig. 8F4a illustrates the workings of a wire metalizing gun. For jobs requiring greater deposition rates or large surfaces, guns are available to handle deposition materials in rod form up to about 3/8 inch (9.5 mm) in diameter.

F4a1. ***electric arc wire metalizing*** - has no external heat source. Two wires are used as shown in Fig. 8F4a1. Each wire is electrically charged and the two charges are of opposite polarity. The wires are fed so that their ends come together, creating an arc that heats and melts the wires. A central air jet propels the molten metal to the workpiece. This method provides good bond strengths with less heating of the substrate than with gas flame atomizing. There is no cost for fuel gas or inert gas. However, the method is limited to ductile materials; wires of carbides, oxides and nitrides are not sufficiently ductile. A common application is zinc coating for corrosion resistance.

F4b. ***powder spraying*** - is a variation of wire spraying. It is performed in the same way, except that the material to be applied is fed to the gun in powder rather than wire form. Gravity or pressure may be used to move the powder which is supported in a gas carrier. An advantage of using powder is that materials like ceramics, cermets, carbides and oxides, which are difficult to fabricate into wire, can be fed easily. Additionally, the size of the droplets of coating material is determined by the particle size of the powder rather than the degree of atomization provided by compressed air or gas. With some applications and coating alloys, it is necessary to postheat the workpiece after coating to fuse the coating material. See Fig. 8F4b.

F4c. ***detonation gun spraying*** - involves the repeated detonation of a mixture of fuel gas, oxygen and nitrogen to which coating material in powder form has been introduced. The powder is expelled at high velocity toward the workpiece. The gun has a long barrel to confine and direct the powder. The detonation repeats 4 to 8 times per second. The powder being expelled has an explosive velocity that causes it to be further heated when it strikes the workpiece, providing a strong metallurgical and mechanical bond. Bond strength is considerably better than with conventional powder spraying and porosity is greatly diminished. These factors enhance wear resistance. Because of

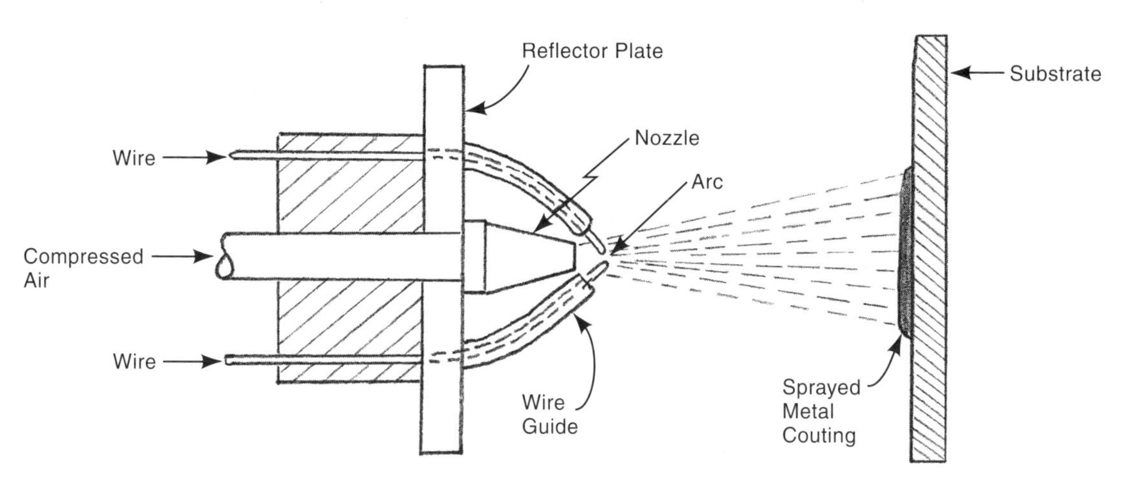

Fig. 8F4a1 Electric arc wire metalizing.

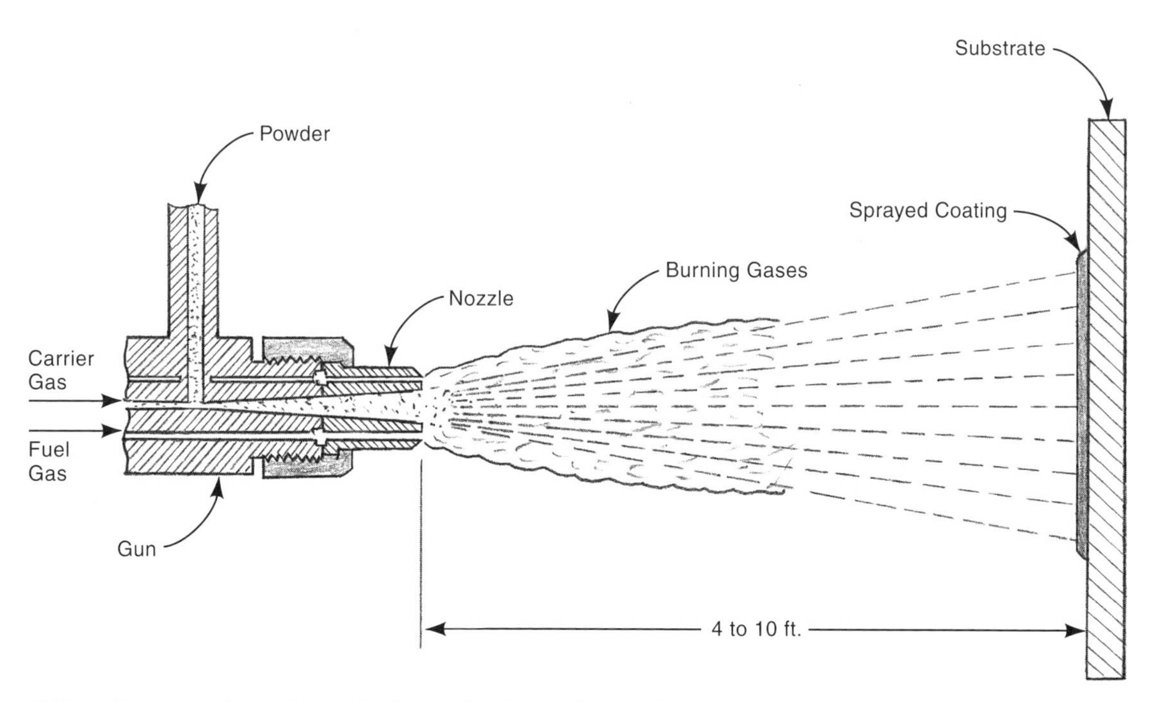

Fig. 8F4b Cross section of a typical gun for thermal spraying powdered metals, ceramics, cermets and carbides to a metal workpiece surface.

the extreme noise of the operation, it is performed within a soundproofed chamber. The operator remains outside the chamber but controls both the workpiece movement and the gun operation. Typical coating depths are 0.005 to 0.010 in (0.13 to 0.25 mm) but can be as much as 0.030 inch (0.8 mm). A cooling system may be used to avoid overheating the workpiece. Tungsten carbide, aluminum oxide and chromium carbide are sprayed with this method which is normally limited to metal substrates because of the force of the explosively driven powder. The prime applications are wear resistant coatings, particularly when conditions are severe. The method is illustrated in Fig. 8F4c.

F4d. ***plasma arc spraying*** - uses a direct current arc, similar to a welding arc, to heat a gas. The gas is constricted at the arc, further increasing its

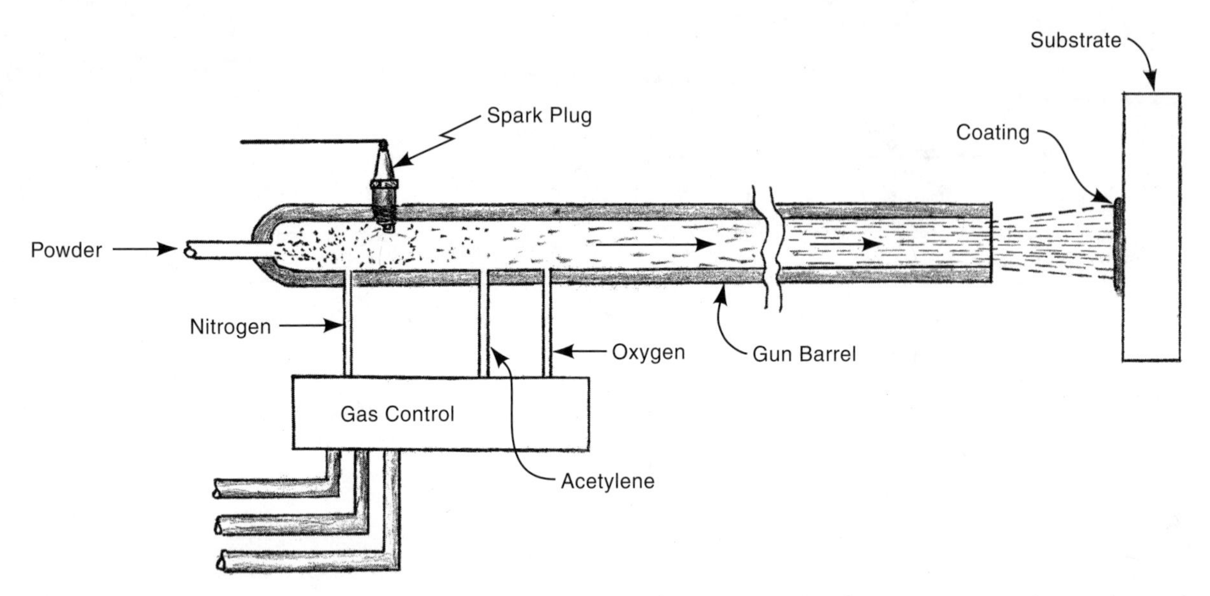

FIG. 8F4c detonation gun powder spraying. Because of extreme noise from repeating detonations, the operation is performed in a soundproof chamber.

temperature to 10,000°F (5500°C) or above. The gas becomes ionized at that temperature and streams from the nozzle of the gun as a plasma. As it exits, powder is added to the stream. The gas used is argon or nitrogen in combination with hydrogen. The powder achieves high velocity as it melts and strikes the workpiece surface, producing a coating with much less porosity than conventional wire or powder spraying. Adhesion is also very good. The process can be used to apply materials with melting points up to 6000°F (3300°C). Ceramics, intermetallics, cermets, carbides, refractory metals and plastic materials as well as other metals and alloys can be deposited. A wide variety of metallic and nonmetallic substrate materials can be coated with the process. Fig. 8F4d illustrates the operation of a typical plasma gun.

F4d1. A process variation, ***transferred plasma-arc spraying***, uses a second arc current

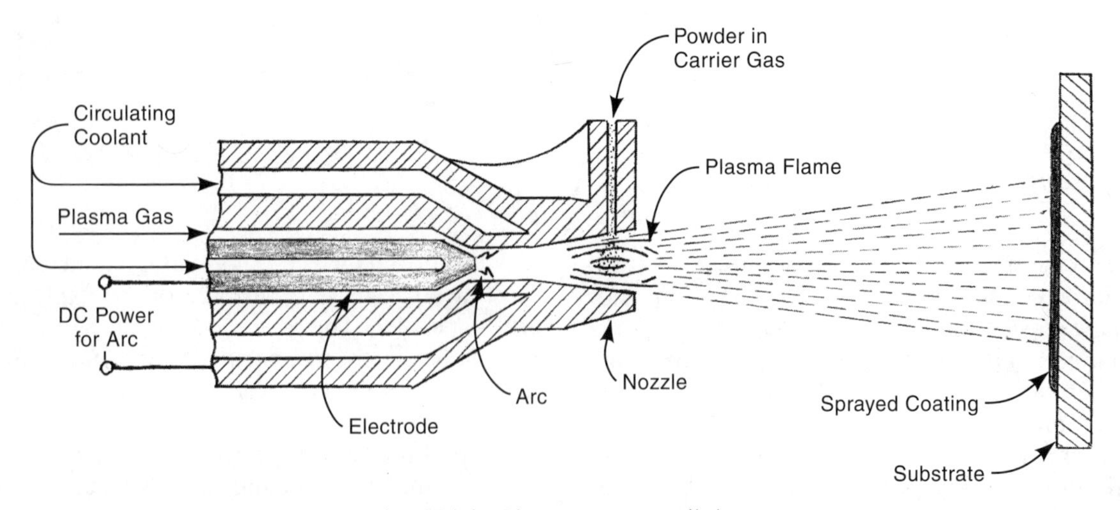

Fig. 8F4d Plasma arc metalizing.

between the spray gun and the workpiece. This arc heats the workpiece, aiding the coating process so that thicker, denser, coatings are feasible with good metallurgical bonding. Less costly powders of larger particle size and wider size distribution can be used. Electrical power requirements are also reduced. However, the workpiece must be conductive and able to withstand some melting. Coatings on plowshares, digging machinery, and valve seats are typical applications.

F4e. ***high velocity oxy-flame coating (HVOF)*** - is another process that uses coating material in powder form. The oxy-fuel mixture (oxygen plus acetylene, propane, hydrogen or MAPP) is ignited in a high pressure chamber where it attains a temperature of about 3800°F (2100°C). (In some applications a mixture of liquid kerosene and air is used.) The resulting gas stream exits through a small-diameter orifice traveling at a supersonic speed of over 1000 ft/s (300 m/s). Powder is injected, usually axially, into the stream which melts the particles and propels them against the workpiece. Dense, well-bonded coatings result; they have superior wear resistance. Metals, cermets and ceramics can be coated with the process, which is now widely used. Coatings for wear resistance are a prime application. Fig. 8F4e shows the HVOF operation.

F5. ***clad metals*** - (See 7C13d.)

G. Heat Treating of Metals

G1. ***annealing*** - is a general term for the softening of metals. It usually involves heating the workpiece to a suitable temperature, holding it at that temperature for a prescribed time and then cooling it at a particular rate. The specific temperature, holding time, cooling rate and other process details depend on the metals involved and the purpose of the annealing operation. Annealing reduces the yield strength and hardness of the workpiece material, removes or reduces internal stresses, reduces segregation, restores ductility, refines grain size and modifies electrical and magnetic properties. Subsequent forming and machining operations can then be performed more easily. With steel, the heating is to a point above the austenitizing temperature, the temperature at which the material's structure starts to change, and holding that temperature until the transformation is complete. The workpiece is then slowly cooled to room temperature. When the term, "annealing", is applied without qualifying adjectives to ferrous metals, it usually refers to full annealing as described below. For the descriptions that follow, as applicable to ferrous metals, refer to the iron-carbon phase diagram in Fig. 8G1 which shows the transition lines between different states in standard terms.

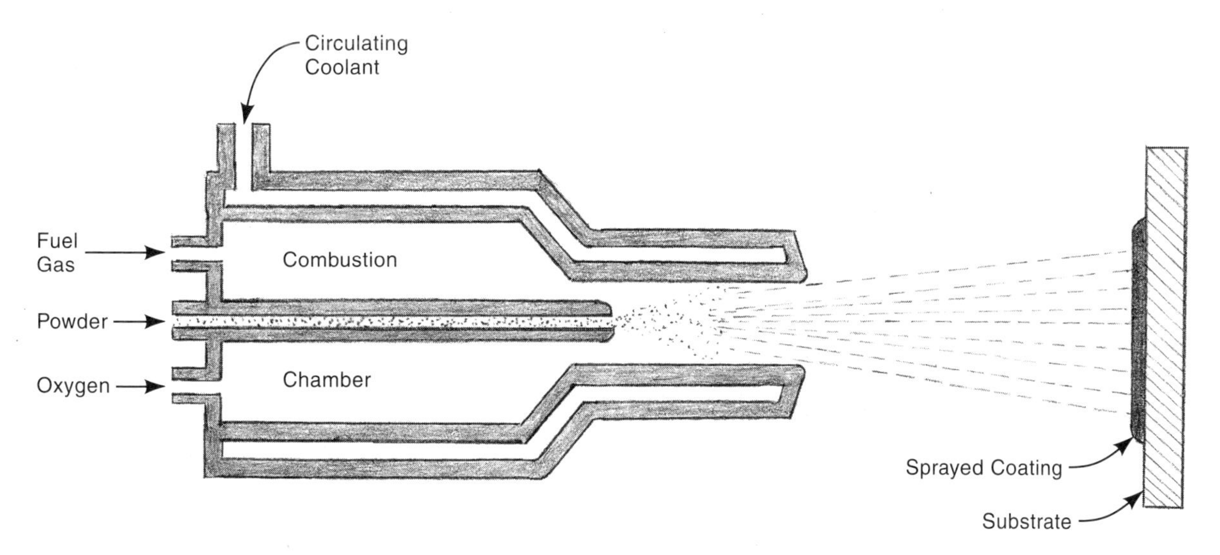

Fig. 8F4e High-velocity oxyfuel (HVOF) powder spraying.

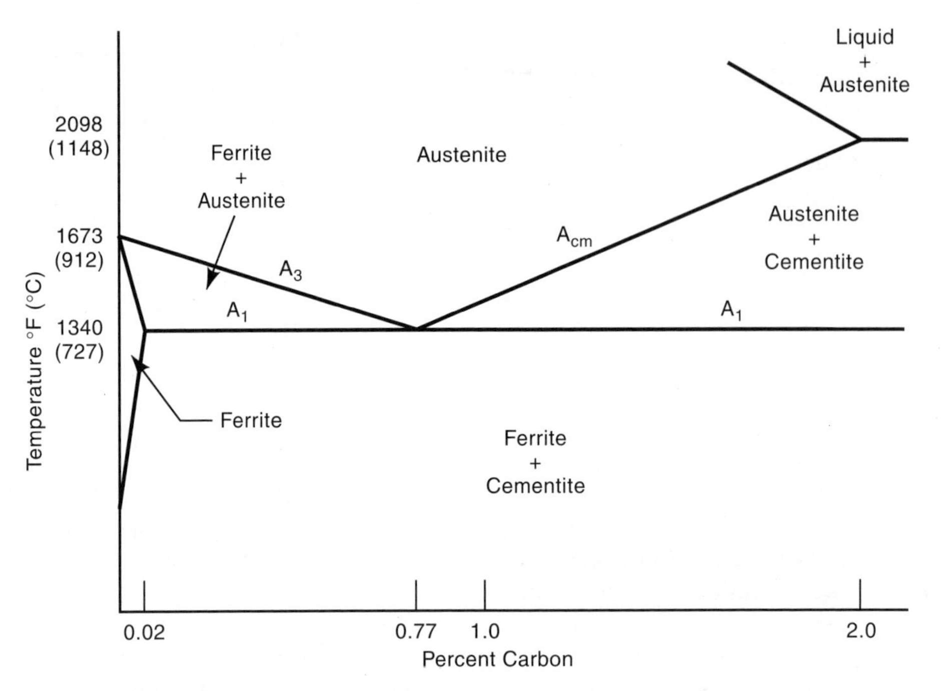

Fig. 8G1 A phase diagram for steels. The diagram, simplified, shows structures that develop in steel at different levels of carbon content and temperature. The lines show the transition temperatures from one type of structure to another.

G2. *annealing processes for steel*

G2a. ***full annealing***[1,2] - For hypoeutectoid steels, those with less than 0.77% carbon, the workpiece is heated to point 50 to 100°F (30 to 55°C) above the critical temperature (A_3 in Fig. 8G1). That temperature is maintained until it is uniform throughout, and the structure is converted to single-phase homogeneous austenite. The workpiece is then cooled slowly in the furnace. The cooling rate must be controlled and slow enough so that the declining temperature is approximately the same throughout the workpiece. The controlled cooling continues until the workpiece temperature is at least 50°F (30°C) below the A_1 line. Fig. 8G2a shows a temperature-time diagram of the process. Full annealing removes existing internal stresses in the workpiece and all traces of the previous structure, replacing it with a crystalline structure primarily of coarse pearlite, providing a softer, ductile metal.

For hypereutectoid alloys, those with greater than 0.77% carbon, the process is basically the same. However, the heating is to a slightly lower temperature, about 50 to 100°F (30 to 55°C) above the A_1 line. This, again, results in a soft, ductile primarily pearlitic structure.

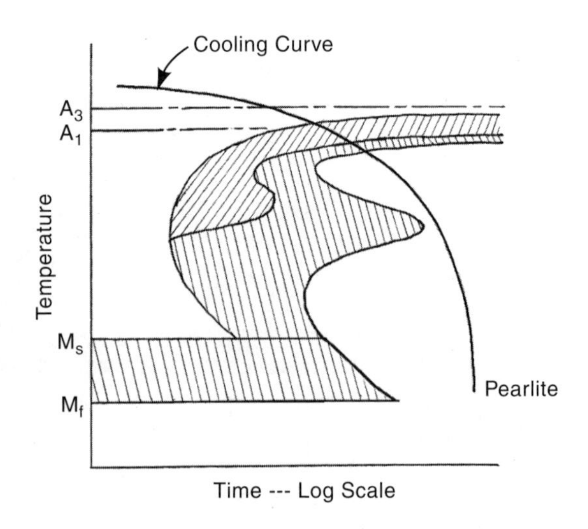

Fig. 8G2a Full annealing of steel. The steel is thoroughly heated to above the A3 temperature level and is then slowly cooled.

G2b. ***isothermal annealing***[2] - Fig. 8G2b illustrates the time-temperature curve for this process. Initial heating and temperature holding are identical to that for full annealing but, once the structure becomes fully austenitic, the metal is quenched rapidly to a temperature below that of the A1 line where the structure changes to a relatively soft ferrite carbide aggregate. The temperature is then held for a period (normally several hours) while the austenite completely transforms to pearlite. After the transformation is complete, the workpiece can be cooled further in any manner. This process shortens the time required compared to full annealing and is less costly. It provides a more uniform structure than full annealing but accurate temperature control is more critical. The fineness of the pearlitic structure depends on the transformation temperature used.

G2c. ***spherodizing*** - is an annealing process that has the purpose of producing a structure in which cementite is in the form of small spheroids dispersed in a ferritic matrix. The process is used to improve the cold formability of steels but is especially useful in providing improved machinability for hypereutectoid and tool steels. There are several ways to produce this structure. In one method, the workpiece is slowly heated to a temperature just below the critical level (A_1 in Fig. 8G1), held there for a prolonged period and then slowly cooled. In another method, the workpiece is alternatively heated and cooled to temperatures just above and just below the critical level (A_1). A third method, used for tool and high alloy steels, is to heat them to a temperature of 1400 to 1500°F (750 to 815°C) or higher, maintain them at that temperature for several hours and then slowly cool them.[1]

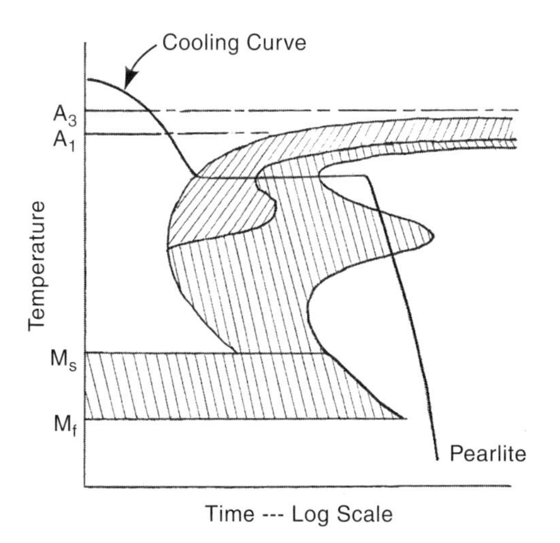

Fig. 8G2b Isothermal annealing of steel provides a more uniform structure than full annealing, but the temperature controls are more critical.

G2d. ***stress relieving*** - is accomplished by heating the workpiece uniformly to a suitable temperature below the A_1 (Fig. 8G1) critical temperature. (This is the temperature at which the material's structure changes.) This temperature is maintained for a predetermined period and the workpiece is then cooled gradually and uniformly to room temperature. The purpose of stress relieving is to relieve the stresses that are developed in the workpiece when forming, machining and other operations are performed on it, without greatly affecting other properties. Rolling, casting, forging, shearing, bending, drawing and particularly welding are sources of residual stresses that may have to be relieved. Uniform heating and cooling during stress relieving are important to avoid inducing new residual stresses.

G2e. ***normalizing*** - The workpiece is heated to a temperature 100°F (55°C) above the A_3 level (Fig. 8G1) if the material is hypoeutectoid (less than 0.77% carbon) or above the A_{cm} level if the material is hypereutectoid (0.77% or greater carbon), and maintained at this temperature until a uniform austenitic structure is formed. The workpiece is then removed from the furnace and allowed to cool in still air. The careful cooling rate of other annealing processes is not maintained. However, the resulting structure is usually fine pearlite with an excess of ferrite or cementite. The exact structure tends to vary within the workpiece and depends on the cooling rate at each point of the workpiece which, in turn, depends on the shape and size of the workpiece. Nevertheless, the process is useful when maximum softening is not required and the cost and time required for the operation need to be kept at a minimum. Normalizing is also used when the part needs to be harder and correspondingly stronger than it would be if annealed. Machining may be more easily performed on some materials when normalized instead of annealed. Air hardening tool steels are not subjected to this process.

G2f. ***tempering*** - is performed on through-hardened steel parts that may be too brittle for use after the hardening operation. The part is heated to a temperature below the critical temperature and is then air cooled. Usually, the part is held at the desired temperature only long enough to ensure that it is uniformly heated. Higher temperatures and longer times at the elevated temperature produce increased softness and ductility in the part. The operation reduces residual stresses in the part from the hardening operation and increases its toughness.

G2g. ***process annealing*** - is performed on workpieces that are to undergo further cold-working operations. Its purpose is to relieve stresses caused by cold-working and to restore ductility when workpiece material has work hardened. The workpiece is heated to a temperature just below the critical level (not as high as in full annealing or normalizing), held at that temperature for a period and then cooled slowly. The process is less costly and produces less scale than full annealing or normalizing. Process parameters may vary from case to case, depending on what is necessary to permit whatever further cold-working is required. Wire parts that require further drawing or upsetting and sheet metal parts requiring deep drawing are commonly process annealed.

G3. ***hardening processes for steel*** - require that the material to be hardened be heated above the austenitizing temperature [100 to 200°F (55 to 110°C) above the A3 line in Fig. 8G1]. The material is held at that temperature and then rapidly cooled (quenched). The length of holding time at that high temperature and the cooling rate depend on the particular steel alloy involved. Hardness is chiefly determined by the carbon content and the depth of the hard zone is affected by alloying elements and the hardening method used. Further heating and cooling may be undertaken after hardening to modify the metal structure. Various methods are used for both surface and through hardening of steel depending on the results desired and the alloy involved.

G3a. ***surface hardening*** - has the purpose of providing a wear resistant surface to a part while retaining a tougher, fracture-resistant core. Two basic methods are used to achieve this structure: 1) selective heating processes that heat only the surface material followed by quenching. The hardness of the finished workpiece surface depends on the carbon content of the workpiece and the method of quenching. With some authorities, the term "surface hardening" refers only to these processes. 2) processes that alter the chemistry of the surface material. Usually, these processes are referred to as "case hardening". Higher initial carbon content is not required in the workpiece with these methods. They may involve the use of hazardous gaseous, liquid or solid materials.

G3a1. ***flame hardening*** - uses an oxy-acetylene flame to heat the surface of the workpiece. The flame source is of high intensity and heating is rapid. It continues to the point where the surface material changes to austenite but the core material remains below the critical temperature. The workpiece is then immediately quenched in water and tempered. By carefully controlling the flame intensity and duration of heating, the depth of heating and eventual hardness can be controlled. The depth of the hardness typically ranges from about 0.030 to 0.250 in (0.75 to 6 mm). The process is often utilized with large workpieces that require wear resistance only in certain areas and where furnace treating would require extra large equipment that may not be available.

G3a2. ***induction hardening*** - uses induction heating to produce high temperatures at the surface, or in those areas requiring hardening. As with other processes requiring induction heating (See 7A2h, 7B4 and 7C2.), an electrical coil must be designed and fabricated to produce the heating effect in the desired location. The process is well suited for surface hardening because the depth and location of the treatment can be controlled by establishing the optimum heating time, current, frequency, power level and coil configuration. (Higher induction frequencies concentrate the heating effect more at the surface and produce shallower hardened cases; lower frequencies produce deeper hardening or even through-hardening.) After the heating cycle, which is quite rapid, the workpiece is immediately quenched. Round and cylindrical workpieces are most easily processed since simple coil shapes can be utilized. Shaft and crankshaft bearing surfaces are frequently surface hardened by induction heating. Vehicle and machine shafts and hydraulic piston rods are typical parts treated by induction hardening for improved wear resistance and fatigue life. When the operation is automated, the shafts pass through the induction coils in a timed cycle and are immediately spray quenched when they exit the coils.

G3a3. ***laser-beam hardening*** - is a surface hardening technique for ferrous materials that uses laser energy to provide localized surface heating. The workpiece is typically coated first with a material with a more light-absorptive surface. Manganese phosphate and zinc phosphate are two commonly used surface coating materials. Laser parameters, such as beam size, power level, and beam movement speed as the surface is scanned, can be set to provide optimum heat input and to limit heating to only the surface of the workpiece. Laser beams for heat treating are usually more broadly focused than those used for metal cutting or welding. However, heating is rapid and the process can be very selective as to surfaces treated. The surface is heated to the austenitizing temperature, and quenching then forms hard martensite. Both water and oil quenching can be used but quenching often results simply from conduction of heat from the surface to the colder interior of the workpiece. Surface hardness up to Rc65 is feasible in steel of 0.40% carbon.[1] The process is readily adaptable to computer control. It is used to improve the wear resistance and fatigue strength of highly-stressed machine components such as gear teeth, cams, and crankshaft surfaces.

G3a4. ***electron-beam hardening*** - is similar to laser-beam surface hardening. In both processes, a high-energy beam is focused accurately to rapidly heat selected surface areas of the ferrous workpiece for hardening. The surface metal is heated to the austenitizing range and is then quenched to form martensite as the heat is conducted throughout the workpiece and the surface cools. In electron-beam heating, a stream of electrons from a heated cathode forms the beam that is directed by means of electromagnetic and electrostatic elements. The process is carried out in a partial or complete vacuum. The need for a vacuum lengthens the process and limits the size of the workpiece that can be processed. Typical hardened depths range from 0.01 to 0.04 in (0.25 to 1 mm)[5]. Electron-beam equipment is costly.

G3a5. ***other surface heating methods for hardening*** - Molten salt or lead baths can be used to heat workpieces rapidly enough so that only the surface - to the desired depth - is heated to the transformation temperature. Another approach is to use an electric arc lamp for selective heating of a workpiece. Higher heat intensities can be obtained with these methods than with flame heating and the heat source can be farther away, leading to less distortion from the process and easier heating of irregular surfaces. Very high arc powers are required.

G3b. ***case hardening*** - hardens a layer (usually a thin layer) of material at the surface of the workpiece. Carbon and/or nitrogen are first diffused into the workpiece surface at an elevated temperature. Although the entire workpiece is heated, the interior of the workpiece, not having received these hardening agents, does not harden but retains its toughness; only the exterior *case* is hardened. Case hardening is suitable for parts made from low-carbon steels. This is in contrast with the surface hardening methods described above which rely on the carbon already in the steel to provide the necessary hardness. There are a number of case hardening processes, all somewhat similar, described below.

G3b1. ***carburizing*** - is the oldest case hardening method, commonly applied to low-carbon steel workpieces. They are heated to above the transformation temperature range while in contact with a carbon-containing material in gaseous, liquid, or solid form. The carbon is absorbed by the workpiece material and the outer surface thus becomes high-carbon steel. Workpieces are then either quenched or slowly cooled and further heat treated, depending on the application and the initial grade of steel. The depth of the case depends on the time and temperature of the carbon absorption operation. The three approaches are illustrated in Fig. 8G3b1.

G3b1a. ***pack carburizing*** - utilizes a solid material as a carbon source. The workpiece is placed in a closed container along with charcoal, coke or other carbonaceous material. Heating takes place over an extended period during which the hot

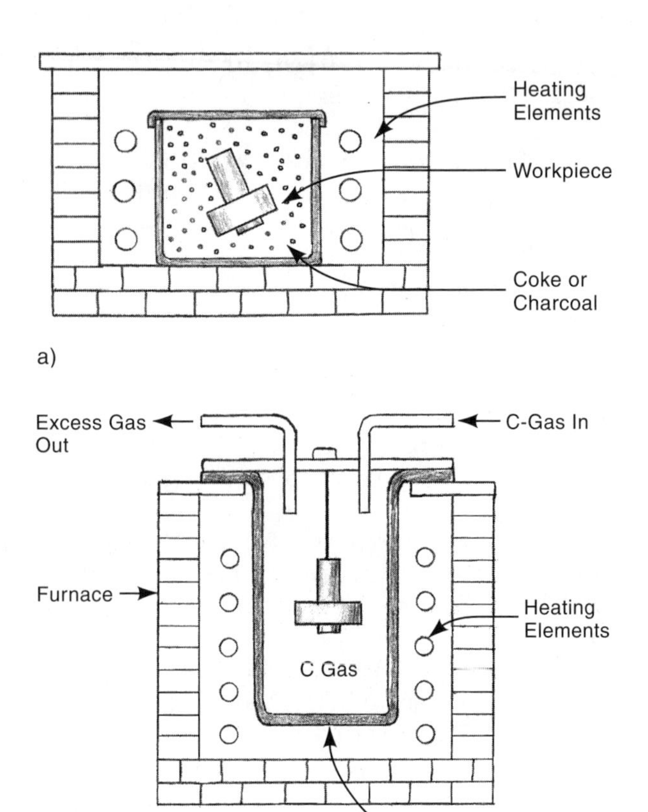

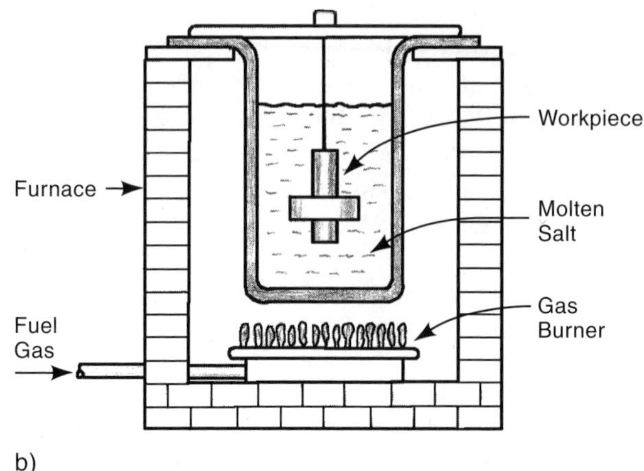

Fig. 8G3b1 The three methods of carburizing: a) pack carburization, b) liquid carburization, c) gas carburization.

carbonaceous material gives off carbon monoxide gas. This reacts with the workpiece metal, releasing carbon that is absorbed by the workpiece surface. Carbon content typically ranges from 0.7 to 1.2% and case depths from about 0.003 to 0.35 in (0.8 to 9 mm) but a depth of 0.06 in (1.5 mm) or less is most common.[1] Pack carburization requires less specialized techniques and less sophisticated equipment than the other two alternatives but is not as well suited for high-production applications.

G3b1b. ***liquid carburizing*** - utilizes a molten salt bath instead of solid material. This provides more uniform heating and easier handling of the workpiece, particularly if automatic operation is involved. Cyanide has been a common salt bath material, providing both carbon and nitrogen infusion, but safety concerns have limited its use. Some non-cyanide materials have been developed as a substitute. Case depths with liquid carburizing are usually thin and the process is used mostly for small and medium-size parts. The process is faster than pack carburizing and is adaptable to workpieces of various shapes and various case depths.

G3b1c. ***gas carburizing*** - is carried out with a carbon-carrying gas instead of liquid or solid material. Usually, natural gas, propane or a mixture of carbon monoxide, nitrogen and hydrogen is used. Gas carburization provides a more controllable process that is also faster and more easily automated since continuous furnaces can be used. However, special precautions must be taken because of the toxicity of carbon monoxide, a critical element of the gases used. Case depths up to 0.4 in (10 mm) are feasible, but shallower case depths, of 0.005 to 0.030 in (0.13 to 0.75 mm), are common. The process is usually used for small parts which can be quenched immediately upon exiting the heating furnace.

G3b2. ***carbonitriding*** - In this process, the workpiece is heated in an atmosphere containing nitrogen and carbon. Ammonia, in a mixture with a carbon-rich gas, is normally used. Heating is to a temperature above the critical range. As the workpiece is held at that temperature, it absorbs nitrogen and carbon. It is then quenched. A wear-resistant case depth of 0.003 to 0.030 in (0.08 to 0.75 mm) results. The process develops less distortion than carburizing. It is useful for less costly steels, providing properties equivalent to those obtained when carburizing more expensive alloy steels.

G3b3. ***cyaniding*** - This process is sometimes referred to as ***liquid carbonitriding***. It involves immersion of the workpiece in a bath of molten cyanide salts to heat them to a temperature at which austenite begins to form . Typical immersion periods are one half to one hour at that temperature. Nitrogen, produced as the cyanide bath decomposes, and carbon enter the surface of the workpiece. Immersion in the bath is immediately followed by oil or water quenching. The case hardened surface thus produced contains iron nitrides and carbides and has high hardness and good wear resistance. The case is typically 0.005 to 0.015 inches (0.13 to 0.38 mm) deep. The process is most commonly used to case harden small parts. Potentially serious hazards attend the use of the cyanide baths because of the poisonous nature of cyanide.

G3b4. ***nitriding*** - consists of heating the workpiece in an atmosphere of ammonia or other gas containing nitrogen. The process is limited to alloy steels that have the capability of absorbing nitrogen. Upon prolonged heating (20 to 100 hrs) at temperatures from 925 to 1050°F (500 to 565°C)(below the transformation range), the workpiece absorbs nitrogen from the gas and forms nitrides which provide the necessary hardness. Special steels have been developed to facilitate nitriding. Quenching is not required.

Hardenable steels to be nitrided are first hardened and tempered. Normalizing may also precede the nitriding operation if workpiece sections are large. Decarburized material must be removed before nitriding and thorough workpiece cleaning is also important before the operation. Nitriding produces very hard cases and little workpiece distortion. Case hardened depths typically range from 0.008 to 0.30 in (0.20 to 7.5 mm). This process also provides improved wear and corrosion resistance and lessened possibility of galling or fatigue failure.

G3b5. ***liquid nitriding*** - involves immersion of the workpiece in a bath of molten cyanide salts instead of a gas atmosphere. The operation takes place at the same temperature (below the transformation range) as when a gas atmosphere is used. It produces a thinner case than gas nitriding, with depths from 0.001 to 0.012 in (0.03 to 0.30 mm) but with results and applications that are otherwise similar. Liquid nitriding provides more nitrogen and less carbon to the workpiece than similar processes, liquid carburizing and cyaniding, which also use baths of molten cyanide salts.

G3c. ***through hardening*** - provides a hardened structure throughout the workpiece. There are a number of processes for through hardening that involve heating the workpiece to a temperature above the critical temperature, holding it at that temperature until it is uniformly heated, and then quenching it, (cooling it rapidly). The quenching medium can be water, oil, or air, depending on the alloy being treated and the size and shape of the part. Tempering normally follows the hardening operation.

G3d. ***martempering*** - is a hardening process that provides uniform hardness throughout the part and a minimum of residual stress and distortion. The heated workpiece is quenched with a hot fluid (molten salts, hot oil, fluidized particle bed, or molten metal), at a temperature above the martensite range. The workpiece is held in the quenching liquid until the temperature throughout is essentially uniform. The workpiece is then cooled at a moderate rate, usually in air, to prevent large temperature differences between its outside and internal portions. Straightening or forming, if required, can be performed while the workpiece is still hot. Conventional tempering then always follows. Fig. 8G3d shows the temperature transformation diagram for martempering.

G3e. ***austempering*** - is a means for hardening a workpiece while retaining ductility and toughness. The steps are as follows: The workpiece is heated to a temperature within the austenitizing range, normally 1450 to 1680°F (790 to 915°C). It is then quenched in a bath - usually molten salt - whose temperature is in the range of 500 to 750°F (260 to 400°C). The workpiece is left in this bath

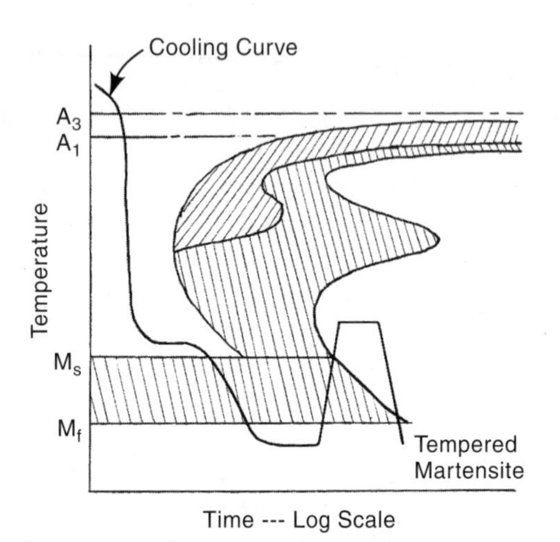

Fig. 8G3d Martempering of steel, a hardening process that provides uniform hardness throughout the part.

long enough for the structure to transform isothermally to bainite. It is then cooled to room temperature. Fig. 8G3e illustrates the time-temperature cycle for austempering[1].

G4. ***solution treating/precipitation hardening (aging or age hardening)*** - is normally a three-step process, most commonly performed in sequence. The operation is prominent with non-ferrous metals. The first step, *solution treatment,* involves heating the workpiece material, which contains some alloying elements, thoroughly to a temperature just below the eutectic melting temperature. At this temperature, alloying constituents become uniformly dispersed in a solid solution. The second step, *quenching,* preserves the solution in a supersaturated solid state because the alloying constituents do not have time to precipitate. In the third step, *precipitation hardening*, which often involves heating the workpiece to a lower temperature for a period of time, alloying constituents precipitate throughout the workpiece in a dispersed arrangement. They form particles at the grain boundaries and slip planes which reduce slippage, producing increased hardness and strength. With some alloys, the precipitation occurs over a period of days at room temperature. The precipitation hardening step is often referred to as aging, age hardening or artificial aging (if performed at an elevated temperature).

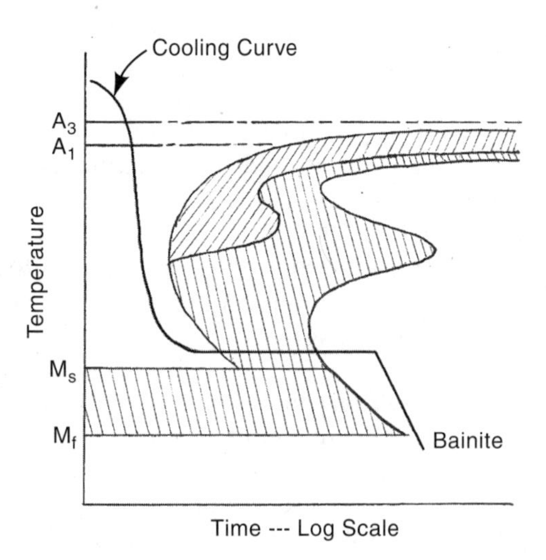

Fig. 8G3e Austempering is a means for providing hardness in a steel part while retaining ductility and toughness.

Solution treating/precipitation hardening or aging are commonly used on aluminum, copper, nickel, magnesium, titanium, zirconium, and their alloys, and on heat-resistant alloys. Fig. 8G4 illustrates a simplified diagram for aluminum-copper alloys and illustrates how temperature levels govern the operation.

Semiaustenitic stainless steels are hardened by this method and are solution treated by first heating, e.g., to 1900°F (1040°C), and then quenching in air to put the material in the solution-treated condition. Precipitation hardening is then performed by heating the workpiece to a subcritical temperature range, i.e., 900 to 1150°F (480 to 620°C). This treatment provides tensile strengths up to 190,000 lbf/in^2 (1310 MPa).

G5. *heat treating processes for non-ferrous metals*

G5a. ***aluminum alloys*** - are ***hardened*** by solution treating followed by quenching, and a precipitation phase of the alloying element. Several of these hardening elements (e.g., copper) are much more soluble at elevated temperatures than at room temperature so the hardening process first involves a heating and soaking phase. The temperature and duration of the heating phase vary with the alloy being treated, the size and thickness of the workpiece, and whether it is wrought, forged or cast. Typical solution treating phases involve heating to

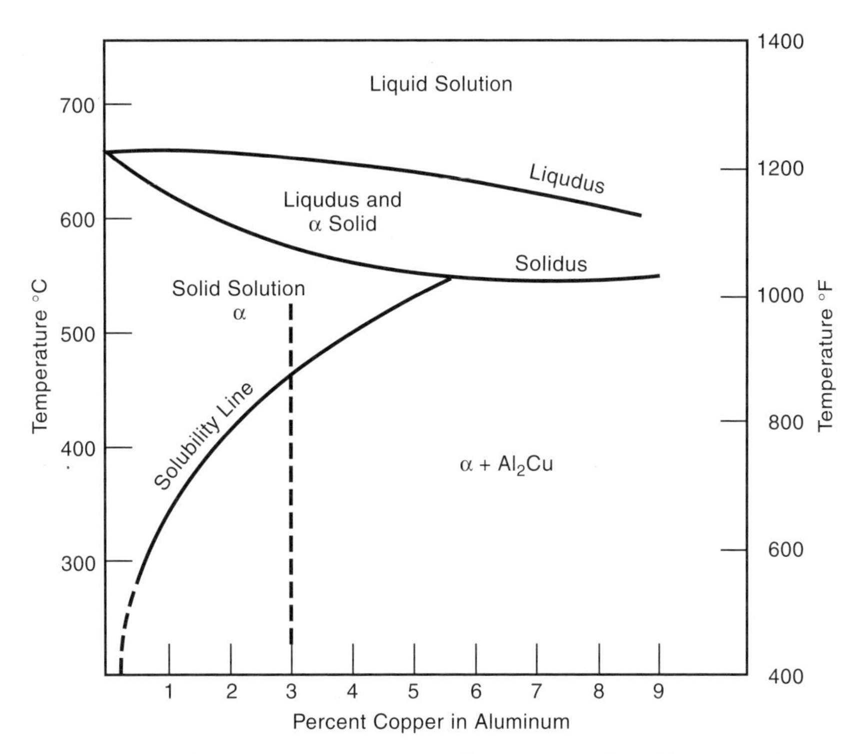

Fig. 8G4 Simplified diagram for aluminum-copper alloys. Note the alloy containing 3% copper. At a temperature above the solubility line at about 900°F, the copper goes into solution. If the workpiece is rapidly quenched to a lower temperature, the copper is held in a supersaturated solution. Then, at 400°F or lower, the copper will gradually precipitate as fine particles which enhance hardness and strength of the alloy.

a temperature between 870 and 1055°F (465 and 570°C) and soaking at that temperature for a period that may be as little as 20 minutes and as much as 75 minutes. (if in an air furnace. Salt bath soaking times are shorter.) Time is required to heat the workpiece uniformly and to allow the hardening element to go into full solution. The soaking temperature must not be so high that eutectic melting takes place at the interface of the aluminum and the alloying element. The allowable temperature range may be quite narrow, typically ±10°F (5.5°C). After soaking, the workpiece is immediately quenched, usually in water below 100°F (38°C). The quick quenching prevents the rapid precipitation of the hardening element, which is not desirable. Instead, precipitation is allowed to take place at room temperature for as much as a month or more, or at an elevated temperature of up to 400°F (200°C) over a period of several hours. This provides stability of the properties resulting from formation of finely dispersed precipitates. (Fig. 8G4 shows the diagram for aluminum-copper alloys.)

Annealing of aluminum alloys requires a heating phase at a temperature that may be as low as 500°F (260°C) or as high as 775°F (410°C) followed by slow cooling to about 500°F (260°C). Times and temperatures depend on the alloy type, temper and initial structure. Stress-relief annealing uses the lower temperatures of this range for shorter periods; full annealing utilizes the higher temperatures and some holding period at the specified temperature before slow cooling. Full annealing requires closer control over both the temperature and the heating period.

G5b. ***copper and copper alloys*** - may undergo the following treatments: annealing, stress relieving, solution treating and precipitation hardening (aging). Both furnace and salt bath methods are used to provide the necessary heating. A protective atmosphere is commonly employed to prevent tarnish of the workpiece. Normally, the atmosphere is either an exothermic (reducing) gas or dissociated ammonia.

Annealing is performed on wrought alloys to produce a desirable combination of ductility and strength for cold forming operations. It is accomplished by heating the workpiece to an elevated temperature for a period of time. Temperatures used range from 500 to 1500°F (260 to 815°C) but most alloys fall into the range of 700 to 1200°F (370 to 650°C). The temperature must be sufficient to cause recrystalization and, if desired, is brought to higher levels to cause grain growth. Heating and cooling rates tend to be relatively unimportant.

Stress relieving reduces the tendency of formed copper alloys to crack from stress corrosion. The intent is to relieve internal stresses without affecting the workpiece properties to an appreciable degree. Treatment involves heating to a particular temperature that depends on the alloy involved, the workpiece shape and the extent of existing stresses. Temperatures range from 355 to 970°F (180 to 520°C) with most alloys falling closer to the middle of that range.

Solution treating/precipitation hardening is the hardening treatment for most copper alloys, though some are hardened by quenching from a high temperature to produce a martensitic-like effect[7]. Except for those alloys, solution treating involves temperatures in the range of 1400 to 1830°F (760 to 1000°C) for 1 to 5 hr., followed by a water quench, and precipitation hardening at a temperature in the range of 575 to 1400°F (300 to 760°C). Specific values depend largely on the alloy involved. Alloys for electrical applications are commonly solution treated if spring action or other hardness is required.

G5c. ***magnesium alloys*** - may receive the following treatments: solution heat treating, precipitation hardening (artificial aging), annealing, and stress relieving. These alloys are treated to improve strength, toughness and other mechanical properties or to condition the metal for further operations.

Annealing removes strain hardening and temper resulting from previous treatments. Wrought workpieces are heated to a temperature of 550 to 850°F (290 to 450°C) for one hour or more and are then fully annealed. (The particular temperature used depends primarily on the alloy involved.) Since most forming of magnesium is performed at an elevated temperature, there is less need for annealing than is required with many other metals.

Stress relieving is carried out to reduce or remove residual stresses that result from forming operations, welding, casting, machining, extrusion or other operations. Typically the workpiece is heated to an elevated temperature somewhat below that required for annealing and holding the workpiece at that temperature for 1/2 to 2 hours. The time and temperature used depend on the alloy and the nature and strength of the residual stresses. For example, forgings have higher residual stresses than rolled sheet and require higher temperatures. Stress relieving temperatures range from 300 to 800°F (150 to 430°C).

Solution heat treating and precipitation hardening (aging) are carried out to improve strength, toughness and shock resistance. The full process involves solution treating at temperatures to a point within the range of 725 to 1060°F (385 to 570°C) for a period of 2 to 72 hours. Aging also takes place at an elevated temperature of 335 to 480°F (170 to 250°C) for from 5 to 48 hours. Time and temperature depend chiefly on the alloy and the results desired. Magnesium castings may be supported during elevated-temperature heating to prevent sagging. Flat parts may also need support and straightening after the treatment.

G5d. ***nickel and nickel alloys*** - are subjected to solution treating, age hardening, annealing, stress relieving and stress equalizing.

Solution treating is not usually required before precipitation hardening but is used in some circumstances to provide special properties such as hardness at high temperatures. When used, it takes place at temperatures of 1525 to 2260°F (830 to 1240°C) to put carbides and age hardening elements into solid solution. Solution treating, when carried out, is normally part of the age hardening treatment. However a high temperature solution treatment at 2100 to 2400°F (1150 to 1315°C) is also part of a solution annealing process.

Age (precipitation) hardening produces a precipitation of submicroscopic particles throughout the base metal, increasing both its strength and hardness. Solution treating beforehand is normally not required as it is for other alloys. Age hardening is commonly performed by heating the workpiece to a temperature of 900 to 1600°F (480 to 870°C) and holding at that temperature for a long period, followed usually by furnace or air cooling. The operation can proceed after hot or cold working, or with the workpiece in an annealed condition. The operation is sometimes performed in a dry hydrogen or ammonia atmosphere or with the workpieces in a sealed box in the furnace.[7]

Annealing requires heating to a prescribed temperature, usually one between 1300 and 2200°F (700 and 1200°C) for a period of time and then cooling at a slow or rapid rate. A recrystallized grain structure results and the workpiece material is softened. Heating and cooling parameters depend primarily on the nickel alloy involved and the degree of hardness before annealing. Annealing is common after cold forming operations that work harden the material. Sometimes, annealing is an in-process operation between successive cold forming stages. Heating may take place in the protected atmosphere of a furnace or in a salt bath.

Stress relieving is a treatment that does not result in recrystalization but it does reduce the stresses in non-age-hardenable alloys that have work hardened. Temperatures used range from 800 to 1600°F (425 to 870°C), depending on the nickel alloy involved, and the degree of work hardening. Careful temperature regulation is required. Cooling is most commonly by air but water quenching is used for Hastelloys.

Stress equalizing is a low-temperature operation. It is used to balance the residual stress after cold working without a significant loss of mechanical strength. The workpiece is heated to a specific temperature between 450 and 900°F (230 and 480°C) for one or more hours and then air cooled. The temperature used depends on the alloy involved. Coil springs and stampings with spring effect are given this treatment.

G5e. ***titanium and titanium alloys*** - are treated to relieve residual stresses developed in fabrication operations, to improve workability or machinability or to increase fatigue strength, creep strength and toughness. Stress relieving, annealing and solution treatment, and aging are the most common operations.

Stress relieving involves heating to a point in the 750 to 1500°F (400 to 815°C) range for a period of time that may be a little as 1/4 hr. and as much as 48 hr. Time and temperature vary with the alloy. Higher temperatures allow shorter holding times. The operation is performed to lessen residual stresses from previous operations without significantly reducing strength and ductility.

Annealing requires somewhat higher temperatures than stress relieving. Titanium and its alloys are heated to a temperature in the range of 1200 to 1650°F (650 to 900°C) for a period of 1/10 to 8 hours and then are cooled in air, in the furnace, or at particularly slow rates. Specific conditions for heating and cooling depend on the alloy and the purpose of the annealing. Annealing provides improved toughness, ductility, machinability, and/or dimensional and structural stability at elevated temperatures.

Solution treating and aging first require that the workpiece be heated to a point within the range, 1275 to 1940°F (690 to 1060°C) for a period of a few minutes up to 2 hrs. Quenching in water, oil, or air follows this step. The workpiece is then aged by reheating it to a temperature of 735 to 1400°F (390 to 760°C) and holding this temperature for a period that may be as short as 2 hr and as long as 100 hr. The properties wanted, the workpiece thickness and the alloy involved affect the choice of treatment parameters.

H. Shot Peening

Shot peening is a process intended to improve the fatigue strength of a workpiece. It produces a residual compressive stress at the surface, a few thousandths of an inch deep, from the effect of many small steel balls that are thrown against the workpiece. The shot balls are impelled by air pressure through a nozzle, or by centrifugal force from a spinning wheel. Sometimes, cut steel wire or glass beads are used instead of steel shot. Masking may be used if the effect is wanted on only part of the exposed portion of the workpiece. Fig. 2J7 in Chapter 2 (Metal Forming) illustrates an application of the process where the compressive stress provided by peening is used to modify the form of the workpiece.

I. Product Marking Methods

I1. ***manual marking*** - Information is often applied to a workpiece or assembly by hand printing or writing, using a pencil, crayon, pen, chalk, or brush. Graphite, paint, ink, stain, dye, or another material that has a contrasting color from the workpiece surface may be used. Such markings are normally used for raw-material grade identification, or in-process or stockroom identification of the workpiece.

I2. ***stamp indented marking*** - Letters, symbols or numbers are pressed into the surface of the workpiece with a hardened die in a punch press, arbor press or hydraulic, pneumatic or solenoid press or individually, with the impact of a hammer. Sometimes, roll dies and equipment are used. Fig. 8I2 illustrates some low-stress-inducing dies for stamp indenting several letters or symbols. A more recent method is to use a programmable stamper that impresses a series of individual dots to form recognizable lettering. Computer control enables the dot pattern to be altered as desired from part to part so that serial numbers and other variable information can be impression stamped.

I3. ***etching (chemical etching)*** - involves the chemical or electrochemical removal of workpiece surface material, usually metal, to create a marking.

Fig. 8I2 Tools for stamp-indented markings. The dot pattern of each die reduces the stresses induced in the workpiece surface from the operation. (*Courtesy Pannier Corporation.*)

Typical depths of etching are 0.003 to 0.012 in (0.08 to 0.3 mm). A stencil is normally used to control the pattern of the etch. Etchant liquid is applied to the workpiece through the stencil with a manual or machine-guided pad. Instrument and ruler graduations are typical applications. When photographic techniques are used, similar to those of photochemical blanking (See 3S4), very fine, detailed etches are feasible. *Electro etching* uses electrolytic current to enhance the operation. The workpiece is connected as one electrode and a pad soaked with electrolyte liquid is the other.

Chemical etching, performed over a wide surface rather than through a stencil, is performed to prepare a surface for a further operation such as adhesive bonding or painting. The etched surface has minute irregularities, which improve adhesion by increasing the total surface area and by creating peaks and valleys that tend to lock the adhesive or paint in place.

When the etched marking is much deeper than that normally used in etching, the terms *chemical engraving* or *photochemical engraving* may be used.

I4. ***engraving*** - produces marks that are much deeper than those from etching. Usually, engraving is a milling operation performed with small end-cutting tools that remove material to produce patterns in the surface of the workpiece. The operation also may be performed with abrasive tools. The markings are normally depressed, but surrounding material can be removed instead, to create a raised image. Nameplates, trophies, plaques and jewelry are commonly engraved. The movement of the cutter can be manually- or machine-controlled. Computer control of the movement of the cutter is now prevalent.

I4a. ***pantograph engraving*** - is engraving guided by a mechanism that allows the operator to trace a known pattern and reproduce it by milling away the surface of the workpiece. The engraved pattern can be an enlargement or reduction-in-size of the master. The operation is used extensively in the production of nameplates, jewelry, trophies and small signs.

I5. ***laser marking and engraving*** - (See section 3O for information on laser machining). Laser energy can be used to make permanent markings by removing part of the workpiece material (engraving), inducing a contrasting color in the workpiece surface or by removing a surface coating that has a color that is different from that of the workpiece material. Laser markings can be computer controlled to provide accurate markings, with flexibility for including variable information from workpiece to workpiece. Changeover from one type of mark (e.g., bar code, pictorial, numerical, or alphabetical) involves only a change of the controlling software program. The operation itself is also quick. Electronic components are frequently laser marked.

I6. ***stenciling*** - uses a sheet device with patterned openings. The sheet is held against the workpiece surface, and ink, paint, dye, stain or other liquid coloring is applied through the openings. The liquid can be applied by any of a variety of devices: brush, roller, spray, or pad, etc. The stencil can be of any sheet material that resists the coloring agent. Paper, sheet metal, fiber or plastic are all used. Stenciling is commonly used to mark shipping cartons. When abrasive blasting is used instead of a coloring agent, glass, ceramic and stone workpieces can be permanently etched.

I7. ***printing*** - A variety of methods are in use to transfer an image to a workpiece, or to a label, tag, or nameplate that will be affixed to the workpiece. In *contact* or *flexographic printing*, an inked die, usually made of rubber, is pressed against the surface to be marked. Non-contact printing utilizes a fine jet to apply an ink marking to the workpiece. Printing is used to mark shipping containers, raw materials for grade identification, labels, nameplates and tags. Chapter 9 describes a number of printing processes, all of which can be used in the printing of labels, tags, decals and nameplates. Many of them also can print directly on the workpiece.

I7a. ***pad printing*** - is a useful method for printing small areas, particularly if the workpiece surface is curved or slightly irregular, thin walled, or sensitive to deformation, since the pressures of pad printing are very low. The method involves the transfer of the image from an etched or engraved flat die to a very soft rubber pad which, in turn, transfers it to the workpiece. Ink is applied to the die plate; excess is removed with a rubber squeegee. When the soft pad contacts the die plate, it picks up the ink held in the recesses of the plate. See Fig. 8I7a. for an illustration of the method. Tooling is quite inexpensive and multiple color images can be printed without waiting for successive applications to dry. This method is used for decorating pens, medical devices, eyeglass frames, electrical devices, computer parts, spools and sporting goods.

I7b. ***screen printing (silk screening)*** - is a refinement of Stenciling. It is often used to add decorative markings such as symbols, product names and trademarks, as well as alphanumeric data to products and components. The stencil-like screen, now usually made from nylon or stainless steel fabric and tautly mounted on a rectangular frame, is coated with a solid material except in openings corresponding to the pattern to be marked. Ink or paint, in paste form, is spread on the screen and the screen is placed on the workpiece. Fixtures are usually used to insure accurate registration of the mark on the workpiece. A rubber squeegee is used to force the ink through the screen openings, manually or mechanically, and onto the surface of the workpiece. The screen printing process is widely used for printing signs, posters, nameplates and other information or decorations directly on products. Paper, paperboard, glass, fabrics, leather, and other materials can be screen printed. Fig. 9D4b in chapter 9 (paper-making and printing) illustrates the process. Successive operations on the same surface with different screens can produce multicolored markings. Photoresist techniques are used to produce the desired pattern in the screen coating.

I8. ***branding*** - involves the use of a heated tool that burns the mark into the surface of the workpiece. The method is effectively limited to non-metallic workpiece materials such as wood, leather, fibre and composition materials. Hot stamping of plastics is similar, but the base material is not burned; instead, it is heated sufficiently so that a film-carried image adheres to it. The method is described in section 4M2.

I9. ***decal marking*** - involves two major steps: 1) the printing, metalizing or otherwise affixing of an image, often multicolored, onto a lacquer or plastic film that is coated with a special adhesive and is supported by paper. 2) the transfer of this image to a workpiece and its release from the paper or

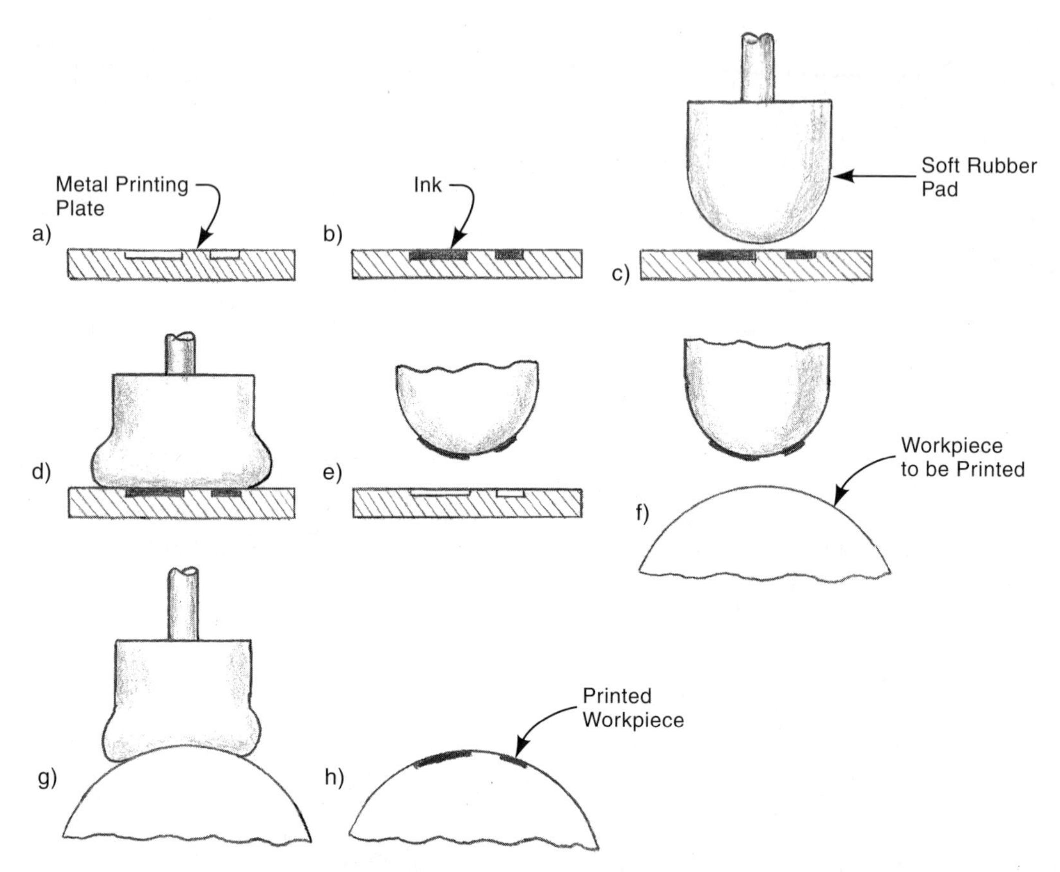

Fig. 817a Schematic illustration of the pad printing method of decorating: a) etched printing plate, b) recesses in printing plate filled with ink, c) soft rubber pad in position to contact the printing plate, d) pad pressed against the plate, e) ink transferred from the plate to the pad, f) pad moved to the workpiece, g) pad pressed against the workpiece, h) ink image transferred from the pad to the workpiece.

plastic backing. High-speed methods can be used to print the decal. Sometimes, the film with the image, the decal, is lifted from the backing and transferred by pressing it with a heated pad or roller against the mounting surface. In another system, the decal is first immersed in water, which loosens the film decal from the backing, whereupon it is slid onto the surface of the workpiece. It is pressed in place and dried and, upon further drying, adheres securely to the surface. Decals are used when the workpiece to be decorated cannot be easily printed directly. They are used to affix trademarks, product names and decorations to a product. Large and small appliances, dishes and other china wear, glass objects and toys are decorated with decals.

I10. ***casting and molding*** - Permanent markings can be incorporated in a cast or molded workpiece by incorporating a suitable engraving or insert in the mold or casting die. The information then becomes an integral part of the surface. The marking can be either raised or depressed. Cast automotive parts often have part numbers and other data marked with this method. The process is also common on molded plastic parts.

I11. ***embossing and coining*** - are explained and illustrated in sections 2D7 and 2D6. These methods are most commonly used for nameplates and tags.

I12. ***nameplates, labels and tags*** - may be used to receive the printed or otherwise marked material. All marking methods may be used, depending on the material involved. Nameplates are normally attached with adhesive, threaded fasteners or rivets.

I13. ***hot stamping*** - is a marking and decorating procedure for plastic parts and is described in section 4M2. The method also is used on the painted surfaces of parts made from other materials.

I14. ***dyeing*** - is accomplished when the workpiece absorbs a liquid colorant. The dye penetrates the workpiece and provides permanent coloring. Usually the workpiece is immersed in the dye, but spraying or brushing may also be used if the conditions warrant. Fabrics, leathers, anodized aluminum and a number of plastics (acrylates, cellulose acetate, nylon and thermosetting polyesters) are commonly dyed. (Also see 10G1.)

I15. ***flocking*** - This treatment is usually applied to fabrics. See 10F3l.

Note: For marking and decorating plastic parts, see 4M.

References

1. *Materials and Processes in Manufacturing, 8th ed.,* Paul DeGarmo, J. T. Black and Ronald A. Kohser, Prentice Hall, Upper Saddle River, NJ, 1997
2. *Manufacturing Processes and Systems, 9th ed.,* Phillip F. Ostwald and Jairo Munoz, Wiley, New York, 1997.
3. *Processes and Materials of Manufacture, 3rd ed.,* Roy A. Lindberg, Allyn and Bacon, Newton, MA, 1083.
4. *Design for Manufacturability Handbook, 2nd ed.,* James G. Bralla, McGraw-Hill, New York, 1999.
5. *ASM Handbook, Vol. 5, Surface Engineering,* ASM International, Materials Park, OH, 1994.
6. *Plastics Engineering Handbook of the Society for the Plastics Industry, 5th ed.,* Van Nostrand Reinhold, New York, 1991, ISBN 0-442-31799-9.
7. *ASM Handbook, Vol. 4, Heat Treating,* ASM International, Metals Park, OH, 1991.

Chapter 9 - Paper, Fiber and Printing Processes

A. Definition, Paper

The term, *paper,* has traditionally referred to (and still refers to) a material composed of thin sheets of matted or felted fibers, usually of cellulose. Paper is used as a base for writing and printing, for packaging and wrapping, as a filter medium and, in heavier sheets, as a raw material for manufacturing furniture and other products and as a building material. The sheets are produced by collecting the fibers on a fine wire screen from a dilute water suspension. The water is removed, the sheet is dried and the fibers bond together. Almost all common paper is made from cellulose fibers, but some specialty papers are made from synthetic or mineral fibers, bonded together by other methods.

B. Paper-making Processes

B1. ***raw materials*** - Wood fibers make up the dominant raw material for paper. Many other fibers are also composed of cellulose and may be used in making paper. Natural fibers from plants other than wood, recycled wastepaper of various kinds and recycled paper board are also used. Fibers from linen, cotton or other scrap rags are ingredients in finer papers and, before the development of wood pulp processes, were a major ingredient in many papers. Wood fibers are obtained from the pulpwood of tree trunks, often from smaller trees not suitable for lumber uses. Scrap and sawdust from the lumber, furniture and other woodworking industries are also used. Woods used include oak, beech, birch, aspen, gum, hemlock, pine, fir and spruce. Bamboo, hemp, jute, wheat or rice straw, esparto grass, ambary and sugar cane fibers are also sometimes employed. Currently used non-cellulose fibers include those made from various polymers.

B2. ***paper making from wood (by machine)*** - involves a sequence of steps: 1) Trees are harvested and the logs are transported to a pulp or paper mill. 2) Logs are cut to a convenient length and are debarked. 3) The wood is converted to pulp by one of several methods. 4) The pulp is refined. 5) The pulp is washed and screened to remove foreign materials, and often is bleached. 6) Some pulps go through a beating process to further process the fibers. 7) Fillers, sizing agents, dyes and adhesives may be added to the pulp to improve properties for the end use of the paper. 8) A slurry of pulp and water is spread on a porous surface; excess water drains off. 9) A continuous web of pulp, supported by a wire mesh or other porous conveyor, passes a series of suction devices and rolls that remove more water from the pulpy sheet, and then press it to a smooth finish. 10) The sheet passes between a large number of steam-heated rolls that dry the sheet. 11) The paper web then may undergo "calendering", that is, pressing it between rolls to provide a smoother "machine" finish. 12) The completed paper is slit and wound onto reels or cut to length and stacked, in both cases to prepare it for further processing into useful products or material. Fig. 9B2 illustrates the operation sequence.

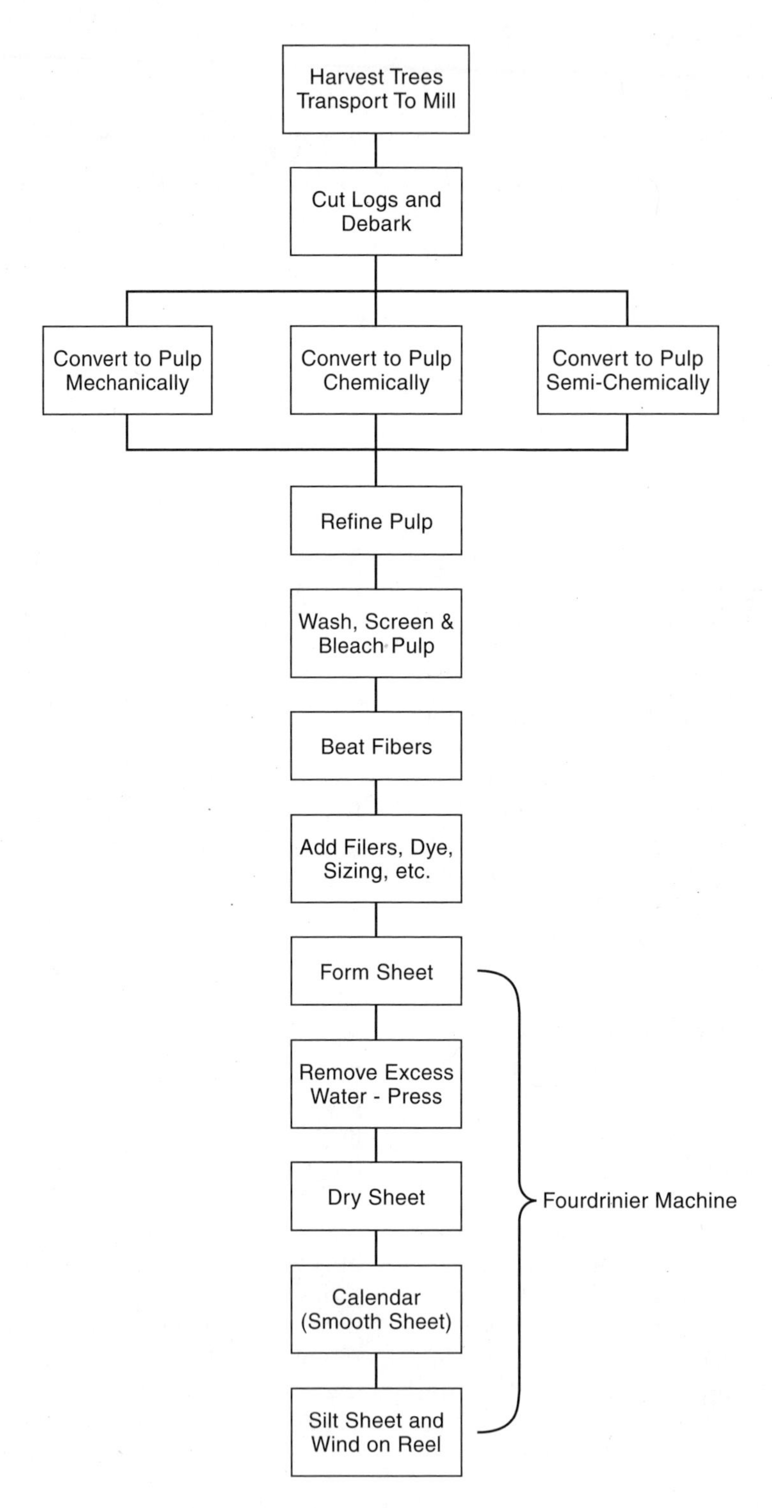

Fig. 9B2 Block diagram of the operation sequence for making paper from wood pulp.

B2a. ***debarking of wood logs*** - is described in 6A1 (Woodworking).

B2b. ***pulping of wood*** - is the separation of wood fibers from one another so that they absorb water, swell and become flexible. There are three basic approaches: mechanical, chemical and semi-chemical.

B2b1. ***mechanical pulping of wood*** - makes use of large, heavy duty, grinding wheels that reduce pulp logs to fiber. Almost the entire content of the logs is converted to pulp. The exception is a small portion of the log that is water soluble. The logs are first cut into *bolts*, having a length of either 2 or 4 feet[4] (0.6 or 1.2 m), the length depending on the size of the grinding equipment. If the moisture content is less than 30 percent, the bolts are soaked in a pond or sprayed to bring the moisture content to 45 to 50 percent. Debarking is the next step, for which the bolts are tumbled in large drums. They are then transferred automatically to the grinding machines which, with large aluminum oxide or silicon carbide grinding wheels, reduce the wood to a pulp. High horsepower (e.g., 10,000) is used along with high pressure from hydraulic cylinders that force the logs against the wheels. Friction develops heat that breaks the bonds of lignin between fibers which are then removed by the abrasive action of the wheels. The method can be used to produce up to about 6 tons of pulp per hour per grinding wheel.

In another method called *thermo-mechanical pulping*, the logs are first cut into chips that are heated and fed between the disks of a refining machine. The action of the rapidly rotating disks breaks the heated chips into wood fibers.

Paper made from mechanically-processed pulp may contain some unwanted ingredients. The pulp does not bleach to a high degree of whiteness and is usually used for lower grades of paper.

B2b2. ***chemical pulping of wood*** - uses chemical means to dissolve the lignin between the wood fibers. Both batch and continuous processes are used. The debarked logs are first cut into chips from 1/2 to one inch (13 to 25 mm) long. The chips are cooked by one of several processes for as long as 12 hours. Chemical action during the cooking dissolves the lignin, the material that holds the fibers together.

One particular method, the *sulfite process*, operates on a batch process to cook the wood chips in a steam-heated pressure vessel called a *digester*. The digesters are large with capacities of 15 or more tons of pulp. They are made of steel with a ceramic tile lining set in acid-resistant cement. Chemical action results from an acid bisulfite solution, created by dissolving sulfur dioxide gas and bisulfite in water. The vessels are first filled with wood chips; then acid is added, displacing any air. Heating is by steam. After the cooking cycle is completed, the chips are blown from the bottom of the digester. The intensity of the action helps separate the wood fibers.

In the *Kraft* or *sulfate process*, the chips are cooked in a solution of sodium sulfide (Na_2S) and caustic soda ($NaOH$). This method has replaced sulfite processing as the predominant chemical means of pulping wood. Sodium sulfate is used as part of the cooking solution though it is not an active ingredient. Sodium sulfide and caustic soda are the active ingredients. The digester is heated with heat exchanger coils or by direct injection of steam. Cooking temperatures are typically 320 to 460°F (160 to 235°C) at pressures of 115 psi (800 kPa) for 1/2 to 2 hours. Both batch and continuous digesters are used but most current production is from continuous equipment. The chemical solutions used in the process are reused, but the recovery process is quite complicated. The directly reused liquid is referred to as "black liquor"; the chemically-recovered sodium sulfide and caustic soda are in a solution referred to as "white liquor". A mixture of both white and black liquor are used in the cooking process. The kraft process produces paper that is particularly strong and durable. Kraft paper is used, for example, for paper grocery bags. For some time, its use was limited to applications where the characteristic brown color was acceptable, but means have been developed to bleach it for other uses.

B2b3. ***semi-chemical pulping of wood*** - consists of cooking with sulfite or alkali to soften the lignin, followed by a mechanical treatment with disk refiners that separate the fibers from one another. Preparatory operations are approximately the same as those followed when providing chips for chemical pulping. The chemical treatment phase uses a pressurized reactor but is somewhat milder than full chemical pulping. A near neutral sodium sulfite solution is common although kraft pulping solution, caustic soda or acid sulfite are also used. After the pulp is thoroughly saturated with solution, it is transferred to a disc refiner consisting of a rotating disc paired with a parallel disc that is stationary or rotating in the opposite direction. The disks have blades or bars on them. The plane of action is perpendicular to the rotating disk's shaft and the pulp passes between the two discs. The device is designed to bend the fibers rather than to break or cut them. Several stages of refining may be used to reduce the softened chips to pulp. Semi-chemical pulp is used to make low-cost printing papers, fluted interior layers of corrugated boxes, and for other applications where intermediate strength and chemical resistance properties are needed.

B2c. ***refining*** - is the disc operation described above as part of the semi-chemical pulping of wood. The same basic process is also used when baled pulp is shipped to another factory that completes the papermaking process. In this case, the pulp is unbaled and, with water, is placed in pulpers, machines with disc refiners. The operation is frequently performed at room temperature and the disc blades are smaller than those used for semi-chemical pulping. Recycled waste paper may also be added. The refiners improve the flexibility of the fibers, increase their surface area and blend the ingredients into a uniform mixture.

B2d. ***removing foreign material*** - The pulp from the chemical and semi-chemical processes listed above is first washed to remove the processing chemicals. Then, debris, "shives" (unfibered wood chips), bark, undigested knots and other foreign materials are removed from the pulp using a series of screens. With mechanically made pulp, a series of rifflers may also be used before screening to remove the heavier foreign pieces. Another method uses centrifugal force and a vacuum to draw pulp through a rotating, drum-shaped, screen in the pulping tank. A third approach is a vortex machine that spins the pulp rapidly using weight rather than particle size as a basis for separating pulp from unwanted materials. The heavy pieces separate from the pulp and fall to the bottom.

B2e. ***bleaching and washing*** - bleaching is a multistage process, especially when kraft paper is processed, since kraft is dark in color and not easily bleached. Four to eight operations are typically involved. Each operation includes a mixing step wherein the bleaching agent is pumped into the pulp mixture. This is followed by a reaction period, that may last from one-half to several hours, and washing to remove lignin and residual chemicals from the pulp. In current bleaching practice[1], the first step is chlorination of the unbleached pulp, effected by mixing chlorine gas with the pulp at a temperature of 70 to 80°F (21 to 27°C). This produces an acid that reacts with the non-carbohydrate components in the pulp. The reaction products are then dissolved by treating the mixture with dilute caustic soda ($NaOH$). They are then washed out of the pulp mixture. The next major treatment is with alkaline hypochlorides to neutralize the solution. Then there is a final wash. Small amounts of chlorine dioxide (ClO_2) may be used in the process to improve the brightness of the bleached pulp.

Washing occurs at several stages in the preparation of pulp for paper making. The sequence of washing operations depends on the type of pulping method used. Its prime purpose is to remove spent cooking liquor or other soluble chemicals from the pulp. In bleaching kraft paper, washing out the spent pulping liquor immediately follows cooking, before the bleaching operation begins.

B2f. ***beating*** - is an operation that compresses and works the fibers. It enables water to penetrate better, causing the fibers to swell and become more flexible. The action separates and frays the fine filaments of the fiber so that they bond together more securely. The result is paper with higher strength, greater density and stiffness, and lower porosity. Previously, this was accomplished with a class of machines known as Hollander beaters. These machines include a heavy roll that revolves against the wood pulp that lies between it and a base plate. The spacing between the roll and plate is gradually reduced, progressively squeezing the fibers and fraying the fibrils. Hollander beaters are presently used only in the production of specialty papers and in smaller-quantity production. Larger mills utilize the refining process described above to achieve the equivalent improvement of the fibers.

B2g. ***making paper from pulp*** - The first step in transforming a slurry of pulp in water into paper involves spreading the slurry evenly on a porous surface. The excess water drains away, leaving a wet mass of cellulose fibers that tend to interlock, forming both mechanical and chemical bonds, and giving strength to the sheet that will result. Further water removal, pressing and drying take place before the sheet of paper is finished. Some finishing operations may follow. The paper may then be slit and sheared into sheets. These operations have been performed for centuries on a hand basis but, except for hobby, laboratory and artwork, all current production takes place on large, complex machines which produce as many as 300 tons of paper per day.

B2g1. ***Fourdrinier and cylindrical paper-making machines*** - transform a dilute slurry of fiber and additives into a roll of paper in several steps. As a first step, the cellulose fibers, mixed liberally with water (more than 99% water), are collected on either a flat forming screen, a cylindrical screen or a suction cylindrical roll. Excess water drains off, leaving a layer of wet pulp. With cylindrical screen machines, the screen-covered cylinder rotates in a vat containing the slurry of pulp. The slurry is flowing, causing the pulp to be captured by the screen while the excess water is removed through the cylinder. The high dilution of the slurry helps insure uniform density and thickness of the sheet. Some cylinder-type machines use suction to draw the pulp to the cylinder as it rotates in the slurry. The wet pulpy sheet is pulled off the cylinder as it completes its rotation and is placed on a wire mesh conveyor.

Fourdrinier machines use a wire mesh conveyor that is in the form of a loop. The wire mesh is normally metal but may be plastic. Strands are fine, typically woven 55 to 85 per in (21 to 33 per cm).[1] The Fourdrinier method is illustrated in Fig. 9B2g1. The slurry of pulp flows from a headbox that distributes it uniformly across and along the wire mesh conveyor as the conveyor passes underneath. The conveyor is flat at that point and as much as 33 ft (10 m) wide. The slurry in the headbox is under pressure to assure proper and uniform flow. Excess water flows through the mesh screen, often aided by an air vacuum below the wire mesh. Rolls that support the wire mesh also absorb water from the pulp. The wire mesh belt may oscillate from side to side to aid in interlocking the fiber ends.

The pattern of the wire mesh is imparted on the paper's upper surface by a "dandy roll", a cylindrical roll covered with a mesh of wires in the pattern desired. Lettering or designs woven into the mesh appear as watermarks on the paper, identifying the manufacturer and the grade of the paper. The roll also flattens the upper surface and improves its finish. A mesh pattern is also imparted on the bottom surface

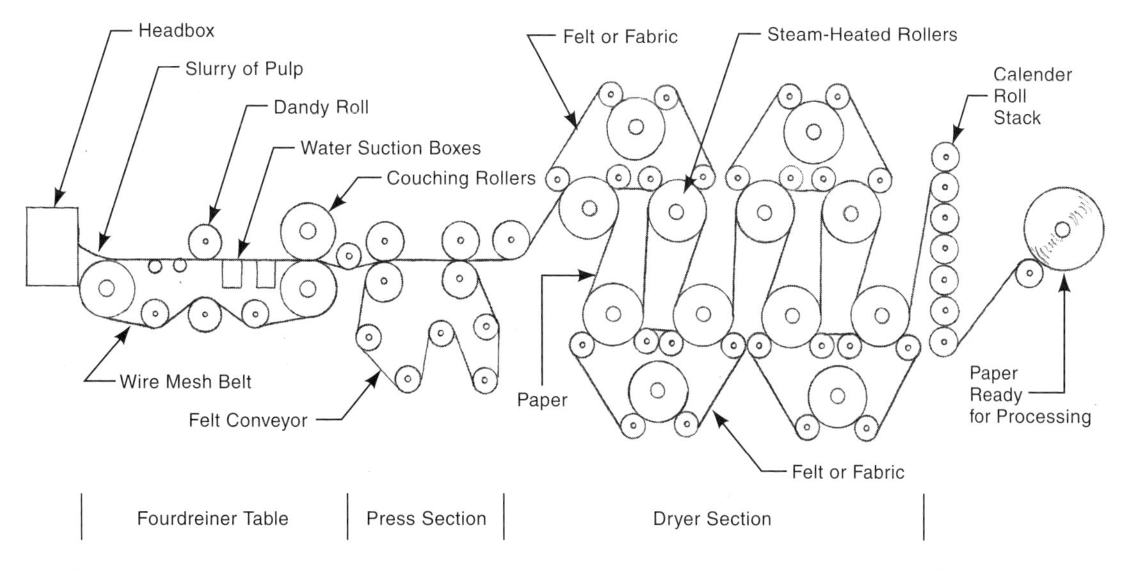

Fig. 9B2g1 Schematic diagram of the manufacture of paper on a fourdrinier machine. A slurry of pulp is spread on a wire mesh, drained and drawn of moisture, pressed and dried through several stages to produce paper.

of the paper from its contact with the mesh conveyor. At the end of the table section of the machine, the paper passes between two opposing felt-covered "couching rolls" that squeeze still more water from the web of paper and force the fibers more closely together. Suction is also used on current couching rolls to aid the water extraction.

The paper is strong enough at this point to be transferred to a felt web that carries it through the press section of the machine. In this section, the paper web passes through a series of rolls that remove additional free water. The rolls are steel with a rubber coating that, with the felt conveyor, prevents the paper from being crushed or having the markings obliterated. The felt, made from a combination of wool and plastic fibers, is porous enough to absorb water removed by the pressure of the rolls. In the press section, the moisture content is reduced but still comprises 60 to 68 percent of the weight of the paper[1].

The next machine operation is drying. The paper web is guided around a series of steam-heated cylindrical rolls and is held in contact with them by dryer felts or dryer fabric. As many as 70 rolls may be incorporated in the dryer section of a typical high-production paper machine. The drying operation removes approximately two tons of water for every ton of paper produced.

Next, the web is "calendered". It is pressed between opposing smooth, chilled rolls that impart a smooth surface finish to the sheet. Several pairs of such rollers in a vertical stack are normally employed.

After drying and calendaring, the paper may be coated with a varnish-like material to give it brightness and gloss. It is then slit and is normally wound on reels for later use, or for additional operations. Alternatively, the paper may be cut to length and stacked at the end of the machine.

Some Fourdrinier machines have two layers of wire mesh conveyor and the sheet is formed between them. This arrangement aids in removing water from the top as well as the bottom of the web. Such machines are called, "twin-wire machines".

B2h. ***finishing (converting)*** - Paper from a paper-making machine normally must undergo one or more further operations to improve its condition or change its configuration for its intended use. There are two kinds of paper converting operations:

Wet converting operations are performed on paper from rolls to improve the paper's printability and opacity or to provide gloss or greater smoothness. Coating, impregnating and laminating are typical wet converting operations. Sizing, the operation that prevents ink from penetrating and spreading in the naturally absorbent cellulosic fibers of paper, is a major finishing operation. Coating and sizing with starch, clay and glue - with additional calendering - provide an excellent surface for printing fine halftone illustrations. Colored paper is made with dyes that are absorbed by the pulp fibers.

Dry converting changes the paper from rolls into a more useful form by slitting, sheeting and stacking, or forming and glueing it into such products as envelopes, bags, boxes, paper plates and cups, tubes, corrugated cartons, writing pads and coasters.

Some specific wet converting operations are the following:

B2h1. ***extrusion coating and laminating of paper*** - is described in Chapter 4, section I4, of this handbook.

B2h2. ***water dispersion coating*** - is performed either as part of the papermaking operation between drying sections of the machine, or as a separate operation later, using previously made paper rolls as raw material. The water dispersion that is used contains some pigment or filler (calcium carbonate, clay, titanium dioxide) to provide uniform distribution of the material and an adhesive binder (latex, starch or a synthetic material) to bond the coating to the paper.[1] The paper, passing through a bath of the dispersion, picks up the desired amount of coating and then passes between drying and smoothing rolls. The coating materials provide better opacity and brightness to the sheet, permitting it to be printed on both sides without any image from the reverse side showing through.

B2h3. ***sizing*** - is material that prevents ink and other aqueous liquids from soaking into the paper. One common sizing material is rosin which is applied to the fiber as a dispersion with soap and water. The amount of rosin, after application, is from 1 to 5 percent of the weight of the fiber. Alum, or aluminum sulfate ($Al_2(SO_4)_3$), causes the rosin to precipitate from the dispersion and adhere to the surfaces of the fibers.

B2h4. ***filling*** - Fillers are incorporated in paper to increase white brightness and surface smoothness, to make the paper more opaque, and to improve its ability to accept ink. Many different compounds can be used: calcium carbonate ($CaCO_3$), kaolin (clay), titanium dioxide (TiO_2), zinc oxide, zinc sulfide, calcium sulfate, talc and barium sulfate. As with rosin sizing, alum may be used to provide an attraction of the filler to the fibers.

B3. ***paper making from rags and other textile fibers*** - Clean rags, textile mill cuttings, garment and other sewing scrap, short fibers from raw cotton processing and flax fibers are used in making finer grades of paper. Operations involved in processing baled rags are as follows: 1) Bales are opened and rags are threshed mechanically. 2) Threshed rags are manually sorted to remove rubber, synthetic fibers, metal, papers and plastic-coated rags. 3) Rags are cut into small pieces. 4) Magnetic rolls remove any extraneous pieces of iron and steel. 5) Rags are cooked in a dilute alkali solution [sodium carbonate (soda ash - Na_2CO_3), or caustic soda (NaOH) with lime (calcium oxide - CaO), and detergents or wetting agents]. Cooking, with steam heat, lasts for 3 to 10 hours to remove grease, wax, fillers and oils. 6) The rags are washed. 7) They are beaten mechanically, as described in paragraph B2f above, to shorten and fray the fibers and increase their tendency to swell when wet, all of which enhance the bonding of fibers to each other in the paper. 8) Fillers, sizing material and colorants are added to the mixture as desired to provide added body and weight to the paper. 9) Additional beating with a Jordan engine, a machine with a pair of concentric conical surfaces containing knives, may take place. One cone rotates inside the other and the slurry of rag fibers flows between them, producing further separation and fraying of the fibers. 10) Subsequent operations to create paper from the fibers are essentially the same as those described above when wood pulp is the raw material. Rag pulp produces "rag bond", the fine papers that are used for important documents, currency, fine writing paper, art papers and other applications where durability and long life are important. Papers are also frequently made from blends of rag and wood fibers.

B4. ***paper making from synthetic fibers*** - Synthetic fibers - nylon, polyesters, rayon, acrylics and glass - and blends of them with conventional pulp - are used to make specialty papers used in filtration and where chemical and water resistance and dimensional stability are important. Synthetic fibers are bonded with adhesives instead of by the natural cohesion of cellulose fibers. Otherwise, they could be processed on conventional paper making machines. Applications also include electrical insulation and, for flexible fabric-like blends, garment linings.

B5. ***paper making from waste paper (paper recycling)*** - Waste paper comes in several categories: corrugated cartons, newsprint, white business paper and mixed paper. Coated papers used in packaging, magazines and advertising applications are more difficult to recycle and are processed separately until the coating material is removed. Such coatings include various plastics, metallic foils, asphalt, synthetic adhesives and certain inks.

There are two basic pulp preparation processes for wastepaper: 1) those that involve ink removal and, 2) those that do not include de-inking. The former are used for papers to be printed again and the latter are used for coarse papers and box board, and account for the great bulk of recycled paper products.

Recycled fiber is also used in the manufacture of paper towels and paper napkins.

B5a. ***waste paper processed with de-inking*** - Paper from bales is inspected and fed into a pulping tank. Caustic soda (NaOH) or other chemicals [soda ash (Na_2CO_3), silicate of soda, phosphates and wetting agents] in hot water (150 to 190°F - 65 to 90°C) are the other contents of the tank. The tank is agitated with blades that circulate the stock and, with the aid of the chemicals, separate the fibers. After pulping, the pulp is screened to remove trash pieces and washed to remove ink and chemicals. Bleaching with hypochlorite may follow if whiteness is a requirement.

B5b. ***waste paper processed without de-inking*** - is processed similarly to that requiring de-inking except that chemicals required to dissolve and disperse the ink are not required.

C. Manufacture of Various Paper Grades

Paper-making processes vary, depending on the grade of paper produced and its application. Key factors are the type of pulp used, the paper making method and machine involved, the additives used and the amount of refining done. The following are some important paper grades with comments on any special aspects of the processes used to make them:

C1. ***bond paper*** - is high quality paper used for currency, insurance policies, legal documents, letterheads, advertising pieces and certificates. These papers are especially strong and stiff, of bright color and high cleanliness and particularly resistant to spreading and over-penetration of ink. Much bond paper is made from rag fibers though some is made from chemically-produced wood pulp.

Papers made from chemically-treated pulp are sometimes called, "wood-free papers". (The term, "wood free" really means free of mechanical pulp.). These high-grade papers are bleached and often coated. Typewriter, computer-printer, copier and envelope paper are normally uncoated. When coated, "wood free" papers are used for magazines, annual corporate reports, sales brochures, and similar application. Coatings include clay, calcium carbonate or titanium dioxide.

C2. ***newsprint*** - is made primarily from machine-ground pulp. A common practice is to use 75% mechanical (machine ground) pulp and 25% chemical pulp. The machine-ground pulp is lowest in cost and has good printing qualities; the chemical pulp is included to increase the strength of the paper. Improvements in the manufacturing process, notably better bleaching, have led to use of this grade of paper for other applications such as catalogs, magazines, and paperback books. These papers take ink well but have a tendency to turn yellow when exposed to light over a period of time.

C3. ***paperboard*** - is heavy board, 0.012 in (0.3 mm) or more in thickness and 0.66 oz/ft^2 (200 g/M^2) or more in weight. It is used for paper boxes, formed food trays, paper plates, corrugated shipping containers, boards used in electrical and building applications, and book cover stock. Paperboard is made from wood pulp, wastepaper and straw pulp or a combination of them[1]. Cylinder machines are used to layer several sheets of pulp together prior to pressing and drying to achieve the desired thickness.

The term, chipboard, is often used for paperboard made from unbleached mixed pulp and used for applications where appearance and strength are not critical. Tablet backs, posterboards, folding cartons and boxes, and backings for photos are common applications. Cereal, cigarette and other boxes are made from chipboard coated with other papers and sometimes impregnated with wax.

C4. ***sanitary papers*** - include paper towels, toilet tissue, paper handkerchiefs, and paper napkins. These products are made from recycled paper or a blend of mechanical pulp, bleached kraft and sulfite pulps. They are processed with a minimum amount of finishing additives except for plastics which are added to increase wet strength. One operation common on these papers is *creping*. This involves running the sheet against a smooth drying roll but stripping the paper from the roll, after partial or complete drying, with a stationary blade that bears against the roll. The sheet folds against itself creating the desired creping effect. Thin facial tissue is dry when removed by the blade; heavier paper toweling and napkins are usually still wet. The light grades are dried on Yankee machines. These have very large steam-heated dryers that eliminate the need for felt transfers of the thin stock. Moist sheets of napkins and toweling are sometimes embossed with textured rollers to provide a decorative pattern.

C5. ***kraft paper*** - is the brown paper used for paper bags and corrugated paper cartons. It is usually made in heavier thicknesses. Kraft is produced from softwood pulp, usually from pine trees using the kraft or sulfate chemical pulp preparation process described above in paragraph B2b2. Softwood has longer fibers than hardwood and they provide greater strength to the paper. Sizing and treatment with plastic resins, when used, decreases the paper's water absorption and increases its wet strength. A common use for kraft paper is grocery bags. Bulk materials such as animal feeds and cement are often shipped in multiwall kraft paper bags. Kraft paper can be bleached and is then used for bags or cartons where more prominent printing or colored decoration is desired. Bleached kraft is used for food packaging. When coated with wax or plastic, it is used for paper cups and milk cartons.

C6. ***vulcanized fiber*** - is a fibrous material made by treating paper pulp, derived from recycled cotton waste, with zinc chloride. The zinc chloride is later bleached out. The material has been called, "the first plastic" and has many uses. It possesses strength and resistance to heat, wear, oils and solvents. It has light weight, good electrical insulation properties, and is economical to produce. Uses include luggage and musical instrument cases, protective helmets for firemen and others, quiet gears, gaskets, knife handles and electrical insulators in transformers, electrical plugs, switches and sockets. However, vulcanized fiber has been replaced in many applications by other plastics.

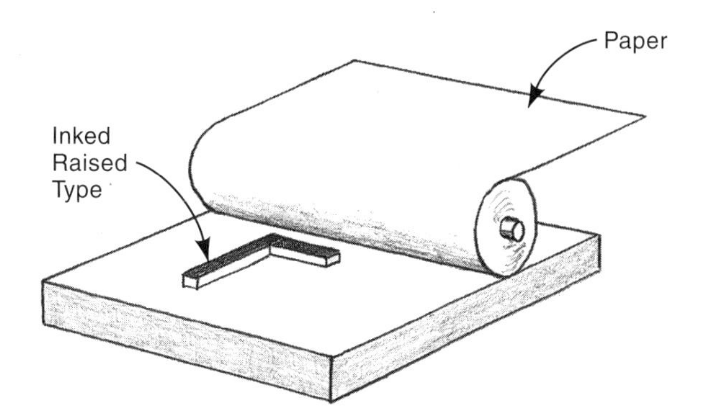

Fig. 9D1 The principle of letterpress printing. Raised type is coated on its top surface with ink that is transferred to the paper when the paper contacts the plate.

D. Printing

Printing is one of a number of processes for producing identical copies of written text and images. Reproduction can take place on paper, plastics, or fabrics, or on various solid objects including product components. Images can be in black or color. There are a number of available processes including the following:

relief printing - letterpress, flexographic. (In relief printing, the printing surface is raised above the non-printing surface.)

planographic printing - offset lithography, screenless lithography, collotype, waterless printing. (In planographic printing, the printing surface and non-printing surface are at the same level.)

intaglio printing - gravure and rotogravure. (In intaglio printing, the printing surface is below the level of the non-printing surface.)

stencil and screen printing

electronic printing - includes laser, other electrostatic, magnetographic, ion or electron deposition, ink-jet, dot-matrix, thermal methods, photocopying and microcapsule methods.

D1. ***relief or letterpress printing*** - is the earliest printing process. It involves the use of type with a raised surface surrounded by a relieved surface. The raised surface is in the shape of type faces, lines and dots. The earliest forms of relief printing, in China over a thousand years ago, used wooden type with the non-printing areas carved away by hand. Gutenberg's presses, of about 1450 AD, the first Western Hemisphere printing, used the same principle. Ink is applied to the raised surface only, and when paper is pressed against the type, the ink is transferred to the paper. Fig. 9D1 illustrates the principle. There are three basic configurations of letterpress machines as shown in Fig. 9D1-1. The platen press is shown in two versions: a1) the rudimentary press as used by Gutenberg, and a2) a production platen press with an automatic inking roll. View b) shows the flatbed cylinder press where sheets are held by an impression cylinder that rolls against a flat bed of type as it moves forward. View c) shows the rotary press where a continuous web of paper is fed between an impression roll and another roll containing a curved printing plate. Rotary presses are used for mass-production situations, but for much mass production, offset printing and other processes have largely supplanted letterpress because of better quality and reduced set up costs.

Letterpresses print from a metal or plastic plate incorporating the raised type, or from a type form, a metal frame containing individual metal type pieces locked into place.

D1a. ***typesetting for letterpress printing*** - may still be performed by hand assembly of individual pieces of movable type, but methods of making a whole page of type as one piece have prevailed for many years. Linotype machines were used to cast type, one line at a time, with justified lines (aligned margins) and the plate consisted of assemblies of lines of the cast type. Current practice uses large metal or plastic type plates made

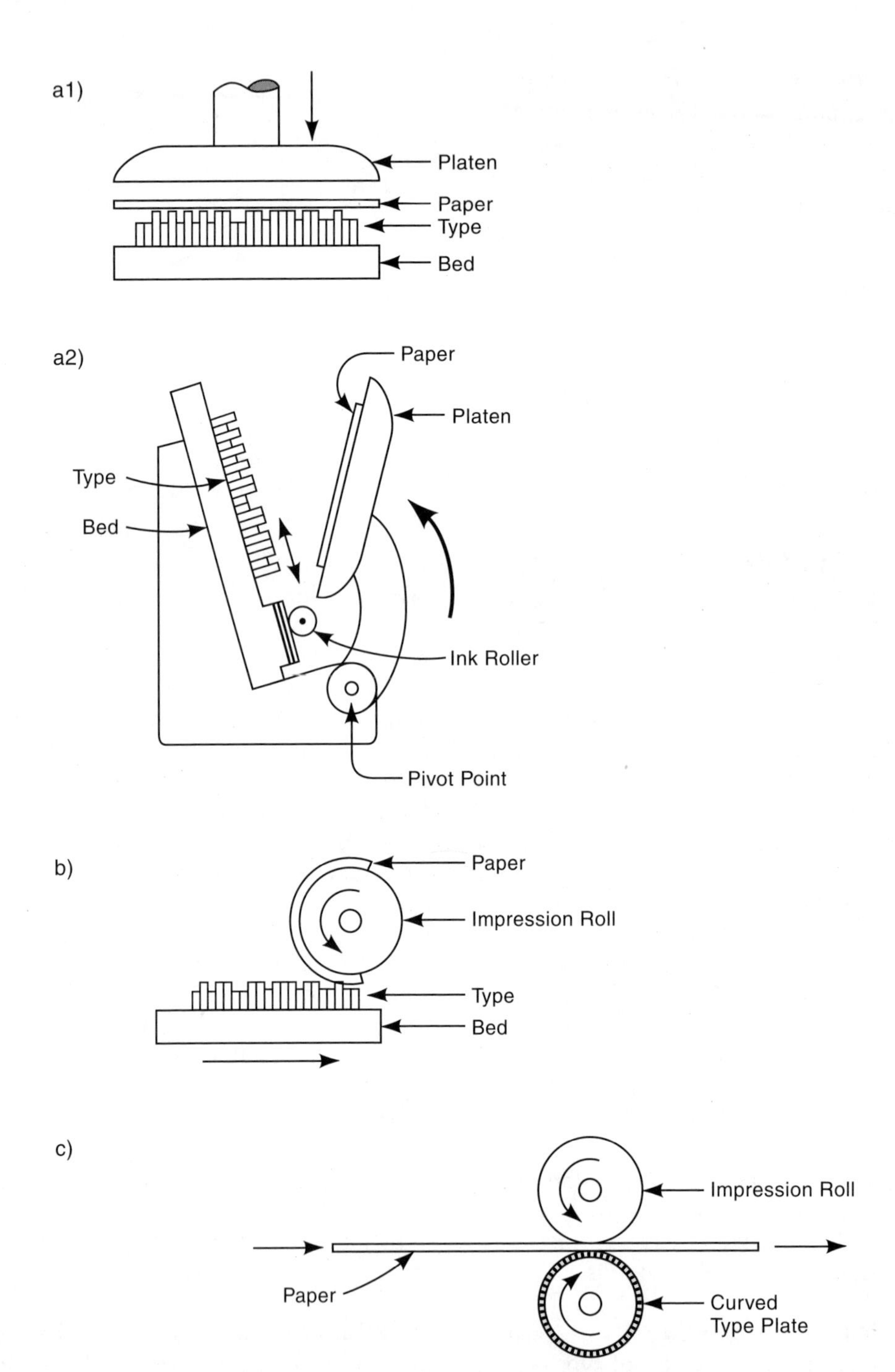

Fig. 9D1-1 Different configurations of letterpress printing presses. a1) and a2) utilize a flat type bed and flat platen to press the paper against the type. The Gutenberg press of a1) is fully manual. a2) is a much later production press with automatic inking. b) shows the principle of a flat type bed with paper pressed against the type by an impression roller; c), the principle of a rotary press, with a curved type plate mounted on a roller opposed by an impression roller that presses the paper against the type.

photochemically. Each page is first designed and composed by computer with the desired type sizes, page composition and illustrations. A film negative of each page is printed on suitable film by a laser printer, under the control of the computer. Then, ultraviolet light is shone through this negative and against a plate made from a photosensitive polymer. Light strikes those areas that are to transfer ink. Where the ultraviolet light contacts the photopolymer plates, the surface is polymerized and hardened. In other areas, the material is soft and is removed by a water or air blast. With metal plates, a similar approach is used, but the equipment applies a chemical etchant to remove metal in areas not exposed to light. The printing plate, thus produced, is mounted on the letterpress.

D1b. ***flexographic printing*** - is a variation of letterpress printing. An engraved printing plate is used and ink is carried on the raised areas. However, the plate is made from rubber or another flexible, resilient, material that enables weaker or irregular surfaces to be printed. Corrugated cartons, labels, unfinished surfaces, cardboard, plastic film, wrapping paper, newspapers and magazines, are all printed with flexographic equipment. Equipment is rotary and web-fed and the relative position of the impression roller and type plate roller is adjusted to apply just enough pressure to get a clear reproduction. Ink is fluid and is transferred from a roll that rotates in an ink container, or from a flow of ink, to a transfer roll and then to the inking roll that contacts the flexible plate. A doctor blade may be used to remove excess ink from the transfer roll.

The printing plate is fabricated by first making a conventional metal plate and using it to form a negative impression in soft plastic or soft cardboard. The soft negative plate then becomes a mold or *mat* for making the flexible rubber or plastic printing plate.

D2. ***planographic printing*** - In this basic process, the printing surfaces of the plate are on the same plane as the non-printing surfaces. Fig. 9D2 illustrates the principle. The basis for lithography, the prime planographic technique, is a printing plate that is receptive to ink in certain areas (type and image areas) and repellent to ink in other areas. The method is based on a 1798 discovery by

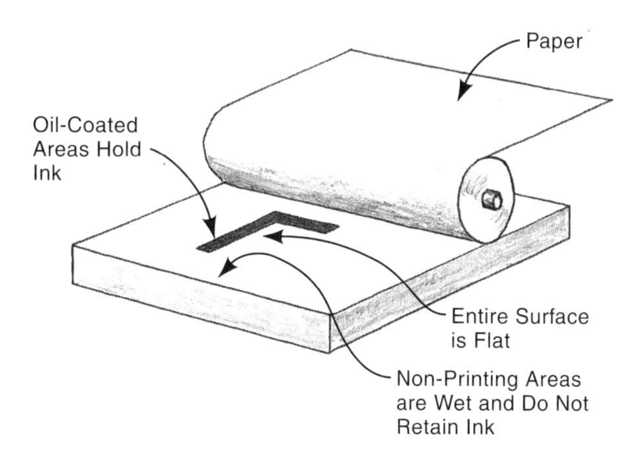

Fig. 9D2 The principle of planographic (lithographic) printing. Lettering, in grease, is applied to the flat printing plate. Ink is then applied to the plate. Ungreased areas repel the ink but the ink is held by the grease and then is transferred to paper when the paper contacts the plate.

Senefelder, in Germany, that wet limestone would repel oil-based printing ink except in areas previously marked with a grease pencil. The grease markings would retain the ink so that a damp sheet of paper pressed against the stone surface would receive the marked image. A number of copies could be made by repeating the wetting, inking and pressing steps.

D2a. ***lithography and offset lithography*** - are planographic processes. Current-day lithography uses a thin aluminum printing plate thinly coated with a light sensitive polymer. (Stainless steel or plastic plates are also used.) The coating undergoes a change in solubility when exposed to intense blue or ultraviolet light. The exposed areas become insoluble to water and receptive to ink. The unexposed areas are washed free of the photopolymer and, if kept wet, are repellent to ink. If the plate is kept wet and inked, paper or other substrates pressed against it will be printed.

The desired text and images can be processed onto the plate by one of several methods. Placing a photographic negative or positive on the plate before exposing the plate is the earlier method, and is still used. A computer-controlled laser beam achieves the same effect on the plate without the time and cost of preparing a film negative.

Rotary presses are used for lithography. The plate is mounted on a cylinder that rotates during printing. Rollers that contact the plate cylinder are used to provide the proper amounts of both wetting and ink to the plate as the plate cylinder rotates. Sometimes the dampening (wetting) is provided by a spray head. Dampening water is actually a mixture containing some alcohol and a small amount of either an acid or alkali. Normally about 98% water, it may contain as much as 20% alcohol, for some applications. Because the ink is highly viscous and thixotropic, and because of the tendency for some water to get into the ink, the ink delivery system may be complex and may include a series of rollers and vibrators.

Though printing directly from the plate cylinder to the paper was widespread, it was discovered that it was advantageous to use an intermediate roller to transfer the ink pattern from the lithographic plate to the paper. This intermediate metal roller, covered with a rubber, is called a *blanket cylinder*, and the process of using it is called, *offset lithography*. The use of the blanket cylinder facilitates printing on metal, plastics and other rigid, smooth surfaces, as well as on surfaces with some irregularities or texture.

The image quality from offset lithography is excellent. The process is used on small, sheet-fed equipment for printing modest quantities of brochures, flyers, labels, stationery, and newsletters. For large-quantity production, web equipment is used and magazines, catalogs, newspapers and books are printed. Color printing using multiple offset units is common. Because of the low cost of plates, the easy set up, the high press speeds and the quality of the printing, offset has replaced letterpress as the most widely used printing process. Fig. 9D2a illustrates web printing on both sides of a sheet using offset lithography.

D2b. ***collotype (collography)*** - is a planographic process, similar to lithography in that the

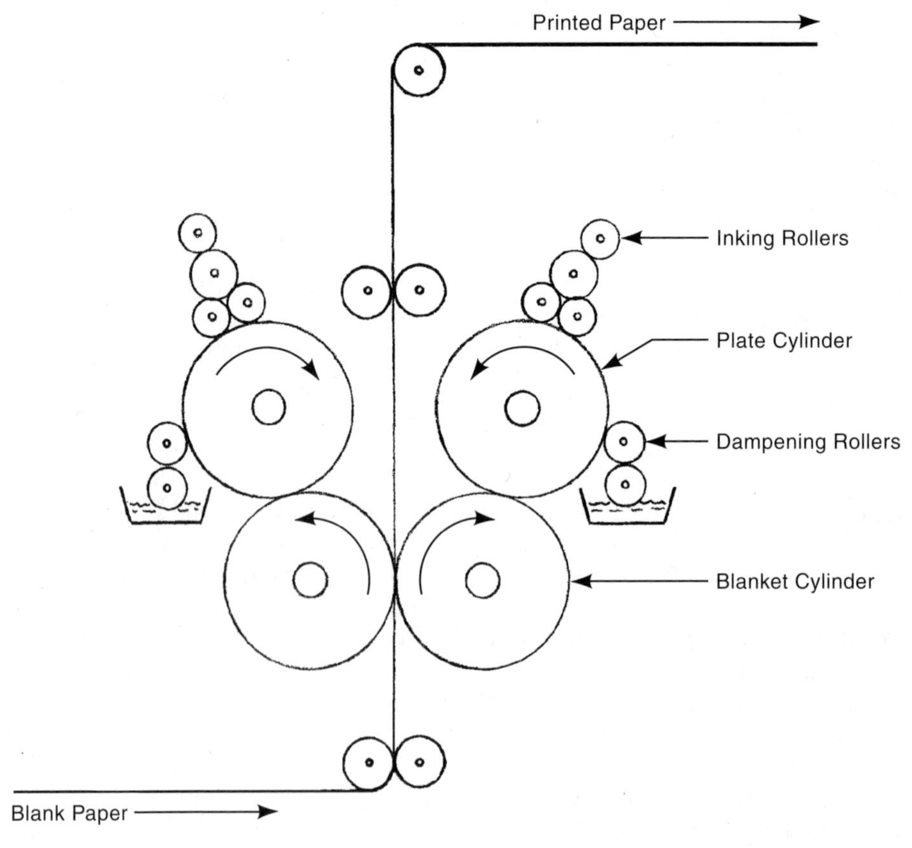

Fig. 9D2a Web printing both sides of a sheet using offset lithography.

flat printing plate accepts ink in some areas and not in others. A coating of light-sensitive gelatin on a metal or glass plate is exposed to light through a negative film of the text and images to be printed. The light, passing through the negative, hardens the gelatin to a variable amount, in proportion to the amount of light that it has received. The plate is soaked in a glycerin-water mixture. The softer parts of the gelatin coating absorb the mixture; the harder parts absorb the least amount. When the plate is mounted in a printing press, the driest parts accept the most ink; the softest, wettest areas accept the least ink. The gelatin retains moisture so no dampening is needed during printing, but humidity may have to be high in the pressroom to maintain the dampness of the nonprinting areas of the plate. The number of copies obtainable from a plate is limited normally to about 2000 but sometimes as many as 5000 can be produced. Printing speed is also low, 200 copies per hour or less. Collotype produces prints of excellent quality and prints photographs without the need for a halftone screen but with a gradation of tones that is close to that achievable with photographs.[4] It is used to make high-quality reproductions of paintings and to print greeting cards, postcards, posters and advertising material, including transparent illustrations.[1] Collotype printing is no longer a common process in the United States.[2]

D3. ***intaglio printing*** - is printing with a system wherein ink is transferred to the paper from surfaces depressed below the basic surface of the printing plate. This differs from letterpress where the inking surface is raised and planographic where there is only one surface of the plate. Fig. 9D3 illustrates the principal of intaglio. Gravure printing is the intaglio method. Rotogravure is a variation of this process that uses a rotating cylindrical printing plate.

D3a. ***gravure printing*** - uses a copper cylinder or plate engraved with the image or text to be printed. Ink is retained in the engraved recesses of varying depth on the surface of the cylinder or plate and is transferred to paper or other material pressed against it. The recesses or *cells* are produced by any of a number of methods: electromechanical or mechanical engraving, photo-chemical etching, laser or electron-beam engraving. The recesses

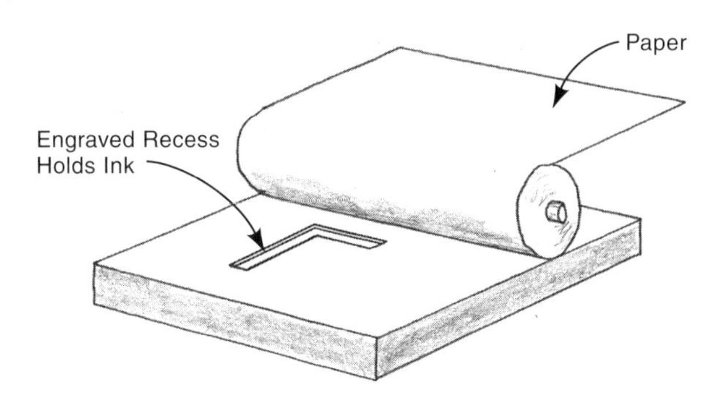

Fig. 9D3 The principle of intaglio (gravure) printing. The ink is held in a depression in the printing plate and is transferred to the paper when it comes into close contact with the plate.

may be very small and very numerous (over 20,000 cells per sq. in) but vary widely in both size and depth. Size variation, however, is within the borders of partitions between the cells. A typical cell is 0.0014 in (35 microns) deep and 0.005in (125 microns) square.[4]

Printing operation using a cylindrical plate are called *rotogravure*. The cylindrical plate rotates through a bath of liquid ink or is subjected to a spray or roller which liberally coats the cylinder's surface. A flexible metal doctor blade, that extends across the full length of the cylinder, then wipes the ink cleanly from the polished surface but not from the recesses. Walls of uniform height between the cells (recesses) support the doctor blade. The walls provide a contact and support surface for the doctor blade and ensure that the exact desired quantity of ink will remain in each cell. In the press operation, a sheet or web of paper (or other material) passes between the cylinder and a rubber covered impression roll, is pressed against the cylinder as it passes between the rolls, and is printed. The paper is given a positive electrical charge and the cylinder is grounded or given a negative charge. These opposite charges attract, aiding in drawing ink to the paper. Fig. 9D3a illustrates gravure printing.

In color printing, a separate cylinder is used for each color. After each color is printed, the paper web passes through a drying section where the solvent in the ink is evaporated. The process produces high-quality printing and color. However, the cost of the engraved cylinders necessitates that the process is limited to large-quantity printing so that these costs can be amortized. Web surface speeds

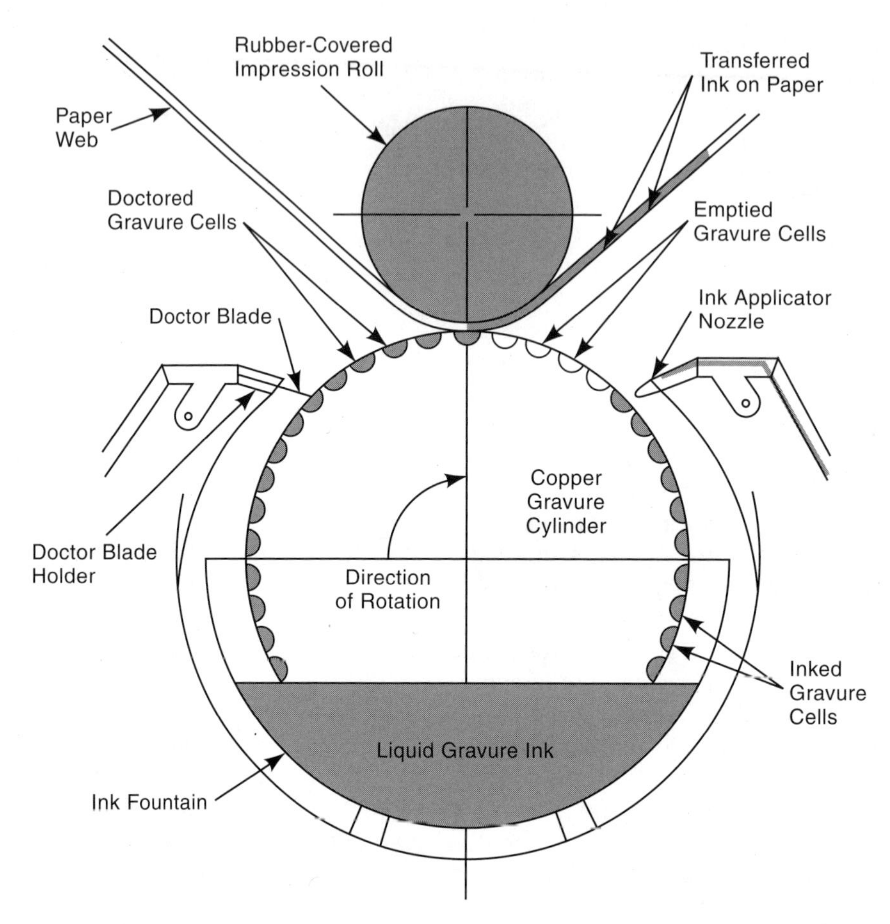

Fig. 9D3a Gravure printing *(from the McGraw-Hill Encyclopedia of Science and Technology, McGraw-Hill, New York. Used with permission.)*

can exceed 3,300 ft (1000 m) per minute. Magazines, catalogs and newspaper supplements are normally printed by rotogravure. The process is also used extensively in packaging and to print such products as shower curtains, wall coverings, table cloths, paper towels and decorative laminates.

Flat engraved plates rather than cylinders are used to print currency, bank notes, stock certificates and postage stamps.

D3b. ***making gravure plates*** - Almost all rotogravure plates for high-production applications are copper-plated steel cylinders that are engraved by computer-controlled electromechanical engraving machines. The engraving heads use cutting tools with fine diamond points. As many as eight such heads may be incorporated in one cylinder engraving machine.

One system (that was previously dominant and is no longer so, but still used) utilizes a special sensitized carbon tissue paper coated on one side with a layer of sensitized gelatin. The gelatin becomes a transfer film. The gelatin on the carbon tissue is sensitized by immersion in a solution of potassium dichromate. It is then contact printed through a gravure screen, a grid of transparent lines and opaque square dots spaced at 150 to 175 per in (60 to 70 per cm). Then the coated tissue is exposed again with a photographic film image of the material to be gravure printed. (The film image is a positive and its illustrations have not been projected through a halftone screen.) During these exposures, the gelatine is hardened by exposure to the light. It becomes hardest in the areas exposed to the most light, in white areas from the film positive and the grid lines. It is less deeply hardened in the

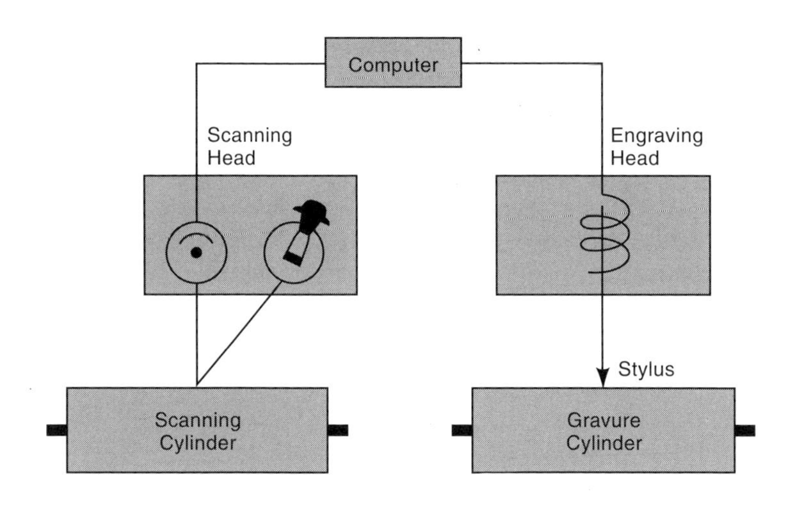

Fig. 9D3b The electromechanical system for making gravure plates. *(from the McGraw-Hill Encyclopedia of Science and Technology, McGraw-Hill, New York. Used with permission.)*

grayscale areas of any illustrations. It is not hardened in areas covered by lines and text. The twice exposed film is then moistened and pressed - with the gelatin side down - into contact with the copper surface of the cylindrical printing plate. The image is developed to produce a gelatin relief resist, and the carbon tissue is stripped off. Then the copper coating of the cylinder is etched with a ferric chloride solution. The depth of the etching corresponds to the degree of hardening of the gelatin resist. Etching is deepest where the gelatin is least hardened from the light exposure. No etching takes place where the clear lines of the screen or bright areas in the image fully expose the gelatin. Due to the exposure of the grid lines, even broad dark image areas consist of a series of separate cells. After etching, the plate may be chromium plated to improve wear resistance in printing.

A different sequence using electro-mechanical engraving is illustrated in Fig. 9D3b. The equipment has three major elements: scanner, computer and engraving head or heads. The scanner scans a photographic print of the material to be printed. The computer processes the scanning data and sends impulses to the engraving heads causing the diamond styli to move and engrave cells into a rotating cylindrical plate. The depth of the engraved cell at any one point corresponds to the intensity of the image at the corresponding point in the scanned photograph. The computer is programmed to provide unengraved grid lines between cells. The engraved cells vary in depth and area corresponding to the image tonal values they represent. Fig. 9D3b-1 illustrates how different tones in the original image are characterized in the printing plate surface.

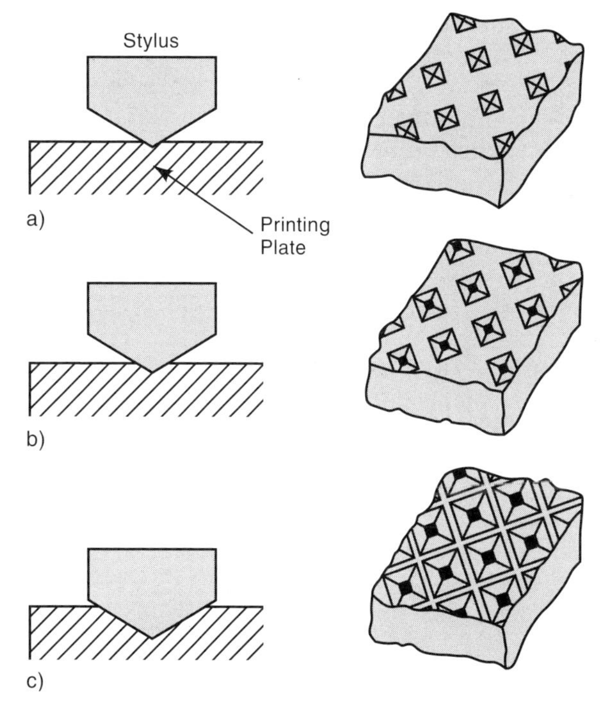

Fig. 9D3b-1 The depth of electromechanical engraved cells, their size and shape determines the different tones in gravure printing plates: a) highlights, b) middle tones, and c) shadows. *(from the McGraw-Hill Encyclopedia of Science and Technology, McGraw-Hill, New York. Used with permission.)*

There are other plate making methods. One process, called direct transfer, uses a light sensitive coating on the cylinder and exposure to light through a positive halftone film. Another method uses photosensitive polymers for the plate instead of copper-plated steel. The dark areas of the image to be printed are etched deeper in the plastic. Nylon is one material used for such plates because it has good resistance to wear from the doctor blade contact. This approach is useful for packaging and other printing runs for quantities under 100,000.[4] Another method uses laser energy instead of diamond-tip styli to engrave the plate. This method retains the scanning and computer-control aspects of the electromechanical procedure.

D4. ***stencil and screen printing (porous printing)*** - differ from the printing methods described above in that the ink (or other colorant) is not applied to a surface that transfers it to the printable substrate, but instead is applied directly to the surface to be printed through openings in a mask. The openings correspond to the image or text to be printed. The ink is applied or forced through these openings and onto the surface that receives the printing. Both methods are probably used more for printing on products or manufactured components, rather than on paper sheets, though they are used extensively for printing notices, posters and signs.

D4a. ***stencil printing*** - uses a sheet device, a stencil, with patterned openings. The stencil is held against the printable sheet (or a workpiece surface) and ink, paint, dye, stain or another paste-like colorant is applied through the openings. The ink can be applied by any of a variety of devices: brush, roller, spray, pad, or squeegee. The stencil can be of any sheet material that resists the coloring agent. Paper, sheet metal, fiber or plastic are all used to make stencils. Stenciling is commonly used to mark shipping cartons. When abrasive blasting is used instead of a coloring agent, glass, ceramic and stone workpieces can be permanently etched with the aid of stencils of suitable material.

D4b. ***screen printing (serigraphy) (silk screening)*** - is a refined method of stenciling. In addition to printing on paper, screen printing is often used to add decorative markings such as symbols, designs, product names, and trademarks as well as alphanumeric data to products and components. The stencil-like screen, now usually made from nylon, polyester or stainless steel fabric instead of silk, and tautly mounted on a rectangular frame, is coated with a solid material except in openings corresponding to the pattern to be marked. Ink or paint, in paste form, is spread on the screen and the screen is placed on the sheet or surface to be printed. Fixtures are usually used to insure accurate registration of the print. A rubber or polyurethane squeegee, operated manually or mechanically, is used to force the ink through the screen openings and onto the surface of the paper or other material to be printed. Fig. 9D4b illustrates the process.

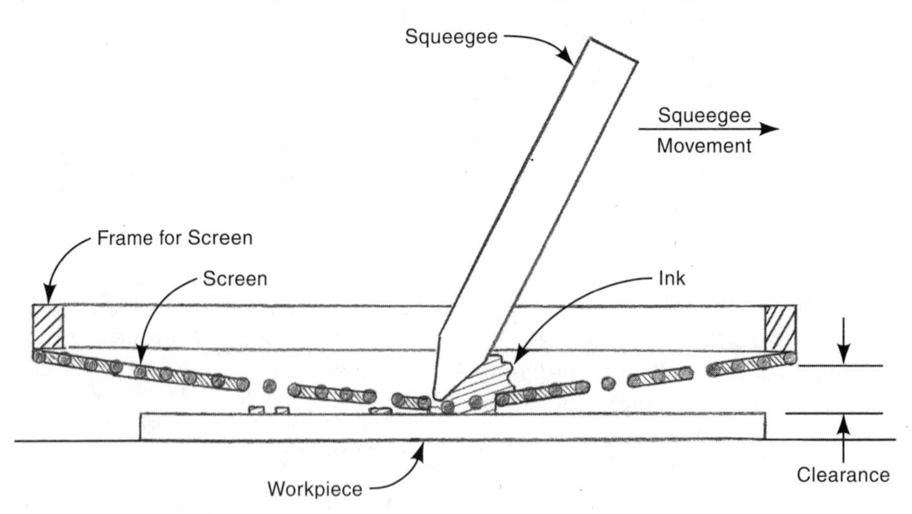

Fig. 9D4b The screen printing process. A flexible squeegee presses ink through the openings of a coated screen, leaving an image on the workpiece - a sheet of paper, other sheet material or product component that is positioned below the screen.

Successive operations on the same surface with different screens can result in multicolored markings.

Screen printing is used on paper but is also applied to fabrics, glass, plastics, leather, wood, ceramic and metal parts, painted surfaces for signs, posters and nameplates. The process is versatile enough to work well with surfaces of various shapes, as well as a variety of materials. Bottles and other glassware, barrels and other containers, t-shirts, hats, banners, appliances and commercial vehicle panels are often decorated by screen printing. The process is also used in the electronics field to deposit conductive, resistive or semiconductive patterns for circuit boards, keypads and membrane switches.

The density of the screen fabric ranges from 25 to more than 500 threads per in (10 to 200 per cm). The higher mesh counts are used when the image to be printed is highly detailed; low mesh counts are used for coarser images and for much textile printing.

There are a number of methods to provide the necessary screen openings. They range from simple manual cutting of openings in stencil material with a knife and then bonding the remaining material to the screen, to hand painting the nonprinting areas, and to several types of photoresist systems. The photoresist techniques are most common. One frequently used system employs a liquid photosensitive polymer that is coated on a screen, allowed to dry and then exposed to strong ultraviolet light that shines through a film positive of the image to be printed. The exposed areas on the screen harden and can be retained while the unexposed areas remain soft and are flushed away. What remains is a screen with openings corresponding to the image to be printed. Other photoresist methods use a polyester sheet that is wetted, applied to the screen, dried, and then processed in the same way as a screen coated with liquid polymer. Gelatin sheets are processed photographically before being adhered to the screen.

Screen printing is frequently performed manually with simple wooden fixtures to position the image correctly on the sheet or workpiece. In such cases, the operator supplies ink to the screen periodically and draws the squeegee manually across the inked screen for each printing. For large-quantity work, presses are available that hold the sheet or workpiece with vacuum or a fixture, position the screen with respect to the work, supply ink to the screen and operate the squeegee automatically. For truly high-production screen printing of web paper, textile fabric or other sheet material, cylindrically mounted screens are used with the squeegee inside the cylinder. As the material passes between the rotating screen and a backup roll, it is printed.

When screen printing involves several colors, each color is normally printed and dried separately, and fixtures or mechanisms ensure that each successive color is in proper registration. Drying may take place by evaporation or by curing with heat or ultraviolet light. Screen printing deposits a heavier coating of ink than normally results from other printing methods. This is desirable for product marking and outdoor applications where more durability is required. Heavier coatings also can add to the intensity of the color.

D5. ***electronic printing methods*** - are printing processes under the control of a computer. In contrast to the various mechanical printing processes described above, electronic printing requires no set of printing plates and no significant set up and run-in to insure proper alignment and proper coloration. Thus the one-time costs per document are low. On the other hand, electronic printing is not nearly as fast as the mechanical processes. However, image information can be changed from copy to copy, though the image recording must be repeated for each copy. Mechanical processes, then, are best suited for large quantity printing and long press runs. Electronic methods are best especially when the information printed is variable from copy to copy or for other short-run situations. Serial numbers, bar codes, product expiration dates, and invoice information are all variable but are suited to computer-controlled electronic printing.

D5a. ***laser printing*** - makes use of electrostatic attraction to make an image on paper. It utilizes a photo-receptor, which is a roller or *drum* coated with photosensitive material. The drum is initially given an overall positive charge by a nearby parallel corona wire or a charged roller. Digital data from a computer activates a laser that shines a narrow, coherent beam of light on the drum, "writing" the material to be printed as the drum revolves. A mechanism in the printer, directs the laser beam as necessary to trace the type fonts and lines of the image. The positive charge on the drum is discharged (removed) from the points contacted by the laser beam, but from only those points. The printer then subjects the drum to toner from a feeder roll or a cloud of toner. Toner is a fine powder with black or other pigments and particles of thermoplastic. The toner is positively charged so that it is attracted to the drum except for those areas that still retain the positive charge. The toner, then, sticks to the drum only at those points traced by the laser. The drum, with the powder image, then rolls across a sheet of paper that is moving on a belt below the drum. A negative charge, applied to the paper beforehand, draws the toner to the paper. The toner transfers to the paper, forming the exact image that was on the drum. The paper's charge is then neutralized by another corona wire. This prevents the paper from adhering to the drum. The paper then passes through the fuser, which is a pair of heated rollers. The heat from the rollers softens the plastic particles in the toner causing them to adhere to the paper and bond the pigment particles together. The paper then leaves the printer and the drum continues to rotate past the discharge lamp, which provides a bright light to erase any charges remaining on the drum. Then the drum surface passes the corona wire that applies the positive charge, and the process is repeated. An on-board microprocessor in the printer controls the operation of the laser, mirrors, lenses, corona wires, and other machine elements that all must work properly in sequence for the printing operation to proceed correctly. Fig. 9D5a illustrates the operation of a typical laser printer. Data fed to the printer comes from a word processor, desk-top publishing or other computer programs.

Laser printing, though quite slow compared to mechanical printing processes described above, is nonetheless very useful for small and moderate quantities. It is very widely used for business correspondence, home computing and other small and moderate quantity work.

D5b. ***copy machine printing*** - Xerox and other photocopier machines also utilize electrostatic attraction to make an image on paper with a sequence similar to that of laser printing. The key element of the copier is a drum (and, in some machines, a belt) coated with photoconductive material such as silicon, germanium or selenium. The drum or belt is given a positive static electrical charge by a nearby parallel corona wire subjected to a high voltage. Another corona wire imparts a positive charge to the copy paper. A white-light lamp inside the copier moves along the paper original, illuminating a narrow band as it passes. A mirror and lens arrangement focuses the reflected light and the image onto the rotating drum below. The reflected and focused light strikes the drum and neutralizes (discharges) the positive charge on the drum. This leaves the drum charged only at those points that correspond to printing, lines, or other dark points in the image being copied. Toner, a combination of fine powdered pigment and a thermoplastic, is a coating on very small beads that are stored in the machine's toner cartridge. The powder has been given a negative charge and the beads have a weak positive charge so that the powder adheres to them. The exposed areas of the drum pass rollers that have been coated with the beads of toner. The toner particles adhere to the drum surface but only to those points that still have a positive charge. (The exposed areas of the drum, corresponding to the background of the image copied, have no charge and do no attract the particles.) The copy paper, still strongly positively charged, then contacts the drum surface. The strong positive charge on the paper is greater than that on the drum and the toner is attracted to the paper and receives the exact pattern of toner that was on the drum. The paper is then heated with tubular quartz lamps and passes between two Teflon-coated rollers that compress the toner and bond it to the copy paper sheet.

D5b1. ***electrophotographic printing systems***$_4$ - There are several production printing systems, similar in principle to photocopying machine printing, that are used to produce moderate-quantity work. They use liquid or solid toner, electrostatic drums and curing of the image by evaporation or heating/cooling. Some use light-emitting diodes as a light source. Most such equipment is capable of color printing, often by using separate printing units in tandem to process paper in web form, but some work by processing the print paper through the unit several times. Systems for printing on both sides of a web in color can have eight units, four for each side of the web. Web speed can range up to about 300 ft/min (90 m/min) of variable images. Sheet feed equipment can produce 4000 variable sheets per hour (1000 sheets per hour with 4-color printing from one drum.)

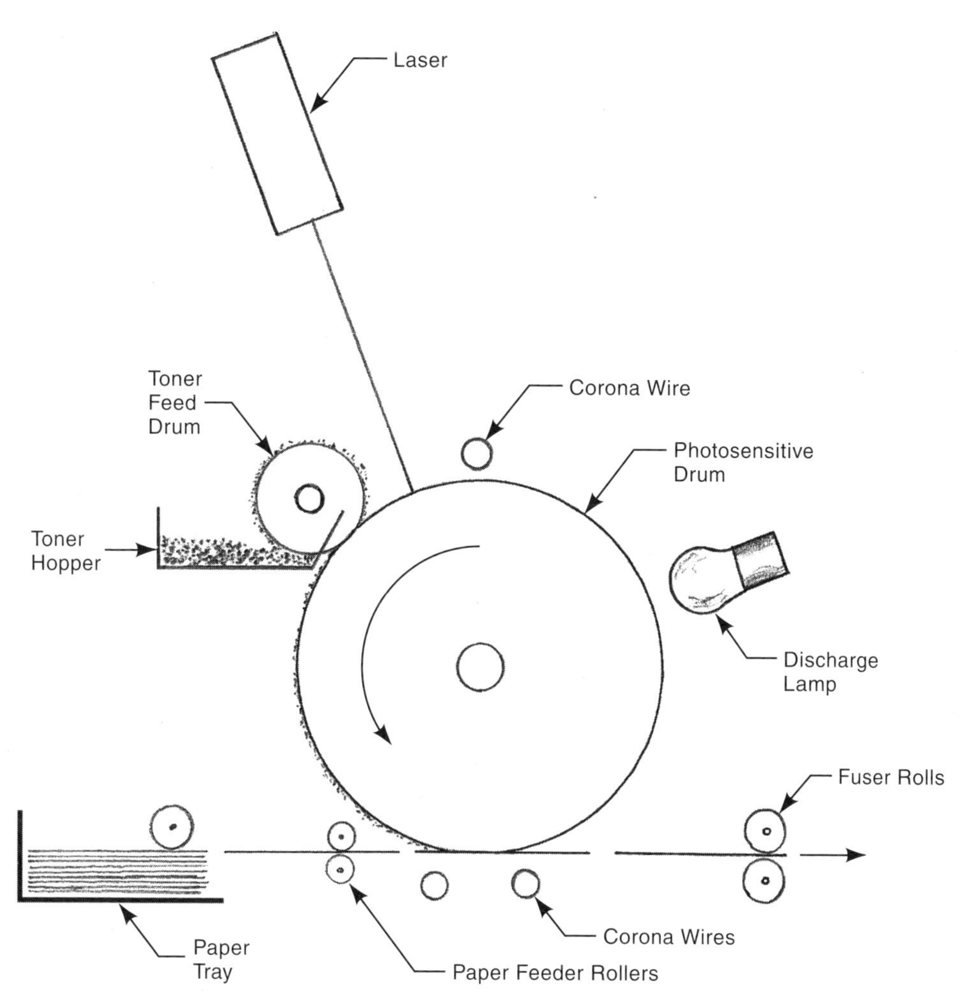

Fig. 9D5a A typical laser printer. The positively charged drum with a photosensitive coating is subjected to a laser beam that neutralizes the positive charge. Toner powder, also with a positive charge, is applied to the drum but adheres only where the positive charge on the drum has been neutralized. The toner is transferred to paper that contacts the drum and is fused to the paper by heated rollers.

D5c. ***ink-jet printing*** - Several varieties of the ink-jet process are in use, but all direct tiny droplets of ink to paper or other substrates without print head contact. There are no printing plates and the operation is computer controlled. Most systems use a bank of ink nozzles, each of which is controlled by the computer program. The computer activates the nozzles individually in quick succession and simultaneously as the nozzle moves across the paper and again as the paper is advanced underneath the nozzles. In some systems, an electric current passing below the firing chamber, heats the ink to produce a bubble of steam. The bubble expands, forcing the ink through the nozzle. When the bubble collapses, a vacuum is created in the print head. This draws more ink into the print head from the ink cartridge. This type of ink-jet printer may have as many as 600 nozzles.

In other printers, a piezo- crystal oscillates, pushing ink from the nozzle when it moves in one direction and, on the return oscillation, drawing more ink into the print head.

Another method uses solid sticks of wax-like ink. Heat melts ink from a stick and it is applied to the printing paper where it solidifies. High resolution and brilliant color can be achieved. The approach is usable on a variety of substrate materials.

The droplets in ink-jet printing are charged electrostatically and are deflected by electrostatic field plates under computer control to form images and characters. Printing in colors is fully feasible and ink-jet printers can produce picture quality equivalent to that of lithographs. Ink jet printing is most common for printing documents from personal computers. The advent of digital photography has opened the use of the ink-jet method for printing photographs. Ink jets are also used for printing address labels and expiration dates on food packages and other variable information such as serial numbers on products. Commercial ink jet equipment is used for billboards and advertising posters for buses, taxis, airports and bus terminals.[4]

D5d. ***magnetographic printing***[4] - uses a carbon-coated metal drum that is selectively magnetized by an array of computer-controlled electromagnetic writing heads. An image is generated when powdered magnetic toner is attracted to the magnetized drum. When the drum makes contact with paper, the toner image is transferred to the paper. The toner then is fused on the paper by a flash fusing device. Colors are dark and opaque except for some limited areas. The image can have infinitely-variable gray scales. After transfer of the toner, the drum is cleaned of toner and the magnetic image is erased. The drum is then ready for transfer of another image. The process can be repeated quite rapidly. The method is used for printing variable information such as that required on some business forms, direct mailings, lottery tickets, tags, labels and bar codes.

D5e. ***ion or electron deposition printing (ionography)***4 - In this process an electron cartridge produces negative electric charges on a heated dielectric aluminum oxide surface. The charges attract a special magnetic toner. The operation is computer controlled and is limited to single color printing or spot-color applications because the pressure of transferring the image and of fusing the toner can distort the printed surface. The system is used for variable printing in forms, invoices, reports, letters, proposals, tickets, tags and checks.

D5f. ***microcapsule printing***[3] - is a method for producing high-quality color reproductions in small quantities. The "printing plate" is paper impregnated with billions of microscopic-sized capsules of liquid dyes based on polymers with photosensitive properties. The plate is exposed to light reflected from the original image. This light hardens the polymer dyes in the micro capsules in proportion to the amount of light received, but capsules not receiving light still retain liquid dye. The paper to be printed and the exposed paper are placed together and run between a pair of pressure rollers. The pressure breaks the micro capsules and the varying amounts of the unhardened dye are deposited on the print paper.

D5g. ***thermal sublimation, dye sublimation, thermal wax and wax transfer printing*** - are a group of processes that use arrays of heating elements under computer control to heat sheets or ribbons of film with dye or wax-based pigments. The dyes or waxes vaporize and are transferred to the printing paper where they cool and solidify. Higher temperatures yield greater amounts of the particular dye where needed. Separate lengths of sheet or ribbon film each contain a dye of a different color: cyan, magenta, yellow and black. The process is repeated with each color on every printing sheet to produce a full color image. These processes are slow because of the repeated passes and the materials are costly. They are used only for limited quantity printing.

D5h. ***dot-matrix printing*** - involves a computer-controlled printer with a head containing an array of very small movable pins. The head moves across the paper and, as it does, one or more of the pins are driven electromagnetically to strike an inked ribbon that transfers a dot of ink to the paper at each point of impact. Enough dots are printed to form letters, numbers, punctuation marks and other characters. Because of their better quality and quiet operation, ink jet and laser printers have largely supplanted dot-matrix printers for personal computer use.

D6. ***sheet and web printing*** - All printing methods function when discrete sheets of paper are printed. In the simplest cases, sheets are manually positioned in a press and removed after the impression is made. In situations where the quantities are sufficient to justify an additional investment in equipment and set up costs for the print run, sheet feeding and removal may be automatic. Sheet methods may be required when heavy paperboard or metal are the substrate materials but otherwise are found only when quantities are small or moderate. For the highest levels of production, it is advantageous to use a continuous strip or web of paper from a roll and run the web through the press (or presses if there are several printing stages). Web printing uses cylindrical plates that roll on the paper as it passes through the press. This provides the fastest production rates. While sheet fed offset presses may print 280 sheets per minute, web presses may be capable of speeds of up to 2000 to 3000 ft/min (600 to 900 m/min). Newspaper web presses can produce more than 50,000 finished newspapers per hour[5]. No time is lost in paper handling between impressions. In addition to the labor savings and faster output rates, the use of a web aids in registering different imprints when color printing or other multiple impressions are involved and enables the paper to be printed more easily on both sides simultaneously. As many as 18 individual press units may be used in tandem on one web press line. Where it is advantageous, web printing equipment can include paper cutting, folding, stacking, stapling, glueing and bundling apparatus at the end of the printing line. Fig. 9D2a, which illustrates offset lithography, and Fig. 9D7, which illustrates multi-color printing, both also show web printing.

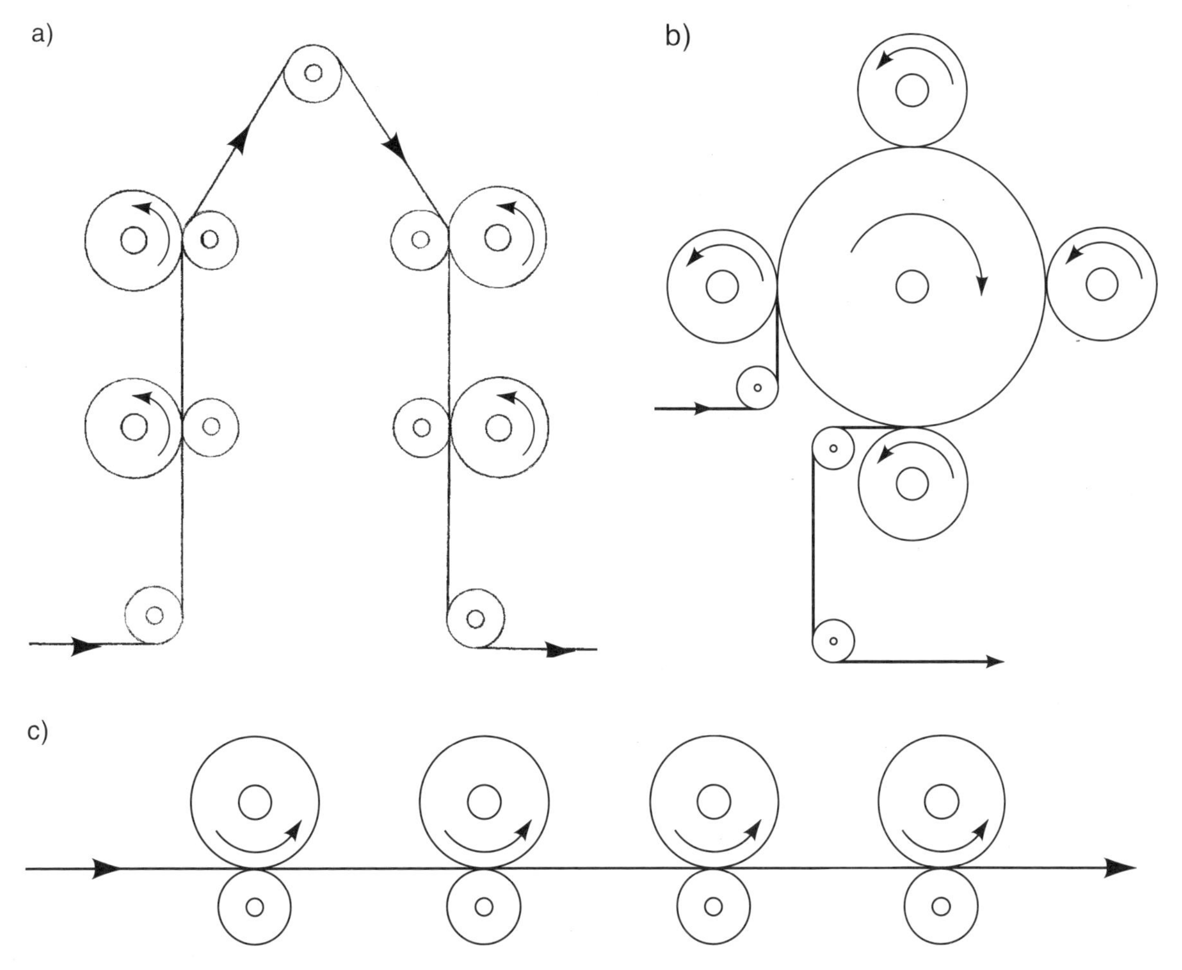

Fig. 9D7 Three arrangements for color printing shown with four printing cylinders, one for each color. Inking rollers are not shown for simplicity of presentation. In all, a web of paper passes between a separate printing cylinder for each color and an impression cylinder. a) stack arrangement of printing cylinders. b) one common large impression cylinder for all colors, c) in-line arrangement.

D7. ***color printing*** - For almost all processes (ink jet and micro-capsule printing are exceptions) color printing involves a series of successive printing operations in which different printing units each lays down only one color on the substrate. Four or more printing plates are used to provide full color. Colors are printed one on top of another in varying amounts. The colored inks are transparent when printed. All other colors are blends of the four. The plates for each color and for black are all different. The four colors are yellow, magenta (a purplish red), cyan (a blue shade) and black, which adds sharpness to the printed image. Yellow, magenta, and cyan are used instead of the three primary colors (red, green and blue) because the system that works best is a subtractive one, wherein each color on the printed sheet reflects color but also absorbs and blocks reflection of other colors. Yellow, magenta and cyan are complementary to red, green and blue.

Often, a greater number than four tandem printing units or successive operations on one press may be used, sometimes with other base colors, to provide a range of colors suitable for particular applications. Such arrangements are more likely when the material to be printed uses a large amount of certain colors, for example, in advertising and packaging printing where a particular trademarked color dominates.

To make the plates, *color separation* is performed, traditionally with color filters used to produce separate photographic images of the work to be printed. Current practice produces the color separations by scanner and computer. When multiple colors are printed, various arrangements of the plate cylinders and impression rolls are used to provide all the colors needed. Fig. 9D7 shows three common arrangements. Color printing requires precise positioning of the successive, single color impressions in order to combine the colors properly.

D8. ***halftone screens*** - Special approaches are needed when a letterpress, lithography, or other press system is used to print photographs or other images with a range of shades from light to dark. The shades may be between black and white or, with color, in various levels between brilliant and pale. A system is needed to create the intermediate tones when the printing system applies only solid colors or no colors. The method that has been used with letterpress and other printing systems for some time is to use a halftone screen to convert such images into ones that are printable. With halftone screens, the picture to be printed is projected through the screen. The screen breaks up areas of intermediate tone into a series of small, closely spaced dots. The darker the gray (or the more intense the color), the larger the dots and the smaller the spaces between them. With pale tones, the dots are smaller and have more space between them. The human eye, at a normal reading distance, is not capable of distinguishing dots spaced at 125 to the inch or closer and interprets the series of small dots as an overall area of uniform tone. The printing plate is thus made from the halftone image instead of the original picture. Fig. 9D8 illustrates how halftones represent different intensities of gray

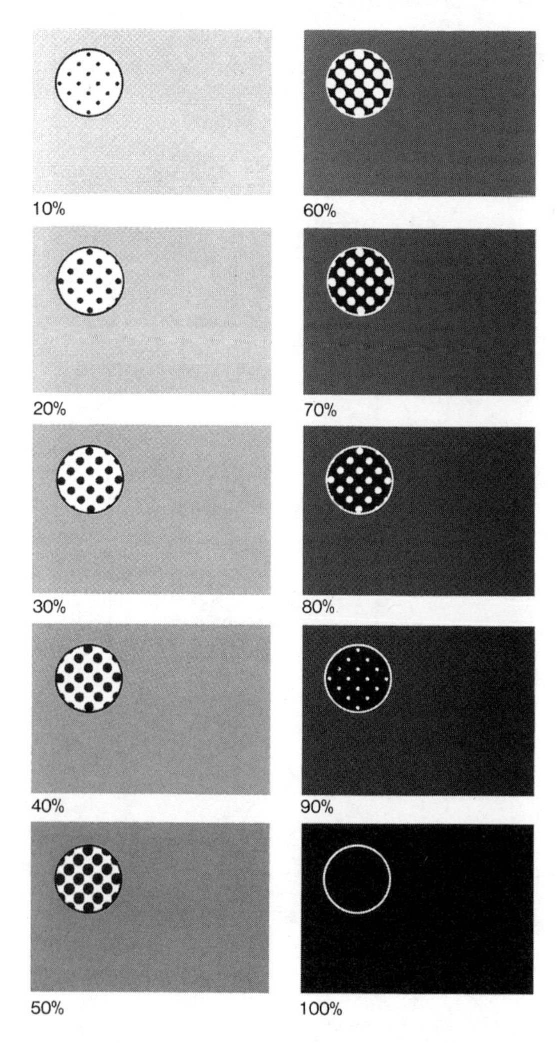

Fig. 9D8 Gradation and magnification of half tones. (*Reprinted with permission from the Pocket Pal, 18th ed., International Paper Company, 2000.*

tone. When electronic image scanning can be employed, the scanner can create the dots and no intermediate screen is required.

D9. ***pad printing*** - is a technique for printing small areas on products and parts. It is based on gravure printing and is described and illustrated in Chapter. 8, paragraph 8I7a of this handbook.

References

1. *The New Encyclopedia Britannica, Macropaedia, Vol. 21,* 1989, Chicago.
2. *The World Book Encyclopedia,* World Book, Inc., 2000, Chicago.
3. *Encarta Encyclopedia (from CD),* Microsoft Corp., 2004, Redmond, Washington.
4. *McGraw-Hill Encyclopedia of Science and Technology, Vol. 14,* 8th ed., McGraw-Hill, New York, 1997.
5. *Academic American Encyclopedia,* Arete' Publishing Co., Princeton, NJ, 1980.
6. *"What Makes Magnetography the Technology of Choice?"* - Nipson Digital Printing Systems, PLC, 2004. Internet: www.nipson.com.

Chapter 10 - Textile Processes

Textiles are fabrics (cloth) and other materials made principally from combinations of fibers. These fibers may be woven, knitted, braided, tufted, or made, by mechanical or chemical bonding, into non-woven fabrics. Yarns, sheets, films, foam materials, furs and leather may also be used in textile products. Garments, sheets, blankets, rugs and carpets, upholstery, drapes and curtains, nets, and various industrial components are important applications of textiles.

A. Textile Fibers

Fibers are long, hair-like, wire-like or thread-like materials whose lengths are 0.2 in (0.5 cm) or more and are greater than 100 times their diameters[2]. They come from plant, animal or mineral sources, or can be synthetic materials. *Textile fibers* are those that can be made into fabrics by the operations described below. Fibers occur or are made into different forms: *staple fibers* are relatively short fibers, normally under 6 in (15 cm) long, *filaments* are long or continuous fibers, *monofilament* is a continuous or long single fiber, usually a thick fiber, *tow* is a bundle of untwisted continuous fibers, *yarn* is a bundle of twisted fibers.[5]

A1. ***natural fibers*** - are those derived from plant, animal and mineral sources. The major ones are cotton, linen, wool and silk. Wool from sheep is the principal fiber produced from animal hair, but camel, llama, alpaca, guanaco, vicuna, rabbit, reindeer and goat (angora and cashmere) hair are also used.[5] Horse and cow hair are sometimes made into felt. Wool is sorted, graded, and scoured before it is processed into yarn.[5] Silk is an important fiber of natural origin, made principally from the cocoon of the silk worm. It is the only filament-length natural fiber. In the natural state, silk fibers are covered with a waxy or glue-like material that is removed by washing in warm water. The fiber is unwound from the cocoons and then spun into threads in a process called *throwing*. The other fibers from animal origin are in short lengths and are combined and spun together to form yarn before being made into fabrics. Broken fibers from silk manufacture are similarly processed.

Cotton is the most important textile fiber from plant sources and, in fact, is the most widely used textile fiber. It comes from the soft hairs that surround the cotton seed. The hairs are separated from the seeds by cotton gin machines, each of which consists primarily of a fixed comb and a rotating cylinder to which saw-like teeth are attached. Raw cotton is fed to the gin and is pulled by the saw teeth through the comb. The seeds, leaves and other debris, that cannot pass through the comb, are left behind. Fig. 10A1 illustrates the operation. (Cotton seeds from the ginning operation are made into cottonseed oil, cattle feed, and fertilizer.) Cotton is used extensively in clothing, household furnishings and industrial products.

Flax fiber, used to make linen cloth, is another important plant fiber. Flax is the designation for a family of plants. Some are grown for their seeds (from which linseed oil is made); others are grown for the fiber that comes from the stems of the plant. Several chemical and mechanical operations are performed to convert the flax into a fiber that can be spun. Hemp, jute, kenaf, ramie, abaca, and sisal

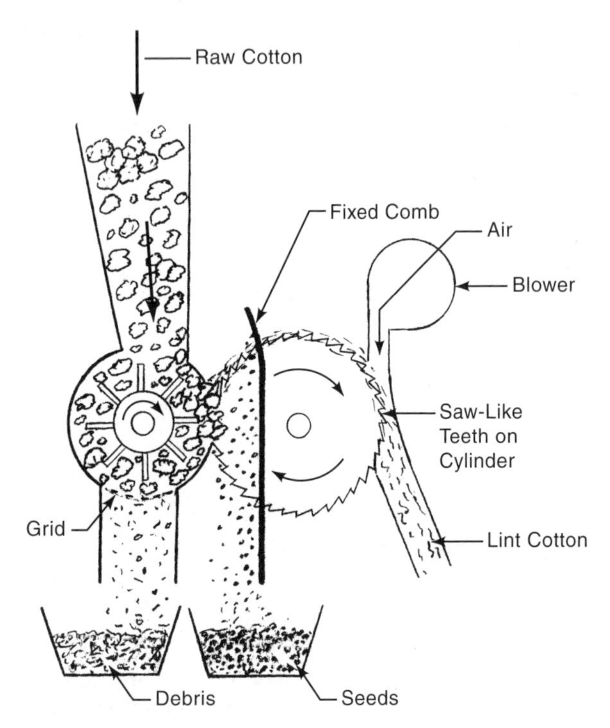

Fig. 10A1 A cotton gin. Raw cotton fed to the machine is thrown against fast-moving saw-like teeth on a cylindrical roller. Some debris in the cotton falls through a grid. The saw teeth pull cotton against the comb but cotton seeds are unable to penetrate the comb and fall from the machine. Cotton lint on the sawteeth is blown from the teeth and into a conveyor pipe.

are plant-sourced fibers that are made into coarser fabrics, rope and other cordage.

Asbestos is a general name of several natural mineral fibers. However, asbestos is no longer used in textiles because of health concerns despite its desirable properties of heat and chemical resistance.

A2. ***manufactured and synthetic fibers*** - Manufactured fibers used in textile manufacture come from both natural and man-made sources. Natural sources are either organic or inorganic. Organic materials include those from plant cellulose or rubber and those from manufactured polymers. Those from polymers, derived primarily from petroleum, coal and natural gas, include polyesters ("Dacron"), acrylics ("Orlon", "Acrylan" and "Dynel"), nylon, polyethylene, polypropylene, polyvinylchloride, polyurethane (spandex or elastane), and synthetic rubbers. Synthetic fibers made from cellulose include rayon, acetate and triacetate. Inorganic fiber materials include metal and glass.

Continuous glass fibers are made by drawing molten glass to very fine diameters. These are used in curtains and draperies and other applications where fire resistance and resistance to deterioration from sunlight and moisture are needed. Glass woven fabrics and unwoven staple fibers of glass and ceramics are also used for plastics reinforcement.

Metal fibers are made by cold drawing metal wires to fine diameters. (See 2B2.) Gold and silver fibers are sometimes used in fabrics for decoration. Conductive metal fibers may be incorporated to dissipate static electrical charges.

Synthetic fibers from thermoplastics are produced by extruding the molten plastic through extrusion dies (*spinnerets*) into a stream of cold air that cools and solidifies the plastic. (The operation is referred to as *melt spinning*.) The spinnerets have many very small die openings. The thermoplastics are similar to those of the same basic materials when they are used for making molded products, but modifications may be made for use in textile fiber applications. Nylon is an important synthetic fiber. After extrusion, fibers are drawn to approximately four times their original length, which aligns the molecules to provide much greater strength[3]. The fibers are often textured prior to use. Fabrics made from nylon are used in hosiery, undergarments, upholstery, draperies, parachutes and carpeting. Polyester is another important synthetic fiber material. Several varieties are used, but PET (polyethylene terephthalate) is the most common. Spun polyester fibers are used in clothing, usually blended with either cotton or wool. They are also used in carpeting. Acrylic fibers are made from polyacrylonitrile copolymerized with other materials. The fiber is textured to provide wool-like properties and is used with wool or nylon, or by itself, in socks, sweaters, blankets and carpets.

Viscose rayon is made from wood pulp that is treated to form a thick liquid. This liquid is extruded into a mild acid bath that converts the filaments back to pure cellulose. (The operation is referred to as *wet spinning*.) Acetate and triacetate fibers are made by treating cellulose with acetic anhydride to

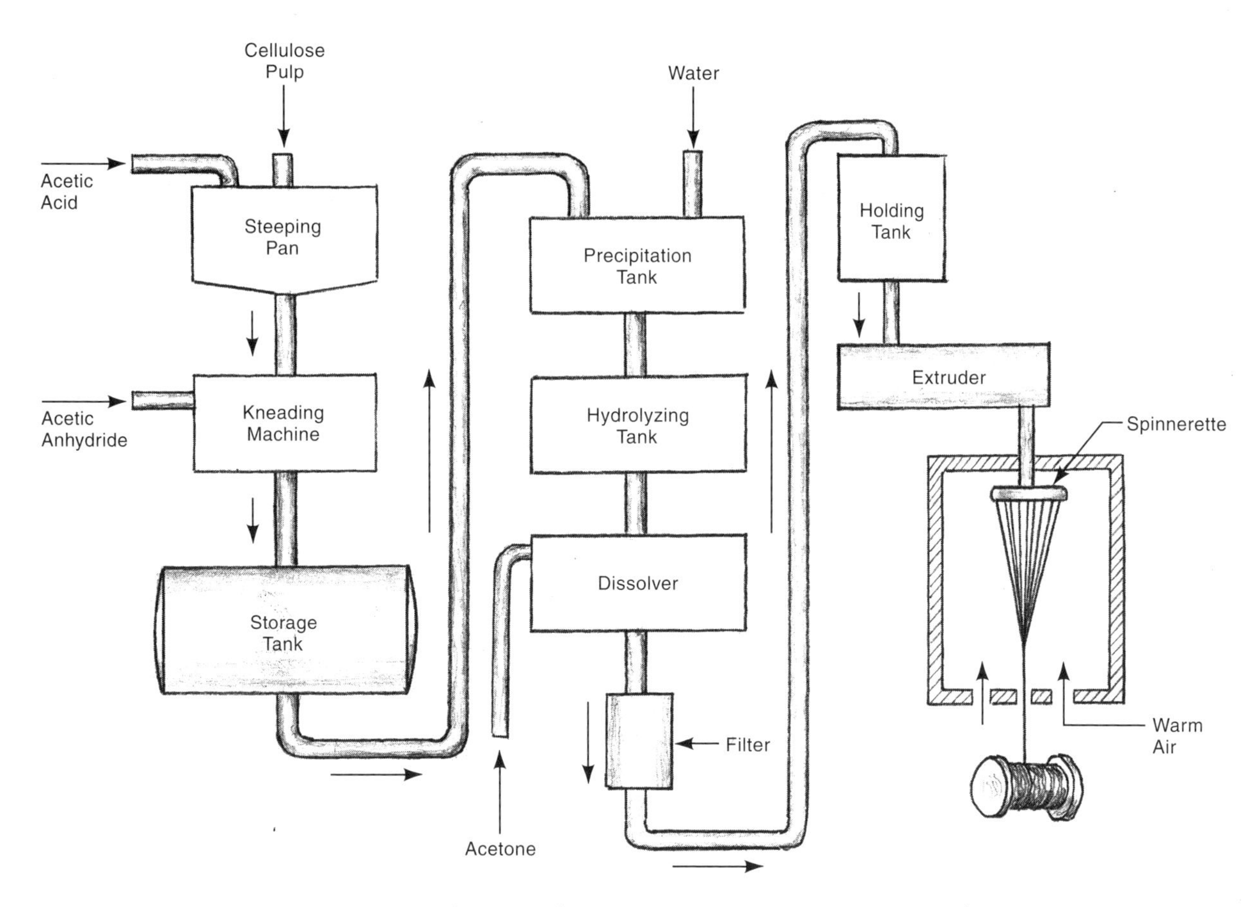

Fig. 10A2 The manufacturing process for acetate fiber.[4]

produce triacetate. To make acetate, wood pulp or cotton linters (short fibers that remain stuck to cotton seeds after ginning) are soaked in acetic acid. A small quantity of sulfuric acid is added. The mixture is then aged at a controlled temperature and mixed with acetic anhydride, producing liquid cellulose acetate. Additional water is added and the cellulose acetate precipitates. After being dried, the cellulose acetate, in the form of flakes, is dissolved in acetone and filtered repeatedly. The acetate is then in the form of a viscous liquid suitable for extruding. It is extruded through a spinneret into a chamber of warm air that evaporates the solvent[4]. (The operation is called *dry spinning*.[3]) Fig. 10A2 illustrates the process.

Carbon and graphite fibers, used as reinforcements in plastic parts, are made by heating acrylic and rayon fibers to a temperature in the range of 1800 to 4500°F (980 to 2480°C) to carbonize the material.

B. Yarn Making (Spinning)

Yarns are continuous strands of fibers that can be woven or knitted into fabrics. The term, "spinning" refers both to the final yarn-making operation that puts a twist in the yarn (B5 below), and also to the entire sequence of operations that convert raw fibers into usable yarns. Yarn making from staple fibers involves picking (opening, sorting, cleaning, blending), carding and combing (separating and aligning), drawing (re-blending), drafting (drawing into a long strand) and spinning (further drawing and twisting)[3]. Silk and synthetic filaments are produced by a less extensive procedure. Current high-production yarn-making operations are performed on integrated machines that perform this entire sequence as one combined operation.

B1. ***picking (including opening and blending)*** - includes the separation of the raw fibers from unwanted material: leaves, twigs, dirt, any remaining seeds, and other foreign items. The fibers are first blended with fibers from different lots or other sources to provide uniformity. (They also may be blended with different fibers to provide improved properties in the final fabric.) When cotton fibers are processed, the raw cotton is run through a cotton ginning operation and then undergoes a cleaning sequence before it is pressed into rectangular bales for shipment to the textile mill. There, the picking starts with a blending machine operation. Bales are opened and cotton from several lots is fed to the machine. The cotton then proceeds to an opening machine that opens tufts of cotton with spiked teeth that pull the fibers apart. Up to three stages of picking follow, after which the cotton is often in the form of a *lay*, a roll of cotton fiber about 40 in (1 m) wide, 1 in (25 mm) thick and weighing about 40 lb (18 kg)[1]. Figs. 10B1a, 10B1b and 10B1c show the blending, opening and picking operations.

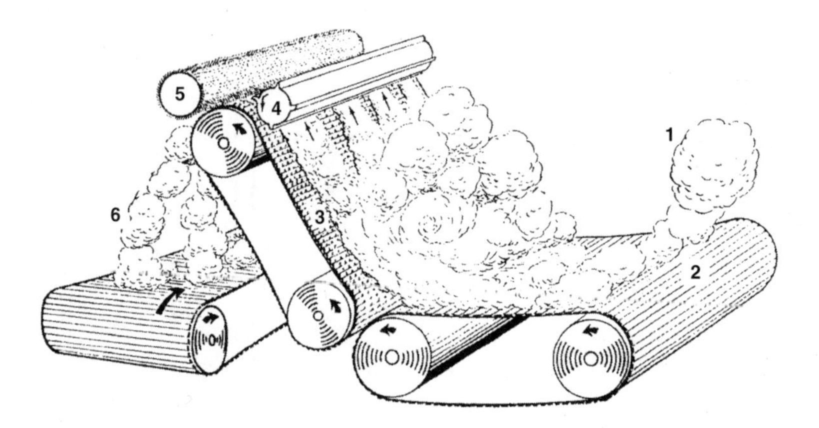

Fig. 10B1a Blending and feeding cotton fibers. Cotton from bales (1), is dropped onto an apron conveyor (2), and moves to another apron conveyor (3), whose surface is covered with spikes. The spikes carry the cotton upward where some of it is knocked off by a ribbed roller (4). The cotton knocked back mixes with cotton carried by the spiked apron. Cotton that passes the knock-back roller is stripped off by another roll (5) and falls (6) to a conveyor that carries it to the next operation. (*Illustration used with permission, Dan River Inc.*)

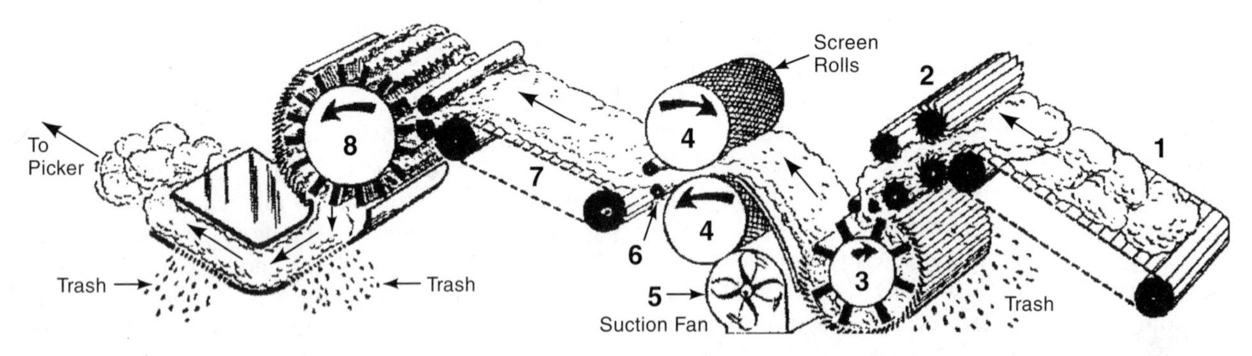

Fig. 10B1b Opening cotton fibers: Cotton from the blending operation falls on an apron conveyor (1) and passes between feeder rolls (2) to a beater cylinder (3). The beater cylinder has rapidly rotating blades that take small tufts of cotton from the feeder rolls, loosen the bunches, remove trash, and move the cotton to the pair of screen rolls (4). The surfaces of these rolls are covered with a screen material. Air is drawn through the screens by a fan (5), pulling the cotton against the screens and forming a web. Small rolls (6), pull the cotton from the screen rolls and deposit it on another conveyor (7), that carries it to another beater (8), that removes more trash. The cotton then moves to the picker operation. (*Illustration used with permission, Dan River Inc.*)

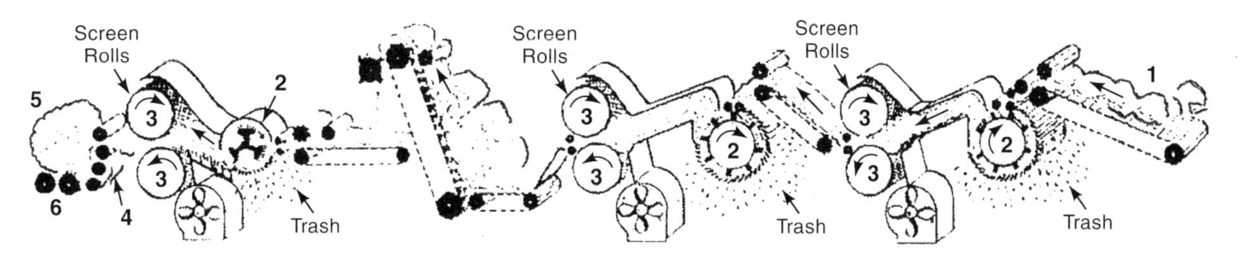

Fig. 10B1c Picking cotton fibers: Cotton from the opening operation falls on an apron conveyor (1) which moves it to the first of a series of beaters (2), and screen rolls (3). The beaters and screen rolls in the series are all similar but are progressively more refined as the cotton moves through the equipment. Each beater removes more trash from the cotton. When it reaches the output section (4), the cotton is in the form of a web or lap that is wound into a *lap roll* (5) by winding rolls (6). The lap roll in then ready to be transported to the carding equipment. *(Illustration used with permission, Dan River Inc.)*

B2. ***carding*** - is a process similar to combing and brushing. It disentangles bunches and locks of fibers and arranges them in a parallel direction. It also further eliminates burrs and other foreign materials and fibers that are too short. The operation is performed on cotton, wool, waste silk, and synthetic staple fibers by a carding machine that consists of a moving conveyor belt with fine wire brushes and a revolving cylinder, also with fine wire hooks or brushes. The fibers from the picking operation are called "picker lap", and are fed between the belt and the cylinder whose motions pull the fibers in the same direction to form a thin web. The web is fed into a funnel-like tube that forms it into a round rope-like body about 3/4 in (2 cm) in diameter. This is called a *sliver* or *card sliver*. The carding operation is illustrated in Fig. 10B2.

B3. ***combing*** - is an additional fiber alignment operation performed on very fine yarns intended for finer fabrics. (Inexpensive and coarser fabrics are made from slivers processed without this

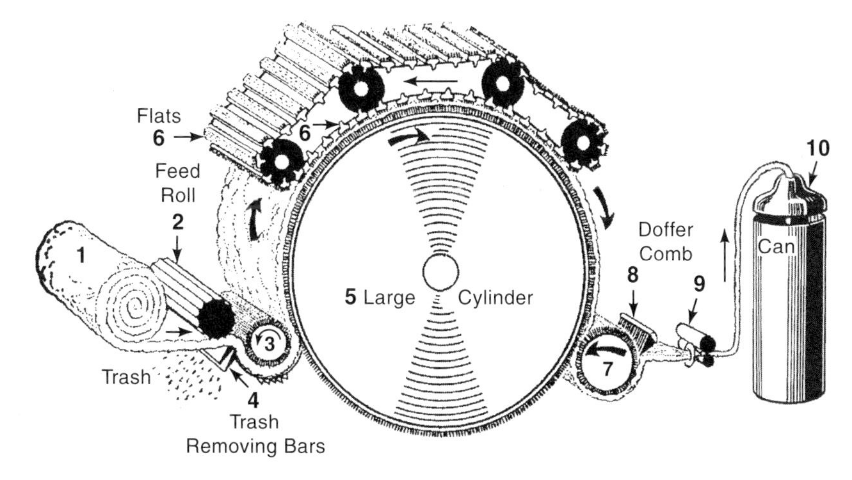

Fig. 10B2 Carding cotton fibers: The lap (1) from the picking operation is unrolled and fed by the feed roll (2), to the *lickerin roll* (3), which has wire shaped like sawteeth. The lickerin roll moves the lap against cleaner bars (4), that remove trash, and passes it to the large cylinder (5). The surface of the large cylinder holds the cotton with thousands of fine wires. The *flats* (6), with more fine wires, move in the direction opposite to that of the large cylinder. The cotton remains on the large cylinder until it reaches the *doffer cylinder* (7), which removes it from the large cylinder. A *doffer comb* (8), vibrates against the doffer cylinder and removes the cotton from it. The cotton, in a filmy web, passes through condenser rolls (9), and into a can through a coiler head (10). The subsequent operation is either combing or drawing. *(Illustration used with permission, Dan River Inc.)*

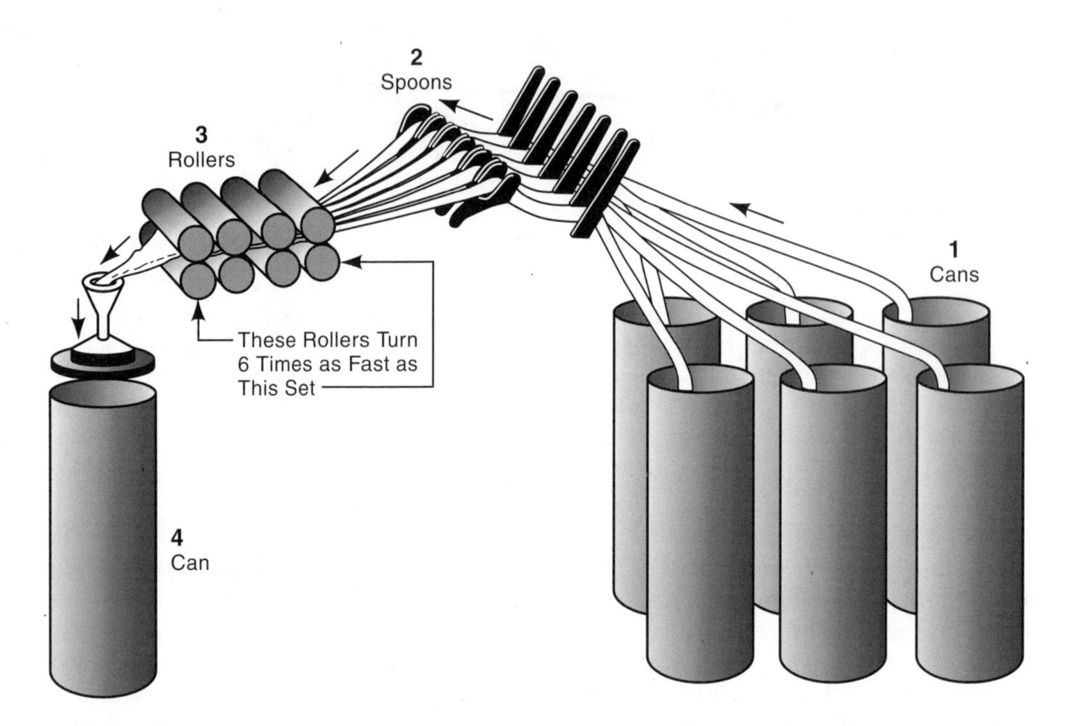

Fig. 10B4 Drawing: Cans (1), filled with slivers from the carding operation, feed the slivers to the drawing frame. The slivers pass through *spoons* (2), that guide the slivers and stop the equipment if any should break. The rollers (3), turn successively faster as the slivers move through them, reducing the size of the slivers and increasing their length approximately sixfold. At this point, the slivers are combined into one which is deposited into a can (4), by a coiler head. The sliver fibers are much more parallel, and the combined sliver is much more uniform after the operation, which is usually repeated for further improvement of the cotton slivers. (*Based on an illustration from Dan River, Inc. Used with permission.*)

further refining.) Fine-tooth combs are applied to the sliver from carding, separating out the shorter fibers, called *noils*, and aligning the longer fibers to a higher level of parallelism. The resulting strand is called a *comb sliver*. With its long fibers, the comb sliver provides a smoother, more even yarn.

B4. ***drawing (drafting), (re-blending)*** - After carding and, if performed, combing, several slivers are combined into one strand that is drawn to be longer and thinner. Drawing frames have several pairs of rollers through which the slivers pass. Each successive pair of rollers runs at a higher speed than the preceding pair so that the sliver is pulled longer and thinner as it moves through the drawing frame. The operation is repeated through several stages. The drawing operations produce a product called *roving* which has less irregularities than the original sliver. Afterward, the finer sliver is given a slight twist and is wound on bobbins. Fig. 10B4 illustrates the drawing operation.

B5. ***spinning (twisting)*** - further draws out and twists fibers to join them together in a continuous yarn or thread. The work is performed on a spinning frame after drawing. The twist is important in providing sufficient strength to the yarn because twisting causes the filaments to interlock further with one another. The roving passes first through another set of drafting rolls, resulting in lengthened yarn of the desired thickness.

There are three kinds of spinning frames: ring spinning, open-end (rotor) spinning, and air-jet spinning. With the common ring spinner, the lengthened yarn is fed onto a bobbin or spool on a rotating spindle. The winding is controlled by a traveler feed that moves on a ring around the

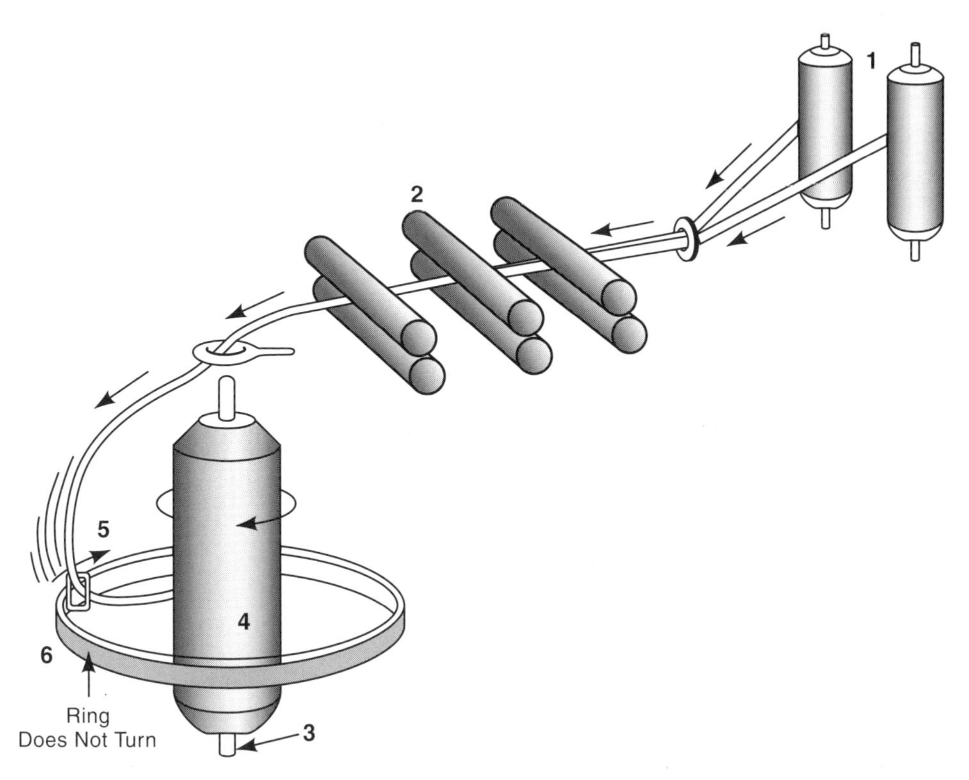

Fig. 10B5 Ring spinning. Spun sliver from the drawing operations, which is then called *roving*, and is wound on bobbins (1), and is fed through another series of drawing rollers (2), that further draw the strand to its final desired thickness. A larger bobbin (4) on a rotating spindle (3), turns at a constant speed. The speed of the final pair of drawing rollers is set a the speed that delivers the yarn so that it is twisted by the desired amount as it is wound on the bobbin. The yarn is guided by the traveler (5), which slides around the bobbin on the ring (6). Because of some drag on the traveler, the yarn winds on the bobbin at the same rate of speed as it is delivered by the final pair of rollers. (*Illustration used with permission, Dan River Inc.*)

spindle but at a slower speed than that of the spindle. The result is a twisting of the yarn. The yarn guide oscillates axially during winding to distribute the yarn neatly on the bobbin. The yarn can then be used to weave or knit textile fabrics or to make thread, cord or rope. Staple yarns, made from shorter fibers require more twist to provide a sufficiently strong yarn; filaments have less need to be tightly twisted. For any fiber, yarns with a smaller amount of twist produce fabrics with a softer surface; yarns with considerable twist, hard-twisted yarns, provide a fabric with a more wear resistant surface and better resistance to wrinkles and dirt, but with a greater tendency to shrinkage. Hosiery and crepe fabrics are made from hard-twisted yarns.[5] Fig. 10B5 illustrates ring spinning.

B6. ***spinning synthetic fibers*** - The term "spinning" is also used to refer to the extrusion process of making synthetic fibers by forcing a liquid or semi-liquid polymer (or modified polymer, e.g., rayon) through small holes in an extrusion die, called a spinneret, and then cooling, drying or coagulating the resulting filaments. The fibers are then drawn to a greater length to align the molecules. This increases their strength. The monofilament fibers may be used directly as-is, or may be cut into shorter lengths, crimped into irregular shapes and spun with methods similar to those used with natural fibers. These steps are taken to give the synthetic yarns the same feel and appearance as natural yarns when they are made into thread, garments and other textile products. (Section A2, above, describes wet and dry spinning methods of making rayon and acetate fibers.)

C. Weaving

Weaving is the interlacing of yarns in a regular order to create a fabric . The operation is performed in a machine called a *loom*. Two sets of yarns are interlaced, almost always at right angles to each other. One, called the *warp*, runs lengthwise in the loom; the other, called the *filling*, *weft* or *woof*, runs crosswise. Woven fabric is normally much longer in the warp direction than it is wide, that is, in the weft direction. Warp yarns are fed from large reels called *creels* or *beams*. Typically, these hold about 4500 separate pieces of yarn, each about 500 yards (450 m) long.[3] The filling yarns are fed from bobbins, called quills, carried in shuttles (hollow projectiles) that are moved back and forth across the warp yarns, passing over some and under others. The shuttle is designed so that the yarn it carries can unwind freely as the shuttle moves. Each length of yarn, fed from the shuttle as it moves across the loom, is called a *pick*. The yarn folds over itself at the end of each pick and forms another pick as the shuttle returns. When the yarn in a particular shuttle is exhausted, current production looms have automatic devices that exchange the empty quill with a full one.

Looms perform the following functions: 1) raising selected warp yarns, or *ends*, with suitable *harnesses*, consisting of frames of *heddles*, with taut vertical wires and eyelets, or strips with openings in the middle. There is one heddle for each end that is threaded through the eyelet. The heddles guide and separate the warp yarns, raising some of them to make room for the shuttle during the pick. This action is called, *shedding,* and the space between the warp yarns is called the *shed*. Simple weaves require only two harnesses; complex weave patterns may require as many as 40[3]. 2) *picking*, laying a length of the filling or weft yarn between warp yarns from the shuttle (a hollow projectile that holds weft yarn inside) as it moves across the shed. 3) *battening* or *beating in*, forcing the filling yarn from the pick against the just-formed cloth next to the previous pick. This step is necessary because the shuttle requires some space in its movement across the loom and it is not possible to deposit the pick closely against the previous picks. Battening is done with the *reed*, which is a grating of parallel vertical wires between the warp yarns. 4) *taking up*, winding the cloth, as it is formed, onto a take up reel, the *cloth beam*. 5) As the cloth is taken up, warp yarn is released from the warp beam. This action is called *letting off*. Fig. 10C illustrates major loom operations.

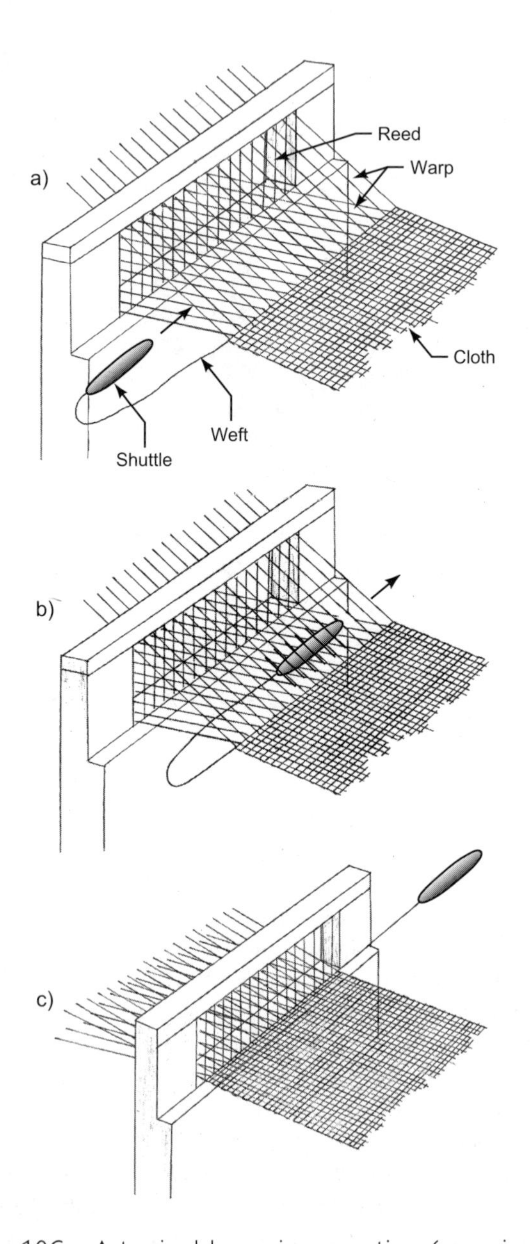

Fig. 10C A typical loom in operation (weaving): a) shedding, raising some warp yarns to make room for the shuttle, b) picking, laying the weft (filler) yarn across and between warp yarns, c) beating in, pushing the reed against the last filler yarn against the woven cloth.

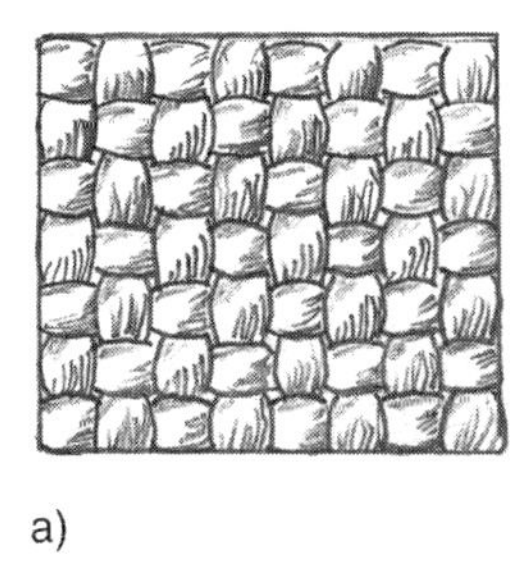

a)

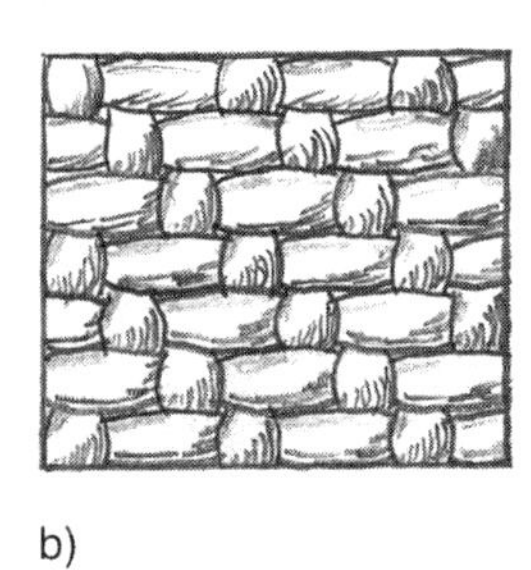

b)

c)

Fig. 10C-1 Three basic weave patterns. a) *plain weave*, also called *taffeta*. Filling yarns pass over and under alternate warp yarns. Other plain weaves are *broadcloth, muslin, batiste, percale, seersucker, organdy, voile*, and *tweed*. b) *twill weave*. Filling yarns pass over two warp yarns and under a third, and repeat the sequence for the width of the fabric. The next filling yarn repeats the sequence but shifts one warp yarn sideways, creating a diagonal pattern. *Herringbone, serge, jersey, foulard, gabardine, worsted cheviot*, and *drill* are twill weaves. c) *satin* weave. Filler yarns pass over a number of warp yarns, four in this illustration, and under the fifth. *Damask, sateen*, and *crepe satin* are satin weaves. Exposed yarns reflect light and give the weave its sheen.

In the simplest weaving arrangement, alternate warp yarns are over or under the shuttle as it moves in one direction and the warp yarn positions are reversed for the return stroke of the shuttle. This weave can be made on a loom with only two harnesses. [See view a) of Fig. 10C-1.] In other arrangements, several warp yarns may be moved upward or downward together, or several filling picks may take place before the warp yarns change position. In still other cases, the warp yarns are raised or lowered with respect to the picks in some predetermined sequence, creating a pattern in the appearance of the weave. These patterns may affect the feel and strength of the woven fabric. Such weaves may require looms with five or more harnesses.

The warp yarns may be coated with a temporary sizing for protection against damage during the operation. The process of applying this coating by taking yarn from a large rack, called a *creel*, passing it through comb guides and through a bath of starch, and winding it on a warp beam, is called *beaming* or *slashing*.

Weaving is the most widely used method for making cloth. It is simple, inexpensive, suitable for high-quality fabrics, and adaptable to special effects. Garments and household and industrial fabrics are made with the method. Fig. 10C-1 illustrates three of the most basic weave patterns. Fig. 10C-2 shows the major components of a simple loom.

C1. ***Jacquard loom weaving*** - Jacquard-type looms are looms with an automatic, selective method for *shedding*, the lifting of certain warp yarns for each cycle of the loom. The mechanism permits the use of continuously varying shedding patterns, to create corresponding patterns in the woven cloth. Complex patterns, including pictures, can be woven into the cloth. The original Jacquard process used a series of perforated cards to control the operation. Needle-like components, connected to hooks that controlled the heddles, passed through holes in the cards, and raised the warp yarns. Each heddle moved independently of the others. Where there were no holes, the needles did not move through and the heddles were not raised. The cards were moved with each cycle of the loom, creating a variable weaving pattern corresponding to the hole patterns in the cards. Current Jacquard looms use sophisticated electronic means to control the pattern of shedding. Jacquard loom weaving is used in making upholstery and drapery fabrics, in table linens and in some garments. Damask, brocade, brocatelle, matelasse and tapestry fabrics are made on Jacquard looms.

C2. ***automatic bobbin changing*** - Because of space and size limitations, the amount of filling yarn that can be carried in a shuttle is also limited. Bobbins in shuttles must be replaced when the yarn is exhausted. As noted above, this operation has been automated. Automatic bobbin loaders sense when the filling yarn is exhausted, remove empty bobbins and insert full ones when the shuttle is momentarily stationary. The operation does not reduce the speed of the loom.

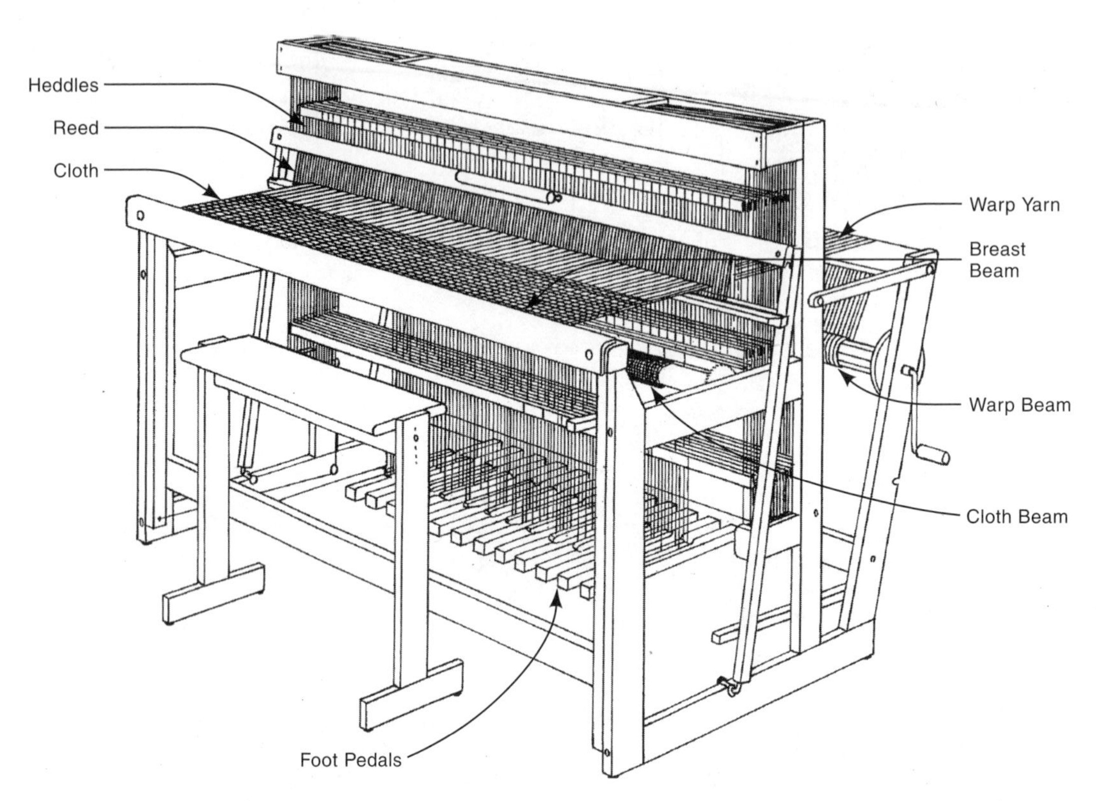

Fig. 10C-2 The major components of a loom illustrated by a hand-operated unit used for craft work. (*Courtesy Louet Sales, Prescott, Ontario, Canada.*)

C3. ***shuttle-less looms*** - Many current production looms do not use shuttles. In some looms, air or water streams propel the end of the filling yarn for each pick. In others, dummy shuttles pull the filling yarn but do not carry a bobbin. The rapier method uses an arm or tape-like machine element that grasps the filling yarn and pulls it across the web of warp yarns. One arm usually feeds the yarn halfway across the loom and an arm on the other side grasps the end of the filling yarn and pulls it the rest of the way across. Newer looms simply propel the end of the filling yarn across the loom by inertia. All these arrangements provide quieter operation, reduced wear, elimination of the need to protectively coat the warp yarn, and increased weaving production.

C4. ***pile weaving*** - is usually a plain weave in which either the filler or the warp yarn is drawn from the fabric to form loops between the intersecting yarns. The loops provide a thickness to the cloth. Turkish toweling is made from pile weaves with the loops uncut. Velvet is a pile fabric, but the loops are cut. In another method, special looms weave two fabrics face-to-face simultaneously. They are connected together by pile yarns. When the pile yarns are cut, two fabrics result, each with a pile. The process is less costly than weaving individual fabrics with a pile which must then be cut. Woven rugs and carpets are pile fabrics.

D. Knitting

Knitting is fabric- or garment-making by forming a series of interlocking loops in a continuous yarn or a set of yarns. In production situations, the work is carried out through the movement of hooked needles. (Hand knitting is normally performed with straight needles.) Each row of loops is

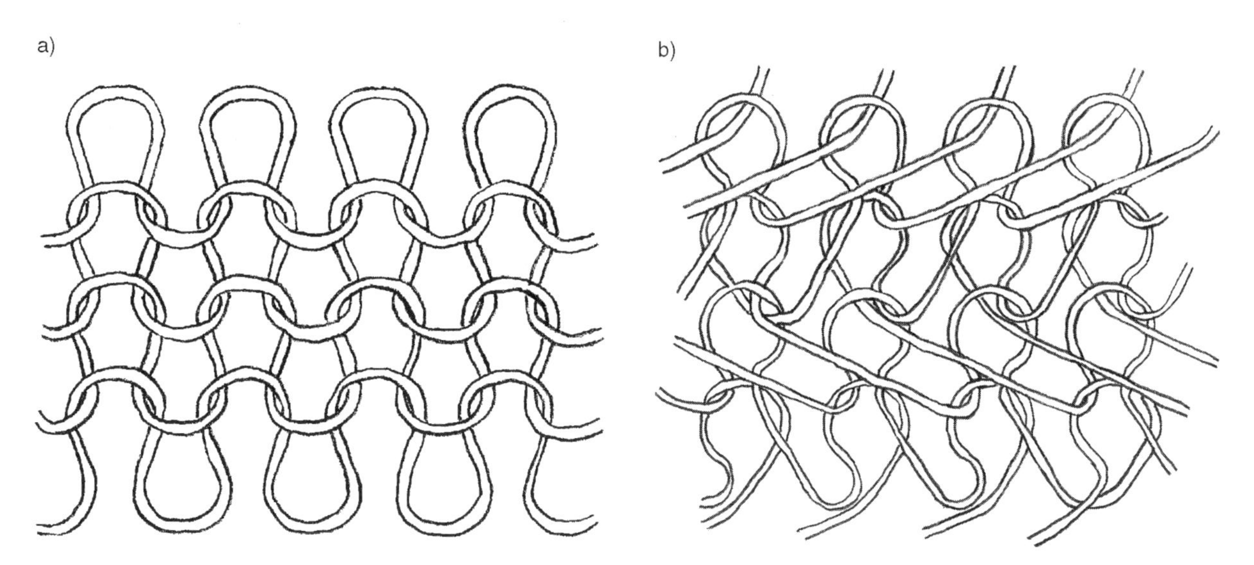

Fig. 10D Two examples of knit fabrics made by interlocking continuous strands of yarn: a) a plain knit made on a weft or filler knitting machine. The path of each crosswise yarn is called a course. b) a single-warp tricot knit.

vertically interlocked with the preceding row. With a sufficient number of loops, the yarn becomes a fabric. Knitted fabrics have the advantage of stretchability, a property not possessed by woven fabrics. Stretching can be in any direction even if the yarn used has little elasticity. Fig. 10D illustrates two types of knitted fabric. Mechanized production knitting utilizes a series of needles commonly operated by cams.

There are two basic types of knitting, weft or filler knitting and warp knitting. Weft knitting is somewhat more common. In weft knitting, the courses (crosswise rows of loops) are composed of continuous yarns. Weft knitting can be done by hand or machine but production weft knitting is a machine operation. The individual yarn is fed to one or more needles at a time. In warp knitting, the wales (predominantly vertical columns of loops) are continuous.[3] Separate yarns are fed to each needle. The warp knitting operation is always produced by machine.

Knitted fabrics can be either flat or tubular in form. Warp knits are usually flat; weft or filling knits are most often tubular.[1]

Two types of hooked needles are used in production knitting machines, the bearded or spring needle and the latch needle. They are illustrated in Fig. 10D-1. With both designs, the needles draw new loops through the previous loops that they have retained. Once the needle head and new loop have gone through the old loop, the old loop is cast off. The latch needle is most often used. It operates more automatically than the bearded needle which requires other machine elements to present the loop

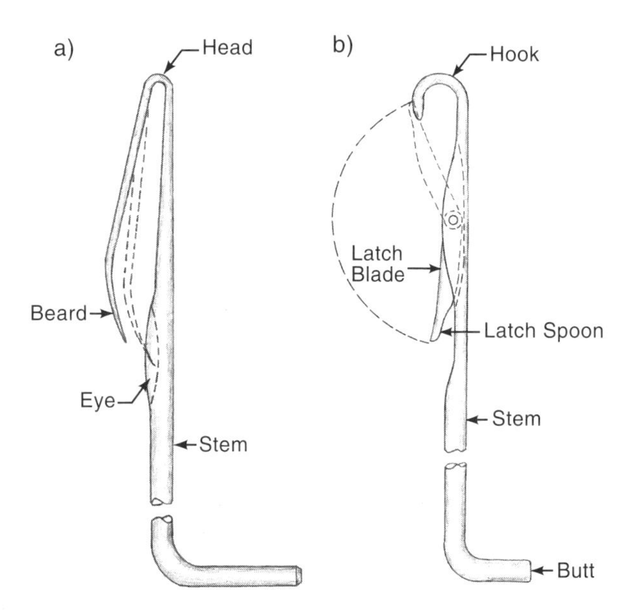

Fig. 10D-1 Production knitting needles. a) the bearded spring needle used for fine knitted fabrics and b) the more common latch needle.

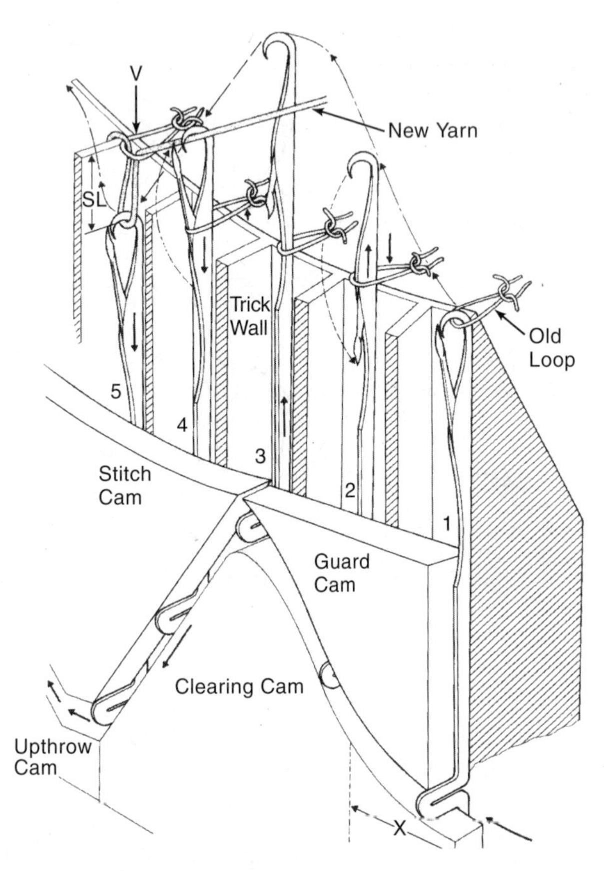

Fig. 10D1 Stitch formation in weft knitting with latch needles in a circular machine. (*from Knitting Technology by David J. Spencer. Reprinted by permission of Elsevier Science.*)

and close the hook. Fig. 10D1 illustrates stitch formation with a latched needle in a circular machine.

D1. ***weft or filling knitting*** - can be produced on either flat or circular knitting machines. In weft knitting, one continuous yarn runs crosswise in the fabric and makes up all the loops in one course. The needles either act in succession or the yarn is fed in succession, so that loop formation and interlocking is not simultaneous. Fig. 10D, view (a), illustrates a basic weft knit jersey cloth. Fig. 10D1 illustrates weft knitting with latch needles and shows that the multiple, evenly-spaced, needles have hooks with latches at the end. The needles are moved upward or downward by cams. As each needle rises, the needle hook loops over the yarn which it hooks on the down stroke, and the yarn is held in place by the needle latch. At the bottom of the needle stroke, a previous loop slips off the needle, and the new loop is held in place with the latch. On the next cycle, the loop is released from the latch as the needle rises, another loop is formed and the process is repeated.[3] Fig. 10D1-1 shows six stages of weft knitting with bearded needles.

Several different stitches can be formed in weft knitting. In the *knit* stitch, the loop is drawn from the back and passed through the front of the preceding loop to the front of the cloth. In the *purl* stitch, the loop is drawn from the front through the back of the preceding loop to the back of the cloth. In the *miss* stitch, no loop is formed. In the *tuck* stitch, two courses on one wale are looped over a third. The stitches, and various combinations of them, make all the patterns of knit and double knit cloth. Distinct patterns can be made from combinations of the knit and purl stitches since the knit tends to advance and the purl to recede.[1] Double knits are made by machine only, using two yarns and two sets of needles. These knits use a variation of the rib and interlock stitches, drawing loops from both directions.[1] Jersey is a common knitted cloth, made from only knit or only purl stitches.

Circular weft knitting machines are used to make hosiery, underwear and simulated furs. They can knit shaped garments. Jacquard effects are possible, and are now generally controlled electronically. Flat knitting machines can also produce shapes by increasing or decreasing loops. Full-fashioned garments can be made on flat knitting machines.

D2. ***warp knitting*** - is usually accomplished on flat machines but can also be tubular. Warp knitting differs from weft knitting in that each needle has its own yarn. The yarns are fed from a large reel or warp beam as in weaving with a loom. The yarns, then, generally run lengthwise in the fabric. The needles all move together and form parallel rows of loops simultaneously. The loops are interlocked on a zigzag or vertical path. The yarn section is held on one end by the previous loop and at the other end by the yarn guide. The yarn is trapped within the hook of the needle as it descends. With latch needles, the hook is closed as the needle descends. This allows the previous loop to slip off the hook while a new loop is held.[3] If bearded needles are used, a yarn guide, called a *sinker*, positions the yarn across ascending needles and then retracts as the needles descend. Fig. 10D2 illustrates warp knitting.

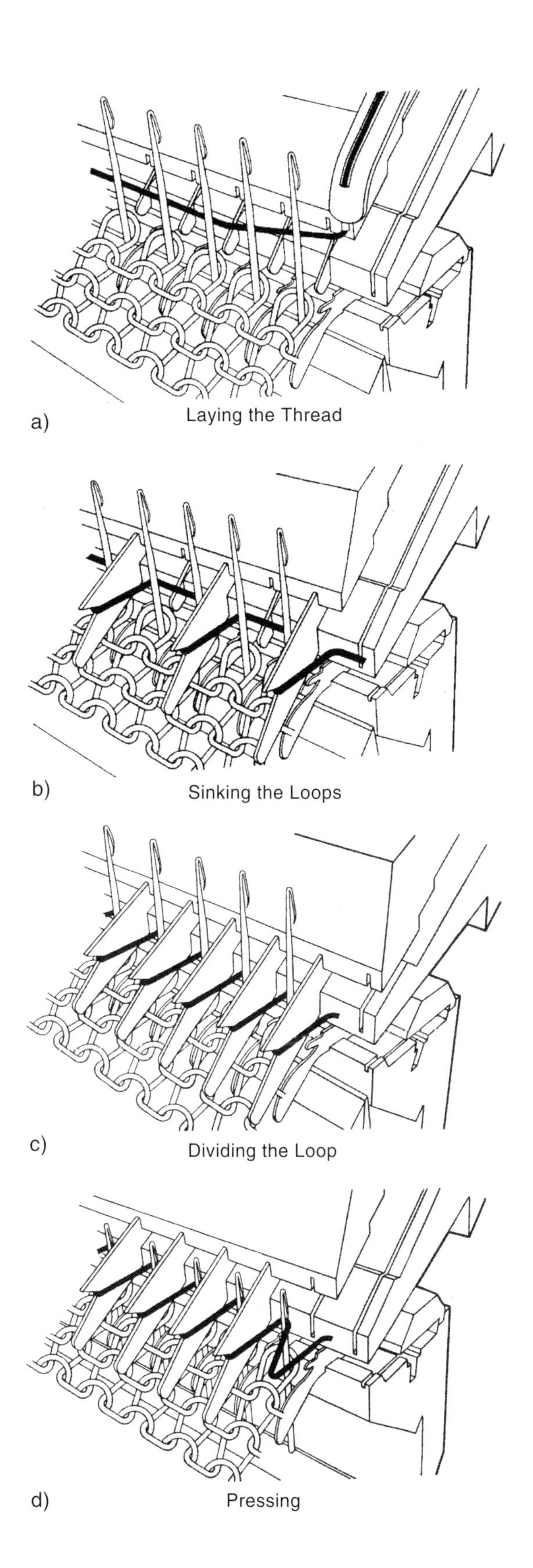

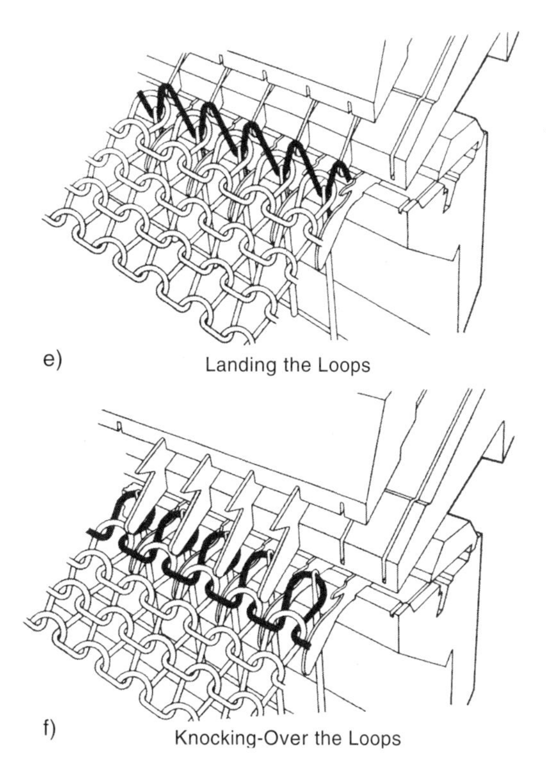

Fig. 10D1-1 Six stages of stitch formation in weft knitting with bearded needles. In this example, all stitches in one row are formed at the same time. (*from Knitting Technology by David J. Spencer. Reprinted by permission of Elsevier Science.*)

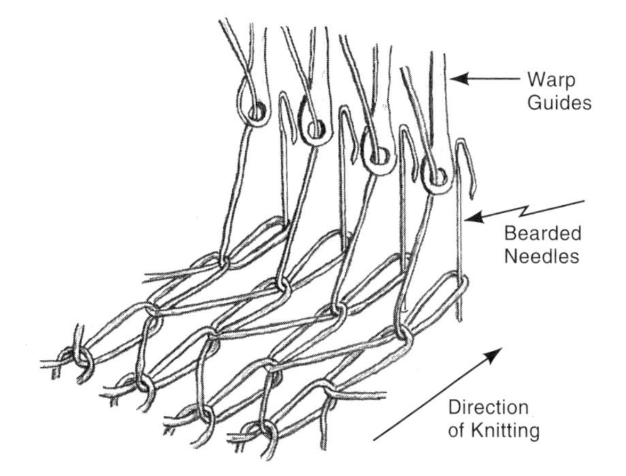

Fig. 10D2 Warp knitting with bearded needles. Loop forming is performed simultaneously with separate yarns fed through warp guides.

Warp knitting is a versatile process, but standard warp knitting machines make just three basic stitch variations: open loop, closed loop or no loop. Various fabric patterns are created from different combinations of these stitches. One simple pattern produces tricot knit, which consists of a zigzag pattern of closed loops of parallel wales. Tricot fabrics are run-resistant. Other warp-knit patterns are simplex, milanese and raschel. Milanese knitting produces run-resistant fabrics with a diagonal rib pattern. Several sets of yarn are used. The raschel knit is made with latched needles rather than the spring beard needles used for other knits. One or two sets of latch needles are used. Raschel knit fabrics are used frequently for underwear.

Warp knitting is used to produce fabric for dresses, lingerie, upholstery and draperies. among other products.

E. Non-woven Fabrics

Non-woven fabrics are fabrics made by bonding or interlocking individual, randomly oriented fibers together, rather than by interlacing continuous yarns. The fibers are held together as the result of mechanical, chemical, thermal or solvent methods or a combination of them. First, a matting of fibers is prepared, usually by taking webs of fibers after they have undergone carding and laying the webs on each other. They are laid with a parallel or crisscross pattern until a mat of sufficient thickness has accumulated. Another method uses equipment that blows loose fibers against a perforated drum. A vacuum inside the drum causes the fibers to adhere to it and the fiber deposition continues until the desired thickness is obtained. Then the mat is passed to another drum with teeth that break up the web and further randomize the direction of the fibers. The fibers are then blown onto another perforated drum. The final mat has a uniform thickness and fully random orientation of fibers. Fibers of wool have a scaley surface and have enough irregularity so that they can be compressed directly into a useful non-woven fabric. Wool felt relying on this irregularity has been made for many years. Another principal mechanical method utilizes a needle-punch machine with barbed and hooked needles that repeatedly penetrate a mat of fibers, interlocking and interlacing them. Non-woven blankets and felts of many fibers are made this way.

Chemical, thermal or solvent methods involve adhesive bonding of the fibers. In one method, thermoplastic fibers are blended with a base fiber. The melting or softening point of the thermoplastic is below that of the base fiber. The fibers are distributed to form a blanket or web that is passed through a pair of heated rollers. The heat of the rollers softens or melts the thermoplastic to the point that it flows and bonds the fibers together. Another method uses an adhesive sprayed on or applied as a foam or powder to a blanket of fibers. The adhesive can be a thermoplastic, solvent-based, or thermosetting material. Latex and acrylic-based adhesives are often used. In all these cases, heat from pressure rollers, or other sources, cures the adhesive into a solid material that holds the base fibers together.

Some non-woven fabrics are reinforced with sewing machine stitches

The wet-lay method of making non-woven fabrics utilizes modified paper-making equipment. Synthetic fibers are used alone or in combination with wood pulp. An adhesive or other binder may be included.

Non-woven fabrics are used for filters, garment lining, shoe insoles, pennants, industrial fabrics, and padding. Felt hats are made from felt made from the fur of rabbits, muskrats, beavers and nutria. Felt is used for the surface of billiard tables.

F. Finishing

Finishing processes include a variety of operations to make a textile fabric more suitable for its application. Finishing operations can be chemical, mechanical, or a combination of the two. They include treatments to improve the appearance or touch of the fabric and processes intended to improve its performance. Before finishing, woven or knit cloths are sometimes referred to as "gray goods". Companies that finish gray goods are referred to as "converters". (In the trade, dyeing and printing of fabrics are classified separately, not as finishing operations.)

F1. ***preparation***[1] - for finishing operations includes the removal of impurities from the initial fibers and irregularities, foreign matter, and defects, from the weave or knit. The objective is to produce clean and absorbent material, ready for finishing and coloring. *Burling and mending* are hand operations, primarily applicable to cloth made with natural fibers, to remove any burrs, stray yarn, knots, slubs and foreign matter. Tears, holes and broken yarns are mended.

Scouring is the removal of sizing applied to warp yarns as part of the weaving operation or any dirt, oil or lint that may have resulted from the operation. Synthetic fibers are washed with water and mild detergents. Natural fibers may require a more aggressive treatment with a strong detergent and an alkali with heat. Removal of sizing starch also requires aggressive steps. However, other special procedures are required with protein fibers (wool and silk) because of their sensitivity to alkali and strong detergents.

Mercerizing is an operation applied primarily to cotton fabrics or yarns but also to linens. It swells the fibers, shortens their length, and improves their appearance (smoothness and luster), and strength. High-quality mercerized cotton yarns have a silk-like luster. The operation also greatly improves the affinity of the fibers to dyeing. The process involves immersion of the yarn or fabric, under tension, in a cold 15 to 20 percent solution of caustic soda (NaOH) in water, followed by neutralization in an acid and thorough washing.

Drying is another preparatory operation, performed to remove excess accumulated moisture in the fabric from previous operations. Centrifuge and vacuum-chamber methods are used as an initial drying step, followed by running the fabric through a heated drying oven and then over a series of heated cylinders.

Bleaching before dyeing is also considered a preparatory operation. (See F2 below)

Singeing (gassing) - burns off any fuzz, yarn ends or projecting fibers from a yarn or fabric and makes its surface smoother. The process is performed extensively on cotton and frequently on rayon, but not on wool, silk or synthetic fibers. The yarn or fabric is passed rapidly through a gas flame or over a heated copper plate. By limiting the time of contact with the heat source, only extraneous fibers are singed. However, the operation is usually followed by a wetting step to extinguish any smoldering. If sizing precedes singeing, cotton cloth, after singeing, is run through an enzyme solution, squeezed, and allowed to stand for a period to digest and drain sizing starch.

F2. ***bleaching*** - whitens the fabric by removing the natural colors of the fibers and any stains from previous operations. The process used varies with the fibers involved but is usually a chemical process involving oxidation. Reduction using hydrogenation is the other common process. A treatment with heated hydrogen peroxide is usually used on cotton and other cellulose fibers. Cotton fabrics are often scoured and bleached in sequence in the same operation. A typical sequence for cotton involves putting the material through a steam chamber to remove sizing, washing it and impregnating it with a mild caustic soda solution, and then holding it in a "J-box" container for a period of an hour or more. The material is then washed and impregnated with a 2 percent hydrogen peroxide solution and put into another J-box for another hour. After washing, the cotton is fully bleached.[4]

Sulfur dioxide is the usual bleaching agent for wool and other animal fibers. The process involves prolonged boiling under pressure with a mild solution of caustic soda, soap and sodium silicate. The fibers are then washed with cold water, scoured and neutralized, washed again and pulled into a 2 percent sodium hypochlorite solution and into a J-box for a period. Finally, they are run through a weak sulfur dioxide solution, washed and dried.[4]

Synthetic fibers may be treated either by oxidation or reduction, depending on the material involved. They require less preparation than natural fibers but are not as easy to bleach. Sodium chlorite is used on nylons and chlorine or peroxide on polyesters and acrylics.

Sunlight has been used for many years as a bleaching agent for linens. Hydrogen peroxide treatment often follows the sunlight treatment.

F3. *finishing to improve appearance*

F3a. ***napping*** - is a brushing process that lifts the loose, short fibers, primarily from the weft yarns, into a down or nap. (The process is different from pile weaving which produces loops during weaving to provide a third dimension to the fabric. In napping, the raised fibers are only a surface effect.) Napped cloths have a warmer feel. The operation is performed by passing the fabric over rollers covered with fine wires. The wires lift the short fibers to the surface. The process is applicable to woven or knitted fabrics of spun yarns including wool, cotton, silk and rayon. Suede cloth, flannelette and wool flannel are napped fabrics.

F3b. ***shearing*** - is performed with rotary cutters, to trim a raised nap to a uniform height. It is also carried out on pile fabrics, often by a machine with spiral blades mounted on a cylinder. Automatic brushing follows shearing to remove the sheared ends of the fibers and yarns.

F3c. ***brushing***[1] - can be used to raise a nap on woven and knitted fabrics. It also is used to remove loose fibers and short fiber ends from smooth fabrics. Another use is the removal of cut fibers after shearing. The operation is carried out with bristle-covered cylindrical rollers which rotate and advance across the fabric.

F3d. ***beetling*** - involves the beating of dampened linen or cotton fabric with wooden mallets as the fabric is tightly wrapped over steel cylinders. The operation is automatic and produces a fabric with a permanently harder, flatter, highly lustrous surface and less porous weave. It makes cotton fabrics more linen-like and is used on table linens but not on linens used for garments.

F3e. ***decating*** - involves the application of heat and pressure to the surface of wool and other fabrics to set the nap, even the grain, develop luster and provide a softer hand. In the wet decating method, the material is tightly wrapped on a perforated roller and immersed in a trough of hot water. In dry decating, steam is used instead of hot water. Decating is used to improve the luster and color of rayon fabric and to make color unevenness, where it exists, more uniform.

F3f. ***calendering*** - is another process that applies heat and pressure to smooth the surface of a fabric, making it flatter and more glossy. Calendering is carried out as a final finishing step, especially when a flat, smooth surface is desired. The fabric is passed between two or more heated rollers. The degree of heat and pressure controls the amount of luster developed. Calenders may have as many a seven rollers, four of steel and three of a non-metallic material. The steel rollers are heated by steam or a gas flame. Fabric moves through the rollers at about 450 ft (135 m) per minute with a pressing force normally of 40 to 60 tons but occasionally as much as 100 tons. Sometimes, one of the steel rollers with a polished surface is geared to rotate with a higher surface speed than that of the fabric so that a burnishing action supplements the pressing. Such calenders are called *friction calenders*. The effects of calendering are normally not permanent. The process is applied to cottons, silks, rayons and synthetic fabrics. The operation is called *pressing* when applied to wool.

F3g. ***creping*** - Crepe is a fabric with a finely ridged or crinkled surface. The crepe effect is most permanently produced by weaving with hard twisted yarns but can also be produced in a fabric by causing it to shrink in certain areas but not in others. The process uses caustic soda (NaOH) applied by roller in a particular pattern. The areas treated with the caustic soda shrink; the other areas pucker. Another method uses a resist such as wax to block the contact of caustic soda in certain areas when the fabric is immersed in a caustic soda solution. Several different patterns of crepe can be produced, depending on the pattern of application of the caustic soda. When silk is creped, sulfuric acid is used instead of caustic soda. After a few minutes, the acid is rinsed off and the silk is neutralized with a weak alkali. The crepe effect can be produced temporarily by passing the fabric through a pair of rollers that have a pattern of indentations in their surfaces. Steam is applied during the rolling and the rollers are heated. After passing through the rollers, the fabric has the crepe pattern of the rollers formed in it. With synthetic fabrics made from thermoplastics (e.g., nylon, acrylics, polyethylene and polypropylene), the creping effect produced by heated, patterned rollers is permanent.

F3h. ***embossing*** - is essentially the same as the embossing operation applied to plastic or metal sheets. The fabric is run between a pair of matched, heated rollers that each have a design on their surface. Where the design is raised on one roller, it is recessed in the mating location of the opposing roller. The design is thus pressed into the fabric as a raised design. If the fabric is made with thermoplastic fibers, and the rollers are at the right temperature, the design will be permanently embossed. (The redult is the same as that with the creping operation described above, but with a different pattern.) If the fibers are not thermoplastic, the design can be made permanent if the fabric is treated before the operation with a thermoplastic resin. The decorative effects can be produced on plain fabrics to simulate the appearance of fabrics with woven decorations.[4]

F3i. ***optical brightening***[1] - is effected by a dyeing process. Optical brighteners (optical bleaches) are dyes that contain colorless fluorescent materials. These change the way that the dyed fabric reflects light by reflecting more blue light, giving the fabric a brighter appearance.

F3j. ***tentering*** - is a process that can be carried out at various stages of finishing but is commonly a final operation. The fabric, usually wet from some other operation, enters a frame and conveyor mechanism where it is gripped at its selvages (edges). The grippers are on moving chains and the entire fabric is eventually held in the machine. As the chain conveyor moves, the fabric is subjected to dry heat from a blast of air. At the same time, grippers gradually move outward to a specified setting. The fabric is thus brought back to its original width. Devices on the machine ensure that the weft and warp yarns of the fabric remain square to one another.[4]

F3k. ***crabbing*** - is a process for wool that has an objective similar to that of tentering. The crabbing process differs for wool in that the fabric is fed over hot rollers, then into cold water and then to a pressing station. The fabric is stretched or loosened as necessary to restore the proper width and relationship of the weft and warp yarns. The process helps prevent the development of uneven shrinkage or creases in subsequent operations.

F3l. ***flocking***[6] - is the deposition of short fibers onto an adhesive-coated surface. One method of applying flocking is to fling fiber dust mechanically onto an adhesive surface that is made to vibrate. Electrostatic charges may be applied to the fibers to align them and attract them to the electrically grounded substrate. Another method feeds the fibers into a pneumatic tube and to a nozzle from which they are sprayed onto the substrate. This method also uses electrostatic charges on the flock to aid in their alignment and attraction to the substrate. Nylon, rayon, cotton and polyester fibers are used with lengths up to about 1/8 in (3 mm). Flocking provides a soft, pleasant, fabric-like feel to solid objects and less-soft fabrics. Automobile interiors and toys are two notable applications. Carpeting is also sometimes flocked. See Fig. 10F6f.

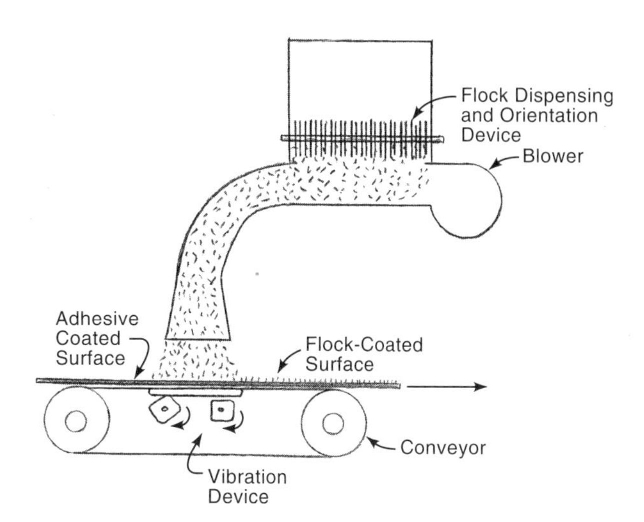

Fig. 10F6f Flocking of a sheet or fabric with a mechanical system. When three-dimensional objects are flocked, an electrostatic charge may be added to the flocking material and the work-piece may be electrically grounded to aid in the attraction and orientation of the flocking material.

F4. *finishing to improve feel (tactile properties) of the fabric*

F4a. ***sizing (stiffening)*** - Starch, gelatin, clay, glue or casein are often applied to cotton yarn and cotton cloth. The yarn or fabric is immersed in a starch solution and then dried. The starch forms a film around the yarns or individual fibers and adds weight, stiffness, smoothness and luster to the fabric. It fills the openings between yarns in the fabric. The finish is not permanent. Some cotton garments are given a starch treatment to help in keeping their appearance and feel fresh until they are sold.

F4b. ***weighting*** - is a process for adding weight and body to a fabric to improve its drape and hand. With silk, its weight can be increased by treating the fabric with tin salts. The salts permeate the yarns and become a part of the fabric, adding weight to it. The results are not permanent but the process can be repeated. Wool and napped cotton can be weighted by adding flocks, extremely short fibers, to the fabric. The flocks are driven into the fabric by air pressure.

F4c. ***fulling, (felting or milling)***[1] - is a process applied to wool fabrics to increase their thickness and compactness. The wool is heated, moistened and subjected to friction under pressure until a shrinkage of 10 to 25 percent takes place. Shrinkage is in both directions and the finished fabric is smooth and resembles felt because of its tight compaction. Sometimes, chemicals are added to aid in bonding the fibers together.

F4d. ***softening***[1] - Sulfonated oils, sulfated tallow, glycerine, dextrin or sulfated alcohols are applied to fabrics to make them softer with a more desirable feel and, often, more absorbent.

F5. ***finishing to improve performance*** - includes various finishes to make the finished fabric more usable, easier to maintain and more resistant to adverse environmental conditions.

F5a. ***anti-shrinkage treatment*** - Several techniques are used, depending on the fabric involved.

Wools are stabilized with a number of methods. With the London shrinking method, the fabric is held between wet blankets for about 20 hours. The moisture in the blankets penetrates into the wool fibers and they shrink. The fabric is then slowly dried and subjected to high pressure which stabilizes it. Another method involves chlorination. The wool is treated with a dilute solution of sodium or calcium hypochlorite. This causes some fusion within the yarn which inhibits shrinkage. The process is used in woolen socks, sweaters and underwear. Another technique is to coat the yarns with a thermosetting plastic resin. When the resin is cured, it tends to prevent the fibers from shrinking.

Cotton is treated by pre-shrinking the fabric after weaving. The process is also called relaxation or compression shrinkage. The cotton is moistened by spraying with water and then pressed against heated rollers that are coated with a thick layer of felt or rubber. In sophisticated systems, the cotton cloth is first tested for its natural shrinkage; production quantities are then shrunk to that degree. The cloth is moistened with water and live steam and held firmly against a wool blanket that is under controlled tension. The tension of the blanket and of the cotton cloth is relaxed to the desired measurements. The cotton is then run over a heated drum to dry it.

Rayon fabrics are also stabilized by a resin treatment that locks the fibers together. Another rayon treatment is to induce cross linking of the rayon molecules by acetal chemical treatment. Polyester and nylon fabrics are stabilized by heat setting the finished fabric.

F5b. ***durable press (permanent press) (wash and wear)*** - The processes used involve plastic resin impregnation of the fabric either before or after it is made into a garment. Melamine or epoxy are commonly used plastics. The plastic resin, in liquid form, impregnates the fabric, which is then dried. The finished garment is pressed and then heated further to fully cure the resin. Successful treatment makes the garment crease and wrinkle resistant and provides smooth seams, shape and pleat retention even after the garment is washed and tumble dried repeatedly. Garments woven with thermoplastic yarns, in whole or in part, have heat setting characteristics from the thermoplastic (nylon, acrylic, polyester). When these yarns are pressed and heated in the desired shape, they can retain the shape through repeated launderings.

F5c. ***antistatic treatment***[1] - is applicable to fabrics made from synthetic fibers such as, nylon, acrylic or polyester. The main effect of static electricity on garments made from these fibers is a tendency to cling. The treatment involves coating the fibers with an anti-static agent that conducts away any electrostatic charges that might occur. Several commercial antistatic agents are available.

F5d. ***treatment for soil and stain release properties*** - consist of applying a coating of fluorocarbon plastic to the yarn fibers. The procedure is similar to that used to apply plastic resins to fabrics in a permanent press treatment. The fluorocarbon plastic coating is slippery and stains and soils do not adhere to nor impregnate the fibers. The treatment of upholstery fabric for protection against soiling is a common application.

F5e. ***water repelling and water proofing treatments*** - are achieved by coating the fabric with waxes, varnishes or enamels, bituminous coatings, metallic salts or silicones. Heat treatment with special quaternary ammonium compounds produces water repellency that withstands normal cleaning processes.[7] Whether the fabric is water repellent or waterproof depends on both the amount of coating applied and the nature of the coating. The heavier the coating, the more the drape and hand of the fabric will be adversely affected, but the longer the fabric will shed water. Water repellent fabrics, however, will eventually allow rain or other water to penetrate if the exposure is long and severe enough. Waterproofing involves a full coating of material so that there are no openings between the woven yarns and protection against rain or water flow will last indefinitely. Vinyl plastic and rubber (both natural and synthetic) are materials used for such coatings. Firemen's raincoats are examples of protective garments receiving this treatment. With water-repellent coatings, garments are still porous to air flow and are therefore more comfortable to wear in most conditions than waterproof garments. They also have a more normal appearance and drape. Silicone compounds are used for many water repelling applications. Another approach is to weave the fabric with 100% synthetic yarn of a material that is, itself, water repellent. Nylon and polyester are two examples.

F5f. ***other treatments*** - to provide antibacterial and antifungal properties, flame retardance, anti-moth protection, and slip resistance can all be provided. The yarn, fabric, or the completed product, is coated with or immersed in the appropriate treatment solution. The solution is usually absorbed in the fibers and gives the product the necessary property. However, some of these treatments provide only temporary protection and some may have an adverse effect on the strength or drape properties of the fabric. Fire-retardant treatments often involve immersion in a chemical or mixture containing phosphorus, nitrogen, antimony, chlorine or bromine.[3] Cellulose fibers can be given a temporary flame-retardant treatment with ammonium salts or borax and boric acid.[7] They also can be mildew-proofed with a number of compounds including acrylonitrile, salicylanilide, organic mercury compounds and chlorinated phenols.[7]

G. Coloring

The natural color of textile fibers is only infrequently acceptable as a color for the finished textile product. Coloring the fabric with one solid color or with a decorative pattern is the norm. Dyeing and printing are the basic processes used to impart desired colors to a textile product. Both dyes and pigments are used.

G1. ***dyeing*** - involves the immersion of the fiber, yarn, cloth or finished product in a solution, usually aqueous, containing the dye. (Most fabrics can be dyed at any stage, but manufacturers generally prefer to dye as late as possible in the manufacturing sequence to minimize the risk of being overstocked with the wrong color material.) The dye saturates the fibers and is fixed by heating, aging or steaming the fabric. The dyeing operation may be performed on either a batch or continuous basis. Washing of the fabric normally follows dyeing to remove loose dye materials. Some fibers receive preparatory operations before dyeing.

Not all dyes are suitable for all fabrics. Successful dyeing requires that the dye be compatible with the fiber to be colored. It also must be suitable for the particular application. Color fastness varies with the environment that the fabric will face. Fastness to light is important in some applications, for example, in draperies or curtains. Color fastness in laundering is necessary for clothing or napkins which frequently become soiled. The earliest dyes came from natural sources, plants or minerals. Present-day dyes more often are synthetic, being derived from coal tar or petrochemicals. Other chemicals in addition to the dye itself are usually part of the dying process. They promote penetration and uniformity of distribution of the dye. There are a number of dyeing methods: stock dyeing (dyeing loose fibers in a vat), top dyeing, yarn dyeing, piece dyeing (dyeing after weaving or knitting), cross dyeing and solution pigmenting or dope dyeing. The different dye types currently in use include: acid, mordant, sulfur, azoic, vat, disperse, substantive and reactive dyes. Combinations of dyes may be used to get proper color effects or if several kinds of fibers are included in the material. Heating is a common part of most dyeing processes to aid in the transfer of the dye molecules to the fiber and to make the fiber more receptive to the dye. When the dyeing equipment is built to permit pressurized dyeing, temperatures as high as 265°F (130°C) may be used.[1] The dye bath is often agitated to help ensure that all portions of the material are contacted by the dye. Sometimes, hydrocarbons rather than water are used as the carrier of the dye. They have the advantages of faster wetting of the fiber and lower dyeing and drying temperatures.

Dyeing equipment is normally fabricated from stainless steel because dye solutions are often acidic. The equipment consists of a vessel to hold the dye liquid and the material, a means for agitating or circulating the liquid or moving the material and heating and cooling capability. In batch dyeing processes, the material to be dyed is often held in perforated containers through which the dye liquid circulates.

G1a. s***tock dyeing***[4] - is the dyeing of unspun fibers in a batch operation. A large vat is filled with loose fibers and the heated liquid dye is circulated through the mass at a high rate in order to contact all the fibers as thoroughly as possible. Even though the fibers absorb the dye thoroughly there may be some fibers with only partial penetration of dye. Nevertheless, after blending and spinning, a high quality of dyeing results and the resulting yarn has a uniform color. The operation is commonly applied to wool.

G1b. ***top dyeing***[4] - is the dyeing of the rope-like gathering of wool fibers that result from the combing operation. The rope or "top" is wound on a perforated drum and the dye is circulated through the drum holes into the rope. This method produces highly uniform dyeing.

G1c. ***yarn dyeing***[4] - is dyeing done after the fiber has been spun into yarn. It is sometimes referred to as *skein dyeing* when the yarn is coiled in a skein or on a reel. The skein or reel is immersed in the dye bath. It is left there long enough for the dye to thoroughly penetrate the fibers of the yarn. When yarn is wound on a reel, the central spindle may be perforated and the dye liquid is pumped through the holes into the yarn. Yarn is dyed as a yarn instead of a fabric (*piece dyeing*) so that it can be woven with other yarns of different colors. Plaids, checks, and stripes can be woven this way.

G1d. ***piece dyeing***[4] - is the dyeing of a woven or knitted fabric. This is the most common dyeing method, because, when the dyeing is deferred to a later stage in manufacturing, there is less chance for the manufacturer to be burdened with material of the wrong color when customer preferences change. Piece dyeing is performed on either a batch or continuous basis. The continuous basis is used when the quantities are large. Rolls of cloth are fed into the dye bath and rerolled after dyeing. Smaller quantities, and particularly wool fabrics, are dyed on a batch basis on perforated rolls. The operation, when performed under pressure, allows higher temperatures to be used and this shortens the dyeing cycle.

G1e. ***cross dyeing***[4] - is a combination process. Stock or yarn dyeing first takes place for some of the yarns that are used to make a fabric. The fabric is then woven with the desired pattern using both dyed and undyed yarns. After weaving, the fabric is piece dyed. The previously gray yarn in the fabric is given a new color and the previously colored yarns may be changed somewhat in shade but will blend with those that are newly dyed.

Some fabrics, woven of two different fibers, can be cross dyed in a bath that contains two dyes, one with affinity for each fiber. Fabrics containing both viscose and acetate rayons have been dyed this way. Special color effects can be produced.

Another approach, when the fabric has yarns of different fibers, is to use two separate dye baths, each one with affinity for one of the yarns.

G1f. ***solution pigmenting (dope dyeing)***[4] - is applicable to synthetic fibers extruded from spinneret dies. The dyeing or coloring is achieved by mixing a pigment or other colorant in the fiber material before it is extruded. This is the same method used to color non-textile thermoplastic extrusions. The method is applicable to fiber made from solid thermoplastics (nylon, acrylic, polyester, polypropylene and polyethylene) and to viscose rayon and acetate. (For rayon and acetate fibers, the materials are wood fiber or cellulose solutions rather than solid thermoplastics.) The color extends throughout the fiber and color fastness is excellent.

G2. ***printing*** - is involved when a fabric is decorated with a pattern of color or colors. It differs from dyeing in that the entire fabric is not made into the same color. Printing is very common in producing decorative fabrics used for clothing, furniture, draperies, carpets, tablecloths and other applications where some kind of decoration is wanted. A variety of methods are used, depending on the amount of fabric to be printed and the pattern wanted. The principal printing methods are: 1) block or relief printing, 2) intaglio printing, sometimes called engraved, roller or gravure printing, 3) duplex printing, 4) discharge printing, 5) heat-transfer printing, 6) screen or stencil printing, 7) resist printing, 8) warp printing, and 9) pigment printing.

The printing ink is usually applied from a thickened paste of dye or pigment. After printing, the fabric is dried to retain the sharp image through further handling. The application of steam heats the ink and causes it to migrate deeper into the fabric while the thickening agent prevents it from spreading. Most important, the ink is set or fixed, usually by the heat, and excess ink may be removed by washing the fabric. The final step is drying the fabric.

G2a. ***block printing*** - is relief printing (See 9D1) applied to textiles. Traditionally, it has been done by hand with wooden blocks carved to put a design in relief. The block, coated with an ink or dye is pressed against the cloth. This step is often repeated in different places on the cloth to produce a design or pattern. Applications are made with different carved blocks and different colors, depending on the pattern and colors wanted. The method now is limited to hand-crafted products. Relief printing with equipment similar to that used with paper is the current production outgrowth of block printing. This relief printing may also be referred to as surface, block, or kiss printing.

G2b. ***engraved printing (intaglio printing)*** - The intaglio printing process, gravure or rotogravure, described in sections 9D3, 9D3a and 9D3b, is also utilized for high-production printing of fabrics and is the most common production printing method for fabrics. When fabrics are printed , the impression roller, the roller that provides backup pressure for the fabric, is covered with a thick, resilient blanket or an endless such blanket that passes against its surface. Another endless cloth, called the backing fabric or *back gray* passes between the blanket and the fabric to be printed. The purpose of the backing fabric is to absorb any excess printing ink and protect the blanket from staining. The fabric to be printed, in web form, passes between the back gray and the printing cylinder. If multiple colors are to be printed, there may be several cylinders, one for each color, all of which bear against one large impression cylinder. This approach is illustrated in Fig. 9D7, view b, and is also known as *roller printing*. After printing, the web of fabric passes through a drying and steaming oven to fix the color. In high-production printing situations, the back gray and blanket webs are automatically cleaned of any excess ink that they absorb.

G2b1. ***duplex printing*** - is roller printing performed on both sides of a fabric. Separate printing and impression rollers are used, one on each side of the fabric. Care is taken to align the printing on both sides. The fabric, then, appears the same as one in which the colors are woven-in rather than printed.

G2c. ***discharge printing*** - is a process in which the cloth is first dyed with some background color. This color is then selectively discharged or removed in some areas with a printing process that applies chemical reagents or reducing agents in paste form. The printed area then becomes white or some other color, depending on the chemical agent and other colorants in the paste. Caustic soda or sodium hydrosulfate are typical reagents. Steaming and washing follow to remove the reagents.[4]

G2d. ***heat-transfer printing*** - is very similar to transfer coating (hot transfer coating) of plastics parts described in 4M2a except that only dyes and no plastic film is transferred to the work. The method was first used with polyester fabrics. The desired pattern is first printed on paper in one or more colors with thermoplastic dyes that are compatible with the fabric. Then the fabric and paper are run together through heated rollers with the printed side of the paper facing the fabric. The heat softens the dyes and transfers them to the fabric. Detailed patterns, including halftone pictures can be printed on fabrics with this method.

G2e. ***stencil and screen printing*** - are explained in sections 8I6, 8I7b and 9D4a, 9D4b respectively, but are both used for decorating fabrics as well as for printing on paper and on other surfaces. Screen printing is the more common method. With either process, the cloth to be printed is normally laid flat on a work table and held by clamping, pinning, or some other method while the image is applied. Separate screens or stencils are used for each color. A common manual screen printing application is the decoration of T-shirts. However, for large scale printing of fabric in web form, automatic screen-printing equipment is available. This equipment advances the web of fabric, positions the screen, dispenses ink to the screen, moves the squeegee to apply the ink and lifts the screen to allow the fabric to move to the next position. Screen printing is sometimes carried out with a rotating cylindrical screen that contacts a web of fabric as the fabric passes against it. This method is sometimes referred to as rotary screen printing. The operation is economical for printing large quantities of fabric.

G2f. ***resist printing*** - The fabric is first printed with a paste material, called a resist, a resinous material that blocks the fabric from accepting dye. The fabric is then dyed but the dye is absorbed only in the resist-free areas. The resist material is then removed chemically, leaving the fabric colored in the selected pattern. A manual variation of resist printing has been used for years in Indonesia and other southeast Asian countries to make *batik fabric*. A design is hand painted on fabric with paraffin wax or beeswax. The fabric is then dyed; the dye penetrates only those areas that are not wax coated. The wax is then removed by heat or solvent, leaving dyed and un-dyed areas on the fabric. The process must be repeated for each color.

G2g. ***warp printing*** - is printing, usually by roller printing methods, on the warp yarns before they are woven into a cloth. Weft yarns that are finer than normal are then used in weaving the fabric. The effect is a soft or shadowy presentation of the pattern on the finished fabric. This approach is sometimes used on cretonnes and other upholstery and drapery fabrics.

G2h. ***pigment printing*** - uses an insoluble pigment instead of a thickened dye. The pigment is mixed with a plastic binder that holds the pigment in place on the fabric after printing. The plastic resin is cured through the application of dry heat. Washing is not required after printing. The process is widely used. However, the fabric may have a harsh feel and the printing may eventually wear away from use of the printed goods.[3]

H. Manufacture of Clothing and Other Sewn Products ("Needle Trades")

Clothing is the largest category of sewn textile products, but footwear, leather products, upholstery, curtains and draperies, towels, bedding, flags, parachutes, table linens, tents, industrial filters, and other industrial components made of fabric all require the same basic sequence of operations when produced in quantity. These products rely on stitching as the predominant means of fastening pieces together to make the finished product. Not only cloth, but leather and plastic sheeting may be used as base materials. Rug manufacture, described below, also usually involves some sewing operations. Although garments and other sewn products are still made by tailors, seamstresses and other skilled persons who each make the entire product, production processes for these products involve a division of labor. The work consists of a series of operations, each performed by a separate person, using production-type equipment. Individual workpieces or bundles of them are moved by conveyors, chutes or other means from operation to operation. The following are the operations involved for most sewn products.

H1. ***spreading/stacking*** - In production operations, the individual pieces making up the items to be sewn are not cut individually from the fabric but are cut from a stack of many layers of fabric. All layers of the stack are cut at the same time. The stack is made by spreading layers of the fabric on a work surface. Short stacks may be spread by hand, but, if production quantities are involved, the fabric is spread from a wheeled carrier, a spreading machine. The machine carries a bolt or bolts of cloth and spreads one layer after another smoothly on the stack as the machine moves back and forth along the work bench. For uniformity of the pieces that will be cut from the stack later, it is important that the layers do not have either too much or too little tension.

Spreading machines can be primarily manual in operation, semi-automatic or fully automatic. Fully automatic machines traverse under their own power. Electronic sensors are sometimes used to superimpose each layer precisely on the stack, particularly when there is a pattern to the fabric that has a relationship to the pieces to be cut. To control tension, many spreading machines are designed with devices that unroll the bolt of cloth as the carrier moves, so that the cloth is unwound at the same speed as the carrier moves.

There are several ways in which the fabric can be placed in the stack. Often, when both surfaces of the fabric are the same, or when cut pieces can be symmetrical, the machine folds the fabric face-to-face at the ends of the stack and spreads the fabric when moving in either direction. When the fabric cannot be spread face-to-face, the machine, equipped with a turntable for the bolt of cloth, rotates the turntable at the end of each layer (after the fabric is cut), and deposits the new layer with the same face up. Fabrics with decorations, patterns or nap that must always face the same direction are spread only when the machine is going one direction. Some products require the fabric to be spread face-to-face but only in one direction. In either case, the fabric is cut at the end of the stack and the carrier or machine returns to the starting point without spreading. However, some machines have a double-deck feature so that one-way, face-to-face stacking can take from separate bolts of cloth as the carrier moves in either direction. Layers are spread alternately from each bolt. Sometimes, double-deck machines are used to spread an outer fabric and a lining alternately from the same spreading machine. After cutting, the outer fabric and lining are handled together.

In most spreading, the alignment of the layers is controlled at one of the edges, but often, when there is variation in the fabric width, the layers are aligned at the center.

H2. ***marking*** - The *marker* is an arrangement of outlines of all the individual pieces of the garment or other product to be sewn. Its purpose is to provide a guide for cutting pieces of the proper size and shape from a stack of fabric. The marking operation involves the arrangement of these patterns for individual pieces in such a way as to minimize the wastage of material between pieces. The marker may be printed or traced on a sheet of paper, a layer of inexpensive fabric, or a layer of the fabric to be cut. Making a good marker manually to maximize the yield of the material is a time consuming process and requires considerable skill. Traditionally, markers have been made by tracing full-size cardboard or fiberboard patterns on a sheet of paper that is the same width and length as the spread fabric for a production lot of the product. The marker is fastened to the stack of fabric by staples, double sided sticky tape, or by an adhesive. It then serves as a guide for the manual cutting of the individual pieces from the stack. The marker may also be developed first in miniature with accurately scaled-down patterns and then printed enlarged on the marker sheet. Perforation of the pattern outlines on the paper is sometimes used instead of ink printing. Then the perforated sheet serves as a stencil for marking the top layer of fabric. Chalk or other powder is dusted through the stencil perforations. The stencil thus produced can be used on subsequent lots of the same product.

Fig. 10H2 illustrates a typical marker, this one for several sizes of overcoat.

In well-equipped production facilities, marker preparation is now done by computer. This saves time and optimizes utilization of the fabric. The computer equipment then prints a full-scale copy of the marker on paper. When computer-controlled cutting equipment is used, a separate marker sheet is not required. The marker pattern is contained in the computer's memory instead of on a sheet of paper.

H3. ***cutting (chopping or knifing)*** - Two basic approaches are used in cutting the stack of fabric into pieces of suitable size and shape to be sewn into finished products. *Pattern chopping* is the cutting of the fabric into the exact shape required for sewing. *Block chopping* is a kind of rough cut that produces pieces close to the final shape, but which require some trimming cuts to bring them to the exact shape needed. Several methods are available to cut the stack of spread fabric into pieces. The methods available include several with manually-controlled cutters (rotary blade machines, reciprocating blade machines, and continuous band knives, similar to band saws in principle). Whether circular or straight, the knife blades may have straight, wave-like, or saw-tooth edges, depending

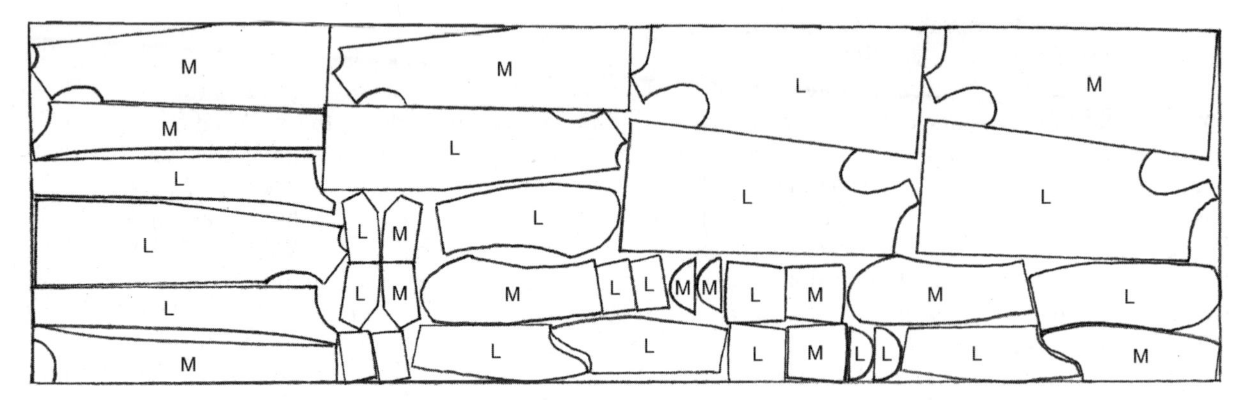

Fig. 10H2 A marker for production of overcoats of two different sizes, medium (M) and large (L).

on the material to be cut. More automatic methods include clicker die press systems and computer-controlled laser cutting equipment.

Round knife (rotary blade) and *oscillating knife* (reciprocating blade) machines are manually operated but have electric power to provide the blade motion. The cutter operator guides the knife by hand through the stack of cloth as it cuts, following the pattern lines on the marker. The result is a stack of cut pieces all the same shape and size, often with a paper piece from the marker at the top of the stack.

When a *continuous blade* (band knife) machine is used, it follows block chopping by a manual cutting machine. The block-chopped stack of pieces is brought to a nearby, more-or-less stationary, continuous blade machine to be trimmed to the precise dimensions and shape required. The stack is manually guided against the blade. (The band knife can be made narrower than a reciprocating blade and thus can cut sharper corners. When the shape of the pieces to be cut includes intricate curves and sharp corners, band knife cutting is most suitable.)

Press cutting uses metal dies (clicker or steel-rule dies) made to the shape of the pieces to be cut. The stack of material to be cut is moved to the press bed, the die is positioned on the stack and the press ram descends, forcing the die through the stack. The sharp edge of the die cuts the pieces from each layer of the stack in one press stroke. A special die is used for each piece to be cut. Steel rule die cutting is described in paragraph 2C4a of this handbook. The cost of making, handling and storing the dies limits this method to large-scale production of pieces of a certain shape and size. Parts for footwear, purses and similar items are cut with this method.

Computer controlled cutting - Three computer-controlled cutting methods are available for cutting fabric pieces. They are the vertical-blade, water-jet and laser-beam machine methods. All three methods provide cutting without the need for a paper or other separate marker sheet. The pattern of cutting is controlled by data in the computer's memory. Vertical-blade computerized cutting uses a cutting head that contains a reciprocating knife blade. The head is positioned by an X-Y mechanism and the blade penetrates the stack and cuts along the perimeter of each piece to be cut. The machine uses a vacuum table to hold the stack for cutting. A plastic film at the top of the fabric stack maintains the vacuum. The water jet method also cuts a stack of fabric, using an extremely fine, extremely high-speed jet of water. The method, which is used for other machining and cut out work, is described in paragraph M, Chapter. 3, of this handbook. Computer-controlled laser cutters operate in the same manner as laser cutting in the metalworking industries. (See Chapter. 3, paragraph O.) The laser beam traces the outline of the pieces to be cut and burns or vaporizes the fabric to separate the pieces wanted.

These computer-controlled cutting processes are fast and accurate, eliminating variations inherent in manual methods. Another advantage is that no investment is required in physical patterns or cutting dies. All the information needed to cut the pieces for a particular product, is contained in the computer data storage system. Laser cutting is particularly useful for lower quantity production including one-of-a-kind products.

Stacks of cut sections of the product may be drilled or notched to guide the subsequent sewing operations. There are two drilling methods: awl needle drilling and hypodermic drilling. *Awl drilling* (needle drilling) uses a solid tool that rotates as it is pressed into the stack of fabric. The awl may have elements that cut a hole in the fabric stack or may simply penetrate the stack spreading or severing the yarn. The purpose of *hypodermic drilling* is to leave a mark on the fabric rather than a small hole. The needle is hollow and, as it is withdrawn after penetrating the stack of fabric, it leaves a dye mark on the fabric. The dye is either a fluorescent type that is not visible in normal light but is detectable under ultraviolet light or an ink that is visible in normal light but disappears when the sewn fabric is pressed.

Tickets are affixed to each stack to identify lay and lot numbers and ensure that each final product is made from correctly matching material.

H4. ***sewing*** - The basic sewing operation in the production of garments and other textile products is the manually controlled sewing machine. "Its function is to form a chain of interlocking loops (or links) of thread around small sections of fabric." [10] The sewing machine makes each stitch and moves the fabric into position for the next stitch in the series. Industrial machines are all powered by electric motors and some reach speeds as high as 8000 stitches per minute. World-wide, there are thousands of different models of machine, many made for special purposes, such as overedging, embroidery, chain stitching, continuous seaming, and blind-stitching, as well as for certain operations such as pocket sewing, button and button-hole sewing. However, almost all have the common characteristic of relying on a human operator to obtain and position the fabric pieces, direct their movement through the sewing head, control the starting, the speed and stopping of the stitching, the movement aside of the sewn component, and the replenishment of thread. From this highly manual system, there has been an evolutionary movement towards more and more automatic operation of portions of sewing operations and also of complete operations. Most progress has been made with those operations that are more highly repetitive. The simplest automation is in machines that perform an automatic sewing cycle on fabric pieces that are manually placed in position and manually set aside after the operation. Examples are button or buttonhole sewing, bar tacking, dart sewing and pocket sewing. These machines are sometimes referred to as *stop motion machines*.[10] Next on the degree of sophistication are semiautomatic machines that perform such operations and move the sewn assembly aside afterwards. The most sophisticated machines are those that take fabric pieces from a hopper or magazine, place them in the sewing machine, perform the operations automatically, and set the assembly aside. The operator's duties are to load the hoppers and monitor the machine operation. More recently-developed machines perform these operations under computer control and can sew with variations of size and shape as dictated by the program.

Attachments for sewing machines improve the quality and add to the productivity of certain operations. The attachments usually consist of fixtures and guides to direct the fabric to the correct position and mechanisms to perform certain other tasks. They are frequently used in production sewing. Examples include hemming fixtures, seam guides, needle positioners (which control the height of the needle when the machine stops), stitching templates, thread trimmers, knives, positioners, pipers, gatherers, binders, rufflers and shirrers and devices called stackers that remove and set aside the sewn pieces. They may do this by sliding, lifting, or inverting the piece.[1] Other devices move the fabric automatically to the correct position for sewing and to a new position after the first sewing operation is completed.

Some of the operations performed by semiautomatic and more fully automatic machines are the following: buttonholing, button sewing with or without automatic feeding of the buttons, tacking, welting, dart stitching, contour or profile sewing in which curved seams are sewn automatically, and pocket setting. Backtracking and angular profiles can be sewn automatically on some machines. Sometimes, two or more sewing machines are arranged in series so that the first machine performs a sewing operation and the material is moved automatically to the second machine for another sewing operation. For products such as sheets, pillowcases and table cloths, hemming may be performed on both edges of fabric fed to two machines from rolls, and it passes through the machines for continuous sewing and cutting off. Buttonholing and button sewing machines may be made to repeat these operations at prescribed spacing on a garment. These machines are programmable so that the number of buttons and their spacing. can be changed when different garments or different sizes are sewn.

The sewing machine is a complex mechanical device that performs many functions with respect to the thread and the fabric, each at a precise instant in the sewing cycle. Each machine includes devices to maintain tension in the thread most of the time and other devices to put slack in the thread when it is needed, machine components to move the fabric between stitches and other elements to hold the fabric motionless when necessary. The sewing machine needle is precisely shaped with a hole near the point to carry the thread and a groove to contain the thread when the needle passes through the fabric. The shuttle used in some stitches must oscillate or rotate but is not solidly connected to any shafts because the thread must pass around it. Surfaces of all thread handling elements of the machine must be very smooth so as not to catch the thread during its movement.

There are two basic sewing machine stitches: *chain stitches*, which are made with one thread, fed from the top, that interlocks with itself and *lock stitches*, which use one thread fed from the top and another in a bobbin in the machine to interlace with the top thread.

Chain stitches are made by a hook-like element called a looper that is beneath the bed of the sewing machine. When a sewing needle and thread penetrate the fabric and start an upward return stroke, there is a small amount of slack in the thread. The looper catches the slack loop and moves it to a position where the next needle stroke passes through it. As this operation continues, a series or chain of stitches is formed, all with the same thread. Chain stitches are simpler and faster to sew than lock stitches and have more stretchability but have the possibility of unraveling if the thread breaks and one of the thread ends is pulled in a certain way. Fig. 10H4(a) shows a simple chain stitch.

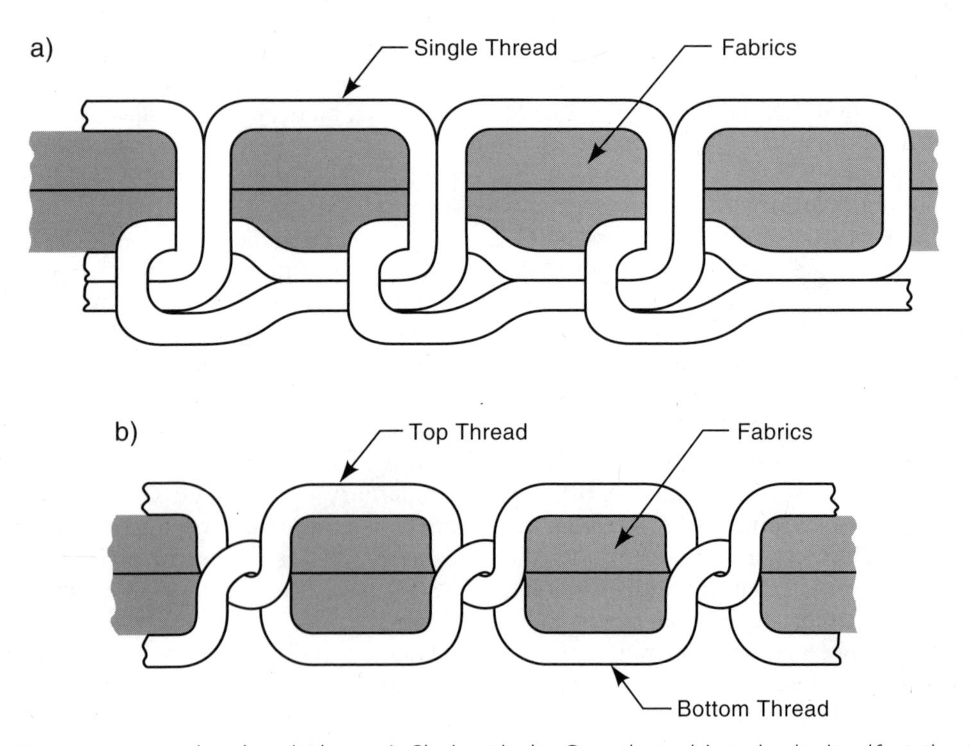

Fig. 10H4 Two common simple stitches: a) Chain stitch. One thread interlocks itself at the bottom of the fabrics being sewn. b) Lock stitch. The top thread interlocks the bottom thread. (Note: Stitches are shown very open to clearly illustrate how the threads interlock.)

With lock stitches, a moving hook in the machine bed catches the slack in the needle thread, just as a looper does, but moves it so that the shuttle, which carries a bobbin of thread, passes through the thread loop. This produces a bottom thread, below the surface of the fabric, that interlocks with the needle thread from the top of the fabric. Fig. 10H4(b) shows a simple lock stitch. Lock stitches essentially remain in place even if the thread should break.

H4a. ***stitch-less joining (seam bonding) (heat sealing)*** - Some fabrics are joined without using sewn stitches. Instead, the separate pieces are bonded together along a seam. Two basic methods are used: fabrics are either fused together, or held by an adhesive that is added. In fusing, the fabric itself is made wholly or partly from thermoplastic synthetic yarn or is coated, at least in the seam area, with a thermoplastic. Heat and pressure are applied sufficiently along the seam from a wheel or die so that the fibers soften and the fabrics fuse together. Heating for stitch-less joining can be by direct contact, infra red radiation, or high-frequency induction. Most machines used for this operation are equipped with a pair of rotating wheels that apply pressure and usually heat to the fabric as it passes between them. The machines operate like sewing machines in that the joining is along a narrow seam.

When an adhesive is added to the fabric for the joining operation, if it is a hot melt adhesive, heat and pressure are applied to soften the adhesive and cause the fabric layers to bond. Other adhesives may be cured with ultraviolet energy, evaporation, or polymerization, often with less intensive heating.

These stitch-less seams are used in garments, footwear and other products, particularly when plastic film is used instead of, or is laminated with, a woven or knitted fabric.

H5. ***pressing*** - provides smooth surfaces, pleats, creases, and other effects with a combination of heat and pressure on the sewn product. Contoured shapes may also be made by pressing, often in areas of garments and footwear. Manual ironing is one method of pressing, using irons like household irons, sometimes supplemented with devices to aid in handling the iron. *Mangles* are also often used. (Mangles are machines with a pair of heated rollers. The cloth or garment is pressed by passing it between the rollers.) For higher levels of production, presses are used. With all these methods, there is normally a timed cycle during which the pressure and heat penetrate and act on the fabric. A *buck press* is a machine with matched, somewhat contoured sections, between which the fabric product is placed. Steam provides both heat and moisture to the fabric being pressed. *Blocking* is forming on a one-piece die with pressure applied to the fabric with hand tools or steam. Heat is applied by steam, gas, or electrical resistance. Moisture may also be added and a vacuums may be used to hold the fabric against the contours of the die. The contoured sections are made for a specific portion of the garment or other product. Surface texture as well as shape can be changed by pressing operations. Hats, collars, and cuffs can be formed on presses with matched dies. *Curing* consists of heating a sewn component or product in an oven. The prime application is the setting of previously pressed creases in durable press, permanent press or wash and wear garments.[1] Steam chambers are sometimes used to remove wrinkles or creases from fabrics.

H6. ***folding, labeling, wrapping/packing*** - The completion of garments and other sewn textile products includes, in the final phases, these operations that allow the products to be labeled, inventoried and protected, so they can be handled, shipped and presented to customers with proper identification and attractive appearance. Folding prior to packaging is normally a manual operation but, in quantity production, is facilitated by equipment that holds the product and simplifies folding. A variety of boxes and cartons are available, depending on the product. Most of the boxes are shipped to the sewing factory in folded, flat form. Plastic bags, sealed, stapled or taped, are increasingly common to protect finished products until they are sold. Vacuum packaging is increasingly common and is often highly mechanized in machines that reduce moisture in each product by passing it through a heated chamber, encase it in plastic bags, apply a vacuum to remove air from the bag, make a compact package and then seal the bag. Product labels are included inside the bag or adhered to its outer surface.

I. Rug and Carpet Making

(The terms rug and carpet will be used interchangeably in this book. However, rugs often mean floor coverings that do not cover the whole floor and are usually not fastened to the floor; carpets are usually coverings of the entire floor of rooms and may be fastened to the floor.) Handmade rugs are still produced in many parts of the world. There are two prime classifications of handmade rugs: flat-woven and knotted-pile. Knotted-pile rugs are made by fastening pile tufts to an open-weave backing fabric. Short pieces of yarn are wrapped or tied around tight warp threads. The loose ends of the knots form the pile of the rug. Manufactured rugs may be woven, tufted, or knitted. Weaving is the slowest and most expensive production method of rug making; tufting and needlepunching are the least expensive and quickest. The pile of pile rugs may be in loop form or cut or there may be a combination of cut and uncut pile. The yarn of cut pile may be twisted or not-twisted; not-twisted pile provides a softer plush texture. Cut pile of varying heights give a "carved" or sculptured effect. Flat rugs have no pile and can often be used with either side up.

Wool is the major traditional carpet material but nylon, acrylics, polyester, and polypropylene are now widely used. Cotton and rayon are also used, especially for scatter rugs, bath mats and automobile carpets.[5] Backing fabrics are woven from cotton, jute, and hemp, among other fibers. Many rugs are coated on the bottom surface with a non-slip plastic or latex binder or a layer of foam rubber or plastic.

I1. ***weaving rugs and carpets*** - Notable machine-woven carpets are the Axminster, Wilton, Brussels, chenille and velvet.

Axminster carpets have a pile yarn that is inserted mechanically during weaving and bound but not knotted. The looms used require great skill and extensive time for setting up, but production is quick once the set up is complete. Colored yarns for weft strands are fitted into spools and put in an overhead storage unit. Axminster weaving is very versatile and a wide variety of patterns and colors is produced. The pile is normally made from loosely twisted yarn because the weave is not tight, simulating hand-knotted carpet The pile in Axminster carpets is almost always cut.

Wilton carpets are made on special Jacquard looms. A series of perforated cards controls the feeding of yarns which are incorporated as piles into the carpet surface. The surface yarns are held in reels referred to as *frames*. The yarn in each frame is wound on spools and is all of the same color. Piles are made by looping the desired yarn (color) for a particular location on the carpet over a wire. With single-frame Wilton, pile yarns run lengthwise, in the direction of the warp yarns and the wires run crosswise. Each wire has a very sharp knife at its end. After the yarn is looped over the wire, the wire is withdrawn and the knife cuts the loop, creating a cut pile. However, pile loops of Wilton carpets are not necessarily cut. Colors not wanted on the surface of the carpet are woven so as to be below the surface. The yarns not on the surface add resiliency and body to the carpet. The Jacquard equipment enables accurate production of intricate color patterns. Wilton carpets can be of high quality. The best Wilton carpets are those made with more frames (more colors) and a high density of tufts per unit of area.

Brussels carpets are similar to Wilton except that the wire used to make the surface loops are not equipped for cutting. The loops are left as the wires are withdrawn. Jacquard looms are used.

Chenille carpets are woven in two separate loom operations. In the first operation, cotton warp threads are used and the weft yarn is the kind wanted for the carpet surface. It may be either hard- or soft-twisted. The woven fabric is fed to a machine that cuts it into strips. The strips are held together by the cotton warp yarn and are pressed into a V-shaped cross section. The ends are fastened together, forming a long fuzzy yarn with an appearance similar to that of a caterpillar's body. This fuzzy yarn is then used as the filler yarn in a second weaving operation, producing a carpet with a pile that can be quite deep and soft, if desired, when the fuzzy strips are made wide with a soft yarn. Chenille carpets may be expensive but can be of high quality.

Velvet carpet uses the simplest weaving method. The pile for velvet carpets is made by looping the yarn that forms the pile over wire strips. These are removed as each row of loops is completed, and a sharp blade at the end of each wire cuts the loops into piles as the wire is withdrawn. Only one yarn is used and, if a pattern or design is wanted in the carpet, it is achieved by printing after weaving. Velvet carpets are inexpensive. They can be made on ordinary looms.[4] Many different surface sculptured effects can be achieved by varying the cutting of the pile.

Much carpeting with cut pile is now woven in the same manner as other pile fabrics. Two carpets are woven at the same time, face to face with a small spacing, and the pile comes from yarns that extend into, and are woven into, both fabrics. When these yarns are cut, two cut pile carpets result.

Broadloom rugs are woven on looms that are wide enough to weave the rugs to the desired width in one piece.[4]

I2. ***tufting rugs*** - is a very common manufacturing method for pile rugs. 90% of American-made carpeting is tufted. The first step is the weaving of a fabric backing or the use of a non-woven fabric as a backing. Traditionally, backings have been woven of jute, kraft-paper cord, or cotton. More recently, woven or non-woven polypropylene has been used. Polyurethane foam may be bonded to the backing. The backing fabric is passed through a tufting machine that inserts the tufting yarns into the fabric by punching it with thousands of tufting needles. The needles are similar to sewing machine needles in that each has an eye to carry the yarn through the fabric. Each needle has its own yarn supply. Pulling the yarn through the fabric and withdrawing slightly creates a small loop of yarn on the opposite side of the backing fabric. (which will become the upper, visible surface of the carpet). A looper (hook) engages this yarn loop to hold the yarn when the needle withdraws. There is a looper for each needle. As the needles repeat the punching they leave loops (tufts) of pile yarn on what will be the top side of the backing fabric, and a stitch of yarn on the underside. The tufts are then fixed in place by coating the backing fabric and the stitch with latex. Often, another layer of fabric (usually jute) is also added, along with more latex, and a foam rubber or vinyl backing. This layer adds stiffness, dimensional stability, and, with the foam backing, some cushioning to the carpet. The tufts may be cut during tufting for some styles of carpet, but shearing after tufting and backing gives a very smooth surface to the carpet. Fig. 10I2a illustrates the tufting of loop pile. Fig. 10I2b illustrates tufting of cut pile.

Tufted rugs and carpets are customarily made in 12 ft (3.6m) or 15 ft (4.5 m) widths. The patterns of tufts can take many forms of both color and texture. Bas relief patterns are commonly incorporated. The tufted loops can be left as loops, partially cut ("cut and loop") or fully sheared (plush). Shearing is common with fully-cut carpets; otherwise, cut piles from opposite sides of the loopers will have slightly different heights.

I3. ***making knitted rugs*** - Knitted rugs are made with special machines that have three sets of knitting needles. The face, backing and pile yarns are all knitted together at one time, with a method similar to that of hand knitting. The backing is then given additional body with a coating of latex. Frequently, a second backing is added for additional stiffness and dimensional stability. The pile loops may be cut but uncut loops are more common. Multiple color patterns can be knitted with these machines.

I4. ***making needlepunch carpets*** - These carpets are unwoven except for a pre-woven fabric core. A face of unwoven fibers is fastened to that core. The face is formed by punching many barbed needles through the core and a blanket or web of unwoven fibers. The needle punching permanently fastens the blanket fibers and woven core together. The surface has no pile, but the carpet is dense, thick and heavy. It can be printed with different designs and colors. A rubber cushion backing may also be fastened. When needlepunch carpet is made from polypropylene fiber, it can be used for outdoor applications. Major uses are in the kitchen, the bathroom and on patios.

I5. ***making hooked rugs*** - Hooked rugs are hand made by cutting and twisting outworn woolen or cotton cloth or other rags into thin strips and pulling them, with metal hooks, through a woven backing cloth. The backing cloth is woven from burlap, cotton or linen. The strips are pulled to form slightly raised loops and they may then be sewn together. The method permits the inclusion of colorful patterns and pictorial decorations in the rugs. Hooked rugs are used as throw mats in smaller areas of houses.

I6. ***making braided rugs*** - Colorful strips of outworn piece goods are braided together and wound

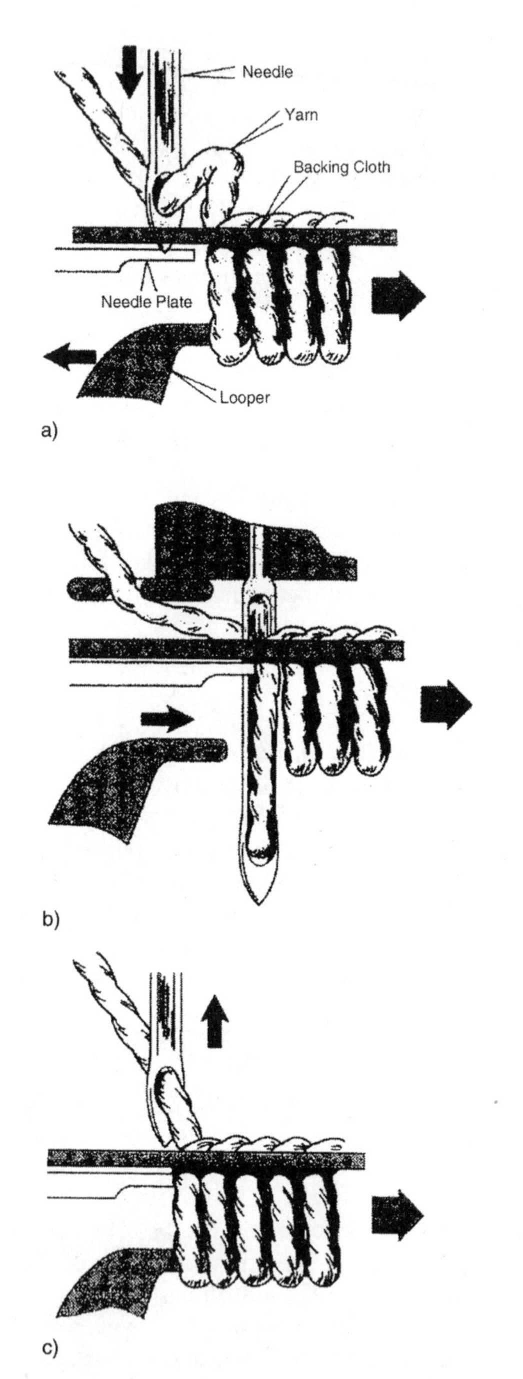

Fig. 10I2a Tufting rugs with loop pile. Note that the operation is done with the carpet upside down. In view a), the needle descends through the backing cloth as the looper moves out of the way. When the needle has fully penetrated the yarn, the looper advances (view b) and catches the yarn, forming a loop as the needle withdraws out of the cloth (view c). (*Courtesy Cobble Tufting Machine Co.*)

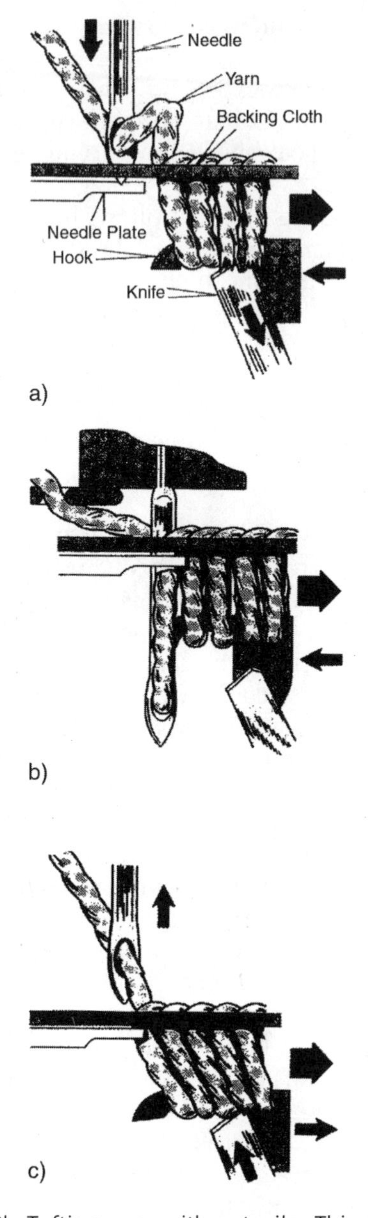

Fig. 10I2b Tufting rugs with cut pile. This operation is also done with the carpet upside down. Note that the looper faces the opposite direction than that used when loop pile is made (Fig. 10I2a). In view a), the looper has just entered the new loop. Loops already on the looper slide along the looper's length as the looper and fabric move in opposite directions. The knife, which has cut the rearmost pile by acting against the looper like a scissor blade, is moving downward. In b), the looper is picking up a new loop and the needle is about to withdraw upward. In c), the needle is moving upward, the looper is moving backward and the knife is moving upward to cut another loop. (*Courtesy Cobble Tufting Machine Co.*)

in concentric flat spiral ovals or circles. The ovals are stitched together.

I7. ***making oriental rugs***[9] - Traditional oriental rugs are made by hand, usually on vertical looms. Working from the bottom up, the worker weaves a backing cloth and inserts short yarns, loops them over the warp yarns, and ties them to create piles. The starting pile yarns are 2 to 3 inches (5 to 7.5 cm) in length. Special Persian or Turkish knots are used. Pile densities range from 50 to 400 per square inch (8 to 62 per sq cm). The more dense arrangements are more valuable. After the piles are completely inserted, they are sheared, and the warp threads at the ends are tied into fringes. Wool yarns are the principal pile material, although goat or camel hair and some silk may also be used. The backing is woven from cotton or sometimes from wool yarns. Pile yarns are dyed with dyes made from natural sources—flowers, roots and insects.

I8. ***making needlepoint rugs*** - is another hand-made approach. Each intersection of a woven canvas backing cloth is covered with individual stitches of wool yarn.

References

1. *The New Encyclopedia Britannica,* 15th edition, 1989, vol. 21 - Industries, Textile.
2. *Materials Handbook,* 14th edition, G.S. Brady, H.R. Clauser, J.A. Vaccari, McGraw-Hill, New York, 1997, ISBN 0-07-007084-9.
3. *McGraw-Hill Encyclopedia of Science and Technology,* 8th ed., McGraw-Hill, New York, 1997.
4. *Textiles: Fiber to Fabric,* B.P. Corbman, D. Potter, McGraw-Hill, New York, 1967.
5. *Encarta Encyclopedia,* Microsoft, 2003.
6. *Plastics Engineering Handbook of the Society for the Plastics Industry,* 5th edition, Van Nostrand Reinhold, New York, 1991, ISBN 0-442-31799-9.
7. *Shreve's Chemical Process Industries,* 5th ed., G.T.Austin, McGraw-Hill, New York, 1984
8. *Knitting Technology,* David J. Spencer, Pergamon Press, Oxford, England, 1983, ISBN 0-08-024762-8.
9. *Grolier Encyclopedia,* Grolier Electronic Publishing, Inc., 2002, Danbury CT.
10. *Apparel Manufacturing Handbook,* Jacob Solinger, Van Nostrand Reinhold Co, New York, 1980, ISBN : 0-442-21904-0.

Chapter 11 - Chemical Processes

A. Batch Processes in General

In batch processing, a fixed quantity of the materials to be processed is placed in the process equipment (the reactor, still, mixer, crystallizer, or other processing device) and the operation proceeds to completion with no material being removed until all the material in the equipment (the batch) is completed. There may be a number of sequential operations, performed in several pieces of equipment, but all material in the batch is processed together and kept together until all operations are completed. Batch processing is used in research, development, and experimental work, since batches can be small and variations in conditions can be tried easily on different batches. Batch processing is also used in production when quantities are not very large or, at least, not large enough to justify the cost of developing and building continuous process equipment. Batch production is most common when specialty chemicals, pharmaceuticals, or foods are produced. In the production of many products, both the materials and the processing steps and the product made are the same from batch to batch. These batches are designated *cyclical batches*. In other cases, with *multi-grade batches*, there may be minor changes in the materials from batch to batch, but the processing conditions are identical. With still another variation, *flexible batches*, both the materials and the processing steps may vary somewhat from batch to batch[5]. For high-quantity production, as exists for commodity products, however, and compared to continuous processing, batch processing may require larger equipment and correspondingly larger amounts of work-in-process. Labor costs are usually higher, due to the handling of material and the need to set, adjust, and monitor operating parameters for each batch.

B. Continuous Processes in General

In continuous processes, raw materials are fed steadily into the processing equipment and the finished product or material continuously exits the equipment. Automatic equipment monitors and adjusts flow rates, pressures, temperatures, and other process parameters. For high-quantity production, continuous processing is far more economical, per unit of production, than batch processing. Equipment size and work-in-process quantities are both smaller for a given level of production. However, there is a development cost involved in establishing a continuous process, and an investment is required for the necessary equipment and controls. These factors demand high quantity production; otherwise, the capital costs of the process cannot be amortized. Process control of conditions must be exacting, but computer control can make this feasible.

When a process is primarily continuous, but there are some elements that occur periodically rather than continually, such as the renewal of a catalyst, or removal of some material, the process is considered to be *semibatch*.

C. Separation Processes

Separation precesses are those that change the proportions of materials in a mixture. The starting material is a solution or other homogeneous mixture, or a heterogeneous mixture such as a suspension of solids in a liquid, or a mixture of various solids. Separation processes include one or more of the following operations: distillation (various methods), absorption-stripping, extraction and leaching, expression, crystallization, precipitation, fluid-particle separation methods, various solids separation methods, adsorption and ion exchange, electrolytic processes, electro thermal processes, and drying.

C1. ***distillation*** - is a means of separating the constituents of a mixed liquid by taking advantage of their different boiling points. It is also used to separate liquids from non-volatile solids. Distillation is the most common industrial method for separating liquid mixtures. The mixture is heated to the temperature at which the most volatile component vaporizes. However, that temperature is not high enough to cause all the other liquids in the mixture to vaporize. The vapor is conducted away from the heating chamber and cooled to a low enough temperature to cause it to condense. The vapor that condenses becomes a single-liquid material provided its boiling point is sufficiently different from that of the other materials in the liquid mixture.

If the liquids to be separated have boiling temperatures that are close together, the removal procedure may require two or more stages. With such a procedure, the condensate from the first stage is heated again to the boiling point of the liquid to be removed and the process is repeated. In some cases, many stages of repeat distillation may be needed. One example is the distillation of whiskey and brandy, where the initial fermented mixture may contain only 5 or 10 percent alcohol and the final liquor is perhaps 90 proof (45% alcohol). Several stages of distillation are needed for such a separation because the boiling point of alcohol, 173°F (78°C), is quite close to the 212°F (100°C) boiling temperature of water. If the mixture contains several different components, the distillation process can be repeated at different temperatures until all desired liquids are removed. Another common application is the distillation of fresh water from sea water. This is normally a multiple stage operation. Petroleum processing involves distillation to separate various fractions including naphtha, kerosene and lubricating oil from crude oil. Petrochemicals and natural gas are commonly processed or purified by distillation.

Distillation is also used when a gas is cooled enough to assume a liquid form and then is heated or allowed to warm, to boil off constituent gases. The distillation of air is one example of this approach. It is used to separate air into nitrogen, oxygen, carbon dioxide, argon, and other minor constituents.

Batch distillation is applicable when quantities are limited, in laboratory work and in larger-quantity production when solid materials, that are difficult to handle or clean from the apparatus, are processed. Materials that yield tars, resins, and other materials that clog the equipment, also lead to batch operation. With batch distillation, a particular amount of feed liquid is charged into a distillation vessel, and sufficient heat is applied to vaporize the desired fraction from the liquid. A still may be connected into the system and reflux may be used. (Reflux is recycled distillate. See fractional distillation below.)

The more-volatile fraction removed from the original liquid is called the *distillate*. The less-volatile fraction from the bottom of the distillation column is referred to as *residue or bottoms*. Distillation equipment consists of a still (retort), the container in which the liquid is heated, a condenser to cool the vapor, causing it to become liquid, and a receiver to hold the distilled liquid (distillate). Fig. 11C1 illustrates a simple distillation arrangement.

Variations in the basic distillation process include fractional, vacuum, multiple effect, and steam distillation.

C1a. ***fractional distillation (rectification, fractionation, or enrichment)*** - is a method that gives the effect of several stages of distillation in one continuous operation. It is useful when the liquids have boiling temperatures close to each other, so that simple distillation does not produce products of satisfactory purity. The process uses a high, insulated, distillation column with a series of horizontal or almost horizontal plates or trays. There is a heat source at the bottom of the column and a cooled distillation chamber is fed from the top of the column. Hot vapor from the bottom of the column rises up the column and either passes through

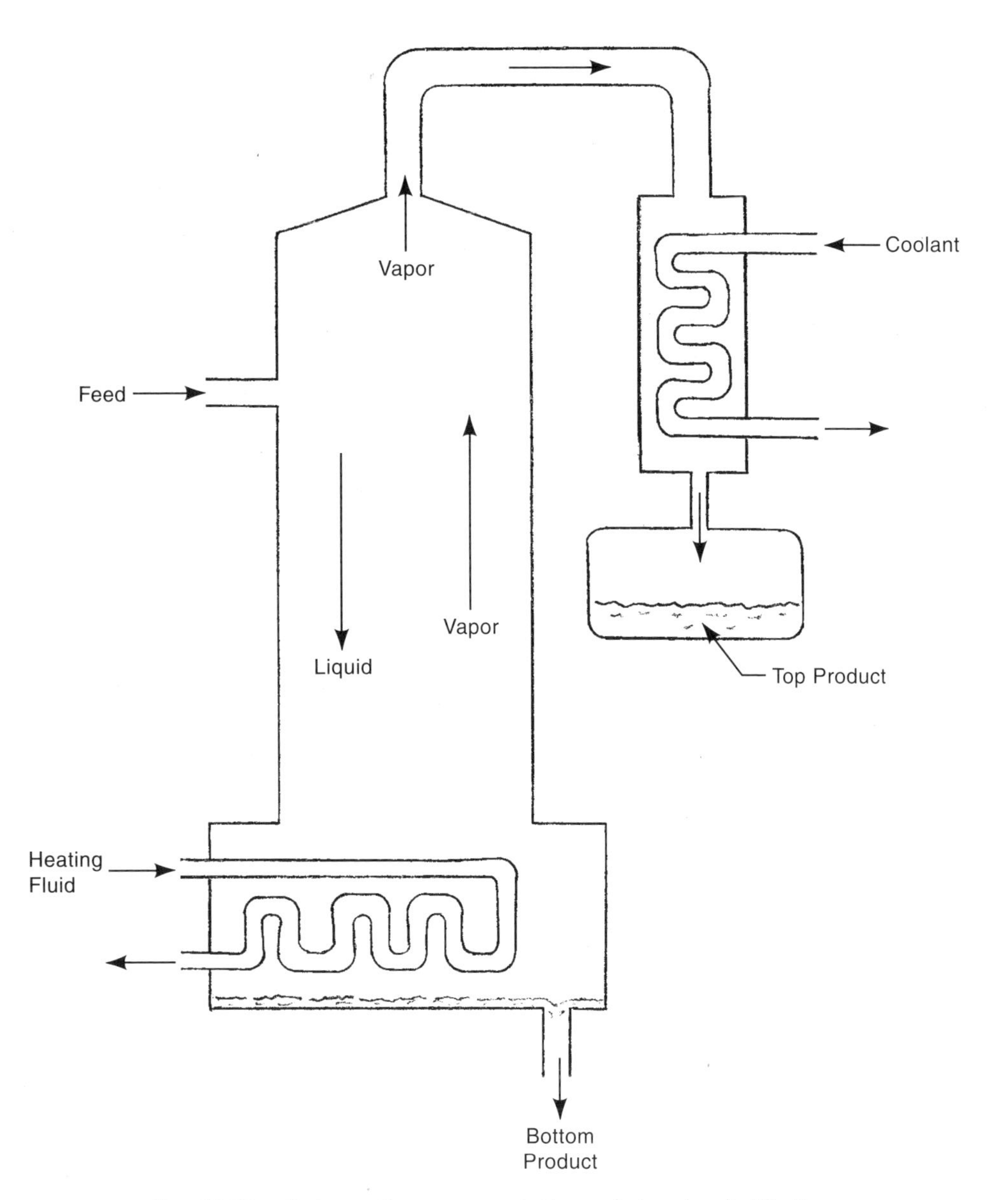

Fig. 11C1 Schematic representation of simple distillation.

perforations in the plates or zig zags past the plates, which are fastened alternatively to opposite sides of the column walls. The hot vapors partially condense on the plate surfaces. A portion of the distillate, referred to as a *reflux*, is also piped from the condenser to the top of the column. As it drips down the column from plate to plate, and as condensate on the plates drips down, it contacts the vapor which is rising in the column. There is an interaction between the liquid and the vapor. Some of the vapor condenses and some of the liquid vaporizes. As the distillate and reflux move down the column, they lose more and more of their more volatile fractions, and as the vapor rises, the cooler liquid dripping down absorbs more and more of the less volatile fractions. In the bubble tower, the most common arrangement, the plates are angled successively at opposite angles rather than being mounted horizontally so that both the liquid and the vapor pass over the full length of plate. The purified distillate, ("top product"), is collected from the condenser in a reflux accumulator and the remaining liquid ("bottom product") is collected from the reboiler at the bottom of the column. Fractional distillation can produce 95 percent alcohol (industrial grade alcohol) in one operation. The process is used for both simple and complex mixtures. Petroleum processing uses fractional distillation, and several different fractions are drawn off at different levels of the column. In the separation of isotopes, as many as 500 plates may be in the column.[1] The process is used to obtain oxygen from liquid air. Columns range from a few inches to 40 ft (12 m) in diameter and heights from 10 to 200 ft (3 to 60 m). Pressures range from a few millimeters of mercury to 3000 psi (21 MPa) and temperatures from –300 to +700°F (–180 to +370°C)[4]. In complex multiple-product columns, some feed material may enter the column at various levels and vapor of different products may be removed at several levels. Fig. 11C1a illustrates fractional distillation. Some authorities use the term *fractional distillation* to refer to any

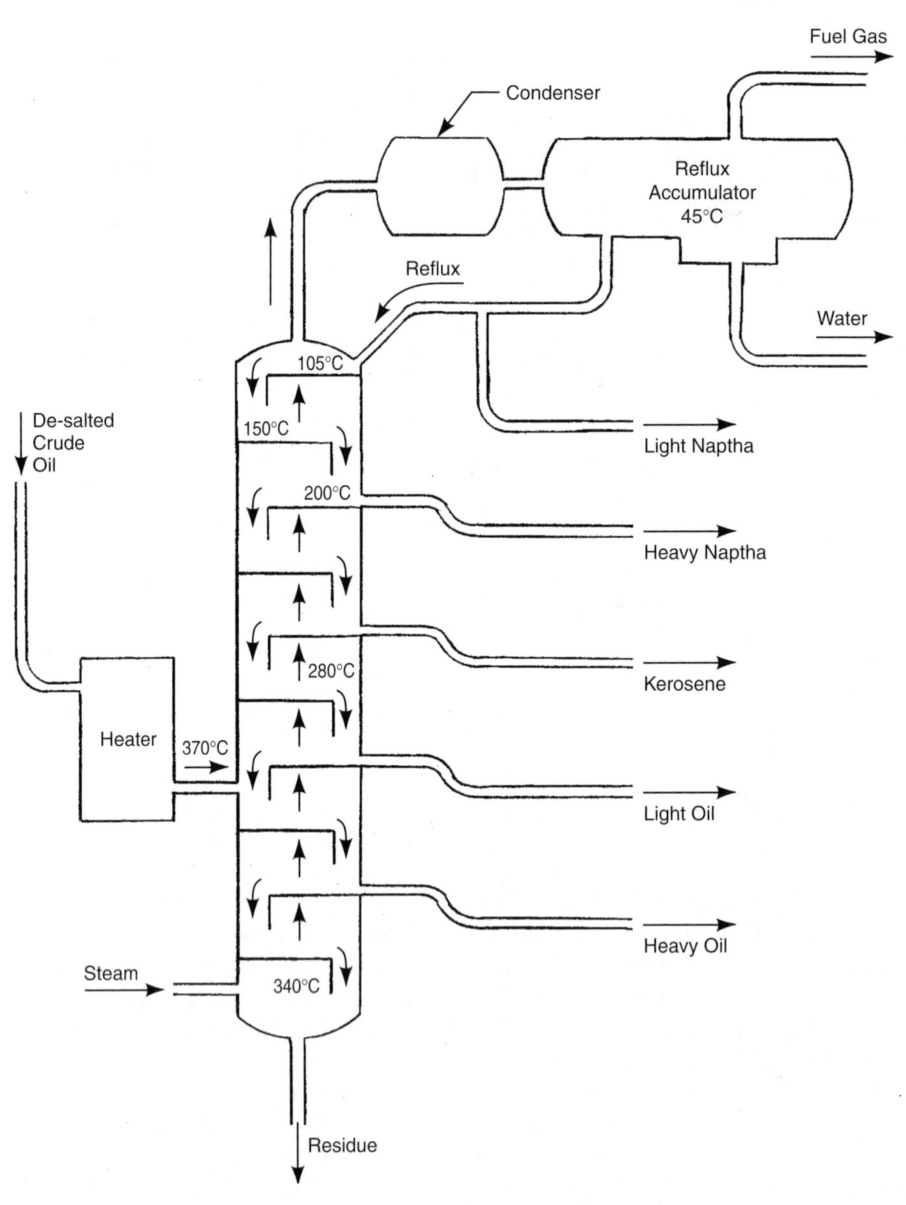

Fig. 11C1a Fractional distillation of crude oil.

process that performs the separation of several components, and *rectification* to the process with the counterflow described above.

An equivalent process is the packed column, which uses a packing of small, complex-shaped metal, glass or ceramic objects, instead of horizontal plates, to provide ample surfaces for the vapor and reflux to interact.

C1b. ***vacuum distillation*** - allows distillation to take place at temperatures below the normal boiling points of the liquids involved. A partial vacuum in the distillation chamber causes the liquid to boil at a temperature lower than it would at atmospheric pressure. The greater the vacuum, the lower the boiling point. When there is a full vacuum, the process is called *molecular distillation*. Vacuum distillation is used to purify substances that would be adversely affected by normal distillation heat, and for other substances when the distillation temperature at normal atmospheric pressure would be inconveniently high. Vitamin purification is one application. In one arrangement, the process is performed in a chamber with the material to be processed in one container that is heated. Another container is the receptacle for the vapor and is chilled. The vapor leaves the heated container and condenses in the chilled one with little loss. Vacuum distillation is common in making lubricants from petroleum.

C1c. ***flash distillation (equilibrium distillation)*** - uses relatively simple equipment illustrated schematically in Fig. 11C1c. The feed liquid, a mixture of two or more liquids, is heated and put under pressure. Then it is fed through a pressure reducing valve and into a vessel (*flash drum*) where, because of the reduced pressure, much of

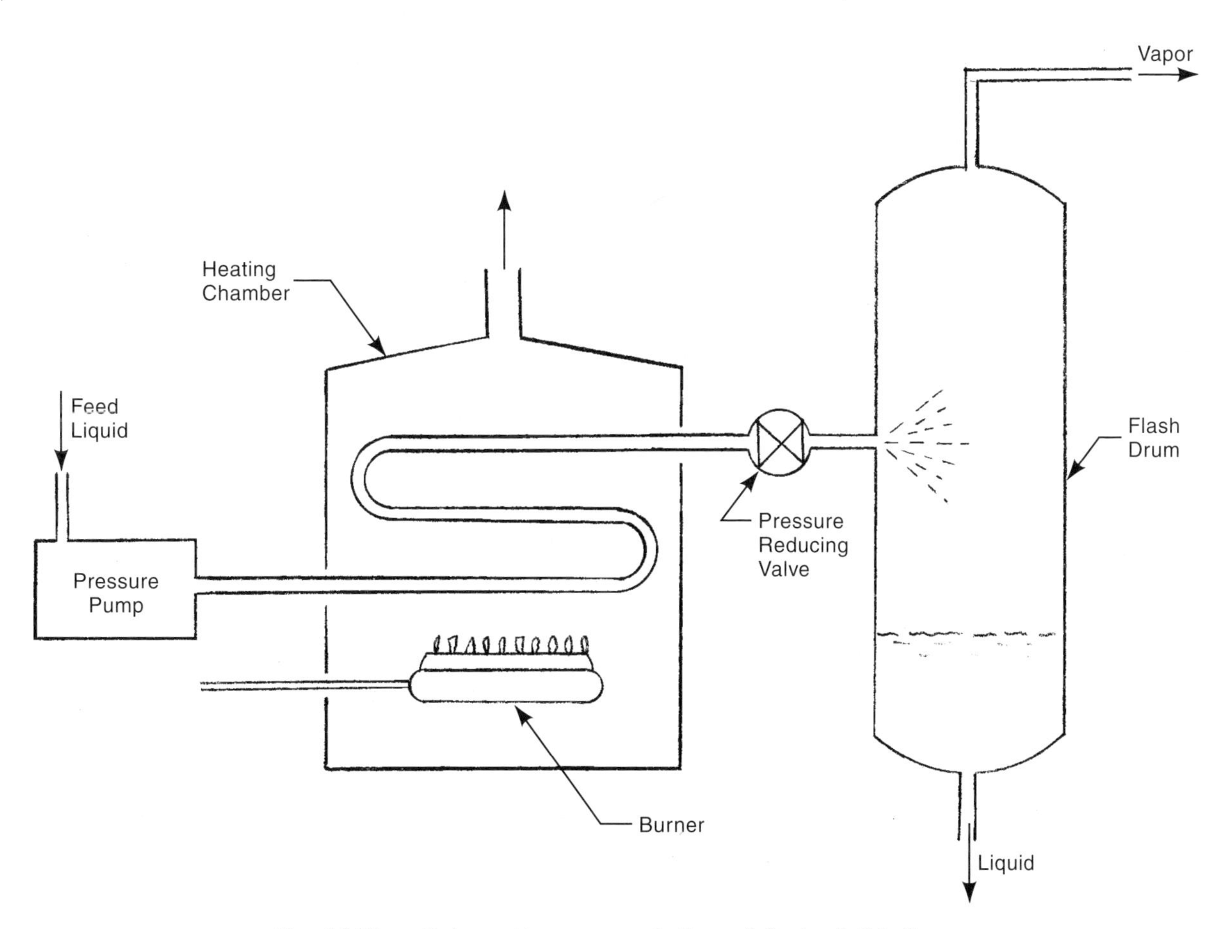

Fig. 11C1c Schematic representation of flash distillation.

the volatile fraction becomes vapor. The flash drum is large enough so that the vapor and liquid separate. The vapor and the residual liquid are then in equilibrium. Vapor is piped from the drum, usually into a condenser. Flash distillation is a common operation, used in cracking petroleum to produce gasoline, kerosene, lubricating oils and asphalt. The method is also used in the production of alcohol, the desalinization of water, and the manufacture of perfume spray.

C1d. ***multiple-effect distillation (multistage flash evaporation)*** - uses a series of vacuum chambers, each with a lower pressure than the one that precedes it. The method does not require heating, though it is used to facilitate the operation. The feed liquid passes from chamber to chamber, and vaporizes further in each. This method is used extensively for desalting water. When salt water is processed, the cooling for condensing the fresh water vapor is supplied by sea water that is thereby preheated by the process. A desalting plant for sea water may have as many as 40 stages of evaporation.

C1e. ***steam distillation*** - is another method for separating materials at a temperature below the normal boiling point of the liquid to be distilled. Steam reduces the partial vapor pressure of the liquid, enabling distillation to take place at a lower temperature. The method is used to remove essential oils from plants. The plant being processed is crushed and/or chopped to open the oil-bearing plant cells. The resulting material is mixed with water or suspended from a grid above the water. Steam is introduced to the water, causing the water to boil and the essential oils to volatilize. The water prevents the plant oils from being overheated, and aids in heat transfer. The oils are carried away by the steam vapor to a condensing chamber. The condensate is a mixture of oil and water and the two are separated easily by gravity. The method is also used to separate high-boiling-point substances from nonvolatile impurities or to remove volatile impurities from substances with still higher boiling points. The steam distillation process is limited to substances that are immiscible with water. However, it is not necessary that the diluting vapor be steam. Theoretically, any liquid that is immiscible with the product to be recovered can be used.

Cinnamon, camphor, anise, clove and peppermint oils are produced with steam distillation. Perfumes and fine organic chemicals are purified by this method.

C1f. ***sublimation*** - In sublimation, a solid material passes directly to a vapor without going through a liquid phase. Upon cooling at the same or low enough pressure, the vapor becomes a solid again without going through a liquid phase. Sublimation is favored if the pressure is low and the temperature is high. Heat must be added to the material to be sublimated to cause the phase change to take place. Sublimation is used to purify a substance or, with condensation of the vapor, to separate and save a fraction of the original material. The process is the same as distillation of a liquid except that special steps are taken to avoid clogging the equipment with solid material. Iodine, naphthalene, and sulfur are purified by sublimation. Freeze drying of food is a sublimation process. The frozen moisture in the food is removed in a high vacuum without becoming liquid.

C1g. ***destructive distillation*** - is decomposition of a material due to heat in the absence of air. In the process, the products of the decomposition are separated and collected as in regular distillation. The material to be distilled is heated, in an inert atmosphere, to a temperature high enough to cause chemical decomposition. Destructive distillation is used to make coal gas, benzene, naphthalene, phenol, coke, and various hydrocarbon materials from coal, in the cracking of crude oil to make gasoline, and for making alcohol, acetone, acetic acid, pine oil, wood tar and charcoal from wood. Oil shale is another material processed by this method to reduce the viscosity and increase the hydrogen content of the oil contained in the shale[4]. In the destructive distillation of coal, temperatures range from 930 to 1830°F (500 to 1000°C), the range depending on what products are wanted. Higher temperatures produce more gaseous products and lower temperatures more liquids. Wood is no longer destructively distilled for liquid chemicals but charcoal for metallurgical processes is made by this method[4].

C2. ***absorption-stripping*** - is a two-step process for removing a minor constituent (and sometimes more than one minor constituent) dissolved in a feed gas stream and then for recovering that constituent in a concentrated form. The purpose is either to purify the gas or to recover the minor constituent for other use. Absorption is generally used to separate materials with high boiling temperatures from gases. Two separate towers are used, as illustrated in Fig. 11C2.

The first step, *gas absorption*, is the operation that dissolves soluble components of a gas mixture in a liquid when the gas and liquid come into contact. The operation utilizes an absorption tower. The feed gas is most frequently fed into the tower from the bottom. As it ascends in the tower, the gas encounters a liquid solvent that absorbs the unwanted constituent. The liquid solvent is fed from the top

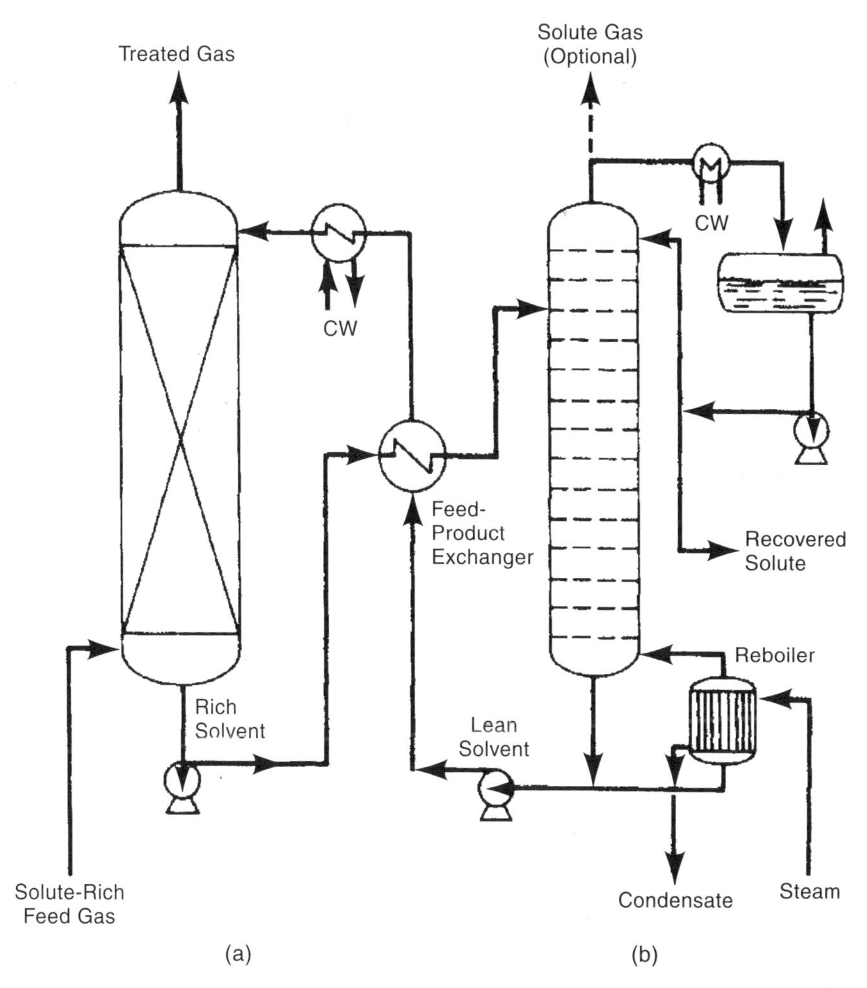

Fig. 11C2 The illustration shows the operation of an absorber-stripper. The feed gas, rich in a solute, enters the absorption tower, (a). There it contacts the solvent, which is flowing downward in the tower. Solute is transferred from the gas phase in the feed gas to the liquid phase in the solvent. This purifies the feed gas. The solvent, containing the solute, is pumped to the top of the stripping tower, (b), where it flows downward over shelves and trays attached to the walls of the tower. There, the solvent contacts a second gas that preferentially absorbs the liquid that was a solute in the feed gas, stripping the solvent of that liquid. The solvent is then recycled back to the absorption column. The solvent gas with its new gaseous solute is condensed or otherwise processed to separate it from the solute vapor. *(Reproduced with permission from Perry's Chemical Engineering Handbook, R.H. Perry and D.W. Green, McGraw-Hill, 1997.)*

and flows downward into and over a series of devices that put the feed gas and solvent in contact. (The devices may consist of a packing of irregularly shaped small plastic, ceramic or metal pieces, sprayers, or cross-flow plates or trays. They all have the function of increasing the surface area of the liquid in contact with the gas.) Contact between the gas and the liquid causes the gaseous solute to be absorbed into the solvent in liquid phase, leaving the feed gas free of solute. The treated gas leaves the absorption tower at the top; the liquid solvent, containing absorbed gas, exits from the bottom of the tower. Absorption towers are sometimes operated at pressures higher than atmospheric to increase the transfer rate and capacity of the tower. For the same reason, the operation is often carried out at a lower temperature, just above the freezing temperature of the solvent. Absorption is a common separation method in chemical manufacture, but is second in usage to fractional distillation. However, fractional distillation requires considerably more energy than gas absorption. One major application of gas absorption is the removal of carbon dioxide and hydrogen sulfide from natural gas. Another is the separation of ammonia from an air-ammonia mixture, using water as the solvent. A third is the removal of benzene, toluene, and xylenes from coke-oven by-product gases[4]. Gas absorption is also used to remove valuable components from the vapors that result from crude oil processing. The ideal solvent for absorption use has a high solubility for the gas to be absorbed and a low solubility for basic gas of the feed stream and other gases that are to be retained.

The second step, *stripping* or *desorption* or *sparging*, is the inverse of absorption. Volatile components in a liquid mixture are absorbed into a gas when the gas and liquid come into contact. The operation utilizes the second tower. The liquid solvent from the first step, containing the unwanted constituent, is pumped to a stripping tower where it flows downward over or through a series of devices that ensure contact between the liquid and another gas that has an affinity for the unwanted component and that is flowing upward in the tower. Conditions in the tower are maintained to facilitate absorption, so that the unwanted solute becomes a component of the solvent gas mixture. In many cases, the stripped solvent then is pumped to the first tower for reuse and this re-processing is one of the purposes of the operation. Some solvent may be vaporized to provide the stripping gas in the second tower. The other purpose of stripping may be to recover the minor constituent gas in a more concentrated state.

Absorption is also used for the removal of trace quantities of impurities such as xylenes from air or other gases. The operation then is called *scrubbing*. Stripping also removes volatile organic substances from water[2]. Nitrogen, steam or air are among the stripping gases used, depending on the materials involved in the process. Some systems remove multiple constituents from a gas, for example, several hydrocarbons from air. Other systems may involve chemical reactions during the gas absorption, after the solute has changed to the liquid phase, to remove the solute from the gas mixture more completely. Applications of this sequence are the removal of hydrogen sulfide from natural gas with aqueous ethanolamine, the scrubbing of gas discharges from coke ovens with dilute sulfuric acid to recover ammonia and the scrubbing of flue gases from coal-fired electrical generating stations to remove sulfur dioxide by reacting it with a solution of sodium carbonate[6].

C3. ***extraction and leaching*** - Extraction is an indirect separation process for mixed liquids that is sometimes used when distillation or evaporation is not feasible due to a narrow difference of boiling points, or if one of the materials would be degraded by the temperature required for distillation. The method utilizes a solvent that preferentially dissolves one or more components of a solid or liquid mixture. The targeted constituent material (*solute*) is transferred from the feed liquid to a solvent that is immiscible with the feed liquid (or, if the feed material is a solid, is not a solvent for its basic material). In some operations, notably the recovery of metals from ores, there is a chemical interaction between the solvent and the solute.

With liquid-liquid extraction, using *tower extractors* without agitation, the heavier liquid - either the feed liquid or the solvent - is fed from the top of a column or tower and the lighter liquid from the bottom. The heavier liquid flows downward over a series of baffles or sieve plates, or through packing designed to ensure the maximum contact between the liquids, and maximum transfer to the solvent of the constituent material. The lighter liquid may be sprayed into the heavier liquid to promote contact. Sometimes mechanical mixing is used to ensure thorough contact. If the feed liquid and solvent are not separated by the column or tower, they can be decanted, and separated by settling, into two liquids. After the solvent has dissolved the material to be separated from the feed material, it is removed from the solvent, usually by distillation. The extraction process is used only when the solute is easier to separate from the solvent than from the feed liquid, when the solvent can be recovered and when the loss of solvent is minimal. Distillation is usually simpler and less costly than extraction and is preferred if it can be used. Fig. 11C3 shows a schematic representation of extraction and distillation.

There are other extraction approaches. Some towers are equipped with reciprocating trays or rotary agitators. Others use valves that promote pulsing action. In the *mixer-settler* method, the feed liquid and solvent are agitated by turbine or propeller mixers. Then the liquids are allowed to separate into layers by gravity. Extraction can be carried out on either a batch or continuous basis, depending on the production quantity involved. Exctraction is used in separating caffeine from coffee, and some aromas and flavors from foods using supercritical carbon dioxide as a solvent. It is also used to remove waxes, sulfur compounds, and other unwanted materials, from high quality lubricating oil and in the production of penicillin and other antibiotics, acetic acid, and caprolactam (monomer for nylon-6). Extraction is also used in the production of copper, uranium, cobalt, nickel and vanadium, and in the recovery of tar acids from crude coal tar oil[4]. A further application is the removal of organic pollutants such as phenol, resorcinol, and cresol, from industrial waste water using a hydrocarbon solvent.

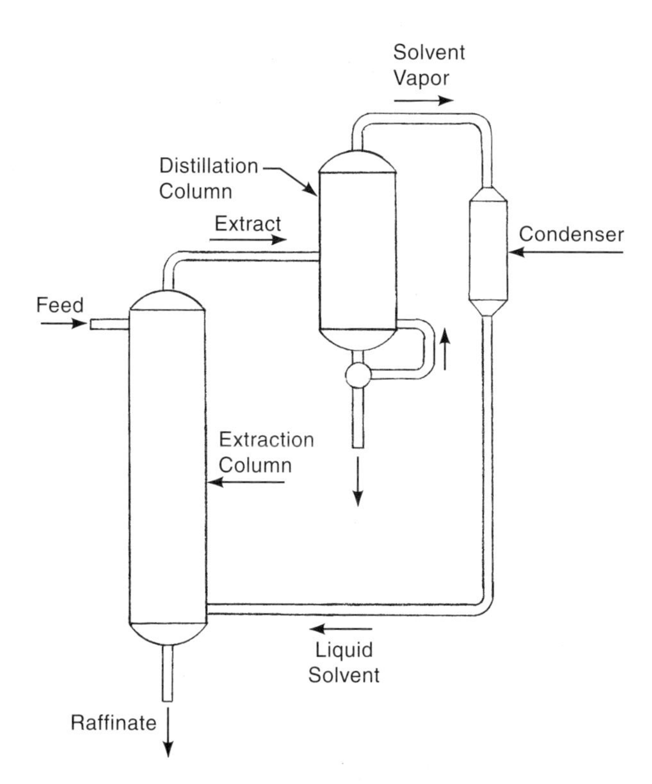

Fig. 11C3 The extraction process. A mixed liquid or solid is fed to the extraction column where one or more of its constituents are dissolved in a solvent. The solvent is piped to a distillation column and condenser where the dissolved constituent is separated from the solvent and discharged. The solvent is then reused.

Sometimes, several stages of extraction are used to ensure maximum separation. The modified feed material exiting from the extraction equipment is known as *raffinate* and may undergo stripping (See C2 above) to remove any solvent it contains. The solvent, after it has dissolved the desired material is called the *extract.* The extract from one stage of extraction may be used as feed material for a subsequent stage. This method is called *crosscurrent extraction* and can have two or more stages. Sometimes, in *fractional extraction*, different solvents are used in two stages.

When the feed material is solid, the terms *leaching* or *solid extraction* may be used to designate the removal of a soluble constituent by selectively dissolving it in a solvent[5]. The balance of the treated material is not dissolved. The feed material may undergo particle size reduction to provide more area of contact between the feed material and the solvent. One of two basic methods is used to apply the solvent to the solid material. Either the solvent is allowed to percolate through the feed material, or the material is dispersed into the liquid. With percolation methods, the feed material is contained in a chamber with a porous or screened bottom. Solvent sprayed into the chamber flows through the solid material, dissolving the wanted fraction, and exits through the porous bottom. In continuous operations, this operation may be carried out by conveying the feed material past a series of solvent spray heads. At the end of the series, the spent solids are discharged from the end of the conveyor. The flow of the solvent is counter to that of the solid material. The freshest solvent is applied to the nearly spent solid material, and is recycled to spray more solid material closer to the head of the conveyor. The richest solvent is sprayed on the material just entering the system. This approach makes the most efficient use of the solvent. Counterflow of solvent is also used in processes in which the feed solid is immersed in the solvent. Fig. 11C3-1 shows such an arrangement.

The second principal method, *dispersed-solid leaching*, involves thorough agitation of the feed material and solvent mixture. The agitation scatters and circulates the solid material and insures thorough contact between solid material and solvent. Then the mixture is allowed to settle to allow the immiscible materials to separate. Centrifuges may be used to hasten the settling.

Leaching is used in the processing of copper and other ores, the extraction of oil from soybeans, the extraction of sugar from sugar beets by using water as the solvent, the production of tannins from tree bark, and in processing Chilean nitrate. A long-standing example of leaching in hydro metallurgy is the dissolution of alumina from bauxite ore with a caustic pressure method[4]. Copper and uranium are leached with a microbiological leaching liquid containing the bacterium thiobacillus ferrooxidans[4].

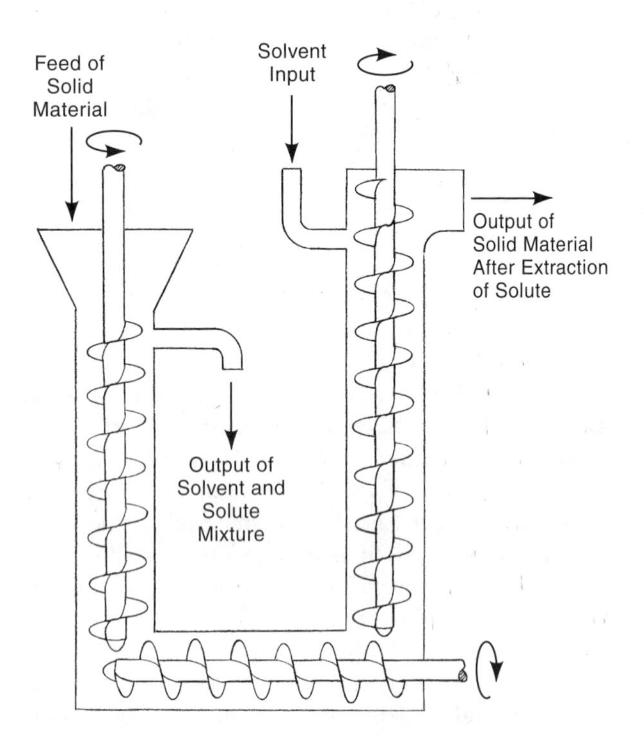

Fig. 11C3-1 Counterflow of solvent and solid material in one method of solid extraction.

In all extraction and leaching processes, the final step is the separation of the solvent from the dissolved material, the *solute*. They are separated by distillation, evaporation, precipitation, membrane separation or crystallization. The solvent, then, normally is reused.

C4. ***expression*** - consists of pressing out, by various methods, liquids contained in a solid material. Expression is frequently used to obtain fragrances or flavors from fruit or other plant material. Its purpose may be to dry the solid material, but most often it is to remove the liquid for other uses. Expression is much more energy-efficient than liquid removal by evaporation or solvent extraction or leaching. Both batch and continuous process methods are in use, depending on production volumes. Batch methods may be manual or machine-based and, if machine-based, may be automatic. Continuous process methods produce a constant stream of the desired liquid. One manual method for removing oil from fruit involves cutting the fruit, trimming off the peel, soaking the peel in water for several hours and then pressing the peel against a sponge. The sponge absorbs the oil. When sufficiently filled, the sponge is squeezed to remove the oil. This method is used for obtaining lemon oil. Machine-based batch methods utilize a container with a porous or perforated bottom or wall. Pressure from a ram, or from compressed air, exerted through a rubber bladder, compresses the feed material so that liquids are expelled.

Continuous process methods, currently in use, utilize one of several available press types[5]. Screw presses use a helical screw, somewhat similar to those used in plastics extrusion and injection molding machines. The feed material is forced into a confined space as the screw rotates. Screened or perforated walls around the screw allow passage of the expressed liquid. Disc presses uses two opposed rotating discs with perforated surfaces. The discs are tilted, so that there is less space between them at the bottom. Feed material, between the discs, is compressed as it works its way to the bottom where the discs are closer together. Perforations in the discs allow the liquid to flow. Roll presses use perforated or screened rollers. Feed material is compressed between the rollers, releasing the contained liquids. Belt presses use screens or perforated belts. As the feed material is transported between two belts, it passes through narrower zones and is compressed, so that the liquid it holds is released to pass through the belts.

Olive, castor, sesame and other seed oils are all obtained by expression, but this approach has many other applications. They include: processing of sugar cane and sugar beets, manufacture of perfume, extraction of pectin and oil from citrus peel, processing of starch and corn silage, dewatering of paper mill or sewage sludge, manufacture of calcium chemical compounds, and processing of magnesium hydroxide, fillers for paper and plastic, and concentrates of metal ores[5].

C5. ***crystallization*** - A dissolved material in solution can often be separated from the solvent by crystallization, the formation of solid, homogeneous particles. If a saturated solution is cooled, the solvent's ability to retain the dissolved material in solution is usually diminished. When the solution is progressively cooled, the solution will become supersaturated and, if nuclei or seed particles are present, the solute will start to deposit on them and form crystals. (Seed particles may have to be added.) If enough material forms a crystal, the solution will no longer be saturated, but further cooling causes it to be supersaturated again and the crystals will increase in size. The crystals can be removed from the mixture when they become large enough. The process must proceed slowly if the crystallized material is to be free of the solvent and other materials that may be present. If the solution contains more than one dissolved material, careful control of the temperature may make it possible to form crystals of one material while the other or others remain in solution. Then, afterward, one or more of the other solutes is crystallized. This process is called *fractional crystallization*. In other cases, supersaturation for crystallization is brought about by evaporation of the solvent rather than by cooling the solution, by both cooling and evaporation, or by adding a substance to the solvent that reduces its solvent property for the material to be crystallized. This third method is used frequently in the pharmaceutical industry when the solute is dissolved in water and alcohol is added to the solution to cause it to become supersaturated. Some crystallizers use a partial vacuum to produce evaporation of the solvent and coincidentally, adiabatic cooling of the feed solution. Crystallization is an economical way to purify some materials because only one step is required, and energy requirements are low compared with distillation and other methods of purification[5]. Typical purities for many crystallized materials range between 99.5 and 99.8 percent[5]. Higher purities can be obtained if the feed material is partially purified before crystallization. Crystallization is also used to put many specialty chemicals, that are to be sold, in an attractive and practical form.

Both batch and continuous crystallizing equipment are employed. Batch processes are used for smaller quantities, and when the crystals need to be within a very narrow size range. All equipment has some means of creating and maintaining a supersaturated solution. Depending on the nature of the solution materials, this may involve cooling the solution, evaporation of the solvent, or a combination of both. Vacuum crystallizers are common. They use a vacuum to produce evaporation and take advantage of evaporative cooling to supersaturate the feed solution. Mixers provide agitation, which equalizes the temperature and concentration of the solution and keeps crystals in suspension. Fig. 11C5 illustrates a Brodie countercurrent, continuous crystallizer in which cooling of a hot concentrated solution is used to create crystals. The feed solution enters near the bottom of the system and flows upward through a system of joined tubes that are jacketed with cooling channels. As the feed solution moves through the system, it is cooled and crystals form. The crystals are conveyed in the opposite direction, and eventually spill into a purification column. As the crystals move through the system toward the column, they continually make contact with a more concentrated feed liquid. Fig. 11C5-1 shows a batch crystallizer.

Crystallization also takes place when a melted material solidifies, when solid material forms from a vapor, or when an existing solid changes to a

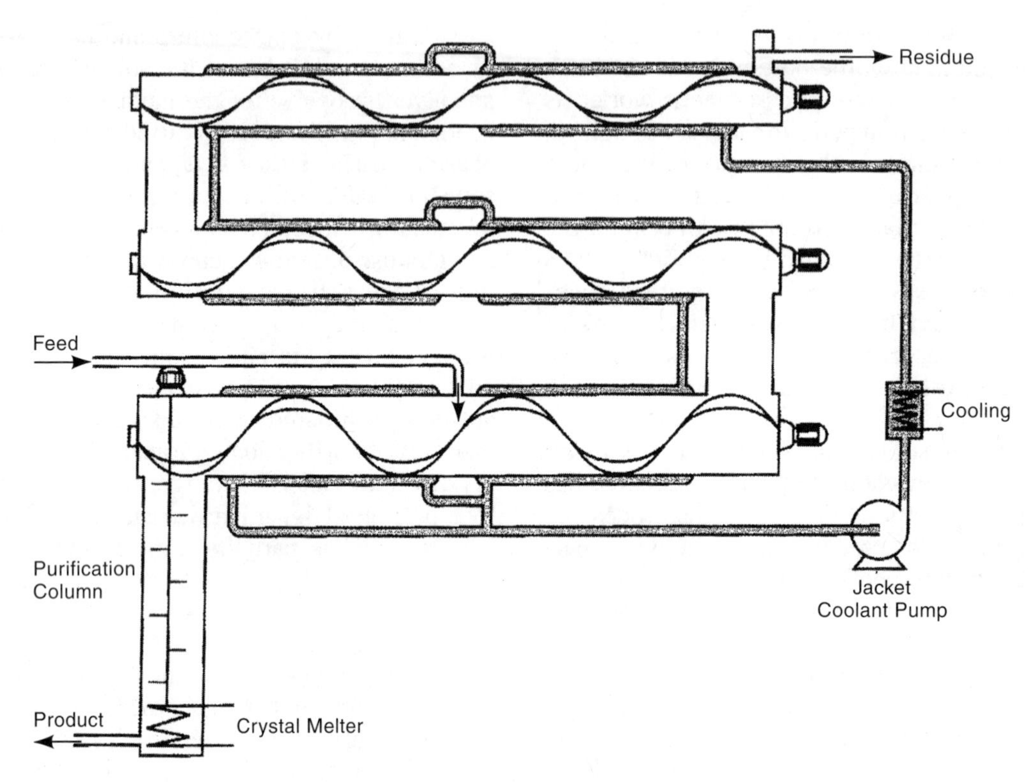

Fig. 11C5 The Brodie countercurrent cooling crystallizer. (*Courtesy Burns and McDonnell Engineering, Kansas City, MO.*)

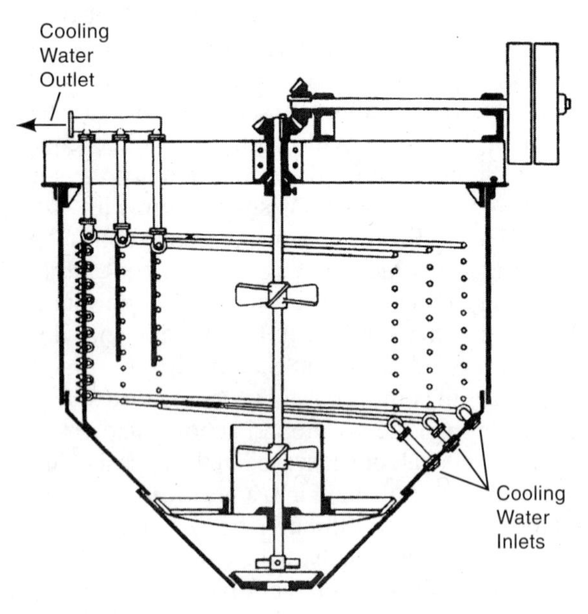

Fig. 11C5-1 A batch crystallizer with cooling coils and agitation. (*by permission, from Introduction to Chemical Engineering, W. L. Badger and J. T. Banchero, McGraw-Hill.*)

different phase. Crystallization from these phases is also used to provide purer material. Crystallization from the melt is used in *zone melting* to purify metals, and in petroleum processing to remove waxes from lubricants[6].

C6. ***precipitation*** - is the formation of a more visible or recoverable state of material from solution within another material. The formation can take place as a result of either a physical or chemical change, which reduces the solubility of the dissolved material or the dissolving power of the solvent. A common industrial application is the formation of a solid material that settles out, and is centrifuged or filtered from a liquid. This result is often achieved by adding a compound which causes a reaction in the solution that converts the material to be separated to an insoluble state. An example in nature is the formation of rain or fog in the atmosphere, when the air is cooled below its dew point. In chemical industries, a liquid material can be similarly precipitated as a condensate from a gas.

Sometimes, with metal alloys, a precipitate of a solid phase of one metal occurs within the solid phase of another metal. The hardness or tensile strength of the metal may be thus increased[1].

Crystallization, described above, meets the above criteria but the distinction with crystallization is that the solid substance yielded by the process has an organized structure.

Precipitation has been used as an important means of separating metals from aqueous solutions. (See K1a1 below.) It is also carried out as part of a chemical analysis (gravimetric analysis) of the starting solution, but instrument methods have generally replaced the precipitation techniques for such analyses.[3] In biochemical applications, salts, solvents or polymers may be added to the feed solution to induce precipitation of wanted materials, or impurities and contaminants[5].

C7. ***fluid-particle separation*** - There are a number of methods for separating solid particles from gaseous or liquid streams. They include: sieving or screening, filtration, sedimentation, centrifugation and cycloning, flotation, evaporation, magnetic and electrostatic separation, membrane separation, the use of scrubbers and the drying of solid materials. (Note: Some of these methods are more applicable to separating fractions of solid materials from other solid materials and are described in C8 below.) Fig. 11C7 charts the applicability of particle separation methods with respect to the usual range of particle sizes.

C7a. ***filtration*** - is the separation of solid particles from a fluid-solids suspension by passing the mixture through a porous barrier. The objective of the operation is either to capture the solid material

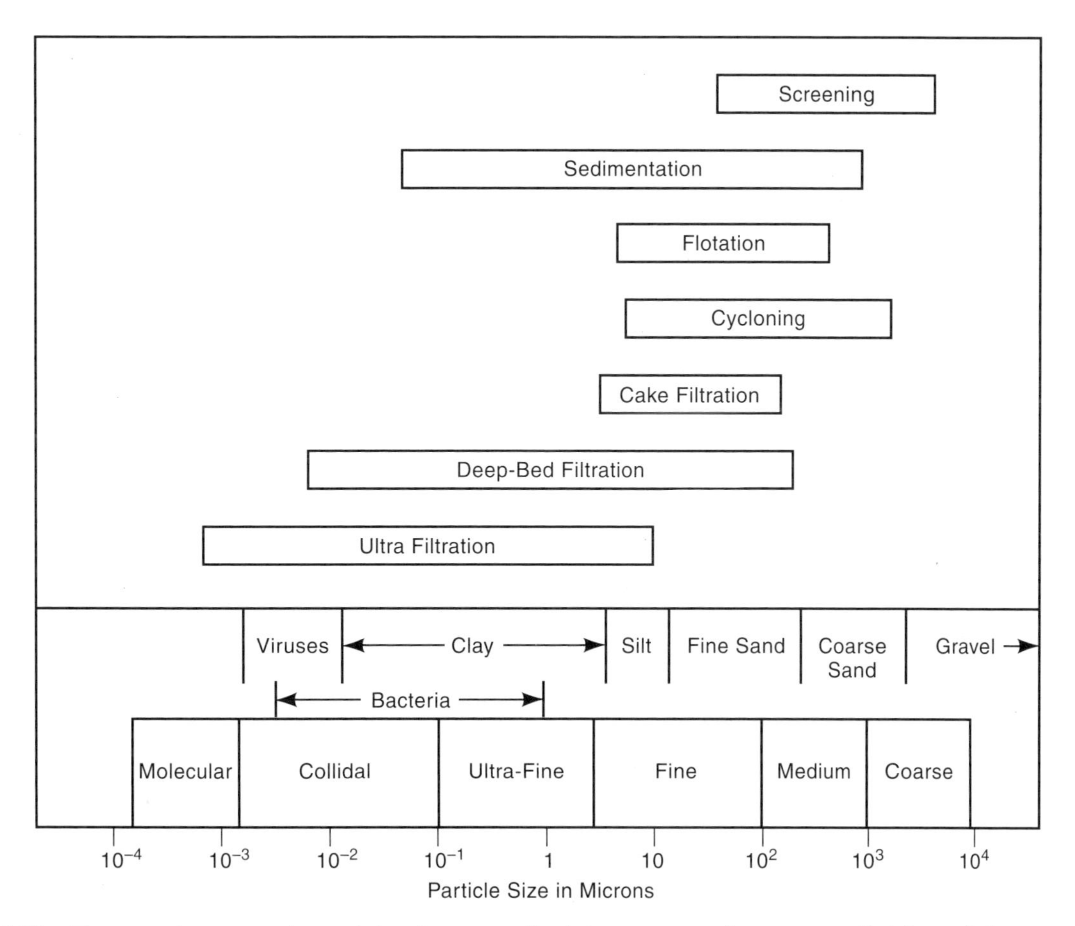

Fig. 11C7 The usual range of particle sizes applicable to several common fluid-particle separation methods. (*Based on data from Dorf*[2].)

from the fluid or to clarify the fluid, or both. The barrier retains most of the solid particles that were part of the mixture. Both gaseous and liquid mixtures are filtered. The fluid flows through the filtering medium because of a pressure differential created across the medium. The differential is supplied by gravity, by pressure applied on the feed side of the filter or by vacuum on the discharge side, or both, or by centrifugal force. The fluid that passes through the filter is called *filtrate*. Normally, the separation of the particles from the fluid is not perfect; filtering cost is higher as the degree of separation becomes higher.

C7b. ***types of filters*** - When the solid particles are blocked at the surface of the filter material, and form a layer or cake of filtered material that increases in thickness as the filtration proceeds, the process is called *cake filtration*. The cake, itself, becomes the prime filter medium. (If the purpose is to capture the solid material, *cake filtration* is always the process.)

When the solids are blocked within the filter medium instead of at its surface, the process is called *deep-bed filtration, depth filtration, filter-medium filtration* or *clarifying filtration*.

Crossflow filters are configured so that the feed slurry moves across the face of the filter, preventing a build-up of solid material on the filter, but allowing some of the feed liquid to pass through the filter.

Filtration can be either a batch or continuous process. When continuous, means must be included to periodically remove the cake, if any, or otherwise remove solid material that clogs the filter. Backwashing is one common method of unclogging deep-bed filters. A clear liquid is pumped through the filter in a reverse direction.

Many materials are used as filter media. They include woven and non-woven textile fabrics, metal fabrics and screens, pressed felt, and cotton batting, filter papers, rigid but porous fabrications of metals, graphite or ceramics sintered from a powdered form of the material, animal and plastic membranes, and granular beds of various solid materials. Sand, for example, is widely used as a filter medium in water supply systems. Woven canvas cloth is a common industrial filter medium. Cloths made from synthetic fibers are used when high chemical resistance is needed[8]. Diatomaceous earth (DE) is often used as a filter aid. When mixed with a slurry, DE provides increased porosity to inhibit clogging of cake[8].

Except for bag filtration, described below, cake filtration is used almost solely for separating solids from liquids. It may be operated with an added pressure differential. The cake must be removed or the filter base back-washed periodically. Because of this, most cake filters operate on a batch basis. Cake filters are normally used when the feed liquid contains 1% or more of solids[4].

Clarifying filters have openings that may be larger than the particles filtered, but have means of trapping, within the filter, the particles to be removed from the feed fluid. These filters are generally used when the portion of solid material in the feed fluid is small, less than 1% but, often, 1/10 percent or less. They are used to clean gases, to clarify beverage liquids and to remove unwanted solids from lubricating oils, pharmaceuticals, beverages, foods, electroplating solutions, and fuel oil[8]. Clarifying filters may include pads of cotton, felt, cellulose pulp, and glass fiber, all used in filtering gases. Beds of sand are used for filtering water. Beds of various other granular materials may also be used.

Fig 11C7b illustrates the principles of three kinds of filtration.

Gas filtration is used to remove dust and other particles from a gas, either because of the contamination or inconvenience they cause or to recover the particles for some useful purpose. Paragraph 11C7f describes the bag filter method. Two other methods are granular-bed filtration which uses beds of carbon particles, sand or other materials as a filter medium and air filters that use a mesh of loosely compacted fibers, often coated with a viscous material. These filters trap particles of dust and other solids which are suspended in air or another gas. They are usually used with dust concentrations of 5 grains per 1000 cu ft (5 grains per 28 cu m) or less. (Also see cyclone separation in paragraph C7e1 and electrostatic separation in paragraph C8e below.)

Filter presses are machines with sets of plates stacked together. The plates are concave or separated by spacers to provide cavities for the feed liquid. Each cavity contains a cloth or other filter medium, an inlet for feed liquid and an outlet for the filtrate. The inlet and outlet openings are joined so that there is one feed inlet and one filtration outlet from the stack. Use of multiple plates provides a large filtration area.

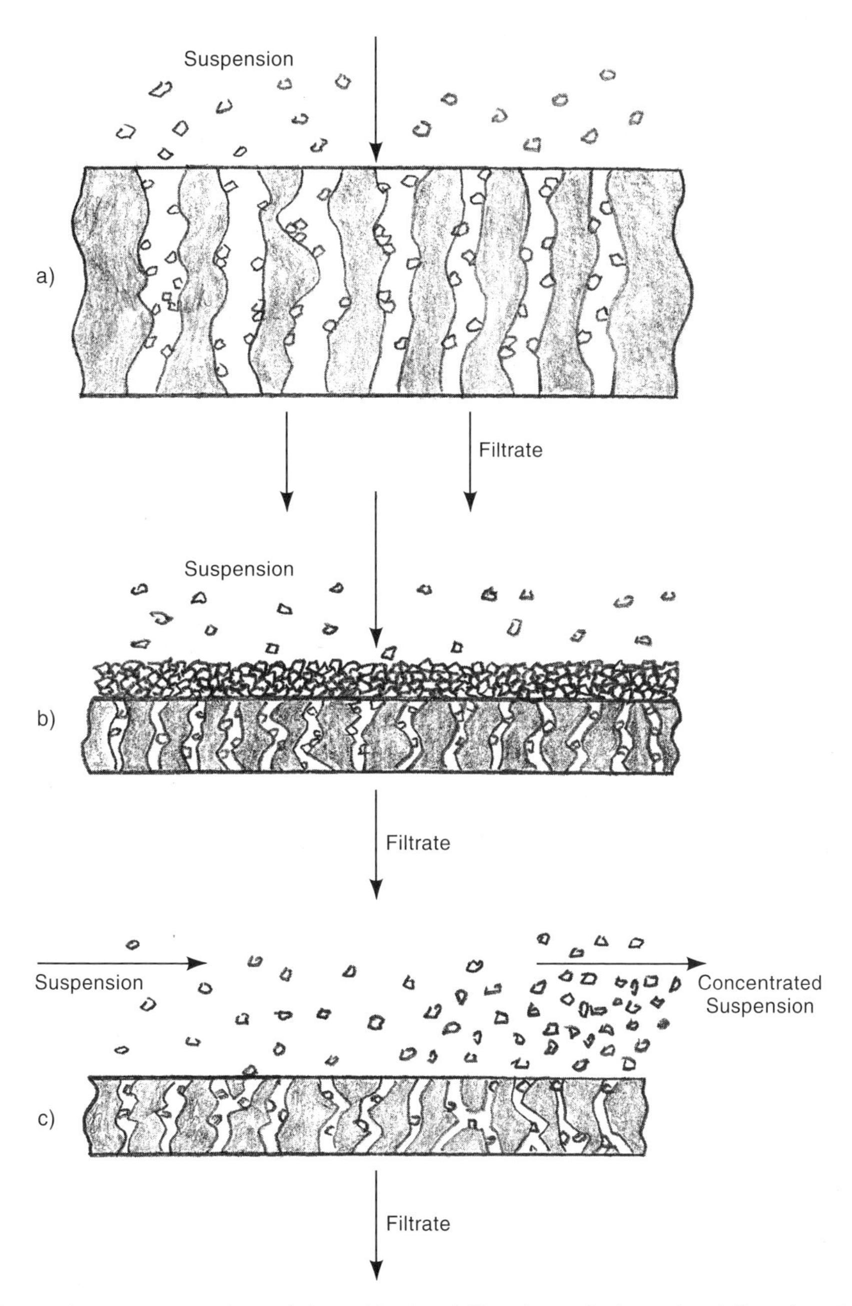

Fig. 11C7b Schematic representation of three kinds of filtration: a) deep-bed filtration, b) cake filtration, c) crossflow filtration.

Pressure is usually applied to the feed slurry. Most filter presses are equipped for back-flow washing, but may have to be opened for removal of dense filter cakes. Many presses are equipped to do these operations automatically[8]. Fig. 11C7b-1 shows a typical horizontal filter press.

Centrifugal filtering makes use of the tremendously higher forces that centrifugation provides, and can be used for materials that form a porous cake. (Centrifugal processing for other operations, as well as filtering, is discussed below in 11C7e.) One major centrifugal filter application is in sugar

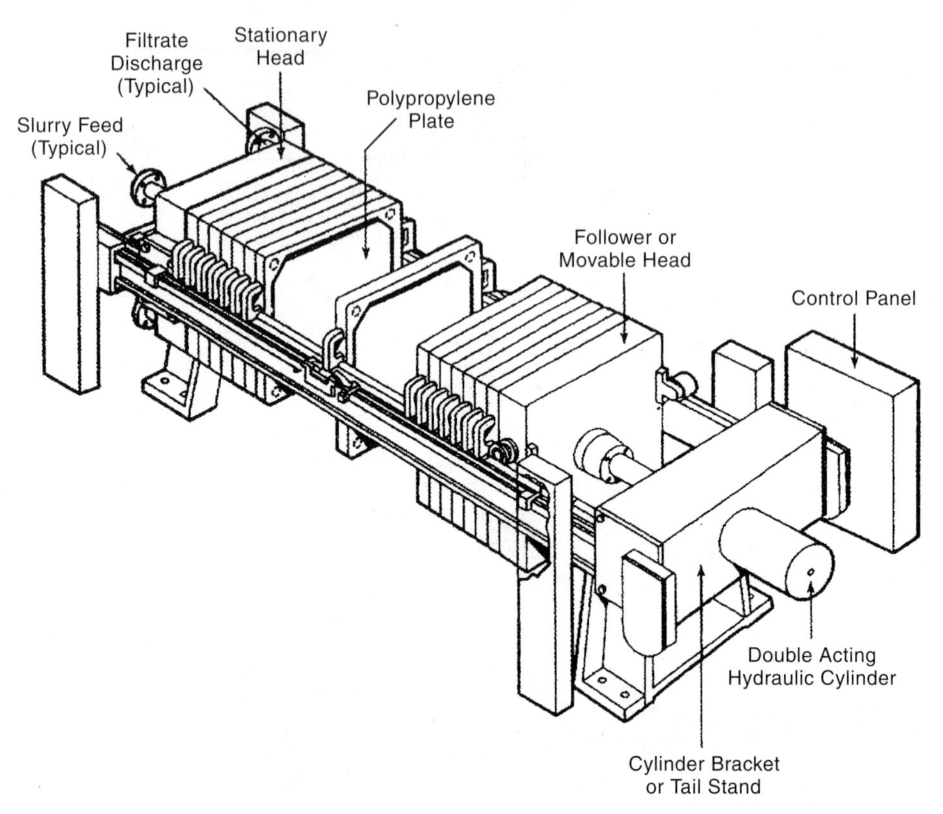

Fig. 11C7b-1 A filter press equipped for automatic operation, with a horizontal stack of filters that act in parallel. (*Courtesy Shriver Filters, Dorr-Oliver USA Inc.*)

processing where the method is used to separate sugar crystals from a slurry of crystals and a sugar solution.

Filtering is extensively used in chemical, food, beverage, and pharmaceutical processing, both in laboratory and production work. Mineral processing, water processing, and sewage disposal are other applications. The manufacture of phosphoric acid includes filtration. It is also used to remove waxes from other products during petroleum refining. Fig. 11C7b-2 illustrates a typical continuous filtration machine that uses a belt filter cloth to support a cake filter. Fig. 11C7b-3 shows the principle of operation of a vacuum rotary-drum filter, a commonly used continuous method.

C7c. ***membrane separation (including ultra filtration)*** - uses a thin semi-permeable barrier, most commonly a plastic sheet, to separate a mixture of miscible fluids. The feed stream is

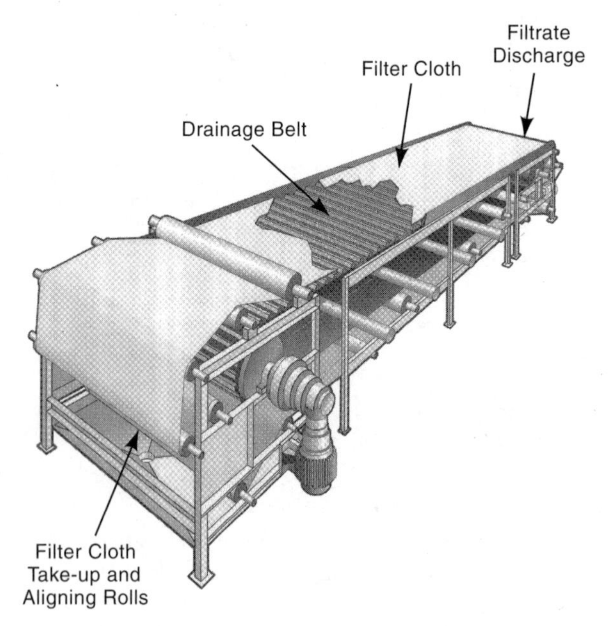

Fig. 11C7b-2 A horizontal continuous belt filtration machine. (*Courtesy Door-Oliver Eimco USA, Inc.*)

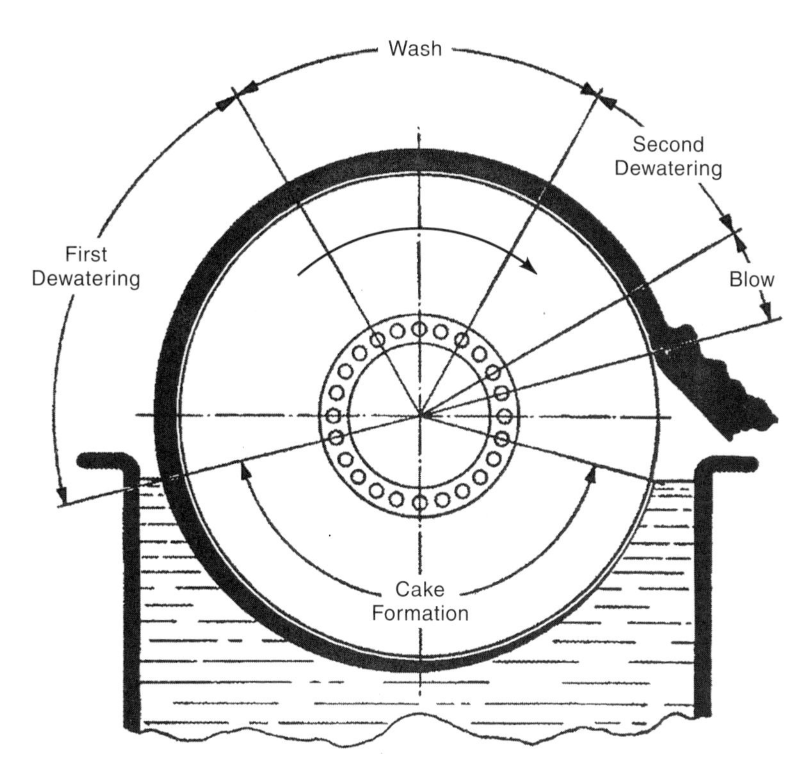

Fig. 11C7b-3 The principle of operation of a typical rotary drum filter. The drum rotates on a horizontal axis and is partially immersed in a container of feed liquid. The drum rotates slowly, clockwise in this view. The drum surface is porous and is covered with a filter medium. Segments of the drum, which would appear pie shaped in this section view, are isolated from one another so that a vacuum or pressure may be applied. Filtrate flows into the portion of the drum that is immersed in the feed liquid, sometimes aided by a vacuum, and filter cake forms on the surface. Filtrate is pumped away. As the drum rotates, each segment moves out of the liquid. Sprays wash the cake that has formed. The segment is still under vacuum and the wash water is diverted from the feed liquid. As the drum rotates, the vacuum is replaced by positive pressure which blows against the filter cake to loosen it so that it can be removed from the drum by a doctor knife. The segment is then ready to reenter the feed liquid and receive more filtrate. (*by permission, from Introduction to Chemical Engineering, W. L. Badger and J. T. Banchero, McGraw-Hill.*)

introduced to one side of the membrane, usually under pressure, and the permeate (filtrate) collects on the other side. Although their non-porosity is in some dispute[5], membranes, said to be non-porous, are used for separation of gases and other ultra fine materials. Membranes can be very selective in allowing passage of one or more materials in a feed stream but not the other components. Fig. 11C7c shows membrane separation schematically. In practice, membrane separation equipment uses stacked arrays of membranes, with feed and filtrate channels to increase the membrane area. Fig. 11C7c-1 shows several arrangements.

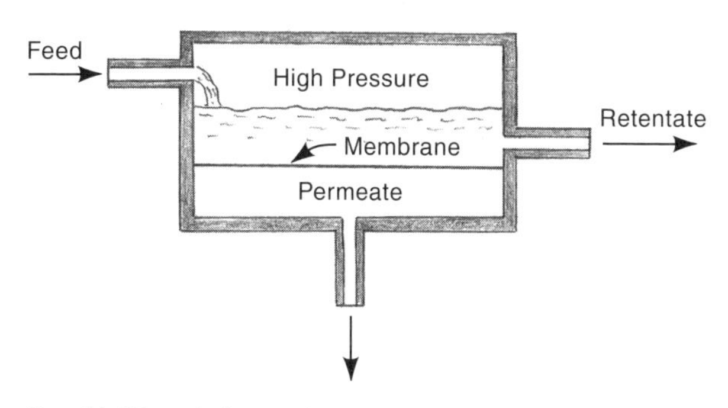

Fig. 11C7c Schematic representation of membrane separation.

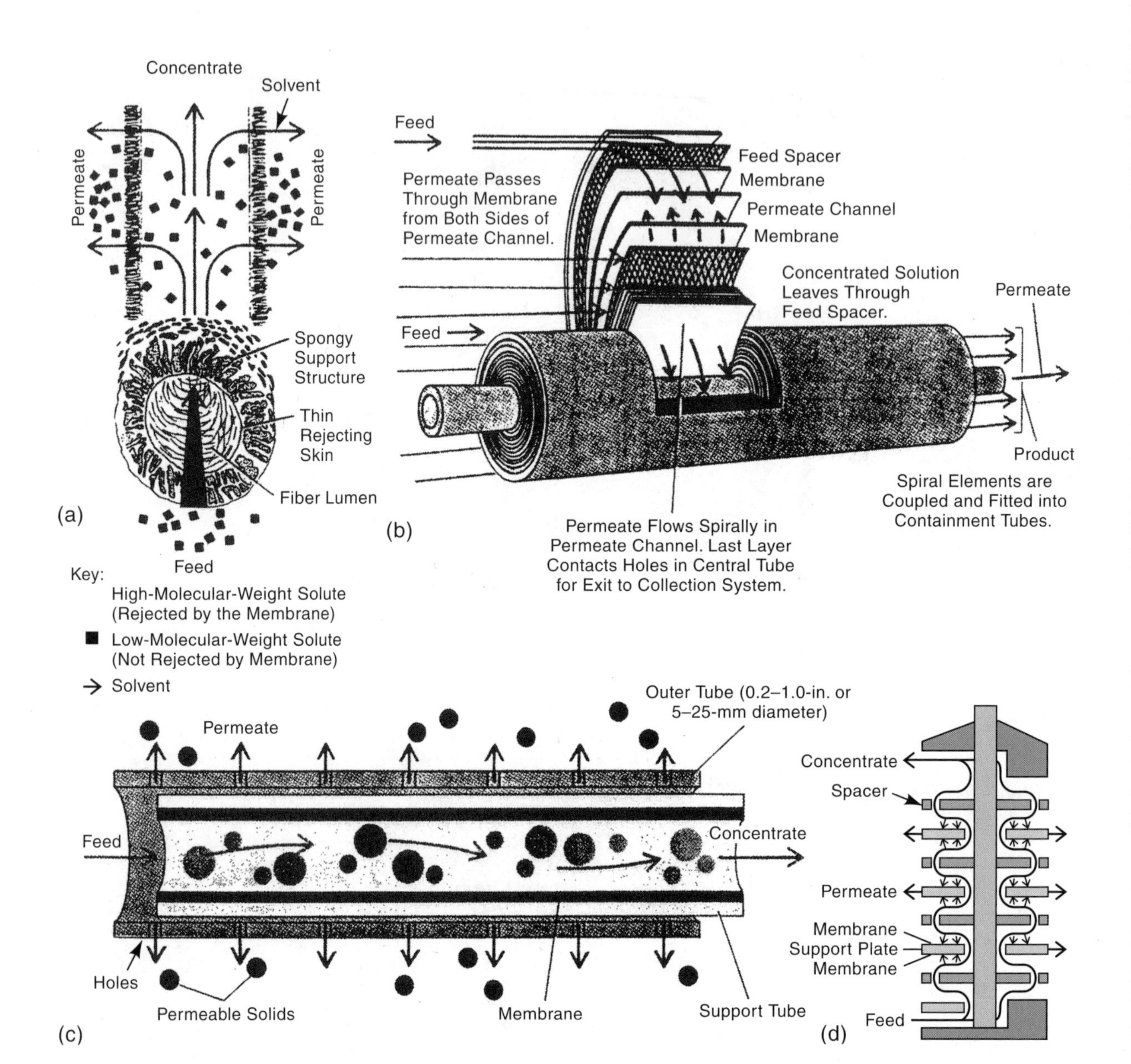

Fig. 11C7c-1 Several arrangements for membrane separation: a) hollow fibers, b) spiral-wound, c) tubular, d) plate system showing internal flow. (*Reproduced with permission from the McGraw-Hill Encyclopedia of Science and Technology, 8th Ed., McGraw-Hill, New York, 1997.*)

Ultra filtration is the separation of very fine particles (macromolecules and colloidal suspensions) from a liquid containing dissolved low molecular weight species. Filtration is with a membrane filter having extremely small openings. Pressure differences of 10 to 200 psi (70 to 1400 KPa) provide the force to move the liquid through the membrane[4]. The feed liquid normally contacts the filter by cross flow. (See view c, of Fig. 11C7b above.)

Dialysis is another membrane separation technique that prevents passage of higher molecular weight solutes and particles (molecular weights over 1000) through the membrane while allowing lower molecular weight solutes and ions to pass. Regenerated cellulose is the most common membrane material. A concentration difference across the membrane provides the driving force[4]. Dialysis is used in removing waste products from the blood of medical patients with kidney disease, in the food industry to desalt cheese whey solids and in microbiology to recover enzymes from cultures[2].

In *reverse osmosis*, the pressure differences across the membrane are as much as 800 psi (5500 kPa) and the permeate is water or another liquid. Almost all solutes and suspended particles are retained in the concentrated feed liquid.[4] (Also see reverse osmosis and ultra-filtration of foods in 12F3.)

Membrane separation is widely used in desalting sea water, purifying water, separating nitrogen from air, production of chlorine, and for separating many organic vapors and gases. In the process, which uses a non-porous membrane, molecules dissolve and diffuse through the membrane, although transient gaps may open within the membrane. In addition to plastics, metals, ceramics, polymer solutions and liquids may be used as membranes. (Liquid membranes are contained in porous media or are immiscible in the fluid being processed.) The shape of the membrane may be tubular or spiral wound from sheet as well as flat.

C7d. ***sedimentation*** - is a means of effecting partial separation of suspended solid particles in a gas or liquid by utilizing gravity to settle the particles. With liquids, the purpose may be to clarify the liquid for its later use (*clarification*) or to separate or concentrate the particles for later use (*thickening*). Several factors affect how the operation is carried out. They include: the degree of concentration of the particles, the viscosity of the liquid, the degree of sedimentation needed, the desired throughput rate, the density of the particles, their size, and their cohesiveness.

Sedimentation using liquids is normally performed in a shallow, round tank of large area, with a bottom that slopes slightly toward the center. Feed piping delivers the liquid or slurry of input material to the tank. Solid material is drained from a central discharge gate that can be opened or closed. The tank includes a raking mechanism on a large arm that slowly traverses the entire area of the tank. Blades on the arm sweep the solid material toward the central discharge gate. Clear liquid spills over the edges of the tank. Large scale equipment may employ tanks as large as 500 feet (150 m) in diameter. With tanks of that size, the raking mechanism may make only two rotations per hour. Fig. 11C7d illustrates one of these devices, which are called *thickeners*. Flocculants are often added to the input material to facilitate the sedimentation. They cause fine particles to agglomerate after which they settle much more quickly.

Thickeners are commonly employed in mining operations. Other applications are cement manufacture, sewage treatment, water purification and the production of magnesium from sea water[8].

Clarification usually involves dilute suspensions. One common application is municipal water or waste processing. Clarifiers and thickeners are very similar but clarifiers often can be built with a lighter structure.

C7e. ***centrifugation*** - applies centrifugal force to separate solids from liquids or gases, and immiscible liquids from each other, as well as to aid filtration as discussed above. Centrifugation tremendously increases the separative force of gravity, greatly speeding up the separation process. Large amounts of material can be processed in little space. The material to be processed is put into a centrifuge, a device incorporating a round container which rapidly rotates. Centrifugal force drives the heavier, denser component of the mixture outward to the container walls. The strongest centrifugal forces, over 100,000 times gravity, are obtained in very small centrifuges rotating very rapidly. Sedimentation of a liquid slurry can be performed with a centrifuge using a solid bowl; perforated bowls are used for filtration. The porous or perforated wall allows liquids to escape while solid material remains in the rotating container. Centrifugation is used in many chemical processes, in removing water from jet engine and diesel fuel, in sugar refining, in separating cream from milk, in rendering oil from animal and vegetable sources, in clarifying liquids, and in the treatment of municipal sewage. Isotopes are separated with centrifuges, and blood cells are separated from whole blood. Centrifuges are of several types, some of which are designed for continuous operation. Others are semi-continuous, providing for continuous removal of liquid but with periodic stops for removal of solids. Because centrifugal forces are much stronger than gravity, some machines can be arranged with a horizontal axis of rotation. Flocculants and coagulants may be added to the feed slurry. Fig. 11C7e illustrates the principle of centrifugation.

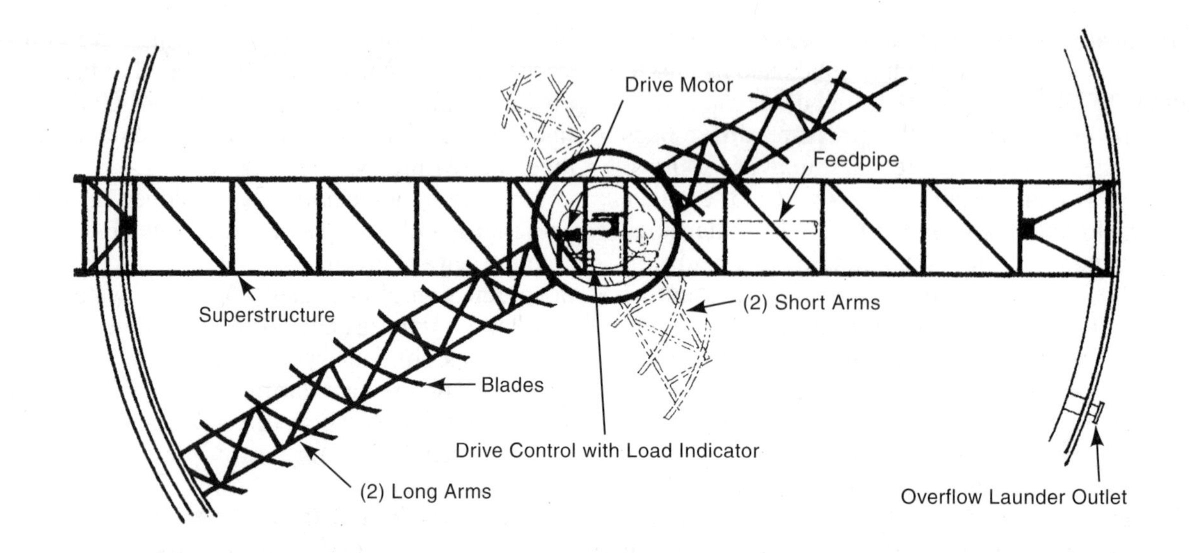

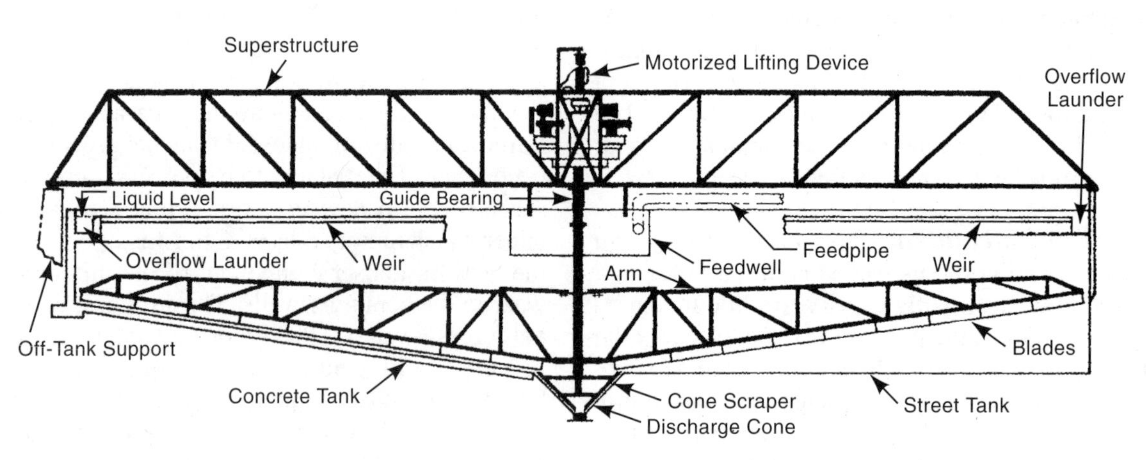

Fig. 11C7d Large thickener with mechanism supported by a bridge. (*Courtesy Dorr-Oliver Eimco USA Inc.*)

Tubular-bowl centrifuges employ long, narrow, cylindrical bowls which rotate very rapidly inside a steel container. Typical bowls are 3 to 6 in (7.5 to 15 cm) in diameter and 30 in (75 cm) in length, rotating at 15,000 RPM[10]. Feed liquid enters at the center of the bowl bottom. Suitably located discharge ports at the top of the bowl, one for light material near the center of rotation, another for heavier material near the bowl walls, carry the processed material to separate discharge spouts. Fig. 11C7e-1 illustrates a tubular-bowl centrifuge for separating two immiscible liquids.

Disc-bowl centrifuges are wider and squatter than the tubular bowl type. They include a series of conical surfaces, called discs, that are stacked with fixed spacing around a central axis of the centrifuge. The discs rotate with the bowl and also have holes in alignment with holes in the discs above and below. Feed liquid enters the centrifuge at the top of the central axis. The holes in the discs allow the feed material to flow between the discs. The heavier liquid collects and flows outward on the underside of the discs from centrifugal force. The lighter liquid flows toward the center on the upper surfaces of the discs. There is shearing at the interface of the two liquid fractions and the distance each drop of liquid must flow to enter the applicable outgoing flow is very small. This improves the effectiveness of the centrifuge in separating the two fractions including those with emulsions. Typical

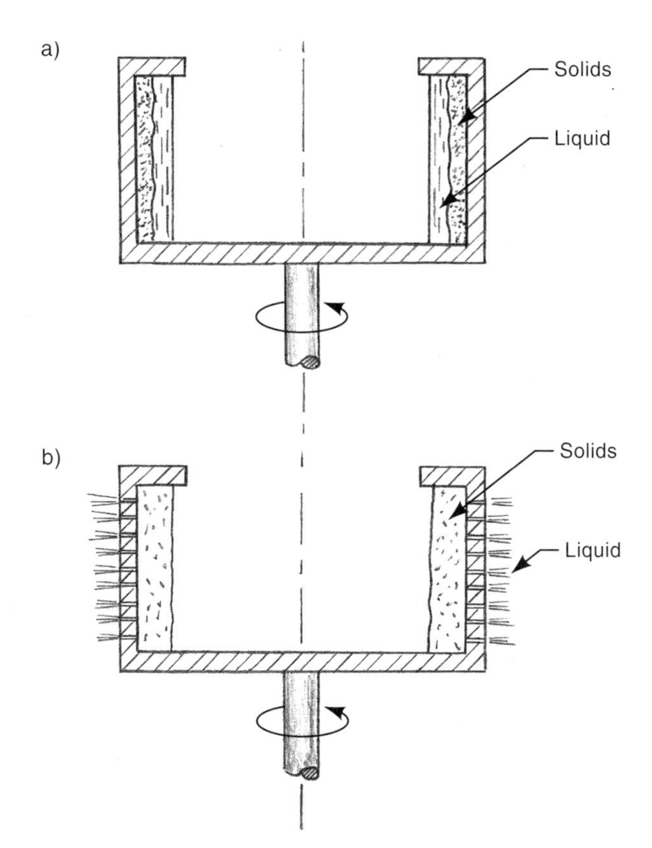

Fig. 11C7e The principle of centrifuging showing the effect of centrifugal force when the device is in operation. In view a), sedimentation takes place in the impervious bowl, creating separate layers of solid and liquid material which can be removed separately. In view b), centrifugal force drives the solid material to the bowl wall where it is retained and forms a cake filter. The bowl is porous, allowing the filtered liquid from the slurry to escape.

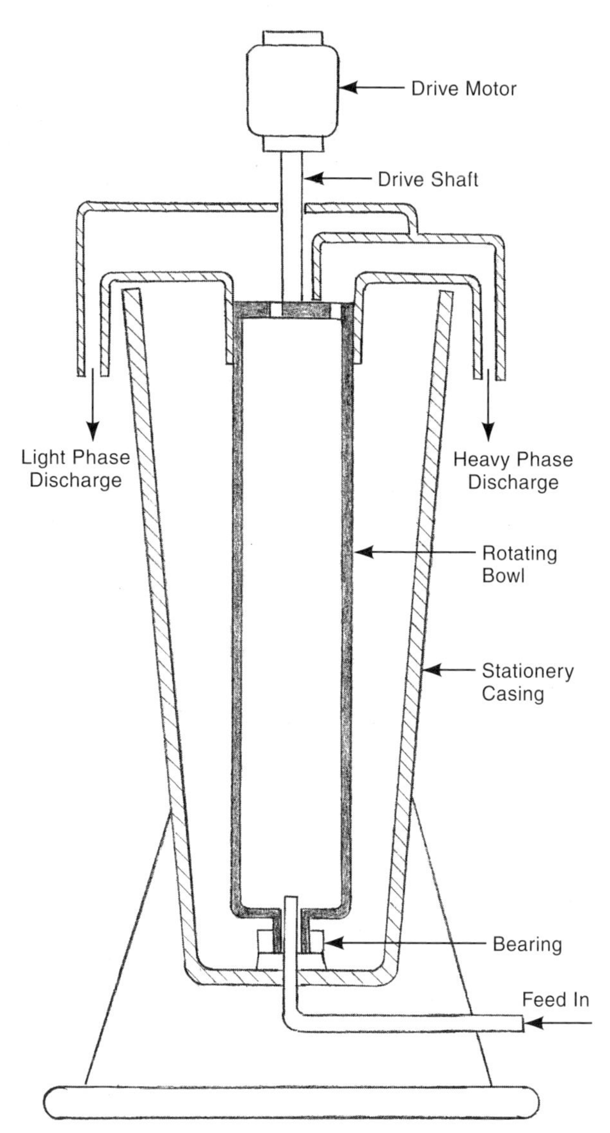

Fig. 11C7e-1 Tubular-bowl centrifuge for separating two immiscible liquids. The heavier liquid gravitates to the walls of the rapidly rotating bowl and leaves the centrifuge from the spout on the right-hand side. The lighter liquid remains closer to the center of rotation and leaves the centrifuge on the left-hand side. Feed liquid enters the centrifuge from the bottom.

bowl diameters are 8 to 40 in (20 to 100 cm). Spacing between discs ranges from about 0.020 to 0.050 in (0.5 to 1.3 mm). A disc-bowl centrifuge is illustrated in Fig. 11C7e-2.

Fig. 11C7e-3 illustrates a typical horizontal industrial centrifuge that separates liquids into two different products in a continuous operation.

Small-diameter, high-speed centrifuges are used in the filtration of varnishes and vegetable oils, the removal of wax from lubricating oils, and the filtration of crude-oil emulsions. Centrifuges are used to remove small amounts of solids from beverages, inks, lubricating oil and other liquids, particularly slimy solids that would clog a filter[8].

The term, *centrifugal molecular distillation* is often used when centrifugal force is applied to separate two gases. An instrument called a vortex is

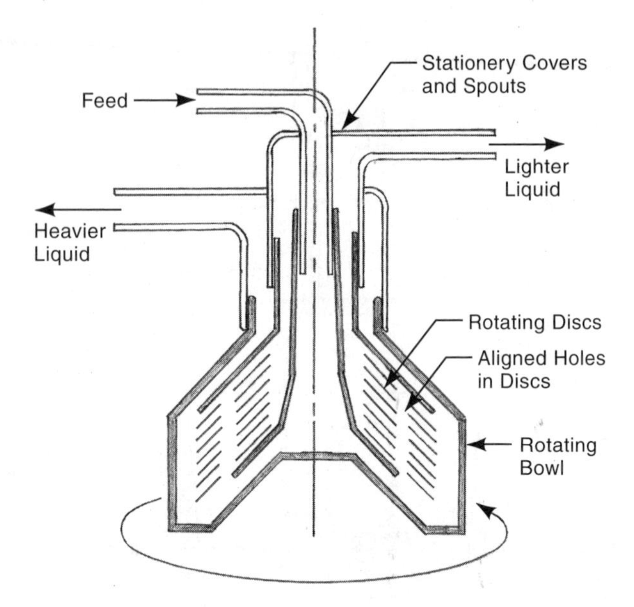

Fig. 11C7e-2 The disc-bowl centrifuge incorporates closely spaced, cone-shaped, rapidly rotating "discs" which effectively separate emulsions and other immiscible liquids. The dense liquid flows outward on the underside surfaces of the disc; the lighter phase flows inward on the top disc surfaces.

used. It consists of a centrifuge that spins rapidly, causing the heavier gas to migrate to the outer portion and the lighter gas to remain in the center. The two gases are withdrawn through suitably placed outlets. The method is used in uranium enrichment processing to separate the U-235 isotope from U-238 by centrifugal molecular distillation of uranium hexafluoride gas. The products of the first separation may be cycled through several additional vortex stages to achieve the desired degree of purity in the final products. Many stages are required for uranium enrichment.

C7e1. ***cyclone separation, cycloning*** - Another centrifugal device, the cyclone separator or *hydrochlone*, does not use a rotating container. Instead, the fluid to be processed is released in a high-velocity stream tangentially to the walls of a round stationary container. The container may also be conical. Centrifugal forces are not as large as with a centrifuge but cyclones are widely used to remove liquids and solid particles from gases and, with smaller units, solid particles from liquids. Separation can be effected from differences in particle size or from differences in particle density, or both. (In popular usage, the terms cyclone and hydrochlone are interchangeable; in strict usage, hydrochlone applies to those devices used with liquid/solid mixtures.) Fig. 11C7e1 illustrates a typical cyclone separator. Cyclone separation applications include separation of heavy and coarse materials from fine dust, classification of pigments and other powdered solids, removal of carbon from gypsum, in the production of alumina, and removal of wood chips and sawdust from woodworking operations. The approach does not always provide

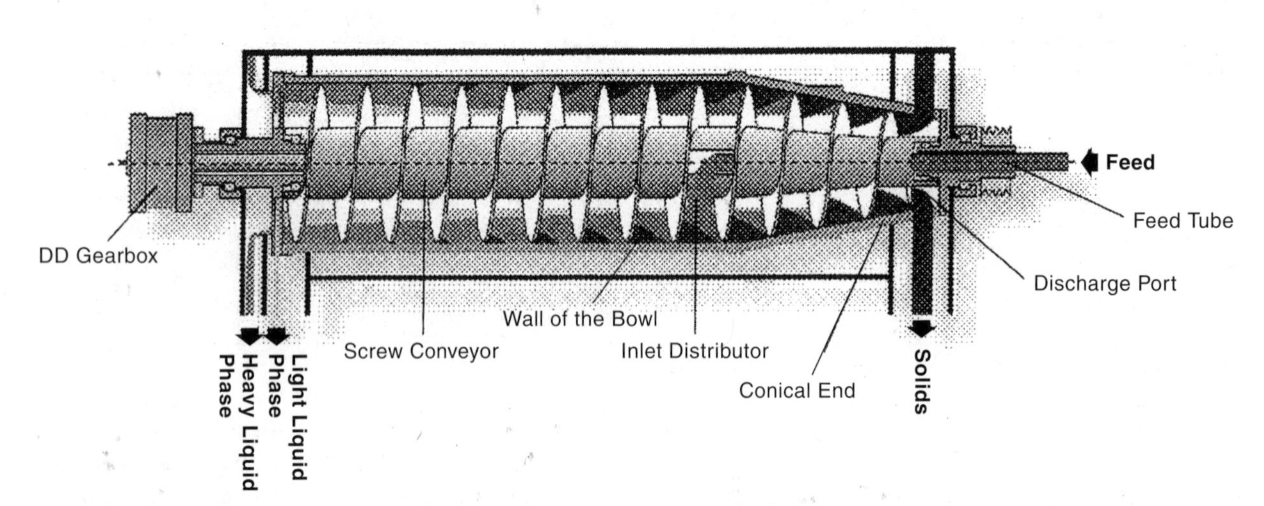

Fig. 11C7e-3 Continuous horizontal centrifuging to separate a liquid slurry into its liquid and solid components. (*Copyright Alfa Laval Inc. All rights reserved. Used with permission.*)

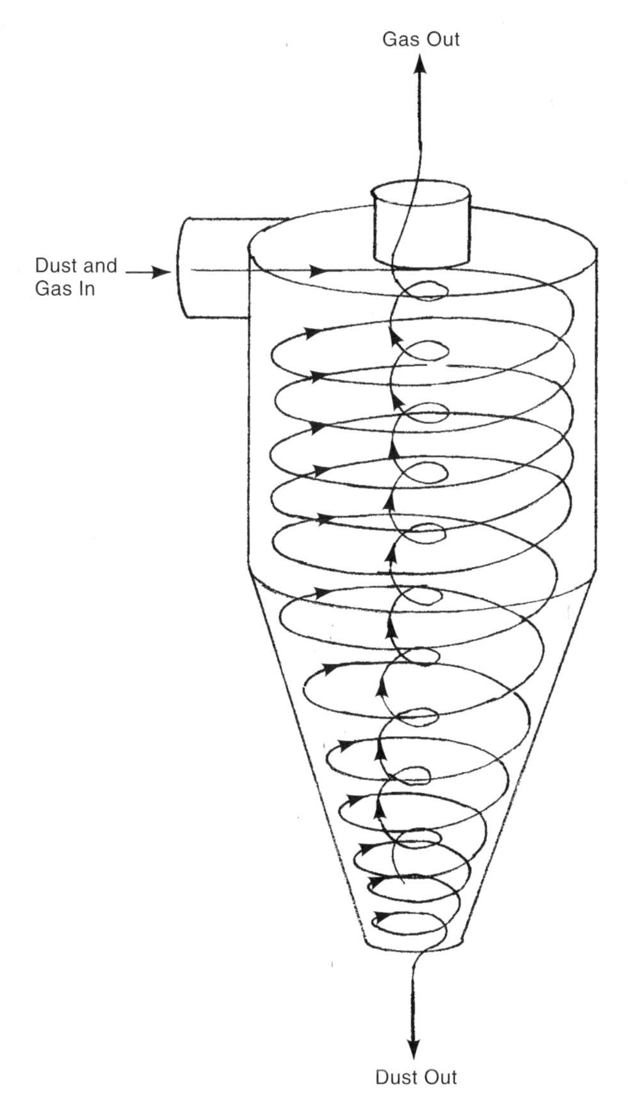

Fig. 11C7e1 A cyclone separator. The gas-particle material enters the device at high velocity and continues on a circular path inside the container. Solid particles tend to remain closer to the walls of the container and fall to the bottom while the lighter gas exits from the central vent.

extremely high separation efficiencies and works best for heavier and coarser dust. Fine dust is better removed with the similar *air separators* which combine centrifugal force with a flow of additional air. This air picks up the finer dust particles but the machine is designed so that the speed of flow of the air containing fine dust is then slowed. The dust then falls from the slow moving air into a discharge chute.

C7f. ***bag filtering*** - An array of bag filters in an enclosure, commonly referred to as a *baghouse*, is often used to separate fine dust from air or other gases, after cyclone separators have been used to separate coarser dust. Bag filters may also be used in conjunction with air separators. A baghouse is illustrated in Fig. 11C7f. Long, narrow filter bags, made from woven cloth, felt or a membrane, are drawn taut in a sealed chamber. Suction fans feed the dust-bearing air into the chamber and through the bags which hang upside down with their bottoms open to the intake gas. The dust laden gas is blown upward through the bags. The dust collects in the bags while the gas passes through the bags and exits from the top of the baghouse. The layer of dust, as well as the filter fabric, acts as the filter. In most equipment with a filter cloth, the openings in the cloth are larger than the dust particles so the filter allows some dust to escape until the dust layer is formed. Thereafter, the filtering action is very efficient except for particles smaller than 0.01 micron (0.0000004 in) in diameter. At regular intervals, the suction fan stops, the exhaust manifold is closed and outside air is allowed into the chamber from outside the bags. This reverse air flow loosens the dust inside the bags. The bags are then agitated and the dust falls into a hopper below. Then, suction and collection of the dust in the bags resume. In this way, the dust collection is intermittent but the operation of the equipment is continuous and automatic. Some installations use several bag compartments side by side, so that bags in some units can be emptied while other units continue in operation. This equipment is a common part of dust collection apparatus in industry, when the operations are smokey or dusty. The maximum operating temperature of the filter medium limits the temperature of the gas that can be filtered, but glass and polymer filter materials raise allowable temperature to the 450 to 550°F (230 to 290°C) range[5]. In some installations, the air or gas to be processed is passed through cooling coils before entering the baghouse.

C7g. ***evaporation*** - is the conversion of a substance from a liquid to a gas. The change is effected by the application of heat. In some instances a vacuum may also be applied. Evaporation is commonly

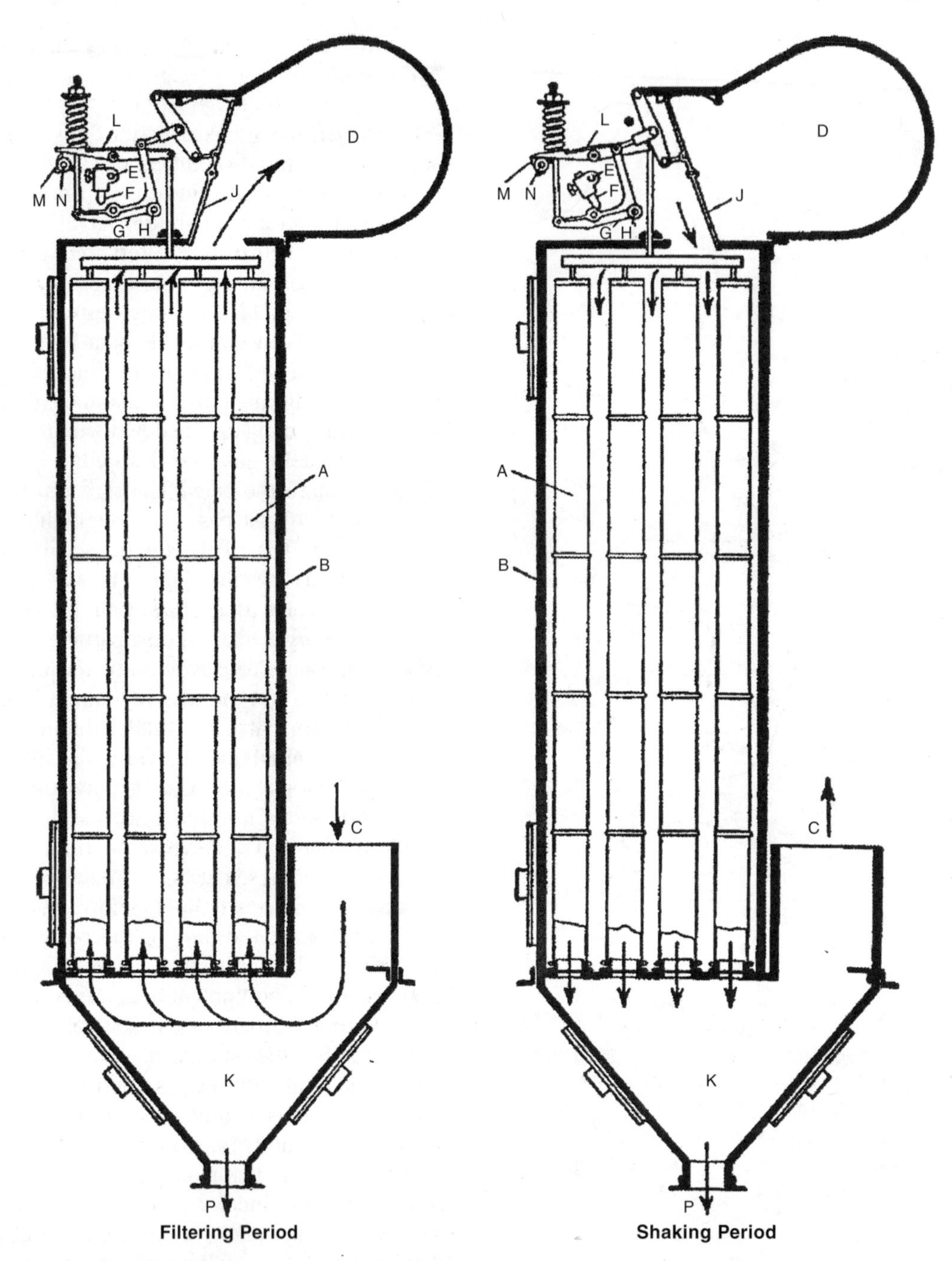

Fig. 11C7f A bag filter for removing dust and other particles from a gas. The left-hand view shows the filter in operation; the right-hand view shows the changed baffle position during the shaking period. A, filter bags; B, casing; C, inlet connection; D, discharge manifold; E, slow-speed shaft; F, cam; G, bell-crank lever; H, bell-crank-lever pivot; J, damper; K, dust hopper; L, shaking lever; M, shaking cams; N, cam shaft; P, product discharge. (*by permission, from Introduction to Chemical Engineering, W. L. Badger and J. T. Banchero, McGraw-Hill.*)

used to make a solution more concentrated for later distillation, or for another operation. The production of orange juice concentrate is an example of a process that includes evaporation. The other common goal of evaporation is to separate a solid material from a solvent liquid, for example, the recovery of salt from brine. In common usage, the term, *distillation*, is used when there are one or more liquid constituents to be removed from a solution by selectively vaporizing and condensing it, whereas *evaporation* is used to separate a solid or, as indicated, to concentrate a solution. Evaporation may be part of a crystallization operation. Distillation includes condensation; evaporation does not necessarily involve that step. In the majority of evaporations, the solvent to be evaporated is water. Sodium sulfate ($NaHSO_4$) and caustic soda ($NaOH$) are two materials produced from evaporated brine[6]. Evaporation is frequent in the food industry, to reduce the volume of a product for easier shipment and storage. Concentration of the juice from sugar cane, prior to crystallization of the sugar, is a common evaporation operation.

Evaporators provide heat by one of a number of methods, further discussed in paragraph N below. Usually the heating medium is separated from the feed liquid by tubular elements, double walls, jackets, or flat plates. Steam or hot gas is the normal heating medium. The heating medium is sometimes brought into direct contact with the feed liquid. Heating by solar radiation is another method. Forced circulation of the feed liquid past heating surfaces is common. *Multiple-effect evaporators* use several stages of heating from one original heat source. In a triple-stage ("triple-effect") evaporator, the vapor from the first stage ("first effect") supplies heat for the second effect and then the vapor from the second effect heats the third effect. This sequence provides better energy efficiency than a single-effect unit. The flow of the feed liquid may be either forward or reversed in the system. Fig. 11C7g shows a simple batch kettle evaporator with a steam jacket. Fig. 11C7g-1 illustrates several alternative approaches in evaporator design.

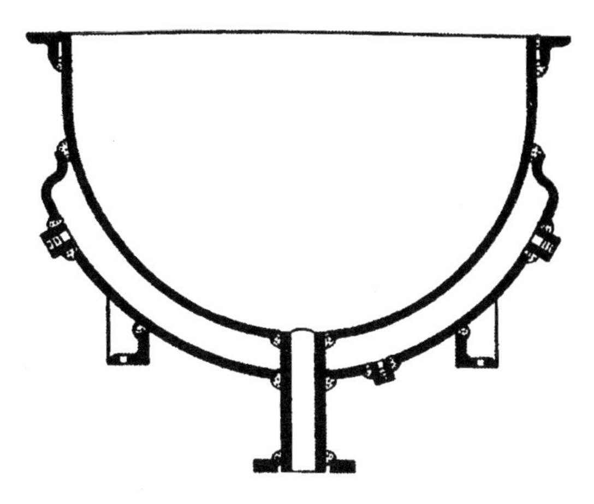

Fig. 11C7g A simple kettle evaporator with a steam jacket used for batch evaporation operations. *(by permission, from Introduction to Chemical Engineering, W. L. Badger and J. T. Banchero, McGraw-Hill.)*

C7h. ***scrubbers*** - or ***wet collectors,*** remove solids or liquids from gases using water or another liquid to capture the particles of the unwanted material. There are many varieties of scrubbers. With some types, a common method is to spray the liquid into the gas so that many tiny droplets of the liquid capture the particles of solid material that it carries. A common arrangement is to provide an upward stream of air or gas containing unwanted material, and a downward spray of water or other capturing liquid from a series of nozzles. Fig. 11C7h illustrates another scrubbing method, a *multi-washer scrubber*. The gas to be scrubbed enters the device near the bottom and moves in an upward spiral inside the walls of the device. The gas passes through curtains of liquid that have come from the top of the device and over deflector cones. Dust or other solid particles contact and are captured by the liquid, which exits the device at the conical bottom along with the trapped particles. The gas is diverted to pass a series of vanes and then through an entrainment separator and out of the top of the scrubbing device.

The venturi scrubber is another common type. The gas is introduced downward into a funnel-shaped container, the liquid distributor. Liquid is introduced tangentially at the top of the same container and coats its walls, where it is contacted by the gas. Both gas and liquid flow downward in the funnel-like container to the bottom where there is a venturi throat that has an adjustable opening. Both liquid and gas flow through the venturi and are squeezed together, thoroughly mixing them. The adjustment at the venturi throat (usually not really a venturi)

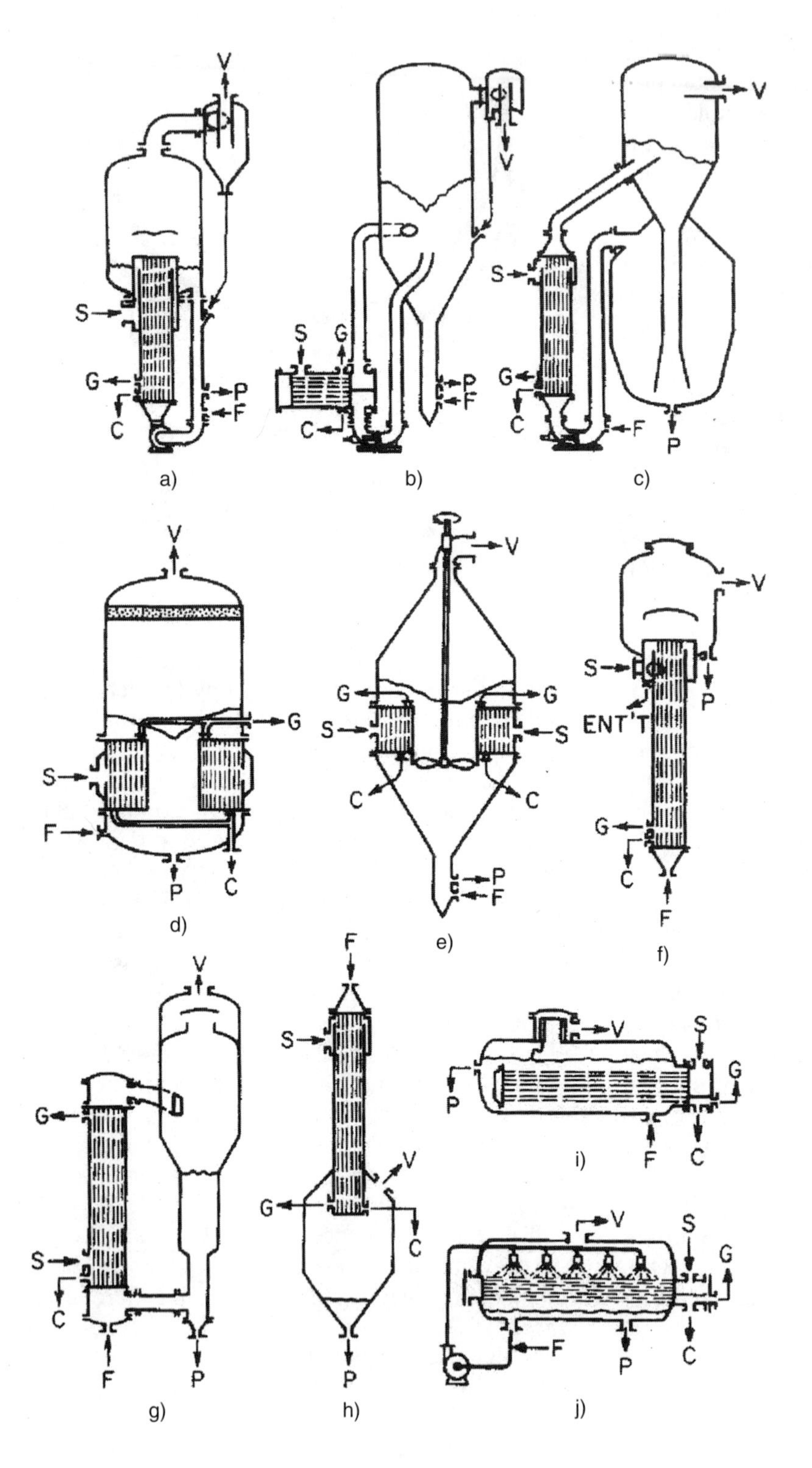

Fig. 11C7g-1 Various evaporator types. a) is a forced circulation evaporator. b) is the much more common submerged tube forced circulation type frequently used for production of salt. c), Oslo-type forced circulation evaporator used in crystallization. Pumps in a), b) and c) ensure circulation of the liquid to be processed in the equipment. d) is a short-tube vertical unit commonly used to evaporate cane sugar juice. (e) is a propeller evaporator. The propeller increases capacity. f), g), and h) are long-tube vertical evaporators that provide the lowest cost evaporation capacity. g) is the recirculating type evaporator with rising film, used for condensed milk. h) is a falling-film type of evaporator. Both g) and h) are used extensively for fruit juices and other heat-sensitive liquids. i) and j) are horizontal-tube evaporators, used for sea water evaporation. C=condensate, F=feed, G=vent, P=product, S=steam, V=vapor, ENTT=separated entrainment outlet. (*Reproduced with permission from Perry's Chemical Engineering Handbook, R.H. Perry and D.W. Green, McGraw-Hill, 1997.*)

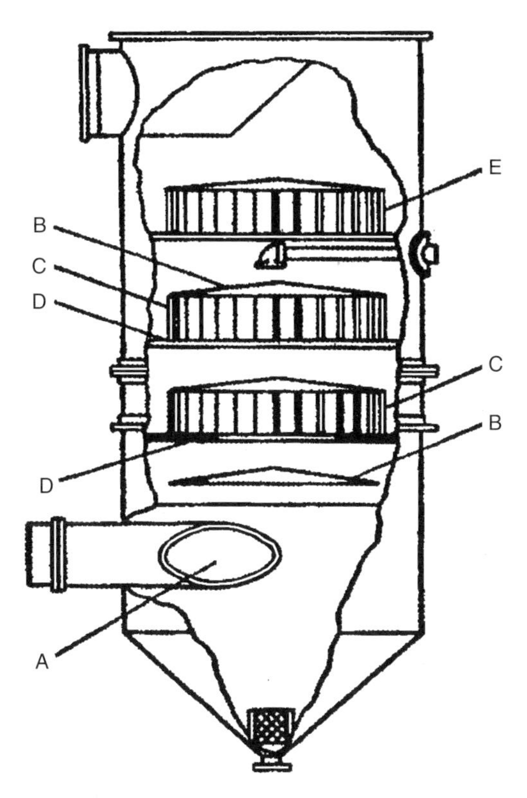

Fig. 11C7h A multi-washer scrubber (dust collector). Gas or air containing dust enters the gas inlet at the bottom of the device and moves upward confronting several stages of liquid curtain. The liquid captures the dust and carries it to the outlet at the bottom. A, gas inlet, B, deflector cones, C, vanes, D, shelves, E, entrainment separator (*Courtesy CMI-Schneible.*)

controls the pressure drop in the device. The liquid and gas are then directed to an entrainment separator, which is usually of the cyclone type. If a cyclone separator is used, the liquid with entrapped particles is discharged at the bottom of the cyclone and the gas at the top. Other scrubbers are similar to gas absorption towers. Still others, called mechanical scrubbers, use blowers or fans to provide the liquid spray that contacts the feeder gas.

Scrubbers are used frequently to remove dust and smoke from the exhaust air of various industrial processes. They have the advantage of fire prevention that bag filters and other types of dust collection do not have. They are used in paper mills, sugar mills and asphalt plants.

C8. ***separation of solids*** - Various methods are in use for separating particles of solid materials with methods that make a separation based on differences in size, density and other properties. Common methods include screening, flotation, dense media separation and magnetic and electrostatic methods.

Another method, air classification, is described in section E5 of Chapter 13.

C8a. ***screening (sieving)*** - is a common method for separating material into the size ranges wanted. Woven wire screens, perforated sheet steel or welded meshes of metal rods or wire are used. The screen openings are of uniform size and act as "go" gages, permitting the passage of gases, liquids and smaller pieces and preventing the passage of larger pieces. Material, in particulate form, is dropped or dumped on the screen, the screen is agitated, and the force of gravity normally drives the particles, if small enough, through the screen. The undersize pieces that pass through the screen are called *fines*, the oversize pieces that do not are call *tails*. The method is applicable for both small and large particles. Small pieces, below 2 in in size, are normally separated by flat screens which oscillate or vibrate to move the particles to screen openings. Most screening, however, takes place with particles ranging from 0.004 to 0.4 in (100 microns to 10 millimeters). Screen movement is produced by electric vibrators, eccentric drives or unbalanced rotating weights. Screens are often stacked so that particulate materials of several size ranges can be removed from the feed at different levels. To dislodge jammed particles in the screen, some equipment includes rubber balls below each screen. The balls bounce from the screen's movement and strike the underside of the screen. If pieces are 2 in or larger in size, the most common screening method utilizes a cylindrically shaped screen, open at both ends. The cylinder is tilted from the horizontal and rotated about its axis. Material, fed at the upper end, works its way through the cylinder and out at the lower end, if pieces are larger than the screen openings. If they are smaller, the pieces pass through the screen. The cylinder may have zones of different mesh openings; usually the finer meshes are first followed by progressively larger openings. Fig. 11C8a shows the cylindrical screen arrangement.

Screening may be performed for many reasons: removal of solid particles from a gas mixture,

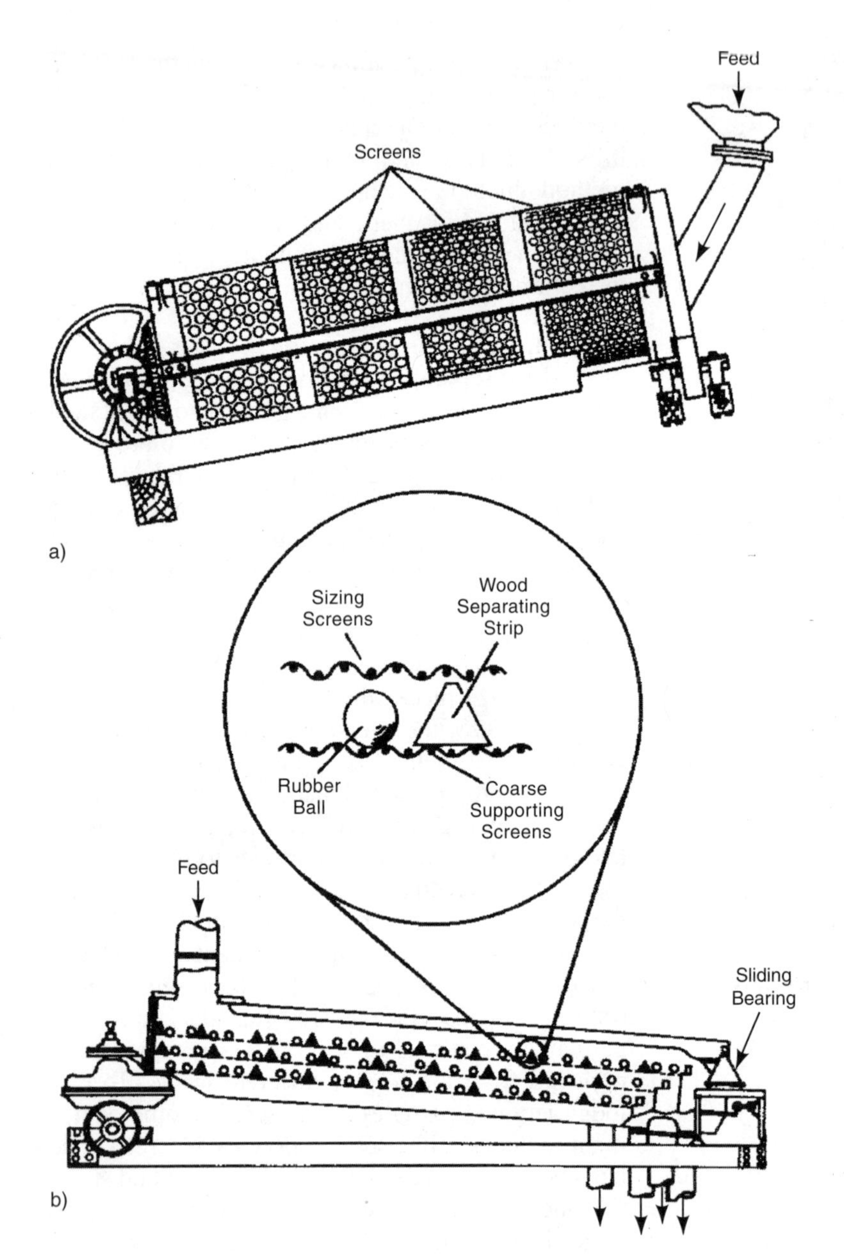

Fig. 11C8a Two types of screen systems for separating solid pieces or particles by size: a) rotating cylindrical (trommel) screen. b) oscillating flat (gyratory) screen. (*Courtesy McGraw-Hill. From McGraw-Hill Encyclopedia of Science and Technology, 1998, New York*).

removal of water or other liquids from a mass of solid material, removal of trash or other extraneous material from processed material, removal of oversize or undersize particles from solid particulate material so that it can be further processed, division of material into batches of certain minimum and maximum particle sizes, and modification of the size distribution of particles comprising a lot of particulate or powdered material.

C8b. ***flotation*** - takes advantages of differences in wettability of materials to effect separation. It is an important method for concentration of metal ores. The process is usable when materials can be reduced to particles smaller than 150 microns. If the material tends to be hydrophobic, that is, not readily wetted by water, it can be separated from materials that are hydrophilic, readily wetted by water. In an aqueous slurry of a mixture of both kinds of particles with suitable chemicals, air bubbles introduced to the slurry will tend to attract the hydrophobic particles but not the hydrophilic particles. The air bubbles will rise to the surface of the liquid, taking the hydrophobic particles with them. This forms a layer of froth on the surface of the liquid slurry and, when the froth is removed, the particulate material is removed with it. The term *froth flotation* is often used. Fig. 11C8b illustrates the process.

Flotation is most frequently used in mineral applications in concentrating the ores of copper, lead, zinc, molybdenum and nickel[5]. Other materials processed are fluorspar, barite, glass sand, iron oxide, pyrite, manganese ore, clay, feldspar, mica, spodumene, bastnasite, calcite, garnet and kyanite[5]. The process is used in coal cleaning to separate the coal from shale and pyrite. Another common use is the separation of silicate minerals.

The material to be processed must be conditioned first. The solid material must be reduced to suitable particle size. Then, oil or other surface-activity modifiers are usually added to the slurry to adjust the wettability of the solution and to disperse the solids in the slurry. Collectors, consisting of organic chemicals that, when coated on the material to be floated, change them from hydrophilic to hydrophobic, are sometimes used. Other additives promote the formation of a stable froth.

Several methods are available instead of air injection to create the bubbles that carry the hydrophobic particles to the liquid surface. They include electrolytic flotation, where a direct electrical current between electrodes creates bubbles of oxygen and hydrogen. Dissolved air systems can also be used. They introduce air to the liquid under pressure and then release the pressure, causing fine bubbles to develop in the slurry. Mechanical flotation cells use mixing impellers and baffles to disperse air in small bubbles in the slurry. *Skin flotation* involves the use of a thin layer or oil on the surface of the water. The feed material is immersed in the water; and the oil aids in the flotation of the wanted material, while the other materials sink. This approach is used in diamond mining and the processing of phosphates.

C8c. ***dense-media separation***[5] - is also known as *heavy-media separation* or *sink-float separation*. Materials to be separated are immersed in a liquid

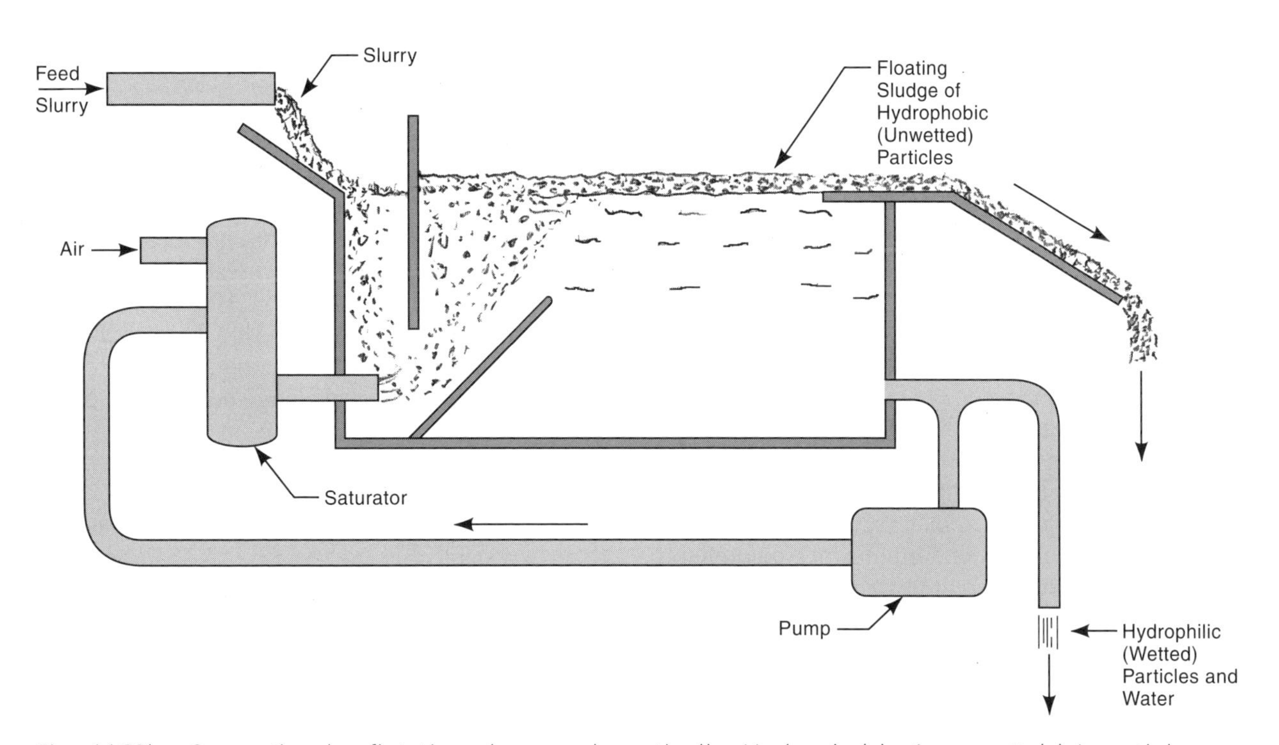

Fig. 11C8b Separation by flotation shown schematically. Hydrophobic (non-wettable) particles are carried to the surface of a water slurry by air bubbles and are skimmed off while hydrophilic (wettable) particles remain dispersed in the water.

of high specific gravity. The specific gravity of the liquid used is less than that of the more dense fraction of the feed material, but more than that of the less dense fraction. Particles of lighter, less dense material float, while particles of heavier, more dense material sink.

The liquid used is developed for the particular application. Normally, it is made by adding a dispersion in water of fine-grained particles of a dense material. Magnetite, arsenopyrite, or ferro silicon, are three minerals commonly used. Mild agitation keeps the particles in suspension, and the flotation effect of the liquid depends on its overall specific gravity, including both the water and the contained particles. The magnetite or ferro silicon can be removed by magnetic methods. The method is in common use in the separation of components of mineral ores, but the most notable use is the removal of foreign materials from coal.

C8d. ***magnetic separation*** - has been used for many years as a method for removing magnetic materials from mixtures. The method is not limited to ferromagnetic substances such as iron, nickel, and cobalt which can be permanently magnetized. It is also used for other materials that can be either attracted (paramagnetic materials) or repelled (diamagnetic materials) by a magnetic field and to separate materials that have different degrees of magnetic attraction. Commonly processed materials include: tramp iron, which is separated from ores, and iron ores, which are concentrated by the process, particularly when magnetite is a major ore ingredient. Silicates and ores of manganese, titanium, and tungsten that have some iron content are separated magnetically. Magnetic separation is also used to remove some magnetic contaminants from foodstuffs. Factors that affect the operation are particle size and the nature of the other materials in the mixture that may impede the movement of the magnetically attracted or repelled material. When there is a strong magnetic attraction, lift magnet equipment may be used. With this equipment, material on a belt conveyor passes under a magnetic belt that lifts the magnetic particles from the mixture on the conveyor and moves them to a discharge point as shown in Fig. 11C8d. Another common device is the magnetic pulley or drum which Edison invented for the processing of nickel ore[5]. It is shown in Fig. 11C8d-1. Some devices use several stages of separation, in

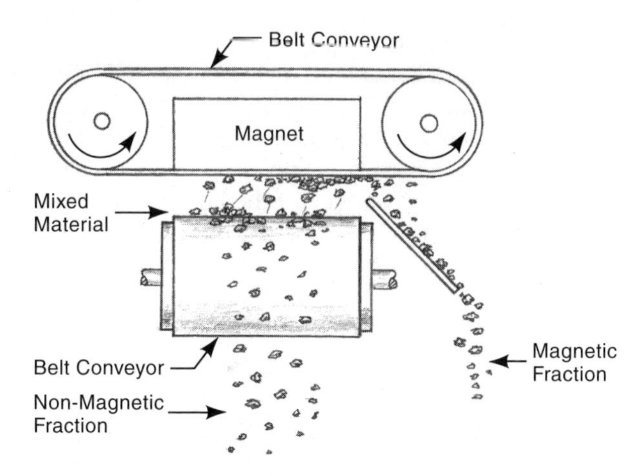

Fig. 11C8d The magnetic separation of materials.

which material remaining from one stage is run through another set of separators, perhaps several times, to enrich the magnetic concentrate and further purify the non-magnetic material. Superconducting magnets are sometimes used with paramagnetic materials or where conditions justify the extra separative forces they can provide. Magnetite processing is one such application; another is the treatment of kaolin (aluminum silicate clay)[5].

C8e. ***electrostatic separation***[5] - is based on the principle that conductive and non-conductive materials, in particle form, behave differently when they are electrically charged and then subjected to the influence of an electrical field. The electrostatic force between the charged particles and the electrical field moves the charged particles in one direction. However, conductive materials lose their charge and are then not affected by the electrical field. Non-conductive materials tend to retain an electrical charge and are subject to

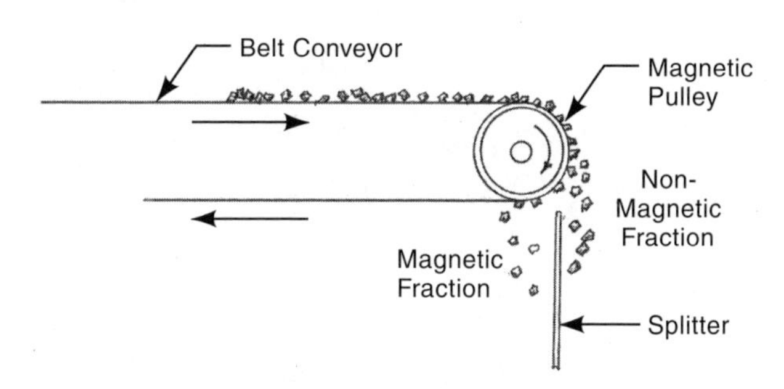

Fig. 11C8d-1 Edison's device for magnetic separation of nickel ore.

movement. This phenomena can be used to separate the conductive from the non-conductive materials. The force must be strong enough to overcome gravity, inertia, friction, cohesion and other drag forces. The materials are first charged by electrical contact, induction or ion radiation. Favorable separation forces are limited to particles smaller than about 4 mm (0.16 in) and larger than about 0.075 mm (0.003 in) if the particles are granular, but somewhat larger particles can be processed if they have long and thin shapes. Sometimes it is possible to separate material of different sizes.

The process can also be used to separate two non-conductive materials that assume opposite charges when they are charged by contact and then make contact with each other. The positively charged particles are attracted to the cathode of the separation system and the negatively-charged particles are attracted to the anode. Fig. 11C8e illustrates this method of electrostatic separation.

Electrostatic separation is widely used in the processing of mineral ores including the processing of iron ore concentrates and heavy mineral sands containing zircon, rutile and monazite[3]. Another common application is the recycling of plastics when nonferrous metals are to be separated from the plastic material or when one plastic material is to be separated from another.

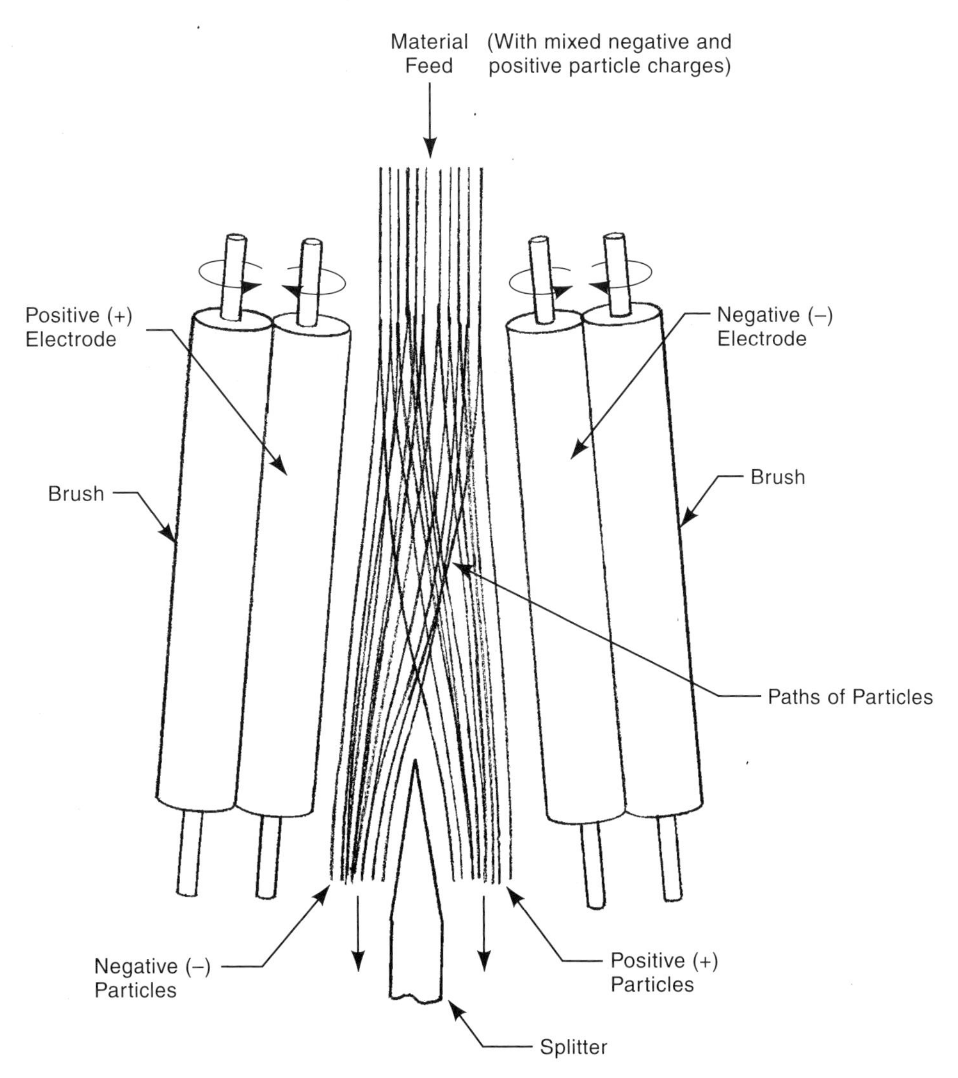

Fig. 11C8e Electrostatic separation of non-conductive materials.

In *electrostatic precipitation*[4], solid particles arc removed from gas streams electrostatically. In one method, the gas travels upward inside a vertical tube. A wire, suspended in the center of the tube is charged with a direct current potential of from 10 to 100 KV. This results in a small corona discharge in the area surrounding the wire. Particles suspended in the gas stream are charged by the corona and tend to flow toward the tubing wall. If the particles are liquid, they usually drip downward and are collected at the bottom of the tube. If the particles are solid, they may have to be scraped periodically from the tube walls, or loosened by vibration so that they fall to the bottom of the tube.

Electrostatic precipitation is used to remove smoke particles from exhaust gases, contaminating materials from chemical process plant gases, impurities from heating or air-conditioning air in buildings, acid fumes from chemical plants and petroleum refineries, and to recover tin, copper, and other metal oxides or other materials of value from exhaust gases.

C9. ***adsorption and ion exchange*** - are processes wherein molecules of gas, liquid or a solute, on contact with the surface of a solid with which they have an affinity, adhere to that surface. This effect can be harnessed as a means to separate mixed materials. (The molecules that adhere are referred to as the *adsorbate;* the material they adhere to is referred to as the *adsorbent.*) Examples of adsorption are the bonding of molecules of a reactant material to the surface of a catalyst, and the bonding of molecules of a contaminant in a gas or liquid to the surface of activated charcoal. The movement of charged electrons to a charged anode is another example, as is the segregation of surfactant molecules at the surface of a liquid. The principle is used in vacuum pumps that remove molecules from materials in the gas phase, and is an important factor in adhesion and lubrication. In the production of helium, adsorption is used to purify the gas to high levels. Adsorption is also used as a means of recovery of solvents used in painting, printing, and other industrial operations. Sulfurous gases and other pollutants or odorous compounds are removed from air in buildings by adsorption. In petroleum processing, it removes heavy materials from gases. The use of activated charcoal in military and industrial gas masks is another important adsorption application. *Adsorption* differs from *absorption* in that the molecules are not sucked up in bulk to fill the pores of another material, but only coat its surface in an extremely thin layer. The amount of gas or liquid material adsorbed is small. To offset this limitation, adsorbents are usually highly-porous solid materials with very large surface areas per unit of volume. They are natural or synthetic materials with a microcrystalline or amorphous structure. Activated carbon is the most common adsorbent. Others are molecular sieves, silica gel and activated alumina[5], polymeric adsorbents and zeolites[2]. Adsorption is used to dry compressed air by passing it through calcium chloride and, in the household, when baking soda or charcoal are placed in a refrigerator to remove unwanted odors[9].

A common method of carrying out adsorption is with fixed-bed adsorbers[8]. The adsorbent is placed in a bed 1 to 4 ft (0.3 to 1.2 m) thick that is supported by a perforated plate or heavy screen, inside a container. The feed gas is fed from above the bed and passes through it. The cleaned gas exits the container at the bottom. Periodically the operation is stopped and the adsorbent is regenerated with hot inert gas or steam. Dual containers of this type are normally used together, so the operation can proceed in one container while the other is undergoing regeneration of the adsorbent. The removal of absorbed molecules from the surface of the adsorbent is called *desorption.* Energy is required to effect desorption, hence the use of hot gas or steam. The adsorbent is cooled before reuse.

Other adsorption applications include: drying of gases and liquids, separation of carbon dioxide-methane, carbon monoxide-hydrogen, fructose-glucose, xylene-cresol, and olefin-paraffin mixtures, the production of ammonia synthesis gas, ozone enrichment, water purification with carbon, and the purification and recovery of enzymes, antibiotics, proteins, and vitamins[2].

In the *Ion exchange* process, solid material containing exchangeable cations or anions is brought into contact with an electrolyte solution in order to change the composition of the solution[8]. The process usually uses a solid polymer, either as a porous solid or a gel that dissolves some fluid-phase solvent[5]. Ions in the fluid being processed (usually an aqueous solution), replace other ions in the solid that are dissimilar but that have the same charge[5]. The process is used in water softening and demineralization.

In softening, sodium ions in the water are exchanged for calcium ions. In demineralization, both cations and anions are removed. The process also is used for obtaining metals from dilute solutions[8].

C10. ***electrolytic processes (electrolysis)*** - use a direct electrical current in a conductive liquid - an electrolyte - to bring about a chemical change in the components of the liquid. The process is carried out in an electrolytic cell that contains a chemical solution that has both negatively- and positively-charged ions. Positive and negative electrodes are immersed in the solution and are connected to a source of direct current. The current flows from the negative electrode (cathode) through the solution to the positive electrode (anode). The negatively-charged components of the solution are attracted to the anode and the positively-charged components are attracted to the cathode of the cell. When they contact the anode, the negatively-charged components give up their electrons and are reduced to a neutral element or molecule. The process can also be used to transform a substance in an electrode, when it gives up its electrons. Electrolysis is the basis of electroplating, described in section 8C1, electro polishing (8B3) and electroforming (2L2). In the chemical and metals industries, electrolysis is used for *electrowinning* (K1c1 below), the extraction of metals from ores, or *electrorefining* (K1c2 below), the purification of metals. Also see *aluminum* and *magnesium*, which are produced by electrowinning the molten salts of these metals. Other uses of electrolysis are the production of sodium and chlorine gas from fused sodium chloride (salt), sodium chlorate from a treated salt brine, chlorine and hydrogen from hydrochloric acid, oxygen and hydrogen from water, caustic soda and chlorine from salt brine and the refining of copper, nickel, cobalt, lead, tin and zinc. Fluorine, peroxysulfate and permanganate are other products made with this process[4].

C11. ***electro thermal processes***[4] - are simply processes that require extremely high heat levels, higher than those attainable from furnaces heated by combustion. Combustion furnaces are limited to about 3100°F (1700°C). Furnaces heated by electric methods, however, are capable of temperatures up to about 8100° F (4500°C). There are four types of electric furnaces: arc, induction, resistance and plasma.

Arc furnaces obtain their heat from an electric arc that passes between two or more carbon or graphite electrodes or between electrodes and the material charged in the furnace. The electrodes may or may not be consumed by the process. Arc furnaces are used extensively in steel production and alloy processing.

Induction furnaces rely on resistance to the electrical eddy currents set up in a conductive object that is exposed to an alternating magnetic field. Water-cooled copper coils surround the object or material to be heated. An electrical alternating current with a frequency of 60 to 500,000 Hz is passed through the coil, inducing eddy currents in the material surrounded by the coil. Commercial induction furnaces usually operate at frequencies below 6000 Hz. Resistance to these currents develops the necessary heat in the material. The principle is the same as that used in induction welding and brazing. Induction heating is also used in steel making and alloy processing.

Resistance furnaces use the electrical resistance of the material to be heated, or of another material to provide the necessary heat. The current, in this case, is due to direct connections rather than induction. When the material to be heated supplies the resistance to the current, the furnace is described as *direct-heated*. When another material is used to provide the resistance, the furnace is *indirect-heated*. Resistance furnaces are used to produce silicon carbide and graphite electrodes.

Plasma arc furnaces use an electric arc, but a gas is passed through the arc to generate the plasma. The heat of the arc ionizes the gas, which then passes to the chamber where the electrothermal process takes place. Constrictions in the flow nozzle of the gas cause it to develop temperatures even higher than that of the arc. Almost any gas can be used. The principle is the same as that used in plasma-arc welding (7C1f).

Arc and resistance methods are the most common electrothermal heating processes. They are used in the reduction of tungsten, molybdenum, and other ores in a hydrogen gas environment and in production of phosphors.

C12. ***drying*** - usually refers to the removal of small amounts of water or other liquid from solid, or nearly solid, material. (The removal of moisture from liquids is usually classified as distillation and the drying of gases as adsorption.) There are many drying methods, depending primarily on the form and nature of the material to be dried which may be in paste, granular, bulk, sheet, or fiber form. Both batch and continuous methods are used. Batch dryers have lower capacities and longer drying cycles. Heat is applied to the material to be dried by one of several methods, vaporizing the liquid in it or on it and carrying away the vapor. The following heating methods may be used: 1) direct application, - the most common method - when hot air or other gas contacts the wet solid, 2) indirect application, in contact dryers or conduction dryers, where the heat is transferred to the wet solid through a tube or wall that is impervious to the hot fluid, usually condensing steam, 3) by infrared radiation from electric lamps, resistance elements, or flame-heated ceramic elements, 4) by dielectric effects when the wet solid is placed in a high frequency electric field. However, dielectric heating is not widely used industrially. Many drying processes provide drying energy from a combination of both conduction and radiation from one heating source. The material to be dried may be turned over or agitated during the drying cycle to provide uniformity of heat transfer. Drying is a very common operation in the chemical industries. In addition to its use for chemicals in process, it is employed in the manufacture or processing of minerals, wood, biological materials, detergents, and waste materials. Drying is frequently the last step in the production of crystalline powders, and other products including foods and pharmaceuticals.

Cabinet, tray or compartment dryers are used when production quantities are suitable for batch methods and the material is in powder, granular, crystalline, or paste form, or is otherwise suitable for loading on trays or other containers that can be placed in a drying compartment. Dye materials and other high-value materials may be dried with this method. Foodstuffs, yarns and other textile materials are also processed. The drying compartments are normally well insulated and equipped with slots for trays, a heating device and fans or blowers for circulating the heated air. Baffles or vanes may be included to direct the heated air to all material in the chamber at a suitable velocity. A vacuum may be applied to provide more rapid drying, but normally is used when it is necessary to limit the drying temperature in order to avoid damage to the material being processed. A vacuum is also used when the material must be kept from contact with air, or when the liquid removed is worth salvaging. Pharmaceutical materials are sometimes dried in vacuum compartments. Other dryers use agitation to expose the feed material to the source of heat. Some drying equipment of this type use steam jackets around the compartment as the source of heat. Equipment with agitators is used when the material is sticky and not as easy to handle as the loose material that is dried with a tray system. Fig. 12H (Chapter 12) illustrates a cabinet/tray/compartment dryer.

Other batch systems use ovens or other drying chambers into which the material or component to be dried is placed. Racks, hangers, movable trucks, and other holding and handling devices may be used. Lumber, painted parts, textile skeins, and hides are materials dried with this kind of arrangement.

Tunnel dryers are used instead of compartment dryers when the quantity of material to be dried is large enough to justify continuous processing. Tunnel dryers use conveyors or truck carriers to move the material through a long, heated tunnel or enclosure. This approach is common when the material must be dried slowly, and when it is bulky. Bricks, ceramics, textile skeins, hides, and lumber are often dried with this approach.

Granular, flaky, and fibrous solids are processed in tunnel dryers with mesh conveyors that allow heated air to pass through the conveyor and upward through the material as it is conveyed through the tunnel. Drying rates with these conveying-screen dryers are faster than with batch compartment dryers because the material can make contact with a greater flow of heated air. The principle is illustrated in Fig. 11C12. Materials processed in these dryers include: cotton and rayon fibers, silica gel, cellulose acetate, starch, pigments, insecticides, dyes and calcium carbonate[4].

Rotary dryers are used for bulk materials that are sufficiently dry so that they can be fed into the device. The dryer consists of a cylindrical shell that rotates on its axis and is set at a slight angle to horizontal. Material is fed into the higher end and discharged, after drying, at the lower end. Fig. 11C12-1 shows such a dryer. Flights mounted on the internal walls of the cylindrical shell help distribute the material so that all of it contacts the heat.

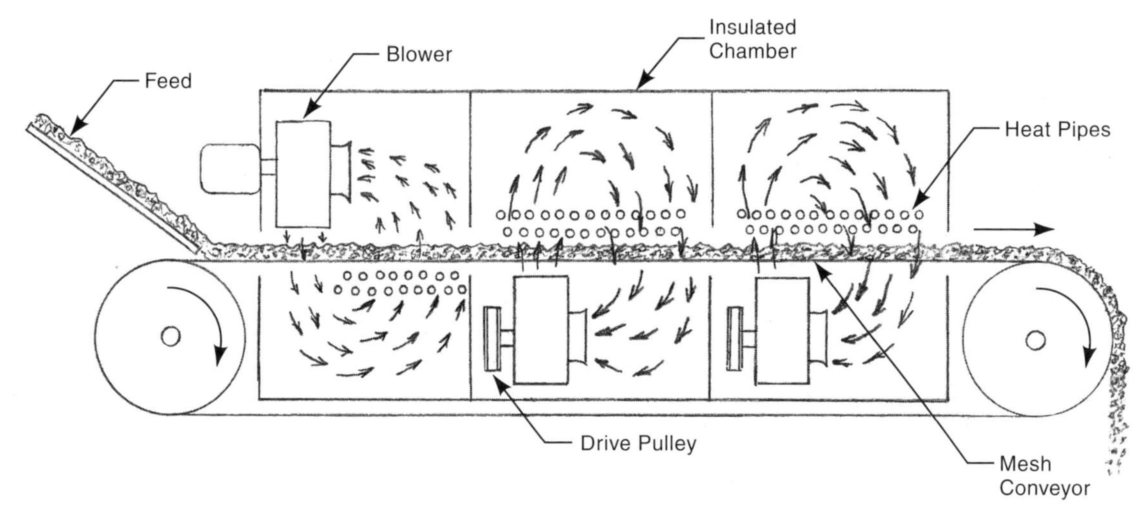

Fig. 11C12 The principal of conveying screen dryers.

Heated air is fed into the shell, usually from the discharge end, to provide drying action. Additional heat may be applied externally to raise the temperature of the shell walls.

Pneumatic-conveyor flash dryers (dispersion dryers) use a high-velocity stream of hot gas to both convey and dry granular, free-flowing, solid materials. Coal, sludges, sodium chloride, filter cakes (broken up beforehand) and whey, are dried with this method. The gas temperature may be as high as 1400°F (760°C) because the material is only briefly in contact with it. Some dried material is often recycled with feed material to help keep it dispersed.

Turbo dryers have vertically-oriented cylindrical or polygonal containers within which a series of horizontal trays slowly rotate. Heated air enters from the bottom and flows upward around a series of baffles, past finned reheaters, and over each tray. Material enters from the top and falls downward onto the upper trays. A series of scrapers and levelers maintain a thin layer of material on each tray and push the

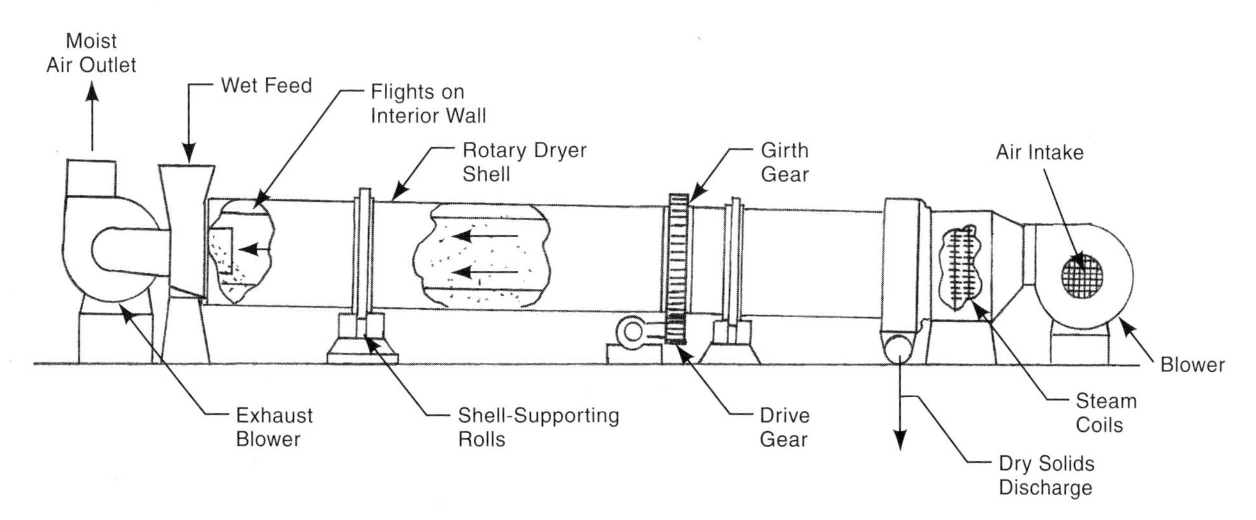

Fig. 11C12-1 A rotary dryer with countercurrent heated-air flow. Wet material is fed to the upper end and is discharged, after drying, at the lower end. Heated air, to provide drying action, is fed from the lower end.

material to openings where it falls to the next lower tray level. Gradually, the material progresses to the bottom of the container and by that time, it is dry. Turbo dryers are used for fragile materials.

Fluidized bed dryers are sometimes used with particulate material. Hot gases are introduced at the bottom of a vertically oriented cylindrical vessel. Material to be dried is introduced at the top. Drying gas flow is sufficient to keep the material fluidized and drying is usually quite rapid. Dried material is removed from the bottom of the cylinder, and drying gas exits from the top. If necessary, a dust collector may be incorporated in the equipment to remove fine particles of material from the drying gas. Fig. 12H4 (in Chapter 12) illustrates continuous fluidized bed drying of foodstuffs.

The *double-drum dryer* shown in Fig. 11C12-2 is used to dry liquid material to the solid state. The liquid - solution, slurry, or paste - is fed to a pool in the space between the heated metal drums. A thin wet film adheres to each drum's surface. As the drums slowly revolve, the film dries and the dried material is scraped off by doctor blades. It falls to a conveyor below. The material is on the heated drum surface for only a short time so there is little chance of overheating.

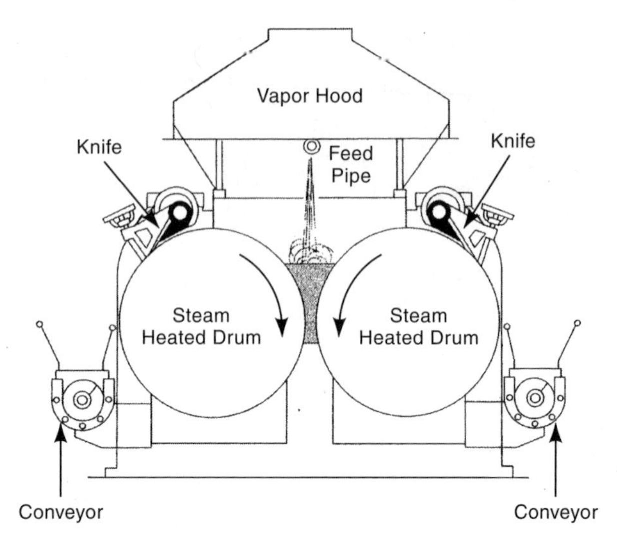

Fig. 11C12-2 A double-drum dryer with center feed of liquid material. Doctor knife blades scrape the dried material from the drum surfaces. (*Courtesy Buffalo Technologies Corp.*)

Vacuum rotary dryers are used for large batches of materials that must be dried in the absence of air or where the solvent is to be recovered. The material is placed in a horizontal cylindrical chamber that

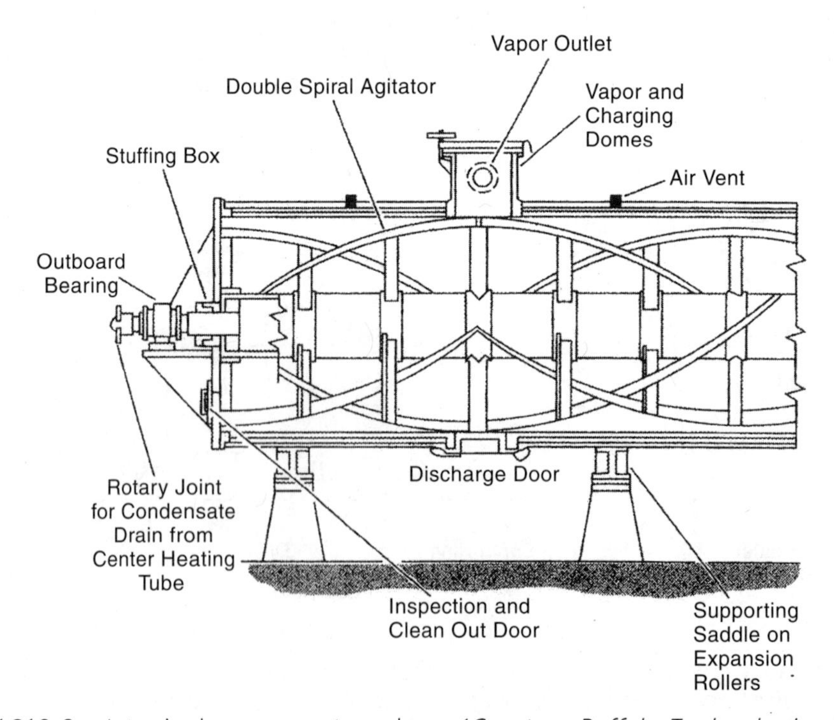

Fig. 11C12-3 A typical vacuum rotary dryer. (*Courtesy Buffalo Technologies Corp.*)

rotates during the operation. Spiral blades in the chamber agitate the material to be dried. Heating comes from steam, hot water or other medium circulating in a jacket surrounding the chamber. Fig. 11C12-3 illustrates this type of equipment.

Pressing and centrifuging are two other methods, not involving thermal vaporization, that are used in drying or partially drying materials.

C12a. ***spray drying liquid materials*** - Liquid materials, including suspensions of small solid particles, can be sprayed with very fine droplets into a stream of hot air or hot gas (up to 1400°F – 760°C). In some devices, the hot air is introduced tangentially into a vertical cylindrical container, and circulates in the container in a spiral direction. The liquid is sprayed into that container where it contacts the hot air. The liquid evaporates and dried particles fall to the bottom of the container. In one arrangement, the hot air enters at the bottom of the container and the excess escapes upward and through a central discharge duct at the top. Small particles may escape with the hot air discharge and are separated by cyclones or bag filters. Other equipment designs use different arrangements for the input, flow and exit of material and hot gas. In many designs, the dried particles and hot gas both exit from the bottom of the container. Water solutions and slurries are commonly spray dried. One of three atomization methods is used to create the spray, depending on the nature of the liquid. Hydraulic pressure or air atomization are used for lower viscosity liquids and a spinning disc is used for viscous liquids. However, slurries or thick liquids above about 1500 centipoises do not atomize well. Spray drying is rapid. Typical drying times are a matter of seconds. Another advantage is that the dried material tends to form spherical granules of uniform size. The method has become widely used. Major applications are the drying of foodstuffs: milk and milk products, coffee extract, and fruit juices. Certain chemical catalysts are also processed by spray drying. Fig. 11C12a shows two arrangements of spray dryers.

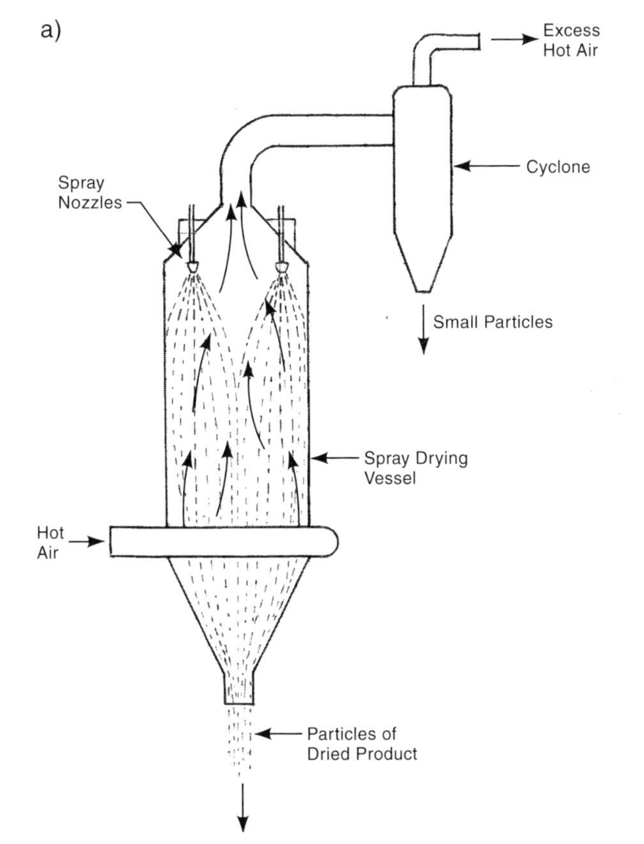

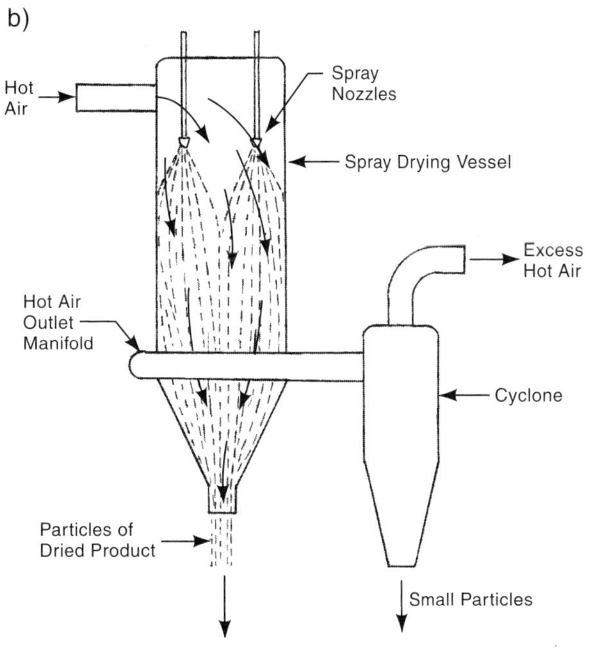

Fig. 11C12a Spray drying in typical arrangements, shown schematically, a) with countercurrent flow of hot drying gas, b) with concurrent flow of hot gas.

Infrared radiation from heat lamps is another method for drying liquids. Its major application is drying paint after it has been sprayed, but this use is limited to thin films.

C12b. ***freeze drying*** - is used with heat sensitive materials such as some foods, vitamins and pharmaceuticals. It is described in 12H5.

D. Size Reduction

Reductions in size of pieces or particles of solid materials is required with many ores, coal, stone, slag, concrete aggregate, clays, kaolins, carbonates and sulfates, mica, agricultural grains, and vegetables. Size reduction occurs in the manufacture of refractory brick, pigments, various other chemicals, soap, fertilizer, Portland cement, flour, cereals and other foods, vegetable by-products, starch, various metals, and the processing of waste materials. It may be done to facilitate a chemical reaction, or some physical operation such as the separation of materials, to facilitate handling or measurement or to put a material in a form needed by customers. Pulverizing to a fine powder requires size reduction in several steps. Size reduction can be classified in three or more stages: 1) primary crushing—12 to 27 in (0.3 to 0.7 m) pieces reduced to 4 to 9 in (10 to 23 cm), 2) secondary crushing—4 to 9 in (10 to 23 cm) reduced to 1/2 to 1 in (13 to 25 mm) in one or two stages, 3) pulverizing—1/2 to 1 in (13 to 25 mm) reduced to 60 – 325 mesh[4]. Table 11D summarizes this classification and indicates typical size reduction methods for each stage. Some methods suitable for one of the stages may be suitable for another stage, but no method is suitable for all stages. Size reduction methods can also be classified by the means used. The three most common methods are: crushing, impact, and attrition (rubbing). Cutting is also sometimes used. Size and hardness of the feed pieces are important factors in the choice of method.

D1. ***crushing*** - is most often effected by a slow application of strong force. A *jaw crusher* has a swinging plate or "jaw" which is connected to a double toggle that is moved by an eccentric on a large flywheel. Each forward movement of the jaw exerts heavy pressure on the material between the jaws and breaks up the large pieces. Crushed pieces fall between the jaws. Large lumps may be hit several times as they work their way down and out of the machine. See Fig. 11D1. *Gyratory crushers* use a cone-shaped pestle that moves eccentrically in a bowl-shaped hopper. Crushed pieces fall to the bottom and exit through the hopper. A gyratory crusher is illustrated in Fig. 11D1-1. These machines are widely used in the first-step crushing of rock materials. Both jaw and gyratory crushers - of an appropriate size - may be used for primary and secondary crushing. Gyratory crushers with wide-angle cone-shaped pestles are often called *cone crushers.* Cone crushers are used more for secondary crushing. Roll crushers, described below, are also primarily used for secondary crushing, except with coal and other friable materials.

Table 11D Size Reduction Classifications and the Common Methods Used

Classification	Feed stock size	Product Size	Reduction ratio	Method Used
Primary crushing	12 to 27 in. (30 to 69 cm.)	4 to 9 in. (10 to 23 cm)	3:1	Jaw, gyratory and cone crushing
Secondary crushing 1 or 2 stages	4 to 9 in. (10 to 23 cm.)	1/2 to 1 in (13 to 25 mm.)	9:1	Hammer mill, jaw, gyratory, cone, smooth roll and toothed rolls
Pulverizing	1/2 to 1 in	60 to 325 mesh	60:1	Ball and tube, rod, hammer, attrition, ball race and roller mills

This table is based on data from the McGraw-Hill Encyclopedia of Science and Technology, 8th ed., 1997, McGraw-Hill, New York.

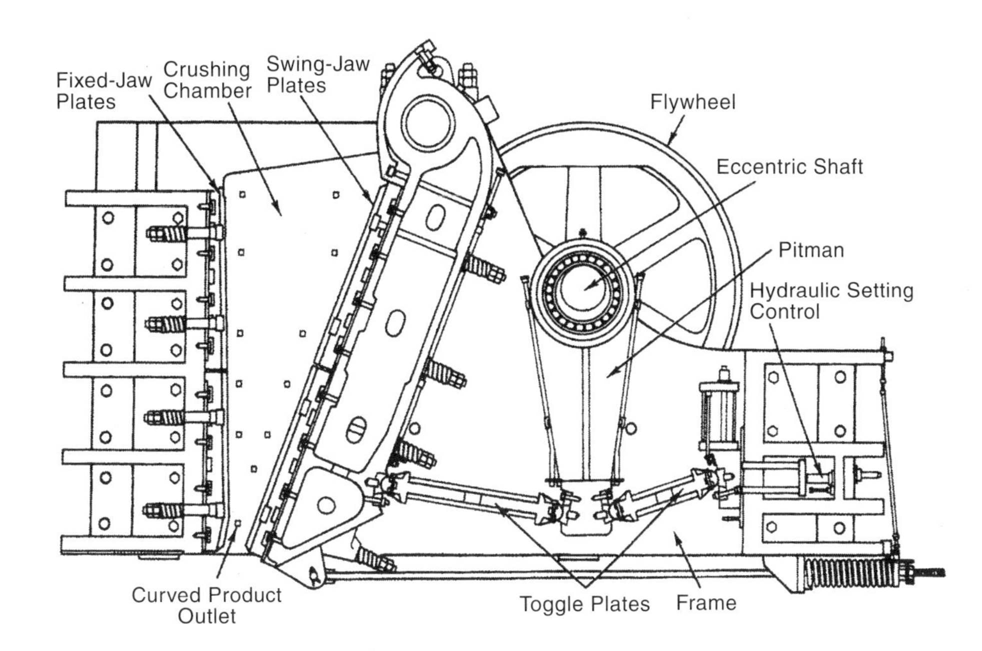

Fig. 11D1 A Blake-type jaw crusher. (*Courtesy Metso Minerals OY.*)

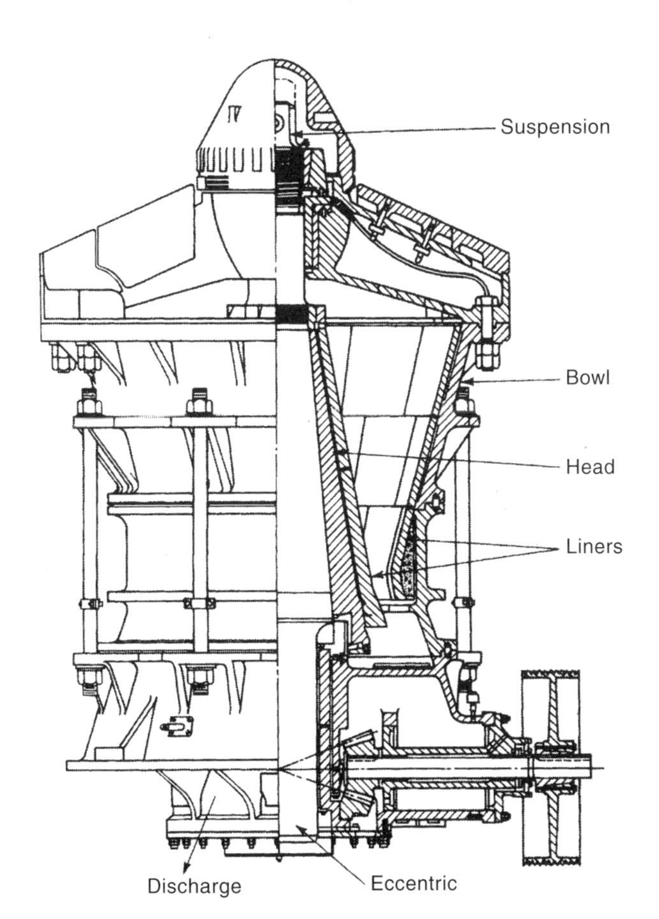

Fig. 11D1-1 A gyratory crusher. (*Courtesy Metso Minerals OY.*)

D2. ***secondary crushing*** - is performed with roller mills (roll crushers) and hammer mills, as well as jaw, gyratory, and cone crushers, engineered for secondary crushing. Secondary crushing follows primary crushing that is normally performed with jaw or gyratory machines.

Roll crushers have either one or two rolls, sometimes with teeth. Material to be crushed moves by gravity, either between the two parallel rolls or between a single roll and the hopper wall. The rolls may be smooth, corrugated, or toothed. Toothed rolls are used for crushing coal, smooth rolls for rock and ore. Toothed-roll crushers can handle a variety of materials and sizes except very hard materials. Fig. 11D2 illustrates a typical roll crusher.

Fig. 11D2-1 illustrates the principle of *impact or hammer milling*. In this machine, hammers mounted on a rotating spindle strike pieces of feed material as they are fed into the hopper. Impact forces rather than crushing forces cause the pieces to break. The broken pieces fall through an opening or a series of openings in the hopper bottom; larger pieces are struck again by the hammers until they break. This kind of equipment is used commonly to crush coal and limestone, but can reduce the piece size of many materials including pastes and clay, tree bark, leather, steel machining chips and hard rock[8].

D3. ***pulverizing*** - is carried out by both impact (rapid blows against the pieces to be pulverized)

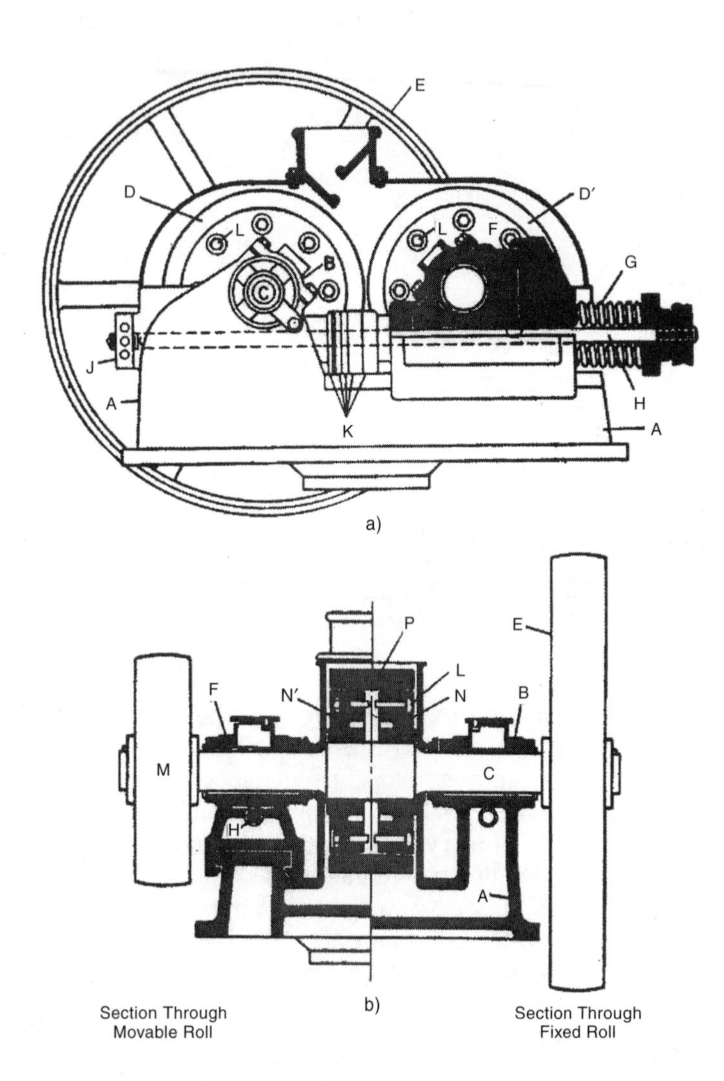

Fig. 11D2 A typical roll crusher: a), elevation view, b), section view. A, frame; B, fixed bearing; C, fixed-roll shaft; D, fixed roll; D′, movable roll; E, main drive pulley; F, movable bearing; G, spring; H, tie rod; J, adjusting nut; K, shims; L, tie bolts; M, movable-roll drive pulley; N, N′, main roll castings: P, roll tire. (*by permission, from Introduction to Chemical Engineering, W. L. Badger and J. T. Banchero, McGraw-Hill.*)

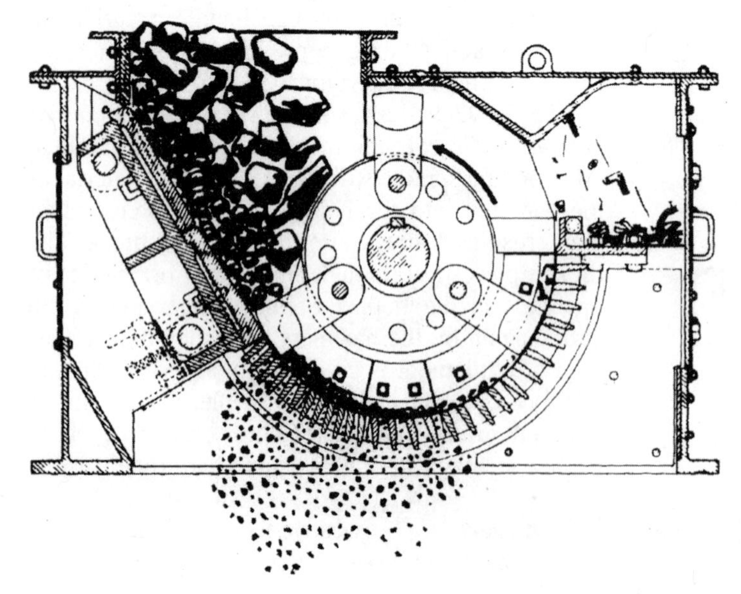

Fig. 11D2-1 A hammer mill (impact mill). Size reduction results primarily from impact but there is also some attrition (rubbing action) of pieces that do not fall through the grating bars. (*Courtesy Pennsylvania Crusher Corp.*)

and attrition (abrasive or rubbing action). Crushing forces are also employed. *Hammer mills, ball mills, rod mills, tube mills, ball race,* and *roller pulverizers* are all used. In some pulverizers, the material passes through the equipment once with no removal of pulverized material during the operation. These machines are called open-circuit pulverizers. In closed-circuit pulverizers, the processed material is conveyed to a classifier and oversize pieces are returned to the pulverizer for reprocessing. Fig. 11D3 illustrates a *ball-tube mill*, sometimes called a *tumbling mill*, with three stages of pulverization. The machine shown is an open-circuit pulverizer. As the tube rotates, the balls and material tumble, and the balls impact the pieces of material, breaking up and abrading them. Similar equipment, with single-stages and short cylinders, are called *ball mills*. Sometimes steel rods are used instead of balls and the machines are called *rod mills*. Sometimes, when steel would contaminate the material, quartz balls and cylinder liners are used. In some applications, notably cement making and some ore processing, water is added to the feed. The finished material exits the machine as a slurry[4].

Roller pulverizers use rolls bearing against a surface to crush and pulverize material. Centrifugal effect, or spring pressure, provides the necessary force when the rollers roll along the walls of a cylindrical vessel. Such machines are sometimes called *centrifugal grinders*. *Roller mills* use pairs of cylindrical rollers rotating toward each other but at different speeds so that material placed between the rolls experiences a shearing action. This kind of equipment is used to grind grain into flour. Other pulverizers use a series of steel balls, rolling in a circular raceway, to pulverize by crushing and attrition. Spring pressure provides the necessary force. Pulverized material discharges from the circumference of the circular raceway. A *ball-and-race pulverizer* is illustrated in Fig. 11D3-1. Cement rock is pulverized with such equipment. *Pan crushers*, similar to the muller illustrated in Fig. 1B6a (in Chapter 1) for the processing of foundry sand, are also used to reduce particle size for medium-hard and softer materials such as clays, shales, cinders, and barites[5]. *Stirred media mills* use balls, stones, or sand, as media with the material to be pulverized. Such mills have a central paddle wheel or an

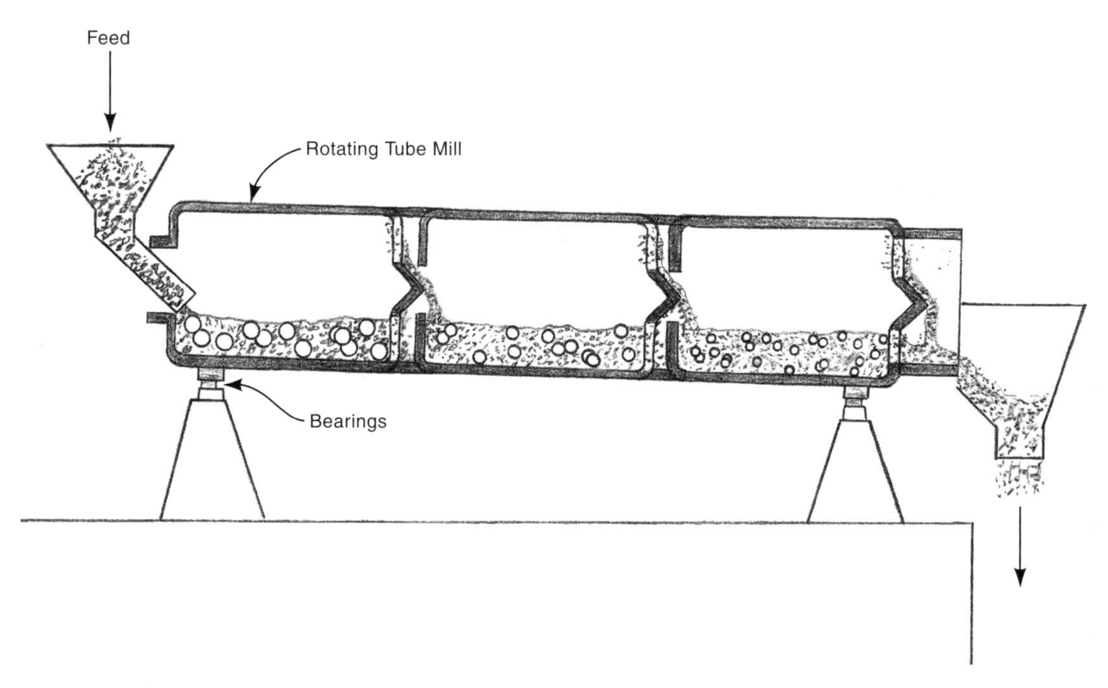

Fig. 11D3 A tube mill pulverizer with three compartments, each containing balls of a different size. As the tube rotates, the balls strike the material and break it into small particles. The material feeds progressively through the three compartments.

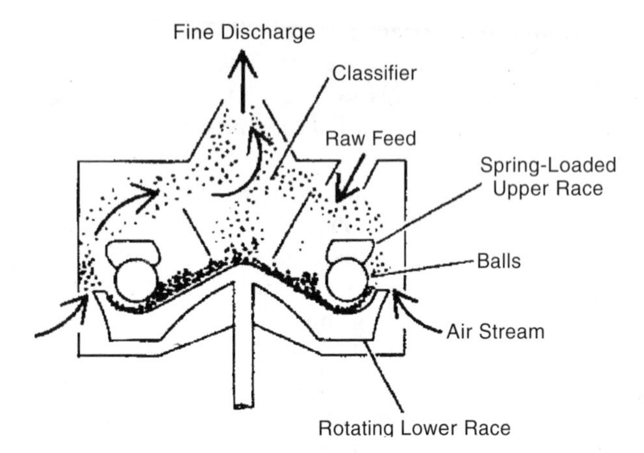

Fig. 11D3-1 A ball and race pulverizer. Coarse raw material is ground by crushing and attrition between balls and races. The air stream conveys the ground material to the discharge port. (*From the McGraw-Hill Encyclopedia of Science and Technology, 1998, New York. Reproduced with permission of McGraw-Hill.*)

armature with discs to stir the media and material, usually as a wet slurry. *Vibratory mills*, with media, are also used to pulverize material. They are primarily used, in a dry state, to mill hard materials[5].

E. Size Enlargement

Several methods are used to gather small particles into larger masses. The operation takes place in some chemical processes to improve handling, prevent dust, and improve processibility. When particles are caused to adhere together, the process is termed *agglomeration*. There are two basic methods for achieving agglomeration: 1) *agitation methods*, which move or circulate the particles in the feed mixture so that they come in contact with one another and adhere. Adherence may result from molecular, electrostatic, or mechanical interlocking[5]. There are several agitation methods: fluidized bed, mixing of various types, and tumbling. (Agitation methods may be referred to as *granulation*.), 2) *compression methods*, wherein particles are forced together. In both basic methods, additives may be introduced to the feed mixture to promote the adhesion of particles to one another. Additives include: binders, wetting agents, surfactants, and lubricants. Other additives may be used to provide such properties as color or flavor (with foods and pharmaceuticals) to the agglomerated product. Binders can be solid or liquid. Blending, milling, mixing, drying, and size classification operations may all precede the agglomeration step. Heat also may be applied to aid the adherence of particles, particularly for polymers, or other materials with sufficiently low softening points. Fine iron ore particles are moistened to help them adhere together and then sintered (See 2L1d) to fuse the particles together for further processing.

Compression agglomeration is typified by tablets, pellets, and briquettes, but extrusion and rolling may be used if the need is for continuous lengths of sheet or some cross-sectional shape. Powder metal parts manufacture, as described in section 2L1, is an agglomeration process. For tablet-making and other mass-produced products, the common arrangement uses a turntable machine. As the table rotates, powdered material fills cavities in the table, excess material is wiped off, rams compress the material into tablet form and the tablets are ejected.

F. Fermentation

Fermentation is the decomposition of organic matter in the absence of air or oxygen. It is the result of life processes of microorganisms, yeast, bacteria, or molds. Fermentation is generally accompanied by the evolution of gas. The most notable example is the fermentation of sugar that converts it into alcohol and carbon dioxide. This is the basic operation involved in the manufacture of beer, wine, brandy, and other alcoholic beverages, and in raising bread. It is described in section 12J4 (Chapter 12). Ethanol as a gasoline additive and for other uses is made from fermentation of grain. Lactic acid, butyl alcohol, acetone, synthetic insulin, monosodium glutamate, and acetic acid are made from various bacteria. Mold fermentation is used to produce enzymes, gluconic and citric acids, some antibiotics including penicillin, riboflavin (vitamin B_2) and vitamin B_{12}.

G. Mixing Methods

Mixing methods are employed to provide a homogeneous blend of material and uniform distribution of any added ingredients. Mixing may also take place to disperse heating or cooling, to promote crystallization or dissolution, to create emulsions, and to facilitate chemical reactions. Liquids, loose solids, and gases are likely to undergo mixing operations. A variety of mixing equipment is available and a choice is made, depending on the nature of the material, the purpose of the operation, the quantity to be mixed, and whether the mixing is part of a batch or a continuous process. Mixing gases with gases is the easiest task. Mixing liquids with other liquids or with gases is common, and not quite as straightforward, but is quite feasible. Mixing liquids with solids, if the portion of solids is small, is often carried out with the same equipment used for mixing liquids with liquids. Mixing solids with other solids may present the most difficult mixing problems.

G1. ***mixing gases with gases*** - is seldom difficult and is usually carried out by injecting one of the gases at high speed into a vessel containing the other gas. An impeller (Fig. 11G3) may be used to insure thorough blending.

G2. ***mixing gases with liquids*** - This operation is performed with a number of different methods, depending on the application, the materials involved, and the temperature and pressure that can be used. The gas is injected into a vessel containing the liquid and a mixer. Impellers (Fig. 11G3) or other sparging devices may be used to disperse the bubbles of the injected gas. These devices can include multiple and special nozzles, a series of angled plates in the path of the bubbles, or other

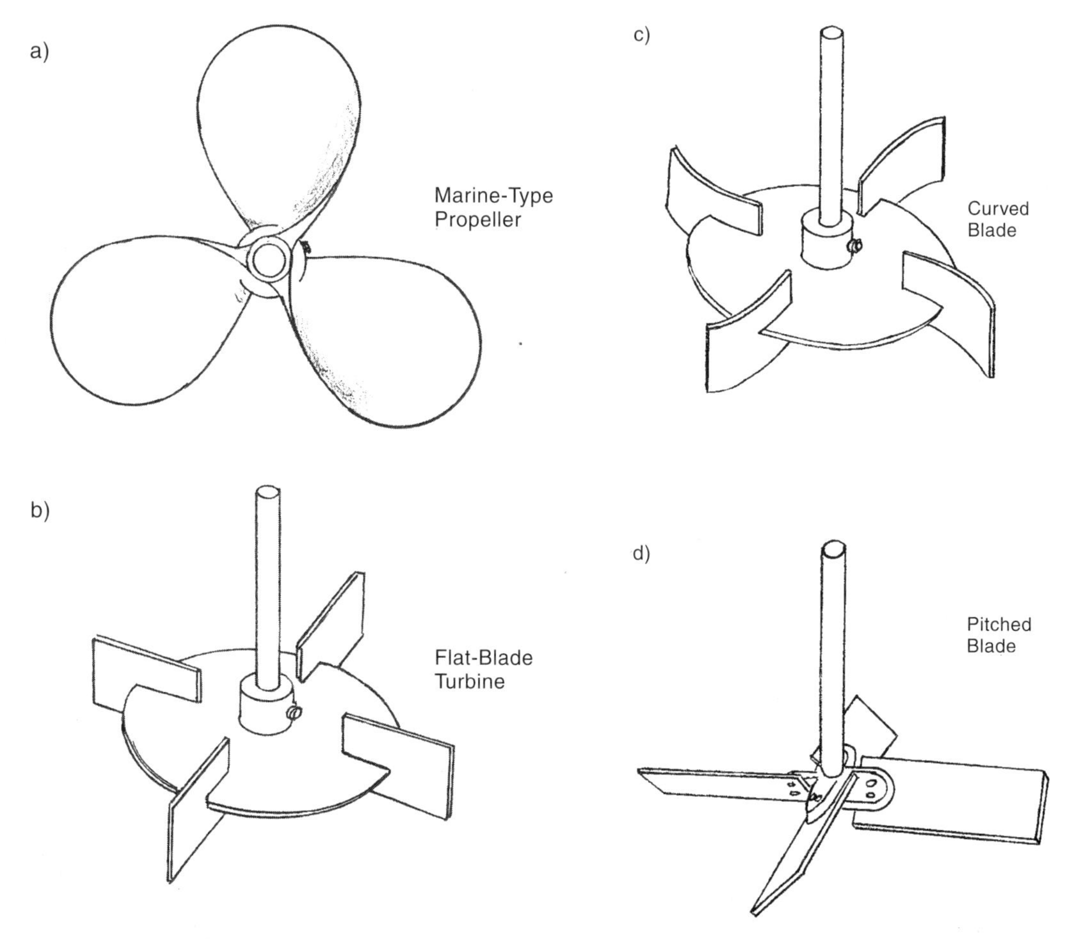

Fig. 11G3 Typical impeller-type mixer blades: a) marine-type propeller, b) flat-blade turbine, c) curved-blade turbine, d) pitched blade turbine.

devices to make the gas bubbles as small as possible, as well as widely dispersed. The objective is to have the largest bubble surface possible per volume of gas introduced to the liquid.

Applications of gas-liquid mixing include operations involving the hydrogenation of oils, aeration of waste water, fermentation operations, the removal of impurities from gases, carbonation of water and the chlorinating, fluoridating or brominating of liquids.

G3. ***mixing liquids with liquids*** - Mixing may be different for immiscible liquids than it is for those that are miscible. Mixing depends on creating turbulence between the two liquids. This can be accomplished with a jet of faster-moving liquid entering a liquid that is stationary or moving slowly. Commonly used are propeller and turbine mixers as illustrated in Fig. 11G3. Turbine blades may be flat or curved as shown in views b) and c). Propellers, such as in a), have become more important. Propeller and flat-turbine mixers include containers with baffles that promote intermixing of the liquids. Baffles are attached to the container walls, except when some solids are to be kept in suspension. In that case, the baffles are spaced a short distance from the container walls. Propeller mixers are sometimes placed at an angle, or off-center in the container. Liquid flow from turbines is radial while that from propellers is axial. Fig. 11G3-1 illustrates these effects.

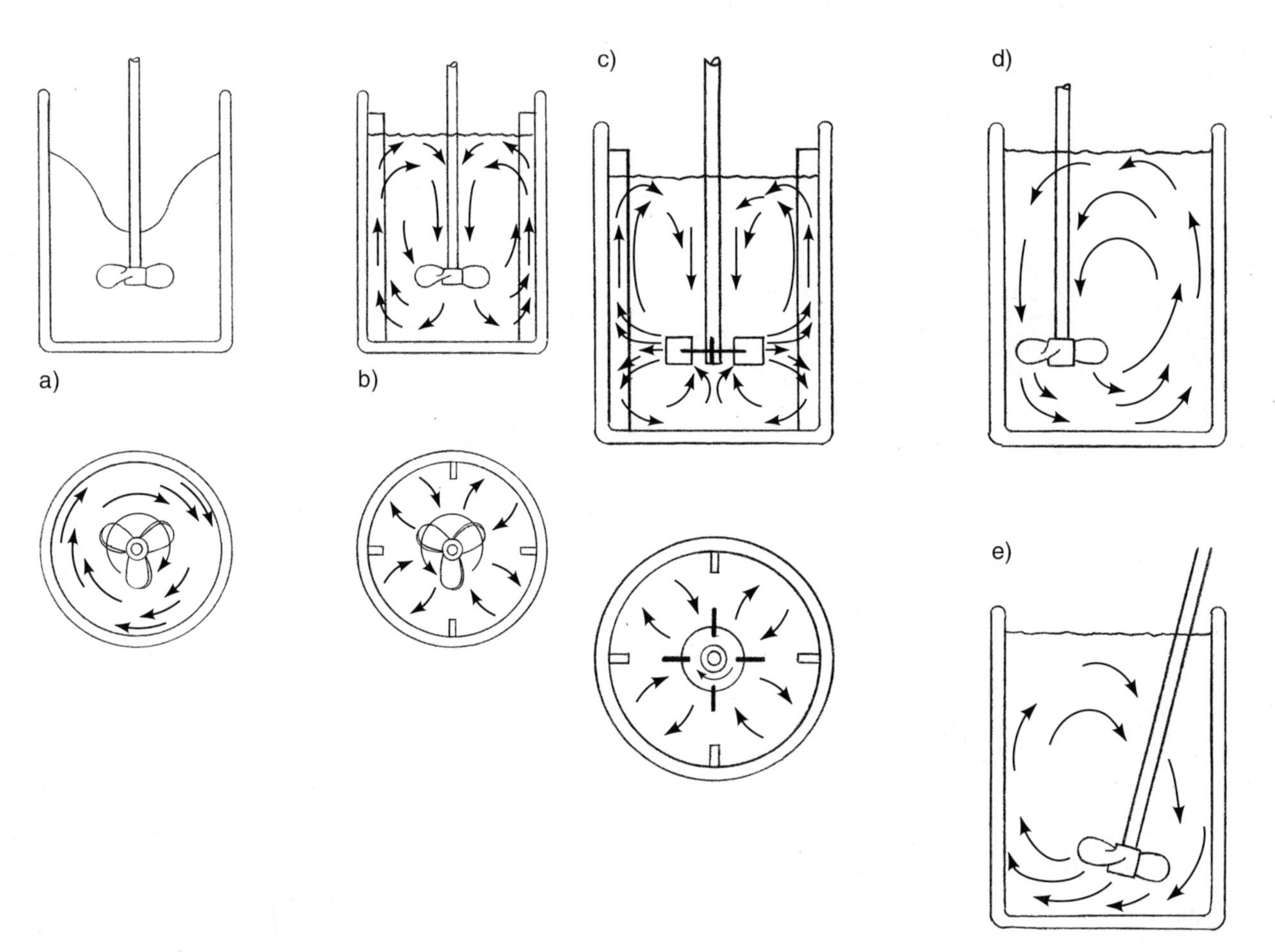

Fig. 11G3-1 Several mixing arrangements for liquids: a) propeller mixer in an unbaffled container, b) the flow of liquids in a baffled container with a propeller mixer, c) the flow of liquids in a baffled container with a turbine mixer, d) a propeller mixer off-center in the mixing container, e) a propeller mixer set at an angle in the mixing container.

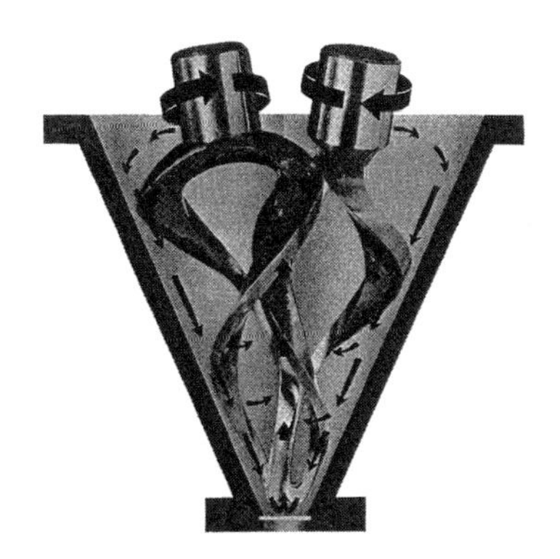

Fig. 11G3-2 A ribbon turbine with double spiral, for high-viscosity liquids. (*Reproduced with permission from the McGraw-Hill Encyclopedia of Science and Technology, 1998.*)

High-viscosity liquids require different mixing methods than those used for low-viscosity liquids. Higher-viscosity necessitates more extensive apparatus and greater power to ensure intermixing because of the reduced flow of such liquids. Fig. 11G3-2 shows a ribbon turbine with double spiral mixer used for mixing high-viscosity liquids. Fig. 11G3-3 shows a helical mixer, also used for these materials.

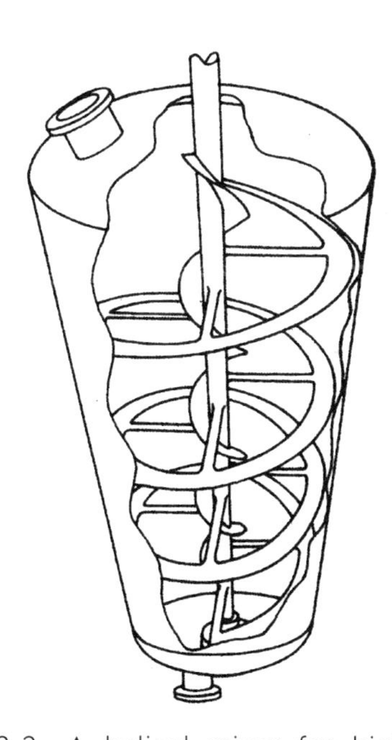

Fig. 11G3-3 A helical mixer for high-viscosity liquids. (*Reproduced with permission from Perry's Chemical Engineering Handbook, R.H. Perry and D.W. Green, McGraw-Hill, 1997.*)

One issue in mixing liquids is that the energy introduced during mixing will increase the temperature of the material. This can be a problem if the material is sensitive or unstable.

G4. ***liquids-solids mixing*** - The mixing method selected depends on the particular materials involved and their proportions in the mixture. Liquid mixing devices with propellers or turbines can be used if the solid material can be suspended in the liquid. Suspension can occur when the particles are not too large and the amount of solids is not too great. Low viscosity of the liquid is also required. When liquids are added to solid materials much of the solids-mixing equipment pictured in Fig. 11G5 can be used. The specific type of equipment used depends on how the solid material is affected by the liquid, and how much liquid is added. Some liquids may add lubricity to the mixture; others may cause caking. Ribbon [Fig. 11G3-2 and Fig. 11G5(e)], screw [Fig. 11G5(f)], rotor, muller [Figs. 11G5(g) and 11G5(h)], Banbury (Fig. 4A4c in Chapter 4) and kneading (Fig. 11G5-2) mixers are most suitable for condition where caking may occur.

G5. ***solids mixing*** - A variety of equipment is available for mixing solid materials. Which device is chosen depends on the nature of the material and the purpose of the operation. Though blending of materials is the most common purpose, mixing operations may be involved in heating, cooling, coating, conveying, polymerizing or reacting[5]. Several types of mixers are illustrated in Fig. 11G5. The tumbling mixers, shown in (a), (b), (c) and (d) are used for gentle blending of abrasive materials and dense powders.[5] Mixers (a) and (b) are adapted to lighter, dry solids. Ribbon mixers, sometimes referred to as dry mixers, are capable of handling a variety of materials from light weight dry powders or granules to materials that are sticky or fibrous. These mixers have a trough with a semi-cylindrical bottom, and a shaft that includes a series of spiral, ribbon-like mixing elements, as well as some straight stick-like elements at an angle to the shaft. The spirals lead in both directions. As the shafts rotate, the material to be mixed is moved back and forth and is intermixed. The mullers shown in (g) and (h) can break apart

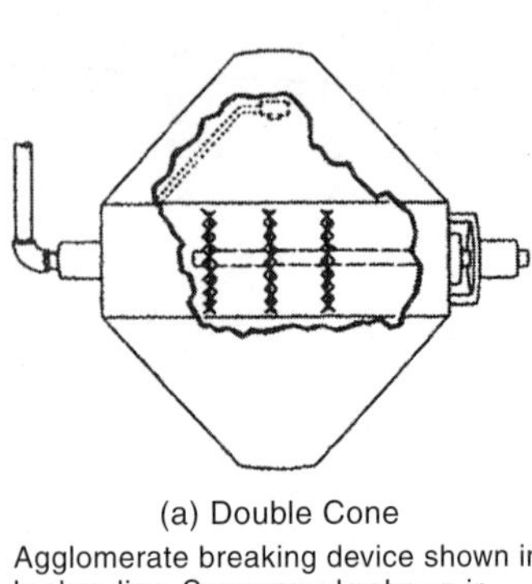

(a) Double Cone
Agglomerate breaking device shown in broken line. Spray nozzle shown in dotted line. Tumblers of this type available plain or with either or both of the above features.

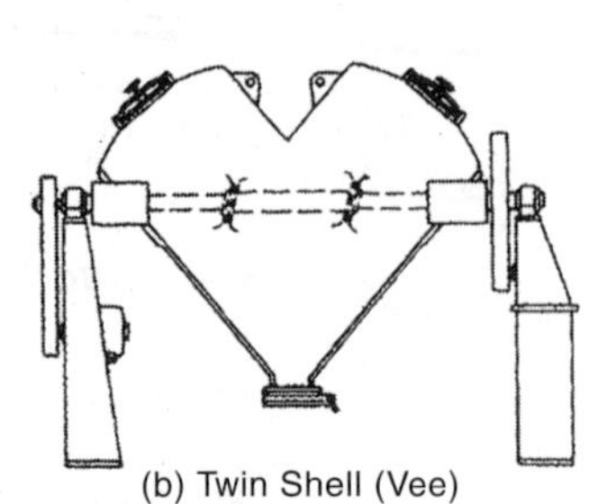

(b) Twin Shell (Vee)
Agglomerate breaking and liquid feeding device shown in broken line. Where no liquid feeding is necessary, a pin-type agglomerate breaking device is used. Tumblers of this type are available plain or with any of the above features.

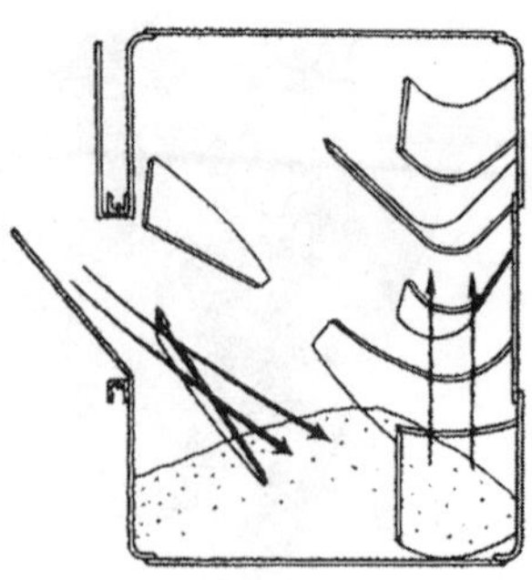

(c) Horizontal Drum (with baffles)

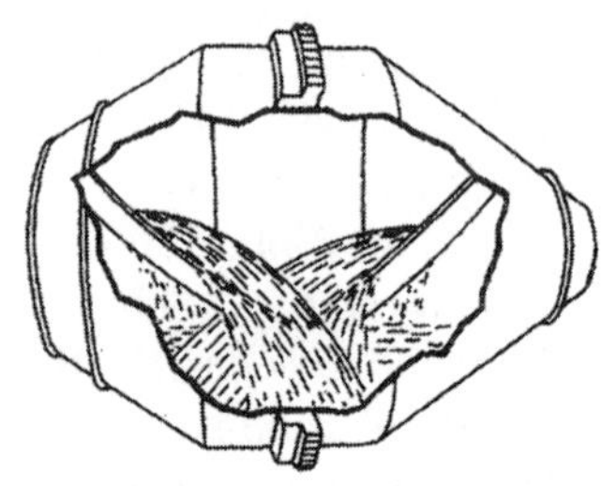

(d) Double-Cone revolving around long axis (with baffles)

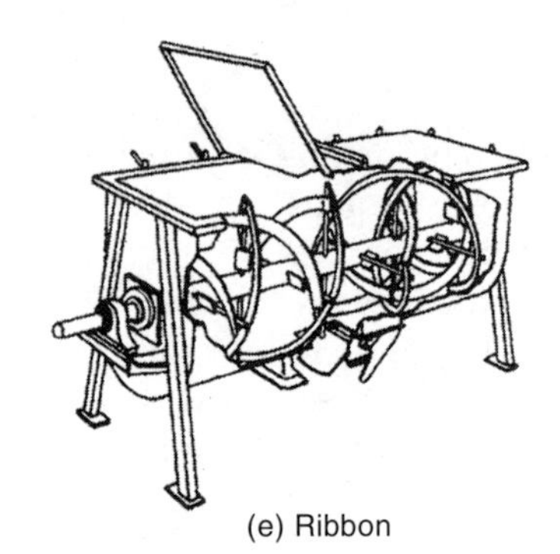

(e) Ribbon

(f) Vertical Screw (orbiting type)

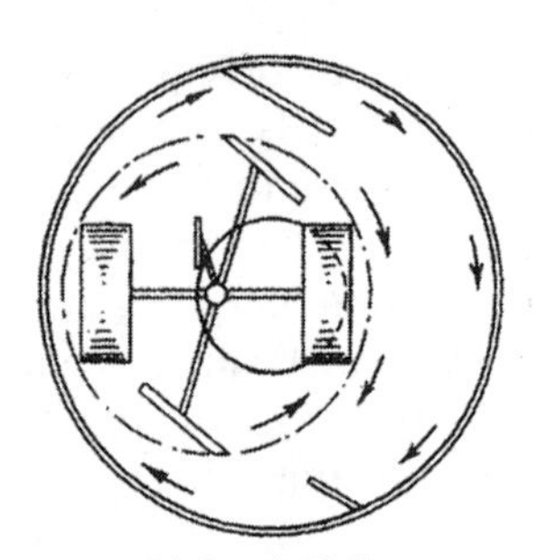

(g) Batch Muller
Three types are available:
(1) pan is stationary and muller turret rotates;
(2) muller turret is stationary and pan rotates;
(3) pan rotates clockwise, muller turret rotates counterclockwise.
Type 3 is illustrated above

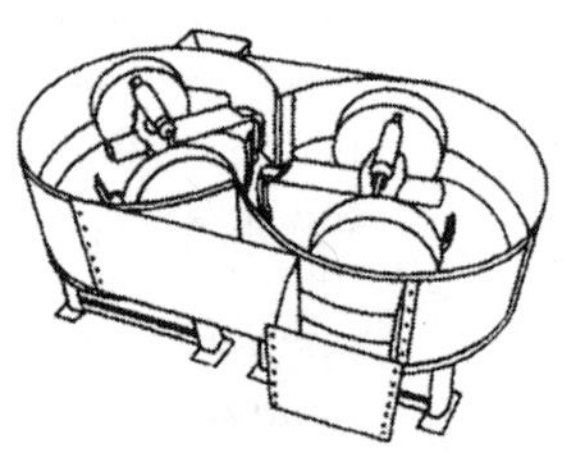

(h) Continuous Muller (stationary shell)

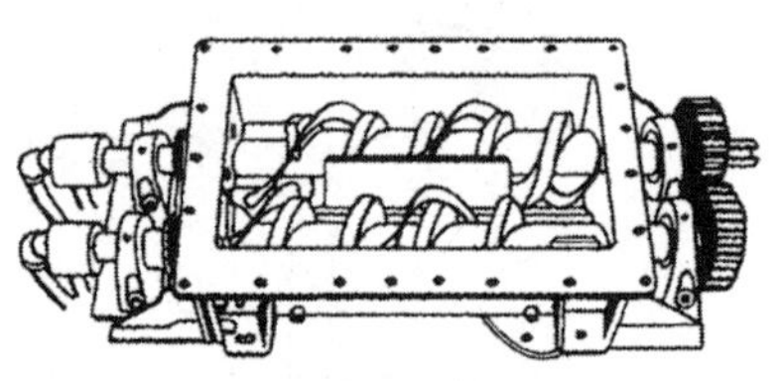

(f) Twin Rotor (adapted to heat transfer-jacketed body and hollow screws)

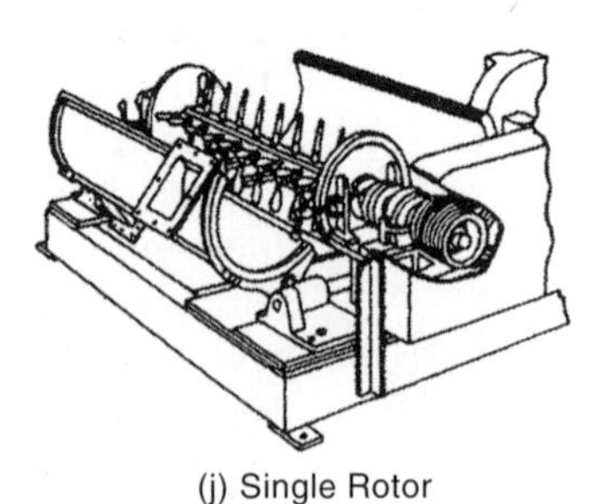

(j) Single Rotor

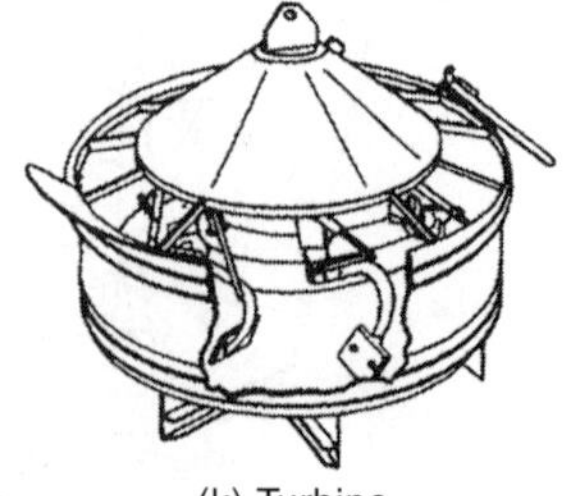

(k) Turbine

Fig. 11G5 Several types of solids mixing machines. (*Reproduced with permission from Perry's Chemical Engineering Handbook, R.H. Perry and D.W. Green, McGraw-Hill, 1997.*)

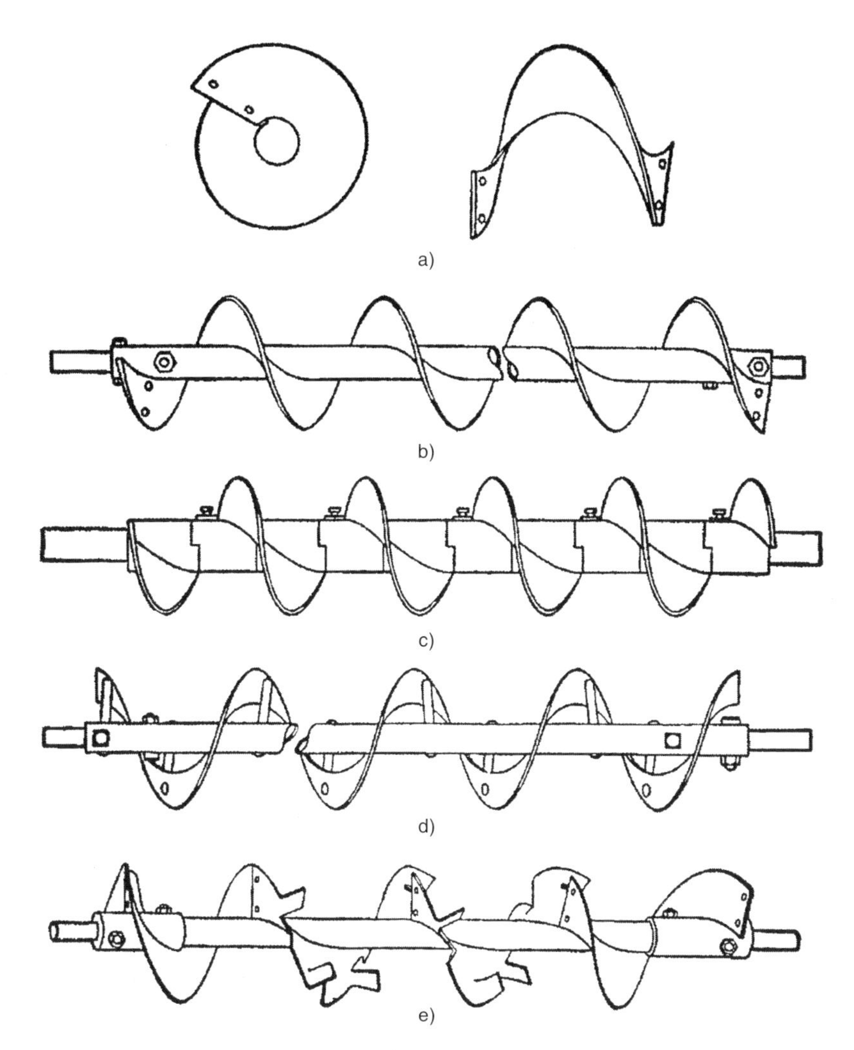

Fig. 11G5-1 Various screw conveyor flights: a) sectional; b), helicoid; c), cast iron; d), ribbon; e), cut flights. (*by permission, from Introduction to Chemical Engineering, W. L. Badger and J. T. Banchero, McGraw-Hill.*)

aggregated material, and make a mixture more dense, but are not so suitable for sticky materials. Foundry sand processing is a common application for these mixers. Turbine mixers, shown in (k) in Fig. 11G5, have a series of legs with plowshares or moldboards that spin through a circular trough. These mixers are suitable for both dry materials that flow well, and lightly-wetted materials that do not flow so well, and also for liquid-solid mixtures[5]. Dry, fine, powder or granules, can also be mixed with screw conveyor-mixers when the material is conveyed. See Fig. 11G5-1. Mixers of the screw type are also incorporated in plastics injection molding and extrusion machines. These machines, illustrated in Chapter 4, are used to mix color concentrates and other additives with the basic plastic resin as part of an extrusion or molding operation.

Stiff, viscous materials are mixed with a kneading machine such as that shown in Fig. 11G5-2. Mixing takes place in an open-top container, with a semi-cylindrical bottom. Two Z-shaped knifes rotate on horizontal shafts in such an orientation that the mass from one knife is picked up and folded by the other knife. As the operation continues, the materials are blended together. The machine pictured also includes a screw conveyor at the bottom of the container. During kneading, the screw conveyor runs in reverse and aids in keeping material within reach of the kneading blades. After mixing is completed, the screw reverses and empties

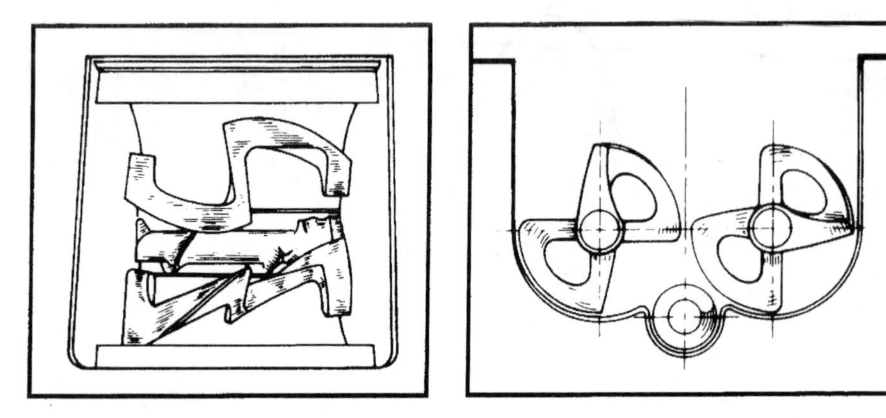

Fig. 11G5-2 A kneading machine with an added extrusion screw. The extrusion screw runs in reverse during the mixing cycle, moving material to the kneading blades. When mixing is completed, the screw changes its direction of rotation and empties the mixing chamber of the mixed material. (*Courtesy Charles Ross and Son Company.*)

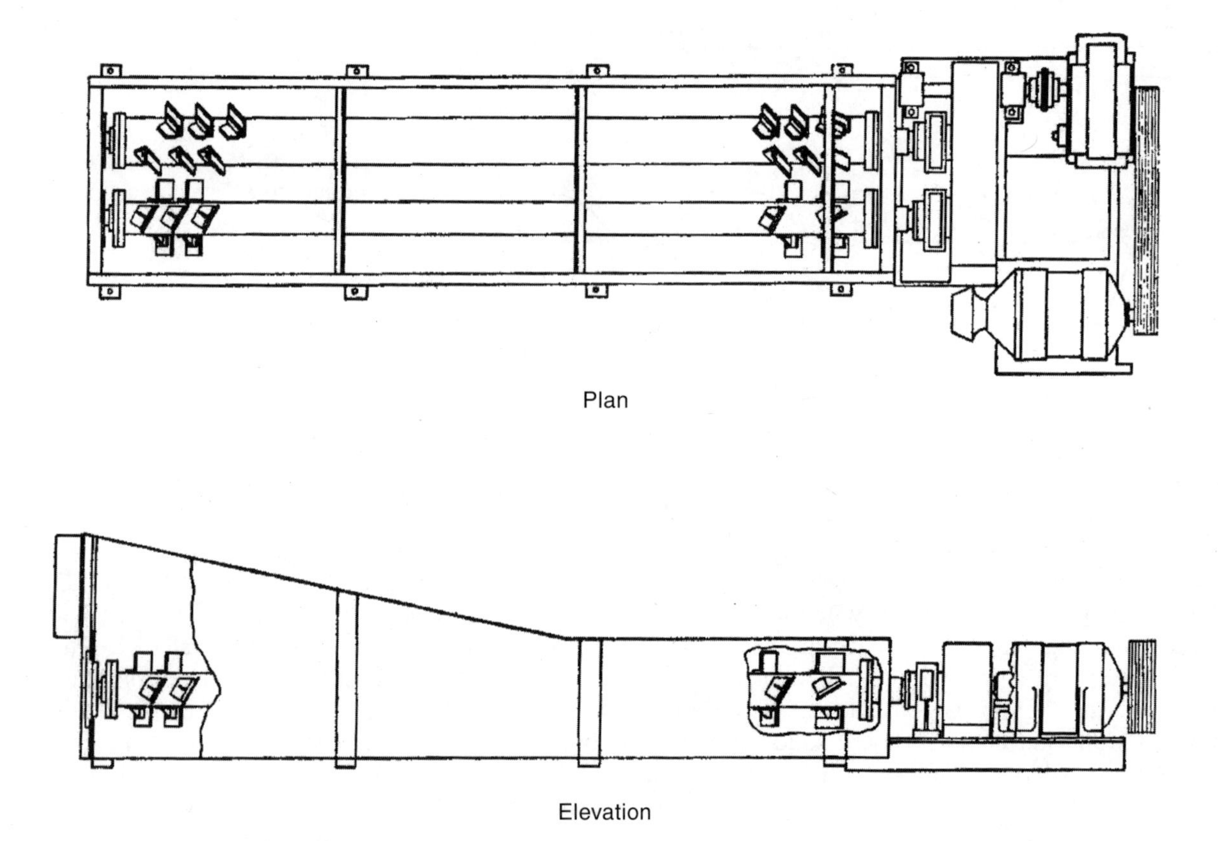

Fig. 11G5-3 A pug mill, used extensively for mixing clay. (*by permission, from Introduction to Chemical Engineering, W. L. Badger and J. T. Banchero, McGraw-Hill*)

the machine. Heating or cooling jackets may be incorporated in the container. Kneading machines operate on a batch basis and are common in the mixing of dough for baked goods. The machines are also used to mix polyester and polyurethane plastics, hot melt adhesives, butyl sealants, metal powders and pharmaceuticals.

Banbury mixers have some similarities to kneading mixers and are used in many solid chemical-mixing operations. They are adapted for mixing rubber, plastics, and other cohesive solids and pastes. Banbury mixers are explained and illustrated in paragraph 4A4c in Chapter 4 in connection with the compounding of plastics and rubber. Other plastics compounding methods, including dry propeller mixing, v-barrel mixing, and other methods discussed in section 4A4, are also applicable to chemical materials.

The pug mixer, illustrated in Fig. 11G5-3, has two rotating shafts in a trough-like container. Each shaft carries a series of inclined blades or pins which are placed so that they overlap the blades or pins on the other shaft. Material is passed back and forth as the shafts rotate. This machine is used extensively for mixing clay.

H. Petroleum Refining and Petrochemicals

Petroleum - crude oil - is a mixture of solid, liquid and gaseous hydrocarbons found underground or under the sea bottom in rock deposits. Refining of petroleum can provide many useful materials including: fuels, lubricants, solvents, and petrochemicals used in the manufacture of plastics and other useful products. Refining involves three basic steps: separation, conversion and chemical treatment. Refineries are normally designed for specific crude oils and particular products, but must be flexible to adapt to variations or changes in market demand for particular products and materials.

H1. ***separation*** - is the first step in petroleum refining. It divides the crude oil into constituents, most of which are further processed with different methods that convert them to salable products. The prime method used to separate the various fractions (hydrocarbon groups) in petroleum is *fractional distillation.*

H1a. ***fractional distillation of crude petroleum*** - (See paragraph C1a which discusses fractional distillation in general.) Crude petroleum is first heated to about 250°F (120°C) and treated with fresh water and an electrical field to remove salt. The crude petroleum is then further heated, in heat exchangers and a furnace, to a temperature above the boiling point of most of the fractions it contains, normally between 600 and 725°F (315 and 385°C). The resulting mixture of gases and liquids is passed to a fractionating tower, sometimes called a bubble tower. This is a high, vertical, cylindrical chamber. As the gases rise in this tower, they cool and condense. Condensation takes place at different levels, depending on the condensation temperature of the fraction. The condensates collect in trays attached to the tower walls at different levels, and are piped off to separate storage tanks for further processing. The cylindrical tower may be as high as 150 ft (45 m), and have as many as 40 fractionating trays. The heavier fractions condense first. Lighter fractions, like gasoline, which boils at 104 to 392°F (40 to 200°C), naphtha, jet-fuel, and kerosene condense at higher levels. Some of the condensed liquid is pumped to the top of the tower as reflux. As it descends in the tower, the liquid collects some of the lighter vapors and absorbs some of the heavier ones. Some vapor fractions do not condense in the tower, and they are piped off at the top of the tower into a vapor recovery condenser. Other fractions, that do not reach their boiling point in the heating chamber, remain in that chamber and are collected at the bottom. They consist of heavier hydrocarbons such as asphalt, and heavy oils. Steam is introduced in the bottom of the column to aid in the separation of the heavier fractions. The fractions produced in fractional distillation of petroleum are known as *straight run products*. The output of the tower includes light and heavy naphtha, kerosene, and light and heavy oil.

When intended to be used for the production of solvents and high-purity petrochemicals, the fractionating column is of smaller diameter, and has many more fractionating trays, as many as 100 or more. More reflux is also used. This approach is called *superfractionation*.

H1b. ***vacuum distillation of petroleum fractions*** - Vacuum distillation, described above in C1b, is used to further distill heavier fractions from the fractionating column, to produce various lubricants. Vacuum distillation is used to avoid heating these heavier fractions to the point where cracking would occur. The usual pressure in this operation is 50 to 100 mm of mercury (6.7 to 13.3 KPa), allowing a lubricating oil to boil at 480 to 660°F (250 to 350°C) instead of at a temperature of over 600°F (315°C) at normal atmospheric pressure.[4] Residues from this vacuum distillation may be used as feedstock for cracking other products, or may be blended to produce fuel oil or asphalt.

H1c. ***absorption/stripping of petroleum factions*** - These methods, described above in C2 , are used to recover light components from the vapors that exit from the top of the crude oil fractionating column. Propane, propylene, butylene, and butane are commonly processed. These vapors are bubbled through, and are absorbed by, heavy naphtha or kerosene in an absorption tower. Other light gases—hydrogen, methane, ethylene, and ethane, are not absorbed. The absorption tower is pressurized to 100 to 150 psi (700 to 1000 KPa). The solvent with its absorbate is then processed in a stripping column with heat. This releases the absorbate that is then condensed to become liquified petroleum gas (LPG).

H1d. ***solvent extraction*** - is used to further separate some of the products of fractional distillation. A solvent such as benzene, phenol, or furfural is mixed with the liquid output from one level of the distillation tower. The solvent may dissolve some of the fractions or cause them to solidify, so they can be separated from the liquid. A common application is the processing of lubricating oils by removing heavy aromatic components. This processing yields an oil with a wider range of temperatures at which the oil's viscosity will remain within the desired range.

H1e. ***crystallization of petroleum fraction contaminants*** - is a third separation method. (See crystallization in C5 above.) Crystallization is used to remove wax and other semisolid substances from the heavier fractions. A common application occurs with lubricating oils which must be free from wax. The oil is mixed with a solvent (usually a mixture of methyl ethyl ketone and benzene, where both solvents have a function in the process). The mixture is cooled to about –5°F (–20°C) sufficiently low to cause the wax to crystallize. The liquid is then filtered with rotating cylindrical filters. The wax is deposited on the cylindrical surface (that is covered with woven filter fabric), and is removed with metal scrapers after the surface oil is rinsed off with the solvent. The solvent-oil mixture is then separated by distillation and the solvent is reused.

H2. ***conversion*** - produces more valuable products from some of the less valuable fractions that result from the separation processes. There are two prime conversion processes: *cracking* and *combining*.

H2a. ***cracking*** - produces a lighter product, primarily gasoline, from a heavier petroleum fraction. It also produces gases: ethane, methane, propylene and propane that are raw materials in the manufacture of plastics, synthetic rubber, detergents, textiles and agricultural chemicals. There are two cracking processes: *thermal cracking* and *catalytic cracking*.

H2a1. ***thermal cracking*** - In this process, heavy petroleum fractions are subjected to intense heat and pressure. This breaks the molecular bonds of the large molecules of heavier fractions, forming smaller and simpler molecules. These molecules then may spontaneously undergo further changes, or combine with other molecules to form molecules of naphtha, gasoline, and other lighter petroleum products. Typical thermal cracking pressures are 100 to 1000 psi (700 to 7000 kPa), and typical temperatures are 850 to 1000°F (450 to 540°C)[4]. However, thermal cracking (without catalysts) is seldom used since the use of catalysts increases the yield of desired high-octane products and produces less of the undesirable compounds such as asphalt and coke-forming constituents.

H2a2. ***catalytic cracking*** - uses a catalyst to facilitate the breakdown of the large molecules. Zeolites (types of clay), and other minerals or molecular-sieve materials are the catalysts used. The heated, heavy fractions come in contact with the catalysts and are converted to lighter materials. Catalytic cracking enables pressures in the cracking chamber to be lower, reduces the amount of energy needed, and produces higher octane (smoother burning) gasoline. Typical catalytic cracking temperatures are 900 to 1020°F (480 to 550°C) with pressures of 10 to 20 psi (70 to 140 kPa). In the fluid-catalytic process, the catalyst is in fine-particle form that acts as a liquid in the cracking unit. It is suspended in a flow of feed liquid vapor. After cracking the catalyst particles are separated from the cracked fluid by cyclone separation. In the fixed-bed process, the feed liquid is passed through a stationary bed of solid catalyst particles. With either process, the catalyst particles become coated with carbon during the cracking operation, and the carbon is removed with steam and by burning. The heat of burning prepares the catalyst for reuse. The product from the cracking reactor is processed by fractional distillation, yielding mostly cracked naphtha, which is blended with other hydrocarbons to make gasoline. The gaseous products of the distillation column consist of propylene and butylene which have petrochemical applications. The balance is consists chiefly of fuel gas and other gaseous hydrocarbons.[3]

When hydrogen is added to the cracking chamber, it reacts with the materials present and the yield of desirable products is increased. The process variation is called *hydrogenation*, and is the reverse of cracking. Smaller, simpler molecules of gaseous fractions are put together, to form longer, more-complex molecules of more-useful products. Three common combining processes are polymerization, alkylation and reforming.

H2b. ***polymerization*** - occurs when light, gaseous fractions, resulting from cracking, are subjected to high pressure and temperature while in contact with a catalyst. The molecules combine, forming longer, more-complex molecules, and the resulting materials are called polymers. Propylene and butylene fractions are commonly polymerized for use as components of high-octane gasoline and other products. Pressures of 400 to 1100 psi (2800 to 7600 kPa) and temperatures of 350 to 450°F (175 to 230°C) are employed in the polymerization chamber. The catalyst usually is phosphoric acid carried on pellets of a porous sedimentary rock.

H2c. ***alkylation*** - produces a fraction known as "alkylate", which is useful in producing certain fuels. Alkylate has a high octane rating and is used to improve unleaded gasoline. The alkylation process is similar to polymerization and is exothermic. It involves a chemical reaction between a hydrocarbon (usually isobutane) and an olefin (ethylene, propylene, butylene, or amylene). The olefin feedstock is obtained from the gases produced during catalytic cracking; isobutane comes from refinery gases. The reaction between them takes place with an acid catalyst (hydrofluoric or sulfuric acid) at a controlled temperature of 35 to 45°F (2 to 7°C) in the sulfuric acid process, and between 75 and 115°F (24 and 46°C) with the hydrofluoric acid process. The product of the reaction is a high-octane branched-chain hydrocarbon.

H2d. ***reforming***[6] - is a means of raising the octane rating of gasoline. It involves changing molecules of gasoline and naphtha to aromatic and branched-chain molecules that have high octane ratings. The molecular changes can substitute for the addition of lead to gasoline, since that is no longer acceptable because of environmental issues. The reforming process is a combination of isomerization and cracking. *Catalytic reforming* is the dominant process. One major method uses naphtha and hydrogen, which are mixed and fed to a preheater and then to four reactors in sequence. Heating takes place before the material is introduced to each reactor. The reactors contain alumina and a small amount of platinum. They operate at pressures of 220 to 1000 psi (1500 to 7000 kPa) and temperatures from 300 to 950°F (150 to 510°C). A series of complex reactions takes place and the resulting product is cooled and then fractionated. The final reformed material is used as either an anti-knock element in gasoline or further fractionated to produce benzene, toluene and xylene. These aromatic materials are used in the production of synthetic rubber, plastics and food preservatives.

H3. ***chemical and other treatments*** - of petroleum products include removal of sulfur and other impurities, addition of various additives, and blending of several fractions with others to improve product performance.

H3a ***sulfur removal (hydrogen treatment)*** - is accomplished by mixing hydrogen gas with the fraction involved, heating the mixture to vaporize the oil fraction, and then passing the mixture over a catalyst. Nickel, tungsten, or a mixture of molybdenum and cobalt oxides on an aluminum support may be used as the catalyst. The operation takes place at a temperature normally between 500 and 800°F (260 and 425°C) and a pressure of 200 to 1000 psi (1400 to 7000 kPa). Hydrogen sulfide is formed and is removed from the mixture by solvent extraction. (See H1d.) The hydrogen sulfide is used as feed material for the production of high-purity sulfur.

H3b ***additive addition*** - Certain chemical compounds may be mixed with a raw petroleum product to improve its performance. Small amounts may produce a major improvement in fuel or lubricant performance. An example would be a detergent added to a lubricant or gasoline. Others are tetra-ethyl lead added to gasoline to raise its octane rating, anti-icing agents, organophosphates, to reduce deposits, and antioxidants as stabilizers. The use of some of these additives has been stopped or reduced for environmental reasons.

H3c. ***blending*** - Petroleum refinery fractions are blended together to produce a better overall product. One example is lubricating oils, which may be blended with different fractions to produce an oil with the desired viscosity. Gasoline blending is much more complicated, and as many as 15 different hydrocarbon fractions can potentially be included in finished gasoline. Octane is a key characteristic that is set by blending different fractions. Other factors that may be adjusted are color, stability, boiling points, vapor pressure, and sulfur, olefin, and aromatics content. The cost of the different hydrocarbons, as well as their properties, must be considered.

I. Chemical Reactions

A chemical reactor is often a complex device because it has to contain the materials that are involved before, during, and after the reaction, where temperature and pressure changes may take place and where the properties of the materials will change. Materials must be brought to, processed, and removed from the reactor. They may be in gaseous, liquid, or solid form. Mixing of the reactants normally is part of the reaction process. Catalysts may be required, some fixed, others mixed with the reactants. Equipment very often operates at higher-than-ambient temperatures but, with other reactions, operates at lower temperatures. High pressures may be used, though some reactions require a vacuum. Production chemical reactions, like other chemical operations, may be performed on either a batch or continuous basis. The choice of either one depends primarily on the quantity of reactant to be produced.

Batch processes are described in A., above. No material is either added or removed from the reaction vessel during the reaction. Batch processing is useful in prototype and small-quantity production, in pharmaceutical manufacture, and in fermenting operations. Fig. 11I3, view (a), shows a typical batch reactor.

Semibatch processes are those where one reactant is fixed in the container and another is fed in small increments, or continuously. The term also applies when one of the products of the reaction is removed continuously, or in small increments, from the reaction vessel. Semibatch processing is also used in small-quantity production.

Continuous processes using chemical reactions include the following:

Continuously stirred tank reactors (CSTR) run steadily during the reaction. Feed materials are added continuously, and reaction products are removed continuously from the reactor. This method is used when the reactants must be mixed or agitated to promote the reaction. The reactor often consists of a series of individual vessels. Material is fed into the first vessel where the reaction begins. Partly-reacted material is fed to a second vessel, whose output is fed to a third vessel, and so on. Mixing and reaction take place in each vessel. At the end of the series, fully-reacted product exits from the last vessel. Some approaches use a similar series of partial reactions but arrange for it to take place on shelves or trays of a single vessel. As the material flows from level to level of the shelves, the reaction proceeds until the final product exits the vessel at the bottom.

Plug flow reactors (PFR) consist of one long reactor, or many short reactors in a tube bank.[2] All materials travel the full length of the reactor. The reaction rate changes as the materials move down the length of the reactor. There is no axial mixing.[5] Gas-phase reactions are usually involved. PFRs are used in large quantity production, including those operations requiring high-temperature reactions.

Tubular *packed bed reactors (PBR)* are tubular reactors packed with a solid catalyst. PBRs are used primarily in heterogeneous gas-phase reactions when a catalyst is involved. The operation is continuous.

Another common type is the *tubular flow reactor (TFR)*. These units have parallel pipes or tubes inside a cylindrical vessel. Reactant materials are fed from one end and the reacted product is withdrawn from the other end. Heating fluids flow in the spaces surrounding the pipes while the reactants flow inside the pipes, or vice versa. TFRs are useful for products that require heat exchange during the reaction.[5]

Chemical reactions are also sometimes classified as *homogeneous reactions* and *heterogeneous reactions*. Homogeneous reactions are those that occur in a single phase, that is, reactions between substances that are in the same phase - gas, liquid, or solid. The most common are reactions between gases, and between liquids or substances dissolved in liquids. Heterogeneous reactions are those that involve two or more phases—gas/liquid, liquid/solid, gas/solid, or two immiscible liquids.

I1. ***gas-phase reactions*** - The gases are mixed by injecting a stream of one gas into the other gas or gases. However, gas/gas reactions are less common than gas/liquid or gas/solid reactions. A gas may be liquified before the reaction to provide reactivity or higher reaction speed, because some reactions that take place in the liquid phase do not take place or are slow in the gaseous state. Gas-phase reactions usually require an elevated temperature, and, often, elevated pressure. Raising temperature and pressure speeds the rate of reaction of gaseous materials. Higher temperature and pressure increase the contact and collision of molecules. According to the collision theory, reactions result when molecules or atoms of the reacting materials contact each other or collide.[3] With gas-phase reactions, heat may be supplied in a number of ways. The reaction vessel may be heated by flame or steam. Alternatively, heated brick linings may provide contact heat, or heated granules of sand may be fed into the reaction vessel. After cooling during the reaction, such pieces are fed out of the vessel, reheated and returned. Sometimes, heat is provided by burning a portion of one of the reactant gases with a small amount of air or oxygen. Exothermic reactions may require cooling of the reaction vessel. Examples of gas-phase reactions are found in the production of olefin plastics, benzene, acetylene, and in the manufacture of ammonia from nitrogen and hydrogen gases.[5] Fig. 11I1 shows two gas reactors schematically.

I2. ***gas/liquid reactions*** - Fig. 11I2 shows several types of reactor arrangements for gas/liquid reactions. Common gas/liquid reactions are carried out to remove small amounts of unwanted constituents from air, hydrogen, hydrocarbons, and other gases, to modify liquids by hydrogenation, halogenation, oxidation, nitration, and alkylation, in the manufacture of nitric, sulfuric, and adipic acids and phosphates, and for certain biotechnical processes such as fermentation, protein production, and the treatment of sludges[5].

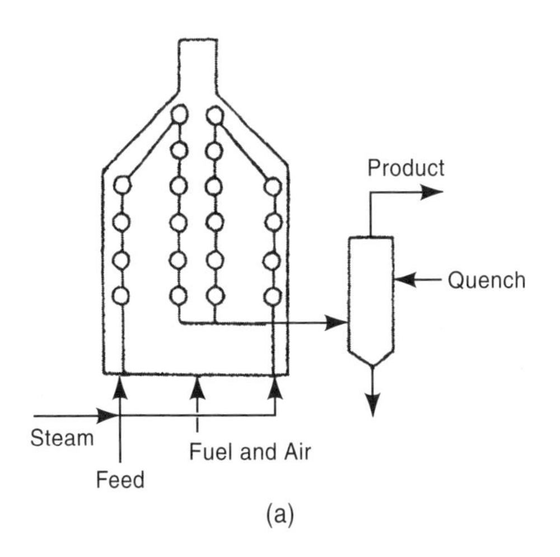

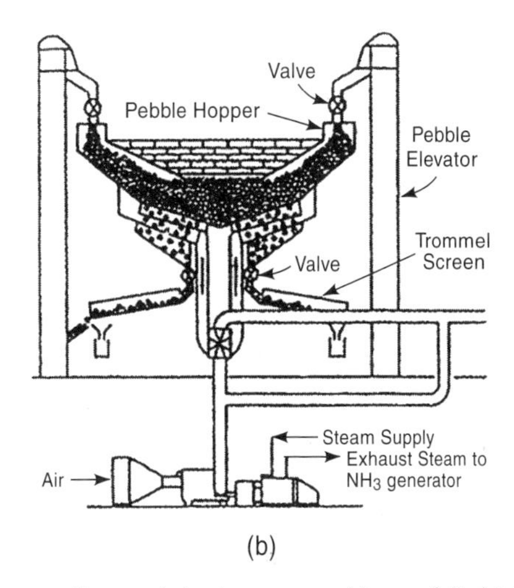

Fig. 11I1 Equipment for two gas-phase, non-catalytic reactions: (a) steam cracking of light hydrocarbons in a tubular fired heater, (b) pebble heater for the fixation of nitrogen from air. *(Reproduced with permission from Perry's Chemical Engineering Handbook, R.H. Perry and D.W. Green, McGraw-Hill, 1997.)*

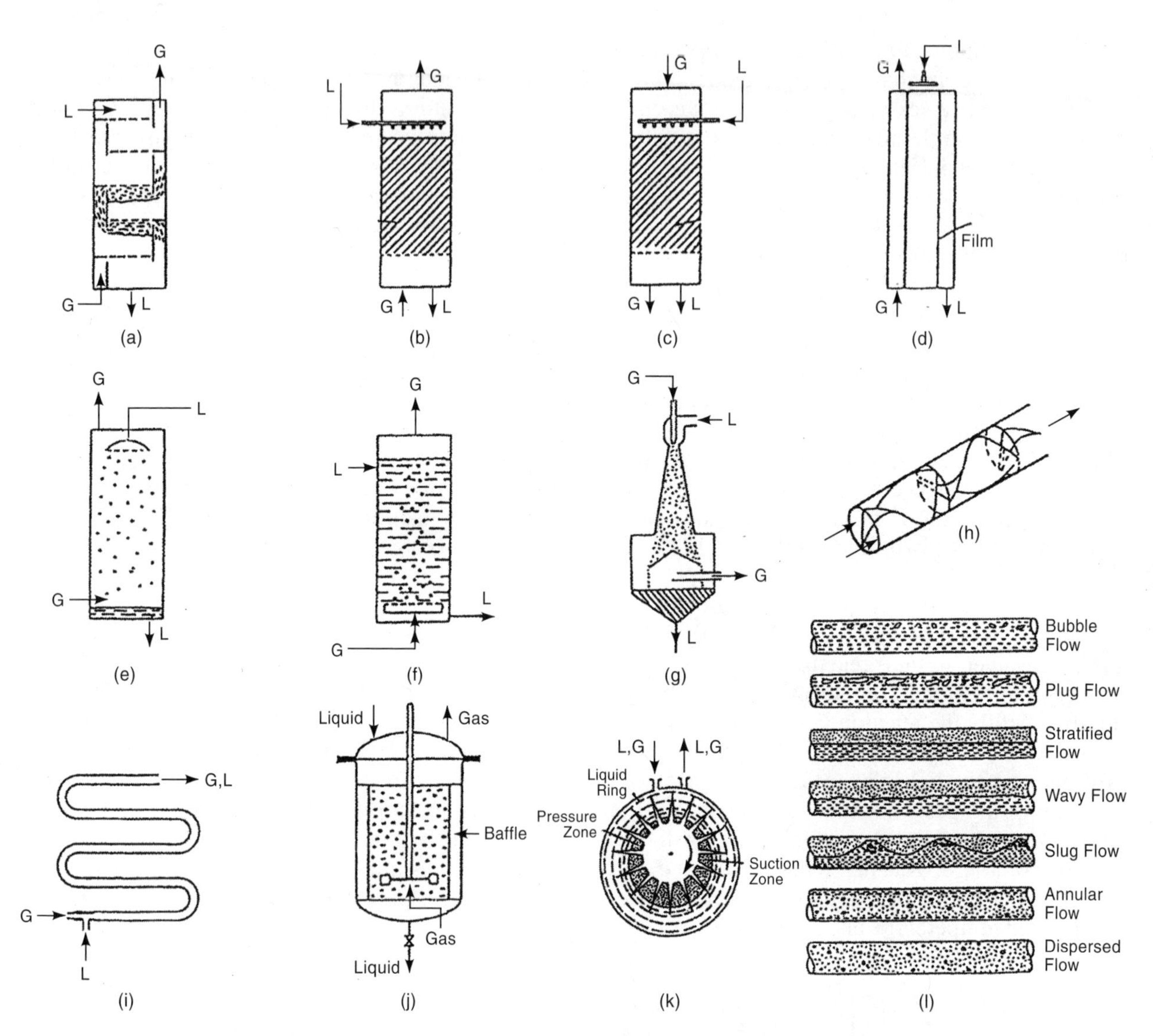

Fig. 11I2 Types of industrial gas/liquid reactors: (G = gas; L = liquid.) (a) tray tower, (b) packed, counter current, (c) packed, parallel current, (d) falling liquid film, (e) spray tower, (f) bubble tower, (g) venturi mixer, (h) static in-line mixer, (i) tubular flow, (j) stirred tank, (k) centrifugal pump, (l) two-phase flow in horizontal tubes. *(Reproduced with permission from Perry's Chemical Engineering Handbook, R.H. Perry and D.W. Green, McGraw-Hill, 1997.)*

In the hydrogenation of oils, the oil and a catalyst (usually nickel in flake form), are placed in a sealed vessel, the air is evacuated and hydrogen gas is fed into the vessel and mixed with the oil by impellers. For edible oils, temperatures are raised but limited to about 355°F (180°C). Pressure in the vessel will range from one to ten atmospheres[5]. The catalyst is filtered from the finished product. Other gas/liquid reactions involve gaseous oxygen, hydrogen, carbon dioxide and carbon monoxide.

I3. ***liquid/liquid reactions*** - are somewhat common. They often involve the use of mechanically-agitated tanks with heat-transfer capability. Other equipment uses the difference in specific gravity of the liquids to provide a flow in the tank and contact with the reactant liquid. Fig. 11I3 illustrates several

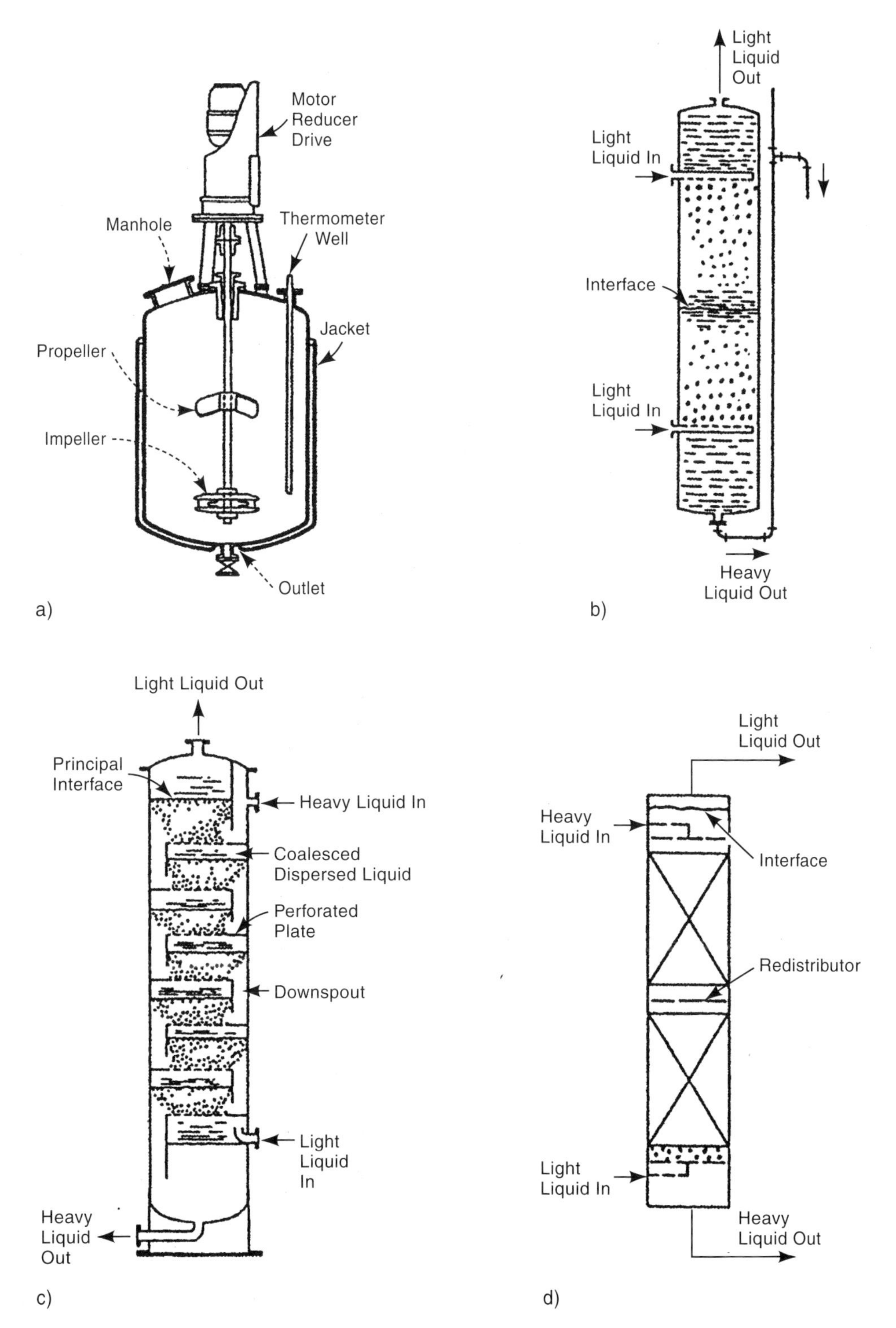

Fig. 11.13 Equipment for liquid/liquid reactors: a) batch stirred sulfonator, b) spray tower with both phases dispersed, c) sieve ray tower with light phase dispersed. d) two-section packed tower with light phase dispersed. (*Reproduced with permission from Perry's Chemical Engineering Handbook, R.H. Perry and D.W. Green, McGraw-Hill, 1997.*)

reactors for liquid/liquid materials. Applications of liquid/liquid reactions are found in soap making with alkali, explosives manufacture from the nitration of aromatics, and gasoline production.[5] Esterification (See L.) and the manufacture of resorcinol from benzene and sulfuric acid are other examples.

I4. ***gas/liquid/solid reactions*** - Common examples of these reactions include the reactions of acids and metals, and the changes that take place when electrolytic cells and batteries operate.[3] In production, these reactions often take place with solids in granule or particle form. The solid may be a reactant or a catalyst. If the solid is a granular catalyst, the surfaces of the granules are normally porous. Several methods are available to provide contact between the solid and the liquid and gas. One method uses a *trickle bed*, a fixed bed of solid material. The gas and liquid flow downward from gravity through the bed of granules, typically about 1/8 inch (3 mm) in diameter. This approach is used in the hydrosulfurization of petroleum oils.[5] Another trickle bed application is the removal of gaseous impurities from the atmosphere in pollution control. The flow of gas is then upward through the bed of solid material.

Another method, the *flooded fixed bed reaction*, uses an upward flow of gas and liquid through a bed of solid particles. A screen at the top of the reaction vessel may be used to block the escape of particles. This approach is used in some hydrogenation operations.[5]

Suspended catalyst beds use smaller solid particles that are either fluidized by upward movement of gas bubbles or as a suspension in a fluid mixture. The solid particles, after the reaction, must be removed from the fluid product by filtering. One application of fluidized reactors is the treatment of heavy petroleum fractions; another is the treatment of waste liquid. Air and liquid are passed through a vertical column of sand and bacterial growth takes place on the sand grains. Still another example is the liquefaction of coal, in which hydrogen is reacted with solid coal in a water slurry. Fig. 11I4 shows a typical three-phase fluidized bed reactor.

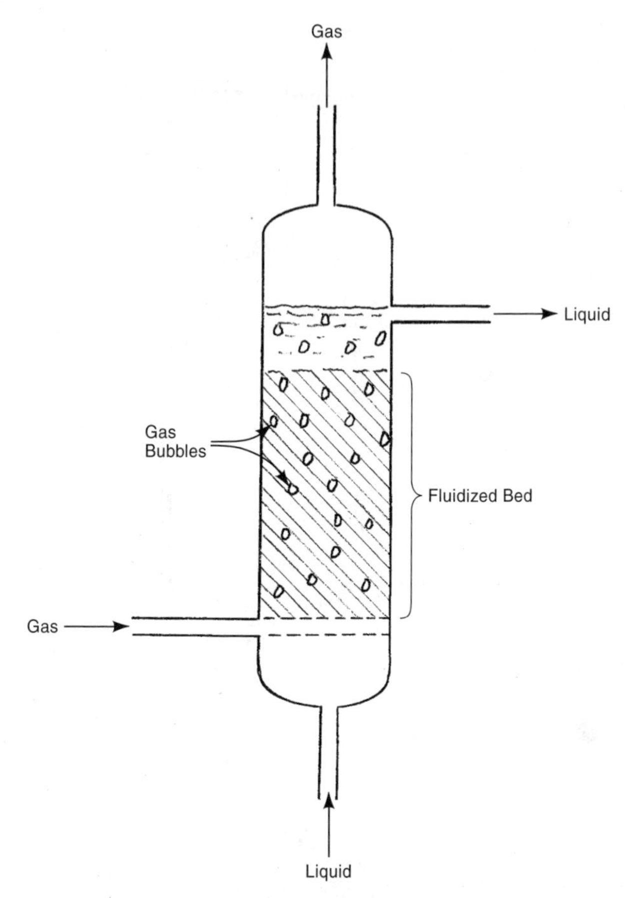

Fig. 11I4 A typical three-phase fluidized bed for reacting gas and liquid with a solid catalyst.

I5. ***solids/solids and solids/gas reactions*** - There are many proprietary processes that react solid materials. These processes often take place at the elevated temperatures that result from the contact of the reactants with combustion gases. Many common solid reactions involve the thermal decomposition of some material. Some common solid reactions are the following mentioned in Perry:[5]

Cement manufacture—the reaction of limestone with clay.

Boron carbide manufacture from boron oxide and carbon.

Calcium silicate produced from lime and silica.

Calcium carbide from lime and carbon.

Leblanc process for making soda ash (Na_2CO_3) from common salt.

In many of these reactions involving solids, the material is in particulate form. There is usually some movement of the material in the reaction vessel to expose it to heat and the reactant gas, if one is involved. Fig. 11I5 illustrates some equipment used in solids reactions.

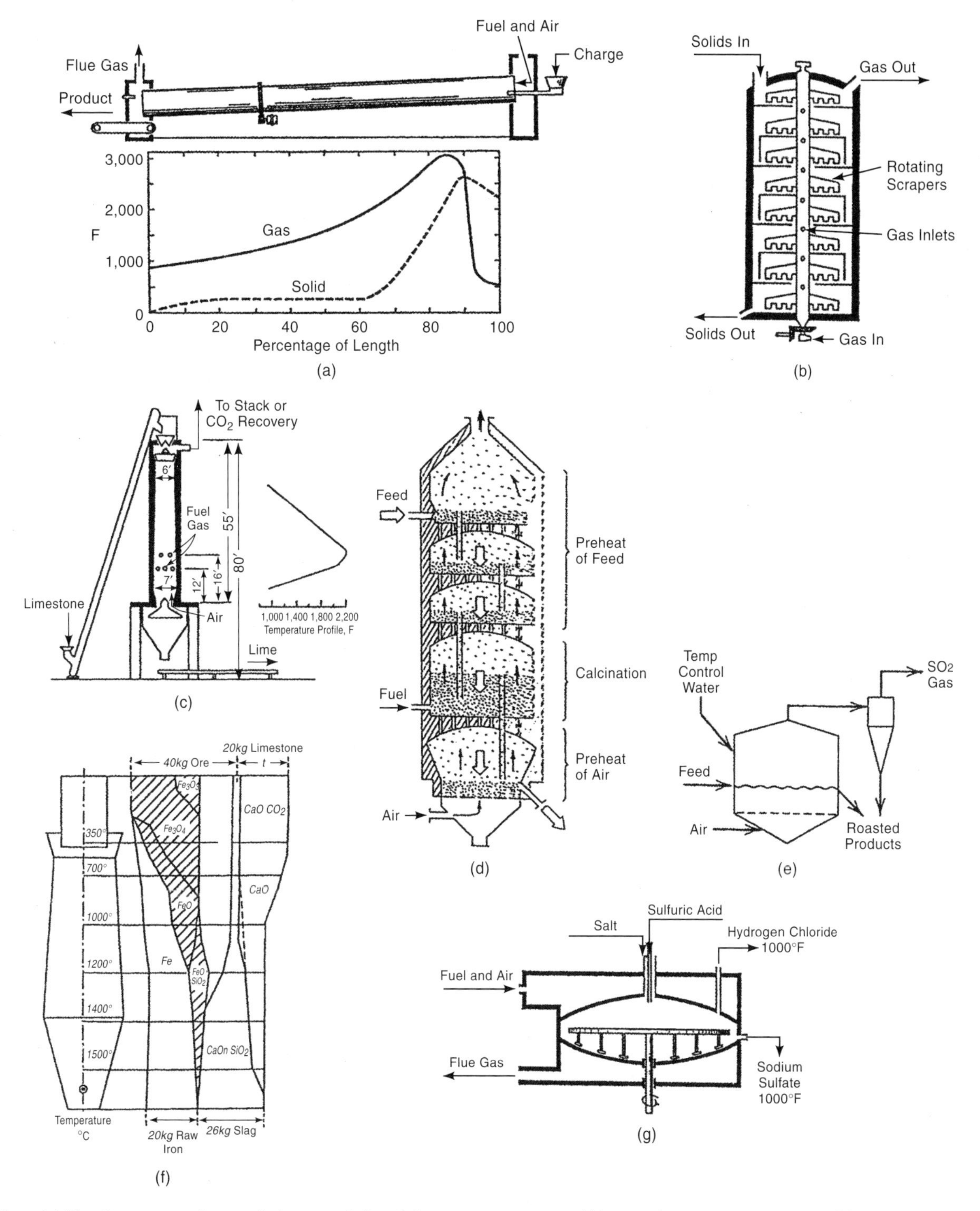

Fig. 11.15 Reactors for solid materials: (a) rotary cement kiln and temperature profiles in the kiln, (b) multiple hearth reactor, (c) vertical kiln for lime burning, (d) five-stage fluidized bed lime burner, (e) fluidized bed for roasting iron sulfides, (f) conditions in a vertical moving bed (blast furnace) for reduction of iron oxides, (g) mechanical salt cake furnace. (*Reproduced with permission from Perry's Chemical Engineering Handbook, R.H. Perry and D.W. Green, McGraw-Hill, 1997.*)

Some applications[5] of the process of reacting solids with gases take place in the mining industry, where sulfide ores are converted to sulfates or oxides that are more easily reduced to metals. Others are the conversion of Fe_2O_3 to Fe_3O_4 in a reducing atmosphere, the chlorination of ores of aluminum, uranium, titanium, and zirconium, the production of hydrogen gas from the reaction of iron and steam, the manufacture of blue gas by reacting steam with carbon, and the production of calcium cyanamide from calcium carbide and nitrogen. Nitriding, using steam for surface hardening is another application.

I6. ***reactions with catalysts*** - involve the addition of a substance (the catalyst) to the reactions. The catalyst either does not change chemically during the reaction or is regenerated at the end of the reaction and thus suffers no permanent effect. Catalysts can be gaseous, liquid, or solid. The reaction with the catalyst is much faster than it would be otherwise. It is believed that an intermediate compound is formed, which reacts with the other reactants present to produce the final product, and at a lower activation energy level than that required for direct reaction of the reactants[2]. Examples of catalysts used in commercial chemical production are the following:

Nitrous oxides, used as catalysts in the oxidation of sulfur dioxide in the production of sulfuric acid.

Sulfuric acid, used as a catalyst in the production of diethyl ether from ethyl alcohol.

Cobalt carbonyl, used as a catalyst in petroleum refining to combine carbon monoxide and hydrogen to olefins to form aldehydes and alcohols. A number of other catalysts are employed in the many operations involved in refining of petroleum into fuel, lubricants and petrochemical products.

Iron, used as a catalyst in the production of ammonia by reacting hydrogen and nitrogen.

Platinum, used as a catalyst to oxidize ammonia in the production of nitric acid.

J. Heat-transfer Methods

Heating is an integral part of many chemical operations; cooling is much less frequent but still may be an essential processing element. The most important type of heat-transfer device is the tubular heat exchanger which consists of tubing inside a vessel so that fluid can circulate in the tubing without directly contacting material in the vessel. Most devices have multiple heat exchange tubes in each chamber or shell. Such devices are also called *shell-and-tube heat* exchangers. Fig. 11J shows a typical tubular heat exchanger, equipped with baffles to ensure more thorough contact of the liquid in the vessel with heat exchanger surfaces. Fig. 11J-1 shows a simple double-pipe heat exchanger. Fig. 11J-2 shows a plate-type heat exchanger. This type provides more efficient heat transfer between the fluids processed. Counterflow (opposite direction flow of the two liquids) is used in all three devices.

In the most common arrangement of tubular heat exchangers, one material is to be heated and it is contained in the vessel while the heat transfer

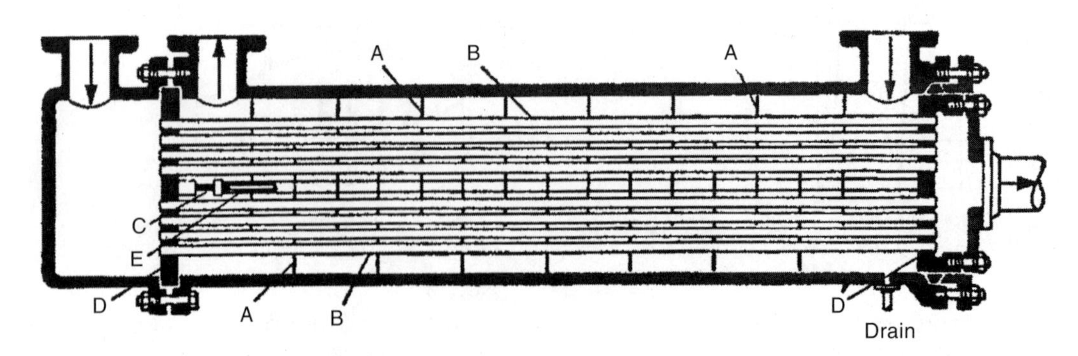

Fig. 11J Liquid-to-liquid tubular heat exchanger: A, baffles; B, heat-exchanger tubes; C, guide rod; D and D', tube sheets; E, spacer tubes. (*by permission, from Introduction to Chemical Engineering, W. L. Badger and J. T. Banchero, McGraw-Hill.*)

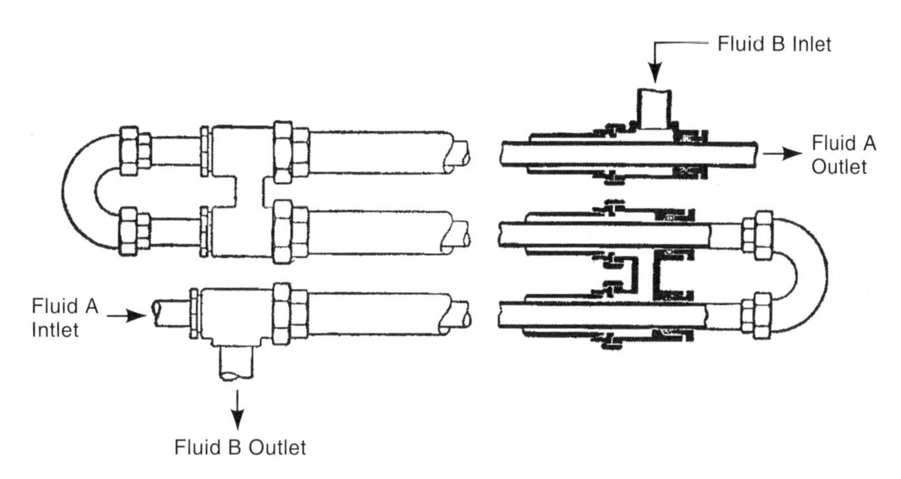

Fig. 11J-1 Double-pipe heat exchanger. Fluids in the inner and outer tubes flow in opposite directions to maximize the heat transfer.(*by permission, from Introduction to Chemical Engineering, W. L. Badger and J. T. Banchero, McGraw-Hill.*)

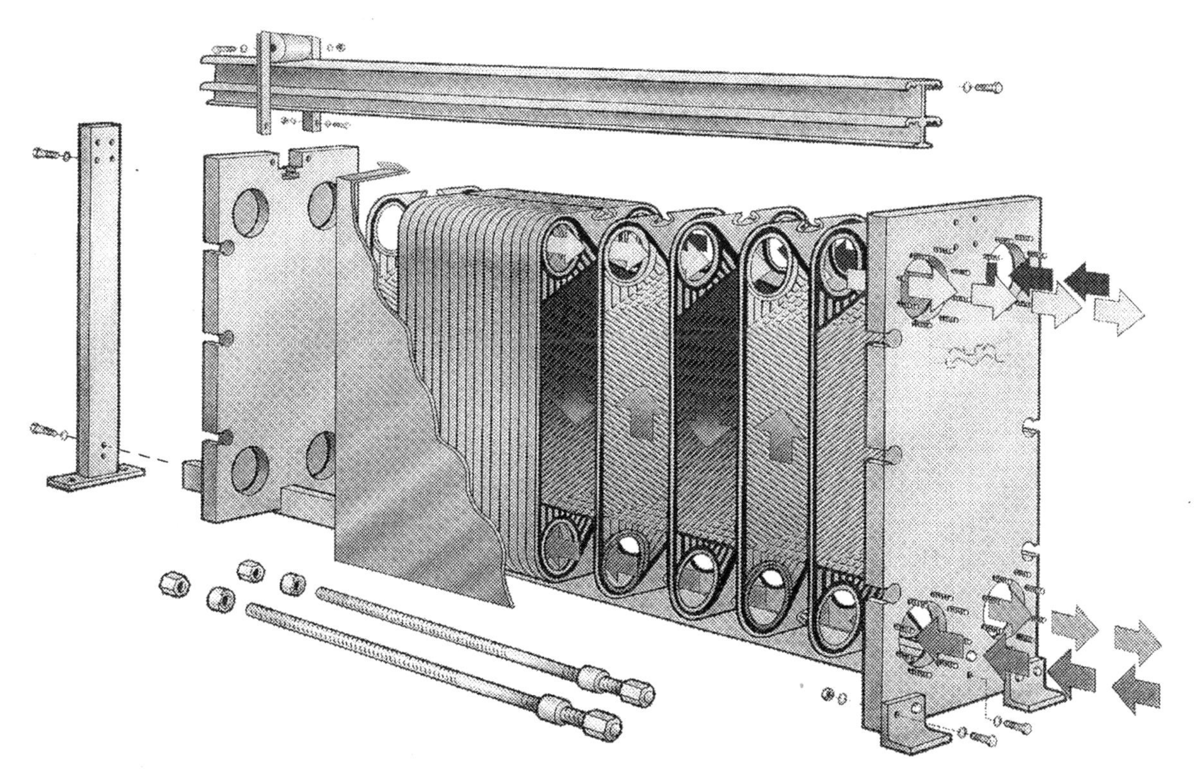

Fig. 11J-2 A plate-type heat exchanger. This type provides more efficient heat transfer than the more traditional shell and tube type. It is used where heat recovery is more important, where corrosive fluids are involved, and where space is more limited. (*Copyright Alfa Laval Inc. All rights reserved. Used with permission.*)

fluid - steam, hot gas, air, a liquid metal, fused salt, water or other liquid - circulates in the tubing, but these positions of the material and heat transfer fluid may be reversed. Heat is transferred to the material in the vessel primarily by conduction through the tube walls. Fins on heat exchanger tubes increase the heating area that can be in contact with a gas or liquid to be heated. Multiple passes of liquid across heat exchanger tubes is also frequently used. Chilled fluid may circulate in the tubes if the operation requires cooling. Refrigeration equipment may be used to chill the fluid. Chemical reactions and phase changes may accompany the heat transfer. Often the reaction or phase change takes place right in the heat exchange device. Examples are evaporators, condensers, stills, and polymerizers.

Applications of heat transfer include the following: transfer of heat from metals in a nuclear reactor to water, cooling of internal combustion engines, turbines and compressors, operation of air conditioning and refrigeration equipment and in many chemical processing and food processing operations where evaporation, condensing, drying and other phase changes are involved.

Some operations involve heating one fluid and cooling another in the same heat exchanger. This takes place in petroleum refining, when one fraction from distillation is cooled while warming another. Another example is the preheating of combustion air supplied to a furnace and the cooling of flue gases. In these examples, the two fluids can perform the function with one another when one circulates in the outer chamber of the apparatus and the other circulates in the tubing.

Paragraphs C11 (electro thermal processes) and C12 (drying) describe some heating methods. Also see Fig. 11C1, 11C7g and 11C7g-1.

J1. ***heat exchange for solid materials*** - Transfer of heat to or from solid material is more difficult than between fluids because the solid material can not be put in contact with the heat transfer surface so easily. However, solid materials, particularly when they are in powder, pellet, granule, or smaller piece form are frequently either cooled or heated as part of some chemical process. Jacketed kettles are more frequently used with solid materials than are tube heat exchangers. Radiant heating is also used. Solids are heated for drying or removal of other volatile constituents, for fusing, solidification, oxidation, or some other chemical reaction. One use is the cooling and solidification of liquid materials. For this purpose, batch kettles are used, with agitators for the liquid as it solidifies; continuous process solidification equipment uses vibrators or sheet metal conveyor belts on the surface of a cooling bath, with sidewalls on the conveyor to keep the coolant from contaminating the material to be solidified. Another method uses rotating drums with coolant sprayed or contained inside, and material to be cooled fed to the outside surface as the drum rotates. These types of equipment are used to solidify sulfur, grease, resins, wax, soap, chlorides and some insecticides. Food applications include cheese, gelatin, margarine, and gums.[5]

Equipment to fuse solids from liquid or powdered material can be of several basic types. They include horizontal-tank types that mix and move the material by screw conveyor inside a jacketed cylindrical vessel. These units are used to melt or cook dry solids, to dry solid materials containing liquids and in the rendering of fats from meat scraps. Vertical, agitated, and jacketed kettles are used for batch quantities. A mixer/agitator insures that all material is brought into contact with the heated walls of the kettle. A third type is a roll mill. In one variation, powdered material is fed from a hopper between the heated rolls. As the rolls rotate, the material is softened by the heat, kneaded, and mixed. When fully blended and fused, it is removed from the roll surface with peeling knives. This type of equipment is used in the compounding of rubber and plastics.

J2. ***cooling with air*** - is used to reduce the temperature of liquids and to condense vapors. In large-scale production situations, the cooling equipment is normally placed outdoors and uses atmospheric air as the cooling medium. These outdoor cooling devices can involve quite large structures and, if so, are referred to as *cooling towers.* Two prime types of cooling towers are in common use, *evaporative cooling towers,* where the liquid (water) to be cooled is brought into direct contact with the cooler air, and *dry (or nonevaporative) cooling towers* that have no direct contact between the liquid to be cooled and the ambient air.

With wet cooling - *evaporative*- towers, there is both sensible heat transfer from the contact with the water and the air, and evaporative cooling as some of the water changes to vapor. The water is sprayed downward into the air stream or onto a series of surfaces called *fill* or *packing*, which slow the downward flow rate of the water and provide much more contact between the cooling air and the water. Cooling from evaporation normally accounts for 65 to 75% of the cooling action.[4] The air is heated by the water it contacts, allowing it to absorb a greater amount of vapor. The amount of water lost from the system, due to evaporation, is not great. However, there is some loss from drift and blowdown.[4] The wet cooling approach is more effective in cooling than the dry or non-evaporative system, but the lost water must be replaced and there is a visible cloud of moisture from the system. Wind, natural draft, and/or fans may be used to move air through the tower. Fig. 11J2 illustrates an evaporative cooling tower. Fig. 11J2-1 shows one kind of fill used to increase the amount of water surface in contact with the cooling air.

Non-evaporative coolers employ heat exchangers, most commonly with bundles of parallel tubes to carry the water or other fluid to be cooled. Each tube has fins extending from the outer surface. Spiral-wound fins are common. Fans blow air streams over a group of parallel tubes and move the cooling air upward. Heat added to the air by the finned tubes increases its upward movement. The equipment may also rely on wind and natural draft for movement of the cooling air. Nonevaporative cooling towers are used if the liquid to be cooled is hazardous, or is expensive to replace. They are also used in arid regions where the cooling water is less available and must be conserved.

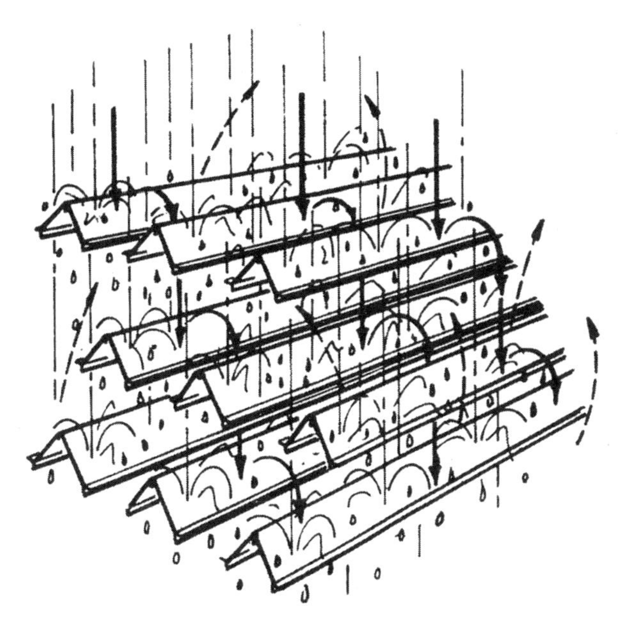

Fig. 11J2-1 A splash-type of fill (or packing) used to increase the amount of contact between water to be cooled and the flow of cooling air. (*Courtesy Marley Cooling Technologies.*)

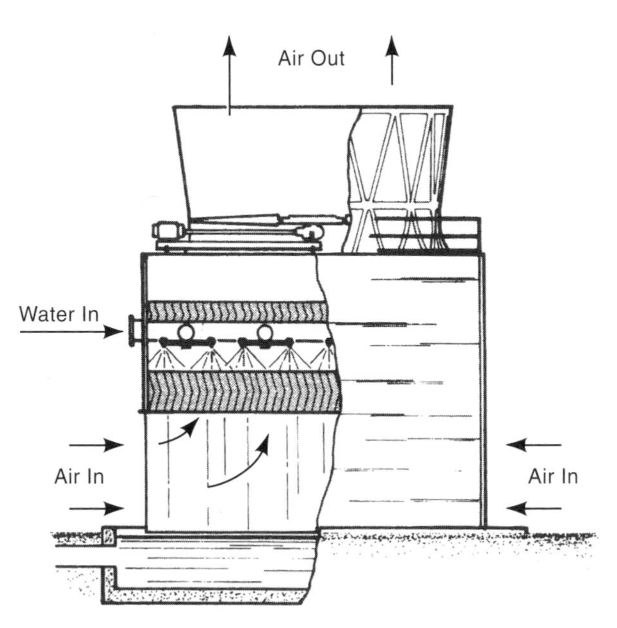

Fig. 11J2 Sketch of an evaporative (wet) cooling tower with counterflow of cooling air. (*Courtesy Marley Cooling Technologies.*)

Some cooling towers are wet/dry, that is, they have both evaporative and non-evaporative sections in combination. The heated water first passes through a non-evaporative section, and then flows, usually by gravity, through an evaporative section. Cooling air from outside the equipment flows through both sections in separate streams. These combination systems lose less water than evaporative systems, and cool better than purely dry systems. Fig. 11J2-2 shows a wet/dry system designed to conserve cooling water.

Cooling towers are also classified by the flow path of the cooling air. Counterflow towers use a vertical upward flow of air that contacts the downward flow of the water to be cooled. Crossflow towers use a horizontal movement of outside air through at least one stage of fill, and then upward movement to exit the tower.

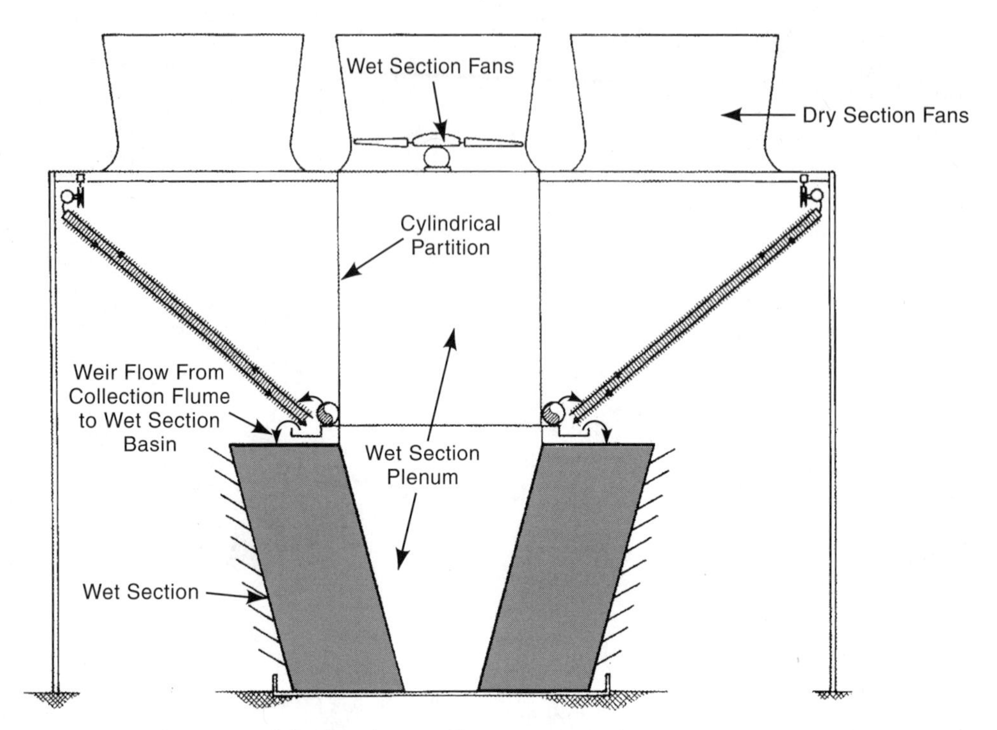

Fig. 11J2-2 A schematic view of a wet/dry cooling tower designed to conserve the amount of cooling water lost to evaporation. (*Courtesy Marley Cooling Technologies.*)

Air cooling devices are used to condense steam, to reduce the temperature of cooling water from industrial processes, and to act as a heat sink for some processes that involve heating and cooling. After giving up its heat to the air, the water is returned to the industrial process to be reused for continuing cooling.

K. Extraction and Recovery of Metals from Ores

This group of processes is sometimes referred to as extractive metallurgy. It includes two basic phases: *concentration* or *mineral dressing*, during which the metallic ore is separated from gangue (stones,·rock, and other unneeded earth material) and *recovery, refining* or *process metallurgy*, during which the ore is then converted to a usable metal. Concentration may first involve a certain amount of size reduction - crushing and/or pulverizing as summarized in section D above. Then, physical separation of the ore from the gangue takes place using one or more of the methods described in section C8 above. These operations include sieving, flotation, dense media separation, drying, and magnetic and electrostatic separation. Sometimes, a specialized process for the particular ore is used. Some ores are then made into larger particles by sintering or pelletizing to facilitate the subsequent operations. Following the concentration, the chemical refining steps can be carried out. These may be quite complex and often involve multiple operations. Chemical, pyrometallurgical, or electrolytic processes, or a combination of them, may be used.

K1. ***recovery/refining processes*** - The metallic compound in an ore is often converted to another compound that is more easily treated before it is reduced to the metallic state. Sulfide ores are converted to oxides, sulfates or chlorides; carbonates are converted to oxides, and some oxide ores are converted to sulfates or chlorides. Three basic types of processes are commonly used to accomplish these changes. Some of these processes, involving aqueous solutions to dissolve the ore, are often referred to under the heading, *hydro metallurgy*. Others utilize heat to convert the ore or intermediate compound to another compound. These processes are sometimes referred to under the heading, *pyrometallurgy*.

Final refining may involve a series of chemical and other operations before metal of sufficient purity results. When electrolytic operations are involved, the term *electrometallurgy* is sometimes applied.

K1a. ***hydro metallurgy*** - is a variety of processes that utilize aqueous solutions and include two basic steps: The first basic step, *leaching*, treats the metallic ore to create a solution of a salt of the metal. The second basic step, which may involve several operations, recovers the metal from its salt solution. The first step involves dissolution or leaching of the metal from the ore. Dilute sulfuric acid is the most common leaching agent. The second basic step recovers the metal from the solution. A sequence of operations may be needed to recover the metal, including purification and concentration of the solution or both, precipitation with a suitable reagent, carbon adsorption, ion exchange, solvent extraction, and electrolysis. Hydro metallurgy is used in the production of many metals, including almost all non-ferrous metals. Most gold is recovered with these processes. Silver and copper are two other notable examples of metals recovered, at least in part, by hydro metallurgy. Others are aluminum, nickel, zinc, cobalt, uranium, molybdenum, tungsten, and beryllium. The processes are applicable to low-grade ores and are adaptable to automatic continuous methods but may be energy-intensive and may produce solutions and residues that must be properly disposed of. (However, the disposal of unusable bi-products is sometimes easier than when pyrometallurgy is used.)

K1a1. ***precipitation in hydro metallurgy*** - (See C6 above) Precipitation in hydro metallurgy can be effected with one of a variety of methods. One method reacts the metal salt solution with a gas, creating a compound that can be reduced more easily than the salt solution. Nickel is refined by treating nickel sulfate solution with hydrogen sulfide gas. Nickel sulfide precipitates from the solution. Another method reacts a more active metal with the one in solution. Copper metal precipitates from a solution containing copper ions when copper cementation iron is made to react with the copper solution. This method is used to precipitate weak copper solutions. A third precipitation method, used in uranium refining, involves changing the acidity of the salt solution of the metal. A concentrated leach solution of uranium is treated by adding sodium hydroxide to raise the ph. Sodium diuranate (yellow cake) precipitates from the solution.

K1b. ***pyrometallurgy processes*** - involve heating the metallic ore to a very high temperature, in either a kiln, hearth furnace, rotary kiln or fluidized bed reactor. (See Fig. 11I5.) The heat has both physical and chemical effects on the ore. Operations include oxidation, calcining, sulfating, chlorination, reduction (smelting), and roasting. Pyrometallurgy is an important means of metal production. In the simplest treatment, a reducing or oxidizing agent is heated with the ore, producing liquid or gaseous metal and a by-product, liquid slag, or some gas. With sulfide ores, the reagent is usually oxygen and the gaseous by-product is sulfur dioxide. Any other metals in the ore form oxides. Pyrometallurgy processing temperatures range from about 300°F (150°C) to about 2900°F (1600°C). Reactions are rapid because of the high temperatures so large production quantities are possible with each furnace. Separation of metal from other materials is made easier because of the liquid or gaseous state of the metal. Examples of pyrometallurgical processes are pig-iron blast furnaces which reduce iron oxide to pig iron, lead blast furnaces that reduce lead oxide to lead metal, and smelting that produces copper matt or converts copper sulfide to blister copper. The complete process for any metal may include both hydrometallurgy and pyrometallurgy and sometimes also electrometallurgy. The latter is usually at the end or near the end of the production sequence but the pyrometallurgy may be either near the beginning or later in the sequence, depending on the ore and metal involved. It should be noted that scrap metals are an important ingredient in much metal production and are melted with metal extracted from ores. Steel, aluminum, copper, zinc and lead production all frequently involve the use of scrap.

K1b1. ***roasting*** - is an effective process for treating sulfite and carbonate ores. The ore is heated in air or an oxygen-enriched atmosphere to a temperature below the melting point of the metal. Temperatures are normally in the range of 1470 to 1725°F (800 to 940°C)[4]. Sulfides in the ore are converted to oxides with sulfur dioxide gas as the by-product. Some operations are performed to partially remove sulfur from the ore; in others, the ore is fully converted to an oxide. With carbonates, carbon dioxide is driven off by roasting, and a metallic oxide remains. In the production of zinc, zinc sulfide concentrates are roasted to fully convert them to oxides. The oxides are then processed further by sulfuric acid leaching. Roasting is still important in cobalt production and is used in some nickel and lead processing. Previously it was also used as a pretreatment before the smelting of copper. The roasting operation is performed in fluidized bed reactors with air entering from the bottom of the reactor to fluidize the particles of ore. The reaction is exothermic, providing enough heat to sustain the reaction as more ore is fed to the reactor and processed ore is discharged.

K1b2. ***smelting*** - is the most prevalent reducing process. A metallic ore, often after concentration, is mixed with a reducing agent and with a flux, and is heated to a high temperature. All constituents of the ore or concentrate are melted. The reducing agent combines with the gangue and forms a liquid slag which floats on the liquid metal, and both can be poured off separately. Smelting is used in the extraction of copper, nickel, lead, and other metals from their ores. Pig iron processing in blast furnaces is another smelting process. Common reducing agents are carbon or coke, natural gas, carbon monoxide gas, ferro silicon and iron (used in the production of magnesium), aluminum (used in the reduction of calcium ores), and magnesium (for titanium, zirconium, and hafnium).

K1c. ***electrometallurgy*** - uses electrolysis to separate metals. There are two processes: electro winning and electrorefining. (Also see C10 above.)

K1c1. ***electrowinning (electrolytic deposition)*** - uses insoluble metal electrodes in a solution containing the metallic compound. The solution can be aqueous or a molten salt, and it ionizes as an electric current passes between the electrodes. The metal ions are attracted to and are deposited on the cathode. The non-metal constituent is deposited on the anode. When the cathode is full, the metal is stripped from it. The process produces metals of high purity, but requires the use of solutions of high metal content. In some cases, the initial metal salt solution is concentrated beforehand by stripping it with another solvent. This concentration takes place in the production of copper from low-grade ores. The low-concentration leaching solution is treated with an organic solvent, immiscible in water, that strips the copper from the solution. Then another stripping operation with another solvent takes place and the resulting solution is suitable for electrowinning. Electrowinning is used in the production of nickel, zinc, manganese, antimony, silver, gold, and cobalt from aqueous solutions. It is used to produce aluminum, magnesium, calcium, beryllium, barium, sodium and potassium from molten salts.

K1c2. ***electrorefining*** - is a metal purification process similar to electrowinning. A metal electrode to be purified is immersed in a salt solution of the same metal and becomes the anode of an electrolytic circuit. Direct electric current is applied and electrolytic action deposits pure metal at the cathode. The method is economical and effective in removing foreign materials from the metal. It is used extensively in copper refining with an electrolyte of copper sulfate and sulfuric acid. Nickel, zinc, lead, silver and gold are other metals which are commonly electro refined.

K2. ***other chemical processes*** - Adsorption with activated carbon is used in gold processing to strip solutions of gold cyanide. The resulting solution is then stripped with another solution that is used in electrowinning the gold. Another operation that may be used in metal refining is the fractional distillation of impure metals to remove minor alloying metals. Zinc is fractionally distilled to remove lead or cadmium. The nickel carbonyl process is used to purify the nickel metal. Impure nickel is reacted with carbon monoxide gas to form nickel carbonyl gas. This gas can then be decomposed to yield nickel of high purity. Chlorination is an operation used in the extraction of some non-ferrous metals. Magnesium oxides or hydroxides are converted to chlorides, to feed electrolytic refining in the production of magnesium metal. Chlorination is also used in nickel processing, and with some refractory metals that do not reduce easily from oxides. Chlorination can be carried out with fine particle ore in fluidized bed reactors using chlorine gas.

K3. ***alloying processes*** - Alloying is normally an uncomplicated operation. Often, it means simply putting the proper weight of ingots of each constituent metal in the melting pot. When, the metals have melted, mixers are inserted in the melt to insure a homogenous alloy. Heat for melting is provided by one of several methods: gas or oil burners, electric arc, electrical induction. The main constituent is commonly melted first and alloying metals are then added and melted. Contamination of the melt must be avoided. Slag may be used to provide protection from the atmosphere to minimize oxidation of the melt, but vacuum furnaces are sometimes used for the same reason.

L. Esterification

Esterification is the chemical combination of an alcohol and an acid, with the elimination of water, to form an ester. The operation is similar to the production of an inorganic salt, except that the organic radical of the alcohol replaces the acid hydrogen of the acid. The alcohol and acid to be reacted are heated, together with a small amount of sulfuric acid. As the reaction products are formed, they are removed, usually by distillation.[6] Esters made from carboxylic acids are the most common. Ethyl acetate is made with this process using some excess of alcohol. The process takes place in a column and a ternary azeotropic alcohol (70% ethanol, 20% ester, 10% water) is used. Isopropyl, butyl and amyl acetates are made from this process with acetic acid. These acetates are useful as lacquer solvents. Other esters are used as plasticizers or in perfume. Vinyl, acrylic, and allyl alcohol esters are important in the manufacture of plastics. Polymethyl methacrylate ("Lucite", "Plexiglas"), polyethylene terephthalate ("Mylar") and textile fibers involved in "Dacron" and "Fortrel" are examples of products made using esterification. Esters of nitric acid are used in the manufacture of some solvents, medicines, perfumes, explosives, and monomers for plastics manufacture. Phosphate esters are used as flame retardants, insecticides and additives for gasoline and oil.

References

1. *Encarta Encyclopedia (from CD),* Microsoft Corp., Redmond Washington, 2004.
2. *The Engineering Handbook,* R.C.Dorf, CRC Press—IEEE Press, 1996, ISBN 0-8493-8344-7.
3. *The New Encyclopedia Britannica,* Macropaedia, Chicago.
4. *The McGraw-Hill Encyclopedia of Science and Technology, 8th ed.,* McGraw-Hill, New York, 1997.
5. *Perry's Chemical Engineer's Handbook, 7th ed.,* R. H. Perry and D. W. Green, McGraw-Hill, New York, 1997.
6. *Shreve's Chemical Process Industries, 5th ed.,* G.T. Austin, McGraw-Hill, New York, 1984.
7. *Introduction to Chemical Engineering,* W.L. Badger, J.T. Banchero, McGraw-Hill, New York, 1955.
8. *Unit Operations of Chemical Engineering, 6th ed.,* W.L. McCabe, J.C. Smith, P. Harriott, McGraw-Hill, New York, 2001. ISBN 0-07-039366-4.
9. *Elementary Principles of Chemical Processes, 3rd ed.,* R.M. Felder, R.W. Roiusseau, John Wiley and Sons, New York, 2000. ISBN 0-471-53478-1.
10. *Food Engineering Operations,* J.G. Brennan and other, Elsevier, London, 1969.

Chapter 12 - Food Processes

Note: Many of the food processes described below are very similar to processes with the same name in the chemical industry, and described in Chapter 11. When foods are involved, special cleanliness and sanitation provisions prevail, state and federal regulations are involved, and the equipment may be adapted to maintain a temperature that prevents degradation or unwanted cooking of the foodstuff. Regular cleaning and sanitizing procedures are required. Stainless steel and other materials that withstand cleaning and sanitizing are used and smooth, continuous surfaces are essential. Otherwise, many food processes parallel those used in chemical manufacture.

Food Processing - Food is processed to make it more digestible, more nutritious, more appealing in appearance and flavor, but most importantly, to make it safe to consume and to preserve it for future consumption. Food preservation has always been important because the time and place of food harvesting or slaughter usually does not coincide with the time and place of food need and consumption. Processes involving both heating and cooling are used to aid in preservation. Heating processes include preheating, blanching, cooking, baking, canning, sterilization, evaporation, and dehydration. Cooling processes include refrigeration, freezing, freeze drying, and freeze concentration. Other processes involve the addition of chemicals and other substances. Drying, fermentation, and irradiation are additional food processing methods in wide use.

A. Cleaning Raw Food Materials

Cleaning is used to remove contaminants from food material at the start of processing. Contaminants may include any of the following: soil, sand, pebbles, oil and grease, twigs, grass, leaves, stalks, other foliage, husks, pits, metal particles, hairs, insects and insect eggs or parts, pieces of string or rope, hairs, excreta, micro-organisms and herbicide or insecticide chemical spray residues or fertilizer residues. Two basic approaches are used for the removal of such materials from the food: dry cleaning methods and wet cleaning methods. Dry cleaning includes brushing, sieving or screening, abrasion, air flow (aspiration or winnowing), and magnetic separation. Wet cleaning methods include: spraying or soaking, filtering, sedimentation, flotation, pressure wash (fluming), and ultrasonic cleaning. Very often, a cleaning sequence includes both dry and wet cleaning operations.

Screening (sieving) is used to remove contaminants that are of a different size than the food material. (Screening is described in Chapter 11, section C8a.) Rotary drum screens are used to remove oversize material from powdered or granular foodstuffs such as sugar, flour or salt or to remove small contaminants such as dust, seeds, or grit from cereals.

Abrasion involves agitation of the food's raw material so that movement and impacts between food particles and between food and the moving parts of the equipment loosen contaminants that have adhered to the food. Screening or other operations to separate the contaminants from the food are then required.

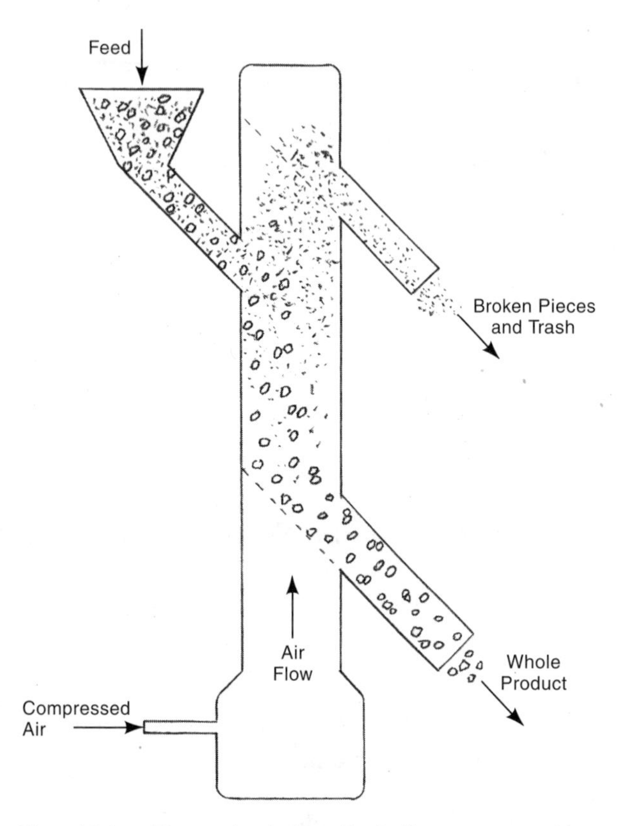

Fig. 12A The principle of air-flow separation. Broken pieces, dust, dirt and trash are more apt to be carried upward by the air flow and exit at the upper chute while the whole product settles in the chamber and is discharged at the lower exit chute. The illustration shows a single unit; production equipment may have two or three stages of separation.

Air-flow separation utilizes the differences in buoyancy of food and contaminant particles suspended in an upward air flow. A typical air-flow separator is shown schematically in Fig. 12A.

Magnetic separation is described in Chapter 11, section C8d.

Soaking and spraying are common cleaning methods for vegetables and fruits. Soaking is especially useful for root vegetables and other heavily-contaminated vegetables and is often a preliminary step before other cleaning. Dirt adhering to the vegetable is loosened by soaking so that some contaminants drop off. The operation is aided if there is flow or agitation of the soaking water and if it is heated, though the temperature must be low enough so that the food is not affected. Detergents may also be used, but they necessitate later rinsing. Spraying provides a more aggressive washing method. Fig. 12A-1 shows two production arrangements, one using a roller conveyor and the other a rotating drum to hold the food and move it under high-pressure spray heads.

Flotation methods may involve simple placement of the food material in water tanks to allow the heavier contaminants to settle to the bottom. When differences in buoyancy are not great enough, flowing water may be used in channels with weirs. The weirs catch the heavier material; the lighter material flows out of the channel with the water. Fig. 12A-2 illustrates such a device, used to remove dirt, plant debris, and stones from beans, peas, and dried fruits.[6] Froth flotation, and other methods described in Chapter 11, section C8b, may also be used with foodstuffs.

Ultrasonic cleaning, described in Chapter 8, section A2b, may be used to remove wax or grease from fruits, or soils from vegetables and eggs.

After wet cleaning, it may be necessary to remove excess water from the food material. One method is to place the material on agitated screens for a period, sometimes with an air blast to assist drying. When moisture removal requirements are strict, centrifuging (Chapter 11, section C7e) is a useful method.

B. Sorting and Grading of Foods

Sorting and grading are separation operations. Food pieces of like size, weight, shape, color, and quality characteristics, are removed from a mixed lot and collected in uniform or nearly-uniform groups. These operations often are performed manually, especially in small-quantity situations. For high production, automatic equipment is commonly utilized. Much current equipment uses machine-vision sensors, weight sensors, or other sensors to characterize each piece that passes it on a conveyor. The datum is sent to a computer processor that compares it with a standard and, depending on the comparison, may actuate a diversion mechanism that routes the piece to the proper channel, perhaps to a chute of rejected items. Sometimes, purely mechanical or non-computerized electromechanical devices, such as the size sorting systems illustrated in Fig. 12B, are used to separate items by size, weight or shape.

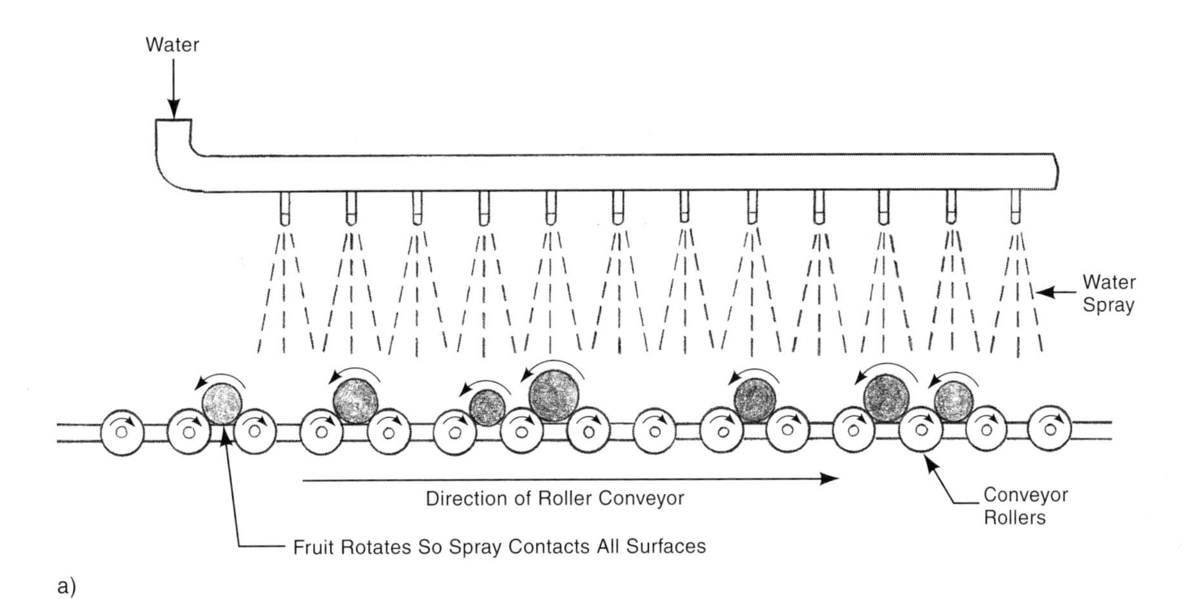

Fig. 12A-1 Spray washing of foods: a) moving conveyor to carry fruits or vegetables under a water spray head. The conveyor rollers are geared to revolve as the conveyor moves, causing the food pieces to rotate, and exposing all surfaces to the spray. b) view of the cross section of a spray drum washer. The drum is tilted and the dirty food pieces are fed from the upper end. As the drum rotates the pieces tumble and are exposed on all sides to the water spray. The dirt, trash and water leave the drum through slotted openings that are shaped so that trash or broken pieces, small enough to enter the slots, are not blocked from falling aside. The washed food pieces leave the drum at the lower end.

C. Size Reduction

Size reduction of pieces of food material may be needed for a number of reasons: to provide the desired size of the product pieces, to remove a shell or husk; to aid in a subsequent operation such as expression, dissolution, extraction, cooking, mixing or drying. Three basic means of reducing size are compression, impaction and attrition. Common methods are roll crushing, hammer milling, ball milling and disc attrition milling. In many cases, several stages of size reduction and several different methods may be involved, the number depending on the size of the pieces of feed material and the final particle size needed. The choice of method depends on the amount of size reduction, its purpose, and the hardness, structure, temperature sensitivity, and moisture content of the feed material.

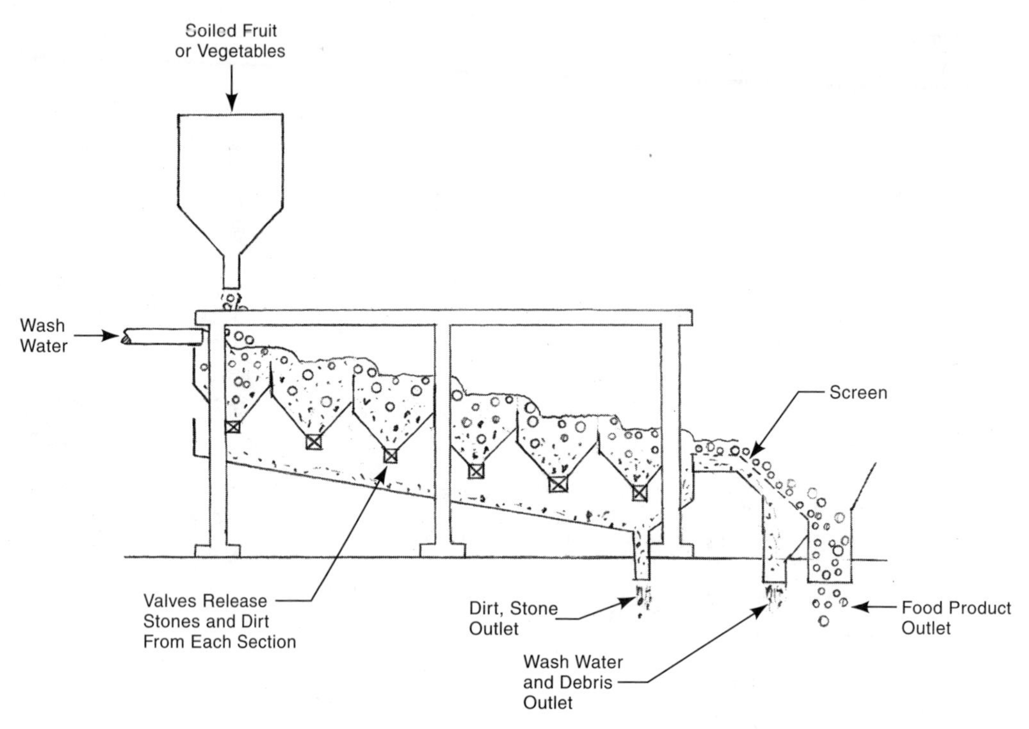

Fig. 12A-2 A flotation separation device to remove dirt, stones and debris from vegetables and fruits.

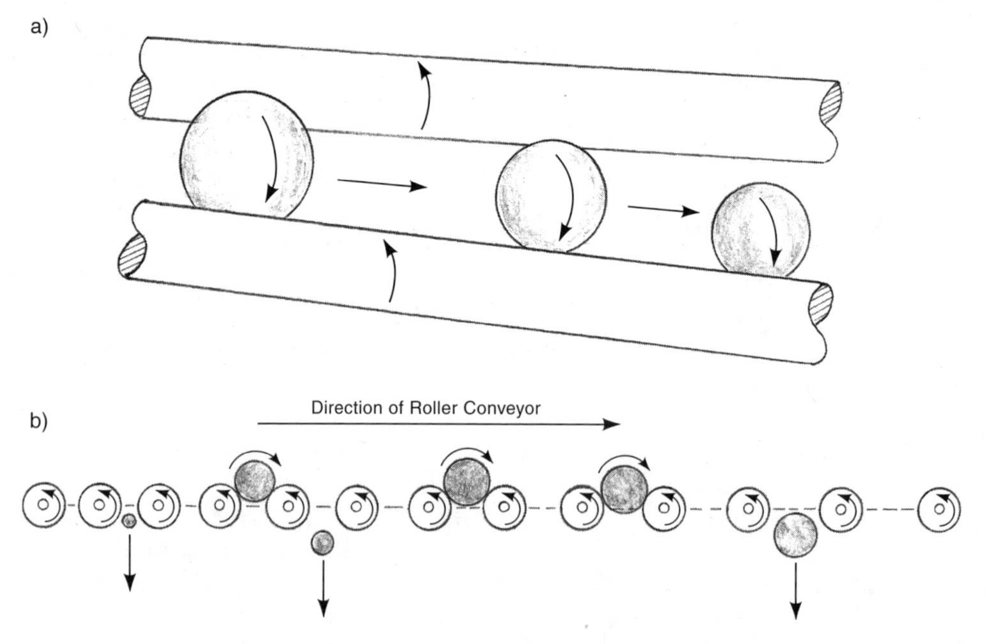

Fig. 12B Two systems for sorting fruit by size: a) apples or other near-round fruit move to the right (down hill) as they rotate. The gap between the rollers gradually increases so that the fruit falls between the rollers when it reaches a point where the gap is large enough. In view b), rollers of a roller conveyor gradually increase in spacing as the fruit moves to the right. When the gap is large enough, the fruit falls through the conveyor.

C1. ***roll crushing*** - utilizes two parallel steel rolls with a small space between them. Both rolls rotate on their axes towards the space that separates them. Food fed into that space is drawn into it and crushed by the weight and strength of the rolls. The roll surfaces may be smooth or grooved. If the rolls are operated at different rotational speeds, some shearing force is also applied. Fig. 11D2 in chapt. 11 illustrates roll crushing of minerals using the same principles and very similar equipment to that used for food materials. Roll crushing is used in the milling of wheat and the refining of chocolate.[6]

C2. ***hammer milling*** - is identical in method to impact milling of minerals, as illustrated in Fig. 11D2-1 (for large mineral pieces). Whether operating on minerals or food, the impact of hammers, attached with a pivot mount on a revolving disc, breaks the material into smaller pieces. Pieces in the mill may be struck repeatedly. The broken particles, when small enough to pass through the screen at the bottom of the impact chamber, are discharged from the machine. These machines have general application in the food industry, being used for various vegetable materials, especially those that are fibrous or sticky, and for crystalline solids. The grinding of sugars, dried milk, pepper, and spices are common applications.[6]

C3. ***ball milling (tumble milling)*** - is illustrated for minerals in Fig. 11D3. Steel balls in a slowly rotating cylinder are lifted and then fall against food material in the cylinder. About 50% of the content of the cylinder is material and 50% the steel balls. Both impact and shearing forces are generated, depending on how the balls strike the material and other balls. The speed of rotation of the cylinder is set to produce the best motion of the balls, which range in diameter from 1 to 6 in (25 to 150 mm). For some materials, particularly those that tend to be sticky, rods are used instead of balls. The rods are as long as the full length of the cylinder. Primarily, shearing forces are exerted on the food material.

C4. ***disc attrition milling*** - exerts high shear forces on the feed material. The food is fed between two rotating discs with horizontal axes, or between one rotating disc and a stationary disc surface. The discs are grooved and rotated at a high speed. The use of two rotating discs provides a higher surface speed because the discs rotate in opposite directions. The size of the gap between the discs is adjustable depending on the material processed. Fig. 12C4 shows both the single-rotating-disc and double-rotating-disc setups. This method is widely used in milling corn and rice, and in making cereal.[6]

C5. ***milling grain*** - is the operation that produces a finely ground meal (flour) from wheat and other grains. Whole wheat flour is milled from the entire wheat kernel including the outer shell (bran), the interior material (endosperm) and the innermost material (embryo or germ). White flour milling involves separation of the edible endosperm (which becomes white flour) from the bran and the germ. There are several steps in the milling process for white wheat flour: The seeds or kernels of the grain to be milled are first cleaned and separated from chaff, dirt, straw, sticks, pebbles, and other seeds. The grain is passed through a series of screens or perforated cylinders that separate the kernels from items of other sizes. This step may also classify the kernels by their size. The kernels are then scoured by passing them through a cylinder lined with emery. They are then tempered to adjust their moisture content to the optimum level by being moistened and washed if too dry, or dried gradually if too damp.

Grinding then takes place by passing the kernels through paired rollers. The rollers are grooved and rotate at different speeds. There is little actual grinding. When this step is complete the grain is reduced to three materials: 1) flour, the ground endosperm or nutritive tissue of the wheat kernel, some of which is in large nodules; 2) bran, the broken husks; and 3) middlings, a mixture of husk fragments and endosperm. These materials are separated by sifting. Air classification may also be used. The nodules of endosperm are again fed through a pair of rollers and then through a series of pairs of rollers with successively closer spacing. The final series of rollers are smooth surfaced. Sieving and further classification, known as "bolting" takes place between rolling operations. Bolting can be carried out by various methods including: plansifting which utilizes several sifting screens arranged one above the other; the use of a reel covered with bolting fabric that retains the middlings and lets the flour pass; air

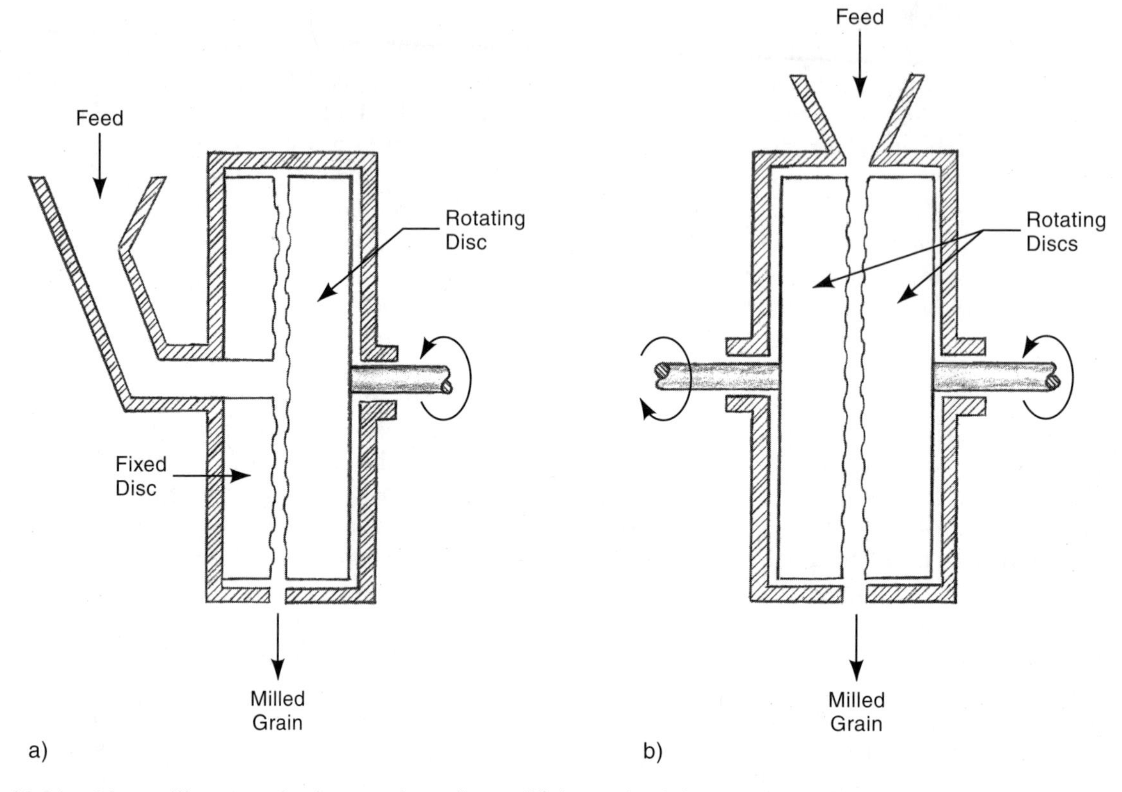

Fig. 12C4 Disc mills: a) a single-rotating-disc mill, b) a double-rotating-disc mill. The two discs rotate in opposite directions to provide faster milling.

classification; and centrifugal methods that also use reels. These processes further remove bran particles from the flour. Ground flour of various particle sizes is blended to provide the desired particle distribution in the finished flour. The blended flour is weighed into sacks or other containers. Methods for grains other than wheat are similar.

C5a. ***post-milling treatments*** - Bread flour may undergo several treatments after milling to enhance its nutritive value, to change its appearance or to improve its baking quality:

C5a1. ***altering protein content*** - Gluten, the protein of wheat flour, can be removed from the starch portion by washing the flour. A dough of flour and water is allowed to stand in water for a short period. This, and further washing, dissolves the starch and leaves the gluten. The gluten then can be added to other flour to enhance its protein content. Such flour is used for high-protein breads. The added gluten, in addition to its nutritive value, holds the flour grains together as the bread rises. Flour for cakes, biscuits, and cookies, that are softer, typically does not need as high a protein content.

C5a2. ***bleaching flour*** - The yellow color of some flour, undesirable for some consumers, is removed by bleaching. Natural bleaching occurs if the flour is allowed to age for several weeks. The aging whitens the flour and improves its baking qualities. Bleaching with chlorine dioxide gas, nitrogen trichloride, benzoyl peroxide, or nitrogen tetroxide, the most commonly used bleaching agents, provides equivalent effects much more quickly and is commonly used in flour production. Potassium bromate and benzoyl peroxide are other flour bleaching agents.

C5a3. ***enriching flour*** - involves the addition of thiamine (vitamin B_1), riboflavin (vitamin B_2), niacin and iron to white flour to provide vitamin content similar to that of whole flour.

D. Mixing

Various mixing methods are described and illustrated in Chapter 11. Most of the reasons for mixing - to provide a homogeneous blend of material, to distribute added ingredients uniformly, to disperse heating or cooling, to promote crystallization or dissolution, and to create emulsions - apply to foodstuffs as well or better than they do to non-food chemicals. The methods described in Chapter 11 apply well to mixing foodstuffs. These methods include: mixing liquids of low or medium viscosity with turbine or propeller mixers, as well as using the same equipment for mixing gases and liquids. These methods are described in sections 11G2 and 11G3 and are illustrated by Fig. 11G3 and 11G3-1. Mixing higher-viscosity liquids is also described in section 11G3 and illustrated in Fig. 11G3-2 and 11G3-3. Mixing liquids and solids together is described in section 11G4. Equipment for such mixing is illustrated in much of Fig. 11G5 which shows devices for mixing solids including those to which some liquid has been added. If the solids-liquid mixture tends to cake, i.e., if the liquid causes adhesion between solid particles in the mixture, the ribbon, screw and rotor muller mixers that are shown in Fig. 11G3-2 and 11G5(e), 11G5(f) and 11G5(h) are commonly used. For heavily viscous material, such as bread dough, the kneading mixer pictured in Fig. 11G5-2 is employed. The material is sheared, stretched, and folded to intermix different components. Another machine used for food mixing is the pan mixer shown in, two versions, in Fig. 12D. Screw-based extruders also may be effective in mixing viscous, dough-like material. See J5 below.

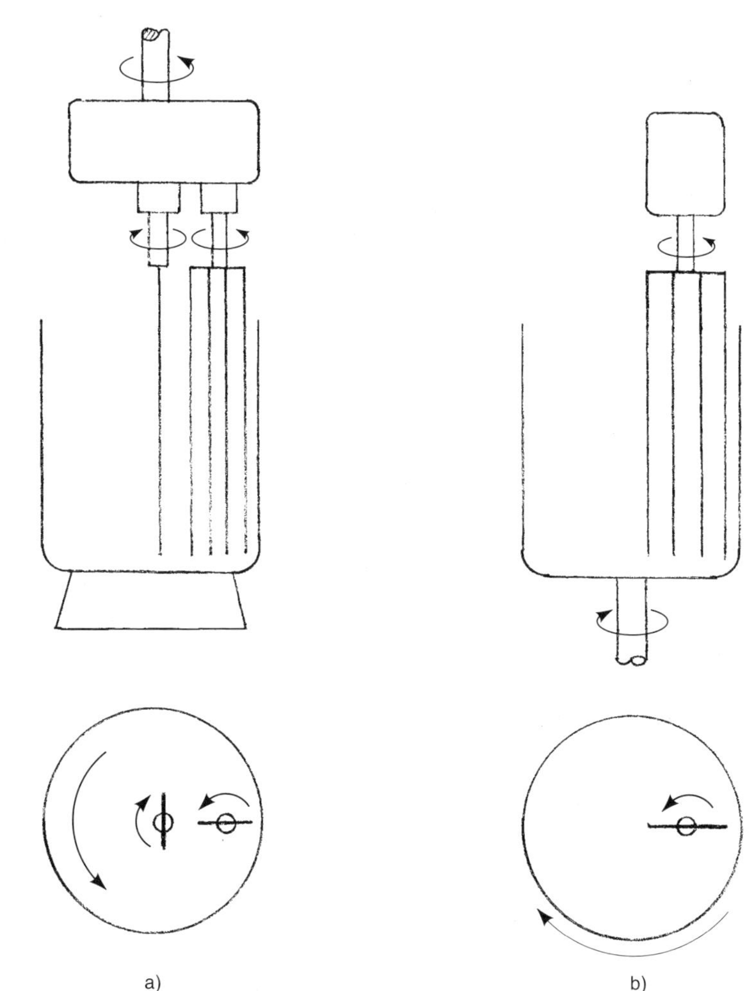

Fig. 12D Two varieties of pan mixer: a) stationary mixing container with two counter-rotating blades which also rotate on a central spindle; b) rotating mixing container with a single rotating mixing blade.

E. Separation Processes

These processes remove one or more ingredients from a mixture of food materials. The major methods are: filtration, expression, centrifugation, crystallization and extraction.

E1. ***filtration*** - is the separation of solid particles from a fluid by passing the mixture through a porous medium that allows passage of only the fluid. It is explained in Chapter 11, sections C7a and C7b. Major method variations are also described. The spectrum of filtration of food materials ranges from the removal of particles larger than 0.004 in (100 microns) in removing curds and casein from milk, to ultrafiltration and reverse osmosis with semipermeable membranes by which soluble starch, salt and sugar molecules are removed from water. Fig. 11C7b shows several types of filters; Fig. 11C7b-1 illustrates a typical horizontal filter press. The presses are used in many food industry applications including the filtering of oil extracted from seeds, and for removing bleaching compounds when the oils are bleached. They are also used in breweries for removing mash and yeast from brewery products and in wineries for the removal of yeast and bacteria from wine.[6] Fig. 11C7b-2 shows a continuous filtration machine and Fig. 11C7b-3, a rotary drum vacuum filter. The rotary drum vacuum filters are widely used in sugar production to filter juice from sugar cane and sugar beets. They are also used to filter gluten suspensions, to remove water from starch, and to clarify fruit juice[6]. Centrifugal filtration is discussed in section 11C7e and illustrated in Fig. 11C7e.

Three types of filtration, described in Chapter 11, are utilized in the food industry for: 1) filtering a slurry of solid material when the slurry contains a significant amount of solid particles - more than one or two percent. This process involves cake filtration as described in 11C7b. The solid material forms a cake on the feed side of the filter, and either the solid or liquid components, or both, may be utilized. 2) clarifying filtration of liquids containing only small amounts of solids where the objective is to obtain a clear liquid for use in food . 3) filtration to remove very fine particles, primarily microorganisms from a food liquid.[6] This is referred to as ultrafiltration or membrane separation. Ultrafiltration is discussed further in F3 below.

Other examples of filtration in the food industry include the clarification by filtration of vinegar, fruit juices, syrups, table oils, jellies and brines[6]. Filter aids (eg., flocculants) may be utilized in these operations. Applications of membrane filtration include vegetable oil refining where gums are removed and hexane is recovered, clarification of dextrose and purification of starch from corn refining, removal of yeast in breweries and distilleries, the desalting of molasses in sugar refining, the production of protein concentrates from various seeds, in the processing of animal byproducts in meat packing and in making dehydrated eggs.[1]

E2. ***expression*** - is the removal of a liquid from a solid material through the application of pressure. It is an important method in food processing for removing oils and juices from fruits, nuts and vegetables. Expression is also has applications in the removal of liquids from various materials in the chemical industry and is described in Chapter 11, section C4.

Common food industry expression operations involve the removal of oils and fats from peanuts, coconuts, soy beans, sunflower seeds, cotton seeds, olives, and rendered fish and meat scraps. The extraction of juice from fruits and vegetables, including sugar cane and sorghum, is a major operation. Expression is used in the preparation of juices for sale to retail customers and to those who use it for ingredients in soft drinks, ice cream and other food products. Other uses are dewatering or solvent removal from filter cakes, removal of whey from cheese, and removal of fat from cocoa. The manufacture of wine involves expression of the juice of grapes. Some preliminary steps may be involved in expression of food liquids to break the cell walls of the fruit or vegetable being processed. Heating may be required. The expression operation may be performed on a batch basis, particularly with smaller production quantities or where a batch identification may be important (eg., some wine making). Continuous methods are used in other instances, particularly when production quantities are large. All expression equipment has some kind of perforated or porous material to hold back solid particles from the expressed liquid. Fig. 12E2 illustrates two batch

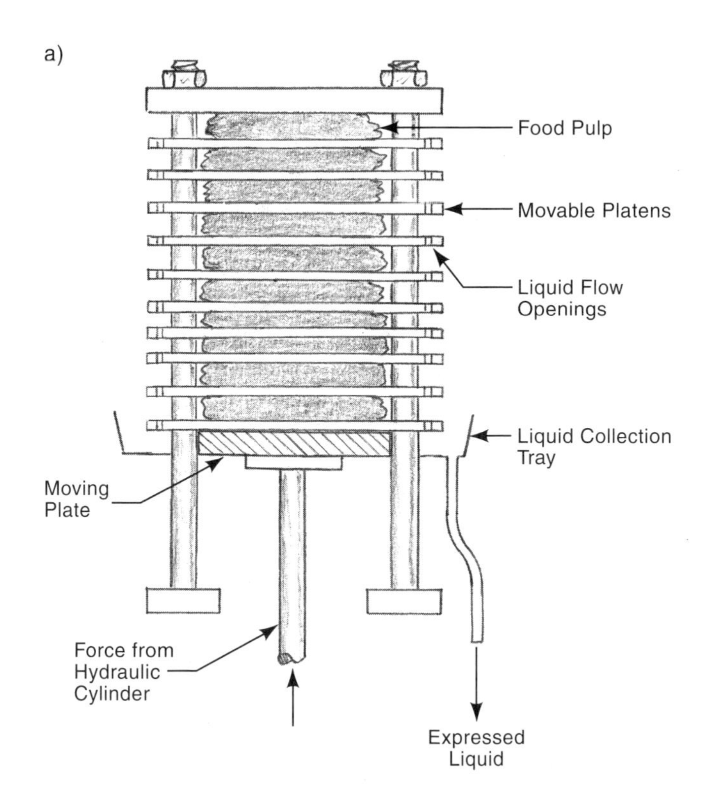

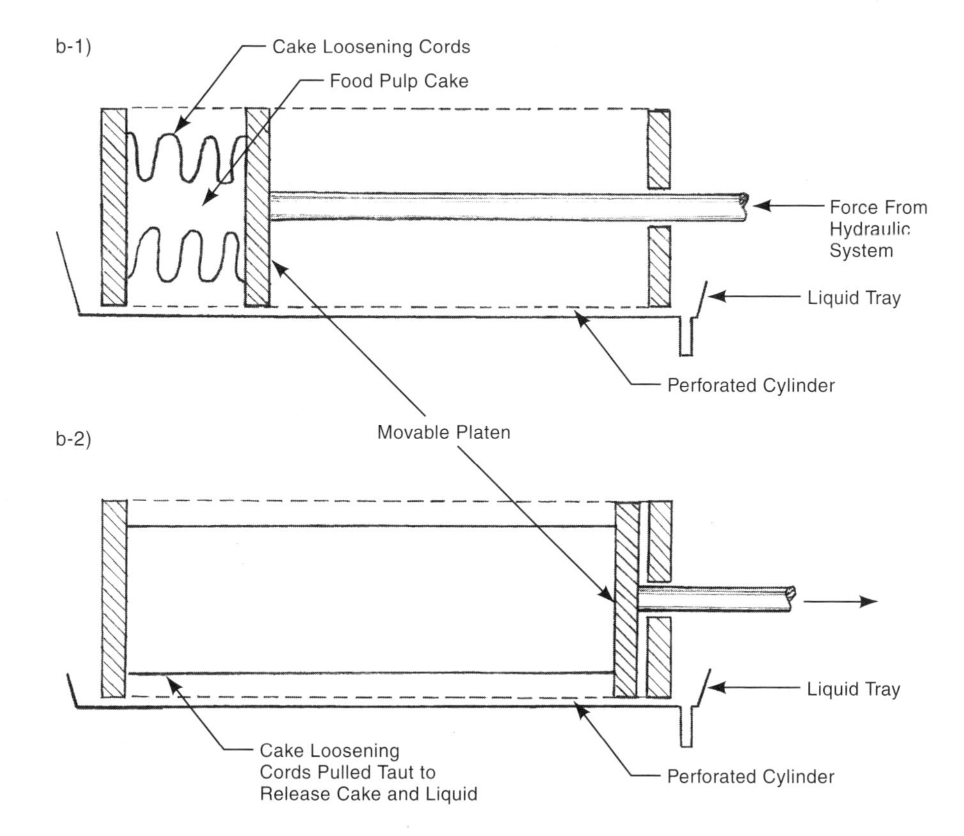

Fig. 12E2 Two batch methods for expression: a) plate press, b) cage press. b-1) closed position when food pulp is compressed. b-2) open position to release food pulp cake and liquid.

methods where hydraulic pressure is applied to the food pulp by pressing it between two platens. In the first illustration, a plate press is shown which includes a vertical stack of horizontal, grooved pressure plates. The second illustration shows a cage press that squeezes the food pulp between two vertical pressure platens. Fig. 12E2-1 shows two continuous methods, Including a continuous three-roll press used, among other applications, for expressing juice from sugar cane. Fig. 12E2-1 also shows a continuous screw press that is used for extraction of both oil and juice. Another continuous method involves a perforated moving belt holding the food material ("cake") and covered by a similar belt moving at the same speed. The two belts pass between two closely-spaced rollers which provide the extraction pressure. Centrifuging is also used to supplement expression in some olive oil and wine processing.

E3. ***centrifugation*** - is widely used in processing foods. The principles of operation are explained and specific types of centrifuges are described in Chapter 11, section C7e. Fig. 11C7e illustrates the operation schematically. In food processing (and in much chemical processing) there are four general uses of centrifugation: 1) separation of immiscible fluids, 2) centrifugal clarification of liquids, 3) desludging and 4) centrifugal filtration.

Fluid separation is used when there is a mixture of two immiscible liquids. The more dense liquid will gravitate to the walls of the centrifuge while the lighter liquid migrates to the space closer to the center of rotation. The movements take place strongly enough to separate the two liquids as long as the difference in density of the two is 3% or greater. The operation can take place continuously if the feed liquid is fed to the center of the centrifuge or to the zone where the two liquids tend to meet; the dense material is piped from the centrifuge through outlets next to the centrifuge wall; the lighter material is piped from a location closer to the center. This method is used to separate cream from milk and to strip small amounts of water that may be mixed with edible oils[6].

Centrifugal clarification is used when there is a small quantity, "a few percent or less"[6] of solids in a liquid, and when the solids have a higher density than the liquid. The solid particles gravitate to the walls of the centrifuge bowl leaving the liquid clear near the center of the bowl. Outlets are provided to permit both to flow into separate containers. The feed liquid enters the centrifuge at the bottom center.

Desludging is the removal of solids from a suspension when the portion of solids is greater than that processible by centrifugal clarification, ie., greater than 5 or 6% by weight[6]. When this high a percentage of solids is present, larger or more exit paths must be provided so that a substantial amount of solid particles is removed as the operation proceeds.

Centrifugal filtration, shown in Fig. 11C7e, uses centrifugal force rather than normal gravity or pressure on the feed slurry to drive the feed liquid through the filter. The centrifuge bowl is perforated on the sides and lined with filter material. However, the major filtration effect comes from the cake of solids that lines the bowl walls. Filtration of vegetable oils is one application of this method.

Tubular-bowl centrifuges are used in removing water from oil or fats from fish, vegetable, or animal sources, and in the clarification of syrups, cider, and fruit juices. Disc-bowl centrifuges are used to separate cream from milk, to refine edible oils and fats, and in the clarification of citrus oils and fruit juices. Other devices are used in separating coffee, tea, and cocoa slurries, removing fish oils and producing fish meal. The clarification of wort; beer, and ale; the dewatering of starches of corn, wheat, and rice; the recovery of vegetable and animal protein; the recovery of yeasts; the removal of small amounts of solids from beverages; the rendering of oil from vegetable or animal sources, and the refining of raw sugar to wash, dry, and recover sugar crystals, are all applications of tubular-bowl centrifuges[6].

E4. ***crystallization*** - is described in Chapter 11, section C5. Two methods, and the necessary equipment, are illustrated in Fig. 11C5 and 11C5-1. In food processing, crystallization may have two purposes: 1) to change a liquid product or ingredient into a more usable solid form, and 2) to separate two materials in a solution by crystallizing one of them. Either the material that crystallizes or the remaining liquid may be the desired material. Freeze concentration is a crystallization process used extensively in food processing. Water is removed from a number of liquid foods by reducing the temperature to the point where ice crystals form. The crystals are then separated from the liquid by centrifugation or filtering.

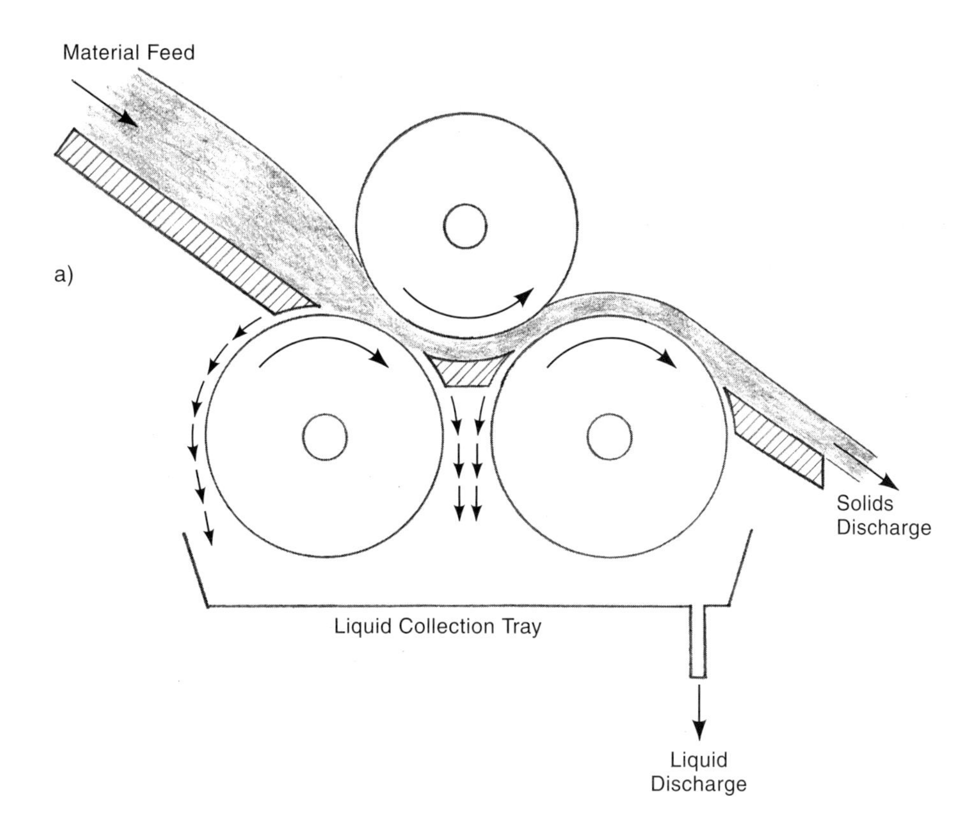

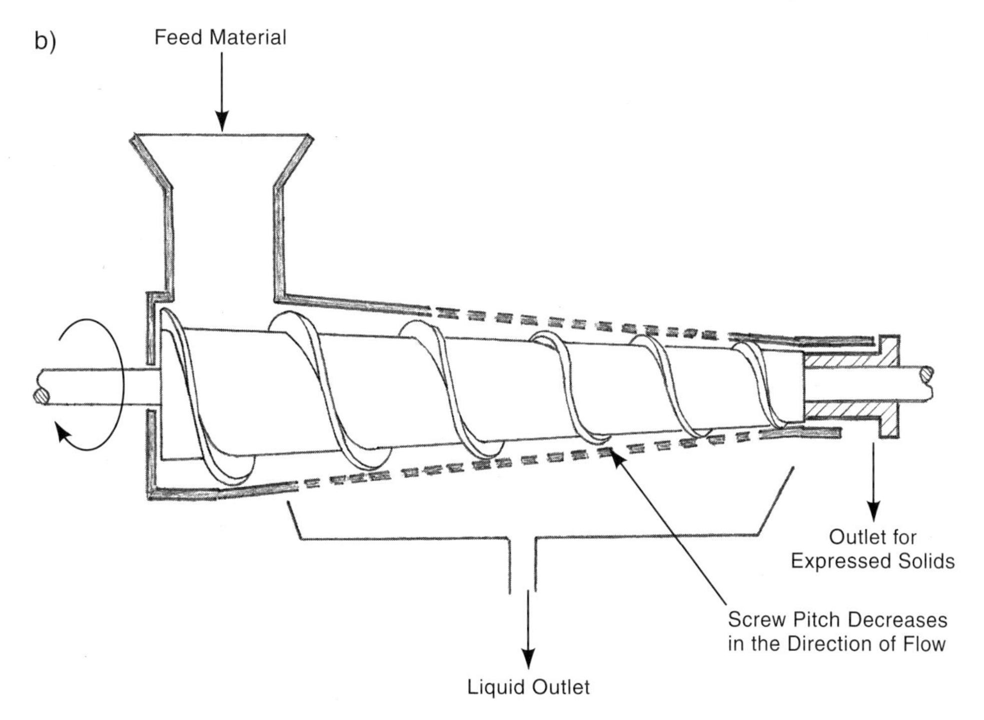

Fig. 12E2-1 Two continuous methods for expression: a) continuous three-roll press and b) screw press.

Another example of crystallizing to improve a liquid is the treatment of salad oils, in which higher melting-point glycerides in the oil are removed by crystallization. This prevents the oil from developing a cloudy appearance should it be stored at the low temperature at which these glycerides would crystalize. Oils are sometimes separated from fats by first dissolving the feed liquid in a suitable solvent from which the glyceride crystals are more easily formed and more completely separated. Acetone and hexane are two solvents sometimes used.[6]

Solid compounds made by crystallization and used in food processing include sugar (sucrose), lactose, salt, monosodium glutamate, and citric acid. With all these compounds, there is also a liquid residue of the feed liquid that is separated from the crystals. Crystallization is also used to produce a solid material where no separation from a residue liquid is involved. Examples of this second situation are the production of ice cream and other frozen foods, sweetened condensed milk, butter, chocolate, margarine, and some fondants and candies. Many of these products require the crystals to be quite small, to ensure smoothness in the finished product. To produce the small crystals, the material to be crystallized is cooled rapidly to a temperature range where the crystallization occurs rapidly, but the crystals are broken up or kept small by stirring the mixture or scraping the surface of the heat exchanger used to provide the cooling. This breaks up the existing crystals and provides more nuclei for additional crystallization. Manufacture of ice cream, margarine, and butter all involve this approach. Fig. 12E4 provides a simplified illustration of equipment used for the scraped-surface method.

E5. ***extraction*** - is reviewed at some length in Chapter 11, section C3. It is applicable to foods as well as chemicals, and is carried out when some component of a food material can be removed from it by dissolving the component in a solvent. The process requires a solvent that selectively dissolves the target component but not the other components of the food material. The method can

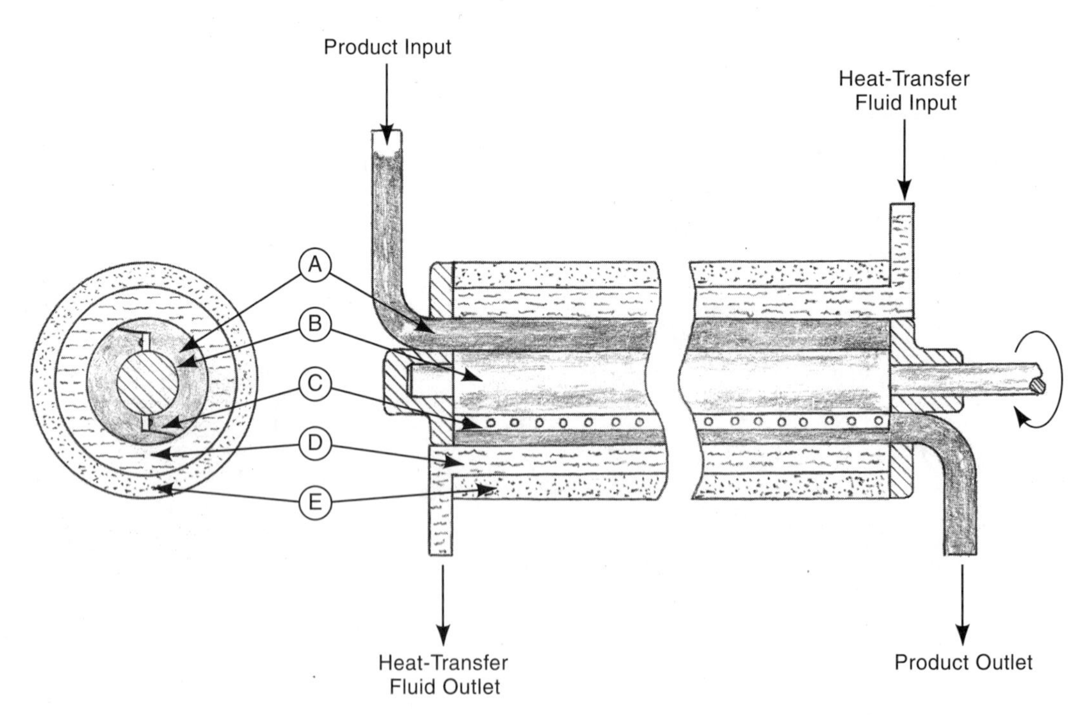

Fig. 12E4 Scraped-surface heat exchanger in two cross-sectional views: (A) product, (B) rotating shaft, C) scraper blades, (D) heat-transfer fluid, (E) insulation.

be used to extract a liquids from another liquid and some material from a solid. In the latter case, extraction is sometimes referred to as leaching. The process can be carried out on either a batch, or continuous basis. Solvents used for extraction in food processing include water, hexane, and other organic solvents and supercritical fluids, notably carbon dioxide. Common food processing extraction operations include: removal of cholesterol from milk fat with carbon dioxide, removal of oleic acid from cottonseed oil with propane, olive oil from pressed olives with hexane, sucrose from sugar beets with water, caffeine from green coffee beans with carbon dioxide, lemon oil from lemon peel with carbon dioxide, and alpha acids from hops with carbon dioxide[1]. After extraction, the solvent and the solute must be separated. Distillation, precipitation, membrane separation, evaporation, or crystallization are the methods used. The solvent can then be reused. Fig 11C3 shows extraction schematically, and Fig. 11C3-1 illustrates extraction from a solid feed material.

F. Concentration

Concentration of liquid foods is an important operation. It is performed to make storage and shipment more convenient and compact, to reduce packaging volume and cost, to induce texture and flavor improvements, and to facilitate further operations such as drying and crystallization. Evaporation, freeze concentration, reverse osmosis, and ultrafiltration are the principal methods used.

F1. ***evaporation*** - the conversion of a liquid to a gas, is reviewed in Chapter 11, section C7g. The methods described are applicable to the evaporation of food liquids. With foods, the liquid to be evaporated is normally water. Fig. 11C7g shows a simple kettle evaporator that could be used for evaporation of batches of food liquids. However, in production situations, evaporation takes place in boiler-like devices, where the liquid product to be evaporated flows through or alongside a series of heat-transfer tubes (shell and tube devices), or plates in a chamber. Steam or other heated vapor provides the necessary heat. As the liquid in the device absorbs heat from the steam-heated surfaces, some of the water content of the food is evaporated. Fig. 11C7g-1 shows several different arrangements for production evaporating equipment. Plate evaporators, similar to the plate heat exchange unit illustrated in Fig. 11J-2, are used in food liquid evaporation.

With short-tube evaporators, the tube arrangement may be vertical or horizontal, and the feed inlet may be either above or below the heat exchange tubes. Boiling and density differences create circulation of the feed liquid through the array of tubes and the heavier, concentrated liquid tends to settle to the bottom of the chamber and be discharged, while the vapor rises and leaves the chamber from the top. Fig. 11C7g-1, view d, illustrates this operation. Often, with this type of equipment, successive stages of evaporation are used. Partially-concentrated liquid from one evaporator is pumped into a second unit and, sometimes, the concentrate from the second evaporator is fed to a third unit. Later stage equipment is run at lower pressures and temperatures so that heated vapor evaporated from the food liquid provides heat for further evaporation. Short-tube evaporators are used in refining cane and beet sugar to concentrate the syrup prior to crystallization. Sometimes, with the addition of circulating mixers, crystallization takes place within the evaporator. These evaporators are used in the concentration of fruit juices and malt extract.

Long-tube vertical evaporators are extensively used in the food industry. These devices have tubular heat exchangers that are as much as 40 ft (12 m) in length. With the rising film evaporator, the preheated feed liquid enters at the bottom of the vessel and flows upward inside the heat exchanger tubes. The heat of the tubes vaporizes some of the liquid, and both the bubbles of vapor and the more concentrated liquid flow upward. At the top of the tube array, the mixture of liquid and vapor is drawn off to a cyclone separator that divides the mixture into vapor and concentrated liquid components. The vapor is drawn off and the liquid falls to the bottom of the cyclone where it exits the system. In the falling film evaporator, the feed liquid enters at the top of the tube array, flows downward through the tubes and partially vaporizes due to the heat transferred by the tubes' inner surfaces. The mixture of liquid and vapor is drawn off and separated. Fig. 11C7g-1, view h, illustrates the functioning of a long-tube, falling-film system. The falling-film method is advantageous for viscous liquids. It can be operated under a vacuum, which is necessary for foods that are sensitive to high temperatures.

Some common food evaporation operations are the production of orange juice concentrate and condensed milk. Concentration of sugar cane juice, prior to the crystallization of the sugar from the concentrated solution, is another application. Evaporation is used in the concentration of fruit juices for jellies, jams, and candies, and the concentration of milk prior to spray drying in making powdered milk. Tomato paste and other vegetable pastes and purees are made by evaporating water from the juices.

F2. ***freeze concentration*** - reduces the water content of liquid food by lowering the temperature of the liquid until ice crystals form. The ice crystals tend to be free of the non-water constituents of the liquid and, when they are removed from the liquid by filtration or centrifugation, a more concentrated liquid results. The liquid food is normally processed in a refrigerated scraped-surface heat exchanger. Small ice crystals form from the water content of the liquid. The liquid, with small ice crystals, then passes to a "ripening chamber". Ripening forms larger crystals and these are separated from the liquid. With the aid of centrifuging, concentrations from a 50 to 60% reduction of the original liquid can be made[1]. The process is more expensive than evaporation and yields a limited amount of concentration but it avoids heat degradation of the food. It also does not cause the loss of volatile aromas and flavors that can occur when evaporation is used to effect concentration. Another factor is the small loss of the concentrated liquid on the surfaces and, to a minor degree, internally in the ice crystals that are formed. Some processes use multiple stages of freezing including washing of the crystals with melt water to recover more of the desired liquid. Multiple stages of crystallization and centrifugation may be used in some freeze concentration operations to remove the freezable component more completely.

Freeze concentration is used to increase the alcoholic content of beverages. One example is the production of "ice beer". The method is also used to adjust the alcoholic content of wine batches, to concentrate vinegar, milk and fruit juices and to concentrate liquid foods prior to freeze drying. Tea, coffee and aroma extracts are made using this method.[8]

Historic non-industrial uses of freeze concentration have included the enhancement of the alcoholic content of hard apple cider and the use of freezing by American Indians to make maple syrup from maple tree sap.

F3. ***reverse osmosis and ultrafiltration*** - *Reverse osmosis* is a membrane separation process related to the membrane separation processes described in Chapter 11, section C7c. It uses pressure to force water through permeable membranes leaving a more concentrated food liquid behind. When two liquid solutions are separated by a semipermeable membrane, without added pressure, natural osmotic pressure of solutions will lead the solvent to pass through the membrane in the direction that dilutes the more concentrated solution. However, when sufficient pressure is applied to the more concentrated solution to overcome the natural osmotic pressure difference, the osmosis is reversed and the solvent passes through the membrane from the more concentrated to the less concentrated solution. Thus, the more concentrated solution becomes further concentrated. Fig. 12F3 illustrates the principles of osmosis and reverse osmosis. An advantage of the process is that no heat is required so that temperature-sensitive food liquids are not adversely affected. The process is used to concentrate milk, whey, and some extracts. Apple juice and maple sap are sometimes preconcentrated by this method prior to evaporation[1].

Ultrafiltration is very similar to reverse osmosis. It differs in that it is usable when the dissolved molecules are relatively large (molecular weights greater than 500[7] and up to 300,000 and typical sizes of 0.002 to 0.2 microns[9]), and pressures are relatively low. Large molecules in solution do not pass through the membrane, but small molecules in solution pass through with the water. A crossflow arrangement for the solution is usual to avoid clogging the membrane. Polysulfone or cellulose acetate plastics are the common membrane materials. Ultrafiltration is used in cheese making to concentrate protein.

The limit of concentration of reverse osmosis and ultrafiltration is about a 20% reduction in solvent content.

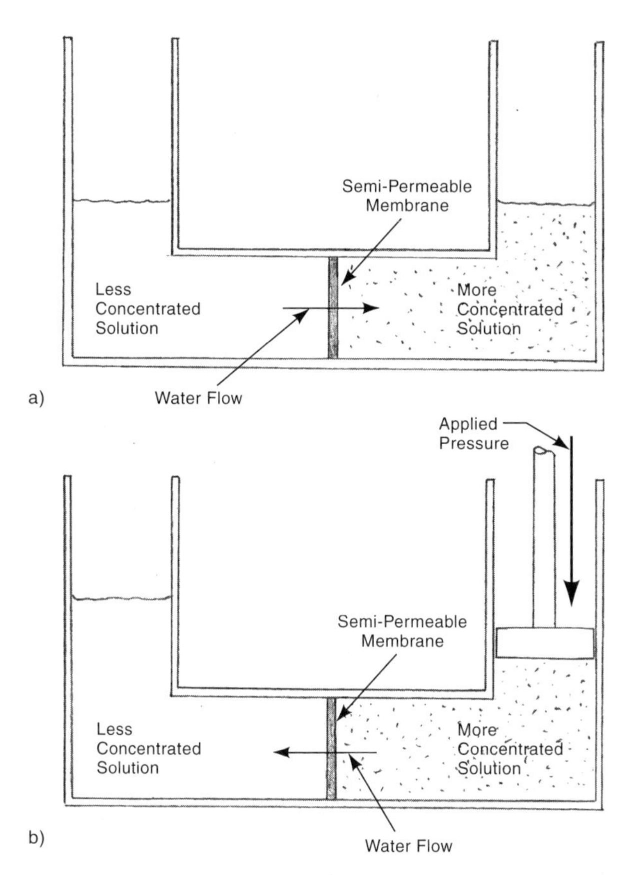

Fig. 12F3 Principles of osmosis and reverse osmosis: a) Natural osmotic pressure forces the solvent (water) through the membrane toward the more-concentrated solution, causing it to be less concentrated, b) When the more-concentrated solution is pressurized enough to overcome and exceed the natural osmotic pressure, osmosis is reversed, and the water is forced through the membrane toward the less concentrated solution, causing the solution from which it flows to become more concentrated.

G. Thermal Processing of Foods

Thermal processing involves heating the food or food material to insure that it is in a healthful condition, to prevent spoilage, and to aid in further processing to make the food more digestible or pleasing in taste.

Heat may be supplied by the combustion of solid, liquid, or gaseous fuels, by electrical methods including resistance heating, dielectric heating, or microwave heating. Radiation, conduction and convection may be involved in transferring heat to the food material.

In heat processing to prevent spoilage, the purpose is to kill any resident food-spoiling organisms. There is some loss of flavor and nutritional value when foods are heated to the extent needed, so careful controls are usually required to gain the necessary sterilization while not causing too much food degradation. The food's packaging and the expected later storage conditions must be considered. For some foods, the operation is performed in a pressurized environment. Cooling immediately after the heating stage is common in order to avoid quality deterioration.

Thermal processing operations include: blanching, pasteurizing, heat sterilization for canning and aseptic processing, cooking (various methods), and baking. In all these processes, heat is applied to the food material to bring it to the necessary temperature for a specified period of time. Several different methods may be used, depending on the nature and condition of the food and the results wanted.

G1. ***blanching*** - is an important operation that involves immersion of vegetables and some fruits in hot water or steam, prior to canning, freezing, or other food preservation operations. Blanching inactivates enzymes that may impair flavor, color and nutrient value during frozen storage or other processing. It also shrinks and wilts the product, which may facilitate proper filling of the container used. Some additional cleaning benefit may also be achieved. The food, however, can lose some heat-sensitive or water-soluble vitamins or nutrients. If the operation is not performed on foods that are heated insufficiently in later processing (for example, foods that are processed by low-temperature vacuum drying or freeze drying), the foods may develop undesirable flavors, colors or odors. The operation is frequently performed on carrots, peas, spinach, and beets before canning. It is also performed before freezing or dehydrating. Blanching temperature is generally at the water boiling point or slightly lower. Foods are heated rapidly and held to this temperature for a period of from about 50 seconds to 10 minutes. They are then cooled rapidly or otherwise processed. In industrial food processes, the operation is performed in equipment consisting either of a rotating perforated drum, a pipe flume or a trough. A screw or other conveyor may be used to move the vegetables or fruit through the equipment.

G2. ***pasteurization*** - is a process for destroying molds, yeasts, and vegetative microorganisms. It is a relatively mild process, less severe than sterilization. Bacterial spores are not inactivated nor are all enzymes that could cause food spoilage in later storage. However, the temperatures are low enough so that flavor, chemical composition or nutritional properties are not harmed. A typical heating cycle for milk is a temperature of 145°F (63°C) for 30 minutes. Another milk pasteurization cycle is heating to a higher temperature, 162°F (72°C) for 15 seconds. Time and temperatures vary with the food, the likely microorganisms present and the use of the food. The pH of the food is a factor since it influences what microorganisms are present. Cooling should take place immediately following the heating phase so that the flavor and nutritional values of the food are not impaired. Still a further process variation for milk is ultra-high-temperature pasteurization (UHT), that subjects the milk (or cream) to a temperature of 280 to 302°F (138 to 150°C) for two seconds or more. If packaged in hermetically-sealed, sterile containers, milk thus treated can be stored without refrigeration for several months.[2] The pasteurization process can be performed on either a batch or continuous basis. Batch equipment usually uses hot water or steam at atmospheric pressure as a source of heat. Agitation of the liquid being processed, eg., milk, is used to insure uniform heating. Plate-type heat exchangers are often employed because such a design provides rapid heat transfer. (See Fig. 11J-2.) These heat exchangers employ multiple plates separated by gaskets. The food liquid and the heating liquid occupy alternate spaces between the plates. Continuous-flow pasteurizers use regenerative systems to minimize energy costs. In this approach, heated pasteurized milk transfers heat to the cool incoming milk. Other separate stages of continuous-flow equipment provide further heating, holding at the pasteurization temperature, and cooling. In addition to milk and cream, fruit juices, canned fruits, beer, wine, ice-cream, and liquid eggs are pasteurized. Bread and cakes may be pasteurized using microwaves as the heat source.

G3. ***heat sterilization*** - is a normal part of the *canning* process further described below. Its purpose is to kill pathogenic organisms and provide necessary storage life for the canned food. The food, which is already packed and sealed in metal cans or containers of glass or plastic-foil laminate, is heated, usually by steam or hot water. Both batch and continuous systems are in use. In a continuous system, the cans or other containers are conveyed through heating and cooling chambers. Batch systems frequently operate under computer control. If a plastic-foil pouch is involved, the operation takes place under pressure to counteract the forces of heat expansion that might otherwise cause the pouch to rupture. Glass containers may be similarly protected. The temperature used in the operation depends on the pH (acidity) of the food. Typical temperatures for low-acid foods are in the range of 220 to 250°F (105 to 120°C). Low-acid foods (eg., crabmeat, olives, eggs, milk, corn, chicken, codfish and beef) require more aggressive heat treatment than high acid foods (eg., lemon juice, cranberry juice, relish, pickles grapefruit, apples and sauerkraut). For fruits and other high-acid foods, the temperatures are typically from 180 to 212°F (82 to 100°C). Cooling immediately follows the heat phase to avoid changes in the taste or nutritional value of the food. Careful control of both temperature and time is maintained in the process to avoid both overheating and underheating.

There are other systems that use higher temperatures and shorter periods under heat. These systems are referred to as HTST (high-temperature, short-time) processes, and typically have two-phase heating cycles. The first phase may involve heating to a temperature of about 150 to 185°F (65 to 85°C) for 5 to 10 minutes; the second phase heats to about 260 to 300°F (125 to 150°C) for only 3 to 30 seconds. The first phase is intended to inactivate enzymes; the second to inactivate microorganisms. The HTST approach is limited to liquid and near-liquid foods. One such process is applicable to soups, purees, yogurt, sour cream and other milk products. The food is sterilized in a heat exchanger before canning; the cans and lids are separately sterilized and the can is filled and closed inside a sterile chamber. Superheated steam provides the heat.

G4. ***canning*** - utilizes heat sterilization to preserve food indefinitely. The food, which has been carefully prepared to prevent contamination, is placed in a container that is sealed and then subjected to the elevated temperature of sterilization.

After a period of time, the container is cooled. The sealed container prevents microorganisms from contacting, spoiling and otherwise contaminating the canned food.

The common tin can (made from steel with a thin interior coating of tin or enamel, depending on the food contained) is inexpensive and durable, and provides a seal against re-exposure to microorganisms. It is the most common container for canning but glass and, more recently, plastics, are also used. Almost all common foods—fruits, vegetables, meats, and sea foods - are canned commercially.

Food preparation prior to canning involves a number of possible operations. They include cleaning, grading, sorting, husking, washing, peeling, cutting, sectioning, coring, pitting, slicing, trimming, soaking, evaporating, and blanching. Which operations are performed depends on the nature of the food being processed. Much of this work is done by automatic machinery. When some of the operations are performed manually, they are usually done on conveyor lines that are synchronized with the automatic sterilizing and canning equipment. Cleaning is an important operation. It may involve high pressure water sprays, water immersion and the use of various special scrubbing apparatus, depending on the nature of the food product. Sometimes air blasts are employed to remove foreign material prior to washing.

The canning process, after food preparation, includes the following sequence of container operations: washing the cans, preheating the food, filling the cans, exhausting air from the cans, capping, sealing and heat-processing the cans and their contents, cooling, labeling, packing in multiple-unit cartons and storage. In commercial canning, these operations are usually highly automated. The food to be canned is usually preheated to insure uniform initial temperature before further heat processing and to aid in producing a vacuum when the can is sealed. Liquid or semi-liquid foods are preheated in tubular heat exchangers that may be equipped with screw conveyors. Steam is the usual heating medium.

Various automatic machines used in the canning sequence are usually interconnected by conveyors so that the cans travel from machine to machine through the entire process without human handling. For large scale operations, the machine pace is quite rapid. Filling is normally automatic but may be manual or partly manual, most notably with fruits and some larger vegetables, to insure filling with the proper amount. After filling, cans may be evacuated of any trapped air or gases, by heating the filled cans in hot water baths or by steam in exhaust chambers. Thermal expansion of the food leaves little or no room for atmosphere in the can as it is sealed. Sealing follows and some sealing machines are equipped to draw any trapped air or gas from the can, if necessary. Lids are placed automatically by the can sealing equipment and the can and cover edges are rolled together with a double seam. There is a thin layer of flexible material near the can edge which acts as a gasket. Some sealing is also performed in a chamber containing an inert gas - nitrogen, carbon dioxide or helium.

The sealed can then undergoes heat sterilization of the contents to specified temperature and holding time as described in G3 above. One continuous method uses horizontal cylindrical pressurized retorts. The cans are conveyed into the retort through a rotary transfer valve that maintains the pressure inside the retort and prevents loss of steam. The cans roll on a spiral track on the interior of the cylinder, helping to insure that the heat is well distributed to their contents. A full, high-production arrangement may have several stages of this equipment in series so that the cans are heated and then cooled under the proper temperature and pressure. Fig. 12G4 illustrates how such machines operate.

Following the heating operation, the cans are cooled in air or water to about 100°F (38°C) and then are labeled and packed. Contraction of the contents of the cans during cooling after sterilization produces a partial vacuum in the cans. Labeling is an automatic operation performed at high speed. Packing in cartons is also normally a machine operation but is manual in some situations.

G5. ***aseptic processing*** - involves both sterilization and packaging carried out at the same time. A sterile product is placed in a sterile container in such a way that microorganisms cannot reenter the product. Microorganism action can be prevented for several years if the process is carried out correctly. The process may involve metal canning but is also usable with various plastic- and paper-based containers. Milk and juice are processed by this method and can be stored at room temperature. The operation is usually carried out with high speed

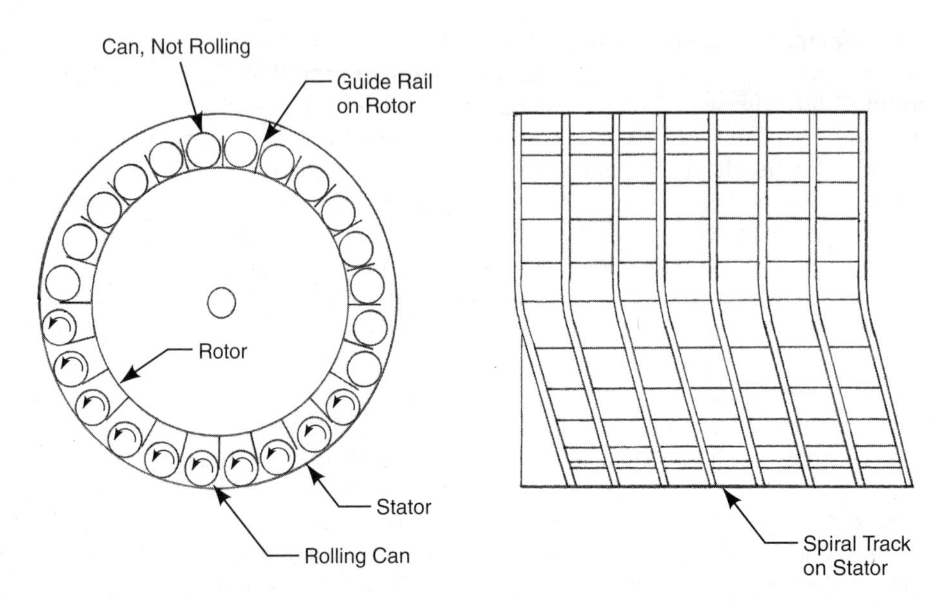

Fig. 12G4 End and side interior views showing the principle of operation of an automatic continuous rotary canning sterilizer. The cans are carried inside a heated, pressurized cylindrical vessel. As they are moved by rails on a rotating wheel, while in the lower portion of the cylindrical vessel, they roll against the cylinder walls, helping to mix the cans' contents. When they are rolling, they are shifted by spiral tracks to the next position and gradually from one end of the sterilizer to the other.

specialized machinery that sterilizes the product and the container. The machinery forms, fills and seals the container, all under sterile conditions, although, sometimes, sterilization of the product takes place at another location.

Sterilization is often by the UHT (ultrahigh-temperature/short time) method that sterilizes the food product and the containers at higher-than-normal-sterilizing temperatures for a short time span. When the container is glass or metal, the container is sterilized with superheated steam at 300°F (150°C) or higher. The food is preheated and then sterilized by heating to a similar high temperature for a short specified holding time followed by rapid cooling. The container is filled and sealed. All these operations take place in a sterile chamber. Glass containers must be heated and cooled at slow enough rates to avoid cracking. Plastic and other non-heat-resistant containers are sterilized with hydrogen peroxide and then dried with hot air and radiant heat. Filling takes place in a sterilized filling station, usually sterilized with hydrogen peroxide for a period of time. Superheated steam is another sterilizing medium. After one of these methods, the machine is dried with sterilized air. Sterile air, made sterile by filtering through microfilters that screen out microorganisms continually bathes the mechanisms and maintains a positive pressure in the machine so that contaminants cannot enter the machine area. The process is also used in the storage and handling of bulk quantities of sterilized food ingredients. Containers up to the size of 55 gallon drums, and even railroad tank cars filled with tomato paste, fruit purees and other liquid concentrates may be used. Special aseptic valves and handling equipment are employed when the sterilized ingredients are transferred from one storage tank or container to another.

G6. ***cooking*** - is the preparation of food for consumption by means of heat. Heating is undertaken to improve the texture, flavor, tenderness, form, or appearance of food or to kill pathogenic microorganisms. There are many methods by which the heating takes place: immersion in a heated or boiling liquid (water, wine, stock) for boiling, poaching, or stewing; immersion in oil or hot fats for sautéing, frying, or deep-fat frying (French frying); steaming; and baking, roasting and broiling, all of which utilize dry heat. Batch baking and roasting take place in an enclosed oven and surround the food with uniform air temperature. With broiling the heating is more intense, usually on one side at a time and, with meat, it sears the surfaces and seals in the juices. Heat for cooking may be applied by any number of ways: flame from a gaseous, liquid or solid fuel; electrical resistance; microwave. The heat is transferred to the food by radiation, convection, conduction or induction (microwave).

To kill pathogens, meat should be heated throughout to a temperature of at least 160°F (70°C) for at least two minutes.[1]

In deep fat frying, the oil temperature is typically 340 to 360°F (170 to 180°C). Typical deep-fried foods, in addition to French-fried potatoes, are doughnuts, potato chips, and fried noodles.

In mass-production applications, much cooking is done on a continuous basis. A conveyor moves the foodstuffs continuously through an oven or other cooking device. Continuous oven systems may include areas that, by design, have different temperatures and humidities, in order to produce the desired properties in baked food. By controlling the conveyor speed, the size of the heating zone and the temperatures, the desired amount of cooking can be assured.

Batch operations are more common because many food products are made in smaller quantities or to order. Batch ovens are sometimes used with a partial vacuum when it is important not to raise the foodstuff to too high a temperature. In such situations, the dwell time in the oven is increased.

G7. ***baking*** - the cooking of flour-bearing foods is an important operation. In foods raised with yeast or baking powder, baking provides heat for the chemical reactions involved. Baking also provides browning, oxidation, volatilization, starch gelatinization, esterification and other physical and chemical reactions that create bread, cake, biscuits, muffins and other baked goods from a paste mixture of flour and other ingredients. Typical baking temperatures are in the range of 240-300°F (115 to 150°C). The food is heated from contact with heated air and radiation from heated oven walls. High-production baking is done on a continuous basis in a tunnel oven, a long chamber as long as 200 ft (60 m) in length, open at both ends. A metal belt conveyor moves the foodstuff through the tunnel which contains several independent heated sections. Heat is usually supplied by gas flames inside or outside the sections, but steam, oil flame, electricity or electronic methods may also may be used. Some oven sections may be humidified and humidity-controlled to ensure proper moisture content in the baked goods. By the time the products reach the end of the tunnel, they are properly baked. Baked goods of controlled shape are conveyed through the oven in pans. Products with toppings, fillings, and icings. have these items added at the proper time by dispensing machines that apply the material as the product passes to or from the tunnel oven. There is a cooling cycle after baking. Cooling is generally at a slow rate using room temperature air rather than refrigeration. This allows the moisture in the product to distribute itself evenly and prevents later condensation of moisture on the crust, which would lead to more rapid spoilage. Packaging usually follows immediately or soon after cooling. Since baked products are distributed with the expectation of short shelf lives, packaging is usually simple, consisting of plastic or paper wrappings, or simple chipboard and cellophane boxes. Much of the packaging is done automatically for high-volume consumer baked goods.

H. Dehydration and Drying

Dehydration and drying are two food preservation techniques that remove moisture from food. They differ from each other in the purpose of the operation. *Dehydration* is a technique that removes moisture from food in such a manner that, when water is re-added, the food will be returned to approximately its pre-dried state. In contrast, *drying* involves removal of moisture for preservation of the foodstuff without the expectation of restoration. Dried fruit and other foods will not return to their predried condition if water is added to them, but dehydrated soup and coffee are reconstructed when water is added. Both methods have the same effect of inhibiting the growth of microorganisms and thereby preserving the food. Both these methods allow longer storage of the food processed, simplified package requirements and reduced weights for shipping and handling.

Methods for removing moisture from foodstuffs range from what probably is the oldest method, sun drying (which is still used), to more sophisticated current methods. Some of these methods are out lined above in the material on concentration. (See sect. 12F.) The common method of removing moisture from solid foods involves subjecting the food to a flow of hot air. Cabinet, tunnel, and kiln dryers are three prevalent types of equipment used. A cabinet/compartment/tray dryer is shown in Fig. 12H. These units are useful for drying vegetables and fruits and various other foods when quantities are not great. Tunnel, rotary drum and fluidized-bed dryers are other methods used. Fig 11C12 shows a tunnel dryer with a mesh conveyor that allows passage of heated air to the foodstuff. These devices are useful for larger-scale production as are the rotary drum dryers illustrated in Fig. 11C12-1. (A different drying method using one or two drums is described below in section H3.) Liquid foods, such as coffee, milk and fruit juice, are dehydrated by spray drying or vacuum drying.

The purpose of both drying and dehydration is to remove sufficient moisture so that microorganism growth does not occur, for the prevention of enzyme actions, and for retarding or preventing other chemical reactions. After dehydration, the food is packaged in moisture-proof packages. The amount of moisture in the food after drying or dehydration varies with the food, the use intended, and the packaging. 5% moisture content is a typical value. Vegetables, fruits, meats and fish may be processed by drying. Milk, including whole milk, skim milk and buttermilk, soup, eggs, yeast, pasta, potatoes, tea and coffee are well suited to dehydration. Typically, these foods, after dehydration, occupy only 1/15 the space required originally or after they are reconstituted[2]. Drying and dehydration methods that shorten the drying time help avoid degradation of texture and flavor as a result of the operation. (Also see section C12 in chapt 11.)

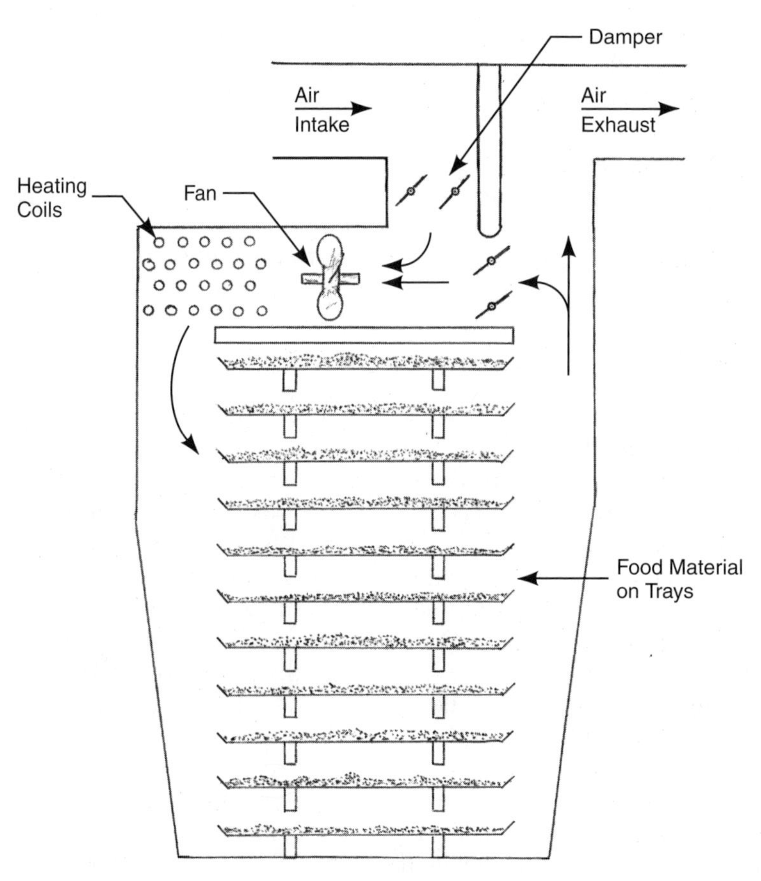

Fig. 12H Compartment/tray dryer.

H1. ***vacuum drying*** - Vacuum is used in drying food that is sensitive to higher heat levels because the vacuum lowers the temperature necessary for vaporization of the moisture in the food. The process is used with fruits and vegetables. It differs from freeze drying in a vacuum in that heat is applied to the food material. With the vacuum shelf dryer method, hollow shelves inside the vacuum chamber circulate a heating fluid. A vacuum of 1 to 70 Torr is used.[6] Careful control is required to avoid overheating the food. The major application is concentrating fruit juices.[6] Other drying methods may be carried out under vacuum conditions in order to lower the temperature required for drying and thus lessen the possibility of adverse effects.

H2. ***spray drying*** - of liquids, involves spraying the liquid, with a fine atomization, into a heated chamber or onto a steam-heated rotating drum, where the water is evaporated. Spray droplets are fine and evaporation time is usually very fast, under 30 seconds, producing a dry powder. The flow of hot gas (usually air) provides drying and movement of the product. The short drying time helps prevent degradation. Two methods are in use to provide the atomization of the liquid: pressure-nozzle atomization and rotating disk atomization.[3] With the pressure nozzle approach, pressures are typically 250 to 8000 psi (1.7 to 55 MPa). In the other method, liquid is fed to the center of a disk rotating at 1700 to 50,000 rpm on a vertical axis. Centrifugal force throws the liquid outward into fine droplets. The disk method is suitable for slurries and pastes, which could clog a pressure nozzle. When the disk method is used, a wide chamber is employed; with the nozzle method, the chamber is tall and narrow. With both systems, the dried particles settle to the bottom of the spray chamber and are removed by conveyor or gravity. Section C12a of Chapter 11 describes pressure nozzle spray drying and Fig. 11C12a illustrates spray drying in a heated chamber with both concurrent (same direction as spray) and countercurrent air flow. The heated air temperature is typically in the range of 400 to 590°F (205 to 310°C. Powdered milk, processed cheese, cream, eggs, fruit juices, whey, yeast extracts, and instant coffee and tea have been made by spray drying.

H3. ***drum drying*** - is used for some powdered foods such as milk and packaged mashed potatoes. A paste, slurry, or solution of the food material to be dried is fed to the outside surfaces of a horizontal, steam-heated metal cylinder that is rotated on its axis. The material is applied as a thin, uniform layer on the drum. The thickness of the layer depends on the food involved and its tolerance for heat. As the drum rotates, the heat from the steam is conducted to the layer, evaporating much of its moisture. At one position, after the drum has rotated half to three quarters of a revolution and as it continues to rotate, a knife blade scrapes the drum and removes the dried layer of food. Sometimes a vacuum is maintained in the chamber where the operation takes place so that the necessary drying temperature can be reduced. Fig. 11C12-2 in chapter 11 illustrates a dual drum-dryer which uses two, side-by-side drums rotating in opposite directions. The food material is fed to both drums from a central reservoir between them. There are several other arrangements with one or two drums. Drum drying is used for applesauce and other fruit purees, dry soups, precooked breakfast cereal, baby food, and bananas. Other food materials processed by drum drying are dried skim milk, malted milk, malt extract, potato flakes and yeast.

H4. ***fluidized-bed drying*** - is applicable to a variety of chemicals as well as foods and is also described in Chapter 11, section C12. A fluidized-bed dryer is pictured in Fig. 12H4. These dryers can be used for smaller-particle foods such as peas, potato granules, diced meat, sugar, salt, flour, coffee, and cocoa. The process is economical and is commonly used in food processing. Air is the usual drying gas and is heated by electricity, steam or by a combustion boiler. Bed heights range from about 1 to 50 ft (0.3 to 15 m). Food materials dried with fluidized beds include potato granules and mashed potatoes, quick-cooking rice and peas, rye and other grains.

H5. ***freeze drying (freeze dehydration)*** - involves freezing the food and removing the moisture by sublimation of the ice. The process takes place under a high vacuum at 26°F (–3°C). Freeze drying preserves flavor, nutrients, and texture better than simple dehydration, and is used for dried soup mixes, instant coffee and some other foods. Meats and other high-protein foods retain their

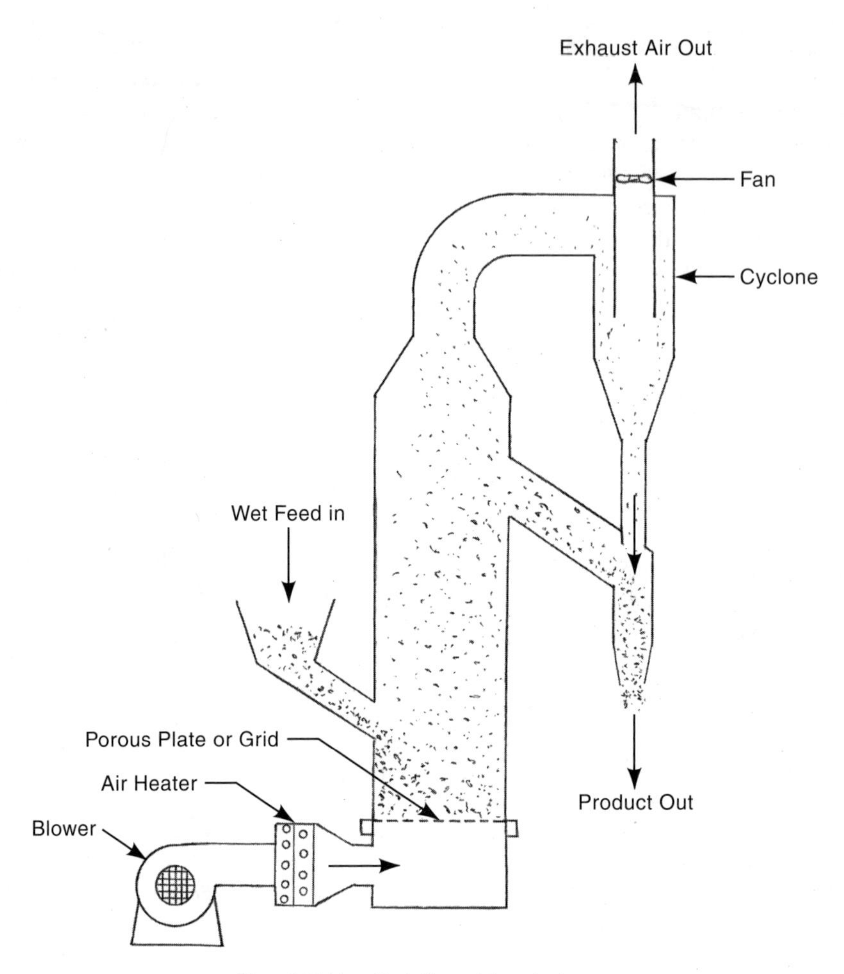

Fig. 12H4 Fluidized bed dryer.

nutritive and palatable properties only if freeze dried. After rehydration, freeze-dried meat closely resembles fresh meat[2]. Compartment dryers as described above and illustrated in Fig. 12H are often utilized with a vacuum but the material is first quick-frozen and spread in thin layers. Its temperature is kept below the freezing point, but some heating may be applied to provide the heat of sublimation[8]. Freeze dryers include, in addition to a vacuum chamber, a duct system to carry vapors away from the vacuum chamber, and a means to remove them from the system. This removal of vapors may involve condensation on a cold surface, adsorption on a solid desiccant, or absorption in a liquid. Both batch and continuous systems are used, depending on the production quantities.

Tunnel freeze dryers are used in some high-production conditions. Trays of food are loaded into the vacuum freeze tunnel at one end through an air lock. The trays are conveyed in the tunnel through several stages of refrigeration and drying and are discharged at the other end.

Handling of food through the freeze drying process requires some special attention. Lean meat, bone, and fat, each have a different moisture content and it may be necessary to process them separately. Most fat should be trimmed from lean meat before freeze drying. Fruit skins may have to be slit or removed. Some fruits are cored and diced. Vegetables may be peeled, washed, trimmed, and cut and blanched before freeze drying. Apples and pears are treated after peeling and before freeze drying with a sodium sulfite solution to prevent unfavorable color changes[8].

I. Cooling for Preservation

Refrigeration and other cooling is a common method of food preservation. Fruits and vegetables are usually cooled after harvesting and meats immediately after slaughtering. Temperatures below 40°F (4°C) do not make significant nutritional or taste changes in the food but retard the growth of microorganisms that cause food spoilage and slow the rate at which fruits and vegetables respirate and lose carbohydrates. Refrigeration is useful throughout the processing of food, its transport to market, and its storage by the customer before consumption. Refrigeration is particularly successful in extending the useful life of meats. Vegetables and fruits with high water content (eg., melons, tomatoes, cucumbers, bananas and pineapple) receive less benefit. The operation is common during many food processes to maintain the food quality before the preservation process is complete. Conventional refrigeration in cooled chambers powered by commercial refrigeration equipment and immersion of the food in chilled water are two approaches used.

I1. ***freezing food*** - prevents food spoilage by suppressing the growth of microorganisms in the food. It is a common and effective method for preserving many foods. When properly done, freezing maintains the natural quality of the food. Quick freezing is used because it forms smaller crystals from the moisture present in the food, and thus reduces cell damage which can change the texture and consistency of the food. Vegetables are blanched before freezing to ensure that enzymes present are inactive and to avoid degrading the flavor. Several different methods are employed to effect the temperature reduction. Some of these use freezing air; some use liquid freezing agents; others use contact with solid cold surfaces.

The following is the sequence for freezing sweet corn, a vegetable frequently preserved by freezing. Corn maintains high quality best if processed within a few hours of harvesting. Harvesting, in production situations, is mechanical. De-husking and removal of the "silk" are the first operations, performed with automatic equipment. Then the ears are washed thoroughly and blanched in steam for 6 to 11 minutes. Cooling follows immediately. If it is to be sold as cut corn, the corn may be blanched twice - once partially, before the kernels are cut from the cob, and then again after the cob is removed. The kernels are separated from any residue of silk or husk by flotation or washing. With both methods, the corn kernels settle to the bottom while the undesirable materials float away. The kernels are quick-frozen by the fluidized bed process and are packaged.[2]

I1a. ***freezing with refrigerated air***[1] - In these systems, air is chilled from contact with the cold coils of a mechanical refrigeration system and then is directed at the foodstuff to be frozen. The air temperature with such systems is typically zero to −22°F (−18 to −30°C) but may be much colder.

The freezing method often used with smaller foods such as shrimp, peas, and beans. is to blow the cold air upward through a mesh conveyor that carries the food. The air blast is strong enough to suspend the food pieces, circulating them and ensuring that cooling air strikes all surfaces. The individual pieces then do not stick together and the freezing is quick, preventing large ice crystals from forming. The suspension of the food pieces is similar to the suspension of particles in a *fluidized bed* and the term is sometimes used to describe this method. Freezing with refrigerated air is also usable with larger pieces of irregular shape that are not as suitable for contact freezing.

In the air blast method, which is used when foods are packaged before freezing, the packages are conveyed through a stream of cold air that is directed at the package at a high velocity, between 100 and 3500 ft/min (30 and 1000 m/min). The air temperature is usually maintained between −22 and −40°F (−30 and −40°C). The high velocity of the air blast and good contact between the package surface and the food provide quick freezing. A lengthy insulated conveyor may be required with packaged or larger food items. A spiral conveyor arrangement is sometimes used for such systems in order to have a more compact system.

A third air freezing method uses the cooling air as the medium of a refrigeration system. The air is compressed, then cooled, then allowed to expand before being directed at the foodstuff to be frozen. The air, then, may have a temperature as low as −250°F (−150°C). This method is classified as *cryogenic freezing*.

I1b. ***freezing with a liquid medium*** - Contact with a cold liquid provides more rapid freezing than is possible with air because of the better heat transfer rate with liquids. Liquid-medium freezing is particularly adaptable to larger food pieces, such as corn on the cob that, because of its bulk, may not freeze quickly enough from refrigerated air. There are several viable approaches:

One approach, used in canned orange juice, is to spray the cans with a chilled liquid (calcium chloride brine). Another approach is to immerse the cans in a bath of the liquid. Temperatures of –20°F (–29°C) are achieved by this method.

Another *cryogenic freezing* method uses liquid nitrogen or liquid carbon dioxide. They are sprayed directly on the food, freezing it very rapidly. The boiling temperature of liquid nitrogen at sea level is –320°F (–195°C). Liquid carbon dioxide boils at –109°F (–78°C) but when sprayed on a food, it changes to solid or snow-like dry ice.

I1c. ***indirect-contact freezing*** - Still a third approach presses packages of the food to be frozen between two chilled metal plates that have channels carrying a refrigerant. This not only provides the chilling needed but maintains dimensional control to counteract swelling that otherwise takes place during freezing. The approach is useful for frozen foods sold in rectangular packages of standard size and shape. The plates, in turn, are chilled by brine or refrigerant that circulates in the attached channels, hence the use of the term, indirect. In most systems there is a series of horizontal plates in a vertical stack. Food packages are placed between the plates, which are pressed together with light hydraulic pressure to insure good contact during the freezing operation. The plates can also be made in non-flat shapes or as cavities that impart a particular shape to the frozen food. Ice cream on a stick is an example. There are also arrangements with a series of vertical plates, between which the food is placed before the plates are pressed together. This method is used for freezing fish and meat, including whole fish at sea. After the freezing is complete, and the plates are spread slightly, the plates may be heated briefly to release the frozen food and to defrost and clean the plate surfaces before the next cycle.

I2. ***dehydrofreezing*** - involves partial drying of the food, to reduce its weight (about 50%[6]) and bulk, before freezing it until it is used. The method utilizes conventional drying and freezing processes and is used to reduce storage space requirements before further processing of fruits and vegetables. Food processed this way can be reconstituted more quickly and easily than foods processed with only drying.

J. Other Operations

Food may be subjected to a variety of processes in addition to those described above. Most have the purpose of aiding preservation, but they also may be used to change the flavor, physical characteristics or other properties of the food. Most, but not all, involve some heating or cooling as part of the sequence of operations. These processes include irradiation, homogenization, hydrogenation, fermentation, extrusion, biopreservation, pickling, salting, candying, glazing and adding sugar, incorporating other additives, high-intensity pulsed electric field processing and ultra-high-pressure processing.

J1. ***irradiation*** - In this operation, the food is moved or conveyed to a shielded chamber where it is exposed to ionizing radiation. One of two forms of radiation is commonly used; one uses gamma-rays (from cobalt 60); the other is from a machine-generated source (an electron accelerating machine). Irradiation kills molds, all bacteria, including harmless species, parasites, and insects, and delays ripening. It also delays sprouting of onions and potatoes. The process has been used for some time on wheat and wheat flour, potatoes, spices, pork, onions, tomatoes, mushrooms, strawberries, and poultry. Its use with beef in the United States was approved in 2000. It has the advantages of requiring much less energy than heat sterilization or freezing, and only minimally raising the temperature of the food being processed. On the other hand, enzymes in the food may not be deactivated by the radiation and they may lead to chemical changes and a small loss of nutritional value. The process may prolong the storage life of the food processed and may eliminate the need for refrigerated storage.

J2. ***homogenization*** - breaks up fat globules so that they remain dispersed in a liquid, notably, milk. A high pressure pump forces the liquid through very small openings at high velocity. This breaks up the fat globules, reducing them to small size, increasing their number and the total fat surface area. The globules may also impinge on a surface that faces opposite to the direction of flow. This further disrupts and divides the globules which then remain dispersed in the liquid. Pressures of 2000 to 2500 lbf/in^2 (14 to 17 MPa) are used in the process. The pressure is adjustable and is regulated to control the fat particle size. Two stages of homogenization are sometimes used, with separate pumps and homogenization valves. Milk is heated before the operation to inactivate lipase activity and to facilitate the break up of the fat globules.

J3. ***hydrogenation*** - is a hardening and raising of the melting temperature of fats and oils. It is accomplished by the addition of hydrogen in the presence of a nickel catalyst. Hydrogenation converts unsaturated radicals of fatty glycerides into more highly or more completely saturated glycerides.[2] The operation takes place in tall, cylindrical hydrogenation columns at a temperature of about 375°F (190°C) and pressure of 30 to 100 psi (200 to 700 kPa). The operation is usually performed on a batch basis and takes about one hour. Hydrogenation extends the shelf-life of fat-containing foods and, for many, also has a favorable effect on flavor and odor. However, it has been indicted as an unfavorable agent in cardiovascular health. Hydrogenation is a vital part of the manufacture of margarine and shortening. The fats in most packaged foods such as cookies, crackers, other snack foods and peanut butter, etc. are at least partially hydrogenated.

J4. ***fermentation*** - is a microbiological decomposition of an organic material, usually a foodstuff, accompanied by the release of a gas. It is the process by which sugars (eg., fructose and glucose) are converted to ethanol. The operation takes place in an atmosphere that excludes air. Yeast, mold, or bacteria are added to the liquid to be fermented to provide the microorganisms that perform the operation. The organic matter decomposes and reduction products remain. Alcohol and lactic acid, both useful products, are examples. The temperature of the liquid and the concentration of the nutrient are controlled to optimize the operation. Carbon dioxide in gaseous form is a normal byproduct.

Microbiological action in the presence of air also is referred to as fermentation although it does not meet the classical definition of the term. The result of the action is an incomplete oxidation. The production of vinegar (acetic acid) from alcohol and of citric acid from sugar are examples.

Applications of fermentation include the production of alcohol and glycerol from the yeast action on sugars, the production of butyl alcohol, acetone, lactic acid, monosodium glutamate and acetic acid from various bacteria and the production of citric acid, gluconic acid, antibiotics and vitamins B-12 and B-2 from mold fermentation. Cheese, wine, yogurt, bread, soy sauce and sauerkraut also are produced with the aid of fermentation.

J5. ***extrusion of foods*** - is an operation quite similar in principle and equipment to the extrusion of plastics, as described in Chapter 4, section I and Fig. 4I, or of metals, as described in Chapter 2, section 2A3. In most cases with foods, however, heat that is applied to the material, and the heating that results from the friction between the screw and cylinder of the extruder, cook the food mixture. This is in contrast to the extrusion of plastics and metals where the purpose of heating is to allow the material to flow through the extruder. The process has the high productivity that is characteristic of the extrusion of non-food materials, and is applicable to the mass production of foodstuffs. With food extruders, the material to be extruded is normally in paste form with a moisture content typically from 10 to 35%. The food is fed from a hopper to a rotating feed screw or twin screws and forced through a shaped die opening. The single screw extruder is more common but the twin-screw variety has greater ability to handle stiff mixtures and provides better control over operating conditions. The screw barrel is normally heated with steam or by electrical resistance. The cross section of the extrudate has the same shape as the die opening. The extrudate, after exiting from the die, is usually cut to some length by a rotary knife. It is also cooled and dried after extrusion. Heat generated or added during the operation may raise the food temperature to as high as 300 to 390°F (150 to 200°C) for a short time before it passes through the die. This temperature deactivates enzymes and microorganisms that could later cause undesirable changes in the food[1]. The high temperature also may cause the extrudate to expand after it leaves the die from vaporization of water in the mixture, creating a puffed or cellular structure for the food product. The product may then be coated with sugar, color, vitamins, minerals or oil.

Food products that are extruded include: wheat dough for pasta, cooked cereal dough for breakfast cereals, corn grits, soybeans, gelatinized corn flour, sausage mixtures, and soy dough. Other products made with extrusion are snack foods, baby foods, beverage and soup bases, candies, animal and pet feeds, vegetable protein and flat breads.[7]

J6. ***food additives*** - Various substances are added in small amounts to foods during preparation for specific purposes: to prevent spoilage or deterioration and to maintain freshness, to enhance appearance, flavor, texture, and nutritional values, or to aid in further processing. Additives used include such compounds as sodium benzoate, calcium propionate, benzoic acid, sorbic acid, citric acid, acetic acid (vinegar), sulfur dioxide, sulfites, and other materials to prevent microorganisms from attacking the food, vitamins, minerals, and other nutrient supplements, antioxidants to prevent browning, sugar, salt, and MSG for flavor enhancement, nitrites and nitrates for pickling, bleaching agents, stabilizers and thickening agents, and various colorants, both natural and synthetic. Anti-sprouting agents may be added to root crops (potatoes, onions, beets, carrots). The use of food additives is controlled in the USA by the Food and Drug Administration.

Bio-preservatives are non-toxic natural substances that have the power to kill harmful microbes that cause food spoilage and danger to consumers. Sugar, salt and pickling solutions all have this effect. Bacteriocins are proteins formed by bacteria that are present in foods. The bacteriocins have the property of killing other microorganisms that are similar to themselves but not other beneficial bacteria. They have been used in preserving pasteurized egg products. Sodium benzoate and other benzoates are common chemical preservatives. They normally comprise no more than 0.1 percent of the total content. Benzoates require an acid medium, common in fruits, to be effective. Sulfur dioxide and sulfites are also common, are effective against molds, and are used in fruits and vegetables. Low quantities are used in wine. Nitrates and nitrites are used in the curing of some meats.

Table 12J6 summarizes many of the reasons for including additives and agents in food and identifies specific ones that are commonly used.

J7. ***pickling*** - is a food preservation procedure that primarily involves immersion of the foodstuff in an acid (usually vinegar). The acid prevents growth of undesirable bacteria. There are many different

Table 12J6 Food additives

Reasons for Including Food Additives and the Agents Commonly Used to Achieve the Desired Effect*

Reason	Agent Used
Prevent growth of microorganisms	sodium benzoate, calcium propionate, potassium sorbate, sulfites
Prevent caking and lumping	calcium stearate, sodium aluminosilicate, cornstarch
Prevent browning and rancidity with antioxidants	BHA, BHT, ascorbic acid, ethoxyquin
Change color	beet powder, caramel, B-carotene, FD&C yellow 5 & 6, red 3, blue 1
Cure and pickle, prevent bacterial toxin from forming, impart flavor and color to meats, reduce spoilage	sodium nitrate, salt, sodium metaphosphate
Emulsify oil-water mixtures	monoglycerides, diglycerides, lecithin, monostearate
Enhance flavor	monosodium glutamate (MSG), disodium inosinate, disodium guanylate
Restore or enhance nutritive properties	various vitamins, iron, amino acids, essential fatty acids, various minerals
Stabilize acidity or alkalinity (pH) to desired level	sodium bicarbonate, vinegar (for acetic acid), hydrochloric acid
Prevent formation of molds or yeasts	sorbic acid, propionic acid, benzoic acid, ethyl formate

Note: Information for this table was taken from reference 1 and other sources.

variations of workable pickling methods, and they each give different results in terms of degree of preservation, flavor, and suitability for certain foods. Most pickling methods also involve the use of salt to provide a brine, which also has anti-microorganism properties. Preservation of food with a brine and spices, without an acid, is also called pickling. Pickling methods fall into one of two basic groupings: 1) Pickling that does not involve fermentation. This is may be called *quick pickling* or *fresh-pack pickling*, and, 2) pickling that includes fermentation. When fermentation is involved, the food usually has better preservation but the pickling process is more complex. Lactic bacteria, which are present on almost all vegetables, act on the starches and sugars in the food to be pickled to create lactic acid, which has food preservation properties similar to those of the acetic acid in vinegar. The process changes the flavor, appearance and texture of the food. Various condiments are also often used to create the particular distinctive flavor that is wanted in the pickled food.

The process of pickling with fermentation to produce lactic acid is also known as *lacto-fermentation*. With either quick pickling or fermentation methods, salt included in the pickling solution strengthens the microorganism protection against some undesirable microorganisms but, if fermentation is desired, it does not prevent it from taking place. Pickling does not preserve food for as long as canning and freezing. Normal usable life of refrigerated pickled foods is several months.

Fresh-pack (non-fermented) pickles are typically made by immersing cucumbers about 12 hours in a brine containing 2 to 3 percent salt, about 12 percent vinegar and spices. The mixture is then heated to a temperature of about 165°F (74°C) for 15 minutes, and immediately cooled. The process for fermented pickles is more lengthy. The cucumbers are placed in a brine with a salt concentration of 8 to 10 percent for one week. The salt concentration is increased 1 percent each week until the concentration is 16 percent. The material, at this stage, is referred to as salt stock and contains fermented pickles. It may be kept for years without spoilage but is not sold to consumers at this stage. The color has changed from green to yellow-green or olive and the interior has become translucent. For consumer use, the pickles are processed to leach out salt with water at a temperature of 110 to 130°F (43 to 54°C) for 10 to 14 hours. The process is repeated at least twice[2]. Tumeric and a final rinse may be added to improve the color and provide firmness. Other processing may take place, depending on the type of pickles to be produced. Sour pickles are made by processing the salt stock with weak vinegar. Sweet pickles are produced by adding sugar, spices and vinegar to processed stock. Processed dill pickles are prepared by adding dill and other spices to the acidified salt solution. Natural dill pickles are made slightly differently. Instead of using salt stock, they are made from fresh cucumbers immersed in a brine with dill and other spices. They are not usually packaged in vinegar but are perishable and are pasteurized and repackaged[3].

Sauerkraut is a notable example of a vegetable produced by fermentation pickling. It is made from cabbage. The cabbage is first shredded to open the cell structure, then salted to about 2.5% and stored in closed crocks. General microorganism activity begins but when the oxygen in the crock is used up, the lactic-acid-producing bacteria take over, producing lactic acid that neutralizes some of the undesirable products of earlier activity. The final sauerkraut typically has a lactic acid content of 1.7%, providing the distinctive flavor.

Olives are also processed by lacto-fermentation and the fermentation takes 6 to 10 months. Green olives are treated with lye (NaOH) before fermentation, and black olives after fermentation, to remove the compounds that cause raw olives to have a bitter taste. Other vegetables processed with lacto-fermentation are beets, cucumber, turnips, green tomatoes, peppers, and lettuces. In Asia, lacto-fermented vegetables are eaten frequently. The vegetables involved are cabbage, turnip, eggplant, cucumber, onion, squash, and carrot. The process begins with placing the vegetable in salt, removing some of the water it contains, and then effecting a slow fermentation with lactic bacteria.

Meats are pickled by treating them with pickling solution in one of three ways: soaking them in the solution, injecting them with the solution, or coating them with a mixture of dry pickling ingredients. Pickling ingredients for meats include salt (sodium chloride), sodium nitrate, sodium nitrite, sugar, and vinegar or citric acid. When dry ingredients are used, the operation may be referred to as *dry curing* or simply, *curing*. Cured meats may be smoked after curing. Meats that are pickled or cured include bacon, sausage, corned beef, pastrami and ham.

Fish, (notably salmon and herring) as well as eggs are processed by pickling. Pickling fish involves the use of vinegar of 5% acetic acid, salt and the desired spices. Pickled fruits include watermelon rind, mango, plum, lemon and kumquat.

J8. ***salting*** - is a method of food preservation. A high concentration of salt on the surface and within food material prevents bacterial growth. The salt prevents bacteria from absorbing additional water needed for the growth that spoils food. In fact, the bacterial cells lose water because of osmotic pressure of the salt solution. This kills the cells. Salt may be applied by one or more of three methods: dry granular salt may be rubbed on the surface of the food or completely cover the food pieces in a container, a brine may be injected into the food, or the food may be soaked in brine. Days or weeks may be required for the salt or brine to fully permeate the food tissues. Spices and flavorings may be included with the salt.

Salt is an important ingredient in pickling as noted above and some pickling methods depend solely or almost solely on salt to provide the wanted food preservation. Fish, pork, eggs and beef (corned beef and pastrami), are commonly processed with salt. Salt preservation is also used to preserve vegetables and fruits, notably mango, peppers, cauliflower, mushrooms, lemon, tamarind, gourd, chilies, and goose berries.

J9. ***sugar curing/sugar addition*** - Sugar is added to many products as a preservative in addition to its flavoring function. If the sugar content of a food is at least 65 percent by weight, the effect on the food bacteria is the same as that from salt, in that water is drawn from the bacterial cells and their growth is halted. It takes much more sugar than salt to get the same microbiological effect. Acidic foods, eg., fruits, are particularly suited to this approach. The sugar can be applied by rubbing it on the surface of the food, or by soaking the food in a sugar solution, or by injecting a sugar solution. Both sugar and salt, and other flavorings and preservatives are sometimes included in one mixture that is used. Ham, bacon, dried beef, and smoked turkey are often treated with this kind of mixture. Jams, jellies, marmalades, and fruit butters all gain storage properties through having a high sugar content.

J10. ***candying and glazing*** - of fruits involves increasing their sugar content to the point noted above where microorganisms that could cause spoilage are inhibited. Candying is effected by immersing the fruit into warm sugar syrup in multiple steps, where the syrup is progressively higher in sugar concentration. This step-by-step approach prevents the fruit from becoming leathery and tough. After the impregnation is completed, the fruit is washed, dried, and packaged. Glazed fruit is processed the same way but at the end is dipped into syrup that is then dried to provide a coating. Sometimes, granulated sugar is used as the final coating. In high production situations, candying, and most of the accompanying operations are carried out on automatic production equipment. Cherries, citrus rinds, many other fruits, flowers and herb leaves are commonly processed with this method.

J11. ***high-intensity pulsed electric field processing (PEF)***[1,10] - is a non-thermal food preservation process that uses electrical energy rather than heat. The electrical energy is in the form of high-voltage (typically 20 to 80 kV/ccm[10]) pulses that pass through the food. The food is placed between and in contact with two electrodes that carry the pulses. The pulses last only a small fraction of a second, enough to affect the cell structure of pathogens and micro-organisms that cause spoilage. The cell membranes of the microorganisms are irreversibly damaged and the organisms are rendered inactive. Heat is not the cause of the cell changes though some heat is generated by the electrical field in the food material. (The food to be processed remains at or near ambient temperature.) There is very little adverse effect on the flavor, texture, or nutritional value of the food. Energy requirements are considerably less than those of thermal processes. No additives are involved, and shelf life of the processed food in lengthened. The PEF process is primarily limited to liquid foods, although bread and brewer's yeast have been treated. Milk, liquid whole eggs, fruit juices and soups are processed with this method.

The equipment used consists of a closed system to prevent contamination from outside sources and external apparatus to provide the electrical pulses and control of the operation. The system is illustrated in Fig. 12J11. The closed portion includes pump, treatment chamber, cooling system and provision for storage and packaging of the finished product. The electrical charge is provided by

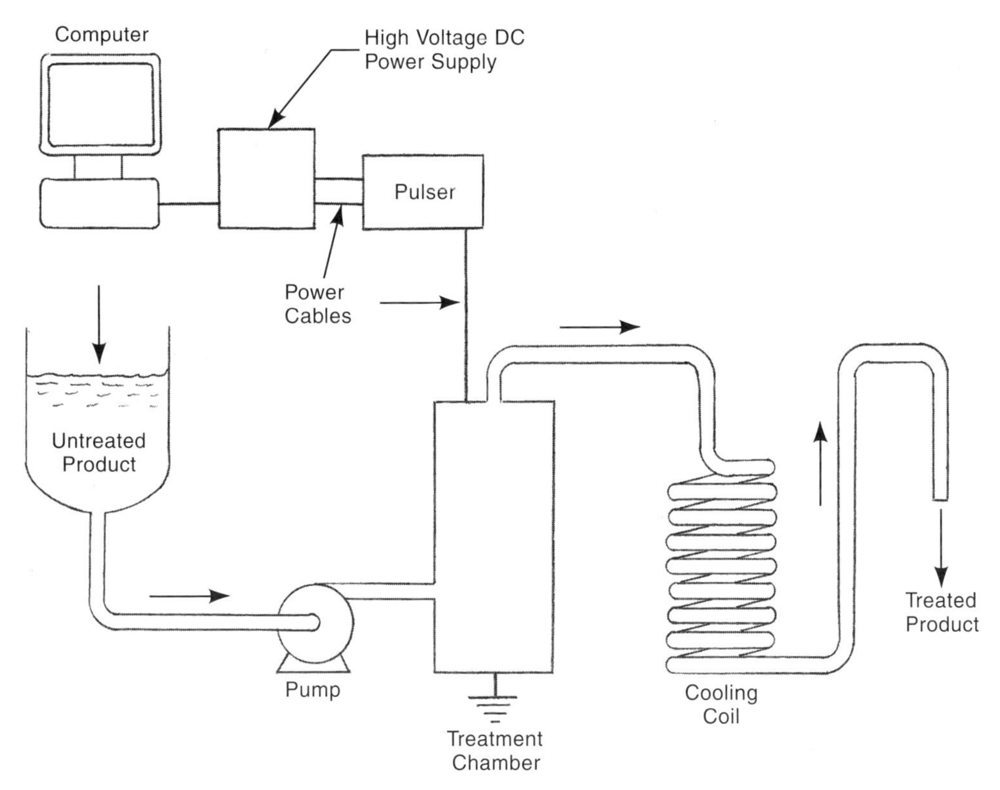

Fig. 12J11 Flow diagram of a continuous pulsed electric field (PEF) processing system.

either 60-cycle power from utility lines, stepped up in voltage and converted to DC, or from a high frequency input to a capacitor that stores the energy and discharges it repeatedly to the food.

J13. ***high pressure processing (or ultra-high pressure processing)*** - uses pressure as a food preservation method. Subjecting the food to pressures in the range of 30,000 to 150,000 psi (200 to 1000 MPa) inactivates micro-organisms and some enzymes. (Although not intended to be a thermal process, heat is generated when the food is compressed. The pressure, rather than the heat, is the prime factor in affecting the microorganisms and enzymes.) Changes in the flavor, texture, nutrients, and color are minimized, compared with those of thermal processes. The operation can be performed in as little as two minutes or may require as much as two hours, but is more commonly carried out in 10 to 30 minutes. Most current operations are on a batch basis, though semi-continuous methods are also in use. The food is placed in a sealed plastic-film package, and loaded into a steel vessel with a pressure transmission fluid, usually a liquid. The vessel is closed, pressure is applied and maintained, and then released. The vessel is opened and the packages of food are removed. The effect on the food is the same, no matter how big a batch is processed at one time. Current applications include pasteurization of fruit juices, fruit jellies, sauces, purees, guacamole and yogurt, improvement of raw ham, reduction of bitterness in grapefruit juice, stopping the fermentation of rice wine, and aiding in the sugar impregnation of tropical fruits[1]. The process is used to kill unwanted bacteria in raw oysters.

K. Meat Packing

Meat packing is the conversion of cattle, hogs, sheep, poultry, and other animals to edible meat food. The process involves a series of operations including stunning, slaughtering, bleeding, eviscerating, skinning, grading, chilling, butchering, the making of sausage and other processed meat products, and packing. The 20th century saw advances in this industry in sanitation, humane treatment of animals, and mechanization of operations. Meat is not the only product. Bones are made into fertilizer, adhesives, animal feed, and pharmaceutical material. Horns and hoofs are sold for other purposes; fat is rendered for use as lard or commercial grease.[4]

K1. ***stunning*** - renders the animal insensitive to the slaughter. When done correctly, the animal becomes unconscious immediately and feels no pain when slaughtered. Several methods are in use: *captive bolt stunning, electrical stunning* and *carbon dioxide anesthesia*. Aside from the humane aspect of preventing animal suffering, proper stunning also provides more assurance of safety for the workers involved, and avoidance of some meat quality problems. Before stunning, the animal may be restrained in a way that does not agitate the animal but does permit more accurate placement of the stunning device. Different types of restraints are used for different animals.

For beef cattle, the common method of stunning is to use a "captive bolt" gun. Using a blank explosive charge or compressed air, a bolt-like projectile is fired at the head of the animal. It penetrates the skull of the animal and has the same effect as a live bullet. The animal immediately becomes unconscious. The bolt, however, retracts after the animal is shot and is reset for the next use. This approach is safer than using conventional bullets and is also used for sheep, goats, pigs, camels and horses.

Electrical stunning utilizes a moderately low-voltage current, applied by electrodes held by tongs at each side of the brain, to induce an electroplectic shock in the brain similar to a grand mal epileptic seizure. The duration of current flow is up to about 10 seconds for sheep, goats, pigs, and turkeys, and about 5 seconds for chickens. Typical voltages are 125 for pigs, sheep, and goats with current at about 1 to 1.25 amps. For poultry, lower values of both can be used. Some methods use higher voltages and shorter stunning periods. The method is less easily applied to beef cattle and not widely used for cattle at present.

With poultry and some genetic strains of pigs, another method is anesthetizing with a gas. Carbon dioxide, in concentrations of 65 to 85% in air, is used. The method requires more technically sophisticated equipment and is used principally at larger meat packing installations. In one method the animals are lowered into a chamber with the high CO_2 concentration, where they become unconscious.

K2. ***slaughtering*** - after the animal is stunned, it is attached by the leg to an overhead moving chain conveyor. The jugular vein and carotid arteries of the animal are cut so that it bleeds to death very quickly. Then, as the carcass moves along the conveyor, it is skinned with an electric knife, beheaded and opened. Viscera are removed and the carcass is split into two sides. The carcass is then cooled to temperatures near freezing to retard the growth of organisms that cause spoilage of the meat.

With hog slaughtering, the steps are similar except that the carcass is not beheaded nor split initially. The first step on the conveyor line, after bleeding, is to move the carcasses through scalding vats to machines that remove hair from the skin. The next processes include eviscerating, washing, and trimming[4]. Thereafter, butchering and other slaughterhouse processing that is common with beef, follow.

Lambs are slaughtered much like cattle. The skin (with wool) is removed, and the carcass is refrigerated before it is split.

K3. ***butchering*** - Carcasses are typically chilled for 24 to 48 hours before grading and processing. With beef cattle, much of the butchering formerly was done at wholesale or retail locations with only primal cuts made at the slaughtering location. That system has changed, partially because of improved packaging at central locations and better and faster means of refrigerated transport. Benefits are also realized from the better facilities for utilization of offal, the non-edible portions of the animals: the bones, hides, intestines, hoofs, etc. when the processing is done at a central location. Specialty edible items such as tongue, kidneys, tail, and fats are also better processed at the packing house rather than at wholesale or retail locations. Tools for butchering include band saws, deboning equipment, and a variety of special cutting knives. Cuts intended for shipment to retail outlets are trimmed, deboned, if applicable, wrapped (often by vacuum packaging in plastic bags that are moisture and gas impermeable) and packed in corrugated boxes

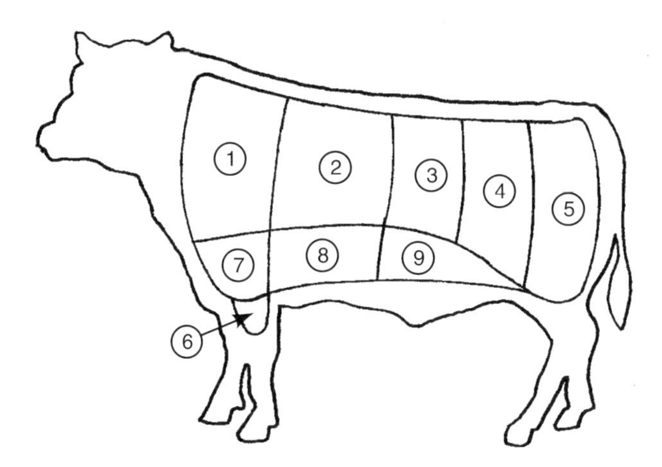

1. **Chuck**
   Pot Roast
   Eye Steak
   Short Ribs

2. **Rib**
   Rib Roast and Steak
   Ribeye Roast and Steak
   Back Ribs

3. **Short Ribs**
   Top Loin (Strip Steak)
   T-bone Steak
   Porterhouse Steak
   Tenderloin Roast and Steak

4. **Sirloin**
   Top Sirloin Steak
   Tri-tip Roast and Steak

5. **Round**
   Top Round Steak
   Round Tip Roast and Steak
   Bottom Round Roast
   Eye Round Roast and Steak

6. **Shank**
   Shank Cross Cut

7. **Brisket**
   Brisket (Whole)
   Brisket (Flat Cut Boneless)

8. **Plate**
   Skirt Steak

9. **Flank**
   Flank Steak

Fig. 12K3 The sources in beef cattle of common cuts of meat.

before shipment. These procedures extend the shelf life of the fresh meat. Fig. 12K3 shows the source in the animal of common cuts of beef.

With hogs, butchering for specific cuts of meat has been more commonly done at the slaughter house for many years. The carcass is cut into legs, loins, shoulders, and picnic hams, etc. Further operations, such as curing and smoking are more common with meat from hogs. About 65% of pork is in the form of processed meat, such as bacon, ham and sausage.

After butchering, meat packing operations may include one or more of the following: sausage making, canning, cooking, tenderizing with enzymes, fat rendering, manufacture of sectioned and formed products, and production of restructured meat products. Hams and bacon are cured by injecting brine curing solution into the meat. Many cuts may also be smoked. These operations provide long shelf lives and reduced shipping and inventory costs. Meat is shipped after slaughtering and butchering to distribution and retail points by refrigerated carriers, usually by truck.

L. Bottling

Bottling is the preferred packaging method for many liquid materials and food products: Soft drinks, beer, wine, distilled spirits, spring water, juices, milk, sauces, syrups, cooking oils, and salad dressings are packaged extensively in bottles. Glass has been the predominant bottle material, but plastics have made and are continuing to make major inroads in the bottle market. In the mass-production situations that are so often applicable to foods, bottling is a highly automatic, high-speed, series of connected operations. With high- production machines of the type used for popular beverages, bottles are removed from shipping cartons with an *uncaser* machine, or are unscrambled, uprighted as necessary, and fed to a machine that washes, sterilizes, and drip-dries them, and then conveys them to a filling station. (Washing, sterilizing, and drying involves inverting the bottles to allow gravity to remove foreign matter and cleaning liquids, after which the bottles are turned upright again.) The filling station is a large table, reminiscent of a carousel; it carries many bottles, each held near the outside edge. As the table rotates, the bottles are filled from inserted tubes. The filling tubes move with the bottles and remain inserted until the bottles are full. After filling, bottles are conveyed aside and empty bottles are moved to engage the circular filling table. The full bottles are conveyed to another rotary table where caps, corks, or other closures are dispensed and positioned at the bottle top. Again, the dispenser and capping equipment rotates with the bottles. At the next rotating station, the bottle tops are crimped, screwed on, pressed in or otherwise processed to effect a seal of the bottle's contents. Many bottles then get a second, tamper-resistant closure in addition to the basic cap. Labels are applied at the next station with adhesive-bonded paper or plastic labels. Some printing, for example, for date codes, may take place on the line. The finished bottles are then accumulated and packed into corrugated cartons or other shipping containers by automatic packing machines. All these operations take place on one large, mechanized bottling line. The movement of the bottles through this equipment is usually continuous. Bottles are fed automatically onto and later off such tables. For many bottled products, the bottling machine operates at a speed of several hundred bottles per minute and, in some machines, as much as a thousand per minute.

The other bottling approach usually involves more of a straight line arrangement with a stop-and-go movement of the bottles. The bottles are stopped, usually in groups, for filling, capping and other operations, and then move again to the next station. This type of arrangement is more common when the production quantities are modest. Fig. 12L illustrates a typical straight line bottling machine.

Fig. 12L A typical automatic bottling system. Bottles are fed to the line from the unscrambling table at the left where they are placed manually. They are conveyed to the filling station, where eight bottles are held and filled simultaneously. The bottles then are conveyed to the capping machine where caps are positioned and tightened, then to an accumulation table where they are removed and packed. Machines for larger production quantities may have automatic feeding and packing machines added to the system. Courtesy *Inline Fillings Systems, Inc., Venice* FL.

M. Other Packaging

Jams, jellies, sauces, condiments, spreads, and semi-solid and solid food products are often packaged into jars, using operations and equipment very similar to those of bottling. Other types of food packages: plastic tubs, paperboard boxes, plastic pouches, formed plastic or aluminum foil trays, dishes or bowls, vacuum shrink packages, and bag-in-a-box packages, are all processed in equipment that, in principle, is similar to bottling equipment. Normal production volume justifies highly mechanized systems. The package moves by conveyor from station to station; each station has some function that advances the product package to its final condition. Packages are fed automatically to the system, opened and sterilized as necessary, are filled, closed, sealed, given identification, and sent to and incorporated in a final outer package. There are differences as well. Non-liquid food is more apt to be dispensed to containers by weight rather than volume; plastic packages are more apt to be preprinted rather than labeled after filling; refrigeration or freezing is more common after the package is sealed, unlike canned or bottled foods. The equipment may be highly specialized; that is, it may be designed and developed specifically for some particular food and package design. In some plants, particularly with more complex and lower-quantity items, the packaging line may incorporate some manual operations interspersed with those that are fully automatic.

Final packaging provides mechanical protection for the product, usually in corrugated fiber containers. Sometimes, stacks of primary or secondary packages are stretch-wrapped. Canned beverages are frequently packed in multiples of 6, 12 or 24 in paperboard carry-out boxes. These operations can be manual but, in high-production situations, are performed automatically. Carton filling and sealing machines collect the primary or secondary food packages in stacks or blocks, insert them in the final containers from the top or side, and fold and glue the closing flaps. Shrink film wrapping has become a common method for holding food cartons on pallets for storage and shipment.

N. Storage

Storage for foodstuffs provides protection from many unfavorable conditions or agents including: microbiological attack, insect infestation, adverse chemical reaction, moisture or humidity damage or corrosion of packaging, and damage to the food by excessively-high or -low temperatures. Adverse effects that can take place in foods during storage include changes in flavor, color, texture, and nutritional values. The impermeability and durability of the packaging materials may be an important factor. The best storage normally occurs in closed buildings that are temperature and humidity controlled. Optimum storage temperatures and humidities depend on the nature of the food and the packaging. Storage of raw or in-process food materials has the same requirements as for finished products. Storage lockers with cold or below-freezing temperatures may be required for some materials. Liquids are stored in tanks which also may keep the contents at below-ambient temperatures. Tanks, piping, pumps, and valves must be thoroughly cleaned between batches.

Proper storage methods include close control and minimum variation of the following factors: 1) temperature. A constant, usually low, temperature is preferred for most foods. Temperature changes, including changes below the freezing point for frozen foods, may result in quality deterioration. 2) humidity. Foods high in water content should be kept in a high-humidity environment; dehydrated foods need just the opposite. 3) atmosphere. Apples and some other fresh fruits and vegetables are stored in a protective gas atmosphere. Otherwise, oxygen may react with the food, leading to color changes and rancidity. 4) light levels. Light striking some fresh fruits and vegetables may cause undesirable changes. 5) stock rotation. The storage system should insure usage of the oldest material first.

O. Food Equipment Cleaning and Sanitizing

Proper food processing demands that the equipment used is clean (free of food soil and any other matter that could be nutritive for microorganisms) and sanitized (to kill any organisms that may be present. Pathogenic organisms can include bacteria and other microorganisms, viruses, molds, and parasites.) Requirements are most severe for surfaces that directly contact the food being processed but non-contacting surfaces in the proximity of the equipment must also be kept free of sources of contamination. Walls, ceilings, light fixtures, equipment frames, and other exterior portions of food processing equipment must not be allowed to contaminate the working surfaces of the equipment.

The standard order of operations for cleaning and sanitizing of food contact surfaces is: 1) rinsing, 2) cleaning, 3) rinsing again, perhaps twice, 4) sanitizing. The use of detergent chemicals for cleaning is standard practice.

Cleaning may occur with the equipment fully assembled ("clean in place"), partially disassembled or fully disassembled, depending on the nature of the probable soil, the difficulty of removing it, and the design of the food processing equipment. Cleaning in place has become the preferred method, for labor-cost and other reasons, if the food equipment can be engineered to provide it. Installed spray heads must adequately reach all surfaces. Valves, pumps, heat exchangers, and mixing devices must be designed so that material is not trapped, and that sprays or agitated soaking reaches all surfaces. When spraying can be made automatic, hotter water, steam, and cleaners with high alkalinity or acidity can be used since human exposure to these hazards is greatly reduced.

The cleaning/sanitizing methods and their frequency must be established and defined for each food operation. In batch operations, cleaning frequently is scheduled after each batch. Cleaning may also be scheduled on a per-work-shift or per-day basis, or on some other schedule. These operations, in some factories, are performed by separate crews from those that operate the equipment.

O1. ***equipment rinsing*** - or pre-rinsing is a first step, often with water of ambient temperature, to loosen and remove as much food material as possible and to soften any that is not removed, so that the cleaning step that follows is more effective. Rinsing also takes place after cleaning to remove the residue of the detergent or other cleaning compounds. Much rinsing is done with manually-held hose and spray heads, but some equipment is equipped with fixed spray heads. Other equipment is disassembled and smaller parts are placed in soaking pans for cleaning and rinsing.

O2. ***equipment cleaning*** - is the complete removal of food remnants, using a hot spray or soak, but with a detergent solution instead of water. Brushing with powered or non-powered brushes may be part of the operation. Either acidic or alkaline detergents - and sometimes both - are used, depending on the nature of the residue material to be removed. Alkaline cleaners commonly utilize sodium hydroxide along with surfactants, dispersants, and water conditioning agents.[1] Alkaline cleaners are generally required for removal of proteins, but are also effective with fats, carbohydrates and sugars.[1] Acid detergents are used for dairy products. They typically contain phosphoric or nitric acid or both. Steam cleaning is also used. Cleaning is more effective with stronger detergent solutions, longer contact time of the detergent, higher temperatures, and agitation of the solution. High pressure sprays of up to 500 psi (3.5 MPa) are sometimes used. Smaller cleaning components are often washed in a tub and reassembled to the equipment after rinsing. They may also be processed through a semiautomatic cleaning tank that provides both chemical and mechanical cleaning.

O3. ***equipment sanitizing*** - There are two basic sanitizing methods: thermal sanitizing using hot water or steam, and chemical sanitizing with certain antiseptic solutions. Chemical sanitizing involves spraying or immersing the equipment in the sanitizing solution. Common sanitizing materials include: chlorine compounds (eg., hypochlorous acid - HOCl - or chlorine dioxide - ClO_2), iodine (iodophors), quaternary ammonium compounds, acid-anionic solutions, fatty acid sanitizers, and peroxides (hydrogen peroxide - HP - or peroxacetic acid - PAA).

Hot water sanitizing of small parts normally requires immersion in hot water (170°F – 77°C) for at least 30 seconds followed by a rinse with water of 180°F (82°C).

O4. ***other factors*** - It is important that the cleaned and sanitized surfaces be dry as well as clean after the procedure because moisture is a factor in bacterial growth. Brushes and other equipment used in cleaning and sanitizing must also be clean and dried after the procedure.

References

1. *McGraw-Hill Encyclopedia of Science and Technology, Vol. 7*, 8th edition, 1997, ISBN 0-07-911504.

2. *The New Encyclopedia Britannica Macropaedia, Vol. 19, 15th ed.* Encyclopedia Britannica, Inc., Chicago.
3. *Van Nostrand's Scientific Encyclopedia, 8th ed.*, 1995, Van Nostrand Reinhold, New York.
4. *Encarta Encyclopedia*, Microsoft, 1999.
5. *Food Engineering Fundamentals*, L.C. Batty, S.L. Follkman, 1983, John Wiley, New York, ISBN-0-471-05694-4.
6. *Food Engineering Operations* - J.G.Brennan, J.R. Butters, N.D. Cowell, A.E. Lilly, 1969, Elsevier, London.
7. *Fundamentals of Food Engineering* - S.E. Charm, AVI Publishing, Westport, CT, 1978, ISBN 0-87055-278-3.
8. *Handbook of Food Engineering* - D.R. Heldman, D.B. Lund, Marcel Dekker, New York, 1992, ISBN 0-8247-8463-4.
9. *Van Nostrand's Scientific Encyclopedia, 9th ed.*, G. Considine, ed., John Wiley & Sons, NY 2002, ISBN 0-471-33230-5.
10. *Pulsed Electric Fields*, U.S. Food and Drug Administration, Center for Food Safety and Applied Nutrition, June, 2000.
11. *Introduction to Food Engineering, 2nd ed.*, Singh, R.P. AND Heldman, D.R., 1993, Academic Press, San Diego.
12. *Fundamentals of Food Process Engineering, 2nd ed.*, Toledo, R.T., 1991, Van Nostrand Rheinhold, New York.

Chapter 13 - Processes for Electronic Products

A. Printed Circuit Boards (PCBs)

Printed circuit boards are also called *printed wiring boards (PWBs)*. They are thin boards made from non-conducting materials upon which metallic circuitry has been deposited. Almost all common electronic products utilize the printed circuit board as the means for connecting and holding circuit devices (integrated circuits, transistors, diodes, other semiconductor devices, resistors, capacitors, inductors, sensors, displays, connectors, and switches). The electrical connections between these devices are in the form of metallic coatings on the board in a pattern to make wiring paths between the devices. Soldering is used to make permanent electrical connections between the devices and the circuit paths and provides a combination of high reliability and uniformity, at low cost. Photochemical techniques are heavily utilized in the manufacture of the wiring paths. Printed circuit boards have proven to be economical, highly reliable, and suitable for providing complex circuits in a small space, at light weight. Although intended for high production applications, the boards are frequently used for low-quantity and prototype production. They are used in television sets, high fidelity systems, radios, computers and computer peripheral devices, military, airborne and industrial equipment. Fig. 13A shows a typical printed circuit board.

A1. ***making bare printed circuit boards*** (boards with no devices assembled to them) - The printed circuit board is normally a flat board made from thermosetting plastic reinforced with glass fiber. See Chapter 4, section G8, for information on the methods used to manufacture laminated boards for various purposes, including printed circuit boards for electronics applications. (Also see A3 below for multilayer boards.) The most common board material is epoxy plastic, reinforced with glass fabric. Some boards use higher temperature epoxy or polyamide resins that are able to resist the solder temperatures involved in lead-free soldering. Kevlar and polytetrafluoroethylene (Teflon®) are used in some military and highly-sophisticated applications. Low-priced boards for some consumer products are made from paper-phenolic plastic laminates. The initial laminations are large and are normally sheared by the manufacturer into smaller panels that are still large enough to hold multiple circuit boards. The panels remain as one piece through the manufacturing operations and are separated into individual boards after processing.

The circuit board differs from other laminated plastic boards in that it normally contains layers of copper foil on the top and bottom surfaces, (only on the top for the simplest, one-sided boards). The copper foil is converted to metallic paths that act as conductors in an electronic circuit, replacing the wiring that otherwise would be required. These wiring pathways are commonly called *traces*. When various electronic devices are mounted on the board, either on one side or both sides, they are connected to the traces with solder joints. Two methods for putting traces on boards are the subtractive method and the additive method. (See A1b and A1c below.)

Fig. 13A A panel of several typical printed circuit boards with resistors, integrated circuits, capacitors, connectors and other devices soldered in position. It is often more convenient and more economical to process a panel of several boards through the assembly and soldering operations and then separate them rather than processing the boards individually. *(Courtesy Universal Instruments.)*

To make the more complex multilayer boards that have an internal layer of traces, a board with traces on both sides is laminated to a non-conductive layer of reinforced plastic and to another board, which may have traces on one or both sides. Controlled pressure, elevated temperature, and time, bond the boards together. Boards with traces on both sides or internally in the board are connected with *vias,* holes drilled through the board and electroplated with copper to provide a metallic electrical connection between the circuit paths involved. Vias may go through all board layers or only some layers, depending on the circuit design. (See Fig. 13A1.) The first plating of the vias with copper is carried out with an electroless process after the drilled holes are deburred. The next step is electroplating with additional copper. The full manufacturing sequence for the bare board often exceeds 100 separate operations.

A1a. ***resists and photoresists*** - are chemically resistant materials that are vital in the production of printed circuit boards and integrated circuits (IC). These materials can form temporary thin-film barriers so that subsequent processing operations, such as plating or etching, can be limited to only the circuit paths or, in other cases, to only the non-circuit areas of the board or IC. Photoresists are materials that change their properties when exposed to visible light, ultraviolet light, x-rays or electron beams. Some photoresist materials consist of a soft, gelatinous plastic that can be polymerized and hardened by exposure to the radiation. Others, based on an alkaline-soluble resin but including a photoactive compound, are initially hard but are softened and made more soluble after exposure. By projecting light or other radiation through an image mask and onto a layer of the photoresist, the desired pattern can be created to provide wiring circuitry on the board or device. The advantage of the photoresist is that it can provide a fine-pitch spacing of circuit elements. Most photoresists are in the form of a solid dry film. The film is laid on the panel and bonded to it with heated rollers. The film may also be applied as a liquid coating that is then dried.

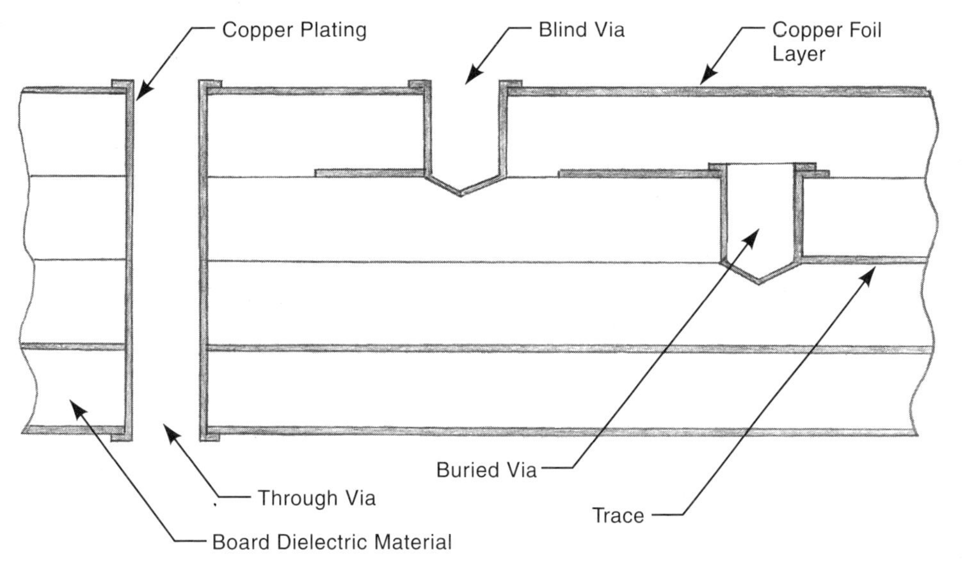

Fig. 13A1 Cross section of three kinds of vias: through vias, blind vias, and buried vias.

Another method utilizes electrophoresis (See 8D9.). Other resists consist of ink, paint or metal plating that does not have photosensitive properties. Inks and paint are applied by silk screening or stenciling to get the desired circuit pattern. This method is faster and less costly than using photoresists but is less precise.

Resist materials are constantly being developed with better resolution (for fine pitch circuit elements), higher purity, better resistance to etchants, and higher sensitivity to low levels of radiation.

A1b. ***subtractive process for making wiring patterns on the board*** - This approach is used on all but a small percentage of circuit boards. The circuit board used with this method incorporates a layer of copper foil in the laminate, on either one or both sides of the board. The copper layer is the base for circuit traces that will contain additional copper and tin/lead solder, at least on the pads, lands and holes. If there are vias (connecting holes between layers of copper foil), they are drilled, deburred and plated with copper to make the connection. Two methods are in use that selectively remove parts of the copper foil to leave copper circuit paths on the board. Both methods involve the application of a protective mask of resist material on the copper layer. Both methods may use either a resist applied in a pattern or a photoresist that is patterned after it is applied.

In the first method, if a screen-printed resist is used, it covers the circuit paths with the resist material and does not cover the areas between the circuit paths. If a photoresist is used, the entire surface of the board is covered with a dry film of photoresist material in a thin layer, 0.001 to 0.002 in (0.025 to 0.050 mm) thick. The board is then subjected to a radiation (usually ultraviolet light) through a mask. The photoresist material reacts to this radiation, leaving harder and stronger film over the planned circuit paths and a softer and weaker film over the board spaces between the circuit paths. The softer photoresist is then washed away, leaving the circuit paths covered with the film. The board then undergoes an acid bath that etches away the unwanted copper foil, that is, the foil not covered by resist material, leaving the circuitry and its photoresist cover in place. The photoresist now can be removed and this is done with a procedure that utilizes a solvent. Additional copper of 0.001 to 0.002 in (0.025 to 0.050 mm) thickness is then electroplated on the copper circuitry. A solder mask, as described in A2d below, is usually applied at this point. The board surfaces not covered with the solder mask are then coated with tin-lead solder by a dip or wave machine. Excess solder is removed by a flow of hot air (*hot air leveling*) or hot oil (*hydro squeegee*) to a thickness of 0.0012 in (0.03 mm) or less. This procedure provides surfaces amenable to later solder connections.

In the second method, the resist or photoresist is applied and processed as above except that the areas protected by the resist are the areas between the circuit paths rather than the circuit paths themselves. A washing operation then removes the unhardened resist that covers the intended circuit paths (traces). The next step is electroplating with copper, adding 0.001 to 0.002 in (0.025 to 0.050 mm) to the copper thickness of the circuit paths. Then, another metal, usually tin/lead solder, is electroplated over the copper. Sometimes, nickel, pure tin, or a nickel/tin alloy is used instead of tin/lead. The tin/lead protects the circuits from copper corrosion that could impair later soldering and protects the copper traces from an upcoming operation that involves etching. The resist covering the spaces between circuit paths is then removed by spraying the board with a solvent. The board then is immersed in an acid bath that etches away the unwanted copper foil in the non-circuit areas, but the circuit paths are protected from the etchant by the solder layer. The solder surface may then be chemically stripped from the copper so that a solder mask can be applied, because these masks do not adhere well to solder surfaces. (See A2d.) Then, after the solder mask is applied, tin-lead solder is coated on unmasked areas by dip or wave soldering and hot air leveling or by using a hydro squeegee (hot oil leveling). This operation sequence, using a photoresist, is shown in Fig. 13A1b.

A1c. ***additive method*** - is used much less than the subtractive method because it is more complex and incurs higher costs. However, it has been found to be useful in making the small boards used in packaging multiple integrated circuits, i.e., when several chips are combined in the same protective package. (See *multiple integrated circuit packages* in N1 below.) In this approach, a thin coating of a dielectric material is deposited on a substrate of silicon or ceramic and a layer of copper, or another conductor, in the wiring pattern, is then plated onto the substrate. The wiring pattern is produced by thin-film photolithography methods. (Patterned etching of full films may also be employed for multiple packages.) The package may include repeated thin-film layers of dielectric and conductive materials. Tantalum, titanium, tantalum nitride and gold are also used. Copper posts, plated with additional copper, are used to connect conductors on different layers. Another dielectric layer is then deposited and the posts are exposed. The next interconnect layer is plated and makes contact with the posts. Additional layers of dielectric and conductor and additional copper posts are added. Thin-film resistors and capacitors may be incorporated in the circuit. Circuit paths and spaces between them may be as small as 0.002 in (50 um).[9]

The additive process is also used with thick film (See K3a6.) layers in hybrid circuits, multichip modules, passive discrete devices, and in other applications where functional requirements and space availability permit larger circuit features.

A1d. ***making photo masters and masks*** - for screens, resists and photoresists. The process starts with the design of the circuit and its arrangement on the printed circuit board. Circuit board design is a computer-aided-design procedure. Very sophisticated programs are available for use in board design, and have enough capability to be called by some, "Design Automation".[7] Design limitations are entered into the computer for such factors as circuit path width and spacing, desired board size and shape, desired and necessary locations of critical components and external connections, necessary spacing of devices, size, shape and number of connections of "chip" packages. The circuit designer enters a schematic design for the circuit, basically a list of all the connections. The computer then takes over and attempts to make the required interconnections while meeting all the other necessary parameters. The program notifies the designer of any conditions that cannot be met and some changes or compromises may have to be made by the designer. Before arriving at a final board configuration, the computer program tries many alternatives and, ideally, arrives at an optimum design. As he examines the computer's solutions, the designer may see alternatives not previously conceived of and may modify his instructions to the computer. When a final solution is reached and approved by the designer, the computer can feed data on the circuit dimensions to a

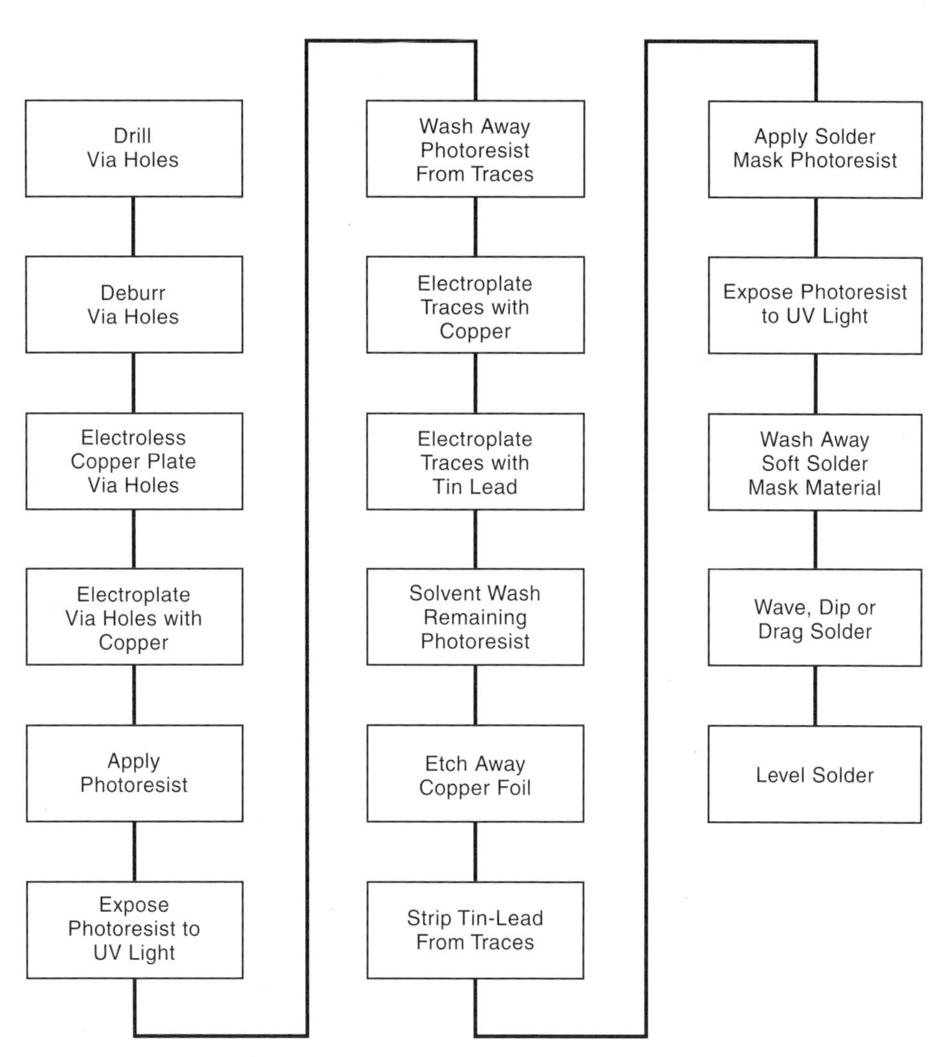

Fig. 13A1b The operation sequence for providing circuit paths to a double-sided board by a common subtractive method. The sequence includes the operations involved in making via holes. It assumes that photoresists are used for masking the board for electroplating of circuit paths (vias) and for applying a solder mask. A plating of tin-lead supplies a mask for the etching operation that removes copper foil from unwanted areas of the board.

laser plotting machine that creates a highly accurate master photographic image of the circuit on film or sensitized glass. This image may be reduced in size by photographic techniques to make photoresist masks, designs for screen printing, fixtures and numerical control programs for hole drilling and other operations on the board.

A1e. ***screen printing the masks*** - follows the same basic method as when silk screening is used to decorate and identify other products. In addition to its use in applying resists, screen printing is used to print product identification and assembly information on the circuit board. It is also used to coat circuit boards with solder mask. See 8I7b and 9D4b for descriptions and illustrations of the screen printing method. Much screen printing of circuit boards is done with automatic equipment. Screen printing of resists is the most critical and demanding of the many applications of the technique.

A1f. ***stripping resists from printed circuit boards*** - The usual stripping agents for resists on printed circuit boards are alkaline aqueous solutions. A common agent is a 2% solution of sodium hydroxide (NaOH). It is applied in a dip tank at a temperature of 125–140°F (52–60°C). The boards are then rinsed with a water spray. These agents work well with resists that are formulated to withstand acidic solutions, such as those involved in electroplating and other acidic treatments, but can be removed with alkaline strippers. Some manual or automatic scrubbing may also be part of the operation. Also used, especially when a stronger stripping agent is required, are semiaqueous solutions that contain, in addition to sodium hydroxide or other alkaline, 10% or less of butyl carbitol, triethanolamine or butyl cellusolve. Flammable hydrocarbons and chlorinated solvents, previously used for stripping resists from printed circuit boards, are no longer in common use because of safety and environmental concerns. Spent alkaline strippers are neutralized with acid before disposal.

A2. ***other board operations***

A2a. ***hole making in boards*** - In addition to the via holes mentioned above (for connections to the other side of a board or to internal circuit pathways), holes are drilled or punched to receive through-hole devices and for use as tooling holes to provide positive location of the board for further operations. Holes are normally drilled on CNC (computer-numerically-controlled) machines and through-holes are drilled with the boards stacked. Once the stack is positioned on the drill table, the operation proceeds automatically. Machine brush deburring follows drilling. The CNC machines often used also have the capability of changing drill bits when they wear or break.

Like via holes, insertion holes for through-hole devices may also be plated to provide large surfaces for soldering as well as connections to the circuit paths.

Tooling holes are usually drilled near the edges of the stack of panels, usually prior to other operations on the stack. Pins, inserted in these holes, hold the panels together while other holes are drilled. The pins extend below the bottom panel and fit into matching holes in the drilling table or fixture to locate the stack accurately for further drilling. A layer of disposable material (of phenolic, paper, or aluminum foil) may be placed on the top and bottom of the stack. They provide an entry layer on top of stack and an exit layer on bottom to reduce burrs and wipe chips from the drill bit as it is withdrawn.

Another drilling operation may take place after the board wiring is completed, to clean any unplated tooling or mounting holes of extraneous mask or plating material.

A2b. ***contact finger plating*** - Contact fingers are metal-surfaced tabs that fit into connector sockets. When the tabs are to be finish plated, plater's tape is applied to mask the board's other circuitry and tin lead solder is stripped chemically from the fingers. The copper surfaces of the fingers are then scrubbed and rinsed and they are electroplated with a barrier coat of nickel, followed by a layer of gold, the preferred metal for surfaces of such contacts. Although these operations can be performed manually and individually, in current practice, all the operations after taping are performed on automatic plating machines developed for the application.

A2c. ***solder fusing*** - Fusing and leveling of electroplated solder pathways ensures that the surfaces are truly wetted by the solder, ensuring full solderability. It also corrects any plating defects that may exist. The operation involves fluxing the circuit board, and heating it by one of a number of methods: infra-red radiation, immersion in hot oil or another hot liquid, immersion in liquid solder, or vapor-phase heating. The molten solder is then leveled mechanically, by pressurized liquid, or hot-air knives. The most advanced fusing arrangement is with a conveyor that moves the boards through infra-red ovens and cooling, cleaning and drying chambers.

A2d. ***solder masks*** - Solder resists (masks) are placed on the circuit board to ensure that only exposed areas of the board are coated with solder during wave, drag, or dip soldering. They prevent solder bridges from forming across dielectric areas of the board between traces, pads, lands, and holes. The masks are made from epoxy or other thermosetting plastic materials that remain on the board and provide protection to it, and increased insulation between circuit paths and components. Screening and photographic methods are used to apply the mask material. With the screening method, the thermosetting material is screened on and cured with ultraviolet light or heat. Photomasks provide a more accurate but more expensive approach and are used for finer-pitched boards. The wet or dry photographic film is applied to the board and processed with a light exposure, development of the film, and removal of the unwanted portion. Wet film is applied by spraying, dipping, curtain coating or roller coating. Dry film is laminated to the board with vacuum equipment.

Temporary solder masks may be applied by similar methods to portions of the board to shield them during wave soldering. Dummy plugs, tape, or precut shapes are sometimes temporarily assembled to the board for the same purpose.

A third type of coating for unassembled boards is a temporary solderable protective film coating. This type of coating protects circuit and pad areas from contamination with dust or dirt and from tarnish during storage before soldering. The coating typically is removed automatically by the heat of soldering or the activity or solvent action of the soldering flux.

A2e. ***separating boards (depanelling)*** - Individual circuit boards are separated from a panel of several boards by CNC (computer-numerically-controlled) routing machines. Routing cutters of 1/8 in (3 mm) diameter are commonly used. Beveling or chamfering, to put a tapered edge on contact fingers, is an accompanying operation to routing. It is performed by an angle-ground or tilted CNC routing tool. Slots and grooves are sometimes machined into the board with the same equipment. An alternative board separation method, less common, is blanking but it entails the expense of making a blanking die for each board design.

A2f. ***silkscreen identification*** - various identifying and instructional information is printed on the circuit board as one of the final bare-board operations. Conventional screen printing techniques are used with epoxy ink followed by drying or curing.

A3. ***multilayer boards*** - are made by laminating double-sided boards together with internal layers of board material. If two double-sided boards are combined with an internal dielectric board layer, a four-layered board will result. This construction is called *cap sheet lamination*. Another method for a four-layer board, the *foil lamination* construction, uses on double-sided board in the middle, covered on top and bottom with a dielectric board layer, and then, on each of them, an external layer of copper foil. Fig. 13A3 shows both arrangements. Boards with 16 or more circuit layers can be created if enough layers of lamination and enough boards are combined. The internal wiring traces are completed before lamination. Also before lamination, sheets of prepreg (uncured reinforced plastic dielectric material), and copper foil, if used, are sheared or purchased to the panel size needed and are cleaned. They and the boards may be drilled for tooling holes to maintain alignment during the balance of the process. The pitch width is normally narrow for multilayer boards since they are used in more sophisticated equipment with more complex and more concentrated circuitry.

The complete procedure for cap sheet board construction is as follows: 1) In several steps, the copper-foil board surfaces are treated with resist, and non-circuit areas of copper are etched away, leaving copper circuit traces, as is done with regular single- or double-sided boards. 2) surface oxidation - These circuit traces are subjected to heated oxidizing chemicals to provide a black copper-oxide surface which improves adhesion of the laminate. 3) The boards are rinsed, and then baked to remove absorbed moisture. 4) lamination - The boards are laid in a carefully aligned stack with layers of *prepreg*, uncured reinforced plastic lamination sheets. (Epoxy/glass sheets are most common.) Temporary “caul” plates, often with alignment pins, hold the stack in alignment. 5) curing - The stack is placed in a press that has heated platens and the stack is kept under pressure until the lamination plastic has cured. This may be done with the stack, and possibly the press, in a vacuum, to reduce the amount of pressure needed and reduce slippage. 6) cooling - The stack is cooled under pressure in another press. 7) stress relieving - The stack

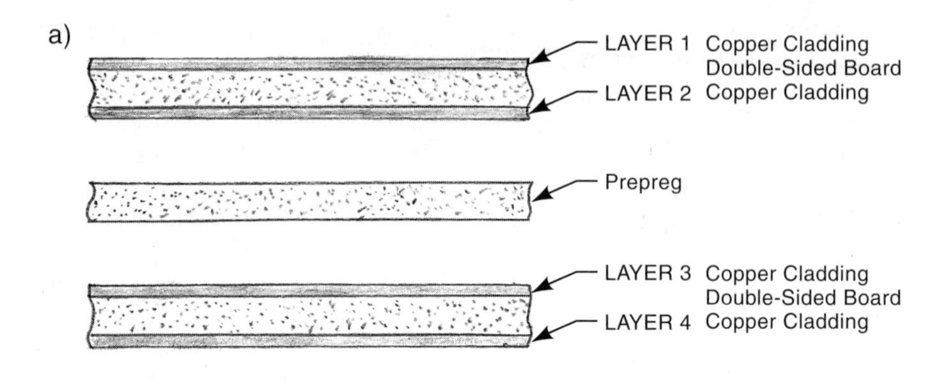

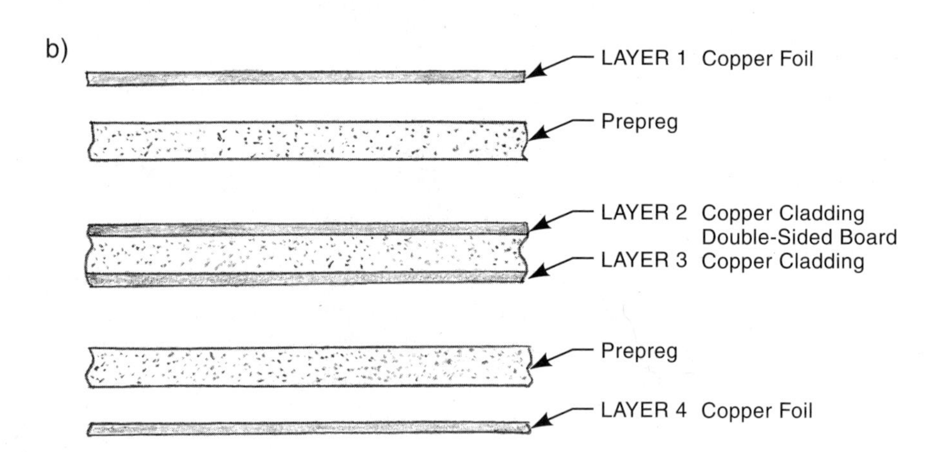

Fig. 13A3 Two ways to construct a four-layer, printed circuit board. With both methods, sheets of *prepreg*, partially cured sheets of epoxy resin reinforced with glass cloth, are used. View a) shows cap sheet lamination, using two double-sided boards (with circuit paths delineated) and one sheet of prepreg. View b) shows one double-sided board with circuit paths, two prepreg sheets and two sheets of copper foil. In both examples, the sheets are bonded together with pressure and sufficient heat to fully cure the epoxy. Multilayer boards with fewer or more layers are similarly constructed.

may be baked in an oven for several hours at about 325°F (160°C) to reduce internal stresses and avoid warpage. 8) Caul plates are removed and any plastic flash is removed. Edges are trimmed, if necessary. 9) Drilling for vias can now take place. Drilled holes are deburred. 10) Drilled via holes are copper plated by the electroless method. 11) Solder masking and electroplating of solder for external surfaces, as with double-sided boards, follows.

The procedure for boards with the foil lamination construction is similar. Internal double-sided boards are processed to produce copper circuit traces having black oxide surfaces, with the operation sequence outlined above. Boards are rinsed, baked, and stacked with layers of prepreg between the conductive surfaces and the layers of cleaned copper foil on the top and bottom of the stack. The stack is cured, cooled, and stress relieved as outlined above. Tooling holes are added if not already in place and via holes are drilled, deburred and electroless copper plated. The copper foil surfaces of the multilayer board are converted to wiring patterns with the methods outlined above, including plating. Solder masking is applied and solder plating, as with double-layer boards, follows.

Multilayer boards are also made using the additive approach.

A4. ***making flexible printed circuit boards*** - These boards use heavy flexible film as a base instead of a rigid, glass-reinforced board. Originally used simply for carrying multiple leads between components that have some motion between them, these boards now contain complex circuits including those that are double sided and, sometimes, multiple layered. Three materials are prominent in construction of flexible boards. All are thermosetting plastics with high temperature resistance: polyimide (Kapton®) plastic film is used in the most critical applications and has the highest temperature resistance and highest cost; liquid crystal polymer (LCP) film has similar characteristics at a somewhat lower cost; and polyester film is used in less critical applications where low costs are more important. Electronic devices are surface mounted on these films and are connected primarily with reflowed solder, though some boards use conductive epoxy instead. Traces on the boards are of copper. Copper is provided by foil that is bonded to the film base, and traces are produced in the copper by the subtractive process. (See A1b.) Traces may be coated with an organic coating to preserve solderability, by tin electrolytically, nickel by electroless plating, or silver by immersion-dispersion. Boards are often given a protective elastomer coating (*conformal coating*. See M.) after assembly is complete. This coating is cured by ultra-violet energy that passes through the flexible base film and cures the coating on both sides.

B. Wiring and Populating Boards

Fig. 13B outlines the operation sequence for populating boards (assembling devices to them) for both those using through-hole connections and those using surface-mount construction.

B1. ***populating boards with through-hole connections*** - See Fig. 13B1, view a) for an illustration of through-hole connections. Assembling

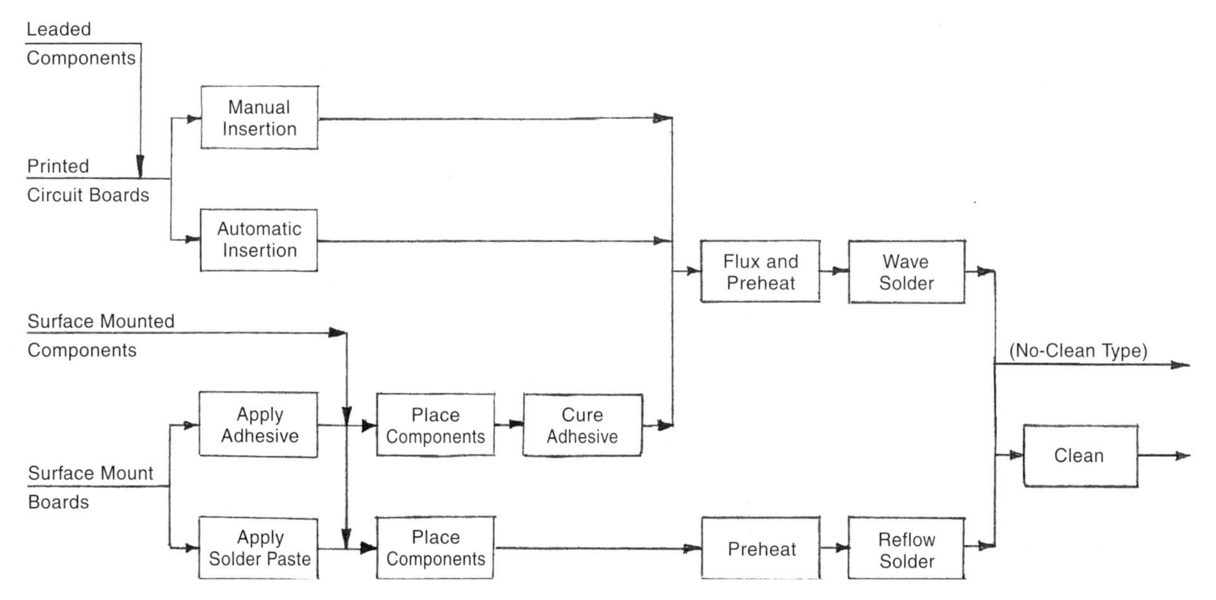

Fig. 13B The operation sequence for populating printed circuit boards, both those using leaded components (those with leads for through-hole connections) and boards populated with surface-mounted components.

a)

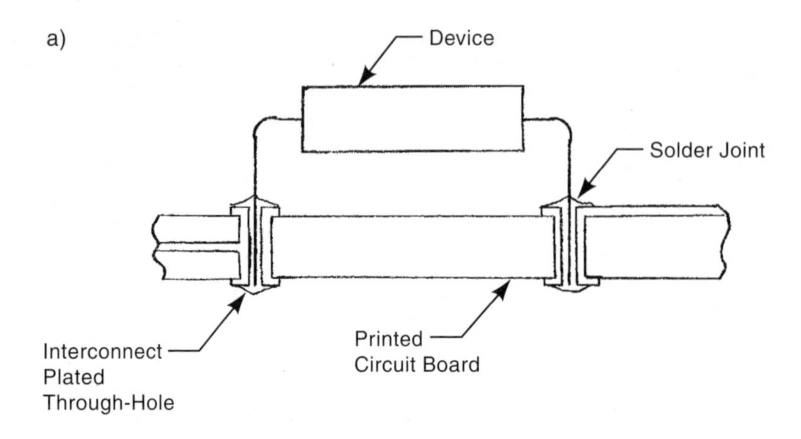

b)

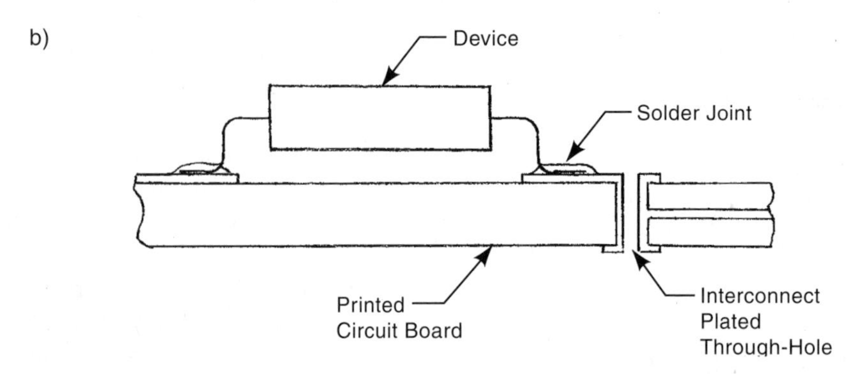

Fig. 13B1 Two common types of connections of electronic devices to printed circuit boards: a) through-hole attachment, b) surface-mount attachment.

through-hole components to circuit boards can be done by hand, with the aid of semi-automatic machines or, when production quantities make it justifiable, by high-speed, fully-automatic, insertion equipment. The semi-automatic machines have computer control that is programmed to move the board for each device to be assembled so that the correct through-holes are in a central position. The operator then inserts each device manually at the same central location on the machine. The computer display advises the operator which device is to be inserted next and displays any special insertion instructions. The leads extending through the board are cut to the proper length and clinched automatically on the underside of the board to retain the component in position, and surfaces are provided for solder connection on the underside. Fig. 13B1-1 illustrates such a machine. When fully automatic equipment is used, the insertion of the lead wires through the board is done by the equipment and the same automatic trimming and crimping steps are part of the operation. The first stage of such machines consists of equipment that bends and cuts the lead wires of each component. The component wires are then inserted by the machine into the corresponding holes in the circuit board; the lead ends are trimmed and clinched to the board. Automatic assembly of components with leads is common in household electronic equipment such as television sets, where circuit boards are not required to be highly compact and sufficient hole diameter allows for easy insertion of lead wires. Versatility of automatic equipment is provided by computer control. When manual insertion is used, a punch press operation, with the proper tooling, cuts and bends the leads of the components to fit the hole spacing of the circuit board.

B2. ***assembling surface mounted components*** - Surface mounting involves the attachment of

Fig. 13B1-1 A workplace for semi-automatic assembly of leaded components to printed circuit boards. The machine moves each circuit board in X and Y directions to bring the assembly point for each device to the same central location. The computer identifies, for the operator, which component is to be inserted and shines a light on the board location where the component is to go. After insertion, the component leads are automatically cut to length and clinched. *(Photo courtesy Contact Systems, Danbury, CT.)*

electronic components to a circuit board without using through holes or other terminals for the connection. Instead, the leads or other contacts of the components are soldered to conductive pads on the surface of the board. See Fig. 13B1, view b), for an illustration. The components are normally placed by specialized, high speed, pick and place equipment; manual placement is difficult to the accuracy required for fine-pitch surface mount technology. Components are fed to such machines pre-attached to tape (The tape is perforated like a movie film to ensure proper registration.), in magazines, on trays and, in some cases, with simpler devices, in loose bulk. Components to be fed from tape, trays, chutes, or some other system where they are in uniform orientation, may be put into that orientation in a preliminary operation with vibratory or other orienting equipment. With fine pitch surface-mounted boards, extremely accurate orientation and placement of components is vital, and the equipment available provides this. Usually, solder paste is applied to the connecting pads and is tacky enough to hold the devices in place until the solder paste is melted and solidified ("reflowed"). When wave, drag or other soldering methods are to be used for surface-mounted devices on the underside, an adhesive is applied to the board where the body of the device will rest. When adhesive is used to secure the device before (wave) soldering, the adhesive placement must also be accurate to make sure that adhesive does not cover parts of the pads to be soldered. Adhesive or solder paste is applied by stenciling, silk screening, use of syringes, or special dispensing equipment that normally is fully automatic if production quantities are high. Fig. 13B2 shows a circuit board assembly system for surface

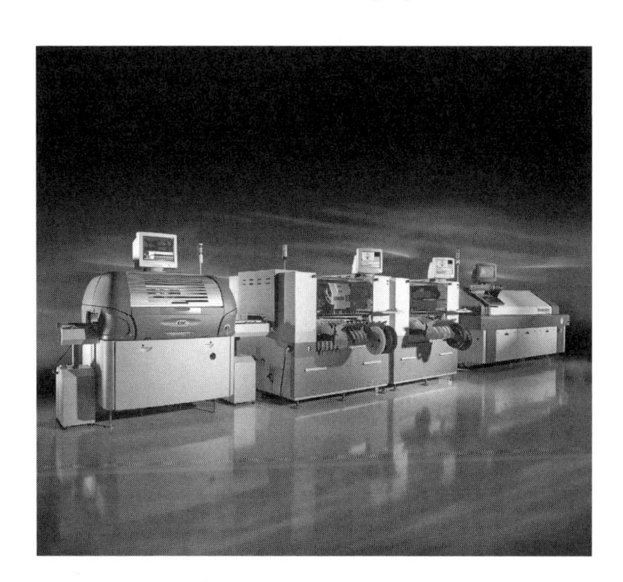

Fig. 13B2 An automatic assembly line for printed circuit boards with surface-mounted devices. All four machines are fully automatic, and the first three use machine vision to control the correct placement of parts and materials. The first of the four machines (at left) is a screen printer, which deposits solder paste on each pad that is to receive a lead from the devices to be assembled. The second and third machines pick and place circuit devices from paper or plastic tape fed from reels visible in front of the machines. The fourth machine is a convection reflow oven. It supplies heat with the proper heating profile to melt the solder paste, electrically connecting and mechanically fastening the devices to the board. *(Courtesy Universal Instruments.)*

mounted devices with two pick and place machines. Fig. 13B2-1 shows the placement of a device on a board.

The operational sequence for assembling and soldering a board with surface mount technology using solder paste is as follows: 1) preparation/cleaning of the board before assembly, 2) application of solder paste to the pads to which components will be soldered, 3) placement of electronic devices on the board, 4) melting or "reflow" of the solder paste to secure and to electrically connect the installed components. 5) cleaning of the board of flux residue, if essential. No-clean pastes, not requiring this operation, are preferred for all but the most critical applications.

Fig. 13B2-1 Placement of devices to the surface of a board by a pick and place machine. This machine uses a four-spindle head. Four devices are brought to the board at the same time and are positioned to the board one at a time. The component in the foreground is an integrated circuit to be fastened to the board by solder from a ball grid array of solder spheres. There is one solder sphere for each pad shown below the device. The spheres, converted by melting to bumps of solder, are on pads at the bottom of the IC and are not visible in the picture. Devices are held on the "nozzle" by vacuum. The spindle lowers the device to the board and the vacuum is released with an "air kiss" - a slight puff of air to seat the device on the soldering pads. The component in the background, waiting to be placed on the board, is a plastic leadless chip carrier ("PLCC") that will be held on the board with solder paste. *(Courtesy Universal Instruments.)*

B2a. ***dispensing adhesives*** - Adhesives used to hold surface mounted devices are dispensed with equipment very similar to that used to dispense solder paste. Screen or stencil printing, syringes and dispensing guns are employed and, for high production applications, the operation is highly automatic. Adhesives are used to hold surface mounted devices on the underside of a double-sided board during reflow soldering. Without an adhesive, the underside devices may loosen and fall as the solder paste flows and loses its tackiness.

B2b. ***using solder paste*** - The viscosity and tackiness of standard paste are high enough for it to stay in place when applied to a printed circuit board and for it to hold surface mounted devices that are placed in it. The paste is dispensed to the board by one of several methods described below. The electronic devices are then placed on the board so that their leads rest on the pads that have been coated with paste. When the board is heated - with one or more of a number of methods - the solder melts or "reflows", connecting the devices electrically to the board and also fastening them mechanically. Solder paste is also used with through-hole devices. (Also see Chapter 7, section A1c, for other solder paste information.)

B2b1. ***syringe dispensing of solder paste (pressure dispensing)*** - can be a manual operation but, in production situations, dispensing solder paste is an automatic operation and the dispensing action is usually pneumatically powered. The common method, with automatic equipment, is to locate the circuit board on an X-Y table under computer control. Z-axis motions may also be made for complex circuit boards. The table shifts, and the dispensing is actuated when the syringe is over the pads that are to receive the solder paste. Solder paste for this method is normally supplied from cartridges supplied by the paste manufacturer. Some equipment is made with multiple dispensing nozzles to provide higher production rates.

Nozzle or needle openings are typically 0.016 to 0.063 in (0.4 to 1.6) mm in diameter although there is interest in smaller sizes.[4] The nozzle is usually held at an angle of 40 to 75 degrees from horizontal.

When performed manually for prototypes or limited-quantity production, the operator positions a syringe over the pads by hand and dispenses a controlled amount of solder paste. With equipment designed for manual operation, the amount of paste dispensed at each point can be preset to provide consistent application.

B2b2. ***pin transfer dispensing of solder paste*** - uses a number of vertical pins on an upper plate. The pins are mounted so that they align with the pads on the circuit board that are to receive solder paste. The pins are first dipped into the paste to a predetermined depth. They are then moved to the circuit board and brought into contact with it so that each pin point contacts a soldering pad. This causes solder paste to be deposited on each pad. The method is suitable for accurate placement of an amount of paste on each pad Fig. 13B2b2 illustrates the process schematically. By making the pins spring-loaded, this approach can be used for boards that are curved or have soldering surfaces at different levels.

B2b3. ***screen printing of solder paste*** - Conventional screening techniques have been adapted to apply solder paste to circuit boards, using solder paste in place of the ink used with conventional screen printing. A fine-mesh woven screen, held tautly in a metal frame is coated with a plastic material in certain areas and positioned against the circuit board. Solder paste is deposited on the screen and a rubber-like squeegee (usually of polyurethane, of 60 to 70 durometer) is drawn either by screen printing equipment or manually, across the screen. In areas where it is desired to deposit paste on the circuit board, openings in the coating allow the paste to be pushed through the screen and deposited on the board. Much of the screen is coated and does not allow the paste to pass through it. The depth of the deposited paste depends on the diameter of the wires of the screen and the thickness of the screen coating. Screen

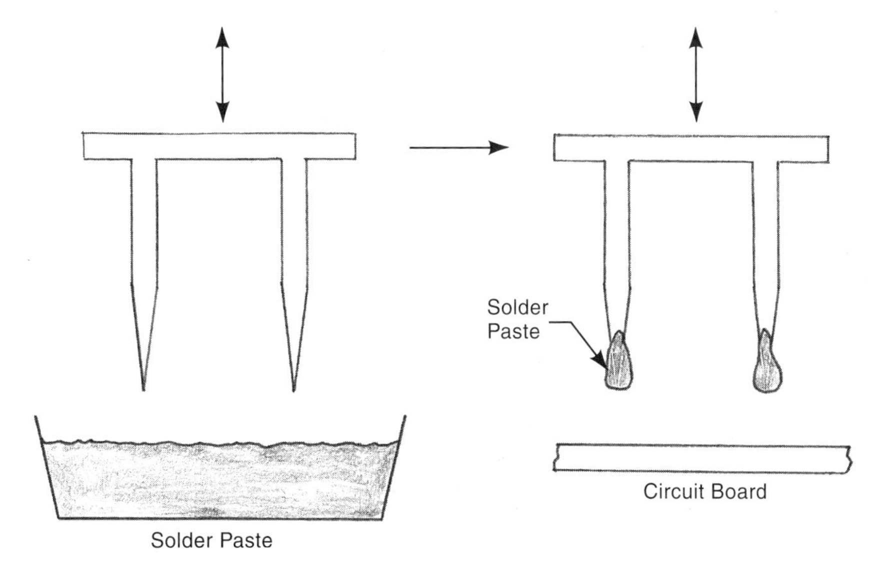

Fig. 13B2b2 A schematic representation of the pin transfer method of applying solder paste to the locations on a circuit board where solder connections are to be made. The pins, spaced exactly in the pattern of soldering pads on the board, are dipped together in a container of solder paste, picking up an amount on each pin end. When the pin ends touch the circuit board, the desired amount of solder paste is deposited on the board.

coatings are typically 0.008 in (0.2 mm) or less in thickness but can be as thick as 0.020 in (0.5 mm). The location of the paste on the board can be well controlled, provided the screen and board are in good registration with one another and the screen does not stretch. The screen is usually constructed of stainless steel to control stretching and provide wear resistance. The screen is usually mounted about 1/16 in (1.6 mm) above the board during the operation to prevent the screen from sticking to the board. The screen coating is applied by either dipping and drying or by manual placement of a film of coating material on the screen. In either method, the coating is "developed" by exposure to ultraviolet light which cures the coating except in areas that are to be open. The undeveloped coating is then removed. Photographic techniques are used to control the locations of the screen openings accurately. The screen printing process is a quick and consistent method for applying solder paste and flux. The approach is limited to flat circuit boards. The operation can be fully automatic in high-production situations where both the board handling and paste screening are fully mechanized. The operation is semi-automatic when board handling is manual but paste dispensing is automatic. It is fully manual with a fixtured board location, a hinged screen and squeegee movement by hand. This approach is used for low quantity and prototype work. Automatic equipment may use machine-vision apparatus to ensure that the registration of the paste on the board's solder pads is exact.

B2b4. ***stencil dispensing of solder paste*** - has become the process of overwhelming choice. It is similar to screen dispensing except that a mask made of sheet metal is used instead of a screen, but occasionally the mask openings are combined with a stainless steel mesh for reinforcement. As in screen printing, a rubber-like squeegee forces the solder paste through the stencil openings and onto the desired locations on the circuit board. The thickness of the sheet metal determines the thickness of the solder paste deposit. Typical thickness are in the range of 0.005 to 0.012 in (130 to 300 μm). Stencils are either cut from sheet metal or formed by electroforming (See 2L2.). Cutting stencils from sheet metal involves either photochemical blanking (3S4), or laser cutting (3O). The metals commonly used are stainless steel for cut stencils and nickel for electroformed stencils. Photochemical blanking normally involves the use of two resist coatings, one on each side of the stencil sheet, exposed and developed with opposite-side images in alignment with each other. The chemical etching takes place from both surfaces. The etching operation is followed by electropolishing (See 8B3.) to remove a sharp edge, if it exists where the etching for each side intersects, and to smooth the whole surface of the stencil. Laser-cut stencils may also be electropolished. Electroformed stencils are made by first coating, imaging, and developing a photo-resist on a substrate. Then electroforming takes place in the openings of the photoresist to create a stencil that is removed from the substrate. As with screen dispensing, stencil dispensing is limited to flat circuit boards, and stencils are less forgiving than silk screens of minor variations in the board surfaces. However, stencil printing is usually more consistent, and alignment with the board is easier to maintain. Stencils are better adapted than screens to fine pitch spacings, especially the laser-cut and electroformed stencils. Metal stencils also have a considerably longer life though they are more costly than screens.

B2b5. ***submerged disk dispensing of solder paste*** - is used for dispensing solid or dashed lines of paste on a board. A rotating disk is positioned so that its edge is immersed in solder paste. As it rotates, the disk carries an amount of paste which, in turn, is deposited on a board that passes above it. The principle is shown in Fig. 13B2b5. A thickness control knife removes excess solder paste before the disk contacts the workpiece. The disk is usually made of non-metallic material and can be of whatever width is needed for the solder paste deposit.

B2b6. ***dip coating of solder paste*** - is used occasionally in applying solder paste to the leads of components such as capacitors. The lead is immersed in solder paste in a tray or other container. Reflow follows assembly of the component to the board.

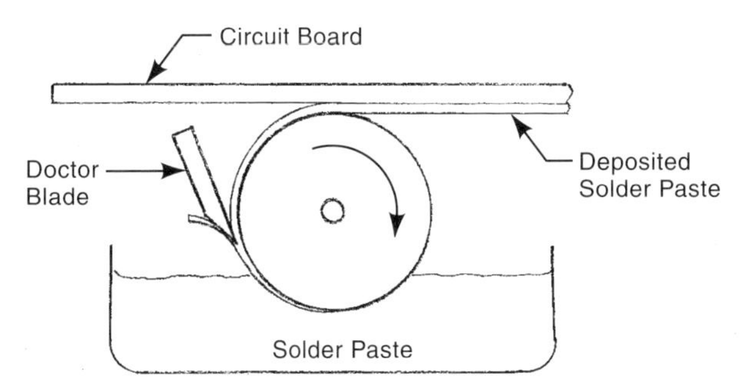

Fig. 13B2b5 The principle of submerged-disk dispensing of solder paste to a printed circuit board.

B2b7. ***roller coating of solder paste*** - to leads of capacitors or other devices is sometimes employed. A nap-covered roller, similar to a paint roller, is used. One common application is the coating of "nail head" ends of capacitor leads by manufacturers of such devices. The capacitors are arranged in a carrier so that they are all have the same orientation. A roller, coated with solder paste from a reservoir, is rolled over the row of "nail heads". The capacitors can then be assembled to circuit boards or other devices and the solder paste reflowed.[4]

B2c. ***using solder preforms*** - Solder preforms are solid shapes of solder manufactured by various methods of metal forming and cutting. Typical shapes are washers, disks, rectangular or square shims, rings, and other wire forms, spheres, sleeves and other special shapes. Typical preform-making operations are extruding, either with or without a flux core, rolling, ring forming and blanking. (These operations are described in general terms in Chapter 3.) Flux coating of preforms, when used, usually involves a spraying operation. Flux-cored preforms are usually made from flux-cored wire solder which may be flattened by being rolled into a ribbon and then blanked to the shape desired. The advantage of a preform is that it can supply the exact amount of solder needed for the joint and, if there is a flux core or coating, the exact amount of flux required. The amount of solder and flux supplied is also constant from assembly to assembly. High-quality, very uniform solder joints are possible. In many instances, the preforms can be placed with automatic equipment. Further economies result from the fact that all joints on a board or on a device or groups of them can be heated simultaneously.

B3. ***cleaning prior to soldering*** - Cleanliness of the surfaces to be joined is an important prerequisite of good solderability. Most cleaning of circuit boards and other electronic components is undertaken to remove foreign matter that may accumulate on the components from prior operations, handling and storage. The foreign matter includes dust, oils, wax, chips from machining the circuit board, and tarnish. Two methods are in predominant use for in-line cleaning, prior to fluxing and soldering: 1) vapor degreasing and 2) water washing. With both methods, care must be taken not to dislodge components assembled to the boards. Vapor degreasing using conventional equipment (See Chapter 8, section A2a3.), is quite satisfactory because it leaves the boards dry as well as clean and does not involve forces that may dislodge devices assembled to the board. An airknife may be used before the next operation (fluxing), to remove any trapped solvent and to cool the board to facilitate foam fluxing. Vapor degreasing has disadvantages from health and environmental standpoints, however. Water washing is more difficult because the force of water agitation or spray can dislodge components on the board. However, satisfactory results can be obtained. Saponifiers and other additives are included in the washing solution.

Ultrasonic agitation may be utilized in both kinds of cleaning if the board is immersed in a liquid. Drying after washing is essential. Baking may be required if the soldering operation immediately follows cleaning.

Components that require more aggressive cleaning because of chemical contamination, corrosion, or to remove protective coatings, may undergo vapor blasting, acid treatment, brushing or scouring. Great care must be taken in such operations to avoid damage to softer materials or dislodging assembled components.

B4. ***prebaking before soldering*** - is an operation that often is not required, but is used when it is necessary to remove volatile materials from the circuit board. These materials include trapped solvent or moisture from a cleaning operation, the volatile materials in the laminated board itself, and moisture that may be absorbed in the board or assembled components during storage and assembly. Baking usually takes place with temperatures ranging from 180 to 250°F (80 to 120°C) for periods of 1.5 to 16 hours. Lower temperatures and shorter times can be used with vacuum ovens (with approximately 1 torr of vacuum).

C. Soldering Processes

C1. ***flux application*** - immediately precedes soldering. The flux removes tarnish from the surface to be soldered and provides some mild cleaning action. Flux keeps the surface clean until the soldering is completed by protecting the surface from oxidation that would otherwise occur during heating. It removes tarnish that may develop on the surface during soldering, and aids in wetting the surface with solder. Acid and rosin fluxes are used. Rosin is a natural product that has acidic properties when heated, but which is relatively inert at room temperature. Typically, the joint surface is coated with flux before being put in contact with solder, though, sometimes, flux-cored wire solder is used in manual soldering and in some automatic operations. Solder paste contains flux and thus it does not have to be applied beforehand. Flux is applied to printed circuit boards as foam, by spray, or by wave application, similar to that used to apply molten solder.

C1a. ***dip fluxing*** - can be used to apply flux to the ends of leads and the edges of circuit devices that are to be selectively fluxed. The operation may be manual for repair or small quantity work, but is sometimes made automatic by conveying the parts so that the solderable portion passes through liquid or paste flux. A wiping or brushing stage may follow to remove any excess.

C1b. ***brush application of flux*** - is primarily a manual method used in touch up or hand soldering of some special components, or to selectively apply flux to critical areas of a circuit board. It can be mechanized in a set up that utilizes a rotary brush that is partly immersed in liquid flux; the brush also contacts the underside of circuit boards as they pass on a conveyor.

C1c. ***foam flux application*** - A porous stone, immersed nozzle, or other diffuser in a bath of liquid flux, emits an airstream that is divided into many elements which form small bubbles in the liquid, creating foam. The foam is contained in an open chimney-like enclosure, and rises to contact the circuit boards that pass above the enclosure. The flux adheres to the circuit board. The liquid flux includes a solvent or vehicle that dissolves rosin or other flux material. The solids content may be quite low, but still sufficient for good fluxing action. Foam fluxing normally is part of an automatic soldering sequence, with the circuit boards being conveyed through the fluxing station. Sometimes, a row of stationary brushes is installed to remove excess flux as the circuit board passes from the flux foaming station.

C1d. ***spray application of flux*** - can provide accurate dispensing of flux on circuit boards, though the need for overspray removal may complicate the operation. Three basic methods are used to create the spray of a flux-containing liquid: (1) *nozzle spray* uses air or gas atomization and one or more spray nozzles to apply the flux. (2) *rotary screen application* uses a rotating cylindrically-shaped screen, whose axis is at right angles to the path of the circuit board. The cylinder is partially submerged into the liquid flux; its rotation carries the flux with it and an air blower forces the flux up against the circuit board. A sensor turns on the air flow only when a circuit board is in position. A mask ensures that the spray is confined to the desired area. This method provides uniform flux application over the width of the board. 3) *ultrasonic application* - Liquid flux is brought into contact with an ultrasonic horn whose rapid vibration atomizes the flux into fine particles. These are then directed to the circuit board by an air or gas jet. An additional air curtain may be used to further distribute the spray, and the ultrasonic horn may move when selective flux application is desired. Another method uses an X-Y table under computer control to move the circuit board, so that the flux is applied only to those board areas where soldering will take place.

In all these spray methods, a mask may be used between the spray source and the circuit board to confine the sprayed flux to areas where it is needed. Fig. 13C1d shows the rotary screen method schematically.

C1e. ***wave fluxing*** - uses the same principle as wave soldering to provide a ridge or standing

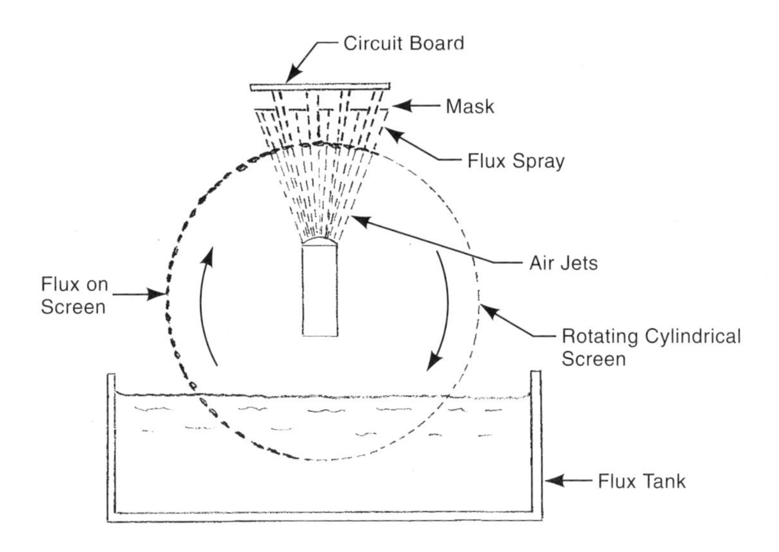

Fig. 13C1d The principle of the rotary screen flux spray. The cylindrical screen rotates in a bath of liquid flux and picks up some flux in its openings. Droplets of the flux are blown by air jets from the screen to the printed circuit board above. A mask prevents flux from hitting the board where it is not needed.

wave of flux that the circuit board can contact as it is conveyed through the fluxing station. The process can be run at a faster speed than foam fluxing and is used in high-speed automatic soldering lines. A wiping operation by brush or airknife in line with the fluxing may be used to remove excess flux after the board has passed through the wave.

C1f. ***roller fluxing*** - uses rollers that are often similar to those used for paint application. The method can be mechanized so that flux is applied as a circuit board passes on a conveyor. Flux application is uniform and the quantity applied can be controlled well. Roller fluxing is commonly used to coat both sides of a circuit board prior to infrared oven fusing.[3]

C1g. ***cored solder fluxing*** - utilizes solder in wire form where the center portion of the wire consists of flux rather than solder metal. The flux is in solid or paste form but melts as heat is applied to the joint and the wire solder. The flux melts before the solder, and flows to coat the joint surfaces. The method ensures that just the right amount of flux is deposited on the work. In electronics applications, cored solder is used chiefly for manual repair or touch-up soldering.

C1g1. ***making flux cored wire solder*** - is a multi-operation process. The first operation is casting the desired alloy in permanent steel molds to form extrusion slugs which are fed to an extrusion machine (2A3). In the machine, hydraulic or mechanical force pushes the solid solder alloy through a die that forms it into a hollow tube, typically of about 1/4 in (6 mm) in diameter. The extruder has a pot of molten flux with piping to feed it to the extrusion die in such a way that the flux flows into the hollow of the tube. The coarse hollow tube containing flux is accumulated on a reel. This reel then provides feed material to a drawing machine (2B2). In that machine, the hollow tube is pulled through a series of drawing dies, each with an opening slightly smaller than the diameter of the incoming tubing. As the tubing moves through the drawing machine, it is reduced in diameter, becoming wire solder. One or two drawing machines, each with perhaps a dozen progressively smaller dies, may be involved in the process, the number depending on the final size wanted in the wire solder. As the wire is reduced in diameter, it retains a center section of flux that has solidified as the wire has cooled. The finished wire solder is wound onto reels or small spools. The finished wire diameter may be as small as 1/64 in (0.4 mm), but 1/32 in, 1/16 in and 1/8 in (0.8 mm, 1.6 mm and 3 mm) are the more common sizes. *Solid* (non-flux bearing) *wire solder* is made in the same way except that there is no flux feeding system at the extruder and the extrudate is solid instead of hollow.

C2. ***preheating*** - circuit boards is part of a wave-soldering sequence, immediately preceding the soldering. It has the following purposes: evaporate volatile flux elements, melt solid flux constituents, start activation of the flux, reduce thermal shock to components when they contact molten solder, and reduce the time necessary for the circuit board to be in contact with the molten solder. Heating is effected by either radiation, convection or, quite rarely, hot plate conduction. Infrared devices are often used as a source of heat. Depending on the board type, heat may be directed at both the top and bottom of the boards. Quartz lamps with reflectors are often used with switches that activate them when circuit boards approach the preheating zone. Sometimes the heaters are controlled by sensors that read the temperature of the circuit board. Typical temperatures of the circuit board, when heated, are 180 to 270°F (80 to 130°C) depending on the type and thickness of the board.

C3. ***dip soldering*** - involves immersion of the solder joint into liquid (molten) solder. The solder pot then provides both the heat and the joint material. Usually, flux is applied to the joint beforehand by dipping, spraying, brushing or some other method. However, a molten flux layer may be used on the surface of the molten solder as a means of applying flux. The dipping process is mostly used to "tin" the ends of wire leads before soldering and has also been used to solder the through-hole connections of printed circuit boards. However, the process is inferior to wave soldering and is currently not widely used. The board is fluxed and moved so that the surface to be soldered contacts the surface of the molten solder, which heats and supplies solder to the joints. Surplus flux from the board must be moved aside so that it does not block the solder from wetting each joint. A rolling motion of the circuit board helps release flux and trapped gases. The surface of the molten solder must be skimmed before the workpiece is dipped; otherwise, dross (oxides) on the surface will interfere with the soldering operation. An oil layer may be maintained on the surface of the solder to retard or prevent dross formation.

C4. ***drag soldering*** - is a variation of dip soldering in which the circuit board is carried by conveyor and is pulled along the surface of the molten solder. The process is less frequently used, at present, having given way to wave soldering. The full operation includes a fluxing station, a preheating/drying station and a soldering station. A skimmer in front of the moving circuit board pushes aside dross that forms on the solder surface. A slight incline and a rocking motion may be used to release trapped flux and gases. Contact time with the solder is typically 5 to 8 seconds[2], longer than that involved when wave soldering is employed. Fig. 13C4 shows the process.

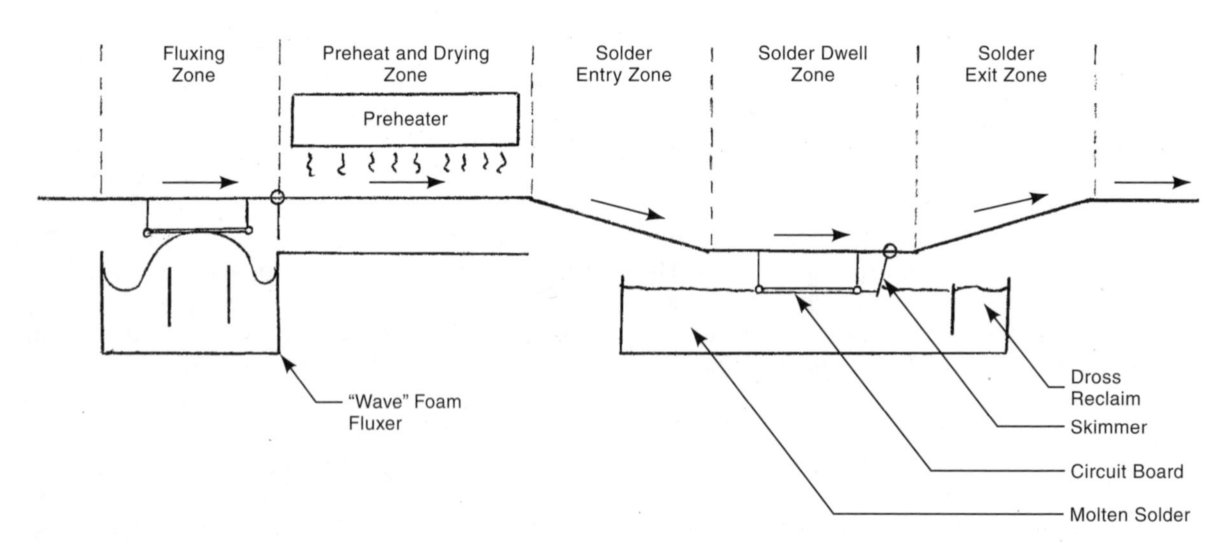

Fig. 13C4 A schematic view of the equipment and sequence of operations involved in drag soldering printed circuit boards.

C5. ***wave soldering*** - is a variation of dip soldering in that the workpiece, a printed circuit board, held in a conveyor, momentarily contacts liquid solder that both heats and feeds molten solder to the joints to be connected. Wave soldering differs from dip soldering in that the liquid solder is lifted as a standing wave, instead of remaining as a level surface in the container. This simplifies the immersion process by limiting the area of contact with molten solder to only part of the circuit board at one time, reducing the chance of overheating electronic components. The solder contacts only the underside of the board. The molten solder is pumped from an intake below the surface so that the wave is essentially free of dross. An oil coating on the molten solder, or an inert atmosphere blanket, may be used to control dross formation. The operation heats and wets (coats) the joints of both devices connected in through-holes and those that are surface mounted. It coats the wire paths on the board and fills plated through-holes in the board. The wave solder operation is illustrated schematically in Fig. 13C5.

The sequence of operations in wave soldering a circuit board in a typical wave soldering machine is as follows:

1) The circuit board, with electronic devices attached but not solder-connected, is fluxed by spraying, contact with flux foam, or by contact with a wave of liquid flux, generated with a method similar to that used in wave soldering. The step applies a coating of flux to the underside of the board.
2) The board is preheated to a temperature of 180 to 270° (80 to 130°) to drive volatile ingredients from the flux, to start the activation of the flux and to reduce thermal shock when the board contacts the molten solder.
3) Wave soldering then takes place while the board is still warm. Each portion of the circuit board contacts the molten solder for about 5 sec. Exposed metal surfaces, including leads or contacts of the attached devices, are wetted with solder. It also fills plated through-holes in the board. A hot airknife (air temperature above the solder melting point) may be used to remove excess solder in the form of solder bridges, solder balls, or other excess deposits.
4) The board is cooled at a controlled rate consistent with the sensitivity of the components to temperature changes. Blower-driven air may be used.
5) Board cleaning, for high-reliability applications, where no-clean flux is not used, is an important operation and immediately follows circuit board soldering. If no-clean flux is used, the cleaning operation is not required.

The wave soldering process is the prime mass-production method used to make electrical connections on printed circuit boards. When wave soldering

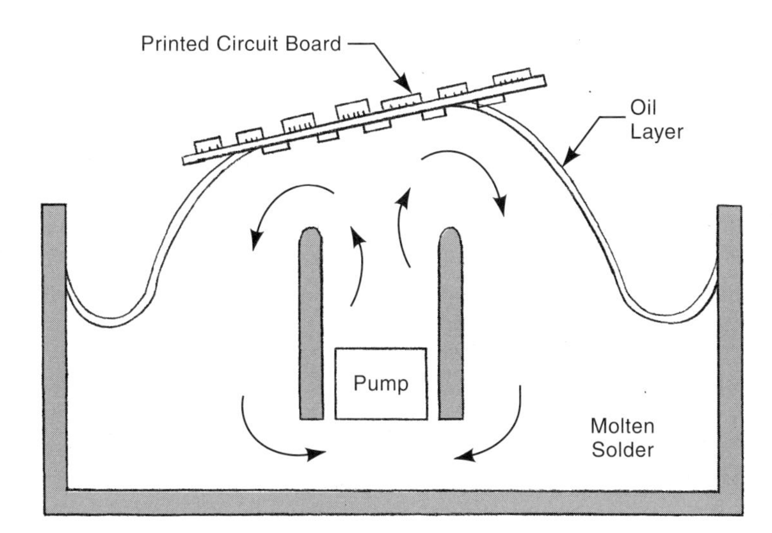

Fig. 13C5 Schematic representation of the wave-soldering process. The printed circuit board, with components attached, is moved against a "wave" of molten solder. Pads, leads, and wire paths on the board are wetted by and coated with solder. An oil layer on the molten solder inhibits the formation of dross.

is used to solder one side of surface-mounted boards (with adhesively bonded components), there is a tendency for air and flux volatiles to become trapped behind components during wave soldering, preventing the molten solder from making sufficient contact with the joint areas. To counteract this, some wave soldering machines have two waves that contact the circuit board. The first is a turbulent wave. Its turbulence gives it a solder flow in multiple directions, having the effect of displacing the trapped gases and allowing solder to come into contact with otherwise inaccessible areas. A second laminar wave of solder provides the desirable joint smoothing.

A dry nitrogen atmosphere is sometimes used in the wave soldering station, especially when no-clean or less active fluxes are used.

Computer control may be incorporated in the wave soldering line. It typically includes both process control functions and management information reporting of output and quality statistics. Process control functions cover such factors as temperature of flux, solder and cleaning agents, flow rates, solder level, liquid levels, airflow and pressure, conveyor speed and flux characteristics.

C6. ***reflow methods*** - "Reflow" refers to the melting of solder that has been pre-placed in the joint. The term originally referred to remelting solder that had been melted earlier to tin a surface. Currently, the term is used whether the surface was pretinned or not. Suitable solder and flux, normally in solder paste or as a fluxed solder preform, are placed in the location where a solder joint is to be made. Heat is applied; the solder melts and forms a joint. The approach is most common with printed circuit boards where circuit devices have been placed in dabs of solder paste.

The term reflow is also used to denote the melting of a plated solder coating to provide better wetting of the surface and an improved metallic structure of the solder.

Proper reflow almost always involves a preheating operation to release solvents from the flux, to start the activation of the flux, to improve the tackiness of the solder paste, if solder paste is involved, and to provide more gradual heating of the whole assembly to prevent thermal shock.[4] A number of different methods are available to provide the heat for reflow.

C6a. ***oven heating for reflow***[4] - The use of oven heat to melt solder paste or solder preforms, and produce permanent connections, is the most common reflow method. Batch ovens can be used, but in much electronics production of printed circuit boards, a conveyorized system is employed. In current practice, the dominant approach utilizes ovens with forced convection. Ovens using infrared radiation have become a secondary method. The atmosphere in the oven may be air or nitrogen (to limit oxidation) or a mixture of nitrogen with 12 to 15 percent hydrogen. Inclusion of the hydrogen has been shown to improve the wetting action of the solder. These special atmospheres are especially applicable when the board is assembled with "no clean" (low residue) flux.[1] The air or nitrogen is blown through electrically-heated manifolds or finned-rod heating elements and diffuser plates above and below the conveyor track and around the circuit boards on the conveyor. Such convection flow provides more uniform heating of the circuit board than is achieved in systems that rely primarily on radiation, or less-forced convection.

Typical stages of heating are as follows: 1) Preheating with gradual heat buildup to about 200 to 250°F (95 to 120°C). This avoids heat shock to the board and its assembled devices and drives volatiles from the flux in a controlled manner. 2) "Preflow", in which the next oven section is a holding zone that brings the board assembly to a uniform temperature just below the melting point of the solder. This stage activates the flux and completes the drive-off of unneeded volatile materials. 3) In a third section of the oven, the temperature is quickly brought up to a level 30 to 55°F (15 to 30°C) above the solder melting point. The time spent by the circuit board at this temperature is very short. 4) Cool-down, an immediate reduction to the solidus temperature of the solder alloy, takes place and there is then a gradual reduction of the board to room temperature. Sophisticated ovens may have as many as 10 heating or cooling zones to achieve the desired heating sequence. An objective of all forced convection ovens is to achieve a minimum "delta T", the temperature variation between the hottest and coolest points on the circuit boards. The major objective is to provide sufficient heat for solder flow without overheating temperature-sensitive elements of the board. To achieve these objectives, special time-temperature sequences are developed for each board, depending on its size, the number of devices, the size and size contrast of the devices on board, and the sensitivity of the components to excessive temperatures. The special program for each board design is often stored in a microprocessor that is part of the oven system. This allows a successful heating profile to be repeated in subsequent production lots. Control systems for some ovens sense the temperature of the boards being processed as well as the temperature of the air or gas oven atmosphere. Variables under control of the system, in addition to the temperature of each heating element, are the speed of the several convection blowers and the speed of the conveyor.

Batch forced-convection ovens are used for limited-quantity production. These ovens also may have capability of storing special heating and cooling profiles in the control system so that each board processed can have an optimum sequence.

Infrared oven systems are still used in many instances. When they are conveyorized, the conveyor carries the circuit boards past an array of infrared lamps. There are two basic approaches. The first uses middle infrared with a bandwidth of 2.5 to 5.0 microns; the other uses near-infrared with a bandwidth of 0.72 to 2.5 microns.

Systems based on middle-range infrared energy use convection as well as radiation. Ceramic elements, coated and insulated, and operating at a lower temperature than that obtained from near-infrared, emit the infrared energy. Radiation at this wavelength is diffuse and somewhat less affected than with near-infrared, by the color of the absorbing surface. The heating is penetrating and easier to make more uniform throughout the board and its components. Some ovens have several heating zones so that a desired heating profile can be utilized. Preheating, and gradual heat-up, are used in this approach with venting to allow flux volatiles to escape. The preheating zones are followed by several additional zones where full heating takes place and then, often, a cooling zone. Top and bottom radiation sources are normally used. Much middle-range equipment includes means for heating and forced circulation of the atmosphere in the oven and this heated atmosphere provides approximately half the heating effect.

The near-infrared system uses multiple non-focused tungsten filament or quartz lamps arranged to radiate heat to the passing circuit boards, often to both the top and bottom surfaces of the board. With the near-infrared lamps, little energy is transferred to the atmosphere in the oven; the bulk of the energy is absorbed by the board and its components which can be heated quite quickly. Care must be taken to avoid having critical surfaces to be heated in the shadow of other objects on the board. Temperature control of the lamps in sophisticated infrared equipment is exercised by computer control.

The infrared systems are less costly than vapor-phase reflow and the sophisticated full-forced convection systems, and are adapted to easy and fast control, so that the heating profile can be changed as necessary and tuned to fit the particular board assembly being processed.

C6b. ***vapor-phase soldering*** - is a method used to reflow solder paste deposits or solder preforms, usually on electronic printed circuit boards. The boards, with solder paste or preforms applied at each joint, are placed in or conveyed into a chamber that contains the saturated vapor of a fluorocarbon liquid. As the vapor condenses on the board, it transfers the latent heat of vaporization to the board, heating the joints uniformly and rapidly. The solder paste melts at each joint. The advantage of the process is that it avoids overheating, because the condensation temperature of the fluorocarbon vapor is a constant, precise temperature, usually about 55°F (30°C) higher than the melting temperature of the solder paste. (Different vapors are used for different solder alloys). Thus, the process avoids damage to critical electronic components on the circuit board. Chemically-inert fluids with low solvent properties are used to avoid adverse effects on the circuit. The equipment used is similar to a vapor degreaser. The fluid is boiled in a container below the work. Vapor rises to the level of the work and condenses on the cooler circuit boards, raising their temperature to the condensation temperature of the liquid. Because heating is rapid with the process, preheating is often used before vapor-phase reflow soldering to prevent flux spattering from sudden volatilization of solvents, and to reduce thermal shock to components. Preheating and cool-down stations are provided in some vapor-phase soldering equipment to reduce the time that the workpieces are exposed to temperatures above the melting point of the solder. Vapor-phase soldering is used to reflow solder connections of surface mounted components on boards with fine pitch lead spacings. Because of the inherent temperature precision of this method, its close repeatability and the fact that the oven atmosphere is oxygen free, vapor-phase soldering is particularly useful with circuit boards having components of high value, or with particular susceptibility to degradation at high temperatures. The process can be made automatic with suitable conveyors and skilled operators are not usually required. A disadvantage of the process, however, is that many of the vapors used are based on chlorinated fluorocarbons and their use is contrary to the provisions of the Montreal protocol. This disadvantage has greatly reduced the use of the process. It is illustrated in Fig. 13C6b.

C6c. ***laser soldering*** - is a reflow method that uses laser energy to heat each solder joint, one at a time, melting the solder and making the desired electrical connection. The laser is a beam of coherent light, closely focused and directed at each joint for a precise amount of time. (See 3O and 7C5 for applications of more powerful laser energy.) A computer controls the movement of the workpiece or laser beam from joint to joint and controls the dwell-time of the beam at each joint. Several different laser systems can be used. Nd:YAG lasers have the advantage of higher thermal efficiencies because less energy is reflected. CO_2 lasers operate at a higher wavelength and have a less concentrated spot of light.

The advantage of laser soldering is that the heat is highly concentrated; each individual circuit board connection can be heated without affecting the connected components (except that ceramic components may require more care). The rapid heating and cooling of the joint metal reduces undesirable intermetallics, and provides desirable ductility and fatigue resistance. However, reflectance of the joint material reduces the heating effectiveness of the laser beam, although soldering flux can be used to reduce or prevent this effect. Additionally, laser equipment is costly and output rates with the one-by-one heating may be slower than other reflow methods, although two or three joints per second has been reported to be feasible.[6] Spattering may occur when solder paste is heated rapidly by laser. The laser beam must be enclosed for safety reasons.

There are two basic laser soldering systems: *blind laser soldering* and *intelligent laser soldering.*

Blind laser soldering is open-loop laser soldering; there is no feed back. (See section 3U and Fig. 3U-1.)

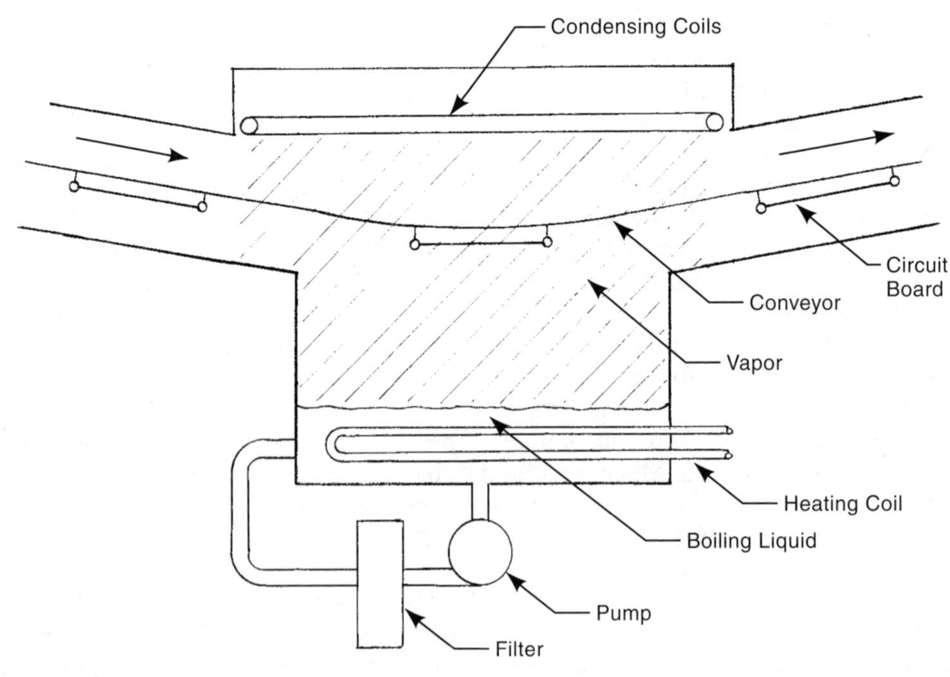

Fig. 13C6b Schematic view of a continuous processing setup for vapor-phase reflow soldering.

The laser heating time, the diameter and power of the beam, and other aspects of the machine's settings, are programmed beforehand for each joint. Differences in joint mass, contamination, or reflectance from board to board, unless programmed beforehand, do not change the heating cycle. This approach is useful when conditions are very predictable and constant, as is the case in many mass-production situations, but is not so suitable for most printed circuit board reflow soldering.

Intelligent laser soldering is a closed-loop process. The equipment incorporates an infrared detector, mounted to be concentric with the laser beam. The detector senses when the joint metal changes from solid to liquid and the computer control then reduces the laser power for a few milliseconds of dwell and then shuts off the power, allowing the solder to cool and solidify. The combination of detector and control obviates the need for joint inspection that would be necessary with blind laser soldering. Reflow soldering of fine pitch printed circuit boards or those with tape automated bonding of components is a major application.

C7. ***hot gas soldering*** - is primarily a manual rework procedure with hand-held heat guns. Air is the most common gas used, but nitrogen or nitrogen-hydrogen mixtures can be employed when it is important to limit oxidation. Heating of the gas is usually by electrical resistance. The method is used to repair defective solder joints or replace defective components on printed circuit boards. The process can be used in other applications such as the soldering of small electronic devices, where only a small area needs to be heated. In printed circuit board operations, care must be taken not to overheat adjacent electronic components. Appropriate nozzles can limit the area heated and baffles can be used to protect critical components nearby. Gas flow rate and temperature are also controlled to avoid overheating in the vicinity of the work.

C8. ***soldering iron soldering*** - is also primarily a rework procedure. Production soldering uses more-automatic methods. Soldering irons are very common, however, for repairs, touch-up, and limited quantity or prototype work. Most irons are heated by electrical resistance. The tips or "irons" (usually copper with an iron or nickel coating) are large enough to serve as heat reservoirs and the current is always "on". Soldering irons transfer heat to the joint by conduction when the iron is brought into contact with the joint surface. Soldering guns are soldering irons with small tips that are part of a secondary transformer coil. The tips heat very rapidly as the trigger is pulled and do not heat otherwise. Another advantage of these guns is that the small tips that can be inserted easily into narrow spaces.

C9. ***using lead-free solders*** - As public and governmental awareness has grown of the potential safety hazards of lead-containing materials, there has been a movement toward the elimination of lead from soldering alloys. Prior to this movement, the most common solders for electronic products contained 37 or 40% lead. These lead solders are relatively inexpensive, reliable, and easily recycled from discarded circuit boards. However, the industry is now switching to solders with typical compositions containing chiefly tin, alloyed with 3 to 4% silver and about 1/2% copper. One commonly-used alloy is SAC305 with 96.5% tin, 3% silver and 0.5% copper, popularized by Japanese companies. These solders require peak temperatures for wave or reflow soldering of 455 to 500 °F (235 to 260°C) compared with 406 to 455°F (208 to 235°C) for tin-lead solders. The methods employed in making and assembling circuit boards with these lead-free solders are basically the same as with lead-bearing solders but require tighter process controls. The lead-free solders are more costly, primarily because of their silver content. Problems can arise with these alloys, especially in applications with severe thermal cycling. Problem areas are surface finish, solder joint integrity, thermal damage to boards, components, and connectors, and in testing, cleaning and rework.[1] Solutions to the problems involve changes to less temperature-sensitive materials and components, minor changes in tooling or methods, and more careful monitoring of process conditions.

D. Cleaning After Soldering

Cleaning printed circuit boards is carried out to remove leftover flux, dust, and various other minor soils. The purpose is to prevent electrical problems from current leakage, to prevent later corrosion, to promote adhesion of later coatings and to facilitate inspection and testing. The operation is necessary except when applications allow the use of no-clean fluxes. Contamination of the circuit board with flux residue and other soils can adversely affect product reliability. As with cleaning before soldering, two basic methods currently predominate: solvent cleaning and water washing.

D1. ***solvent cleaning*** - with vapor degreasers has been the most common solvent cleaning method but is now being limited by health concerns and regulations attendant on the use of non-flammable solvents. Both batch and continuous systems are used, depending chiefly on the quantity of boards to be processed. Cold dipping and brushing are also used, but to a lesser degree. Vapor degreasing is described in Chapter 8, section A2a3. The process has one powerful advantage in that the use of vapor, from boiling the solvent, ensures that the solvent making contact with the solder board is always clean. Dirt remains in the sump of the degreaser.

Thorough batch vapor degreasing of printed circuit boards has the following steps:[6] 1) The board is placed in the vapor zone of the degreaser, above the sump where the solvent boils. The board must be cooler than the vapor, which then condenses on it, flushing dirt from the board. The dirty solvent drips back to the sump. The board should be as cool as possible, because a greater temperature differential with the vapor increases the amount of condensation and flushing that will take place. The temperature of the board gradually rises to be the same as that of the vapor and condensation ceases. 2) The board is then immersed in an auxiliary tank containing cool solvent produced when vapors make contact with the cooling coils of the degreaser. This liquid may have a slight amount of dirt from the board condensate, but not enough to be harmful, considering the subsequent operations. The immersion further flushes the board and cools it. Ultrasonic agitation may be applied to the solvent in this auxiliary tank to aid in loosening soils on the circuit board. 3) The board is again suspended in the vapor until condensation stops. 4) The board is sprayed with clean, cool, distilled solvent from another auxiliary tank. This provides further flushing of the board and cools it again. 5) The board is once again suspended in the vapor until the condensation and dripping ceases. The board is then hot, clean and dry.

Continuous or in-line vapor degreasing has the same steps as those outlined above. However, the spraying step (step 4) usually can be made more vigorous with automatic in-line equipment. There is also less vapor escape with the in-line approach which can be quite important in view of safety and environmental concerns about the escape of vapors of chlorinated and fluorinated solvents.

D2. ***water washing*** - There are a number of water-based cleaning methods that may be used individually or in sequence. Which one is used depends on the nature of the soils to be removed, the application to which the circuit board is to be put and the tolerance of that application for small amounts of contaminants, the board design, and the environment it will face. All the water-washing methods have the advantage of avoiding the environmental problems and costs involved in the use of chlorinated solvents. Waste water is economical to process and dispose of. *Water without additives* is sometimes used as a cleaning agent when the only contaminations to be removed are soluble in water. Organic fluxes and water-soluble oils are two examples. Sometimes, boards are repeatedly flushed with the same water, with the last flush of any particular board being made with the freshest water, and the first flush being made with used water. Thus, repeated flushes are made with successively cleaner, fresher water. This system leaves the board cleanest at the end of the sequence and is referred to as *countercurrent rinsing*. Fig. 13D2 illustrates countercurrent rinsing schematically. The water may be heated, since this aids in the dissolution of materials to be removed. Sometimes *water with neutralizing agents* is used. These agents react with flux acid and other acidic dirts, neutralizing them and making them easier to remove. Neutralizing agents also aid in the removal of metallic salts. *Water with surfactants* (detergents) is effective in removing non-polar contaminants, including oils, waxes, rosins and grease. *Water with saponifiers* (alkaline materials that react with rosin and oils to produce a soap-like material that is washable) are also effective in removing rosin and non-polar contaminants. Sometime a foam suppressor is added to the solution. Water rinses, often those that are multi-stage and countercurrent, follow cleaning steps that use additives with the water.

The water and water-augmented liquid used in these approaches is sprayed, flushed, and agitated, usually at a temperature in the range 120 to 180°F (49 to 82°C)[4]. Pressure spraying can loosen soils that

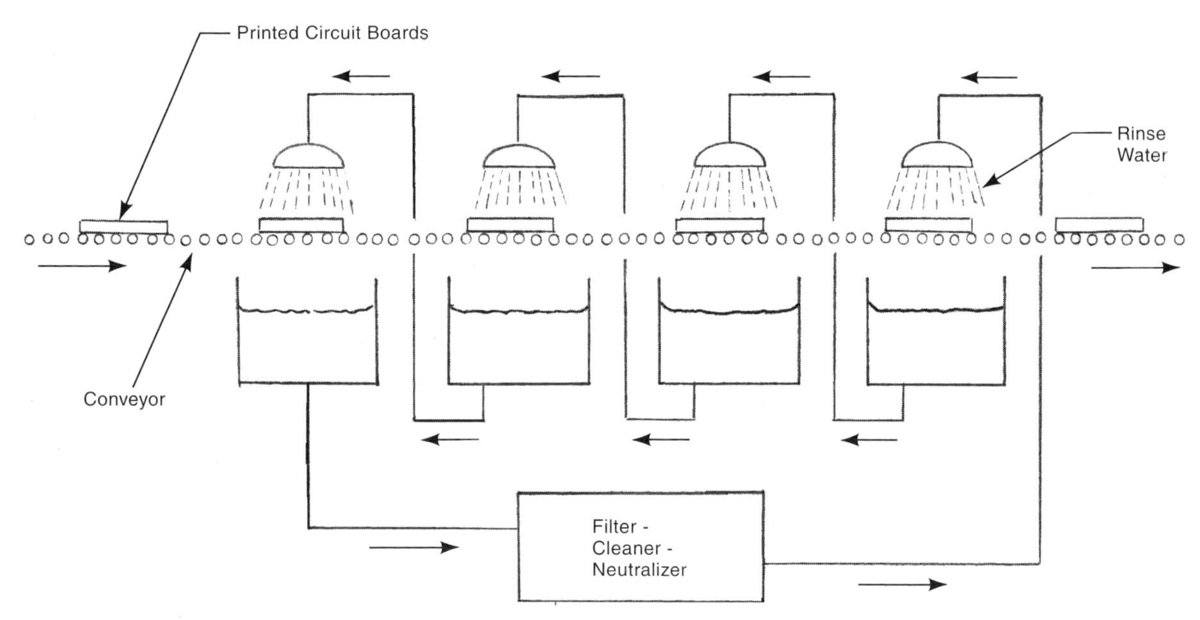

Fig. 13D2 Countercurrent rinsing shown schematically. Circuit boards on the conveyor are subjected to four rinses. The first rinse, with the board at its dirtiest, is with water that has been through three rinse cycles. The second rinse is with water that has been through two rinse cycles. The last rinse, with the board at its cleanest, is with the cleanest rinse water.

are in protected spaces. Spraying at an angle of 45 to 60 degrees has been found to be more effective than vertical spraying. Ultrasonic cavitation (See 8A2b.) can be utilized when the board is immersed, to help loosen adhering soils (though ultrasonics are used less with fine pitch assemblies where there is concern that the cavitation could damage fine-wire leads). Water washing can be a batch or continuous process, depending on the production volume. Batch washers resemble kitchen dishwashers in their arrangement and operation.[2] For higher-level production, these operations can be performed on a conveyorized, continuous basis. Air knives are often used to remove water from the boards after the washing cycle and sometimes also as an in-process step to more thoroughly remove soils from the board when the wash water contains soils. Radiant heat in a final station may be used to enhance the drying of the rinse water. A simple water-washing operation sequence is shown schematically in Fig. 13D2-1.

D3. ***semi-aqueous cleaning*** - uses solvents that do not have the disadvantage of a significant ozone-depletion property, or other serious environmental drawbacks. However, they are flammable. The most common solvents used are terpene (commonly extracted from orange peel) or an alcohol. Terpene has a flash point of 160°F (70°C). Explosion-proof equipment with fire prevention properties is used. The semi-aqueous cleaning process usually has two stages. The first stage involves washing the workpiece with the organic solvent (terpene or alcohol), to remove soils that are solvent-soluble, including rosin flux and non-polar materials. An inert atmosphere may be used during this step for fire prevention. The second stage is a water wash that removes traces of the solvent. A surfactant is added to the water. This stage also removes any water-soluble soils that were not removed by the organic solvent. A nitrogen knife, like an air knife, may be used after the solvent wash, and an air knife after the water wash. If a nitrogen atmosphere is not used in the solvent wash stage, the flushing with the solvent is done by immersion or another method that avoids creating a solvent mist that, in air, could be explosive. Terpene drained from the solvent-wash stage and that recovered from the wash water is treated and reused. The solvent is separated from the water by differences in specific gravity. The cleaning system may also have a final water rinsing stage followed by a drying stage. Fig. 13D3 schematically illustrates a semiaqueous cleaning system for circuit boards.

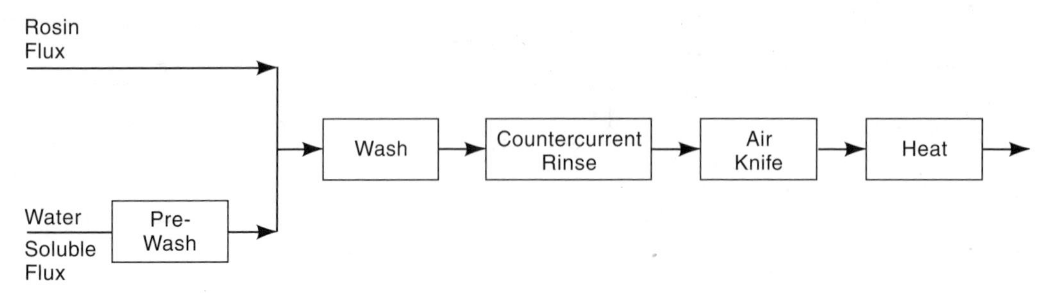

Fig. 13D2-1 Typical water washing operation sequences for circuit boards after soldering: Boards with water soluble fluxes are given a pre-wash which may be with a neutralized water or the least fresh water from the first countercurrent rinse station. The wash water may contain surfactants and/or saponifiers. Rinsing is often done in several stages with a countercurrent sequence. The air knife removes rinse water retained on the boards and the last station provides heating to dry the boards. Any or all of the wash or rinse stations may use heated water.

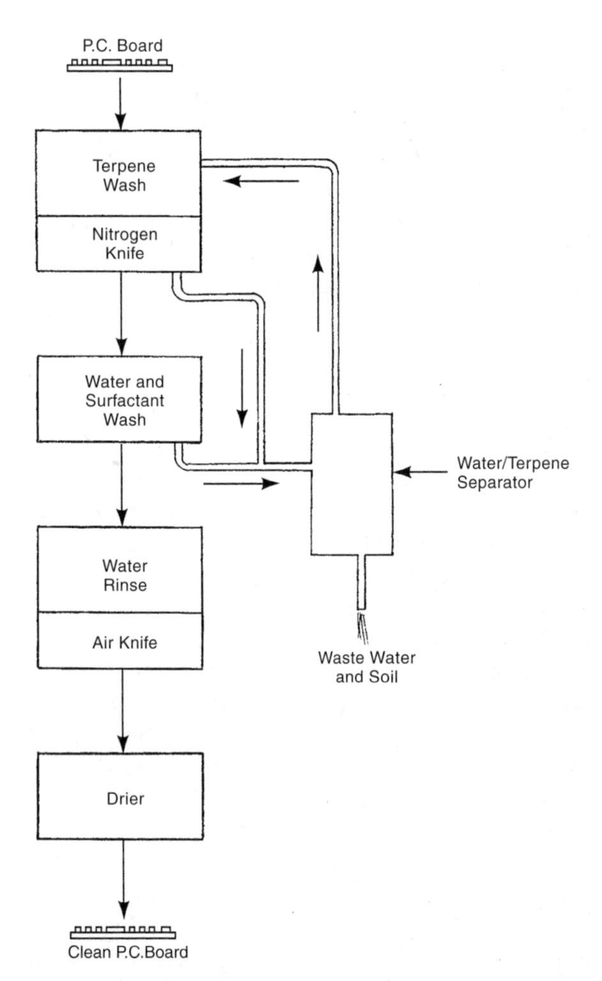

Fig. 13D3 Semi-aqueous cleaning of flux from printed circuit boards using terpene solvent.

E. Making Solder Paste

Solder paste is a homogeneous, stable mixture of flux, pre-alloyed solder powder and other ingredients (thickeners, tackifiers, plasticizers, thinners) into a single material. Solder paste is used extensively in the manufacture of printed circuit boards that include surface- mounted devices. Metal content for electronics applications is typically in the vicinity of 85 to 90 percent metal content, by weight. Solder paste is tacky enough to hold electronic devices in place until the solder is melted ("reflowed"). The paste can be deposited before devices are assembled to the board, to control the position of the devices and the size and shape of the solder fillet on each joint. Formulations with various properties are available. The necessary ingredients are mixed on a batch basis, because the usual order quantities and the many varieties produced make batch-type production most practical. A key to satisfactory solder paste is the use of solder powder with particles of the proper shapes, sizes and distribution of sizes. Several methods are in current use for producing solder powder with the necessary characteristics:

E1. ***making solder powder by gas atomization*** - is the major manufacturing method for solder powders used in solder paste. In the process, a molten solder alloy flows by gravity through a narrow orifice into an enclosed chamber. A nozzle directs a flow of nitrogen gas at this stream of molten solder, breaking it up into fine droplets of molten metal. As these droplets gradually settle to the bottom of the deep chamber, the solder cools and solidifies, changing the droplets to mostly spherical particles of solid solder. Fig. 13E1 illustrates the process. The size of the nozzle orifice, its angle and placement, the gas temperature and velocity, and the flow rate of the solder alloy into the chamber, all affect the size and shape of the particles or powder.[4] The chamber is filled with nitrogen to minimize the formation of surface oxides on the solder particles.

E2. ***spinning disk powder making*** - In this method, molten solder flowing through an orifice onto a spinning disk or cup, is flung outward by cylindrical force, to form fine droplets. These gradually settle to the bottom of the container in which

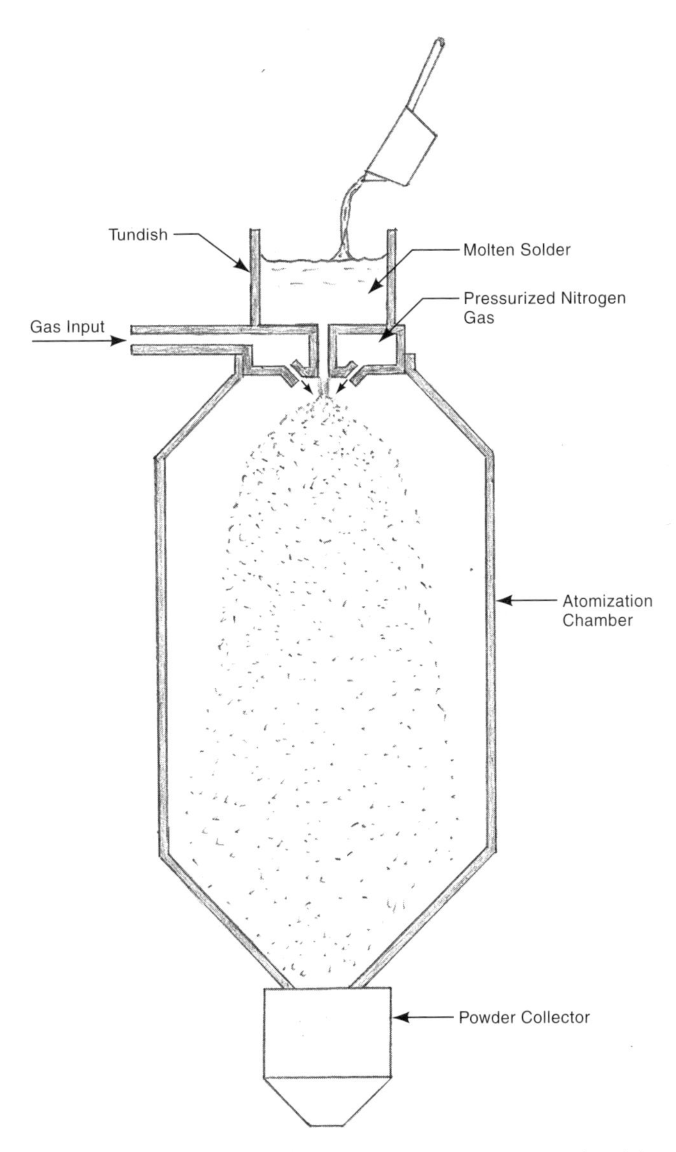

Fig. 13E1 Gas atomization of molten solder in the production of solder powder to make solder paste. For high production installations, the equipment may extend several stories in height. A blast of nitrogen gas against a thin stream of molten solder breaks it up into small droplets that form into spherical shapes and solidify as they fall to the bottom of the chamber.

the operation takes place, cooling and solidifying as they fall. They are collected as they reach the bottom of the container. However, the larger particles, having more mass per unit of surface area, tend to travel farther before they settle to the bottom of the chamber. Therefore, there is a certain amount of sorting that accompanies the process. The speed, diameter, and shape of the disk or cup, as well as the temperature and flow rate of the solder, and the temperature of the chamber, all affect the size and shape of the resulting solder particles.[4] The atmosphere in the chamber is inert, usually of nitrogen, to minimize oxides that form on the surface of the particles during the operation. The chamber is deep enough to ensure that the solder particles have fully solidified before they reach the bottom. Fig. 13E2 illustrates the spinning disk method.

E3. ***ultrasonic method of powder making*** - In this method, the molten solder is directed to flow against an ultrasonic horn, a metal device vibrating at a frequency above that of audible sound. As the horn vibrates, it throws off the solder alloy in fine droplets. These gradually settle to the bottom of the chamber in which the operation takes place, cooling as they settle, and solidifying into more or less spherical particles. The rate of flow with this method is somewhat less than with a spinning disk or with gas atomization, but multiple ultrasonic devices can be used. The powder produced tends to have a narrow range of sizes. As with the other methods described above, the atmosphere in the chamber is inert and the chamber is deep enough to ensure that particles settling to the bottom have fully solidified.

E4. ***screen classification of powder*** - With a vertical gas atomization chamber, the powder settles to the bottom of the chamber. It consists of solid particles of a wide range of sizes. Most particles are quite spherical, but when particles

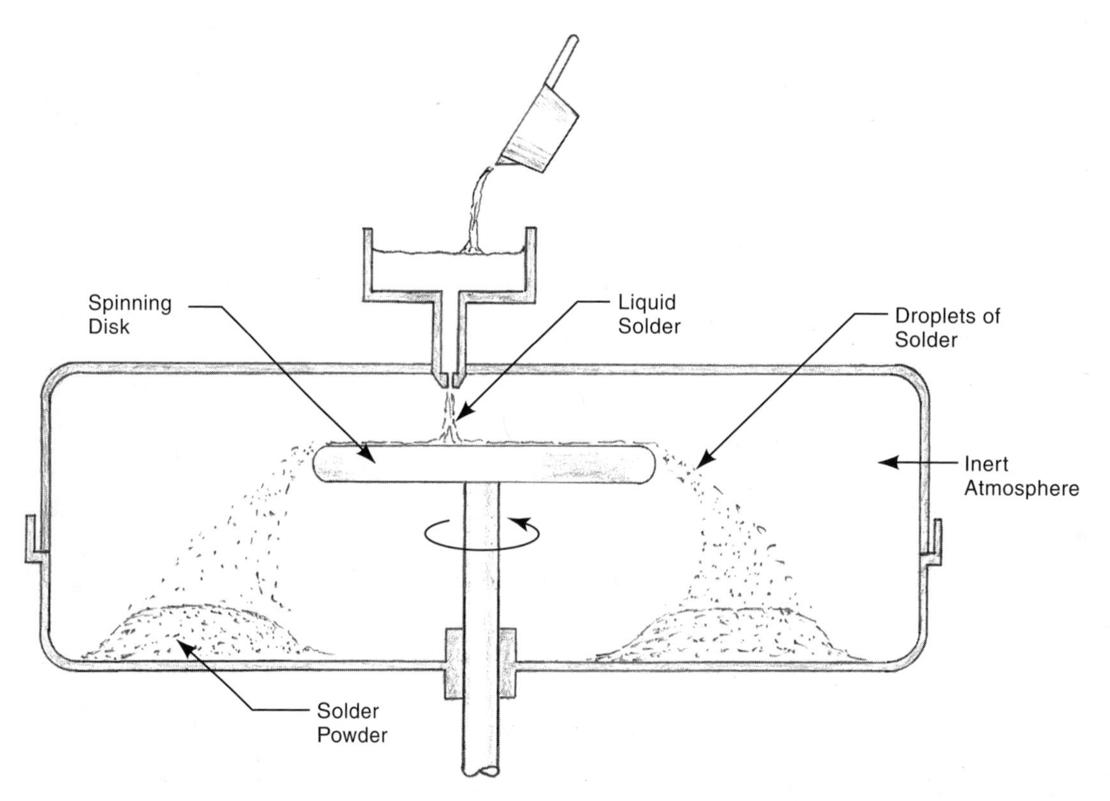

Fig. 13E2 Spinning disk atomization of solder to make powder for solder paste.

collide before solidification, irregular shapes can result. The usual practice is to have an opening at the bottom of the collection chamber through which the powder flows into a series of classification screens. Each of these screens passes particles smaller than the screen mesh openings and retains those particles that are larger than the screen mesh openings. Screens are agitated to provide repeated opportunities for each particle to pass through. Each successive screen has a finer mesh than the one above it. Material at each screen that does not pass through, migrates to one end of the screen and is drawn off into separate containers. Thus, the manufacturer is able to have a stock of powder in groups, with each group having a narrow range of particle sizes. When the solder paste is made, several size ranges of powder may be blended to provide the desired properties to the paste. Screen sizes are designated by the number of openings per inch.

E5. ***air classification of powder*** - is less precise as a classification method for particles above 35 μm but is very discriminating for particles less than 20 μm. In one method, powder of mixed particle sizes is blown into the air in a horizontal direction in a wide chamber. The larger particles have a greater ratio of mass to surface area and thereby have more resistance to the accelerating force of the air jet. Hence, they exit the blower area at a slower speed. These larger particles fall sooner to the bottom of the chamber. The smaller particles have a lesser ratio of mass to surface area, are accelerated more, and are more easily carried by the air flow. Hence they fall to the bottom of the chamber later, after they have traveled a greater distance. In-between sizes, fall in between these locations, approximately in keeping with their particle size. Therefore, there is a graduation of particle sizes in the pile of powder at the bottom of the chamber, with the larger particles closer to the blower and the finer particles farther from it. The graduation of size from large to fine, however, is far from uniform. The operation, therefore, is normally repeated several times, starting with powder mostly within the size range wanted. After each air classification cycle, some out-of-range particles end up out of the target area, and the powder in the target area more closely conforms to the size limits wanted. However, such increased handling results in surface and shape changes to the particles and impairs the performance of solder powder when it is made into paste.

E6. ***inspection of powder*** - The simplest method for checking solder powders used in solder paste is to spread a thin layer of the powder on a glass slide and then examine this layer with a microscope. With this approach, the size distribution of the particles and their shape can be monitored in a qualitative way.

To gain a quantitative breakdown of the portion of the particles in various size ranges, the standard approach is to run a sample from the lot into a stack of small screens with progressively smaller screen openings from top to bottom. With the largest screen opening on the top, the particles larger than the opening are trapped on the screen and smaller particles fall through it to the next screen. The same thing happens at the next screen, which has slightly smaller screen openings. The largest of the particles are trapped; the balance pass through the screen. This process can be repeated with as many different size screens as desired, all placed in one vertical stack. The amount of powder left on each screen indicates the portion of the lot that is in that particular size range.

E7. ***mixing solder paste*** - is a batch operation. Solder powder, flux, plasticizers, tackifiers, thickeners, or thinners, are blended in mixers designed for the high density (because of tin and lead content) and high viscosity of the paste. Some of the mixers described in section 11G5 for stiff, viscous materials can be used for solder paste. Fig. 13E7 illustrates a machine particularly suitable for mixing solder paste. Machines which pass the paste between parallel rollers may also be employed as part of the mixing operational sequence.

E8. ***inspection of paste*** - A number of tests can be made on the solder paste to verify its properties: The ability of the paste to be dispensed - its *rheology* - is controlled most commonly by measuring the *viscosity* for which several viscosity-measuring instruments are available. The most prominent variety uses a rotatable spindle into which a small diameter rod with a cross piece ("T-bar") is inserted. The T-bar is lowered into a container of just-remixed solder paste that is at a specified and

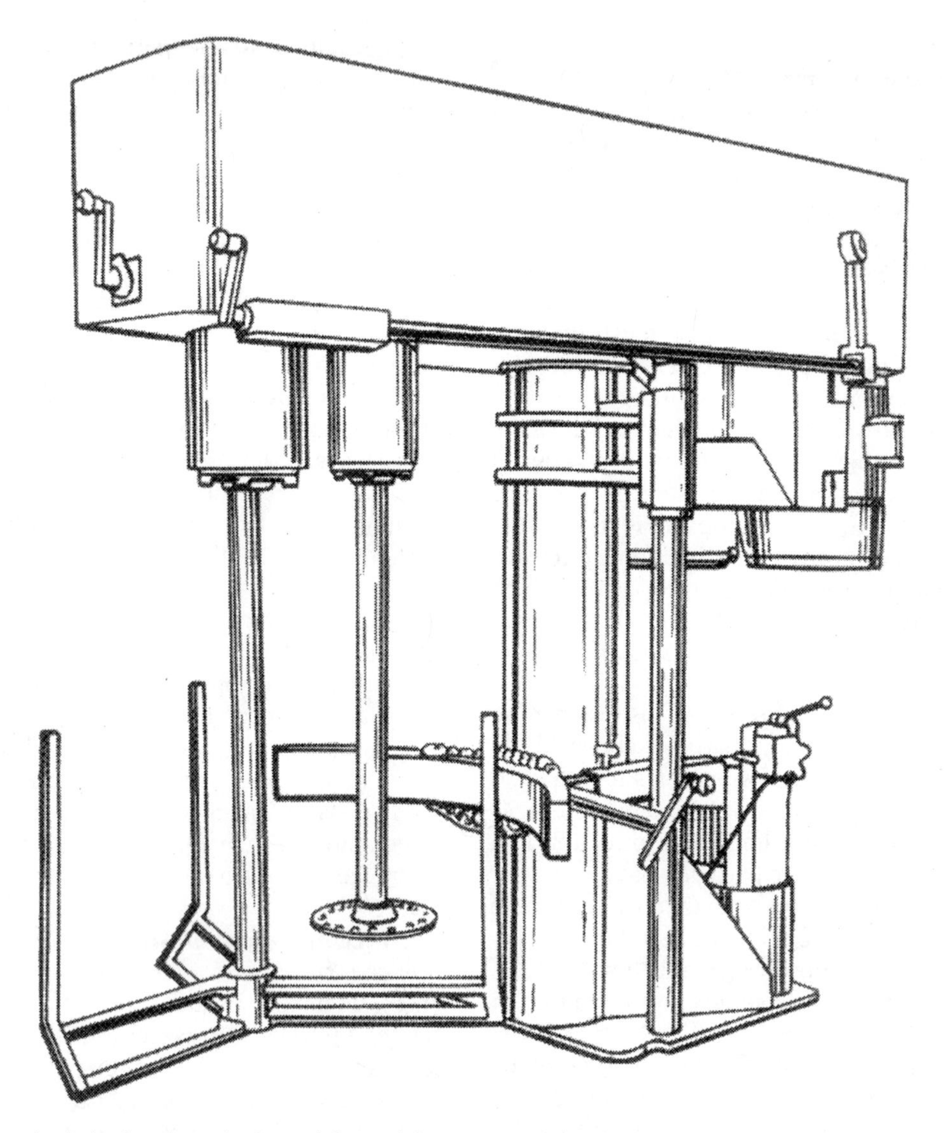

Fig. 13E7 A dual shaft mixer designed for solder paste. One shaft rotates at high speed and disperses the ingredients; the other shaft rotates at low speed and moves and blends the mixture. (*Courtesy Myers Engineering, Inc.*)

uniform temperature. The T-bar follows a helical path so that the bar is always meeting resistance from the paste as the bar rotates. The instrument measures the resistance to the rotation and translates this into a digital viscosity reading.

The *metal content* of the paste is normally checked by weighing a sample of paste, heating it to melt and coalesce the solder into one wafer, washing away the flux, and weighing the resultant metal wafer. The ratio of the two weights indicates the percentage of metal.

Flux conformance to specifications is determined by immersing the paste in a suitable solvent, filtering out the metal powder, evaporating the solvent and performing various analytical tests on the residue.

Fineness of grind of the metal particles in the paste is measured with a gauge based on those in use in the paint industry to measure paint pigments. A sample of solder paste is placed on the gauge, which is a hardened steel block having two tapered grooves in the surface. The grooves range in depth at the deep end of about 185 microns (0.007 in) to zero at the other end. The paste is placed at the deep end of the grooves and a scraping blade draws it along the length of the channel so that it remains on the gauge only in the grooves. The depth of the grooves at the point where the line of paste in the grooves ends indicates the size of the finest metal particles in the paste.

Tackiness testing - verifies that the paste has the necessary tackiness to hold surface mounted devices placed on a circuit board until solder reflow takes place. A motorized commercial testing device is used. A sample of paste is placed on the surface of a glass slide which is then stored for a length of time equivalent to that involved in production conditions. The slide is then placed on the work surface of the testing device. A probe in the device descends into the paste at a controlled rate with a specified amount of force. The probe is then withdrawn and the pulling force needed to withdraw it is measured. The magnitude of this force gives a quantitative indication of the holding power of the paste for mounted devices. Another device, sometimes used, measures the shear resistance of the paste, and therefore its resistance to the movement of devices on the board before solder reflow.

Slump tests - measure the increase in area from gravitational forces of a deposit of solder paste after the solder has been applied to a surface. Standardized test patterns of paste are applied to a surface, by screening or stenciling, and their dimensions are then observed and, if desired, measured for a change in spread. Excessive slump causes problems in holding the mounted components and can also predict solder ball and other problems in reflow soldering.

Performance tests[3] - Several tests can be used to verify that the solder paste performs satisfactorily when used on the components to be soldered: 1) *compatibility tests* verify that the solder paste is suitable for the joint surface materials under the expected production conditions. A small amount of paste (50% or less of the expected production amount) is placed on the joint surface, which is then heated to reflow the solder. The solidified spot of solder is then inspected. If good wetting is evident, the materials are compatible. 2) *solder ball test.* A small spot of solder paste is screened onto a ceramic test surface. The ceramic is heated on a hot plate sufficiently to reflow the solder. The spot of solder is then examined. If the solder forms one large spot, it is ideal. If there are more than three separate spots (solder balls), the paste is not acceptable. The flux area surrounding the solder spots should also be examined. Black particles in the flux indicate unreduced fine solder powder and the paste is not suitable for critical applications.

F. Ball Grid Arrays

Ball grid arrays utilize spheres of solder to provide the necessary material to electrically connect large integrated circuits (ICs) to printed circuit boards. These ICs are enclosed in packages that have solder pads on the underside. Connections cover the entire underside of the package, allowing space for many connections and insuring short paths for connections within the package. The spheres range from about 0.008 to 0.035 in (0.20 to 0.89 mm) in diameter and are held to close tolerances for diameter, sphericity, and surface smoothness. The spheres are placed in position by accurately applying a tacky flux to each contact pad, then causing the solder spheres to adhere to the pads (one to each pad), by vibrating them in bulk against the package. (The package is inverted for this operation.) A machine-vision device verifies from the reflectivity of the sphere's surface that there is a sphere in each position where it is needed. The tacky solder flux holds the spheres in position until the IC is heated to reflow the spheres enough to wet the pads and leave a bump of solder at each pad. The package then can be positioned on a printed circuit board that has an array of connection pads corresponding to those on the IC package. When the board is reflowed, the solder bumps complete the connections from the IC to the board. Fig. 13B2-1 shows a chip package connected to the circuit board by this method.

G. Fluxes for Electronics

Fluxes perform the same functions with respect to the joint surfaces of printed circuit boards as they do when used for mechanical and other soldering and as noted in 7A3. Rosin fluxes have long been common for electrical uses but now are employed less often as the major flux element on printed circuit boards. The availability of other suitable no-clean fluxes, testing difficulties when rosin-based, no-clean fluxes and automatic test probe equipment are used, and workers' allergic sensitivity to rosin, have all led to a reduction of rosin usage. No-clean fluxes have become dominant because of environmental factors and disposal costs related to both solvent and water-based cleaning effluents, the high density of SMT-type boards that are more difficult to clean, and the availability of effective no-clean fluxes. These do not have to be removed after the soldering operation. They typically are free of ionic materials though they contain organic acid activators (which are solids), a solvent (either water or isopropyl alcohol), viscosity modifiers (eg., methyl cellulose), surfactants, and other additives. No-clean fluxes are mixed with the same standard apparatus as are used for conventional fluxes. No cleaning or even rinsing of circuit boards is required when no-clean fluxes are properly specified.

H. Tinning

Leads, contacts, traces, pads, and other solder joint areas are coated - tinned - with a solder alloy prior to the soldering operation in order to facilitate final soldering, lessen the need for strong fluxes that may attack the circuit board, and provide longer storage life. Tinned surfaces have superior storage life and solderability than electroplated coatings. Tinning involves the following steps which may be manual or automatic:[3] 1) surfaces to be tinned are degreased, 2) If necessary, surfaces are microetched with acid, 3) flux is applied, 4) the component is preheated, 5) the surface is dipped or otherwise brought into contact with molten solder, 6) The surface is held in contact with the molten solder until full wetting takes place.7) The workpiece is withdrawn from the solder, 8) cooling takes place, 9) flux residue is cleaned from the tinned surface and adjacent surfaces as necessary. 10) The tinned surfaces are inspected. Tinning by dipping is an economical method for precoating surfaces with solder, but the amount of solder in the coating is subject to variations.

I. Quality Control and Inspection Operations

I1. ***visual inspection of joints*** - is a manual inspection for the following characteristics: 1) degree of wetting of the surfaces to be joined. 2) contours of the joint fillet (indicates the volume of solder in the joint), 3) evidence, if any, of thermal damage to the surrounding area or components, 4) cleanliness of the areas around the solder joint and, 5) consideration of design requirements and special conditions affecting the joint, if any.[2] Visual inspection is an effective method for detecting faults with solder joints. Low-power magnification is sometimes used to aid the operation, especially with fine-pitch assemblies.*[1]

I2. ***incoming inspection*** (of materials and boards before soldering connections) - The following characteristics are checked in incoming components and bare circuit boards: 1) solderability, tested by a performance test or conformance to materials and finish specifications, 2) finish of component and board terminal surfaces to resist tarnish, 3) confirmation that components will be able to withstand the heat of soldering, 4) resistance of the boards and components to the cleaning materials to be used, 5) adequacy of packaging, 6) quality of plating of conductive and terminal surfaces, 7) whether the condition of board coatings is proper, 8) correctness of dimensions of boards, traces, holes, leads, pads, etc. 9) cleanliness of terminal surfaces.

Soldering fluxes are checked for specific gravity or density, color and clarity, ionic content and for the specified chemical analysis.

I3. ***solderability testing*** - is performed by dip soldering the joint area, after proper fluxing, of the component to be tested. This step is followed by a careful visual inspection of the joint area to verify that it has been properly wetted. Sometimes, it is desirable to perform such a test with a weaker flux than will be used in production so that, if the condition of the component is marginal, the problem will be detected. Instruments are available that facilitate the testing operation. One such instrument, called a wetting balance, measures the

* Many inspection details are covered in the IPC 610-C standard. (IPC, Northbrook, IL, www.IPC.org.)

flotation of a sample joint immersed in molten solder. As wetting of the joint proceeds, its flotation decreases. Measurement of this effect over a time span gives worthwhile data on the solderability of the surface tested.

J. Repair and Touch-up

Soldering operations involving circuit boards do not have a 100% yield, so there is always some need to repair or touch-up solder joints after the main soldering operation. The task may be as simple as reheating a poor solder joint, or may involve the removal and replacement of a defective device on the circuit board. Repair and touch-up is normally performed manually, with electric soldering irons or soldering guns (See C8.), hot gas guns (C7), cored wire solder or solder paste. Cleaning of the joint area may be required before the operation and is also necessary after the operation. Flux is usually applied with a small brush. Care must be taken so that neither the components nor the board are damaged from excessive heating. Heat sinks may be put in place temporarily, adjacent to heat-sensitive components. Sometimes, excess solder must be removed and this can be done with hollow-tipped irons using suction or braided and fluxed copper wicks that draw the excess solder by capillary action.

The preferred method for cleaning the repaired area is by application of an uncontaminated cleaning fluid, followed by manual brushing of the soldered joint, and suction of the contaminated liquid into another container. Typically, this three step sequence will be repeated several times until no flux residue remains in the area around the joint. Apparatus for performing this cleaning sequence is commercially available. However, purely manual cleaning methods with solvents, rags, etc. may be used.

K. Integrated Circuits (ICs) (Microcircuits or Chips)

Integrated circuits are electronic circuits in micro- miniature size, existing on a single piece of silicon, germanium, gallium arsenide, or inert material (glass or ceramic) containing up to tens of millions of transistors and other devices (diodes, resistors, capacitors). These devices are formed in the semiconductor substrate, or as part of film layers added to it. Twenty or more layers of circuitry may be involved and the devices are all permanently interconnected. Circuit elements and devices on each chip are extremely small and wiring paths are as narrow as 5 millionths of an inch (0.13 microns), or less. (Process and design improvements are continually being made. The Semiconductor Industry Association has projected circuit dimensions of 0.05 microns, 50 nanometers, or 2 millionths of an inch by 2012.)[12] ICs are produced in mass-production quantities with extensive, highly sophisticated, extremely precise and extremely clean manufacturing processes, often involving 600 or more steps before each chip is completed. Chips vary in size but a common surface area is 0.24 sq. in (1.5 sq cm).

Integrated circuits are the brains of computers and other electronic devices including televisions, radios, stereo equipment, cellular and regular phones, instruments, control devices, military navigation equipment and firearms, aircraft, spacecraft, missiles, medical devices, digital watches, automotive diagnostic devices, traffic control, environmental monitoring, industrial process controls, video games and appliance controls.

The integrated circuit manufacturing process must deal with circuit features that have the very smallest dimensions and are subject to very subtle electrical and chemical effects. In order to prevent manufacturing defects, the entire fabrication sequence must be free from contamination by extraneous particles and chemicals. Clean room conditions, with the highest order of freedom from extraneous particles, are maintained by use of fine filters for ambient air and by limiting garments and room equipment to lint-free types. Clothing must prevent the release of contaminants from workers. Contamination from solvents and other chemicals, tools, equipment and production supplies must also be carefully controlled throughout the process.

IC manufacture includes the following major stages: material preparation, single crystal making, wafer preparation, wafer fabrication (including the incorporation of circuitry), and packaging.

K1. ***material preparation - making ultra-pure silicon***[5] - Quartzite (chiefly silicon dioxide, SiO_2), coke, coal and wood chips (to supply carbon), are placed in the crucible of a submerged electric-arc furnace (See 1A2.). Silicon carbide is formed and it reacts with the silicon dioxide to form liquid metallic silicon that settles to the bottom of the crucible and carbon dioxide gas that is allowed to escape. The liquid silicon, which is about 98% pure,[5] is cooled and solidifies. It is pulverized and brought into a fluidized bed with hydrogen chloride. The two materials react, forming trichlorosilane and hydrogen. Chlorides are also formed from the impurities in the silicon. The trichlorosilane is entered into a heated chamber with a controlled hydrogen atmosphere. This reduces the trichlorosilane to 99.99999% pure silicon[1] in the form of polycrystalline rods. The reaction is:

$2SiHCl_3$ (gas) + $2H_2$ (gas) ——–> 2Si (solid) + 6HCl (gas).

Fig. 13K1 shows the sequence of operations used to manufacture pure silicon in wafer form, and to prepare the wafers for use in integrated circuit manufacture.

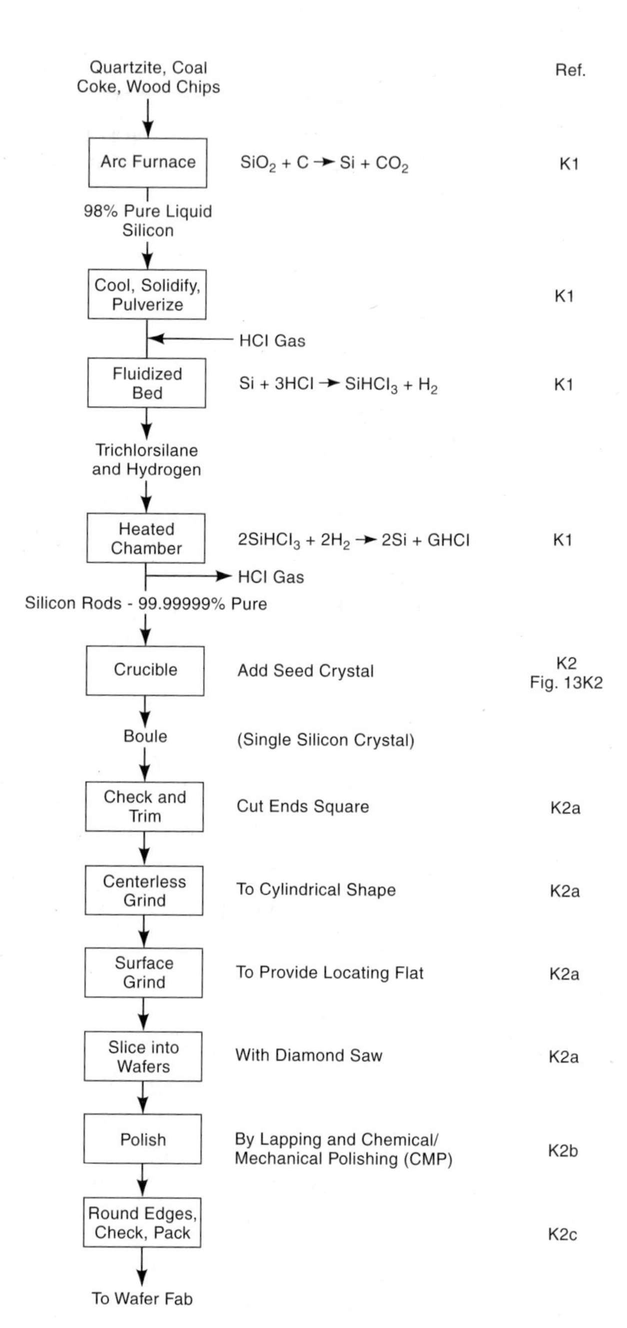

Fig. 13K1 The operation sequence for making single-crystal silicon wafers from raw materials, and preparing the wafers for the wafer fab operations.

K2. ***making a single crystal of silicon*** - Most single crystal silicon is grown by the Czochralski (CZ) method. The rods of electronic grade silicon are broken up, loaded into a large crucible and melted at a temperature of 2580°F (1415°C). Radio frequency induction or radiant heating is used. A seed crystal, attached to a vertical shaft, is lowered so that it just contacts the melt. This starts the growth of a crystal in the melt. The crystal continues to grow where it contacts the melt following the same orientation as the field. The vertical shaft and crystal rotate slowly in one direction, and the crucible in the opposite direction, to prevent any inhomogeneities in the melt from being incorporated in the crystal. When it has developed to the desired diameter, the vertical shaft is gradually drawn upward as it rotates, causing the developing crystal to assume a near-cylindrical shape. The pulling movement of the shaft is computer controlled, based primarily on the weight of the crystal as determined by a sensor. Eventually, a single large crystal, called a *boule*, is developed. It may be as large as 12 in (300 mm) in diameter and several feet in length. The operation may be performed in a vacuum or inert gas to minimize oxygen absorption by the boule. The crystal is doped during this phase. (See K3c below.) by adding already doped pieces of polysilicon to the melt. The doping lowers the resistivity of the crystal and, depending on the dopant used, leads it to become either a P-type (electron poor) or N-type (electron rich) semiconductor. Rotation of the crystal as it grows helps ensure uniform distribution of the dopant.

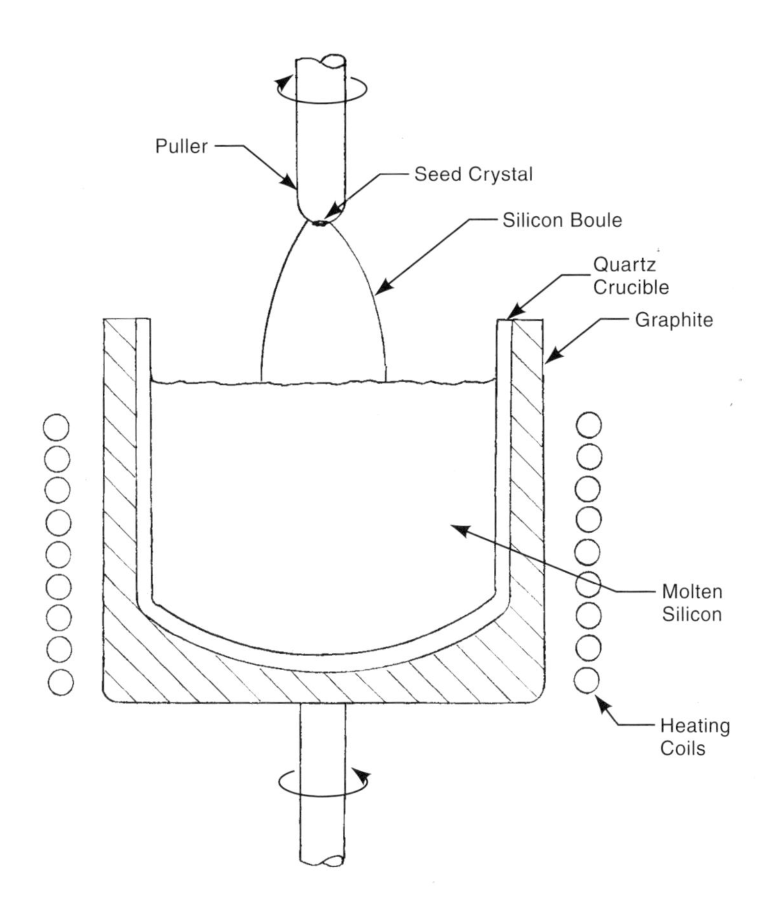

Fig. 13K2 In the Czochralski method for growing large, single-crystals of silicon, the seed crystal is immersed in molten silicon, and then gradually withdrawn as the crystal forms. The result is a large crystal that takes an essentially cylindrical shape because of the rotation and upward movement of the puller. Temperature and visual sensors are connected to a computer system that controls the operation.

Fig. 13K2 shows the crystal forming equipment schematically. When gallium arsenide crystals are made, the process is modified somewhat to prevent evaporation of the arsenic in the melt.

K2a. ***slicing into wafers*** - The boule, after it has reached the desired diameter and length, is removed from the crucible. It is checked for physical defects by etching, for proper doping by making resistivity measurements, and for crystal orientation. Defective parts of the boule, if any, are removed. The ends, of lesser diameter, are sawed off. The boule is then ground on a centerless grinder (3C1b) to produce a uniformly-cylindrical shape of the desired diameter. It is surface ground to produce one or more flat surfaces along its length. The major flat is located along one of the major crystal planes. The flat or flats facilitate location of later circuit elements. The boule is then sawed - with circular or moving-wire blades coated with diamond abrasive - into thin slices to form flat, round wafers. Circular saws are ring-shaped and do the cutting on the inside edge. This enables a thinner saw blade to be used, to reduce the kerf loss from sawing.

K2b. ***polishing the wafers*** - The surface of the wafers must be extremely flat and smooth, far flatter and smoother than the surface resulting from sawing. A two-step process is used. First, the wafers are polished with a lapping operation similar to that employed for lapped metal parts used in mechanical equipment. (See 3J2.) The lapping removes sawing irregularities. Then, a special chemical/mechanical polishing operation (CMP) takes place. The wafers are held in rotating holders, and are put in contact with a polyurethane pad rotating in the opposite direction. The polyurethane may be a solid material or a coating on felt. A slurry of glass particles and ammonium or potassium hydroxide is fed onto the pad. The hydroxide reacts with the silicon to form a thin surface layer of silicon dioxide which is removed by the buffing abrasive action of the pad and the glass particles. High spots are removed most aggressively and the result is an extremely flat surface. Careful control of all process variables and conditions is essential.[12]

K2c. ***other wafer preparation operations***[12] - The edges of the wafers are rounded to minimize the possibility of edge chipping or other damage to the wafer during further processing. Rounding is accomplished by a grinding operation, followed by chemical processing. Another operation that may take place consists of sandblasting the underside of the wafer. This is to *getter* the surface, to create mild damage to the crystal structure of the wafer and prevent ionic contamination, if it occurs, from damaging the circuitry on the top side of the wafer. Another operation, quite common for 300 mm (12 in) diameter wafers, is to polish the reverse side of the wafer to ensure the highest level of flatness.

Before being released for chip-making operations, the wafers are carefully examined with automatic inspection machines or special lights to uncover any surface defects or contamination. Before shipment to customers, they are carefully packed in clean rooms in non-static protective material.

K3. ***wafer fab*** - is the name given to the series of operations that create semiconductor devices on and in the wafer surface. The result of wafer fabrication is a flat round silicon disc, up to 12 in (300 mm) in diameter, with hundreds of individual integrated circuits (ICs) on its surface. Wafer fab for complex microprocessors may require more than 500 operations. In the current state of the art, these operations may all take place automatically with special machines and often with the aid of robots, so that human handling is not required. Four classes of operations are involved in the wafer fab process. They are: 1) *layering* - processes that add thin layers of material to the wafer surface. The layers can be insulators, conductors, or semiconductors, and are either grown by chemical reaction with the existing surface material, or are deposited as new materials. The layers are made by thermal oxidation or nitridation, chemical vapor deposition, vacuum evaporation and deposition, or sputtering. 2) *patterning* - a series of processes that removes portions of surface material so that a layer on the wafer incorporates microcircuit elements. Patterning operations, also known as lithography, photolithography, or microlithography, include mask making, resist application, exposure and developing of the resist, etching and resist removal. 3) *doping* - operations that change the electrical conductivity, usually in localized areas of the wafer surface. Thermal diffusion and ion implantation are two doping methods. 4) *heat treatments* - heating operations that make physical changes in the wafer material. There are several different heat treatment operations, one of which is referred to as annealing.

These operations are repeated many times for up to about 20 layers before the wafer is converted to a quantity of chips (integrated circuits not yet packaged). Individual steps of wafer fab are described at more length below. Fig. 13K3

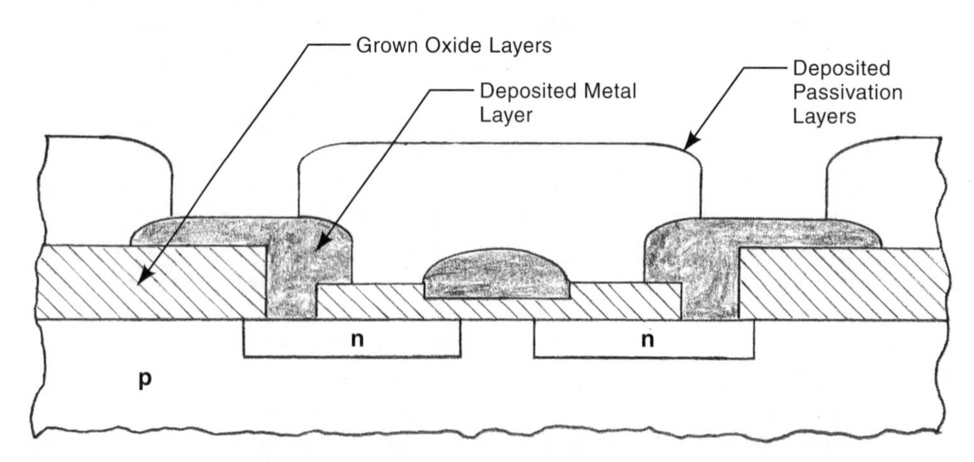

Fig. 13K3 Cross section of a MOS transistor, showing the patterns of both the grown and deposited layers.

illustrates the patterned layer construction of a typical MOS (Metal Oxide Semiconductor) transistor as it exists in an IC on the wafer. Fig. 13K3-1 outlines the operation sequence required to make such a transistor as part of the IC.

K3a. ***layering*** - There are two basically different ways of providing layers in integrated circuits. Grown layers are made by chemical reaction with the existing surface material to create a surface layer of a different compound. Oxidation and nitridation are two processes used to grow new layers by chemical reaction. Deposited layers are thin or thick films that are added to the existing surface. Deposition methods include CVD (chemical vapor deposition), vacuum deposition

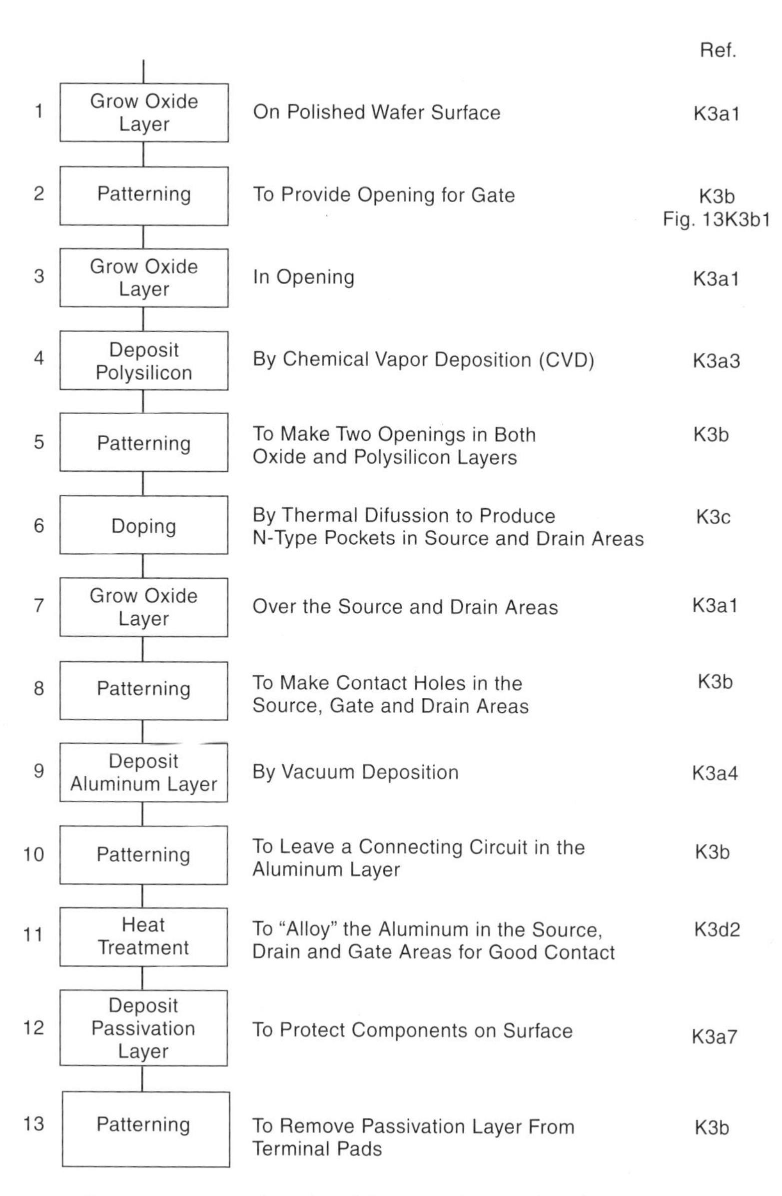

Fig. 13K3-1 The operation sequence involved in creating a simple MOS-silicon gate transistor as part of an integrated circuit. Note that the patterning operation, which is repeated five times in this sequence, requires, in itself, five or more separate operations. (*This chart based on data in Microchip Fabrication by P. Van Zant.*[12])

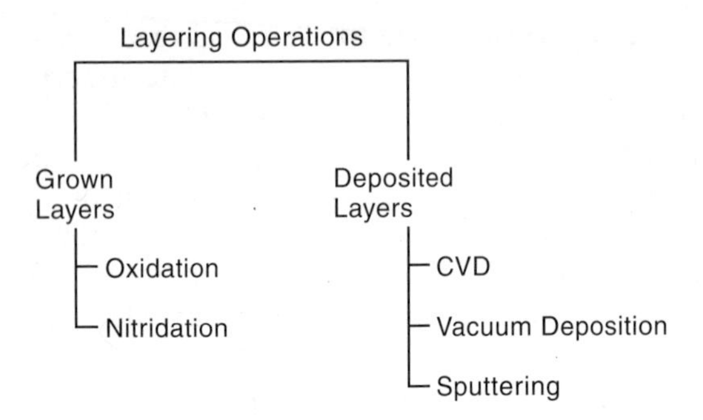

Fig. 13K3a Methods of creating layers in integrated circuits.

(vacuum evaporation and deposition) and sputtering. See Fig. 13K3a.

K3a1. ***oxidation of silicon*** - creates a grown layer of silicon dioxide. This oxide layer has many possible uses. It is a dielectric and is used to electrically separate the circuits on different layers, to provide a dielectric in circuits, and to passivate the surface of the wafer, that is, to provide a protective coating to guard semiconductor surfaces against physical damage, chemical contaminants, and dirt particles. Coating the wafers with oxide film is part of the process for producing doping masks which allow the selective changes to be made in limited areas of the silicon substrate surface. This layer is one of several kinds of film that may be produced on the wafer and is the first film. The first step is to oxidize the entire surface of the wafer, converting silicon at the surface to a silicon dioxide film. This creates a mask over the whole wafer surface. Part of the mask normally is then removed in the areas where it is not wanted by using lithography. (See K3b1 below.)

The oxidation operation is carried out by blanketing the wafer surface with a stream of pure oxygen or steam in a hot furnace. Commonly used temperatures are in the range of 1500 to 2200°F (810 to 1200°C) and a pressure of 20 atmospheres in the furnace, that is tubular with quartz walls. The operation creates a layer of silicon dioxide, (SiO_2), that can function as an effective mask for later operations. The use of steam (*wet oxidation*) produces faster results and, in practice, steam is provided by introducing pure hydrogen and oxygen to the furnace and igniting the hydrogen. This method avoids any problems of contamination that may come from the use of steam from water and minimizes the chance of liquid water getting into contact with the substrate surface, which can cause some unevenness of oxidation. The oxide surface provides protection from unwanted reactions with the silicon surface in subsequent operations. Silicon, without this protection, is very reactive and could bond to other substances and cause electrical misfunctions.

Subsequent operations to produce an oxide layer are carried out as the circuitry and semiconductor devices are fashioned on the chip. At certain points, particularly between layers of circuitry on each chip and after lithography, etching and doping and other thin film operations, a layer is required for insulation and as a mask during doping. These layers are achieved by chemical vapor deposition (CVD) of silicon dioxide or silicon nitride.

K3a2. ***nitridation*** - Some transistors utilize a thin gate oxide of 100Å or less in thickness. Silicon nitride (Si_3N_4) is preferred to silicon dioxide in these cases. Silicon nitride is grown on silicon substrates by *thermal nitridation*, which involves exposing the surface to gaseous ammonia (NH_3) at a temperature in the range of 1740 to 2190°F (950 to 1200°C).

K3a3. ***chemical vapor deposition (CVD)*** - Chemical vapor deposition is a gas-phase process discussed in Chapter 8, section F3b. Most integrated circuit films are produced by a CVD method. The material is deposited from a chemical reaction of gases, producing a vapor of molecules or atoms that form a layer on the wafer surface. There are many different variations of CVD processes. Heat is usually applied to the reaction chamber. The operation may take place at atmospheric pressure or at a reduced pressure. CVD, for IC wafers, is used for deposition of silicon dioxide, silicon nitride, epitaxial silicon and polysilicon. The CVD of wafers is preceded by a cleaning operation, sometimes including etching. The process is usually carried out in a horizontal cylindrical reactor. Heating is generally by induction but radiant heating is also used. A typical temperature is 1830°F (1000°C). Silicon tetrachloride ($SiCl_4$) or silicon hydride (silane) (SiH_4) are the gases used to provide silicon atoms. The growth rate for the silicon film is normally about one micron (0.00004 in) per minute.[11]

K3a3a. ***epitaxy*** - is the process of growing a single crystal, layer by layer, on the flat surface of another single crystal. In wafer fab, epitaxy is the growing of layers of silicon on the silicon substrate of the wafer. The method is used when the silicon to be deposited has a different impurity content than the substrate silicon. The deposited material is usually the same as the substrate material except for the presence of a different amount or type of dopant. Layers that are lightly doped can be deposited on layers of the opposite conductivity that are more heavily doped. The dopant is another semiconductor material such as gallium arsenide, cadmium telluride (CdTe), germanium, or lead telluride (PbTe). Insulators that may be deposited include sodium chloride (NaCl - rock salt) and magnesium oxide (MgO). The substrate and deposited material involved must have nearly the same atomic spacing in their crystal structure for the process to work well. Epitaxial coatings permit closer spacing of components in the IC. Several methods are available for deposition of epitaxial layers: CVD, molecular beam epitaxy (MBE), and liquid-phase epitaxy (LPE). Epitaxy is used in the manufacture of bipolar and CMOS transistors and some resistors. The epitaxial coating is deposited on either the entire wafer surface or only in selected areas in openings in the silicon dioxide or silicon nitride films.

Epitaxial coatings are most commonly deposited by chemical vapor deposition (CVD). With carefully-developed and closely-controlled process parameters, the deposited material follows the same crystal arrangements as the substrate. Silicon tetrachloride, with a small amount of HCl gas for etching the silicon substrate, is the gas most frequently used to supply the silicon. Silane and dichlorosilane (SiH_2Cl_2) are also used. Several stages of thorough gas-phase cleaning of the wafer surface before deposition are vital to the operation.

Another form of epitaxy is molecular-beam epitaxy (MBE) in which a beam of molecules and atoms is directed at the flat surface of the substrate in a chamber of low pressure. Cells containing very pure amounts of the coating material are placed in the chamber where they are subjected to an electron beam. This causes them to be heated until some of the material evaporates and travels through the cell opening to the wafer, where it is deposited in an epitaxial layer. This approach can be used to produce a layer of gallium arsenide, and multiple layers of different materials at a lower temperature than that required for CVD. However, the process is slow.

Liquid-phase epitaxy (LPE) uses a liquid solution instead of a gas or molecular stream to grow crystals on a substrate. The method involves the immersion of the substrate in a saturated solution of the coating material. Gallium arsenide, gallium aluminum arsenide, and gallium phosphide are grown with this technique.

K3a4. ***vacuum deposition*** - is much like the vacuum metalizing process described in section 8F3. The material is evaporated in a vacuum, often by electron beam, and the vapor condenses as a film on the wafer surface. The process is also referred to as *evaporation*. It is used for the deposition of thin film metallic conductors. Aluminum is the most frequently used metal. Gold can also be deposited with this method. Fixturing that moves the wafer during evaporation is commonly utilized, to ensure uniform metal deposition on the wafer. The process is used for integrated circuits having broader wiring paths. It is also used to deposit gold on the back sides of wafers to facilitate adhesion of the chips in packages.

K3a5. ***sputtering*** - Sputtering is another vacuum method and is used for metals, alloys, semiconductor materials, and dielectrics, including glass. High-melting-point metals, such as tungsten is deposited by sputtering which is also known as Physical vapor deposition (PVD). The operation is performed in a vacuum, and is described in section 8F3a and illustrated schematically in Fig. 8F3a. The deposition material is taken from a wide source and therefore covers steps in the substrate surface. Adherence of the film to the substrate is superior to that achieved with vacuum deposition.

K3a6. ***adding thick films*** - involves the printing and then firing of a coating on a substrate material. The film material may provide conductance, resistance, or a dielectric on wafer surfaces, hybrid circuits, or multichip substrate materials. The materials can be ceramic, glass, quartz, sapphire, or metal coated with porcelain enamel. Film thicknesses are typically 0.0005 to 0.0015 in (13 to 38 microns).[9] Resistors, capacitors and inductors can be formed on wafers or other substrates of dielectric materials. These components are used, along with other electronic devices, particularly integrated circuits, in hybrid microcircuits.[1] The thick films are also used in the fabrication of multichip modules, resistors, potentiometers, magnetic devices, circuit protection devices, electroluminescent devices and membrane switches. Thick films are normally in paste form and contain three basic ingredients: a functional material (resistor, conductor, dielectric), a binder (glass powder), and a vehicle (solvents, plasticizers, etc.) The paste is applied by screen or stencil printing . After printing, the paste is allowed to settle for 5 to 15 minutes at room temperature and is then oven dried at 210 to 300°F (100 to 150°C) for 10 to 15 min. Firing temperatures are typically 930 to 1850°F (500 to 1000°C).[1] Various layers are added, depending on the function involved and circuit devices needed. Hybrid circuits may have a mixture of thick and thin film layers.

Thick film capacitors[9] are made in one of two ways: The first is by printing conductive material to form the base electrode and its termination, then depositing a dielectric material and firing it, and then depositing conductive material for the other electrode and its connection. The dielectric film, commonly barium titanate or titanium dioxide in a vitreous mixture, is fired at about 1560°F (850°C). The second method is to print the metal electrodes on opposite sides of the ceramic substrate. In both methods, capacitance may be adjusted to more precise values, if needed, by laser trimming the electrode material. When trimming is required, the amount of material printed provides slightly more capacitance than needed and the trimming of appendages or small parallel trimming capacitors reduces the capacitance of the unit.

Thick film resistors[9] are made by first printing the resistor terminations on the substrate with a conductive ink (usually a metallic paste). The substrate, with terminations, is then fired at 1470 to 1700°F (800 to 930°C). Then the resistance material, also in paste form, is printed on the substrate and dried at about 300°F (150°C). If a resistor network is involved, several different resistance materials are normally used in the network. After drying, the device is fired again at about 1560°F (850°C) if made from mixtures of precious metals, metal oxides, and glass binder, or at up to 600°F (315°C) if carbon. The finished resistors may be trimmed by laser to provide more precise resistance values. When trimming is needed, the amount of resistance material printed is slightly more than needed and the trimming reduces the width of the material, increasing its resistance.

Thick film inductors[9] are made with thick films by printing a spiral pattern of conductive inks on the substrate. Because of size limitations on circuit devices, this method of inductor making is limited to circuits operating at high frequencies, 10MHz or higher.

K3a7. ***adding protective layers*** - Layers of silicon dioxide or silicon nitride are added to provide insulation between devices on the integrated surface and to provide protection to the existing layers. Protection may be needed from chemical action or handling damage. The insulating layers are grown by thermal oxidation (See K3a1.), nitridation (K3a2), or chemical vapor deposition. Adding oxide or nitride layers for protection after the chip is fabricated is called *passivation*. It provides protection for the chip during testing, packaging and use.

K3b. ***patterning*** - is the series of operations that incorporates the circuit layout from a photomask or reticle into the surfaces of the wafer. The purpose of the process is to provide the correct locations and spaces for fabricating circuit devices and the necessary wiring paths to connect them.

K3b1. ***lithography*** - is a means for etching patterns in integrated circuit surfaces corresponding to the elements of the integrated circuit. The size and location of the elements is measured in microns. Photolithography uses photographic techniques and ultraviolet light to provide such small size and precise positioning. One key element of photolithography is the use of a photoresist. Electron-beam lithography and X-ray lithography can provide even more precise positioning but have some process disadvantages. The resist film has two basic vital properties: 1) It changes its solubility (becoming either less soluble or more soluble, depending on its material) when exposed to light or other radiation and, 2) It resists the attack of an etchant that will remove substrate material. The steps of photolithography are illustrated in Fig. 13K3b1 and are as follows: 1) Preparation of the wafer substrate with an oxide layer. 2) *Resist application* - A thin film of photoresist material is applied to the substrate that is to be processed, 3) *Exposure* - Optical light, ultraviolet light, or other radiation is projected through a transparent mask plate. (The mask plate, made of glass or other transparent material, already has the circuit pattern printed on it.) Electron beam radiation produces the finest resolution, but is very slow.[11] The areas of the substrate that received the radiation are either softened (positive resist) or hardened (negative resist), depending on the material involved. The image on the substrate is reduced in size considerably from that on the mask using special optical reducing lenses. Circuit path widths of less than 100 nanometers (0.1 micron) have been achieved. The exposure machine exposes only a small portion of the wafer and then steps to the next position and exposes again. With each exposure, the mask's circuit pattern is duplicated on the wafer surface. The process is repeated until the entire wafer surface is exposed. 4) *Development* - The substrate is washed with a solvent that dissolves the softer photoresist material. (With a *positive resist*, the softer material is that which was exposed to the radiation; with a *negative resist*, it is the material that did not receive the radiation.) 5) *Etching* - An etchant is applied to the substrate. Part of the substrate is removed from those areas not protected by the photoresist; the covered areas are unaffected. 6) *Resist removal* - The resist material is removed, leaving the substrate with a surface etched with the pattern that existed on the mask plate.

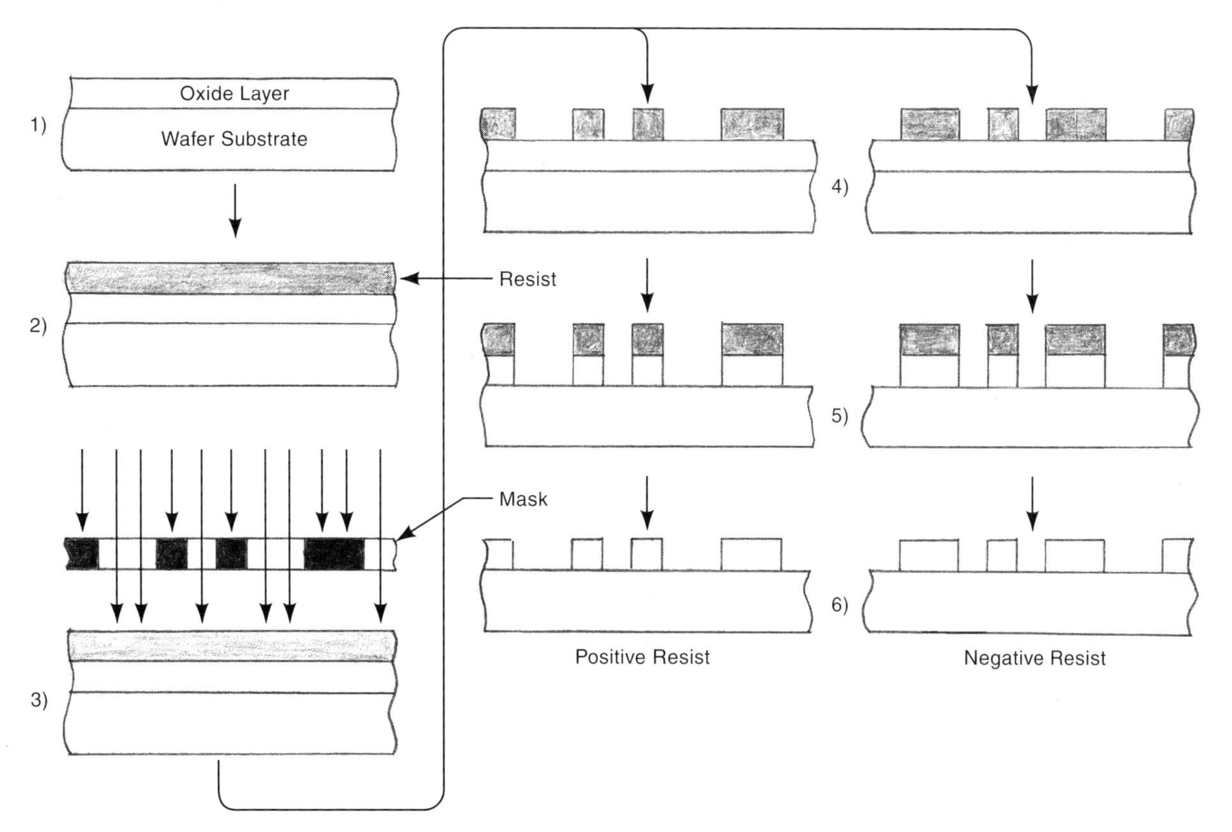

Fig. 13K3b1 The sequence of the lithography process: 1) wafer with a film layer is made ready for the operation, 2) resist is applied to the surface, 3) resist is subjected to radiation of ultraviolet light, x-rays or electron beams, through a mask, 4) the more soluble portion of the resist is removed. With a positive resist, the exposed portion becomes more soluble; with a negative resist, the exposed portion is made less soluble. 5) the layer underlying the resist is etched away, 6) the resist layer is then removed.

K3b1a. ***making masks*** - A unique photomask, which delineates the circuit layout, is made for each layer of the chip. The operation starts with a circuit design for the complete integrated circuit. The design identifies the various transistors, resistors, diodes, capacitors, and connecting wiring to be included in the layer, and the electrical specifications of all components. This circuit design is translated by the chip designer, with the aid of a computer-aided design (CAD) program, into a dimensioned layout for each layer of the chip. The next step is to make a reticle for each layer. The reticle is an enlarged master copy of the layout on a borosilicate or quartz glass panel, with the layout delineated in a layer of chromium on the panel. The pattern in the chromium is made by an operation sequence similar to that used on the chips. Resist material is applied to the chromium-coated panel. A computer- controlled laser beam or electron beam traces the circuit pattern of this design on the panel, exposing the resist. The resist is developed and unexposed portions are flushed away with a solvent. The uncovered portions of the chromium layer are then etched and removed and the photoresist is removed from the finished reticle. Fig. 13K3b1a shows the operation sequence. The reticle, a single copy of the circuit layout, then can be used with a step-and-repeat sequence to project images of the circuit on the resist-coated surface of the wafer. The alternative method is to use the reticle as a means to make a multiple image photomask that, in one operation, projects all the layout images on the wafer. To make the photomask, the reticle is brought into contact with the resist-coated blank mask in a contact printer. The blank glass and chromium mask are also in the contact printer. UV light is then used to transfer the image from the reticle to the resist on the mask. This is repeated until all necessary images on the mask have been exposed. The balance of the patterning operations are performed on the mask until it has the necessary number of layout images delineated by its chromium coating. The resist is then removed. Fig. 13K3b1a-1 illustrates how the reticle and masks are used.

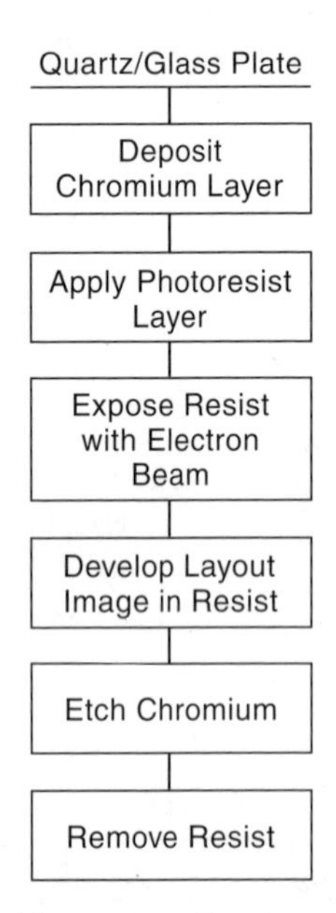

Fig. 13K3B1a The operation sequence for the preparation of a reticle.

K3b1b. ***applying resist (photoresist film)*** - Application of the photoresist involves three steps: 1) priming the wafer surface. 2) applying the resist in a uniform coating, 3) soft baking the photoresist.

Priming is carried out to ensure that the resist adheres properly to the wafer surface. The usual priming material is hexamethyldisilazane (HMDS) in xylene solvent. It is applied by a number of methods: immersion of the wafer, by applying the primer to a rapidly spinning wafer on a turntable, by vapor deposition at atmospheric pressure in a chamber, or by vapor degreasing with HMDS, or with a third vapor method. In this third method, wafers are first heated in an oven in a nitrogen atmosphere to completely dry the wafer; then, a vacuum is pulled in the oven followed by admission of HMDS vapors to the oven. The vapors coat the wafer, providing good performance as a primer for the resist coating with low usage of HMDS.

There are several variations of resist coating process, but the most common methods all use a spinning turntable. The centrifugal force generated by rapid spinning spreads the liquid polymer resist to a uniform thickness, especially near the edges of the wafers, where there otherwise would be an edge-bead buildup. An edge buildup of resist would interfere with the accuracy of the photoresist. Manual, semiautomatic, and fully automatic variations of the spinning method are in use. Some methods combine the operations of primer

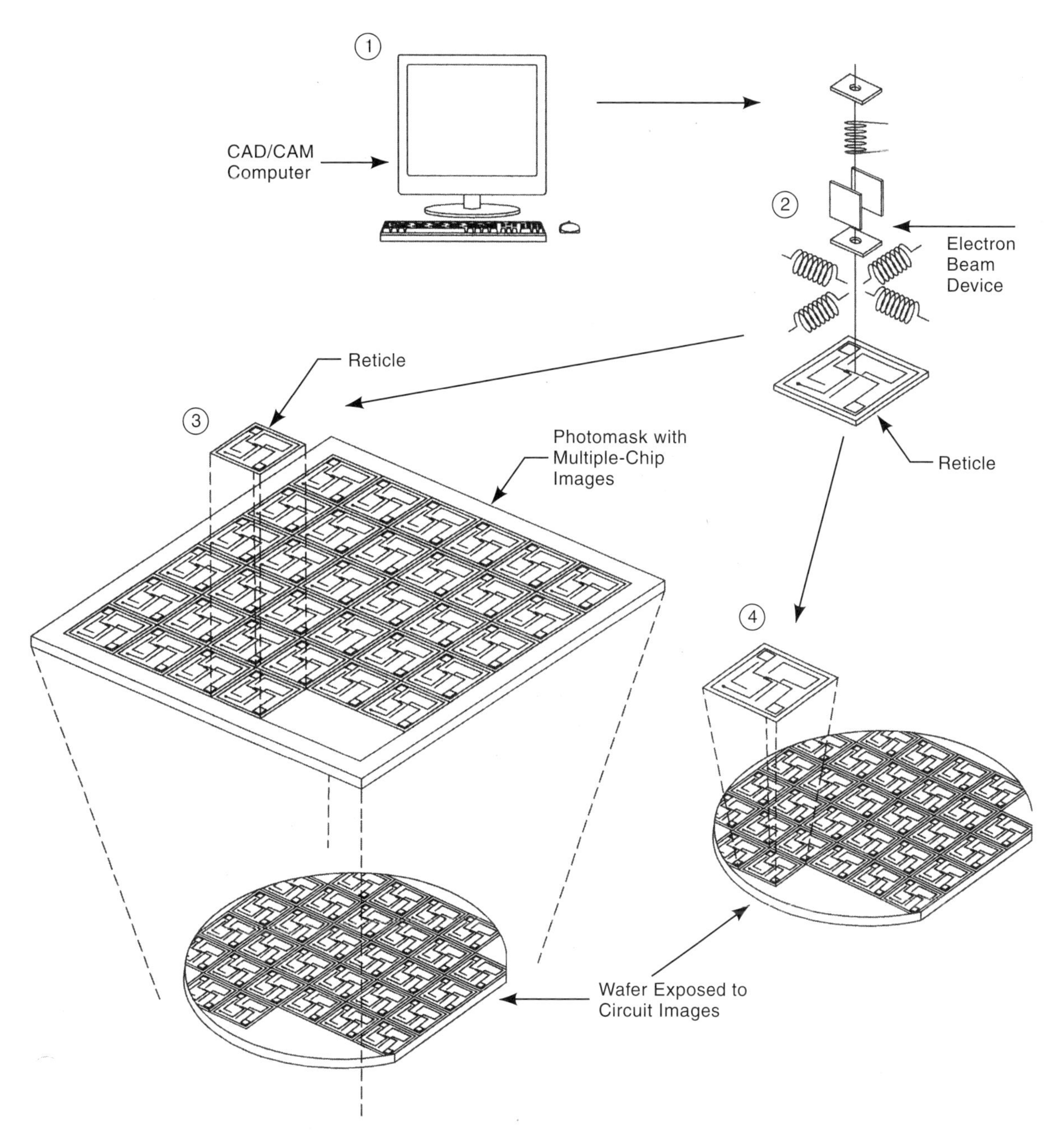

Fig. 13K3b1a-1 Fabrication and use of a reticle and a mask to expose the wafer surface with the design layout for one layer of the chip circuit. 1) The process starts with a CAD dimensioned design layout of one layer of the chip. 2) That design layout is fed to an electron beam or laser beam scanner, which exposes a resist surface on a quartz-glass reticle to the design. 3) In one method, the reticle design is transferred to a multiple-image quartz- glass mask by a series of contact printings. The mask then has multiple images of the chip circuit layout. These images are projected all at one time from the mask to the wafer to expose a resist on the wafer surface. 4) In the other method, the image on the reticle is projected directly to the wafer with a step and repeat exposure and no multiple photomask is used.

application and resist application on the same spinning chuck. The views in Fig. 13K3b1b schematically represent an automatic arrangement after each wafer is positioned on a vacuum spinning chuck where it is held during nitrogen blow off, primer dispensing and spinning, and resist dispensing and spinning, followed by automatic unloading for transfer to the baking operation.

The baking operation, called *soft bake*, has the purpose of evaporating part of the solvents in the photoresist. The provides better adhesion of the resist and better quality exposure of the resist pattern. Conduction (hot plate), convection oven, radiant oven, and microwave heating systems are all used in various soft bake systems. Some systems also use vacuum assistance to aid the evaporation.

K3b1c. ***expose and develop the resist*** - A glass mask or reticle, carrying the desired circuit pattern for each chip, is used in a precision optical device to project an image of the circuit on one small area of the wafer. In the most common method, the projection operation is repeated for each chip ("die") area on the wafer, with precise alignment, one die at a time, until all dies on the wafer surface have been exposed. (Another method, now less used, except in less densely packed chips, is to make a multiple mask that has a copy of the circuit for each die location on the

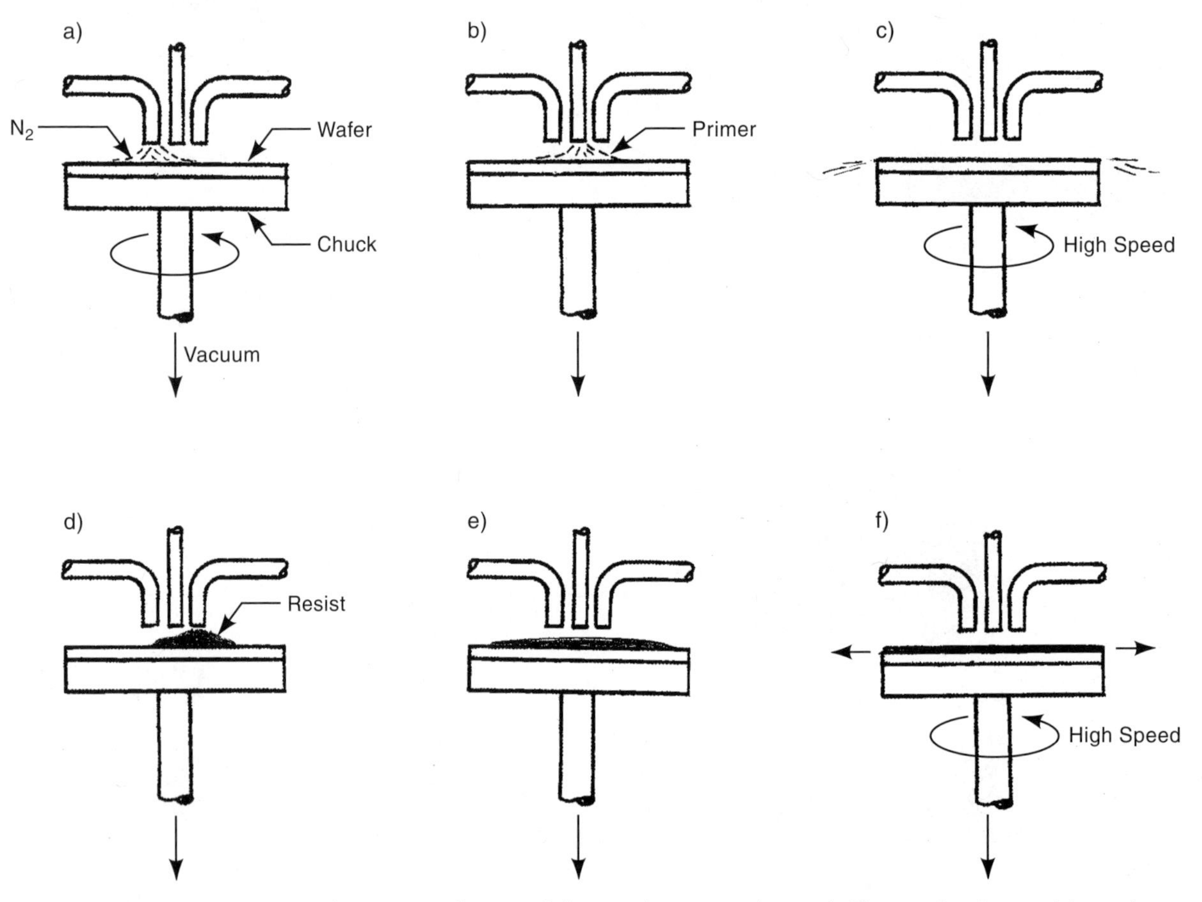

Fig. 13K3b1b One automatic system for applying resist to wafers: a) The wafer is positioned on a chuck, on which it spins, as nitrogen gas is blown against it to remove any dust or other foreign material. b) a primer is then dispensed onto the wafer, c) the wafer is spun at a high RPM to spread and dry the primer, d) resist is dispensed onto the wafer, e) the resist spreads and f) the wafer is spun rapidly to throw off excess resist and ensure that the remaining resist forms a thin, uniform layer. The wafer is then removed from the chuck and transferred to a baking station.

wafer. Then, all dies on the wafer are exposed at the same time.) Radiation from a light, x-ray, or electron beam is projected through the mask. Five steps are involved in these operations: 1) exact alignment of the mask or reticle to the wafer. 2) exposure of the wafer to radiation through the mask. 3) development of the resist, i.e., removal of unpolymerized (softer) resist material. 4) hard baking the resist to maximize its adhesion to the wafer surface. 5) inspection of the resist and wafer for proper alignment and freedom from defects. With a positive resist, the areas of the resist coating that are exposed to the radiation are thus rendered soluble and are dissolved and removed chemically. The open spaces in the resist are available for receiving the next operation which may be etching, doping or deposit of a film of another material.

The *alignment* of the mask or reticle with the wafer is an extremely critical element because of the minute size of the many millions of devices that may be crammed into each chip. Alignment marks are made on the wafers and masks to guide the operation. Several types of alignment devices are employed. One type, used for simpler chips with larger feature dimensions, puts the mask and the resist-coated wafer in direct contact. An optical microscope system is used to adjust the wafer position with respect to the mask until the two are correctly aligned. Another optical system uses light projected through the mask with a scanning technique. A more advanced system uses a stepper mechanism to move the reticle, rather than a full mask. A laser-based automatic device provides the alignment. The laser is projected through the alignment mark on the reticle to the corresponding mark on the wafer. The reflected beam is fed to a computer that adjusts the reticle position until the alignment is correct. Then exposure takes place, the stepper moves the wafer to the next die position and alignment and exposure take place again. The process is repeated until all the dies on the wafer have been covered. Still another system utilizes electron beam alignment and direct exposure of the wafer with a computer-controlled electron beam instead of one projected through a reticle or mask. X-ray alignment systems are under development for the most densely packed chips. In the course of making complete chips on a wafer, different alignment and exposure systems may be employed on the several layers (up to about 20) of each chip. Even the most advanced chips have some layers that do not need extremely small spaces for circuit devices; these layers are more economically exposed with the less advanced alignment and exposure systems.

The *development* of the resist layer takes place by the chemical dissolution of the unpolymerized areas. Negative resists, those that polymerize upon exposure to light or other radiation, are typically dissolved with xylene and rinsed with *n*-butylacetate. Positive resists become more soluble after exposure, and the exposed areas are dissolved with sodium hydroxide (NaOH) or tetramethyl ammonium hydroxide (TMAH) and then rinsed with water. Developer and rinse are applied by spray methods. The wafer is attached by suction to the horizontal surface of a rotating chuck, and the developer, followed by the rinse liquid, are sprayed from above. With positive resists, the developer is heated and first "puddled" on the wafer for a period of time. Then more developer is sprayed on. After rinsing, for both types of resist, the wafer is spun rapidly by the chuck to drive off the liquids and aid drying.

After development, the resist coating on the wafer undergoes *hard baking*, a heating step that evaporates more of the solvent in the resist with the purpose of enhancing its adhesion to the wafer surface. Typical baking temperatures are 270 to 390°F (130 to 200°C) for 30 minutes. Inspection of the resist, with automatic equipment, follows before the wafer is moved to the etching operation.

K3b2. ***etching*** - After the resist is developed and the developed portion removed, the substrate (or film on the substrate) may be etched in those areas not covered by resist. The purpose is to transfer the pattern in the resist to the surface layer of the wafer. Etching is used to create circuit patterns in layers of aluminum or other conductors and to create a mask in an oxide film for high temperature diffusion. The etchant must be a material that dissolves the desired material effectively but does not dissolve the remaining resist mask or other needed microelectronic materials.[5] There are two basic etching methods: *wet etching* and *dry etching*.

K3b2a. ***wet etching with liquid etchants (wet chemical etching)*** - The etchant is applied by immersion of the wafer in etchant solution or by spraying. Liquid etching has limitations in the size of the circuit paths or features that can be etched, and the thickness of the layer etched. Etched features must be no less than 3 or 4 microns (1.2 to 1.6×10^{-4} in)[9] wide. Fig. 13K3b2a illustrates schematically the common effects of the wet etching operation on the microcircuit. The choice of etchant can be made from a variety of possible chemicals, and depends largely on the material to be etched and the adjacent and underlying materials that are not to be etched. Dilute hydrofluoric acid is a common choice because it dissolves silicon dioxide, the usual layer to be etched, but does not attack silicon.

In *immersion etching*, the wafers are placed in a tank of etchant for a specified period, then rinsed twice, and dried by spinning. The wafer is rinsed immediately after quenching in a vessel that flushes the rinse solution away from the wafer to ensure that there is no further contact with the etchant. Careful control of the time of contact with the etchant is important. Surfactants are usually added to the etchant solution to aid in wetting. Agitation and heat are also used to ensure contact with fresh etchant, remove trapped air and any gas bubbles that are formed by the etching reaction, and to promote uniformity of etching.

Wet *etching by spraying* has replaced much immersion etching. Spray application has the advantage of using less etchant and ensuring that only fresh etchant is brought into contact with the wafer. In one spraying method, the wafer is held, by vacuum, on a rotating chuck on a vertical shaft. The wafer stays on the chuck for application of the etchant, for rinsing and for spin drying. (The rotational speed is increased for drying.) Separate nozzles are used for etchant and rinse water, and the sequential steps take place immediately after one another.

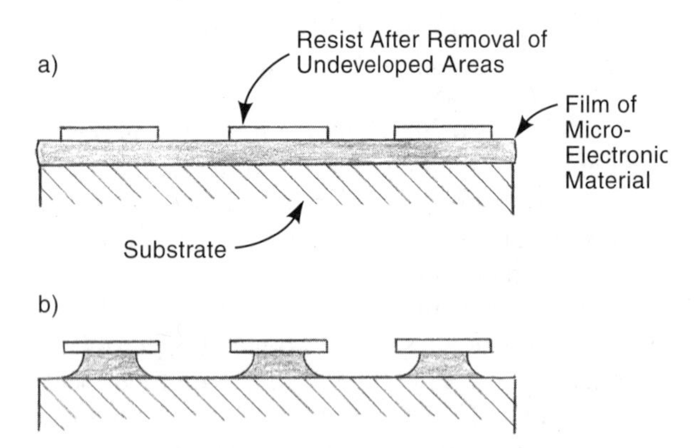

Fig. 13K3b2a. How wet chemical etching, one step in the sequence of making discrete devices from a film of microelectronic material, affects the integrated circuit. a) before etching the film but after removal of undeveloped areas of resist. b) after etching away unneeded microelectronic material. Note the undercuts made at the edges of the protected areas, since the etchant acts horizontally as well as vertically.

K3b2b. ***dry etching*** - includes plasma etching, ion beam etching and reactive ion etching. Two varieties of plasma etching are: planer etching and barrel etching.

The control and environmental disadvantages of wet etching, coupled with the inherent limits to the minimum line width achievable, have led to extensive use of plasma etching, particularly when fine-line circuitry is required.

K3b2b1. ***plasma etching*** - The plasma, a low pressure body of neutral ionized gas, is produced for integrated circuit etching by applying radio-frequency energy to gas contained in a vacuum chamber of approximately 10^{-3} atmospheres (10^2 Pa).[9] The ions, molecular fragments,[5] and electrons in the plasma, all with high energy, react with the film or substrate material of the wafer, to etch the material. If the compounds formed are volatile, they will evaporate and be carried away. Fig. 13K3b2b1 shows the workings of a planer reactor set up for plasma etching with reactive ions. This method provides favorably uniform etching.[5] The wafers are arranged horizontally between two planar electrodes that are closely spaced. Radiation sensors detect changes in the radiation given off by the wafers during the operation and these changes are used to control the amount of etching by triggering shut off devices in the plasma system. The gas used in the operation depends on the material to be etched since it must react with that material. The process, then, provides etching from both energy and chemical effects. Aluminum conductors are etched with chlorine plasma; nitride, oxide, and silicon are usually etched with gases containing fluorine.[5]

K3b2b2. ***ion beam etching (sputter etching or ion milling)*** - is a physical process in contrast to the chemical nature of plasma etching.

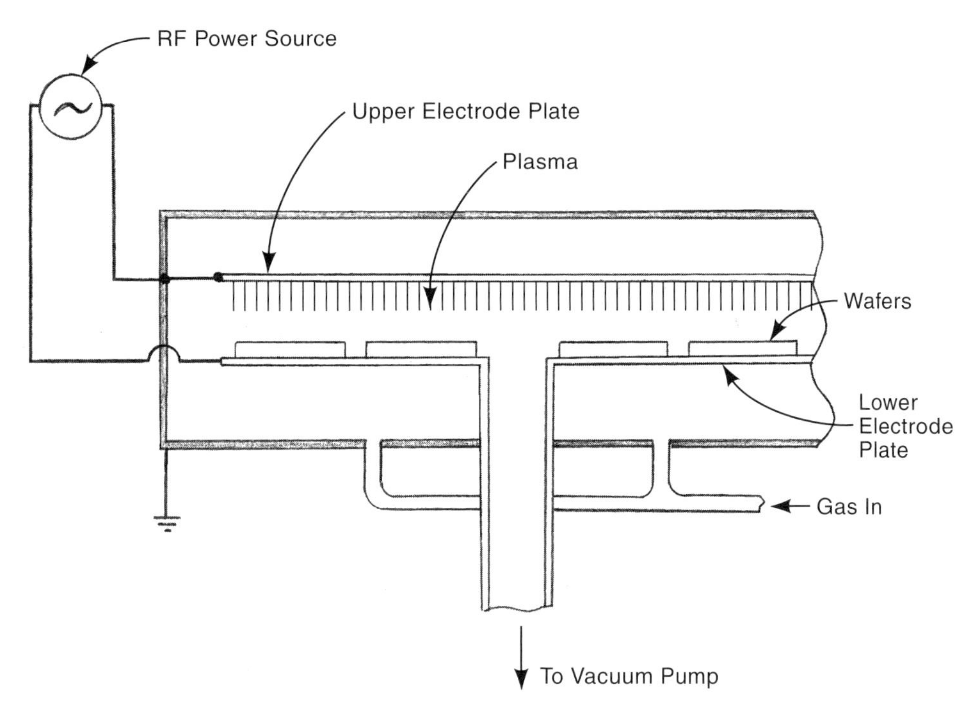

Fig 13K3b2b1 Reactive ion plasma etching.

The process involves the placement of wafers in a vacuum chamber in which they are subjected to a stream of ionized argon gas. The argon is ionized by a stream of high-energy electrons from cathode/ anode electrodes. The wafers are negatively grounded so that the ions are attracted to them, and their speed is accelerated as they approach. When they strike the wafer, the ions have enough energy to dislodge material from the wafer surface. Etched areas have good definition, but the ion beam is nonselective and there are radiation effects. Fig. 13K3b2b2 illustrates the process schematically.

K3b2b3. ***reactive ion etching (RIE)*** - is a combination process involving elements of both plasma and ion beam etching. It is particularly suited to etching layers of silicon dioxide over layers of silicon. The process has higher selectivity ratios than plasma etching and has become common for critical ICs. (The selectivity ratio is the ratio of etching rate of the target material compared with that of other materials.)

K3b3. ***stripping photoresist from wafers*** - The photoresist film is removed after etching, since it has served its purpose of limiting the etching to the desired areas. The film is also removed after ion implantation. Wet chemical stripping is the most

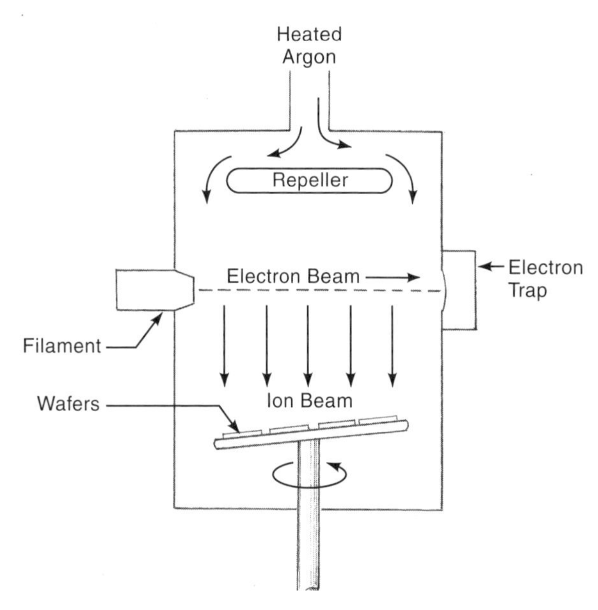

Fig. 13K3b2b2 Ion beam etching shown schematically.

common method but an oxygen plasma is also used. The wet methods are more economical, are effective in removing metallic ions and do not expose the circuit elements to radiation. The choice of chemical strippers depends on the nature of the resist material to be stripped and the wafer materials that will be contacted by it. When the materials involved, other than polysilicon, silicon dioxide, or silicon nitride, are non-metallic, solutions of sulfuric acid with an oxidant (hydrogen peroxide or ammonium persulfate) are used. About 10% of nitric acid in the mixture may also be used as an oxidant. With these strippers, the resist is dissolved by oxidation.

When metalized surfaces are involved, several proprietary formulations are available. Strippers using sulfonic (organic) acid and a chlorinated hydrocarbon solvent are alternatives. These solutions have detergent action and are applied with heat of 195 to 250°F (90 to 120°C), often in two steps, each followed by thorough rinsing and then a final drying operation. Several solvent and solvent/amine strippers are available for positive-type resists. These are often used with heat. Special other strippers have also been developed for particular conditions.

Dry stripping is performed with a plasma. Oxygen (O_2) plasma, created in a chamber containing wafers, oxidizes the resists to carbon dioxide and water. The term, *ashing*, is used to refer to such stripping. Plasma stripping is effective in removing resists that have been hardened by ion implantation but does not remove metallic ions. In some situations, both plasma and wet stripping are used. The plasma removes hardened resist and is followed by wet stripping.

K3c. ***doping (dopant defusion)*** - selectively changes the electrical conductivity of the semiconductor materials (silicon, germanium, gallium arsenide). The operation introduces impurities to the lattice structure of the semiconductor. In producing a silicon semiconductor, two different impurities are used and each is introduced in an adjacent area of the silicon surface. Boron, phosphorus, arsenic and antimony, are dopants for silicon. Two major methods for achieving this condition are *thermal diffusion* at high temperatures and *ion implantation*.

In the *thermal diffusion* method, wafers are first cleaned and then acid etched to remove any oxidation that may have grown on the surface. Then the areas to be doped are exposed to the dopants, while the silicon wafer is heated in a tube furnace to a temperature between 1500 and 2200°F (815 to 1200°C). A nitrogen atmosphere is maintained around the wafers when they are loaded into the furnace and when they are unloaded. The high temperature causes vacant spaces to develop in the crystal structure of the silicon, and the dopant material, if sufficiently concentrated, migrates to these open spaces. The rate of dopant diffusion increases with increased temperature. The source of the impurities is either a gas, a liquid vapor, or a deposit of oxide that contains the impurities. A film dopant mask, with openings, isolates the areas where the dopant material can be absorbed. Following diffusion, the wafers are heated to a higher temperature for a period to distribute the dopant deeper into the wafer. This step is called *drive-in oxidation* and it produces an oxide surface at the doped areas. Figs. 13K3c, 13K3c-1 and 13K3c-2 illustrate equipment used for thermal diffusion doping.

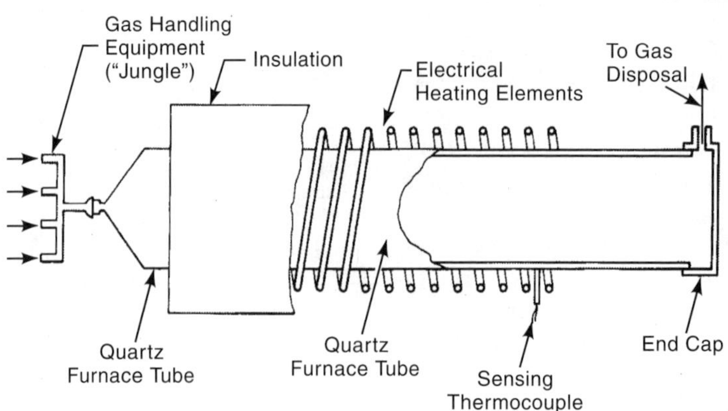

Fig. 13K3c A furnace used in the diffusion doping of integrated-circuit wafers. (*Reproduced with permission from Microlelectric Processing - An Introduction to the Manufacture of Integrated Circuits, by W. Scott Ruska, McGraw-Hill, 1987.*)

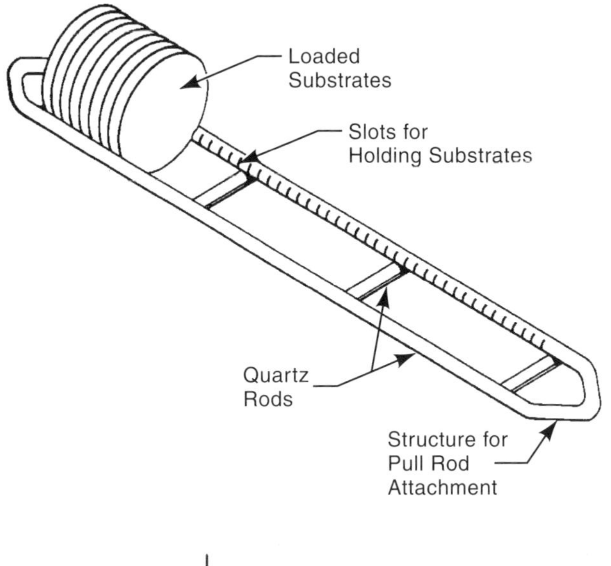

Fig. 13K3c-1 A quartz "boat" for holding integrated circuit wafers during diffusion doping. (*Reproduced with permission from Microlelectric Processing - An Introduction to the Manufacture of Integrated Circuits, by W. Scott Ruska, McGraw-Hill, 1987.*)

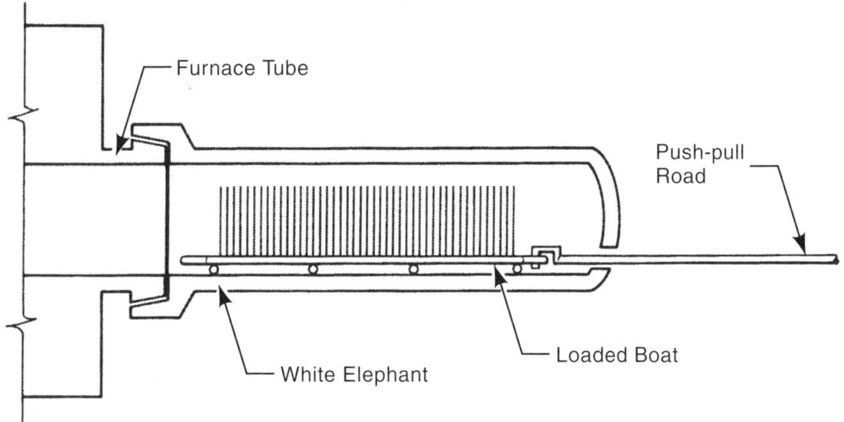

Fig. 13K3c-2 Insertion of a loaded diffusion boat into a diffusion-doping furnace. Note the use of the push pull rod and the temporary furnace extension called a "white elephant". (*Reproduced with permission from Microlelectric Processing - An Introduction to the Manufacture of Integrated Circuits, by W. Scott Ruska, McGraw-Hill, 1987.*)

Ion implantation - For the most advanced integrated circuits with extremely close spacing of devices, closer control of the depth, width, and degree of concentration of the dopant, is required. Ion implantation, which provides better control of these factors, may be used as a doping method rather than thermal diffusion. Ion implantation utilizes intense beams of high-energy ions of the dopant material, at an energy level of 10 to 500 keV. The dopant source is usually a gas, though solid materials are also used. Electrons from oppositely charged electrodes turn the gas molecules into ions. The operation is performed in a vacuum, with specialized equipment such as a Van de Graaff generator.[9] The beams are able to be purified, accelerated, and focused at a precise spot on the wafer with this technique, and the amount of dopant can be closely controlled. It is deposited to a depth of from a few hundred angstroms to several microns.[10] A scanning system is usually used to cover the full area to be doped. Because the beam energy level is sufficient to cause damage to the silicon lattice, annealing after the operation is performed as described below. Annealing removes much of the damage to the structure[9] and diffuses the dopants to the desired locations in the silicon.

K3d. ***heat treating*** - The wafer is heated to produce effects or changes in the wafer material. Annealing and "alloying" are two heat-treating operations. Wafers are also heated for other reasons, including soft and hard dehydration baking, to remove solvents from photo resist film and provide more accurate patterning.

K3d1. ***annealing*** - is a heating operation performed after doping by ion implantation. The implantation disrupts the crystal structure of the wafer. However, heating the wafer to a temperature between 1100 and 1770°F (590 and 970°C), restores the structure and activates the dopant. This temperature, however, is below the level that would cause the dopant to diffuse laterally. The operation is performed in a tube furnace in a hydrogen atmosphere and requires about 15 to 30 minutes.

Another method for annealing is rapid thermal processing (RTP) which provides better protection against spreading the dopants. The cycle takes place in a chamber with a gas inlet and outlet, with radiant heat sources above the wafer, and sometimes below. The operation is automatic. Tungsten halogen lamps are most common as heat sources, but graphite, microwave, and plasma-arc heaters may also be used. The surface of the wafer is brought up to the annealing temperature in less than a minute and then is rapidly cooled. The wafer body remains at a much lower temperature. The necessary annealing of the surface crystal damage takes place, but the doping is not permitted to diffuse.

K3d2. ***alloying*** - is a heat treatment operation that takes place after metallic layers have been deposited and patterned into the wafer. Its purpose is to ensure good electrical contact between the metal circuitry and the locations of semiconductors and other devices on the wafer surface. The operation involves heating the wafer in a nitrogen atmosphere to a temperature of approximately 840°F (450°C). The term "alloying" has been used to identify this operation.

K4. ***wafer testing and sorting*** - After all the circuitry has been incorporated in the wafer, each individual die is tested to verify the circuit function and that the die meets all other design specifications. This testing takes place before the individual dies are separated from the wafer. The wafer is held in a test fixture that aids in the alignment of many narrow test probes, one of which contacts each connection pad of the die. The testing sequence, measurement of results, and decision of acceptability or not, are all under computer control. The computer notes, in its memory, which dies on the wafer are acceptable and which are not. See Fig. 13K4.

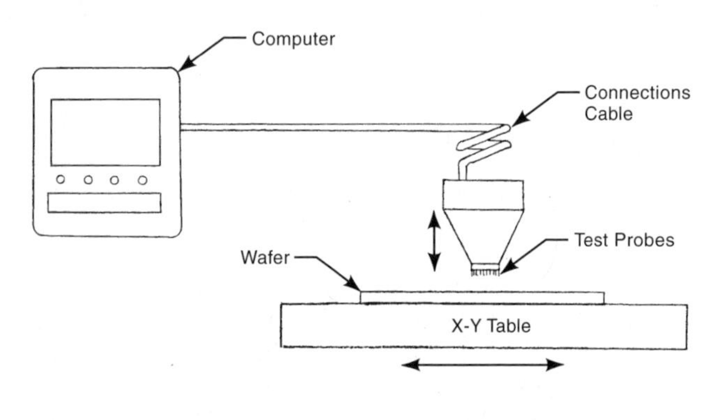

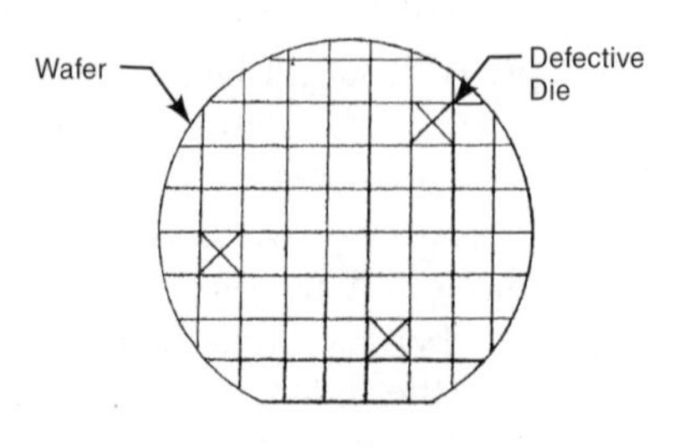

Fig. 13K4 The wafer sorting operation. Test probes contact the connection pads of each die and a computerized functional test is made of the die. The x-y table then moves another die into position for testing. The process is automatic. Any defective dies are identified through the computer network and are separated from the usable dies at the later dicing operation.

K5. ***packaging (assembly) of chips*** - The purpose of packaging (after the die is separated from the wafer) is to provide strong leads for easy connection of the chip to a circuit, to protect the chip from physical damage and exposure to reactive environments, and to dissipate heat. The package consists of a base surface (sometimes a recessed surface) to which the chip is attached, external leads or pads to connect the IC to a circuit board, internal leads or pads to which the chip can be electrically connected, wires or other means of connecting the chip to these internal leads, and an enclosure to provide protection and heat dissipation. The enclosure may be made of metal, ceramic, or plastic. Fig. 13K5 illustrates some typical IC packages. Prior to packaging, the dies are prepared to be assembled into a package. The operations include backside preparation and electrical testing, completed while the dies are still part of the wafer. The good individual dies are marked and separated from the wafer in the dicing operation. Each one is then mounted in and attached to a ceramic, plastic or metal package or carrier. The operations that follow consist of bonding the connecting wires, assembling, closing and sealing the packaging, plating and trimming the leads, and marking and final testing of the packaged IC. The packages may be individual or may contain several chips. A chip may also be mounted on the substrate of a hybrid circuit or directly on the circuit board as a chip-on-board (COB). The package protects the chip while allowing all electrical connections to be made to it. Each bonding pad of the chip must be connected to a lead of the package or to a circuit board. Very fine wires between the chip and leads of the package are bonded, welded, or soldered. The external leads of the package are designed to facilitate insertion into the circuit board or, if surface mount technology is involved, to facilitate a solder connection to the board. The completed packaged ICs are tested, marked and packed for shipment. At the customer's plant the packaged chips are assembled on, and

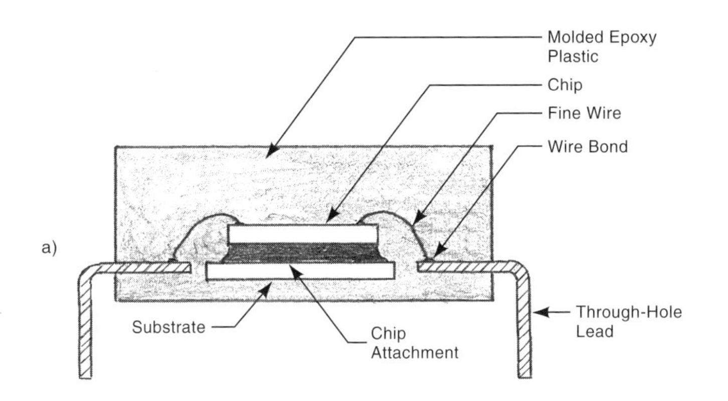

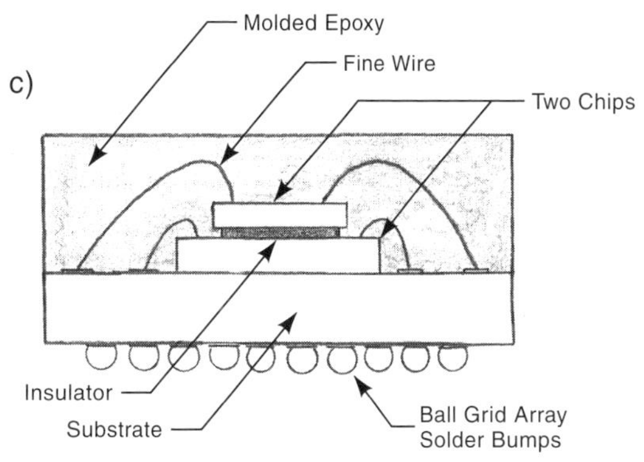

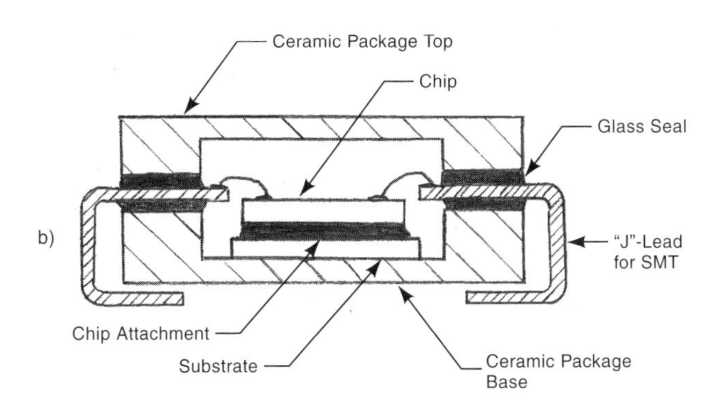

Fig. 13K5 Typical IC packages showing chip, bonding wires, leads, and enclosure. a) a plastic package. The enclosure is a one-piece molding of epoxy. The leads, in this example, are made for through-hole connection to a circuit board. b) a ceramic package in two pieces, sealed with low-melting-temperature glass. This package uses J-leads for connection by surface mounting (SMT) to a circuit board. c) A single package of two integrated circuits. A ball grid array is used to connect the leads to the circuit board.

connected to, printed circuit boards that are used in computers or thousands of other electronic products.

K5a. ***backside preparation***[12] - Some chips that are to be used in thin packages or those that have some backside junctions or damage are reduced in thickness in the wafer state. This is done by surface grinding the bottom of the wafer, followed by chemical-mechanical polishing. Chemical-mechanical polishing is performed in the same way as it is done to prepare the plane flat wafers for processing. Some thinning is also carried out by chemical etching of the bottom side. The thickness, after thinning, is typically 0.008 to 0.020 in (0.2 to 0.5 mm). Gold is sometimes applied to the bottom side of the wafer by vacuum coating or sputtering. When used, the gold becomes part of a gold-silicon eutectic solder to fasten the chip to the package substrate.

K5b. ***dicing (die separation)*** - Individual dies are separated from wafers by one of two methods: sawing, or scribing and breaking. When sawing is used, the cut between the dies is made with diamond-bladed circular saws. Sawing is the preferred method, especially with thicker wafers. With the other technique, lines are scribed between the dies with a diamond- pointed tool, and the individual dies are broken free at the scribe lines by stressing the wafer with a cylindrical roller. Sometimes, a combination method is used: Saw cuts are made part way through the wafer and the dies are then separated at the cut line by roller stressing the wafer. After separation, the dies are referred to as *chips*.

K5c. ***chip insertion and fastening to the package*** - Chips separated from the wafer are placed, with others, on a carrier tray called a plate. This operation may be manual or automatic but, either way, the die is held with a vacuum wand. When the operation is manual, the defective chips are marked so the operator does not move them; when the operation is automatic, computer-controlled pick and place equipment is used, and is fed data from the wafer final testing operation, so that only good chips are placed on the plate. The plates are transported to the package assembly area. From the plate, each chip is picked and placed in the chip attachment area of the designated package. The operation may be manual but, for larger quantity production, it is performed by computer controlled pick and place equipment. Two methods are used to fasten the chip in the designated location: 1) soldering with gold-silicon eutectic or other solder or, 2) bonding with epoxy adhesive.

When the eutectic chip attachment method is used, a small preform of gold-silicon alloy may be placed in the chip-attachment area of the package before the chip is placed. The package is heated to the melting temperature of the eutectic alloy, about 720°F (380°C).

"Scrubbing" takes place after the alloy has melted. This consists of motion between the chip and the package to ensure that the eutectic alloy is distributed uniformly across the joint and that the chip and package are close together. The joint is then cooled until the eutectic alloy solidifies. When automatic placement of the chip is used, the equipment carries out both the placement and scrubbing and controls the heating and cooling cycles. Eutectic attachment provides good heat dissipation properties and is used for high-quality, high reliability applications.

When epoxy attachment is used, a measured amount of thermally and electronically conductive liquid epoxy adhesive is dispensed or screen printed in the chip attachment area. The chip is positioned and pressed into place to provide a uniform adhesive layer between the chip and the package. Oven heat is used to set the epoxy. With this attachment method, the operation may be either manual or automatic. Epoxy attachment is less costly than the gold-silicon eutectic system. The epoxy can constitute an insulation layer or, if silver filled, can serve as both an electrical ground connection and a heat transfer channel.

K5d. ***wire bonding*** - is a means of providing an electrical connection between a wire and a contact surface. It is most used in connecting a fine wire of gold or aluminum between an integrated circuit ("chip") and the electrical leads of the package that houses the chip. The method is also sometimes used to connect fine (0.001 in - 25μm) diameter gold wire to printed circuit boards.[9] The most common use, however, is to connect the chip to the inner ends of the leads that later connect the finished IC to a circuit board. A characteristic of the bonding method is that the connection is made without bringing the metals involved to the melting point. When gold wire is used, the end of the wire to be attached is first made into a ball from the heat of an electrical spark or small flame. There are two common bonding methods: *thermocompression* bonding and *thermosonic* or ultrasonic bonding. With either method, automatic machines are commonly used. In thermocompression bonding, the metallurgical bond at the molecular level is made by heating the metal elements that are in contact without melting them, and then applying pressure. In the ultrasonic method, both pressure and ultrasonic vibration are used to break up surface layers of the materials and achieve a bond. Fig. 13K5d illustrates the thermocompression method. The thermocompression method is the faster of the two. Aluminum bonding wire is usually used because of favorable metallurgical effects. Aluminum is also joined by ultrasonic bonding.

K5e. ***closing and sealing the package*** - After the wiring connections are made between the chip and the leads of the package, a metal, ceramic, or plastic enclosure is assembled over the chip and wires, and is sealed. Seals may be hermetic, with metal or ceramic enclosures, or near-hermetic with plastic enclosures. Hermetic seals are used in many aerospace and military applications when the package may be exposed to harsh environments. Near-hermetic seals are adequate for most consumer-product and commercial applications. Metal enclosures are welded or soldered to the substrate of the package. Resistance welding techniques are used when the joint is welded. Projection welding and seam welding are the methods used. (See Chapter 7. sections C6c and C6b.) Resistance welding has the advantage of having the joint heat highly concentrated so that temperature sensitive materials in the package are not harmed. Soldered joints often use preforms of gold-tin alloy and are melted in a furnace with a nitrogen atmosphere at a temperature of 610 to 680°F (320 to 360°). Ceramic enclosures and lids are normally sealed with glass of low melting point. A furnace temperature of approximately 750°F (400°C) with a dry, clean-air atmosphere is used. The resulting packages are known as cerdip or cerflats depending on the package configuration. Molded epoxy enclosures are predominantly used in many applications. They are produced by transfer molding where the connected chip and lead frame assembly constitute an insert in the mold. The liquid or near-liquid epoxy flows around the inserted parts, fills the mold, and solidifies. Further curing of the epoxy in an oven may take place after the molding operation.

K5f. ***lead plating and trimming***[12] - After the package is assembled and sealed, the external leads are coated with solder, tin, or gold to promote solderability when the IC is connected to the circuit board and to provide corrosion protection. Solder is applied by dipping or by passing the package through wave soldering equipment. Tin and gold are applied by electroplating. Leads, that are sometimes attached together for ease of handling and assembly to the package and to ensure correct spacing, are also trimmed to remove the unneeded attachment and to provide the proper length of lead.

K5g. ***marking and final testing***[12] - The finished IC is marked by laser etching, or by ink-jet or offset printing. The ink is cured by exposure to ultraviolet light, or by oven or room-temperature drying. The marking provides product identification and specifications, the date of manufacture, lot number, and name or location of the factory.

Final testing has the purpose of separating defective or failure-susceptible packages. It includes a series of electrical tests of the circuitry and then deliberate exposure to potential environmental hazards. Prior to the test, the packages are subjected to a prolonged oven bake - typically at a temperature of 300°F (150°C) for 24 hours. This step is designed to stabilize the package and chip and to drive off any volatile materials. The electrical tests include a verification that the IC meets electrical specifications, and an operational test to ensure that it functions as it should. Sometimes , a "burn-in" stress test is conducted. This is intended to induce failure in the test if the chip has a weakness that otherwise could lead to early failure in its use. Tests of package integrity and resistance against environmental threats is also carried out. These tests include exposure to high and low temperatures and to high acceleration forces in a centrifuge. Tests are computer controlled and the loading and unloading of packages into the test apparatus may be either automatic or manual.

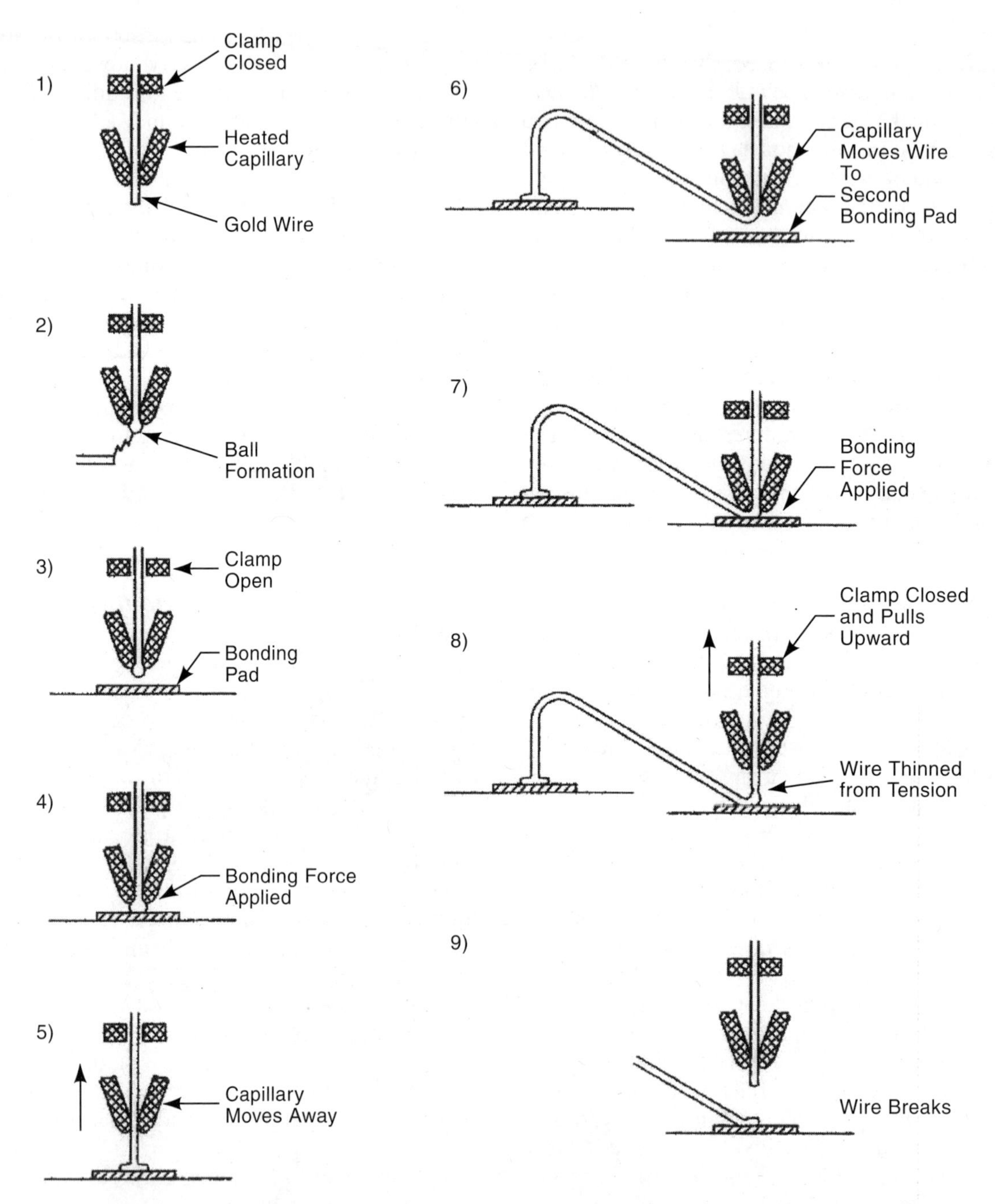

Fig. 13K5d Wire bonding by the thermocompression method, showing formation of ball and wedge bonds: 1) The wire is heated by the hot hollow capillary. 2) A ball is formed at the end of the wire by an electric arc. 3) The ball is lowered to the bonding pad. 4) Force is applied to create the bond. 5) The capillary moves away on the wire; the bonded end remains. 6) The capillary moves the wire to the second bonding pad; enroute it heats the wire. 7) Bonding force is applied to bond the wire to the pad. 8) The clamp closes and pulls the bonded wire; the wire thins from the tension. 9) The wire breaks and the capillary starts to move the wire to the next bonding pad to repeat the cycle. When thermosonic bonding is used, ultrasonic vibration is applied to the wire by the capillary.

K6. *other methods of connecting the integrated circuit to the board*

K6a. ***chip on board (COB) technology*** - refers to the procedure of attaching integrated circuit leads directly to the circuit board. This is often preferred to having the chip manufacturer package the chip in an enclosure, with internal connections to external input-output leads that must then be soldered to the circuit board. In one COB approach, called chip-and-wire-technology, the tiny wires are excluded from the chip as received and are bonded by the board assembler to both the chip and the circuit board. In other approaches, the fine wires are bonded to the chip by the chip manufacturer, and the board assembler then bonds the other ends to the circuit board. In either approach, a glob of epoxy resin over the chip, and the circuit board below it, provide the protection to the chip that would otherwise be provided by the chip package. This approach is used, among other applications, in automotive equipment. Fig. 13K6a illustrates the concept.

K6b. ***conductive adhesive connections*** - are made from one-part, quick setting epoxy, containing small flakes of silver to provide conductive paths. These connections are not used extensively because of higher cost and limitations of impact strength. They are used when there is some flexing of the circuit board and have the benefit that high temperature heating is not necessary to effect the electrical connection. The epoxy is dispensed to connecting tabs of the circuit board where it will bond to copper or solder alloys. Component devices are then positioned and the epoxy is cured, typically at around 265°F (130°C).

K6c. ***tape automated bonding (TAB)*** - is a means of electrically connecting chips' to circuit boards, other substrates or packages. Instead of using fine gold or aluminum wires to connect chips, TAB utilizes thin, flat metal conductors that are mounted on a tape of polyimide film. The tape facilitates the positioning of the conductors to pads on the chip. The copper conductors are either deposited as a layer on the tape by using sputtering or vacuum deposition, or made from a layer of copper foil. The copper lead pattern in the layer is produced by photolithographic processes or by mechanical blanking. The copper conductors are then plated with nickel for corrosion protection, and with gold to facilitate bonding. A window opening is made in the tape for bonding the inner leads to the chip, and another for bonding or soldering the outer leads. The windows are made either by blanking before the foil is added, or by chemical blanking after the leads are in place. Fig. 13K6c illustrates the tape with leads affixed, and shows the means of bonding the leads to a chip. Bonding to the chip is the first connection. Thermocompression bonding is used to connect the inner leads to gold bumps on the pads of the chip. One tool, called a thermode, provides the heat and pressure for all chip connections simultaneously. The operation is highly automatic and quick. After bonding of the tape to the chip, the chip is normally covered with a liquid resin encapsulant that, when the resin hardens, provides protection to the chip and the connections. Another advantage of TAB is that the tape connecting leads are very thin, helping provide a low profile for the IC, if that is needed. The other ends of the tape leads are connected to the circuit board or other substrate by one of several methods: reflow soldering, conductive

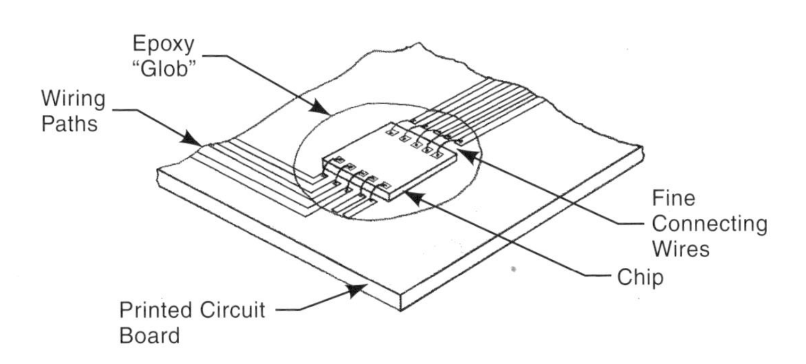

Fig. 13K6a Chip-on-board (COB) packaging of an integrated-circuit chip. A glob of epoxy plastic, material very similar to that used to mold plastic packages, covers the chip and its wiring connections to the circuit board. The plastic provides physical and chemical protection to the integrated circuit. Note: The epoxy glob is opaque, but is shown here as transparent to illustrate the items it covers.

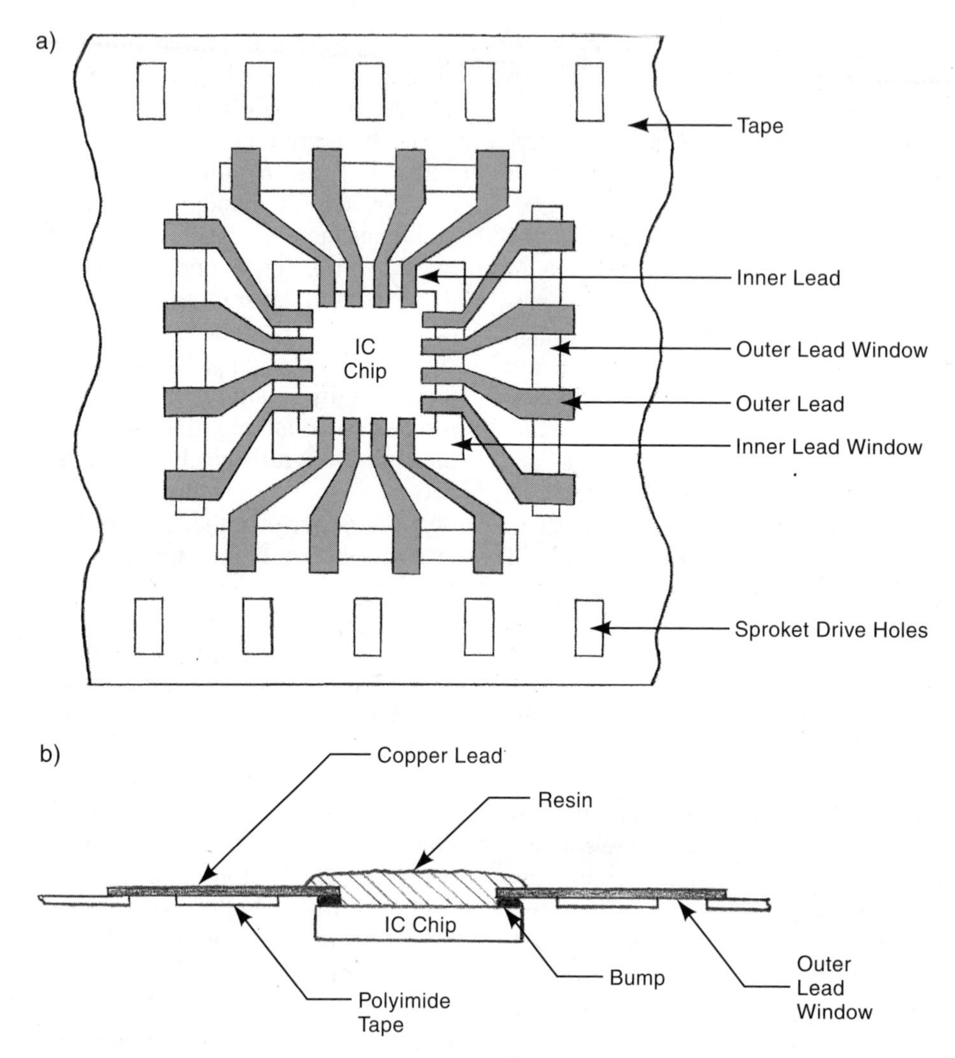

Fig. 13K6c Tape automated bonding (TAB) of an integrated circuit chip: a) a plan view of the copper leads affixed to a polyimide tape with window openings for connection to a chip and later for connection to a package substrate or circuit board. b) a cross section of the chip bonded electrically to the copper leads with a protective resin coating over the chip and the connections.

adhesives, or thermo-compression bonding. A portion of the plastic tape backing is retained.

K6d. ***flip-chips*** - constitute a method for connecting integrated circuits to printed circuit boards with no wire leads and very short connections. The design facilitates the operation of high speed devices such as PC microprocessors. With this method, the chip is first connected to the substrate of the IC package. Solder bumps are provided on the top of the chip by plating, or by stenciling solder paste on the connection pads and then reflowing. The chip is inverted ("flipped") and positioned on the package substrate. Pads on the substrate match the spacing and position of the bumps on the chip. The chip is then reflow soldered again to connect the chip circuits to the substrate. The balance of the package then can be assembled to the chip. There is a small space between the pads on the chip and the pads on the substrate so that the solder connection is almost spherical in shape. This provides some flexibility in the connection. The substrate has internal wiring

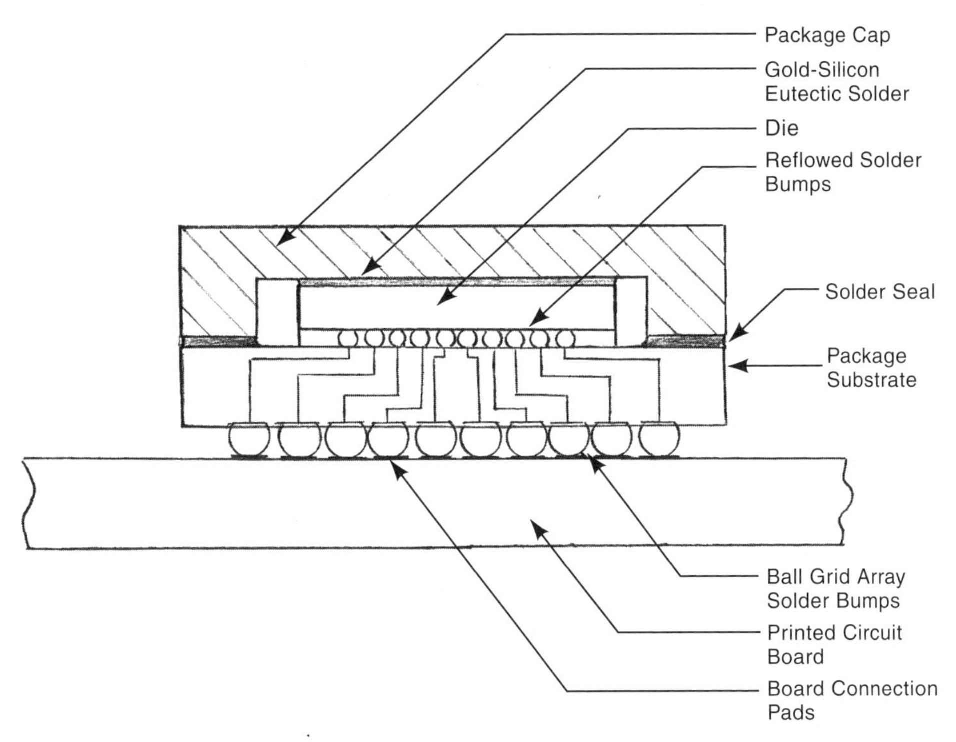

Fig. 13K6d Example of a flip-chip connected to a circuit board with a solder sphere ball grid array.

channels that connect to underside wiring pads, which are larger and more widely spaced than the pads on the chip and the top of the substrate. These pads are used to connect the packaged chip to the circuit board, using the ball grid array method. Solder spheres are positioned and held in place on the circuit board with a tacky flux. When these spheres are reflowed, a very short connection with minimum capacitance effects is achieved. Fig. 13K6d illustrates the completed arrangement.

For added reliability of the solder joints, some flip chips receive *underfilling*. This involves the addition of a plastic encapsulant to the solder joint area. The encapsulant helps to distribute stresses caused by thermal cycling of a group of solder joints. Epoxy or silicone resins in low-viscosity form are dispensed to the joint from hollow needles and flow into the space between the solder bump connections by capillary action. Jet dispensing is also used.

Curing temperatures are in the 280 to 360°F (140 to 180°C) range.

L. Making Discrete Devices

Transistors, diodes, resistors, capacitors, inductors, and transformers are all included, where necessary, in integrated circuits. However, there are circuits in electronic products that need power handling capabilities, or other capacities greater or different than those normally included in integrated circuits. Output stages of electronic equipment and power supplies are two kinds of circuits that require discrete devices. Methods of producing the devices are discussed in the following:

L1. ***making resistors*** - Resistors are of several varieties: composition, film, wire-wound, or integrated circuit.

Composition resistors are molded into a cylindrical shape from a mixture of finely ground carbon powder, a powdered insulator material and a resin binder. The resistance value depends on the ratio of the carbon to the insulating material, as well as the diameter of the resistor. Copper wire leads are attached to the ends, and then a plastic or ceramic jacket is molded over the cylinder. The jacket is marked with the resistance value. Other composition resistors are made from mixtures of tin oxide, antimony, and glass.

Film resistors consist of a thin film of carbon, metal, or metal oxide deposited on a cylinder of ceramic or glass. A helical groove is cut into the film coating to narrow and lengthen it. The amount of resistance depends on the nature (resistivity) of the film, its thickness, and the pitch and width of the helical cut. The value of resistance may be measured during the cutting, and the cutting may be stopped when the desired resistance value is reached. Hence the resistance value of this kind of resistor can be made to be quite accurate. After the desired resistance value is set, the unit is encapsulated in silicone-treated glass and marked or color coded to indicate the value. Some film resistors are made with a serpentine pattern on a flat surface, rather than on a cylindrical surface. This minimizes inductive effects for resistors used in high frequency circuits. Fig. 13L1 illustrates a typical thin film resistor. The manufacture of thick film resistors is described above in section K3a6.

Wire wound resistors are made by winding resistance wire on a cylindrical ceramic form. After winding, the unit is inserted in a ceramic or metal container, or coated with vitreous enamel. These resistors have inductive and capacitive properties that limit their use to direct-current or low-frequency applications.

Adjustable film or wire wound resistors are made by incorporating a sliding contact in the construction and not coating the path where the contact slides. The length of the resistance material in the circuit controls the amount of electrical resistance.

Resistors in integrated circuits are formed from layers of the circuit. In one method, grown oxide layers are masked to provide the necessary shape, size, and location of the resistor, and are changed in resistivity by doping the masked area. Doping is accomplished by diffusion or ion implantation. Diffusion methods follow a series of steps very similar to those used to create transistors in the chip. By limiting the cross-sectional area of the resistance bar, high levels of resistance can be created. The other method of making resistors is to deposit a thin film of resistance material, nichrome, cermet (CrSiO), or tantalum and then laser trimming the film layer.

L2. ***making capacitors*** - Electronic capacitors have two or more electrodes separated by some dielectric. Numerous dielectrics can be used, from air with a relative permittivity of 1, to some ceramics with relative permittivities exceeding 1000. Air dielectrics are used in variable capacitors, and the ceramic, barium titanate, is the dielectric most frequently used in fixed capacitors in electronic circuits. In between these, are capacitors that use plastic films, glass, or kraft paper. Barium titanate ($BaTiO_3$) is made by mixing barium carbonate and titanium dioxide and firing the mixture. The two materials must be of high purity, have a sub-micron

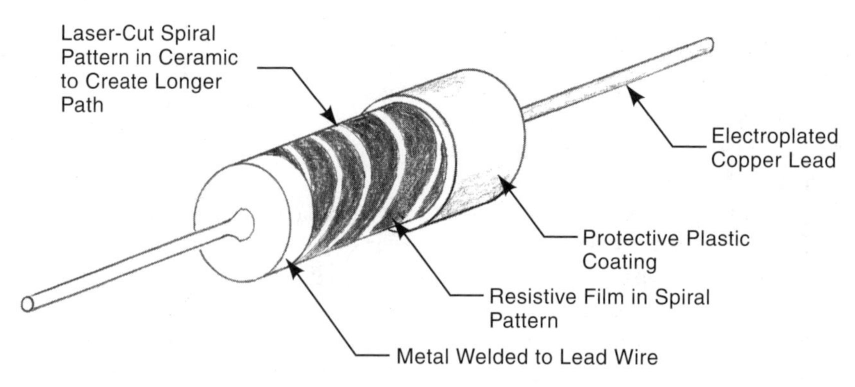

Fig. 13L1 A typical film resistor shown in a cutaway view. The ceramic core is coated with a film of resistive material. A laser cut creates a long spiral electrical path in the film.

particle size, and be mixed in a closely-controlled ratio. For disk capacitors, the resulting material, in dry powder form, is pressed into the shape wanted, or is made into a thin tape and blanked to shape. It is then fired at a temperature in the range of 2280 to 2460°F (1250 to 1350°C). Silver, in paste form, is screen printed on the surfaces to form electrodes, and is bonded to the surface by heating to a temperature of 1380°F (750°C). Leads are then soldered to the electrodes and the disks are encapsulated in epoxy or wax.

Multilayer capacitors can be made with desirably thin layers of dielectric, without suffering a strength loss. They have higher capacitance than the disk type, have small size, and operate well at high frequencies. They are used frequently in surface-mount circuit boards. Dielectric thicknesses can often be less than 0.00022 in (5 microns). Palladium, or a palladium/silver alloy, is used for the electrodes and the dielectric and electrodes are assembled together in layers before firing. (The palladium electrodes have a higher melting point than the firing temperature.) For less critical applications or when lower cost is important, ceramics with added glass or fluxes, and electrodes of nickel or copper may be used. These electrode materials are less costly and the assembled components can be fired together in a reducing atmosphere at temperatures of about 2000°F (1100°C).

Another common type of capacitor uses polyester, polystyrene, or polypropylene film as a dielectric, sometimes combined with a layer of kraft paper. The electrode is aluminum or zinc, coated on the plastic film by vacuum deposition. The combination is wound to a cylindrical roll shape. Connection wires are soldered in place and the cylinder is encapsulated in plastic. Fig. 13L2 shows the construction of typical multilayer capacitors and the construction of the wound-roll type.

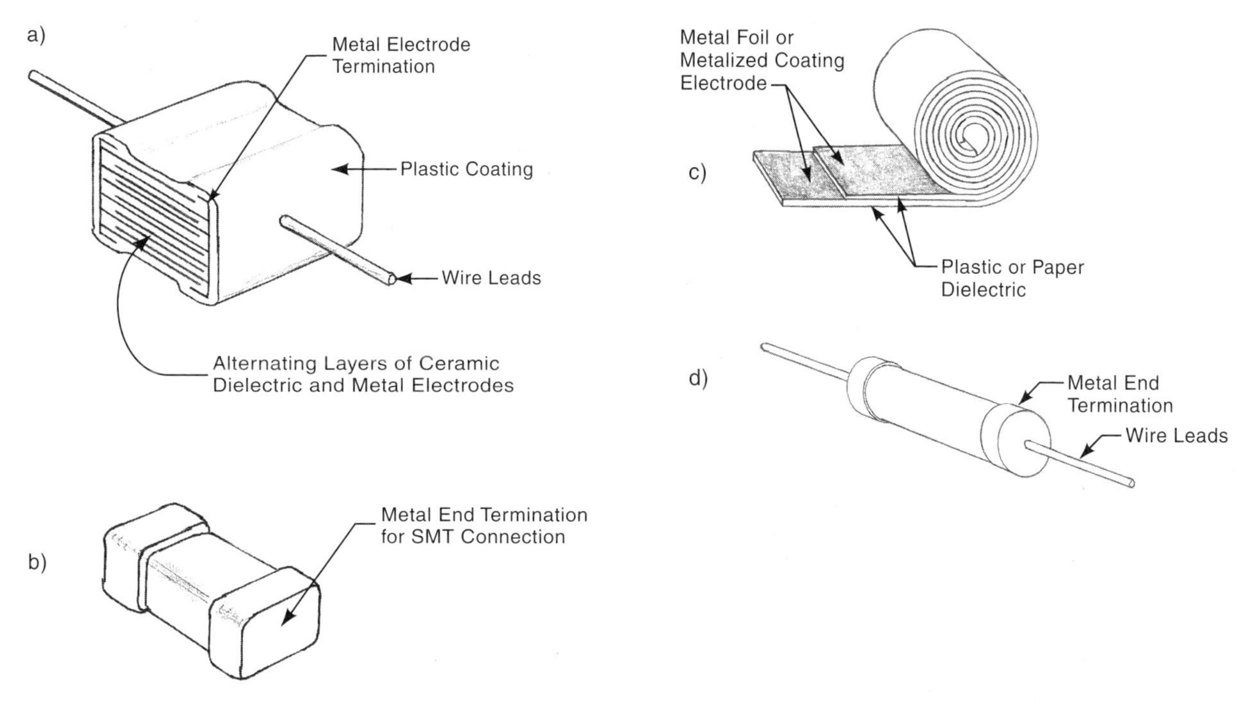

Fig. 13L2 Electronic capacitors -a) sectional view of the construction of a typical multilayer electronic capacitor with a ceramic dielectric, showing alternating layers of electrodes and dielectric connected to external leads. b) a similar capacitor to that in (a) except that, instead of wire leads, metal end terminations are used that are suitable for connection to surface mount printed circuit boards. c) the construction of spiral-wound capacitor with dielectric of plastic film or paper or layers of both, with metalized coatings or foil electrodes. d) the capacitor of view©) with leads connected to alternating layers of electrodes and with plastic encapsulation.

Thin-film capacitors are made by vacuum deposition of conductive layers and dielectric layers through masks that limit the areas of deposition. Dielectric materials used include tantalum pentoxide, silicon dioxide and monoxide, magnesium fluoride, and zinc sulfide. Capacitance is adjusted by laser cutting as is done with thick-film methods.

The manufacture of thick film capacitors is described above in section K3a6.

L3. ***making inductors (chokes, choke coils)*** - These are electrical or electronic devices that tend to oppose rapid changes in current intensity. They consist of coils of conductor wire, wound in a round or rectangular shape. Depending on the application and frequency, the coils may be wound on a magnetic core, a dielectric core, or have only an air core. Magnetic cores may be made of magnetic steel, ferrite, or pressed powdered metals. One common material is 50% iron and 50% nickel. Ferrites are ceramics consisting primarily of iron oxide with nickel, manganese, cobalt, or zinc oxides or carbonates.[12] Coil winding is a semi-automatic operation. The coil wire is wound directly on the core, on a bobbin, or on a mandrel whose cross section is the shape wanted in the coil. One end of the wire is attached to the mandrel that is then rotated, pulling the wire from a supply reel. As the wire is wound and builds on the mandrel, it is fed in such a way as to maintain even layers. Computer control can provide variable spacing of the wire turns, and can stack multiple layers in a desired shape or pattern. Sophisticated motor and braking apparatus can wind and stop quickly, to a desired fraction of a revolution.[12] When the correct number of turns have been made, the wire is cut, stripped, and connected to the terminals of the inductor.

Cores of powdered iron, nickel and other metals are made from finely-ground powder mixed with an insulating powder, pressed to the desired shape, and annealed to provide the desired level of magnetic properties. Ferrite cores are also made from finely-milled metal oxides, mixed with organic binders, dried, and pressed into the desired shape, fired in a controlled atmosphere and then tumbled to improve the surface and edge finish. Both metal and ferrite powder cores may be painted before assembly to the wire coil. After the coil is wound and connected, it is common practice to encapsulate the inductor in soft rubber covered with epoxy, or solely with epoxy. The inductor is then color coded and otherwise marked for identification. Fig. 13L3 illustrates a typical inductor for surface mounting on a printed circuit board.

The manufacture of thick film inductors is noted above in section K3a6.

L4. ***making transformers*** - Transformers convey energy from one circuit to another, usually stepping

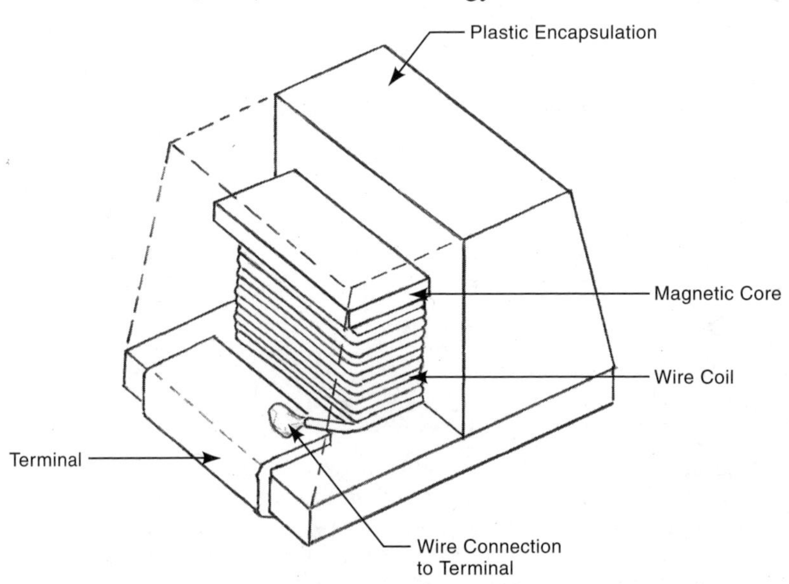

Fig. 13L3 Cut-away view of an inductor suitable for mounting on a surface-mount-type printed circuit board.

AC voltage up or down at the same time. Two major components are the coils of wire, each wound with a specific number of turns. (The ratio of number of turns is the ratio of the voltage change between the circuits.) Another major component is the metal core that is magnetic and is normally made from an iron alloy. The usual method is to build the core from stacks of sheet iron (*laminations*) of the proper shape - a shape of the letter "E" is common with single-phase power - with the open side of the E closed in with straight strips that are laid alternately with the E's. The magnetic permeability of the iron is important and the magnetic lines are carried around the coil in two directions. Iron oxide or another material provides the necessary insulation between the laminations. When the E-configuration is used, both coils are assembled over the center leg of the E. Fig. 13L4 illustrates this construction. When a "C" shape is used, separate coils are assembled on each side of the C. Coils are wound to the prescribed number of turns, first on semi-automatic machines that wrap the wire on bobbins or around mandrels of the proper size and shape. (See *inductors*.) The coil wire is insulated with a clear varnish and sometimes is square in cross-section. Protective insulating tape may be wrapped over the coils and between separate windings to provide protection and added insulation between core and windings or separate windings. The coils and laminations are then assembled together. Except for high mass-production transformers, lamination stacking and coil assembly to the laminations are manual operations, but some robotic handling has been employed. The end wires of the coils are stripped and connected to transformer terminals. The transformer is usually contained in a metal or plastic enclosure, and may be fully imbedded in a plastic material that adds insulation between windings of the coils and between the two coils. Larger power transformers are enclosed in a tank of oil that provides both insulation and cooling effect, especially if the oil is pumped to circulate. Transformers are most frequently found in power supplies for electrical and electronic equipment. They are also used extensively in audio and radio-frequency apparatus to transfer signals from one circuit to another, at the higher frequencies with an air core instead of magnetic laminations. In electronic circuits, many transformers have been replaced with transistor circuits.

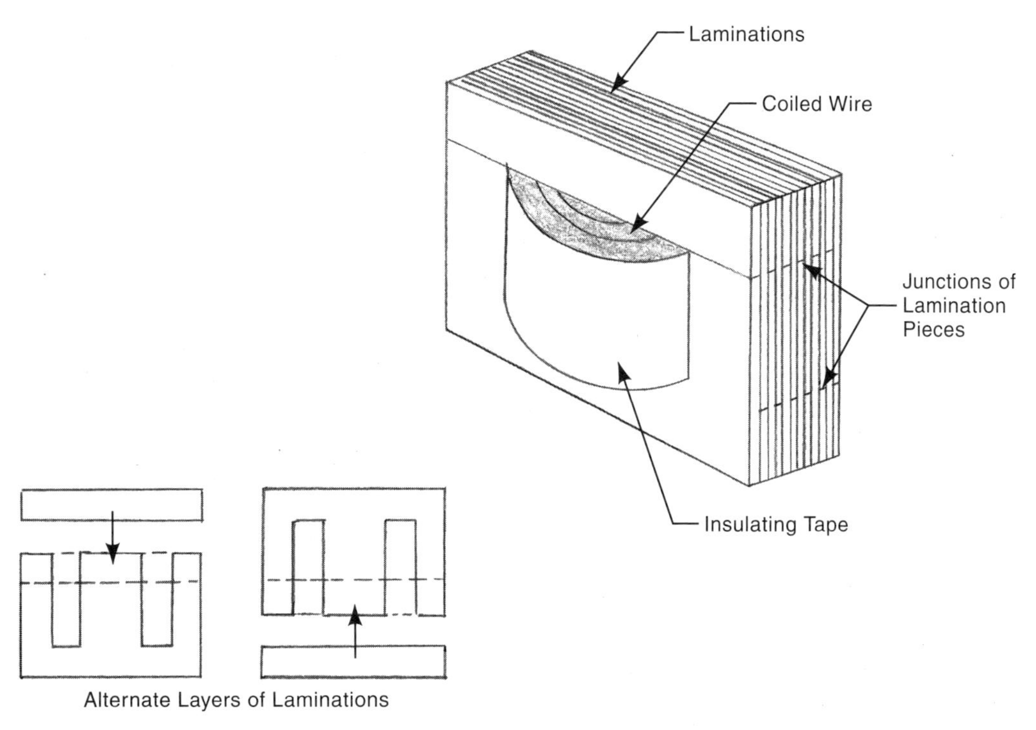

Fig. 13L4 A typical audio transformer before it is encapsulated.

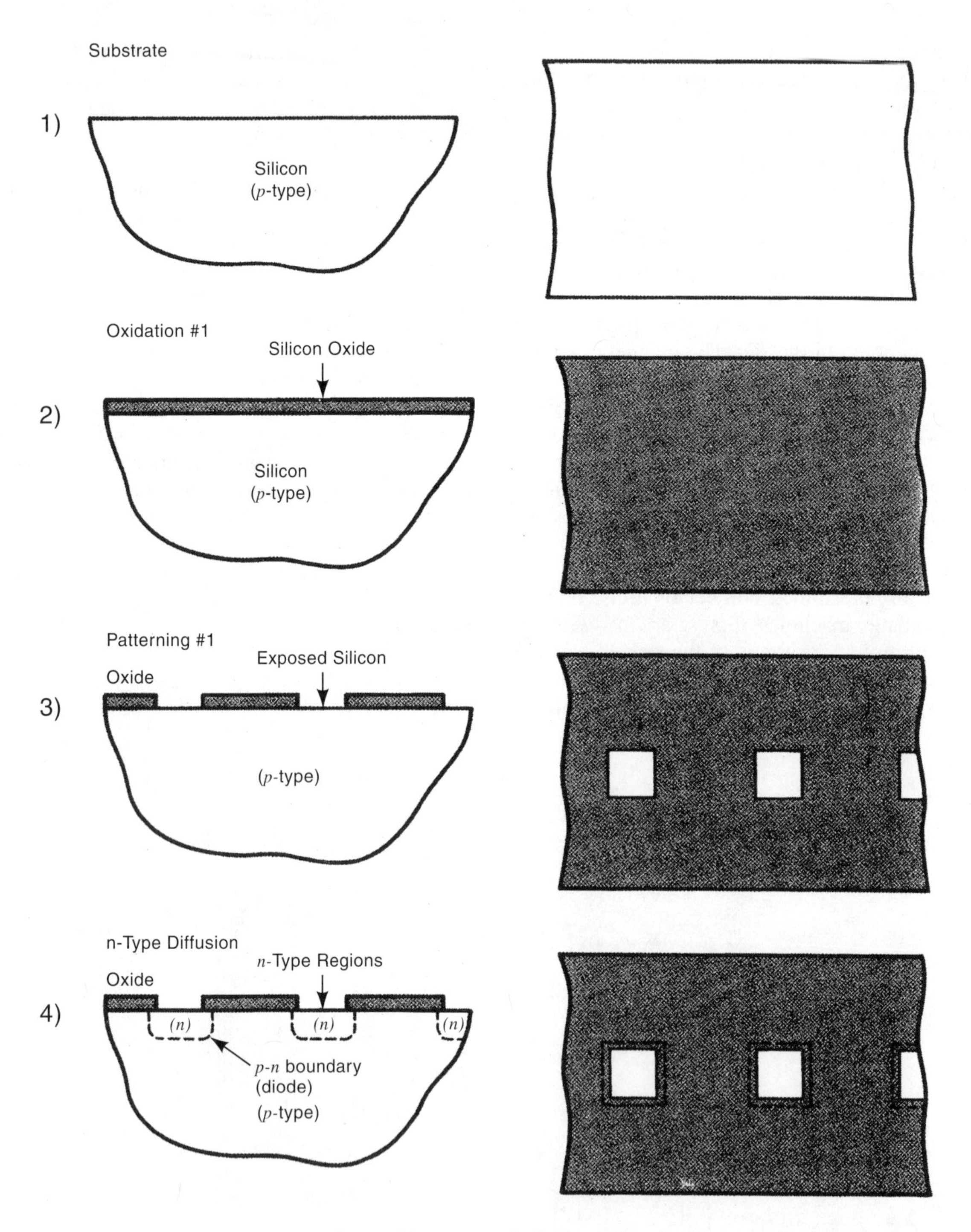

Fig. 13L5 The operation sequence for making a set of diodes. The sketches on the left represent cross-sections of the diodes as they are fabricated on a silicon substrate. The right hand sketches show the corresponding top views. Note that only a part of the substrate and its coatings are shown of the operation that produces a large quantity of diodes simultaneously. (*Reproduced with permission from Microlelectric Processing - An Introduction to the Manufacture of Integrated Circuits, by W. Scott Ruska, McGraw-Hill, 1987.*)

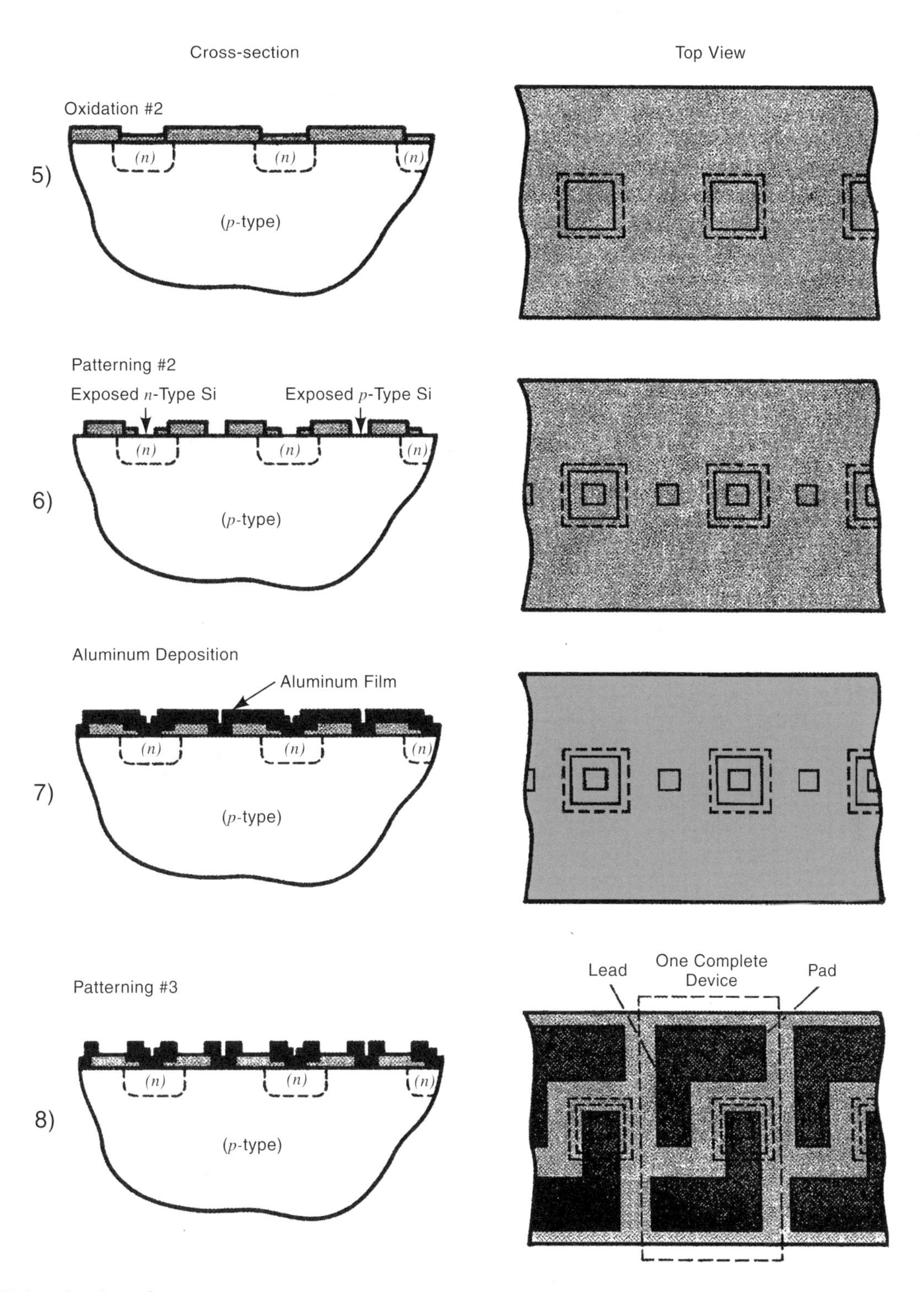

Fig. 13L5 *Continued*

L5. ***making discrete transistors and diodes*** - Transistors and diodes are solid-state electronic devices made from single-crystal semiconductor materials, usually silicon or other semiconductor materials containing gallium, aluminum, and arsenic. These devices are made with the same techniques used in the manufacture of integrated circuits as described above. Layers of silicon dioxide are grown. Layers of other materials are vapor deposited. Openings are made in the layers by photolithographic techniques and etching. Entire silicon layers and exposed areas are doped with impurities to establish the way electric current will flow. Open areas are doped by either diffusion or ion-implantation methods. Deposits of conductor, semiconductor, or dielectric layers, are limited by photolithographic techniques or by printing or stencilling methods. One difference between integrated circuit device manufacture and discrete device manufacture, however, is that the discrete transistors are found in applications where the electrical current and voltage levels may be higher and where larger circuit elements are needed to conduct the current. Additionally, the extremely small size of devices in integrated circuits is not normally required for discrete devices. Like integrated circuits, discrete transistors and diodes are made on wafers, are tested, cut from the wafer, are connected to terminals and assembled in a protective metal, ceramic, glass, or plastic package. The package is provided with external leads, pads, or other terminals for connection to the circuit where it is used. Fig. 13L5 illustrates the manufacturing sequence for production of discrete diodes, the operations for which are the same as for transistors except that diodes are inherently simpler, having two major components (base-collector or emitter-base) instead of three (base-emitter-collector).

M. Conformal Coatings

Conformal coatings are applied to an entire printed circuit board after all traces and devices on the board are soldered. The coatings provide protection to the circuits from the contaminating effects of moisture, perspiration, dust, and ambient atmospheres. Most such coatings can be penetrated for repair soldering if that should be necessary at some point. The coating is applied after the post-soldering flux removal and cleaning to preserve the clean state of the board. The application method is vapor deposition, spraying, dipping, or brushing.

N. Other Chip Configurations

N1. ***multiple integrated circuit packages (multichip devices, assemblies, modules [MCM], system in a package (SIP) or packages)*** - involve the assembly, in one protective package, of several integrated circuit chips (unpackaged) and, optionally, some other components connected on one substrate. This type of package is used when high speed of operation of the circuit is important. Normal connections from chips to other devices and board circuits exhibit capacitance effects that slow the rate of current flow. By putting several interconnected chips in one package, and on one common substrate, the connecting paths between them are made much shorter and the operations of the chips can be faster. Higher frequencies, better performance, and a more compact arrangement are also achieved. The cost, however, is higher than if the chips were mounted in individual packages on a circuit board. Connecting circuits are made using thin or thick films of metal or other conductive materials. This wiring is put on the ceramic or silicon substrate by using the additive method (described in A1c above) to produce a multilayer substrate. Resistors and capacitors may be formed and connected to the chips as well. As many as five chips may be included in the module. These are usually mounted on the substrate with an epoxy adhesive. The conductive films employed may be made from titanium, palladium, tantalum nitride, and electroplated gold. Electrical connections are made by wire bonding. The final package, containing both substrate and chips, may consist of a sealed metal, ceramic, or silicon rubber capsule, with external leads for connection to the circuit board. Fig. 13K5, view c, shows a simple arrangement of two chips.

References

1. *ASM Electronic Materials Handbook, Vol. 1,* ASM International, Materials Park, OH, 1989.
2. *Soldering Handbook for Printed Circuits and Surface Mounting, 2nd ed.,* Howard H. Manko, Chapman and Hall, NY, 1995, ISBN 0-442-01206-3.
3. *Solders and Soldering, 3rd ed.,* Howard H. Manko, McGraw-Hill, NY, 1992, ISBN 0-07-039970-0.
4. *Solder Paste Technology,* Principles and Applications, Colin C. Johnson and Joseph Kevra, TAB Professional and Reference Books, Blue Ridge Summit, PA, 1989, ISBN 0-8206-3203-4.
5. *Microelectric Processing—An Introduction to the Manufacture of Integrated Circuits,* W. Scott Ruska, McGraw-Hill, New York, 1987, ISBN 0-07-054280-5.
6. *Solders and Soldering, 4th ed.* Howard H. Manko, McGraw-Hill, NY, 2001, ISBN 0-07-134417-9.
7. *Handbook of Printed Circuit Board Manufacturing,* Raymond H. Clark, Van Nostrand Reinhold Co., New York, 1985.
8. *Printed Circuit Basics—Michael Flatt,* Miller Freeman, San Francisco, 1992.
9. *McGraw-Hill Encyclopedia of Science and Technology, Vol. 9,* McGraw-Hill, New York, 2002.
10. *Encyclopedia Britannica,* 2002.
11. *Academic American Encyclopedia,* 1998.
12. *Microchip Fabrication, 4th ed.*—Peter Van Zant, McGraw-Hill, NY, 2000, ISBN 0-07-135636-3.
13. *Processing Difficult Lead-free Boards,* paper by Alan Rae, VP Technology, Cookson Electronics.
14. *Electronic Components,* Victor Meeldijk, John Wiley, New York, 1996.

Chapter 14 - Advanced Manufacturing Methods

A. Rapid Prototyping (RP) Methods

Rapid prototyping is a means of producing component parts or accurate replicas of them in a short lead time. These parts are made with special automatic non-traditional fabrication methods from sophisticated computerized designs, without the use of special dies, molds, jigs or other tooling. All common rapid prototyping methods build the parts with a layer-by-layer approach, under computer control. Parts thus made can be tested and evaluated much sooner than would be possible if traditional production methods were employed.

Rapid fabrication methods can be placed in one of three basic categories, depending on the form of material used: liquid-based systems, solid-based systems and powder-based systems. Regardless of the specific process used, there are six major steps in the process for rapid prototypes: 1) The component is designed using a CAD (Computer Aided Design) program with 3D modeling capability. 2) The design data are converted to the STL ("Stereo lithography") or similar format. The data are transmitted to the shop-floor computer that controls the rapid prototyping machine. 3) The design data are transformed in the shop computer to represent a series of "slices" of the solid model. 4) The shop-floor computer converts the design data into operating instructions for the rapid prototyping machine. 5) The prototype part is fabricated by the machine and, 6) Necessary post-processing operations are performed. This sequence is summarized in Fig. 14A.

Rapid prototypes have the following uses: 1) as visual concept models - to visualize and verify appearance, fit and design features. 2) as casting patterns. Wax or plastic prototypes can be used as patterns for the manufacture of investment, plaster or sand-mold cast metal prototypes. 3) for use as

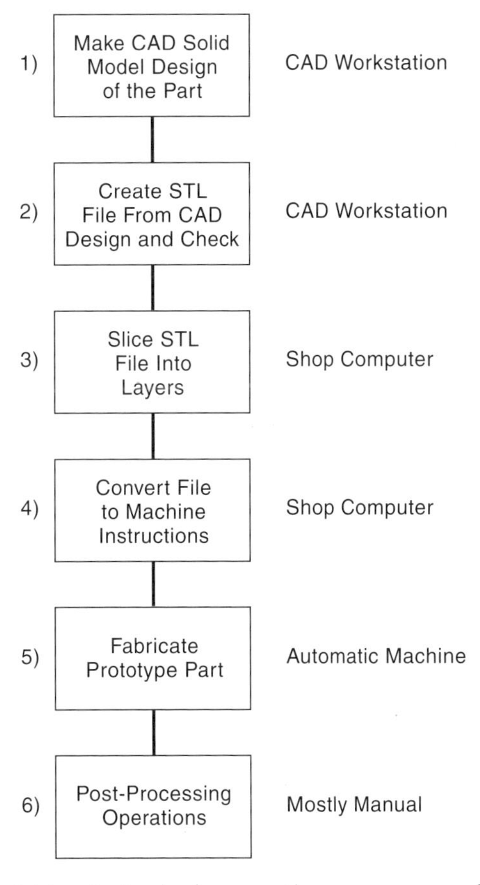

Fig. 14A A typical operation sequence for a rapid prototype part, made by any of a number of RP methods.

patterns for producing non-consumable short-run tooling of epoxy, rubber or other materials for plastic molding of prototypes or short-run production parts. 4) as functional prototypes. Sometimes, rapid prototypes can be made sufficiently strong that they can be used in operational testing of the product that uses the part.

A1. ***initial step: computer aided design (CAD)*** - Rapid prototyping begins with a computer aided design (CAD) of the part to be made. The design must incorporate an advanced 3D geometrical solid modeling of the part, though some surface model files can be used. The CAD file for the design must define a fully enclosed volume. Outside and inside surfaces and boundaries must be completely specified. This level of CAD data for the part to be made is a prerequisite for all rapid prototyping systems.

A2. ***the STL file*** - Another computer program converts the solid model data to an STL (as in stereolithography) or similar file. The STL file processes the CAD data and converts it to a description of the surfaces of the part as a series of triangles. The triangles approximate surfaces that have curvature by using many small triangular facets at slight angles to each other. The x, y and z coordinates for each vertex of each triangle are part of the file. There is also an indication of the direction of a perpendicular line to each triangular surface. (This indication positively differentiates the outside and inside surfaces of the part.) The complex procedures of designing a solid model and expressing the surface as a series of triangles are subject to some human and system errors. Common practice is to check the STL file carefully for gaps, cracks, holes, and other defects in the part's surface, because faults must be eliminated before a satisfactory prototype can be made. Checking is done manually, usually with the assistance of a special program developed to assist the operation. Any necessary corrections are then made. The STL format is the most frequently used format for rapid prototyping and has become a de facto standard. Some systems, however, use other formats. IGES (Initial Graphics Exchange Specifications) uses more complex algorithms to define surfaces but it provides a precise representation of their geometry. Several other formats, including SLC (Stereolithography Contour), are used in some RP systems.

A3. ***the SLI file*** - is usually created in the shop computer, the one that controls the rapid prototyping equipment on the factory floor. The shop-computer program takes the STL data and divides it to portray the part as a series of horizontal slices from 0.001 to 0.020 in (0.025 to 0.50 mm) in thickness. This is the SLI (for SLIce) file. The layers it represents, when stacked together in the proper sequence, very closely follow the shape of the prototype part. All the common RP systems use this layer-additive approach in making prototype parts. The thinner the slices in the program and the greater the number of layers in the prototype, the more closely it describes the shape of the part and the smoother the surface it produces.

A4. ***liquid-based rapid prototyping systems***

A4a. ***the stereolithography (SLA) system*** - The use of this system requires a 3-dimensional CAD solid model file for the part to be produced and the use of the STL program to prepare a file for the part, as discussed above. Then, a shop-floor computer, the control computer for the stereolithography machine, checks and verifies the correctness of the data. The computer then creates an SLI file which slices the model into a series of horizontal cross sections from 0.004 to 0.020 in (0.10 to 0.50 mm) thick, puts the slices in a graphical form that permits them to be visually verified by the designer or machine operator, and, most important, puts the data in a form that operates and controls the SLA machine.

The process involves the solidification of a liquid plastic resin, a photopolymer, through polymerization caused by focused laser radiation. The molecules of the resin, a monomer, link together with the aid of the laser energy and change to a solid polymer. The laser beam is actuated and moved by computer control based on the data in the CAD, STL and SLI files. The operation takes place in a tank containing the liquid and a supporting table for the prototype. The height of the supporting table can be controlled accurately. The operation starts with the fabrication of special supports for the part involved. The supports are usually needed to hold portions of the part being fabricated before it is substantial enough to support itself or hold itself together. The table and supporting surfaces are placed just below the surface of the liquid resin as the process begins. A leveling wiper moves across the table to ensure that there is a complete thin coating of liquid photopolymer. The laser beam scans the liquid surface in accordance with a computer-programmed path, changing the surface resin from a liquid to a solid. The focused laser spot is usually about 0.008 in (200 microns) in diameter. The beam is moved with mirrors so that all points to be solidified are exposed to the laser. The first solidified material becomes the bottom layer of the prototype. After the first layer is complete, the support surfaces are lowered slightly and become covered with another layer of photopolymer. The leveling wiper again transverses across the surface. The laser traces a path of radiation to solidify the next layer. The process continues, solidifying the liquid, layer by layer, to create, from the bottom up, a solid plastic prototype part. Fig. 14A4a illustrates the process schematically.

Postprocessing operations follow fabrication of the complete prototype part. There are three basic operations: 1) final curing of the photopolymer resin, 2) separation of the part from any support structure and 3) any miscellaneous finishing operations that may be required. The final curing involves placement of the workpiece in post-cure apparatus that bathes the workpiece in ultraviolet radiation. This thoroughly cures the resin to full strength, completing any cross-linking that was not accomplished by the laser beam. This operation may take one hour. The next step is to remove any supports that were needed during the layering. This is usually done manually. The third step is to carry out any surface finishing or refinement operations that may be needed. These operations include sanding, glass bead blasting, polishing, electroplating or painting as appropriate and drilling, milling or thread tapping, if required.

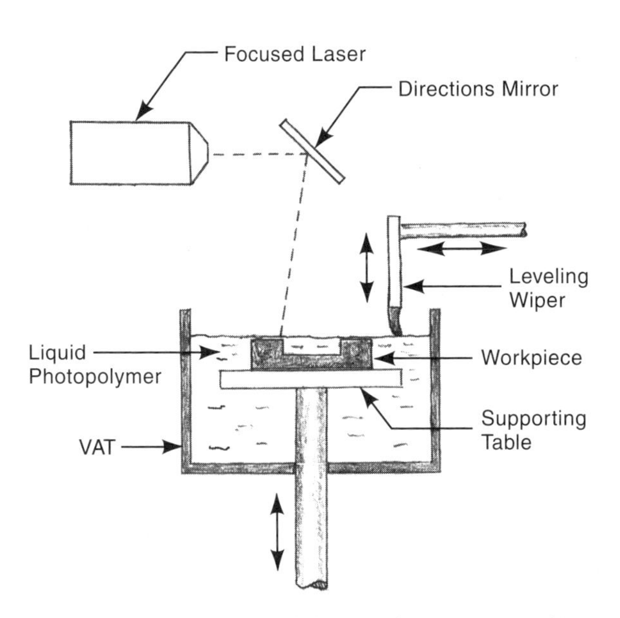

Fig. 14A4a The stereolithography process (SLA). Liquid photopolymer is polymerized by a laser beam in the areas that delineate a horizontal slice of the prototype part to be made. The worktable is lowered, more liquid photopolymer is wiped onto the workpiece and the process is repeated, layer (slice) by layer, until the prototype is completed.

A4b. ***QuickCast*** - is an RP method for making cast metal prototypes. The first step uses epoxy resins in a stereolithographic process similar to that described above. However, instead of acting as the sample part itself, the prototype is used as a pattern in casting a metal prototype. Casting is by the investment casting process. (The investment casting process is described in items G, G1 and G2 of Chapter 1). Patterns made by SL in QuickCast have a "quasi-hollow" structure, i.e., they are hollow except for a light internal structure inside the thin, external shell. This structure provides sufficient stiffness to the pattern and promotes its better accuracy. The hollow spaces lessen the stresses and distortion of the part due to the shrinkage that occurs during polymerization. They also greatly reduce the amount of thermal expansion of the pattern that takes place later, when the ceramic investment casting mold is heated. This expansion would otherwise cause the ceramic mold to crack, rendering it unusable.

After the pattern is formed, it emerges from the SL equipment filled with liquid resin in the spaces between structural members. The part is then placed where the liquid resin can drain. To allow this resin to escape from each internal cell, a special hatch arrangement was developed for the reinforcement structure. The pattern then undergoes post-treatment. This includes removal of any attached support structure, filling any pinholes in its surface, and filling the openings from which the resin has drained. Oven postcuring then takes place to ensure that all the material is fully polymerized. The investment casting process then can proceed, using the shell mold method. If not incorporated in the pattern, material to form sprue and runners is added. The pattern is then coated with or dipped into a slurry of finely-ground ceramic material that is then dried thoroughly. Dipping and drying are repeated through a total of about 6 or 7 cycles to build up a coating of the necessary thickness. The mold is then heated to a high temperature to burn out the pattern and fuse the ceramic of the mold. Molten metal is poured into the mold and allowed to cool and solidify. After the casting solidifies, the ceramic mold is broken away from the cast part, the casting is cleaned of loose ceramic, sprues and runners are removed and other post casting refinements may be carried out. The resulting part is a metal prototype of the CAD design, suitable for functional, environmental and life testing as well as for testing of fit with mating parts and testing of strength, appearance, and all the other properties it requires. A wide variety of ferrous and non-ferrous metals can be cast. The process utilizes epoxy polymers that have been developed to minimize shrinkage during polymerization and distortions afterwards due to creep.

The QuckCast process is used as a rapid tooling method if the patterns made are those of mold halves instead of a prototype part. The two cast-metal mold halves then can be put together and used to injection mold prototype (and production) parts in whatever plastic material is specified for the part. See *Rapid Tooling* below.

A4c. ***solid ground curing (SGC)*** - is another process that uses liquid photopolymer resins but it does not use laser energy to effect polymerization. Instead, it creates a temporary photomask for each layer using a process similar to xerography. In the process, black toner powder is held on a glass plate in areas of the plate that have been electrostatically charged by an ionographic process. (Where the plate is not charged, the toner is easily removed; where it is charged, the toner is retained.) The photomask then is carefully aligned with the workpiece location and used to direct ultraviolet radiation from a high-powered, collimated UV lamp to a layer of liquid photopolymer resin. The radiation cures the polymer in all exposed areas but not in areas blacked out on the photomask. An airknife then removes the uncured liquid resin. The curing step is followed by application of a coating of molten wax to the entire surface but primarily in areas surrounding or open between cured resin areas. The wax supplies support for the prototype, including outlying features, as it is being fabricated. A cooling plate descends on the wax to cool and solidify it and, when the plate is removed, a mechanical face milling cutter passes across the surface and removes excess polymer and wax to provide a flat surface of exactly the correct thickness. (The hardened polymer layer, before machining, is thicker than the final desired thickness.) A vacuum cleaner removes the chips created by the machining. The elevator mechanism then lowers the workpiece slightly below the surface of liquid photopolymer, causing it to be covered again with the liquid. The process of making a photomask, coating the surface with liquid polymer, removing excess, curing the polymer, coating with wax, cooling, machining, and vacuuming it, follows for the next layer. Layers are added until the prototype is completely formed.

The SGC process begins, as with other RP systems, with a CAD model of the part to be made. As with other systems, the CAD data is put into the STL or another similar format and is "sliced" after transmittal to the RP machine's computer. The computer software, in this case, is the manufacturer's DFE (for Digital Front End) program. The program guides the ionographic process, using electrostatic toner, to prepare the mask plate for each layer of the part. When each use of the mask is completed, the electrostatic toner is removed and the glass plate is reused for the next layer. After the mas-making and layering processes are complete for all layers of the prototype part, the finished prototype is removed from the machine. The prototype is surrounded by wax that must be removed. This is done with a high-temperature wash of water and citric acid. A disadvantage of SGC is that it requires the presence of a full-time operator. An advantage of the process is that multiple prototypes can be made simultaneously in the same operation, so long as there is room in the equipment. The capacity of current units is 71 in wide × 165 in long × 114 in high (1.8 × 4.2 × 2.9 m high). Fig. 14A4c illustrates the process sequence.

The process is being used to make patterns for investment casting metal prototypes, for short-run injection molding of plastic parts in silicon rubber, epoxy, or other molds, as well as for making plastic prototypes.

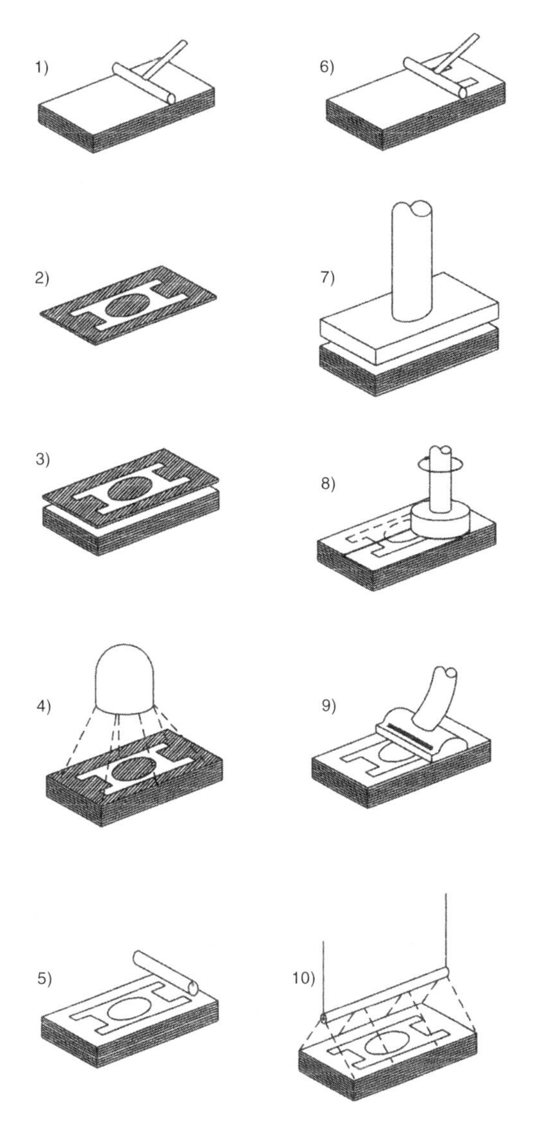

Fig. 14A4c The process sequence for the Solid Ground Curing process (SGC):
1) A coating of liquid photopolymer is applied to the workpiece surface. 2) A glass photomask is made for the next workpiece layer. 3) The photomask is aligned with the workpiece. 4) Collimated ultraviolet light is shined on the workpiece through the mask. 5) Uncured resin is blown away with an airknife. 6) The workpiece is coated with liquid wax. 7) The wax is solidified with the aid of a cooling plate. 8) The surface of the workpiece is machined flat with a face milling cutter. 9) A vacuum cleaner removes loose chips and the workpiece is lowered slightly. 10) Final curing takes place with the aid of another ultraviolet light. The process sequence is repeated for each layer of the prototype workpiece until it is fully formed.

A4d. ***solid creation system (SCS)*** - has been developed in Japan and is used mostly there. It employs an operation sequence very similar to that involved in stereolithography: 1) develop a CAD model (normally solid) of the part to be made, 2) slice the CAD design into layers and generate supports needed, 3) with an ultra-violet laser, scan the surface of a liquid photopolymer to polymerize and solidify a liquid resin, 4) lower the solidified material slightly, 5) repeat the laser scanning and table lowering for each layer of the part, 6) when all layers are completed, carry out post-processing. This includes removal of support structures and, usually, post curing of the resin to ensure full polymerization.

A4e. ***solid object ultraviolet-laser printer (SOUP)*** - has the same basic sequence as SCS except that the laser beam is moved and focused by a galvanometer mirror. This arrangement provides faster scanning and faster production. Proprietary software is used for slicing and generating supporting structures, as needed. Epoxy resin is used. This method is used in Japan to make evaluation prototypes and patterns for casting.

A4f. ***soliform system*** by Teijin Seiki - is also similar to stereolithography (SLA), using an acrylic-urethane liquid resin. The main application has been to make short-run, low- cost, tooling by using the prototype to cast molds of metal-bearing epoxy, or of low-temperature-fuse metal. These molds have been used for injection molding, casting, or vacuum forming of parts in other plastics. ABS, polypropylene, and polycarbonate parts have been made from such tooling.

A4g. ***MEIKO system*** - was developed for prototyping of jewelry and other parts of a similar small size. It uses conventional liquid resin processing, similar to stereolithography, but with proprietary CAD and CAM programs, particularly adapted to typical sizes and shapes of precious metal rings and related items. An x-y plotter system is used to direct the laser solidification process. The photopolymer prototypes are used for design evaluation and as patterns for investment casting of production products.

A4h. ***E-Darts system*** - is a low-priced system that operates on Windows computers and can make rapid prototypes of a small size - within an 8 × 8 × 8 in (20 × 20 × 20 cm) tank. Its forming method is similar to stereolithography. A unique feature of the system is that the bottom of the tank containing liquid photopolymer is transparent. The laser beam originates below the tank and is aimed upward through the tank bottom to the liquid resin. The prototype is formed on the underside of the machine's base which moves upward as each layer is completed.

A5. *solid-based systems*

A5a. ***laminated object manufacturing (LOM)*** - as the name indicates, uses sheets of thin material of varying shapes, stacked together, to create an object with a three dimensional shape. The outline of each layer is first cut with a CO_2 laser operating at an infrared wavelength. Then the sheet layer is bonded to previous layers with a temperature and pressure-sensitive adhesive. As the sheets are stacked, the full three-dimensional shape of the part emerges.

As with other RP processes, the operation sequence for LOM begins with a CAD model design of the part and an STL computer file. The slices are delineated by a software program, LOMSlice, which also controls the operation of the equipment. The computer file includes outline shapes and dimensions for each layer. Fig. 14A5a illustrates the operation. The material (normally paper with adhesive coating) is fed to the work area from a continuous feed roller. The layers are cut from the sheet stock by the laser under computer control with an X-Y positioning system for the laser beam. The cut piece, and a cut surrounding sheet, are both bonded to the existing laminated stack with the aid of a heated cylinder that rolls with pressure across the stack of sheets. The sheets surrounding the part at each layer form the support structure for temporarily-separated or weakly-supported parts of the prototype. The mounting platform descends a small amount with each layer to accommodate the additional thickness of the stack. The unwanted sheet material, outside the outlines of the part, is removed after the part has been built up. To aid in its removal, the unwanted material is diced (cut into small pieces) by the laser after each layer is laid on the stack. After the last sheet is laminated, the material is removed from the platform.

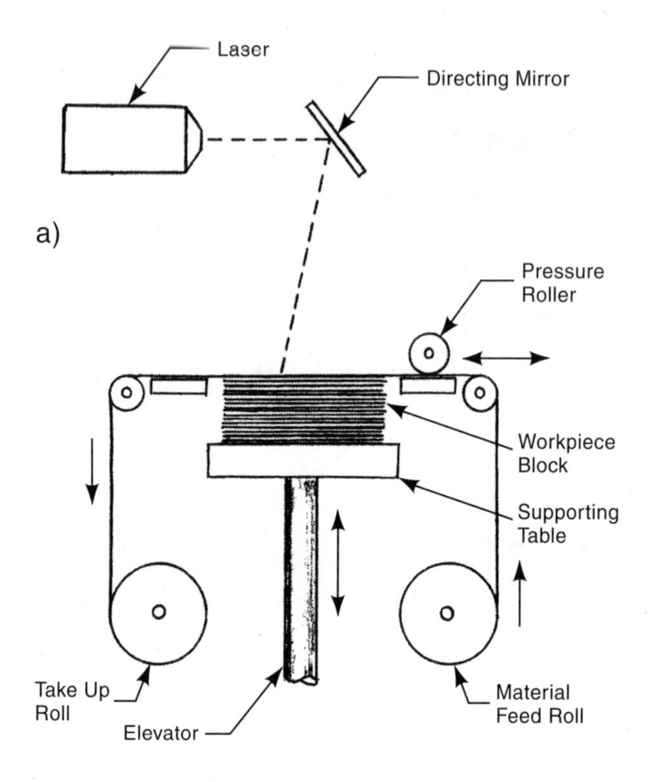

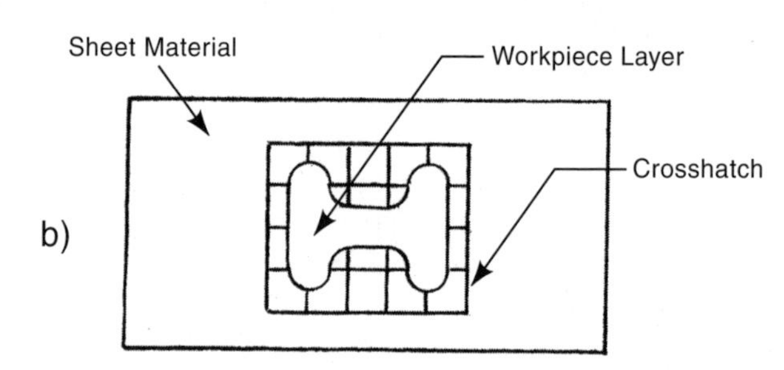

Fig. 14A5a Schematic illustration of Laminated Object Manufacturing (LOM). View a) shows the equipment in operation, where adhesive-coated sheet material is fed to the machine on which the outline of the workpiece is cut. A pressure roller presses each layer against the stack and the sheet material advances for the next layer. The elevator lowers the supporting table slightly as each layer is added. When the final layer is completed, a solid block of material containing the workpiece (prototype part) can be removed from the machine. View b) shows one layer with the outline of the workpiece cut in the sheet by a laser, and cross-hatching around the workpiece. The cross-hatching makes it easier to remove the supporting material that surrounds the finished prototype.

The material, at that point, is in the form of a rectangular block with the part inside. The surrounding material can be removed manually with the aid of a hammer, putty knife and wood carving tools. Other post processing operations may include: sanding, polishing, painting, and sealing with a urethane, epoxy, or silicon spray.[2] Machining operations such as drilling, turning, or milling, may be performed if necessary.

In addition to paper, the laminations can be made from plastics, metals, composites and ceramics. Paper is least expensive and, when laminated, has properties similar to those of plywood. A major application of the LOM method is the fabrication of foundry patterns for sand-mold-cast or investment-cast parts. The LOM patterns look very much like the wood patterns used in the foundry industry. They are also used to make short-run injection molds of silicone rubber or sprayed-on metal. These prototypes also are employed in the design verification functions now common for various kinds of rapid prototypes. LOM-made dies are also used as thermoforming molds for modest-quantity production. LOM can also be used to make patterns of two-part mold halves, similar to the method described above for the QuickCast system. These patterns then can be used to make investment cast molds for production of the part involved, or can be used directly for low-pressure injection molding of a limited quantity of parts.

A5b. ***fused deposition modeling (FDM)*** - uses a thermoplastic as the raw material. The material is heated in a dispensing head to just above its melting temperature and is extruded as a thin ribbon which is deposited on the work surface, and then on a previously deposited layer, in a computer-controlled pattern. The discharge head moves in X and Y directions to cover the entire layer to be deposited. The dispensing is started and stopped as needed to meet the design of the part being produced. The material, after being deposited, quickly cools and solidifies, forming a hardened layer. Previous layers are maintained at a temperature just below, but very near, the melting point of the material so that there is good adhesion with the next layer deposited. The work surface is lowered slightly after each layer so that the discharge nozzle has a constant spacing relative to the existing surface. Typical layer thicknesses are 0.005 to 0.014 in (0.13 to 0.35 mm). Layer thickness is changed by changing the delivery speed of the discharge head. The ribbon width can range from about 0.010 to 0.040 in (0.25 to 1 mm). If supports are needed to hold the prototype while it is in process, the operating program for the FDM equipment designs and generates them as part of the deposition process, using wax from a second computer-controlled dispensing head as the supporting material. Layers are added until the workpiece is completely formed. Then the workpiece is removed from the equipment and the supporting wax structures are separated from the prototype part. The process can be used with polyethylene, polypropylene, polyamide, ABS, MABS (methyl-methacrylate acrylonitrile butadiene styrene), and investment casting grade wax materials.

The process may not be suitable for producing fine holes in prototypes and surface irregularities may result if the discharge head is stopped. Shrinkage occurs as the deposited plastic material solidifies and this may cause some distortion. ABS prototypes can be made with near the strength of injection molded parts so that they can be tested as functional parts. Fig. 14A5b illustrates the FDM process schematically.

A5c. ***paper lamination technology (PLT)*** - is similar to LOM, but no laser is used to cut the paper layers. Instead, a computer-controlled

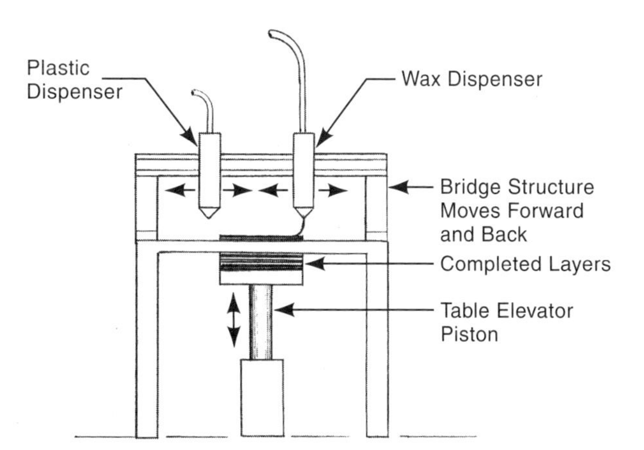

Fig. 14A5b. Fused Deposition Modeling (FDM) shown schematically. A dispensing head extrudes a thin ribbon of thermoplastic in a computer-controlled pattern to create one layer of a prototype. Another dispenser extrudes wax for any necessary supporting structure.

mechanical cutter is used. The process has six steps after the CAD design model, slicing data and machine instructions are completed: 1) print the shape of the layer on paper with plastic resin powder. The printing process is similar to Xerography; 2) align and hot-press the paper on the work surface or on the stack of previous sheets; 3) cut the top sheet to the correct outline; 4) repeat steps 1 through 3 until the full height of the prototype is achieved; 5) remove supporting and unneeded portions of the stack; 6) post-process the workpiece by sanding, cutting or coating as necessary for the surface, shape and application needed. The PLT process has been used in Japan primarily for conceptual modeling.

A5d. ***multi-jet modeling (MJM)*** - is a system that applies wax or thermoplastic material with a method similar to ink-jet computer printing. The printing head dispenses the heated plastic from up to several hundred individual jets at one time. It moves in X and Y directions to cover the full surface of each layer of the prototype, but chiefly in the x direction because the printing head is 8 in (20 cm) wide. The deposited plastic cools and solidifies, forming a layer of the prototype. After each layer, the work surface is lowered and the next layer is deposited. The system operates from proprietary software and is designed for use by engineers for checking design concepts early in the design sequence. The process is quick, clean, automatic and suitable for use in engineering offices. It is illustrated schematically in Fig. 14A5d.

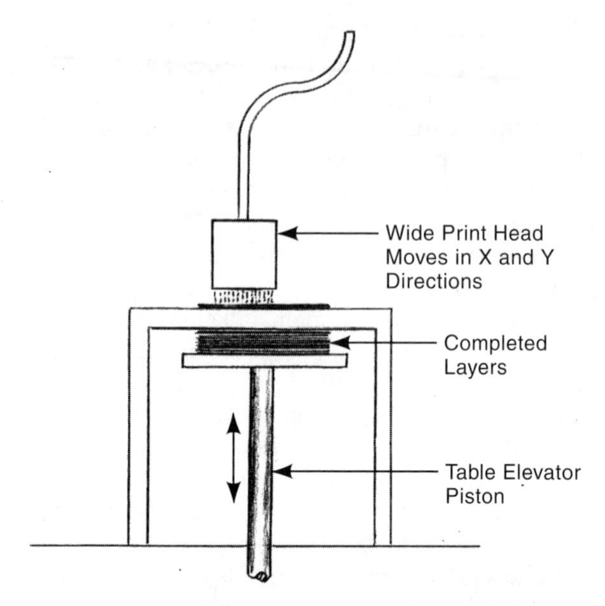

Fig. 14A5d The Multi-Jet Modeling (MJM) system shown schematically. A wide print head, similar to those used on ink-jet printers but considerably wider, moves in x and y directions and deposits wax or thermoplastic material in layers to create a prototype part.

A5e. ***ModelMaker and PatternMaster*** - are two other systems that use ink-jet-like application of thermoplastics to provide concept-verification prototypes early in the design process. Two print heads are employed, one with a thermoplastic, the other with wax. A cutter passes over each finished layer to provide a planar surface, and controlled-layer thickness before the next layer is applied.

A5f. ***slicing solid manufacturing (SSM), melted extrusion modeling (MEM) and multi-functional RPM systems (M-RPM)*** - are Chinese-developed systems and equipment. SSM is similar to LOM, using laser cutting to provide the shape of each layer. The MEM method is similar to FDM. The M-RPM system provides both methods on one machine base, and it can produce prototypes from either layers of sheets or from extrusion of a thermoplastic or wax filament.

A6. *powder-based systems*

A6a. ***selective laser sintering (SLS)*** - uses the energy of a high-power CO_2 laser to selectively fuse particles of powder together. Layers of fused powder are deposited on previous layers to create a part of the desired size and shape. The process is therefore, layer-additive, but it differs from stereolithography primarily in that the starting material is a solid, in powder form, instead of a liquid. The process starts with a CAD modeling design of the part and, as in stereolithography, this design is converted into STL files, which are then delineated as slices of the prototype part. The SLS equipment deposits powder on the work surface and, with a roller, spreads, smooths, and levels the deposit to the correct thin layer. Then, the focused CO_2 laser beam, directed and moved by a pair of mirrors, traces the outline of the part and applies energy to heat the powdered plastic to the fusion point.

The powder is also heated beforehand and maintained at a temperature just below the fusion point, so that the energy load on the laser is minimized. The laser operation takes place in a nitrogen atmosphere to avoid an explosion or combustion hazard with the powder and oxygen contamination of the bonding. Powder that is not sintered is left in place to act as a support for those portions of the workpiece that require it. The elevator then descends slightly. The operation of spreading and sintering the plastic powder is repeated, layer-by-layer until the entire part is completed. Materials used for the operation include PVC (polyvinyl chloride), a thermoplastic elastomer, polycarbonate, nylon, wax, phenolic coated ceramic (for sand casting molds and cores) and plastic coated metal powders.

The surface of SLS prototypes can be quite rough, due to the nature of the process. One factor, with amorphous materials (PVC and polycarbonate) is that powder particles adhere at their points of contact, leaving openings at other points between the particles. These openings also result in densities sometimes considerably less than those that would have resulted from regular molding of parts in the same material. The strength of the prototypes is correspondingly reduced. Prototypes made with the crystalline materials (nylon and wax) do not have this problem, but have the disadvantage of greater shrinkage during solidification and corresponding distortion, if the shape of the part is complex.

When the part is finished and removed from the equipment, it is accompanied by a surrounding cake of powder which must be removed. (The powder cake has an advantageous supporting effect on the prototype as it is formed, but some additional supporting structure may be required, in some cases, as in stereolithography.) Removal of the powder cake is done manually with hand tools: spatulas, dental-like tools, brushes and air nozzles.[1] The surface is then often sanded and may be coated with wax or another material to provide a smoother surface.

SLS prototypes can be used for limited functional testing, as patterns for investment casting and, with metal-based powders or resin-coated sand, the workpiece can be a mold rather than a prototype part. The mold, if made from metal-based powders, may be used for limited or moderate production of plastic parts. When resin-coated sand is used, the workpiece produced is also a mold. The process bonds the sand grains together to produce a sand mold which is used for casting the metal prototype part. Fig. 14A6a illustrates selective layer sintering schematically.

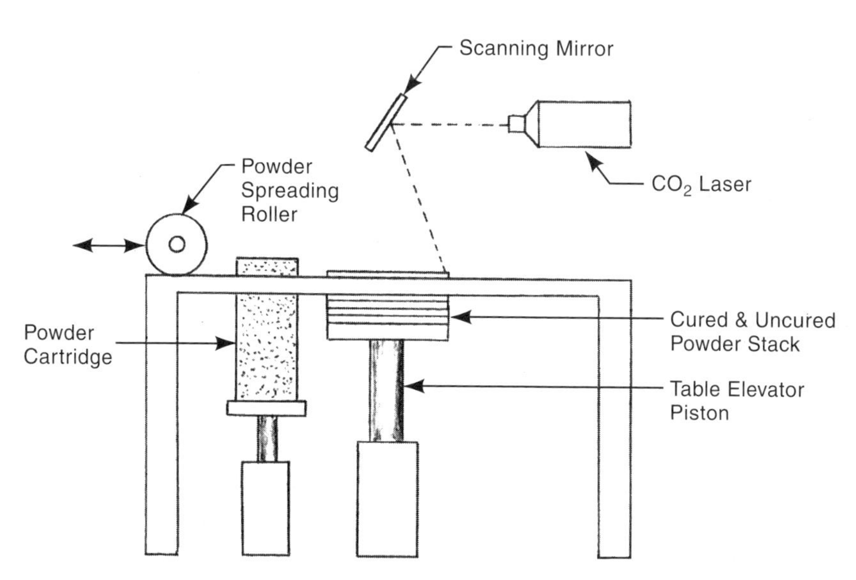

Fig. 14A6a Selective Layer Sintering (SLS) shown schematically. The operation takes place in a nitrogen atmosphere.

A6b. ***EOSINT*** - is a European-developed system similar to SLS described above. With EOSINT, CAD design data are processed by proprietary software to convert them to the layer format that the company's machines use. Layers of powder are deposited and fused together by laser. A wide variety of powdered materials can be processed including polyamides of fine or coarse grain size – with or without glass filler – polystyrene, wax, bronze, steel, and phenolic-coated foundry sand. Available machines are dedicated to only one type of material. The process can be used to make concept or functional prototypes, casting patterns and, with metal powders as indicated below in B8, directly for tooling for injection molding of plastic parts.

A6c. ***three dimensional printing (3DP)*** - uses powder bonded with an adhesive solution that is dispensed from print heads similar to those used in computer ink-jet printers. The objective of the process is to produce inexpensive concept prototypes quickly. The powder used is either a starch-based or plaster-based material. This powder is spread to a uniform thickness on the 3DP machine's work surface. Then the binder solution is jetted onto the appropriate portions of the powder by four print heads. When each layer is completed, the machine table is lowered, another layer of powder is spread and the print heads apply adhesive again. This sequence continues until the full height of the prototype part is achieved. Excess uncured powder is then vacuumed off and the part is removed. To gain additional strength, the part may be dipped in wax or infiltrated with a liquid plastic resin. The prototype may be colored by using colored binders in the adhesive. Sanding and painting also may be done as postprocessing operations.

B. Rapid Tooling

Making "rapid tooling" usually involves the fabrication of injection molding dies by rapid prototyping methods plus some other operations. The dies are normally usable for only limited quantity production and have other limitations. However, they are suitable when quantities needed are modest but greater than common prototype amounts. Rapid tooling also may involve making dies for die casting, patterns for sand casting or forms for thermoforming sheet plastic parts. One of two basic approaches is used. The first is simply to make a master part by RP methods and then use that part as a pattern for making short run injection molds by one of several methods. The second approach is to use RP methods to make patterns of two mold halves for the part. These patterns then are used to make the two halves of an injection mold of steel or other metal by investment casting. The CAD design of the two mold halves (called "two-part negative tooling") is made automatically from the design of the part by software in the CAD computer. Then data for the mold halves rather than the part itself is fed to the RP-making equipment. The QuickCast system, described above, has been used to make production molds by this method. Fig. 14B outlines the steps involved in making metal molds using rapid prototyping methods. For lower-quantity injection molding production, "two-part negative molds" made of epoxy or LOM laminated material can be used, after post-fabrication finishing, directly as injection molds for wax or plastics of lower melting temperatures.

RP patterns of the part can also be used in casting a silicone rubber mold for the part. Silicone rubber molds are most suitable for casting plastics that are in liquid form when the mold is filled. These plastics are usually catalyzed thermosetting types. Another method, sometimes useful for short-run injection molding dies, is to spray or plate a metal coating on a prototype part and then support the thin metal form thus made with a back-up of epoxy or other material to make two mold halves. Another approach is to cast an epoxy mold from the prototype. These short-run molds can often be used for low-pressure injection molding of wax patterns or plastic parts if the shape is not too intricate.

In making injection molds, particularly those for high-production applications, steps must be taken to provide functions that are in addition to the simple reproduction of the part's contours. Dimensions must be precise so that the molded part can fit the space allocated to it in the product. Gaining the necessary precision in a rapid tool requires correct allowances for shrinkage of the RP material when it solidifies, for shrinkage of the mold material, metal or plastic, if it is cast, and for the shrinkage of the production part from the mold.

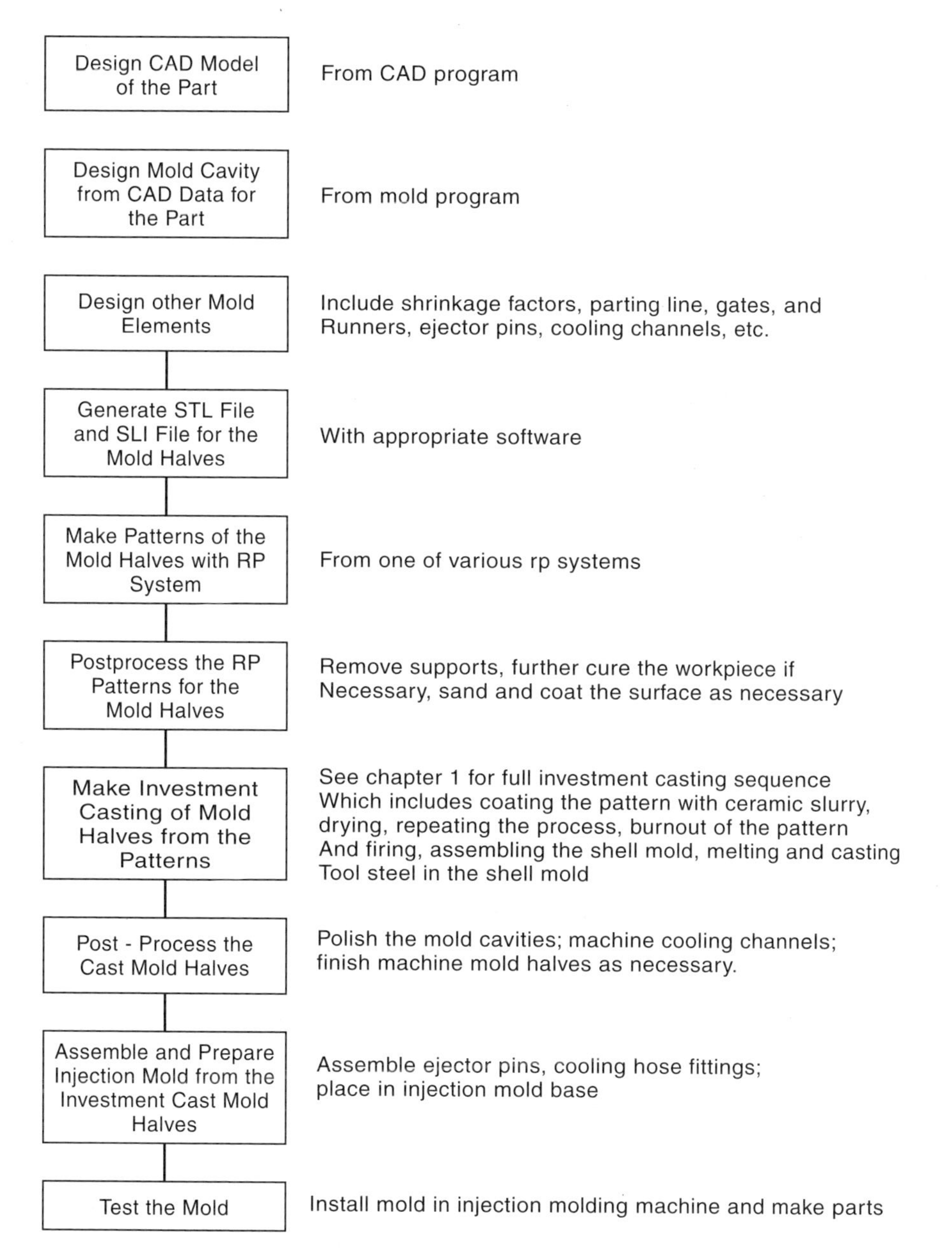

Fig. 14B The operation sequence for "rapid tooling" fabrication of metal molds for injection molding of plastics.

Fastening holes must match; the surface finish must be correct, especially if the part is an external, visible one; cooling channels, ejection pins, sprues and gates must be incorporated in the mold. Software, developed especially for rapid tooling use, can incorporate some of these features in the solid model for the mold but, even then, additional conventional machining and toolmaking operations are necessary. Some of these features can be skipped if the mold is for low production use and much slower production rates and manual mold handling and part handling can be tolerated.

The SLS process, described above in A6a, can be used to make a sand mold for casting a metal part, made by conventional foundry methods. Resin-coated sand is used as the working material with the CAD data reflecting the female mold shape for the part in question, In this approach, the tooling, the sand mold, is usable for only one part.

B1. ***direct shell production casting (DSPC)*** - produces ceramic molds for casting metal parts without the need for making patterns. The system includes software to convert a conventional CAD design of the part to a design for the mold cavities that will produce the part. Sprues, gates, runners and risers in the mold are part of the mold design. The mold is then made by the machine, which combines ceramic powder (fine aluminum oxide) with a colloidal silica binder in a layer-by-layer process. The alumina is spread on a work surface and leveled with a roller. A print head, similar to those used in ink-jet printing, deposits droplets of the binder from a series of jet openings on the print head to the appropriate places on the layer of ceramic powder. An additional layer of powder is spread and leveled, and more colloidal silica is selectively jetted onto it by the print head. When the process is completed, the workpiece is separated from the surrounding loose alumina powder and is post processed as necessary. It is then fired in a kiln. The DSPC process, illustrated in Fig. 14B1, is used exclusively for mold making.

B2. ***Prometal 3D printing process*** - uses steel powder and an ink-jet printed binding agent to make, layer by layer, a "green" powder metal part. The part is then furnace sintered and infiltrated with molten bronze in an infiltration furnace. The resulting part is then of full density. It is post-processed with any necessary machining, polishing and chromium or nickel plating. The process is used in making molds for injection molding and blow molding and dies for extrusion of plastics.

B3. ***RapidTool***[4] - is a method that forms injection molds from resin coated metal powder. A selective laser sintering process (SLS) (See A6a above) is

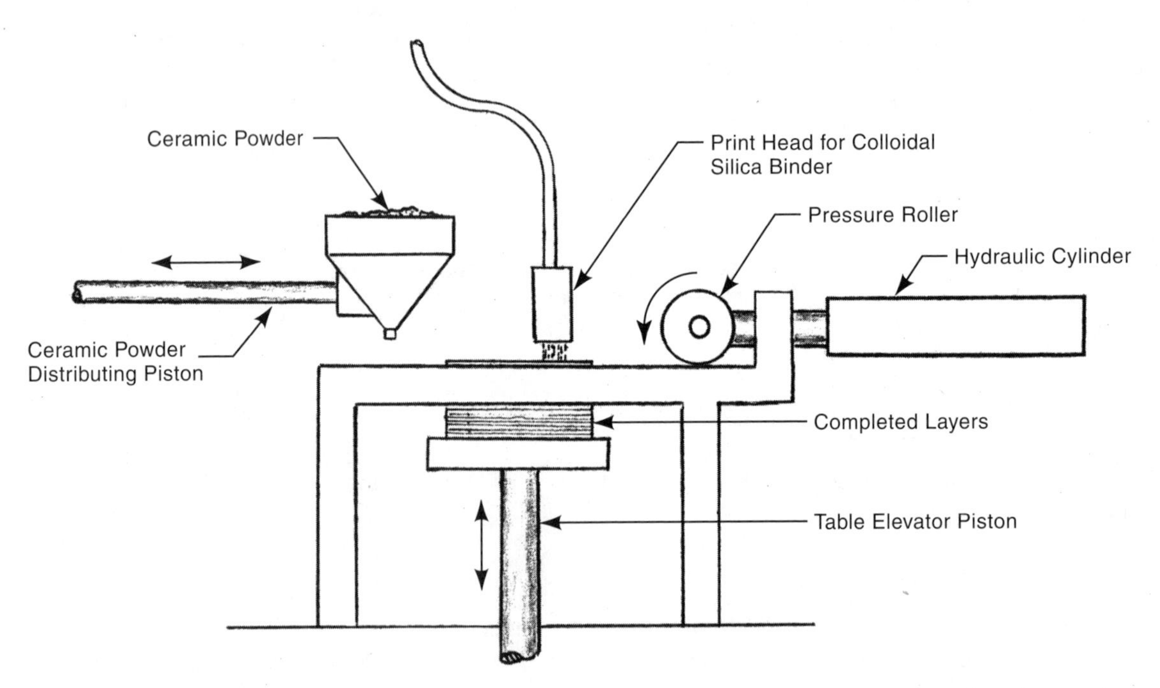

Fig. 14B1 A schematic representation of the DSPC process for making ceramic molds using rapid prototyping techniques. Ceramic powder, deposited in layers and roller pressed to be level, is bonded with colloidal silica. After all layers are completed, the mold halves are removed from the machine, loose powder is removed and the mold halves are fired in a kiln.

used to bond the powder particles together into female mold halves. The mold halves are in a green state at this point and are then sintered and infiltrated with copper. The resulting mold is usable with conventional injection molding equipment and, although less durable than conventional hardened steel molds, can be used in moderate quantity production.

B4. ***laminated metal tooling*** - If the Laminated Object Method (LOM) (See A5a) is used with sheets of steel or another metal instead of paper or other non-metallic material, it can be used to form the halves of an injection molding die. The metal sheets are cut by laser, router or water jet, and are stacked and bonded or bolted.

B5. ***direct AIM***[4] - The stereolithography process, when used with a thermosetting plastic material and with CAD data for the female mold shape, can be used to make molds for injection molding other plastics. Such molds, however, are usable for only small quantity production and low injection pressures because of their lack of mechanical strength.

B6. ***SL composite tooling***[4] - is another system for making plastic molds for injection molding of plastics. The system differs from the Direct AIM method in that plain resin is used only for a thin shell of the mold cavity. This shell is backed up with epoxy plastic filled with aluminum powder and aluminum shot. The back-up materials provide increased mold strength and heat conductivity, both of which increase mold function and life.

B7. ***3D Keltool***[4] - is used to make solid injection molds from powdered steel. The process is lengthy, starting with a prototype part made by conventional stereolithography. The part is finished by sanding and polishing, and then is used as a pattern to cast a mold from liquid silicon rubber. This mold is then used to cast a silicon rubber replica of the part. This replica is then placed in a mold-size box that is then filled with powdered tool steel, coated with a binder material. The powder-binder mixture is cured and removed from the box as solid mold halves. The halves are sintered and infiltrated with 30% copper to fill voids between steel particles and can be finished for use as an injection mold.

B8. ***direct metal laser sintering (DMLS)***[4] - is an EOSINT process (See A6b) using metal powder and a laser of very high power. The laser sinters the metal particles together. Both bronze and steel alloys are used. The process has been used to produce mold inserts as well as metal parts.

C. Manufacturing Cells (Group Technology)(Family of Parts Concept)

The use of manufacturing cells represents a particular kind of factory layout for manufacturing equipment. The essence of the concept is that, when manufacturing a particular part or family of similar parts in substantial quantities, the equipment for successive operations on the part should be grouped together. The operator - or team of operators who make that part (or the family of parts) - operate all the equipment in the cell. This arrangement is in contrast to a more traditional factory layout that groups like equipment together in departments. Then parts move from department to department for each operation. The advantage of the cell layout is that lines of communication and transportation are made very short. The factory's through-put rate is speeded; work-in-process inventory is greatly reduced; when problems arise at one operation, their effect on subsequent operations is immediately recognized. Route sheets to control movement of parts through the factory are greatly simplified or not needed. Operators learn what is needed at each operation to avoid problems at subsequent operations and have the satisfaction of seeing the results of their efforts. Quality tends to improve. The disadvantage is that there is apt to be lesser utilization of some equipment. There is also the necessity of having operators learn the operation of several kinds of equipment rather than specializing in one type. The concept is illustrated in Fig. 14C for one particular family of machined parts. Fig. 14G8 shows a more advanced system where a robot does the handling between operations. When a circular layout is used, the robot can tend all machines in the cell.

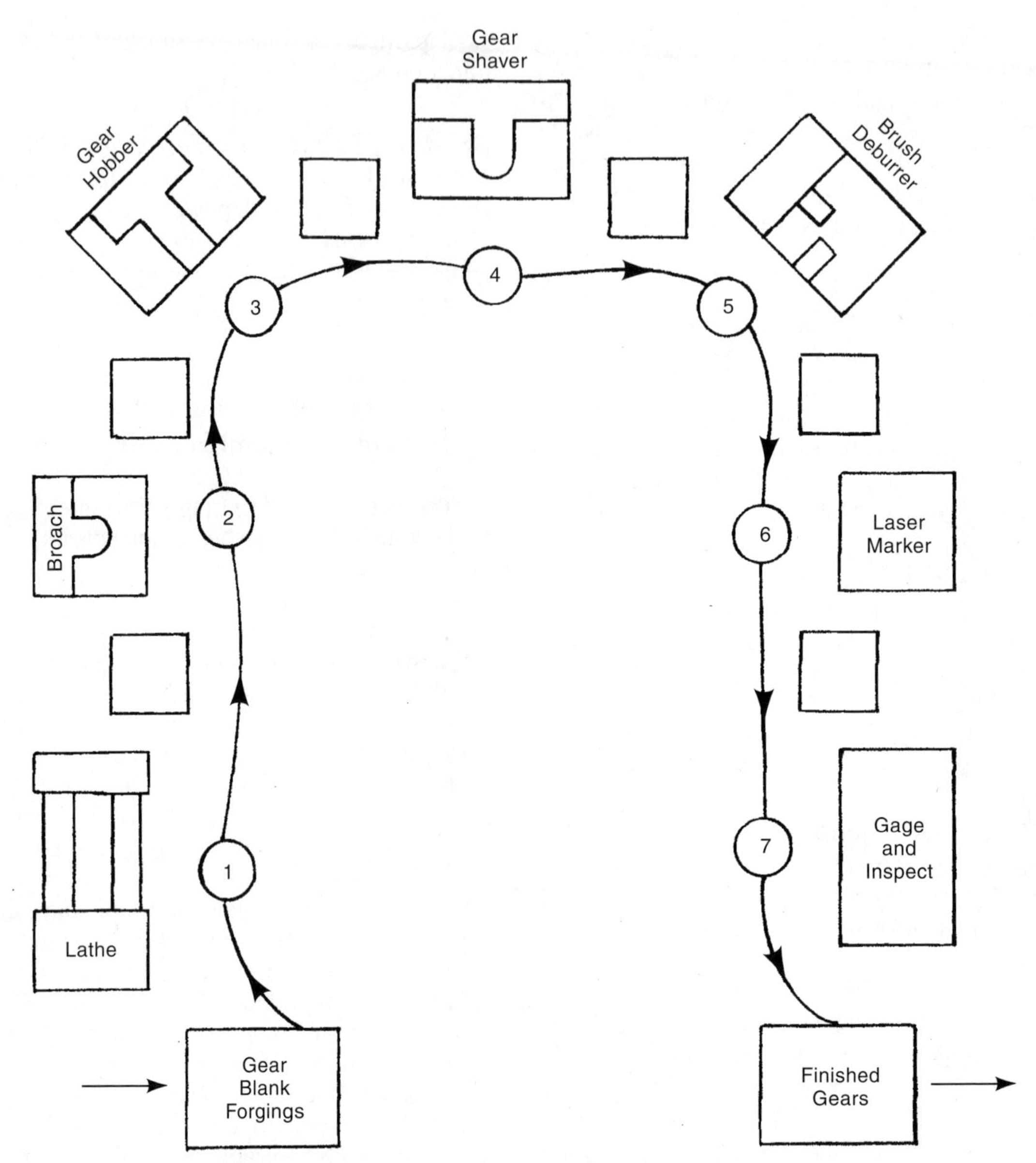

Fig. 14C The concept of a manufacturing cell illustrated with a cell arranged to make a family of spur gears of various pitches and diameters from forged blanks. This cell is shown with one operator for seven operations, based on the use of automatic equipment. A similar arrangement may be made using more operators and possibly duplicate machines for some workstations depending on the production level needed, and the operation time at each workstation. The gear blanks are first faced on both sides, turned, drilled and reamed at station one. At station 2, a keyway is broached in the center hole. At station 3, the gear teeth are hobbed and at station 4, they are finish shaved. Station 5 is a brush deburrer. At station 6, the gear is laser-marked to identify its model or part number, and with other descriptive data. At station 7, various gages and inspection devices are used to ensure that quality is of the prescribed level. The small tables between machines act as decouplers/Kanbans, containing small amounts of work-in-process, to illustrate which machines may need operator attention to maintain production flow. Ideally, these tables are empty and all work-in-process is undergoing machine operations.

D. Advanced Inspection Devices

D1. ***coordinate measuring machines (CMMs)*** - utilize diffraction gratings to gauge the position of a measuring probe with high accuracy. Typical systems include, in addition to the measuring machine, a sensing probe, a control and computing system and measuring software. Various kinds of probes may be used. The probe is moveable with very low friction because of nearly friction-free linear bearings in the machine. The workpiece to be measured is placed on the granite table and is then stationary. The electronic touch probe is guided to move around the workpiece and make contact with it where measurements are wanted. Probe movements can be in the x, y, or z direction. Movement of the probe can be manual, by CNC, or by a programmable controller. The readout console displays the position of the probe simultaneously in terms of the three coordinates. Different probe shapes can be used, depending on the element that is being measured. For example, if dimensions involving the center axes of holes are to be measured, a tapered plug is used as the probe. The measurement accuracy results from use of the moire' fringe pattern from two glass scales placed together on the machine at a slight angle from each other. The fringe pattern between the scales is detected by photocells and converted to electrical pulses. Measurement accuracies within 2 to 4 ten-thousandths of an inch in a span of 10 to 30 inches are common[5]. The probe is often mounted on a bridge-like structure, but may be cantilever mounted or fastened to an articulated arm. The machine is usually installed in a room in which temperature and humidity are controlled. Coordinate, profile, and angular measurements can be made. Some machines are equipped with more than one probe or with a machine vision device or laser scanner (see below), in addition to a contact probe. Such machines are referred to as *multisensor systems*. Fig. 14D1 illustrates a typical coordinate measuring machine.

Fig. 14D1 Dimensional inspection of a machine component with a coordinate measuring machine. (*Courtesy Sheffield Measurement, Inc.*)

D2. ***machine vision*** - in industry, utilizes a pictorial image of the workpiece as part of a system of quality, machine or process control. Control is accomplished by capturing the image by electronic methods and then using digital data of the image, as the data is processed by a computer, to provide a display, to sort good and bad parts or to actuate control mechanisms. Machine vision systems are versatile and have many non-manufacturing applications in the fields of medical diagnosis, surveillance, zip-code mail sorting, traffic control, and bar code reading.

A machine vision system includes means for proper illumination of the workpiece or scene to be pictured, a camera or cameras, an analog to digital converter (though some systems process digital data directly from the camera), sufficient computer stages to process the data, computer software for the application, and actuating equipment that responds to the digital signal from the computer. Proper lighting is a critical factor and there are various lighting arrangements: backlighting, low-angle lighting or diffuse lighting to emphasize the

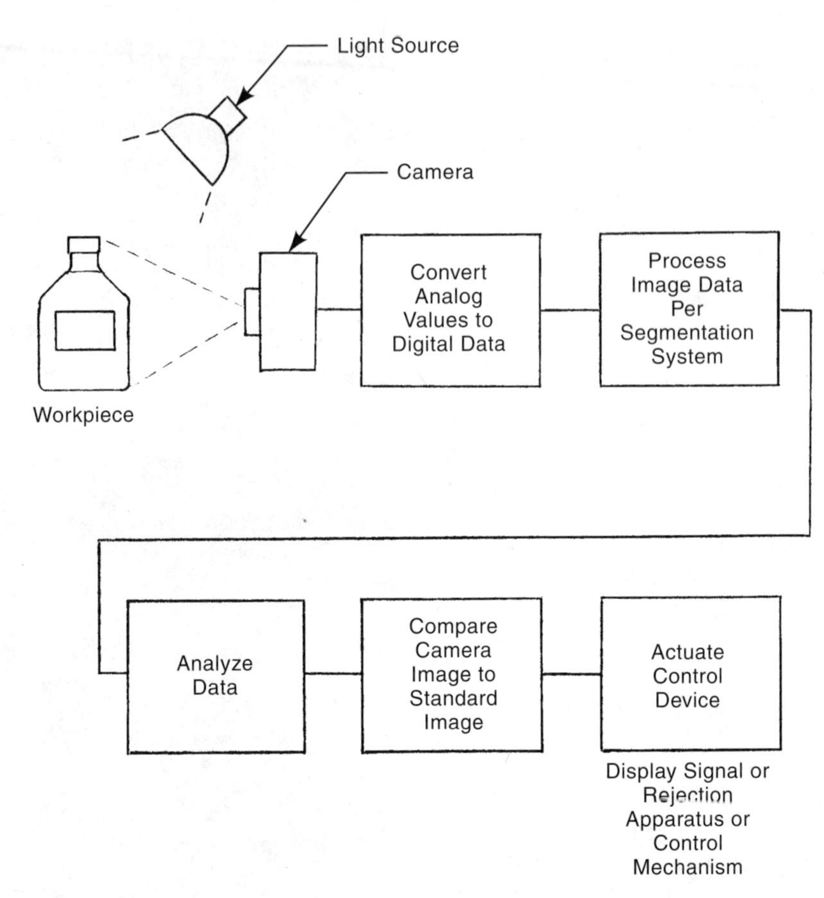

Fig. 14D2 Major elements of an industrial machine vision system used for inspection or control.

essential image elements. Fig. 14D2 illustrates the elements of a typical industrial vision system. Image processing removes unwanted detail. What detail is wanted and what is not wanted is a result of segmentation, the procedure of choosing the elements that must be measured to achieve the desired results. Segmentation reduces the amount of data that must be computer processed, speeding up the operation of the system. One common form of segmentation is to concentrate on the "edges" of the image, the changes in contrast that are characteristic of the edges of the workpiece or the edges of a shadow of a three-dimensional workpiece. (This localized analysis is one reason why proper illumination is critical.) Unexpected edges may be evidence of a surface flaw or other defect in the workpiece. Another segmentation approach is to scan with the camera only a narrow strip across the workpiece instead of its full area, if that is enough to make the desired measurement. Another is to analyze only a small critical portion of the whole image. Even with segmentation, because of the large amount of data in each picture image, high-capacity computer systems are required. The camera includes a lens system to focus the light and a CCD (charged coupled device) integrated circuit that receives the camera image (instead of photographic film that would be used in a film camera). Each pixel (picture element of the CCD sensor) in the camera senses the brightness of light focused on it. The brightness level at each significant pixel is converted to a binary digital number that is processed by the system's computer. Most machine vision cameras are monochromatic, so the brightness value is from a grayscale. Color filters may be used on the camera, however, or color cameras may be used, when color is an important factor in the particular application. All inspection and machine control operations involve computer analysis of the image data. This analysis may involve such things as counting the number of pixels between two lines or edges, determining the radii of some curved edge, counting the number of pixels in the entire image, etc. After such analysis, these measurement data are compared with the data stored in the computer that defines a standard pattern or value for the characteristic being measured. Depending on the difference between the two sets of values, machine or process controls may be adjusted or parts may be rejected.

The advantages of machine vision systems compared with human visual inspection or control are the much greater consistency and reliability of measurement that are inherent in the machine system. Also, machine vision does not suffer from fatigue, can operate in adverse working conditions and can often be operated at higher speeds than a human counterpart.

When used for component inspection, machine vision can provide several different types of information: 1) part identification, 2) the presence or absence of a component or certain features, 3) shape verification, 4) measurement of length, width, area, hole diameter and hole position 5) inspection of surface finish including surface flaw detection. 6) quantity verification when multiple components are involved.

One machine control application is the guiding of robots. Machine vision on a robot can guide the robot to locate the part, then identify it, then direct the robot's gripper to the proper position to grasp the part correctly, and then, after it is grasped and moved to the desired location, orient the part to fit the receiving space. Other applications of machine vision are inspections made throughout the electronics industry for both circuit board and integrated circuit manufacture. Specific examples are inspection of circuit path widths, completeness of population of boards and soundness of solder joints. Machine vision provides feedback control data in wire bonding and die slicing operations. In the paper, textile and plastics industries, machine vision monitors the soundness of continuous webs of rapidly-moving material during processing, including the completeness of coatings applied. In high-production printing, it monitors the registration of different colors on the printed document. In the food industry, machine vision identifies and rejects undersize, oversize or misshapen products in the packaging of cookies, candy bars, and similar products. For many products, it verifies that labels are properly in place on containers. In the painting of various products, it confirms the proper gloss, color, coverage and freedom from runs or other defects. Machine vision finds flaws in individual parts made with highly automatic glass, plastic and metal manufacturing processes, and monitors the quality of manufactured containers.

D3. ***laser scanning*** - is a method used for dimensional inspections. In one common technique, the workpiece is placed between a low-power scanning laser beam and a photodetector. Scanning is achieved by directing the beam to the axis of a rotating mirror which reflects the image to different points on a collimating lens. The axis of the mirror is at the focal point of the lens. The lens directs the parallel beams to a collecting lens that focuses the beams on a photodetector. A workpiece is placed between the two lenses. The collecting lens receives the beams that pass the workpiece but not those that strike the workpiece and cast a shadow on the collecting lens. The dimensions of the shadow, and thus the workpiece, are calculated by the timing of the spaced laser beams on each side of the shadow. A microprocessor makes the necessary calculations and displays the width or other workpiece dimension of interest. Fig. 14D3 illustrates the working principle. Since there is no contact between the workpiece and measurement tools, the procedure is useful for in-process measurements. Workpieces can be measured while they are in motion on a machine or conveyor. Multiple parts can be measured simultaneously by the one beam. Bench-mounted laser "micrometers" are also used. Accuracies of ±10μin (0.25μm) are achievable for dimensions of 2 in (50 mm) or less. For larger dimensions, measurement tolerances are correspondingly larger.

In another system, the scanned laser beam is received by an array of photodiodes instead of a single photocell. Depending on the location of the edges of the workpiece, some diodes receive the laser beam and some do not. From these differences in the signals received, the dimensions of the part can be determined.

Another measurement aims a laser beam at a curved surface and records the reflections.

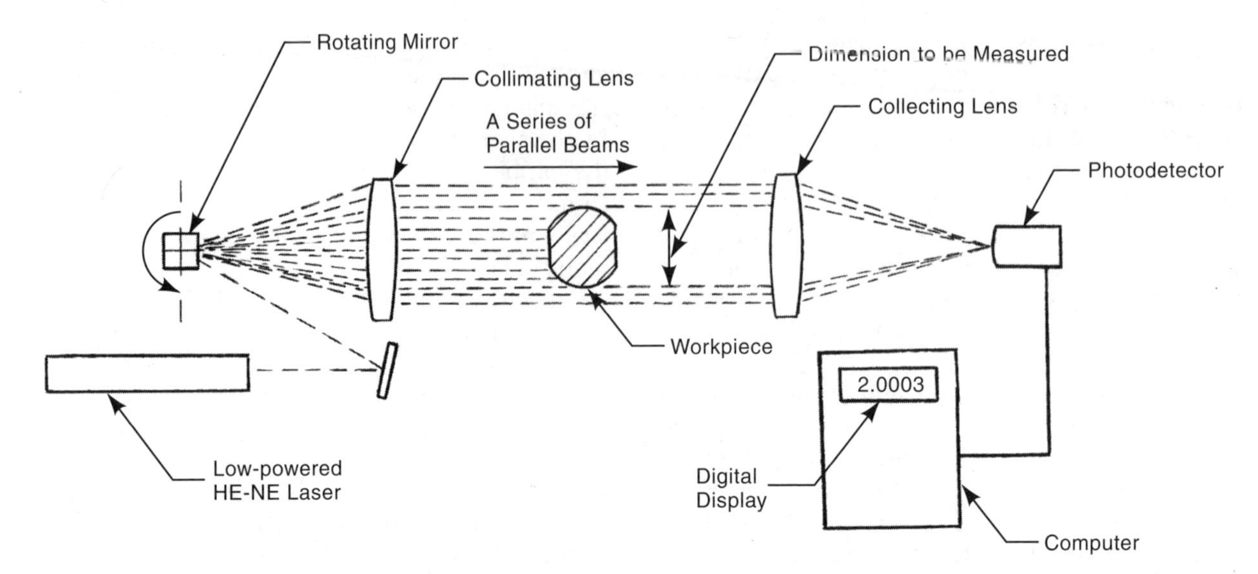

Fig. 14D3 The working principle of laser scanning for dimensional measurement. The dimensional measurement is based on the elapsed time between the detection by the photodetector of the laser beam from each side of the workpiece.

Thousands of points per second are scanned and the resulting data describes the curved surface in detail. The unit's computer compares these data with the specified dimensions and indicates whether or not the shape meets specifications.

Still another system uses interferometry. In this method, a laser beam is split into two separate beams by a partially silvered mirror. One beam travels to a fixed mirror and the other to a movable mirror on the workpiece to be measured. The beams are combined and the visual pattern of the interference of the two beams is fed to a photoelectric detector and digital counter. The number of light fringes indicates the displacement of the movable mirror. The method is very accurate and is used in measuring the accuracy of, and in calibrating movements of machine tool elements.

E. Automatic Guided Vehicle (AGV) Systems

An automatic guided vehicle is a device for moving unit loads of materials from one place to another, within a facility, with no accompanying human operator. Vehicles are battery powered and an on-board computer controls the movement. Guidance is provided by one or more of a number of methods. One method uses an electrical inductance wire embedded in the floor. The vehicle has a sensor system that follows the wire. Other methods use optically-read paint or tape markings on the floor or electronically-read magnetic markings on the floor. Some systems use a scanning laser and reflective markers to determine the vehicle's position by triangulation. An inertial guidance system is used on other vehicles with an on-board gyroscope and odometer. Current systems have the capability to control movement over a number of different routes and destinations. Some have data-entry devices so that the path can be modified by factory-floor personnel when necessary.

The vehicle may be loaded and unloaded by human operators, by robots, or by arrangements involving powered conveyors. Automated guided vehicles (AGV) are considered by some to be robots. Both materials, tools, and major components such as car or truck engines, are moved with AGV systems. The vehicles normally move at a rate that is slower than a person's walking speed and can stop very quickly if necessary. They include easily-found emergency stop buttons and/or obstacle sensors and

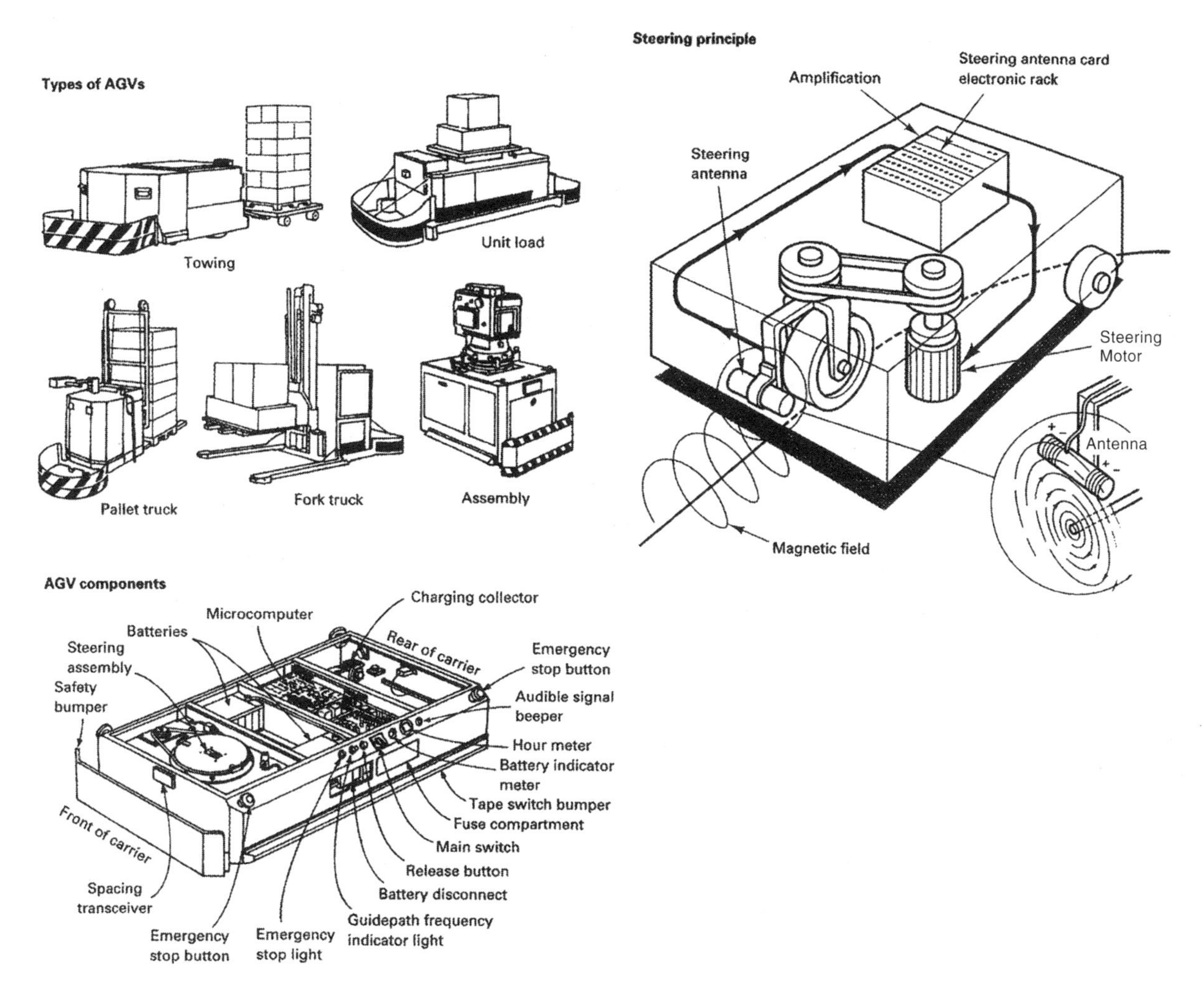

Fig. 14E Automatic guided vehicles, AGV, are used to move materials, parts, products, tooling and supplies within a factory, without an on-board operator. One guidance system using a wire buried in the floor as a guidepath is illustrated. *From Materials and Processes in Manufacturing, 8th ed., E. Paul DeGarmo and others © 1997. Reprinted with permission of John Wiley & Sons, Inc.*

circuitry to stop the vehicle immediately if some object is in the vehicle's path.

There are three main types of AGVs: those that carry the unit load right on the vehicle, those that tow one or more trailers or other non-powered vehicles, and those equipped with lifting forks. The latter type are used to transport pallets or skids of material. They replace, for at least some operations, human-operated forklift trucks. Many unit-load vehicles are made to transport one particular kind of material, for example, paper rolls in a newspaper printing plant or coils of sheet steel in a stamping plant.

Fig. 14E provides an illustration of the AGV concept.

F. Automated Storage/Retrieval (AS/R) Systems

These systems use automatic devices which, under computer control, place standard loads of material on storage racks or remove them from the racks. The location in the racks is random, depending on where there is open space. The system uses a standard load of material - a pallet load, a handling tray load (for small to medium-sized parts) or a tote box load. The racks have openings that match the size of the standard load. The system is under complete computer control; the computer remembers where each item is stored and, when an item is to be retrieved from stock, directs the retrieval device to the oldest unit of that item in the racks. There are four basic components in the AS/R system: 1) the storage/retrieval, stock-handling, machine which usually runs on a track at the storage rack and operates automatically from computer signals, 2) the storage racks, usually one to three unit loads deep at each opening, and built to a height of as much as 70 ft. (21 m). 3) a computer to control the system and keep stock records and 4) a conveyor or other means to handle the stored items to and from the storage /retrieval device. Fig. 14F illustrates a typical system.

AS/R systems are used for a wide variety of materials. They are employed in manufacturing for factory-floor storage of work-in-process, tools and spare parts. In warehousing and distribution, these systems are used for order picking of finished products and for storage of raw materials and component parts. Applications include grocery warehouses, university and corporate research libraries for low-usage books and periodicals, chemical blending operations where drums or raw materials and finished products are stored, textile, sheet metal and printing operations where rolls of coiled material are stored, and storage of bakery pans, office records and production molds, fixtures, tools and dies when they are not in use.

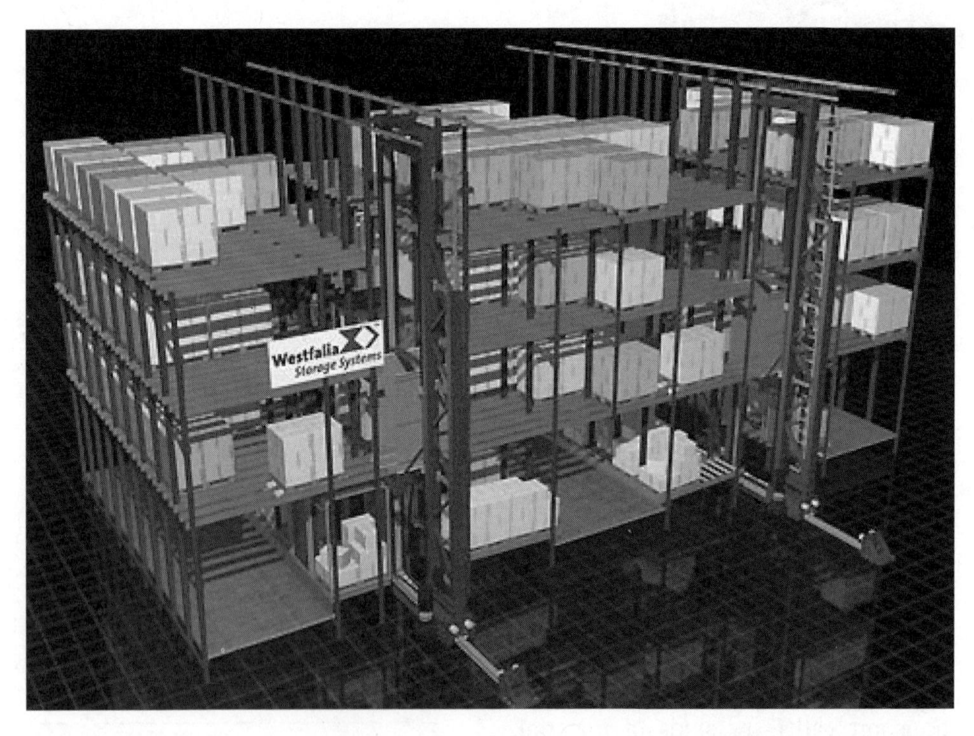

Fig. 14F A complete automatic storage and retrieval system, operated under computer control. Cartons of completed products are automatically placed in open locations in the storage racks. When shipping orders are processed, the computer directs the unloading device to the oldest carton in storage of the product to be retrieved. The carton is automatically moved to the particular loading station where it is needed. (*Courtesy Westfalia Storage Systems*)

G. Use of Robots in Manufacturing Operations

Industrial robots are mechanical devices that can be programmed to perform a variety of tasks of moving and manipulating materials, parts, tools or other devices automatically. Often having an appearance similar to that of a human arm and hand, robots typically have the following major elements: 1) the manipulator - the structure and linkages that provide movement. Robots may have as many as 6 axes of movement, described sometimes as "six degrees of freedom", 2) the end-effector - a hand-like gripper or some device that performs a useful operation, 3) a controller - the apparatus that starts, stops and controls the action of the robot, stores data and communicates with other data devices or with persons, 4) the power supply - for operation of the robot. This may be from hydraulic or pneumatic sources or from electrical servo or stepper motors. Electric power is the most common, 5) sensors (in many robots) - that detect position, contact, force, torque, resistance, or may have vision capabilities. The signals or data are transmitted back to the controller. 6) active devices - Robots may be equipped at the end of the arm with active devices such as spray guns, welding tools, drills, routers, grinders, buffing wheels, or other tools. Fig. 14G illustrates the major elements of a typical industrial robot.

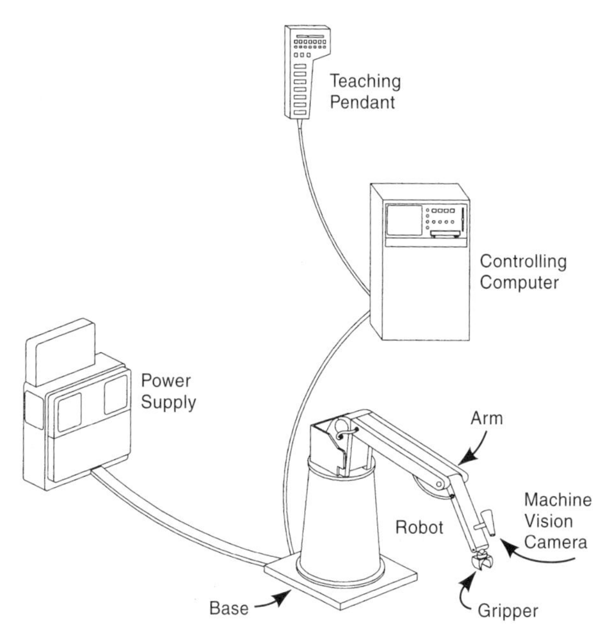

Fig. 14G The major elements of a typical industrial robot.

Some robots are equipped with a "teaching pendant", a hand controller that can be used by a human operator to move the robot's arm and gripper. The pendant records the movements, their rate of speed and the end points. The sequence of motions is retained in the controller's memory and can be "played back" to repeat the desired series of motions necessary to perform the operation. Some robots have point-to-point control over the end effector's movement; others employ continuous path control. Some have fixed motion paths; others provide variation in the movement of the arm depending on information sent to the controller by sensors.

G1. ***areas of robot applicability*** - Robots have been most common in operations where the working conditions are unpleasant, difficult, or unsafe for human operators. Hot forgings and castings are easily handled by robots. Heat, fumes or odors, dirt, dust and solvents are unpleasant, unhealthy, or hazardous for human operators but are not a problem for robots. The other justification for robots comes from the labor savings that they provide. However, the cost of accompanying equipment for feeding, orienting and transporting parts for robotic application is a cost that must be considered in addition to the cost of the robots themselves. The operation must have sufficient volume to justify the investment. Robots are particularly useful in situations where there are frequent product changes; they provide "flexible automation" because many operational sequences can be stored in the memory of their control computers. Because of the sensory capability of present robot-based systems, especially in units with machine vision, justification for use in operations with variable conditions is more feasible. Still, robots are less likely to be found in assembly operation where a great number of parts are involved. The flexibility, compactness and skill of the human operator in those situations is less easy to replace.

G2. ***robots in foundries*** - Because of heat, the heavy weight of many workpieces and the other unpleasant or hazardous aspects of the working environment, robot use is often easily justifiable in foundries. Robots are most common in high-production shops where the opportunities for labor cost reduction may be more substantial.

G2a. ***in die casting*** - The heat, dirt, hazards and repetitiveness of the operation make robotic handling particularly attractive in die casting. An early, and still common, application was the unloading of castings from die casting machines. Spraying lubricant on the die casting die, and ladling molten metal to the machine are other applications. Additional operations are cooling the casting by dipping or spraying, trimming to remove flash, gates and runners, and placement of inserts in the die.[3] Trimming involves the placement of the casting in a suitable press die, activation of the trimming press and the removal and setting aside of the trimmed part. Deburring with robot-held rotating tools has been employed. The tool is moved along the parting line of the casting. An advantage of this method is that it does not require the use of a dedicated die for each casting.

G2b. ***in sand-mold casting*** - Robots are used less in sand-mold foundries than in die casting despite the heat and unpleasant working conditions but there are many possible applications. Robots are used for pouring molten metal into molds, for core handling and core gluing - if needed, core deflashing, spraying refractory coatings on molds, moving molds to and from baking ovens, venting molds and handling of hot castings at shake-out. Other robotic operations are the removal of gates and risers, using robot-held flame or plasma cutters, and the deflashing of castings after shake-out. Dross skimming of molten metal is another application in aluminum foundries. Deflashing may be simply done by breaking off the excess material; otherwise grinding with a robot-held rotating grinder is the robotic method.

G2c. ***in investment casting*** - The process of making shell molds for investment casting requires repeated dipping of the wax pattern in slurries or fluidized beds of ceramic material and sand. Each dipping step is followed by a drying stage. There is a final firing operation to melt the wax pattern and fuse the ceramic. The repeated dipping and the handling to and from the drying and firing operations are carried out robotically instead of manually, in some foundries, with both cost and quality benefits.

G3. ***robots in forging*** - Robots are being used in higher-production shops to load forging billets into furnaces, to move heated billets from furnaces to presses or drop hammers, to move workkpieces from one die station to another, and to move the forgings from forging presses to trimming presses, drawing benches, pallets, or conveyors. They are also used to apply lubricant to both workpieces and dies. Since the workpieces are very hot, the use of robots can be justified for elimination of unpleasant workplace conditions as well as for productivity improvements.

G4. ***robots in metal stamping*** - are used for unloading workpieces from punch presses (fed with coil stock), loading workpieces in presses for secondary stamping operations (for example, unloading a workpiece from a deep drawing press and positioning it in a trimming press die) and for press-to-press transfer of workpieces in stamping lines. Robots are justifiable because of the elimination of exposure of human operators to safety hazards at the presses. When production levels are sufficiently high to justify high speed, dedicated, transfer equipment, robots are not fast enough to compete. However, when there are a variety of parts to process on the stamping line or when quantities of each are more moderate, robots are more apt to be specified.

G5. ***robots in injection molding and other plastics molding*** - Unloading molded parts is the most common robotic operation with plastics. Usually, injection molding is involved but robots are also used with compression and transfer molding, particularly when the parts are large. Another robotic application in the molding of plastics is the placement of metal inserts in molds. In layup molding of reinforced fiberglass parts, the mixture of glass fibers and polyester plastic is sprayed into open female molds by robots. Other robotic operations are trimming parts after molding, drilling, buffing, palletizing and packaging. Fig. 14G5

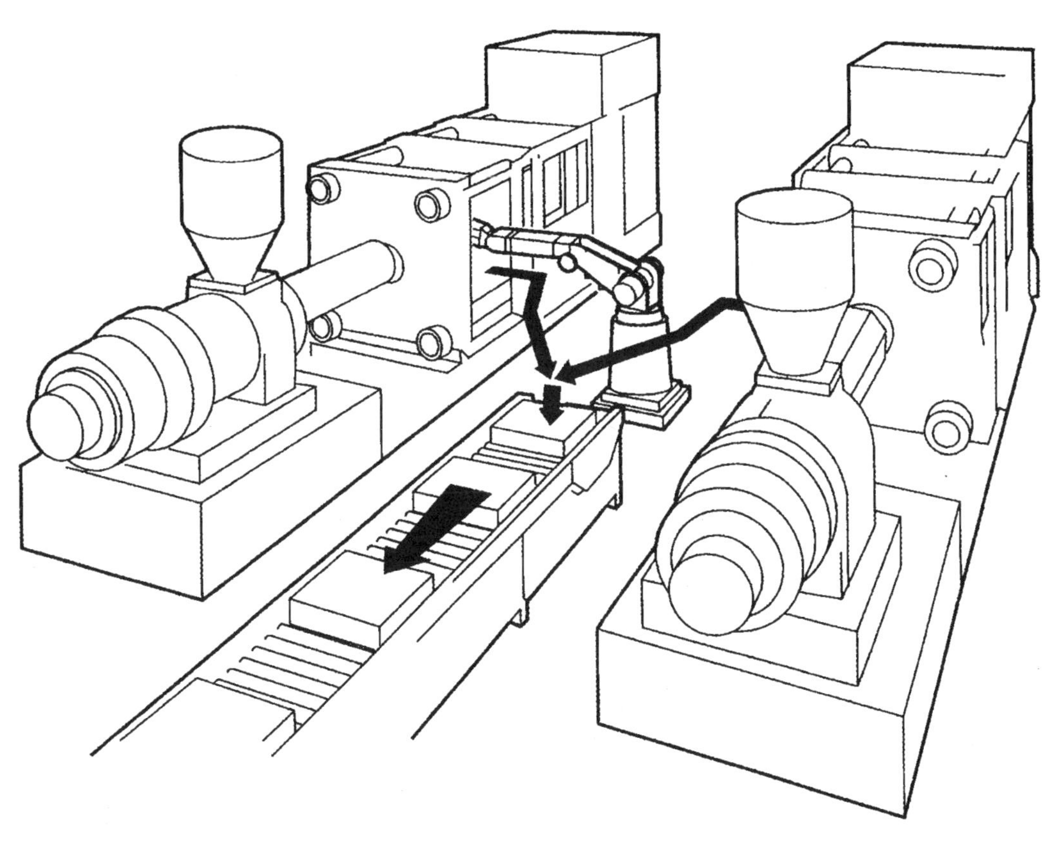

Fig. 14G5 A robot unloading molded parts from two injection molding machines. *(Courtesy Milicron, Inc.)*

illustrates an arrangement where one robot unloads two injection molding machines.

G6. ***robots in welding*** - More robots are involved in welding than in any other industrial operation and the most common robotic welding operation is spot welding. Robotic spot welding is the standard method for automotive sheet metal body components. Several kinds of arc welding and, more recently, laser welding are also carried out robotically. In all robotic welding operations, the guns or opposed welding electrodes are positioned, held and actuated by the robotic devices. With spot welding, the robot (usually six-axis) moves the electrodes from spot to spot on the fixtured sheet metal assembly.

Robots performing arc welding use noncontact seam trackers[3]. The full robotic arc welding system includes, in addition to the robot, a suitable welding gun, a positioner to hold the workpiece in a controlled location, grippers to hold the workpiece, a control system for movement of both the robot and the positioner, arc control equipment, power supply, shielding gas supply, and adaptive control that utilizes feedback from the joint as the operation progresses. Robots with teaching pendants can be programmed to make the weld along complex spatial paths. Submerged arc welding can be carried out robotically, and some robotic laser welding has been used in the automotive industry instead of arc or spot welding to fasten roof panels to other components, for floor pan and truck front-end assemblies, and for frame members[3].

Since welding robots do not get fatigued or distracted, they typically achieve a much higher percentage of time in operation, and much better repeatability than welding tools operated manually. The robots also relieve human operators of the need to carry out an awkward, not fully healthful, operation. Vision systems track and control the welds.

G7. ***robots in painting, sealing, coating*** - Spray painting by robots provides greater consistency and uniformity of coating than that controlled by a human spray painter.[3] Robotic painting is very common in the automobile industry for auto bodies and is also used on appliances, furniture and other commercial components. Paint, both liquid and powdered, is sprayed robotically. Primer, top coat, stain, mold release, porcelain enamels or other materials are similarly applied. When painting parts on a conveyor, the robot and the conveyor are synchronized. Electrostatic attraction typically accompanies the robotic system. Robots are programmed to spray deep pockets where electrostatic attraction is not effective.

Sealants and adhesives are also applied robotically, but rather than using a spray, the robot deposits a bead or spots of sealant material along a prescribed path on the workpiece. Robotic application provides consistent, uniform dispensing with a better utilization of sealant material and freedom from operator exposure to solvents and other possibly harmful materials. The automotive, appliance, aerospace and furniture industries use robotic application of these materials[3].

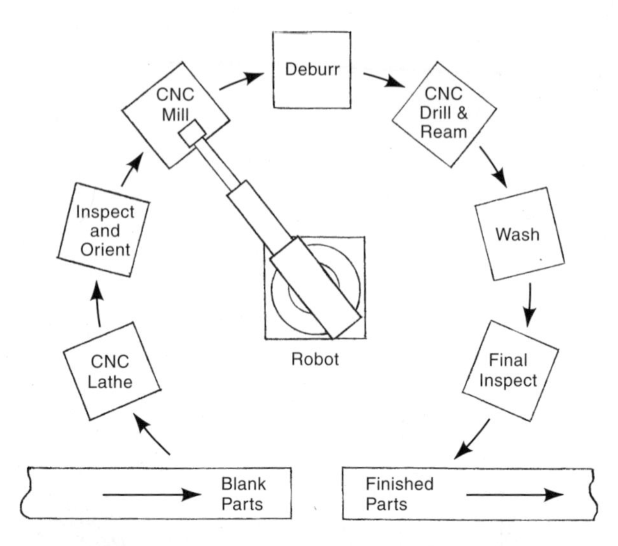

Fig. 14G8 A manufacturing cell with a circular layout so that a single robot can move workpieces between the conveyors and the production machines.

G8. ***robots in material handling*** - are common. Less-sophisticated robots having less precise placement capabilities often can be used when moving workpieces from one workstation to another. In manufacturing cells with a circular layout, one robot may move workpieces to and from seven workstations as illustrated in Fig. 14G8. Other applications include moving workpieces from pallets to machines (and the opposite operation - palletizing workpieces after an operation), removing parts from cases, bins, carousels, and conveyors; positioning parts in fixtures on machines, placing parts in kits; and placing parts or materials on conveyors. Gantry robots, which are conventional robots mounted on the underside of a bridge crane, have a much larger envelope (service area) and are used in material handling applications.

G9. ***robots in mechanical assembly*** - The penetration of robots into assembly operations is less than in the welding, painting and material handling mentioned above. One exception is electronic assembly as noted below and in Chapter 13. Other notable robotic assembly operations are the installation of light bulbs in automotive instrument panels, the installation of auto windshields as noted below, and the spray application of adhesives. Assembly of small electric motors is another application.[3] Video tape recorders have been assembled robotically by Sony Corp and Polaroid camera shutters and other precision components have been assembled by robots.[3] Where necessary or justifiable, robotic equipment can handle and assemble difficult parts such as springs, crooked wires, and compliant plastic parts.[3] Fig. 14G10 shows a robotic pick and place machine assembling parts of an electronic remote key for an automobile.

G10. ***robots in electronics*** - Although they are not normally called robots in the industry, robotic or robot-like machines are used in the placement of circuit devices on printed circuit boards. These units are normally referred to as pick and place machines, or placement machines, and are used for both the surface-mount type of devices and those with through-hole connecting leads. The machines are also used in test and inspection operations, and to load and unload machines involved in parts fabrication. These machines have a limited number of axes of motion compared to the sophisticated 6-axis robots that have pivoting actions similar to those of the human upper and lower arm, wrist and

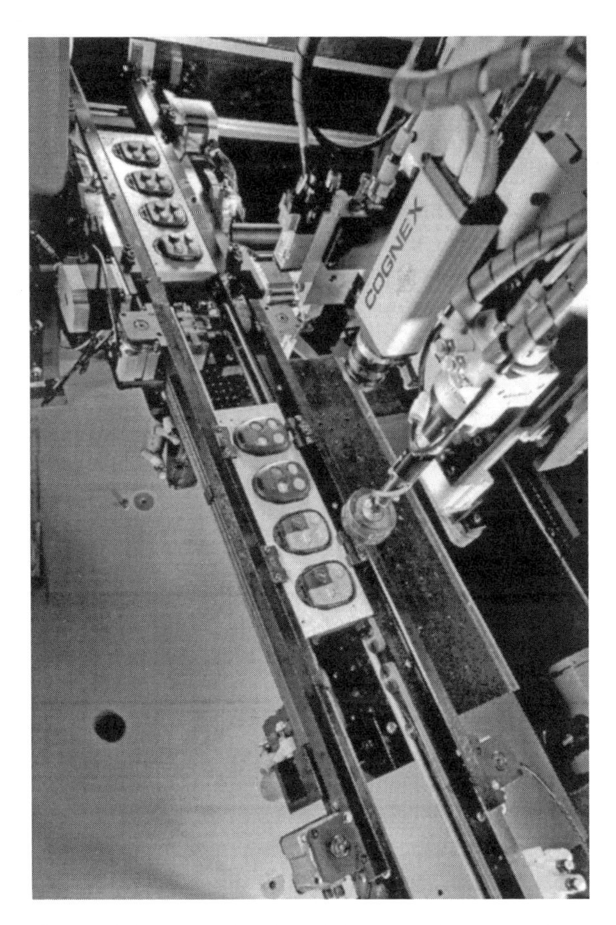

Fig. 14G10 A robotic pick and place machine assembles parts to an automotive remote key pad. *(Courtesy Universal Instruments.)*

hand. Most pick and place machines are mounted on a gantry-type frame which supplies motion in the x and y directions. The gripping device is given Z-axis motion (up and down) and there is also a gripping motion (or use of vacuum) to hold the devices to be placed. Computer control and programmability help these machines conform to the usual definition of robots. Similar machines are also used to assemble major parts of cellular telephones automatically.

G11. ***robots in quality control*** - The major use of robots in quality control is to move and position workpieces in testing and measurement devices and, after the operation, to move the workpiece to where it is needed. In some applications, the robot may actively participate in the inspection or test. In one test of cellular phones, the robot places each phone in a test cradle, presses specific numbered keys and, using machine vision, verifies that the phone's display is correct. In a more conventional arrangement, robots are used to move compact discs between a series of eight testing units, each of which performs a particular test on the discs. Robots with machine vision can make some visual verifications or evaluations during the process of moving a workpiece from one workplace to another.

G12 ***robots in machining*** - Loading and unloading workpieces from machine tools is the prime robotic application in machining. Fig. 14G12 shows a typical application. However, robotically held and used power-tools are used in a number of applications where the precision requirements are less strict or where fixtures can help control the accuracy of the cut. Fixtures may be required because robotic arms do not possess the rigidity of machine tools, nor the accuracy of CNC positioning

Fig. 14G12 Robotic handling of a large workpiece (an aluminum vehicle wheel) between machining operations. *(Courtesy Alcoa Inc.)*

mechanisms. Powered cutting tools held and operated by robots include: drills, reamers, taps, countersinking and counterboring tools, routers, rotary files, grinders, polishing and buffing wheels. These tools are power driven, and are used for drilling and related or similar operations, deburring or surface improvement. When deburring, the cutting tool is usually a rotary file, made of carbide or other hard material. The robot is equipped with a force compensation device to offset deflections caused by variations in the size of the burr to be removed. Other robotic machining involves the use of flame or plasma cutting torches or a laser cutter held by the robot.

Machining operations with the robot handling the cutting tool are most common on large parts where the operation otherwise would be performed by a worker with a hand-held power tool. Aircraft, trucks, space vehicles, vessels, railroad cars and locomotives are examples of products where these operations may be feasible, where it is easier to move the tool to the work instead of vice versa.

G13. ***robots in heat treatment*** - are used to handle parts to be processed, primarily in loading and unloading heat treatment furnaces, salt baths or washing and drying equipment. Robots also are used to immerse workpieces in quenching baths.

G14. ***robots in some specific industries***

G14a. ***in automobile assembly*** - The auto industry has been a leader in the use of robots. The single most prevalent use of robots in automobile assembly is in spot welding body stampings. Other significant uses are arc welding, body painting and coating, dispensing and placement of adhesives and sealers, and loading, unloading and transfer of workpieces.

Assembly of glass windshields to auto bodies is one interesting operation. The windshield is first picked up with vacuum cups and positioned in a fixture. A robot dispenses an adhesive to the edges of the glass in the fixture. A laser system measures the position of the auto body in relation to known reference points and the transport robot uses those data, after the windshield has been picked up again, to position it accurately in the auto body opening.

Body stampings are moved to the assembly line by material handling robots.

Painting of auto bodies includes the use of long-arm robots that can apply paint uniformly to large body panels.

Fully automatic assembly of automobiles and subassemblies is inhibited, per Nof[3], by the very large number of different parts involved, limitations of accuracy and tactile sense of robots, and space limitations.

G14b. ***in appliances*** - One example of robotic assembly in appliance manufacture is the use of a six-axis robot to insert an extruded profile rubber seal in the doors of dishwashers. Robots are also used to unload and load wire coils used in transformers.

G14c. ***in the food industry*** - Robots are used in the food industry for many tasks: to handle poultry products and prawns, to candle eggs, inspect pouches of ready-to-eat foods, sort mushrooms and oysters, grade and cut meat, and process fish. They place airline food, tableware and condiments on trays, using machine vision. Robots assemble assortments of chocolate candies, placing each item in its assigned location. They load machines for packaging wrapped candies. Robots are also used in decorating cakes and chocolate candies. They transfer baked items (bread, cookies, doughnuts and cakes) from oven conveyors and place them on packaging lines with each item prearranged for position in the packages. Gripping is achieved by vacuum. When equipped with machine vision, the robots inspect the products as they are handled.

In meat packing plants, robots pick up frozen fish fillets, ground beef patties, sausages and poultry pieces from freezer conveyors and stack them in packaging containers. Machine vision directs the pick up location for the robots and inspects pieces for correct size and shape. The robots stack the proper quantities of pieces in each container and place each piece in the correct position.[3]

G14d. ***in glass making*** - robots are used to charge molds for molded glass components and to handle sheet glass and molded parts. The heat factor in these operations presents difficult working conditions if human operators are used, but does not impede a properly designed robot.

G14e. ***in chemical industries*** - Robots are used for various material handling applications in chemical processing. They are employed for reactor clean up, particularly when the work would be hazardous for human operators.

G14f. ***in woodworking*** - Robotic handling is sometimes used with furniture components. Other applications are drilling and routing or milling of workpieces by robots and some assembly operations. Robots place components in assembly fixtures and press dowels or similar parts into workpieces for their assembly into furniture. However, the use of robots in woodworking is less widespread than in a number of other industries.[3]

G14g. ***in other industries*** - Robots pack assortments of pills in blister packages using machine vision. Other similar applications are palletizing of containers of various products and packing of bagged materials into shipping containers.

References

1. *Stereolithography and other RP&M Technologies,* Paul F. Jacobs, SME, ASME Press, New York, 1996.
2. *Rapid Prototyping, Principles and Applications in Manufacturing,* Chua Chee Kai and Leong Kah Fai, John Wiley and Sons, Inc., New York, 1997.
3. *Handbook of Industrial Robotics,* Shimon Y. Nof, Ed., John Wiley & Sons, Inc. 1999, ISBN 0-471-17783-0.
4. *Rapid Prototyping, Principles and Applications, 2nd ed.* Chua C.K., Leong K.F. and Lim C.S. World Scientific Publishing Co., Singapore 2003.
5. *Materials and Processes in Manufacturing, 8th ed.,* E.P. DeGarmo, J.T. Black, R.A. Kohser, Prentice Hall, Saddle River, NJ, 1997.
6. *Van Nostrand's Scientific Encyclopedia, 7th ed.,* Considine, D.M., Van Nostrand Reinhold, New York, 1989.

Section II

How Products, Components and Materials Are Made

A

abrasives - are materials used to polish or machine metals, wood, stone, glass, and other materials by the cutting action of the small grains of the material. (Abrasive machining is described and illustrated in 3C). There are two kinds of abrasives, natural and man-made. Natural abrasives include aluminum oxide ("corundum"), emery (impure aluminum oxide), diamond, sand, crushed garnet, quartz, tripoli, talc, and pumice. These materials are mined, crushed, classified by size and shape, and usually bonded together in a grinding wheel, stone or block, or bonded to paper or cloth. Man-made abrasives are silicon carbide (SiC) "Carborundum", aluminum oxide (Al_2O_3) "Alumina", boron nitride (BN), and boron carbide (B_4C) made by various chemical processes.

Silicon carbide is made from pure sand, coke, sawdust, and salt. The mixture of these ingredients is placed in a long, trough-like furnace and heated by an electric current from graphite electrodes. Temperatures up to 4400°F (2400°C) cause a complex chemical reaction that yields SiC and carbon monoxide. Crushing of the silicon carbide to yield small grains follows the furnace operation.

Aluminum oxide is made from bauxite, the ore of aluminum. The calcined bauxite (heated to drive off unwanted materials) is melted in an electric arc furnace. Aluminum oxide is also made with the Bayer process wherein bauxite is mixed with sodium hydroxide and seeded so that aluminum hydroxide precipitates. The aluminum hydroxide is heated to drive off the water and produce granular alumina.[1]

Also see *grinding wheels* and *sandpaper*.

ABS plastics - are a family of plastic alloys that are terpolymers of acrylonitrile, butadiene and styrene. ABS is also SAN (styrene acrylonitrile copolymer) with butadiene-derived rubber dispersed in it. ABS is most commonly prepared from 50% or more styrene monomer. (See *polystyrene*.) The styrene is a clear and colorless liquid at room temperature. It is produced from the dehydrogenation of ethylbenzene, a product of ethylene and benzene, both petroleum derivatives Acrylonitrile is made chiefly from propylene (obtained frorm petroleum refining) by treating it with air and ammonia in a fluidized bed catalytic reactor.[4] Butadiene is a colorless gas used in the production of neoprene and nylon and other materials. It is produced, along with ethylene, from the steam cracking (11H2a and Fig. 11I1) of naphtha and oil obtained from petroleum. One of three polymerization processes may be used to produce the ABS from these materials. They are emulsion polymerization (4A3d), bulk polymerization (4A3a), or suspension polymerization (4A3c).

ABS plastics are used for telephones, helmets, luggage, computer housings and other housings, pump impellers, pipe and pipe fittings, toys and often, plated automotive grills, door handles, window cranks and other components that are electroplated to resemble metal.

acetal plastics - are polyoxymethylene (POM) and have the -CH_2O- unit repeated in their backbone[15]. Acetals are made from the addition polymerization (4A2a) of purified gaseous formaldehyde. Formaldehyde is produced from the oxidation of methanol in the vapor phase.

Three trades names for acetal plastics are Delrin (homopolymer), Celcon and Ultraform (copolymers). Acetals are used for many mechanical parts including gears, bearings, conveyor links, faucet parts, stereo cassette parts, zippers, food processor blades, automobile door handles and seat belt parts.

acetate fibers and fabrics - See chapter 10, particularly section 10A2.

acetone - is made as a byproduct in the production of phenol from cumene hydroperoxide. It is also made by dehydrogenating isopropyl alcohol with a catalyst. Acetone is an industrial solvent, used in the production of rayon, plastics, smokeless powder, lacquers, and lacquer solvent.

acrylic plastics - are a family of plastics derived from acrylic acid. The most common is polymethyl methacrylate (PMMA), made from the polymerization of methyl methacrylate, $CH_2{=}C(CH_3)COOCH_3$. Methyl methacrylate is an ester that results from the reaction of methacrylic acid and methyl alcohol. Another method reacts sodium cyanide and acetone to yield acetone cyanhydrin and then reacts this with methyl alcohol to produce methyl methacrylate. Polymerization (4A2 and 4A3) takes place with heat, light, and organic peroxides as catalysts. Bulk polymerization (4A3a) is commonly used.

Other acrylics are made by combining methyl acrylate or acrylonitrile with methyl methacrylate to produce, after polymerization, copolymers of PMMA. The acrylonitrile is made from ethylene or acetylene gas derived during petroleum refining.

These materials, known by several trade names such as Plexiglas and Lucite, have favorable optical properties and resistance to weathering. Common applications are automotive tail lights, window panes, outdoor signs, aircraft windows, small airplane canopies, watch crystals and various lenses. With mineral fillers, acrylics are used for counter tops and sinks. Acrylic emulsions are used in paints and textile finishing. “Orlon”, “Acrylan” and “Dynel” yarns and fabrics are made from acrylic plastics.

adhesives - are materials that hold other materials together by surface attachment. There are many, many adhesives, some derived from natural sources and some from synthetic resins. The following are some noteworthy adhesives, the means by which they are produced, and their particular applications:

animal and fish glues - are made from waste material not suitable for food use. Animal glues are made from collagen, a protein found in bones, sinews and hides. These and other scraps are cut into small pieces, degreased to remove oils and fats, and then treated with lime, plumped, and washed. Usable material, a gelatin, results, and is removed with hot water. The mixture is filtered; water is evaporated and the residue is chilled, flaked and packaged for customer use.[4] It is also sold in liquid form. These glues are used in woodworking, book binding, sandpaper manufacture, and other applications involving paper. Other glues of animal origin are made from casein, a milk protein, and from blood albumin, treated with an aqueous alkaline solution. Casein glues are used in woodworking; blood albumin glues are used in the manufacture of plywood.

alloy adhesives - are made from combinations of two or more different chemistries, for example, rubber-based and thermoplastic combinations or thermosetting and thermoplastic alloys. These combinations of materials can provide better strength or other properties for some applications than either of the constituents can provide individually.

cyanoacrylate glues (super glues)[13] - are made from ethyl cyanoacetate which is mixed with formaldehyde (methylene oxide - HCHO) in a heated vessel. (The ethyl cyanoacetate is made in a series of steps from acetic acid and other materials.) A condensation polymerization (4A2b) reaction takes place, yielding cyanoacetate polymer and water. The water is evaporated and removed from the vessel, and the vessel is further heated to 305°F (150°C) to crack the polymer into gaseous monomers. These are piped to a condenser and collected, in liquid form, in another vessel. One or two more stages of distillation (11C1) may take place to purify the cyanoacrylate monomer. Additives are mixed with the monomer to inhibit too-early polymerization and to set viscosity at the desired level. The monomer is then packed, in a moisture-free environment, in plastic tubes for distribution and sale. The monomer, in the presence of a small amount of moisture from the atmosphere or moisture or an alkaline on the surfaces to be bonded, will repolymerize into a strong bonding adhesive. Cyanoacrylic adhesives are used in medical, dentistry, and construction applications, as well as for numerous household repairs and projects.

electrically-conductive adhesives - used in printed circuit board manufacture, are conventional epoxy or other thermosetting plastic adhesives, with a conductive filler: carbon powder or small flakes of gold, silver, copper or nickel. (See 13K6b.)

epoxy - is usually a two-part adhesive consisting of a thermosetting resin and a catalyst which is an amine or other curing agent. When the resin and catalyst, which are usually viscous liquids, are mixed, a thermosetting reaction takes place and the mixed material becomes a solid. The resin is usually made by reacting epichlorohydrin with phenol compounds. The epichlorohydrin is made from allyl chloride. Varieties of epoxy formulations are made with somewhat different methods and have a range of properties. Epoxies are used in bonding metals and other non-porous materials, in structural applications, in the aircraft industry - for composite construction and other applications - as coatings, and where electrical insulation is needed.

hot-melt adhesives - are made from thermoplastics that soften or liquify when heat is applied and solidify when they cool to room temperature. They are usually made from polyolefins, polyamides or polyesters, sometimes modified with waxes and other ingredients. They are used in making laminates, and in carpeting, packaging and book binding.

pressure sensitive adhesives - are often mixtures of phenolic and a nitrile rubber in a solvent.

pyroxylin cements - eg., "Duco" are solutions of cellulose acetate or nitrocellulose in a hydrocarbon solvent. When the solvent evaporates the adhesive is solid. These adhesives are used for household cements for wood and paper and in shoe sole bonding.

rubber-based adhesives - Natural rubber, butyl, neoprene, SBR nitrile, and polysulfide synthetic rubbers are widely used as adhesives. Many of these rubber-based adhesives are simply rubber dissolved in a solvent. SBR rubber is made from acrylonitrile and butadiene monomers. EPDM is made from ethylene, propylene and diene monomers. Several silicone polymeric compounds are rubber-like and are used as adhesives, though their best applications are as a potting material for electronic devices, particularly where high voltages and high temperatures are involved. Rubber-based adhesives are used extensively as sealants, where ability to withstand moisture, solar radiation, and vibration are more important than bond strength. They are of relatively low strength and are used to bond paper and other similar materials.

vegetable or plant-based adhesives - Tapioca paste is one basis for such adhesives and is used for gluing paper including envelopes, labels and postage stamps that are made to adhere by wetting the adhesive surface. Other vegetable glues are made from agar, a colloid derived from marine plants, gum arabic, from the acacia tree and from algin, derived from seaweed. Mucilage is a vegetable glue made from water-soluble gums. Starch-based adhesives made from corn, potatoes, and rice, are used for mounting wallpaper and in the manufacture of corrugated cartons.

white glue - used as a household adhesive, is a water emulsion of polyvinyl acetate, made by reacting acetylene gas and acetic acid with a catalyst. The resulting material is then polymerized and mixed with water. White glue is principally used to join paper and wood.

advanced ceramic materials, (high technology ceramics), (modern ceramics), (fine ceramics) - See 5B4a.

air bags (for automotive passenger protection) - Air bag systems have the following major components: air bags themselves, inflation devices for the bags, a pre-tensioning system for seat belts, and the control system. The control system incorporates multiple crash sensors that respond to the abrupt deceleration of the vehicle and are input devices to the computer system of the car. The air bags, called "cushions" are made from high-strength nylon fabric that has been coated with silicone plastic for lubricity (so that the folded bag can inflate in milliseconds). Bags are sewn with CNC sewing machines. Tethers are sewn inside the bags to control their shape for maximum cushioning. The fabric is cut with large openings to allow the inflation gas (nitrogen) to escape at the desired rate on impact. A uniquely-numbered bar code tag is sewn into each bag for later traceability. Sewn bags are run through a metal detector to insure that no sewing needle has broken off. The bag is manually attached to a compressed air source for an inflation test, and is then assembled to a metal frame with rivets or bolts. The frame also contains the inflation device which is contained in a sealed, drawn metal can with perforations, which allow the inflation gas to escape. The inflation device includes tablets of propellant that inflate the bag and an initiator, a electrical rapid-heating device with a small initiator charge that activates the inflation tablets. The bag is folded very carefully in a prescribed pattern. The operation is manual with the aid of fixtures. The airbag cover, the visible part of the car's dashboard or steering wheel, may be attached. A sampling of completed airbag assemblies is subjected to deployment testing.

air conditioners - are complex assemblies of many parts including two heat exchangers (one, the condenser, the other, the evaporator), an expansion valve (capillary tube), a compressor and drive motor sealed in a housing, fans or blowers for each heat exchanger with one or two electric motors to drive them, sheet metal or plastic shrouds for the fans, pulleys, drive belts, brackets, a supporting base, air filter screens, control devices, electric wiring for power to the motors and for control of the system, a volatile fluid refrigerant, a catch-pan and drain for condensate and a housing for the assembly including a front panel with movable louvers, an additional louvered panel and controls. The sizes of commercial air conditioning systems range from the small unitary units that fit into window openings to super-large units that may operate on a college campus or commercial complex for a number of buildings. The unitary type, described here, is made in substantial quantities using mass-production methods. Fig. A1 shows the components of a typical room air conditioning unit of this type.

Heat exchangers in these air conditioners are usually made from lengths of copper tubing bent into banks of connected tubing and press fitted into fins of sheet aluminum. The tubing is received from suppliers in coil form, is straightened (2K), cut to the several lengths needed and bent (2H2), as needed, including to the U-shapes needed in the heat exchangers. Tube ends are expanded (2H2i) or reduced in diameter (swaging - 2F1) for fitting to mating tubes. The fins are blanked (2C4) from coiled aluminum sheet stock. They include holes for the copper tubing and may be embossed (2D7) with

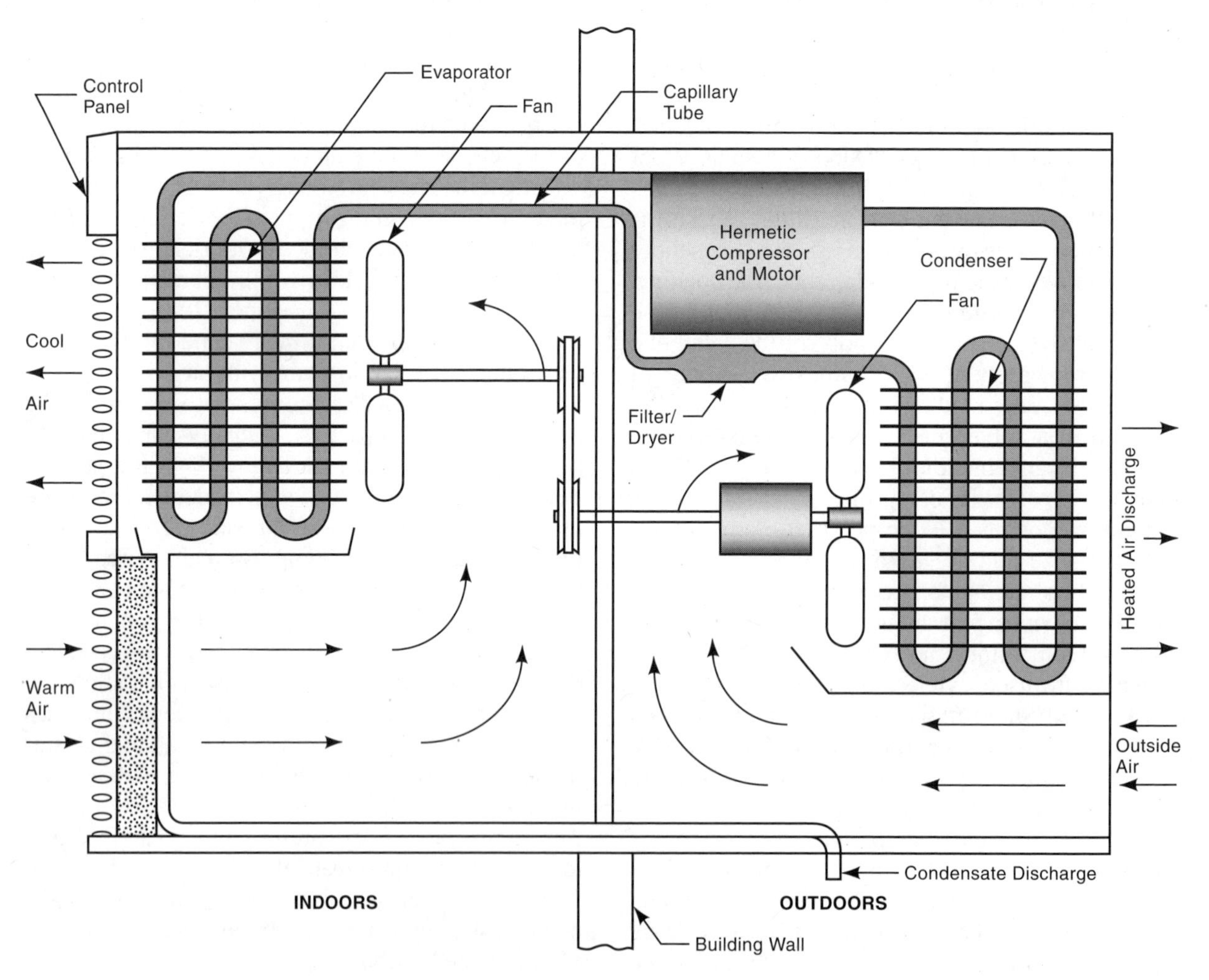

Fig. A1 The major components of a typical room air conditioner.

a pattern to increase the fin area within the air conditioner. Special machines press the copper tubes through stacks of fins. Various means are used to insure a tight fit between the fins and tubes. One involves the punching of extruded (flanged) (2D8) holes, thus providing spacers between fins as well as more contact area. Another method involves expanding the copper tubes from internal pressure after the fins are assembled. These heat exchangers are brazed (7B) to the U-shaped tubes and other tubes that carry the refrigerant to and from the heat exchangers and to and from the compressor, expansion valve, etc.

The compressor is usually a piston type made from cast iron, steel, and die cast (1F) aluminum parts, machined with various milling (3D), turning (3A1a), boring (3B5) and surfacing (3C and 3J) operations. Sometimes, rotary compressors of the vane type are used. Cylinders are cast iron and are bored and honed (3J1) on the inner surfaces. The usual compressor is the hermetic (sealed) type, with the compressor and its drive motor both encased in a steel, can-like, housing. The shaft for the electric drive motor also serves as the shaft for the compressor. The hermetic housing consists of two deep-drawn (2D5) parts that fit together. Sealing is permanent and is accomplished by arc welding (7C1d) the seam between the two parts. The only openings in the housing are leakproof and provide for the electrical connections to the drive motor and inlet and outlet connections for the refrigerant. There is no need for a seal on a moving shaft inside the housing. Complete compressor units with their drive motors are normally supplied to the air conditioner company by vendors who specialize in that type of component.

Electric motors, air filters, pulleys, belts, the expansion valve (capillary tube), sensors, control devices, knobs and fasteners are also purchased from companies who specialize in those particular components. Fan blades and squirrel-cage blowers are usually injection molded (4C) in one piece and press fitted on a splined fan motor shaft or held by a set screw. However, they may be made from galvanized sheet steel with appropriate punch press tooling and screw machine parts (3A2c) for hubs. Shaft bearings are of the pre-lubricated powder metal variety and most often are purchased from specialist suppliers. Fan shrouds and blower housings are also injection molded but may also be made from galvanized sheet steel. They are complex in shape and may have appendages that have structural, supporting functions and may separate the condensing section from the evaporating section of the unit. Other injection molded parts are the control panel parts, louvers, and the frame for the air filter screen. The screen is made by weaving wire or strands of high-melting-point plastic fiber into screening that, when cut to size, becomes an insert in the mold when the frame is molded (4C6), usually of ABS or impact polystyrene. Knobs and the control panel are hot stamped (4M2) after molding.

Brackets, supports, mounting base, and housings are all most commonly formed from sheet steel using blanking, piercing and forming operations though some are injection molded of plastic. Much of the sheet stock is galvanized (8F2) beforehand and supplied in coil form. Sheet metal operations, for high production air conditioner facilities heavily utilize progressive and compound high production dies. Larger, more specialized commercial application units, made in smaller quantities, utilize sheet metal parts made more often with shear (2C1), press brake (2D1a) and CNC turret punch equipment (2C5a). Sheet metal parts, when fastened together are joined by resistance welding (7C6) or, with high production window units, by projection welding (7C6c). External housing parts are cleaned and spray painted on automatic electrostatic painting (8D7) lines. Some parts are powder coated (8D8a) instead of painted. The catch pan for condensate is injection molded or, in some cases, is blanked and formed from galvanized steel sheet. If metal, a drain fitting and pan corners are soft soldered (7A2b and 7A1a) in place. A vinyl drain tube is attached with a spring clamp.

Final assembly of the components including torch and induction brazing of tubing connections, charging and testing of the unit are performed on an assembly line basis (7F2). See Figs. A2 and 7F2-2. The housing parts are fastened to the internal structure with sheet metal screws in pre-molded or pre-punched holes. Operation and leak testing are key parts of the final assembly sequence. Completed units are packed in individual corrugated shipping containers for storage and transportation to retail outlets.

aircraft, (airplanes) - (Note: Much of the following description covers the manufacture of large "transport airplanes", commercial airliners, and similar business aircraft. However, smaller com-

Fig. A2 An assembly line for the production of residential air conditioners. (*Courtesy Carrier Corporation.*)

mercial and private planes require most of the same manufacturing operations, with commensurate care, attention to detail, and quality control.) Large airliners are extremely complex assemblies of structural and functional components and external skin for the body, wings, stabilizer, elevator, rudder, and engine nacelle or cowling. Other key components are the propulsion engines; the landing gear system, including retraction, extension, steering and braking mechanisms; engine and flight control apparatus, including autopilot, instruments, navigation, communication and radar gear; windshields, anti-ice and rain protection systems, windows, doors, electrical power, lights, wheel brakes, fuel tanks, fuel distribution and fuel pump system, communication and radar gear; flight control surfaces, electrical wiring, hydraulic and pneumatic systems including tubing and flexible hoses, pumps for fuel and lubrication; fire protection system, emergency equipment, cargo handling system, passenger entertainment system, galleys, lavatories, water and waste systems; heating and ventilation, cabin pressurization and air-conditioning equipment; passenger and crew seating and interior finishing; and a multitude of sensors and gages to support the listed airplane systems. Commercial airliners may have as many as five million individual parts. Military aircraft also include armament devices and projectiles. Private planes, though they have less auxiliary facilities and equipment, are still highly complex products. Most of the above parts are contained in five principal components in current planes, each of which consists of at least one major subassembly: 1) the fuselage (body), 2) the propulsion devices - jet, turboprop or piston engines, 3) the wings, 4) the tail assembly, 5) the landing gear.

The tremendous quantity and range of variety of components needed in a modern aircraft, their necessarily high quality standards and the specialized skills and manufacturing processes needed, requires that the manufacturing tasks be spread beyond one factory and one company and even one country. Many components, such as engines, instruments, electric motors, electronic and hydraulic devices, and hardware, are made by vendor companies, sometimes in other countries.

Aluminum, in alloy form, is the major metal utilized in aircraft components because of its light weight. Aluminum sheet has formed the skin of modern jet aircraft, and aluminum forgings and machined parts have been used extensively. However, there is a revolution in process in the construction of commercial aircraft, replacing aluminum riveted structures with those made from reinforced plastic composites. (Where aluminum is being retained, much rivet fastening of the skin to structural members is being replaced by friction stir welding (7C13i). See Fig. A4.) The plastic composites consist primarily of carbon fibers in a matrix of high-strength epoxy thermosetting plastic. Fiberglass and kevlar fibers are also used and polyester and phenolic may comprise the matrix for some components. This composite construction enables designs that have less weight, greater strength, more corrosion resistance and a lower parts count than an aluminum structure, though at higher cost. As an example of the importance of composites, the materials content of the Boeing 787 airliner is as follows:

plastic composites - 50%
aluminum - 20%
titanium - 15%
steel - 10%
other materials - 5%

The aluminum, aluminum-magnesium alloys, titanium, steel, and other metal alloys that are used involve, in many cases, the most sophisticated and advanced alloys available. Military aircraft have even higher standards because of the high-temperature effects of high-speed airflow, the need for greater strength to withstand maneuverability stresses, and the shocks from use of armaments.

The development of composite plastic parts started with members that were less structurally critical such as nose cones. Smaller parts such as trim tabs and tail control parts were developed later. After these applications have been proven successful, the composite materials have been used increasingly in the fabrication of larger components. The entirc empennage (tail surfaces) and floor beams of the Boeing 777 are composite and on the Boeing 787, wing and fuselage are also of composite construction. Fuselage "barrel" sections of approximately 22 ft (7 m) length and 19 ft (6 m) diameter, including structural stringers and skin are made of carbon reinforced epoxy construction.

The aircraft materials are formed into component parts by both traditional and advanced machining, forming, and joining techniques, as described in Chapters 2, 3 and 7. Critical structural metal parts are forged. Some of the more advanced and sophisticated metalworking techniques such as explosive forming, electrical discharge machining, chemical machining, electron beam welding, friction stir welding, and diffusion and adhesive bonding, are utilized in aircraft structural components because the requirements for strength, light weight and reliability are so severe. Conventional machining and joining techniques are also prevalent. Computer controlled manufacturing equipment (3U2) is extensively used because of its precision and reliability. The plastic components often involve a sandwich structure with foam plastics or honeycomb cores. (See *sailplanes* for more details on using composite construction in aircraft.) Filament winding (4G7) is employed for reinforced plastic construction of hollow components such as tanks and ducts. Pultrusion (4G11) is used for some composite structural parts. Special fixtures and jigs are used to control the assembly of structural components and many other subassemblies to insure their accuracy, correctness of assembly and fit to other components. Electronic systems (avionics) make maximum use of integrated-circuit technology. Quality control steps are numerous and are vital in aircraft component manufacture.

Final assembly for commercial aircraft and other airplanes produced in some quantity is on an assembly line basis. Fig. A3 shows such an assembly line. However, because of small production quantities and the extremely large number of components in the aircraft, movement of the planes from station to station is less frequent, and the work content at each station is greater than with lines for smaller and simpler products. The lines, however, are increasingly being established on a continuously moving basis with quicker throughput, fueled by a Japanese "kanban" approach that delivers components to the line just when needed, and avoids high stocks of parts. The parts fed to the final assembly line are often large assemblies in themselves, structural subassemblies, or major components such as engines, landing gear, doors, and navigation and communication equipment that may have been made elsewhere. Many of the subassemblies that are added at the final assembly line are, themselves, often assembled on lines, though simpler subassemblies and one-of-a-kind subassemblies may be put together at fixed workstations rather than on moving lines. For larger airliners, the fuselage is made up of a number of sections or "barrels", each a subassembly. Fixtures guide the manufacture of these subassemblies. In some plants, the fixtures are placed in a vertical position for convenience of use. All subassemblies are carefully inspected, gaged, and, where applicable, tested before they are moved to the final assembly line. Movement to a position on the assembly lines often requires overhead cranes.

Fig. A3 A "flow" final assembly line for aircraft used for business and personal flying. After the components specified are installed at each station, all the planes move to the next station for additional assembly work. (*Courtesy New Piper Aircraft.*)

The main airplane assembly, the combination of all these components, is held, in its earlier stages, in a large steel holding and supporting fixture. In addition to the joining of the large structural segments, various internal components and parts involved in the many systems of the plane are assembled to the main fuselage assembly. When the prescribed work has been completed at each station, the assembly is moved to the next station. Moving is accomplished by equipping the holding fixture with wheels, by using overhead cranes or, with the largest planes, by a factory-floor air-cushion technique. As each plane moves along the assembly line, it accumulates more and more major subassemblies. The parts and subassemblies for the various systems that involve wiring, hydraulic lines, ducting, control cables, and other interior and external parts are gradually added. Major subassemblies - the engine or engines, nose section, wings, landing gear, and doors, are also added and fastened in place, primarily with threaded fasteners or rivets.

The fuselage is normally the first major component to be put together on the final assembly line. For large commercial airliners, the fuselage consists of several major subassemblies: a nose assembly, a forward cabin assembly, a mid cabin assembly and a rear cabin assembly. (Smaller aircraft do not require this many subassemblies.) When these fuselage sections arrive at the final assembly line, they are essentially structurally complete. The mid cabin ("mid cabin barrel") subassembly is the first to be placed on the line. The rear bulkhead and wing attachment members are among the components assembled. After it is placed on the final assembly line, the forward cabin barrel assembly is moved into position and attached to the mid cabin assembly. The major assemblies are mated together using laser and other measuring equipment, indexing tools and holding fixtures to insure precision alignment of the fuselage sections. When one fuselage section is in alignment with another, they are joined together using rivets or other permanent fasteners. Sevral different joint structures are employed to fasten the large fuselage sections to one another. In one system, butt joints are used and they are reinforced with a circumferential lap joint ring. The ring extends around the junction area on the outside of the fuselage, and is fastened with "hi-lock" fasteners (high-strength, high-alloy fasteners engineered specifically for aircraft applications) to the two sections, and to internal stringers that connect them. Other designs use lap joints between fuselage sections. One manufacturing technique employed in the industry for joining fuselage sections uses a semi-automated riveting tool that is rotated around the fuselage circumference when the fuselage sections are being riveted together. Special equipment can punch the rivet hole, insert the rivet and clinch it, all in one operation.

The nose assembly with windshield, cabin door and pilots' cabin, assembled on a branch line, is next brought to the fuselage line and attached. It is positioned with the aid of a large locating fixture that has elements to ensure proper alignment of the sections. The rear cabin assembly, made off-line with the aid of an assembly jig, is then added to the fuselage, with the aid of the same large assembly fixture. The resulting fuselage, with these components assembled, becomes the base unit for final aircraft assembly.

Jet or turboprop engines are attached with supporting structures either to the fuselage or to the wings. (See *jet engines* for the manufacture of jet and turboprop engines.) Nacelle assemblies enclose the engines and provide air inlets to the engines, exhaust nozzles for the jets, and diversion channels to provide heating and pressurization air for the cabin and for deicing the leading edges of the wings. The highly-formed sheet metal nacelle parts are made with press operations or roll forming and are pre-assembled and attached to the engine structure after engines and their accessory fuel, ignition and control apparatus are in place.

Wings are assembled on the wing line. With aluminum wing construction, wing spars and ribs are fastened together and the skin pieces, like those on the fuselage are drilled, countersunk, and riveted to ribs and other frame members. The wing skins are chemically machined (3S1) to reduce thickness and weight in non-critical areas. Riveting is performed by automatic machines while the components are held in wing assembly fixtures. Some planes have a center wing section in addition to the left and right wings. The leading edge slats, ailerons, spoilers and trailing edge flaps have been fabricated with the aid of manufacturing fixtures, and are installed on the wing, typically after the wing has been joined to the fuselage. If the wings are made in three parts - left, center and right sections - these are bolted together in a fixture before the wing is assembled to the fuselage. Wings are fastened to the fuselage where mating structural members - wing spars and body frame members - come together. Large, high-strength metal pins connect the structural members. Various other high-strength fasteners: pins, special bolts and rivets, are all used to secure wings and structural components together in building an airplane.

The tail assembly or empennage consists of a fixed vertical stabilizer or fin with a hinged rudder to control yaw (aircraft turning), and a horizontal stabilizer with two hinged elevators to provide pitch control (up and down movement) of the aircraft. These components include spars, stringers, ribs and sheet aluminum skin, that are assembled with rivets, other fasteners, friction spin welding and/or adhesive bonding with the aid of assembly fixtures.

Landing gear assemblies, with retractable wheels at the wings and the nose of the plane, are assembled on separate lines and transported to the final assembly line for installation after the wings are attached to the fuselage. When the landing gear, wheels and tires are installed, the airplane can be moved on the assembly line on its own wheels.

The control systems are installed late in the assembly process on the final line after the large-size components have been attached. Rudder, elevators, flaps, ailerons and trim tabs may be installed on the final line rather than as part of the wing or tail subassemblies. The hydraulic and electrical apparatus involved in the flight control system are also installed at this point.

There are many other components and systems in the aircraft that have important elements installed after major components are in place. These include the windshield, avionic systems, cabin pressurization system, anti-ice system, emergency oxygen system, cargo handling system, etc. Interior work, such as installation of passenger entertainment systems, passenger cabin sidewalls, seats, galleys, lavatories, flooring, overhead luggage bins and cabin partitions, is performed at workstations near the end of the line, partly because these items are often customized for each customer. Exterior painting of the airplane is the final step in the manufacturing sequence.

Testing is particularly important in the aircraft industry because of safety issues and the possible effects of the failure of a basic aircraft system or of its subsystems. All subsystems are tested before assembly and the completely-assembled aircraft is tested as much as possible inside the final assembly building. Computer simulation techniques are used where possible to save costs and, more important, to uncover any defects at an early stage. Tests performed in the final assembly building include: avionics systems testing, leak tests on the fuel system, and pressurization checks to confirm the pressure integrity of the airframe, among many other tests. Then, after the plane departs from the final assembly building, extensive ground and in-flight testing takes place. Rigorous tests of many of the plane's systems through different operation modes are carried out. Corrections and adjustments are made, if necessary, before the new airplane is delivered to the customer.

Figs. A4, A5, and A6 show components of business or private aircraft in process of manufacture.

alcohol, denatured - is ethanol mixed with small amounts of unpleasant substances (camphor, wood alcohol, benzene, pine oil and kerosene) to prevent it from being used as a beverage. Its usefulness for industrial applications is not affected.

alcohol, ethyl (ethanol) - also known as grain alcohol, C_2H_5OH, is the alcohol contained in alcoholic beverages - wine, beer, whiskey, brandy, gin, etc. It is produced from the fermentation (12F) of sugars or

Fig. A4 Part of the cabin structure of a business jet aircraft. The stringers and ribs are fastened to the aluminum skin by friction stir welding (See 7C13i.) instead of with rivets. (*Photo courtesy Eclipse Aviation.*)

Fig. A5 The rear section of a business jet fuselage being assembled. Note that much of the aluminum skin is not fastened with rivets to interior stringers and ribs. Instead, it is fastened by friction stir welding. (*Photo courtesy Eclipse Aviation.*)

Fig. A6 One side of the major part of the cabin of a four-passenger private aircraft being moved into position for additional assembly. Note the light weight of the composite structure, permitting easy handling in the factory in addition to the flight advantages of such a structure. (*Courtesy Cirrus Design Corporation.*)

starches. Corn is the most common raw material. Black strap molasses is another. In Brazil, much is made from sugar carne. The yeast enzyme, zymase is used to convert these sugars and starches into ethanol. Carbon dioxide is a bi-product. After fermentation, the liquid is only about 7 to 12 percent ethanol. A series of distillations (11C1) increases the content to as much as 95 percent ethanol. Animal feed is a by-product of ethanol production from corn. Ethanol is also made for commercial applications by other methods. One process involves the hydration of ethylene derived from petroleum. Another process utilizes acetaldehyde made from acetylene. Ethanol's industrial uses include it application as an ingredient in lacquer, perfumes, synthetic rubber, explosives and many organic chemicals. Ethanol is used in automotive antifreeze, as the fluid in thermometers, and mixed with gasoline to produce gasohol.

alcoholic beverages - See *distilled spirits* or the listing for the specific beverage: ale, beer, wine, brandy, whiskey, rum, and vodka.

alcohol, isopropyl, (isopropanol or rubbing alcohol) - is a petrochemical, made from propylene gas. The propylene is treated with sulfuric acid followed by hydrolysis and distillation (11C1). Isopropyl alcohol is also a by-product of some fermentations (11F and 12J4). In addition to its medical use, it is a solvent for oils, resins, alkaloids and gums and is used in the manufacture of antiseptic solutions, soap, and acetone.

alcohol, methyl - also known as ***methanol*** or wood alcohol, CH_3OH, was traditionally made from the destructive distillation of wood (11C1g). Currently, methanol is most commonly made by reacting carbon monoxide and hydrogen with the aid of a catalyst: (See "reactions" in entry 11I.)

$$CO + 2H_2 \rightarrow CH_3OH.$$

Methanol is extremely poisonous for either drinking or inhaling. It is used as an antifreeze and as a solvent for lacquer, gums and other materials. Derivatives of it are used in the synthesis of plastics, drugs, dyes and perfumes. Methanol can also be used as a high-octane, clean-burning fuel.

ale - is a variant of beer produced with water containing calcium sulfate and from top-fermenting yeast. (Beer is made from a bottom-fermenting yeast.) Like beer, ale contains both malt and hops. The fermenting temperature for ale is higher than that used in the lager beer process. Ale is somewhat bitter, full-bodied and has a stronger hop flavor and higher alcoholic content than beer. See *beer*.

alloys - are blends of two or more elements, normally metals. (Plastics are also alloyed and some metallic alloys include carbon and other nonmetals).The usual manufacturing method for metal alloys is to melt, in one vessel, measured quantities of all the ingredients and then to stir or otherwise mix the molten mixture thoroughly. (See 11K3.) One or more of the ingredient metals may be melted separately before mixing. The initial melt may include already alloyed metals, particularly if scrap material is included in the original melt. If so, calculated amounts of pure metals are included in order to achieve the desired metal ratios in the alloy produced. Other alloying methods include powder metallurgy, which is the mixing of solid metal powders, pressing them and sintering the mixture, by heating it to a temperature just below the melting point. Another alloying method is ion implantation in a vacuum chamber.

Almost all commercial applications of metals involve the use of alloys because alloying ingredients can greatly improve the properties of a basic metal.

aluminum - is produced from the aluminum ore, bauxite, in two steps: 1) refining the bauxite. This produces alumina (Al_2O_3), and separates it from the oxides of iron, silica, and titanium that are in the bauxite. 2) producing aluminum from alumina by smelting.

Refining bauxite uses the *Bayer process.* The bauxite, crushed to powder form, is mixed with a solution of sodium hydroxide (caustic soda). The mixture is heated to a temperature of 300 to 480°F (150 to 250°C) under pressure for about 1/2 hour. The alumina dissolves in the caustic soda, forming a solution of sodium aluminate. The other materials in the bauxite remain in solid form and are filtered from the solution in a series of tanks with cloth filters. The solution is then treated in precipitation tanks by adding crystals of aluminum hydroxide. After several days, most of the alumina in the solution precipitates and collects on the crystals. When precipitation is complete, the solution is filtered (11C7a) to separate the liquid and the aluminum hydroxide crystals. The crystals are then heated at 2000 to 2200°F (1090 to 1200°C), driving the water from the hydroxide and leaving alumina in fine white powder form. Fig. A7 illustrates the processing of bauxite to yield alumina.

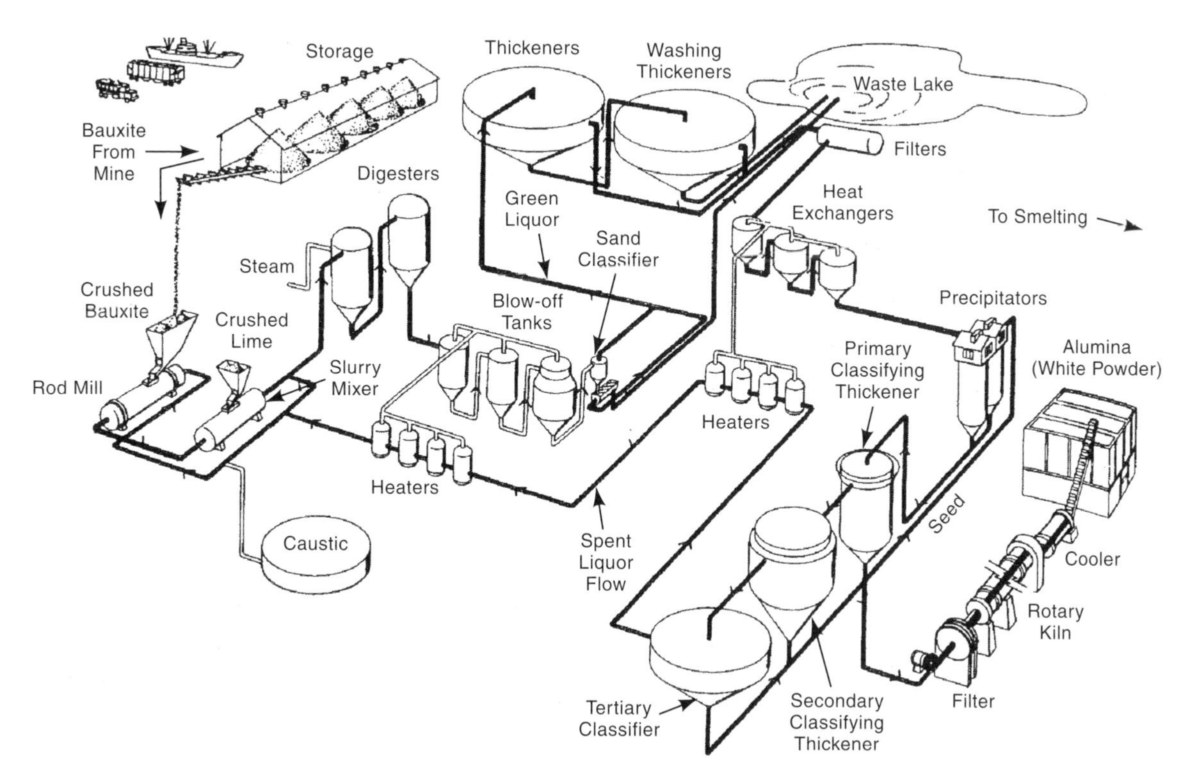

Fig. A7 The process for making alumina from bauxite. *(Courtesy Alcoa Inc.)*

In the Hall-Heroult smelting process, the alumina is dissolved in a chemical bath primarily of molten sodium aluminum chloride ("cryolite") that also contains aluminum fluoride and calcium fluoride. The tanks used are steel with carbon liners. The solution is heated to 1740°F (950°C). Carbon anodes, connected to a power source, are lowered into the solution. The lining of the tank is connected to the power source and becomes the cathode of a direct electrolytic circuit. Resistance to the passage of electrical current through the bath generates heat that keeps the bath molten. The electrolytic action separates the alumina into aluminum and oxygen. The oxygen combines with the carbon of the anode to form carbon dioxide gas. The aluminum, in liquid form, collects at the cathode, which is the carbon tank lining at the bottom of the tank. The process is continuous and many electrolytic tanks may exist in one factory. Periodically the molten aluminum is drawn off from each tank into crucibles from which it is poured into ingot molds. Alumina is periodically added to the bath to replace the aluminum that has been drawn off. The carbon anodes are eroded by the process and they are also replaced as necessary as the continuous operation proceeds. The cost of the electricity needed for the electrolysis is a major expense in the production of aluminum, leading to the location of aluminum smelters near dams that produce inexpensive hydroelectric power. Fig. A8 illustrates the electrolytic process and the casting of aluminum ingots.

Since recycling of scrap aluminum requires far less electrical energy than the processes for making aluminum from bauxite, there is a powerful cost advantage in utilizing recycled material. Much present-day aluminum comes from melting and

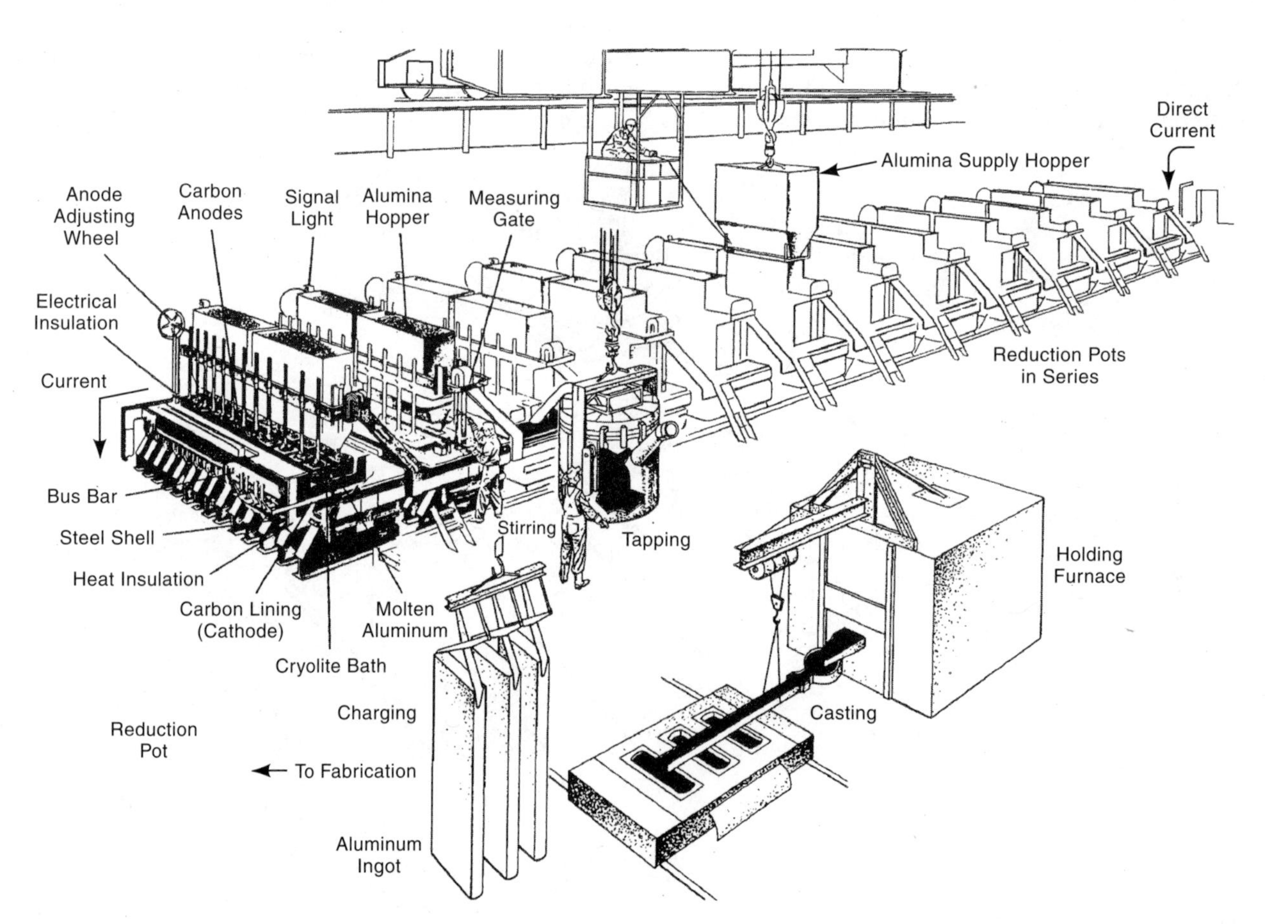

Fig. A8 The production of aluminum metal by electrolytic refining of alumina. *(Courtesy Alcoa Inc.)*

refining scrap aluminum, including aluminum beverage cans.

Aluminum's light weight makes it useful for applications where weight reduction is beneficial. Components for aircraft, railroad cars and automobiles are major applications. However, the greatest tonnage of aluminum is found in architectural and building applications: doors, windows, screens, downspouts, gutters, siding, and building panels. Aluminum has been used increasingly for electrical wires, despite its lower conductivity than copper because of its lighter weight and lower cost. Beverage can, cookware and foil applications are also significant.

aluminum foil - is made with one of two processes. In one, "reroll stock", large slabs of essentially unalloyed aluminum are annealed and then run through a rolling mill. In the other method, the foil rolling mill is arranged in line with the ingot casting equipment and the material then does not require annealing.

The rolling mill utilizes a pair of smooth-surfaced steel work rolls on parallel axes with parallel backup rolls providing support to each work roll against the high forces involved. The aluminum stock is passed several times through the rollers, which reduce the stock thickness and spread it to a longer length. Lubricants are added to the metal surface to facilitate the operation, and annealing may be required between rolling operations when the aluminum work-hardens. In each successive rolling operation, the thickness is reduced and the length is increased. Roll rotational speed is increased at successive stages. After the final rolling to foil thickness, the material is slit into the widths desired. It may then be coiled and packaged for use as foil or may be laminated or coated for use in other packaging or for other special uses.

aluminum oxide - See *abrasives*.

ammonia - NH_3, a gas, is made by the Haber-Bosch process, which uses direct synthesis of hydrogen and nitrogen at high pressures and temperatures. Ammonia also is produced as a by-product of coke production. It is used as a fertilizer both directly when liquified and applied to soil, and as a basic ingredient in the manufacture of solid fertilizers. Other applications include heat-treating steel, use as a refrigerant, and in the manufacture of explosives, plastics, and use in many chemical and other industrial processes.

antifreeze - The major antifreeze ingredient for automobiles and in other internal combustion engines, and for other freeze-prevention uses, is ethylene glycol. It has the desirable properties of low freezing point, high boiling point, and water solubility. In vehicle cooling systems, it is mixed with water and anti-corrosion and anti-foam agents. A 25% solution will lower the freezing point of water to −5°F (−21°C). Ethylene glycol is made from ethylene that is reacted with oxygen (often from air) to form ethylene oxide. This is purified and converted, by hydration, to ethylene glycol. Unreacted ethylene oxide, and other products of the reaction, are separated or reacted. A phosphate, nitrate or other agent may be added to provide anticorrosive properties.

Methyl and ethyl alcohol and propylene glycol are also sometimes used as antifreezes.

antimony - Sb, is found primarily as a sulfide in the ore, stibnite (Sb_2S_3). Two methods are in use to derive metallic antimony from this ore. In one method, the ore and scrap iron are melted in a furnace, causing the iron to react with the sulfur, forming iron sulfide in liquid form. It floats on top of the antimony, which also melts to the liquid state and can be removed from the bottom of the furnace. In another method, the stibnite is roasted to form antimony trioxide, Sb_2O_3. This material is then reduced to metallic antimony by heating it with carbon. Antimony is used as an alloying metal in the lead used in storage battery plates, with lead and tin as type metal and in babbitt metal used in bearings.

anti-shrink fabrics (cloth) - See 10F5a.

apple sauce - Apples are peeled, cored, and chopped. They are cooked in water, sometimes with the addition of sugar. The cooked material is passed through a fine-mesh screen, often through one of cylindrical shape. Paddles within the cylinder push the mass through the screen to a canning operation. Any seeds or large particles that do not pass through the screen fall to the bottom of the cylinder.

argon - is obtained from the liquification and fractional distillation (11C1a) of air. The first fractional distillation yields a mixture of argon, nitrogen and oxygen. The oxygen is removed by reaction with hydrogen and the hydrogen and remaining nitrogen are removed by a further distillation operation. Argon is used as a inert shielding gas for arc welding and various industrial processes including the production of silicon and germanium for semiconductors. Argon is used to fill fluorescent lighting tubes and other electric light bulbs.

asbestos/asbestos board - Asbestos is a name for not just one material, but for a variety of minerals that occur in nature as fibers. Fiber length, however, may be short. One of the more commonly mined minerals for asbestos is chrysotile, $Mg_6Si_4O_{11}(OH)_6$. Quebec is a major source; it is also mined in South Africa. The fibers in the ore are separated by several methods: air suction, crushing and vibrating screens. Fibers of about 3/8 in (10 mm) in length or longer are suitable for spinning and twisting into yarn for textile applications where temperature resistance is important. Shorter fibers are molded with plastic resins or portland cement into boards, pipe, brake linings, gaskets or with asphalt, plastics or rubber into felt. Use of asbestos is now restricted because of the adverse health effects of inhalation of the fibers.

asphalt - is a variety of naturally occurring bitumen and is also a by-product of petroleum refining. The material ranges from a highly viscous liquid to a solid substance. Venezuela, Trinidad and California have natural supplies of asphalt. When made from petroleum, asphalt is the by-product when more volatile components are removed from petroleum. A type of asphalt also results as residue after the distillation of coal. In natural deposits, it is believed to be a product of the decomposition of organic marine materials that later form petroleum. It is used for road pavements when mixed with sand and gravel aggregates. It is also used for roofing, pipe coating, sealants in water tanks, canals and reservoirs, floor tiles, paints and laminates.

aspirin - is acetylsalicylic acid, a solid. It is made from the action of acetic anhydride on salicylic acid, a substance that occurs in the bark of the willow tree and many other plants. The acetylsalicylic acid, corn starch, water, and a lubricant are the raw materials for aspirin tablets. The corn starch and (cold) water are placed in the same vessel and are stirred as the water is heated. When blended, the acetylsalicylic acid and the lubricant are added and are mixed thoroughly to blend all the ingredients and expel air. The blend is processed through a device that forms slugs of about an inch in diameter. These are forced through a screen to remove lumps and air pockets. The mixture is then blended and mixed gently with additional lubricant. Next it is fed to a tableting machine that feeds the mixture to small tablet-sized cavities on a rotary indexing table. As the table rotates, each cavity passes under a compression station where a punch descends and presses the material in the cavity to a solid tablet. At another station, as the table rotates further, the tablet is ejected from the cavity. (The action of the tableting machine is similar to that of the machines that compress metal powders when making powder metal parts (2L1c). The completed tablets are inserted into bottles by automatic machines that also insert cotton packing, attach a cap, attached a label to the bottle and insert the finished bottle in individual boxes, if used, and then into a shipping carton.

athletic shoes - See *shoes, athletic.*

automobile engines - are assemblies of many precision cast, stamped, forged, and machined parts, some of which are electroplated or painted. Except for specialty situations where only a very limited number of a particular engine is built, assembly takes place on an assembly line. Some portions of the assembly operation may be robotic or mechanized with special equipment. (See 7F3b and 7F3c.)

The basic engine block is normally an iron casting made in sand molds. (See *engine blocks.*) It is then machined extensively by milling, drilling, boring, reaming, grinding, and honing. The crankshaft is either forged (2A4) or cast (1B), and is turned (3A1a) and ground (3C1). Connecting rods are usually forged, bored (3B5) and honed (3J1). Pistons are sand cast or permanent-mold cast (1D1) of aluminum and turned on special machines. Valves are forged, turned, and ground. Camshafts are forged or cast, and turned and ground on special machines. Manifolds are cast and machined by milling (3D) and other operations. Machine screws to fasten parts together are usually cold headed (2I2) with rolled threads. (See *screw threads.*) Many other parts, made in the engine factory or purchased, are included in the assembly. These include parts that may be stamped from sheet metal (2C and 2D), die cast (1F), or molded from plastics. They also include spark plugs, electrical wiring, oil and air filters. bearings, seals, insulators, electronic ignition and fuel metering parts, carburetors, fuel injection parts, coils, drive belts, and pulleys. After assembly, the engine is tested for correct operation and power at a test stand. If satisfactory, it is moved to the final assembly line for installation in an automobile. Due to the high production volumes that typically accompany automotive production, many of the parts making operations are highly automatic and engineered specifically for the component in question. Special machines and transfer lines are often part of the parts-making operations. (See 3X and 3Y.)

automobile bodies - Auto body parts are made from sheet steel although, increasingly, fiberglass reinforced polyester plastic and formed thermoplastic sheet parts (4D) are finding their way into current designs. With the sheet steel parts, blanking, forming, and deep drawing operations are performed (See 2C4, 2D4, 2D5). These operations are performed on high-production equipment with compound dies (2E3) and progressive dies (2E1), where applicable, with robotic unloading of the stamped parts (14G4). Body parts are fastened together by resistance welding (7C6) and some arc welding (7C1), most of it robotic (14G6). Weld joints are made smooth by application of high-lead body solder, sanded smooth. The welded body assembly is dipped in a cleaning bath and then given a zinc phosphate treatment (8E2) to aid in corrosion resistance. Plastic sealers are applied in locations where moisture can be trapped. The complete metal body assembly is then painted. The first coat is often applied with the electrophoretic method (8D9), dipping the body into a vat of water-based paint. The selected color is often applied with robotically-manipulated, electrostatic paint guns (14G7), with some manual spray application to selected or difficult-to-cover areas. A final clear coating is applied similarly, and is buffed and polished after it dries. Sound deadening materials are applied in some areas with rubber-based adhesives. A polyurethane coating is applied to the bottom surfaces to provide protection against flying stones, gravel and other debris. After painting, doors, deck lids, hood, trim, windows, doors, bumpers, interior panels, the dashboard with instruments, seats, lights, radios, speakers, carpeting, and various hardware items are assembled to the body as part of the final auto assembly operation. The body is then conveyed to the main assembly line where it is assembled to the other components that make up the car. (Also see 14G14a.)

automobile chassis - the steel frame that supports the car, is used in many automobiles. However, the more common auto designs now incorporate a unitized body. With the unitized design, extra members are added to the body to enable it to support the weight of the vehicle and to withstand road shocks. The supporting members then, are in the body assembly rather than part of a separate chassis. Where a separate chassis is used, it is made from heavy gauge sheet steel that is blanked, formed, and hole-punched. It is assembled and arc welded with other similarly-made chassis components into a strong and rigid assembly. Even with a unitized body, however, there normally is a sub frame, similar to the earlier chassis but only in the front of the vehicle, to support the engine, transmission, and front suspension. In many designs there also is a small rear frame to support the rear axle, differential, and suspension. These frames are also made of heavy gage steel stampings, welded together. The net effect of the unitized body construction is a reduction in vehicle weight.

automobiles - are highly complex assemblies, consisting of about 14,000 or more parts, in many subassemblies and systems from many different suppliers. Automobiles are produced in large quantities. The production process for automobiles consists of the manufacture of all the individual parts, including their finishing with heat treatments, plating and painting, if used, their assembly into various mechanical subassemblies, followed by the combination of all these subassemblies and parts into a finished vehicle. Since the days of Henry Ford's Model-T, line assembly methods (7F2) have been used for the final assembly of automobiles. In principal, this is still true, but the present-day assembly line is far more automatic and the products produced on it are much more variable, from car to car, than the early lines of Ford. Automatic and robotic equipment is interspersed with human assemblers. Many of the components assembled on the line are subassemblies that were, themselves, manually assembled on lines with some interspersed robotic and automatic assembly stations. Examples of these subassemblies are the chassis, body, bumpers, fuel pumps, piping and tank, radiator, suspension system, seats, engine, transmission, drive shaft, rear axle, wheel assemblies, instruments and instrument panel assembly, steering system, brake system assemblies, electrical wiring, battery, generator or alternator, starter, headlights and interior lighting system, as well as auxiliary equipments such as air conditioning, radio, stereo, and cruise control. Fig. A9 shows the assembly of dashboards and accompanying components. All these subassemblies and many parts are delivered to the point on the assembly line where they are needed.

Some subassemblies are put together completely with dedicated (special purpose), high production equipment, others with a combination of robotic and dedicated equipment, with or without manual assembly of some components. Robotic operation is common for such operations as welding, painting, windshield assembly, and placement of heavy components like the engine, transmission and body assembly. (See Chapters 14 and 14G14a.) Fully automatic assembly with dedicated equipment is most common with components such as spark plugs, hydraulic brake cylinders, shock absorbers and other subassemblies that are used in multiples in the car. (7F3b)

Fig. A9 A subassembly line for automotive dashboards. The fixture in the foreground holds the dashboard in an upside-down position, in which wire cable bundles, instruments, air bags, controls, a central console and other components are attached as the fixture moves along the line. *Photo courtesy of General Motors.*

The assembly line starts with the attachment of the chassis to the assembly line conveyor (if the car being assembled has chassis rather than "unibody" construction). As the chassis moves down the line, components such as wheels, suspension systems, steering and braking components, and gas tanks, are added. Major stations on the final assembly line involve the joining of the chassis and body (the "body drop"), if that type of construction is used, and the assembly of the engine, transmission and drive train to the body or chassis.

Much of the assembly involves permanent joining of the constituent parts by welding, brazing, soldering or adhesive bonding. The body, chassis, muffler and fuel tank are components that are permanently assembled, mainly by welding. Permanent assembly also makes for a quieter operation of the car. Other assembly is carried out with mechanical fasteners or other fastening methods that can be reversed so that the assembly can be taken apart for maintenance or repair during the life of the vehicle.

Final operations on the line involve the assembly of trim, and the addition of a spare tire, fuel, and antifreeze. Then the assembled vehicle leaves the conveyor; the engine is started, and lights, horn and accessories are tested. Adjustments are made; defects are repaired and, if the vehicle meets the quality standards, price and shipping labels are attached. Fig. A10 shows a typical automobile final assembly line in operation.

automobile windshields - consist of curved pieces of safety glass. The glass is made with the methods described in chapt. 5, section A1, with raw materials including potassium, magnesium, and aluminum oxides, in addition to the more common materials, to provide hardness and other properties. The molten glass is fed to float glass equipment (5A3f) to produce a large, flat glass sheet. Each sheet is cut into smaller, windshield-size sheets. These are then bent to the desired curvature by heating them and draping them over a form of refractory material. Gravity, and the softness of the

Fig A10 An automobile final line: Cars placed sideways on this portion of the line move past workstations where interior, trunk, hood and door components are installed. The conveyor, referred to as the "skillet", sits above the floor and operators can adjust its height for optimum ergonomic conditions. (*Photo courtesy of General Motors.*)

heated sheets, causes them to take the shape of the form (5A5a). The bent sheet is tempered (5A4b), cleaned and assembled with an internal layer of plastic and a second layer of glass. (See *safety glass*.) These three assembled pieces are placed in an autoclave, which provides pressure to force the three layers together and heat to bond the plastic to the glass surfaces. The finished windshield then undergoes a plastic injection molding operation where it becomes an insert in an injection mold and a plastic frame is molded around it. (Insert molding is described in Chapter 4, section C6.) The windshield is then ready for shipment to the automobile assembly factory.

B

bacon - American bacon is made from hog bellies, Canadian bacon from pork loin and European bacon from the ham and shoulder. The meat is rubbed with several possible ingredients: salt, sugar, sodium nitrite, sodium phosphate and sodium erythorbate, or they are dissolved in water and the meat is soaked in the resulting brine. Another approach, common for commercial bacon, is to inject the brine uniformly into the meat with a multiple-needle machine. This step is followed by smoking at a temperature between 130 and 140°F (55 and 60°C) for a period of 2 to 10 days. Chilling, forming into slabs, slicing and packing operations follow.

bags, paper - The typical brown paper grocery bag is made from kraft paper (See 9C5), on special machines. The machines work on rolls of wide paper, to print, slit, cut and fold the paper, glue the seams, and stack the completed flattened bags.

bags, plastic - The initial manufacturing process is film extrusion (4I5) to form the bag body. Zipper-type closures are extruded (4I1) in two pieces, cut to length with shears, and joined to the bag sides with radio frequency sealing (4L9). Radio frequency sealing also joins the two sides of the bag at the bottom. Because of the high volume production of such bags, producers have developed automatic production equipment that performs all these operations and packages the bags in chipboard boxes, all without the bags being touched by human hands.

baking powder - is most commonly a dry mixture of sodium bicarbonate (NaHCO3) with one or more agents to completely decompose it, and a drying agent such as corn starch or flour. A major decomposing agent is monocalcium phosphate. Under heat and moisture, the baking powder decomposes to produce carbon dioxide for leavening of bread and other baked goods. Baker's yeast performs the same function in leavened baked goods, but works more slowly and by fermentation rather than decomposition (See bread and 12G7).

ball bearings - have four main parts: inner and outer grooved raceways, rolling balls and a retainer or cage to hold the balls in the raceways. Except for the cages, the parts are usually made from 52100 high-carbon, high-chromium steel, or 440C stainless steel if corrosion resistance is needed.[9] The steel raceways are machined on lathes or screw machines (3A2c) from heavy-wall steel tubing, heat treated for hardness (8G3c), and finish-ground both internally and externally (3C1 and 3C2). The rolling surfaces for the balls are then lapped (3J2) to a near-mirror finish. The balls are made from heavy wire. The first operation is shearing the wire and forming a ball shape in cold heading (2I2) machines. Flash from cold heading is removed in special automatic machines where the balls roll repeatedly between two grooved cast iron disks. The balls are hardened and tempered and then ground to nearly perfect sphericity and accurate diameter in special grinding machines. They are then lapped to a fine finish. The retainers are steel stampings made with progressive dies (2E1), or may be injection molded (4C) from a plastic. The components are assembled, lubricated and tested before shipment.

ball grid arrays - See 13F.

ballpoint pens - are of many different designs. The simplest designs are produced by the million, with dedicated automatic production machines, and are assembled "untouched by human hands". The components of a simple, low-cost pen are as follows: a ball that rotates during writing, (It has a textured surface to aid in retaining and spreading ink to the paper.), a brass part that holds the rotating ball, an ink reservoir tube, a spring to retract the ballpoint when it is not in use, an external body for the pen, a pushbutton device at the top of the pen to extend the point when needed, a cap for the top of the pen with a center hole and bearing surface for the pushbutton device (The cap usually includes an integral clip to hold the pen in a shirt pocket.). There is also a small plastic part that fits inside the body and holds the ballpoint in the extended position after it is moved by the pushbutton. Ink, in paste form, is held in the ink reservoir.

The ball points are made of tungsten carbide using powder metal methods (2L1). The brass part that holds the ball point is made by cold heading (2I2) or by screw machining (3A2c). This part contains a recess for the ball and, after the ball is placed, is crimped at the end with just enough deformation to hold the ball but not prevent its rotation. The ball assembly is press fitted into the extruded (4I1) polyethylene ink reservoir. Ink is a blend of pigment, lubricant, surfactant and thickener, mixed on a batch basis. Heating or cooling may be part of the mixing operation, depending on the ingredients used. The body of the pen, the top and the clip, the pushbutton and the inside plastic part are all injection molded (4C1). The spring is a compression type, wound on conventional spring-winding machines of steel spring wire. (See *springs*.) The ink reservoir tube is deformed a small amount at one point to provide an end bearing surface for the spring. Deformation is done with a heated press forming die. The ink reservoir is filled by a special machine and the parts are then assembled automatically on special dedicated machines (7F3 and 7F3b).

balls, athletic - See *baseballs, footballs* and *golf-balls* for representative construction and methods.

banknotes - See *paper money*.

bar codes - are binary renditions of a series of numbers. They are used for identification and are printed on products, packaging, or labels by a variety of methods depending on the quantity to be printed and the surface to be printed on. Short-run or individual quantities are printed by regular computer ink jet or laser printers. Larger quantities, as required for supermarket products, are printed by various high-quantity methods discussed in Chapter 9. Some companies are in the business of supplying preprinted bar code labels. They use conventional printing processes as well as some specially adapted for high speed printing of labels, including those with sequential numbering or other variable information. Machines are also available to print bar codes directly on a product package or carton rather than on a label.

In a typical bar code, the presence or absence of a bar and the bar width are interpreted by an optical scanner to represent a series of binary digits. The digits are then fed to a computer for processing. For identification of common products, such as those sold in grocery or other retail stores, a Universal Product Code (UPC) is used where the first five digits represent the manufacturer or supplier of the product while the second five digits identify the specific product or component. Similar standard coding systems are in use in different countries. Bar codes are used in retail pricing, for inventory control at the retail, wholesale and manufacturing level, to track lots or individual pieces in a factory and to track books in a library and parcels during shipment.

baseballs - Spherical centers, about 0.8 in (2 cm) in diameter, cut from cork and machined for roundness and size, are wrapped or molded with a layer of black rubber and then with a layer of red rubber to a diameter of about 1.3 in (3.3 cm). A thin layer of adhesive is then applied to the surface, and multiple layers of wool yarn are wrapped by machine around the rubber. Constant tension is maintained on the yarn during winding. When it has been wrapped to the specified diameter, the ball is wrapped again, this time with a layer of cotton yarn that gives the finished ball a smoother finish. The yarn-wrapped balls are then dipped into a bath of latex. This bonds the yarns together. Two cover pieces of a figure-eight shape are blanked from leather, given a coat of latex adhesive, and stapled to the wound ball. The seams are stitched by hand around the ball, using a two-needle method and a standard stitch pattern of 108 stitches. (Machine sewing for this operation has not been developed.) Fig. B1 shows the construction of baseballs used in the major leagues. The staples are removed and the sewn balls are rolled in a machine for a few seconds to press down any raised stitches. The balls are

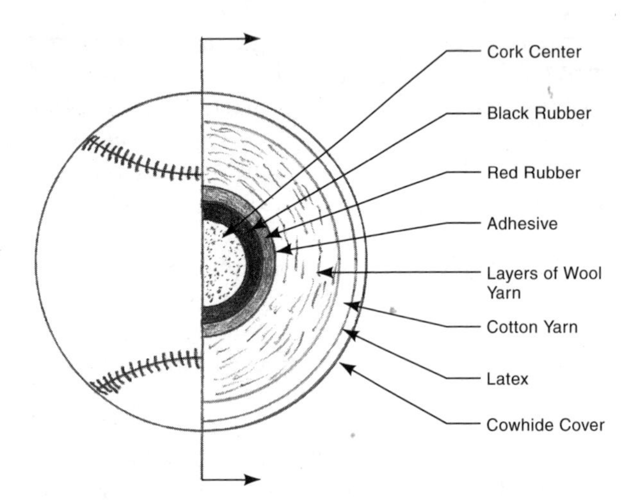

Fig. B1 Section view of the construction of a major-league baseball.

inspected, confirmed to meet the specified weight of 5 to 5.25 oz (142 to 149 g) and circumference of 9 to 9.25 in (22.8 to 23.5 cm) and stamped with identifying designations.

bathtubs, cast iron - are made from gray iron cast in sand molds (1B). After casting and shakeout, the tubs are snagged (1B8g) and may be annealed (8G1). Drain holes are drilled (3B1) and some surfaces are machined (3D). Then all surfaces that will be visible are prepared for application of vitreous enamel (8F1). The first step is grit blasting (1B8c) to remove sand, rust, scale, and dirt. Then the surfaces are smoothed with manual polishing wheels (8B1 and 8B1a). The enamel application is by powder spray or fluidized bed coating, wet spray, wet dip, or flow coating. Powder application is followed by drying and firing. Then a second coat is applied and fused by firing. After quality checking, the bathtub is crated for shipment. Cast iron bathtubs are heavy but solid, and have very good sound deadening properties. Also see 5A5i and *enamels, vitreous.*

bathtubs, plastic - can be made by several different methods. Frequently, fiberglass reinforced thermosetting polyester is used, with the spray-up (4G2) or hand lay-up (4G1) methods. The molding operation preceded by a spray application of a gel coat of the polyester to insure a smooth surface. Another fiberglass method is compression molding with sheet molding compound (4G10). Trimming and hole punching follow these molding operations. Oven heating of the molded tubs may take place to further cure the polyester resin. When high-tonnage, large-platten, injection molding machines are available, tubs can be injection molded (4C1). Common materials used in injection molding are ABS or acrylic thermoplastics, often incorporating glass or other reinforcing fibers, fillers and pigments.

bathtubs, steel - are made from drawing quality, low-carbon sheet steel. One or more deep drawing operations (2D5b) may be required and the workpiece may be annealed (8G1) between draws. Holes for drain fittings are punched with suitable dies after deep drawing. The tub is then annealed to remove stresses from the drawing operations, is shot blasted (8A1b) to remove any scale or other soils, is vapor degreased (8A2a3), and then further degreased with an alkali rinse (8A2d). It is then acid pickled (8A2f) to prepare the surface for vitreous enameling. The enameling and finishing operations proceed as described above for cast iron bathtubs.

bats, baseball - Pennsylvania and New York ash is the wood used for making professional baseball bats. The wood is selected for straight grain and freedom from knots, and is cut into 40 in (1 m) lengths. These are split into pieces of the approximate width and thickness needed for bats and are called "splits". The splits are rough turned (3A1a) to round cylindrical shape and are then known as "billets". They are checked again for proper grain and are usually bundled for shipment to another factory for completion.

At the new location, the billets are dried outside for a period of 6 to 24 months to lower the internal moisture content to the proper level. They are then lathe turned (3A1b) and sanded to near-final bat shape. When ordered for a specific player's preference, they are final turned on a CNC lathe (3T1) programmed for that player's specification. They are sanded, branded (8I8) with the trademark and player's name, stained, if necessary, varnished, and packed for shipment to the team.

batteries, flashlight (dry cells) - The traditional flashlight dry cell is made of a zinc cup (as a cathode) carrying zinc chloride, ammonium chloride, and graphite in moist paste form as an electrolyte, and manganese dioxide as a depolarizer, with a carbon rod in the center as the anode. Starch or flour is used as a gelling agent in the electrolyte. The zinc cup is deep drawn (2D5b) from sheet stock. The carbon anode is compression molded (4B1) from carbon

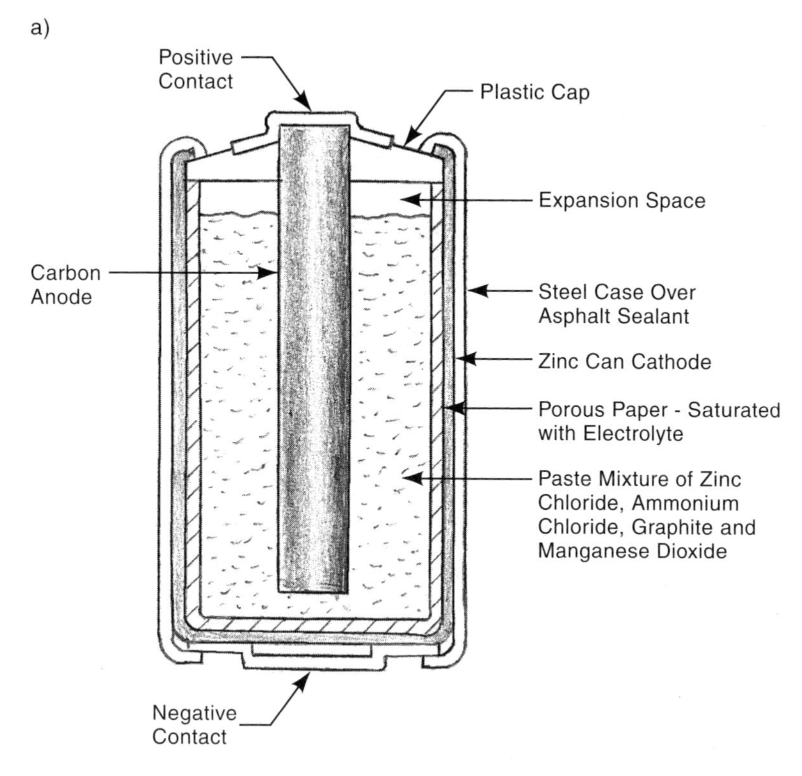

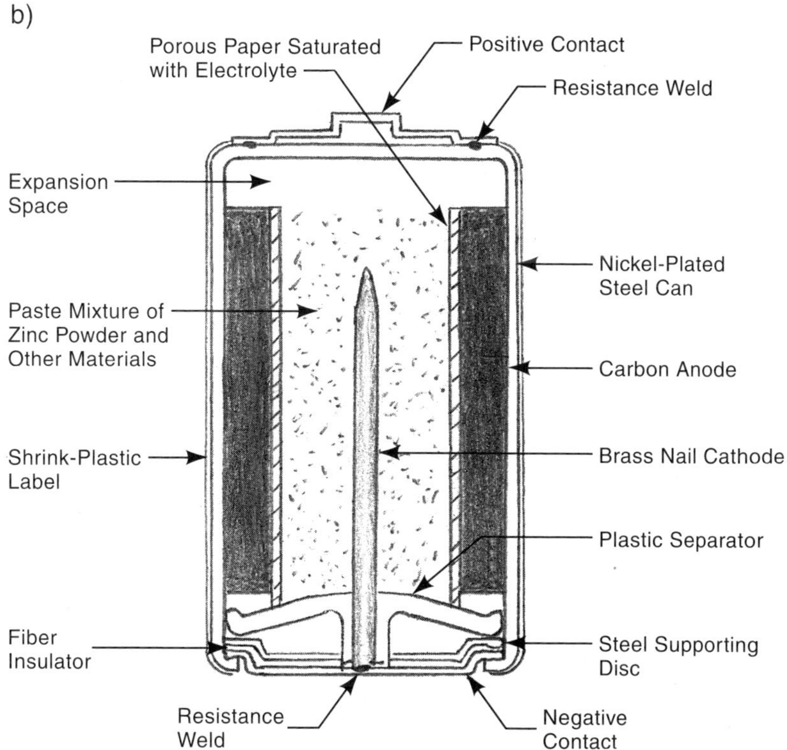

Fig. B2 a) Section view of a typical carbon-zinc battery (dry cell). b) Section view of a typical alkaline battery.

powder and a bonding agent. The electrolyte and depolarizer are made into a paste from the powdered material and water. The zinc can is lined with cardboard which is soaked in ammonium chloride and zinc chloride solutions. The assembly is sealed to prevent escape of the electrolyte. An injection molded (4C1) plastic top cover isolates the anode from the zinc cup and holds a brass or nickel-plated steel electrical contact cap to the carbon anode. Space is allowed between the top cover and the material in the can to allow for expansion. The can is crimped to the top cover. For additional sealing, another top cap and a bottom cap are blanked and formed (2C4 and 2D2) from nickel electroplated (8C1) steel stock. A steel covering for the can, coated internally with asphalt, may be used to further seal the entire battery so that, if the zinc can is penetrated by the electrolytic action that feeds on it, there will be no external leakage. The steel covering also holds the top and bottom caps in place. The sheet steel for the cover is lithographed (9D2a) with product and brand identification before it is blanked. The electrolyte mixture is normally made in a batch mixing process (11G4), but all assembly and sealing operations, including forming of the cover and making rolled cover joints, are carried out in dedicated automatic equipment. Some carbon-zinc batteries are made with an inside-out construction where the zinc cathode is an internal element and a plastic coated paperboard (9C3 and 4I4) container (inside a steel cover) holds the electrolyte. Alkaline, lithium, and rechargeable batteries use different combinations of electrolyte, electrode materials and containers.

In alkaline batteries, "alkaline" refers to the electrolyte, which contains potassium hydroxide. The cathode consists of a brass nail that is surrounded with a paste of zinc powder, potassium hydroxide and other materials. The drawn-steel can that contains the battery, lined with a carbon-containing sleeve, becomes the anode. A layer of porous paper, wet with the potassium hydroxide electrolyte, separates the anode and cathode materials.

The anode liner of the can is a mixture of carbon black (graphite), manganese dioxide, potassium hydroxide solution and starch or flour. The liner is pressed or extruded into a cylindrical shape and inserted into the nickel-plated steel can that contains the battery. A suitably shaped disc is welded to the can end to provide the anode terminal.

The brass nail head is resistance welded to a plated steel disc that becomes the cathode terminal of the battery. An injection-molded plastic part, and a paper fiber disc, electrically isolate the cathode terminal and the nail from the can surface. The plastic part is pressed into the steel container. Expansion space is provided at the opposite end of the steel can. The open end of the can is crimped and sealed over a steel supporting disc and a plastic sealant. A vinyl shrink label is slipped over the can and shrunk to fit tightly. All assembly operations are performed on automatic equipment.

Fig. B2 illustrates both the carbon-zinc and alkaline batteries. These batteries (dry cells) are also used to power toys, portable radios, cameras, tape recorders, electric razors, television remote controllers and many other electrical and electronic devices.

The mercury cell, used for small electronic devices such as hearing aids and wristwatches, has a zinc cathode, an anode of mercuric oxide, and an electrolyte of potassium hydroxide.

bauxite - See *aluminum.*

beams, plastic, reinforced - Plastic structural members also include channels, angles and squares, and are normally made with glass or other fiber reinforcement to provide the necessary strength. The pultrusion process (4G11) is used, and parts are then cut to length by abrasive saw (3G5). These members are used where light weight and corrosion resistance are important.

beer - The major ingredient in most beers is barley, and the first step in making barley-containing beers is to convert the barley to barley malt. To do this, barley grain is softened by soaking it in water until it starts to sprout. Then, it is dried in a kiln. The grain is then milled between parallel rolls that break the brittle, modified starch into small pieces without breaking up the husks. Malting contributes to the desired flavor and provides needed enzymes. Barley malt, thus produced, is mixed with water, hops, and yeast, and usually with other grains such as rice and corn. Hops add bitterness to the flavor and add control to the later fermentation. The water dissolves starches and other molecules and enzymes that result from malting. The enzymes start to act on the other ingredients.

The malt-grains-water mixture is heated in a process called mashing, performed in a vat called a mash tun. Processes vary with different ingredients and different breweries, but they always involve heating the mixture, often to temperatures around 150°F (65°C). This temperature is held is for an hour or more. This heating, through the action of the enzymes, converts the starches in the grains to sugar and other carbohydrates. The mash liquid is known as wort. The wort is filtered slowly to remove husks and other solid materials. The resulting liquid is boiled to sterilize it, stop enzyme activity and, with added hops, enhance the beer's flavor. The next step is fermentation (12J4), changing the sugar into alcohol. When the desired degree of fermentation has taken place, the beer is filtered to remove yeast residue, and aged for several weeks to further improve its flavor. It is then filtered (12E1) again, possibly pasteurized to kill any residual microorganisms, and bottled, canned, or put into kegs. (For some "draft" canned beers, microfiltering replaces, pasteurizing.) Fig. B3 illustrates the brewing sequence.

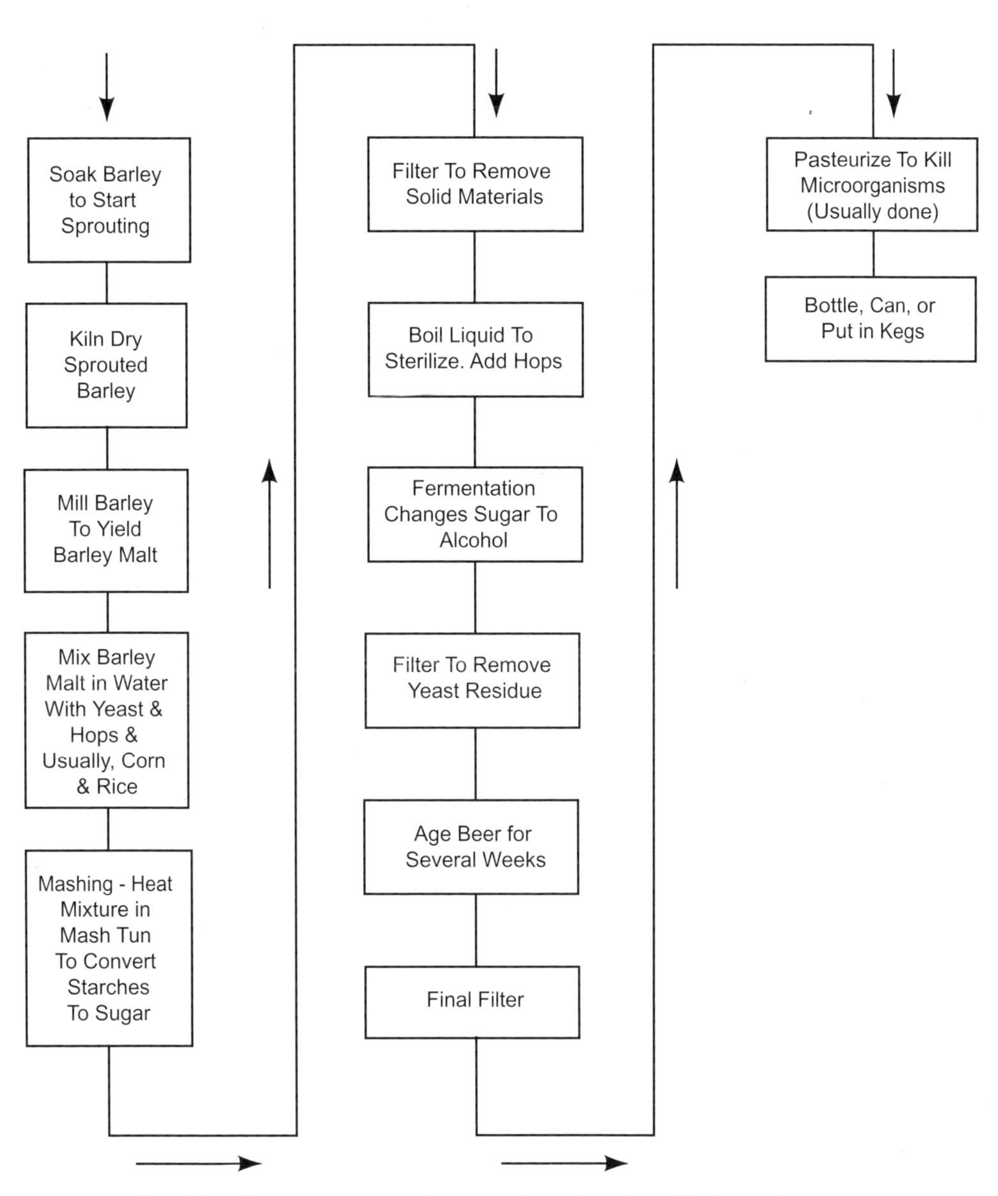

Fig. B3 The sequence of operations involved in beer brewing.

bells[3] - Bells of the traditional shape are cast from bell metal, a bronze alloy, normally four parts copper and one part tin, although zinc, lead and silver are sometimes used. Sand molds are used (1B). The first step is to make the core, the mold for the inside surface of the bell, which is done by rotating the sand and shaping it with a sickle board. Then a clay pattern of the bell is formed over the core sand. Any lettering or design pattern for the exterior of the bell is made from wax and attached to the clay pattern. Lettering is in relief (raised). The next step is to make the cope, the mold for the outer surface of the bell. The cope is built up over the pattern. A thin mixture of clay is used for the surface, particularly where there is lettering and the balance of the cope is made from a thicker sand/clay mixture. The molds are then dried and baked. The heat from baking melts the wax lettering but leaves cavities for it on the cope half of the mold. The cope is lifted off and the clay pattern is removed. The cope and core halves are then reassembled and fastened. Melted bronze then can be poured into the cavity. After the casting has solidified and cooled, the mold is removed; the bell casting is snagged (1B8g) and sand blasted (1B8c), and sometimes machine trimmed in a lathe (3A1a). Trimming can adjust the pitch of the bell. For large bells, vertical lathes are used.

bent wood components - See 6D.

bicycles - Most bicycle frames are made from arc welded (7C1) assemblies of welded steel tubing. (Racing bicycles and others of light weight may be made from aluminum tubing, or tubing made from pultruded plastic reinforced with carbon or other fibers). Metal tubing components are bent and formed with conventional tube processing methods (2H2). Some drilling, boring, and reaming (Chapter 3) operations are performed. The head frame and frame member containing the sprocket shaft bearings are bored (3B5) and reamed (3B4), after welding. Some frame parts are formed into curved and partially-flattened shapes by press operations (Chapter 2). After the frame is welded, it is heat treated to anneal the welds (8G1). Handle bars are bent from steel tubing. Sprockets are blanked from steel plate (2C4) and, like handle bars and some other bicycle parts, are polished (8B1), buffed (8B1a), and electroplated (8C1). Wheel hubs and wheel shafts and most fasteners are made on automatic screw machines (3A2c). Pedal arms and various gear shifting parts are forgings (2A4). Wheel rims are contour roll-formed (2F7) and three-roll bent (2H2f) to the correct cross sectional and circular shapes before the ends are welded. Chain guards, wheel and seat parts and other smaller parts are stamped from sheet steel (2C, 2D, 2E). All parts are cleaned (8A) and either painted or are polished and chrome electroplated. Frame painting is often electrostatic (8D7) with an enamel coat followed by clear coating. Wheel spokes are made from drawn (2B2) and plated steel wire, straightened, cut to length and formed at the ends, as necessary. Wheels with spokes and hub are assembled (7F1) and adjusted to run true. Chains are assemblies of stampings and screw machine parts. Seats, tires, ball bearing assemblies, and threaded fasteners are supplied from companies that specialize in these components. The bicycle is put together on an assembly line (7F2) where parts are added as the bike moves down the line. Decals and nameplates are also added. For compactness in shipping, handle bars, pedals, and seats are usually not part of final assembly, but are often installed by the dealer. After final testing and inspection, the components and the main assembly are wrapped and inserted in corrugated cartons prior to shipment.

bills, dollar, and other paper money - See *paper money*.

bleach - Household-strength chlorine bleach is commonly an aqueous solution of 5.25% sodium hypochlorite, NaOCl. The sodium hypochlorite is made from caustic soda (NaOH), chlorine and water. The chlorine and caustic soda are made by electrolysis (11C10) of a salt solution (See *chlorine.)* and initially kept separate. The chlorine, in liquid or gaseous form, and the caustic soda, dissolved in water, are reacted together, either in a large batch reactor or in continuous process equipment, to produce the sodium hypochlorite. The solution is cooled, filtered (11C7a), and bottled (12L) for shipment. In lower concentrations, sodium hypochlorite is used to disinfect water to make it potable. Hydrogen peroxide, another liquid sometimes used as a household bleach, is also used as a disinfectant. Calcium hypochlorite, chlorine, sodium hypochlorite and hydrogen peroxide are all used as industrial bleaches. Major applications are in the paper and textile industries.

blue jeans - are made with conventional garment-making methods (10H) from blue-colored denim fabric. See *denim*.

boats, small, plastic - Small plastic boats, if not made with fiberglass (See below.), are most often thermoformed (4D). Various mechanical assembly operations (7F) to add such items as seats, oar-locks, and other hardware, are performed after the shell is formed.

boats, fiberglass - Fiberglass reinforced boats are made by the hand lay-up (4G1) or spray-up (4G2) methods. Mechanical assembly (7F) of hardware and seats follows.

bolts (machine screws, cap screws, set screws) - can be machined (cut) on lathes and screw machines but almost all standard commercial bolts and machine screws are made by forming rather than machining. Round stock of a formable grade of steel (or other metal if steel is not used), usually coiled, is the raw material. The first operation is cold heading (2I2) which cuts a blank piece to length and imparts the basic shape. The head is formed to its hexagonal, square, or round shape, with slots or recesses, if used. Then the bolt blank is tumbled (8B2) and fed automatically (usually by vibratory feeder) to a thread rolling machine where the screw threads are formed (3E7). The screws are then cleaned (8A), barrel electroplated with zinc (8C and 8C3) and given a chromate conversion coating (8E3). These coatings improve appearance and corrosion resistance. Sometimes, black oxide (8E4) or other surface treatments may be applied.

books[3] - Production of books requires a series of operations in addition to the writing, editing, format designing, and related manuscript preparation operations. Current practice makes extensive use of computer-based techniques, not only for these activities, but also for plate making for printing and, in some cases, where quantities of the finished book are limited, for the printing also. These methods are in contrast to the traditional system wherein a text manuscript and illustrations were combined by manual methods to make a master copy of each page with the desired type fonts and layouts. These are sometimes called "pasteups" or "mechanicals". In the traditional system, the master is photographed and used to make a plate for offset lithographic printing, as described in section 9D2a. Offset is the printing process normally used for books and both sides of the paper are printed in the same operation. 8, 16, or 32 pages are typically grouped together to print multiple pages at a time in an arrangement that allows shearing and folding equipment to complete a section of the book. These groups of pages are called "signatures", and are collected with others to form the complete book. The folded signatures are fastened together by gluing or stitching along the spine. A reinforcing gauze is often wrapped around the spine to add strength. End sheets for the front and back of the book are attached. The book block is then trimmed in a machine that shears the pages to the desired size.

If the book is to have a hard cover (called a "case"), it is made from heavy cardboard, covered with cloth or a special paper. The cover is printed, hot stamped, or embossed, or processed by a combination of these operations. The process of adding the cover is called casing-in, and may include applying a headband, a cloth band applied at the top and bottom of the book block spine, rounding the cover, and further application of adhesives (usually of the hot melt type). The case is applied by a machine which feeds it from a hopper, applies adhesive to the endsheets and presses them in place. The hinges for the covers are hot formed into the cover after it is attached. Lastly, the paper jacket may be added to the book and it is placed into a shipping carton with others.

When electronic processes are used, word processing or desktop publishing programs can provide a page master with illustrations, tables or other non-text material, if any. Plates can be made directly from the computer, and can include grayscale photographs without the need to make a photo master of each page. If illustrations or text are in color, color separations can be prepared by computer. The computer can also develop and control the contents and sequence of each "signature", and the assembly of signatures into a book block. Direct digital printing, without plates, can be economic for as many as 500 copies of the book, but are perhaps most useful for print-on-demand, where no book inventory is maintained but copies are printed by computer printing methods as orders are received. The cost of printing may be greater but there are no inventory carrying costs nor risks of having to pay for disposal of unneeded copies.

bond paper - see 9C1 and the various paper making processes described in Chapter 9.

bottled drinks - See 12L and *soft drinks.*

bottles, glass - Both glass and plastic bottles are manufactured by very similar blow-molding methods. Glass bottles are blown by one of several variations of the blow molding processes. Which method is used depends on the size, shape, quantity needed and other factors. See section A2b in Chapter 5 for a description of the different methods, particularly the blow-blow, press-blow, and rotary mold processes, and accompanying operations such as trimming and annealing.

bottles, plastic - These are blow molded. Smaller bottles tend to be injection blow molded (4F2), while larger bottles are extrusion blow molded (4F1). Bottles containing carbonated soft drinks may be multi-walled, blow molded from coextruded parisons (4F4, 4I2), or made by stretch blow molding (4F3).

bowling balls - are molded of mineral-filled thermosetting plastics in two molding stages. A core piece, known as the "weight block," is molded first. Its material varies, depending on the weight desired in the finished ball, but its size and shape are essentially fixed. Sometimes, light weight foam is incorporated in the core when a lighter ball is needed. After the core is molded, it is trimmed, and a hole is drilled in one place for a locating pin that is part of the mold for the shell, the outer portion of the bowling ball. The shell material is engineered specifically for the wear, appearance and control properties wanted in the finished ball. After the shell has hardened, the ball is removed from the mold. The hole needed for the locating pin is then filled with another plastic, but in a contrasting color from the rest of the shell. The color difference aids the operator, who will later drill finger holes, in locating them in the correct position. The ball is then positioned in a lathe that makes a trimming cut to remove flash, ensure roundness, and smooth the surface. A series of sanding and polishing operations further smooths the ball's surface. Trade name, manufacturer, and model number are engraved in the surface and the recessed lettering is filled with paint. Inspection, testing, packing and shipping to the retail dealer follow. At the dealer's location, finger holes are custom drilled to fit the hand of the intended user. Bowling balls have a thick enough shell to permit later surface smoothing and polishing if the ball becomes scratched or cut.

bowls, glass - are pressed (5A2a) and annealed (5A4a).

boxes, corrugated - See *cartons, corrugated.*

brake linings - A brake lining must have the following properties: high coefficient of friction, slow rate of wear of the lining coupled with minimal wear of the brake drums or discs it bears against, low noise generation during braking, fade resistance, ability to perform when wet and ability to withstand high temperatures and to dissipate heat. A combination of materials is used to achieve these properties. Asbestos previously was a key fiber used but has been eliminated in production brake systems because of its health risk to those who service vehicle brakes. Now, the list of materials used is almost always proprietary with each manufacturer. As many as 15 or more different materials may be used in one brake lining and a manufacturer may make different varieties of lining from a list of 35 or more materials. The following are materials that can be found in present day products: carbon or graphite, sintered metal, metal fibers (steel, copper, brass, titanium), polymer fibers (aramid, acrylic and cellulose), various ceramic fibers, glass fibers, clay and other mineral fillers, and a phenolic resin or rubber matrix. Depending on the use, the lining may have a preponderance of one class of material, eg., metal, ceramic or polymer fibers. Ceramics have come into increasing use in recent years. A typical lining is made from a grouping of fibers, sometimes in a woven or non-woven fabric mixed with fillers and some modifiers, bonded together by compression molding with phenolic resins. Metal fibers provide heat conduction as well as friction. The final shape may include chamfers and slots, the latter to provide noise improvement, cooling and space for dust from wear of the lining.

brandy - is distilled from wine or from the marc - the pulp residue from grapes or other fruit that remains after pressing or straining. See *distilled spirits.*

brass - is an alloy of copper and zinc. There are various brasses with different levels of these two major alloying ingredients. To produce brass, zinc and copper are melted together in a reverberatory, crucible, or cupola furnace. The alloy is cast into ingots which are cold worked into sheets, bars, rods, wire or other shapes. Depending on the alloy and hardness, brass is exceptionally well suited for machining or forming operations. It is used extensively in screw machine products, various stampings, and castings. Door hardware, electrical contacts, lamps and light fixtures are common products that include parts made from brass alloys.

bread - can be made from many different recipes. The simplest bread uses one grain (wheat, rye, oats, corn, barley, buckwheat or millet), milled into flour and mixed with water. The resulting dough is shaped and cooked. Matzos, tortillas and chapaties (India), are examples of such breads. These breads are flat or unleavened. Leavened breads have an ingredient that makes the bread rise by releasing carbon dioxide gas which, together with the steam that is formed, creates larger bubbles in the dough, giving it a lighter structure. Leavening agents are baking powder, baking soda, and yeast. Yeast is a microorganism that feeds on the carbohydrates in the flour, turning them into alcohol and carbon dioxide (See fermentation - 12J4.) With baking soda or baking powder, the carbon dioxide gas is released by a chemical reaction with the ingredients and a separate rising operation is not needed. Wheat flour is an important ingredient in raised breads because it includes gluten, a protein that has the flexibility needed to hold pockets of the gas released when the bread rises; other grains, except rye, have little or no gluten.

In addition to leavening agents, most breads have other ingredients to increase their flavor, texture and nutritional value. Eggs, milk, multiple grains, sugar, fats and oils (shortening), salt and various flavorings may be included. The fats and oils give the bread a finer, softer consistency. Various seeds and nuts and raisins may be added, for additional flavor and nutrition. Spices may also be added. Most cooking of bread is done by baking, though steaming and frying are sometimes used.

Breads containing yeast are commonly made in seven basic steps: 1) *mixing* the flour with water or milk, yeast and shortening, salt, sugar and other ingredients. 2) *kneading* of the dough after it becomes too viscous for stirring. (See 11G5 and Fig. 11G5-2.) 3) *rising*. The dough is set aside in a moist and warm environment before baking to allow fermentation to take place and the dough to rise. Typically, the dough is allowed to rise to twice its original volume. 4) *further mixing* by kneading and punching to break up large gas pockets and ensure that the yeast has contacted all the carbohydrates. 5) *further rising* 6) *shaping*. The dough is divided into loaf-sized and shaped amounts and placed in pans. 7) *final rising*. 8) *baking*. Fig. B4 illustrates this commercial bread making processes, which includes removal from the pans, slicing, and packaging after baking. High production bakeries make bread as a continuous process.

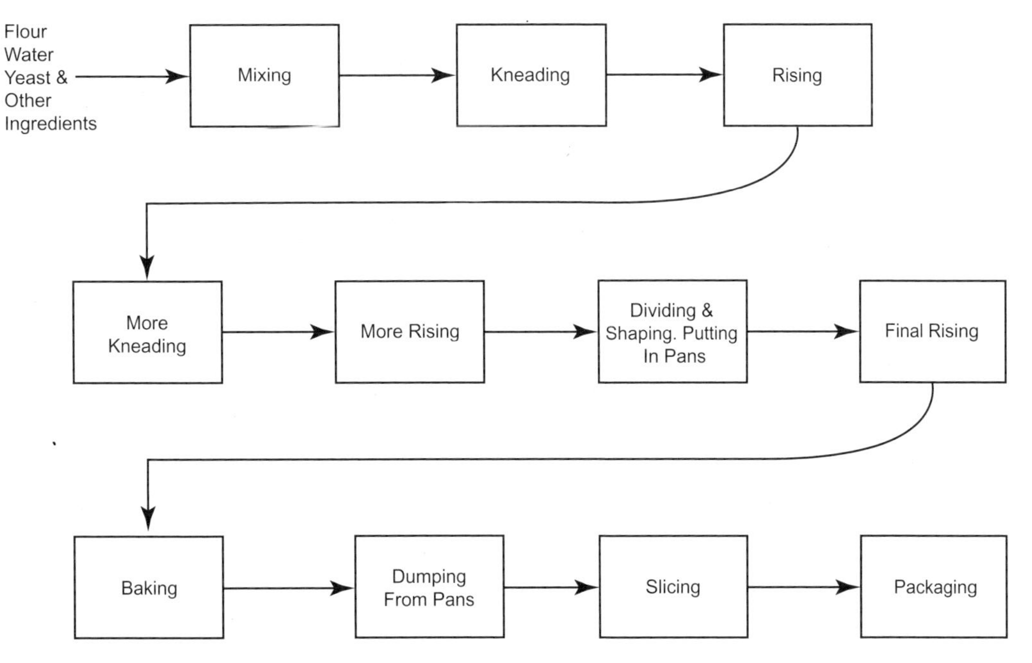

Fig. B4 The sequence of operations involved in commercial bread making.

breakfast cereal - See cereals, breakfast.

bricks[4] - are made from clay (most commonly red burning clay). The clay may be dried, screened, and crushed, to provide the proper particle size distribution. In the stiff-mud process, it is mixed with enough water to bring the total water content to 12 to 15%. This provides adequate workability. The mixture is extruded to the desired cross section, cut to length and, if to be used for face brick, may be repressed to provide a more uniform shape and smoother surface. The bricks are dried in ovens or in the atmosphere, and then fired in a kiln at 1600 to 1850°F (870 to 1010°C). Fig. B5 illustrates the full brick-making process.

bronze - was originally any one of a number of alloys made from copper and tin, but the term has also been applied to some other brass-colored alloys, especially if they contain tin. Other minor ingredients may be used, but copper and tin are still the principal alloying metals. Zinc, lead, and other metals may also be included. Additions of phosphorus and aluminum add strength. Bronze alloys are hard (harder than pure iron), strong and corrosion resistant. They are easily lubricated and used extensively in bearings and other machine parts. Pump parts, valves and marine propellers are other applications. Bronze parts are commonly made by casting or forging and machining.

brushes - There are many kinds of brushes. *Toothbrushes*, cleaning and *scrub brushes*, *paint brushes*, wire brushes and revolving brushes are most common, but manufacturing methods for almost all are similar. All manually-operated brushes have a handle of some kind. Handles traditionally were made from wood but most now are injection molded (4C1) of plastics, some from structural foam plastics (4C3). Wooden handles are made from various hardwoods, depending

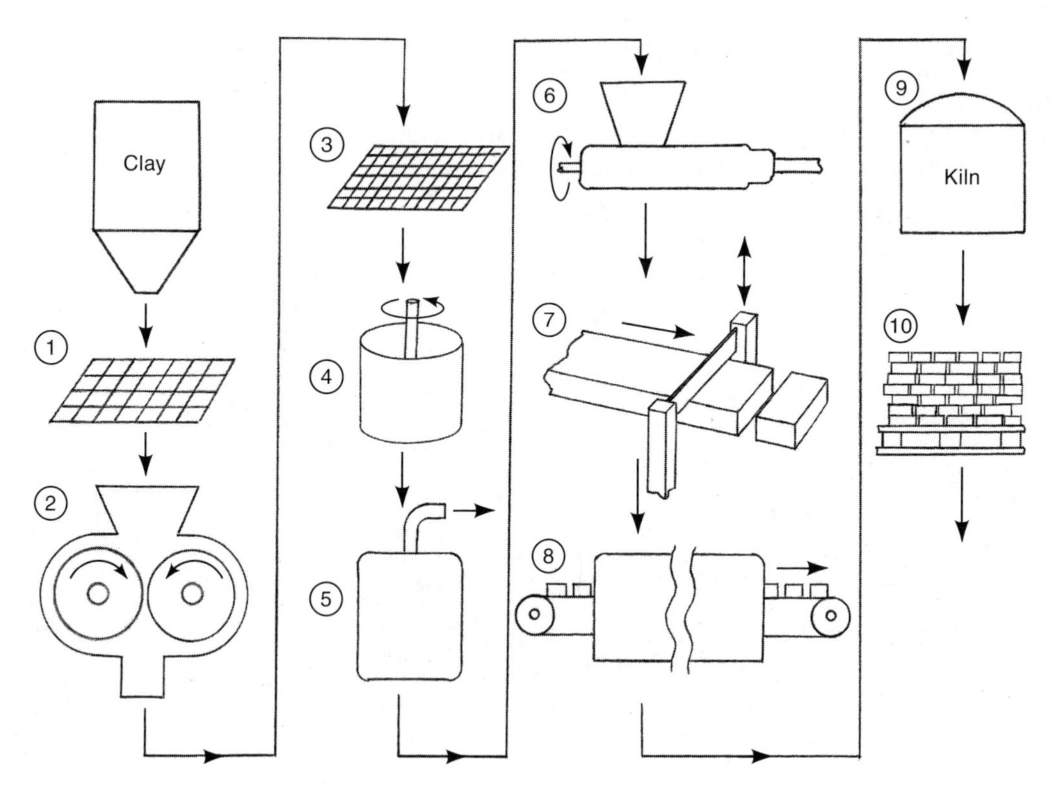

Fig. B5 - The sequence of operations involved in brick making: 1) initial screening of clay, 2) crushing of large clay pieces, 3) further screening 4) mixing of powdered clay with water to plastic consistency, 5) vacuum degassing, 6) extrusion of clay paste to form a continuous block, 7) cutting the block into bricks, 8) oven drying slowly to remove moisture while preventing cracking, 9) kiln firing of bricks at a high temperature, 10) cooling of fired bricks prior to storage and shipment.

partially on the application. Polypropylene is a common plastic handle material; cellulose propionate is used for many toothbrushes; elastomers are used for some handles. Bristles for the best paint brushes have traditionally come from the back hairs of semi-wild hogs grown in cool climates, not from hogs raised for meat. Artificial bristles are now widely used. Nylon is the major material. Other bristle materials are polyethylene, polypropylene, polystyrene and other man-made fibers, and tampico, bassine, palmetto, palmyra, and other natural vegetable fibers. Metal bristles are made from steel, stainless steel, brass, copper, or aluminum wire. (See *wire, mechanical.*) Plastic bristles are extruded (4I) from multiple-orifice dies. Bristles may be crimped along their length to provide greater flexibility and better holding power when they are held in place with an adhesive or plastic. Bristles are cut to length, gathered in bundles, held together at one end by metal staples, adhesive or, if thermoplastic, by being heated to the softening point at one end so that the individual bristles bond together. Many brushes use bristles of twice the length needed so they can be folded in the middle at the handle end over a metal staple. The bundles of bristles are punched into holes in the brush handles with enough force to drive the staple into the wood or plastic handles deep enough for secure holding. This method is common for tooth brushes. Other brushes use adhesive to hold the bundles of bristles. With many brushes, the bristles are trimmed after insertion. Trimming provides uniform bristle height and special shapes to the bunched bristles as is desired in some toothbrushes and other brushes. In high-production situations, all this handling and assembling is highly mechanized with special machinery so that little manual labor is involved. Packaging is also automatic.

bulbs, light - See *lightbulbs.*

bulletproof glass - See *glass, bulletproof (bullet resistant).*

bullet-proof vests - are made from high-tensile strength Kevlar, Twaron, Spectra or Bynema fibers.(See *Kevlar*. Twaron is a related material.) Spectra is an ultra-high-molecular-weight polyehylene-based fiber. Bynema is similar to Spectra. Kevlar vests are made from multiple layers of woven (10C) fabric. Spectra vests are made from a non-woven fabric that has parallel yarns of the material held together with a plastic resin. From 8 to 30 layers of the Spectra fabric are layered and sewn together, with alternate layers at right angles to each other, and with each pair separated from the others with a layer of polyethylene film. With both materials, vests are sewn of polyester/cotton or nylon fabric and the Spectra or Kevlar panels are sewn inside or placed in pocket-like pouches. (See 10H.) Other pouches may be placed in critical locations to hold ceramic or metal plates for extra protection.

bullets (small arms ammunition or cartridges) - have four major components: 1) The bullet (projectile), a cylindrically-shaped lead slug with a conical point, coated with copper. 2) A cartridge case, formed from brass. 3) a primer, a central cup, loaded with a percussion explosive, eg., mercury fulminate, and press fitted at the base of the cartridge case. This primer is a detinator for 4), the gun powder, also contained in the case. The case, primer, explosives and bullet are all assembled automatically in current production practice. The brass case is crimped over the projectile to hold all the parts together. Fig. B6 shows the complete cartridge assembly.

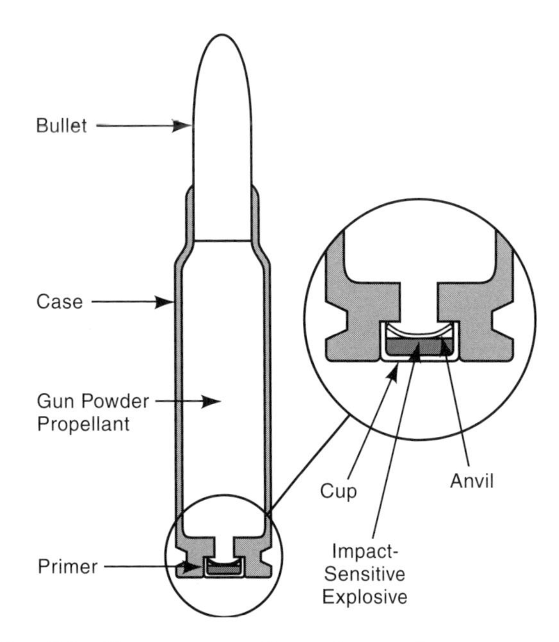

Fig. B6 The cross-section of a small arms ammunition cartridge.

The bullets are cast in permanent molds (1D1) or cold-formed (2I2) from a lead alloy, are swaged (2F1, 2D12) to the final size and shape, are cleaned in a multi-step operation with aqueous cleaners, and then barrel plated (8C, 8C3) with a thick coating of copper. The bullets then undergo "final striking" in a rotary die punch press to make the shape more exact and the dimensions more precise. Some bullets are formed from copper or brass and are lead filled afterwards.

The case is made from cartridge brass sheet that is deep drawn (2D5, 2D5b) in a series of steps. There are up to five deep drawing operations to convert flat sheet to the long cylindrical shape needed. The material is annealed (8G5b), washed, and lubricated between the draws. Then a cold heading operation (2I2) forms the primer pocket and the rim or groove needed for ejection of the spent cartridge from the gun after it is fired. There are further forming operation to reduce the diameter of the case near the mouth, trim it at the mouth and anneal it again so that it can be crimped around the bullet. An alternative method for forming the case instead of deep drawing from sheet involves upset extrusion (2I4) from a slug of brass.

The primer consists of three parts: a cup of about 3/16 in (5 mm) in diameter drawn from brass or copper sheet, an *anvil*, a small stamping that fits inside the cup, and a percussion explosive that is squeezed between the anvil and the cup bottom when the gun's firing pin bends the bottom of the cup bottom inward. Filling the cup with percussion explosive is a dangerous operation and is often performed by robots or other automatic equipment without handling by human operators. The complete operation includes placing a layer of foil or paper between the explosive and the anvil, inserting the anvil and sealing and drying the explosive charge.

The assembly sequence for the cartridge is preceded by tumble polishing (8B2) and resizing the case. The primer is then pressied into the case bottom, the case is filled with the exact amount of gun powder needed, the bullet is inserted into the case, and the area where they overlap is crimped to make a secure assembly. The primer and bullet are fed automatically to the assembly machine from hoppers, and the gun powder is metered from a supply hopper. The final step is to pack the finished cartridges into a box, usually 50 in each box, for shipment to retail outlets and customers. Sensors verify the completeness and correctness of the assembly operation and a sample of each lot is test shot to verify function and accuracy.

bungee cords - consist of a core of natural or synthetic rubber, a braided cotton or nylon covering and hooks or other metal terminations at the ends. The rubber is blended, usually with some reclaimed material, heated and extruded (4I) into ribbons 0.09 to 0.12 in (2.3 to 3.0 mm) thick and 0.25 in (6.3 mm) wide. The ribbons are dusted with talc or finely powdered soapstone to provide a lubricant so they do not stick together. Several such ribbons are bundled together, stretched to reduce the diameter slightly, and fed into a braiding machine, which provides the fabric cover. In some constructions, two layers of braided covering are used. The braided rubber cords are cut to length and assembled to metal end hooks that have been formed from painted or plastic-coated heavy steel wire. The cord is doubled over each hook and tightly wound with another steel wire in a special assembling machine to hold the hook securely. The number of rubber strands in the core determines the rated tensile strength of the cord.

burlap - The coarse fabric is woven (10C) from jute or jute-like natural fibers. India is a major source for the fibers and for much of the burlap cloth. Burlap is sewn into bags and used for wall coverings, as backing for linoleum and in upholstered furniture.

butter - is made from the milk fat in cream. The milk is first processed in a cream separator that divides it by gravity into skimmed milk and cream. The cream, of 35 to 45% fat, is chilled and held to a low temperature. It is then strongly agitated (churned) until some of the fat globules break down, releasing liquid fat which then bonds other globules together. These globules form a somewhat solid mass accompanied by liquid buttermilk. The buttermilk is removed; the coagulated mass is washed (to remove milk curd and other non-fat residue) and is further worked to distribute the remaining moisture uniformly. At this point, the fat content is approximately 98%. During this working operation, salt and coloring are normally added. Current large-quantity production systems carry out the foregoing operations on a continuous basis in contrast to earlier production operations that were batch by batch. The butter is then chilled, extruded and cut into sticks which are wrapped and packaged with automatic equipment.

buttons - There are three basic methods for making clothing buttons in production quantities: 1) cutting the buttons from sheet plastic, normally polyester. 2) injection molding (4C1). 3) compression molding (4).

Injection and compression molding methods produce good, workable buttons, but neither process provides the pearlescence often wanted in buttons.

The process of cutting from polyester sheet is akin to the historical method where buttons were cut and shaped from a variety of natural materials: sea shells, bone, horn, wood, shells of Brazilian nuts, brass, pewter, precious metals, tortoise shell, and ivory. Some buttons are still made from these materials for costly garments. The method of cutting from sheet polyester has the following steps: 1) Liquid polyester plastic, a dye or other colorant, liquid wax, and a catalyst, are mixed together. 2) The mixture is slowly poured into a large cylindrical vessel that rotates on a horizontal axis. The liquid spreads from centrifugal force and lines the interior surface of the cylinder. The plastic polymerizes and becomes solid but initially is still quite soft. The wax has migrated to the inner and outer surfaces of the polyester. 3) The cylindrical vessel is stopped; the sheet is cut, carefully removed, flattened, and placed on a conveyor belt. The wax is removed. 4) The sheet is carried on the conveyor to a punch press. The press, with a blanking die, cuts the sheet into a series of round button-size discs (2C4). 5) The discs, still hot and soft, are transferred to a nylon bag and immersed in hot, salty water, where the polymerization continues and the blanks harden. The water and the blanks cool and the blanks are then transferred in the bag to a cold water tank and, after a period, to a centrifugal dryer. 6) The blanks are transferred to machining equipment where each button, securely held in a collet, is machined with a form cutter to the particular shape needed, for example, with beveled edges and a concave or patterned front. The machine also drills two or four thread holes in each button. The buttons may also be reversed to bevel the back edge. This machining is carried out on special automatic, multiple-station turret machines. 7) Following machining, the buttons are tumbled with an abrasive to remove tool marks, burrs, and sharp edges, and to provide the surface finish wanted. 8) The buttons are then washed, dried, checked for defects, and packed for shipment.

C

cabinets, wood - See Chapter 6.

cams[17] - are rotating or reciprocating machine elements that create a prescribed motion in other machine elements. Two typical cams are illustrated in Fig. C1. Steel and ductile cast iron are primary materials for cams though some cams designed for less heavy usage in consumer products (for example, making decorative stitches in household sewing machines, or switching washing machine cycles) are molded from phenolic and other plastics. Most metal cams are machined by milling (3D) and grinding (3C) but some are made by powder metallurgy (2L1), or by sheet metal blanking (See fine blanking, 2C9 and Fig. 2C9-2). When machining is utilized, the blank may be a simple unshaped piece, or may be a forging (2A4) or casting (1B)(1G) that is made to near the final net size and shape to minimize the amount of machining necessary. All cam machining and grinding requires fixtures and cutter-path or depth-of-cut controls to achieve the proper contour. The most accurate cam machining involves CNC control (3U2) that utilizes polynomial, circular, or linear interpolation between discrete location points. Other cam machining utilizes special machines that employ master cams to control the path of the milling cutter (3U6) or grinding wheel. Most cams require some heat treatment for wear resistance. Case hardening (8G3b) or other surface hardening methods (8G3a) are normally used, though many steel cams, especially if small, are through-hardened. After heat treating, cams are usually ground to correct any distortion resulting from the heat treatment and to provide a smoother surface. Sometimes, EDM wire machining (3I1b) or other is used on through-hardened blanks, eliminating the need for prior near-net-shape forming and for milling. "Hard machining" techniques can also be used to mill cam blanks prehardened to Rc 55 or less. Some cams are polished with a soft wheel and fine abrasive (8B1) to provide the desired surface smoothness.

candied fruit - See 12J10.

candy - "sweets" in Britain, are confections of many varieties that all have one basic similarity - the

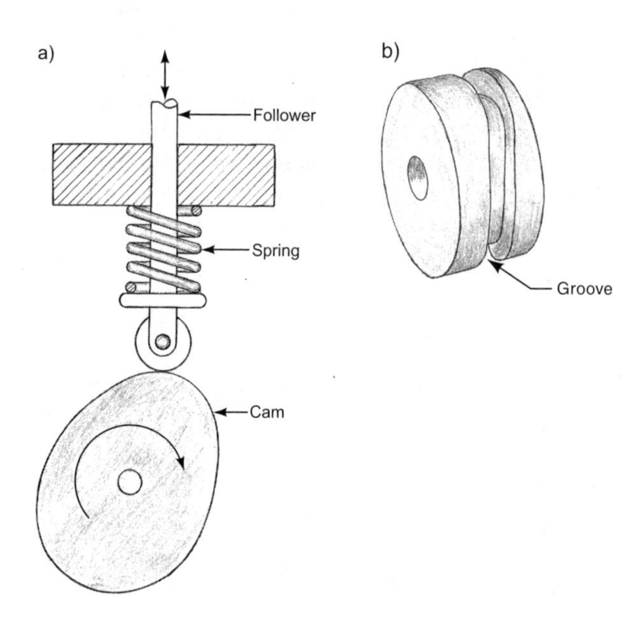

Fig. C1 Two typical cams: a) an open radial or plate cam with a roller follower. The cam moves the follower in one direction. Motion in the other direction is provided by the spring but is limited by the cam. b) a cylindrical or barrel cam with a grooved cam surface. The cam moves the follower in both directions.

main ingredient is sugar. Sucrose (from sugar canc and sugar beets), invert sugar (a combination of glucose and fructose), corn syrup, honey, maple syrup or some other sweetener may be used. The second ingredient is a flavoring such as chocolate, vanilla, fruit flavors, peppermint, licorice, spices and herbs. Fillings such as peanut butter, peanuts and other nuts, fruit, cereals, flour, milk products, starch and gelatin may be included. Coloring agents, salt, fat and some preservative may also be added. Candies can be considered to exist in five basic varieties: chocolate, hard, chewy, whipped and panned. Most candies of these varieties are made by mixing the main ingredients with water and heating the mixture to a high enough temperature to melt and blend the ingredients and boil the mixture. The length of time candy is boiled, the temperature to which it is cooled after boiling and the way is processed after cooling, affect its hardness and other properties.

Chocolate candy gets its flavor and texture from chocolate liquor and cocoa butter. (See *chocolate.*) Cocoa butter is the fat of the cocoa bean and chocolate liquor is a combination of cocoa butter and the shelled and finely-ground cocoa seeds. In making chocolate candies, the melted candy mixture is either poured into molds or poured over cookie or other candy filler material and allowed to cool and harden.

Hard candies are made from a mixture that includes sugar, corn syrup and a small amount of water. (The corn syrup, in addition to its sweetening properties, aids in controlling crystallization.) The solution is boiled to a temperature of about 300°F (150°C), to remove almost all the water. Flavorings are added, and the mixture is cooled and dumped on a cooling table. The mass is then worked by pulling and rolling it into long rods. The rods may be cut into sticks or pressed into other shapes. Candy canes are made from pulled red and white strands twisted together and bent to shape.

Chewy candies usually include jelling or softening agents such as gelatine, starch, flour, milk or fats and have a higher moisture content than hard candies. The paste formed after boiling is either formed in molds or cut from flat masses to make caramels, gums, jellies, toffees and other softer candies.

Whipped candies are mixed vigorously to entrap air and produce a smooth texture. Egg whites and gelatine are included in the mixture and act as whipping agents. Marshmallows and nougats are made with this operation.

Some panned candies are made by placing small center pieces in a rotating pan and spraying with cooked syrups, which form layers on the center piece. Jelly beans (See *jelly beans*) and various chocolate covered candies are madc by this method.

Confectioners' glaze provides a smooth and harder shell to several different kinds of candies, preventing sticking, functioning as a sealer and extending shelf life. It is used on jelly beans, malted milk balls, "M and M"- type chocolate candies, gum balls and licorice candy. It consists of food grade shellac in an alcohol solution, 2 to 6 lb of lac per gallon of alcohol. When applied to a candy, the liquid is sprayed while the candy tumbles in a rotating container. About 4-8 oz is applied per 100 lbs of product. Cool, dry air is blown through the tumbling material to evaporate the alcohol and dry the product. Carnauba wax may be added to the final coating to provide a shiny finish.

Commercial branded candies are made in very large quantities and the mixing, boiling, cooling, crystallization, casting, *enrobing* (poured coating), other processing operations, wrapping and packaging are all highly mechanized.

canned food - See 12G4.

cans, metal - Aluminum cans, used primarily for beverages, are made on fully automatic equipment where deep drawing (2D5b) is the key operation. Sheets of a special aluminum alloy (with 1% Mg, 1% Mn, 0.4% Fe, 0.2% Si, and 0.15% Cu) are blanked to produce flat discs. The discs are about 5 $^1/_2$ in (14 cm) in diameter for a standard U.S. 12-ounce can. The discs are deep drawn to a cup shape, about 3.5 inches (9 cm) in diameter and 1.3 in (3.3 cm) deep. Two more drawing steps increase the height to about 5 in (13 cm) and reduce the diameter to its final value. The bottom of the can is formed to its special shape to increase stiffness, and the top is trimmed to remove the wavy edge that results from deep drawing. The can is then cleaned and printed in several colors with the desired logo and contents identification. The top of the can is then necked in and a small outward flange is left in place. The can is now ready for filling.

The top and pull tab are made from separate aluminum sheets of a different thickness and alloy. The lid is blanked, scored for the opening and formed to include a small projection at the center that forms a rivet to fasten the pull tab. The blanked (2C4) and formed (2D2) pull tab is positioned in place by automatic equipment and the rivet is set.

After the can is filled with the beverage, the flange from the can top is folded over with an extension of the lid and they are crimped together to seal the can. Fig. C2 illustrates the forming sequence.

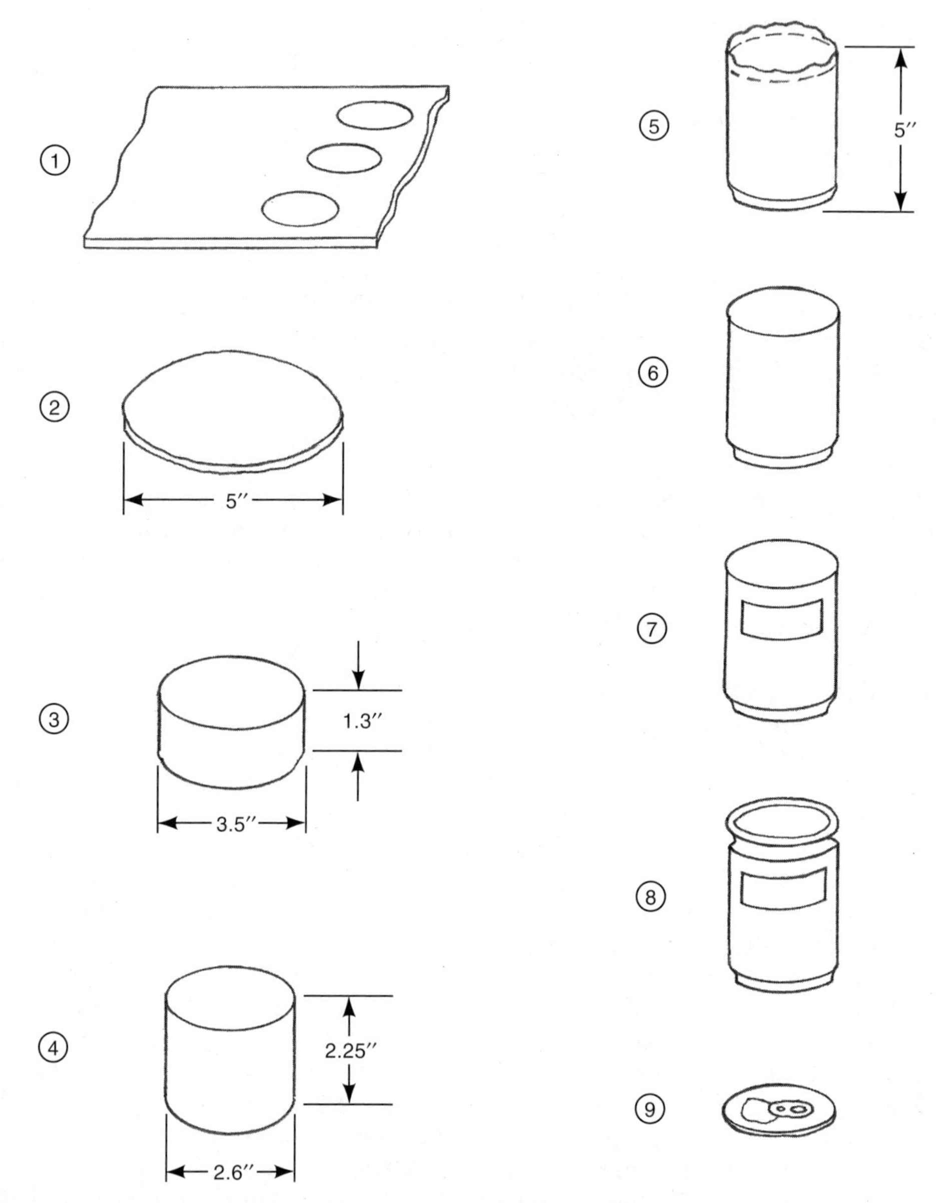

Fig. C2 The manufacturing operation sequence for making aluminum beverage cans: 1) & 2) Discs are blanked from sheet aluminum. 3) The disc is deep drawn to a cup shape. 4) A second drawing makes the cup deeper and narrower. 5) The third drawing reduces the can's diameter to its final value and extends its length. Forming of the bottom is part of the operation. A wavy edge exists at the top of the can. 6) The wavy edge is trimmed off. 7) The can is printed in several colors with identifying information. 8) The can top is formed leaving a flange for attachment and crimping of the cover. 9) The cover is blanked, formed and assembled as a separate part, which is assembled to the can and crimped for a seal after the can is filled.

capacitors, electronic - See 13L2.

carbon, activated - is usually made from peat or sawdust that is mixed with a chemical agent and dried. The mixture is then heated to 1600°F (870°C) with steam or carbon dioxide to carbonize it. The chemical agent is then removed with water, leaving carbon particles without other hydrocarbons and a very greatly increased surface area. Activated carbon is used in gas masks, various kinds of filters for purification and odor control and, in industry, for the recovery of vapors.

carbon black - is produced from the incomplete combustion of gas or liquid fuel. The product is collected on a metal surface. Most current production is from the oil furnace process that uses, as fuel, residue from catalytic cracking of petroleum. The fuel is burned with insufficient air for complete combustion but the heat produced pyrolizes unburned oil droplets into carbon black particles. The black smoke is first cooled with a water spray and then passed through precipitators, cyclones, and bag filters, to separate out combustion gases, cooling water and any undesirable grit. The soot is then pelletized for convenience of handling and to prevent excessive dust. Carbon black has its main use in rubber tire manufacture. It is also used as a black pigment in inks, paint and for coloring plastics.

carbon fibers[4] - are made with several processes. High-modulus fibers are made from rayon fibers that are charred at 390 to 660°F (200 to 350°C), carbonized at (1830 to 3630°F (1000 to 2000°C) and then heat treated and stretched at 5400°F (3000°C). Fibers with lower properties but better process yields are made similarly from polyacrylonitrile (PAN) fibers. Another process melts coal tar or petroleum pitch, spins the liquid into yarn in an oxygen atmosphere and treats the fibers at 5400°F (3000°C). Carbon fibers provide very light and very strong reinforcement in composite plastic parts used in sailplanes, powered airplanes and other products. Fishing rods, skis and archery bows are molded with carbon fiber reinforcement.

carpets - are rugs or other heavy fabric pieces that are normally fastened to the floor and cover the floor completely. In large or long rooms, several pieces may be joined together when the carpet is laid. The term "carpet" is also applied to oriental or Persian rugs made by hand in western and central Asia, in the Caucasus and in North Africa. In Chapter 10, see I7. These rugs have been imported into Europe from the 16th century, but were produced centuries earlier in Asia for local use. The tufts are inserted, woven and tied by hand to heavier warp yarns. Tufts are normally of wool, colored with natural dyes. In Chapter 10, see I1 for Axminster, Brussels, chenille, velvet and Wilton types. See I2 for pile rugs, I3 for knitted, I4 for needle punch, I5 for hooked, I6 for braided and I8 for needle point types.

carrying cases, power tools and instruments - Inexpensive carrying cases for power tools and similar items are made by a number of different methods, depending on the manufacturer and the size and weight of the device to be carried. The cases are normally molded in plastic with an internal cavity or dividers to fit the shape of the product they carry. Injection blow molding (4F2) is one method and it is often utilized to make the case exterior, the interior nest for the tool, the hinges, the clasps and the handle, all as one part and in one operation. However, to create more robust hinges, especially for power tools of heavier weight, the carrying case is often blow molded in two basic parts, connected by hinge elements molded into each half. Clasps, and sometimes a handle, are injection molded (4C1) as separate pieces. The two halves snap together at the hinges and the clasps and the separate handle, if used, also snap-fit together. Some manufacturers injection mold the two halves of the case and, when injection molding is used, the parts for inside the case have interior wall dividers to locate and hold the product instead of the surface cavities formed in the inner wall when blow molding is used.

cars - See *automobiles.*

cast iron - is normally made in a cupola (1A1) with pig iron, steel and cast iron scrap, limestone and coke.

CDs - See *compact discs.*

cellophane - is a transparent viscose film extruded from a solution similar to that used to make rayon. (See 10A2.) The extrudate enters a bath of buffered sulfuric acid and this is followed by a warm water bath for rinsing. The film is then immersed in a bath of sodium sulfide to remove the sulfur, then bleached with a hypochlorite solution, washed and treated with glycerol to provide plasticity. One surface is usually coated with nitrocellulose lacquer or a plastic to provide moisture proofing.

cellular glass - See 5A7c.

cellulose acetate plastics - are transparent thermoplastics. Cellulose, made from wood pulp or cotton linters is reacted with acetic acid or acetic anhydride in the presence of sulfuric acid[2]. Variations of the resulting material include triacetate, tetracetate, pentacetate, or mixtures of these. Other variations are cellulose acetate butyrate, made by esterification of cellulose with acetic acid and butyric acid with the aid of a catalyst and cellulose acetate propionate, similarly made. These plastics are used as ingredients in lacquer, for photographic film, for the injection molding of eyeglass frames, tool handles, toothbrushes and hairbrushes, packaging tubes for small items, and various transparent items. These plastics are also cast into sheet, thermoformed and used in coatings and insulation. Cellulose acetate butyrate is used in outdoor signs.

cement, Portland - Raw materials are: 1) limestone, chalk, oyster shells or another mineral rich in calcium, 2) clay or another mineral rich in silica, 3) gypsum, 4 to 5% of the mix, to slow the hardening process, 4) blast furnace slag, iron ore, waste bauxite and/or sand, which may be added in small quantities to adjust the mix. The materials are crushed and finely ground, mixed, and placed in a rotary kiln. Production kilns may be as large as 17 ft (5 m) in diameter and 650 ft (200m) long. The material mix is burned, ie., heated to a temperature of about 2700°F (1500°C), to form clinker. The kilns are slightly inclined so that the raw materials, fed at the high end, gradually work their way to the lower end after about six hours. Clinker results from a number of reactions that take place during burning including the evaporation of water and evolution of carbon dioxide. The clinker is milled to a fine powder and is stored for bagging or bulk shipment to customers.

The cement is chemically unstable and when water is added, the structure rearranges and cures, first as a jelly-like mass and then, gradually, as a hard material. Portland cement, mixed with sand and aggregates and often reinforced with steel rods or bars, is widely used in road, bridge and building construction.

ceramic materials, advanced, (high technology ceramics), (modern ceramics), (fine ceramics) - See 5B4a.

ceramics - See 5B.

cereals, breakfast

corn flakes - Corn grits (corn kernels without the bran and germ) are pressure cooked with water, malt, sugar, salt and certain flavorings. Pressure is approximately 18 psi (124 kPa) and cooking proceeds for one to two hours. The cooked grits are dried and tempered to about 20% uniform moisture content. They may then be pelletized to incorporate other ingredients and grains, depending on the formulation. The pellets are rolled into flat shapes (flakes) between pairs of stainless steel rollers. Preheating may precede the rolling operation. The rolled shapes are then toasted in air at 525°F (275°C) or higher to dehydrate, crisp, brown and blister the flakes. An additional operation may be spraying with a mixture of vitamins, minerals, flavorings or sugar. Flakes are then dried to about 3% moisture content and packaged.[3]

wheat flakes[23] - differ from corn flakes in that the whole wheat kernels are used whereas, with corn flakes, only the endosperm is used. Before the kernels are cooked, a process called *bumping*, is carried out. The process has two steps: first the kernels are lightly steamed; then they are run through a pair of opposed rollers that partly crush the kernels. These steps allow all parts of the kernel to be cooked; otherwise the bran would block the interior. *Cooking* is next with the addition fine granulated sugar at about 10% of the weight of the bumped wheat, malt syrup, and salt, each at about 2% of the wheat weight, and water sufficient to yield a moisture content in the cooked mixture of about 30%, including condensed steam. The ingredients are mixed and then cooked with steam for about 1/2 hour at ambient pressure. The mixture tends to form lumps and these are broken up in special *lump breaking* machines that push the lumps through a comb-like structure. Several stages of progressively finer breaking are usually used, with screening operations in between the stages to remove small lumps from the mix. When breaking is completed, the mixture is also cooled during this operation. The finished lumps ("grits") range from 1/8 to 1/2 in (0.3 to 1.3 cm) in diameter. The grits are then dried to a moisture content of 16 to 18%, cooled, and tempered. Normally, after drying, the outer portion of each grit has less moisture than the center portion. *Tempering* involves holding the grits for a period of time in bins so that the internal moisture content becomes more uniform within each one. The grits then are *rolled* and *toasted* with methods similar to those used with corn flakes. The final moisture content is 1 to 3%.

oatmeal[23] - The groat portion (kernel with hull removed) of the grain is the part processed for oatmeal. The first operation on the grain is *separation* to remove foreign matter such as trash from the fields by sieving, and finer material, such as chaff dust, by aspiration. Then the grain undergoes *cleaning* to further remove dust, weed seeds, stems and oat grains that are undesirable for milling. This operation is performed with several pieces of equipment that utilize width and length disc separators, screens, gravity separators, and aspirators. The resulting product is then *hulled*. The hull (external shell) is removed by an impact process that uses centrifugal force to throw the kernels off a spinning disc and against the wall of a cylindrical housing. The hulls tend to break off from the groat. They both fall to the bottom of the housing and are separated from the groat by table separators that use vibration to move the materials along a path. Differences in density, surface smoothness and shape cause the groats to separate from the hulls and any partially-separated kernels. (The latter are returned to the hulling operation for reprocessing.) Then the groats are cut into 3 to 5 pieces by a special machine, and are passed through a steamer to heat the pieces and bring their moisture content to within 10 to 12%. The groats are then *rolled* to create flakes of oatmeal. Rolling involves passing the pieces between two cast-iron rollers. This further conditions them and produces a flake-like shape. They are then packaged for sale as quick rolled oats for oatmeal. (Groats that are not cut before steaming are rolled to flakes of greater thickness to produce old-fashioned rolled oats without the quick cooking feature.)

shredded wheat cereals[23] - are made primarily from soft white wheat, using the whole kernel. The kernel is first separated from all foreign material: dust, sticks, other grains, stems, dirt, and stones. It is then cooked in excess water for about 1/2 hour at a temperature just below the boiling point. Cooking is stopped when the center of the endosperm changes to a translucent gray color. At that point, the moisture content of the kernel is 45 to 50%, and it is ready for further processing. The kernels are moved to cooling equipment where circulating air reduces their temperature to ambient. They are then tempered to equalize the internal moisture content and develop the firmer consistency needed for shredding. Shredding is accomplished by feeding the kernels between two parallel, closely fitting rollers, one with a smooth surface, the other with a series of parallel, closely spaced, circumferential grooves. The kernels are squeezed into the grooves, filling them with cooked wheat. A series of comb teeth fit into the grooves on the exit side of the rollers and force the strings of cooked wheat to fall on an accompanying conveyor, producing one layer of "biscuits" of shredded wheat. Additional rollers along and above the conveyor produce additional layers of shreds. There may be as many as 20 pairs of rollers to deposit 20 layers of shreds, depending on the size and thickness wanted in the finished product. Different groove sizes, spacings and shapes and various roller arrangements may be used. Cutters after the last roller pair score and partially sever the shreds across the conveyor. The stacked layers of shreds are then fed on the conveyor to an oven for baking. The initial oven zone is at a high temperature to insure initial heating of the biscuits, and is followed by lower temperature zones. The biscuits lose moisture, rise in thickness, develop the desired color, and are further dried. Final moisture content is about 4% in contrast to the 45% present when the biscuits enter the oven. The final severing of the individual biscuits from the continuous layers is easily accomplished where the layers were scored.

other shredded cereals[23] - The same shredding method used for shredded wheat can be employed in making other cereals such as "Chex" squares that feature bite-sized pieces. The method is used with corn, rice, oats, and wheat or combinations of them. Whole kernels are used in some cereals and parts of kernels and flours are also used. The nature of the grain ingredients depends on the characteristics wanted in the product and the method of cooking before shredding. Other ingredients that may be used are malt, salt, sugar, colorings, starches, vitamin and mineral enhancements, and preservatives, as are used with other cereals. These ingredients are mixed and cooked by one of two methods: pressure cooking, very similar to that used with corn flakes and other flakes, or extrusion cooking. Extrusion cooking utilizes equipment that mixes food materials and forces them through an orifice of a prescribed shape and size. The extruders, like those used for plastics extrusion (4I), utilize one or two longitudinal screws that churn the material as well as forcing it through the orifice. Heat is generated from the friction of mixing and is also applied from external electric resistance bands, both of which provide cooking. The moisture content of the material, after cooking, ranges from 25 to 32%. (This content is less than that needed for shredded whole wheat). For pressure-cooked material, tempering takes place to allow equalization of the moisture content, but this is not necessary with extrusion cooked material. Shredding then takes place as with shredded wheat but the rollers usually have cooling channels to offset the heat generated by the drier mixed mass of grain and more cross grooves are incorporated in the rollers. Fewer layers of shreds are deposited on the conveyor. Scoring or cutting of the cereal squares then takes place. The squares are conveyed to a baking oven which may be of a continuous conveyor type if the shreds are only scored, or a fluidized-bed toaster if the squares are severed from one another. For either type, the baking takes only one to four minutes and the moisture content of the finished product is between 1.5 and 3%. For corn and rice squares, the oven is configured to provide some puffing of the product. The oven has two sections, the first for partial drying, and the second at a higher temperature of 550–650°F (290–340°C) that causes the internal moisture in the shreds to turn to steam and expand the shreds.

oven-puffed cereals[23] - are normally made from rice or corn or mixtures of the two. When rice is used, it is first pressure cooked in water with sugar, salt, and malt. Cooking time is about one hour at a pressure of 15 to 18 psi (100 to 125 kPa). The mixture, then with about 28% moisture, is cooled and any agglomerated pieces are broken up. It is dried to about 17% moisture content and tempered for up to about 8 hours to allow the internal moisture level to be uniform. Following tempering, the particles are slightly flattened in a bumping operation and again dried, this time to about 10% moisture level. The particles are then fed into an oven that reaches a temperature of 550–650°F (290 to 340°C) at the end of the heating cycle, which lasts about 90 seconds and toasts and expands the cereal. Fluidized bed or rotary ovens are used. The cereal is then cooled, treated with preservatives and vitamins, if specified, and packaged.

chairs, upholstered - See 6H.

chairs, wooden - See Chapter 6.

charcoal - is made by the destructive distillation of hardwoods. (See 11C1g.) Charcoal is used for activated charcoal in filters and absorbent applications, as fuel, and in the carburizing of steel.

cheese - Cheese is formed from bacterial action on milk. The following sequence, for cheddar cheese, is typical of those involved with dryer cheeses: The milk is pasteurized. A culture of bacteria is added and the milk is heated to approximately 88°F (31°C), (Slightly different temperatures are used for other cheeses) for a period of 1/2 to 1 hr. Rennet is added to accelerate the bacterial process and is allowed to work for another 1/2 hr. By this time, the mixture has separated into curds (semi solid material) and whey (liquid). The coagulated curds are cut with a tool of multiple wires called a "curd knife" and the mixture is allowed to sit or "heal" for about 10 min. Then the curds are allowed to float an additional time, perhaps 1/2 hour, as they tend to loose moisture and become harder. The mixture is heated to about 101°F (38°C) and stirred to keep the particles of curd floating. When the curds are of the right "feel", they are allowed to settle to the bottom of the tank. The whey is then drained off and the curds remain at the bottom of the tank for about 2 and 1/2 hr. The curd is then cut into small pieces and salt is added. The salt has three functions: 1) It stops the bacterial action, 2) it aids in drying the curds, and 3) it adds flavor. The curds are then placed in boxes in which they are pressed to remove additional whey (moisture). The resulting blocks are then aged for a period of from several months to about 2 and 1/2 years at a temperature of 38 to 40°F (3 to 4°C). The longer the aging period, the sharper the cheddar flavor. Softer, more moist cheeses have a shorter process sequence with less processing of the curds and either no aging or a far shorter aging period.

cheese, Swiss[1] - is made initially with methods similar to those for other cheeses, but then undergoes a secondary fermentation that produces carbon dioxide gas, which creates the openings or "eyes" in the cheese body. The secondary fermentation takes place when the cheese is removed from refrigerated curing after two weeks and moved to a room with a temperature of 68 to 75°F (20 to 24°C). A bacterium, *propionibacterium shermanii*, feeds on the lactates in the cheese and carbon dioxide gas is generated. The action is allowed to proceed for 3 to 6 weeks, after which there are well-rounded eyes in the cheese. The cheese is then returned to a cold room at 45°F (7°C) for 4 to 12 months of aging.

chewing gum[3] - consists of a matrix of a natural or synthetic gum or insoluble latex, together with sugar, softeners, flavors, food coloring, and other additives. Chicle, from the sap of a Central American tree, has been used as the gum base, but almost all current products utilize synthetic resins, gums and waxes because they give more consistent texture and longer lasting release of flavor. Polyvinyl acetate, polyisoprene rubber, polyethylene rubber and resin esters, are used. Pieces of the base material are soaked to soften them and are mixed and heated with corn syrup, sugar (or other sweeteners), flavoring (eg., mint, licorice or fruit), and a small amount of glycerine. The mixture is discharged from the heating chamber as blocks of material weighing 8 to 10 lbs (3.6 to 4.5 kg). These blocks are cooled and passed through a series of paired rollers that gradually reduce the thickness to that of the finished stick of gum. Powdered sugar may be used during rolling to prevent sticking. The final pair of rollers also scores the sheets to gum stick size. The sheets are placed on trays and conditioned for 24 to 48 hours in a chamber at a cool, controlled temperature, and moderate humidity. The sheets are then fed to packaging equipment that breaks the sheets into sticks, wraps them, assembles them into packages, wraps them again, and packs them in cartons. These are shipped in multiple units to warehouses for distribution. The final operations are fully automatic and, in high production facilities, the above blending, treating and packaging of ingredients is on a continuous-flow basis.

Bubble gum is made similarly from a different matrix gum and is normally not rolled into sticks but, instead, is extruded into a rope-like shape and then cut into pieces. The pieces are wrapped, packaged and shipped.

chinaware - is ceramic table ware or whiteware (consisting of such items as dishes, plates, cups, platters, bowls, and pitchers) of a high grade, made from ceramic materials that provide translucence in the finished product. Porcelain and bone china are in the same category. Bone china uses animal bone ash as an ingredient, in addition to the clays used in chinaware. Bone provides greater strength to the product. Stoneware and earthenware usually refer to similar products that are not translucent, that utilize a less pure or less refined clay as raw material and are fired at lower temperatures. Pottery also usually refers to a non-translucent product although the term is also a general one for all kinds of ceramic ware. The manufacturing process for these ceramics is basically the same as with chinaware.

The following is the operation sequence involved in manufacturing boneware dishes or plates, cups, bowls, and mugs:

1. The raw materials: bone ash, kaolin (a pure grade of hydrated aluminum silicate clay), and china stone (another clay), feldspar, and flint, all finely ground, are mixed with water to form a slurry. The slurry is filtered to remove air and water and the resulting mix is extruded into cylindrical pieces called pugs. (See 5B4, 5B4a and 5B7)
2. jiggering (See 5B9): The pugs are pushed through another extruder to remove any remaining entrapped air and are sliced into round clay discs by a special machine. The discs are each placed on a rotating plaster mold of a jiggering machine. An overhead shaped tool descends on the rotating disc and forms the clay disc into the dish shape. Excess clay is squeezed outward and is removed by a cutting tool. The plaster mold absorbs some of the moisture in the clay disc. These operations are all automatic in current high-production facilities.
3. drying (See 5B13) - The formed dish is first dried while on the plaster mold. Then it is removed from the mold and moved to another dryer. After drying to about 0.5% moisture content (from 20% before jiggering) the dishes move to a finishing machine where damp sponges smooth the edges. Before firing, dishes and other chinaware at this stage is referred to as greenware.
4. firing (See 5B16) - The dishes are placed on metal forms called setters that preserve the dish's shape during firing. Setters with dishes are stacked in racks that are placed on a conveyor that moves through the kiln. The kiln temperature is approximately 2290°F (1254°C) and the conveyor moves slowly enough so that each dish is in the kiln for a 9-hour period.
5. vibratory polishing - (See 8B2 and 3K11.) After firing, the dishes are vibratory tumbled to smooth the surface finish.

6. glazing (See 5B15) - The dish is sprayed with glaze from multiple guns to insure coverage, is wiped on the bottom to prevent adhesion to the rack, and is conveyed through the glaze kiln. The temperature in the kiln is about 2020°F (1100°C), and the dwell time is 8.5 hours including 3 hours of cool down.
7. decoration - Decorations on dishes are added, for high-production lots, from decals that are applied by machine or by hand (8I9). Some decoration, for specialty lots of high grade products, is applied by hand brushing by skilled workers. Pad printing (8I7a) is another method. With all these methods, the ink applied is either vitreous or metallic and is melted to adhere to the dish. The typical kiln temperatures for these materials ranges from about 1400 to 1600°F (760 to 870°F) for a period of 1.5 to 2.5 hours.
8. Final inspection and packaging for shipment follow.

Many chinaware products are formed by casting rather than jiggering. Slip casting is the usual method. The raw materials, after mixing, are kept in a liquid state and are called slip. The slip is poured into plaster molds of two or more pieces. The molds are somewhat porous and are similar to plaster molds used in metal casting (1C5). The plaster absorbs water from the slip, leaving a leathery coating in the mold after about 10 or 15 minutes. The excess slip is then poured from the mold; the mold is carefully taken apart and the cast piece is removed. It is still somewhat soft at this point, and is wiped with a damp sponge and other hand tools to remove mold seam marks and to smooth the surfaces. Handles, if used, are made by casting, and are fastened by hand. The cast greenware is then ready for firing as described above.

chipboard (wafer board) - See 6F3.

chips, electronic - See 13K.

chlorine - is made primarily by the electrolysis of salt brine. From salt (NaCl), and water (H_2O), three materials are created: caustic soda (NaOH), hydrogen gas and chlorine gas. The operation is simple except that steps must be taken to keep the caustic soda separate from the chlorine. Membranes and mercury cells are two methods that have been used to accomplish this. Chlorine is also produced as a bi-product when metallic sodium is made by the electrolysis of molten sodium chloride. Chlorine is used in the production of solvents, plastics, synthetic rubber, bleaches and dyes.

chocolate[16] - typical chocolate candy production involves the following major steps: 1) pods that grow, in tropical countries, on the cacao tree, are picked. 2) The pods are cut open and the beans that they contain are removed. 3) The beans are placed in trays and fermented in the sun for several days. 4) The beans are sorted, washed, and blended with beans of other varieties. 5) The beans are roasted. 6) Winnowing machines are used to break and remove the shells. 7) The nibs that remain are ground into a liquid paste containing cocoa butter. 8) The cocoa butter is removed from some of the paste, leaving cocoa cake, which becomes cocoa powder after grinding and is used for chocolate flavoring in other food products. 9) The cocoa butter is mixed with other ingredients including sugar and milk powder (if milk chocolate is being made), and is blended thoroughly. 10) The blending, or heating raises the temperature to 120 to 190°F (50 to 90°C) which melts the cocoa butter and, with nuts, crispy rice, raisins or whatever is added, it is poured into molds of the shape desired. 11) After cooling, the resulting chocolate candy is wrapped and packaged with automatic machines.

chokes, choke coils, inductors (electronic) - See 13L3.

chromium[11] - is produced chiefly from the ore chromite, which contains significant amounts of both iron oxide and chromium oxide. There are several processes in use to obtain metallic chromium from this ore. In one common process, the ore is roasted with soda ash to convert the chromium oxide to sodium dichromate, ($Na_2Cr_2O_7$). This is then reduced to green chromic oxide (Cr_2O_3) by heating it with coke. The chromic oxide is mixed thoroughly with aluminum and the mixture is heated to initiate the reduction of the chromic oxide. This reduction is accompanied by the release of considerable heat, sufficient to melt the chromium. Metal of 97% to 99% purity is produced. Other methods of chromium production use an electric arc furnace for the oxide reduction or electrolysis of chromic acid to yield high purity chromium. Chromium is used as a key ingredient in stainless and alloy steels, in cutting tools, for electroplating and in various chemical compounds where pigments and refractories are two of the major applications.

cider - is made by pressing the juice from apples. The apples are picked, inspected, washed, and rinsed. They are then placed in a machine that breaks and crushes them, reducing them to a pulp. The pulp is pressed in a vertical, top-down press to extract the juice. The usual arrangement in the press is to spread the pulp on mats made from cloth, coconut fiber, or other filter material that retains the solid pulp material but allows the juice to flow. The filter material covers both the top and bottom of the spread pulp and several layers of mats are normally used in each pressing operation. The juice is collected and, by present practice, is pasteurized by heating it to a temperature of 160°F (70°C) for 6 to 10 seconds and then immediately chilling it to 40°F (4°C). It is then bottled. Subjecting the juice to ultraviolet light is sometimes used as an alternative to heat pasteurization. Before bottling, the juice may be tested for sugar (with a hydrometer to measure specific gravity or by a hand-held refractometer that measures the percentage of sugar directly) and for pH. The juice may then be blended or adjusted as required.

hard (alcoholic) cider is made by approximately the same series of initial operations though it may not be washed as thoroughly since various contaminants on the apple skins may aid in the fermentation process. For fermentation, the liquid is traditionally placed in wooden barrels with added yeast and kept in a dry, light area. Fermentation takes several months at a temperature of 45°F (7°C) to 60°F (16°C). Carbon dioxide gas is produced by the fermentation process and must be allowed to escape from the barrels. Sediment is removed before the cider is bottled.

clad metals - See 7C13d.

cloth, anti-shrink - See 10F5a.

cloth, knitted - See 10A, 10B and 10D.

cloth, non-woven - See 10A, 10B and 10E.

cloth, woven - See 10A, 10B and 10C.

clothing - See 10H.

coal gas - See *gas, manufactured.*

coffee - is the liquid made from the bean of a tropical evergreen tree. There are several varieties of such trees, the most important being the Arabian coffee tree. Most commercial coffees are blends, made from the beans of a number of different tree species, from different locations, or from different growing techniques. The fruits from the tree each contain two small green seeds, the coffee beans. The beans are softened in water, fermented, washed and dried in the sun or in heated rotating cylinders. They are sorted mechanically or by hand to remove extraneous material and unsatisfactory beans. After being blended with other beans, the mixture is roasted. Roasting takes place in rotating horizontal drums, heated to a temperature from about 380°F (193°C) to about 425°F (218°C). The higher the temperature, the darker the roast. In roasting, new compounds are formed that produce the characteristic coffee aroma and flavor. The coffee is ground before use, either at the retail store for the customer or in roller or plate-type grinding mills, and packed in sealed containers before shipment to the retailer.

coffee, decaffeinated - Coffee beans are exposed to steam to raise the moisture content and to bring dissolved caffeine to the surface. The caffeine is extracted (See extraction, 11C3) with supercritical carbon dioxide as the solvent. (Methylene chloride was previously commonly used.) The solution is washed from the beans with steam. The beans are then dried. The caffeine is separated from the solvent and used as an ingredient in other caffeine bearing foods and medicines.

coffee, instant - regular brewed coffee, made with hot water, is brewed in several cycles to a specified strength. It is then dehydrated. The prime method used is spray drying (12H2), which produces a granular powder that is soluble in water to make coffee. Freeze drying (12H5) and vacuum drying (12H1) may also be used to make the powder.

coils, electrical - See 13L3.

coins - U.S. dollar, half dollar, quarter, and dime coins are made from clad sheet metal that has copper sandwiched between two layers of nickel alloy (25% nickel and 75% copper). The clad sheet is made by pressure rolling the three clean, annealed sheets together under high pressure. Pennies are made from zinc-copper alloy electroplated with copper. Nickels are made from a copper-nickel alloy. All these metals are produced by suppliers to the U.S. mints and are delivered to the mints as coiled sheet stock, fully annealed, for blanking and coining. Round, coin-size discs are blanked (2C4) from sheets of this stock with multiple blanking dies. The unused portions of the sheet stock are recycled.

The blank discs are annealed, washed, pickled, tumble burnished and inspected by machine vision devices (14D2). The discs are then fed automatically to presses that coin (2D6) the design on both surfaces. Coins with ribbed edges may have these formed by the coining dies or by another press operation after coining. The coins are again inspected and are machine-counted into cloth bags that are stitched closed before storage and shipment.

Coin designs are first rendered as a flat sheet drawing. From these drawings, large size bas relief masters are made by hand sculpturing. The approved master is used to make plaster castings. Rubber molds are made and from these plaster castings, and epoxy masters are cast in the rubber molds. From these masters, coin-size "hubs" or hobs are machined in tool steel by special pantograph milling machines that trace the epoxy masters. The hubs are finished, polished, and heat treated and then used to hub (pressure form) the working dies used in the coining operation.

combs - are made by injection molding. See 4C1.

compact discs - are made from a polycarbonate-based material with a special process that is primarily injection molding (4C1), with an element of compression molding (4B). The injection mold is designed so that slight movement is possible between the mold halves when the mold is closed. After injection, when the material is still soft, force is applied to the mold halves and the mold closes an additional fraction of an inch. This motion impresses in the surface of the disc the fine surface features (pits and lands) that convey the disc's digital information. The disc is then vacuum coated (8F3) with a very thin reflective layer of aluminum (though silver or gold are sometimes used.) In the disc player, the reflection of laser light from this surface provides digital data that is converted to the audio or visual output of the player. A thin coating of clear acrylic plastic is applied to the disc for scratch protection and identifying information is printed on the disc's top surface.

The injection/compression mold for the disc is created from a tape recording of the audio or video material to be incorporated in the disc. From the output of the tape, the recording is changed to a digital format. A glass disc, coated with a bonding agent and a layer of photoresist material, is exposed to laser light that fluctuates in response to the digital data from the tape. When the photoresist layer is developed and etched (See 13A1a and 13K3b1), a pattern of pits and lands is created in the glass surface. The pattern conforms to the sound or visual input from the tape recording. The disc is then vacuum coated with a thin layer of silver and then with a thicker electroformed (2L2) layer of nickel. The nickel layer is removed from the glass disc and its bottom surface now has a negative impression of the master pattern of pits and lands on the disc. This nickel layer could be used as a mold to make production CDs but, in practice, is used as a submaster to make a number of mold inserts using a two-step electroforming process that makes another positive master disc and a number of negative copies of it. These negative copies, suitably reinforced and fitted into precision molding equipment, are used to mold the compact discs sold to the public.

composite structural lumber - See 6F7.

concrete blocks - are made from portland cement, sand, water, and some of the following: gravel, crushed stone, cinders, expanded slag, and vermiculite. Typical ratios of cement, sand and coarse aggregate range from 1:1:3 to 1:3:6 (in the listed order, by volume). These ingredients are mixed thoroughly to the desired consistency and the mixture is introduced to metal molds that are in the shape of the blocks to be produced. When filled, the molds are vibrated to compact the concrete mixture and to eliminate voids. The molds are removed when the cement mixture has set sufficiently. The blocks are then allowed to fully cure over a period of at least seven days.

condensers, electronic - See 13L2 and 13K3a6.

confectioner's glaze - See *candy*.

contact lenses - See *lenses, contact*.

containers, plastic - The injection molding process (4C1) predominates for manufacture of all kinds of containers. Large containers may be molded by one of the structural foam processes (4C3). Blow molding (4F) is used for bottles and similarly-shaped containers.

cooking utensils - are made from a number of different materials: stainless or carbon steel, aluminum and copper in sheet form, cast iron, ceramic materials, and glass. (Glass and ceramic are limited to oven use or for covers. Uneven heating at the stovetop could cause cracking or breaking.) Many different finishes may be applied, and various kinds of handles are used. Stainless steels are usually of the variety with 18 percent nickel and 10 percent chromium.

Cast iron utensils, notably frying pans and cooking kettles, are sand-mold cast (1B). They are typically finished with a black oxide (8E4) oil finish. Aluminum may be cast also, but most aluminum pans are made from sheet material. Cast utensils usually have one or more handles, or brackets for handles as part of the casting.

Sheet metal methods involve blanking (2C4) from coil or sheet stock, and a series of deep drawing (2D5b) and forming (2D2) operations, with annealing (8G2 or 8G5b) between various stages of deep drawing. Some utensils are made by spinning (2F6). The utensils made from sheet material have separate handles.

Ceramic oven utensils are made with the usual ceramic fabrication methods(5B).

Glass utensils and covers are made primarily by pressing (5A2a) borosilicate glass that withstands the temperatures of cooking without fracturing. The glass is also known as Pyrex and is made from 60 to 80% silicon dioxide, 5 to 20% boric acid, 5% fluxes, 2% stabilizers, and small quantities of other materials. These constituents are mixed and melted at a temperature of 2900°F (1600°C), held at a high temperature, for up to 24 hours, formed to shape at a somewhat lower temperature and annealed (5A4a) after forming.

Separate handles are made of various materials and manufacturing methods. One type of handle is compression, transfer, or injection molded of phenolic resin. There may be a sheet metal (stainless steel) reinforcement near the fastening point and a base piece that is welded or riveted to the sheet metal utensil for fastening the handle in place. Handles may also be formed from sheet metal. Some sheet metal handle parts are polished and chromium electroplated.

Handles are fastened with rivets, screws, or by resistance welding or brazing. Filets may be formed where the handle meets the pan.

Surface finishes vary, but there are common approaches for utensils of like material. Many pans are coated with Teflon or other non-stick materials on internal surfaces. Both external and internal surfaces may be porcelain enameled (8F1) but pans with Teflon coatings on the inside surface and porcelain on the exterior are first porcelain coated, and then Teflon coated.

Teflon coating is limited to aluminum surfaces and involves a number of steps: Washing, rinsing, and hydrochloric acid etching of the surface are the first operations. Application of a primer to the surface by spraying and then allowing it to dry are next. Then the Teflon is applied in two coats, allowed to dry in an oven and then fused by gradually heating the utensil to about 800°F (425°C) for about 5 minutes, followed by slow cooling.

Some stainless steel pans are electroplated with copper in the area that contacts the stove flame or heating elements in order to transfer heat more quickly. Some sheet metal pans are polished to a near-mirror finish, usually only on the exterior surfaces. Pans are often polished while rotating in a lathe chuck. Satin finishes, most often put on the bottom or interior, can then be given a symmetrical circular pattern. Aluminum pans may be anodized (8E1) to provide a hard, attractive finish.

Covers are supplied with pans, except some frying pans. Covers are often made from material that matches the pan material except that glass is commonly used for pans of other materials. Metal covers are blanked and formed with several operations, sometimes with in-process annealing to offset work hardening.

Brand and manufacturer information may be impression stamped (8I2) into the bottom of the utensil. (Also see *handles, cooking utensils.*)

copper - Worldwide, about 15 different ores are sources of copper metal. Most contain less than 2% copper but are economically processed because they also yield gold, silver, nickel, and other valuable metals. Chile is the largest source of copper; but much is also produced in the United States. Because of the low metal concentration in the ores, the first step is concentration. This process usually involves crushing (11D, 11D1, 11D2), and ball milling (11D3), followed by flotation separation (11C8b) to remove unwanted minerals and bring the copper content to about 25%. The major North American and English ore is chalcopyrite or copper pyrite. After concentration, this ore is processed by smelting (11K1b2) with sulfur that reacts with the copper, yielding a matte of CuS_2, and FeS and impurities. The matt is melted in a reverberatory furnace (1A4), air is blown through it, converting the sulfur to sulfur dioxide gas and yielding iron oxide and blister copper. Blister copper contains 1 to 4 percent other metals, arsenic, sulfur, and other impurities. It is further furnace refined or electrolytically refined (11C10 and 11K1c2) to produce commercial copper. Fig. C3 shows this sequence.

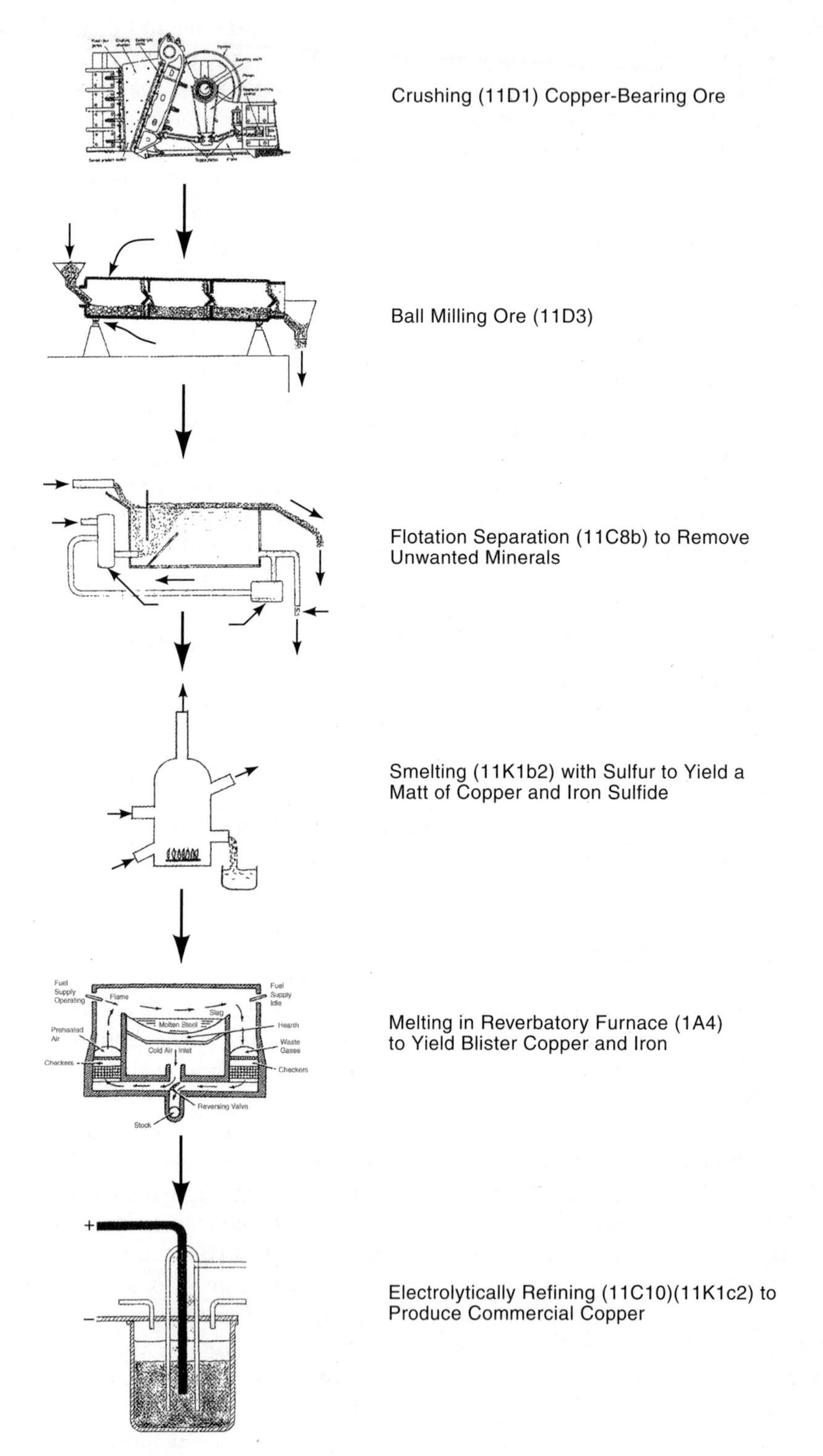

Fig. C3 The manufacturing operation sequence for refining chalcopyrite (copper pyrite) ore into copper.

Another copper ore is chalcocite containing cuprous sulfide, CuS_2, which is processed similarly. Native metallic copper is crushed and washed, cast, and refined.

The desirable properties of copper are its very high electrical conductivity, its malleability, and its resistance to corrosion. Electrical wires, switch and connector parts, pipe and coins, are made of copper and copper alloys. Major alloys are brasses and bronzes.

corn flakes - See *cereals, breakfast.*

corn, frozen - See 12I1.

corn oil - is obtained by expression (See 12E2 and 11C4.) of the germ of the corn kernels. Hulls and kernels of corn are separated with a bath of warm water containing sulfur dioxide. Attrition mills then break the germ from the kernel and the two are separated by flotation. The kernel is washed and dried and the oil is expressed with worm presses. The oil is screened and filtered. Corn oil is used for salad dressings and as a raw material for some soaps.

corrugated cartons - are made from kraft paper (9C5). The first major operation is to produce corrugated board and carton blanks when the board is cut to overall size. The operation takes place on a corrugating machine that typically extends about 300 feet (90 m) in length. A typical corrugating machine is illustrated in Fig. C4. Kraft paper is fed from multi-ton rolls that, with current equipment, can be 98 or more inches (2.5 m) wide. Three sheets of kraft paper are fed to the machine when single-wall corrugated board is produced; one of the three sheets is corrugated; the other two are called outside liners. (See the standard board construction illustrated in Fig. C5). When a double-wall board is to be made, five sheets are fed through the machine. Corrugated inner sheets are formed between a pair of grooved, mating rollers that are heated to a temperature of 350°F (177°C). The corrugations are bonded to the liner sheets with a thin (0.008 in - 0.2 mm) layer of starch-based adhesive, applied to the corrugations by glue rollers. In addition to starch, the adhesive contains caustic soda and borax. The flat liner sheets are heated to about 200°F (93°C) before being glued to the flute tips (ridges) of the corrugations. The sheets are pressed together and the adhesive sets almost instantly, but the bonded board then passes through a machine section for additional pressing between two heated belts for further curing. The board is then slit and cut to length as necessary to make blanks for the cartons to be produced. The blanks are stacked automatically and conveyed to an in-process storage area. Production from machines like this is as high as 1600 ft (485 m) of board per minute.

From the storage area, the boards are fed to machines called, "flex-folder-gluers". When production quantities are high, these machines function automatically. They print identification on the outside board liner sheets, cut slots in the board for carton flaps, score the board where it is to be folded, apply an adhesive to the board near one edge and fold the board so that the four sides of the carton are closed

Fig. C4 A typical corrugating machine that makes either one or two layer boards, slits and cuts them to size for carton blanks and stacks the blanks for future use. The complete machine extends over 280 ft. (90 m) in length.

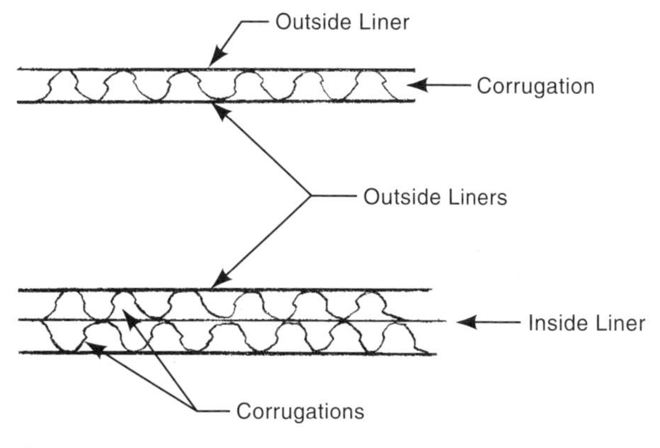

Fig. C5 Construction and terminology of single and double corrugated board.

and the carton is flat. (Note: Cartons are shipped from the factory in flat form. Forming to the box shape takes place where the cartons are filled.) The flat cartons are stacked in bundles of 25, wrapped with a plastic strap and conveyed to a palletizing station. When cartons are produced in smaller quantities, these operations may be performed on several machines with a larger content of manual work.

When the cartons are to be finished in a color other than the standard brown of kraft paper, they are made from rolls of either bleached kraft (white color) or kraft coated with the desired color.

cotton fabric - Cotton fibers come from the soft hairs that surround the cotton seed. The hairs are separated from the seeds by cotton gin machines that consist of a fixed comb and a cylinder with saw-like teeth attached. Raw cotton is fed to the gin and is pulled by the saw teeth through the comb. The seeds, which cannot pass through the comb, are left behind. This operation is commonly performed at a different location than that of the spinning and weaving operations. After ginning, the cotton is cleaned and baled and sold to a textile mill. There, the cotton is processed through the yarn-making steps described in section 10B. These operations include blending, picking, carding, combing (sometimes), drawing, and spinning. The cotton is then normally woven on looms (10C) or knitted (10D), finished (10F), and made into garments or other products (10H).

cottonseed oil - See *margarine.*

crayons (wax) - are made from paraffin wax, colored with solid powder pigments. The wax is heated to the melting point (about 240°F - 115°C) and mixed with the pigment (a batch for each crayon color) so that the color is evenly distributed. The crayons are cast in open flat bed molds. Each mold has many deep pockets for the crayons. When the wax in the pockets has cooled and hardened, excess wax in the flat bed is scraped off and removed. This operation is mechanized. The crayons are expelled from the cavities by ejector pins and are fed to a labelling machine that wraps a pre-printed label on each crayon. They are checked for completeness of casting and for proper label position and are fed to a packing machine. The machine has separate feed magazines for each color in the package, 12 feed magazines for an package of 12 assorted-color crayons. The machine fills the packages, closes it, and inserts them in shipping containers.

crepe fabric - See 10F3g.

crystal, lead glass - See 5A1.

cups, plastic, disposable - Three major processes are used in the production of common plastic drinking cups. The most common method is thermoforming. Because of the large depth-to-width ratio of these cups, they are produced by either the plug-assist method (4D4), the slip-ring method (4D6), or the vacuum snap-back method (4D5). Injection molding (4C1) is also very common. Insulated foam-plastic cups are made with the expanded polystyrene foam bead molding process (4C4 and 4C4b).

cups, paper - See *drinking cups, paper.*

cut glass - See 5A5b.

D

denim - is a twill-woven (10C and Fig. 10C-1) cotton fabric, usually with the warp yarns dyed blue (10G1). (The filling threads are white.) Denim is usually all-cotton, but some is made with a cotton-synthetic fiber yarn. The twill weave provides good durability. Denim is used for jeans, overalls, trousers, and other garments. Denim yarn or fabric is sometimes sized (10F4a).

detergents[4] - There are many different types of detergents, including soap. Detergents work with water as a soil remover by acting as an emulsifier and surfactant, breaking up oily films and helping the water to penetrate particles of soil. The most common synthetic detergents are the anionic type, which are usually the sodium salts of organic sulfates or sulfonates.[4] Frequently used raw materials are fatty alcohols (from tallow - animal and vegetable fats) or alkylbenzene (from petroleum). These materials are reacted with sulfur trioxide gas, made either by burning sulfur or by vaporizing sulfuric acid anhydride. After the reaction with sulfur trioxide, (SO_3), the fatty alcohol becomes fatty alkyl hydrogen sulfate and the alkylbenzene becomes alkylbenzenesulfonate. These are then treated with caustic soda, (NaOH), to neutralize the acidic compounds present. Other ingredients that may be used are sulfuric acid, oleum ($H_2SO_4 \cdot SO_3$), sodium silicate as a corrosion inhibitor, sodium tripolyphosphate and water. There are other miscellaneous additives that are used to improve brightening or bleaching and to provide a pleasant odor. An illustration of the detergent-making process and equipment, in greatly simplified form, is shown in Fig. D1. Other types of detergents, less used, are nonionic, ampholytic and cationic types.

diamonds, synthetic - are produced by several companies with proprietary processes. Most synthetic diamonds are made from graphite, which is subjected to a pressure of 750,000 psi (5 GPa) at temperatures of approximately 2700°F (1500°C)

Per Shreve[4], the General Electric process involves heating and pressurization, apparently of graphite, for a number of minutes in a catalyst of a molten "group VII metal alloy," apparently a nickel alloy. Temperatures are approximately 2000 to 2700K and pressures from about 9400 to 16,700 psi (65 to 115 MPa) are applied. Different types and sizes of diamond crystals are formed from different pressures, temperatures, time, catalysts and solvents. Crude pieces are cleaned and graded by size and shape.

Per Brady[2], the early GE process, developed in the 1950's, subjected graphite to pressures of 800,000 to 1,800,000 psi (5.5 to 12.4 GPa) at temperatures of 2200 to 4400°F (1200 to 2430°C) with a molten metal catalyst of aluminum, cobalt, nickel or another metal. These formed a film around the growing diamond crystals. More recently, GE is reported to use chemical vapor deposition with methane gas, enriched with additional carbon, to deposit a sheet of polycrystaline diamond. The sheet is then crushed and used as source material in the high pressure, high temperature process. DuPont makes synthetic diamonds with an underground explosive process. It produces pressures of 2,000,000 to 7,000,000 psi (14 to 48 GPa). Then, a series of chemical and mechanical steps extract a powdered diamond which is cleaned, sorted and shaped. Particle sizes range from 3.9 to 2,300 micro inches (0.1 to 60 microns).

Synthetic diamonds are used to provide cutting action on saws, in grinding and polishing wheels for hard substances, and for honing and lapping operations.

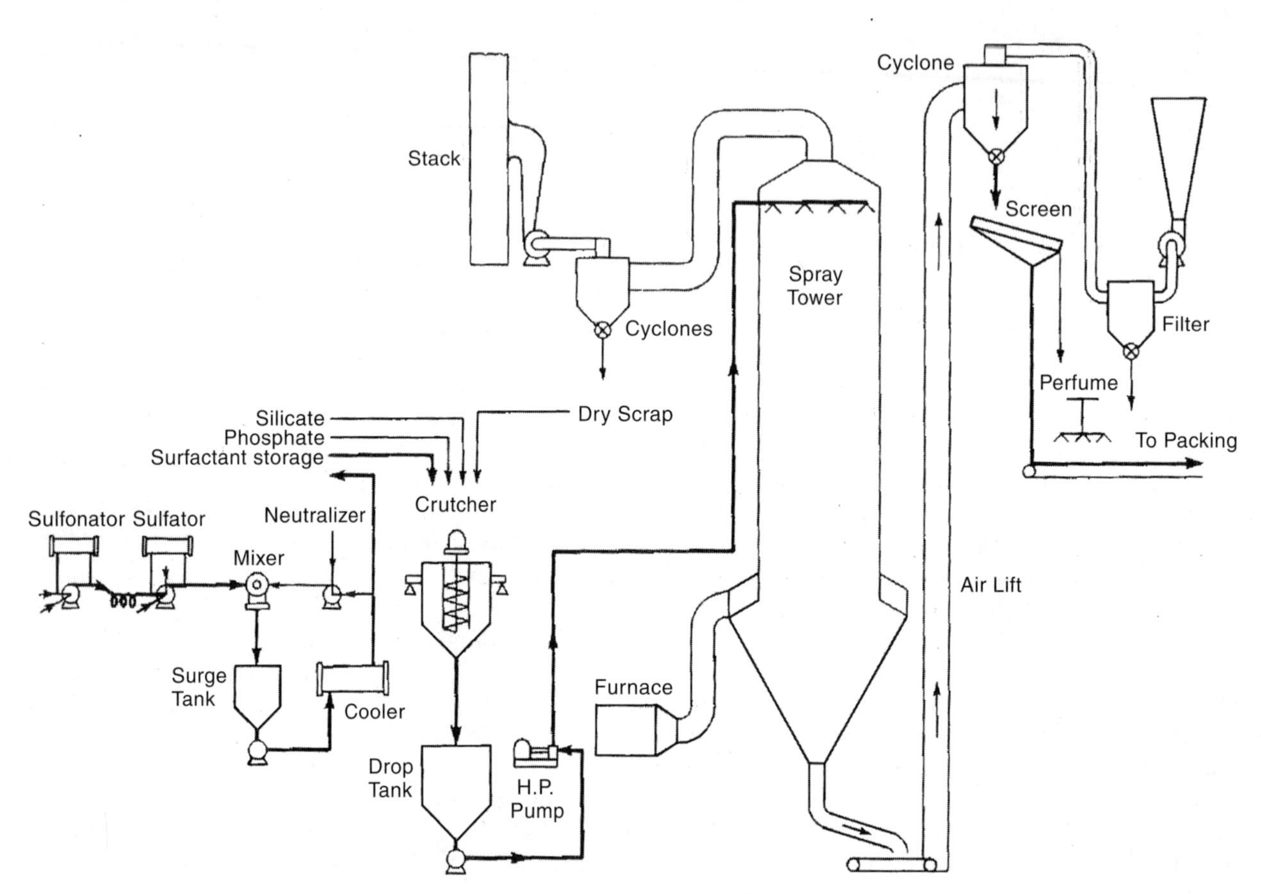

Fig. D1 A simplified flowchart for the continuous production of a heavy-duty detergent in granular form. (*The Procter and Gamble Company.*)

Diamond coatings for military and aerospace applications and cutting tools are made in low-pressure, very high temperature chambers by chemical vapor deposition (CVD) (See 8F3b.) from hydrocarbon gases with added energy from plasma, microwave or other sources. These films have excellent heat transfer ability, transparency to infrared and visible light and high electrical resistivity in addition to diamond hardness.

diesel fuel - See *fuel, diesel.*

digital video discs (DVDs) - A significant amount of preparation is required before a motion picture can be reproduced on a digital video disc. The film is first photographed, frame by frame with a digital camera. The digital data thus obtained is recorded on tape. The tape is processed by a video encoder to compress it, that is, minimize the amount of digital data required to define the images and the space that they will require on the disc. The motion picture then goes through the "authoring" process where a skilled technician customizes it for presentation, adding subtitles as necessary and perfecting the stereo sound. The complete presentation is checked and, if satisfactory, the finished digital tape is sent to the disc factory for production by the following steps: 1) Laser equipment (8I5 and 3O) is used to transfer the data from the tape to a thin, polished glass disc which has previously been coated with a photosensitive material. 2) The glass disc (master disc) is developed chemically, leaving minute pits, 0.00002 to 0.00008 in (0.5 to 2 microns) wide where the coating was activated by the laser. These pits have a digital pattern that corresponds to the picture, subtitles, and sound of the motion picture. 3) The glass master disc is immersed in a bath that deposits a thin coating of nickel by electroless plating (8C2). Then, a further, much heavier, coating of nickel is added using electroforming (2L2) and electroplating (8C1) processes. 4) The nickel layer is removed from the glass master and may be assembled as part of a mold for injection molding of plastic. Alternatively, it may serve as a sub-master for further electroforming to make inserts for production molds. The mold, then, has high spots that correspond to the minute pits in the glass master. 5) Polycarbonate plastic discs are injection molded (4C1) in this mold. The discs have minute pits equivalent to those in the glass master. 6) The plastic discs are given a very thin coating of aluminum or gold by sputtering (8F3a). 7) The disc is assembled and adhesively bonded to another disc which may or may not have data on its other surface. 8) The disc is labeled by offset printing (9D2a). 9) Discs are assembled into packages with accompanying leaflets. These are shrink wrapped and inserted with other copies of the disc into shipping cartons, which are sent to warehouses or stores.

dinner plates - See *chinaware.*

diodes and transistors - See 13L5 and Fig. 13L5.

dishes, china - See *chinaware.*

dishes, glass - are made by glass pressing (5A2a) and annealing (5A4a).

dishes, plastic - picnic dishes are normally either vacuum formed (4D1) or injection molded (4C1). (Those with dividers are more likely to be injection molded.) Other plastic dishes are compression molded (4B1) of urea or melamine. Decoration of such dishes is normally by in-mold methods (4M5).

distilled spirits (distilled liquors) - include whiskey, brandy, gin, vodka, rum and other liquors. Brandy is a liquor distilled from wine; applejack is distilled from hard cider. Distilled spirits are produced with two major operations, fermentation (12J4), wherein sugars and other carbohydrates are converted to alcohol, and distillation (11C1), in which the ethyl alcohol content in the beverage is increased. The first steps in whisky-making, before distillation, are the same or similar to those of beer brewing. The making of distilled spirits starts with the collection of a fruit, a sugar-bearing plant, or a grain or other starch-bearing plant material. Foods that are used include grapes, peaches, pears, apples, apricots, sugarcane, sugar beets, honey, molasses and juice from a cactus, These foods are all sources of sugar. Wheat, rye, corn, rice, barley and some roots are sources of starch. These foods are converted to juice by squeezing, crushing, or cooking in water, or by a combination of these methods. The sugary juices are available for fermentation, but juices from starchy materials must first have their starch (carbohydrate) converted to sugar. This is done by introducing an enzyme to the liquid. The enzyme formed when barley grain begins to sprout is one that is used. (See *beer.*)

Yeast is another enzyme that enters the process and it is used to convert the sugar in the mixture to ethyl alcohol during fermentation. Fermentation produces drinks of low or moderate alcohol content (up to 12%). When the fermentation is complete, distillation can then be used to increase the alcoholic content. The complete process normally has other steps: filtrations to separate the liquid from solids, heating or cooling of the liquid to facilitate some of the operations, the addition of some flavoring or coloring material, and the blending of several different liquors to provide a standardized or desired flavor. (Distillation requires considerable heat energy.) Aging is an important operation for some liquors, notably whiskies, brandies, and some rums. It is done in oak casks, sometimes casks that are charred on the inside. Flavor and alcoholic content of the liquor may change during aging, depending on the cask, how much it has been used previously, and the temperature and humidity of the storage area. The final operation is bottling (12L). In modern high-production distilleries, this is highly automatic, including placement in shipping containers. Fig. D2 shows the full operation sequence for the production of whisky and gin.

downspouts, roof - most frequently are made from either vinyl plastic or sheet aluminum. Vinyl downspouts are made by profile extrusion (4I1) of a mineral- and pigment-filled PVC material. Elbow fittings to direct the water flow at ground level and fittings to connect the downspout to roof gutters are injection molded (4C1) of essentially the same material. Holding brackets are also injection molded.

Aluminum downspouts are contour roll formed (2F7) from coiled aluminum sheet stock, slit to the width needed, and painted (before forming).

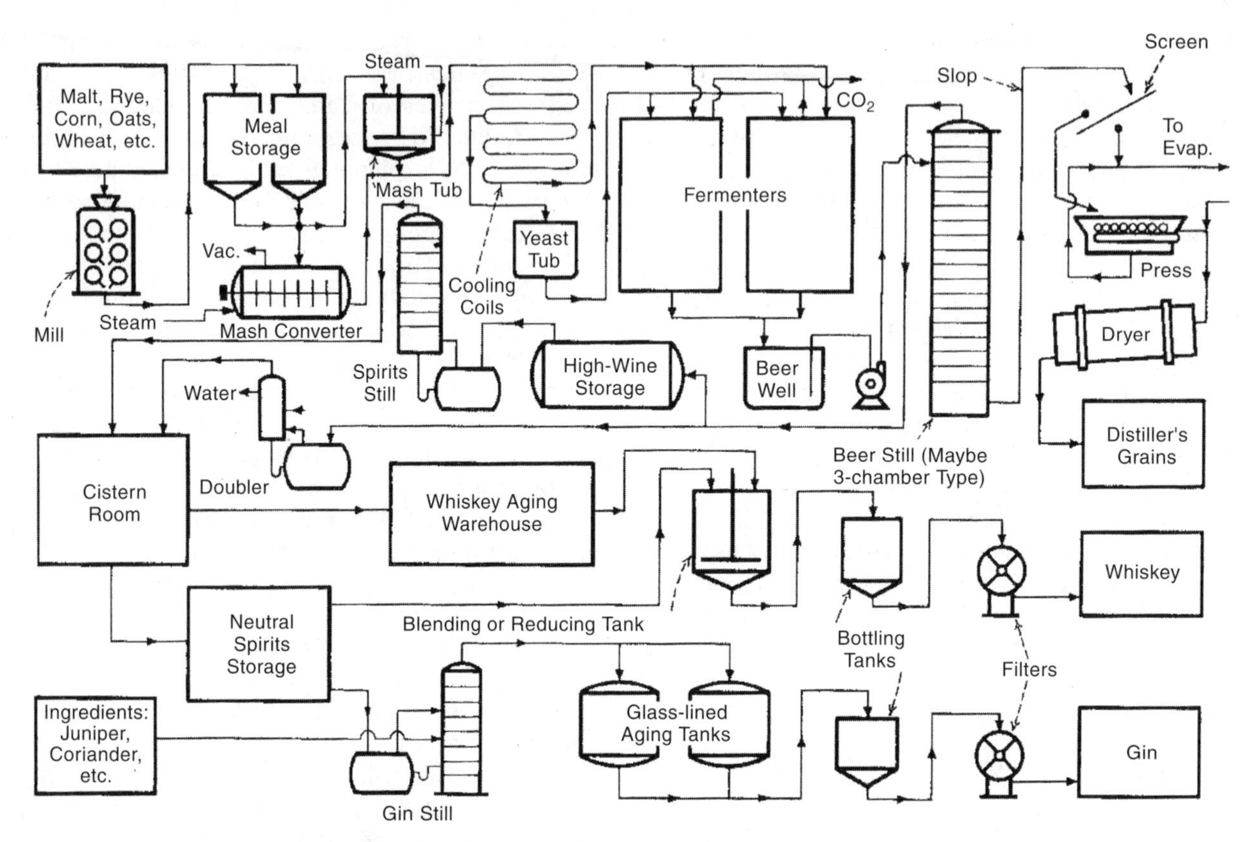

Fig. D2 The operations involved in the manufacture of distilled spirits (*from Shreve's Chemical Process Industries, 5th ed., G.T. Austin, McGraw-Hill, New York, 1984. Used with permission.*)

Painting is commonly on both sides and the edges by immersion of the sheet stock in a paint bath. As the sheet emerges from the bath, excess paint is blown off with air knives. Another method, if a different color is used inside the part, is roller coating (8D2 and Fig. 4K1b-view b). The contour roll-forming operation bends the sheet around to a square cross section, and then bends and closes a lock seam joint to hold the edges together, forming a square conduit. Holding brackets for aluminum downspouts are made by blanking and forming parts from the same precoated aluminum stock and riveting two such pieces together. This operation is performed on special press equipment. Elbows are made from short lengths of downspout, which is wrinkle bent to near 90 degrees. The operation is preformed on a special press set up. The tooling provides preliminary deformation and, as the part is bent, a series of wrinkles form on the inside of the bend.

drill bits - are made from high-carbon, tool-steel drill rod. It is heat-treated to a hardness of Rc 62 and centerless ground to a diametral tolerance of ±0.0005 in (0.013 mm). (See 3C1b.) Flutes are ground into the rod in a spiral pattern with special form grinding machines using creep feed grinding. (See 3C3c.) Larger drills are flute ground with wheels dressed to grind the flutes, grind relief in the flutes and leave a narrow land, all in the same pass. The drill is then given a black oxide treatment (8E4) and the point is ground and sharpened in a special cam-operated grinding machine. If the drill is large, the shank may be tempered to reduce its hardness and increase its toughness. Size and other identification is then usually marked on the drill by acid etching, but may, instead, be pressure imprinted before heat treatment and centerless grinding.

drinking cups, foam plastic - are made with the expanded polystyrene foam bead molding process (4C4b).

drinking glasses - are made by glass pressing (5A2a) or, sometimes, machine blowing (5A2b3) often followed by annealing (5A4a).

drive screws - See *screws.*

drums, 55 gallon - Plastic 55 gallon drums are extrusion blow molded (4F1).

dry cells - See *batteries, flashlight.*

dry ice - is the solid form of carbon dioxide gas. Carbon dioxide, from a number of possible sources, is purified and then liquified by compression and refrigeration to a temperature of −71°F (−57°C) or lower. The pressure is then reduced to the atmospheric level, allowing expansion and adiabatic cooling, causing the liquid to partially solidify into snow-like flakes. These are compressed together to form a cake which is then cut into blocks of the desired size.

ductwork, steel - is usually custom made with job shop forming methods, from galvanized sheet steel or sheet aluminum. Squaring-shear cutting (2C1a), notching (2C5b), nibbling (2C2) and turret punching (2C5a) all may be used to make sheets of the desired size and shape. The sheets are then formed, normally with job-shop bending and forming methods. Operations and equipment used includes press brake bending (2D1a), flanging (2D8), beading (2D9), hemming and seaming (2D10), edge curling (2D11) and three-roll forming (2H2f). Some ready-cut and preformed pieces, available at hardware stores and home centers, are made from coiled sheet stock on contour roll forming equipment (2F7).

ductwork, plastic - Plastic ductwork for automotive applications is extrusion blow molded (4F1).

DVD's - See *digital video discs.*

dyed fabrics - See 10G1.

dyes - are intensively colored complex organic chemicals that are used to color other materials and products. The term "dye" pertains particularly to colorants that are in liquid form. Although dye chemicals occur in nature, almost all current production is of synthetic dyes made by chemical processes. One exception is logwood, derived from the heartwood of a Central-American tree and used to dye some fabrics dark black. Synthetic dyes are made from coal tar or petroleum but coal tar predominates. The sequence for production of dyes includes the use of aromatic hydrocarbons such as benzene, toluene, naphthalene, anthracene, xylene, and some paraffins made from the distillation of petroleum[4]. These compounds are made into chemical intermediates that are used as materials for dye production and, in many cases, for other uses. The intermediates, also made with petroleum refining methods, include styrene, ethylbenzene, cumene, phenol, p-xylene, cyclohexane, and numerous other materials. They are chemically converted to dyes by a variety of operations that depend on the particular chemistry of the dye produced. Azo dyes, which have an ·N:N· linkage, constitute a large portion of dye production. They are produced by diazotization of primary arylamines which is followed by reaction with phenols, aromatic amines, and enolizable ketones[2].

The most important use of dyes is to impart colors to textiles. (See Dyeing, 10G1.) Other products that are dyed are paper, leather, food and cosmetics.

dynamite - a solid explosive made by absorbing nitroglycerine (a liquid) in a "dope" of wood flour, sawdust, starch, wheat flour, or other similar materials or a combination of some of them. The solid dynamite permits safer and easier handling than that achieved with liquid nitroglycerine which is unstable and apt to explode during handling. The nitroglycerine ($C_3H_5(NO_3)_3$ is made from the nitration of glycerol with a mixture of sulfuric and nitric acids followed by the removal of the acid from the mixture. Sodium nitrate or ammonium nitrate may be added to the mixture to increase its explosive power. A small amount of calcium carbonate or zinc oxide may also be added to the mixture. When these ingredients are mixed, the material is ready for packaging. This is done by pressing the mixture in a paper tube, and sealing it with wax. The assembly constitutes a "stick" of dynamite. Various sizes are made but a diameter of 1 1/4 in (32 mm) and length of 7 7/8 in (20 cm) are common. Because of the danger from an unwanted explosion, many special arrangements for the production equipment, plant layout, storage facilities, and work procedures are made for the above operations. Dynamite has largely been replaced by other explosives. Nitroglycerine also has a medical use to induce dilations of blood vessels.

E

edible oils - See *oils, edible.*

elastomers - are plastics with rubber-like properties. See 4O2.

electrical wire - See *wire, electrical.*

electricity - Most electricity is produced by devices, called generators, that convert mechanical energy into electrical energy. Whenever a wire passes through a magnetic field, current is induced in the wire. An electrical generator is a device made to optimize this effect so that the electrical energy developed can be utilized. Generators contain coils of electrical wires, attached to a shaft. When the shaft and coils rotate, the coils pass through a magnetic field and electric current flows in the wires. The mechanical energy that turns the generator can come from one of a variety of sources. With hydroelectric power, it comes from water turbines spun by the force of water falling or under pressure from a dam. With most commercial generating plants, the power is supplied by steam or gas turbines. When steam turbines are used, the energy for converting water to steam comes from the combustion of coal, oil or natural gas. It also comes, in many power plants, from nuclear sources. A pressurized water reactor system to generate electricity from steam is shown in Fig. E1. Diesel and gasoline engines are also used, usually in smaller installations. Automotive generators or alternators work in the same way to provide electricity to charge the car's storage battery.

Storage batteries, dry cells and other electrical batteries generate electricity by converting chemical energy into electrical energy. Chemical reactions release electrons that flow from the battery terminals as electric current.

electric light bulbs - See *light bulbs* and *fluorescent lights.*

electric motors - See *motors, electrical.*

electric transformers - See 13L4.

electrical wire - See *wire, electrical.*

enamel, vitreous[4] (porcelain enamel) - is a colored glass coating, formulated with low enough melting temperature so that it can be used as a protective and decorative coating for metal and ceramic appliances, other products and artwork, including glass artwork. (When used with ceramics or glass, the enamel is referred to as a *glaze.*) The materials used to make vitreous enamels are clay, quartz or feldspar, fluxes (to reduce the melting temperature), metal oxide coloring and opacity agents, and other agents. The raw materials are mixed, pulverized and ground so that the particles are small enough to pass through a 200 mesh screen. See F1, porcelain enameling, in Chapter 8 for a summary of the application methods.

Vitreous enamel coatings are common on stoves, bathtubs, cooking utensils, outdoor signs, architectural panels, chemical process equipment, automobile exhaust parts, jewelry, and vases, bowls, cups, and other ceramics or glassware. (See 5A5i for its use as a decoration on glass objects.)

enclosures, shower - Plastic shower enclosures are most commonly made by the spray-up method with fiberglass reinforced polyester (4G2).

engineered lumber - See 6F7.

engine blocks - The major component of an internal combustion engine is the engine block, normally a sand cast (1B) part, made, in high production conditions, by the flaskless method (1B3g). Engine block castings are extensively machined. The surfaces for the cylinder heads are milled (3D) or

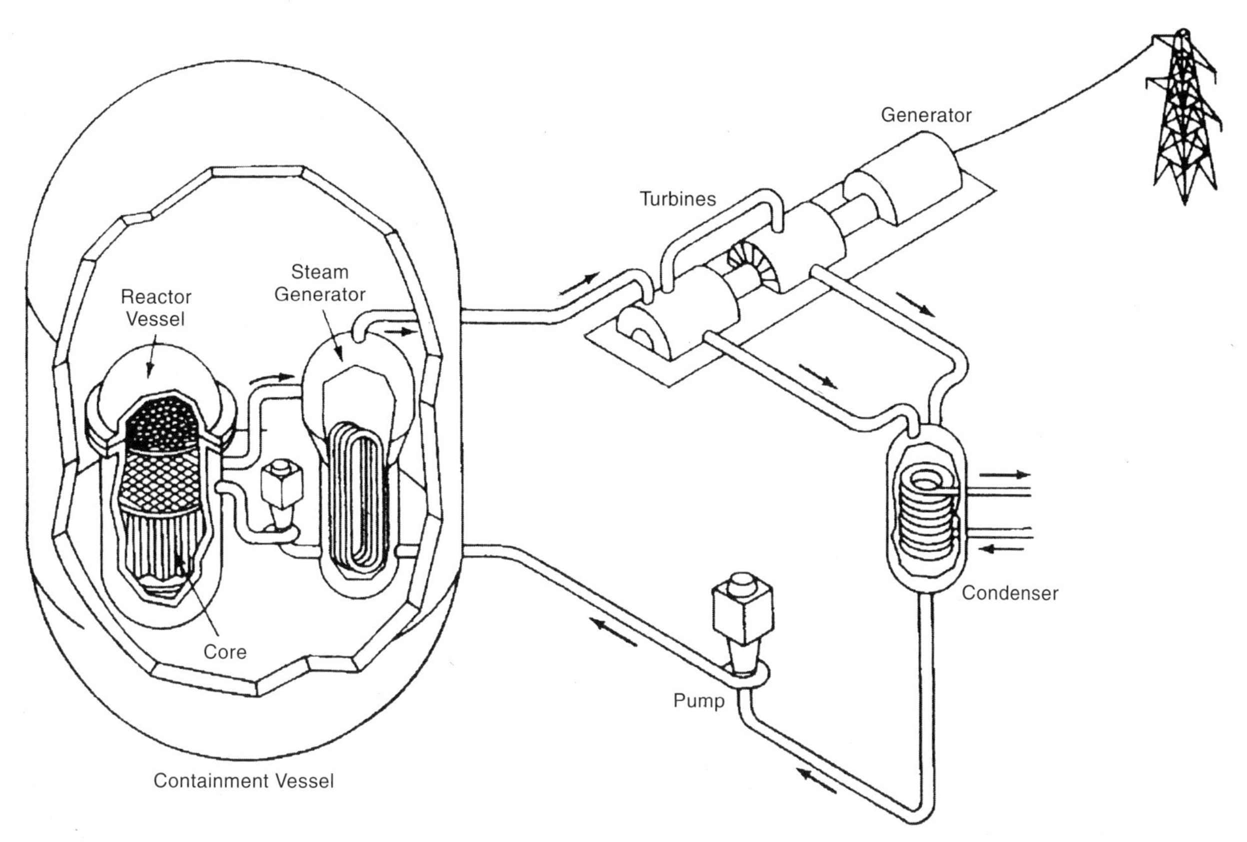

Fig. E1 Schematic illustration of a pressurized-water nuclear reactor. Water flows through the reactor vessel and, though it becomes very hot, does not turn to steam because it is under high pressure. It also does not leave the containing vessel but is circulated through a heat exchanger. In the heat exchanger, it transfers its heat to water in the other circuit of the exchanger. That water turns to steam and drives a steam turbine that, in turn, drives an electrical generator. (*Illustration from U.S. Department of Energy, Publication DOE/NE-0029, 1982.*)

broached (3F). The cored cylinder holes are bored (3B5) and honed (3J1). There are other milling, drilling (3B1), reaming (3B4), boring, and tapping (3E2) operations. In high-production situations, which usually exist in the automobile industry, these machining operations are performed by special purpose equipment (3X) on transfer-lines (3Y). At lower production levels machining centers (3T) are used.

envelopes - are all made on machines designed specifically for envelope making. Paper in rolls is fed to one end of the machines and travels rapidly through the various stages of the operation. Completed envelopes, packed in boxes, exit from the other end. The machines cut the paper to the size and shape (usually a rhombic shape), apply glue to the overlapping areas, fold the paper where necessary, seal the joints, apply gum adhesive to the flap that is to be closed by the user, print postal information and the return address as specified and pack the completed envelopes in boxes. Most machines are modular so that printing, packing, transparent window, metal clasp assembly, and other elements, can be added or removed. (Some machines print on the paper before cutting; others print on the formed envelope.) Most also can be set up to make envelopes of different types and sizes. Speeds of production equipment range up to a maximum of about 500 envelopes per minute for smaller, simpler machines and as high as 1500 per minute for the highest speed, most advanced machines. Such machines are of extended length and require a major capital investment.

epoxy[4] - The basic resin is most commonly made from a polymerization reaction of bisphenol A (made from phenol and acetone), with epichlorohydrin. (Epichlorohydrin is made from allyl chloride.) To cure the resin into a solid, it is mixed with a curing agent. Amines, acid anhydrides, or mercaptans are used. They cause the epoxy molecule chains to lengthen and cross-link. Epoxies are used as adhesives, coatings, potting compounds for electrical devices and in laminated boards for printed circuits. (Also see *epoxy* under *adhesives.*)

essential oils - See *oils, essential.*

etched glass - See 5A5f.

ethanol - See *alcohol, ethyl.*

ethylene (ethene) - a gas, is an important raw material. It is constituent of natural gas and occurs in petroleum. It is most commonly made by fractional distillation (11H1a) and cracking (11H2a) of petroleum. Ethylene is also derived from natural gas. In one method, ethane, another constituent of natural gas, is steam cracked at a temperature of 1550°F (840°C) and a pressure of 24 psi (165 kPa) to produce ethylene. Ethylene is the raw material for polyethylene plastics and in making trichloroethylene, a powerful solvent and degreasing material.

explosives - See *dynamite.*

eyeglasses - Eyeglasses with plastic frames are made with one of two methods:

The most prevalent method for the frames is injection molding (4C1) where both the part that holds the lenses, the front, and the two pieces that hook around the user's ears, the temples, are injection molded to almost the final size and shape needed. The temples may be molded with metal wire inserts for reinforcement (4C6). Hinges are normally formed and machined from a brass alloy in three pieces per hinge, two half-hinges and a connecting screw. Special equipment is used. Little or no machining is required of the molded components, but tumbling with abrasive (8B2) is used to remove mold parting line flash and to enhance the smoothness and polish of the frame parts. Hinge parts are assembled and bonded to the plastic parts by ultrasonic insertion (4N1b).

The other method is much more labor intensive and its use is limited to higher-priced eyeglass frames. The fronts and temples are blanked from a heated sheet of plastic, approximately 1/3 in (8 mm) thick, with steel rule dies (2C4a) or other blanking dies. The plastic usually is cellulose acetate. (See *cellulose acetate plastics.*) The undercut groove that holds the lenses in the frame is milled with a router-like cutter. The standard groove width is 1/6 in (4.2 mm). The edges of the pieces are rounded and smoothed by routing machines and by hand operations. With the aid of a fixture, nose pieces are bonded to the insides of the front piece and the adhesive is allowed to cure for 24 hours. Then further grinding smooths the area of the joint. At another machine and fixture, a slot is machined into each end of the frame at the proper angle. Half of a metal hinge for each temple is positioned in each slot and bonded with ultrasonic welding (4L4). The two temples are heated and a steel reinforcing wire is pressed into the core area of each to provide reinforcement. Then the temples are cooled and machined at the frame ends to fit the other half of the hinges. The other half of each hinge is inserted ultrasonically. The hinges are covered with protective caps and the parts are then thoroughly tumbled with pumice and other media to further round and smooth their edges. Tumbling continues for 24 hours. The frames are heated again and bent in a press operation to the curved shape needed. Then the frames and temples are tumbled again repeatedly with progressively finer abrasive and finally, with wax, to polish their surfaces. The temples or fronts may be hot stamped (4M2) on an inside surface for product identification. Finally, the parts are packaged for shipment to dealers who install lenses and assemble the temples to the fronts.

For manufacturing information on eyeglass lenses, see *lenses.*

F

fabrics, anti-shrink - See 10F5a.

fabrics, dyed - See 10G1.

fabrics, knitted - See 10A, 10B and 10D.

fabrics, flocked - See 10F3l.

fabrics, non-woven - See 10A, 10B and 10E.

fabrics, permanent press (wash and wear) - See 10F5b.

fabrics, printed - See 10G2.

fabrics, stain release - See 10F5d.

fabrics, woven - See 10A, 10B and 10C.

felt - a non woven fabric made from wool, fur, other hair fibers or synthetic fibers. Felt was previously used extensively as a hat making material and is currently used for industrial purposes and where padding is needed. (See 10E, *non-woven fabrics.*)

felt-tip marking pens - See *marking pens, felt tipped.*

fertilizer - Currently manufactured fertilizers have three major ingredients: nitrogen, phosphorous and potash (potassium carbonate - K_2CO_3 or $K_2CO_3{:}H_2O$). Superphosphate is one fertilizer that contains approximately 12% of each of these three ingredients. Superphosphate is made from phosphate rock that is treated with nitric or sulfuric acid, then reacted with nitrogen to neutralize the acid and add nitrogen. Potash salts are then added. Trace quantities of other elements, needed by plants in very small quantities, may also be added.

Ammonia, in liquid form is used as a direct-application, nitrogen fertilizer. Ammonium nitrate, ammonium sulfate, ammonium phosphate, sodium nitrate, potassium nitrate and urea are other nitrogen compounds used in fertilizer. Phosphate rock from Florida is a major source of phosphate, but animal bones are a minor source and basic slag from steel mills that use phosphatic ores are also sources. Other potassium compounds may be used in place of potassium carbonate. Potassium chloride (KCl), also called potash, is used in fertilizer and is mined in combination with sodium chloride as sylvinite. The potassium chloride is separated out by making use of its changes in solubility in sodium chloride at different temperatures.

fiberboard, low density (insulation board) - See 6F6.

fiberboard, medium density - See 6F6.

fiberglass insulation - fiberglass wool (5A6d) is sold as matt for insulation purposes. It is often adhesively bonded to asphalt-coated kraft paper (9C5) which serves as a vapor barrier.

fibers, glass - are made by a number of different methods, depending on the use intended for the fibers. See 5A6, 5A6a, 5A6b, 5A6c and 5A6d. For optical fibers for general light transmission, see 5A6e, for data transmission, see 5A6f.

fibers, textile - See 10A.

fibers, optical - See 5A6, 5A6e and 5A6f.

fibers, synthetic - See 10A2 and 10B6.

fiber, vulcanized - See 9C6.

film, photographic - previously made from cellulose acetate, photographic film is now made from polyester which is stronger and thinner. In liquid form the polyester is poured onto a highly polished metal belt. The solvent carrier evaporates leaving a thin plastic film. The film is then coated with light sensitive chemical compounds: silver halides, organic photoconductors, diazo compounds, amorphous selenium and zinc oxide.[4] Several layers of light-sensitive materials may be used if the film is made for color photography.

film, plastic - (See section 4I5.)

filters - (See filtration, 11C7b.) are made from many materials. All have openings of some size to permit the passage of a desired gas or liquid, but which are not large enough to allow passage of solid materials of a specified size. Common filter materials are: woven fabrics, non-woven fabrics, felt, cotton batting, filter papers, sintered metals, ceramics, glass or graphite, animal membranes, plastic membranes, granular beds, usually of sand or diatomaceous earth (powdered chalk-like rock) and the solid particles that are in the liquid to be filtered.

Woven fabrics, commonly of canvas, synthetic yarns, or metal wire, are made with common weaving methods (10C). Metal screens are similarly made.

Non-woven fabrics and felt are made with methods described in 10E.

Some filters are made by sintering processes in which powdered material is pressed to shape and then heated to bond the particles together. Small spaces between particles provide a path for the filtrate. These filters include those of metal made with powder metal processes (2L1), fritted glassware (5A8a and 5A8c), and ceramic filters, made from ceramic powder (5B). With all these materials, careful control of particle size, the compression of the powder and the degree of sintering are necessary in order to provide the needed size of path openings for the filtrate. In glass filters, the sintered material may consist of glass fibers rather than glass powder. Ceramic filters are used in metal casting and for diesel engine exhaust.

Pores in ceramic filters can be produced by one of the important methods: foaming, the plastic sponge method, and extrusion.[13] In the foaming method, the ceramic material is mixed with an organic foaming agent. The mixture is pressed to shape and processed to cause the agent to release a gas. The gas bubbles create openings in the mixture. After drying and firing, the workpiece has a permanent porous structure and is suitable for use as a filter. Extrusion (5B7) is used to make filters with a honeycomb-like structure having small openings. After the extrudate is cut to size, dried and fired, the openings allow passage of the filtrate but not solid particles larger than the openings. In the plastic sponge method, the following manufacturing sequence is used: A suitable sponge of plastic material (polyurethane, a cellulosic polymer or PVC) is selected. The sponge is immersed in a slurry of fine ceramic particles and water. (The sponge is compressed both before and after immersion in the slurry, so that all pores are filled with the right amount of the ceramic slurry.) The filled sponge is then oven-heated and dried, and then is further heated to a temperature high enough burn off the plastic sponge material. This leaves an all-ceramic material with many pores and passages in the space formerly occupied by the sponge. The workpiece is then fired to fuse the ceramic particles.

Membrane filters are used in ultra-filtration, dialysis, and reverse osmosis. See 11C7c. Membrane materials include cellulose acetate, polyamides, polysulfone, polytetrafluorethylene, polyvinylidene fluoride, and polypropylene. Micropores are induced into the membrane by stretching, or by a thermal process[3].

Filter paper is made from longer fibers than are incorporated in standard papers. Sometimes, other fibers such as glass are used for reinforcement. (See Chapter 9 for details of paper manufacturing.)

finger-jointed lumber - See 6F7.

fireworks - The principal ingredient in fireworks is black gunpowder, which contains potassium nitrate (saltpeter), charcoal, and sulfur. The potassium nitrate supplies oxygen for the explosive reaction and the charcoal is consumed. The sulfur does not burn but facilitates the combustion. Metal salts may be added to provide the colors red, green, yellow, and blue. Powdered or flaked aluminum and magnesium add brilliance. Aerial display fireworks have a charge for firing the shell, another charge to burst it open, and a third charge to provide the sound. Shells are cylindrical or spherical in shape and are made from multiple layers of heavy kraft paper, bonded with wet paste, shaped when softened by the moisture, and allowed to dry. For aerial display fireworks, numbers of small containers, "breaks", are incorporated into the large shells; they provide secondary bursts of visual and sound effects in the display. String is used to bind and hold the breaks. Safety is essential in fireworks manufacture. Sparks, friction, and inadvertent mixing of materials are carefully avoided. The mixing of the chemicals and their assembly into shells or containers is largely a manual operation because of concerns about sparking and frictional heat from powered equipment. The operations are performed in small one-room buildings which are widely spaced on the factory grounds for safety reasons. Even lubricating oil can cause problems by reacting with the explosive materials.

fishing rods (poles) - consist of the rod itself, a mounting seat for the reel, a handle, and a series of guides for the fishing line. The rods are made with a composite construction similar to that used in composite golf club shafts. (See *golf clubs*.) The predominant manufacturing method consists of wrapping steel mandrels with epoxy-impregnated carbon and glass fiber fabrics, curing the resin, and then removing the mandrels. (See Fig. F1.) Other methods are pulforming (4G12) a variant of pultrusion (4G11) with carbon or glass fiber reinforcement or a combination of both. E-glass fiber is used more in fishing rods than in golf clubs because it provides more flexibility than other glasses. A minority of rods have boron reinforcement in an E-glass/boron composite. After the composite material has cured, it is sanded and painted with methods very similar to those used for composite golf clubs.

There are a number of different designs for reel seats. They are often molded from plastic filled with graphite fiber. Another method is to machine the seats from aluminum using a number of operations, primarily turning (3A1). The seat is hollow for part of its length to accept the rod, and is usually threaded at the upper end. A tubular part, a clamp ring, fits over the rod and has a mating screw thread. The clamp ring often doubles as a second handle. When it is loosened by turning on the screw thread, it releases the reel; tightening it holds the reel in place. This method permits attachment and removal of the reel without the need for tools. If the reel seat is aluminum, it usually is anodized after being polished. In some designs, there is no reel seat, per se, and the reels are simply clamped to the plain

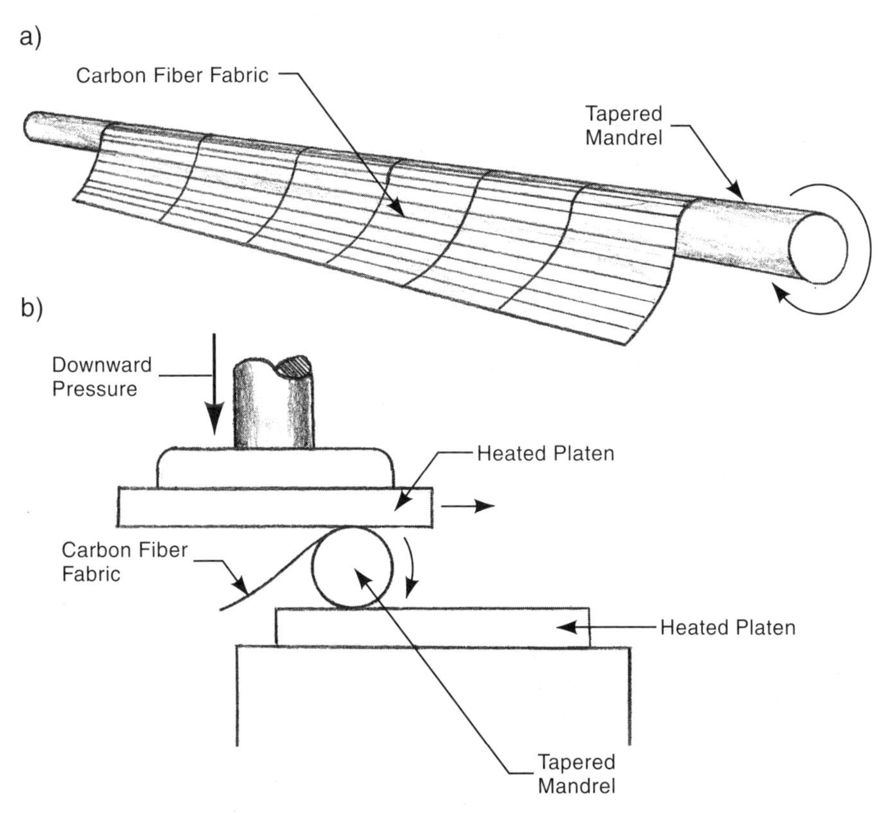

Fig. F1 The wrapping method of fabricating fishing poles and golf clubs from composite material. a) carbon or glass or combined carbon/glass fiber fabric, impregnated with epoxy, is wrapped around a tapered steel mandrel. b) The mandrel is further wrapped and is subjected to heat and pressure to polymerize the epoxy and create a solid workpiece. Removal of the mandrel and further operations result in a finished shaft.

rod or to the handle with a die-cast clamping part and screw fasteners.

Handles or "grips" are commonly molded from EVA plastic or hypalon synthetic rubber. (Hypalon is chlorosulfonated polyethylene, made by reacting polyethylene with sulfur and chlorine.) More traditional grips are made from wood or a series of cork rings, or are wound on the rod with cork tape or tape made from hypalon filled with granulated cork.

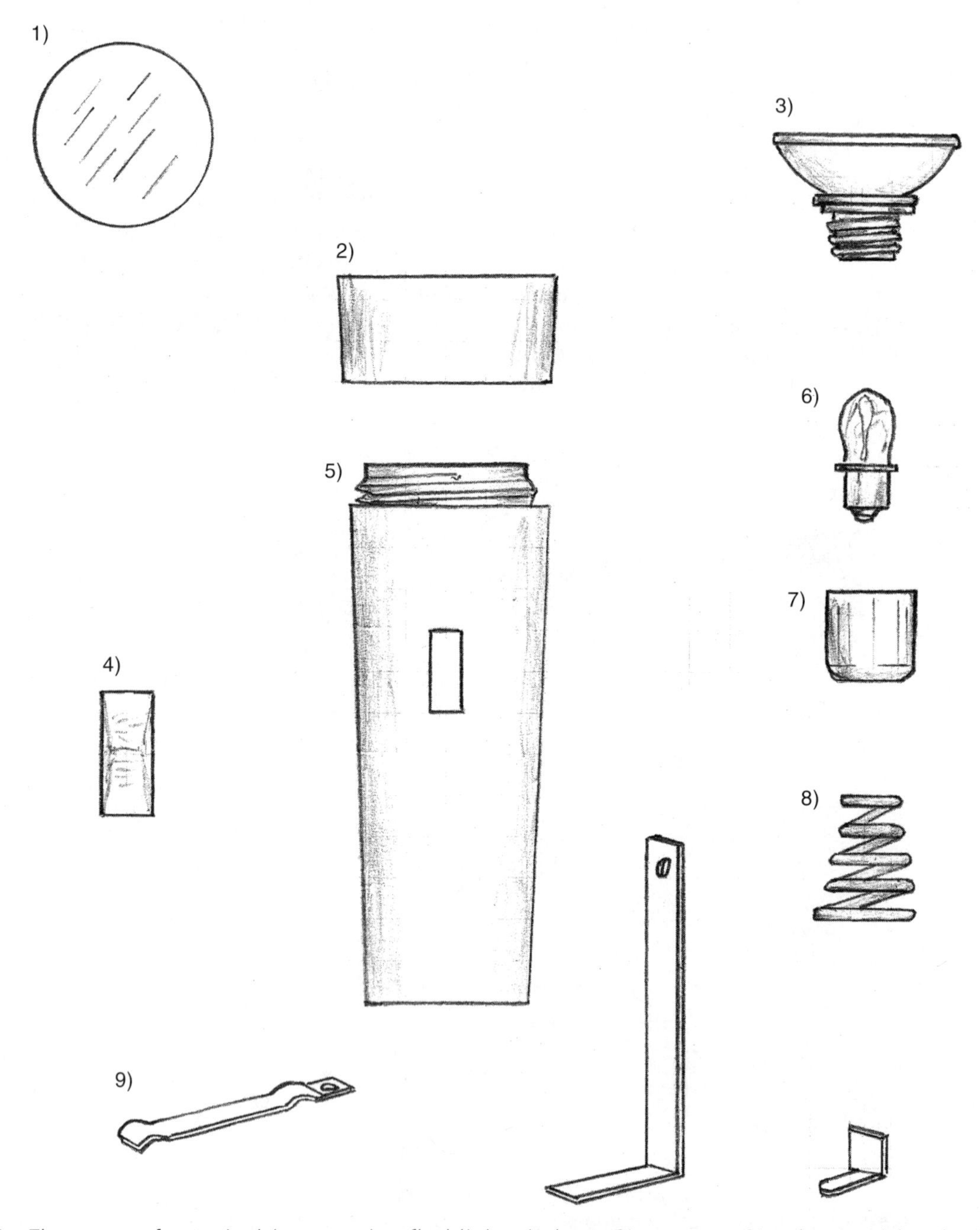

Fig. F2 The parts of a typical inexpensive flashlight: 1) lens, 2) top cap 3) subassembly of reflector, lamp connector and lamp base, 4) sliding switch, 5) body, 6) lamp, 7) lamp holder, 8) battery spring, 9) electrical connecting and contact strips.

The handle and reel seat are assembled at the large end of the blank rod. Both parts are held in place with epoxy adhesive. Some rods, especially the long ones, are made up of shorter pieces that fasten together, end-to-end, with the aid of metal ferrules - made on screw machines (3A2c) - that are bonded to the shorter pieces.

There is additional work on the rods to attach guides for the fishing line. The guides are fastened with epoxy and impregnated cord that is wrapped around the rod and over the wire parts of the guides. Wrapping to attach the guides may be completely manual or may be facilitated by rotating the rod in a lathe. After wrapping with the cord, additional epoxy is brushed over the cord. The epoxy is then heated to cause it to flow into a smooth surface. Guides for the line are made in one of a variety of materials and methods: from stainless steel wire formed by 4-slide operations (2G2); sometimes, from welded combinations of two formed parts; from carbon steel wire similarly formed and hard-chromium plated (8C1); or from molded eyelets of aluminum oxide or other ceramics (5B), held by the steel wire parts. A final clear coat on the rod covers and protects decorative effects and the model and brand identification.

flashlights - The simplest flashlights have the following parts: body, lens, reflector, lamp base, lamp (light bulb), top cap, lamp connector, lamp holder, sliding switch button, lamp contact spring, connecting strip from bottom to switch, connecting strip from switch to lamp, contact strip in the lamp holder, and a helical spring to maintain battery contact. Fig. F3 illustrates these parts.

Most flashlight bodies are injection molded (4C1) of a thermoplastic. Others are molded of thermoplastic elastomers (synthetic rubber compounds). Other parts made by injection molding are the top cap, reflector, switch button, and lamp holder. The lens is made from extruded transparent plastic material which is blanked (2C4) to a round shape. The electrical connectors and contacts are made from brass alloy strip stock, and are cut and formed as necessary on four-slide machines (2G2). Contacts and switch parts with spring properties are made from half-hard stock and may be of beryllium copper alloy. They may be zinc plated at contact points. The reflector is vacuum metallized (8F3) to provide reflectance. The light bulb is made with standard incandescent lamp methods. (See *light bulbs, incandescent*.) The battery spring is wound by conventional spring winding methods. (See *springs*.) The lamp holder and lamp connector are pressed together and then bonded to the reflector by ultrasonic bonding (4L4) or adhesive bonding (4L7) The two connector strips are connected at the switch button with an eyelet. Other parts are all assembled easily using molded screw threads to hold parts together. The sliding switch button is heat upset from inside the body to hold it in place while allowing it to slide.

flatware (tableware)(silverware) - The usual metals used for making knives, forks and spoons are stainless steel, sterling silver, or brass alloys, electroplated with silver. A commonly-used grade of stainless steel contains 8% nickel and 18% chromium. Sterling silver contains 92.5% silver and 7.5% copper alloy. Table knives in a sterling silver set usually have stainless steel blades and silver handles.

The operations for many pieces begin with a heavy sheet of the base material used. The sheets are blanked (2C4) into pieces that approximate the eventual outline of the item. These pieces are rolled (2B1) a number of times to reduce the thickness to near that of the final product. Thickness is usually different in different areas, e.g., the handle area is often heavier than in the working area, so the rolling is selective. Several passes of rolling may be needed and annealing is usually required between operations because of work hardening of the material. The annealing process used and the amount of annealing depends on the alloy involved, but it involves heating the workpiece to a high temperature and either cooling it gradually or quenching it back to near room temperature.

After rolling, the workpieces are blanked again, this time to the final outline of the knife, fork or spoon, except for forks, the tines of which are connected and held in place by a thin web. Coining (2D6) is carried out for spoons where the handle pattern and identifying data on the underside are formed. In making forks, the tines are first formed in another coining operation but a thin connecting piece is still left at the end of the tines until after the handle is coined. After the coining of the handle, the piece connecting the tines is trimmed off in another press operation. For knives, the handles are made from two halves of thinner stock that are blanked, formed (2D2) and embossed (2D7), or coined with the pattern. Then the two halves are brazed (7B) together to provide a hollow handle. The handle is

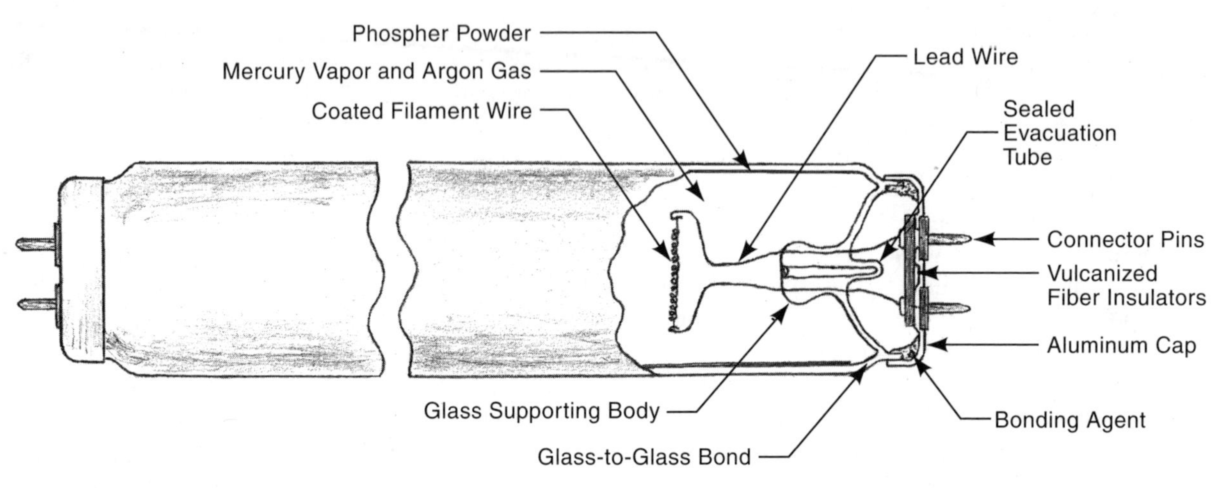

Fig. F3 A typical fluorescent light tube.

polished (8B1) and buffed (8B1a) to remove evidence of the brazed joint and to achieve the finish wanted. The knife blade, usually made from stainless steel, is also blanked, coined, trimmed, and sharpened, and is assembled to the handle and fastened with a strong adhesive. (If the spoons and forks have hollow handles, the operations performed are essentially the same as those for knives.)

Forming the bowl shape is the next operation for spoons and requires several cold forming press operations and a trimming operation for the flash that results. All pieces are then polished to remove burrs and buffed to the surface finish required. Pieces to be silver plated are then cleaned (8A) and electroplated (8C1). Final inspection is the last step before packing, for all pieces.

flexible printed circuit boards - See 13A4.

flocked fabrics - See 10F3l.

floor covering, vinyl - See *vinyl flooring.*

flour - The most common flour is finely ground wheat meal. See "milling grain" (12C5 and 12C5a).

flour, bleached - See entry 12C5a2.

flour, enriched - See entry 12C5a3.

fluorescent lights - convert ultraviolet radiation between two electrodes in a long tube to visible light. They include the following components: a glass tube with internal surfaces coated with phosphor powder, an electrode assembly with sealed end caps at each end of the tube and mercury vapor and argon gas that occupy the open space in the tube. Fig. F3 illustrates the construction of a typical fluorescent light. The electrode assemblies at the ends of the tube consist of a filament wire, two lead wires to support and connect the filament, a glass supporting body with the lead wires molded in, two brass connector pins made from fine brass tubing, three parts made of vulcanized fiber or phenolic plastic that hold and insulate the two connector pins, an anodized aluminum cap, and a sealant-bonding-supporting agent.

The glass tube is made with the methods described in section 5A2c. Identifying information is printed or etched on the tube. The connector pins are made from brass seamless tubing drawn to a small diameter. The electrode assemblies at each end of the tube are assembled to the tube in two stages, each with a separate subassembly. The filament subassembly consists of a molded glass supporting body incorporating the two copper lead wires and the filament wire. The filament wire is coated with an alkaline-earth oxide. It is held in place on the lead wires by crimping. The other subassembly for each electrode consists of an aluminum sheet metal end cap, the two brass connector pins and three pieces of vulcanized fiber that hold and insulate the pins. The aluminum cap is blanked, formed and punched to provide room for the vulcanized fiber parts and pins. The fiber parts are blanked and pierced from sheet material. The pins are assembled to the holders and fastened by upsetting the open ends, as with tubular rivets.

Each glass filament assembly is fastened to the fluorescent tube by heating and softening the glass supporting body and the tube ends, and pressing the two parts together. Before filament assemblies are attached at both ends, phosphor powder is introduced to the tube. The phosphor powder is usually zinc silicate or magnesium tungstate. Before both ends are sealed, air is evacuated from the tube and replaced with a mixture of mercury vapor and argon. The cap and pin subassemblies are put in place with the connecting wires inserted in the open ends of the connector pins. The tubular pins are crimped at the ends to round the ends and to hold the connecting wires in place. A bonding and supporting adhesive is incorporated in the assembly and the aluminum cups are swaged against the glass tubes to insure a tight fit.

This final assembly takes place on special rotary indexing table equipment in which the glass tubes are held in a vertical position as the table rotates. At each station, some element of the assembly takes place. At one station, the powdered phosphor is discharged into the top of each tube where it coats the inner surface as it falls to the bottom. At other stations, the gas is injected and the top and bottom end filaments are installed, the tube is sealed and caps are assembled and bonded to the tubes. At the last position, the tube is tested.

When the fluorescent tube is used in lighting applications, the other components involved are a socket for each end of the tube, wiring to an electrical plug or to connection points in the building, a light fixture incorporating reflectors and a holder for the fluorescent tube, sockets for the tube end connection plugs, starter, ballast transformer and an on-off switch.

flux-cored wire solder - See 13C1g1.

foam plastics - See 4C3.

food wrap, ("saran wrap") - See *saran.*

footballs - are made from cowhide as the external material with vinyl sheet, cotton fabric, polyurethane rubber, lacing, and stitching yarn. (See *leather* and *leather goods*) The tanned cowhide is split/shaved to less thickness; the outer surface is given a pebble texture, and some are decorated with the manufacturer's logo, team identification and other information. The leather is cut into elliptical pieces with steel-rule or similar dies. Four such pieces are sewn together, along with a vinyl and cotton lining. This work is done with the ball inside-out. The ball is then turned right-side-out and a polyurethane bladder is inserted into position through a small opening. (The bladder is made by heat sealing four elliptical sheets and a valve together. The ball is then pre-inflated to stretch the leather and straighten the seams. Adjustments in shape or seam straightness are made along with any needed repairs. The ball is partly deflated and the opening is laced by hand.

footwear - See *shoes.*

Formica®, Micarta® (rigid plastic laminates) - See 6F8 and Fig. 6F8.

fragrances (perfumes) - Fragrances for perfumes, cosmetics, soaps and other products are obtained from natural sources (flowers, fruits, plants and animals) by methods of extraction (11C3), distillation (usually with steam) (11C1e) and expression (11C4). However, synthetic and semisynthetic fragrances are increasingly being used. Synthetics are now the major ingredients for perfumes, and are produced by various chemical processes from natural materials and from coal tar hydrocarbons, alcohols and other organic chemicals. Some of the synthetics are chemically identical to ingredients from natural sources. Others mimic a natural fragrance or provide an entirely new odor. Most perfumes are a blend of a number of different fragrances. As many as 100 different ingredients may be used in the finest perfumes. Some of the ingredients in perfumes are not fragrances, but are fixatives that enhance or give longer life to other odors or are agents to aid in blending of other materials. Some fixatives and fragrances have unpleasant odors when extracted from their source but, when extremely diluted, provide pleasant effects. Often several ingredients, both natural and synthetic, when blended together, imitate a natural fragrance.

Some important natural ingredients from plants are citrus oils (from skins), lavender (from the flower lavandula vera), attar of rose (from the flower of the attar or damask rose), gardenia, oak moss , cinnamon (from bark), anise (from seeds), mint (from leaves and stems), thyme and sage (from leaves), citronella (from the leaves of lemon grass), orris (from roots of the Florentine iris), vetiver (from the roots of a tropical grass), and geranium oil (from the flowers or leaves of the plant). Animal-sourced fragrances, all fixatives, are musk (a secretion from the musk deer), ambergris (from the spermaceti whale), civet (found under the tail of the African civet cat), and castor (from beavers). Some notable synthetic fragrances are indole, C_8H_7N, made from phenylthydrazine and pyruvic acid, but first found in some flower oils, muscone (synthesized musk), acetophenone ($C_6H_5COOH_3$) from benzene and acetyl chloride, and hydroquinone dimethyl ether ($C_8H_{10}O_2$).

Perfumes are alcohol solutions of the fragrant oils and solids. They typically contain 10 to 25% of the concentrated ingredients. Cologne (eau de cologne or toilet water) usually contains 2 to 6% of the concentrate. Fragrances are used in lotions, face powder and other cosmetics, deodorants, hair dressings, shaving creams, toothpastes and medicines. Industrial uses include paints, artificial leathers (to give a leather odor), cleaning materials, and some product packaging.

freeze dried food - See 12H5.

frozen food - See 12I1, 12I1a, 12I1b, 12I1c and 12I2.

fuel, diesel - is a product of the fractional distillation of crude oil (11H1a). It is a heavier hydrocarbon than gasoline which is also a product of the same fractional distillation equipment. Distillates having boiling points from 350 to 650°F (177 to 343°C) provide diesel fuel[1]. Different fractions within that range are used for different types of diesel engines. The smaller, higher speed engines with frequent changes of speed and load as used in vehicles, utilize the lighter, more volatile distillate with the lower boiling point. Larger engines with more uniform speeds, e.g., electricity generation, use the heavier grades.

fuel, jet - is another product of the fractional distillation of crude petroleum. It is a lighter hydrocarbon than the diesel fuel described above but is heavier than gasoline. See 11H1a.

furniture, upholstered - See 6H.

furniture, wooden - See Chapter 6.

G

garlic - See *spices.*

garments (clothing) - As described in detail in chapter 10, sections H, H1, H2, H3, H4, H5 and H6, garment making has the following major sequence of operations: 1) Fabric for the garment is spread on a long table in a high stack. 2) The top layer of the stack is marked with chalk or with an attached paper layer, to outline the parts to be cut from the fabric, including garment panels of various sizes and other parts such as pockets. 3) The stack is cut along the marks to provide stacks of garment panels and parts. Bundles of mating parts are prepared for sewing. 4) Parts are sewn together. In mass production, each operator (or automatic machine) performs only one or a few operations and the bundle of sewn parts is conveyed or taken to the next operation where another panel or part is added by sewing. This sequence of operations continues until all sewing is completed. 5) The finished garment is pressed. 6) The garment is inspected, folded, labeled, and wrapped for shipment.

gaskets, packings and seals - are made from sheet materials, cordage and molded parts. A number of softer metals may be used including lead, tin, zinc, copper, aluminum, and low-carbon steel. Nonmetals used include rubber, both natural and synthetic, paper and other fiber sheets - especially vulcanized fibre - cork, asbestos, various plastics, carbon or graphite fibers, glass and aramid (e.g., Kevlar) fibers. Composite materials are common. Metal and nonmetal materials may be combined in a gasket or seal. Plastics may be combined with fibers to hold the gasket together and provide a filler and seal between fibers. Various manufacturing methods are used depending on the material involved and the shape of the seal or gasket. Fibers may be coated with other materials for corrosion protection or better sealing. Fiber materials are sometimes twisted and braided into a rope-like material. Thermoplastics and some rubbers are injection molded. Thermosetting plastics are usually processed by compression or transfer molding. Many composite materials are blended and calendared to form a sheet and are then blanked with steel rule or conventional blanking dies to the shape needed. Profile extrusions of rubber or plastics are used for some long, narrow seals.

gas, manufactured - There are a number of processes for making fuel gas from coal, some of them dating to the 17th and 18th centuries, when gas was used for illumination. One closely-related current process makes *producer gas* by partially burning coal in a closed furnace having an atmosphere of air and steam. The resulting gaseous product contains carbon monoxide, hydrogen and nitrogen. Although the heating value of this gas is lower than that of natural gas, it does have some industrial applications in heating and as an intermediate material when some chemicals are manufactured.

Coal gas is made from the destructive distillation (11C1g) of coal. The full process may also include some methanation and gas cleaning operations. The resulting gaseous product primarily contains hydrogen, methane and carbon monoxide. In some operations, steam is injected into the furnace to react with the hot coke, increasing the yield of combustible gas. Coke and coal tar are by-products of the operation. Most processes involve passing steam and air through a bed of hot coal or coke.

gas, liquified petroleum (LPG) - usually has propane or butane as its major constituent with some pentane, though the composition varies depending on the source and the intended use and use location, since propane is more suitable for northern climates and butane for the southern. LPG is a product of the refining of petroleum and is compressed to a liquid and stored in metal vessels for ease of storage, transportation and handling. A primary use is for cooking and heating in areas where pipelined natural gas is not available, but it is also used as a vehicle fuel and a source of heat for grain drying and tobacco curing.[4]

gasoline - is a product of petroleum refining (11H). Gasoline is one of the lighter products of fractional distillation (11H1a) of crude oil. Additional gasoline is extracted from crude oil by cracking (11H2a) heavier oil products of the fractional distillation.

gears - are many different kinds of gears with different requirements that necessitate different manufacturing methods. Fig. G1 illustrates various kinds of gears. Fig. 14C shows a manufacturing cell layout for gear machining.

gear milling - is one common method for machining gears. A milling cutter, with the teeth ground to the shape of the desired gear teeth and their spacing, is fed across the gear blank. Gear teeth are machined one at a time. The gear blank is stationary during the cutting but is indexed between cuts. See 3D and 3D5. All kinds of gears except spiral bevel gears are feasible with this process, but internal gears are only sometimes feasible. Fig. G2 illustrates milling of a simple spur gear.

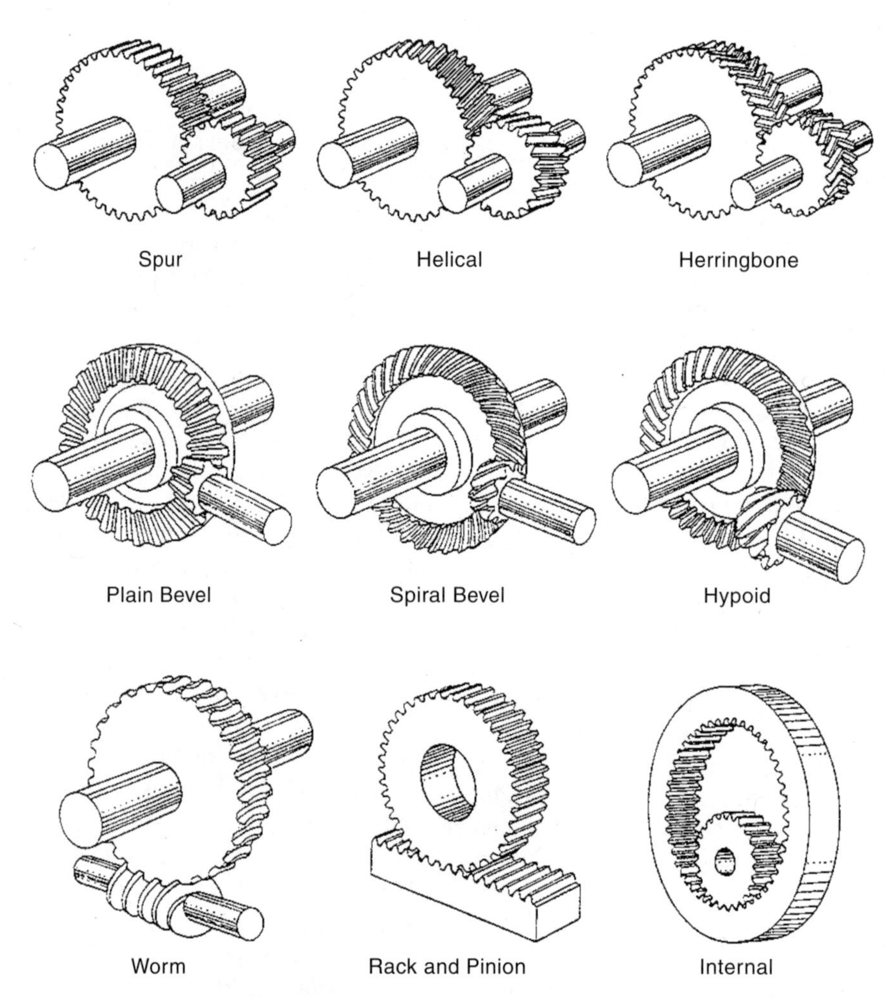

Fig. G1 Various kinds of gears. (*From Maintenance Engineering Handbook, 3rd ed., L. Higgins and L. Morrow, McGraw-Hill, New York, 1985, used with permission.*)

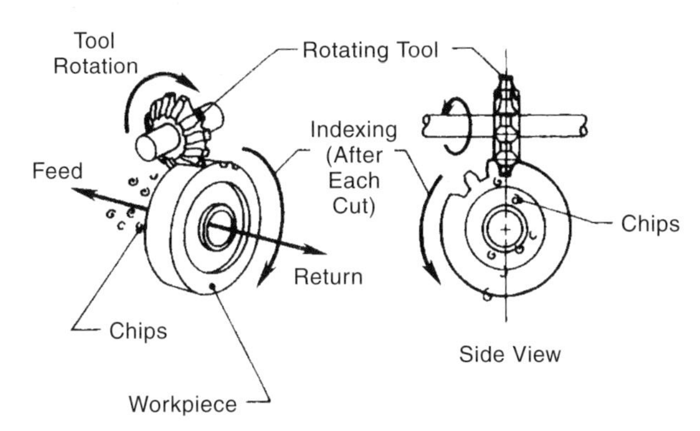

Fig. G2 Gear milling a simple spur gear. The form milling cutter machines one gear tooth at a time as it feeds across the gear blank. (*From Todd, Allen and Alting, Manufacturing Processes Reference Guide, Industrial Press, 1994.*)

gear hobbing - is a specialized gear machining process. The hob is a rotating cutter which resembles a worm gear with material removed to provide cutting edges. It is mounted on a spindle that is geared to another spindle which holds the gear blank at approximately a right angle to the hob. (The right angle is modified to the extent of the helix angle of the hob.) As the two spindles rotate, the cutting teeth advance into the gear blank, cutting the gear teeth. The shape of the gear tooth is gradually generated as the cutting of each tooth proceeds. The gear blank does not have to be indexed between teeth because of the worm-gear shape of the hob. The process is rapid and produces high quality gears. It is applicable to spur, helical, and worm gears. Bevel and internal gears are not feasible with this method. Fig. G3 illustrates the process schematically.

gear shaping - A special shaping machine, similar in principle to those described in sections 3P and 3R, has the cutting tool and the gear blank geared together. As the cutter, which resembles a gear, moves back and forth, both the cutter and the blank rotate. Material is removed on each forward stroke of the cutter and as the operation progresses, the desired tooth shape is generated in each gear tooth machined in the workpiece. The process can be used to machine both external and internal spur and helical gears, although, with helical gears, an additional guide to produce the helical motion of the cutter is required on the machine. The process is not suitable for bevel gear machining. Fig. G4 illustrates the process.

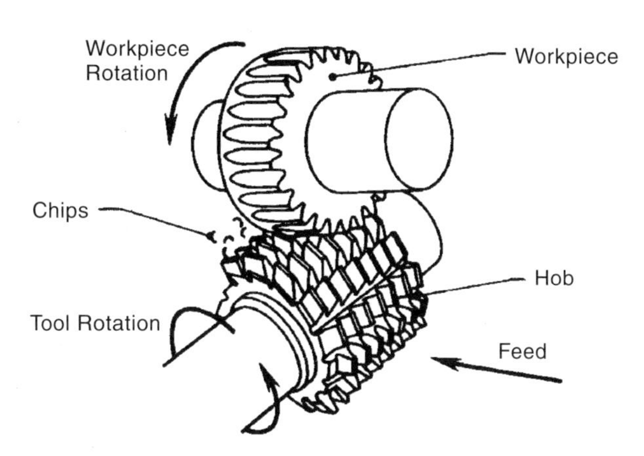

Fig. G3 Machining a spur gear by gear hobbing. The rotation of the hob (cutting tool) and the gear blank (workpiece) is synchronized and continuous. (*From Todd, Allen and Alting, Manufacturing Processes Reference Guide, Industrial Press, 1994.*)

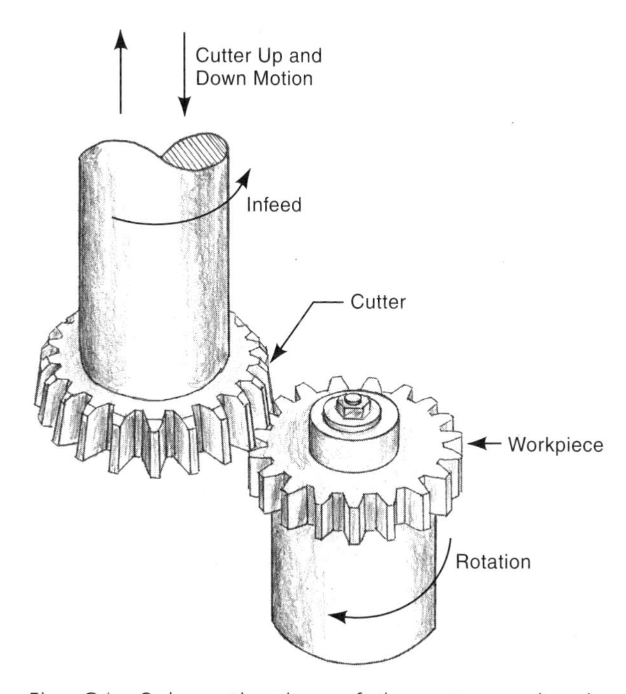

Fig. G4 Schematic view of the cutter action in gear shaping. The cutter moves up and down as both the cutter and workpiece rotate, feeding the cutter into the gear blank. The illustration shows a spur gear but the method is applicable for helical gear shaping when the cutter moves at an angle to the axis of the gear.

gear broaching - (See section 3F.) Gears can be completely machined in one pass with this method. It is more commonly used for internal gears than external gears because tooling for the latter is quite bulky and expensive. The broaching tool contains formed cutting teeth that produce complete gear tooth forms in the workpiece. Helical as well as spur gears can be machined with this method. With helical gears, there is rotational motion between the broach and the workpiece during the cut to provide the helical shape.

shear cutting of gears - This process is quite similar to both shaping and broaching. All the spaces between teeth of the gear are cut at one time with individual cutters mounted in a circular tool holder. There are multiple passes of the cutting tools. The tool motion is reciprocating. Each time the tool crosses the workpiece in the cutting direction, each of the cutting tools is advanced into the work. The toolholder incorporates a double cone system to advance the cutters. After the final pass, all gear teeth are fully formed. Tooling is expensive so the process is applicable chiefly to high-production situations but is fairly rapid. Both internal and external spur gears can be produced with this method, but the system is used more for external gears. Helical gears are not feasible.

straight bevel gear planing - is a method that is applicable to straight bevel gears but not helical bevel gears. It is essentially an application of shaping as described in section 3Q, and has similarity to gear shaping described above. However, except for indexing between gear teeth, the gear blank is held stationary during the process. The cutting tool, ground on the sides to the profile of a single gear tooth space, moves forward for a cutting stroke, then retracts and repeats the reciprocating motion. Because of the bevel gear shape, the path of the cutter is radial toward the center of rotation of the bevel gear, producing the taper that the bevel gear requires. The process is most applicable to larger, coarse pitch bevel gears.

two-tool planer method - is similar to straight bevel gear planing except that there are two cutting tools, one for each side of the gear tooth. The tools are given a rolling motion during the cutting stroke, to generate the proper tooth profile. Unlike straight bevel gear planing, the method is applicable to both fine and coarse-pitch bevel gears. With fine-pitch gears, both roughing and finishing cutters are mounted on the tool holder. With coarse-pitch gears, the common practice is to use two operations, rough machining and finish machining of the gear.

dual rotating-cutter method - This method is a faster production method than the two-tool planer method for straight bevel gears. Two rotating form milling cutters, which have interlocking cutters, machine both sides of the gear teeth simultaneously. They follow an angled path to provide the proper tooth taper and have a rolling motion to generate the tooth profile. The cutters roll back after each tooth is cut, and the gear blank indexes for cutting the next tooth.

revacycle process - is a broaching process that appears, at first glance, to be a milling process. It is rapid for the production of straight bevel gears. The cutter is a large circular broach, similar in appearance to a milling cutter but each successive tooth is slightly larger than the one that precedes it. The cutter revolves only once as it passes across the gear blank, starting at the inside edge. The successively larger teeth put a taper in the teeth and the space between them. Each pass of the cutter machines one gear tooth space. After each cut, the gear blank indexes to present the next gear tooth position to the cutter.

spiral gear planing generator cutting - is illustrated in Fig. G5. The process is a variation of straight bevel-gear planing. The single tooth cutter, which has a reciprocating motion, tracers a helical

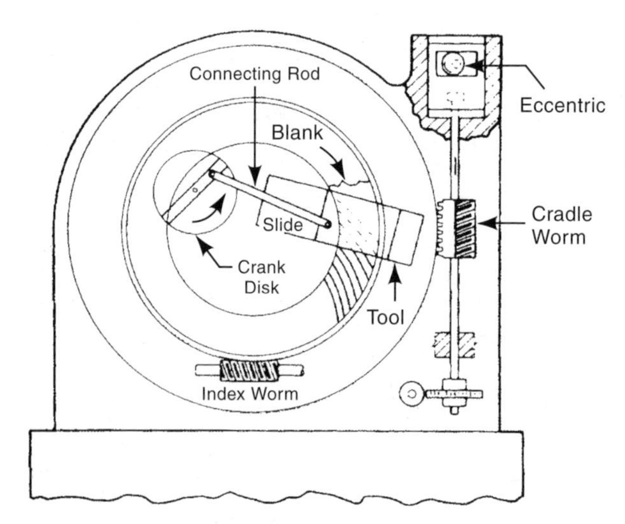

Fig. G5 Spiral gear-planing generator cutting (*from D.W. Dudley, Gear Handbook, McGraw-Hill, New York.*)

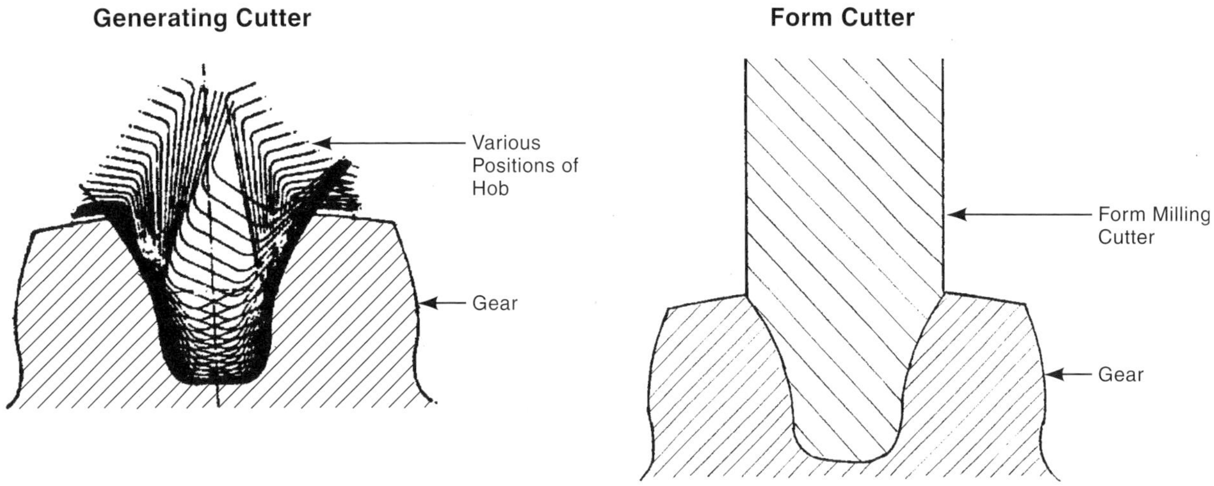

Generating Cutter **Form Cutter**

Fig. G6 The action of a generating-gear-tooth cutter as contrasted with a form cutter. *(The drawing of the generating cutter is based in part on an illustration from D. W. Dudley. Gear Handbook, McGraw-Hill, New York, 1962.)*

path on the gear blank during the forward stroke. The path is helical because the gear blank is rotated at a controlled rate during the cutting stroke. The gear blank's rotation is continuous so that the cutter removes material from successive gear tooth spaces with each stroke. The gear tooth shape is generated with each successive pass in each tooth space. One side of each gear tooth is machined, and the equipment is then reset for cutting the other side of the next tooth. The process is used for spiral bevel, hypoid and zerol gears.

face mill cutting - utilizes face-milling cutters somewhat similar to those pictured in Figs. 3D1 and 3D (view j) except that, instead of having cutters to mill a flat surface, the cutter teeth are form cutters in the shape or near-shape of the space between the gear teeth. They cut circular paths across the face of a gear blank to create bevel, hypoid, or zerol bevel gears. Two different arrangements are used. In one arrangement, the shape of the cutting tooth is exactly the shape of the space between the gear teeth. In the other arrangement there is a rolling motion between the face mill and the workpiece during cutting so that successive cutters generate the gear tooth shape. Two passes may be needed on each tooth, one for each side of each tooth.

worm gear methods - Worm gears are made by one of three methods: 1) single point turning on a lathe, similar to thread turning. 2), hobbing (See above) and 3), thread milling (3E5), all of which can produce the helical shape. With all three methods, the shape of the machined teeth are the form of gear teeth with helicoidal sides, whereas turning and milling of regular screw threads produces straight angled sides.

gear shaving - is a gear finishing method, used to remove fine surface irregularities that can occur with gear machining processes. The process utilizes an accurate gear-shaped cutter that can mesh with the gear to be machined. The cutter gear has slots or gashes across the width of each gear tooth to provide a number of cutting edges. The gear and cutter are run together, but the cutter is slightly helical and the axes of the two are set at an angle of about 15 degrees. This arrangement produces a sliding action where the cutter and the gear teeth mesh, so that the sharp edges of the cutter remove minute amounts of metal, providing superior accuracy and finish on the gear teeth. The process can be used for both straight spur and helical gears and is illustrated in Fig. G7.

gear grinding - is normally used to refine the dimensions, tooth profile, and surface finish of gears that have been heat treated and may have undergone some heat-treatment distortion. However, grinding is also sometimes used to machine fine pitch gears from solid stock. See 3C3a. Various methods are used, but they all resemble milling, with a grinding wheel replacing the milling cutter. In some set ups, the grinding wheel is dressed to the desired

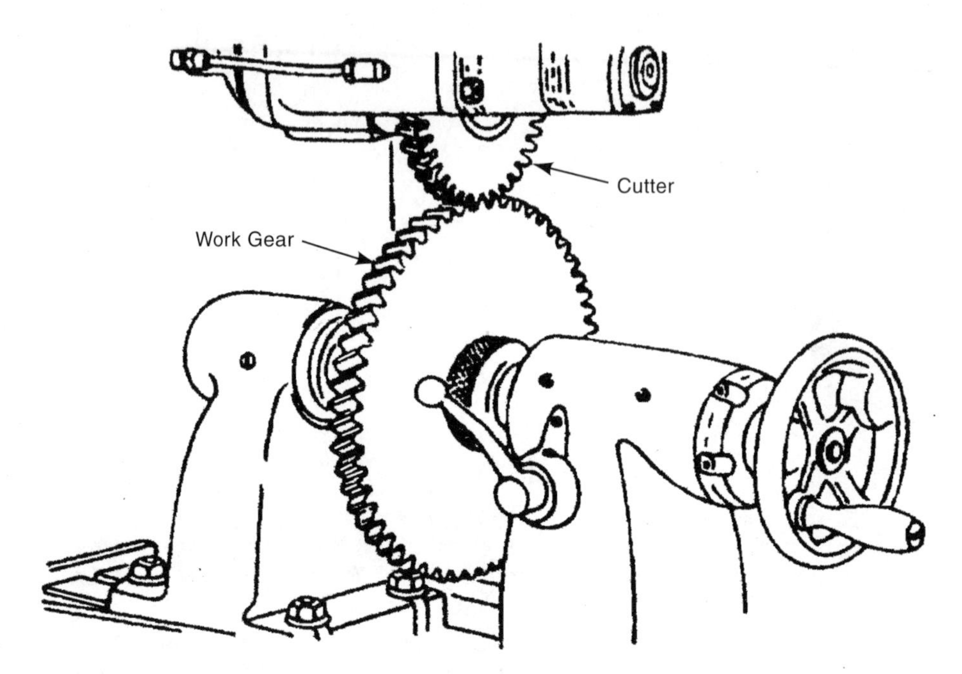

Fig. G7 In the gear shaving operation, a gear-shaped cutter and the gear workpiece are run together. The cutter is shaped so that there is some sliding action as the two rotate. The action shaves a very small amount of metal from the gear surface, producing a higher quality involute gear tooth form. (*Reprinted with permission of the Society of Manufacturing Engineers, Tool and Manufacturing Engineering Handbook, 4th ed., Vol. 1 Machining, copyright 1983.*)

tooth profile; in others, the wheel generates the tooth profile through multiple passes. Grinding is suitable for almost all gear sizes and shapes, the exceptions being very large gears, where suitably-sized grinding equipment is not available, and small internal gears that are too small to permit the insertion of a grinding spindle.

gear honing - is another gear finish-machining method. It is similar to gear shaving in that there is a tool in the shape of a gear that engages the workpiece gear. However, the tool is made of plastic and is impregnated with an abrasive. The tool teeth are helical, so that when they engage and drive the workpiece gear, there is some transverse sliding action against each tooth. Driving takes place for a period of time in both directions. During the sliding action, the abrasive particles work against any minor irregularities in the gear surface, honing the surfaces to be more smooth and more accurate. Gear honing is used after heat treatment and can take the place of grinding if the heat treatment distortion is minimal.

gear lapping - is another abrasive process for finishing gears. Like honing and shaving, it uses a gear-shaped tool. Abrasive compound is placed between the workpiece gear and the tool and, when the two are run together, minor imperfections at the surface are reduced by the abrasive. Sometimes, a mating gear is used instead of a specially-made, gear-shaped tool. If the mating gear or gear-shaped tool are not helical, some sideways reciprocating motion may be introduced to provide better abrasive action across the width of each gear tooth.

gear burnishing - also uses a mating gear-shaped tool to run with the workpiece gear. The gear-shaped tool is hardened, ground, and polished, to accurate dimensions and tooth form, and a smooth surface finish. The pressure applied by the burnishing tool forces down any raised surface imperfections and improves the surface finish. Sometimes two or three burnishing gears are run simultaneously on opposing sides of the workpiece gear.

gear casting - sand mold casting (1B), plaster mold casting (1C5), permanent mold casting (1D1), investment casting (1G), and die casting (1F) methods are all used for the production of gears, given a pattern or mold of the proper shape.

All these processes have some limitations for gear making, due primarily to insufficient accuracy, strength, or hardness of castable materials, and limitations of castable shapes. Sand casting is now more widely used for making gear blanks for later machining than for as-cast gears because of the roughness and inaccuracy of the as-cast tooth surfaces. Plaster and permanent-mold casting can be used for lightly loaded, slower moving spur gears, and for other applications where tooth accuracy is not critical. Investment casting can also be used for helical and bevel gears but the process is limited to smaller-size gears. Die cast gears are widely used in commercial products where strength and tolerances less stringent than those achievable with machined steel gears, are sufficient. Die cast gears normally require trimming operations to remove mold flash.

gear forming methods - Almost all metal forming processes can be used to form gears of one type or another.

extrusion - (2A3) is useful for pinions and other small straight spur gears, especially those of coarse pitch. Normally a secondary drawing operation is required after extrusion to achieve the desired accuracy. The extrusion is cut to the desired length and center holes are drilled and bored. Lathes or screw machines, equipped with the necessary collets, are used.

cold drawing - (2B2) may follow extrusion to refine the surface and accuracy of the tooth form or may be used to form gear teeth in round stock. As with extrusion, the round bar with formed gear teeth is cut off to the width of each gear and center holes are drilled and bored.

forging - (2A4) is a common process for making gear blanks but, under some conditions, may be used to make gears without subsequent machining. To do so involves machining the blank before forging, and processing the blank through both rough and finish stages of forging. This method is chiefly applicable to straight bevel gears and face gears. Accuracy is less than that found in machined gears.

stamping (blanking) - Gears can be made by the stamping process (2C4). Material is blanked to the silhouette of the gear. The limitation is that sheet materials must be used, limiting the face width of the gears. However, when the blanked sheets are stacked and fastened together by riveting, press-fitting or welding, a sufficiently wide gear can be made. Fine pitch gears cannot be blanked from thick materials. Another problem with stamped gears is the "drawdown" and "breakaway" that occur with normal stamping processes, resulting in a portion of the edge being rough and not of the optimum dimensions. A secondary shaving operation can alleviate this or fine blanking (See below.) can be used. The use of stamping is limited to production of straight spur gears.

fine blanking - (2C9) The normal fine blanking process can produce good quality gears from sheet metal, usually by stacking and fastening several fine-blanked pieces together. The process avoids the drawdown and breakaway disadvantages of regular blanking and can produce gears of somewhat finer pitch than those producible by conventional blanking.

powder metallurgy - Normal powder metallurgy methods (2L1) can be used to produce gears of high accuracy. Spur, helical, bevel, and face gears can be made. Gear size is limited by the press force required and the size of presses available.

plastic gear molding - Injection molding (4C) is frequently used to make gears of injection-moldable plastics. Glass or other reinforcements may be incorporated in the plastic before molding to provide increased strength. The method is identical to that used to mold other parts, requiring only a mold with the necessary dimensional precision. The method is rapid and inexpensive, once a mold is available. Similarly, compression molding (4B1) and transfer molding (4B2), though less common, can be utilized with standard techniques.

gelatin[4] - is an animal product. There are two types, type A, made from skins, and type B, made from bones. Type A is produced by causing skins (after washing) to swell by soaking them in a solution of hydrochloric, hydrophoric or sulfuric acid for 10 to 30 hours and then extracting the gelatin in a series of hot water soakings. Four or five stages of soaking are used at progressively-hotter temperatures. Each stage extracts additional gelatin from the skins. The extraction liquid is filtered, degreased, deionized, and concentrated with the aid of a vacuum. The rich liquid is then chilled and dried on screens. The solid gelatin thus produced is ground and blended with material from other batches. Most type A gelatin is used for food. Other uses are pharmaceutical capsules, the production of photographic film, and as an emulsifier. Type B gelatin is made with an alkali process over a period of several months or by a method where the bone is dissolved in hydrochloric acid. The resulting material is washed to remove the acid and the residue is ossein which is about 65% gelatin[2].

gemstones - See *jewelry.*

gin - See *distilled spirits.*

ginger - See *spices.*

girders, steel - are normally hot rolled (2A1) to shape by passing a heated billet, bloom or ingot between pairs of heated rollers. Repeated passes are made until the desired cross-section (usually an I-beam) is achieved. Pickling (immersion in warm, dilute sulfuric acid) to remove scale, and oiling usually follow the hot rolling.

glass - See 5A.

glass bottles - are made by glass blowing (5A2b).

glass, bulletproof (bullet resistant) - is a special form of safety glass (See *safety glass.*) made by laminating layers of glass with layers of clear, flexible polyvinyl butyl plastic. Alternating layers are used with at least four panes and, the more layers, the better the bullet resistance. Total thickness may be 3 inches or more[27]. Thinner panes, with a thickness of one inch or more, and with at least four laminated panes, are not bulletproof but provide resistance to handgun fire. Another approach is to use polycarbonate as the internal plastic layer or layers.

glass, cellular (foam glass) - See 5A7c.

glass ceramics - See5A7a.

glass containers - See 5A2b1 and 5A2b3.

glass, cut - See 5A5b.

glasses, drinking - See *drinking glasses.*

glasses, eye - See *eyeglasses.*

glass, etched - See 5A5f.

glass, pyrex - See *cooking utensils.*

glass fibers - are made by several different methods, depending on the use intended for the fibers. See 5A6, 5A6a, 5A6b, 5A6c and 5A6d. For optical fibers for general light transmission, see 5A6e, for data transmission, see 5A6f.

glass filters - See 5A8c.

glass, foam (cellular glass) - See 5A7c.

glass jars - are made, in production quantities, by machine blowing (5A2b3).

glass lenses - See *lenses.*

glass microspheres - See 5A7d.

glass, photosensitive - See 5A7b.

glass pitchers - See 5A2b1.

glass, plate - See 5A3e.

glass, safety - See *safety glass.*

glass thermometers - see *thermometers, glass.*

glass tubing - See 5A2c.

glassware, laboratory (scientific) - See 5A2b1, 5A2b2.

glass vases - See 5A2b1.

glass, window - See flat glass processes, 5A3.

glass wool - See *glass fibers.*

glazing compound - See *putty.*

glove compartments, automotive - These are customarily made of plastics and can be extrusion blow molded (4F1), injection molded (4C) or made by one of the deep draw thermoforming methods (4D).

gloves - of a fabric or leather are sewn using the same methods as are employed with other garments and sewn products and described in Chapter 10, section H. (Fabric making for various products, including gloves is described in Chapter 10, sections A through G.) Protective gloves such as medical gloves, are made by dip molding latex or vinyl plastisol (4K2 and Fig. 4K2) or are made by heat sealing (10H4a) two plastic sheets together with a suitable die and then cutting the gloves free from the sheets with a suitable blanking die. Sometimes, fabric work gloves, after sewing, are given a protective vinyl coating as described in 4K2a.

glue - See *adhesives.*

glued-laminate lumber ("glulam") - See 6F7.

gold - Most gold is recovered by the cyanide process. The gold is found in the natural state (not as an oxide or as part of another chemical compound). However, the amount of gold per ton of gold-bearing ore is normally very small, and chemical means are used to recover it. The crushed ore is treated with an alkaline cyanide solution in the presence of air. The gold is converted to liquid sodium cyanoaurite. Sodium hydroxide is also formed from reaction of the sodium with oxygen in the atmosphere. The operation may be performed by spraying a dilute solution of the sodium cyanide on a heap of ore or, when the gold content is higher, the operation is performed in large agitated vats. The gold-bearing liquid is separated from ore solids by filtration. Gold is recovered from the solution by electrowinning (See 11K1c1.) or by reduction of the solution with zinc. Another recovery method is to process the ore with mercury, which captures the gold. With alluvial ores, simple panning may be used to separate particles of gold from accompanying sand and gravel.

golf balls - may have any of several different constructions. Most golf balls are of two parts; others have three components; a minority have four or more. The two-part balls provide longer shots, but less spin and control; the three-part balls travel less distance but can be given more spin and better control. Two-part balls consist of an inner spherical core, molded of a polybutadiene compound or of another elastomeric compound of high durometer. The core is tumbled (8B2) to remove mold flash. A cover of ionomer plastic is injection molded (4C1) over it. Retractable pins in the mold hold the core in a central position during molding. They retract after sufficient material is in the mold to hold the core in a central position but while there is still enough material flow to fill the holes left by the pins. The mold cavity surface is patterned to provide the dimpled surface for the balls.

Three-part balls use a smaller core, wrapped with layers of rubber thread that is stretched to 10 times its normal length as it is wrapped. The cover is then molded over the rubber wrap but, in this case, it is molded in two halves which are then heated, placed over the core, and pressed together to fuse. This approach is used to prevent distortion in the rubber band winding of the core.

Four-part balls may have a layer of other material between the rubber wrap and the cover. Some four-part balls have more than one such layer. Some balls - usually those of three or more parts - have covers of polyurethane or balata instead of ionomer. Balata is a thermosetting material of little or no elasticity, made from the sap of a South American tree. Covers of balata are compression molded (4B1a) in two halves. Polyurethane covers may use the thermosetting variety of polyurethane, cast (4H) over a layer of ionomer.

All balls are tumble polished to remove mold flash. Each is then placed on two holding pins that spin it while two or more coats of enamel are sprayed on (8D6). The paint is dried and the ball is hot stamped (4M2) with brand identification and variety information. Quality control and packaging follow. All operations are highly automatic.

golf clubs - are made by a variety of methods from a number of different materials, depending on the use of the clubs and the design developed by the manufacturer. Clubs called "woods", traditionally made with a wooden head, are now hollow and made of metal. "Irons" have solid metal heads. Stainless steel and titanium are the metals normally used for both types but irons are also made from carbon steel, beryllium copper, or beryllium nickel. Putter heads may also be made from bronze or aluminum. Tungsten is often used to provide additional weight in a desired location. Shafts may be of steel, stainless steel, aluminum, titanium, or now most often of a composite of epoxy resin and a reinforcing fiber, usually of carbon but sometimes of boron. 17-4 and 431 are two commonly-used stainless steels and Ti- 6Al - 4V is the most common titanium alloy but other alloys of both metals are also utilized.

Metal shafts are made from tubing which is tapered in 7 or 8 steps along the length by drawing (2B2) in successively smaller dies for short sections or by rotary swaging (2F1). Metal shafts are hardened (8G3c), tempered (8G2f), straightened (2K), and polished (8B1) after tapering and, if carbon steel, are chromium electroplated (8C1). Composite shafts are made by filament winding (4G7) with pre-impregnated fibers ("prepreg"), or by a roll-wrapping process that utilizes unidirectional sheets of pre-impregnated carbon fabric. Sometimes pultrusion (4G11), or pulforming (4G12), a variation of pultrusion, is used or, in other designs, a combination of wrapping and filament winding. Filament winding is carried out with fibers at a low angle to the axis of the shafts to provide the optimum reinforcement. When the wrapping method is used, the prepreg is wrapped around a mandrel and covered with a shrink wrap. After curing, the shrink wrap is removed. Fig. F1 illustrates the wrapping method. After the composite material is cured, the shaft is sanded on a centerless grinding 3c1b machine with wide-belt abrasives. The shafts are coated with a highly-filled urethane enamel using a dip (8D4) and circular squeegee method. Further sanding follows with a finer abrasive and, when the shaft is fully sanded, and of the desired smooth surface, it is dip painted with enamel and then a clear coat. Labeling by pad printing (8I7a), screen printing (8I7b), or transfer labeling takes place before the clear coat is applied.

Metal heads for irons are made with one of two processes, forging (2A4) or investment casting (1G). Metal heads for woods are normally investment cast but are sometimes made from forgings. Some woods, in order to reduce weight, have the top portion of the head injection molded (4C1) from graphite mixed with ABS plastic. Most metal head parts are surface hardened at high-wear areas after forming, typically by induction (8G3a2) or flame (8G3a1) hardening. Metal heads for woods are hollow and are often filled with foamed polyurethane plastic. Some woods and putters have separate striking surfaces for the ball. For putters, the striking surface may be molded from a semi-resilient grade of polyurethane. For woods and irons, inlaid striking surfaces may be titanium, stainless steel, zirconia ceramic, or a ceramic composite material with a titanium matrix.

Gripping handles for all clubs are usually injection molded of an elastomer or rubber, using the shaft as an insert in the mold (4C6), but some handles are leather that is wrapped on the shaft and adhesively bonded. Rubber or elastomer grips may use material filled with granulated cork for light weight. Leather grips are cowhide or calfskin. Appearance is important in selling the clubs, so finishing operations including polishing and buffing of metal parts, painting of the top portion of the wood heads and finish coating of the shaft with urethane or other varnish are additional steps carried out after the parts are made.

Shafts are assembled by insertion into a socket in the club head, and the two are commonly fastened with adhesive but, with metal shafts, the two may be drilled and pinned together.

graphite[4] - is obtained from both natural and manufactured sources. Manufactured graphite is made from petroleum or retort coke which is calcined to remove volatiles, screened, crushed and ground, and mixed with a coal tar binder. The mixture is cooled and run through an extruder or molding operation to make a workpiece of the desired shape. The workpiece is baked at about 1650°F (900°C) to convert the binder to carbon and put the carbon in amorphous form. Firing in an electric furnace at 4900°F (2700°C) converts the amorphous carbon to graphite. Workpieces are then machined as necessary to make finished graphite electrodes, chemical process equipment parts, rocket nozzles, nuclear energy components, electric motor brushes, and sealing rings. Natural graphite is used as a lubricant, for making pencil leads, and for making crucibles and refractories.

gravure printing plates - See 9D3b.

grease, lubricating - is produced in a variety of consistencies from mineral oil, which is made up of the heavier liquid fractions produced by the fractional distillation of petroleum (11H1a). The mineral oil is then thickened to a grease form by compounding it with soaps of calcium, lithium, aluminum, or sodium, and sometimes, non-soap thickeners. The soaps impart stiffness to the mixture. Other materials may be added in smaller quantities to improve temperature or water resistance, and to inhibit oxidation and corrosion. Grease is also made by rendering the inedible fat of hogs, cattle, or sheep. Greases are used as lubricants when the surfaces to be lubricated are not fully contained and the stiffness of the grease acts to keep it in place.

grinding wheels - are made by bonding abrasive grains together in the shape of a wheel. Various abrasives may be used. (See abrasives.) Bonding agents include ceramics, sodium silicate, shellac, rubber, plastic resins or oxychloride. The type of bonding agent used depends on the intended application of the wheel. Ceramic bonds are used for precision grinding applications. Softer, tougher, more resilient bonds are used for heavier cutting operations such as snagging of castings and for abrasive saw blades.

Ceramic-bonded wheels are made by mixing the abrasive and the ceramic material (clay or feldspar), pressing them together into the approximate wheel shape at high tonnage in steel molds and then firing the wheel at high temperature, e.g., 2300°F (1260°C). The wheel is then trued with steel cutters and ground to the exact shape and dimensions required. Wheels using resin, shellac or rubber bonds are baked at temperatures of 300 to 400°F (150 to 200°C). After size finishing, wheels are tested at high speed for balance and soundness. Honing and sharpening stones, and shaped tumbling abrasives, are made by the same basic methods.

guitars, acoustic - High quality guitars are still made with methods that rely heavily on manual operations. The first step is to produce the sides of the instruments, usually from strips of thin rosewood, cut to the proper width. The sides are softened in a steam chamber and bent to the curved shape in a bending press. Pieces for the top and bottom surfaces are cut to size and hour-glass shape by bandsaw, CNC laser cutter, router, or blanking die. (See 6B3 and 6B7b.) Tops are normally made from spruce or other softwood; backs are made from hardwood, especially rosewood, but often are blanked in two pieces that are glued together with a decorative center strip. Color and grain pattern are matched in the top, back, and side pieces. Straight grain patterns are preferred. Spruce braces are glued to the inside surfaces of both the top and back pieces. The pattern of the braces is important in the acoustic properties of the guitar. The top, back and sides are glued together in a fixture that maintains the proper alignment and applies pressure until the glue sets. The neck of the guitar has to be shaped from mahogany by manual or machine methods and is filed and sanded smooth. The neck is cut to allow attachment of an ebony finger board. The neck is also often bored through its length and a steel rod is inserted for reinforcement. Basswood strips join and reinforce the side pieces. The fingerboard, in turn, has pieces of pearl shell inlaid, and 20 nickel-silver frets glued into slots in the proper places. The neck is trimmed as necessary to fit the guitar body closely but they are not yet glued together. The body and the neck are then finished with many coats of lacquer, except for the fingerboard, which is oiled instead of lacquered. The neck and body are then glued together. When the glue has set, the lacquer finish is polished. The smaller parts, which support and contact the guitar strings, called the saddle, bridge, and nut, are attached. The "tuning machine", that is, the knobs and gears that tighten the strings are fitted to the end of the neck and the strings are then attached. The completed instrument is tested by an accomplished guitarist, not only for tone and pitch, but for smoothness of surfaces, level of frets, appearance, and all other characteristics. Any guitar that has defects is returned to the responsible production department for repair. Then, after a recheck, it is ready for packing and shipping.

gum, chewing - See *chewing gum.*

guns (firearms) - Gun manufacture involves the fabrication and assembly of a considerable number of precision parts, mostly made from steel. Historically, the complex-shaped metal parts have been hot forged to approximate shape and then finished with a series of machining operations involving milling, turning, broaching, drilling, boring, reaming, threading and various grinding operations. In current practice, many of the parts are made by methods that provide a finished part in its final shape and dimensions, with little or no post-forming machining. Investment casting (1G), powder metallurgy (2L1) and metal injection molding (2L3), are three

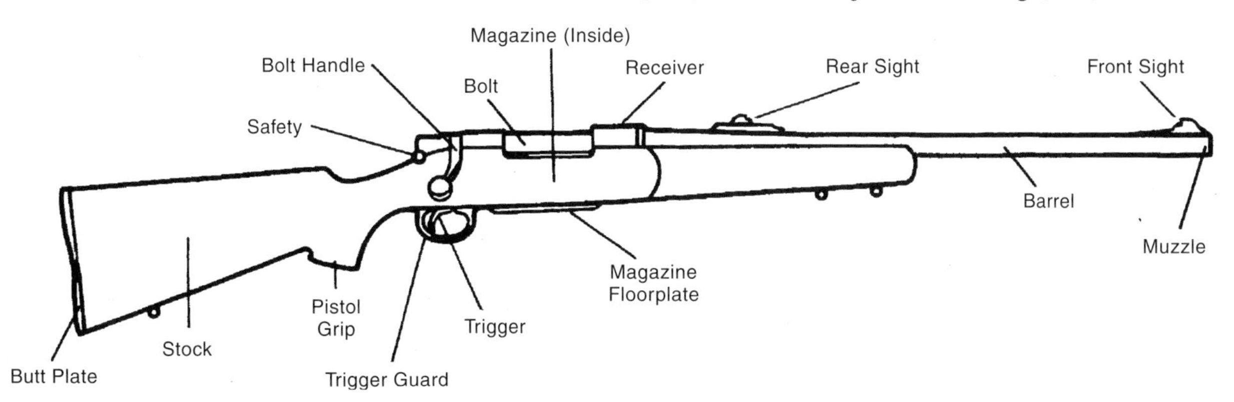

Fig. G8 The parts of a simple bolt-action rifle. *(Courtesy Remington Arms Company, Inc.)*

metal-forming processes used to provide near-net-shape gun components. Some parts are also made by stamping sheet metal. Except for those of stainless steel, all steel parts are normally given a black oxide surface treatment (8E4).

The major components for a typical basic bolt-action rifle are illustrated in Fig. G8. These parts include the barrel, receiver, trigger, and associated parts, trigger guard assembly, bolt assembly, firing pin assembly, front and rear sights, magazine, stock, and buttplate.

The barrel is made from alloy or stainless steel. Operations vary from manufacturer to manufacturer, and may include forging, normalizing, turning, gun drilling, reaming, and rifling (spiral grooves in the barrel to put a spin on bullets as they leave the gun). Gun drilling (3B6) is a critical operation because the hole must be straight and centered in the barrel over about 30 in (76 cm). Fig. G9 shows a common gun drilling machine for making the bore holes in two barrels simultaneously. Rifling of the bore of the barrel is produced by one of several methods:

Fig. G9 A gun drilling machine for rifle barrels, drilling two at a time. Two pieces of steel bar for the barrels (like the one shown leaning on the machine) are placed in the chucks of the machine (in the background in the picture). The drills (shown in the foreground), starting at one end, advance and drill holes the full length of the barrels. The barrels rotate on their axes but the drills are stationary except for advancing into the work. The drill shanks are hollow and cutting oil fed through the drills removes metal chips as the drills advance. (*Courtesy Remington Arms Company, Inc.*)

1) A hook-type single cutter at the end of a long rod makes a series of passes in the bore of the gun barrel. The cutter is fed on a spiral path, and several passes are needed to cut one rifle groove to the proper depth. This is a shaper-like operation and is repeated for each of the 6 grooves that are typically cut. 2) A spiral broach is used to cut all 6 grooves in one pass. 3) Forming rather than cutting the rifling by forcing a carbide "button" through the bore. The button has a smooth non-cutting surface and displaces the metal of the bore to make the spiral grooves and raised spirals between the grooves. 4) Cold forging: The barrel is placed over a mandrel that contains a negative impression of the rifling - and then cold forged against the mandrel with a series of hammer blows in a machine that is a type of rotary swager (2F1).

The receiver of the rifle is machined from a solid block, or finish-machined from an investment casting with a series of CNC machining operations (3T & 3U2) or operations performed by dedicated machines arranged as a manufacturing cell. Broaching may be one operation. The machined part is heat treated (8G3) to a hardness in the Rc 40 range. Other parts are either machined from solid stock or made to near final shape by one of the processes mentioned above. Parts that are investment cast by some manufacturers include bolt handles, receivers, sears, hammers, and triggers.

The Remington model 700 bolt-action rifle includes the following powder metal parts: the floor plate latch, rear safety cam, front and rear spacers of the trigger assembly, safety button, and rear sight aperture. Metal-injection-molded parts include the front sight, rear sight base, rear sight slide, and trigger. The trigger guard is die cast of aluminum. The magazine box and spring are made on four-slide machines. Wooden stocks are made from knot-free walnut and are first machined with CNC milling machines that rotate the stock material and shape it with a milling cutter. Smoothing and finishing of stocks are essentially hand operations. Stocks are varnished after being shaped and surface smoothed. Checkering, if used on the stock, is done by CNC machines. (Checkering is an operation performed on the wood stock of rifles, shotguns, and other shoulder arms, and on the grips of handguns, for decoration and to provide a more frictional surface for holding.) Originally in the industry, and currently for some deluxe guns, checkering is done manually by artisans who cut a series of spaced parallel V-grooves in the wood with hand tools similar to files. Some lower-priced guns have checkering done by impression rather than cutting. Other guns use injection-molded plastic stocks. The butt plate is compression molded (4B1) from phenolic resin. Springs and screw fasteners are normally purchased from subcontractors who specialize in these components.

Final assembly is normally a bench operation. Experienced workers put the entire gun together from finished subassemblies and parts. The firing mechanism is tested without bullets and then is test fired with an excessive charge to confirm the strength of the components. Shooting accuracy tests are made on samples of finished guns. After test firing, the gun is cleaned, polished, inspected, wrapped, boxed, and shipped to wholesalers.

gutters, roof - if metal, are normally made of aluminum and are made by roll forming (2F7). Painting normally precedes forming and is carried out by roller coating (8D2) or spray painting (8D6). End pieces, tees, and elbows, are made by conventional blanking and forming press operations (2C4 and 2D2). Plastic roof gutters are profile extruded (4I1) of mineral-filled vinyl material, normally white in color, from titanium dioxide pigments mixed into the vinyl. Plastic downspouts are also extruded. Plastic elbows, mounting brackets, and connectors are injection molded (4C1) from essentially the same vinyl material.

gypsum plaster - Gypsum is a natural material, calcium sulfate ($CaSO_4A\ 2H_2O$), found throughout the world. When used to make plaster, gypsum is normally heat treated to about 375°F (190°C). This treatment partially dehydrates it and changes the chemical formula to $2CaSO_4\ A\ H_2O$. The material is then called *calcined plaster* or *plaster of Paris.* When water is re-added, the mixture will set to a solid material but, before setting, can be cast or applied and smoothed to form wall and ceiling surfaces. It is used for plaster walls in building interiors, for casting into decorative objects, and for making plasterboard (wallboard, gypsum board or drywall).

gypsum board - See *plasterboard.*

H

ham - is the meat from the rear quarters of the pig. The meat is cured by one or more of several methods: pickling (12J7), salting (12J8), usually by injecting a brine solution that includes salt, other preservatives, and flavorings; sugar curing (12J9); and smoking. Sometimes the ham is soaked in the brine instead of being salted by injection. A third method is to rub on a solid mixture of granules or powders of salt and the other substances. Smoking for a prolonged period may follow.

hammers - Hammer heads are drop forged (2A4b & 2A4b1) from carbon steel, trimmed (2C6), heat treated for through hardness (8G3c), and polished (8B1). The specified hardness per ASME Safety Requirements Standard B-107.41-2004 is R_c 45 to 60 for the striking face and R_c 40 to 55 for the claws. Handles are made as described below.

handles, cooking utensils - There are many different varieties. One of the most common types is compression molded (4B) or injection molded (4C5) of phenolic or another thermosetting plastic. These handles usually have a metal reinforcement or base structure, made from sheet steel by blanking (2C4), forming (2D2), polishing (8B1) and chromium plating (8C1) operations, before being incorporated as inserts in the phenolic moldings. A more recent development is the use of a thermoplastic elastomer instead of the phenolic, for easier gripping. The elastomeric material has temperature resistance to 400°F (204°C). Handles are fastened to the utensil by riveting or by screw fasteners, but spot resistance welding (7C6a) may also be used. Cast iron frying pans and pots usually have the handle as part of the casting. Other utensils, especially those to be finish coated with vitreous enamel, use handles formed from sheet metal, tubing, or solid rods. They are brazed to the utensil. (Also see *cooking utensils.*)

handles, tool - Wooden handles for hammers, axes, rakes, hoes, and similar tools are machined from hickory or ash. Round handles are turned on lathes. Those with special shapes are machined on special equipment. On these machines, the workpiece is held in a spindle and slowly rotated. Milling cutters move in and out as the workpiece rotates to machine an oval or other cross section in the workpiece. As the operation progresses, the cutter also moves axially along the workpiece. Movement of the cutters may be controlled by cams, tracing templates, or a CNC (computer numerical control) system. Plastic handles for hammers are made from glass- or carbon-fiber-reinforced polyester, molded with the pulforming method (4G12). Plastic handles for screwdrivers, files, and other hand tools are injection molded, usually with the metal member of the tool as an insert in the mold (4C6).

hand tools - See *hammers, pliers, wrenches, screwdrivers* and *handles, tool.*

hardboard (including tempered hardboard) - See 6F6.

helium - is produced chiefly from natural gas. The gas is liquified at a low temperature and high pressure to remove other components. (See 11C1a - *fractional distillation.*) Helium is 70–85% pure at this point, with the balance composed primarily of nitrogen, but with some hydrogen, methane and neon. The crude helium gas is then compressed to about 3000 psi (21 MPa) and cooled to about –340°F (–207°C or 66K). At this temperature, it is brought into contact with cooled activated charcoal. Adsorption by charcoal of the other gases yields helium of 99.995% purity[3].

herbicides - There are many different types of herbicide, both in their action and chemical make up. The common 2,4-D type is made by chlorinating phenol to 2,4-dichlorophenol which is then distilled to purify it and convert it to the sodium salt. That material is reacted with sodium monochloroacetate, formed by chlorinating acetic acid, and then reacted with the 2,4-dichlorophenol to form 2,4-D. 2,4-D attacks broad-leaf plants but not grasses. Cresol, a derivative of coal tar or petroleum is a raw material for other herbicides. One herbicide used in aerial spraying of cocaine-producing plants is glysophate. Others are sodium arsenate, which is applied to leaves, and chlorinated benzene, which is added to water to control aquatic weeds.

high-density polyethylene - See *polyethylene, high-density.*

hosiery (stockings and socks) - Seamless hosiery, by far the most common type, is knitted in tubular form on circular warp knitting machines (10D and 10D1). (Also see 10A2 re nylon for hosiery.) Shaping of circular-knit hosiery is achieved by gradually increasing or reducing the size of the knit loops from the top of the hosiery to the toe portion. The part that fits the heel is shaped by a mechanism that drops courses in the knitting sequence. The fit of hosiery to the legs is facilitated by the stretch capability of knitted fabrics and by the use of nylon and spandex yarns that are stretchable. Sewing machine stitching (10H4) is used to close the hosiery at the toe end, and to provide a hem (welt) at the top.

Some hosiery is knitted on flat machines and achieves a good fit throughout its length by reducing the number of knitting loops for the portion that must be smaller to fit the leg or foot of the wearer. However such hosiery has a seam where the edges of the knitted piece are sewn together.

houses, prefabricated (modular houses) - are made with methods that, in many ways, are not strikingly different from those used when constructing a house on-site. The same wood and wood-product materials are used. The differences are that the prefabricated house is built inside a factory building that supplies light, heat, fixtures, cranes, workbenches, and power tools, that make the tasks faster and easier under more ideal working conditions. Framed floor, wall, ceiling, and roof panels can be completed on fixtured workbenches, and then moved by crane - or other means - to be assembled with other panels. Wiring and piping can be installed when conditions are optimum to do so. On the other hand, the need to transport the finished units on roadways limits the size, especially the width (maximum: 16 ft.) and height, that can be prefabricated. Houses within the transportation size limits can be virtually completed - as much as 95% complete - within the factory before shipment. For larger and more complex house designs, modules for sections of the house are prefabricated in the factory, and shipped with other modules to be mounted and fastened together at the house site. Using the modular system, very large and elaborate houses can be constructed but, in such cases, the modules may comprise only 40% of the total house-building operation; the rest of the work is done on site.

Another difference in prefabrication compared with site fabrication is that adhesive bonding of certain components, such as plasterboard and much sub-flooring, can be employed, leading to improved quality and reduced labor time. In the factory, more sophisticated cutting machines, including those that are computer-controlled, can be used to make necessary cuts in frame and panel members, particularly the angled cuts needed with roof structures, roof intersections and dormers. However, no foundation or other masonry portions of the house can be prefabricated. Brick or stone siding, steps, fireplaces, and concrete floors, are made at the house site. At the site, modules are positioned on the prepared foundation, and fastened together, and electrical and piping connections are made. Modules are also weather-sealed together. Final roofing and exterior siding are completed at junction points.

Fig. H1 shows operations in one factory for prefabricated house modules that employs assembly-line techniques. Normally, in this facility, a house module, constructed from individual wood pieces, is completed in five days. Panel assemblies are fed from one side of the assembly line to the modules-in-process on the line. The modules-in-process are advanced along the line at the same time. (The modules are all assembled on wheeled frames.) When a module arrives at the end of the line, it is nearly ready for shipment and final items such as kitchen cabinetry and exterior siding are then installed.

Fig. H1 An 80,000 sq ft (7400 sq m) factory building for the construction of house modules. The module assembly begins in the foreground, starting with the floor panel, and extends to the end of the building. The panel assemblies for floors, walls, ceilings, and roof are assembled in the area on the left and moved to the modules in process at the right. After several panels are added, the modules are moved one or two spaces along the line to be in position to accept further wall panels, then ceiling and roof sections, and other components. Finished modules are shipped from the far end. *(Courtesy Signature Building Systems, Inc., Moosic, PA.)*

housings, appliance - Housings for large appliances such as refrigerators, stoves, dishwashers, clothes washing machines, and dryers, are made from sheet steel except for plastic trim, handles, knobs, and most control panels. These products are made in large quantities. Therefore equipment and tooling designed for high production are economically justified. Mild steel sheet in roll form is fed to large mechanical punch presses for blanking (2C4), punching (2C5), forming (2D2), drawing (2D5) and trimming (2C6) operations with sophisticated dies. Progressive (2E1) and compound dies (2E3) are frequently used. Transfer die lines (2E2) along with robotic handling (14G4), for some operations, may be employed to handle the large stampings involved. These parts are fastened together with projection welds (7C6c) and other resistance welds. Removable housing parts are typically fastened with sheet metal screws and machine screws to projection-welded nuts (7F4a). The housing parts are painted on electrostatic painting (8D7) lines that usually include cleaning (8A) and phosphatizing (8E2) operations.

Plastic parts are normally injection molded (4C1). Hot stamping (4M2) is the common means of providing marking on control panels.

housings, business machine - The machines used in copying and printing centers have housings that tend to be a mixture of sheet metal and plastic parts with plastic parts predominating, except for the very largest machines. The plastic parts are those that are contoured, round-cornered, and smaller; the metal parts are those that are larger, or contribute to the structure of the machine, or are rear or side panels. Control panels are normally plastic. While conventional injection molding (4C1) is widely used, the moderate production quantities of this equipment makes structural foam injection molding an attractive alternative. The purpose of the structural foam is to provide thicker, more-rigid walls for the parts, but a major advantage is that the low-pressure foam process enable less-rigid, far less expensive, molds to be used. Low-pressure injection molding of structural foam (4C3a) and reaction injection molding (4C3b) are such processes. If higher-quantity production is involved, higher tool costs can be justified, and the high pressure process (4C3c), or the gas counterpressure process (4C3d) for structural foam, can be used. Co-injection molding (4C3e) or gas-assisted injection molding (4C3f) are other alternatives. The plastic parts are molded and metal panels are painted to match one another's color.

The metal parts, for the low- or moderate-quantity production, that is common for such products, are made from operations and equipment typical of a sheet metal job-shop: shearing with squaring shears (2C1 and 2C1a), hole making with turret punching equipment (2C5a), that is sometimes computer numerically controlled, with notching and other operations to create the required outline of each part. Bending and some forming are commonly performed with press brakes (2D1a).

Business equipment of the personal computer variety and the personal printers and scanners that accompany them are produced in high quantities and standard, high-production, injection molding is used for many parts. Metal cabinets for PC's can be made with production metal stamping processes. With most desktop units, the major portions of the case, including the cover panel, are made from sheet steel. This also may be done by job-shop shearing, punching and forming equipment but, if the product is to be made in large quantities, high-production compound and progressive die punch-press operations (2E1 and 2E3)are often employed. Reinforcing members, made with sheet metal processing methods, are part of most computer cases. The main components and smaller parts of the case are fastened together with pop rivets (7F4b), resistance welds (7C6), and screw fasteners (7F4a). After assembly the case is spray painted (8D7) or powder coated (8D8), usually electrostatically. The front panel of the case is made from several injection molded (4C1) parts and includes or accommodates switches, slots and drive bays.

hydraulic fluid - for hydraulic power systems ("fluid power" systems), is mostly made through the distillation of petroleum but other common types are made from synthetic lubricants, oil-water emulsions and water-glycol mixtures. Specialty and proprietary hydraulic fluids are made by a number of chemical companies. There are numerous different varieties and many ingredients are usually involved. Ingredients are added to the basic material for special properties or property enhancements. These additives include defoaming agents, lubricants, thinners or flow aids, corrosion inhibitors, and viscosity modifiers.

Organophosphate esters and polyalphaolefin are each the basis for two types of synthetic-lubricant fluids. They are usable over a wide temperature range and are suitable for higher-temperature environments. Polyalphaolefin fluids are made from ethylene, $H_2C{+}CH_2$. Other fluids, developed for higher temperature applications (such as the hydraulics of die casting machines), include diphenyl didodecyl silane and a fluid that has a base of tricresyl phosphate.

Flame-resistant fluids are used in aircraft hydraulic systems. One such fluid is a water-glycol liquid that incorporates additives for thickening, lubrication and corrosion resistance.

Biodegradable, vegetable-based, fluids are being used in some installations for environmental reasons to replace petroleum-based fluids that do not degrade well. The biodegradable fluids are made from estolides, which are fatty acids from sunflower, safflower, or other high-oleic oilseeds.

Hydraulic fluids are used in industrial machinery including cranes and agricultural machinery, aircraft controls, landing gears and brakes, automobile brakes, and transmissions.

hydrochloric acid (HCl) - a major compound used in chemical manufacture, is made primarily as a byproduct of the chlorination of hydrocarbons, both aromatic and aliphatic. It is also made by dissolving gaseous HCl in water, or by reacting sulfuric acid with sodium chloride. HCl gas is made by the direct combination of hydrogen and chlorine at a temperature above 482°F (250°C). The largest use of hydrochloric acid is in the pickling of steel. Other applications are the production of chemicals and pharmaceuticals and food processing.

hydrogen - for commercial quantities, is derived from hydrocarbon fuels. There are a number of processes. In one major method, steam is reacted with natural gas, oil refinery gas, methane, propane, ethane, or other light hydrocarbons. Nickel is the catalyst, and temperatures of the reaction range from 1300 to 1850°F (700 to 1000°C). The reaction yields hydrogen and oxides of carbon. With propane, the reaction is as follows:

$$C_3H_8 + 3H_2O \Rightarrow 7H_2 + 3CO$$

Hydrogen is also obtained from the electrolysis of water containing dissolved potassium hydroxide. Another source is as a by-product when brine solutions undergo electrolysis in the production of alkalies. Liquefaction of the other constituents of fuel gases also yields hydrogen.

I

I-joists, wooden - See 6F7 and Fig. 6F7-1.

ice, dry - See *dry ice.*

ice cream - is a frozen blend of cream or butterfat, milk, sugar, flavorings and sometimes small pieces of fruit, nuts, or other ingredients. Sometimes eggs are also included and the dessert is then called French ice cream or frozen custard. The liquid ingredients (milk, cream, sugar syrup) are added first and are thoroughly mixed. The mixture is heated and solid ingredients: (powdered milk, sugar, dried eggs, and a stabilizer, if used) are added. Agar and gelatin are common stabilizers that give a smooth consistency to the mixture. Pasteurization and homogenization follows heating and mixing. Homogenization breaks up particles of butterfat. Then the mix is refrigerated for a period and solid pieces of fruit, etc. are added, if part of the formulation. The mix is then frozen while still under agitation, to add air and ensure a smooth blend. The near frozen mix is then put into containers or other packages and is "hardened" (frozen solid). Soft ice cream does not receive the hardening step.

Sherbet and sorbet are made from fruit puree. Sherbet includes milk but has much less butterfat than ice cream; sorbet has no butterfat.

ice skates - have two major components: boots to fit the skater, and blades that are assembled to the bottoms of the boots.

Ice skate boots for figure skates are normally leather, but hockey skates are commonly made of a combination of plastic, leather, and fabric components. Both types of boots are made with the same methods used for shoes and other leather boots, with a few exceptions. The fit of the boot to the foot is closer to provide better control over the skates. Reinforcement is added to provide ankle support, and there is some additional padding. The soles and heels of the boots are made thicker, with vulcanized fiber, high durometer rubber, or a similar plastic to provide material for attachment of the blades. Internal steel guards may be installed at the toe and heel ends of the boot.

For figure skates, the blades are made from steel plate of approximately 1/8 in (3.2 mm) thickness. The plate is blanked (2C4) to the shape needed (including the ice pick teeth at the front of figure skates blades). The blades are heat treated for greater hardness and wear resistance (8G3). The sole and heel attachment plates are blanked from thinner stock than the blades and are welded by GMAW (gas-metal arc welding) (7C1d), or another arc-welding method to the blades. The welded assembly is polished (8B1) and chrome plated (8C1), though many made from stainless steel do not require plating. The blade is sharpened by form surface grinding (3C3) to provide a concave surface between the edges. The blades are then assembled to the boots with rivets in punched holes. The boots are polished and laced and the skates are packed in boxes for sale to customers.

inductors (chokes, choke coils) - See 13L3.

ink - is a dye or a fine dispersion of pigments in a liquid or paste. The base may be a drying oil or a plastic resin. Drying agents and thinners may also be used. Some inks are made from liquid dyes in similar vehicles. There are many formulations, depending on the application. Common uses for inks are writing, drawing, printing and marking. The pigment in most black inks is carbon black, but black writing inks usually contain gallotannate of iron.[2] This is made by reacting ferrous sulfate or another iron salt with an aqueous mixture of tannin and gallic acid. The resulting solution is not sufficiently dark when first applied, so is usually supplemented with a blue or black dye. After drying, it becomes darker and insoluble in water.

The simplest printing inks consist of carbon black in linseed oil with an additive to speed drying. More complex printing inks are now increasingly common and contain one or more of various suspended pigments in an oil vehicle with resin, solvent, drier, and adhesive. Drying may be by penetration and oxidation or by evaporation.

in-line skates - are made with a construction that, in many respects, is very similar to that used for ice skates, but in-line skates often tend to be more complex in the boot area. Both types of skates use a boot that is high enough and reinforced to support the ankle of the user. The boot also fits snugly for good control and has some internal padding. Otherwise it is made with methods that are consistent with those used to make other boots. Both ice and in-line skates have a frame that is attached to the sole of the boot to carry the blade or wheels. The boots can be made from fabric, leather, or injection molded plastic parts or various combinations of these. Some have external stiff plastic parts for ankle support, sometimes in one piece for each boot, sometimes in two pivoted pieces. With some skates, the entire boot shell may be molded of urethane plastic. The skate wheels are normally made by suppliers to the skate company and consist of ball bearings, nylon or other plastic wheel hubs and polyurethane or PVC tires. All the plastic parts of in-line skates are injection molded (4C1) with the exception of polyurethane wheel tires, which may be cast (4H). Four or five wheels are mounted in tandem on a metal frame that has an upside-down, U-shaped cross section. The frames are blanked (2C4) and formed (2D2) from sheet steel, aluminum, or titanium, and then chromium plated (8C1). Alternatively, frames may be made from extruded aluminum (2A3), or from injection molded plastic. They are usually fastened to the soles of the boots with rivets or a coupling device. Axle holes are punched or drilled in the frame. Axles are formed by cold heading (2I2), are plated, and held in place with threaded nuts. Many of the skates have a second frame piece at the heel, either of plastic or sheet metal, that holds a plastic block that can be dragged by the skater along the skating surface to act as a brake. The skates are assembled on a production line and packaged for shipment. Fig. I1 illustrates the major components of typical in-line skates.

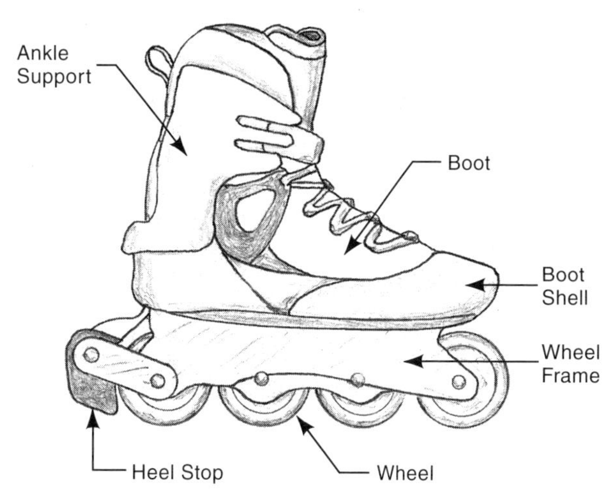

Fig. I1 The key components of typical in-line skates.

insecticides (pesticides) - are made in many formulations, depending on the insects they are to be used against, how they act on the insect (from contact, digestive poison, inhalation poison, or reproduction blockage) and what is to be protected (people, crops, household goods, etc.). Both natural and synthetic active ingredients are used. One common ingredient in insecticides used as digestive poisons is calcium arsenate [$Ca_3(AsO_4)_2$]. Common active ingredients for contact insecticides are rotenone dust, nicotine sulfate solution and sulfur dust.[2] Powdered insecticides are usually mixed with a carrier powder that is inert to the insecticide but allows it to be spread more uniformly. Kaolin clay and limestone, finely ground, are two materials used.

DDT, previously used extensively, but now sparingly because of its toxicity to mammals and its long life, is a chlorinated hydrocarbon, dichlorodiphenyltrichloroethane $C_6H_3Cl_2(C_6H_4 \cdot CH_2CCl_3)$. Since restrictions on the use of DDT were instituted, many other chlorinated hydrocarbons, and other complex organic compounds that break down readily in the environment into nontoxic materials have been employed. They include dieldrin, aldrin, chlordane, endrin and heptachlor. Organophosphates that attack the nervous systems of insects but have a shorter life, are one family of insecticides being used more extensively. Carbamate insecticides, which are esters of carbanilic acid and are derivatives of carbamic acid, NH_2COOH, kill insects and their larva on contact, are less dangerous to humans, have a short active life, and soon break down to non-toxic substances[14]. These insecticides are carried on talc or synthetic clays[2].

Digestive poisons include several arsenic compounds (lead arsenate, copper acetoarsenite - Paris green, and calcium arsenate), and fluorine compounds (sodium fluoride and cryolite). Some contact insecticides derived from natural sources are nicotine (from tobacco), rotenone (from a plant root), and pyrethrum (from the flower of the chrysanthemum). Some inhalation insecticides are hydrogen cyanide, nicotine, methyl bromide, and naphthalene. They are used on plant materials in storage.

instant coffee - See *coffee, instant.*

insulation board (low-density fiberboard) - See 6F6.

integrated circuits - See 13K.

iron - Almost all iron is made by blast furnace. Other methods are direct reduction (producing solid or sponge iron), and direct smelting.

In the blast furnace operation, the product is pig iron, an intermediate material between iron ore and steel. Iron ore, coke, and limestone are loaded into the blast furnace. It is a steel container lined with fire brick and equipped with tuyeres, in the lower part, through which heated air is forced under pressure into the furnace. The combustion of the coke produces carbon monoxide gas which, as it rises in the furnace, reacts with the iron oxides to produce carbon dioxide and iron. The reaction is as follows:

$$Fe_2O_3 + 3CO \rightarrow 3CO_2 + 2Fe$$

Limestone acts as a flux and provides additional carbon monoxide for the reaction. The molten pig iron is drawn from the bottom of the furnace. Slag from the limestone and various impurities float on the surface of the molten iron and are also periodically removed from the furnace. The pig iron is high in carbon and contains silicon, manganese and phosphorous that are removed, if the iron is to be converted to steel during a subsequent process. Fig. I2 shows blast-furnace iron-making as part of the steelmaking sequence.

With direct reduction methods, the iron ore and other materials are heated to a temperature below

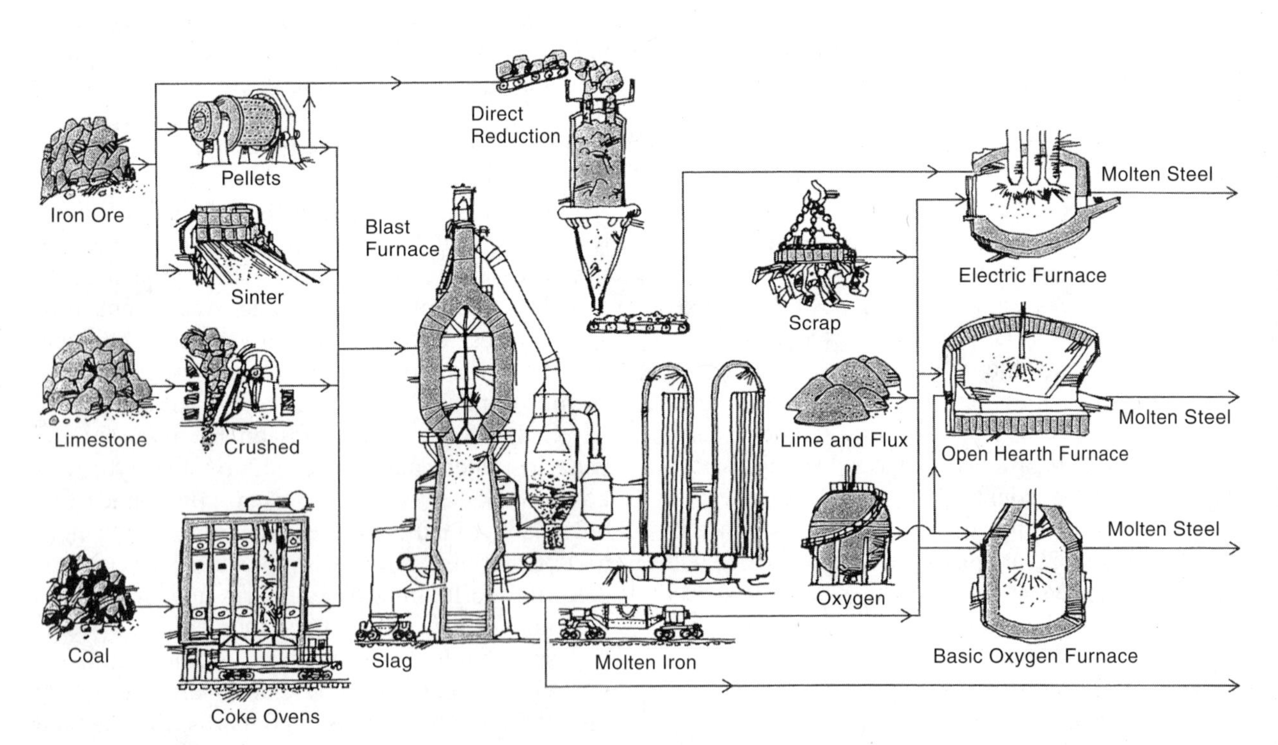

Fig. I2 The position of the iron-making processes, using a blast furnace and direct reduction, in the sequence of operations that produce steel (*Courtesy American Iron and Steel Institute, AISI.*)

the melting point of iron. There are several variations in the process. Some methods require natural gas. Methane, CH_4, can be converted to carbon monoxide, CO, and hydrogen, H_2, which reduce pellets of iron ore. Other approaches provide different reducing agents; some provide different means to bring the reducing agents in contact with the iron ore. Some methods use coal that is partially burned to produce carbon monoxide and other reducing gases. One method uses fluidized beds, another a moving-bed furnace, a third method, a rotary kiln. The iron produced by direct reduction is solid or sponge iron.

High-purity iron, called ingot iron, is made with the basic open-hearth process incorporating a 1- to 4-hour extension of the heating cycle at a temperature of 2900 to 3100°F (1592 to 1704°C). This process is used in applications where high ductility is required.[2]

isopropyl alcohol (isopropanol or rubbing alcohol) - See *alcohol, isopropyl.*

J

jars, glass - See 5A2b3.

jelly - is made from the strained juice of fruit or, in a few recipes, vegetables. Sometimes, several fruits are used. Citrus fruits and apples have high pectin content and may be included in jellies of other fruits. Jams, preserves, and marmalades include fruit pulp or fruit pieces. The fruit is harvested, crushed, separated from stems, leaves, and skin, heated, filtered, pasteurized, and chilled. It is then pumped into kettles and cooked three times. Sugar, pectin, and gelatin, if used, are added before the last cooking. The mixture is fed to a filling machine that meters the correct amount into jars that are then vacuum sealed and capped with metal covers. The jars are automatically labeled and packed by the dozen or more into corrugated cartons for shipment. Jellied candies are made similarly but include starch or agar.

jelly beans - are made from sugar, corn syrup, flavoring, coloring, and confectioner's glaze. Cornstarch is used as a mold material in the manufacturing process. It is deposited in metal trays where it is formed into open mold cavities. The forming is done by a machine with a large number of identical punches that are pressed into the cornstarch. Each tray has about 1200 cavities. The operation sequence is then as follows: 1) The jelly bean ingredients are well mixed and cooked in large vats. 2) The resulting liquid is transferred to a special casting machine, called a "mogul". The mogul deposits a small amount of the liquid mix in each mold cavity. 3) After the cavities are filled, the trays are conveyed to a dry room where the material cools and solidifies. The trays are left in the dry room overnight. The material in each mold will become the core of a jelly bean. 4) The next operation is a "sugar shower" that coats the cores with a layer of granulated sugar. The cores are subjected to both moist steam and the sugar, with enough movement of the cores to cover all sides. 5) The sugar-coated cores are then dried again for a few days. 6) The next operation is called "engrossing" or "panning". The coated cores are placed in a rotating container and sprayed with flavoring, coloring, and fine sugar which are then allowed to cool and dry for about 2 hours. This step is repeated several times - an average of 4, the number depending on the variety and flavor of the bean. 7) The candies are then polished. They are sprayed with confectioner's glaze, which is food-grade shellac in an alcohol solution with some carnauba wax. The candies are tumbled as the alcohol evaporates, and a shell is formed. The tumbling action smooths and polishes the surface of each jelly bean. This takes about 2 hours. 8) The completed candies are inspected for quality and, with some manufacturers, brand identification is stamped on each with a marshmallow-based edible ink. 9) They are usually then mixed with other flavors, packaged by weight by automatic equipment and shipped. The entire process takes 7 to 10 days.

jet engines (gas turbines) - The manufacture of jet and turboprop engines for aircraft and gas turbines for other applications involves the use of difficult-to-process metal alloys, ceramics, composites and other sophisticated materials. Titanium is a common material, often alloyed with nickel and aluminum.

The typical turbofan jet engine has almost 25,000 parts. The major components are: an intake fan; a compressor (with alternate rows of rotating and stationary blades) to raise intake air pressure to 2 to 12 times its original value in two stages, a low-pressure booster stage and a high-pressure stage;

a combustion chamber with up to 20 fuel spray nozzles; a turbine to drive the compressor and a fan, (The temperature in the chamber typically reaches 2000 to 2700°F (1100 to 1500°C) and even higher for short periods at take-off of the airplane); a main central shaft, or two or three concentric shafts, and bearings; and an exhaust duct system. The exhaust system may include a mixer, to combine cool air from the fan with hot exhaust from the combustion chamber. There are also accessories (fuel pump, lubrication pump, instruments, and an electric starter-generator) which are connected by gearing to the main shaft. There is a fuel control system that senses pressures, temperatures, and rotational speed to prevent excessive speed or temperatures. The engine also includes nacelles or other air ducts to channel the air flow from the fan. See Fig. J1 for a simplified illustration of the components of a typical turbofan jet engine.

The air intake fan, which may consist of three sets of blades, is made from a titanium alloy. Individual blades are hollow, but the hollow space contains a titanium honeycomb structure to support the titanium skins. The skins are hot formed, assembled with the honeycomb and welded. The lower ends of the blades are machined to fit a hub or pitch-adjusting components on the main compressor shaft. Some jet engines are now using plastic composite intake fan blades, made from carbon fibers and epoxy.

The compressor sections include a series of notched rotating discs to which the individual compressor blades are attached. They also have stator blades attached to the inner surface of the compressor casing in between the rotor blades. The discs that hold the rotor blades are powder metallurgy forgings (2L1g) of aluminum alloy. Hot isostatic pressing (5B17a) is used to compact and sinter the powder particles[26]. The formed discs are then machined in lathes (3A) and machining centers (3T) and "fir tree" slots are broached (3F) near the edges to accept each blade. The compressor blades are made from titanium alloys by ceramic mold casting (1C1) and machined to a "fir tree" shape by broaching or form milling (3D5) at the bottom, to match the slots in the discs. Threaded fasteners lock each blade in place.

Combustion chamber parts are hot formed from sheet titanium alloys and welded by GTAW (Gas Tungsten Arc Welding)(7C1e) or GMAW (Gas Metal Arc Welding)(7C1d) with argon or argon-helium shielding gas. The combustor may be lined with ceramic materials to provide resistance to the high heat. Plasma-sprayed zirconium is currently used (8F4d).

Turbine discs are also made by powder metal forging and, after machining, are hard faced with refractory metal to provide wear resistance against the hot gases that strike them from the combustion chamber. Turbine blades are made by

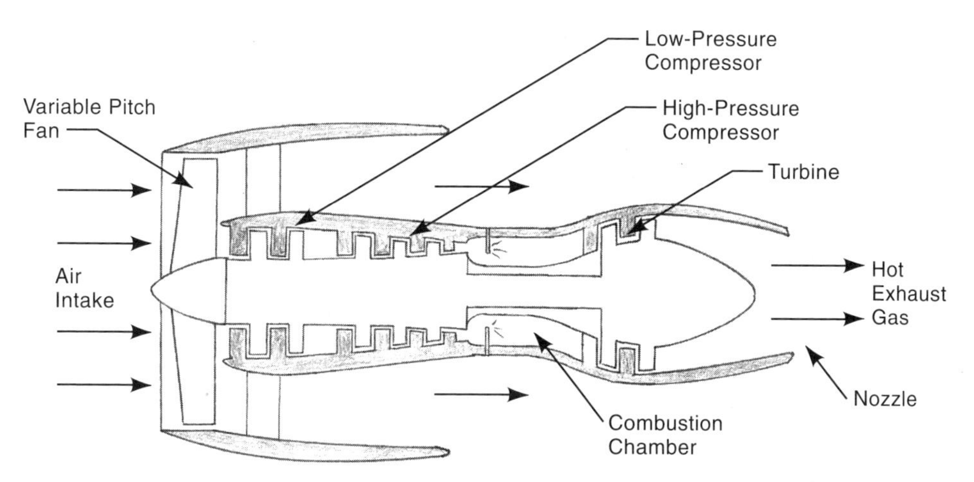

Fig. J1 A simplified cross-sectional view of the components of a typical turbofan jet engine.

investment casting special nickel-aluminum-titanium alloys[26]. The casting process has included directional solidification of the molten metal. This is an oven treatment that controls the cooling to a prescribed rate to align the metal molecules within the blade and improve its strength against the extreme centrifugal forces and hot gases that it faces. However, current practice is to cast the blades with methods that cause the metal to solidify as a single crystal.[27] (See 13K2 for single crystal casting of silicon for semiconductors.). The blades are subjected to hot isostatic pressing to close any porosity that may result from the casting operation. The blades may also be given aluminide diffusion coatings by pack aluminizing for oxidation resistance, or a coating of MCrAlY (M = Co,Ni, or Co/Ni) by PVD (physical vapor deposition - sputtering) (13K3a5) and overcoated with a ceramic coating (zirconia) as a thermal barrier. An intricate pattern of cooling channels for the blades is cast in, but there is extensive machining of the blades after casting. Automatic transfer equipment is used to move the blades from machine to machine and machine loading and unloading is robotic. Electrical discharge machining (3I1) and electrochemical machining (3I2) are normally employed, the latter to drill small, parallel cooling holes in the blades. These holes provide a layer of cooling air on the surface of the blades. The surfaces are honed (3J1) for improved dimensions and surface finish.

The exhaust duct system includes inner ducts and afterburners (tailpipe) cast or formed from titanium or aluminum with a sound-absorbing ceramic honeycomb material. Outer parts are made from Kevlar® and carbon fabric reinforced plastics.

The quality of all parts is carefully verified before the parts are assembled. Castings and stampings are checked for freedom from checks and cracks, using fluorescent penetrant, ultrasonic, and other tests. Dimensional checks are made with coordinate measuring machines (14D1) and shadowgraphs. All rotating parts are dynamically balanced before assembly. (The 20,000 rpm rotating speed of the engine applies very high centrifugal forces and stresses in rotating parts.) These parts are all assembled on a main shaft or concentric shafts which are turned on CNC lathes from tubular stock.

Considerable assembly work is required to convert these components into a working engine. At some manufacturers, the turbine blades are robotically assembled to the turbine hubs but most assembly is manual. The process begins with the making of a series of subassemblies or modules. A typical engine will have seven or more major modules. These usually are assembled in other plant locations or may be made by another company. After these modules are made, they are brought together at the engine assembly area. An assembly fixture is employed and, for the addition of some modules, the engine may be oriented in a vertical position to facilitate the work. Cranes and other material handling devices move the modules into position in the fixture. Major modules are put together before peripherals, the auxiliary equipment such as electrical devices and connecting wiring, hydraulic lines, pumps and valves, etc., are assembled. Full assembly of tubing, wiring accessories and the engine nacelle takes place at the aircraft manufacturer's facility rather than the engine factory. However, sufficient apparatus is installed to permit thorough operational testing of the engine before shipment.

The first tests are static tests in that the engine does not run. All systems (fuel, cooling, instrument, control) are first tested individually to verify that they operate properly. The engine is then operated while it is mounted on a test stand to confirm that it runs correctly, delivers the specified thrust, uses fuel at the specified rate, and is free from vibration, overheating, and other faults.

Machining processes for all the mechanical parts are extensive and must be precise and capable of handling alloys of low machinability. Sophisticated computer-controlled equipment is employed in many areas, and advanced techniques such as electron-beam welding, critical heat-treating processes, laser machining (3O), and electrical discharge machining (3I1), are utilized, along with more conventional machining and grinding.

The turboprop engine is very similar to the turbojet engine, differing in that the power is delivered primarily from the rotation of a propeller rather than the reaction from expelling exhaust gases at high speed. Gas turbines for non-aircraft use are also similar to those for aircraft, from the standpoint of both the design, and the manufacturing processes used. Turbine blades for power generation or vehicles may have turbine rotors made from silicon nitride ceramic with the ceramic process supplemented by hot isostatic pressing.

jet fuel - is one of the light distillates of petroleum refining. The U.S. military uses JP-4, a naphtha-based jet fuel with properties between that of gasoline and kerosene. Kerosene is also used as a jet fuel. (See 11H and *kerosene*.)

jewelry - Major raw materials for precious jewelry are gold, silver, and platinum metal, and various gemstones. The gold used is normally *karat gold*, an alloy of gold and other metals, usually copper, with some zinc. Silver is also alloyed, principally with copper, to provide greater strength. Sterling silver is 92.5 % or more pure silver. Sheet gold is available as a coating on another metal (usually brass or bronze), when strength or cost conditions dictate. This is "gold filled" material or "rolled gold", achieved by mechanical bonding when the materials are pressure rolled together. Jewelry can be made by most of the metal working methods outlined in Chapter 2, but much jewelry making, especially with precious metals and costly gemstones, is for one-of-a-kind or small-quantity fabrication, and is carried out by artisans using manual methods with a variety of bench and hand tools instead of mechanized equipment. Suppliers provide jewelry artisans with pre-alloyed precious metal in forms most suitable: sheets, strips, wire, rods, and ingots for casting. Wire and rods are available in a variety of sizes and cross-sections including square, triangular, half-round and other shapes in addition to round. These shapes are also available in various tempers for the jewelry maker. Suppliers also provide some common findings - preformed shapes of common components such as clasps and stone settings. These are prefabricated from production tooling to reduce the amount of hand labor necessary by the artisan.

Some of the specific operations that may be employed in making jewelry parts from precious metals are: rolling of sheet metal and wire, drawing of wire, making and bending tubing, forging - often with the metal at room temperature and with anvils and hammers, filing, drilling, cutting, texturing the surface, polishing, and patination.

Annealing is often necessary with gold and silver workpieces when they are heavily formed into new shapes. Annealing involves heating the workpiece to the point where a crystalline structure forms, holding it at that temperature for a period, and then allowing it to cool. Sometimes, the metal is quenched as it cools. Acid pickling is usually then necessary to remove scale. Silver can also be hardened when necessary. Both operations are performed in a furnace with a non-oxygen atmosphere.

Much precious metal jewelry is made by investment casting (1G). With original designs, a model of the proposed piece is sculpted from wax by artisans using a fine-pointed, electrically-heated forming tool and cutting tools similar to those used by dentists. After the carving of the wax model of the piece is completed, wax is added to provide pouring channels, runners, gates, and risers, as needed. The model is then the basis for an individual investment casting. With high-production jewelry, the wax model is usually injection molded of wax or polystyrene in a aluminum or, sometimes, a hardened steel mold.

"Soldering" (really brazing - 7B) is used to join metal parts together and the brazing alloys are formulated to match the color of the pieces being joined.

gemstones - are processed from rough stones with a variety of techniques.[22] These include *sawing* with several varieties of saws coated with diamond abrasive. For the initial cutting of large stones, first cuts are made with large-diameter circular saws. Further cuts are then made with smaller-diameter, thinner, circular trim saws. Wire or band saws are used for making curved cuts. The sawing machines are specially designed for gem cutting.

Grinding wheels, made with either silicon carbide or diamond abrasive, are used for rough shaping the stones. Then, wheels with finer abrasive may be used for further shaping and some surface smoothing. These operations use bench-top grinders and are manually controlled.

Lapping (3J2) against a flat surface is used to create flat surfaces on the stone. With a finer abrasive, lapping is also used to polish flat surfaces. A further shaping and pre-polishing operation, usually used for curved surfaces of cabochon gems, is *sanding*. Sanding of gemstones is usually performed on manually-controlled, abrasive-belt grinders.

Polishing (8B1) is a final step for all surfaces. It is performed by pressing the gem surface strongly against a moving surface of felt, leather, cloth, cork or wood, in the presence of a polishing agent. Typical polishing agents are diamond dust, silicon carbide, aluminum oxide, ceric oxide, chalk (calcium carbonate), rouge (red oxide of iron), tripoli (silicon dioxide), tin oxide, and zirconium dioxide. Some agents provide polishing action even though they may be softer than the gemstone being polished. Holes are made in gemstones by *drilling* with a rotating needle, rod, or tube, and an abrasive (diamond, silicon carbide or boron carbide), either loose or embedded in the end of the drill. Drilling machines that use ultrasonic vibration can be faster than those that use rotation to move the drills.

Faceting is a key operation in which angled flat surfaces are made and polished with a lapping technique to emphasize the reflectance, refraction, and color properties of the gemstone. Diamond dust is used as the abrasive and it is pressed ("charged") into the lap surface. The lap surface may be made from copper, tin, methacrylate, or phenolic plastic, ceramic, wood, or wax. Fig. J2 illustrates a typical "precision cut" faceting device. All the cutting, grinding, drilling, and polishing operations may be performed with some liquid coolant, either water or oil, or an emulsion of both. The means used for holding the gemstone when it is cut, ground, polished, or faceted is doping. Doping wax is a hot melt adhesive that holds the gemstone at the end of a hardwood or metal stick. Another gemstone treatment is *tumbling* (8B2) to shape rough gemstones to a smoother, rounder shape and polish them. Rotating or vibratory tumbling barrels are used.

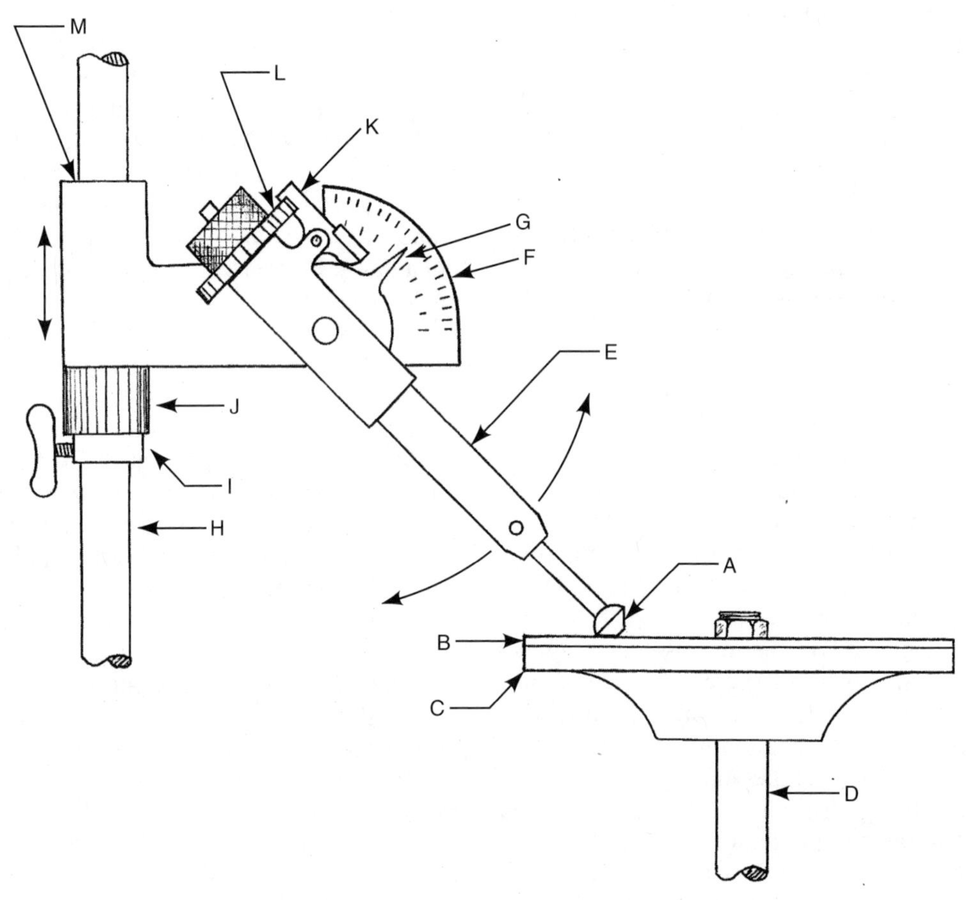

Fig. J2 A precision-cut faceting device for gemstones: A - gemstone being faceted, B - lap, C - master lap, D - rotating lap shaft, E - drop arm that can swing up and down to vary the elevation angle, F - elevation angle quadrant, G - elevation angle pointer, H - supporting shaft, I - locking sleeve for sleeve elevation, J - fine elevation adjustment, K - rotation locking arm, L - notched gear, M - sleeve that is moved up and down as the elevation angle is changed.

K

kerosene - is one of the light distillates of petroleum refining. It distills from the fractional distillation tower after gasoline and at a temperature between 345 and 550°F (174 and 288°C).[2] (See Chapter 11, sections H, H1a, C1 and C1c.) Kerosene is used for jet fuel and in some space heaters. Previously, it was used extensively for home heating and lamps.

Kevlar®[4] - is an aromatic polyamide, (aramid). It is a para isomer that is produced from *p*-phenylenediamine with isophthaloyl chloride. Kevlar has very high strength and is used as a tire cord and, with short fibers, as a substitute for asbestos, for example, in high-temperature gaskets.[2] It is also used as reinforcement in bulletproof vests. To make kevlar fabric, rods of the polymer, polyparaphenylene terephthalamide, are extruded through spinnerets to produce fine fibers which are spun together to make yarns that can be woven.

keyboards, computer - There are several different methods for converting movement of the computer keys to electrical pulses for a computer's processing circuits. Most of these methods use electrical contacts made by bringing together two electrical conductors. Other designs make use of the change in capacitance that occurs when two charged surfaces are brought nearer to each other. Perhaps the most common is the contacting system that uses a domed rubber sheet to push the electrical contacts together. These types generally have the following parts: 1) An injection molded (4C1) plastic (often ABS) upper housing that has spaces for the individual keys. 2) Individual keys, that are dual color injection molded (4C5) to provide visible lettering or numbering. These keys are inserted in openings in the upper housing. 3) A blanked (2C4) and formed (2D2) sheet steel or aluminum base. If steel, the base is electrostatically painted (8D7); if aluminum, it is anodized (8E1). 4) An injection molded elastomer sheet that incorporates a dome-shaped area at each key. (The domes act as springs for the keys.) 5) Two sheets of flexible plastic printed circuit board, separated by a blank sheet with holes corresponding to the contact area below each key. (The blank sheet provides spacing to insure that the contact areas don't touch unless a key is pressed. The circuit boards have no attached integrated circuits or other devices. They are purely to provide circuit paths in the key matrix.) See 13A4 for manufacture of flexible circuit boards. Circuit paths on the top and bottom sheets are screen printed (9D4b) with conductive ink. 6) A small conventional printed circuit board (13A) with one integrated circuit and several other devices attached. This board controls the key function. 7) A cable to connect the keyboard to the computer. Cable leads are soldered to the small circuit board. 8) A number of self-tapping screws to hold the keyboard assembly together. All the parts are assembled together, and the unit is tested and packed.

keys - for locks are normally made from brass sheet stock. Keys are blanked (2C4), embossed (2D7), and coined (2D6) or form-milled (3D5) along the key, to a cross section that mates with the key slot. They are then milled across the key to a profile shape that conforms to the particular lock on which they are to be used. This is a computer controlled operation (3U2) in which the cutter follows the prescribed path. The key may be further processed by electroplating (8C1) and insert injection molding (4C6) to attach a small plastic handle.

keys for computer keyboards - are two-color injection molded (4C5)(See above.)

kidskin - is leather made from the hides of goats, traditionally young goats. It is used for gloves, shoes, pocket books, and linings.[2] See *leather*.

knit fabrics - are fabrics that have yarns fastened to each other by interlocking loops. Knitting is described in 10D.

kraft paper - See 9C5, 9B and 9B2b.

L

lace - originally made only by hand, has been made since the early 19th century by special lace-making machines. The methods are: needlepoint lace-making that uses only one thread, which is made into various stitches with a needle, and bobbin lace-making that uses many separate thread bobbins and the threads are intertwined, braided, or tied. Automatic lace machines are of three types: One type, similar to knitting machines, is identified by the Leaver, Nottingham and Pusher machines; another type is the circular Barmen machine that produces a lace closely similar to handmade lace using the bobbin method; and the third is the Schiffli embroidery machine that uses cloth as a base for the lace stitches, and where the base fabric is destroyed chemically after the operation. The Leaver machine utilizes a pattern control similar to those of Jacquard looms (See Chapter 10, section C1.) Lace fabrics are purely decorative and are used in curtains, table cloths, pillow covers, and for ornamental effects on garments and handkerchiefs.

lacquer - is a paint made by dissolving a plastic resin in one or more solvents. The formulation may also include a plasticizer or softener, and a pigment. Common solvents are anhydrous alcohol, toluol, or benzol. Nitrocellulose, cellulose acetate, and cellulose butyrate are commonly used plastic resins, but acrylics and melamines may also be used. Lacquer dries rapidly from the evaporation of the solvent, leaving a solid thermoplastic coating on the object painted. It is used in furniture making, as a quick-drying coating for many products, and for household use. With pigments, it is sometimes referred to as a lacquer enamel. Automobile enamels that are not lacquers nevertheless are often referred to in the trade as lacquers.

ladders - Ladders are commonly made of extruded (2A3) aluminum or, sometimes, magnesium. The rungs constitute one extrusion, the side frames another. However, the legs on some ladders are made from extrusions of fiberglass-reinforced polyester (4I1). The parts are extruded and then saw-cut to length. Holes are punched (2C5) in the side frames to accommodate the cross bars. It is common to make the rungs hollow with an egg-shaped cross section and to hold them in place by upsetting the ends that otherwise would slightly protrude from the frames. Some diagonal braces between frames and steps are blanked (2C4), and formed (2D2), from steel strip and then zinc plated (8C1). Zinc-plated steel rivets are used to fasten the braces to other members through punched holes. Extension ladders have aluminum latching members made by permanent mold casting (1D1). Other latching parts are steel stampings and standard commercial bolts. Many ladders have injection molded plastic feet placed over the bottom ends and held by rivets. Rivets are upset with hand-operated power tools as the ladders are assembled.

lasers - are of several types: low-power varieties such as diode lasers, that are used as laser pointers, levelers, and target designators; and higher-power, solid-state or gas lasers that are used for manufacturing processes and military applications. Beam quality, wavelength and efficiency, often depend on the nature of the material in which the laser beam is generated. Solid-state lasers, i.e., those that are crystal-based, are often characterized by higher operating efficiencies and gas lasers by broader wavelength range capabilities.

The following is a manufacturing sequence for solid-state, high-power YAG (Yttrium-Aluminum Garnet) lasers:

1. Large, single-crystal boules of yttrium-aluminum garnet ($Y_2Al_3O_3$) are made using the same single-crystal technique used to make silicon semiconductor wafers (See 13K1 and 13K2). The boules are doped with Neodymium, a rare-earth metal. The boules, typically 4 inches (10 cm) in diameter and 8 in (20 cm) long are smaller than those made for semiconductor wafers. Such boules are often supplied to the laser manufacturer by specialty optical crystal suppliers.
2. Undoped YAG segments are often attached to a doped (i.e., Nd:YAG) component. This mitigates thermal effects during the laser's operation. The method used to attach the segments is termed *adhesive-free-bonding* (AFB®), and it produces a permanently fastened, multi-element, component.
3. The bonding surface of the blocks or boule-segments of both doped and undoped YAG are lapped (3J2) and polished, using fine abrasives such as diamond and aluminum oxide-based slurries. The operation is continued until the resulting surfaces are polished to a flatness of 1/10th the wavelength of light and a smoothness of less than 5 angstroms RMS.
4. Under high-level cleanroom conditions, the two YAG blocks are optically contacted together, with the neodymium YAG often as the center portion of a sandwich arrangement. Because of the extreme smoothness and flatness of the parts, molecular attraction holds the parts together as one piece. The AFB process is then finalized by means of a high-temperature heat treatment of the composite element.
5. The resulting AFB boules are then either cut into rectangular blocks (slabs) using saws with diamond coated blades (3G5) or are made into cylindrical rods using diamond coated ultrasonic core-drills (ultrasonic deep-hole trepanning) (3C11 and 3B7).
6. For rod components, final finishing is often done in a center-type grinder (3C1a), using diamond abrasives, to produce precise cylindrical shapes. For slab components, the side surfaces are either as-cut, or are given a final finish by additional lapping processes. For both rods and slabs, the final surfaces may have a ground (matte) finish or be polished, depending on the specific laser design.
7. The two ends of the workpiece are trimmed off with another diamond saw operation and the end surfaces are lapped and polished to a smooth and flat mirror finish. Standard specifications for these surfaces are flatness of 1/10th the wavelength of light, smoothness of less than 10 angstroms RMS, parallelism within less than 60 arc seconds, and perpendicularity to the long axis within less than 5 arc minutes. These two surfaces then define the resonator of a solid-state laser.
8. The part's end faces are then coated, either with a material such as magnesium fluoride, that is anti-reflective ("AR coated") at laser oscillation frequencies, or with a combination of highly reflective and anti-reflective materials ("HR/HT coated"), often using a dichroic coating process involving a combination of material layers of aluminum oxide and titanium oxide.
9. In the case of an AR coated laser component, external mirrors which are either reflective and/or partially reflective at the laser wavelength (i.e., 1064 nanometers for the case of Nd:YAG lasers), are placed on either end of the rod or slab to define the laser cavity and enable multiple-pass laser operation through the YAG component. For the HR/HT laser component, the component end-faces define the laser cavity and no additional outside mirrors are required.
10. After coating, the laser crystal is complete, ready to be assembled with other components to make a laser head, the heart of a laser device. The other components include pumping apparatus such as laser diodes or flash-lamps to stimulate laser operation in the laser crystal, and external optics such as mirrors, beam splitter, lenses, and frequency shifters to produce the desired laser output.
11. After integration with appropriate power supplies and chillers, the laser is complete. Fig. L1 illustrates a typical YAG solid state laser.

latex[2,4] - is the sap of the rubber tree, *Hevea brasiliensis*, native to South America but now grown in Malaysia, Liberia and other countries. Rubber content in latex varies from 20 to 50%.

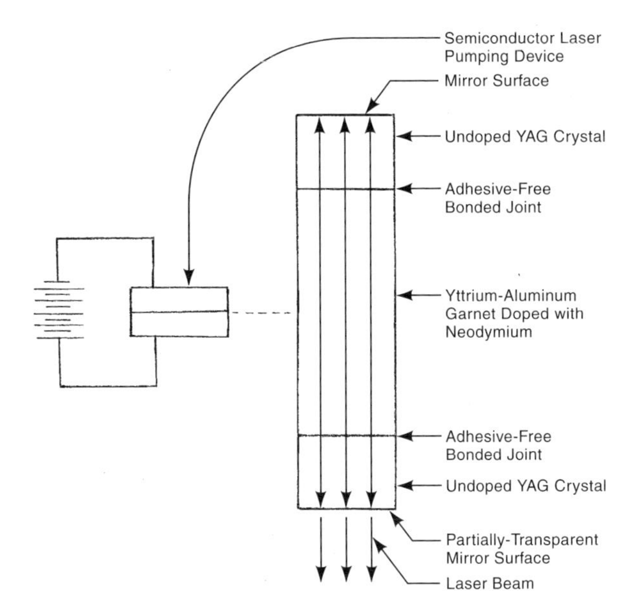

Fig. L1 A schematic view of a yttrium-alumina garnet (YAG) laser pumped by a semiconductor laser.

The sap is collected frequently in small containers to avoid spoilage. Before shipment of latex to a processing factory, the latex is strained and treated with ammonia or sodium sulfate, which acts as an anti-coagulant and preservative. The collecting and processing operations are labor intensive. See *rubber, natural* for information on the process of making rubber from latex. The term, *latex*, also now refers to dispersions of synthetic elastomer particles in water. Latex paints are water dispersions of synthetic rubber or plastics.

lawn mowers - The role of the household lawn mower manufacturer is to put together, on production assembly lines, a large number of components, most of which are received as sub assemblies from suppliers who specialize in components of the particular type. Gasoline or electric motors, injection molded plastic (4C1) motor housings, discharge chutes, wheels, handle bars, handles, control cable devices, batteries and starter motors, if used, drive belts, various screws and other fasteners, and mower housings (usually die-cast from aluminum (See 1F.) or drawn from sheet steel (See 2D5.), and shipping cartons, are some of the major components commonly supplied by other firms. To these components are joined a large number of major and minor parts made by the lawn mower company. These parts may include the rotary blade, axle shafts, small metal stampings and other parts that vary somewhat from company to company. The mower housing and some other metal parts are normally painted in-house with automatic electrostatic equipment (8D7) as the parts are conveyed overhead. The many parts are put together on an assembly line with a conveyor that moves the mower from station to station as it is assembled (7F2). Assembly is assisted by fixtures and power tools and, sometimes, robotic equipment. Riding mowers and other mowers have different parts but the system is essentially the same. Fig. 7F2-1 (in Chapter 7) shows an assembly line for professional riding mowers.

lead - is made from galena, an ore that contains a significant amount of lead sulfide. The ore is concentrated by gravity methods (that are feasible because of the high density of the lead it contains) to a lead content of at least 40%. The concentrate is then roasted, which removes the sulfur, by converting the sulfide to an oxide. The ore is then smelted in a blast furnace with coke, reducing it to metallic form. Other metals in the ore, principally silver but sometimes cadmium, bismuth, and copper as well, are well worth separating and this separation is done by various means. Scrap lead, principally from storage batteries, and then smelted, is another major source of lead metal.

lead glass (lead crystal) - (lead-alkali silicate glasses) are made by substituting lead oxide (PbO) for calcium oxide (CaO) in the batch of materials that is melted in glassmaking. (See 5A and 5A1a.) Glass formulations do not normally have more than 15% calcium oxide, and the percentage of lead oxide in lead glass may also be low, but may range up to 50% or more, so additional lead oxide may be added to the batch. Lead glasses have good working qualities. They are suitable for making engraved artistic glassware, and in other applications where optical properties and high electrical resistance are important and where nuclear shielding is required. Neon sign tubing and electric light bulb parts are other applications.

leather - processing is sometimes divided into two major stages: Stage 1, wet-blue processing that includes pre-tanning operations and tanning, and Stage 2, finishing operations, that make the hide fully ready for conversion into leather products.

The processing sequence for the first stage is as follows: 1) *fleshing* - Hides (kips) are run through a machine to remove excess flesh and fatty tissue from the under side. 2) *soaking* - Hides are immersed in a mildly alkaline, water-detergent solution, and soaked for 3 to 24 hours. This removes dirt and blood from the hide surface and restores moisture if the hide has dried out. 3) *beaming* (hand work) is performed on the hide as necessary. 4) *liming*. The hides are immersed in a bath of calcium hydroxide [$Ca(OH)_2$] and sodium sulfide (Na_2S) for 10 days. 5) Hair is removed from the hides with a *dehairing* machine. 6) Further liming takes place to remove unwanted proteins from the hides which are then de-limed. 7) *bating* - soaking the hides for 20 to 30 minutes at 90°F (32°C) in a solution of tryptic enzymes followed by rinsing. This treatment refines and cleans the grain surface. 8) *tanning* by the chrome process. This is a multi step operation that includes pickling in $Na_2Cr_2O_7$, two salt solution baths, reduction in $Na_2S_2O_3$, settling in borax to set the chrome salts on the leather fibers and rinsing in water. (The alternative tanning procedure, vegetable tanning, requires 2 to 4 months compared with 1 to 3 weeks for chrome tanning.) Tanning makes the hide into a stable material that resists spoilage or deterioration. 9) *sammying* - consists of passing the tanned hides through large rollers under pressure to remove excess moisture.

The sequence for the second stage may include the following operations: 1) *splitting* and *shaving* - The tanned hide is split through its thickness in a machine to produce layers of about 0.40 to 0.80 in (1 to 2 mm) in thickness. Any high spot or fleshy material not wanted is shaved off. 2) *retanning* - a second tanning operation to make the leather softer or firmer, as desired, and to remove any free acids that may remain from the original tanning. 3) *dyeing*, if specified. 4) *fatliquoring* - an oil treatment that ensures flexibility and gives the leather a soft feel. 5) *drying* - either in a vacuum dryer, by air drying, or stretched on a frame and oven dried. The objective is to reduce the moisture content to 15–20%. 6) *staking* - a mechanical treatment to restore flexibility that may be lost during drying. 7) *buffing* - of the surface to smooth the grain of the leather, especially if there are scratches or blemishes. 8) *finishing* - a wax or polymer finish coating may be applied. 9) *embossing*, if specified. 10) *measuring* - the size of the finished leather sheet. Fig. L2 shows this sequence.

leather goods - After the above processing, sheets of leather can serve as the major raw material for many products including: shoes (See *shoes*.), wallets, suitcases, purses, saddles, gloves, belts, key cases, watchbands, cases and containers, and many other items. Most of these products involve some sewing and are made with operations very similar or identical to those outlined in Chapter 10,

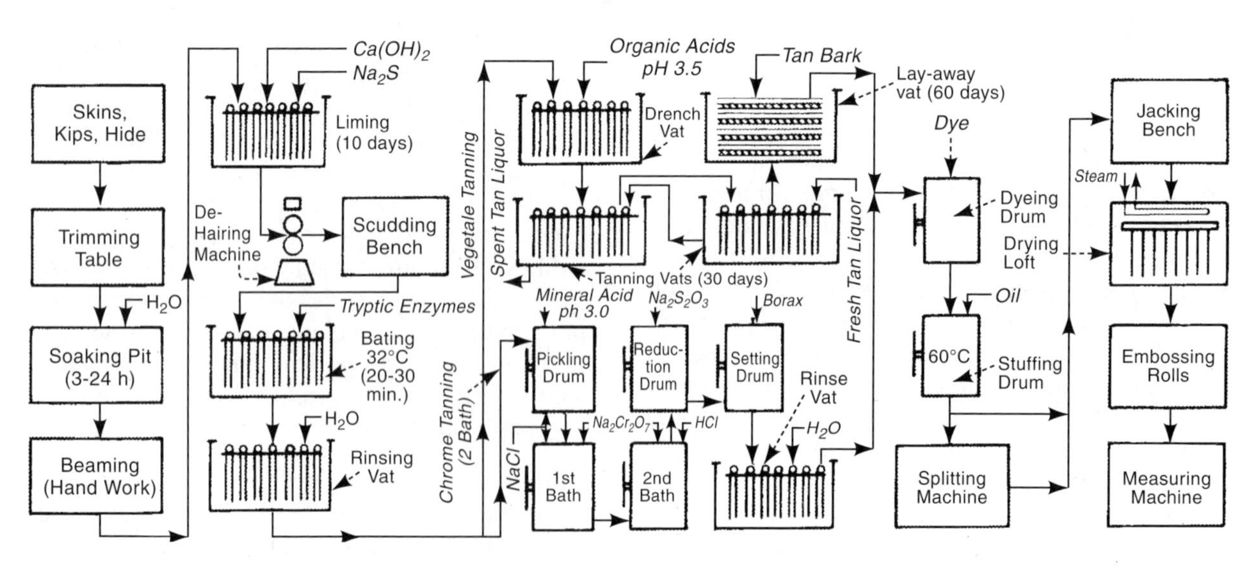

Fig. L2 The sequence of operations followed in manufacturing leather from hides. (*from Shreve's Chemical Process Industries, 5th ed., G.T. Austin, 1984, McGraw-Hill, New York, used with permission.*)

sections H through H4a which describe the production steps involved in making clothing and other sewn products from cloth and other materials (spreading, marking, cutting, sewing, seam bonding). When production quantities are involved, the leather sheets (skins) are stacked, marked at the top of the stack for cutting according to patterns, cut to the marked shape, with cuts through the whole stack at the same time, and sewn together in one or more sewing operations to make the desired product or major parts of it. Stacking sheets of leather for cutting is not as simple as it is with fabrics because skins are irregular and vary in size and shape, and consideration must be given to natural differences in various areas of the skin. Because of these differences, many leather parts are cut individually or in multiples of a few pieces by press stamping with steel rule or similar dies. (See 2C4a.) This approach simplifies the task of avoiding natural irregularities or discontinuities in the leather sheet. Some leather parts undergo skiving (thinning), edge beveling, or V-gouging (before folding, if the leather is thick). These operations are done with hand tools or, if the quantities are sufficient can be mechanized.

Sewing machines for leather are very similar to those for cloth but are designed and made specifically for leather work. Before or after sewing, the leather parts are processed further and may also be formed, decorated, and have special surface treatments to provide the desired appearance. Making optimum use of the leather sheet is even more important than it is with fabrics because of the higher cost of leather, compared with cloth. After cutting and sewing, parts may be further fastened together by riveting (7F4b), adhesive bonding or laminating (7D), or lacing. Before riveting or lacing, holes must be punched into the leather piece and this operation can be carried out with the same steel rule die that cuts the outline of the part, or can be done as a separate operation. Press tooling, simple hand punching tools, or hammer-driven punches made for hand leather work can be used. Rivet setting can be by a punch press, by foot-operated machines, or with hand tools or a hammer. Eyelets, grommets and snaps are also frequently used in leather products and are similarly installed.

Surface changes to leather include dyeing, stamping, modeling, burning and buffing, and may be made before or after leather parts are fastened together. Dyeing normally takes place at the tannery for the whole hide or lot, but parts are often selectively dyed. Dyes consist of oil, alcohol or water-based solutions that can be brushed, sprayed, or wiped on the surface. Stamping puts a design in the surface by compressing some areas. The leather piece is first softened by thoroughly moistening it from the backside. Decorations can be pressed in with special dies made for the particular design, creating an embossed or coined effect. When the design is made manually, with single-point or blunt hand tools, the operation is called modeling. Burning is a similar operation performed with a tool hot enough to singe the surface of the leather. After these operations, an oil-based conditioner may be applied to the parts or the finished product to ensure its flexibility and softness, and to protect it from moisture.

lenses - are made from glass or polycarbonate plastic. Glass lenses are made most often from clear crown optical glass with a refractive index of 1.523.[1] The glass is supplied to the lens manufacturer in plate form, free from imperfections. Many current eyeglass lenses are made from polycarbonate.

The methods used to make lenses for eyeglasses, telescopes, cameras, microscopes, binoculars and projectors, all follow the same process but with greater precision for instruments and many optical devices, and lesser precision for eyeglass lenses. The first step is to cut the glass plate, with a diamond-abrasive circular saw, into pieces of the desired size. These pieces are then either cut or chipped to a round shape, or the glass blank is heated to a softened state and then rolled to a round shape. The blank also may be pressed (5A2a) to a preliminary approximation of the surface curvature needed. Grinding (lapping - 3J2) to the precise surface curvature needed on each side then takes place in special machines, which use shaped lapping tools and an abrasive powder of aluminum oxide, silicon carbide or diamond. The tools for convex and concave lens surfaces are illustrated in Fig. L3. The operation starts with a coarse abrasive and proceeds in steps, with progressively-finer abrasive particles, and two or more successive tools. Both the tools and the lens workpieces rotate and their axes of rotation intersect. The surface generated is normally spherical, i.e., its radius of curvature is constant at all points. Both sides of the lens are ground in this manner. The next step is polishing

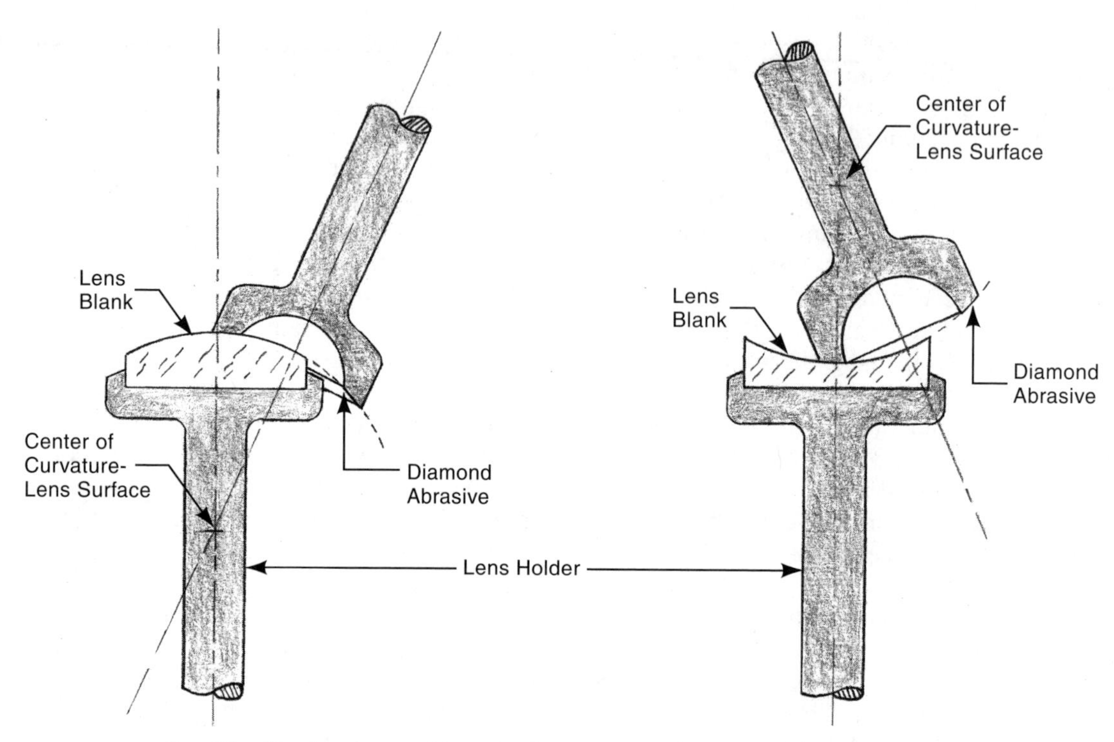

Fig. L3 The tools used to grind convex and concave lens surfaces.

each surface. Polishing is accomplished by mounting a number of lenses on a block, where they are held with pitch, against a rotating wheel covered with cloth, wax, or pitch, and charged with an abrasive of rouge (hydrated iron oxide) or other metallic oxide. The process may take a number of hours. Both sides of the lens are polished and then each lens is ground at the edge to center it in a lens holder. The edge grinding is carried out by holding the lens in a lathe chuck with its optical center on the axis of rotation and using a brass tool and abrasive to effect the necessary cutting. Eyeglass lenses are ground to the shape needed to match the eyeglass frame and ground with beveled edges to secure them in the frame.

Plastic lenses are made with a similar process of grinding and polishing the surfaces and grinding the shape. Plastic lenses for eyeglasses may be dipped into surface treatment liquids to provide ultraviolet blockage, colors, or tints, and scratch resistance. Plastic lenses are lighter than glass, less costly, and more impact resistant.

Compound lenses are assemblies of several lenses in a tubular holder with their optical centers in line. The lenses are ground to complement one another in providing sharper focusing and less aberration of the image. They may be adhesively bonded together or held in accurate fixed positions by the tubular holder. Telescopes, microscopes, and cameras, may have as many as 20 lenses in a compound arrangement when zoom capability is needed and up to about 10 in other designs.

lenses, contact - Soft contact lenses can be made by a number of methods. In one method they are made by casting a liquid, transparent plastic, poly hydroxethyl methacrylate (pHEMA), to form a "button" of approximate lens shape from which the lenses are made. The buttons are initially rigid and are carefully inspected. Those with flaws are rejected. The acceptable ones are machined in a lathe to the correct curvature and size. Diamond-pointed cutting tools are used and both sides are machined. Both sides then may be lapped with methods similar to those used in making glass lenses (See *lenses*). The workpieces are held in the lathes and lapping machines on shaped arbors with suction, wax, or adhesive. The finished shape must not only provide the desired optical correction but must also fit the patient's eye. In some cases, particularly if the shape needed is uncommon, the lens shape may be provided completely from a flat plastic disc by machining and lapping. Other soft lenses are injection molded (4C1), but extreme care and the most sophisticated computer-controlled process is needed to achieve the correct shape, a smooth surface and freedom from internal plastic flow lines. These lenses may also be finished by lathe cutting and then lapping but, increasingly, the injection molding process is providing more finished lenses. Careful inspection is performed at critical stages during the process for all these methods. Magnification and shadow-graph methods are used to verify correct size, curvature, and freedom from internal flaws. Finished lenses are sterilized and boiled in a salt solution for several hours. The plastic absorbs the water and softens to the degree of flexibility needed for use. The lenses are then packed in individual bottles with salt solution approximating the composition of human tears, and are labeled for the left or right eye and with data identifying the size and correction factors.

licorice - is made from an extract of the roots of a group of fabaceae plants, that grow in Southern Europe, the Near East and parts of Asia. The roots are dug up, separated from the rest of the plant, and dried for several months. They are then crushed, ground up and boiled in water. The resulting juice is then strained and evaporated. The remaining material has medical uses, is used as a frothing agent in fire extinguishers and beverages and as a sweetener and flavoring in candy and tobacco. In making licorice candy, flour and starch are added and the mixture is extruded into the familiar candy shapes.

light bulbs, incandescent - are made fully automatically on special machines. Each bulb contains a tungsten filament, nickel-iron connecting wires for the filament, a sealed glass bulb, a glass mount for the filament wires, and an aluminum threaded base. See Fig. L4. A mixture of nitrogen and argon gases that replace air is contained in the finished light bulb. Also involved in many bulbs is a special white, light diffusing, coating powder for the inside of the bulb. The powder is "getter", a mixed material that removes the last of the oxygen from the bulb. A cement/sealant is used to fasten the threaded base to the bulb and mount, and a central insulator to the bottom of the base. The bulb is made by glass blowing. See Chapter 5, section 5A2b3, and

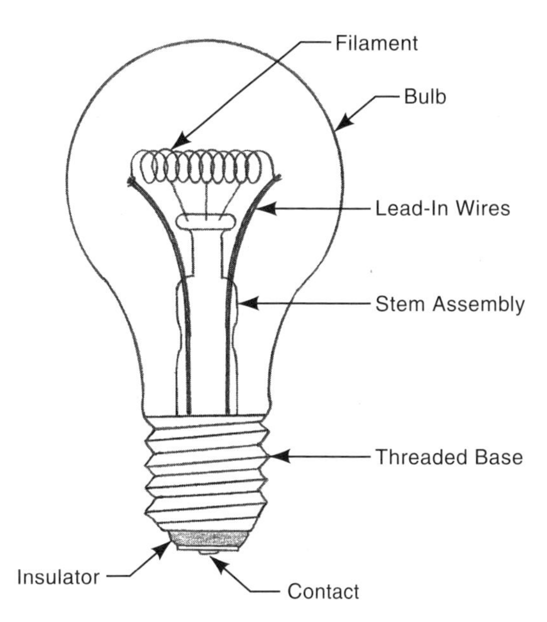

Fig. L4 A typical incandescent light bulb.

particularly 5A2b3f and Fig. 5A2b3f, which cover the specific ribbon-machine blowing method used for making light bulbs. The glass mount that supports the filament is made by machine pressing a gob of glass (5A2a). The tungsten wire filament is wound into a coil shape, and the filament connecting wires are formed in wire forming stations on the automatic light bulb equipment. (See 2G for forming methods.) The formed filament and connecting wires are pressed together to form an electrical contact, and are incorporated in the mount as it is formed. The threaded base is formed from sheet aluminum (or brass) by deep drawing, and special forming operations that create the screw threads. (See 2B2 and 2D5.)

The end of the threaded base includes a press-formed glass ceramic insulator. The glass bulb, after forming, is heated, and the white coating powder is sprayed inside. The mount, cement/sealer, the bulb, and the base are all assembled together by the machine at high heat levels that melt the cement. During assembly, one connecting wire is welded to the threaded base and the other is assembled through the base insulator and is upset by cold heading (2I2) to connect it to a central contact blanked from sheet brass. Identifying information is printed on the top of the bulb. The bulbs are tested before shipment and the first surge of electricity heats the getter and uses up any oxygen in the lamp.

lights, fluorescent - See *fluorescent lights.*

linen - is the name for both the yarn and fabric made from the flax plant, which also provides seeds from which linseed oil is extracted. The stems of the plant yield the fiber that is converted to linen yarn and cloth. Fiber strands from flax are typically 12 to 30 in (30 to 75 cm) in length. Plants are uprooted and piled in the fields to dry. After drying, there are a series of mechanical and chemical operations to prepare the fibers. The operations include retting, drying, scutching (crushing and beating), and hackling. In retting, the plant stems are exposed to natural microbiological action to promote partial decomposition and separation of the fibers. This step is accomplished by spreading stalks on the ground, exposing them to rain, drying, freezing, and thawing. Alternative approaches put the stalks in streams, ponds, or special tanks. After retting, the stalks are scutched mechanically and the fibers are separated from the woody portions. Hackling combs the fibers to separate the long lines from the tow (short fibers). The long line fibers are drafted and doubled to form a rove (slightly twisted sliver of flax fiber). Both the rove and the tow are spun into yarn and further processed into textile products. (See Chapter 10.) The line (long) fibers produce a strong yarn. The tow produce a heavy coarse yarn used for heavier fabrics and for knitwear.

Applications include apparel, tablecloths and other household furnishings. Linen fibers are strong, and lower grades are used for such applications as twine, fire hose, industrial sewing thread used in shoe manufacture and bookbinding, fishnets, and hooked rug backing.

liquid crystal displays (LCDs) - range from simple black and white (light gray) displays such as those used on digital watches or digital thermometers, to large color screens for high-definition television sets. In black and white displays, an electric current, directed to a particular point of the display, causes the molecules of the liquid crystal material to change their orientation so that light no longer passes through and the display appears black at that point. In other points, the display appears white.

Digital displays include at least the following: 1) two layers of sheet glass, 2) a film of silicon dioxide on each of the sheets, 3) a mirror in the back of the display to reflect light back to the user of the display. In computer and TV liquid crystal displays, a light source is used instead of a mirror, 4) two grids of conductors made of indium-tin oxide that provide energizing electric current to points in the display. (The pattern of the two conductors is identical except for the leads connected to them.) The thin-film conductors are applied as a layer on the coated glass sheets. 4) a coated layer of transparent plastic to provide proper alignment of the liquid crystals, (Polarization on one glass sheet is at right angles to that on the other glass sheet.) 5) a layer of nematic liquid crystal material (Nematic liquid crystals have twisted molecules) and, 6) plastic spacers, sealed to the other LCD components, to contain the liquid. Fig. L5 illustrates the construction of a typical simple display. In addition to the display, there is a source of a series of charges directed to the display from sensing, integrated-circuit or computer equipment. Active-matrix LCDs have a more complex structure, with transistors behind each pixel of the display.

The glass plates are made from borosilicate glass because it has a low ion content and stray ions could distort the display image. Some displays use a plastic sheet instead of glass. The plastic is ion-free but has poorer optical properties than glass. The glass sheets are cut to the needed size by diamond sawing or scribing and breaking, and are polished and washed. Polishing (8B1), referred to as *lapping*, is done with a polishing wheel embedded with abrasive. The glass is then coated with a layer of silicon dioxide. Chemical vapor deposition (See 13K3a3.) is the common means for making such coatings. The coating insulates the conductive grid that will be installed on the glass from ions that may be present in the glass. The conductive grid or pattern of indium-tin oxide is evaporated onto the glass surface in a complete, very thin layer by methods similar to those used to make circuits on printed circuit boards and integrated circuits. (See vacuum deposition, 13K3a4 and sputtering, 13K3a5.) Then, using photoresist methods or screen printing to create a mask (See 13A1a through 13A1f.), the unwanted portion of the coating is chemically etched away (13K3b2) to leave the desired display pattern. The conductive pattern is coated with a transparent plastic, usually polyamide (nylon). The coating is then brushed with soft material to create grooves in the plastic that will polarize light that passes through it. The grooves on one sheet thus processed are at right angles to the grooves on the plastic coating on the

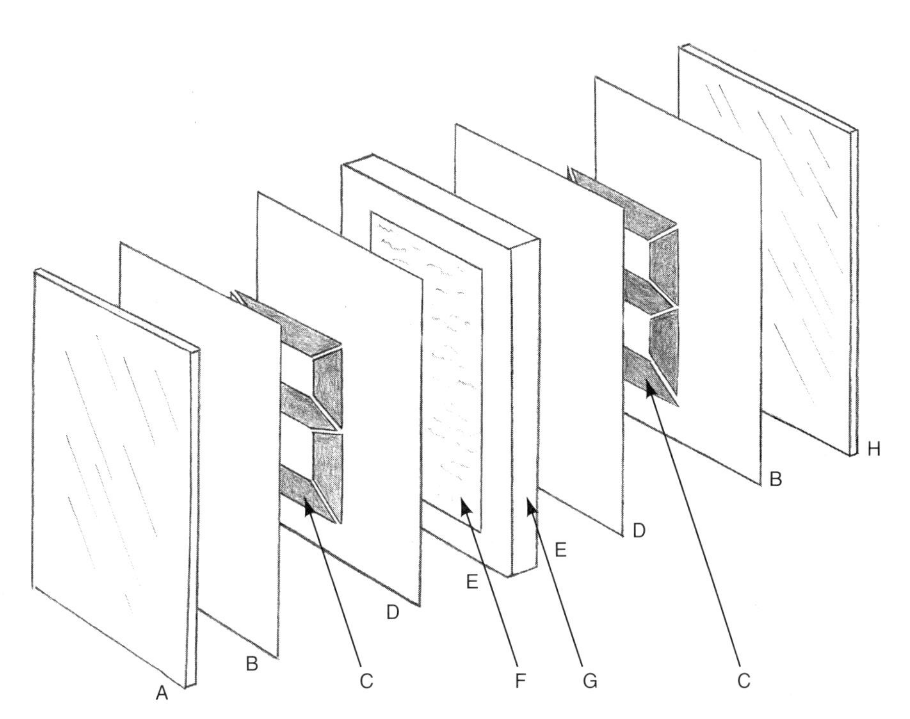

Fig. L5 The components of a typical simple liquid crystal display: A - front glass, B - silicon dioxide coating on one side of glass, C - grid pattern of a very thin layer of indium tin oxide, (7-segment numeral pattern, in this example), D - plastic coating over the grid and glass, E - sealant for the liquid crystal spacer, F - liquid crystal, G - liquid crystal spacer, H - rear glass with mirror coating on the back.

other glass sheet This method is one of several ways to provide alignment of the molecules of the liquid crystal. Other methods involve a different coating material or special application of the silicon dioxide coating of the glass.

When the display is assembled, a sealant is applied to the glass panels to contain the liquid crystal, which is injected after the spacer and glass panels are assembled. A critical dimension is the thickness of the liquid crystal, usually within 0.0002 to 0.0010 in (5 to 25 microns). The finished display is aged and mounted on the circuit board that provides the activating electronics.

Computer and other color displays require a much more intricate manufacturing process, especially with the active-matrix types. Clean-room conditions are required. Light for computer and TV displays is provided by a very narrow fluorescent light with a diffuser sheet to spread the light across the entire screen.

locks, combination - are made mostly from a series of sheet steel and die cast zinc parts. The parts used in a typical combination lock include: the outer casing (in two pieces, a cup and a cover),an inner casing, the shackle (U-shaped solid steel rod), combination discs that can rotate on a central stud (When slots in the discs are in alignment, the lock can be opened.), two studs, a latching device that may have an internal spring and a sliding element, a disc cam, a numbered dial (with knob), and some latching parts and fasteners. The dial and latching parts are die cast (1F). The discs are blanked (2C4) from sheet steel. Casing parts are blanked and drawn (2D5), trimmed, and shackle openings are punched in the side. The studs are screw machine parts (3A2c). The shackle is made from steel rod, cut to length and grooved on one end on a screw machine, notched by a milling machine operation (3D), press-bent cold (2H2c), heat treated for hardness (8G3), and chromium plated (8C1). The outer casing parts are also chromium plated and the dial is spray painted (8D6) (black with the recessed numbers painted white and wiped). Inner parts are zinc barrel plated (8C3) for corrosion resistance. The parts are carefully assembled. Studs are riveted to mating stampings by upsetting one end. A removable tag or label with the combination is assembled to the lock. The cover is crimped in a press operation to permanently close the internal parts. The locks are then blister packaged for retail sale.

low-density fiberboard - See 6F6.

lubricating grease - See *grease, lubricating*.

lumber - is made from logs as follows: 1) Logs are transported from the forest to a sawmill. 2) The logs are placed on a moving deck to move them to a debarking machine. 3) Logs are debarked by a machine with a rotating head equipped with a series of flexibly-held cutters that abrade the bark surface of the log. The log also rotates on its axis and, as the bark is removed, there is axial motion between the cutter and the log. 4) The log is moved to a headsaw, a large-diameter circular saw or a band saw accompanied by a device that holds the log and moves it axially against the saw blade. In one common system for cutting the log, the headsaw makes four cuts, changing the round log to a "cant" of essentially rectangular cross section. 6) The cant is moved to the resaw operation where a bandsaw cuts it into a series of boards. 7) The boards are run through an edger that trims off rounded edges and any bark remaining at the edges, but otherwise produces as wide a board as possible, or one of a prescribed width. The edger has two blades so that the width of the board is cut in one pass. 8) The ends of the boards are trimmed so that the boards are of prescribed length. 9) The boards are sorted by length, quality, thickness, and width. 10) Boards are stacked with like boards, with separating strips between layers, and the stack is transported to a location for air drying. 11) After air drying, which may require several weeks, the boards are kiln dried. 12) The boards are then planed on all four sides to more exact dimensions and smoother surfaces, and then are re-stacked without the separating strips, preparatory to shipment to customers. See Chapter 6 for further information and illustrations.

lumber, pressure treated - is conventional softwood lumber, usually of a high grade, that is pressure impregnated with a liquid consisting, in the United States as of 2004, alkaline copper quat (ACQ), a pesticide with anti-bacteria and anti-fungal properties. The ACQ was adopted by the industry to replace chromium copper arsenate (CCA) because of health concerns. The lumber to be treated is placed in a treatment cylinder and the cylinder is put under a vacuum for a period. This sequence extracts moisture from the wood. Then the ACQ is introduced and pressure is applied, which forces the chemical into the pores of the wood.

M

magnesium - is commonly produced from sea water, which contains magnesium chloride. The first step in the frequently used Dow process is to mix lime (calcium oxide, CaO) with the sea water. The lime reacts with the magnesium chloride to produce magnesium hydroxide and calcium chloride. The magnesium hydroxide precipitates and is filtered from the mixture. It is treated with hydrochloric acid, and the reaction creates magnesium chloride and water. The water is evaporated and the resulting magnesium chloride is melted by heating it to 1310°F (710°C). The next step is electrolysis which converts the molten magnesium chloride into magnesium and chlorine gas. (See 11C10 and Fig. M1.) The magnesium, still in a molten state, is cast into ingots for further processing.

Magnesium, alloyed with copper or aluminum, is used in cast parts for aircraft, lawn mowers, vacuum cleaners and other products where light weight is important.

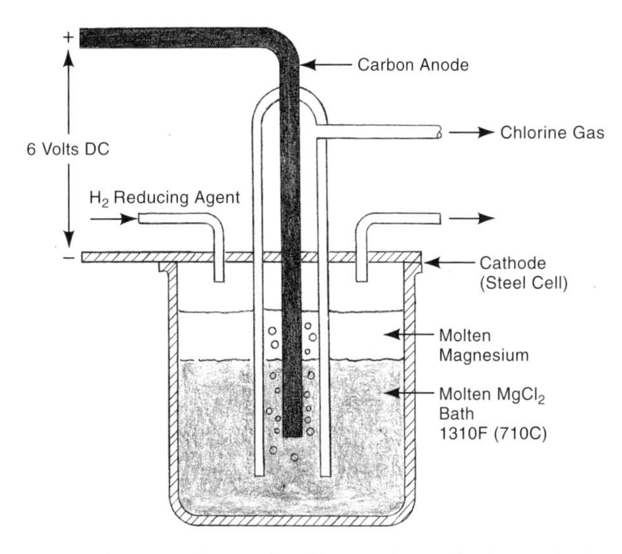

Fig. M1 A schematic illustration of electrolytic refining to produce molten magnesium and chlorine gas from molten magnesium chloride.

magnets - can be made from several materials. There are two basic types of magnets: electromagnets and permanent magnets. Electromagnets obtain their magnetism from current flow in wire coils that surround the magnetic material. The magnetism occurs only when current flows in the coil and the core material is engineered to lose its magnetism when the current stops. Permanent magnets are engineered to retain their magnetism indefinitely and do not require surrounding electrical coils except when they are initially magnetized.

Electromagnets are commonly made from iron and iron alloys. Silicon is one alloying metal. The magnetic core may be made from layers of annealed sheet iron, from powdered iron compressed with a binder into the desired shape, or the core may be cast.

Permanent-magnet materials include various iron alloys, alnico alloys (iron alloyed with aluminum, nickel and cobalt) and a number of ceramic materials. Prominent among them are barium ferrite and strontium ferrite. Others materials are samarium-cobalt and neodymium-iron-boron. The latter makes the strongest permanent magnets. These materials are reduced to a fine powder and are processed into magnets with powder metallurgy techniques (2L1).

The process for making neodymium-iron-boron permanent magnets is as follows:

1. The three metals are melted together in a vacuum and cast into ingots (1D5).
2. The ingots are broken up, crushed (11D1) and pulverized (11D3) in a ball mill into a fine powder.

3. Using powder metallurgy techniques, the powder is compressed in a suitable die to within about 1/8 inch (3 mm) of its final thickness. In one difference from conventional powder metallurgy, the powder is subjected during compaction to a magnetic force in order to align the particles magnetically in the same direction.
4. The "green" workpiece is removed from the die and is sintered (2L1d).
5. The workpiece is heated again to anneal it, remove internal stresses, and thereby strengthen it.
6. The workpiece is machined as necessary (Chapter 3), to bring it to the desired final dimensions.
7. A protective coating is applied.
8. The workpiece, which up to this point is not fully magnetic, is subjected to a very strong magnetic force from a surrounding electromagnet, and the force is held for a period of time. This energizes its permanent magnetic properties in the desired direction.

manganese - occurs in nature in the ores pyrolusite (MgO_2), the principal ore, manganite, and in most iron ores. Manganese is a common alloying metal with steel, and also is alloyed with some non-ferrous metals. For use with steel, ferromanganese is made from manganese ores - that also contain iron - by a smelting process in blast furnaces or electric furnaces. Electrolytic manganese is produced from low-grade ores by an electrochemical process.[2] Another method for producing metallic manganese is to ignite pyrolusite with powdered aluminum. In steel, manganese facilitates hardness and wear resistance. It acts as a deoxidizer in steel making.

maple syrup - is made by concentrating the sap of the maple tree. The sugar maple tree, acer saccharum, has the highest percentage of sugar in its sap, but the sap still must be concentrated to about one fortieth of its volume to produce syrup. The sap is gathered in the early spring when it flows up from the tree roots on warm days. In a typical maple farm, it is obtained from the tree by drilling holes in the trunk and placing spouts called spiles in the holes. Flexible tubing, attached to the spiles, carries the sap to a central processing shed, called the "sugar house". There it is boiled in large covered pans heated by a wood-burning fire or by an oil or gas flame. The water gradually boils off. When the proper degree of evaporation has taken place, the syrup is filtered, usually under pressure, and is bottled.

marbles - are made from glass, both glass made from silica sand and that made from cullet, glass reclaimed from scrap bottles and other scrap material. (See 5A1 - basic glassmaking.) These materials are heated to 2300°F (1260°C) to melt and blend them. Then the molten material is discharged into another container, a flow tank into which glass of a contrasting color is injected. Coloring is added to almost all batches. Iron oxide gives glass a green color; manganese provides purple; cobalt, blue and uranium oxide, chartreuse. Most marbles are made with two colors. The two colors are mixed partially so that there are streaks of both colors in the mix. Cube-shaped gobs of the molten mixture are placed in a shaping machine. In this machine, the cubes of molten glass move between two parallel cylinders that have spiral-oriented slots, similar in appearance to screw threads. As the cylinders rotate, the glass material is rolled in the slots so that, as it moves along the cylinders, it gradually assumes a spherical shape with curved color streaks. When the material has solidified and cooled, inspection takes place, and rejected marbles are returned for remelting. Good marbles roll to a packaging machine where they are packed in bags or boxes.

margarine - is an emulsion of fats and oils in water or milk. Emulsifying agents and other additives may also be used. Equipment similar to that used in butter making is used to churn the mixture.

Cottonseed oil is a common base-material for margarine. The process of making margarine from cottonseeds includes screening (11C8a), cleaning (12A) and delinting the seeds. Delinting is performed with machines similar to cotton gins. The seeds are then split, and the meat and hulls separated by screening and air classification. The meats are rolled into thin flakes. These are cooked for 20 minutes at 230°F (110°C) to open the oil glands, and are then fed to screw presses to express (12E2) the oil. The oil is screened, cooled, and filtered (12E1), and held for further refining. The meat of the seeds is pressed again with hexane solvent and then is subjected to further solvent extraction (12E5). The hexane and oil are separated, and the resulting oil is combined with that from the first expression. Further refining is carried out by degumming the mixture with a small amount of phosphoric acid and processing the mixture by centrifugation (12E3). The liquid is then bleached by adsorption (11C9) using bentonite. Hydrogenation (12J3) follows with a nickel catalyst, after which the product is deodorized, colored, flavored, and churned. Like butter, margarine is cooled, extruded (12J5), cut, wrapped and packaged with automatic machinery.

marking pens, felt tipped - Felt-tipped marking pens differ slightly in design from maker to maker but often consist of seven components: a body, a rear plug for the body, a fiber filler for the pen, a flexible plastic tube to hold the fiber, ink to saturate the fiber, a writing tip, and a cap to fit over the writing tip. The body, rear plug, and cap are injection molded (4C1), normally of polypropylene plastic. The body is printed (9D1b), hot stamped (4M2), or label-wrapped with identifying information. The filler is made from absorbent fibers and is inserted in the extruded (4I1) flexible plastic tube, which is open at both ends. Ink is injected into the fibers with a needle-like injector and the plastic tube helps keep it from drying in the pen. The ink consists of dyes/pigments dissolved in an ethyl-alcohol solvent. However, some highlighting pens use a water-based ink. The tip is made from polyester felt (10E), compressed and bonded tightly with heat in a compression-molding operation (4B1) to form a rigid but porous part. It is then trimmed to an angle for writing. The tube of ink-saturated fiber is inserted in the body from the back end. The writing tip is inserted in the other end of the body and the back cap is pressed into place. Both the back cap and tip fit tightly. The size and fit of these parts insures that the back end of the writing tip always presses against the inked yarn so the ink can flow to the tip by capillary action. All assembly operations are performed on special automatic equipment (7F3b) that assembles them, without human handling, at a rate of about 3500 markers per hour.

matches - are made in special-purpose, highly-automatic equipment. Match sticks are cut from pine or other wood in a series of cutting operations, or are die-cut from paperboard. With wood matches, the initial operations from logs are as described in Chapter 6 (Woodworking), but the final steps in making the sticks are shearing operations. In safety matches, the striking surface is coated with a mixture of ground glass, glue, and red phosphorous. The match heads are coated with antimony sulfide, and contain starches or gums, a glue binder, clay, or diatomaceous earth to regulate the flame. In strike-anywhere matches, the ingredients are all in the match head except for a ground glass friction striking surface. To coat the heads, thousands of matches are dipped at a time and then dried in special machines. Packing in cardboard is the final operation. Friction from striking the match generates heat that starts a reaction between the coatings, creating sufficient heat so that normal combustion of the match can continue.

meat - See 12K.

meat tenderizer - is papain, the dried extract of sap and the fruit of the papaya tree. The sap is dried at a temperature below 158°F (70°C) Because higher temperatures destroy the tenderizing enzymes. The final tenderizers is in either powder or liquid form.

melamine plastic - See *urea and melamine plastic.*

mercury - is derived from the mineral cinnabar, which contains mercuric sulfide, HgS. The mercury is obtained by a distillation process (11C1). The ore is first concentrated by various processes including flotation (11C8b) and screening (11C8a). After concentration, it is heated in the presence of air. The sulfide is converted to sulfur dioxide, and the mercury is freed as a vapor. The mercury vapor is then condensed to produce pure liquid mercury. Mercury is used in mercury vapor lights, in thermometers, in batteries, and in separating gold and silver from their ores. Mercuric sulfide and mercuric oxide are used as red pigments. All mercury compounds are poisonous.

medium density fiberboard - See 6F6.

metal cans - See *cans, metal.*

metal powders - See *powders, metal.*

methane - (CH_4), is the major constituent of natural gas and is retained in the gas when it is used for fuel. Removal of minor unwanted compounds (e.g., water and hydrogen sulfide), and other valuable constituents (propane and butane) from the gas leaves methane. Removal of these other ingredients is achieved by a combination of compression, refrigeration, and adsorption (11C9) operations. Methane is also made by reacting carbon monoxide with hydrogen, by the reaction of water and aluminum carbide (Al_4C_3), or by the destructive distillation (11C1g) of coal. Methane is an important raw material for many plastics, fertilizers, explosives, and various chemicals, including methanol, chloroform, carbon tetrachloride, and carbon black, used in tire making.

microcircuits - See 13K.

microspheres, glass - See 5A7d.

milk, condensed - goes through an evaporation process (12F1) to remove water and increase the solids content. It is usually performed with heat but under a vacuum to reduce the temperature needed. Solids content increases from 8.6% to 45%. In a two-stage process, a temperature of 154°F (68°C) is used in the first evaporator and 115°F (46°) in the second. Condensed milk is used chiefly to replace cream in cooking baked goods, ice cream, cheese and candies. *Evaporated milk* is very similar but usually refers to the product when the milk is heated, concentrated, and sterilized, in individual cans. Sweetened, condensed milk contains sugar which acts as a preservative.

milk, powdered - is produced by spray drying milk (primarily skim milk) in an airstream. (See 12H2.)

milk, skim - Skim milk (now designated in the U.S. as "Fat Free" milk) is whole milk with the cream removed. In production situations, a centrifugal separator is used. (See 12E3 and 11C7e.) The milk is introduced to a bowl that rotates at 6000 to 10,000 rpm, generating centrifugal forces up to 500 times the force of gravity. The bowl contains a series of stacked cones that separate the milk into thin layers. The cream, being lighter, tends to remain in the center and rise to the top of the bowl, while the centrifugal forces drive the heavier skim milk to the periphery and the lower part of the bowl. (See Fig. 11C7e-2.) The milk is usually heated to a temperature of 90°F (32°C) to as high as 160°F (71°C) to facilitate the separation of the two portions.

mineral wool (rock wool) - See5A6d.

mirrors - use flat glass of high quality so that there is no distortion of the reflected image. Float glass (5A3f) or plate glass (5A3e) is used. The pane used is first washed and treated with a stannous chloride solution to activate the glass surface. Then, the back surface is sprayed with a solution consisting of silver nitrate, ammonia, caustic soda or caustic potash, dissolved glucose, and distilled water. This treatment causes a fine crystalline film of solid silver to form on the glass surface. The coating is built up to a thickness of 0.0004 inch (0.01 mm). A similarly-applied coating of copper is added, followed by two coats of lacquer, all for protection of the silver coating. (See 5A5j, metallic coating of glass.)

molybdenum - is made from the minerals molybdenite and wulfenite, and as a by-product from copper refining. Molybdenite, the principal ore, is roasted with air to form molybdenum trioxide (MoO_3). Mixing this with iron oxide and igniting it produces ferro molybdenum, which is used as an alloying agent in steel. Metallic molybdenum can be made as a powder by reacting the trioxide with ammonia to produce ammonium molybdate $(NH_4)_2MoO_4$ and then reducing this with hydrogen. The powder can be formed into useful shapes by powder metallurgy (2L1) techniques, or by arc-melting and casting it in a copper mold. Molybdenum is principally used as an alloying ingredient in steel to provide greater strength, toughness, and corrosion resistance. Applications also include heating elements, electrical contacts, high-temperature aerospace components, and as a flame-resistant coating for other metals[2]. Molybdenum disulfide is a high-pressure, high-temperature, solid lubricant.

monuments - *Commemorative stones* for public areas and *gravestones* are commonly made from granite rock. The manufacturing steps involved are as follows: 1) Stone blocks are removed from the ground by one of two methods: a) explosive charges in deep holes drilled into the rock by diamond drills, or, b) by cutting out blocks of stone using a metal wire coated with diamond abrasive. The wire acts somewhat like a bandsaw against the granite. 2) The stone is then cut into monument stones, close to final size by either a large circular saw coated with diamond abrasive, or by diamond beads on a moving wire. When curved surfaces are needed, the wire traverses a curved path, guided by a template or computer control. 3) The stone workpiece is then polished on the front and back surfaces as required to remove saw marks and provide the desired appearance. This operation is done by large machines with worktables of 20 by 60 to 20 by 100 ft (6 by 18 to 6 by 30 m). A group of stone workpieces, all to be made to the same standard thickness, is placed on the table. An overhead gantry structure, which can move the length of the table, holds a motorized vertical spindle and a disc on which are mounted carborundum (aluminum oxide) or diamond-coated abrasive stones called "shoes". The shoes rub against the stone workpieces, as the gantry moves forward and back, and as the spindle traverses the width of the gantry, so that all surfaces of the monument stones on one side are contacted repeatedly by the shoes. Repeated passes with progressively-finer abrasives are made until the stone achieves the desired thickness and degree of polish. 4) The stones are all turned over and the polishing is repeated for the opposite side. 5) The top surface of the monument is similarly polished if it is flat, and is polished manually with hand-controlled polishing wheels, discs, or belts, if it is curved. (Some monument surfaces are given a rock-like texture by artisans using hand chiseling.) 6) Carving of any images and lettering is done by "sandblasting" the surface with steel shot in the open spaces of a rubber stencil mask that is temporarily glued to the stone surface. (The rubber stencil is made in layers so that wider and deeper sections of the engraving can be made by peeling off part of the mask and sandblasting again.) This operation is partly automatic in that the movement of the stone and the sandblasting nozzle are machine controlled. Deeper carvings may be made by hand by an artisan using a vibrating carbide-tool chisel. 7) After carving, the stone is buffed and washed. 8) Some lettering may be spray painted. 9) The finished monument is crated and shipped to the dealer who installs it at the desired site on a base piece which, in turn, is on the necessary concrete footing. The installation follows a careful, lengthy procedure designed to prevent movement of the stone over a period of time.

motors, electric - Each electric motor, whether used with alternating current (AC) or direct current (DC), consists of an armature (a rotating electrical conductor) or a rotor attached to a central shaft, a stationary magnet or stator, and an air gap - a narrow space - between the rotor and stator. Magnetic effects are achieved either by electromagnets - coils of wire surrounding a stack of laminated sheet steel - or by a permanent-magnet structure. Coils of wire are used in either the armature or stationary field, or both, to create electromagnets. Direct current and universal motors have a commutator - a device to reverse the direction of current when the rotation of the motor armature passes a certain point - and carbon brushes to convey the electric current to the armature. Most AC motors do not have commutators or brushes. Some depend on the regular current reversal to achieve the desired motion. The speed of the motor is then proportional to the frequency of current reversal, and these are synchronous motors. Others for AC are induction motors that have copper bars in slots in the rotor. The bars are connected to rings at the ends. A current is induced in the bars and this produces a magnetic effect in the rotor that reacts with the field magnet, causing the rotor to turn. Fig. M2 illustrates the component parts of a typical universal (AC or DC) motor.

All rotating electric motors have a central shaft that holds the armature or rotor and also holds a pulley, gear, or other means to transfer the motor's rotation and power to other devices. They include bearings for the shaft and an external housing for protection. Almost all have an internal cooling fan. The fan is normally blanked (2C4) and formed (2D2) from sheet stock and is press fitted on the central shaft. The shaft is turned on a screw machine (3A2c) from long lengths of steel bar drawn (2B2) to the proper diameter. Shafts may be milled (3D) to a D-shape or spline, or may be knurled (3A1f) for the necessary torque-handling strength. Bearings are either commercial ball bearings or bronze bushings made by powder metallurgy processes (2L1). The laminations are stamped on high speed blanking presses from thin sheets of a special steel alloy which has desirable magnetic properties.

Commutators for DC and universal motors are cold formed from coiled sheet copper by special blanking and forming press operations. The segments produced are inserted in a mold with the central mandrel. Phenolic plastic is compression molded (4B1) around these inserts to produce the full commutator, which is then machined by turning (3A2 and 3A2d) to provide a smooth surface at the brush contact area. Strips of sheet mica or synthetic insulation may be placed between commutator segments prior to molding. The insulation between the segments is then undercut slightly on special machines to ensure a smooth electrical connection to the brushes. Field coils are wound on coil-winding machines (13L3 and 13L4) on forms of the shape needed to fit the laminations, wrapped with tape at least partially to hold the wires together, and then are assembled to the stack of field laminations. Sheets of vulcanized fiber, or of polyester, are usually placed between the laminations and coils to provide both mechanical protection to the wiring

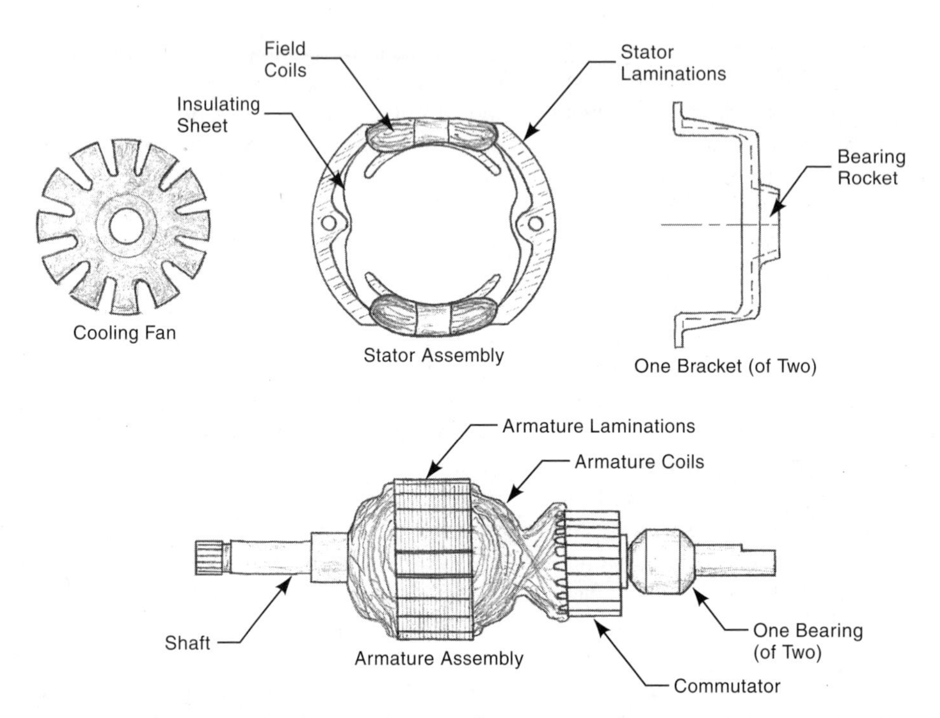

Fig. M2 The principal components of a small universal electric motor (for sewing machines). Two brackets hold the armature assembly in alignment with the stator assembly with a small air-gap between the two. An external plastic or sheet metal housing protects the motor. Only one of the two brackets and two bearings is shown. Also not shown: brushes and brush holders, external wiring, and housing parts.

and electrical insulation. Coils for armatures of DC and universal motors are often wound directly on the armature which, at that stage, has the armature laminations and commutator press fitted on the armature shaft. Small sheets of vulcanized fiber are placed in each slot in the laminations to protect the windings from sharp corners.

Each armature coil wire is connected to the appropriate commutator by crimping and spot welding the connection point. Then the windings are given a thick coating of varnish (usually epoxy) by dipping. The varnish provides supporting strength to the coil against the centrifugal force of rotation, and gives additional protection and insulation. Brushes are pressed and sintered from carbon graphite powder, and are held in square brass sleeves, that are made from strip stock on four-slide machines (2G2). The sleeves provide electrical contact with the brushes. The motor housings are welded assemblies of various blanked and formed sheet steel parts. Brackets and other parts are resistance welded (7C6) to the housing. Sometimes, the housing is deep drawn (2D5b) from steel or die cast (1F) from aluminum. The housing assembly is spray painted (usually electrostatically - 8D7) to the color needed. Some housings are injection molded (4C1) of a thermoplastic resin able to withstand the heat that the motor develops. Modified polyphenylene oxide is one plastic used. When a plastic housing is used, brackets, stamped from sheet steel, are fastened to the field laminations to hold the bearings and rotating parts.

The assembly of motors, particularly the placement of coils, is heavily manual but robotic assembly is employed in some cases. Components attached to the main shaft are normally fastened by press fitting. Electrical connections of the field coils and brush sleeves are usually made by soft soldering (7A). A running test is a final assembly line operation before packing for shipment.

multiple chip packages (electronic) - See 13N1.

musical instruments - See *guitars* and *musical instruments, brass*

musical instruments, brass - include trumpets, cornets, trombones, french horns, baritone horns and tubas, all of which are made from brass with very similar manufacturing methods. (though tubas - Sousaphones - for marching bands are made, in the larger sections, of reinforced plastics for weight reduction.) Except for stainless steel screws, the electroplating of some parts with chromium or nickel alloys, silver, or gold, key finger pads and valve seals, only brass is used in these instruments. Yellow brass (70% copper and 30% zinc) is the most common alloy. Another is gold brass (80% copper, 20% zinc) and a third is silver brass (copper, zinc and nickel). The brass material is supplied to instrument makers in sheet and tubular form. Most production work involves a combination of manual and machine operations. With some instruments, particularly those with special requirements, the operations are highly manual and are performed by skilled artisans. With those made to standard specifications, for example, those instruments intended for beginner or school use, much more use is made of methods involving machines and production tooling. The operations involved for either type include parts making, assembly, final finishing and testing.

Brass tubing for most parts of the instrument is drawn (2B2) to the diameter needed. Tubing that is drawn, however, usually must be annealed (8G5b) and then cleaned by immersion in dilute sulfuric acid to remove surface oxidation that forms during the annealing. Annealing is also required whenever severe working is done on the parts in spinning, flaring, tapering, and other operations. Some tubing pieces must be bent to 45, 90, or 180 degree angles and this bending is done with conventional tube bending equipment, usually with internal ball mandrels to prevent the tubing from collapsing (2H2 and Fig. 2H2a-1) during bending. Some shops use water under a high pressure of 3900 psi (27,000 kPa) instead of a ball mandrel, to prevent the walls from collapsing, and others may fill the tubing with pitch or solder that is melted out after the bending is complete. The bell of the instrument is formed from sheet material in a series of blanking (2C4), forming (2D2), and flaring press operations but, in small artisan shops, may be hammered to the approximate final shape. With either approach, it then undergoes torch brazing (7B2) to fasten the ends together to form a round shape. The edges are overlapped, held in place with a fixture and brazing alloy is introduced to the overlap area. Brass-colored brazing alloys containing silver and phosphorous are used, with the appropriate flux. Some hand hammer-and-anvil shaping takes place before the bell end is given its final form by metal spinning (2F6). Where two ends of tubing are joined, flaring or swaging (2D12) or both are used so that the end of one piece fits into the end of the other. The bead at the open end of the horn is made as part of the spinning operation. A wire is laid inside the bead to facilitate its formation. If the instrument has valves (all do, except trombones), they are made from heavier-walled tubing and are machined to the required dimensions in lathes (3A), as are the valve pistons and finger buttons. Computer-controlled machining (3U2) may be used, especially when rotary valves are involved. Cross holes are drilled in the valve bodies where they join tubular parts. Rotary-tube saws are used as drills, and the parts are held in fixtures during drilling to insure correct location of all holes. Finger rings, hooks, a water-release key, and other smaller parts are usually press formed. Mouthpieces are cast and finish-machined on lathes or made from brass bar stock on screw machines (3A2c) and then either polished or electroplated (8C1), usually with silver but sometimes with gold.

When all parts are completed, the mating surfaces are cleaned with mild abrasive, the parts are inserted in assembly fixtures where they are fastened together by torch brazing (7B2). There may be several stages of assembly where already-brazed subassemblies are joined together, again by brazing. The fully-brazed assembly is then immersed in an acid bath to clean off all scale and residual flux. It then undergoes a complete polishing (8B1) and buffing (8B1a) operation. The entire instrument then is cleaned, final-polished, and either electroplated (with silver or gold) or lacquered. Identification information may be engraved (8I4) or acid etched (8I3) into the surface of the instrument. Then, the final assembly takes place with the mechanical valves, valve pistons and mouthpiece. Lastly, the instrument is tested by a qualified musician and adjusted as necessary. If the test is successful, the instrument is wrapped and packed for shipment to the customer or retail outlet.

N

nail polish - is made from nitrocellulose resins and plasticizers dissolved in an acetone solvent. Dyes and pigments are added for color. Acetone is commonly made by the dehydrogenation of isopropyl alcohol.

nails - are made on special cold heading machines (2I2). Most nails are made from carbon steel, but aluminum, brass, copper, stainless steel, and other alloys may be used. Wire from coils is fed to the machines where it is clamped securely by gripper dies. While the wire is gripped, it is struck on a protruding end by a die with a cavity that forms the nail head. Then, with the wire still gripped, a set of shaped punches strike the wire at another location to form the point of the nail and sever the nail from the wire coil. The grippers then open and the nail is ejected into a chute below. The wire in the coil then advances into the grippers and the operations are repeated to form another nail. This sequence takes place at a rapid rate, as many as 700 times per minute. The formed nails may then be fed into another special machine that forms ribs in the shank of the nail or twists it to provide a spiral shape.

Many varieties of finishing operations then take place, depending on the intended application of the nails. Most are first fed into a tumbling barrel for alkali cleaning to remove the forming lubricant and any small pieces of metal left from the forming of the point. Other nails are galvanized (8F2), or are barrel electroplated (8C3) with zinc or another metal. Some nails are painted to provide a matching color for their application. Others are heated to provide a blue color and some corrosion protection. Still others are coated with a plastic for corrosion protection and the plastic may contain ceramic powder to improve the nail's holding power. Nails intended for masonry uses are heat treated to increase their hardness (8G3c).

All nails are packaged automatically with the quantity controlled by weight. Some go into small boxes that hold a few ounces, or into other boxes of 1, 5, or 10 pounds each.

nameplates - See 8I12, 8I2, 8I4, 8I4a, 8I5, 8I7 and other items under 8I (Product Marking).

naphtha - See 11H1a and 11H2a1.

napkins, paper - See 9C4 and 9B5.

napped fabrics - See 10F3a

natural gas[4] - is recovered from underground and undersea deposits through oil and gas wells. The gas is formed from the decomposition of marine plankton and other organic matter, and from geologic changes over millions of years that cause it to be held underground. It accumulates in pockets, and often accompanies crude oil which is formed from the same or similar organic matter. Its major constituent is methane (CH_4).

Before distribution, natural gas may be treated to condense some of the less-volatile hydrocarbons it may contain, notably propane and butane. These are used for liquified petroleum gas or other products. The remaining gas is purified to remove water and hydrogen sulfide, which are incompatible with pipeline transmission. Carbon dioxide is also removed, because it lowers the heating value of the gas. Water causes corrosion and operating problems with valves and regulators and the formation of unwanted hydrates. Sulfur compounds also cause corrosion and undesirable odors, and air pollution when the gas is used. Water is removed by compression or refrigeration, which cause contained moisture to condense, by adsorption, or by treatment with drying materials. Refrigeration is less used because is usually more expensive. When drying substances are employed, glycols are predominant because of their good drying characteristics and freedom from reaction with, or absorption of, the gas. Silica gel, activated alumina, and other drying agents can also be used. When this type of drying is employed, the equipment used regenerates the drying agent periodically. Hydrogen sulfide and carbon dioxide are removed with a variety of processes that use reagents or solvents. The most common process uses monoethanolamine to absorb the hydrogen sulfide.

The processed gas may be liquified for tanker transportation, or may be transported as a gas in pipelines.

natural rubber - See 4O1.

needlepoint carpets - See 10I8.

needlepunch carpets - See 10I4.

neon signs - are made from glass tubing, filled with neon or argon gas, and subjected to a high voltage/low amperage current, which causes the gas to glow. The fabrication sequence is as follows: Straight glass tubing is heated, where the bends are to take place, to a temperature of 1100 to 1400°F (600 to 760°C) which softens it sufficiently for hand bending. Mild air pressure in the tube prevents it from collapsing during bending. The glass is bent to match the curvature desired in the sign lettering. A full-size template drawing is usually used to guide the placement and angle of the bends. Once a series of bends has been made, electrodes are welded to the ends of the tubing and a small tube is welded to one end. The small tube is connected to a vacuum pump and the air in the tubing is evacuated. A 25,000 volt charge is applied to the electrodes. This creates a white glow in the tube and, more importantly, burns out impurities in the tubing. Neon or argon gas is then introduced to the tubing and the small filler tube is closed. The colors of the sign are controlled by choosing colored glass tubing, and by the selection of either neon or argon gas. Neon is used when the desired color of the sign is red, pink or orange. Argon is used for blue, greens, or whites. Portions of the tube not wanted in the sign lettering are painted an opaque black. Operation of the sign involves application of potentials of 2000 to 5000 volts at low current levels, from a suitable power supply.

neoprene - See 4O2.

newspapers - are printed on a special type of low-cost paper, *newsprint* (9B and 9C2). Before printing takes place, considerable preparation and composition work is needed. Each news story is written by a reporter on a computer. The story, and all other articles with their accompanying illustrations, captions, and headlines, stay in digital electronic form throughout the editorial and composing steps until printing plates are prepared. The data are moved within the paper's offices, and to the printing plant, on a local area network. The story is first reviewed and often modified by a city-desk editor, and then by the paper's news editor. Stories that originate out-of-town from news services arrive at the paper in digital form, by wire. The news editor decides which stories to run in the paper, where they will appear in the paper, what kind of headline should be used, the size of type, the final length, and other details. A make-up editor prints a copy of the story and inserts it into a "dummy" copy of the page, including pictures, tables, headings, advertisements, etc. When this is approved by the editor on duty, it is still in digital form, but is ready for the printing process.

The page is then printed on plastic film by laser. Scanned pictures are included. There may be two stages of plastic film, but the final version includes several pages and is converted into a negative that is used to prepare the final printing plates. Most current newspaper printing is by the offset litho-graphy process (9D2 and 9D2a). Some newspapers use flexographic letterpress printing (9D1a1 and 9D1b).

Newspaper printing takes place on extremely large, rotary web-printing machines. Paper is fed to the presses from large continuous rolls, and as many as eight copies may be printed simultaneously (9D6). Both sides of the paper are printed simultaneously with color utilized for some pictures and type (9D7 and Fig. 9D2a). Equipment is included to cut and fold the pages automatically, as part of the printing line. Speeds are as high as 60,000 copies per hour.

newsprint - See 9C2

nickel - is most commonly made from pentlandite and nickel-bearing pyrrhotite sulfide ores. The ores are first crushed (11D1) and pulverized (11D3) and separated from gangue by flotation (11C8b). The concentrated ore is then smelted to produce a matte (an impure mixture of sulfides) of copper and nickel. In the electrolytic process for separating the two metals, electrolysis first removes the copper; then, with a different electrolyte and voltage, metallic nickel is deposited. In another process (the Mond process), the matte is dissolved in dilute sulfuric acid, removing the copper and leaving impure metallic nickel. The impure nickel is subjected to a carbon monoxide atmosphere, which converts it to nickel carbonyl gas. This gas is heated to 392°F (200°C) where it decomposes, depositing pure nickel. Nickel is used in alloys with steel, notably stainless steel, in Monel copper/nickel alloys, in Inconel and Hastelloy temperature and corrosion resistant alloys, for electroplating and in coinage.

nitrile rubber - See 4O2.

nitrogen - is a common element which, in gaseous form, makes up 78% of the volume of the earth's atmosphere. For commercial use, it is obtained from the fractional distillation of liquid air. Membrane separation (11C7c) is another process used. The boiling point of nitrogen is −320°F (−196°C) sufficiently below the −297°F (−183°C) boiling point of oxygen so that, in fractional distillation, it distills off first and is collected. (See 11C1 and *oxygen*.) Nitrogen gas is used in many chemical processes, where its inert properties are needed. These include its use in shielding gas in welding and brazing, for foaming plastics and rubber, as a diluent for reactive gases, to prevent oxidation during chemical reactions and to prevent explosions and fire. Nitrogen is also used in the manufacture of ammonia gas. Non-gaseous applications include nitric acid, fertilizers, explosives and many organic compounds. Liquid nitrogen is used as a cryogenic refrigerant.

no-clean solder flux - See 13G.

non-woven fabric - See 10E.

nuclear power - See *electricity* and Fig. E1.

numerical controls - See 3U and 3U1.

nutmeg - See *spices*.

nuts, screw - Standard commercial nuts are made on special machines, cold nut formers. Low-carbon steel wire in coils, either round, or drawn to a hexagonal shape, is fed to the machines. A shearing die cuts off a workpiece from the wire, and it is positioned in a holding die on a rotary indexing table. A series of press strokes by different punches, as the table rotates, flattens the workpiece, forms the hexagonal shape as necessary, chamfers the two sides, preforms a chamfer for a center hole, punches the center hole, and then burnishes it with another punch to control the diameter and surface finish. (The diameter is held to +/− 0.0005 in (0.013 mm) to insure a proper thread.) The nut blanks are then discharged from the forming machine. They are tumbled with media, cleaned and fed, via a vibratory feeder, to either a tapping or thread-forming machine. Following tapping, the nuts are normally barrel plated with zinc.

nylon - Nylons are a family of plastics composed of *polyamides* of high molecular weight. The common nylon 6,6 is made from the polymerization reaction of adipic acid and hexamethylene. (Adipic acid is made from a two-step oxidation, from air, of cyclohexane. Hexamethylene is made from butadiene or acrylonitrile.[4]) Nylon 6 is similar in manufacturing and properties to nylon 6,6. Nylon 6 is made from caprolactam, the molecules of which join together by self-condensation to form polycaproamide,[3] or nylon 6.

Nylon is used to make yarn by extrusion through spinnerets to make fine filaments that are gathered and spun together, cold drawn to a smaller diameter, and wound on bobbins. The yarn is especially strong and is used extensively in carpets, hosiery and other wearing apparel, parachutes, auto air bags, and tire cords. Coarser mono-filaments are used for fishing lines, insect screening, tennis rackets, brushes and weed trimmer cord. Nylon is also used extensively for injection-molded parts including: gears, bearings, anti-friction or high temperature machine components, painted exterior auto body parts, and combs. Fig. N1 shows the operation sequence for making nylon resin.

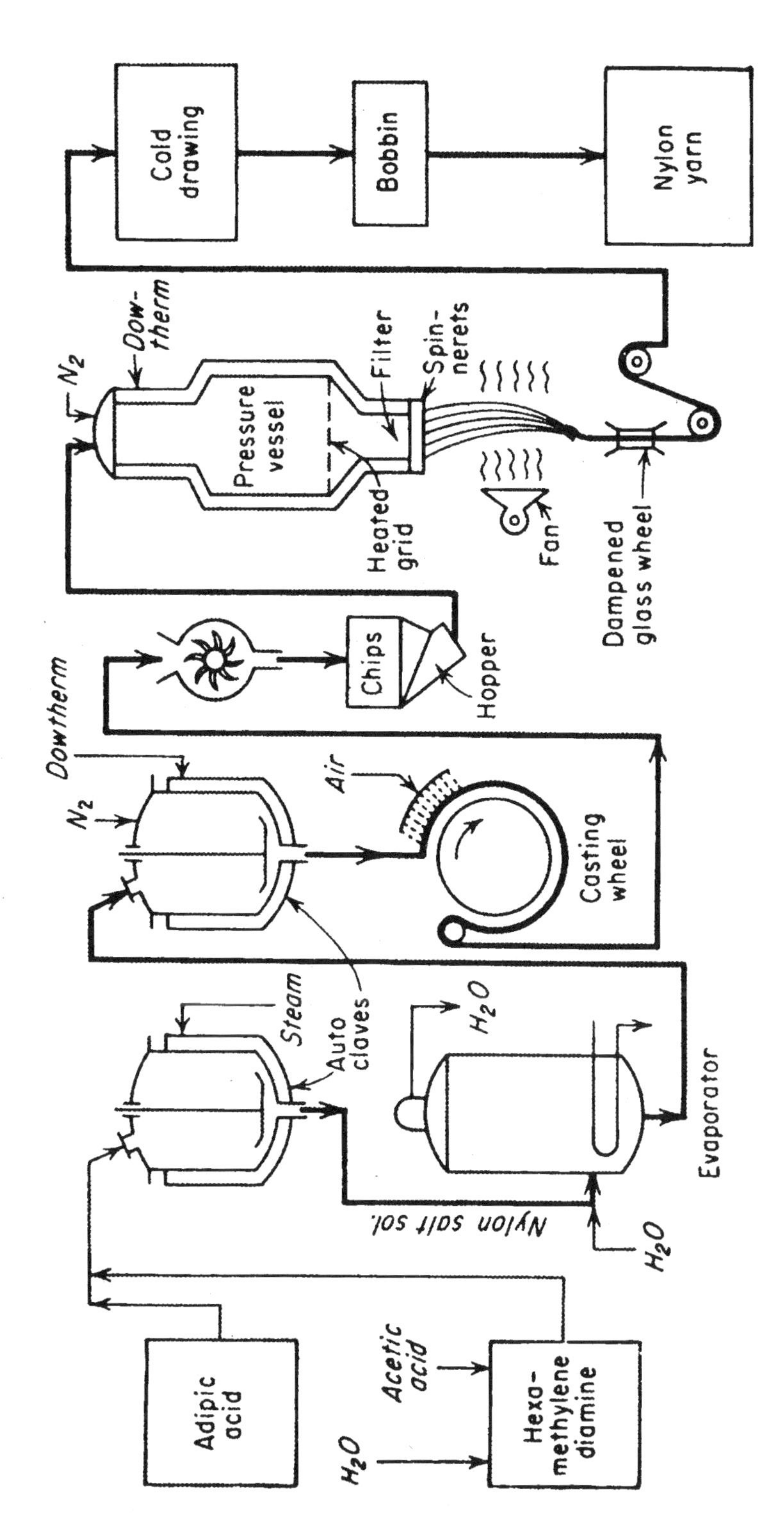

Fig. N1 The sequence of operations involved in making nylon yarn. *(from Shreve's Chemical Process Industries, 5th ed., G.T. Austin, McGraw-Hill, 1984. Used with permission.)*

O

oils, edible - include olive, corn, palm, coconut, peanut, safflower, canola, cottonseed, and soybean oils, and others of vegetable origin. Edible oils and fats of animal origin include cod liver oil, shark liver oil, and other fish oils, lard, and butter. Almost all vegetable oils are obtained by solvent extraction (11C3) or expression (11C4). (Cottonseed and safflower seeds are processed with both methods.) Palm oil is removed by rendering: boiling the fruits in water; the oil rises to the surface. Expressed oils usually are filtered to remove particulate matter. Oils are usually further refined to put them in better condition for use. Treatments may include centrifugation (12E3 & 11C7e) after phosphoric acid treatment to remove gummy material, bleaching by adsorption (11C9), and deodorization with steam. Oils may be hydrogenated (12J3) to make them less liquid and to greatly reduce the possibility of deterioration in storage.

Animal fats are removed from carcasses by several methods including rendering or cooking in water, followed by decanting the oil, which rises to the top. Centrifuging, filtering and bleaching may also be used. (Also see *butter*.)

Fig. O1 illustrates the operation sequence for production of a typical edible oil from seeds.

oils, essential - Essential oils are those, contained in odoriferous plants, that are used for aromas and flavorings. They are utilized to provide a pleasant odor in cosmetics, soap, perfumes, and detergents. These oils are also used to provide flavor to baked products, confections, meats, pickles, soft drinks, candy, and medicines. Common essential oils are wintergreen, peppermint, cinnamon, rose, camphor, turpentine, birch, anise, clove, sage, and nutmeg. Essential oils occur as very small droplets in the cells of plants. They are most commonly extracted from the plant using *steam distillation* (11C1e). *Extraction* (11C3), using volatile fats, is employed to obtain aromatic essential oils from flowers for perfume and cosmetic fragrances. The process is sometimes carried out with cold fat placed in contact with the flower petals. The fat, after 1 to 3 days, absorbs the flower oil and its fragrance. This approach is called *enfluerage*. *Expression* (11C4), a method that presses the plant to force out the oils, is used to remove them from the skins of citrus fruit. The oil is then separated from accompanying water and solid plant particles by filtering (12E1 and 11C7a), decanting, and/or centrifuging (11C7e).

oil, fuel (furnace oil) - is one of the products of fractional distillation of crude oil (11H1a) and the further vacuum distillation (11H1b) of the heavier residues of fractional distillation. Fuel oil is the residue of these processes, a relatively heavy product, and one that is more costly to refine into higher-valued lighter products such as gasoline, kerosene, and diesel oil. Fuel oil is made by blending various fractions of these distillations to achieve the required viscosity for handling, and the desired flash point. Applications are the heating of buildings, and fuel for diesel engines, ships, and industrial plants.

oil, lubricating[4] - is made from the heavy distillates of petroleum processing. Solvent extraction (11H1d) is used frequently to remove unwanted compounds. Asphalts may be removed with propane. Blending (11H3c) with other fractions is employed for some applications. A series of additives may be mixed with the oil. These include detergents, anti-foam agents, antioxidants, viscosity-index improvers, extreme-pressure agents, and anti-scuff agents.

oils, vegetable - See *oils, edible; olive oil; oils, essential.*

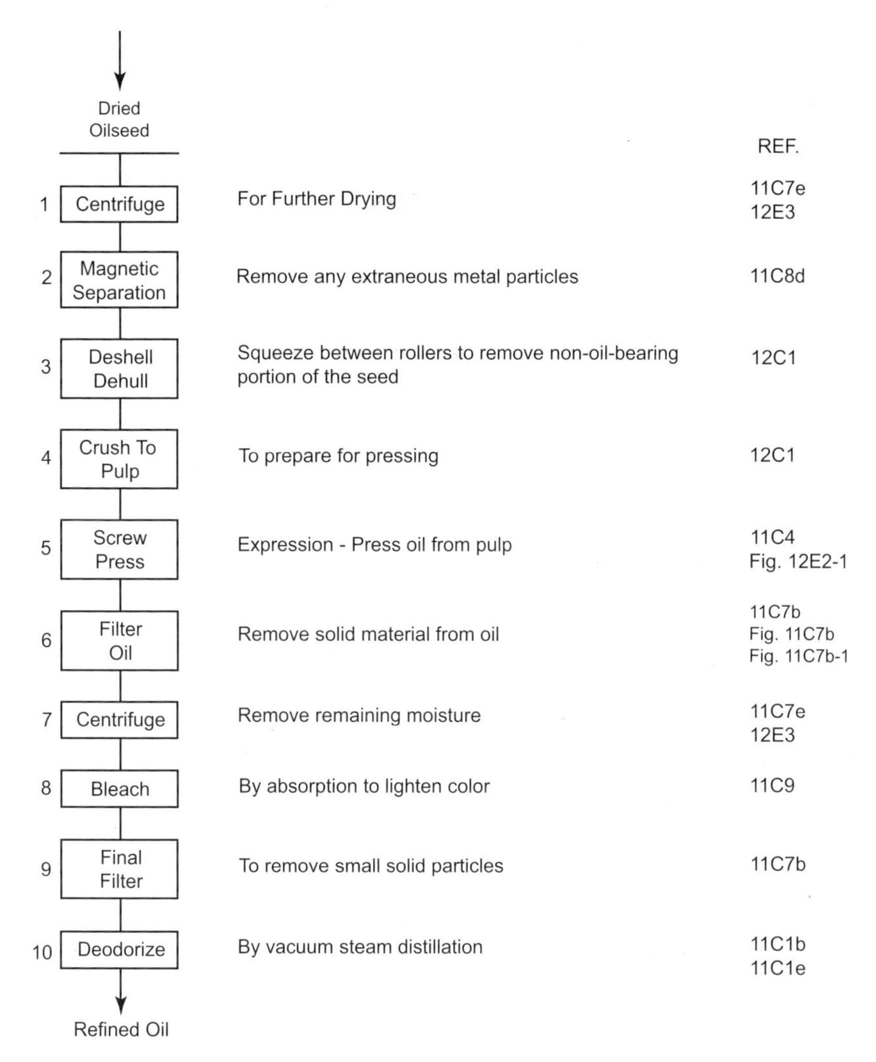

Fig. O1 The sequence of the major operations in the production of edible oil from oilseeds[1].

olive oil - The olives, when ripe, are picked by hand or by machine. Whole olives, with pits, are crushed in a machine that strikes them with metal hammers. The brown paste that results is spread on woven nylon mats. The mats are stacked and the stack is squeezed in a hydraulic press at pressures of approximately 6000 lbf/in^2 (40 MPa). Water and oil flow from the press and are collected and centrifuged (11C7e) to separate the two. The oil is then bottled. Oil from the first pressings that meets standards is classified as virgin olive oil. Oil from later pressings may be classified as "pure" or "edible". Still later pressings provide oil that is processed to remove acidity, odor, and color.

olives, green - Olives, for eating, are harvested when they have reached full size but before they are fully ripe. They are processed by soaking for 9 to 12 hours in a dilute solution of sodium hydroxide (NaoH) that reduces the bitter flavor. (A small amount of the flavor is allowed to remain to provide the characteristic taste.) The olives are rinsed thoroughly and soaked for about 24 hours in fresh water to remove the sodium hydroxide solution. Salt, applied after rinsing in a brine solution, also removes some of the natural bitterness, acts as a preservative and provides the salty flavor.

optical glass fibers - See 5A6, 5A6e and 5a6f.

optical lenses - See *lenses.*

orange juice - Most American orange juice comes from Florida. The common supermarket container of orange juice has undergone the following production steps: 1) Oranges ripen on the tree. Samples are taken and tested for acid/sugar ratio. 2) If ready, oranges are picked from the tree, usually by hand. 3) Oranges are dumped into trucks and transported to the juice factory. 4) The truckload is weighed and a sample is tested for juice content and other attributes. 5) Oranges are classified according to the test results, and are placed in temporary storage. 6) Several lots of oranges are selected for processing together, to provide the juice properties desired. 7) The oranges are washed automatically. 8) They are graded and substandard oranges are removed. 9) They are put in automatic equipment that peels the oranges, extracts the core, and squeezes the juice from the fruit and oil from the peels. The juice, at this point, has a high pulp content. 10) The juice is screened to remove seeds and pulp. 11) Not-from-concentrate juice is pasteurized, chilled and placed in storage tanks for later packaging. Juice to be concentrated is run through vacuum equipment that evaporates most of the water it contains. 12) Concentrated juice is frozen at about 0°F (–18°C) and stored at 18°F (–8°C). 13) Frozen concentrate from several tanks is thawed and blended; oils and essences removed in earlier processing are added back to enhance flavor. 14) The concentrate blend may be packed into cans and sold as frozen concentrate or mixed with water and packed into cardboard containers for sale.

oriented strand board (OSB) - See 6F4.

o-rings - are toroidal-shaped sealing rings, molded from synthetic rubber or other polymers and used as seals in various fluid-handling systems. Common o-ring materials are buna-N, neoprene, polyurethane, EDPM (ethylene propylene), silicone, Teflon® and Viton®. O-rings are most commonly compression or injection molded, but large sizes may be made by splicing the ends of extruded rods or tubing. Sometimes, ring-shaped seals with a square, rather than round, cross-section are made by cutting off extruded tubing into rings. Molded o-rings are normally tumbled at low temperature, after molding, to remove mold flash. The operation is carried out with the aid of liquid or solid carbon dioxide as a refrigerant. The choice of o-ring material depends on the temperature and pressure of the working environment and the nature of the fluids that will contact the seal. Typical o-ring applications include seals in valves, faucets, pipe flanges, compressors, engines, and tube fittings.

Orlon - is a synthetic fiber, somewhat similar in feel to wool. It is made from polymerized acrylonitrile. Orlon is made into garments and filters. See *acrylic plastic* and manufacture of synthetic fibers (10A2).

oxygen[4] - is obtained from the compression and rectification of air. The air is filtered and compressed in centrifugal compressors to about 75 psi (520 kPa). It is cooled moderately to enable moisture to be removed and is further cooled to near the dew point. Moisture is deposited on the walls of the heat exchanger and freezes. As the temperature goes still lower, carbon dioxide gas in the air also is deposited on the walls of the heat exchanger as dry ice. The chilled air, still in the gaseous state, passes through a fixed-bed adsorption unit, which removes any remaining carbon dioxide and any hydrocarbons which it may have held. The air is then fed to a double-column rectifier. This is a two-column fractional distillation device (11C1a.) which separates the air into a low-purity stream of nitrogen that exits the upper column at the top, and a stream of oxygen vapor that exits from the main part of the apparatus. The rectifier includes silica gel adsorption traps that remove carbon dioxide and hydrocarbons from the oxygen as it circulates in the apparatus. Oxygen leaves the rectifier at 99.5% purity. Another manufacturing method involves electrolysis (11C10) of water. Oxygen is used in metal cutting and welding, and in chemical processes in the steel, cement, petrochemical, paper, and glass industries.

P

packages, blister - are made by thermoforming thermoplastic sheet. Several thermoforming processes, as outlined in 4D, are available, but straight vacuum forming (4D1) is the most common method for making blister packages. The plastic sheets may be formed over the product but, more often, are formed from tooling that duplicates the product's shape. In high production systems, special machines, using rotary index tables, are employed. The tables include mechanisms around the periphery that cut off and form the plastic sheet, cut off the preprinted paperboard base, assemble the product to the board and the clear sheet, and wrap the sheet around the board or seal it to the board, all done automatically.

paint[4] - Though the ingredients in present-day paints are usually the products of chemical processes, the final paint preparation is purely a mechanical operation. The ingredients used depend on the type of paint involved, for example whether it is oil-, water- or solvent-based. However, the demarcation that previously existed between lacquers and oil paints no longer exists because of more complex formulations. Paints dry by oxidation or polymerization of the oil or resin they contain, or by evaporation of the solvent vehicle. Each of these reactions can take place at either room or elevated temperatures. The preparation/mixing sequence for a typical paint is shown in Fig. P1 and can be described as follows: 1) accurate quantities of pigments and vehicles (oil, solvent, or water) are fed into a feed container. Quantities of each are controlled by weighing the mixture. 2) The mixture is piped to a mixing vessel where it is thoroughly mixed by a machine similar to those used for mixing dough. 3) The mixture is fed into grinding equipment which may consist of ball mills, roll mixers, high-speed dispensers or a combination of these. Solid particles are reduced in size and thoroughly dispersed. 4) The ground mixture is fed to a tinting

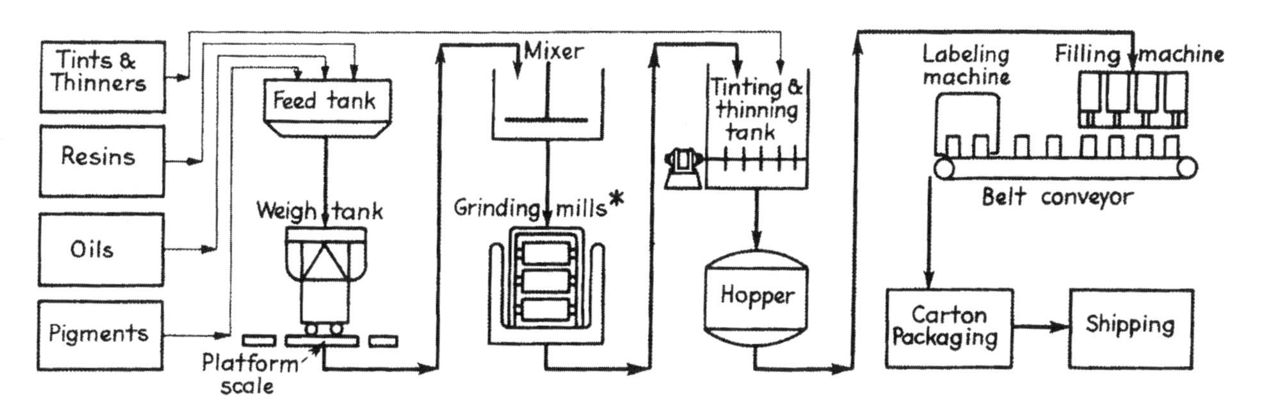

Fig. P1 The operation sequence for paint making. (*Shreve's Chemical Process Industries, 5th ed., G.T. Austin, McGraw-Hill, New York, 1984. Used with permission.*)

and thinning tank, where the color and viscosity are adjusted. 5) The mixture is strained as it leaves the grinding tank and fed to either a holding tank or the feed tank of a filling machine. 6) Cans are filled and closed. 7) labels are applied and cans are packed in cartons for storage and/or shipping.

paint brushes - See *brushes*.

paint removers (paint strippers)[24,6] - are primarily made from strong hydrocarbon solvents, but aqueous caustic solutions are also used. All ingredients in paint removers are hazardous in some respect. The formulation of removers from different suppliers varies considerably. Hydrocarbon solvent paint removers usually are made from one or more of the following: methylene chloride, acetone, methanol, toluene, n-methyl pyrrolidine, and di-basic esters. Methylene chloride is a strong paint stripper and an important ingredient, but is toxic. It is made by reacting methanol with hydrogen chloride to make chloromethane, and then reacting the chloromethane with chlorine. Its most common use is in a formulation with methanol. Another common paint remover formulation includes methylene chloride, toluene, acetone, and methanol.

Caustic solutions of sodium hydroxide (caustic soda or lye) alone, or with ammonium hydroxide (ammonia), are powerful paint removers but are little used in furniture refinishing because they stain wood, change it to wood pulp, and break down some wood glues. Some solvent paint removers have added caustic (usually ammonia), to increase their strength, particularly against tough epoxy or polyester paints.

pallets, plastic - are molded from structural foam plastics (4C3). Low-pressure injection molding (4C3a) is often used. Another method used is dual-sheet forming (twin-sheet forming) (4D13) of high-density polyethylene.

pallets, wood - are made from the less-expensive, lower grades of hardwood lumber that, because of irregularities, discoloration, knots, or being a less desirable species, is not marketable for cabinet-work. However, almost any species may be used. Many pieces are made from the cores of logs, after more desirable boards have been cut. Beech, birch, maple, oak, aspen, and ash are among the Pennsylvania hardwoods that are used. Some softwood pieces may also be utilized. Component pieces are saw-cut to size but are not planed. Pneumatic nailers or staplers are used for fastening.

pans, cooking - See cooking utensils.

paper - See 9A and 9B.

paperboard - has been defined as paper thicker than 0.012 in (0.3 mm) and heavier than 0.66 oz/ft^2 (200 g/M^2). See 9C3.

paper, bond - See 9B and 9C1.

paper clips - are made on four-slide machines (2G2) or special machines similar to them. Wire in coil form is fed to the machines, which cut and bend the wire into the shape of the paper clips at a rate of hundreds of finished clips per minute.

paper handkerchiefs - See 9C4.

paper, kraft - See 9B, 9B2b2, and 9C5.

paper money - is printed by the intaglio (gravure) method (9D3a). Master plates for printing are hand engraved, because hand engraving with fine and coarse dots, dashes, and lines, is more difficult to counterfeit than a computer-generated engraving. Hand engraving master plates is a lengthy, painstaking process. Over 1000 man-hours are required for completion of the master engraving for any one note. The master engraving is hardened and used to make production plates. The hardened engraving is pressed into a soft transfer roll which, when hardened, becomes a master die for making production printing plates. The production plates are assembled in multiples of 32 so that 32 notes are printed with each impression. Sheet-fed rotary presses are used, with output rates of over 8000 sheets per hour. Sheets are inspected and, if satisfactory, are over-printed by letterpress (9D2) with a serial number, the Federal Reserve District seal, and number. Both black and green ink are used. The sheets are cut into individual notes that are collected into stacks of 100 notes, of which 40 are bundled together into "bricks" of 4000 notes. The paper used consists of 25% linen and 75% cotton fibers. Red and blue plastic fibers of various lengths are distributed in the paper as a further protection against counterfeiting.

paper, rag (rag bond) - See 9B and 9B3.

paper, sanitary - See 9B and 9C4.

paper towels - See 9C4.

particle board - See 6F5.

partition glass - See 5A3g.

pasta - is traditionally made from semolina, the flour of durum wheat. [See roll crushing (12C1) and milling (12C5) of grain.] The semolina is mixed with warm water and is kneaded until it becomes a stiff, smooth, dough. The dough is extruded into various shapes, the particular shape being controlled by that of the openings of the extrusion die. Round, string-like shapes make spaghetti or vermicelli, flat shapes make linguini or lasagna, and tubular shapes make macaroni. After extrusion, the material may be further formed into shapes such as shells (conchiglie), twists (rotini), butterflies (farfalle), bent tubes (elbow macaroni), and curls (ricciolini). The dough may also include eggs or juices to provide flavor and coloring (green from spinach juice, red from beet juice). The dough may be rolled and cut instead of extruded. After the pasta shape is made, the dough is slowly dried to reduce its moisture content to about 12 percent, in which condition it can be stored at length without loss of quality. Asian pastas, usually referred to as noodles, are made from other wheat, rice, mung beans, or parts of other plants.

patterns for casting - See 1B7.

peanut butter - is made by grinding the seed kernels of peanuts. Peanuts are harvested, dried and shelled. They are separated from any soil or stones and are roasted (12G6). After roasting, blanching machines (12G1) remove the reddish-brown skins from the kernels. Off-color or un-skinned kernels are sorted out automatically. The peanuts, sugar, salt, and partially-hydrogenated vegetable oil (12J3)(for stabilizing and extending the mixture), are fed to a mixing and grinding mill (12D). The mill grinds the peanuts and mixes them with the other ingredients to a smooth consistency. The peanut butter is then piped to an automatic filling machine that feeds a measured amount into jars on rotary tables (12M), places and tightens a cap, and moves the jars to another rotary table where labels are attached and code numbers are ink-jet printed. The jars are collected in multiples of 12 or 24, placed on trays and stretch wrapped, or put into shipping containers. Trays or containers are stacked on pallets, stretch wrapped again, and are ready for shipment.

pencils, lead - do not contain lead. The lead in a lead pencil is a blend of graphite and clay. (The greater the graphite content, the softer the pencil lead and the darker the imprint from the pencil.) The graphite and clay are mixed with water to form a paste that is extruded into thin rods. These are cut to length, straightened and kiln dried. They are then dipped into molten wax to provide a lubricant coating. Standard pencils are made from cedar or another softwood that provides ease of sharpening of the pencil. The wood is cut into strips of 1/8 inch thickness and pencil length. Several half-round grooves are machined in the strips by special machines, the number of grooves depending on the strip width. The leads are placed in these grooves and another similar strip is glued to the top, enclosing the leads. The strips are cut into narrower strips, each containing one lead. These smaller strips are then machined, usually to a hexagonal shape, with another special machine. The pencils are painted, printed with the manufacturer's name, brand, and type identification. They are lathe turned at one end to fit a ferrule formed from brass tubing, which is assembled along with an eraser, and crimped to hold the eraser in place. The opposite end of the pencil is cut to a square face. These operations are all performed by automatic machines with no manual handling of individual pencils.

pens, ball point - See *ballpoint pens.*

pepper - See *spices.*

perfume - consists of a blend of fragrant oils dissolved in highly-refined ethyl alcohol with some water. (The alcohol is first deodorized.) Perfumes contain 22 or greater percent fragrant oils and 78 or less percent alcohol. Eau de parfum, eau de toilette, and eau de cologne each contain a progressively lesser portion of oil and a greater portion of alcohol. Oils are of two basic types: 1) those containing fragrances originating from plants, particularly flower petals, and 2), those from animal sources. Now, many ingredients are synthetic. Synthetic chemical fragrances reproduce or imitate the odor of the natural substances but not their chemical composition. Plant fragrances may come from the leaves, bark, wood, roots, fruit, and resins as well as the flower blossoms. Two common flower blossoms used for perfume are night blooming jasmine and May rose. It takes 800 lbs (360 Kg) of rose blossoms or 2500 jasmine flowers to make a pound (450 gr) of concentrate. Flower picking is a hand operation. Odors are extracted from blossoms with solvents (11H1d) and distillation (11C1) or by the ancient French method called "enfleurage", where the fragrances are absorbed by trays of fat. Animal-sourced ingredients include musk - from a deer, similar substances from musk-ox, muskrat, and Florida alligator, civet - a secretion from the glands of a civet cat, castor from beavers, and ambergris from spermaceti whales. Animal-sourced ingredients are important in the final product. They are fixatives, liquids of less volatility, providing a permanence to the more fleeting vegetable odors. Fixatives may be of synthetic or vegetable sources as well as from animals. They may or may not add to the odor of the perfume but must blend with it. Typical perfumes consist of a blending of several or many fragrant oils in the alcohol solution. The finished perfume is aged for up to a year to provide a harmonious and stable blend. The aging process cannot be shortened without a loss of quality.

permanent press fabrics - See 10F5b.

pesticides - See *insecticides.*

petrochemicals - See 11H.

petroleum - the heavy flammable liquid found in underground and undersea deposits. Its refinement into useful products is covered in section 11H.

petroleum jelly (petrolatum, Vaseline) - is a product of the fractional distillation of petroleum (11H1a), distilling from the petroleum at 577°F (303°C).[2] It is a greasy semi-solid, clear or slightly yellow in color. It is used as a base of salves, ointments and cosmetics, as a lubricant, and in rubber compounding.

pewter - traditionally, was an alloy of tin and lead. Ancient Roman pewter had 70% tin and 30% lead. The percentage of various ingredients has varied considerably since then. From the 14th century in Europe, pewter was used for plates, bowls, drinking vessels and church chalices. 16th century pewter had as much as 90% tin. Current pewter is largely Britannia metal, an alloy of 89% tin, 7.5% antimony, and 3.5% copper. It has the easy workability and castability of lead-containing pewter, but retains a bright sheen if polished, compared to the lead-bearing variety that takes on a dull, dark-gray patina over time. Britannia metal is also stronger and safer to use with food.

phenolic plastics - are made from the reaction of formaldehyde and phenol, condensation-polymerized (4A2b) with an acid catalyst. In one process, the resole process, an excess of formaldehyde is used with a water solution base catalyst, and the reaction is stopped just after crosslinking occurs. This allows for further crosslinking during molding of the material. In the novolac process, insufficient formaldehyde is used in the polymerization reaction but this is compensated for by blending in compounds that decompose to provide formaldehyde when the compound is heated during a molding operation. When they are to be used as molding compounds, phenolics are filled with wood flour, mineral powders, glass fibers, paper, or chopped fabric. Phenolics are thermosetting and are widely used in compression, transfer, and injection molding to make commercial parts, especially those used in electrical outlets, handles for appliances and pots and pans, switches, as commutator insulation in motors, and for other electrical uses.

photographic film - See *film, photographic.*

pickles - See 12J.

pig iron - an intermediate material between iron ore and steel, is produced in blast furnaces. See *iron* and Fig. I2.

pile rugs (tufted rugs) - See 10I2.

pipe, plastic - See *tubing, plastic.*

pipe and tubing, metal - Formed metal pipe and tubing may be either seamless or welded, and may be made from various metals. Steel is perhaps the most common, but copper is used for much water piping, and other metals are used in various applications. Several methods are usable for pipe and tubing manufacture: 1) Seamless metal pipe and tubing often starts with a round hollow extrusion (2A3) that is drawn (2B2) repeatedly until the desired diameter is achieved. 2) Another process uses the hot piercing method (2A5), which pierces a heated and softened round rod axially with a pointed tool as it is pressure rolled circumferentially, converting the rod to a length of hollow tubing. This also is drawn until the desired diameter is achieved. 3) Hot drawing or cupping (2A2), is another method used to make pipe or tubing. In all these methods, the internal diameter of the blank before drawing is controlled to produce the desired internal diameter after drawing as closely as possible, and drawing is done with a mandrel to control the internal diameter. Fig. 2B2, view b, illustrates this operation. 4) Butt welded pipe and tubing is made by contour roll forming (2F7a) a strip of sheet metal (a *skelp*), into a round tubular shape, and resistance welding the butt joint where the edges come together. Fig. P2, view a, illustrates this. 5) Another method of welding pipe, forge seam welding (7C13b), uses a scarfed joint under pressure to achieve welding. Welding takes place when the tubing is red hot and is fed between external form rollers and an internal mandrel to apply pressure to the joint. View b of Fig. P2 shows this method. (Also see 2A6.) 6) Still other methods use arc welding (7C1), induction welding (7C2) or flash welding (7C10) to join the edges after the skelp has been formed to a tubular shape. Fig. P2, view c, illustrates arc welding. For cast metal pipe, see 1E1.

pipe, cast - See 1E1.

pipe, welded - See *pipe and tubing, metal* above.

plaster - See *gypsum plaster*.

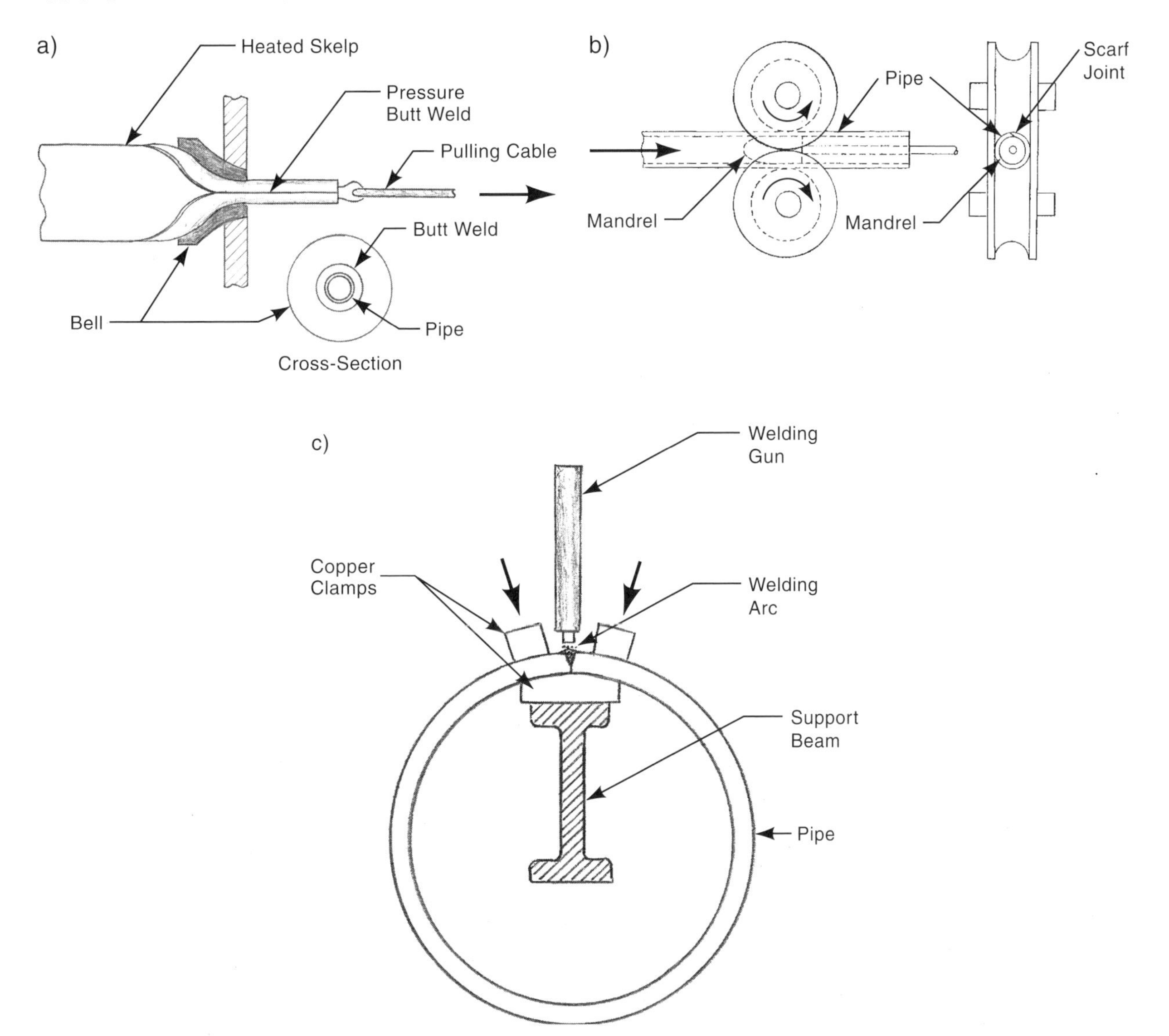

Fig. P2 Schematic illustrations of pipe welding methods. a) pipe with a butt-welded seam being made from strip material. b) forge welding. The pipe fed to the machine is red hot and the rollers and mandrel apply pressure to the scarf joint, causing the edges to fuse together. c) arc welding of a large-diameter pipe.

plasterboard, gypsum board, wallboard, sheet-rock, and drywall - are all common names for board material used for wall and ceiling surfaces in buildings. The board material is made from gypsum (hydrated calcium sulfate - $CaSO_4$ - $2H_2O$), mined in various parts of the world, or obtained as a bi-product of the desulfurization of flue gas in electric power generating stations. Some gypsum is obtained from recycled pieces.

The gypsum is crushed to a powder and heated to remove about 75% of its water content, to produce plaster of Paris. (See *gypsum plaster*.) Plaster of Paris, when mixed with water, solidifies but, before solidifying, can be cast or formed into various shapes. For plasterboard, the gypsum powder (plaster of Paris) is mixed with small portions of various additives (paper pulp, starch, foaming agents, glass fibers, wax, or asphalt) each of which produces or enhances some property of the finished board. The resulting powder is mixed with water and the paste/slurry formed is fed between two layers of paper from large rolls into plasterboard machines which may be as long as 800 ft (244 m). As the plaster and paper coverings move through the machine, rollers on the machine distribute and compact the gypsum plaster and form the board to the desired thickness. Standard thicknesses in the USA are 3/8, 1/2 or 5/8 in (9.5, 12.7 or 15.9 mm). As the material moves through the machine, the plaster sets and the material is sliced to the desired width (usually 4 ft - 1.2 m) and cut to length (usually 8 or 12 ft - 2.4 or 3.7 m). Edges are finished, and the boards move through a long drying oven to complete the curing of the plaster. The completed boards are then stacked and palletized for shipment.

plastics - See 4A or the listing under the name of the particular plastic of interest, e.g., polyethylene.

plastic film - See 4I5.

plastic laminates, rigid ("Formica" or "Micarta") - See 6F8.

plastic wood, wood filler - is made from wood flour, finely ground dried wood, which becomes a filler that is blended with a plastic resin and a solvent (n-Butyl acetate, acetone and/or MEK) for the resin. The product is packaged in tubes or cans. When applied, the solvent evaporates, leaving a solid material that can be machined, sanded and varnished or painted.

plate glass - See 5A3e.

platinum - is found in metallic form as small grains or pebbles in alluvial sand and gravel in several countries. These grains or pebbles usually also contain other metals of the platinum group, alloyed with the platinum.[2] Platinum is separated from the other metals by a very complex aqueous chemical process. It is used as a catalyst in automotive engines and petroleum processing, in jewelry, in dental fillings, as a coating for laboratory dishes, crucibles, and other devices, and in electrical contacts and electrodes.

playground equipment - includes swings, slides, teeter-totters (see-saws), gyms, ladders and climbing frames, platforms, trapezes, and rings. Climbing structures are made either of wood or welded steel tubing. Cedar is used for quality wood equipment because of its appearance and durability. However, pressure-treated spruce is more common. Wooden frames consist of material about 4 × 4 or 2 × 8 inches (10 × 10 or 5 × 20 cm) in cross section, with rounded corners, cut to the size needed and drilled with necessary bolt holes. The frames are called "moulding" by the manufacturers. Metal frames of welded steel tubing (See *pipe and tubing, metal*) are prepunched with holes for connections. The finish is sometimes by galvanizing, but much more commonly is by electrostatic spray painting (8D7) or powder coating (8D8). Tubing pieces are connected with corner fittings made from steel stampings that wrap around the pipe, by aluminum die castings (1F), or by arc-welded (7C1) assemblies, fabricated from slightly larger steel tubing. All fittings incorporate holes to match cross holes in the tubing for fastening bolts. The connector stampings are formed from sheet steel in a progressive-die operation (2E1). Chain used for swings and trapezes is zinc plated and covered with extruded (4I1) flexible vinyl tubing. Rope may also be used for swings. Swing seats are injection or blow molded (4C or 4F) of high-density polyethylene. Slides are blow molded or made by rotational molding (4E) of the same material. Teeter-totters are made from steel tubing with blow- or injection-molded seats. Trapeze bars are made from steel tubing. Rings are injection molded.

The full units are normally customer assembled though some pre-assembly that does not significantly increase carton size may be carried out by the manufacturer. Sometimes parts are decorated with stencils. Injection-molded plastic caps are used to cover ends of tubing.

plexiglas - is a trade name for methyl methacrylate sheets or rods. See Chapter 4, especially 4H1a and 4H1b.

pliers - are hand tools that are usually impression-die forged (2A4b,) or drop forged (2A4b1) from steel of 0.25 to 0.55% carbon, trimmed, normalized (8G2e), and cooled quickly. (Some pliers are made from chrome-vanadium steel or manganese-bearing alloy steel.) The gripping teeth are form milled (3D5), and the pivot hole is drilled. Gripping teeth and cutting blades are heat treated, commonly by induction hardening (8G3a2), followed by tempering (8G2f). Surfaces are polished (8B1). The rivet that connects the two gripper parts is cold headed (2I2). All parts are usually electroplated (8C1) with nickel, zinc, or chromium. Hand grips are often molded on the handles by dip coating in vinyl plastisol (4K2). The parts are then assembled and the connecting rivet is set (7F4b).

plywood - See 6F2.

polycarbonate plastic, PC - is a special variety of polyester resin. It differs from other polyesters in that a derivative of carbonic acid is substituted for adipic, phthalic, or other acid, and a diphenol is substituted for the glycols normally used. A melting process or a phosgenation process is used in its preparation. Bisphenol A is reacted with phosgene, a highly toxic gas or, in another method, polyphenol is reacted with methylene chloride and phosgene.[2] Still another method, developed in Japan, eliminates the need for phosgene by reacting biphenol-A with diphenylcarbonate at a temperature of about 525°F (275°C), with a special catalyst and an excess of a chloride.[2]

Polycarbonate, is used for high-quality, impact resistant glazing, lenses, including lightweight eyeglass lenses, safety helmets, housings, aircraft parts, boat propellers, signs, insulators, other electrical components and compact discs.

polyester plastic[4] - can be either thermoplastic and thermosetting, but is most common as a thermosetting material. It is manufactured in two basic steps: 1) condensation of dibasic acid and a disfunctional alcohol to form a soluble resin. Maleic, phthalic, or itaconic acids can be used, with allyl alcohol or ethylene glycol. 2), the addition of a cross-linking agent to convert the resin to be thermosetting. Polymerization is by condensation reaction (4A2b). Some of the reactants (phthalic anhydride, fumaric acid) are solids; others, such as ethylene glycol, are liquid. The reaction takes place in an insulated glass-lined or stainless steel vessel at 390°F (200°C), and requires up to 20 hours, during which water, inert gas and glycol are continuously removed. The cross-linking operation takes an additional 2 to 4 hours. The liquid resin is then available for storage or shipment, in suitable containers. Polyester resins are used in the fabrication of fiberglass objects such as boats, skis, building panels, fishing rods, and aircraft components. When made into a fiber, such as "Dacron", it is blended with cotton fiber to provide permanent-press qualities to garments. Polyester is also made into useful film, e.g., "Mylar".

polyethylene plastic, PE - is made by addition polymerization (4A2a) of ethylene, $CH_2 = CH_2$. Ethylene is a colorless gas, produced by cracking petroleum. In the polymerization process, the double bonds of the carbon atoms are broken, and replaced with other ethylene molecules in long molecular chains[2]. The several varieties of polyethylene include low density (LDPE), high density (HDPE), linear low density (LLDPE) and ultra-high molecular weight (UHMWPE). (See *polyethylene, high-density*, *polyethylene, low-density* and *polyethylene, ultra-high molecular weight*.) All varieties require somewhat different processes, but there are two basic variations, one that produces branches on the backbone of the long molecule, and one that produces linear backbones with an absence of branches. The process for branched forms uses very high pressures during the polymerization; the process for linear forms uses low pressure. Other differences in manufacturing processes depend both on the variety to be produced and the manufacturer, but are all produced from ethylene with the aid of a catalyst. Most linear varieties also have a copolymer as a secondary ingredient. Common co-monomers used are hexene, butene, 4-methyl-1pentene and octene.[15] Branched varieties are less apt to have copolymers, but some formulations include acetate, acrylic acid, and other co-monomer materials.

Polyethylene plastics are extensively used in injection- and blow-molded products including milk and kitchen chemical bottles, toys, and housewares. Polyethylene film is widely used for packaging and other applications. Polyethylene is extruded into pipe and electrical wire insulation.

polyethylene plastic, high-density, HDPE - is made with low-pressure, low-temperature addition polymerization, from ethylene gas and a comonomer (alpha-olefin), with a catalyst and a hydrocarbon diluent. Two commonly-used catalyst systems are the Phillips type (chromium-based), or titanium compounds with aluminum alkyls.[15] Various additives may be included to prevent oxidation, to provide UV light resistance, and antistatic properties, etc. The materials are fed to a reactor that operates at about 210°F (99°C) at a pressure of 100 to 290 psi (690 to 2000 kPa). Gas that does not polymerize is cooled and recycled into the reactor, while the finished product, in granular form, is drawn from the bottom of the reactor, separated from any accompanying gas, and discharged. Fig. P3 illustrates the operation schematically.

High density polyethylene is used in blowmolding milk bottles, other product containers, fuel tanks and drums. Other uses are injection molding of pails, toys, bottle caps and appliance housings. Pipes, hoses, and wire insulation are extruded.

polyethylene plastic, linear, low density, LLDPE - is made with a low-pressure, low-temperature, polymerization process, similar to that used for HDPE. Ethylene is copolymerized with 1-butene and with lesser amounts of 1-hexene and 1-octene, using catalysts.[1] LLDPE has similar properties and applications as LDPE but is less costly to produce and can be more easily modified by changing copolymers.

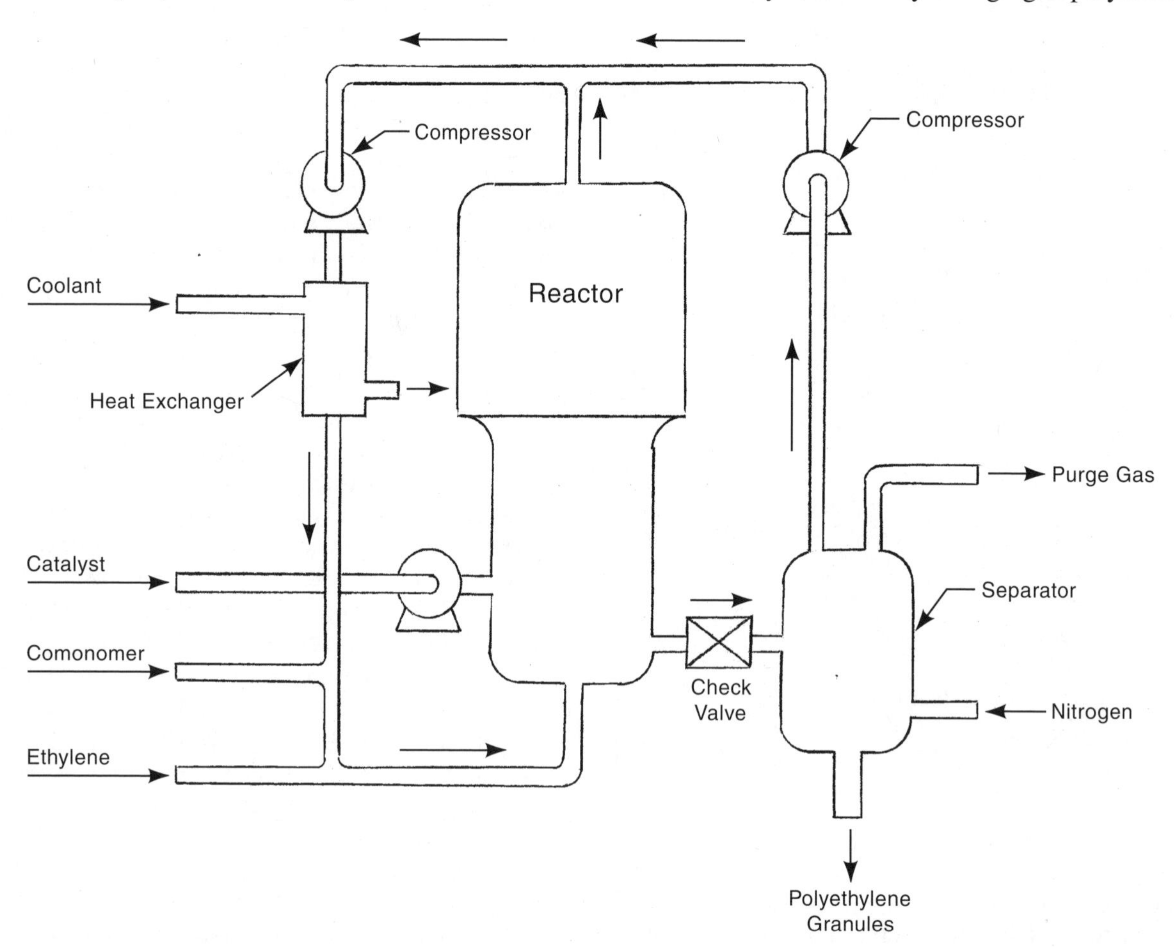

Fig. P3 The polymerization of ethylene gas to produce high-density polyethylene in a low-pressure reactor operating at a temperature of 212°F (100°C) and pressure of 100 to 290 psi (690 to 2000 kPa).

polyethylene plastic, low-density, LDPE - (See *polyethylene* above.) LDPE is produced by addition polymerization of ethylene with very-high-pressure methods. Ethylene gas is first purified to 99.8% by passing it through a demethanizer and then a deethanizer, from which the methane and ethane are removed and recycled. The ethylene is compressed to a pressure of 15,000 to 40,000 psi (100 to 275 MPa), and fed to a reactor with a peroxide catalyst. The reaction temperature is 300 to 500°F (150 to 260°C)[15]. The reaction converts about 30% of the ethylene; the balance is recycled through the reactor. The polyethylene product is extruded, pelletized, and dried.

Low-density polyethylene is extruded as wire and cable insulation, or into film for grocery and trash bags, garment bags, packaging and agricultural applications, or for lamination with paper, cloth, and other materials. LDPE is injection molded, or blow molded, into squeeze bottles, housewares, and toys.

polyethylene plastic, ultra-high-molecular-weight, UHMWPE - (See *polyethylene* above) UHMWPE is a polyethylene with extremely long linear molecules, having a molecular weight of over 3 million. It is produced by a low-pressure process similar to those used to produce linear low-density polyethylene and high-density polyethylene, and with identical reactors.

UHMWPE is used for machinery parts requiring high chemical resistance, low wear, and low friction, and as a fiber in bullet-proof vests and soldiers' helmets. It is very difficult to injection mold or extrude and is usually processed into parts by compression molding, sintering, forging, and machining.

polypropylene plastic, PP - is made by methods very similar to those used in the manufacture of polyethylene, but propylene, ($CH_3CH : CH_2$), is the basic ingredient. The propylene is a product of petroleum fractional distillation (11H1a), absorption/stripping (11H1c) and catalytic cracking (11H2a2). It is polymerized by addition polymerization (4A2a) in the presence of a catalyst using low temperatures and pressures. Sometimes polypropylene is copolymerized with ethylene or another material. There are many process variations, depending on the manufacturer. Polypropylene is a mass-produced product with a large number of producers. It is injection molded into luggage, housewares, toys, medical equipment (because it can withstand sterilization temperatures), and electronic components. PP is also made into fibers for rope and carpeting, and used in making film and coatings and in other applications.

polystyrene plastic, PS - is made from the polymerization of styrene monomer. The styrene, $C_6H_5CH = CH_2$, is clear and colorless, and liquid at room temperature. Polystyrene is made from ethylene and benzene, both of which are derived from petroleum. These are made into ethylbenzene by alkylating benzene with ethylene.[4] Then the ethylbenzene is dehydrogenated to styrene with an aluminum chloride, solid phosphoric acid, or silica-alumina catalyst. The following formula is applicable:

$C_6H_6 + C_2H_4 \rightarrow C_6H_5CH = CH_2 + H_2$. The styrene is then polymerized by addition polymerization (4A2a) with free-radical catalysts, typically in a continuous process.

Polystyrene is widely used in blister packaging, toys, ballpoint pen barrels, and, when expanded into foam, in flotation devices, packaging materials, egg cartons, hot and cold drink cups and styrofoam panels.

polyurethane plastic - is not one compound, but a group of plastics based on polyether or polyester resin.[2] A hydroxyl-terminated polyether or polyester is reacted with a diisocyanate to form a prepolymer with higher molecular weight. This prepolymer is then treated by adding difuntional compounds containing active hydrogens (from glycols, water amino alcohols, or diamines) to extend the molecular chains.[7] Many variations of properties are attainable, including rubbery materials with a wide range of hardness and elasticity.

Polyurethanes are used in applications where strength and resistance to abrasion are important. Rigid polyurethane foam is used for insulation and as a core material for aircraft wings and skis. Flexible polyurethane foam is used for upholstery, mattresses, and clothing liners. Elastomeric polyurethanes are used in roller-skate and skateboard wheels, industrial rollers, shoe soles, forklift truck tires, and medical equipment. Adhesives and spandex fiber are other polyurethane applications.

polyvinyl chloride plastic, PVC, vinyl[4] - is produced from the polymerization of the monomer, vinyl chloride ($CH_2 = CHCl$). Several methods are available for making vinyl chloride. One method employs ethylene ($CH_2:CH_2$), chlorine, copper chloride ($CuCl_2$), catalyst and oxygen (from air). The plastic resin is made from vinyl chloride liquid in a physical polymerization process that involves vigorous stirring of a mixture of vinyl chloride liquid, water, a peroxide catalyst and an emulsifying agent over a several-day period. PVC particles are then removed from the mixture by spray drying or by coagulation from acid addition. To provide greater flexibility, toughness, and chemical resistance, PVC is often made with polyvinyl acetate as copolymer. This is done by mixing monomers of both materials with a solvent and polymerizing them together in an autoclave. Plasticizers, stabilizers, fillers, pigments, and lubricants are also often blended with the basic resin. Rigid PVC is extruded for roof gutters, building siding, window channels and piping, It is blow molded into clear bottles, and injection molded into pipe fittings. Flexible (plasticized) PVC is molded into shoe soles, and made into sheet for rain gear, gloves, upholstery, and floor tile. Adhesives are also made from PVC.

polymers - See 4A or the listing under the name of the particular plastic of interest, e.g., polyethylene.

porcelain - is a ceramic made from mixtures of clay, quartz, feldspar, kaolin and other materials. It is usually translucent and is used for chinaware, pottery, chemical-resistant parts, electrical parts, and dental components. Porcelain is especially hard and requires a temperature of about 2650°F (1450°C) for firing. See Chapter 5. (The term, "Porcelain enamel" is often used for the glass-like coatings of stovetops, cooking utensils, and signs although these are not made from porcelain materials; "Vitreous enamels" would be a preferred term.)

portland cement - See *cement, portland.*

"popcorn" loose-fill packaging - These very light weight shapes are made from expanded polystyrene foam beads. See 4C4 for a description of the operation when used for pre-expanding such beads for later molding into insulated drinking cups and other shapes requiring insulating properties. When the beads are to be used for loose-fill packaging, they are run through the expansion operation two or three times in order to ensure maximum expansion. The beads are conveyed through the steam-heated expansion chamber on a wire mesh conveyor.

potato chips - Potatoes for making potato chips are sorted by size and fed to the chip-making operation on vibrating screens that separate any residual dirt or other foreign matter from the potatoes. They are then conveyed in a water stream to peeling machines. The water stream provides washing action as well as transportation. Peeling is achieved by abrasive action in automatic abrasive-lined barrels. The residue is used for animal feed. The potatoes are cut in half and then fed to automatic rotary slicing machines. These machines use straight knives, mounted at an angle on the sides of a drum-shaped cutter, to slice the potato halves into pieces of 0.040 to 0.065 in (1.0 to 1.6 mm) thickness. The sliced pieces are conveyed to deep fryers and cooked at 250°F (121°C) in partially hydrogenated soybean oil or, for some varieties, in lard. However, increasingly, for health reasons, the deep frying takes place in liquid, non-hydrogenated oil. Paddles in the frying kettles keep the potato slices in motion. After frying, the chips are spread on a conveyor where salt is added and the chips are dried and inspected by machine vision. Chips that are too dark or have dark spots are rejected automatically. The initial moisture content of potatoes is about 75% but this is reduced to only 2% after cooking, though 30% of the initial weight is restored with oil. Finished chips are conveyed by vibratory conveyor to a packaging machine. Packaging is automatic by weight. The chips fall into pockets on the machine, which open when the desired weight is reached, dropping the chips into an open bag. Nitrogen is introduced to the bags to provide a longer shelf life and the bags are sealed.

pottery - See *chinaware* and 5B2.

powders, metal - One of several possible manufacturing methods is employed, the choice depending on the metal used and the application of the powder. One of the most common methods is *atomization*, the spraying of molten metal into a chamber where the droplets cool and solidify as they fall to the bottom of the chamber. One approach, used to achieve the atomization, is to break up a stream of the molten metal with a jet of inert gas, air, water or steam (13E1). Another method uses the discharge of a small stream of metal on a spinning disc (13E2) to break up the metal into small droplets which become powder grains. A third method uses a plate vibrating at ultrasonic frequency (13E3) to get the same effect. Following solidification of the particles, sieving is used to select and classify particles of the desired size.

Another method is to use *chemical reduction*[25], wherein the oxide of the metal is brought in contact with a reducing gas at a temperature below the melting point. With copper and tungsten, a fine powder of the oxides is reduced with hydrogen. For iron powder, iron oxide (from mill scale), is ground to powder form and is reduced with carbon monoxide gas in a furnace at about 1900°F (1040°C). Another method with iron uses a bed of coke and limestone to reduce treated iron ore to porous cakes, which are then ground into powder. For some metals, the starting material for chemical reduction is a liquid solution of a metallic compound. Other metals that are made into powder by chemical reduction are molybdenum, cobalt, nickel and tungsten.

Electrolysis[25] (11C10) is still another powder production process. When iron powder is made by this method, a steel plate forms the anode of an electrolytic cell and stainless steel plate is used as the cathode. Direct current, over a period of hours, deposits iron on the cathode. The iron is stripped off periodically, washed and screened for sizing (11C8a). Since it is initially brittle, it then may undergo annealing (8G2).

powdered milk - See *milk, powdered.*

powder metal parts - See 2L1.

pretzels - are commonly made from wheat flour, vegetable oil - often partially hydrogenated (12J3) - salt, yeast, malt, and baking soda (sodium bicarbonate). There are, however, some differences, depending on the manufacturer and pretzel type. The ingredients, with some water, are mixed in kneading mixers, as pictured in Fig. 11G5-2. (Also see 12D.) After mixing, the dough is discharged into an extruding machine that forces the dough through an oval-shaped die and then cuts the extrudate into loaves called, "dough balls." The dough balls are conveyed to another extruder where they are made into the traditional pretzel shape in one of two ways: In one method, used with the large pretzels, the dough is extruded into a rod or "noodle" whose cross section is the size wanted in the pretzel. A special machine grabs each end of the rod and twists the rod into a pretzel shape. In another method, dough from the dough ball is extruded through a die whose opening matches that of the pretzel shape wanted, which may be a non-traditional shape. The extrudate is sliced to the thickness of the pretzel and the sliced pieces drop to the surface of a conveyor. With either method, the dough, once formed, is conveyed to a salting station where salt crystals are sprayed on and adhere to the soft dough surface. The pretzels then move on the conveyor to the baking oven. There are two stages of baking: In the first, at a temperature of about 550°F (288°C) for 5 to 15 minutes, depending on the size of the pretzels, they are baked mostly on the surface and acquire their characteristic brown color. They are crisp on the surface only. Then the pretzels exit the oven, fall off the end of the conveyor onto another conveyor beneath the first one, and are carried back into the lower portion of the same oven. Here the temperature is about 250°F (120°C). This second conveyor moves more slowly than the first and the pretzels remain in the oven for from 20 to 80 minutes, and become fully crisp throughout. The pretzels are conveyed to a packaging machine that dispenses them into chutes that are part of a weighing apparatus. When the standard weight of a package is discharged into the chute, it opens, discharging the pretzels into an open plastic foil bag. Nitrogen gas is puffed into the bag to provide longer shelf life for the pretzels than is possible with ambient air. The filled bags are inspected and placed in corrugated cartons for shipping. The entire operation, except for the initial dough mixing, is on a continuous basis. In one plant, two operators tending the series of machines, produce 4000 pounds of pretzels per hour.

printed circuit boards - See 13A through 13D.

printed fabrics - See 10G2.

printing plates, gravure - See 9D3b.

propane - See *gas, liquified petroleum.*

prototypes, rapid - See 14A.

putty - is a mixture of finely-divided calcium carbonate powder and 18% boiled linseed oil, sometimes with white lead added. The calcium carbonate powder is made by wet grinding and levigating natural chalk.[4] Other putties also include red lead, rubber, or plastic resins and other inert fillers. The prime applications are to cement panes of glass in wood or metal window frames and to fill nail holes in wood. Other glazing compounds that retain flexibility are made from a number of hydrocarbon solvents, polymers, and fillers. One variety is made from a mixture of acrylonitrile, ethylene glycol, phthalate ester, aceldehyde, formaldehyde, and crystalline silica.

pyrex glass - See *cooking utensils.*

Q

quarter-sawed lumber boards - See 6A2, 6A4 and Fig. 6A2-1.

quartz glass - is a type of glass that is particularly temperature and chemical resistant but difficult to work. It is made from quartz (Si_2O_7) sand with little or no other additives. Quartz glass components are made with essentially the same methods as are used with other glass compositions (Chapter 5), but its high melting point and high viscosity when melted require special care and skill in fabrication. Quartz glass is used in light bulbs, crucibles, tubes and rods in furnaces, in optical glass, and in chemical process and integrated circuit production equipment.

quicklime[2] - is calcium oxide, CaO, and is obtained by heating limestone or oyster shells in a kiln at about 1000°F (540°C). This burns out carbonic acid gas. Quicklime is used in glass manufacturing, water treatment, air pollution control, iron melting, and in a number of chemical processes.

R

rag paper (rag bond) - See 9B3.

rapid prototypes - See 14A.

rapid tooling - See 14B.

rattan furniture - is made from the stems of several species of climbing palm plants of the Calamus and Daemonorops family that grow in Southeast Asia. The stems grow to several hundred feet in length and are quite uniform in diameter. They are cut near the base, pulled from the plant, stripped of leaves and tendrils and cut to lengths suitable for shipping. Rattan is used for canes, umbrella handles, and similar objects as well as furniture. It is also split and used for seating, baskets, and heavy cordage.

rayon - is a general name for a number of artificial silk fibers made from cellulose-based plastics. The cellulose is obtained from soft wood, or from cotton linters, the short fibers that adhere to the cotton seeds. In the viscose process for producing rayon, cellulose is treated with caustic soda (sodium hydroxide), shredded, aged to form alkali cellulose, and then treated with carbon disulfide to form cellulose xanthate. This material is allowed to ripen, is filtered, and then extruded through a spinneret (an extrusion die with many small holes) into an acid solution that hardens the cellulose. See Fig. R1. The filaments are stretched, washed, dried, and packaged.[3] In another process for viscose rayon, the cellulose is dissolved in an ammonia solution of copper sulfate and then extruded through spinnerets as with the viscose process. By stretching the fibers, superfine yarns can be produced, and the resulting fabrics have the appearance of sheer silk. Rayon fabrics are used for vehicle tire reinforcement and, woven with other fibers, for women's dresses and underwear. They are also used for carpets and home furnishings, and surgical materials. Also see 10A2.

refractories - are materials with exceptionally high melting points and with strength at very high temperatures. Materials with melting points above 2880°F (1580°C) can be so classified. The most commonly used refractories are ceramic. Natural refractory materials include kaolin, kyanite, chromite, dolomite, bauxite, zirconia, and magnesite. Magnesite and dolomite are the most important. Artificial refractories include silicon carbide and aluminum oxide. Refractory materials can be formed into various shapes, and each has properties that determine its usage. Refractories are used for furnace linings, melting pots, kilns, and similar applications. They are important materials in metals refining and processing, and in glass manufacturing.

Most refractories are processed in some way before being made into useful forms. From the natural state, they may be washed, crushed, ground, and mixed. There should be a balanced mix of large and fine particles. Screening may be used to get the proper balance of ingredients. Water may be added to aid in compaction for molding, most commonly into firebricks. Refractories can also be made into moist pastes with water for use as mortars or for forming by ramming into place, or into molds. With additional water to make them more liquid, they can be cast. After molding, the molded parts are dried and then burned (calcined) in a kiln.

resistors, electronic - See 13L1.

rice wine - called sake in Japan is made by fermenting rice with the mold, tane' koji[2]. The process starts with steamed rice which is mixed and kneaded with the mold and water, heated, and placed in

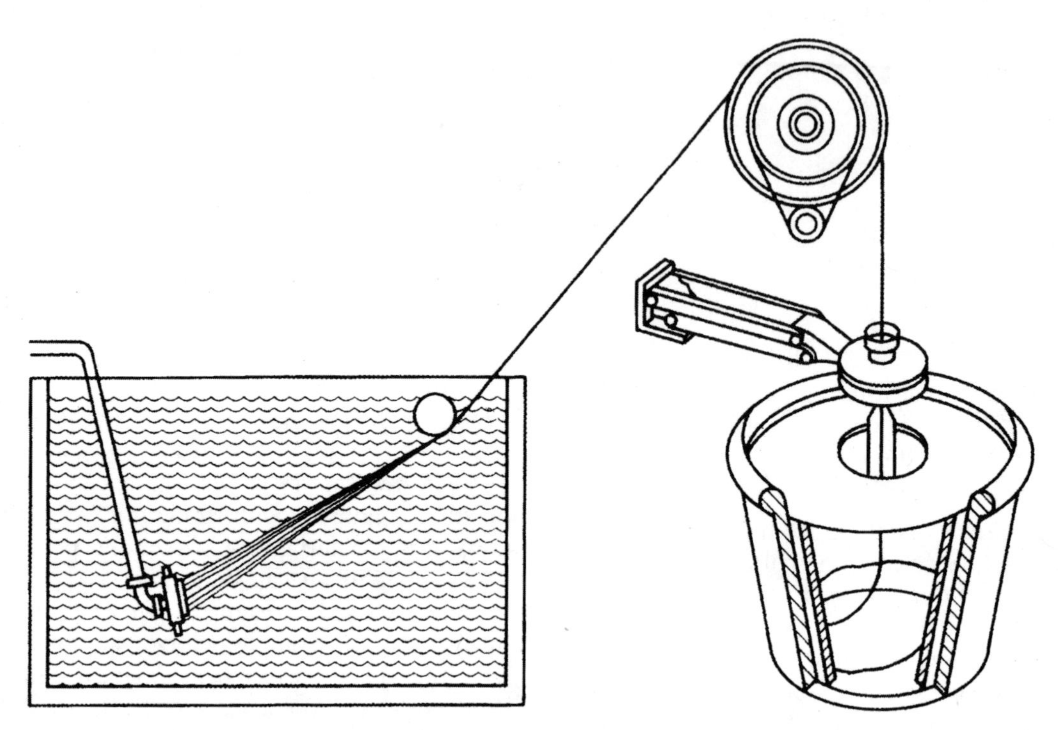

Fig. R1 Spinning viscose rayon yarn in an acid bath. (*from Shreve's Chemical Industries, 5th ed., G. T. Austin, 1984, McGraw-Hill, New York, used with permission.*)

large vats. The mixture is allowed to ferment for about six weeks, after which the liquor then is filtered and bottled. See fermentation in Chapter 12, section J4.

rings (jewelry) - See *jewelry.*

rock wool (mineral wool) - See 5A6d.

roller blades and skates - See *inline skates.*

rope - is made from both natural and synthetic fibers. Hemp is a traditional material used for strong marine ropes. Sisal, flax, and jute are other natural materials. Nylon, polypropylene, polyester, rayon, and polyethylene plastics have become more common in recent years because of their high tensile strength and light weight. Fiberglass is used in some applications where chemical or electrical resistance is important. Many ropes are made from combinations of fibers.

Rope manufacturing requires several steps. The first step involves spinning fibers into yarns by the spinning, carding and combing methods used in the textile industry. (See 10B, 10B2 and 10B3.) In the second step, these yarns are twisted into strands on stranding machines called *formers* or *bunchers.* Then, the strands are twisted into a rope. The most common industrial rope consists of three S-shape strands, twisted together in the direction of opposing twist (Z-shaped twist). This rope is known as hawser-laid or plain rope. Other ropes are made with four strands and are known as shroud-laid rope. Twisting is carried out on special machines that feed the strands from bobbins, pull them through compression tubes with a mechanism called a capstan flyer, and twist them into rope on a revolving flyer.

rubber bands[13, 18] - are made from both natural and synthetic rubber. Synthetic rubbers are used for most rubber bands, but those from natural rubber have the best elasticity. The initial processing of rubber is as described in section 4O1 for natural rubber and 4O2 for synthetic rubber. Compounding (4O3) then takes place to add sulfur for vulcanization, pigments for coloring, and other additives for various properties. This operation typically involves a 400 lb (180 kg) batch. The rubber is milled by being fed between two parallel, opposed rollers that rotate in opposite directions at different speeds. Thc rubber is kneaded and folded over and over repeatedly, and undergoes a change to a smoother, more even consistency. It exits the machine as a 1/2 in (13 mm) thick slab. As it exits, the slab is slit into strips about 8 in (20 cm) wide, by rotating cutters that bear against the cylinders. The strips are lubricated with talc and fed to an extruder that extrudes (4I) the rubber as a long round tube. For different sizes of rubber bands, different sizes of extruder dies are used. A hole in the center portion of the extrusion die discharges air into the tube so that it doesn't collapse. The tubes are fed from the extruder into a long tank containing a liquid salt solution that is maintained at a temperature of 370°F (190°C). The tube remains in the salt solution for a short time but the time is long enough for the heat to vulcanize the rubber. The tubes then pass through a washing bath that removes residual talc and salt. They move to a cutting machine that slices across each tube repeatedly to cut it into narrow bands. The completed rubber bands are then conveyed by a vacuum system to temporary storage, or to packaging machines that load them into cardboard boxes, by weight, for shipment to customers.

rubber, natural - See 4O1.

rubber, synthetic - See 4O2.

rubber, silicone - See *silicones.*

rubber, urethane - See *polyurethane* and 4O2 (*synthetic rubber*).

rubies and sapphires[2] - are different colors of corundum - crystalline alumina, (aluminum oxide, Al_2O_3). The raw stones are found in alluvial deposits containing sand, gravel, silt, and clay, and in corundum deposits. The red color of rubies comes primarily from chromic oxide in the corundum; the blue of sapphires comes from iron oxide and titanium dioxide.

Artificial sapphires and rubies are made by crystallizing pure alumina. To achieve the necessary purity, the alumina is made from calcined ammonium aluminum sulfate and is in powder form. The oxides needed to provide the desired coloring are mixed with the powder. There often is a blending of the color-producing oxides, and colors can be modified by careful heating and cooling, or exposure to strong radiation. In the Verneuil process, the alumina is flame-treated in a hydrogen-oxygen flame, burning downward at 3430°F (1890°C) to fuse and form a single rod or carrot-shaped crystal boule[4]. The boules can be as large as 400 carats but generally average about 200. The boules are cut and polished using saws, grinders, polishers, and lappers, that achieve their cutting action from diamond abrasives. Most artificial sapphires and rubies are used for bearings in watches, clocks, and instruments, for valves, for thread guides in textile machinery, and in other parts requiring wear resistance and dimensional stability. However, gemstones for jewelry can also be made with this process. (For information on gem cutting, see *jewelry*.)

rugs (carpets) - See 10I1 for Axminster, Brussels, chenille, velvet, and Wilton types. See 10I2 for pile rugs, 10I3 for knitted, 10I4 for needle punch, 10I5 for hooked, 10I6 for braided, 10I7 for oriental and 10I8 for needle point types.

rum - is an alcoholic beverage made from molasses, the thick syrup that remains after sugarcane has been processed into sugar. The molasses, with water and often other sugarcane residues, is fermented, changing sugars to alcohol. The fermented mixture is distilled, yielding a clear liquid that is aged in oak casks for a period of one to seven years. Some coloring may be absorbed from the oak casks, but some rums are further darkened by the addition of caramel coloring. (See *distilled spirits*.)

S

safety glass for automobiles - There are two prime manufacturing methods: In the first method, two pieces of thin sheet glass, made by the float process (5A3f), are assembled with a viscous plastic layer (usually polyvinyl butyl) between them, and are pressed together and heated. If the sheets are to be curved, as in windshields, they are put together with talc or another separating agent and bent by gravity (5A5a). They then pass through an annealing lehr (5A4a). The two formed sheets of glass are then assembled with the viscous plastic layer between them. This assembly occurs in an autoclave, where pressure and heat ensure a proper assembly. The edges of the sandwich are sealed with a water-resistant material. If the glass is broken during use, the broken pieces stick to the internal plastic layer, which is tear-resistant. Safety glass of this type, in addition to its use in automobiles, is also used in machinery guards, buildings, television sets, and instruments.

Another form of safety glass is tempered glass (5A4b) which, if broken, breaks into small, somewhat regularly-shaped pieces with no long, sharp, cutting edges. This type of safety glass is used in automobiles in Europe, and in side windows of cars in the USA.

Wire glass (5A3g) may also be considered a variety of safety glass.

sailplanes (gliders) - are unpowered aircraft that maintain their flight by utilizing natural updrafts of air. They are, of necessity, of lightweight construction. Current sailplane structures are made from composite materials consisting of plastic resins reinforced with high strength fibers. The external skin is an integral part of the supporting structure.

The method normally used involves hand lay-up of thermosetting plastics (4G1). Manufacture of each piece of wing or fuselage structure begins with a female mold for the part to be produced. There are four major parts for the wings: a top surface with reinforcements and a bottom surface with reinforcements, for both the left and right sides. The mold surfaces are first spray coated with a gel coat of polyester resin that will form a smooth exterior surface for the part. When the resin has set but is still tacky, an epoxy resin coat is applied by rollers on top of the gel coat and a thin layer of fiberglass is laid on and pressed into the epoxy with rollers. Then a layer of carbon fiber fabric is laid in the mold and bonded to the existing material with a liberal amount of epoxy. (The carbon fabric is the main reinforcing medium.) Then a layer of PVC foam sheet (See *polyvinyl chloride.*), about 1/4 in (6 mm) thick, is carefully laid on the epoxy. The foam sheet has been previously punched with many small holes to provide a means of penetration for the bonding epoxy. Openings between pieces of the foam sheet provide room for a wing spar, ribs, and control devices. More epoxy is added, and the wing spar and several ribs, made earlier with epoxy and carbon fiber, are put in place. Another layer of carbon fiber fabric is added with more epoxy to bond the spar and ribs to the existing structure. The process is largely manual, with the necessary precision controlled by the molds. After all wing materials are in place, a layer of plastic film is applied temporarily over the entire surface of the material in the mold. A vacuum is applied under the film so that atmospheric pressure forces the film to squeeze the epoxy and reinforcement, and they are compacted with no voids between layers. The epoxy is cured overnight with the aid of heat supplied to the mold through channels in which warm water is circulated. After the epoxy has cured, hinges for control surfaces and control apparatus, and wing-mounting hardware are bonded into place. The wing top and bottom, both made with the same hand lay-up process, are bonded together with the aid of a precise fixture and an extensive process that assures that the correct amount of epoxy is used on all bonding surfaces. Fuselage and tail components (stabilizer and vertical fin) are made with similar hand lay-up methods and bonded together. (The fuselage may use Kevlar fiber reinforcement instead of a foam sandwich construction.)

Control hardware and other non-plastic parts are next assembled to the fuselage, wings, and tail. Control surfaces, ailerons, elevators, and rudder, are also installed. Before completion, however, the exterior must be finished and this usually involves spray painting with polyurethane enamel and many hours of wet sanding by hand with progressively finer abrasives. Final assembly is next. The cockpit canopy, seating, various seals, control apparatus, instruments, and electrical connections are installed. The wing halves are mechanically attached to a central spar in the fuselage.

After assembly, a lengthy quality check is made. The sailplane is flight tested in accordance with a prescribed test program. If successful, the sailplane awaits an airworthiness certificate, license, and shipment to the customer. Fig. S1 shows wing molds in use.

salt - (sodium chloride - NaCl), is produced by one of three methods: 1) underground mining of solid salt, producing relatively coarse rock salt. 2) solution mining, where underground salt is dissolved in water to form a saturated brine that is converted by evaporation to granulated salt, often referred to as "table salt". 3) solar evaporation of sea water or salty lake water to produce "solar salt". The purity of rock salt can vary between 95 and 99% NaCl. Solar salt is typically around 99.5% pure and granulated salt has a purity varying between 99.8 and 99.95%.

Underground salt mining is quite similar to coal mining and the equipment used is often the same. A room and pillar system is used with extraction rates varying from 65 to 75%. After the rock salt is blasted free from the solid deposit, it is fed to a series of crushers (11D1) and screeners (11C8a), to produce various grade sizes ranging from 3/8 in (10 mm) down to 12 mesh (1/16 in - 1.5 mm).

In the production of evaporated granulated salt, fresh water is fed to a deep rock salt deposit. The

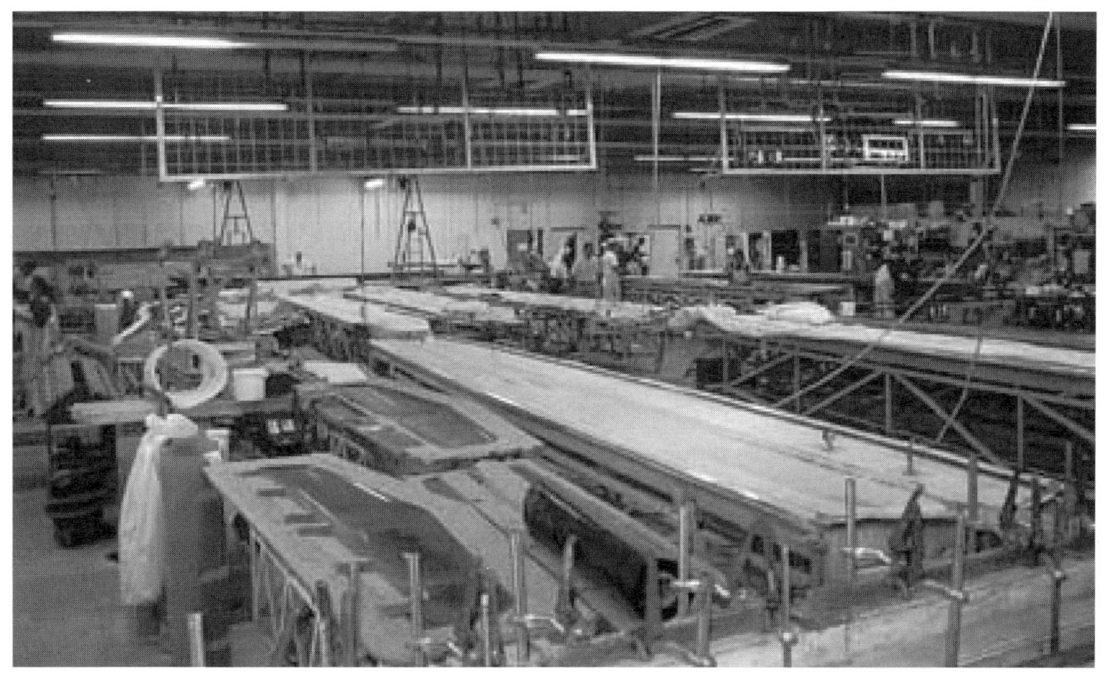

Fig. S1 The factory floor of a sailplane manufacturer showing molds for tail and wing components. (*Courtesy DG Flugzeugbau GmbH.*)

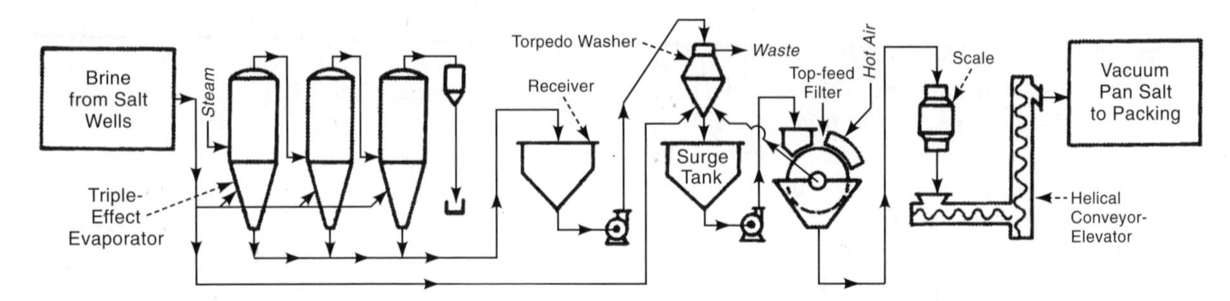

Fig. S2 The vacuum pan system for making salt from brine. *(from Shreve's Chemical Process Industries, 5th ed., G. T. Austin, McGraw-Hill, New York, 1984. Used with permission.)*

rock salt dissolves in the water to the point of saturation, resulting in a brine of about 26% NaCl. This brine is returned to the surface where, after chemical treatment to remove impurities (mostly calcium sulfate), it is fed to a series of multiple-effect evaporator pans, operated under vacuum (See 11C1d and Fig. S2.). The pans are heated, usually by low-pressure steam. Salt in suspension in the brine is removed by filtration (11C7a).

Solar salt is usually produced from seawater, which contains an average of 2.8% NaCl plus many other minerals in various quantities. (Sometimes, denser lake water is used.) Fresh seawater is fed to a series of irregularly-shaped and shallow concentrating ponds, where solar radiation, wind, and heat cause evaporation. The brine increases in concentration of salt and impurities, particularly calcium sulfate (gypsum). The gypsum, and other impurities precipitate out before the salt does. When the brine reaches a concentration of about 26% NaCl, it is fed to a series of shallow, rectangular, crystallizer ponds where the salt crystallizes in a bed that can range up to 8 in (20 cm) or more in thickness. This salt is harvested using mechanical harvesters, and trucks then haul it to a washing plant where other impurities, including organic matter, are washed out. The salt is then conveyed to a storage pile where dewatering occurs. After dewatering, the salt is reclaimed and loaded for bulk shipment.

Much salt is shipped in bulk freighters, particularly if the salt facility or user are near water-shipment facilities. Rail cars and trucks are also used. Bulk salt is heavily used for winter road use and in chemical processing. The amount of salt used in food processing (e.g., for pickling and meat treatment) is much less and only a small percentage of total production, on a weight basis, becomes table salt. Some salt produced by the above methods is packaged into 15, 50 and 80 lb (6.8, 23 and 36 kg) paper or plastic sacks, but most is shipped in bulk.

Table salt (mostly from solution mining), may be crushed and ground to the desired particle size. Magnesium carbonate, calcium carbonate, and sodium ferrocyanide, may be added to ensure that it flows freely, even if the humidity is high. Potassium iodide or sodium iodide may also be added for health reasons. Efficient automated high-speed packaging lines produce the familiar 26 oz. (0.74 kg) cylindrical cardboard containers.

sandpaper - is made from crushed grains of quartz bonded to heavy paper. Sandpaper is sometimes called, *flint paper*. Long lengths of paper are processed at one time. They are first coated with adhesive. Various types of adhesives are used, depending on the application. The quartz grains are then placed on the paper in a single layer using electrostatic attraction in order to position sharp edges upward, where they will contact the work, and to properly space the grains. The adhesive is allowed to dry and another layer is applied and dried. The sheets are then cut to the desired length. When aluminum oxide, embedded in iron oxide, is used as the abrasive, the product is called emery paper, or emery cloth if the backing is cloth rather than paper. Silicon carbide grains are used in some abrasive paper. Sandpaper is used in the finishing of wood, and emery and silicon carbide are used when metals are to be polished.

sanitary paper - See 9B and 9C4.

sanitary ware - such as toilets and lavatories, are ceramic products made from clay mixtures. See section 5B. Slip casting (5B8), (also known as drain casting), is a common manufacturing method for such products because of their often hollow construction. The workpieces are smoothed by hand, with sponges, after casting, dried and fired. Glazing (5B15) is applied to the product after firing, after which there is a lower-temperature firing, followed by inspection and packing.

sapphire, synthetic - See *rubies and sapphires*.

saran - is a trade name for a copolymer of polyvinyl chloride (13%) and vinylidene chloride (87%). The monomers for both materials are mixed together with a catalyst and heated to bring about polymerization. The copolymer is then usually extruded to make film (blown film extrusion - 4I5a), sheets, or fibers. Saran is made into the common kitchen wrap, and filters, insect screening and upholstery.

satellites and spacecraft - are made with manufacturing techniques quite similar to those used in aircraft manufacture. In both cases, light weight and extremely high reliability are paramount. These requirements have led to a substantial use of composite materials. Another major factor is that these devices contain much electronic gear for communication and instrumentation. Honeycomb and other light weight composites are used extensively for structural and body components. One major difference between conventional aircraft and these products is that these devices are assembled under clean room conditions. The assembly room is sealed; entering air passes through fine filters; floors, walls, and ceilings have few seams and are cleaned each day; temperature and humidity are closely controlled. Because of the one-of-a-kind or very limited-quantity production, line assembly is not common and manual operations predominate. There is an intense concentration of inspection and testing operations as the assembly progresses, in keeping with the extreme reliability requirements for these products. Some testing takes place in a vacuum to simulate space conditions. Testing devices are computerized and automatic, and where machining and other operations are involved, computer control is utilized to insure high precision and reliability of dimensions and other characteristics.

satin[2] - was originally a heavy silk fabric woven with a close twill weave. (See Fig. 10C-1.) Now, the same type of weave of other fibers is also called satin. The warp threads are very fine and are prominent; the weft threads are covered. In the most common arrangement, an eight-leaf twill, the weft is intersected and bound down at every eighth pick. Satin fabrics can be dyed in many colors and are used for the lining and trimming of garments. They are also used for dresses, upholstery, and bedspreads.

sausages - are a product of meat processing. Their manufacture involves the use of finely ground meat, usually highly seasoned, which is stuffed in a casing. There are many varieties of sausage, depending on the meat used, other ingredients, the spices involved and the means of preparation. Though sausages may contain almost any meat, poultry, or fish, the most common meats are beef and pork. Often, mixtures of several meats and meat-processing by-products are used. Other ingredients are cereals, soy flour, water, vegetable starch, colorings, and flavorings. Seasonings are widely varied and include salt, garlic, pepper, coriander, nutmeg, vinegar, mace, cloves, and chili peppers. Casings may be made from the intestines or other internal organs of meat animals, from extruded plastics, from fabric coated with paraffin, or from other animal protein reconstituted to tubular form.

The meat contents are chopped and ground, and mixed with other ingredients (12D). The mixture is extruded (12J5) and stuffed into the casings. Cooking (12G6), smoking, and pickling (12J7) are common processing operations that follow.

sauerkraut - is made from cabbage that is fermented (See 12J4). Salt controls the bacterial action. Acid develops during the fermentation and acts as another preservative, as well as contributing to the flavor of the sauerkraut.[2]

screwdrivers - The processes for making conventional screwdrivers for slotted screws and those for Phillips-head types are the same except for the end-shaping operations. The operation sequence begins with coils of carbon-steel wire of a larger diameter than the screwdriver shank. Wire drawing (2B2) is the first operation to reduce the diameter of the wire to the desired dimension. The wire is then annealed (8G1) and straightened (2K2), cut to length, and formed to shape with a variety of press forming, trimming (2C6), milling (3D), broaching (3F), and grinding (3C) operations, depending on the type of screwdriver, the particular manufacturer's methods and the finish level of the model involved. "Wings" in the shank near the upper end are press cold formed to transfer torque from the handle to the shank. Standard screwdrivers are first flattened at the tip in a cold-forming press operation, and then press trimmed and/or machined or ground to refine the outline. Machining involves either milling, broaching, or surface grinding (3C3). The tips of Phillips-head screwdrivers are either machined or hot-formed, and ground to produce the four-grove pointed shape. After press operations and machining, the screwdrivers are through-hardened (8G3c) with oil quenching and tempering (8G2f). The completed shanks are polished (8B) chemically and/or mechanically and electroplated (8C1) with nickel or chromium. Handles are normally injection molded with the shank as an insert in the mold (4C6). Cellulose acetate is the most common plastic used, particularly for higher-grade screwdrivers. Some handles are made from extruded plastic, sometimes with two different-colored materials coextruded (4I2), then machined at the ends and drilled in the axis. The shanks are then pressed into the handles. Handles may be hot stamped after assembly with the manufacturer's identification. A final operation, with most tools, is packaging, usually blister packaging. The operation is fully automatic with dedicated equipment (7F3b). Fig. S3 illustrates the full manufacturing sequence.

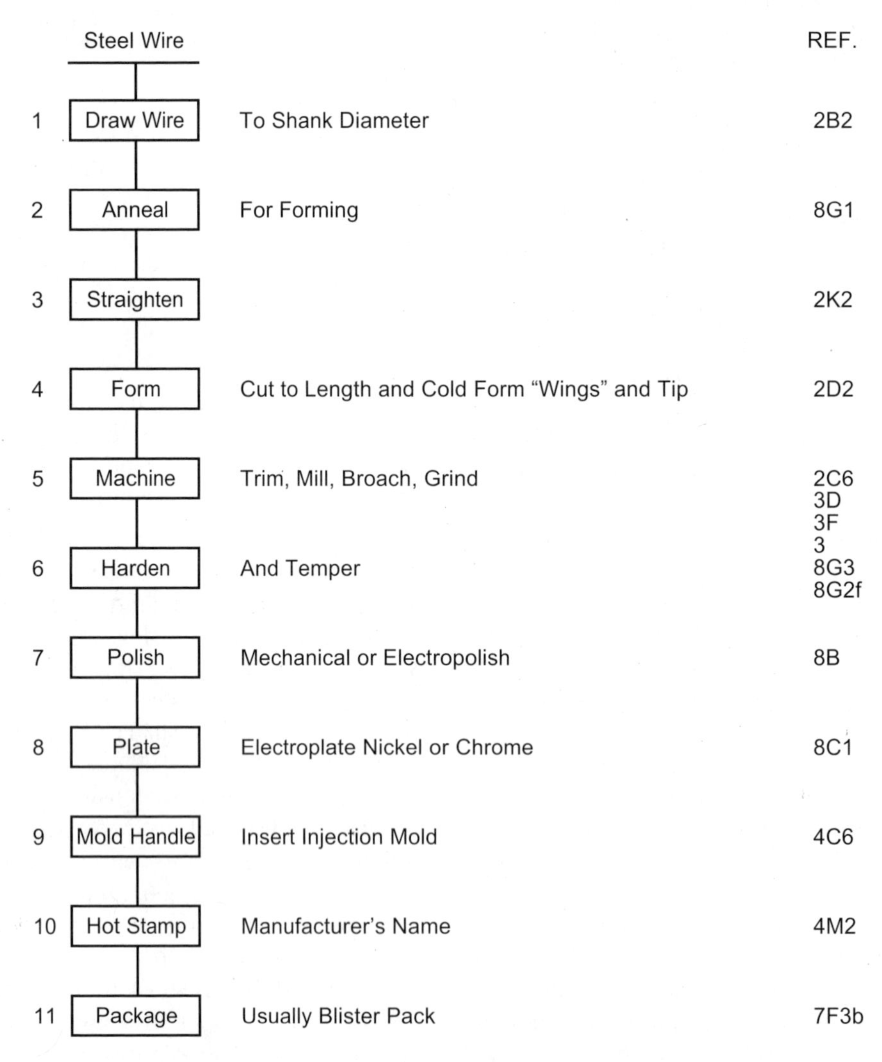

Fig. S3 The sequence of operations for making a screwdriver.

screws, machine and cap - See *bolts*.

screws, wood, drive, and sheet metal - are made by cold heading (2I2) followed by thread rolling (3E7) in a highly automatic operation. Carbon steel wire is fed to the cold-heading machine and, in a series of blows with the necessary tooling, the head is formed including slots or recesses for driving. A taper is formed at the end of the shank. The screw blanks are then tumbled (8B2) and fed by vibratory feeder to a thread rolling machine. Threads are rolled, often including a gimlet point at the end. Sheet metal and drive screws are heat treated for hardness (8G3c). The screws are then cleaned (8A) and tumbled, and usually barrel plated (8C3) with zinc for corrosion protection, and then given a chromate conversion coating (8E3). Other surface treatments such as black oxide (8E4) or ceramic coating may also be employed.

screw threads - See 3E.

semiconductors - are materials that are neither conductors nor insulators of electricity but lie somewhere between. Silicon is the most common of these materials, but germanium, gallium arsenide, selenium, and other compounds are also used. The semiconductor properties are obtained by starting with a material of extremely high purity and then doping it with a very small quantity of another element such as boron, arsenic, antimony, phosphorus, or cobalt. Doping changes the electrical properties of the material, allowing it to become much more conductive. Semiconductors are made as a single crystal of high purity, and the doping can be carried out by one of two methods: thermal diffusion or ion implantation (13K3c). Semiconductors in the form of transistors and diodes in integrated circuits are the basis of present-day computers. Semiconductors are also used in rectifiers, as controllers of electrical current, and as sensors. See 13K and 13L5.

sewing machines - The machines that put stitches in fabric to make garments and other cloth products are complex assemblies of hundreds of parts, most of which are of high precision. The main structural parts, the arm and the bed, have been made for many years of cast iron, using sand molds (1B). These parts are machined with a number of drilling, reaming, boring, milling, and grinding operations (Chapter 3) to provide a base and bearing surfaces for parts that make up the mechanism. Some holes are both rough and finish bored (3B5) to provide precise alignment and location. For household sewing machines, produced in considerable quantity, most of these operations are performed on special machines. The arm and bed are polished (8B1) and filled to provide smooth surfaces, bolted together, electrostatically painted (8D7) and decorated with decals. Household sewing machine arms and beds are more commonly made of die cast (1F) aluminum instead of cast iron and, more recently, reinforced and filled thermosetting plastics are being used (4C2). Both of these alternatives reduce the amount of machining required.

Many metal parts are involved in the sewing mechanism. Some are made by machining small sand castings, die castings, or investment castings (1G). Others are made on screw machines (3A2c). Conventional metal stampings (2C and 2D) or fineblanked stampings (2C9) are made, and sometimes are subjected to finish-machining operations. Other parts are made by powder metallurgy methods (2L1). Gears are machined with gear-making equipment (See *gears*) or are molded (4C1) from polyurethane or other materials. Some critical parts are surface heat treated (8G3a and 8G3b) to provide longer wear resistance. The shuttle, a critical part involved in handling the thread during stitching, requires 60 or more separate operations. These parts, and others that touch the thread, must be polished to a very smooth surface so that stitching takes place reliably. Most such parts are then chromium plated (8C1). All these parts are assembled to produce the stitch-forming and fabric-moving mechanisms. Some external parts are injection molded (4C1) of ABS and other plastics. Current machines now often include electronic controls with linear motors to move the needle when making decorative stitches. Other machines use plastic or metal cams to achieve the same result mechanically. An electric motor drives the mechanism.

The assembly of the machine takes place on a conveyorized assembly line (7F2), typically of 20 or more manual workstations. Final operations involve visual inspection and sewing performance tests for each machine before it is packaged for shipment.

shampoo - See *detergents*. Shampoos consist of one or more detergents.

shellac - is a kind of varnish made by dissolving lac resin, secreted by the lac insect, which is found in India, with denatured alcohol. Shellac is useful in interior wood finishing when a quick-drying, light-colored, hard finish is desired. The resin has dielectric properties and can be compression molded into plastic parts.

shirts - are made with standard garment-making methods. See 10H. Also see *T-shirts*.

shoes - are made from a variety of materials. Leather predominates. The type used depends on the customers' demands and the leathers available. Cowhide and calfskin are most common but pigskin, horsehide and other leathers can be used. For high-style footwear, the hides from exotic wild animals may be employed. Synthetic leathers, fabrics, and some composition materials are often used. Heels and soles are often injection or compression molded (4C1 and 4B1) of natural or synthetic rubber.

The following operations are involved in making footwear from leather or synthetic substitutes for leather[1]: 1) cutting, which includes spreading and marking hides with methods very similar to those used when cutting pieces for cloth garments. Fabric and synthetic materials are spread in stacks and cut, multiple pieces at a time, as described in 10H1, 10H2 and 10H3. Leather hides may also be stacked and cut similarly or with steel rule dies (2c4a). For special, more limited production levels, leather pieces are cut individually with shears. Stacks of parts are bundled for later sewing or gluing. 2) stitching, assembling and sewing the upper portion of the shoe. Sewing of leather components is done on heavy-duty sewing machines similar to conventional industrial machines (10H4), but with larger needles and machine elements, and with coarser, stronger, threads. Some operations, particularly in decorative design work, are cycled automatically under computer control. Some sewing is done with the parts inside-out and, if the leather is stiff, the parts may be soaked in water to soften them so that they can be turned right-side out. 3) stock fitting, to prepare the sole. The sole may consist of only one, or up to three, layers. If more than one layer is used, bonding is the usual fastening method. 4) lasting, an operation that attaches the upper part of the shoe to a wooden form that is shaped like a foot. 5) bottoming, the attachment of the upper part of the shoe to the sole. Bottoming is achieved by one of three methods or a combination of them: sewing, adhesive bonding, or nailing which includes fastening with nails, staples, screws, or pegs. 6) heeling, the attachment of the heel and the shaping of it, if necessary. 7) finishing, the final production operation of polishing, removing the lasts, inserting and fastening inner pads, stamping the brand name, trademark, and size on the sole, 8) treeing - attaching laces, buckles or other closures, and final cleaning and inspection.

Leather boots, including those used for ice skates and in-line roller skates, slippers, and sandals are made by methods and process sequences that are similar to that described above.

shoes, athletic and ***sneakers*** - are all very similar although special athletic shoes are made for different purposes. Sneakers are normally simpler and lighter than shoes, with fewer parts and upper parts made primarily from fabric. Most athletic shoes are made from thicker material: woven fabric (often from plastic fibers), leather, rubber, and soft plastic upper parts. Athletic shoes normally have some stiffening and reinforcement to provide arch or ankle supports, internal plastic foam padding covered with fabric, and extra padding in critical areas. Both sneakers and athletic shoes have molded rubber or elastomer soles. Running shoes have extra support and cushioning on the insoles; basketball shoes are higher around the ankles and have reinforcement and extra cushioning in the ankle area. Athletic shoes for general gym purposes are a result of design compromises, and are usable in several sports, though not optimum for exclusive use in any one. There are also "fashion" athletic shoes, designed with attractive appearance as the main objective. Fig. S4 identifies the components of a typical running shoe.

The athletic shoe manufacturing process starts with the manufacture of the individual parts that will make up the shoe. Leather (often dyed white) and non-molded plastic parts are cut from sheet materials with the same basic methods as are used for leather shoes and garments. (See *shoes* above and Chapter 10.) Sheet materials (fabric, leather, and solid and foam plastic sheets) are stacked in multiple layers, and the stack is cut into parts. Some materials are cut with steel rule dies. The cut parts are bundled for later assembly. Most are sewn to other parts with heavy industrial and shoe-type sewing machines, but some parts are glued together. Soles and some other parts are injection, transfer,

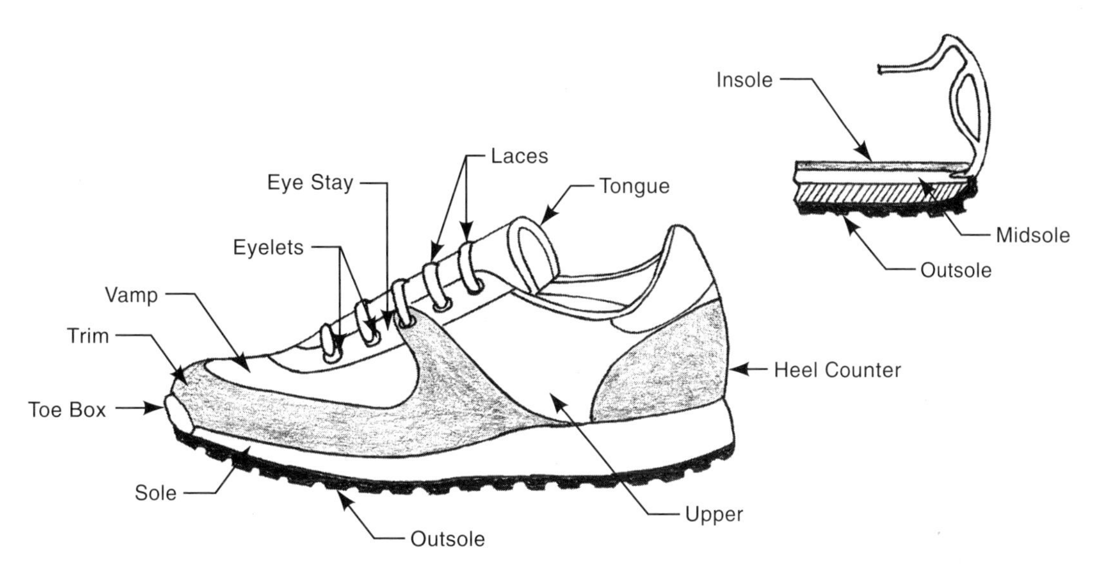

Fig. S4 The components of a typical running shoe.

or compression molded. Lace holes are punched, and eyelets are assembled and upset with eyeletting machines. The upper part of the shoe is sewn and bonded together as a subassembly. The subassembly includes stiffening members and cushioning, and the tongue. The upper is partially sewn to the insole, heated and wrapped around a last. It is mechanically pulled to tightly fit the last. Then, hot-melt adhesive is applied to complete the fastening of the insole and the parts are pressed together. Parts for the midsole, wedge, and outsole, are next bonded together and assembled with more adhesive to the upper subassembly on the last. Hot-melt adhesives are used, and heat to soften the adhesive is applied after the parts are aligned and pressed together. Holding and pressing continue until the adhesive cools. When cooling is complete, the shoe parts are securely bonded. Any excess adhesive is removed and the shoes are taken off the lasts, inspected and packed for shipment.

shortening - is made by the *hydrogenation* (12J3) of edible oils, almost always vegetable oils. Cottonseed, safflower, corn, and soybean oils are used. (See *oils, edible*.) Unsaturated fatty glycerides in these oils are converted to more saturated forms by hydrogenation, providing a more solid, less liquid consistency, improving the shelf life of the fat involved, and providing a more neutral flavor and odor. Some other ingredients may be added to aid moisture absorption, and retard the development of rancidity. Shortening adds tenderness to baked products. If shortening is not used, margarine (produced similarly), butter, or lard may be used to achieve a result that is similar or the same. (Also see *margarine*.)

shrinkproof (shrink resistant) fabrics - See 10E5a.

signs, neon - See *neon signs*.

silicon - See 13K1 for silicon refining.

silicon carbide - See *abrasives*.

silicones (silicone plastic resins, silicone oils, silicon rubber) - have, as their basic raw material, sand (quartzite or silicon dioxide, SiO_2). The silicon dioxide is reduced to metallic silicon as described in 13K1 and then is pulverized. The pulverized silicon and gaseous methyl chloride are reacted in the presence of a copper catalyst to produce a mixture of silane gases. These are condensed into methyl chlorosilane liquid. Fractional distillation of this liquid yields methyl chlorosilane, dimethyl dichlorosilane and trimethyl trichlorosilane. These are hydrolyzed with water to produce cyclic linear polymers. The dimethyl dicholorosilane produces silanol ($Cl_2Si[CH_3]_2$), which is unstable and condenses to form polydimethylsiloxane, the most common siloxane polymer.[1] Hydrogen chloride, a bi-product, is captured and used to make the methyl chloride needed for the process. Whether the polymer is an oil, resin, or elastomer is determined by controlling the size of the individual molecules, and the polymerization of adjacent molecules.[5]

Silicone oils are used as hydraulic fluids, for high-temperature lubricating oils, and to give water repellence to textiles, papers, and other materials. Silicone resins are used for electrical insulation and protective coatings. Silicone rubber is used in O-rings, seals, gaskets, and surgical implants, as a potting material for electronic devices, and for molds used to cast plastics and low-melting-temperature metals.

silicone rubber - See *silicones* above.

silicon single crystals - See 13K2.

silicon wafers (for integrated circuits) - See 13K1 through 13K4.

silk - is a fiber obtained from cocoons produced by silkworms. (The silkworm is not really a worm but is the caterpillar form of the mulberry silk moth.) Sericulture, the raising of silkworms, involves incubation of the eggs from the silkworm moth, raising and feeding the caterpillars and, most important, getting them to spin cocoons. Mulberry trees are planted to provide leaves that serve as food for the silkworms. Cocoons are made from one continuous filament. After the cocoons are harvested, the worms inside are killed (normally by heat), and the silk fiber is obtained from the cocoons by a careful process. The cocoons are placed in boiling water to dissolve sericin, the gummy material that binds the fibers together. The filaments are unwound from the cocoons onto a holder in an operation called, *reeling*. The filaments from 4 to 8 cocoons are joined and twisted. (This is called *throwing*.) Similar twisted filaments are combined to make a thread that is collected on a reel. (See *spinning*, 10B.) The resulting thread, raw silk, usually consists of 48 individual fibers. This thread is composed of unusually long fibers. (Shorter fibers that result from damaged cocoons and certain parts of the cocoons are made into a lower-grade silk that is spun into yarn.) One or more of the threads produced are twisted together again to make a still heavier thread that is suitable for knitting or weaving. (See Chapter 10.) Various treatments and kinds of twisting arrangements for these final threads determine which type of silk fabric will be woven. Further boiling of the yarn or fabric removes the balance of the gummy sericin, leaving the silk smooth, semi-transparent, and lustrous. Silk is used for clothing, lace, draperies, linings, and handbags.

silver - Silver occurs in nature as a metal and in combination with chlorine and sulfur. Most silver is obtained as a bi-product of the refining of copper, lead, and zinc sulfide ores. Antimony is also present in these ores. Silver also occurs in the metallic state in deposits where it is alloyed with gold. Recovery of silver from used photographic film and spent developing solutions is another source of the metal.

The sulfide ores are separated from other materials by flotation. After the copper and lead are removed, silver is usually recovered by furnace roasting the ore to convert the sulfides to sulfates, followed by chemical treatment to precipitate metallic silver[2] . Another method is to treat the ore with liquid mercury. The silver forms an amalgam with the mercury and then is separated out by distillation. Metallic silver that is found alloyed with gold is separated by leaching with a 35% nitric acid solution. Impure silver is most commonly refined electrolytically.

Silver is used in tableware, jewelry, electrical contacts, integrated circuits, electronic circuit boards, brazing alloys, and coins.

silverware - See *flatware.*

single crystals of silicon - See 13K2.

skim milk - See *milk, skim.*

soap - is made from animal or vegetable fat and an alkali. Some of the most-commonly used fats are beef tallow, cottonseed oil, palm oil, coconut oil, corn oil, fish oil, palm oil, olive oil, and soybean oil. Caustic soda (Sodium hydroxide - NaOH) is the most common alkali. Caustic potash (potassium hydroxide) is less common but also used. Other ingredients may be water softeners, fragrances, optical brighteners, colorants, and abrasives. Soap is the sodium or potassium salt of a fatty acid, formed when fats or oils such as these react with the alkali. Current soap manufacture is a continuous process that involves the splitting or hydrolysis of the fat at high pressure and temperature and with the aid of a zinc soap catalyst. This operation produces fatty acids and glycerin. These are later separated. The glycerin is drawn off from the mixture and is a useful by-product. The fatty acid, mainly stearic acid, is purified by distillation in a vacuum and is then reacted with caustic soda (NaOH) to produce sodium stearate - soap - and water. The water is removed from the mixture by decanting and distillation. (Stearic acid, $CH_3(CH_2)_{16}COOH$, is a solid wax-like material that is useful in making greases, paint dryers, rubber, cosmetics, and coatings, as well as soap.) Various special machines mold or extrude, cut and stamp the soap into bars, flakes or powder. Fig. S5 illustrates the soap-making process for aerated (floating) soap. Fig. S6 illustrates the manufacture of soap in milled bars.

sodium carbonate (soda ash)[1] - is an important industrial chemical used in making chemicals, paper, glass, soap, and detergents. It is made from salt brine by the Solvay or ammonia-soda process. Sodium chloride brine is treated with ammonia and carbon dioxide, and yields sodium bicarbonate and ammonium chloride. The sodium bicarbonate is then heated to yield sodium carbonate. The ammonium chloride is treated with lime (calcium oxide) and calcium chloride and ammonia result. The ammonia is reused in the brine treatment portion of the process; the calcium chloride is sold as a winter melting agent for snow and ice.

soft drinks - consist of water and flavorings (juice concentrates, fruit juices, or blends of other flavorings), sweeteners [sucrose (table sugar), high-fructose corn syrup, aspertame, or sacharin], and a number of other ingredients that may or may not be present in a particular drink. They are: carbonation (dissolved carbon dioxide), colorings, acidulants (citric, malic or phosphoric acid) and preservatives

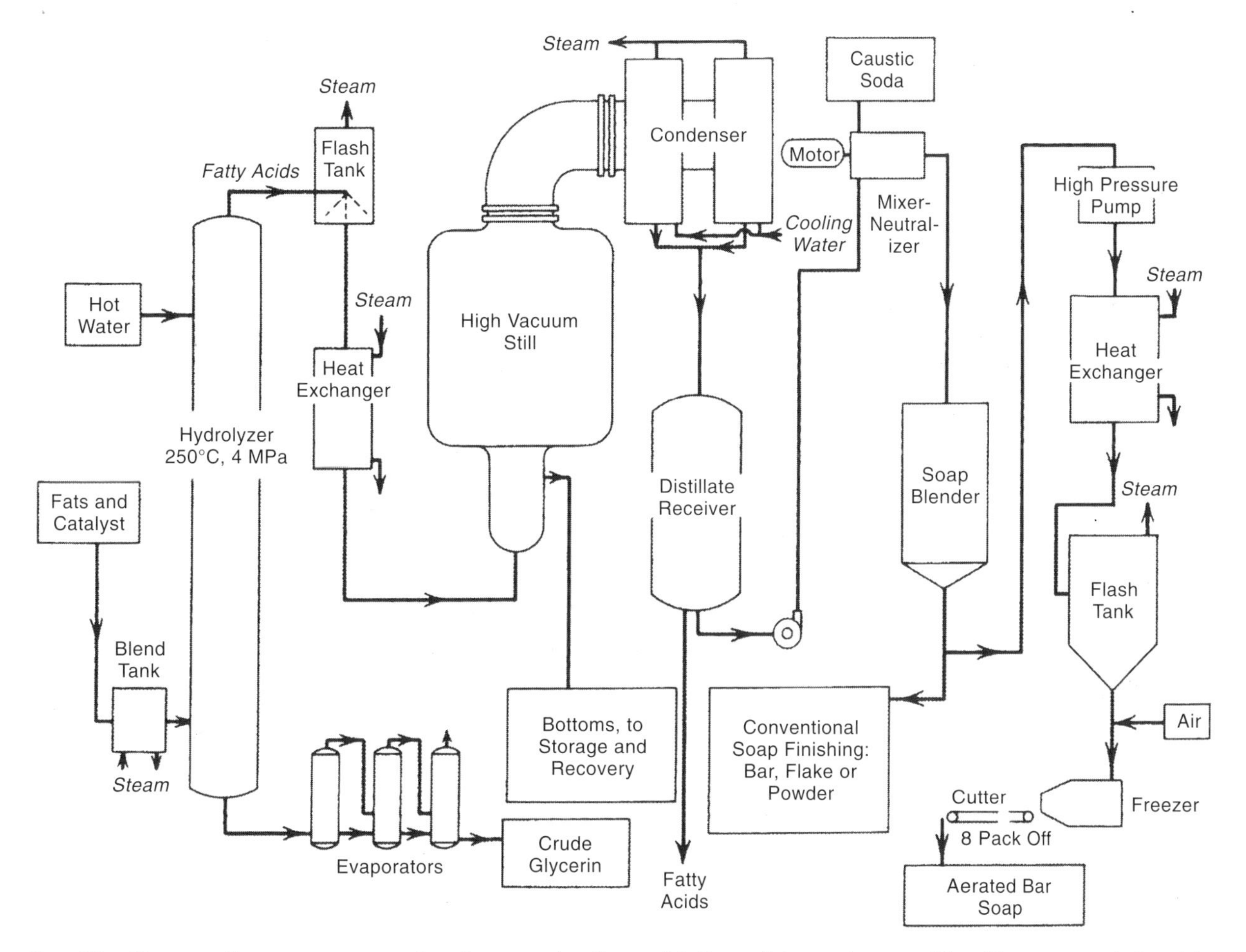

Fig. S5 The continuous process for the production of fatty acids and soap.(*The Proctor and Gamble Company.*)

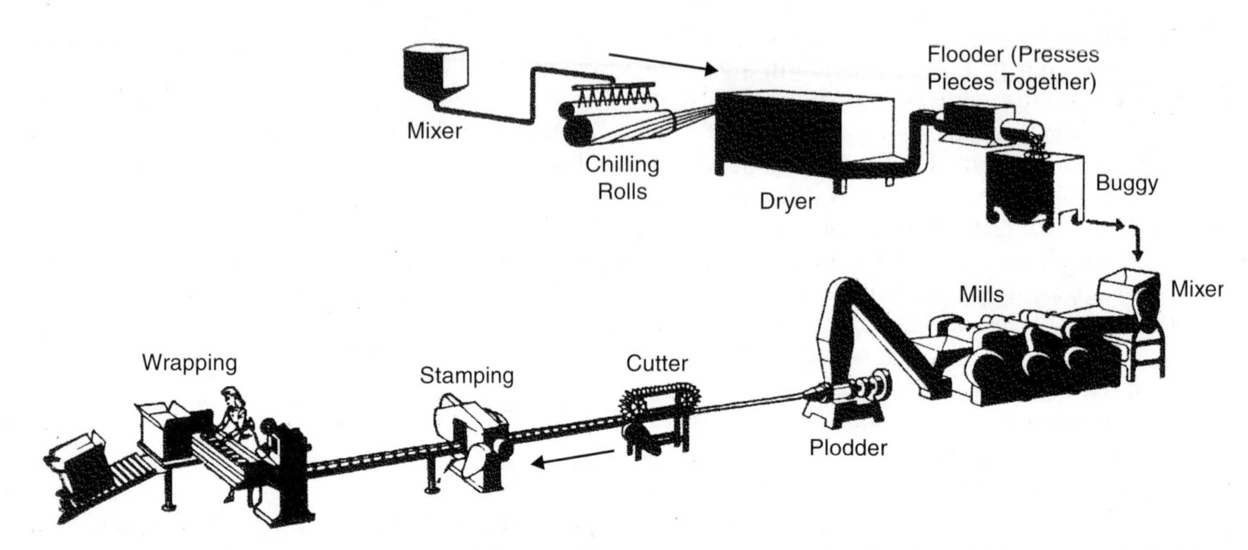

Fig. S6 The production of soap in milled bars. Milled soap undergoes a series of operations involving sets of heavy rolls or mills that knead and mix it, providing a uniform product. *(from Shreve's Chemical Process Industries, 5th ed., G. T. Austin, McGraw-Hill, New York, 1984. Used with permission.)*

(sodium benzoate, potassium sorbate). The water, even if it is good-quality municipal water, is treated before being used. The treatment typically involves chlorination or other oxidant treatment with chlorine, chlorine dioxide, lime, or ferrous sulfate. (Lime serves to remove dissolved calcium and magnesium bicarbonates and aids in the filtering of minute particles, including algae, if present.) The water is then filtered with both sand and carbon filters. The carbon removes any residual color or taste and chlorine.

The flavorings and other ingredients (except sweeteners and water) are often prepared at another location and shipped to the bottling location as one combination ingredient called, *syrup concentrate*. A first step at the bottling plant is to mix this concentrate with the sweetener and some water to produce the *finished syrup*. Once mixed and quality-checked, the finished syrup is mixed with additional water in a ratio of about 5 to 7 parts water to one part finished syrup. Except for carbonation, this is the finished beverage. Carbonation takes place in a reduced-temperature vessel, called a *carbo-cooler*, that is pressurized with carbon dioxide gas. The beverage liquid is sprayed into this vessel and it absorbs the carbon dioxide to provide the characteristic bubbly fizz of soft drinks. This is then the liquid that fills the bottles or cans. Bottling takes place as described in Chapter12, section L. If the drink is canned instead of bottled, the operations involving the preparation, washing, filling, and closing the cans are very similar to those involved when bottling takes place but the heat sterilization, described in Chapter 12, section G4, for canned foods, is not needed with soft drinks.

solar cells (photovoltaic cells) - convert energy from sunlight into electric power. The cells are made from semiconductors and fully conductive materials, employing methods that are similar to those used to produce integrated circuits and other semiconductor devices. However, the semiconductor materials in solar cells occupy large areas whereas most semiconductors in electronic devices are as small as possible. Typical cells are about 4×4 in (10×10 cm) in size. In the solar cell, light energy excites electrons so that they move from one layer to another through a semiconductor. The reaction takes place where an n-semiconductor layer meets a p-semiconductor. Electrons flow across the junction and to metal contacts near the top and bottom of the cell. The semiconductors are doped to provide "n" or "p" (negative or positive) conducting properties. (See doping, 13K3c.) Materials commonly used are silicon in single-crystal form (sawed into thin plates from a boule - See 13K2, 13k2a, 13k2b.), cadmium sulfide (CdS), copper-indium diselenide, gallium or gallium-indium phosphide, cadmium telluride and gallium arsenide. Polycrystalline silicon is also used; it is less costly but less efficient. Most of these materials, when used, are applied as thin film coatings (See 13K3a.) Amorphous silicon is polycrystalline material coated by chemical vapor deposition (13K3a3) on glass or other substrates. This form of silicon is still less efficient but is inexpensive and is used in watches and pocket calculators where it supplies sufficient operating current.

Solar cells are used to power calculators, watches, electronic and electrical devices in remote locations, satellites, and spacecraft. The arrangement of components in a typical solar cell is illustrated in Fig. S7.

solder - normally refers to alloys with low-melting temperatures that are used to make electrical and electronic connections, for joining plumbing pipes, and in other fastening applications. These alloys are often called *soft solders*. Alloys of lead and tin are most common but antimony, bismuth and cadmium are also employed. The lead-tin eutectic, (63% tin, 37% lead), and similar alloys are common in the production of printed circuit boards. They are produced by melting tin and lead together and thoroughly mixing the melt. Lead-free solders, used for connecting water pipes, and increasingly in electronics, are principally tin, alloyed with copper or silver. Alloys with higher melting temperatures, above those of the tin-lead solders are called *hard solders*. These are often silver alloys, "silver solders", alloys of silver, copper, zinc, and phosphorous. (Also see 7A and Chapter 13.)

solder paste - See 13E.

solder preforms - See 7A1b.

solder powder - See 13E1 through 13E6.

spandex[4] - is a flexible stretchable fiber made from polyurethane plastic. The basic material is made by reacting diisocyanates with long-chain glycols (usually polyesters or polyethers). The product of this reaction is then chain-extended or coupled using glycol, a diamine or sometimes water. The resulting polymer is dry spun into fibers. Rigid and flexible segments are alternated in a typical spandex fiber. Used with other fibers, spandex is made into surgical hose, foundation garments, sportswear, swimwear, and other items where elasticity is important.

spark plugs[9] - consist of four basic parts: the outer metal shell, the ceramic insulator, the central electrode and the side electrode. There is also a small terminal nut to secure the high voltage wire to the plug, and a copper ring gasket to seal the plug to the engine cylinder head. See Fig. S8 for a section view of a typical plug. All these parts are made with high-production, continuous-flow, dedicated equipment.

The outer metal shell supports the plug and provides an electrical ground and mechanical screw thread connection to the engine's cylinder head. It is made from low-carbon steel and is first formed by the cold extrusion process (2I4). Knurling (3A1f), thread rolling (3E7), and some finish machining (3A2d) take place. Finishing is by either a black oxide surface treatment (8E4), or electroplating with zinc (8C1).

The ceramic insulator separates the high voltage electric current from grounding elements. It is made from alumina mixed with other ceramics, and is formed by a sequence of ceramics processes: pressure casting (5B8a), drying (5B13), form cylin-

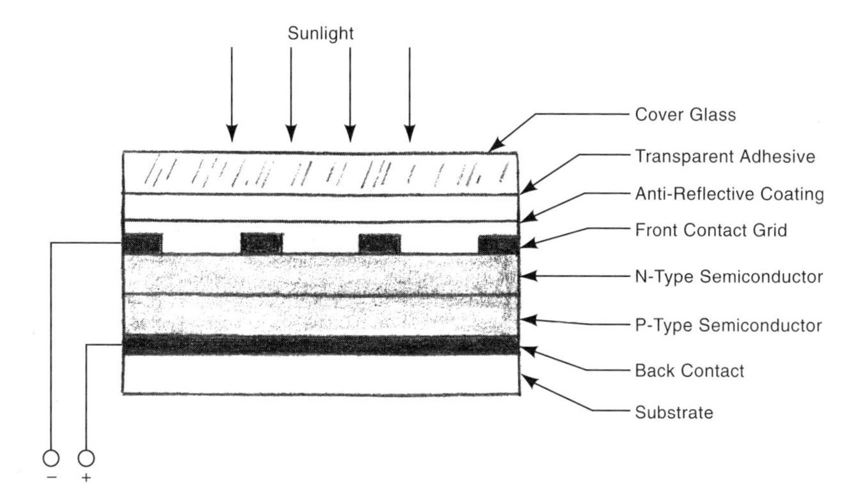

Fig. S7 A typical simple solar cell shown in cross-section. Light falling on the cell passes through the glass or plastic cover, an anti-reflective layer, and past a front contact grid to the semiconductor layers where it creates an electron flow if the top and bottom contacts are part of a circuit.

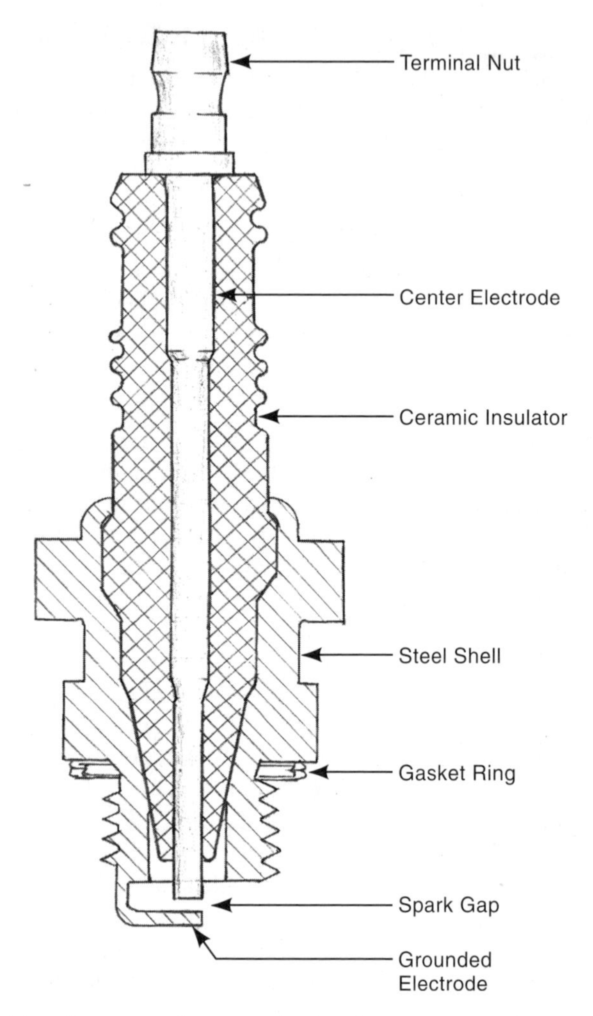

Fig. S8 A typical spark plug for an internal combustion engine.

drical grinding (5B14), firing (5B16), printing, and glazing (5B15).

The central electrode conducts the high voltage charge to the spark gap. It consists of two parts, the electrode at the lower end - made from a high-nickel alloy - and a low carbon wire that forms the top terminal of the plug. Before the two are electrically butt welded together, the top portion is formed to the shape needed by cold heading; then threads are rolled at that end.

The side or grounding electrode, also made of high-nickel alloy wire, provides the other conductor for the spark gap. The wire is fed from a coil, straightened and resistance welded to the outer shell of the plug. It is bent to near the position needed for the spark gap.

Assembly of the plug is also an automatic operation. Key portions of the operation are the press fits of the central terminal/electrode to the center hole of the insulator, and of the insulator to the metal shell. These components must withstand the internal engine compression pressure of 2000 psi (14 MPa). The edge of the metal shell of the plug is crimped around the insulator to complete the seal. The lower end of the electrode is trimmed, and the grounding electrode is further bent to provide the specified spark gap. The fastening nut is threaded onto the central terminal, and the copper ring gasket is assembled to the plug shell and crimped into place. The finished spark plug is then packed into an individual box which, in turn, goes into a carton of plugs for shipment to customers.

spices - are obtained from the flowers, seeds, fruits, leaves, bark, stalks, bulbs or roots of plants, usually from tropical or subtropical regions. Spice seeds and herbs (from plant leaves) are also called spices. Source materials for spices are usually dried and, in for many spices, ground to a fine powder in roll mills. They then may be coated with a water soluble gum or dextrose to preserve their flavor. Popular spices and their sources are as follows:

allspice (pimento) - is made from the dried, unripe berries of a tree, Pimento dioica or Pimento officinalis, which grows in the West Indies.[14]

anise, aniseed - is the seed of the Pimpinella anisum, a member of the carrot family. It has a licorice-like flavor. It is used in baked goods and confections, and the liquors anisette and absinthe.

basil - is the dried leaves of the herb, Ocimum bascilicum, a small bushy plant. Egypt is the main source, followed by the U.S.A. The leaves are dried and sold whole, or ground. Basil is used in tomato sauce, pizza, pestos, cheeses, and Italian seasonings.

bay leaves (sweet laurel) - are the dried leaves of the evergreen bay leaf laurel shrub or tree which grows in California and Turkey. The leaves are dried and used whole in soups, stews, meat, and vegetables as they are cooked. They are not left in the food when it is served.

cayenne pepper - is a very hot pepper that is made from the deep red to orange colored fruit of a small Capsicum (chili pepper) plant. The fruit is dried and then ground.

celery seed - is not from the vegetable celery plant but from the seeds of its relative, Apium graveolens, which have a similar flavor and aroma. The very small brown seeds are dried and sold whole or are ground. Most come from India and China. Celery seed is used in pickling, in salad dressings and in cooking vegetables, soups, breads and tomato items. Celery seed is the main ingredient of celery salt.

cinnamon - is the dried bark of various tropical evergreen laurel trees, Cinnamonum, made into rolls, sticks or powder. The inner bark is pressed, rolled and dried and most is ground to powder form. Sri Lanka is the source of true cinnamon, but most of that used in North America comes from Vietnam, China, Indonesia, and Central America. Cinnamon is used in stewed fruits and other desserts, breakfast rolls, cakes, pies and cookies.

clove - Cloves are the dried, unopened buds of an evergreen tree, Zyzygium aromaticum, that grows in some African countries and Indonesia, Zanzibar and Madagascar. Cloves are sold both whole and ground. They are used in ketchup, Worcestershire sauce, meats, salad dressings, and desserts.

garlic - comes from the bulb of allium sativum, originally from Asia and now grown widely. It is related to the onion.

ginger - comes from the bulb-like underground stems of the herb, Zingiber officinale. It originally grew in the Orient but now is also raised in Nigeria and Jamaica. The root is dried and ground into ginger powder, sold as root ginger, or as an essential oil that is extracted to make ginger ale and other beverages, chutneys, and sauces.[5]

mustard - One strong mustard comes from black mustard, Brassica nigra, a tall plant that grows in Israel. The seed is ground into powder and mixed with lemon juice, wine or vinegar, which preserves the flavor.[5] Another variety, brown mustard, has a less pungent flavor. The common American mustard is made from the still milder seeds of the white mustard plant, ground into a powder, and mixed with vinegar, sugar, and tumeric as a coloring agent. There are other varieties of mustard made from these three types of seeds, made into a powder and combined with other ingredients. Ingredients that may be used include grape, lemon or lime juice, beer, vinegar, cider, wine, salt, and various herbs. Mustard is used in spice mixtures for seafood, meats, and sauerkraut. It is an ingredient in mayonnaise, and salad dressings and in flavoring barbeque sauces.

nutmeg - is the round, brown, wrinkled seed of a fruit similar to an apricot, of a tropical evergreen tree, Myristica fragrans. The tree is native to the Moluccas, but now grows in Grenada.

parsley - the dried leaves of the herb, Petroselinum crispum. Sources are the U.S.A., Canada and Europe. Parsley is used to add color and a pleasing appearance to soups, stews, omelets, and other cooked dishes and to bring out the flavor of other herbs.

pepper - Common black pepper comes from Piperaceae, a vine-like perennial plant, which is grown in many countries, notably Brazil, Indonesia, China, Burma, Vietnam, the Philippines, and Malaysia. The plant yields red berries that are harvested and boiled in water for approximately 10 minutes. They turn black and are dried to become common black peppercorns. They are sold in that form but are also ground into a coarse powder in pepper mills.

rosemary - is the dried leaves of the evergreen, Rosmarinus officinalis. The leaves very small and resemble curved pine needles. They are hand-harvested. Most rosemary is supplied from France, Spain and the former Yugoslavia. It is a powerful spice and is used sparingly. Meat, fish, poultry and vegetables are seasoned with it, especially in Italian and other Mediterranean cooking.

saffron - comes from stigmas of saffron crocuses that are grown in Spain, the Middle East and Italy. It is used to both color and flavor food dishes.[5]

sage - is made from the dried leaves of the plant, Salvia officinalis, a member of the mint family, which is grown in Central Europe. The most common form is "rubbed sage" in which the leaves are given a minimum amount of grinding and then are passed through a coarse sieve. Another form uses the leaves cut into smaller pieces. The spice resembles rosemary and is strongly aromatic. It is used in Italian and Greek meat and poultry dishes. Most supplies come from Southeastern Europe.

sesame seeds - grow as part of the annual plant, sesamum indicum. The seeds are harvested by hand because they would otherwise scatter.

thyme - comes from the leaves and sometimes the whole plant of Thymus vulgaris, a small flowering perennial plant. There are several varieties but that with a strong lemon-like flavor is most common. The leaves are harvested just before the plant flowers. The plant originated in Southern Europe, and now grows in Eastern Europe and the

United States as well, but most U.S. thyme is imported from Spain. The leaves are very small, less than 1/4 inch long, and are dried, and often ground. Thyme is used in cooking meat, fish and poultry.

vanilla - historically comes from the fruits (beans) of a climbing orchid-family plant, Vanilla planifolia, which requires special hand pollinating. The beans require special curing, and the vanilla flavoring is then extracted. Much vanilla is now synthesized from the hydrolysis of wood.[5] (See *vanilla*.)

Spices are utilized to add flavor, aroma and piquancy to foods. Spices are used in making sausages and other processed meat, salad dressings, sauces, prepared mustard, pickles, preserves, salad dressings, cookies, confections, cakes and beverages.[5]

spirits, distilled - See *distilled spirits*.

sporting goods - See *bats, baseball; baseballs; footballs; tennis balls; golf clubs; golf balls; fishing rods*.

springs - Fig. S9 illustrates several kinds of springs. Most are made from high-carbon spring steel, hardened (8G3) and either annealed (8G2g)

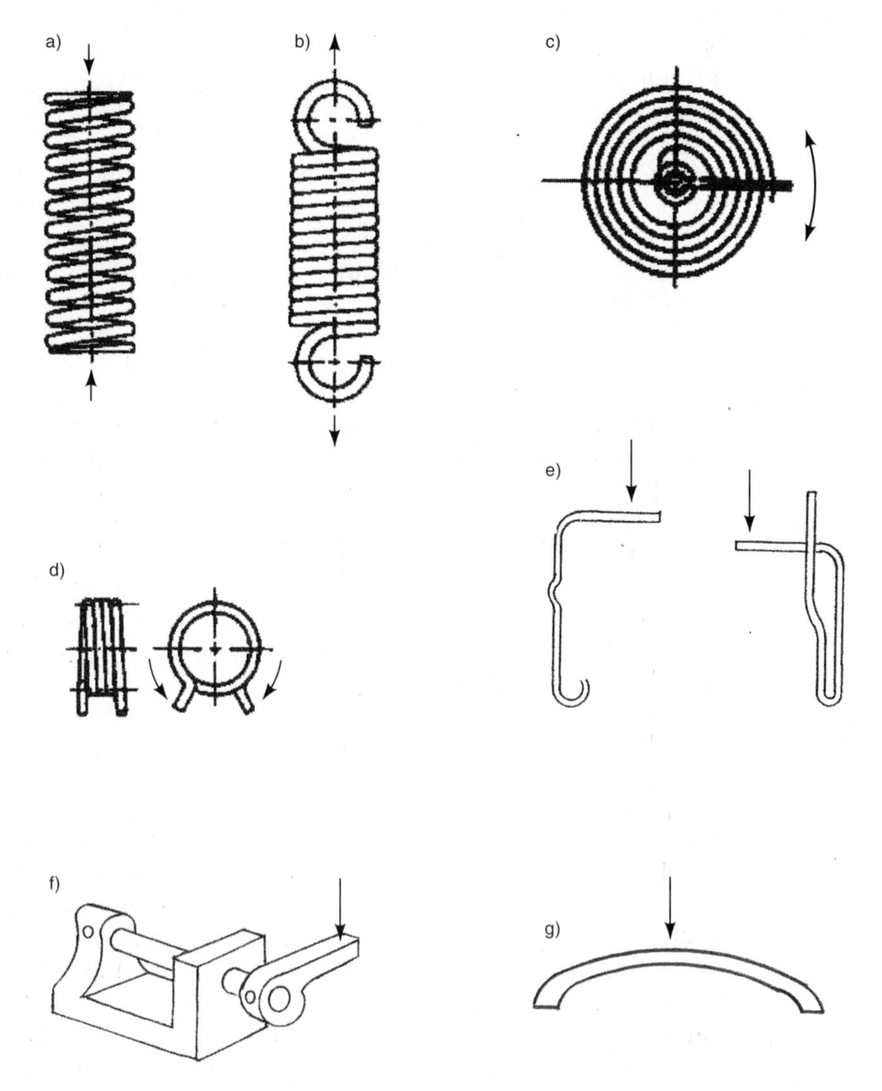

Fig. S9 Different kinds of springs: a) helical compression spring, b) helical extension spring, c) hairspring (for clocks and watches), d) helical torsion spring e) non-helical wire springs, f) torsion bar spring, g) flat spring.

or tempered (8G2f) before forming. Stress relieving (8G2d) may follow forming. Common grades used are 1050 (G10500), 1075 (G10750), and 1095 (G10950). Stainless steel types 301 (S30100), 302 (S30200) and 1.7-7 PH (S17700) are used where corrosion resistance is necessary. Phosphor bronze and beryllium copper are used for springs requiring high electrical conductivity. Springs are used to absorb impact, reduce vibration, provide stored energy for clocks, watches, and door closers, and for weighing with spring scales.

helical springs - In mass production situations, compression, extension and torsion helical springs are wound from coiled spring wire on automatic spring coiling machines. Automatic spring coilers use one or more pairs of feed rolls to feed a predetermined length of straightened wire against a coiling point that imparts a curvature to the wire. A stationary arbor is used to hold the coil during forming. The coil is separated from the feed wire by shearing. For low-quantity production, engine lathes can be used. The wire is fastened to an arbor that is held in the chuck of the lathe. As the arbor rotates, the wire is fed from between two wood blocks on the lathe's cross slide and winds around the arbor. Various hand tools are used to form the ends and cut off the wire.

spiral springs - are wire or strips of sheet material coiled in a spiral pattern in a flat plane. Spiral springs are used to provide the power to drive clocks and watches, and, in high-production, are wound on special machines.

non-coiled wire springs - are made on four-slide machines (2G2). Paper clips are one example, but many products have wire parts with a spring function.

torsion bar springs - are bars set so that they can twist. One end is anchored and the other is free to rotate. Torsion bars are made from suitably heat-treated metal bars by conventional metal forming and machining methods.

flat springs - include cantilever and beam types and, when multiple layers are involved, leaf springs, such as those used in automobiles. These types of spring are blanked and formed from high carbon steel and other metals having a high yield point, with methods almost identical to those used for blanking, piercing and forming low-carbon steel and other metals. With hardened spring materials, higher press forces are required, tooling must be more wear-resistant, and greater clearances between punch and die are normally employed. (See sections 2C and 2D).

stained glass windows - Rolled glass (5A3g), that has been colored by the addition of metal oxides in the batch (5A1a1), is hand-cut to the size and shape desired and assembled with strands extruded (2A3) from lead alloy. (The extruded strands have lengthwise slots to accept the glass.) Junctions of lead strands are soldered with a high-lead solder.

stainless steels - See *steels, stainless.*

stain-release fabrics - See 10F5d.

stamps, postage - In the U.S., stamp designs and other decisions about stamps are made with the help of a Citizens' Stamp Advisory Committee. Artists are hired for the art work needed for new designs. Plate-making and printing are then carried out by the Bureau of Engraving and Printing, or by qualified government vendors. Printing is by either the offset lithography (9D2) or the intaglio (rotogravure) (9D3) processes. Paper in rolls, precoated with adhesive on one side by the paper supplier, is fed to the presses. Perforations are made in line with the printing by suitable stamping dies with one small punch for each perforation. The printed and perforated paper is then cut into sheets as part of the same operation. Other countries use similar methods with varying degrees of automation and may use watermarked or other special papers.

starch[4] - is a common natural substance, a complex carbohydrate, produced by plants, and a major element in the human diet. Chemically, starch is based on the formula $(C_6H_{10}O_5)n$ where n ranges from 250 to over 1000.[4] Most starch produced in the United States comes from corn, but rice, wheat, potatoes, arrowroot, and tapioca are other significant commercial sources. In addition to each use in foodstuffs, starch is used in textiles, paper, adhesives, paints, insecticides, soaps, explosives, and as a basis for corn and other sweeteners. Corn starch manufacture has the following sequence: 1) Corn kernels are cleaned 2) They are soaked in warm water for two days to weaken the hulls and soften the gluten. 3) The softened kernels are processed by equipment that uses two studded steel plates, one rotating and the other stationary, to break the kernels but not crush the corn germs. 4) The corn germs are separated from the hulls in special equipment that uses centrifugal action on the kernels in water. 5) Oil is extracted from the germs (See 11C3). 6) The heavier starch is separated from the gluten by "hydroclone" machines. 7) The starch is dried and is then ready to be sold. The major use for corn starch is the production of corn sweeteners. If used for other food or commercial purposes, starch is processed through other operations that modify it physically or chemically.

steel - Three prime processes are used to produce steel from pig iron: the basic oxygen process, the open hearth process, and the electric furnace process. All these processes burn out the excess carbon and impurities from the pig iron.

The ***basic oxygen process*** - like the earlier Bessemer process - utilizes a pear-shaped furnace that is turned sideways for charging and pouring, and upright for processing the charge. The charge consists of molten or solid pig iron, scrap steel and iron, lime (CaO), and flux. The scrap material acts as a coolant to prevent the furnace temperature from rising too high. After charging, the furnace is tilted upright and an oxygen lance is lowered into it from the top, with the water-cooled tip about 6 ft (2 m) above the level of the charge. Oxygen is blown into the furnace through the lance at very high speed. The oxygen churns the charge and causes the carbon and impurities to burn (be oxidized). A carbon dioxide/nitrogen mixture may be blown from tuyeres at the bottom of the furnace to promote mixing of the charge. Oxygen may be also introduced at low pressure from low-placed tuyeres. Excess carbon escapes from the furnace as carbon monoxide and carbon dioxide gases; the impurities in the charge (phosphorus, silica, and manganese) form slag. The operation is rapid and 275 metric tons of steel can be made in an hour[5]. Fig. S10 illustrates a basic oxygen furnace schematically.

The ***open hearth process*** (1A6) uses regenerative preheating of fuel and air used for its combus-

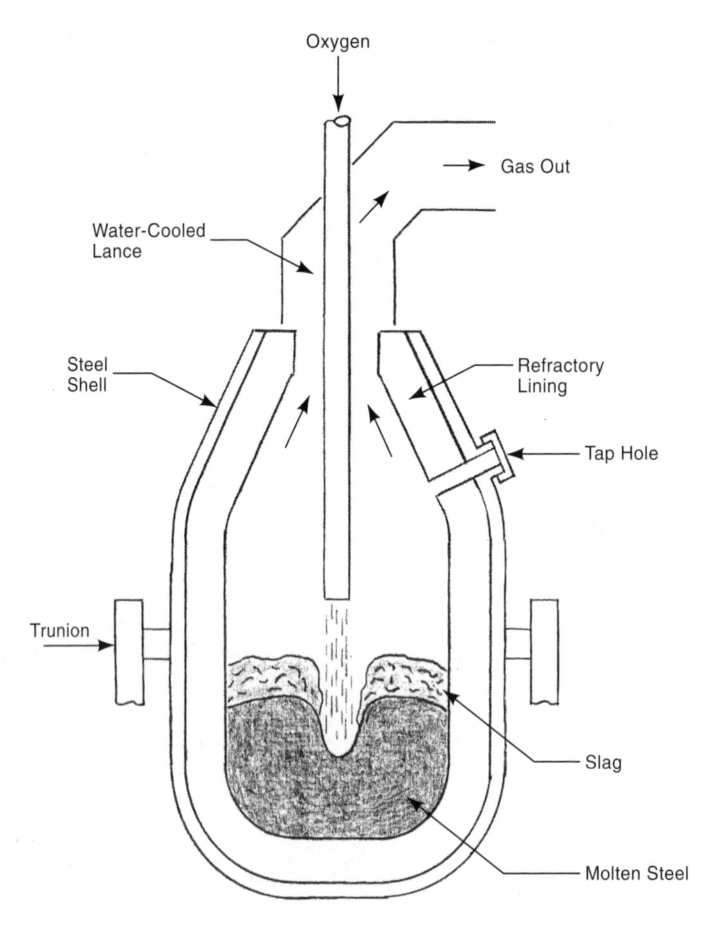

Fig. S10 A basic oxygen furnace for steelmaking.

tion to hold the full furnace charge to a temperature between 2800 and 3000°F (1540 and 1650°C) - well above the melting point of iron (2500°F - 1370°C) - for a prolonged period. The charge consists of pig iron, either solid or molten, or portions of both, scrap steel, iron ore (to provide additional oxygen), and limestone flux and fluorspar to add fluidity to the slag. The carbon in the iron is gradually oxidized, and impurities such as silicon, manganese, phosphorus and sulfur, react with the limestone to form slag. When the carbon content in a measured sample has reached the desired level, the furnace is tapped into a large ladle (set below the furnace level), and the molten steel is poured into molds to make large ingots for further processing. Typical production from open hearth furnaces is about 100 metric tons per 11 hours.[5]

Electric furnace steelmaking uses the electric arc furnace (1A2) to provide the necessary heat for melting and refining. The furnace is closed to the outside atmosphere which, together with the use of electric heating, allows tight control over the refining conditions. Because of the more rigid control, the electric arc process is particularly useful when high-grade alloy or stainless steels, with their more exacting specifications, are to be made. The charge metal usually consists largely or completely of scrap iron and steel, but the scrap is first analyzed for its content and unwanted alloys are excluded from the charge. Small quantities of dry lime and iron ore are included, to assist in the removal of impurities and unwanted carbon. The electric arcs provide the necessary heat but an injection of pure oxygen at the beginning of the process shortens the time required to reach operating temperature. Electric furnace steelmaking from scrap material has become a common process for smaller steel mills making high-quality low-carbon steels.

In all steelmaking processes, deoxidizers may be added to the raw steel to remove dissolved oxygen from the molten metal. Aluminum and ferroalloys react with the oxygen, and the resulting oxides precipitate from the steel. Other treatments may also take place after the molten steel has been tapped from the processing furnace. This step, called *ladle steelmaking*, produces higher-quality steel for specific applications such as deep-drawing.

The finished steel is cast into ingots. (However, methods have been developed for processing the steel into useful forms directly from the molten state.) Ingots are reduced to billets, blooms, or slabs, and then made into semi-finished or finished shapes (I-beams, rails, bars, angles, channels, tees and other cross-sections, plates or sheets) by any of several hot-forming operations. These include hot rolling, which involves passing the steel through sets of rollers that are shaped to produce the cross section wanted. After rolling, the workpieces are usually pickled by immersion in warm, dilute sulfuric acid to remove scale, and are then oiled. Cold drawing and cold rolling may occur afterward to produce cold-finished steel. Further finishing operations may include some machining, further pickling, grit blasting, water washing, and immersion in a slaked lime solution. Stress-relieving, annealing, and normalizing treatments may be carried out, depending on the application intended for the steel.

Fig. I2 (*iron*) shows the entire iron and steel making operational sequence including the three alternative steel-making methods. Also see *ultra-high strength steel.*

steels, stainless - are alloys of iron and 10 to 30% chromium, often with sizable amounts of nickel. Manganese, silicon and molybdenum are also often present. Stainless steels gain their corrosion resistance from a chromium oxide film that forms on their surface and is non-porous and self healing. 12% or more chromium is required for this effect.

After initial melting and alloying in electric or basic oxygen furnaces, stainless steel is refined in another vessel with the main purpose being the reduction of carbon content. An argon-oxygen process is usually used, with a number of different possible ratios of these two gases to oxidize carbon without oxidizing chromium. With this process, high-carbon ferrochromium can be used as a raw material.

Common applications are food processing equipment, table flatware, surgical instruments, aircraft and spacecraft.

structural composite lumber - See 6F7.

structural foam plastics - See 4C3.

styrofoam - is polystyrene expanded to 42 times the original size[2]. It is used, usually in panel form, as a cold-temperature insulation. The panels are made by extrusion, most commonly with a dual, tandem extruder system. The first extruder heats and mixes the polystyrene with a nucleating agent, a fine powder of talc or other material, with small amounts of other materials. The output of the first extruder is passed under pressure to a second extruder, sometimes called a cooling extruder, where the liquid or gaseous blowing agent is injected. The blowing agent is usually a blend. Several key materials that may be used include: pentane, carbon dioxide, butane, and HFC-152a[15]. The blowing agent is kept under compression as it is mixed into the cooling resin. At the nozzle of the second extruder, the blended material passes through a shaped die into the open atmosphere where the blowing agent expands, changing the extrudate to a foam. Often, an annular die is used, creating a tube of foamed polystyrene. The tube is slit and the foamed material is flattened to thick sheet or slab form. Slabs are further cooled and cut to standard lengths.

sugar - Sugar cane is the prime source for sugar, though sugar beets are another source, and sweeteners are made from corn and are widely used in canned beverages. Cane sugar stalks, after they are harvested, are washed, chopped, crushed, and shredded to remove the sweet liquid they contain. Water sprays dissolve additional sugar from the stalks. The liquid thus obtained contains some impurities. These are removed by screening and then heating the liquid and adding calcium hydroxide (lime) that causes the impurities to settle out from the liquid. Carbon dioxide is bubbled through the liquid to remove the excess calcium hydroxide. The settled matter is separated by filtration, but is processed separately to remove the sugar it contains.

The liquid filtrate is then pumped into large evaporator tanks where it is heated sufficiently to boil off most of the water. The liquid, at this point, is thick and syrup-like. Not all the water can be removed at this stage, however, because doing so may overheat and scorch the sugar. The balance of the water is removed by a combination of vacuum and heat. The vacuum reduces the boiling temperature of the syrup enough so that it does not scorch.

Removal of additional water causes the sugar to crystallize. The mixture is placed in a centrifuge to separate the crystals from the remaining syrup. The sugar crystals at this point are yellow-brown in color and are referred to as raw sugar.

The raw sugar is refined further to yield common white table sugar. The crystals are first subjected to a water rinse that flushes off some surface impurities from the crystals, which are then dissolved again in water. The mixture is filtered to remove further impurities, and becomes a colorless, clear liquid. The evaporation is repeated to form white crystals that again are separated centrifugally. The crystals, now solid sugar, are conveyed to drying drums where heated air removes any remaining moisture.

Remaining syrup from the evaporation operations may undergo rinsing, dissolving, filtering, and evaporating several times to produce additional white sugar crystals. Syrup that remains after several cycles is used to make brown sugar. The final syrup is blackstrap molasses.

suits - See manufacture of clothing (10H).

sulfuric acid - has wide industrial use. It is the most heavily-produced product of the chemical industry[1]. A major usage is the production of phosphate fertilizers. It is also used in many chemical processes, in ore processing, petroleum refining, steel pickling, pulp and paper processing, and in making rayon, paints, pigments, and storage batteries. The contact process - with vanadium catalysts and double absorption - is the prime one for sulfuric acid manufacture. Sulfur and air are the raw materials. The sulfur is burned to produce sulfur dioxide, SO_2 and this is oxidized to form sulfur trioxide, SO_3. The reaction is reversible, and takes place with catalysts (iron or vanadium oxide and platinum) at a temperature between 750 and 1020°F (400 and 550°C). These oxidation reactions are exothermic. The heat generated is utilized to facilitate the reactions and for other heating purposes. The sulfur trioxide is passed through liquid sulfuric acid that absorbs it in the reaction: $SO_3 + H_2O \rightarrow H_2SO_4$. Water and dilute sulfuric acid are added continuously to maintain the correct concentration of the finished acid, usually about 95%. Environmental concerns with escaping SO_2 have led to the use of the double absorption process which captures the stack emissions and absorbs SO_2.

superconductors - are materials that can carry electrical current with no significant resistance. They must operate, however, at temperatures well below zero Fahrenheit - near absolute zero - and must be within a critical magnetic field. The metals columbium, lead, iridium, mercury, tantalum, tin, and vanadium, and the alloys niobium-germanium, columbium-tin, columbium-titanium, and lead-molybdenum-sulfur, have such a property and are drawn into wire or otherwise fabricated into other shapes as needed for the application. Cooling is with liquid helium, which exists at 4.2 K (–452°F or –269°C), to maintain the necessary temperatures. Certain ceramics can operate with superconductivity at the higher temperature of less costly liquid nitrogen 77 K (–321°F or –196°C). (Liquid nitrogen is also a more effective cooling agent than liquid helium.) Yttrium-barium-copper-oxide (YBCO) compound is one such ceramic material. In 1988, a thallium-barium-calcium copper oxide with a critical temperature of 125 K (–235°F or -148°C) was discovered.[5] However, these materials are very difficult to fabricate into wire coils or other needed shapes. Making large superconductor devices of these materials necessitates that the whole structure has essentially the same crystal orientation.

There are two classes of devices that use low-temperature superconducting materials: 1) Small-scale devices that use thin film production techniques based on those used for integrated circuits and other semiconductor equipment. Applications for such superconducting components are ultrasensitive magnetometers, switching apparatus and frequency standards. These units operate at the 4 Kelvin temperature range that requires liquid helium cooling. 2) Large-scale devices utilize niobium-tin, niobium aluminum, niobium-germanium, and vanadium-gallium, and operate at higher temperatures of 20 K (- 423°F or -253°C), still requiring helium as a coolant. Medical MRI systems are the major application but other uses are limited and involve fusion research and the storage of electrical energy for use in peak periods. The MRI devices incorporate closed cooling systems that minimize the amount of liquid helium that must be added to the system.

Efforts to join small pieces of YBCO material and maintain a suitable crystal structure have met with some success. When a thin layer of thulium-barium-copper-oxide (TmBCO) is placed between two pieces of YBCO to be joined and the joint is heated to the temperature half-way between the melting points of the two materials (TmBCO melts about 20°C lower than YBCO.) the TmBCO liquifies and follows the structure of the YBCO as it cools and crystallizes. Superconductivities of 95% of that of YBCO are reported for the combined structure.[28]

A recent development is the use of magnesium diboride (MgB_2) which becomes superconductive at 40K (–388°F or –233°C). Wires of this intermetallic material can be made by reacting boron filaments with magnesium vapor at about 1830°F (1000°C). Development is on-going to make practical sheathed electrical conductors. The 40K temperature can be provided by liquid neon, liquid hydrogen or closed-cycle refrigeration.[30]

swiss cheese - See *cheese, Swiss.*

switches, electrical - See *electrical switches.*

synthetic fibers and fabric - See 10A2, 10B6, 10C and 10D.

synthetic lumber (composite lumber) - See 6F9.

synthetic rubber - See 4O2.

T

T-shirts - are made from knitted fabrics (10D) using yarns of cotton or polyester, or a blend of cotton and polyester. The fabric for the body is usually knitted on circular knitting machines so there is no seam on the side of the garment. Neckbands for crew-neck type shirts are also usually knitted on circular machines. Neckbands are usually one-inch rib knits. The steps involved in making T-shirts is the same as that for other garments as described in 10H. Spreading of the fabric (10H1), marking (10H2), cutting (10H3), and sewing (10H4), are all involved. Due to the high quantities usually involved, and the standard, fairly simple construction of T-shirts, these operations tend to be automated. Overedge stitching is common for seams, providing stretchability of seams comparable to that of the fabric. Sleeves are hemmed before the edges are sewn together, and before the sleeves are joined to the garment body. The body is hemmed at the bottom, usually with an overedge stitch. Sleeves, neckband, label and, sometimes, pockets, are attached. Pockets are sewn with semi-automatic pocket setting machines. When decoration is part of the factory production, it is done by screen or stencil printing (10G2e). Finished T-shirts may be steam pressed (10H5) and are then folded, wrapped, and packed (10H6).

tableware, plastic - Plastic knives, forks, and spoons are invariably injection molded (4C1).

For the most common picnic-type of tableware, the material used is polystyrene.

tableware, metal (silverware) - See *flatware*.

tacks - are cold headed (2I2) from low carbon steel wire (or, occasionally, aluminum wire), on single-station heading machines. These machines produce a flat head and cut the shank with a sharp point, using a special pair of cutters. The tacks are given a barrel-plated (8C3) zinc finish followed by a bright chromate sealer, or with a black-oxide finish (8E4). Packing is by weight on automatic equipment. Thumb tacks and round-headed tacks are made on special machines that feed both a strip of sheet steel and a wire from coils. The machines blank and form the strip to make the tack head, and cold head the wire to form the pointed shank. Then they join the two parts at a central hole in the head and upset the shank to permanently rivet the two parts together. Finishing is most commonly by brass or nickel barrel electroplating.

talcum powder, baby powder - is talc, magnesium silicate, finely ground and in the form of $3MgO{\cdot}4SiO_2{\cdot}H_2O$. Talc is a soft mineral, a form of soapstone. In talcum powder, it is mixed with a perfume. Many baby powders are now made with a mixture of talc and cornstarch instead of all talc, or with essentially all cornstarch but with some tricalcium phosphate (an anti-caking agent). In addition to its use in toilet preparations and cosmetics, powdered talc is used as a filler in plastics and paper.

tanks, fuel for automobiles - Blow-molded plastic automotive fuel tanks are a relatively recent development. The extrusion blow-molding process (4F1) is used. Metal fuel tanks are blanked (2C4), deep drawn (2D5b), trimmed (2C6), and seam welded (7C6b) from terneplate sheet - steel coated with a lead/tin alloy. (Because of health concerns from the high lead content, development is underway to substitute tin/zinc or another non-lead alloy.)

tanks, plastic, storage for chemicals - Plastic storage tanks for chemicals may be made with a variety of processes, depending on the design, size and application of the tank. Cylindrical tanks often are made with filament-wound reinforcement of thermosetting resins (See 4G7.) or with a resin-impregnated, tape-placement process. Tanks requiring large open tops, or large covers, are often hot gas welded (4L11) of vinyl panels, which have been made by calendering (4J), or extrusion (4I1). If larger quantities are required, the reaction injection molding process (4C3b), low pressure injection molding (4C3a), or casting (4H2) of structural foam, may be used to make the tank in one piece.

tea - is made from the leaves and buds of the shrub, *camellia sinensis* or *thea sinensis* that grows in a warm, subtropical, humid climate. Young leaves and leaf buds are picked. *Green tea* is made by drying the leaves in sunshine or artificially. They are processed almost immediately to reduce oxidation and prevent fermentation. Green tea is a pale green-yellow color that is slightly bitter and mild. Black tea is made by first only partially drying the leaves, allowing them to ferment and then fully drying them. The leaves are rolled to break them and release the juices. Black tea is a dark amber-colored beverage, full flavored, but not bitter. Both green and black teas contain caffeine and a strong antioxidant and anticarcinogen. There is also a semi-fermented tea called "oolong". Packaging tea in bags is an automatic operation with special equipment.

tea, instant - is made by concentrating the liquor produced in making tea from leaves and leaf waste, and drying this concentrate to form powder. The drying process is freeze-drying (12H5), vacuum-drying (12H1), or spray-drying (12H2).

teflon - is polytetrafluoroethylene plastic (PTFE), which has a structure somewhat similar to that of polyethylene. PTFE is made by polymerizing the monomer, tetrafluoroethylene, a gas at normal temperatures, by free radical vinyl polymerization. The monomer is produced by heating chlorodifluoromethane to temperatures around 1200°F (650°C).

PTFE is heat resistant, very resistant to attack by other chemicals, and has a slippery surface. It is used in gaskets, corrosive-resistant liners for pipes, hoses and containers, bearings, corrosion-resistant valve and pump parts, and slippery coatings for saw blades and cooking utensils.

tennis balls - are made in two pieces of "dog-bone" shape that are compression molded (4B1) of a very high grade of resilient rubber with gas barrier properties. Flash is trimmed from the two molded pieces and the edges are buffed and coated with an adhesive. The two halves are then butted together in a fixture under pressure, and heated to cure the adhesive. A pellet, placed inside the two halves before bonding, releases gas and fully inflates the ball as it is heated to vulcanize the rubber. In another method, the halves are vulcanized before being joined, and the bonding fixture is under pressure, so that the balls emerge from bonding already pressurized. The ball is coated with an adhesive and two pieces of fabric, also in a "dog-bone" shape, that fit together, and are bonded in place to the surface of the ball under pressure. The fabric is a combination of woven cotton with added fibers of nylon and wool, treated to have a felt-like surface. The finished balls are normally packed in aluminum cans under pressure sufficient to preserve their pressurization.

textile fabrics - See Chapter 10.

thermometers - of the traditional type are made primarily from glass tubing and a liquid, either an alcohol mixture, dyed red, or liquid mercury. (The use of mercury has been strongly discouraged in recent years because of environmental concerns, but mercury is still used where accuracy is important.) Standard outside or inside household thermometers, also have plastic or sheet metal mounting plates, suitably marked with the temperature graduations. Medical thermometers have the graduations etched directly on the glass tube. They are also made with a strip of white-colored opal glass as well as clear glass so that the temperature graduations are more visible against the white background. Two tubes are involved: One is the capillary tube; the other is a thin-walled tube from which the reservoir bulb is blown. Both kinds of tubing are drawn by machine (5A2c) and are usually supplied to the thermometer plant by a glass manufacturer.

The glass capillary tubing is drawn to produce a fine diameter bore. Some thermometers use round tubing; others use tubing of a rounded triangular cross-section. The triangular tubes are called the lens type because they magnify the width of the liquid in the tube. Fig. T1 shows the cross-section of this type of tubing. After incoming inspection, the bulb reservoir is made from the thin-walled tubing by heating the end portion of the tube, pinching it to close the bottom opening and blowing to form a bulb shape (glass blowing - 5A2b). The bulb is attached and sealed at the bottom of the capillary

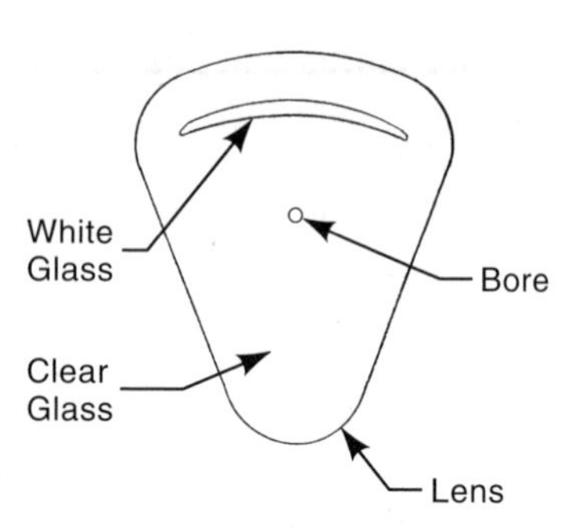

Fig. T1 Enlarged cross-section of a glass capillary tube for a clinical thermometer using a column of mercury to indicate temperature.

tube. The tube is then placed in a vacuum chamber, upside down. When air has been removed from the bulb and bore, the alcohol mixture, or mercury, is fed into the chamber and is allowed to rise in the tube to a prescribed level as the vacuum is gradually removed. The top is then sealed. The tube may undergo aging at this point, depending on its intended application. The tube is then inverted, heated and then cooled, and placed in a bath of controlled temperature, the temperature depending on the use and temperature scale wanted, and a mark is made in the tubing for that temperature. (In some thermometers, the mark is made at the freezing point of water, 32°F (0°C) and also at the boiling point of water, 212°F (100°C). Then, when the scale is put on the glass tubing or the mounting plate, it is located to conform to the reference points.

Scales on the glass are etched by coating the tube with wax, removing the wax where the markings and numbers are to be placed, and dipping the tube into hydrofluoric acid, which is subsequently washed away. Paint is then rubbed into the etched numbers. Printing the scale on plastic mounting plates is done by hot stamping or screen printing, and on metal mounting plates, by printing after they are blanked, formed and painted. Product and manufacturer information may also be printed on the mounting plate. The thermometers are then packaged for shipment and sale.

Oven thermometers use a bimetallic strip in spiral form and an analog dial. Digital thermometers use an electronic sensor and an integrated circuit with an LCD display.

thermoplastics - See entries under the name of the particular plastic.

thread - is twisted yarn, made for sewing operations rather than for weaving or knitting. The manufacturing process for threads is very similar to that for yarn, as described in section 10B. Sewing threads, however, are ply yarns (made from two or more yarns), which have a particularly-balanced tight twist to provide a circular cross section and smooth surface. The smoothness facilitates movement in sewing machines and penetration into the fabric being sewn. Cotton, silk, and nylon are three common thread materials, but many other natural and man-made fibers may also be used.

tiles, ceramic - are made with conventional methods for ceramics. (See 5B.) Most tiles are formed by pressing (5B5). Also see 5B3 and, if tiles are glazed, 5B15.

tiles, floor - See *vinyl flooring.*

tiles, plastic - See *vinyl flooring.*

tin - chiefly comes from the mineral cassiterite where it occurs in the oxide form, SnO_2. To extract the tin, the ore is first crushed and washed. It is then roasted so that the sulfides of copper and iron are oxidized. The ore is washed a second time to further remove impurities. It is reduced by smelting in a reverberatory furnace with coal or coke. This treatment yields molten metallic tin that is collected at the bottom of the furnace. The tin is cast into blocks. When these are remelted, the impurities form a slag that is skimmed off. The remaining tin may be further refined electrolytically. Tin is an alloying metal in soft solder, used extensively in the assembly of printed circuit boards, in alloying bronze, brass and babbitt metal, and as an electroplated or hot-dipped coating.

tires, rubber - Present vehicle tires are usually made with both natural and synthetic rubber (See *rubber, natural* and *rubber, synthetic*.) Sheets of mixed natural and synthetic rubber are reinforced with cords of nylon, rayon and/or steel. The cords are first pre-coated with latex rubber and then are pressed into the rubber sheets. The sheets are fed to a collapsible drum-shaped form and wrapped around it to a prescribed thickness, sandwiching layers of fabric impregnated with rubber. Several layers are laid on the form in the desired pattern. For tubeless tires, an air-proof layer is included. The cables or wires to reinforce the bead are laid in place on each side and are held by folding over the ends of the cord fabric. Then the final layer, an extrudate of rubber that will form the treads, is wrapped around the form and overlapped at the ends. The form is then collapsed and the tire blank is removed and placed in the bottom half of a round mold in a compression molding press (4B1). A butyl rubber bladder is placed inside the tire. The upper half of the mold descends to close the mold, and the bladder is inflated to press the tire blank against the mold walls. Heat is applied through the mold and from steam inside the bladder. The combination of heat and pressure forms the tire treads and molds identifying information on the tire walls. The heat cures the rubber and a formed tire is ejected from the mold. Flash is removed, as necessary, the tire is inspected, a label is affixed, and the tire is ready for shipment. Fig. T2 illustrates a common tire manufacturing sequence.

titanium - is made from the ore, rutile, an oxide of titanium, or ilmenite, an oxide of iron and titanium. The sand ores are screened and separated from unwanted materials, and concentrated by a series of operations that involve gravity and electrostatic and magnetic processes. If ilmenite is the source alloy, a series of hydro- and pyro-metallurgical steps removes the iron and upgrades the ore to a become a synthetic rutile containing more than 90% TiO_2 - titanium oxide. Titanium oxide is an important white pigment and 95% of mined titanium ore ends up as a pigment instead of metal. The oxide is converted to titanium tetrachloride with the carbochlorination process. The converted material is further purified and then reduced with molten magnesium in an inert atmosphere. Titanium is an attractive aerospace structural material because of its favorable strength-to-weight ratio and corrosion resistance. It is also used in chemical processing equipment, prosthetic devices and pump and valve parts.

toilets and other sanitary ware - See *sanitary ware.*

toilet paper - See 9C4.

tooling, rapid - See 14B.

tools, hand - See *hammers, pliers, wrenches and handles, tool.*

toothpaste - usually contains about 12 ingredients including: a mild abrasive - diatomaceous earth, dicalcium phosphate, sodium metaphosphate or calcium pyrophosphate; a binder, gum tragacanth, carrageenin - derived from a red algae, "Irish moss" - or cellulose gum to prevent separation of the ingredients; flavoring - peppermint, spearmint or other oils; water; a humectant, usually glycerine, to provide plasticity and prevent drying; saccharin or cyclamate as sweeteners; a detergent to add cleaning power; sodium salicylate, myrrh or another germicide to inhibit plaque build-up; coloring agents; and stannous fluoride in some toothpastes to prevent tooth cavities. These ingredients are thoroughly mixed and dispensed by automatic machines into tubes that are capped, open at the bottom and inverted

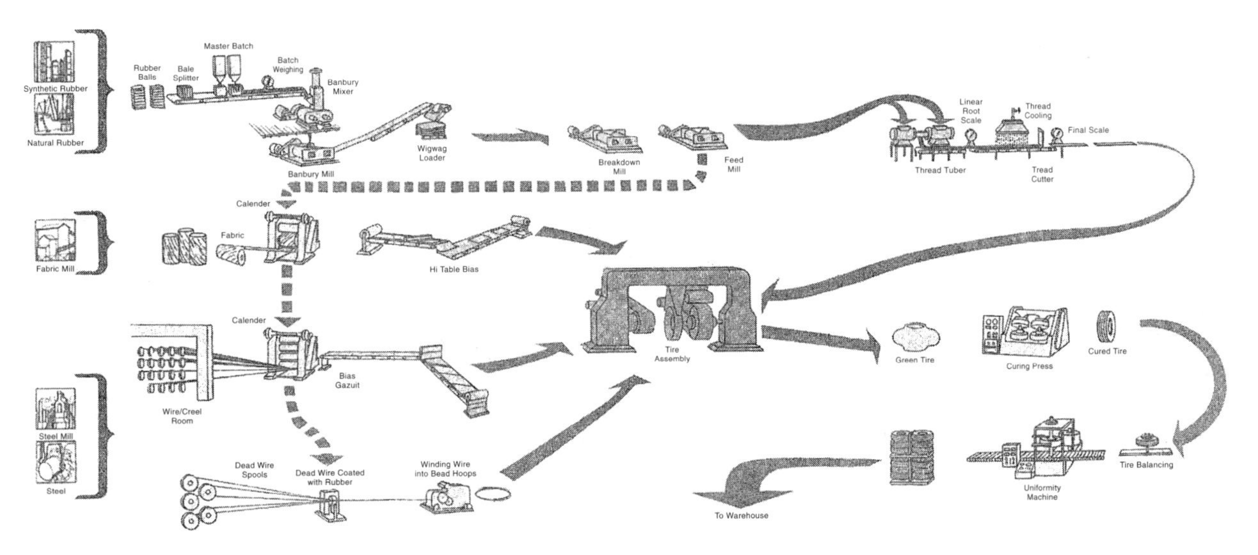

Fig. T2 The radial tire manufacturing process.

for filling. The tubes are injection molded from plastic, color printed and assembled to injection molded caps all by automatic equipment. After filling the tube bottoms are heat sealed by the filling machine. The tubes are inserted in cardboard boxes which, in turn, are placed in shipping cartons.

toothbrushes - See *brushes*.

towels, paper - See 9C4.

trailers - are made for a number of purposes: to transfer products, product components, materials, foods, and other goods, to transport people for travel and vacation, and as temporary offices and living places. The designs of trailers, their structures and their auxiliary equipment, vary considerably, depending on the uses to which the trailers will be put.

Airstream® travel trailers are made, in their external surfaces, with stretch-formed aluminum sheet, using hydraulic machines of 100-ton-capacity for the stretch forming. (See 2F3.) The sheets are trimmed (2C6), and pre-punched (2C5), or drilled (3B1) for rivet holes. Aluminum rivets, similar to those used in aircraft, are used to fasten the sheets to the frame. Body frame ribs and spars are also aluminum, blanked (2C4), and press-formed (2D2) from sheet stock. There are five major subassemblies that together constitute the shell or body of the trailer: two sidewalls, the rear and front end assemblies, and the roof. All are made with aircraft-like aluminum construction, held together with rivets. Chassis are made from box channel and other steel structural members, arc-welded together. After the chassis is welded and painted, it is moved to an assembly area. It is inverted and underside components (axles, a galvanized steel (8F2) protective sheet, gas lines, stabilizer jacks, and a spare tire carrier) are assembled. Then the chassis is turned upright and water tanks, piping, ductwork and under-floor fiberglass insulation are added.

The shell assembly (aluminum body) is made separately. The five shell subassemblies are made independently and a steel fixture is used to guide their attachment to each other. They are held together with more rivets. They are also fastened on the bottom to an internal subfloor made of oriented strand board. (See 6F4.) An aluminum channel provides reinforcement and moisture protection at the junction of the walls and floor, and is held to the shell by riveting. Openings for windows and doors are cut with manually-held routers, guided by fixtures. Window assemblies (windows and frames) are also riveted in place. Similar openings are made in the roof for ventilating fans and a roof-mounted air conditioner. The body shell is then moved and carefully lowered onto the chassis and the two are bolted together.

After the body shell and chassis are fastened together, considerable additional assembly and finishing work is involved. Sealant is applied to all seams. Electrical wiring, the roof air conditioner, doors, and TV antenna are installed. Exterior trim, tail lights, and other lights are installed, and decorations are added. The exterior is then inspected and the partially-completed trailer is subjected to a water test with a high pressure spray. The interior finishing then takes place.

Interior work involves the installation of fiberglass batten insulation in all ceilings and walls, the completion of wiring inside the unit to electrical outlets, switches, lights, TV antenna and phone sockets, and the installation of interior plumbing. An aluminum skin is installed on the interior walls using "pop" rivets (blind rivets - 7F4b) for attachment to the structure of the exterior shell. Then a vinyl liner is glued to the interior aluminum wall. Interior cabinets, furniture, and other items, are installed. These include kitchen cabinets, a dinette, closets, sinks, faucets, counter tops, bathroom shower stall, lavatory and toilet. Kitchen appliances and vinyl and carpet floorings are installed. Then, there is a final cleaning and inspection to ensure that the interior work is of the desired quality, and that all apparatus is in working condition. When all tests and inspections are successfully completed and passed, the trailer is ready for shipment to a dealer for sale to a customer.

Other makes of trailer are assembled from flat panels of fiberglass-reinforced plastics instead of aluminum. Frames utilize both wood and steel members. Steel members are bolted or welded to produce a strong structure; wooden frames are fastened with screws, bolts, and adhesives. Trailers for cargo utilize steel framework, often much stronger and heavier than that used in travel trailers. All have a chassis welded from steel structural components. After the frame is completed, and before or after external panels are fastened, electrical wiring is installed as are duct work and piping, if needed. Fiberglass or plastic foam panel insulation is fitted from the inside. Assembly is largely manual, but is assisted by powered hand tools. The assembly is usually on a line basis, especially for quantity trailer producers and the assembly work, therefore, is divided so that each worker performs a fairly limited task, for which he or she can become highly skilled. Internal prefinished wall and ceiling panels are installed after all items in the walls are completely in place. The sequence of operations for all travel trailers is quite similar to that described above for Airstream units.

trampolines - are assembled from frame parts, springs and a nylon or polypropylene "mat" which serves as the jumping surface. The frame is made from welded and galvanized steel tubing. The tubing is made from coiled strip that is contour roll-formed (2F7) to a round cross section. The edges are heated by induction and butt-welded as the strip moves through the roll former. (See 2A6.) Holes are punched in the tubing for spring attachment, and the ends are swaged (2F1) to be smaller or expanded (2H2i and 2H2j), so that they will fit together, end-to-end. Assembly of the frame takes place after the tubing is formed to a circular shape with a three-roll forming machine (2H2f). Leg lugs are welded to the tubing. Legs for the trampoline are cut to length, swaged at the ends to fit the lugs and together, and are bent to a J-shape. The mat is woven (10C) from nylon or polypropylene yarn and pressed in heated rollers (i.e., *calendered*) to provide a smoother surface and to interlock the yarn. The resulting fabric is cut to a circular shape, hemmed in sewing machines and assembled to grommets that will each hold one end of the springs. Reinforced polyethylene sheet is sewn to a circular bag shape, and then it receives sections of polyurethane foam that provides cushioning from the springs. These components and, typically, 88 springs, are packed in a corrugated carton for the customer. Cartons are weight-checked to ensure that all parts are included. The customer assembles the trampoline at the location where it is used. The springs that surround and hold the mat, and the natural springiness of the mat, provide the "bounce" that characterizes the trampoline.

transformers - See 13L4.

transistors - See 13L5.

trays, plastic - Although many trays are injection molded (4C1) or thermoformed (4D), the common fiberglass-reinforced cafeteria tray is made by matched-metal-mold forming (4G9).

trumpets and other brass musical instruments - See *musical instruments, brass.*

tubing, glass - See 5A2c.

tubing, metal, seamless and welded - See *pipe and tubing, metal.*

tubing, plastic - is made by profile extrusion (4I1) of various plastic resins. Flexible tubing is extruded from plasticized vinyl, polyurethane, polyethylene, nylon, teflon, silicone, neoprene and other plastics and elastomers. Tubing reinforced with braided wire or fiber is made by passing the braid through the extruding die so that both the braid and the plastic matrix exit the die at the same time. Rigid tubing (plastic pipe), commonly is extruded from unplasticized PVC or CPVC with mineral fillers.

tungsten - a heavy metal with a very high melting point, is made from the ore, wolframite, which has the composition, $(FeMn)WO_4$. The ore is concentrated by gravity methods, yielding a concentrate with 60 to 76% tungstic oxide (WO_3)[2]. This material is fused with sodium carbonate (Na_2CO_3) to produce sodium tungstate (Na_2WO_3) The sodium tungstate is dissolved in water but the tungstic oxide precipitates as a yellow powder when hydrochloric acid is added to the solution. The precipitate powder is reduced with hydrogen in an electric furnace, to metallic tungsten. It is then pressed into bars and sintered. Another source for tungsten is depleted uranium. Tungsten is used in tool and alloy steels, magnets, electrical contacts, spark plugs, rocket nozzles, radiation shielding, and other applications, particularly when resistance to high temperatures, wear, radiation, and acids is important.

turbine blades - See jet engines.

turpentine[4] - was traditionally made from the gum (oleoresin) of the pine tree, or pine tree stumps, distilled (11C1) into rosin and oil (or spirit) of turpentine. Current production is by distillation from tall oil, a by-product of kraft paper manufacture. The other product of the distillation is pine oil. Strong odoriferous substances in the turpentine are removed by treatment with sodium hyperchlorite or other mild oxidizing agents. Turpentine is still used as a paint thinner, but the main applications are in chemical processes, particularly as a raw material in the production of synthetic flavors and fragrances.

U

ultra-high-molecular-weight polyethylene - See *polyethylene, ultra-high-molecular-weight.*

ultra-high-strength steels - are steels with tensile strengths above 200,000 psi (1,380MPa). Quite a few alloy steels meet this specification. Steels alloyed with chromium-molybdenum, and chromium-nickel-molybdenum, with suitable heat treatment and sufficient carbon content, can achieve the necessary strength. Several types of stainless steel meet the specification, notably cold-rolled austenitic and semi-austenitic grades. Maraging steels containing less than 10% nickel, plus 10 to 14% chromium, can be formed, machined, and welded, and then, by a simple aging at about 900°F (480°C), can well exceed 200,000 psi in yield strength[2].

umbrellas - normally consist of a fabric canopy stretched over a light metal frame that can be collapsed to make the device compact when not in use. The fabric is typically polyester or nylon. The frame consists of a central shaft, a slide that fits the shaft, of a series of arms called "stretchers," and ribs that support the fabric canopy. The stretchers, which extend the ribs, are attached to the slide at one end, and to the ribs at the other end. Some shafts have a telescoping feature that allows the folded umbrellas to be further shortened for easier carrying or storage. Shafts have a handle at the lower end.

The shaft may be turned from wood but is more commonly made from drawn (2B2) steel tubing, particularly if there is a telescoping action. The stretchers are made from sheet steel, blanked (2C4) and formed (2D2) to the necessary shape with a U-cross-section. The ribs may be similarly formed but are often made from round steel wire. Connections between stretchers and ribs are made by means of brass eyelets in punched or drilled holes. Wire ribs are flattened by a press operation to provide space for such connections. The slide is injection molded of acetal or other plastic and contains slots for attachment of the stretchers with steel wire. There is a similar, non-sliding, injection-molded piece at the top of the shaft for fastening the ribs to the shaft at the top. There is a slot in the shaft and space for a spring, pivot, and triangular catch piece, blanked from sheet steel, to hold the umbrella open. The handle may be injection molded or lathe-turned from wood. Metal parts are electroplated (8C1), usually with a chromium finish. The canopy is sewn from six or eight pie-shaped pieces of fabric that are cut, sewn together, and hemmed at the bottom edge with methods described in 10H. The umbrellas are assembled manually, with machine assistance for eyeletting and other operations. The workstations are usually laid out on an assembly-line basis. A ferrule of metal (made on a screw machine) or plastic (injection molded), is press-fitted and/or bonded to the end of the shaft. Small similar ferrules are pressed onto the ends of the ribs and they contain small holes for thread attachment of the canopy. The last line operation is testing and packaging the finished umbrellas.

undergarments (underwear) - are made with the methods commonly used for other garments and sewn products. (See 10H.) Woven (10C) or knit (10D) cotton is the common fabric, but rayon, nylon, wool, polyester and linen fabrics are also used. Also see *lace* and *spandex.*

unwoven fabric - See 10E.

upholstered furniture - See 6H.

uranium fuel[4] - The fuel most commonly used in nuclear reactors is uranium dioxide, UO_2 in pellet, rod, sphere, plate, or pin form, clad with zirconium, stainless steel, other alloys or aluminum oxide. (Thorium and plutonium are other reactor fuels.) Uranium ore, pitchblend, is mined in several places and contains about 2% uranium oxide, U_3O_8. The ore is milled and the U_3O_8 concentrated to form "yellow cake". This operation is accomplished by sulfuric acid extraction (11C3) followed by solvent concentration. Ion exchange and precipitation with sodium hydroxide may also be used.

The uranium in the yellow cake is then further concentrated and purified to extremely high levels by a complex, multi-step process that includes digestion with nitric acid and steam, solvent extraction of the uranium with tributyl phosphate (TBP), reextraction from the TBP in pumper-decanter pulse-extraction columns, denitration to UO_3, hydrogen reduction to UO_2, and hydrofluorination to yield uranium tetrafluoride, UF_4. Further fluorination with fluorine gas, F_2, in a special furnace at 660°F (350°C) yields uranium hexafluoride, UF_6. The UF_6 contains two isotopes of uranium, U-235 and U-238. The isotope used for fuel must include sufficient U-235. The concentration of U-235 is increased by repeated steps of either centrifugation (11C7e) or gaseous diffusion (See membrane separation - 11C7c). The uranium hexafluoride is reduced to metallic uranium in a batch-type furnace with magnesium, followed by reconversion of the uranium metal to UO_2. (An alternative ammonium diuranate process converts the UF_6 gas to UO_2 in powder form.)

The UO_2 powder is then fabricated into pellets or other fuel-use forms by methods used for processing ceramics: pressing (5B5), sintering (5B16) at 2730 to 3270°F (1500 to 1800°C) and grinding (5B14) to the precise size. The pellets are clad and sealed for use in a reactor.

urea and melamine plastics - are made by reacting urea ($NH_2{\cdot}CO{\cdot}NH_2$) or melamine ($C_3H_6N_6$) with formaldehyde (CH_2O). The resulting plastics are also known as urea-formaldehyde (UF) and melamine formaldehyde (MF). Urea, a major constituent of mammalian urine, is made commercially from the reaction of ammonia and carbon dioxide, both as liquids. The reaction takes place at a high temperature and high pressure to form ammonia carbamate. This decomposes to urea and water. The formaldehyde is made by oxidizing methyl alcohol. Urea-formaldehyde is then the product of the combination of urea and formaldehyde, heated with mild alkalies. A condensation polymerization takes place, yielding the water-soluble resin. The resin is mixed with wood fiber or other fillers, pigments, and other materials to provide a thermosetting molding material. During molding, the heat and pressure cause further reactions to take place and the resin becomes heat and temperature resistant.

Melamine-formaldehyde is made by similar condensation polymerization reactions but with melamine instead of urea reacting with the formaldehyde. The melamine $(N{:}C\ NH_2)_3$ for this reaction is produced by heating dicyandiamide under pressure, or by reacting ammonia and urea at an elevated temperature.

Urea, melamine and phenolic plastics are similar in their properties, processing and applications. All are thermosetting, are blended with fillers for molding, and have good heat and water resistance. Urea-formaldehyde is used for appliance handles and knobs, knife handles, buttons, housings, and table plates. The resin is used as an adhesive in making plywood and waferboard, and as a treatment for wash and wear properties in fabrics. Melamine-formaldehyde is used in molding dinnerware, buttons, and electrical components, and in laminated table and counter tops ("Formica" or "Melmac"). The polymer is used for baked coatings on appliances, metal furniture, and automobiles, and for adhesives.

urethane rubber - Polyurethane plastics, made from the reaction of organic diisocyanates and polyglycols, are rubbery in nature. Two common reactions are TDI (tolylene diisocyanate) and MDI (4,4'-diphenylmethane diisocyanate) reacting with linear polyols of polyether or polyester. Chain extenders, water, glycols, aminoalcohols or diamines, are used to achieve a long-chain product[2]. Urethane elastomers have high abrasive resistance and are usable at elevated temperatures. If the diisocyanate contains free carboxyl and hydroxyl groups, the reaction with polyglycols will produce gas, which will cause the urethane to foam. The result can be a flexible cellular material that is useful for mattresses, upholstery and other applications of foam rubber. Rigid foam can also be made by changing the reacting materials. Among other applications, rigid polyurethane foam is used as insulation in refrigerators, freezers, and buildings. (Also see *polyurethane*.)

utensils, cooking - See *cooking utensils*.

V

vacuum bottles (Thermos® bottles, Dewar flasks) - were previously made only from glass. The vast majority are now made from stainless steel.

When glass bottles are used, two bottles are blown by the blow-blow (5A2b3c) or press-blow (5A2b3d) methods and placed one inside the other. The space that will be between the bottles is coated with silver in the same way as the rear surface of glass mirrors are silvered (See *mirrors*). The inner bottle is inserted in the outer bottle. The top portion of both bottles is then heated and the bottles are finish formed and joined (5A5d) so that there is an air space between them. The space between them is open only at one small hole at the bottom of the outer bottle. The air is then evacuated from the space between the bottles by a vacuum pump connected to the small opening, after which the opening is sealed with epoxy plastic, leaving a vacuum between the bottles. The double-walled bottles are then assembled into blow-molded plastic (4F) containers. A threaded, hollow (blow molded) stopper is provided. In most models, the stopper has an elastomer O-ring or gasket for additional sealing of the bottle's contents. An additional injection molded (4C) threaded cap is provided, to provide further insulation and serve as a drinking cup.

Stainless steel bottles also consist of a bottle-within-a-bottle, with a vacuum space between the two. The two stainless steel bottles are made from welded stainless steel tubing (See *pipe and tubing, metal*) with drawn (2D5) and formed (2D2) end pieces. The inner bottle, at the pouring end, is reduced in diameter and formed with screw threads for the stopper. This is done by a series of operations including swaging (2F1) and forming with annealing (8G1) between operations to remove work hardening. The pouring end of the outer bottle is formed to mate with the inner bottle. The other end of the inner bottle is a deep drawn piece that is arc welded, by the GMAW or GTAW methods (7C1d or 7C1e), to the tubing. The surfaces of the two bottles that will face the vacuum are polished and electroplated (8C1) with copper to provide a barrier to block the passage of radiant heat. The formed pouring end of the outer bottle is fastened to the inner bottle by roll welding (7C13e). Then a drawn bottom piece is welded to the outer bottle. This completes the welding that seals all the joints to the space between the bottles. One small hole is left at the bottom of the outer bottle for air evacuation. Air is evacuated from the space between the paired bottles, by a vacuum pump, and sealed with hard solder (7B). This step is done for one vacuum bottle at a time. A new method places a group of bottles in a large vacuum chamber where, after the air is evacuated, the openings are sealed inside the chamber. Visible surfaces of the outer bottle and spout are polished, usually to a satin finish. Most bottles also have an exterior blow molded plastic container that covers the arc welded joints and provides an air space between the container and the outer stainless steel bottle.

vanilla - is extracted from the vanilla bean, the immature fruit of a tropical vine. The beans are picked before they are fully ripe, and are cured in a treatment that involves repeated exposure to sunlight during the day and sweating at night. The curing produces fermentation that divides the glucoside glucovanillin in the bean into glucose, vanillin, and other aromatic substances. The beans are cut up into small pieces and soaked for three cycles in 35% ethyl alcohol to soften and break down these materials. They are recombined in the desired portions to make vanilla extract. Vanilla flavor is common in ice cream, baked goods, chocolate candy, beverages and other foods.

varnish - is unpigmented enamel, a solution or dispersion of colloidal resins in drying oils and/or solvents that produces a transparent coating. Varnish is frequently used on wooden products to enhance and protect the wood grain finish. See *paint* for the manufacturing process. The resins may be natural or synthetic. Although varnishes contain no pigments, they sometimes contain dyes to change the color shade of the object coated, especially when the object is wooden. Varnishes dry or cure to provide a transparent film that provides protection and enhances the color of the object that has been coated. Like enamels, varnishes dry by oxidation, polymerization, and/or evaporation of the carrier liquids. Varnishes made with alkyd and urethane resins have become prominent because of their improved properties, but phenolic, acrylic, epoxy, and ester gum resins are also widely used. Varnishes are used as coatings on furniture, wood floors, boats, and house siding.

vases - (made by manual blowing) See 5A2b1.

veneer, wood - See 6F1, 6B7b, 6E4, 6D2.

vermouth - is a wine-based alcoholic beverage. It contains a blending of herbs and other flavorings in a solution of alcohol. Juniper, cloves, hyssop, quinine, chamomile, orange peel, coriander and other ingredients may be used in various proprietary formulas. The flavor-carrying elements are steeped in alcohol, either heated or at room temperature. The flavored alcohol is then added to sweet or dry wine, depending on the variety of vermouth, aged for several months in tanks and then bottled.

vials, glass - (made from tubing) See 5A2c.

vinegar - is made from any of a variety of liquid food materials by a fermentation process. The full process involves two steps of fermentation. The food (grapes, apples, sugar, rice, malted barley) is first converted to juice or other liquid, which is fermented to convert the sugar it contains to alcohol in dilute solution (forming wine, beer, hard cider, or other alcoholic beverage). (See 12J4 for fermentation of sugar-bearing liquids into alcohol-bearing liquids.) A second microbiological process, also referred to as fermentation and using acetobacter bacteria, then converts the alcohol into acetic acid. (The acid gives vinegar its sour taste.) Household vinegar contains about 4 percent acetic acid. This second process also utilizes the oxygen in the air to complete the fermentation reaction. In modern production processes, the liquid undergoing the second fermentation may be continuously aerated to accelerate the process. Other organic acids and esters in the original juice remain in the vinegar to give it its particular aroma and flavor. Herbs, spices, garlic, tarragon and onion may be added to the vinegar to enhance or change its flavor.

vinyl flooring - consists of tiles or sheets of floor covering material, with polyvinyl chloride plastic as a major component. Other components include plasticizers (which provide flexibility/resiliency to the vinyl), calcium carbonate (chalk, a whitener and filler), other pigments, stabilizers, and a backing or carrier sheet. The carrier sheet is a heavy paper or felt that supports the vinyl. High-gloss ("no wax") flooring has, in addition, a clear coating of polyurethane on the top surface. Standard sheet widths in the United States are 6 or 12 ft (1.8 or 3.7 m), and the standard tile size is 12 × 12 in (0.3 × 0.3 m). Pre-glued tiles have a coating of pressure-sensitive adhesive on the bottom surface, with a temporary protective paper layer that is removed when the tile is laid.

The manufacturing process involves mixing the vinyl resin, plasticizer, and other ingredients, sometimes with AZO, a foaming agent. (The AZO releases bubbles of nitrogen gas when heated.) The mixture is heated, deposited, and smoothed on a moving sheet of the backing material with a reverse roll coater (4K1b). The coated sheet is passed through an oven of controlled temperature, and is then cooled. The thickness of the foamed vinyl layer is normally from 0.010 to 0.030 in (0.25 to 0.75 mm) depending on the grade of product being produced. Decorative patterns are printed on the surface of the vinyl using the intaglio (gravure) process (9D3 and 9D3a). Then, a second coating of plasticized vinyl, with no filler material, is applied to the printed vinyl, using the same reverse roll-coater method. The printed sheet is again passed through an oven to fuse the two layers, and is again cooled. The second layer is a clear protective layer to maintain the decorative pattern and color even when the vinyl wears from foot traffic. If the product is the "no wax" variety, a third layer, this one of polyurethane, is deposited by roller coating, and the polyurethane is cured by ultraviolet radiation. The sheet is then ready to be cut to the standard widths and wound into rolls.

Vinyl floor tile may be made somewhat differently. Sometimes, no back up sheet is used, but the vinyl plastic incorporates a calcium carbonate filler. The mixture of vinyl, filler, plasticizer, pigments, and other ingredients is heated and then calendared (4J) into sheets. The sheets may be printed and embossed. Various pigments may be included in the vinyl mix to provide an inlaid color. The surface may also be embossed, especially if the inlaid color is stone-like. Another inlay method uses clear vinyl with no filler, but with small pieces of colored vinyl mixed in. If the color pattern is printed, a layer of clear vinyl is added as it is with the sheet material and, if "no wax" properties are wanted, a polyurethane third layer is also added. If a pressure-sensitive adhesive is specified, it is applied by roller coating to the sheet bottom and a protective paper sheet is also applied. The sheets are then cut into individual tiles, with a machine using steel-rule type dies (2C4a). The tiles thus produced are cooled and packed into corrugated boxes of 10 or 12 tiles each, ready for shipment to customers.

vinyl plastic - See *polyvinyl chloride (PVC).*

vinyl plastisol coatings - See 8D10.

vinyl siding (for buildings) - is made by the profile extrusion (4I1) of wide panels of polyvinyl chloride resin, mineral fillers, pigments and other additives. For some siding, a thin layer of acrylic plastic is coextruded (4I2) on the outer surface to provide better color retention and weatherability.

viscose (viscose rayon) - See *rayon* and 10A2.

vitreous enamel coatings - See 8F1.

vodka - is an alcoholic beverage originally made from potatoes but now made mainly from wheat and other cereal grains through fermentation (12J4) and distillation (11C1). Also see *distilled spirits.* The manufacturing process removes much odor, color, and flavor. The liquid, then, is highly neutral and clear in color. It is sometimes flavored with lemon, "buffalo grass", berries, peppercorns or caraway.

vulcanized fiber - See 9C6.

W

wafer board lumber - See 6F3.

wallboard (plasterboard, gypsum board, drywall, sheet rock) - is gypsum plaster, solidified in board form, often with fiber reinforcement, faced on both sides and edges with paper. (See *plasterboard* and *gypsum plaster.*) Wallboard is used for interior ceiling and wall paneling.

wall paper - Present-day wallpapers (wall coverings) are commonly made either from paper coated with vinyl and some other materials, from fabric coated similarly or from solid vinyl sheet. Paper making for wallpaper follows normal papermaking processes from wood pulp as described in section 9B. Coating materials, in addition to vinyl, are latex, kaolin (clay), and titanium dioxide (white pigment). Other ingredients sometimes used are mold-retardants, insecticides, fungicides, and flame retardants. Vinyl sheeting is usually produced by calendaring (4J) and vinyl-coated wallpaper or fabric is produced by plastic coating/laminating (4I4). The vinyl is usually colored in the background color of the printed design that will be made on the sheet. Printing can be by one of four common methods: letter press with rubber type (flexographic - See 9D1b.), rotogravure (9D3a), silk screen (9D4b), and stencil printing (9D4a), performed with rotary printers using photographically-produced stencils. An additional clear vinyl plastic coating may be put on the paper after printing. By printing with an adhesive, and spraying powdered material or yarn, a flocked appearance can be created. Roller coating the reverse side of the paper with an adhesive is now very common. The adhesives are usually made from corn starch or wheat starch. When used, they are activated with a sponge or other means to wet them.

washers (as used with bolts, etc.) - are blanked from coiled sheet metal with high-speed, multiple blanking dies (2C4.) Washers can be made from almost any metal, but hardware-store washers are blanked from mild steel sheet, and are cleaned (8A1d), and bright-zinc barrel-plated (8C3).

watches - Digital watches utilize the oscillations of an electronic circuit containing a quartz crystal. The quartz crystal oscillates electrically at a fixed high frequency, and a series of divider circuits reduces the frequency of the current. Each circuit divides the frequency in half, and switches the signal to another circuit. The final once-per-second electrical pulses activate a digital display through further circuitry connected to the applicable segments of the display. The components of watches with a digital display consist of an assembly of one or two integrated circuits (13K), a small printed circuit board (13A), a digital liquid crystal display, (See *liquid crystal displays.*), an enclosing case of formed metal or molded plastic, a glass lens, two or more push buttons, a small silver oxide, mercury, or lithium cell to provide the electric current, and a wrist strap with a buckle. The crystals used are made from synthetic quartz that is made in a process carried out in an enclosed pressurized chamber. The chamber contains an alkaline silica solution, heated to about 750°F (400°C). Seed crystals are suspended in the liquid. Additional quartz then deposits on the seed crystals to eventually form crystals of usable size. This takes about 75 days[9]. The crystals are purified by running an electric current through them at a high temperature (932°F - 500°C). They then are cut by diamond abrasive saws to the size at which they oscillate at the desired frequency of 100 megaHertz. The crystals are connected to wires at each end and encapsulated under a vacuum. When the printed circuit board is made, the crystal is incorporated as part of it.

Analog electronic watches (those with a dial and hands) have similar electronics but translate the oscillations into the movement of watch hands with a small stepping motor and a train of gears. The gears are mounted on a central bridge or plate and connected to the hands of the watch. The gears and their mounting structure may be made, as they are in the most expensive mechanical watches, from fine-blanked (2C9), or conventionally blanked (2C4) and shaved (2C7), brass-alloy sheets. These parts may be plated with gold or silver. Shafts are made on Swiss-type screw machines (3A2c2). Gear-train parts are assembled with jeweled bearings (See *synthetic sapphires.*) that support the shafts. Precision-machined or stamped gears have been replaced in some watches with those injection molded to exact dimensions from glass-reinforced engineering plastics. (The reinforcement adds dimensional stability.)

Mechanical watches draw their power from the energy stored in their spiral mainspring (See springs.). The oscillation takes place in a balance wheel, controlled by an escapement mechanism, using a balance spring, a club-toothed escapement wheel, an escapement lever, shafts, bearings, and mounting plates. These parts are made with methods similar to those used to make the mechanisms of analog electronic watches. For wear resistance, the club-toothed wheel may be made of steel, heat treated for hardness, and then ground and polished.

water repellant fabrics - See 10F5e.

water, potable - for public water supplies. The water undergoes a number of purification treatments before it is usable. The number and extent of the steps taken depends on the quality of the water supply. Water from reservoirs requires more steps than ground water because it is more apt to contain foreign matter. Larger pieces of suspended or floating material are removed by intake screens at the processing plant. Aeration may follow to remove unwanted odors and tastes, and to solidify some dissolved compounds. Coagulants and flocculants are added and the water is piped to a sedimentation tank (See 11C7d.), where heavier impurities settle out. Filtration, commonly in sand-bed filters, is a vital step to remove particles of organic and inorganic solids and to clarify the water. (See 11C7a.) Carbon adsorption (11C9) removes molecules of impurities. The clarified water may be treated with chlorine to disinfect it against microorganisms, including those that could enter the water later. Additional treatments, sometimes used, involve fluoridation and water softening. Water for chemical processes may require ion-exchange processing or distillation to remove minerals in solution. Water used in nuclear reactors is normally purified to 0.08 ppm or less.

Fresh water is made from sea water or brackish water by a number of distillation processes. The major process is multistage flash distillation, described in sections 11C1c and 11C1d. Dissolved salts can also be removed by membrane separation (11C7c).

wax[4] - Waxes are high-molecular-weight fatty acids, in combination with high-molecular-weight alcohols (in comparison with oils and fats, which are fatty acids in combination with glycerin.) Waxes are of animal, vegetable, or synthetic origin. Animal waxes come from protective material secretions of insects. The most notable such wax is beeswax. Vegetable waxes are coatings on leaves, seeds, stems, and flowers. Synthetic waxes originate from petroleum refining, and from processes that use coal, lignite, or peat as raw materials. Paraffin is a product of petroleum refining; 90% of currently-produced waxes are petroleum-sourced. Beeswax and similar waxes are obtained by boiling nest material in water, or by solvent extraction (11C3) or expression (11C4). Vegetable waxes are obtained by various methods. Carnauba wax, used in automobile, furniture, and floor waxes, is obtained directly from the leaves of a palm tree. The leaves are dried and beaten, and the wax falls from them, after which it is melted and filtered. Another common vegetable wax, candelilla, is obtained from the stems of a plant that grows in the southwestern US and Mexico. The stems of this plant are boiled in dilute sulfuric acid, and the wax they contain floats to the surface and is skimmed off. Extraction with hexane is also used. Paraffin comes from lubricating oil, obtained from the fractional distillation of petroleum. The oil fractions are concentrated by freezing and filtering and the wax is removed by solvent extraction followed by distillation (11C1). Waxes can be made from the hydrogenation of oils or by removing glycerin from oils and replacing it with higher-molecular-weight alcohols.[2] Polyethylene plastic of low-molecular-weight has wax-like properties and is blended with softer waxes to provide more toughness, durability, and gloss.[2]

Waxes are used in the manufacture of polishes, cosmetics, printing inks, food containers, matches, crayons, candles, wax paper, and for waterproofing fabrics and other items.

whiskey - See *distilled spirits* and Fig. D2. Bourbon whiskey is made from a grain mixture with at least 51% corn. Rye whiskey is made with at least 51% rye grain, and Scotch whiskey is made with malted barley. All have some malt that is made by soaking kernels (normally of barley) in warm water, holding them until they sprout, and then drying them in a kiln. The sprouting produces a chemical, "diatase," that converts starches into sugar that, in turn, is converted into ethanol during fermentation. Straight whiskeys are made mostly from one grain; blended whiskeys are made from a broader combination of grains. Whiskey is aged in oak barrels that have charred interiors. The charcoal in the barrels, aided by the oak wood, neutralizes some of the harsh minor constituents of the liquid and adds some wood flavors to the whiskey. The aging process takes place in an air-conditioned environment.

white glue - See *adhesives*.

window panes - are made from drawn or float-process flat glass or, sometimes, plate glass. See 5A3 for a description of the various flat-glass processes. The flat glass is cut to size by scoring the surface at the cutting line with a small sharp metal wheel, or a pointed tool of diamond or tungsten carbide, and then breaking the glass panel along the scored line. Breaking is accomplished by applying a bending force, often with rollers, or by applying localized heat, at the score. In production situations, this operation is mechanized and, often, is fully automatic.

window panes, antique - See 5A3a.

windows - Flat glass for windows is made by one of a number of processes; however, the prime current method used is the float glass process, described in section 5A3f. The flat glass is cut to the size wanted by the method described above (*window panes*). In high production situations, the scored line is made by computer-controlled equipment. The glass panes are assembled in wood, plastic or metal frames. Wood frame members are made with suitable wood milling methods described in Chapter 6. Plastic frames (normally of vinyl) are made by extrusion (4I1) or injection molding (4C1) of the components, and metal frames are made by various metal forming methods (Chapter 2). Frame components are fastened together with adhesives, by welding (of vinyl or metal), or with metal fasteners. The glass and frames are assembled together by manual or robotic methods (Chapter 7), are labeled and packed for shipment.

windshields, automotive - See *automotive windshields*.

wine: Wine is made from the fermentation of the juice of grapes. Typical process steps are as follows:

1. As soon as grapes have ripened, usually in the fall of the year, they are harvested. They are picked by hand or with mechanical harvesters that shake the vines.
2. Juice is squeezed from the grapes with roller crushers or rotary paddle crushers that also remove the stems.
3. With white wine, the juice is separated from the pulp and skins by filtering and centrifuging; with red and rose wine they are left together.
4. The juice is placed in tanks in which fermentation (See 12J4.) will take place. A small amount of sulfur dioxide is added to the tank to act as a disinfectant. It kills naturally occurring yeasts and other organisms that are present on the grapes. A particular yeast is added to the liquid according to the characteristics wanted in the finished wine. (However, in some European wineries, the naturally occurring yeast on the skins of the grapes is relied upon to provide the fermentation.) Fermentation extends over a period of 3 days to two weeks, changing the glucose and fructose in the juice to ethanol and carbon dioxide gas. The temperature of the liquid in the fermenting tank is controlled. Since fermentation releases heat, cooling may be required. For white wine, the temperature is typically 50 to 60°F (10 to 16°C). Red wine is fermented at a higher temperature to help absorb color from the grape skins. The wine is periodically stirred during the fermentation

cycle, especially with red wine, to extract color from the skins. Sometime during fermentation of red and rose wines, the skins and pulp are removed from the juice in the tanks.

5. After fermentation, yeast and any other solid material that may be suspended in the wine is removed. This process is called "racking".
6. The wine is further clarified by filtration or "fining," a process in which bentonite (clay) or other material is used to absorb material suspended in the wine.
7. Aging. The wine is stored at a controlled temperature for a period of 5 to 12 months. Further racking and/or filtration may take place after aging.
8. Bottling, in production situations, is by automatic bottling machines.

It should be noted that air is kept from the liquid during almost the entire wine-making process by using an inert atmosphere or insuring that all vats are full of liquid. This precaution is taken because oxidation impairs the quality of the wine. Fig. W1 summarizes the wine-making process.

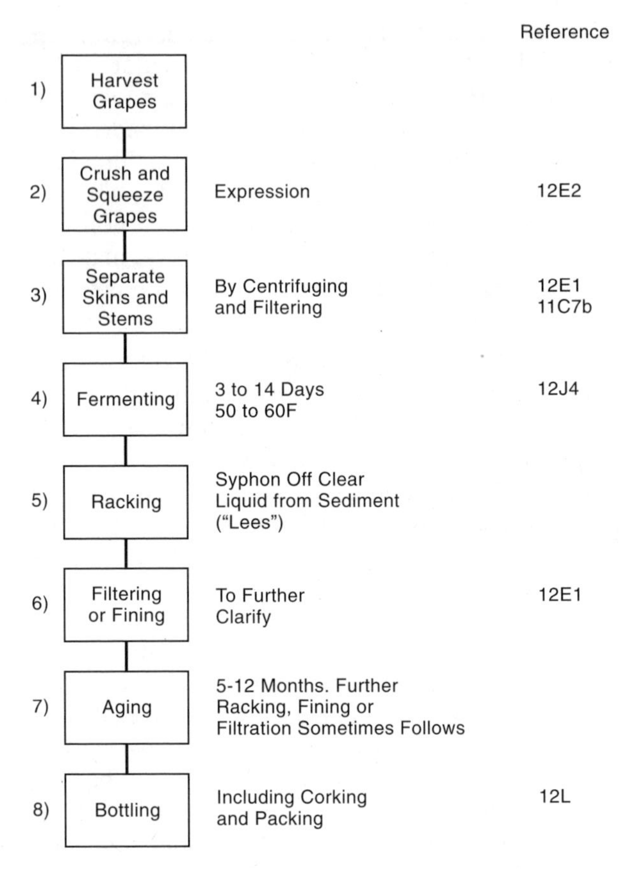

Fig. W1 The operation sequence for the manufacture of white wine. Red wine has a very similar sequence, except that the skins and stems of the grapes are removed after partial or full fermentation rather than before. Also, the occurrence and sequence of racking, fining and filtering operations often varies from the above sequence depending on the type of wine and the winery involved.

wire, electrical - is most commonly made from copper but is sometimes aluminum. Wire making starts with the extrusion of a rod of the metal used. The rod becomes the initial material for a series of wire drawing operations. In wire drawing, the wire is pulled through a series of successively smaller-diameter dies. Ten or more dies may be arranged in series, and one continuous length of wire is pulled through all ten. As the wire passes through the dies it is gradually reduced in diameter and increased in length. (See cold drawing, 2B2 and Fig. 2B2.) When the wire has reached the desired diameter, after one or more drawing operations and if it is to be insulated, the final operation is to extrude a plastic sheath around the wire. (See 4I4a and Fig. 4I4a.) Flexible vinyl is a common insulating material but is only one of several plastics that may be used. For common electrical cable for household and other building wiring, after each wire is insulated, two, three, or four wires are run together through a machine that wraps kraft paper over the wires. Another extrusion of the outer insulation follows, with the wrapped wires as the core. Wire intended for high flexibility is made from multiple strands, drawn to a small diameter and twisted together with other strands to achieve the total diameter needed for the desired current-carrying capacity. Appliance power cords and extension cords are made by this approach.

wire forms - include such items as dish drainers, coat hangers, hooks, clips, oven or refrigerator shelves, kitchen implement parts, store shelves and racks, shopping cart components, and similar open-frame components made from joined pieces of heavy wire. The wire parts for these assemblies are normally made on four-slide machines (2G2). Wire in coils is fed through straighteners and to the machines, which cut off, form, and discharge individual parts, but regular punch press forming (2G1b), similar to that used with sheet stock, is also used. Individual formed pieces are placed in fixtures and fastened together by resistance-welding methods (7C6). In high production situations, several joints are welded simultaneously, with a multiple-electrode arrangement. Assemblies are then cleaned, electroplated (8C1), powder-coated (8D8) or coated with vinyl plastisol (8D10).

wire glass - See 5A3g.

wire, mechanical - is used for fencing, (including barbed wire fencing), cable making, screen and sieve making, in textiles, for various packaging and wrapping purposes, in making needles, nails, pins, hair pins, rivets, many cold-headed parts and for formed racks and holders made from stiff wire. Most of these applications use steel wire, though aluminum, brass, copper, stainless steel, and other metals may be involved.

The basic operation for mechanical wire is wire drawing, as described above for electrical wire and in section 2B2. The initial steel rod may be descaled by pickling (8A2f) before drawing, and is lubricated during the drawing operation with oil or a soap solution. Metals subject to work hardening may be annealed (8G1) or stress relieved (8G2d), between drawing operations.

When reduced to the proper diameter, the wire may be galvanized (8F2) for corrosion protection or may be plated with zinc, nickel alloy, chromium or other metals, or may be painted or coated with vinyl plastisol (8D10) or other plastics (8D8).

Wire ropes and cables are made by twisting together a number of strands of finer wire.

wooden I-joists - See 6F7 and Fig. 6F7-1.

wood veneer - See 6F1.

wool - is a fiber made from the protective fleece of sheep. The operations involved in making these fibers into useful products are as follows:

1. Wool is received in bales or packed bags and is sorted and graded by skilled workers in accordance with the fineness, length, elasticity, and strength of the fibers so that like varieties are kept together or blended as necessary.
2. *Scouring* - The wool is thoroughly washed in a mild alkaline solution with soap to remove foreign material. It may be treated with a solvent to remove excess wool grease.
3. Thc wool is dried to a moisture content of about 12 to 16 percent.
4. Treatment of the wool with various oils takes place to maintain its softness and provide lubrication for subsequent spinning operations.
5. If dyeing is to take place with the raw stock, it is done at this point. (It may instead take place later with the yarn or fabric)
6. *Blending* - Wool of several grades may be blended together at this stage to provide suitable properties for the intended process and application. Some synthetic fibers (acrylics, polyesters or nylon) and/or a small amount of cotton may be added to the mixture. The choice of added fibers and their quantities depends on the properties (strength, drape, body, hand, warmth, and absorbency) wanted in the finished fabric.
7. The wool is carded. (See 10B2.) to produce a yarn suitable for spinning. (*Carding* is a process similar to combing and brushing that disentangles bunches and locks of fibers and eliminates burrs, foreign materials, and fibers that are too short.) There are two varieties of wool yarn and fabrics: worsted and woolen. Regular woolen yarns contain a wider variety of raw wool, with more irregular yarns and a more open and hairy structure. Worsted yarns undergo further operations.

   Separate process paths are followed from this point, depending on whether the finished fabric will be a woolen or a worsted. For woolens, the result of the carding operation is roving, a rope of wool, suitable for spinning. Worsted fabrics require their fibers to undergo the further operations of grilling, combing and drawing. Fig. W2 shows the differences in the operational sequences for woolen and worsted fabrics.

   7a. *Grilling* and *combing* (See 10B3.) - The worsted yarn is processed to remove the shorter fibers (noils), which are used for regular woolens or felts, and arrange the longer fibers in a more parallel orientation in a uniform thick strand called a "wool top". Some further loose impurities are also removed.

   7b. *Drawing* (See 10B4.)-runs the strand through rollers that reduce the strand in diameter, mix the fibers and make them more parallel and create a thinner, twisted yarn called "slubbers".

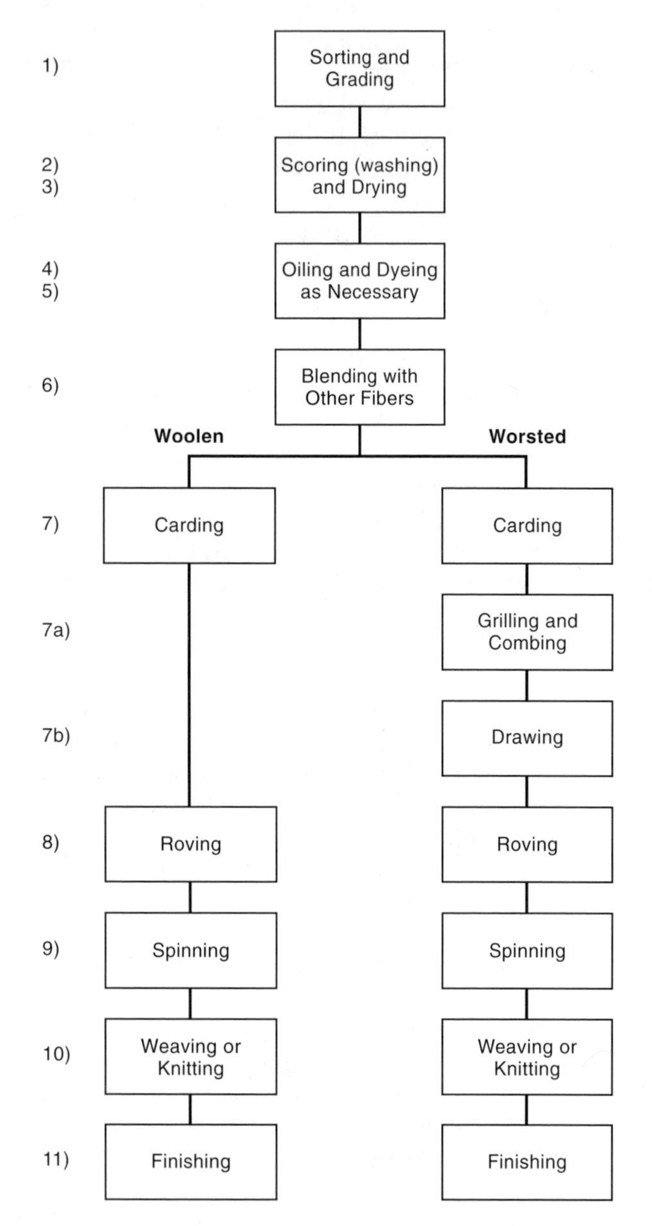

Fig. W2 The operation sequence for manufacture of both woolen and worsted fabrics from wool.

8. For both varieties of wool fabric, *roving*, a light twisting operation, may take place to hold the thin slubbers in place.
9. The *spinning* operation now takes place. (See 10B5.) This operation is performed on a spinning machine that produces yarn ready for conversion to a fabric.
10. Spun yarn of either variety is processed into fabric by *weaving* (10C) or *knitting* (10D). Both woolen and worsted woven fabrics are normally between 54 and 70 in (1.4 and 1.8 m) wide, and 50 to as many as 140 yards (46 to 130 m) long. Both woolen and worsted yarns are knitted by machine into garments such as sweaters or socks, or fabric to be made into coats, dresses, and other garments. Yarns for hand knitting are also produced from wool.
11. Woolen and worsted cloth undergo *finishing* operations (10F) at some time during the manufacturing process, especially after weaving. The purpose of the finishing processes is to improve the appearance and feel of the finished product. There are two predominant finishes: clear finishes and face finishes. The clear finishes, normally given to worsteds, provide an even surface without nap, with prominent colors. The face finishes, in contrast, have a distinct pile or nap, more subdued colors, and a less distinct weave. Dyeing is a common finishing operation that may take place anywhere in the manufacturing process. Printing is another operation that is used with neckties, beachwear, and dress fabrics. Wool is used primarily for clothing. Other wool products are blankets, rugs, carpets, upholstery, and draperies. Industrial felt and cloths are other uses. Wool felt is made from un-spun fibers that, because of the nature of wool fibers, adhere tightly together. Felts undergo a combined mechanical-chemical process involving both heat and moisture. Unwoven felt is used in the hat industry, and in a number of industrial applications as a covering or padding. Woven felt is used in pianos, printing, and in optical and chemical industries.

woolen fabric - is woven or knitted. See *wool* above, Chapter 10 and Fig. W2.

woven fabrics - See 10A, 10B and 10C.

wrenches - are most commonly produced by drop forging (2A4b1), normalizing (8G2e) and a number of other operations. They are made from chrome-vanadium steel or medium carbon steel (0.25 to 0.55% carbon). A minority of wrenches are made by sand-mold casting (1B) or investment casting (1G). Press trimming of flash may follow forging and then barrel or vibratory tumbling is used to improve surface smoothness. Some surfaces, for example, the contacting surfaces of open-end wrenches are machined on milling machines (3D) or on special grinding machines (3C3). The end openings of box wrenches are broached (3F). Deburring, heat treating, usually by induction (8G3a2), and tempering (8G2f), for hardness and wear resistance follow machining of most wrenches. Finish grinding of some surfaces may follow heat treatment. Other finishing is normally by polishing (8B1) and chromium electroplating (8C1) though some wrenches are given a black oxide surface treatment (8E4). Adjustable wrenches have additional parts and additional machining operations to provide adjustability. The adjusting screws for adjustable open-end wrenches are screw machine parts (3A2c), with the crests of the threads knurled.

wrench sockets - Sockets are either impression die forged (2A4b) or investment cast (1G), of chrome-vanadium steel. Broaching of the internal surfaces may also take place. Heat treatment by induction hardening (8G3a2) follows. After the initial operation, the external socket surfaces are polished (8B1) and chromium plated (8C1).

writing paper - The best quality writing paper is rag paper, or paper made from rags, or scrap cuttings of linen or cotton, or from a combination of rags and wood pulp. See 9B and, especially, 9B1 and 9B3.

xylene (xylol) - is usually a mixture of three closely-related volatile liquids that occur together. They all have the formula $C_6H_4(CH_3)_2$ and very close boiling points. They are colorless and used together as paint and ink solvents, in small amounts in aircraft fuel, and individually, as ingredients in dyes and plastics. They are made from the fractional distillation of petroleum (11H1a) or coal tar.

Y

yarn - consists of bundles of fibers twisted or laid together to form continuous strands[2]. Ply or plied yarns are single strand yarns twisted together. Cabled yarn or cord consists of ply yarns twisted together[2]. (See 10B.)

yogurt - is semi-fluid fermented milk. Two yogurt cultures, lactobacillus bulgaricus and streptococus thermophilus, are added to pasturized, homogenized cow's milk. (Goat, sheep, and water buffalo milk is used in some countries.) In commercial production, milk solids are added to provide a custard-like consistency. (In home production, the milk is boiled to concentrate it, and a portion of a previous yogurt batch is added to provide the cultures.) The mixture is placed in a temperature-controlled environment of 110 to 112°F (43 to 44°C) for four or five hours by which time curd forms. Fresh fruit, jam, juices, sweeteners and flavorings may be added before the yogurt is packaged for sale.

Z

zinc - is produced chiefly from two ores, sphalerite (zinc sulfate) and smithsonite (zinc carbonate). The ores are transformed into zinc oxide by heating them to a high temperature in the presence of air. This converts them to zinc oxide. (See roasting - 11K1b1.) The oxide is reduced by heating in the presence of carbon. In one process, enough heat is used in the electric reducing furnace to melt, boil and then distill the zinc. The metal thus produced contains small amounts of some other metals and is called spelter. In another process, the oxide is leached with sulfuric acid. The resulting solution is processed to remove impurities and is then electrolyzed, producing zinc of high purity. (See electrorefining - 11K1c2.)

Zinc is used for plated or dipped coatings on ferrous materials for corrosion protection (galvanizing). It is also used for die casting, particularly of smaller parts, because of its excellent castability, as an alloying ingredient (See *brass*), and in dry-cell batteries.

zippers[13] - have several basic parts; 1) the teeth that interlock to hold the zipper and the attached garment opening closed, 2) the fabric tape that supports the teeth, 3) two stringers, the assemblies of tape and teeth that can mesh together from each side, 4) a slider that joins or separates the teeth of the two stringers, 5) stops, which are fittings that prevent the slider from moving off the end of the stringer, 6) a tab attached to the slider to make it easier to grasp and move. See Fig. Z1.

The tapes are made with conventional textile methods, primarily knitting, with a strong hem or bead where the teeth are attached (Chapter 10). Cotton and polyester are the usual fabrics. The teeth are now mostly made from plastics, commonly nylon, but polyacetal and polyester are also used. Metal teeth are made from brass, steel, aluminum or zinc. Some are painted, plated, or given a black oxide finish to match or complement the product's color. The teeth are formed and attached to the tape in special machines. Zinc teeth can be die cast right onto the tapes in special die casting machines. Some plastic teeth are injection molded onto the tape in similar special injection molding machines. However, machines that form the teeth from either flattened metal or plastic wire are also widely used. These machines usually form the wire into a Y-shape, with the bottom of the Y formed into a cup-like shape or other shape that

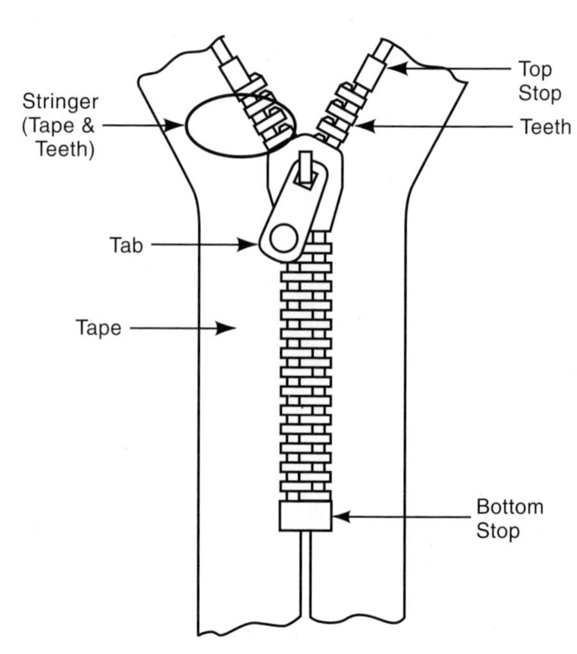

Fig. Z1 a typical zipper assembly.

will engage with and hold the opposing teeth. The tape is fed to the machine and the arms of the Y's are clamped together to hold the teeth in place on the tape. All this forming is done automatically at a high speed, producing continuous tapes of assembled stringers. The slide, stops, and tabs, are made by metal-forming methods, are injection molded, if plastic, or die cast, if zinc. The stringers are fed to machines that join both sides together, apply a wax lubricant to the teeth, and deburr them by wire brushing (if the teeth are metal), apply starch to stiffen the tapes, dry and straighten the tapes, and roll them onto large spools for later processing. Then, later, the tapes are cut to standard lengths, or the special lengths needed, and top and bottom stops, slider, and tabs are assembled and fastened. The completed zipper assemblies are sold in standard lengths to garment, luggage, and other manufacturers, who sew them into their products, and to distributors for eventual sale to individuals.

References

1. Encyclopedia Britannica, Deluxe CD Rom edition, 2002.
2. Materials Handbook, G. S. Brady and H. R. Clauser, McGraw-Hill, New York, 1997.
3. McGraw-Hill Encyclopedia of Science and Technology, McGraw-Hill, New York, 1998.
4. Shreve's Chemical Process Industries, 5th ed., G. T. Austin, McGraw-Hill, New York, 1984.
5. Encarta 2004 Encyclopedia Standard - Microsoft, Inc. Redmond, WA.
6. Perry's Chemical Engineers' Handbook, 7th edition, D. W. Green, ed., McGraw-Hill, N.Y. 1997.
7. The Engineering Handbook, R. C. Dorf, CRC Press - IREE Press, 1996, ISBN 0-8493-8344-7.
8. Grolier Encyclopedia, Grolier Electronic Publishing, Inc., 2002, Danbury, CT.
9. How Products Are Made - N. Schlager, ed., Gale Research, Inc., Detroit, 1994.
10. Design for Manufacturability Handbook, J. G. Bralla, editor, McGraw-Hill, New York, 1999.
11. Encyclopedia Americana, 2000.
12. Electronic Components, Victor Meeldijk, John Wiley, New York, 1996.
13. CDs, Super Glue and Salso, How Everyday Products are Made, (Series 1, 2 and 3), S. Rose and N. Schlager, K. Witman, K. L. Kalasky, M. L. Rein, Gale Research, Detroit, 1995, 1996 and 2003.
14. http://www.encyclopedia.com - from the internet.
15. Plastics Handbook, ed. Modern Plastics, McGraw-Hill, New York, ISBN 042805-0.
16. Design for Excellence, James G. Bralla, McGraw-Hill, New York, 1996.
17. Cam Design and Manufacturing Handbook, Robert L. Norton, Industrial Press, Inc., New York, 2002.
18. Factory Made - How Things Are Manufactured, Leonard Gottlieb, Houghton Mifflin Co., Boston 1978.
19. The Complete Book of Jewelry Making, Carles Godina, Lack Books, Sterling Publishing, New York, 1999.
20. Introduction to Lapidary, Pansy D. Kraus, Chilton Book Co., Radnor, PA 1987.
21. Firearms Encyclopedia, G. C. Nonte, Jr., Popular Science Publishing Co., 1973.
22. Gem Cutting, A lapidary's Manual, 3rd ed., John Sinkankas, Van Nostrand Reinhold, New York, 1984.
23. Breakfast Cereals and How They Are Made, 2nd ed., R. B. Fast and E. F. Caldwell, American Association of Cereal Chemists, St. Paul, 2000.
24. Understanding Wood Finishing, B. Flexner, Rodale Press, Emmaus, Pennsylvania, 1994.
25. The Tool and Manufacturing Engineers Handbook, 4th edition, Vol. II, Forming, Charles Wick editor, Society of Manufacturing Engineers, Dearborn, MI, 1984.
26. How Jet Engines Are Made, Julian Moxon, Facts On File Publications, Bicester, England, 1985.
27. Jet Propulsion, Nicholas Cupsty, Cambridge University Press, United Kingdom, 1997.
28. Superweld - Unlocking the Potential of High-Temperature Superconductors, Energy Science News, internet: http://www.pnl.gov/energyscience/08-00/art2.htm.
29. Glass Engineering Handbook, E. B. Shand, McGraw-Hill, New York.
30. Scientific American, Vol. 292, No. 4, April 2005, "Low-Temperature Superconductivity is Warming Up."

Index

B

C

D

E

F

G

H

I

J

M

Q

R

S

U

V

X

Y

Z